CW01573020

SAE Dictionary of Aerospace Engineering

2nd Edition

SAE Dictionary of Aerospace Engineering

2nd Edition

Joan L. Tomsic
Editor

with contributions by
Charles N. Eastlake
Professor of Aerospace Engineering
Embry-Riddle Aeronautical University

Published by:
Society of Automotive Engineers, Inc.
400 Commonwealth Drive
Warrendale, PA   15096-0001

Library of Congress Cataloging-in-Publication Data

SAE dictionary of aerospace engineering / Joal L. Tomsic, editor;
with contributions by Charles N. Eastlake.—2nd ed.
    p.   cm.
"SAE order no. R-211"—T.p. verso
ISBN 0-7680-0245-1
1. Aerospace engineering—Dictionaries. 2. Aeronautics—
Dictionaries. 3. Astronautics—Dictionaries.   I. Tomsic, Joan L.
II. Eastlake, Charles N. III. Society of Automotive Engineers.
TL509.S24 1998
629.1'03—dc21   98-28541
                                                          CIP

Library of Congress Catalog Card Number 92-64101
ISBN 0-7680-0245-1

SAE Order No. R-211

PREFACE
TO THE 2ND EDITION

Although this dictionary is concerned primarily with the terminology of contemporary aerospace technology, words that are used in other scientific fields, such as computing, astronomy, and geophysics, have also been included. This dictionary should thus prove useful to students and teachers of aerospace engineering, scientists, technologists, and technicians in research and industry.

In such a fast-moving technology as aerospace engineering, new terms are created and others become redundant almost weekly. We have made every effort to give more space to the emerging technologies and to delete the more obsolete terms.

While this dictionary contains over 20,000 terms and definitions, size alone is a poor criterion of merit. A work of reference must stand or fall by its accuracy and authority. The authority and accuracy of this work rests on the decades of experience of SAE engineers who created the SAE standards and the generous contributions from the National Aeronautics and Space Administration (NASA).

Joan L. Tomsic
Editor

PREFACE
TO THE 1ST EDITION

The SAE DICTIONARY OF AEROSPACE ENGINEERING is a technical publication edited for the aerospace engineers who design, test, and manufacture aerospace vehicles, components, and parts. Every term and definition has been extensively reviewed by experts, and will be continuously reviewed in the future to incorporate changes as they occur.

This completely new dictionary is, remarkably, the first such dictionary created for <u>engineers</u> who must communicate clearly and accurately in an increasingly more complex technology. This volume has the quality and comprehensiveness that aerospace engineers have come to expect from SAE publications. The engineering terms and definitions are compiled from the following sources:

- 5,188 terms from SAE AEROSPACE STANDARDS and REPORTS... peer reviewed information from these technical publications.

- 4,038 terms from the NASA THESAURUS...a publication reviewed and updated repeatedly since the 1970s.

- 9,541 general engineering and computer terms from ENGINEERING RESOURCES, INC....a leading developer of engineering terms and definitions that are germane to all engineering disciplines.

In sum, here are nearly 20,000 terms and definitions contained in the <u>first</u> dictionary for aerospace experts who actually design and build aircraft and their related components rather than the aerospace "buff."

Obviously, aerospace engineers and those technicians and clerical people who support them, will have a primary interest in the new dictionary. Aerospace engineering students and writers will also welcome this first technical dictionary in aerospace technology.

William H. Cubberly
Editor of the 1st Edition

HOW TO USE
THIS DICTIONARY

Basic format

The format for a defined entry provides the term in boldface type; if the term has a commonly used abbreviation or acronym, this follows the term in parentheses, also in boldface type. The definitions appear in regular (light) typeface. A term may have more than one definition, in which case the definition is preceded by a number.

Some terms contain a "See" cross reference rather than a definition. In such cases, the cross-referenced term provides the definition and/or explanatory information. Some terms contain a "See also" cross reference in addition to a definition. In such cases, the cross-referenced term provides additional or related information.

In general, SAE Aerospace Standards and Reports definitions appear first. The standard or report in which the term and definition originally appeared is referenced and shown in brackets, such as [ARP1201]. Please note that in this edition, original definitions have in some cases been copyedited to a consistent style and to improve readability. Thus, if an exact original definition is required, the referenced standard should be consulted.

Abbreviations and acronyms related to aerospace are shown as normally used as well as spelled out, except in those cases where the abbreviation or acronym is so well known as to preclude the need for a spelled-out entry.

Alphabetization

The terms are alphabetized on a letter-by-letter basis. Superscripted and subscripted letters are considered in the sequencing of terms; however, word spacing, hyphens, apostrophes, commas, and slashes are ignored. Arabic and Roman numerals, Greek letters, and other miscellaneous symbols are also ignored.

To aid the user in finding words or abbreviations, the first word in the left-hand column and the last word in the right-hand column of each page are shown in boldface type at the top of the appropriate page. When using this aid, however, the user should be aware that the terms in this dictionary are not uniformly listed in either singular or plural form, thus both the singular and plural forms should be checked when looking for a particular word. Further, certain terms are not listed in normal word order, thus alternate word orders should be checked; for example, for "residual stress," one might also check "stress, residual."

a *1. See* amp or ampere. *2.* Arterial symbol, used to designate source of blood gas, or other components of arterial blood, e.g., PaO_2, partial pressure of arterial oxygen tension; $PaCO_2$, partial pressure of arterial CO_2 tension, etc. [ARP171]

A *1. See* Angstrom. *2. See* amp or ampere. *3.* Alveolar symbol, used to designate alveolar gas tensions, e.g., $PACO_2$, alveolar CO_2 tension; PAO_2, alveolar oxygen tension, etc. [ARP171]

A-basis The "A-basis" or "A" mechanical property value is the value above which at least 99% of the population of values are expected to fall, with a confidence of 95%. [AIR4844]

aberration *1.* In astronomy, the apparent angular displacement of the position of a celestial body in the direction of motion of the observer, caused by the combination of the velocity of the observer and the velocity of light. *2.* In optics, a specific deviation from perfect imagery, for example, spherical aberration, coma, astigmatism, curvature of field, and distortion. *3.* Deviation from ideal behavior by a lens, optical system, or optical component. Aberration exists in all optical systems and designers must make trade-offs among the different types.

abhesive *1.* A material that resists adhesion. *2.* A film or coating applied to surfaces to prevent sticking, heat sealing, etc., for example, a parting agent or mold-release agent. [AIR4844]

ablating materials Special materials on the surface of a spacecraft that can be sacrificed during re-entry into the earth's atmosphere. [ARP4386]

ablation *1.* The removal of surface material from a body by vaporization, melting, chipping, or other erosive process. Specifically, the intentional removal of material from a nose cone or spacecraft during high-speed movement through a planetary atmosphere to provide thermal protection to the underlying structure. *2.* An exothermic process involving the degradation, decomposition, and erosion of a material caused by high temperature, pressure, time, percent oxidizing species, and velocity of gas flow. [AIR4844]

ablative materials Materials, especially coating materials, designed to provide thermal protection through the loss of mass to a body in a fluid stream.

ablative plastic A material that absorbs heat through a decomposition process that takes place at or near the surface exposed to the heat. This mechanism essentially provides thermal protection of the subsurface materials and components by sacrificing the surface layer. [AIR4844]

ABL bottle An internal pressure test vessel about 460 mm (18 in.) in diameter and 610 mm (24 in.) long used to determine the quality and properties of the filament-wound material in the vessel. [AIR4844]

abnormal Deviating from normal operation.

abnormal electric-system operation The unexpected loss of control of the electric system. The initiating action of the abnormal electric-system operation is uncontrolled and the exact moment of its occurrence is not anticipated; however, recovery from this operation is a controlled action. Abnormal electric-system operations occur, perhaps, once during a flight as a result of damage, or they may never occur during the life of an aircraft. An example of an abnormal electric-system operation is the faulting of electric power to the structure of an aircraft and its subsequent clearing by fault protective devices. [AS1212]

abnormal limits Limits that accommodate the trip bands of protective equipment in the primary power generating system. [AS1212]

abort *1.* Failure to accomplish a mission or portion of a mission for any reason. [ARP4386] *2.* To cut short or break off an action, operation, or procedure with an aircraft, space vehicle, or the like—especially because of equipment failure, such as to abort a take-off, abort a mission, or abort a launch. *3.* An aircraft, space vehicle, or the like that aborts. *4.* During catapulting, the act of suspending a launch and removing the airplane from the catapult. [AIR1489] *5.* In data processing, the termination of a computer operation before its normal conclusion.

above ground level (AGL) Height of an aircraft, clouds, or the top of an obstruction (for example, building, tower, or bridge) above the surface of the earth in the immediate vicinity, usually expressed in feet. [ARP4107]

abrade To prepare a surface by roughening it by sanding or any other means.

abrasion *1.* The wearing away by friction, especially of hose cover or reinforcement, due to rubbing or vibrating against another hose or permanent fixture in an installation such that it evidences damage, fraying, etc. [ARP1658A] *2.* Removal of surface material by sliding or rolling contact with hard particles of the same substance or another substance. The particles may be loose or may be part of another surface in contact with the first.

abrasion resistance The ability of a material to withstand abrasion. [ARP1931]

abrasive *1.* Particulate matter, usually having sharp edges or points, that can be used to shape and finish workpieces in grinding, honing, lapping, polishing, blasting, or tumbling processes. Depending on the process, abrasives may be loose; formed into solid shapes; glued to paper or cloth; or suspended in a paste, slurry, or air stream. *2.* A material formed into a solid mass, usually fired or sintered, and used to grind or polish workpieces. Common forms are grinding wheels, abrasive discs, honing sticks, cones, and burrs.

ABS *See* acrylonitrile butadiene styrene.

absolute accuracy error The deviation of the analog value at any code from its theoretical value after the full-scale range has been calibrated. Expressed in percent, ppm, or fractions of 1 LSB.

absolute address In the actual machine code address numbering system, an address that indicates the exact storage location at which the referenced operand is to be found or stored. Synonymous with specific address and actual address and related to absolute code.

absolute alarm An alarm caused by the detection of a variable that has exceeded a set of prescribed high- or low-limit conditions.

absolute altimeter *See* terrain clearance indicator.

absolute altitude Distance from an aircraft or spacecraft to the actual surface of a planet or natural satellite.

absolute angle of attack The angle measured between the chord plane of an airfoil and the position that plane would have if the airfoil were producing zero lift; that is, the sum of the geometric angle of attack and the zero-lift angle of attack. Also called an aerodynamic angle of attack. *See* angle of attack, zero-lift angle of attack, and critical angle of attack. [ARP4107]

absolute ceiling The maximum height above sea level in a standard atmosphere at which a given airplane, under specified operating conditions, can maintain horizontal flight. [ARP4107]

absolute code Coding that uses machine instructions with absolute addresses. Synonymous with specific code.

absolute encoder An electronic or electromechanical device that produces a unique digital output (in coded form) for each value of an analog or digital input. In an absolute rotary encoder, for instance, the position following any incremental movement can be determined directly, without reference to the starting position.

absolute feedback In numerical control, assignment of a unique value to each possible position of a machine slide or actuating member.

absolute humidity The weight of water vapor in a gas/water-vapor mixture per unit volume of space occupied. Absolute humidity may be expressed, for example, in grains or pounds per cubic foot.

absolute instrument An instrument that determines the value of a measured quantity in absolute units by making a simple physical measurement.

absolute measurement A measured value expressed in terms of fundamental standards of distance, mass, and time.

absolute pressure *1.* Pressure value expressed using absolute vacuum as a reference. [ARP4386] *2.* Absolute static pressure in psia of the liquid and vapor phase. [AIR1326] *3.* The combined local pressure induced by some source plus the local atmospheric pressure. *4.* Gage pressure plus barometric pressure in the same units.

absolute programming In numerical control, use of a single point of reference for determining all positions and dimensions.

absolute rating A theoretical size designation that is an estimation of the largest particle, by length, that can pass through a filter with a specific rating. [ARP4386]

absolute sealing A level of sealing that requires all seams, slots, holes, and fasteners passing through the seal plane to be sealed. (All integral fuel tanks require absolute sealing.) [AIR4069]

absolute stability A linear system is absolutely stable if there exists a limiting value of the open-loop gain such that the system is stable for all lower values of that gain, and unstable for all higher values.

absolute value error The magnitude of the error disregarding the algebraic sign; or, for a vectorial error, the magnitude of the error disregarding its function.

absolute viscosity A measure of the internal shear properties of fluids expressed as the tangential force per unit area at either of two horizontal planes separated by one unit thickness of a given fluid, one of the planes being fixed and the other moving with unit velocity.

absolute zero Temperature of -273.16°C, -459.69°F, or 0 K at which molecular motion vanishes and a body has no heat energy.

absorb To take in or assimilate (e.g., sound) with little or none of it being transmitted or reflected.

absorbance An optical property expressed as log $(1/T)$, where T is the transmittance.

absorbate A material that is absorbed by another. [AIR4844]

absorbent A material that takes in, or absorbs, another. [AIR4844]

absorptance *1.* The ratio of the radiant flux absorbed by a body to that incident upon it. *2.* The fraction of the incident light absorbed.

absorption *1.* The penetration into the mass of one substance by another. *2.* The process whereby energy is dissipated within a specimen placed in a field of radiant energy. *3.* The capillary or cellular attraction of adherent surfaces to draw off the liquid adhesive film into the substrate. *4.* A process in which one material takes in or absorbs another. [AIR4844]

absorption band A region of the electromagnetic spectrum where a given substance exhibits a high absorption coefficient compared to adjacent regions of the spectrum. *See also* absorption spectra.

absorption coefficient An inherent material property expressed as the fractional loss in radiation intensity per unit mass or per unit thickness determined over an infinitesimal thickness of the given material at a fixed wavelength and band width. *See also* absorptivity.

absorption cooling Refrigeration in which cooling is effected by the expansion of liquid ammonia into a gas and the absorption of the gas by water. The ammonia is reused after the water evaporates.

absorption cross sections In radar, cross sections characterized by the amount of power removed from a beam by absorption of radio energy by a target to the power in the beam incident upon the target.

absorption curve A graph of the variation of transmitted radiation through a fixed sample of material of a given thickness while the wavelength is changed at a uniform rate.

absorption dynamometer A device for measuring mechanical force or power by converting the mechanical energy to heat in a friction mechanism or bank of electrical resistors.

absorption-emission pyrometer An instrument for determining gas temperature by measuring the radiation emitted by a calibrated reference source both before and after the radiation passes through the gas, where it is partly absorbed.

absorption hygrometer An instrument for determining water vapor content of the atmosphere by measuring the amount of water vapor absorbed by a hygroscopic chemical.

absorption meter An instrument for measuring the quantity of light transmitted through a transparent medium by means of a photocell or other light-detecting device.

absorption spectra The arrays of absorption lines and absorption bands that result from the passage of radiant energy from a continuous source through a selectively absorbing medium cooler than the source.

absorption spectroscopy The study of the wavelengths of light absorbed by materials and the relative intensities at which different wavelengths are absorbed. This technique can be used to identify materials and measure their optical densities.

absorption tower A vertical tube in which a gas rising through a falling stream of liquid droplets is partially absorbed by the liquid.

absorptive index *See* absorptivity.

absorptivity The capacity of a material to absorb incident radiant energy, measured as the absorptance of a specimen of material thick enough to be completely opaque, and having an optically smooth surface.

AC or a-c *See* alternating current.

acapnia The complete absence of carbon dioxide (usually in the blood). Commonly used incorrectly for hypocapnia. [ARP171]

acapnic Suffering from acapnia. [ARP171]

accelerate To make or become faster; to cause to happen sooner.

accelerated aging A test in which parameters such as voltage and temperature are increased above normal operating values to obtain observable or measurable deterioration in a relatively short period of time. The plotted results predict expected service life under normal operating conditions. Also called accelerated life test. [ARP1931]

accelerated life test A method of estimating reliability or durability of a product by subjecting it to operating conditions above its maximum ratings.

accelerated simulated mission endurance test An accelerated test derived from a real-time representation of the mission profiles and mission mix for a specific aircraft application for a period of at least 1000 hours. The real-time representation is usually expressed in terms of power-lever angle versus time and engine-inlet temperature and pressure conditions. [AIR4896]

accelerated stall A stall occurring under acceleration, such as in a pullout. Such a stall usually produces more violent motions of the airplane than does a stall occurring in unaccelerated flight. [ARP4107]

accelerated test A test in which the applied stress level is chosen to exceed that stated in the reference conditions in order to shorten the time required to observe the stress responses, or magnify the responses in a given time. To be valid, an accelerated test must not alter the basic modes and/or mechanisms of failure or their relative prevalence. [ARD50010]

accelerate-stop distance A measure of the field length requirement for aircraft operation. The accelerate-stop distance is the sum of the distance required to accelerate the aircraft from a standing start to the critical engine failure speed and the distance required to bring the aircraft to a full stop, assuming the critical engine to fail at the critical engine speed. [AIR1489]

accelerating agent A chemical that hastens the curing of rubber, plastic, cement, or adhesives, and may also improve their properties. Also known as an accelerator.

accelerating electrode An auxiliary electrode in an electron tube that is maintained at an applied potential to accelerate electrons in a beam.

acceleration *1.* The act or process of accelerating, or the state of being accelerated. Negative acceleration is called deceleration. [AIR1489] *2.* The time rate of change of velocity. *3.* The second derivative of a distance function with respect to time.

acceleration cardiovascular effects Reduction in performance capability due to grayout or blackout resulting from high positive G. Analogous negative G effects, such as redout, are also classified as acceleration cardiovascular effects. *See* grayout, blackout, and redout. [ARP4107]

acceleration displacement effects Reduction in performance capability due to the physical displacement or restriction in movement of the operator as a result of $\pm G_z$ (head to toe), $\pm G_y$ (side to side), or $\pm G_x$ (front to back), where G_x, G_y, G_z are the three vector components of the total acceleration to which the operator is being subjected (generally stated in units of the standard value of the gravitational acceleration, at the earth's surface, $g = 32$ ft/s^2). [ARP4107]

acceleration effects Reduction in performance capability due to the effects of acceleration on the cardiovascular system, the vestibular apparatus, or restriction of body movement. [ARP4107]

acceleration factor *See* factor, acceleration.

acceleration orienting effects Reduction in performance capability due to the effects of acceleration on the vestibular apparatus or visual system. [ARP4107]

acceleration time *1.* The span of time it takes a mechanical component of a computer to go from rest to running speed. *2.* The measurement of time for any object to reach a predetermined speed.

acceleration tolerance The capability to withstand the effects of changing velocity over time on biological processes or spatial-orienting mechanisms. [ARP4107]

accelerator *1.* A chemical additive used to hasten a chemical reaction under specific conditions. [ARP1931] *2.* An ingredient used in a sealant formulation to accelerate the rate of cure. [AS7200/1]

accelerometer An instrument for measuring acceleration or an accelerating force, such as gravity. If the instrument includes provisions for producing a recorded output, it is called an accelerograph.

acceptable message segment (AMS) Condition occurring within a bus monitor when the monitor determines that a message segment contains only valid word(s) and valid data words; that the words are contiguous where required; and that the bus monitor is programmed to accept such a valid message segment. [AS4116]

acceptance Acknowledgment by a certification authority that a submission of data, argument, or claim of equivalence satisfies applicable requirements. [ARP4754]

acceptance test *1.* A test conducted on an item under specified conditions by, or on behalf of, the government, using delivered or deliverable items to determine compliance of the item with specified requirements. Includes acceptance of first production units. [ARD50010] *2.* A test performed on a production component to assure the continuance of the quality of that component. [ARP1281A]

acceptance-test tab An extension of a forging to provide material for lot-acceptance testing for subsurface chemical uniformity. [AMS6474]

access Pertaining to the ability to place information into, or retrieve information from, a storage device.

access control *1.* The procedures for providing systematic, unambiguous, orderly, reliable, and generally automatic use of communication lines, channels, and networks for information transfer. *2.* Hardware or software features, operating procedures, or management procedures designed to permit authorized access to a computer system.

accessibility A measure of the relative ease of admission to the various areas of an item for the purpose of operation or maintenance. [ARP4386]

accessible Within the reach of a crew member whose lap belt is on, whose shoulder harness is unlocked, and whose seat is at the position which locates his eyes at the DEP. [ARP4101]

accessible terminal A node in an electronic network that is configured to allow it to be connected to an external circuit.

access method Any of the data-management techniques available to the user for transferring data between main storage and an input/output device.

accessory A part, subassembly, assembly, or component designed for use in conjunction with or to supplement another item. [ARD50010]

access procedures The procedure by which the devices attached to a network gain access to the medium. The access procedure typically includes provisions to guarantee fairness in sharing the network bandwidth between attached devices. The most common access procedures for LANs are CSMA/CD, Token Bus, Token Ring, and Slotted Ring. *See* MAC.

access, random *See* random access.

access time The interval between a request for stored information and the delivery of the information. Often used to refer to the speed of memory.

access unit interface (AUI) The optional interface between a data station using an IEEE 802.3 LAN and a transceiver or modem. The AUI permits transparent connection of a data station to either baseband or broadband media.

accident An occurrence associated with the operation of an aircraft which takes place between the time any person boards the aircraft with the intention of flight and the time when all such persons have disembarked, and in which any person suffers death or serious injury, or in which the aircraft receives substantial damage. [ARP4107]

acclimatization *1.* The adjustments of the human body or other organism to a new environment. *2.* The bodily changes that tend to increase efficiency and reduce energy loss.

accommodation coefficient The ratio of the average energy actually transferred between a surface and impinging gas molecules scattered by the surface to the average energy that would theoretically be transferred if the impinging molecules reached complete thermal equilibrium with the surface before leaving the surface.

accredited standard committee (ASC) A standard committee accredited to ANSI.

accretion disks Rotation disks of matter surrounding an astronomical object, such as a star, galactic nucleus, black hole, which are accumulated gravitationally by the object.

accumulation test A leak-test method using pressurized gas to determine the leak integrity of a component.

accumulator *1.* A device or apparatus that accumulates or stores up; for example, a contrivance in a hydraulic system that stores fluid under pressure (energy). *2.* An electric storage battery. [AIR1489] *3.* The register and associated equipment in the arithmetic unit of the computer in which arithmetic and logical operations are performed. *4.* A chamber or vessel for storing low-side liquid refrigerant in a refrigeration system. Also referred to as a receiver, reflux receiver, or reflux drum.

accumulator, compensating An accumulator that, in addition to its high pressure volume, incorporates low pressure volumetric capacity which will accommodate a like volume of fluid to that discharged from the high pressure chamber. In a compensating accumulator, the sum of the volumes of the high and low pressure chambers remains constant. [ARP4386]

accumulator, flexible separator An accumulator in which the fluid is separated from the compressible medium by means of a flexible bladder or diaphragm. [ARP4386]

accumulator, hydro-pneumatic An accumulator in which the stored operating medium is hydraulic fluid, pressurized by means of compressed gas. [ARP4386]

accuracy *1.* The degree to which the controlled temperature approaches the steady-state setpoint value. [ARP89] *2.* The deviation, or error, by which an actual output varies from an expected ideal or absolute output. Each element in a measurement system contributes to deviation or error, which should be separately specified if they significantly contribute to the degradation of total system accuracy.

accuracy, measured The degree to which an indicated value matches the actual value of a measured variable.

accuracy, total The deviation, or error, by which an actual output varies from an expected ideal or absolute output. Quantitatively, the total accuracy is the difference between the measured value and the most probable value for the same quantity, when the latter is determined from all available data.

ACEE program (Aircraft Energy Efficiency Program) A NASA program started in 1975 to reduce fuel consumption for transport aircraft through the study of structural and aerodynamic energy efficiency as well as engine energy efficiency. Results of this program have included engine-component improvement, new energy-efficient engines, and advanced turbo-propellers.

acetone Commonly used wipe solvent. Used for cleaning composite surfaces prior to bonding and metal surfaces prior to other treatments. [AIR4844]

AC generators (a-c generators) Generators for the production of alternating-current power. Also known as alternators.

achieved airplane performance The performance that an airplane is actually attaining under the particular combination of circumstances and environment present during the time period the operation is being conducted. [AS8004]

achromatic Optical elements that are designed to refract light of different wavelengths at the same angle. Typically, achromatic lenses are made of two or more components of different refractive index, and are designed for use at visible wavelengths only.

acicular alpha In titanium, a product of nucleation and growth or an athermal (martensitic) transformation from beta to the lower temperature allotropic alpha phase. Acicular alpha may be needle-like, lenticular, or of flattened bar morphology in three dimensions. Its typical aspect ratio is about 10:1. [AS1814]

acid Any of a large class of substances whose aqueous solutions are capable of turning litmus indicators red and of reacting with bases or alkalis to form salts.

acid cleaning The process of cleaning the interior surfaces of steam-generating units with an acidic inhibitor to prevent corrosion, and subsequently draining, washing, and neutralizing the acid by a further wash with alkaline water.

acidity Represents the amount of free carbon dioxide mineral acids and salts (especially sulfates of iron and aluminum) which hydrolyze to give hydrogen ions in water. Acidity is reported as milliequivalents per liter of acid; as ppm acidity as calcium carbonate; or as pH, a measure of the hydrogen ion concentration.

acidosis, respiratory A condition of the blood in which there is an increase of the carbonic acid fraction of blood relative to the bicarbonate fraction, with resultant decrease of pH. This occurs whenever elimination of carbon dioxide through pulmonary ventilation is impaired, as in respiratory obstruction, paralysis, pulmonary fibrosis, or other diffusion impairment. [ARP171]

acid wash A chemical solution containing phosphoric acid which is used on metal surfaces to neutralize residues from alkaline cleaners and to simultaneously produce a phosphate coating that protects the surface from rusting and prepares it for painting.

a-c input module I/O module that converts process-switched a-c to logic levels for use in the PC.

ACK *See* acknowledge.

acknowledge (ACK) A message sent between peer entities to indicate that data was properly received.

ACLS *1. See* automatic carrier landing system. *2. See* air cushion landing system.

acme screw actuator A mechanical actuator consisting of an acme threaded shaft and an acme threaded nut and arranged so that a rotary input motion is converted to a linear output. [ARP4386]

Acme screw thread A type of power-transmission thread made in four series: 29° general purpose; 29° stub; 60° stub; and 10° modified square. In the Acme screw thread, the number of threads per inch is not standardized according to shank diameter.

acoustic Related to sound.

acoustical ohm The unit of measure for acoustic resistance, reactance, or impedance. It equals unity when a sound pressure of one microbar produces a volume velocity of one cubic centimeter per second.

acoustic compliance The reciprocal of acoustic stiffness.

acoustic coupler A type of communications device that converts digital signals into audio tones that can be transmitted by telephone.

acoustic delay lines Devices used in a communications link or a computer memory in which the signal is delayed by the propagation of sound waves.

acoustic dispersion Separation of a complex sound wave into its various frequency components, usually due to variation of wave velocity in the medium with sound frequency. Usually expressed in terms of the rate of change of velocity with frequency.

acoustic emission *1.* A measure of the integrity of a material, as determined by sound emission when the material is stressed. [AIR4844] *2.* The stress and pressure waves generated during dynamic processes in materials and used in assessing structural integrity in machined parts.

acoustic excitation The process of inducing vibration in a structure by exposing it to sound waves.

acoustic generator A transducer for converting electrical, mechanical, or some other form of energy into sound waves. Also called a sound generator.

acoustic holography A technique for detecting flaws or regions of inhomogeneity in a part by subjecting it to ultrasonic energy, producing an interference pattern on the free surface of water in an immersion tank, and reading the interference pattern by laser holography to produce an image of the test object.

acoustic impedance *1.* A material property that determines the product of the velocity of sound in a material and the density of the material. Used in determining the reflection characteristics of interfaces. [ARP5089] *2.* The complex quotient obtained by dividing sound pressure on a surface by the flux through the surface.

acoustic inertance A property related to the kinetic energy of a sound medium which equals $Z_a/2f$, where Z_a is the acoustic reactance and f is the sound frequency. The usual units of measure of acoustic inertance are g/cm^4. Also known as acoustic mass.

acoustic interferometer An instrument for measuring either the velocity or frequency of sound pressure in a standing wave established in a liquid or gas medium, between a sound source and reflector, as the reflector is moved or the frequency is varied.

acoustic levitation Method by which molten materials in space are suspended during processing experiments in the low-gravity environment. Also, the use of very intense sound waves to keep a body suspended, thereby eliminating any container contact.

acoustic microscopes Instrument that uses acoustic radiation at microwave frequencies to allow visualization of microscopic detail exhibited in elastic properties of objects.

acoustic reactance The imaginary component of acoustic impedance.

acoustic resistance The real component of acoustic impedance.

acoustic retrofitting Modification, especially of aircraft, to effect noise reduction. Specifically, the introduction of absorber materials and jet-noise silencers.

acoustics *1.* The technology associated with the production, transmission, and utilization of sound. *2.* The science associated with sound and its effects.

acoustic signature In sonar applications, the profile that is characteristic of a particular undersea object (or class of objects); for example, the profile of a school of fish or a sea-bottom formation.

acoustic spectrometer An instrument for analyzing a complex sound wave by determining the volume (intensity) of sound-wave components having different frequencies.

acoustic stiffness A property related to the potential energy of a medium or its boundaries which equals $2fZ_a$, where Z_a is the acoustic reactance and f is sound frequency. The usual units of measure of acoustic stiffness are $dyne/cm^5$.

acoustic streaming Unidirectional flow currents in a fluid that are due to the presence of sound waves.

acoustic velocity The speed of propagation of sound waves. Also known as sound velocity.

acousto-optic Refers to an interaction between an acoustic wave and a light wave passing through the same material. Acousto-optic devices can serve for beam deflection, modulation, signal processing, and Q switching.

acousto-optic glass Glass with a composition designed to maximize the acousto-optic effect.

a-c output module I/O module that converts PC logic levels to output switch action for a-c load control.

AC phase voltage A phase being considered the line-to-neutral circuit at the equipment terminals. [AS1212]

acrylic plastic Any of a family of synthetic resins made by the polymerization of esters of acrylic acid and its derivatives. [AIR4844]

acrylonitrile butadiene styrene (ABS) An unusually tough type of thermoforming plastic.

actinicity The ability of radiation to induce chemical change.

actinometer *1.* The general name for an instrument used to measure the intensity of radiant energy, particularly that of the sun. *2.* An instrument for measuring the flux density of solar radiation.

activation The chemical process of making a surface more receptive to bonding to a coating or an encapsulating material. [AIR4844]

activation analysis A method of determining composition of composite substances, especially the concentration of trace elements, by bombarding the composite substance with neutrons and measuring the wavelengths and intensities of characteristic gamma rays emitted from activated nuclides.

activator A chemical additive used to initiate a chemical reaction in a specific chemical mixture. [ARP1931]

active An adjective that describes: (a) a system or portion of a system that is in control on-line, in contrast to being in standby; or (b) the operational status of a servo control device after a failure, if the device is activated or remains in control. [ARP4386]

active built in test A type of built-in test that is temporarily disruptive to the prime system operation through the injection of test stimuli into the system. [ARD50010]

active bus interface A bus interface that is connected to the bus media, and is currently capable of participating in bus transactions. Also referred to as bus interface. [AS4710]

active control The automatic activation of various control surface functions in aircraft.

active control technology An airplane design concept in which vehicle performance, weight, and economic characteristics are optimized through a reliance on automatic subsystems within the flight control system to augment the stability of the airplane, to reduce the designing loads (through load reduction or redistribution and structural mode damping), and to manage the configuration of the airplane for aerodynamic efficiency. Active control functions include: pitch stability augmentation, lateral and directional stability augmentation (or both), angle-of-attack limiting, wing-load alleviation, maneuver-load control, gust-load alleviation, flutter-mode control, and ride smoothing. [ARP4386]

active device A type of device that responds to signals separate from the fluid stream that powers the output. [ARP4386]

active maintenance time The time during which corrective or preventive maintenance is being done on an item. Active maintenance time is comprised of the following task times: preparation time, fault-location time, item-obtainment time, fault-correction time, adjustment and calibration time, and checkout time. [ARD50010]

active medium The material in a laser that produces the amplified stimulated emission. The name of the laser identifies the active medium, for example, ruby laser or neon laser.

active repair time That portion of downtime during which one or more technicians are working on the system to effect a repair. [AIR4896]

active satellites Satellites that transmit a signal, in contrast to passive satellites.

active transducer A transducer whose output waves are produced by power derived from a source other than any of the actuating waves, but whose output power is controlled by the actuating waves.

activity Ratio of escaping tendency of the component in solution to that at a standard state. The activity of a component is obtained by multiplying the concentration of the component by an activity coefficient. For example, the ion activity is obtained by multiplying the ion concentration by an activity coefficient.

actual address *See* absolute address.

actual burst pressure *See* burst pressure, actual.

actual special sortie *See* special sortie, actual.

actuarial data Refers to the type of information used to define the subcomponents of an engine or system (e.g., serial numbers) and often includes the bookkeeping of data that indicate the life usage on these subcomponents (e.g., operating hours, LCF counts, hot section usage). [ARP1587]

actuarial program *See* program, actuarial.

actuate To put into action or motion.

actuating system A system (mechanical, gas, or otherwise) that supplies and transmits energy for the operation of other mechanisms or systems. [AIR1489]

actuation Operating or releasing a switch by depressing or releasing its plunger or rotating its shaft. [AIR4077]

actuation signal The setpoint minus the controlled variable at a given instant. Same as error.

actuator *1.* The mechanical link that drives the plunger of a basic switch or the shaft of a rotary switch. [AIR4077] *2.* A device for converting fluid energy into mechanical energy. [ARP4386]. *3.* The component of the actuation system that does work or dissipates energy to control a load. Actuator output is achieved by conversion of energy from the power system or load into mechanical work, torque, or force. [ARP4386] *4.* Mechanism to activate process control equipment, e.g., a valve.

actuator, bellows type A fluid-powered device in which the fluid acts upon a flexible convoluted component, the bellows.

actuator, diaphragm A fluid-powered device in which the fluid acts upon a flexible component, the diaphragm.

actuator, double acting An actuator in which power is supplied in either direction.

actuator, electric A device that converts electrical energy into motion.

actuator, electro-hydraulic type A device that converts electrical energy into motion.

actuator housing *See* brake housing.

actuator, hydraulic A fluid device that converts the energy of an incompressible fluid into motion.

actuator, hydro-mechanical A hydraulically actuated mechanism that incorporates mechanical linkage for modification of output. [ARP4386]

actuator-locking An actuator with integral provisions for locking the output rod or shaft in a given position or positions. [AIR1489]

actuator, piston type A fluid-powered device in which the fluid acts upon a movable piston to provide motion to the actuator stem.

actuator, pneumatic A device that converts the energy of a compressible fluid, usually air, into motion.

actuator, power flight control A flight-control actuator using fluid or electrical power to drive the controlled-output member, usually a ram or rotary shaft. [ARP1281A]

actuator, rotary *See* rotary actuator.

actuator, servo An integral servovalve and hydraulic actuator for use in a control system. [ARP4386]

actuator, signal conversion An actuator capable of converting one or more input signals to a single output which is used to drive a power actuator. [ARP1281A]

actuator, single acting An actuator in which the power supply acts in only one direction, e.g., a spring diaphragm actuator.

actuator, vane type *See* vane actuator.

acute fatigue *See* fatigue, acute/transient.

A/D *See* analog-to-digital and converter.

Ada A programming language based on PASCAL, originally developed on behalf of the US Department of Defense for use in embedded computer systems.

adaptation The adjustment, alteration, or modification of an organism, so as to make it more perfectly suited for existence in its environment.

adapter *1.* A mechanical means for connecting some part of one oxygen system to another oxygen system. [ARP171] *2.* Devices or contrivances used or designed primarily to fit or adjust one thing to another. 3. Devices, appliances, or the like used to alter something so as to make it suitable for a use for which it was not originally designed.

adaptive algorithms Algorithms that change based on the past history of inputs and outputs. [AIR4548]

adaptive control A control system in which response to inputs is adjusted based on previous experience. Automatic means are used to change the type and/or the influence of control parameters in such a way as to improve the performance of the control system.

adaptive flight control system A flight control system having the capability to vary its performance parameters in flight and thus adapt to changing flight conditions so that the structural integrity and stability limitations of the vehicle are not exceeded during critical phases of flight. [ARP4386]

adaptive gain control A control technique that changes the gain of a feedback controller based on measured process variables or controller setpoints.

adaptive optics *1.* Real-time optical correction for atmospheric perturbations and other system error sources. *2.* Optical components whose way of reflecting or refracting light can be changed. In practice, the term usually refers to mirrors with surface shapes that can be adjusted.

adaptive tuning *1.* In a control system, a way to change control parameters according to current process conditions. *2.* The identification of process gains and time that can be used to improve the response of a control loop.

ADC *See* analog-to-digital converter.

A-D converter (ADC) A hardware device that converts analog data into digital form. Also called an encoder.

adder A device that forms, as output, the sum of two or more numbers presented as inputs. Often no data-retention feature is included, i.e., the output signal remains only as long as the input signals are present.

adder-subtractor A device whose output is a representation of either the arithmetic sum and/or difference of the quantities represented by its operand inputs.

addition polymerization A chemical reaction in which simple molecules are added together to form long-chain molecules without forming byproducts. [AIR4844]

additive A material or substance added to something else for a specific purpose.

address *1.* An identification, represented by a name, label or number, of a register or location in storage. Addresses also are part of an instruction word, along with commands, tags, and other symbols. *2.* The part of an instruction that specifies an operand for the instruction.

address bus The highway-linking subcomponents of a microcomputer system along which address data is transferred.

address field That part of an instruction or word containing an address or operand.

address format The arrangement of the address parts of an instruction.

addressing The means whereby the originator or control station selects the unit to which it is going to send a message.

address modification The hardware action of computing the effective operand address of an instruction by some sequence of the following operations as prescribed within the instruction: (a) indexing–adding an index to the address; and (b) indirect addressing–using the intermediate computed address to obtain another address from memory.

address register A register in which an address is stored.

add time The time required for one addition, not including the time required to get and return the quantities from storage.

adducts Chemical compounds with weak bonds, e.g., occlusive or van der Waals bonds.

ADF *See* automatic direction finder.

adhere To cause two surfaces to be held together by adhesion. [AIR4844]

adherent *1.* A body that is held to another body, usually by an adhesive. *2.* A detail or part prepared for bonding. [AIR4844]

adhesion *1.* Bonding between two surfaces. Usually refers to localized welding at high points under substantial contact pressures. *2.* Bonding between two surfaces, assisted by an adhesive substance.

adhesion promoter A material applied to a surface for the purpose of chemically enhancing the adhesion of a sealant to the surface. Adhesion promoters for polysulfide integral fuel tank sealant also contain an organic solvent cleaner. [AIR4069]

adhesive Any substance capable of bonding two materials; can be a liquid, film, or paste.

adhesive age The age of the material measured from the date of shipment by the manufacturer to the present date. [AIR4844]

adhesive batch One production mixture of adhesive by a manufacturer. Each batch has a batch number assigned by the manufacturer. [AIR4844]

adhesive bonding *1.* Bonding accomplished by adding an adhesive coating to the surface of wire or cable and curing the adhesive. Examples are bonding to potting material at the cable end of an electrical connector and bonding to silicone pressure seals. *See* potting. [ARP1931] *2.* A commercial process for fastening parts together in an assembly using only glue, cement, resin, or other adhesive.

adhesive failure Rupture of an adhesive bond such that the separation appears to be at the adhesive-adherent interface. [AIR4844]

adhesive film A synthetic resin adhesive, with or without a carrier fabric, usually of the thermosetting type, in the form of a thin film. Used under heat and pressure in the production of bonded structures. [AIR4844]

adhesive flash The cured adhesive squeezed out around the edges of doublers, at butt splices, and at the ends of an assembly. [AIR4844]

adhesive, intermediate temperature setting *See* intermediate temperature setting adhesive. [AIR4844]

adhesive joint The location at which two adherents or substrates are held together with a layer of adhesive. [AIR4844]

adhesive lot One batch of adhesive, or a portion of one batch, submitted for acceptance at one time. [AIR4844]

adhesive sealing The bonding of faying surfaces to each other through the use of a structural adhesive in the form of a tape or a film. [AIR4069]

adhesive strength *1.* The strength of the bond between a sealant and a substrate. It is desired that the adhesive strength be greater that the cohesion of the sealant. Thus, if an attempt were made to peel the sealant from the surface, the sealant would pull apart before the bond between substrate and sealant would fail. [AS7200/1] *2.* The strength of an adhesively bonded joint, usually measured in tension (perpendicular to the plane of the bonded joint) or in shear (parallel to the plane of the joint).

adhesive, supported *See* supported adhesive film.

adhesive system A compatible primer and film adhesive system for bonding metal-to-metal assemblies; or a primer, pour coat, and film adhesive system for bonding sandwich assemblies. [AIR4844]

adhesive, unsupported A film adhesive that has no carrier cloth. [AIR4844]

ADI *See* attitude director indicator.

adiabatic A process in which an energy change is accomplished on or by a fluid without heat transfer to or from the surroundings. [ARP147C]

adiabatic curing Curing concrete or mortar under conditions whereby heat is neither gained nor lost.

adiabatic demagnetization cooling Cooling through the use of paramagnetic salts cooled to the boiling point of helium in a strong magnetic field, then thermally isolated and removed from the field to demagnetize the salts. Temperatures of 10^{-3} K can be attained through this process.

adiabatic engine Any heat engine that produces power without a gain or loss of heat.

adiabatic temperature The theoretical temperature that would be attained by the products of combustion provided the entire chemical energy of the fuel, the sensible heat content of the fuel, and combustion air above the datum temperature were transferred to the products of combustion. This assumes: (a) combustion is complete; (b) there is no heat loss; (c) there is no dissociation of the gaseous compounds formed; and (d) inert gases play no part in the reaction.

adjacent channel In FM/FM telemetry, the modulated signal bandwidth immediately below or above the channel of interest.

adjacent conductor An insulated conductor next to any other insulated conductor. [ARP1931]

adjacent pitch propeller/variable pitch propeller A propeller with a pitch setting that can be changed in the course of field maintenance, but not when the propeller is rotating. [ARP4107]

adjustable restrictor valve A restrictor valve having provisions for external adjustment. [ARP4386]

adjust/align/trim To bring variable elements of an item within specified limits. [AIR4896]

adjustment The process of altering the value of some circuit element or some component of the mechanism of an instrument, controller, or auxiliary device to bring the indication to a desired value—usually a value corresponding to an independently determined value of the measured variable within a specified tolerance.

adjustment error *See* error, adjustment.

adjustment hardware Any hardware designed for adjusting the size of a torso-restraint system to fit the user, including such hardware that may be integral with a buckle, attachment hardware, or retractor. [AS8043]

adjustment, span In instrumentation, the change in slope of the input-output curve.

adjustment, zero In instrumentation, the reading is zero when the value of the measured quantity is zero.

admixture The addition and homogeneous dispersion of discrete components, before cure. [AIR4844]

adsorb To take up (liquid or gas) on the surface of a solid.

adsorbents Materials that take up gases by adsorption.

adsorption *1.* The adhesion of a thin film of liquid or gas to the surface of a solid substance. The solid does not combine chemically with the adsorbed substance. *2.* The concentration of molecules (of an element or compound) at a phase boundary, usually at a solid surface bounding a liquid or gaseous medium containing the molecules.

advanced composites Composite materials applicable to aerospace or other high-performance component construction and made by embedding high-strength and/or high-modulus fibers within an essentially homogeneous matrix. [AIR4844]

advanced development In testing, to: (a) demonstrate the technical feasibility of a design; (b) determine its ability to meet existing performance requirements; (c) secure engineering data for use in further development; and, where appropriate, (d) establish the technical requirements for contract definition. [AIR4896]

advanced filaments Continuous filaments made from high-strength, high-modulus materials for use as constituents of advanced composites. [AIR4844]

advanced range instrumentation aircraft An NKC-135 aircraft configured for reception recording and real-time relay of telemetry data.

advanced technology laboratory An all-pallet payload utilizing the Space Shuttle and the European Spacelab and designed to accommodate 8 to 15 experiments per mission.

advancing blade Any rotor blade or wing on a rotary-wing aircraft in horizontal motion, moving into the relative wind. [ARP4107]

advection The process of transport of an atmospheric property solely by the mass motion of the atmosphere. Also, the rate of change of the value of the advected property at a given point.

advection fog Fog resulting from the movement of warm, humid air over a cold surface, especially a cold ocean surface. Also, sometimes, steam fog, which results from the transport of cold air over relatively warm water. [ARP4107]

adverse yaw *1.* A negative yawing moment caused by a positive rolling surface deflection. [ARP4386] *2.* Yaw in the opposite sense to that of the roll of the aircraft; for example, a yaw to the left with the aircraft rolling to the right. [ARP4107]

advisory Advice and information provided to assist pilots in the safe conduct of flight and aircraft movement. [ARP4107]

advisory indicating system A system that indicates to the pilot or crew member a safe or normal configuration or operation of essential equipment; or that otherwise attracts attention for routine purposes. [ARP1088]

advisory, safe A signal indicating a safe condition. [AIR1161]

advisory, status A signal indicating a status condition only—not necessarily a safe condition. [AIR1161]

advisory system A system that supplies the crew with information and guidance which they can follow only if they have other reasons that reinforce such information and guidance. [ARP4153]

aeolight A type of glow lamp whose intensity of light output varies with an applied signal voltage. Its components include a cold cathode and an envelope filled with a mixture of gases.

aerate *1.* To supply or charge liquid with a gas. *2.* To expose to the circulation of air.

aerator Any device for injecting air into a material or process stream.

aero- Prefix designating air, gas, or the atmosphere.

aeroassist Changing orbit size by utilizing aerobraking, aerocapture, or aeromaneuvering.

aerobatic confidence maneuvers Aerobatics intended to increase pilot skills. [ARP4107]

aerobatic demonstration Aerobatics intended to increase or demonstrate pilot skill, or to demonstrate aircraft capabilities. [ARP4107]

aerobatics Preplanned, precisely executed flight maneuvers in which the aircraft exceeds either 60 deg of bank or 30 deg of pitch. [ARP4107]

aerobiology The study of the distribution of living organisms freely suspended in the atmosphere.

aerobraking Changing orbit size by using the upper atmosphere to create drag.

aerocapture Making use of the atmosphere of a planet or planetary satellite by capturing the object and reducing the orbit size so that it remains in orbit or lands on the body.

aerodynamic An incomplete term popular among nontechnical people for describing a shape with low drag. Preferable term would be aerodynamically clean or well-streamlined.

aerodynamic angle of attack *See* absolute angle of attack.

aerodynamic center Measured along the chord of an airfoil, the reference point about which the pitch moment coefficient does not change with angle of attack. For most airfoil shapes the aerodynamic center is approximately 25% of the chord aft of the leading edge.

aerodynamic coefficient A dimensionless coefficient that quantifies the magnitude of aerodynamic forces and moments, as well as airflow conditions. Lift coefficients and drag coefficients are examples of aerodynamic coefficients.

aerodynamic force The force exerted by a moving, gaseous fluid upon a body completely immersed in it. [AIR1489]

aerodynamic heating The rise in skin temperature of an aircraft due to friction of air at high speed. [ARP4386]

aerodynamic interface plane An instrumentation plane used to define inlet distortion and performance. [ARP1420]

aerodynamic missile A missile that uses aerodynamic forces to maintain its flight path, generally employing propulsion guidance and a winged configuration. [ARP4386]

aerodynamics The science that deals with the motion of air and other gaseous fluids, and the forces acting on bodies when they move through such fluids, or when such fluids move against or around the bodies.

aerodynamic stability *See* stability, aerodynamic.

aeroelasticity The study of the response of structurally elastic bodies to aerodynamic loads.

aeroelastic research wings Wings designed with less-than-normal stiffness, used to test devices that suppress flutter.

aeroembolism The formation or liberation of gases in the blood vessels of the body, as brought on by a too-rapid change from a high, or relatively high, atmospheric pressure to a lower one.

aerograph Any self-recording instrument carried aloft to take meteorological data.

aerology The study of the free atmosphere throughout its vertical extent, as distinguished from studies confined to the layer of the atmosphere adjacent to the earth's surface.

aeromaneuvering Changing orbit size and/or plane by entering the upper atmosphere to create drag and/or lift.

aeromaneuvering orbit to orbit shuttle (AMOOS) Proposed reusable upper stage for the Space Shuttle, superseded by the orbit transfer vehicle.

aerometer An instrument for determining the density of air or other gases.

aeronautical beacon A visual NAVAID displaying flashes of white and/or colored light, to indicate the location of an airport, a heliport, a landmark, a certain point of a Federal airway in mountainous terrain, or an obstruction. [ARP4107]

aeronautics The science of the design, construction, and operation of aircraft.

aeronomy The study of the upper regions of the atmosphere, where ionization, dissociation, and chemical reactions take place.

aerosol *1.* Submicron particles suspended in air, gas, or vapor. *2.* A fog, fume, or smoke. [AIR4783]

aerospace *1.* The region consisting of the earth's atmosphere and outer space. *2.* Adjective referring to flight operations within the atmosphere as well as space flight outside the atmosphere.

aerospace safety The engineering assessment and analysis of systems, subsystems, and functions of spacecraft, missiles, advanced aircraft, and ground support in order to identify hazards associated with such systems, and to design procedures that eliminate those hazards or determine tolerable safety levels.

aerospace technology transfer Technology transfer germane to aircraft and space vehicles, their propulsion, guidance, etc.

aerospace vehicles Vehicles capable of flight within and outside the sensible atmosphere.

aerothermodynamics The study of aerothermodynamic phenomena at sufficiently high gas velocities that thermodynamic properties of gas are important.

aerothermoelasticity The study of the response of elastic structures to the combined effects of aerodynamic heating and loading.

aerozine A rocket fuel consisting of a mixture of hydrazine and unsymmetrical dimethylhydrazine (UDMH).

AFC (control) *See* automatic frequency control.

AFCS *See* automatic flight control system. [ARP4386]

AFCS per-engage synchronization The process of biasing a new input signal to correspond with the AFCS output position prior to switching to the new input signal. [ARP4386]

affective states Subjective feelings of pleasantness or unpleasantness that a person has about aspects of his/her environment, other people, or himself/herself. Affective states are subdivided into emotions and moods depending on their duration and intensity. [ARP4107]

affective states, emotions *1.* Affective states that tend to be disruptive of mental, physiological, or behavioral processes. Emotions are relatively brief in duration but strong in intensity. *2.* A mental state, characterized by strong feeling and accompanied by motor expression, that is related to some object or external situation. [ARP4107]

affective states, mood *1.* A relatively mild emotional state, enduring or recurrent. *2.* An echo of an emotional reaction with or without remembrance of the original stimulus. [ARP4107]

afterbodies *1.* Companion bodies that trail satellites. *2.* Sections or pieces of rockets or spacecraft that enter the atmosphere unprotected behind nose cones or other bodies that are protected for entry. *3.* Afterparts of vehicles.

afterburners *See* afterburning.

afterburning *1.* Irregular burning of fuel left in the firing chamber of a rocket after cutoff. *2.* The function of an afterburner, a device for augmenting the thrust of a jet engine by burning additional fuel in the uncombined oxygen in the gases from the turbine.

afterglow *1.* Broad, high arches of radiance or glow seen occasionally in the western sky above the highest clouds in deepening twilight, caused by the scattering effect of very fine particles of dust suspended in the upper atmosphere. *2.* The transient decay of a plasma after the power has been turned off. *3.* Luminosity that persists in a gas after an electrical discharge passes through it. This phenomenon is sometimes utilized in flow measurement.

AGC (control) *See* automatic gain control.

age-control For an elastomeric material, the designation of a specific maximum period of age after the cure-date that will assure desired conformance characteristics. Age control is based on the premise that elastomers deteriorate upon exposure to ozone, oxygen, heat, sunlight, rain, and other similar environmental factors. [AS1933]

aged beta A beta matrix in which alpha, typically fine, has precipitated as a result of aging. [AS1814]

age exploration A systematic evaluation of an item based on analysis of collected information from in-service experience, development tests, or scientific handbooks. Age exploration assesses the resistance of an item to a deterioration process with respect to increasing age. [ARD50010]

age hardening Raising the strength and hardness of an alloy by heating a supersaturated solid solution at a relatively low temperature to induce precipitation of a finely dispersed second phase. Also known as aging or precipitation hardening.

ageing The effect on materials of exposure to an environment for an interval of time. [AIR4844]

age sensitive The characteristic of an elastomer that makes it subject to deterioration by oxygen, ozone, sunlight, heat, rain, and similar factors present in normal environmental exposure subsequent to vulcanization. [AS1933]

age, threshold The time at which inspection of the condition of an item is required.

agglomerate Two or more particles in intimate contact, which cannot be separated by gentle stirring or by the small shear forces thus generated. [ARP1179]

agglomeration Any process for converting a mass of relatively fine solid material into a mass of larger lumps.

aggregate Natural sand, gravel, or crushed stone that is mixed with cement to make mortar or concrete.

aging *1.* Soaking solution-heat-treated parts at a moderately elevated temperature (or, for some alloys and tempers, soaking at room temperature) to precipitate alloying elements from solid solution and thereby develop strength and corrosion-resistance properties. [MAM2771] *2.* Alteration of the characteristics of a device due to use. *3.* Operating a product before shipping it to stabilize component functions or detect early failures. *4.* Any time-dependent change in properties of a material—especially age hardening at room- or slightly elevated temperatures. *5.* Curing or stabilizing parts or materials by long-term storage outdoors or under closely controlled storage conditions.

agitation Movement of parts or material, or circulation of liquid media around parts or material. [ARP1917]

agitator A device for mixing, stirring, or shaking liquids or liquid-solid mixtures to keep them in motion.

AGL *See* above ground level.

agricultural aircraft Light aircraft specially equipped for agricultural applications such as crop dusting.

agRISTARS project A multiagency program utilizing Landsat remote sensing data to predict crop yields and land use, and to detect pollution.

agrophysical units Geographic areas defined for statistical purposes by AgRISTARS personnel. The boundaries of agrophysical units are based on natural rather than political lines for the purpose of comparing similar agricultural regions.

AGT *See* automated guideway transit vehicles.

AHDL *See* analog hardware description language.

AI *See* artificial intelligence.

AIAA Acronym for American Institute of Aeronautics and Astronautics.

aiding load A force or torque on the actuator, provided by load restoration and/or inertia, that acts in the same direction as the desired direction of load motion. *See also* opposing load. [ARP4386]

aileron A control surface set into or near the trailing edge of an airplane wing to produce roll moments about the longitudinal axis. [ARP4386]

aileron angle The angular displacement of the aileron surface relative to the wing surface. [ARP4386]

aileron control A control for producing a rolling moment on the airplane through the use of ailerons. [ARP4386]

aileron yaw Yawing moment caused by aileron deflection. [ARP4386]

air *1.* A port for application of pressurized air. Applicable to hydropneumatic accumulators and air-pressurized reservoirs. [ARP4386] *2.* The mixture of oxygen, nitrogen, and other gases, which, with varying amounts of water vapor, forms the atmosphere of the earth.

air atomizing oil burner A burner for firing oil in which the oil is atomized by compressed air. The compressed air is forced into and through one or more streams of oil, breaking the oil into a fine spray.

air bearing Gas bearing using air as working fluid.

air-bearing table Table supported on a spherical air bearing, and thus free to tilt, without sensible friction, to any attitude within design constraints.

air bind An air pocket in a pump, conduit, or piping system that prevents liquid from flowing past it. Also called a liquid trap.

air binding The inclusion of air in a space, hindering the flow of some other gas or liquid.

air bleed Air bled from the compressor of a gas turbine engine. [ARP147C]

airborne *1.* Carried in the atmosphere—either by being transported in an aircraft or by being dispersed in the atmosphere. *2.* Pertaining to personnel, equipment, etc. transported by air. *3.* Pertaining to material being or designed to be transported by aircraft, as distinguished from weapons and equipment installed in and remaining a part of the aircraft. *4.* Describes an aircraft from the instant it becomes entirely sustained by air until it ceases to be so sustained. [ARP4386]

airborne integrated reconnaissance system (AIRS) Aerial reconnaissance system incorporating various modes of detection.

airborne radar approach The use of airborne radar for aircraft approach control—the radar cursor technique.

airborne static electric power inverter Equipment, or a combination of equipment, used in aircraft to convert direct current (DC) electric power to 400 Hz alternating current (AC) electric power. [AS8023]

airborne support equipment All equipment required on the aircraft to support the operation and maintenance of the aircraft and all of its airborne equipment. [AIR4896]

airborne windshear alerting system A device or system that identifies the presence of windshear after it is encountered. Alerting devices of this type do not provide guidance information to the pilot. [ARP4109]

airborne windshear autoflight recovery system A system that integrates or couples the autopilot and the autothrottle systems of the aircraft with an airborne windshear flight guidance system. [AIR4102/11]

airborne windshear automatic recovery system A device or system that integrates or couples autoflight systems of the aircraft with an airborne windshear flight guidance system. [ARP4109]

airborne windshear detection and avoidance system A device or system that detects a severe windshear phenomenon far enough in advance of the encounter—in both the takeoff/climbout profile and the approach/landing profile—to allow the pilot to avoid the windshear. [ARP4109]

airborne windshear flight guidance system A system that provides the crew with flight guidance to improve recovery probability in windshear encounter. [AIR4102/11]

airborne windshear situational display A display that presents pertinent windshear information, such as flight path and stall margins; may be available in conjunction with alerting, guidance, and/or detection/avoidance systems. Airborne winshear situational displays may supply windshear severity information to the pilot on a continuous basis; however, they do not provide guidance information. [ARP4109]

airborne windshear system System that alerts the crew of the presence of windshear either before or after the phenomenon is encountered, and/or aids in the avoidance of or escape from the hazard. [AIR4102/11]

air bottle A container suitable for storing a quantity of air under pressure. May be made of steel, titanium, fiberglass, or other suitable material, with or without a plastic or aluminum liner. [ARP906A]

air breathing boosters Boosters that take in oxygen for their engines through inlets, rather than carrying oxygen on board, as in a conventional rocket. Air-breathing boosters are possible substitutes for rocket engines.

air breathing missile A missile with an engine that requires the intake of air for fuel combustion, as in a ramjet or turbojet. In contrast, a rocket missile carries an oxidizer and can operate beyond the atmosphere. [ARP4386]

air-bubbler liquid-level detector A device for indirectly measuring the level of liquid in a vessel—especially a corrosive liquid, viscous liquid, or liquid containing suspended solids. The device consists of a standpipe, open at the bottom and closed at the top, which is connected to an air supply. The air supply pressure is maintained slightly above the maximum head of liquid in the vessel. Air bubbles out of the bottom of the pipe, maintaining the internal pressure equal to the head of liquid in the vessel. Pressure is measured by a simple gage or transducer. Also known as purge-type liquid-level detector.

air-bubbler specific-gravity meter Any of several devices that measure specific gravity by determining differential pressure between two air-purged bubbler columns. The devices ordinarily use either of two principles for determining specific gravity: (a) comparison of sample density with density of a known liquid; or (b) comparison of pressure between two bubbler columns immersed at different depths in the process liquid.

air-bubble void Air entrapment within and between the plies of reinforcement, or within a bond line or encapsulated area. Air-bubble voids are localized, non-interconnected, and spherical in shape. [AIR4844]

air, cabin *1.* Air flowing into the cabin. *2.* Air in the cabin proper. The condition of cabin air is normally determined at the point where the air leaves the cabin. [ARP171]

air, compartment Air in a compartment proper. The condition of compartment air is normally determined at a point where the air leaves the compartment. [ARP147C]

air compressor A machine that raises the pressure of air above atmospheric pressure and normally delivers it to an accumulator or distribution system.

air condenser *1.* A heat exchanger for converting steam to water; the heat-transfer fluid is air. Also known as an air-cooled condenser. *2.* A device for removing oil or water vapors from a compressed-air line.

air conditioned area *See* area, air conditioned.

air conditioning Controlling the atmospheric environment in a confined space by measuring and continually adjusting factors such as temperature, humidity, air motion, and concentrations of dust, gases, odors, pollen, or microorganisms.

air-cooled engine An engine, such as an internal combustion engine, whose waste heat is removed directly by a flowing stream of air—either a stream blown across the external surfaces of the engine, or a stream blown through internal cooling passages.

air-cooled heat exchanger A device for removing heat from a process fluid by passing it through a bank of finned tubes that are cooled by blowing or drawing a stream of air across the tube exteriors.

air, cooling A stream of air used as a heat sink. [ARP171]

aircraft Any machine that can be supported for flight in the air by buoyancy or the effects of the air against its surfaces. [ARP4107]

aircraft accident An occurrence associated with the operation of an aircraft which takes place between the time any person boards the aircraft with the intention of flight and the time when all such persons have disembarked, and in which any person suffers death or serious injury, or in which the aircraft receives substantial damage. [ARP4107]

aircraft approach category A way of grouping aircraft based on a speed of 1.3 times their stall speed, in the landing configuration, at maximum gross landing weight. For example, aircraft whose stall speed (with landing gear, flaps, etc., in proper position for landing) multiplied by 1.3 is less than 91 knots are classified as Category A aircraft; those whose stall speed multiplied by 1.3 is at least 91 knots but less than 121 knots are classified as Category B aircraft; and so on, up to Category E aircraft, whose stall speed multiplied by 1.3 is 166 knots or more. [ARP4107]

aircraft braking environment That environment the aircraft experiences during the landing and braking phase of operations, including: runway length, width, surface texture, slope, crown, and contamination level; wind conditions, temperature and pressure; and touchdown point proficiency. [AS483A]

aircraft construction materials A general term designating the materials used in manufacturing an aircraft.

Aircraft Energy Efficiency (ACEE) program *See* ACEE program.

aircraft environmental control A system that provides an environment that is controlled within specified operational limits of comfort and safety, for humans, animals, and equipment. These limits may include the following: air pressure, temperature, humidity, velocity, and composition; ventilation rate; wall temperature; and audible noise and vibration. The environmental control system encompasses the air-conditioning system but is broader and does not necessarily include air alone as the controlled medium. [AIR746A]

aircraft gas turbine engine Any gas turbine engine used for aircraft propulsion or power generation, including those commonly called turbojet, turbofan, turboprop, or turboshaft type engines; also includes afterburning engines in the nonafterburning mode of operation. [AIR1533]

aircraft ground equipment Equipment required on the ground to support the operation and maintenance of an aircraft and its airborne equipment. *See also* ground support equipment. [ARP4386]

aircraft, incipient skid *See* incipient skid.

aircraft, individual wheel control *See* individual wheel control.

aircraft in service The average number of aircraft used in aircraft operations and normal maintenance during a reporting period. [AIR4896]

aircraft intermediate maintenance department That department of an aviation ship or naval air station responsible for the check, test, repair, or manufacture of aeronautical components, and SE for the supported aircraft. [AIR4896]

aircraft multiplexing A means of increasing the number of signals transmitted between units, while decreasing the number of signal paths. The result is a significant reduction of wire bundles and weight in the aircraft. (Interior communication (signal flow) in an aircraft is conventionally accomplished by point-to-point wiring between the various electronic and electrical equipment.) [AIR1207]

aircraft noise prediction *See* noise prediction (aircraft).

aircraft on ground (AOG) The highest priority designation to process a requirement for a spare part(s) and/or maintenance action. Indicates that an aircraft is unable to continue or be returned to revenue service until the appropriate action is taken. [AIR4896]

aircraft operational period The time interval between the start of preparation for flight and post-flight engine shutdown, with consequent deactivation of the aircraft electric system. [AS1212]

aircraft power supplies Electrical sources for the normal operation of aircraft.

aircraft refueling adapter In the aircraft fuel system, the mating portion of the quick disconnect, by which the hydrant servicer (nozzle) is connected to the aircraft, in the aircraft fuel system. [AIR4782]

aircraft refueling pressure The fuel pressure controlled as near as possible to the aircraft adapter during a refueling operation. [AIR4783]

aircraft required horsepower The total power required by all engines to generate the mission element. [ARP1352]

aircraft runup Final engine check prior to take-off.

aircraft, slip ratio The ratio of reduction in wheel-rotational velocity under the influence of braking forces to the equivalent free-rolling rotational velocity of the unbraked wheel. [AS483A]

aircraft spin A prolonged stall in fixed-wing aircraft, characterized by a sustained spiral descent, usually with the nose down.

aircraft structural integrity program A time-phased set of required actions performed at the optimum time during the life cycle (design through phase-out) of an aircraft system to ensure the structural integrity (strength, rigidity, damage tolerance, durability, and service life capability) of the aircraft. [ARD50010]

aircraft subsystems Lesser systems that are components of major aircraft systems. For example, subsystems of the hydraulic system include landing gear, brakes, wing flaps, nose-wheel steering, and speed brakes. NOTE: The terms "system" and "subsystem" are often used synonymously. [ARP4107]

aircraft systems Major components of the aircraft which operate from a common source of power, provide a common power source to similarly powered components, or perform a major function encompassing lesser functions or components. Examples include hydraulics, electric systems, flight control systems, avionics, engine power systems, fuel systems, and all-weather systems. *See* aircraft subsystems. [ARP4107]

aircraft utilization *See* utilization, aircraft.

aircraft vectoring *1.* The act of directing control of in-flight aircraft through control commands of azimuth headings. *2.* The act of changing direction of the aircraft in pitch and yaw. [ARP4386]

aircraft, wheel speed detectors, generators AC or DC generators containing bearings, and which output an AC or DC voltage signal, respectively, as a function of wheel speed. [AS483A]

aircraft, wheel speed detectors, inductive type System containing a sensor and exciter ring. The sensor is a magnetic field device whose reluctance is varied as a function of wheel speed by a rotating exciter ring, thus resulting in a frequency proportional to wheel speed. In inductive systems, special attention must be given during installation to air-gap tolerances, effect of wheel deflection to provide suitable signal amplitude compared to controller input saturation for reliable wheel-speed detection. [AS483A]

aircraft, wheel speed detectors, inertia type A hub cap or rim-driven, rotating masses on overriding clutches operating switch(es) or hydraulic/pneumatic valve(s). [AS483A]

air cushion *1.* A mechanical device that uses trapped air to absorb shocks or arrest motion without shock. *2.* The partly confined stream of low-pressure, low-velocity air that supports a vehicle known as an air-cushion vehicle, ground-effect machine, or hovercraft, and allows it to travel equally well over water, ice, marshland, or relatively level ground.

air cushion landing systems (ACLS) Landing systems based on the ground effect principle, whereby a stratum of air (in place of landing gear) is utilized as the aircraft ground-contacting medium.

air cycle machine *1.* The basic component of an air-cycle refrigeration system. The device consists of an air-expansion turbine, which is connected by a shaft to a compressor and/or a fan. The power generated in the expansion turbine is used to drive the compressor and/or the fan. [ARP147C] *2.* A turbine-fan assembly, a turbine-compressor assembly, or a turbine-fan-compressor assembly. Air cycle machines may also have multiple turbines. [AIR4073]

air cylinder A cylindrical body for storing compressed air, for compressing air with a piston, or for driving a piston with compressed air.

air deficiency Insufficient air, in an air-fuel mixture, to supply the oxygen theoretically required for complete oxidation of the fuel.

air drag The drag exerted by air particles on a moving object, for example, on an aircraft or rocket moving through the air, or on an earth satellite that comes within the earth's atmosphere during orbit. [ARP4386]

air dry Air in which no water vapor is included. This term is used comparatively, since in nature there is always some water vapor included in air, and such water vapor—being a gas—is dry.

air drying The drying of material at ordinary room temperature without the use of artificial heat. [AIR4844]

air ejector A device for removing air or noncondensable gases from a confined space (e.g., the shell of a steam condenser) by eduction using a fluid jet.

air eliminator A device installed in a fueling system which allows for automatic removal of trapped air. [AIR4783]

air embolism A dangerous condition characterized by abnormal formation of an air bubble inside the body. Most frequently induced by transpulmonic pressure causing air insertion into one or more of the following: (a) arteries and brain, as in arterial and cerebral embolism; (b) lungs, as in mediastinal and subcutaneous emphysema; (c) pneumopericardium; and/or (d) pneumothorax. [ARP171]

air entrainment Artificial infusion of a semi-solid mass, such as concrete or a dense slurry, with minute bubbles of air, especially by mechanical agitation.

airfield, expeditionary An extension of SATS that provides a surfaced runway 4000 feet long and 96 feet wide, and parking/maintenance areas for up to four squadrons of aircraft. The field includes catapults and primary recovery systems identical to SATS, as well as three Fresnel Lens Optical Landing Systems (FLOLS), two M-21 emergency recovery systems, and expanded field lighting and communications systems compatible with the expanded capability. *See also* SATS. [AIR1489]

airfield, forward-area An airfield that must support the operation of liaison, observation, and light-transport type aircraft, including heavy-cargo helicopters, for a period ranging from a few days to three weeks. [AIR1489]

airfield, heavy-load An airfield that must support heavy bomber-type aircraft. The load-carrying capacity of pavements for this type of airfield is equivalent to a main gear load of 265,000 lb on a four-wheel, dual-twin configuration having tire contact areas of 267 sq in. for each wheel, twin spacing of 37 in. center-to-center, and inside wheels of twins spaced 62 in. center-to-center. [AIR1489]

airfield index An index of soil-shearing resistance obtained using an airfield penetrometer. [AIR1780]

airfield, light-load An airfield that must support fighter- and medium-cargo-type aircraft. The load-carrying capacity of pavements for this type of airfield is equivalent to a main gear load of 25,000 lb on a single wheel having a tire contact area of 100 sq in.

airfield, medium load An airfield that must support heavy-cargo-, tanker-, and medium-bomber-type aircraft. The load-carrying capacity of pavements for this type of airfield is equivalent to a main gear load of 100,000 lb on a two-wheel, twin configuration having tire contact areas of 267 sq in. for each wheel and wheel spacing of 37 in. center-to-center. [AIR1489]

airfield penetrometer A probe-type instrument designed to measure soil strength. [AIR1780]

airfield, rear-area An airfield that normally must support the operation of heavy-cargo aircraft, medium-cargo aircraft, and fighter-bomber aircraft for a period of four to six months. Airfields of this class must be constructed, rehabilitated, extended, and maintained by engineer construction battalions and must usually be located in the Zone of Communications or in the Army rear area. The strength characteristics of the rear-area airfield normally govern the landing gear flotation design for heavy-cargo and fighter-bomber aircraft. The controlling rear-area airfield is characterized as a field having the equivalent of a T11 landing-mat surface laying directionally on a 4-CBR subgrade. [AIR1489]

airfield, support-area An airfield that normally must support the operation of medium-cargo aircraft (and, conceivably, certain fighter-bomber aircraft designed for close tactical support) for a period of from two weeks to one month. [AIR489]

airfield, theater of operations Theater-of-operations(TO) airfields are limited-life facilities which represent the maximum construction capability of engineering troop units in the field, considering time limitations imposed by the tactical situation and available construction equipment and surfacing materials. The TO airfield classes are defined as rear-area, support-area, and forward-area airfields; and light-VTOL landing areas. [AIR1489]

airfield, zone of interior Zone of interior (ZI) airfields are permanent facilities constructed in accordance with the criteria given in Air Force Manual 88-6. Pavements may be either rigid (concrete); flexible (bituminous); or a combination thereof. The ZI airfield classes, as defined for Air Force construction, not only represent a range of flotation capabilities but also directly represent the designs on which most existing military airfields are based. Thus, the relations presented have a direct application to existing airfields, in addition to providing a basis for comparison of proposed new aircraft landing gear designs with those of existing aircraft. The ZI airfield classes are defined as heavy-load, medium-load, or light-load airfields. [AIR1489]

air filter A device for removing solid particles, such as dust or pollen, from a stream of air, especially a device in which the airstream is caused to pass through layered porous material, such as cloth, paper, or screening.

airflow correlation A method that uses cell airflow to calculate the thrust corrections required to determine a correlation factor. [ARP741]

air flow factor *See* air flow parameter.

air flow function *See* air flow parameter.

air flow parameter A mathematical expression of gas flow, in units of mass flow per unit time, through a pneumatic starter nozzle under choked-flow conditions. Also called air flow factor, air flow function, and corrected flow. [ARP906A]

airfoil *1.* A surface or body, such as a wing, propeller blade, or rudder, especially designed to obtain a reaction, such as lift or thrust, from the air through which it moves. [ARP4386] *2.* The cross-sectional profile shape of a lifting surface.

airfoil-vane fan A device for creating a stream of moving air by drawing it into a fan casing near the hub and propelling it centrifugally with a rotor whose vanes are curved backward from the direction of rotation.

airframe *1.* The structural components of an airplane, including the framework and skin of various parts. *2.* The framework, envelope, and cabin of an airship. *3.* The assembled principal structural components—less propulsion system, control, electronic equipment, and payload. [ARP4386] *4.* The fuselage, booms, nacelles, cowlings, fairings, airfoil surfaces (including rotors, but excluding propellers and rotating airfoils or engines), and landing gear of an aircraft, and their associated accessories and controls. [ARP4107] *5.* The assembled structural and aerodynamic components of an aircraft or rocket vehicle which support the different systems and subsystems integral to the vehicle.

airframe manufacturer The manufacturer responsible for providing to the galley buyer or user the airframe structure and provisions for aircraft galleys. [AS1426]

airframe mounted accessory drive (AMAD) An accessory drive package mounted in the airframe remote from the engine, but driven by the engine or an APU. Usually shaft- or pneumatically driven. Sometimes called airframe mounted accessory gearbox (AMAG). [ARP906A]

airframe mounted accessory gearbox *See* airframe mounted accessory drive.

air-free The descriptive characteristic of a substance from which air has been removed.

air-fuel ratio The ratio of the weight, or volume, of air to fuel.

air furnace Any furnace whose combustion air is supplied by natural draft, or whose internal atmosphere is predominantly heated air.

air gage A device for precisely measuring physical dimensions by measuring the pressure or flow of air from a nozzle against a workpiece surface, and relating the measurement to distance from the nozzle to the workpiece.

air gap The space between two ferromagnetic elements of a magnetic circuit.

airglow The quasi-steady radiant emission from the upper atmosphere, as distinguished from the sporadic emission of the auroras.

air handling unit An assembly of air conditioning equipment, usually confined within a single enclosure, that treats air prior to distribution and provides the means of propelling the treated air through the distributing ducts.

air-hardening steel A type of tool steel containing sufficient alloying elements to permit it to harden fully on cooling in air from a temperature above its transformation temperature. Also known as self-hardening steel.

air infiltration The leakage of air into a setting or duct.

air law The body of domestic and/or international laws dealing with regulations and liabilities in civil or military aviation.

airline operating profile Information supplied by the airline to the equipment manufacturer for calculation of spare items needed to support the fleet of the airline. [AIR4896]

airline transport pilot (ATP) A pilot who has the privileges of a commercial pilot with an instrument rating, and is certified to serve as pilot in command for passenger-carrying operations of a turbojet airplane, of an airplane having ten seats or more (excluding the pilot seats), or of a multiengine airplane operated by a commuter air carrier. [ARP4107]

air locks *1.* Surface depressions on a molded part, caused by trapped air between the molded surface and the plastic. [AIR4844] *2.* A stoppage or diminishment of flow in a fuel system, hydraulic system, or the like, caused by pockets of air or vapor. *3.* Chambers capable of being hermetically sealed which provide for passage between two places of different pressure, for example, between an altitude chamber and the outside atmosphere. *4.* An intermediate chamber between an environmentally controlled confined space and the outside atmosphere which provides for entry of personnel and materials by sealing a door between the chamber and the confined space, opening a door to the outside to admit personnel or materials, closing and sealing that door, changing environmental conditions in the chamber to match those in the confined space, then opening an interior door to permit entry into the confined space. (The process is reversed when exiting the confined space.) *See also* air bind.

AIRMET Acronym for airman's meteorological information. In-flight weather advisories issued only to amend the area forecast concerning weather phenomena that are of operational interest to all aircraft, and potentially hazardous to aircraft having limited capability because of lack of equipment, instrumentation, or pilot qualifications. AIRMETs cover weather of lesser severity than that covered by SIGMETs or convective SIGMETs. AIRMETs cover moderate icing; moderate turbulence; sustained winds of 30 knots or more at the surface; widespread areas of ceilings less than 1000 ft and/or visibility less than 3 miles; and extensive mountain obscurement. *See* SIGMET and convective SIGMET. [ARP4107]

air meter A device for measuring the flow of air or other gas, in which flow is expressed as weight or volume per unit time.

air moisture The water vapor suspended in the air.

air monitor A warning device that detects airborne radioactivity or chemical contamination and sounds an alarm when the radiation, or gas or vapor level of the contaminant exceeds a preset value.

air motor An engine that produces rotary motion using compressed air or other gas as the working fluid.

air nozzle An air port having direction and appreciable length for directing an air stream.

air permeability A method of measuring the fineness of powdered materials, such as portland cement, by determining the ease with which air passes through a defined mass or volume of the material.

airplane An engine-driven, fixed-wing, heavier-than-air aircraft that is supported in flight by the dynamic reaction of air against its wings. [ARP4107]

airplane reference performance The takeoff performance that a particular airplane can be expected to demonstrate under the expected airfield and environmental conditions, in the absence of failures, and when operated using normal procedures. [AS8004]

air port An opening through which air passes.

airport advisory area The area within ten miles of an airport that does not have a control tower (or where the tower is not in operation) and on which a flight service station (FSS) is located. In such cases, the FSS provides advisory service to arriving and departing aircraft. [ARP4107]

airport marking aids Markings used on runway and taxiway surfaces to identify a specific runway, a runway threshold, a centerline, a holdline, or other designated area. A runway should be marked in accordance with its present usage, for example: (a) visual; (b) nonprecision instrument; and (c) precision instrument. [ARP4107]

airport rotating beacon A visual NAVAID operated at many airports. At civil airports, alternating white and green flashes indicate the location of the airport. At military airports, the beacons flash alternately white and green, but are differentiated from civil beacons by dual-peaked (two quick) white flashes between the green flashes. [ARP4107]

airport security Organization of trained security personnel, surveillance and screening devices, and procedures used for the protection of airport and airline property, aircraft, passengers, employees, and visitors from injury, air piracy, and other unauthorized acts.

air surveillance radar (ASR) Approach control radar used to detect and display the position of an aircraft in the terminal control area. [ARP4107]

airport traffic area Unless specifically designated in FAR Part 93, that airspace within a horizontal radius of five statute miles from the geographical center of any airport at which a control tower is operating. The airport traffic area extends from the surface up to, but not including, an altitude of 3000 ft above the elevation of the airport. [ARP4107]

air-position indicator An aircraft navigation device that integrates headings and speeds to give a continuous, dead-reckoned indication of position with respect to the surrounding air mass.

air preheater A heat exchanger for transferring some of the waste heat in flue gases from a boiler or furnace to incoming air, thereby increasing the efficiency of combustion.

air-puff blower A soot blower that is automatically controlled to deliver intermittent jets or puffs of compressed air to remove ash, refuse, or soot from heat-absorbing surfaces.

air purge The removal of undesired matter by replacement with air.

air, recirculated Air in enclosed spaces or equipment cooling systems which is recirculated by fans or blowers. [ARP171]

air regenerated Air that has been regenerated and made suitable for respiration by removing excess carbon dioxide, water vapor, odor, or other contaminants; and by adding oxygen. [ARP171]

air regulator A device for controlling airflow; for example, a damper to control flow of air through a furnace, or a register to control flow of heated air into a room.

air resistance The opposition to the passage of air through any flow path.

air route traffic control center (ARTCC) A facility established to provide ATC service to aircraft operating on IFR flight plans within controlled airspace, principally during the en route phase of flight. [ARP4107]

AIRS (reconnaissance system) *See* airborne integrated reconnaissance system.

air, saturated *See* saturated air.

airships Propelled and steerable dirigibles dependent on gases for flotation.

airsickness *See* motion sickness.

air slew missiles Solid-propellant rockets utilizing thrust vector control.

airspace *1.* The atmosphere in which aircraft operate, extending upward from the surface of the earth. [ARP4107] *2.* The atmosphere above a particular portion of the earth, usually defined by the boundaries of an area on the surface projected perpendicularly upward.

airspeed The speed of an aircraft relative to its surrounding air mass. *See* airspeed, indicated; airspeed, calibrated; airspeed, equivalent; and airspeed, true. [ARP4386]

airspeed, calibrated Indicated airspeed corrected for instrument installation error. [ARP4386]

airspeed control A function or mode of operation that provides constant airspeed by regulating the throttles in response to signals from an airspeed sensor. [ARP419]

airspeed, equivalent Calibrated airspeed corrected for compressibility error. [ARP4386]

airspeed, indicated The airspeed shown by an airspeed indicator. [ARP4386]

airspeed mach indicator (**AMI**) An aircraft instrument that displays the speed of the vehicle as a ratio of the true airspeed to the speed of sound. [ARP4107]

airspeed sensor A device that provides output signals proportional to airspeed pitot-static pressures. [ARP419]

airspeed, true Equivalent airspeed corrected for error due to air density (altitude and temperature); the true speed of the vehicle through the air. [ARP4386]

air spring A device commonly used instead of a mechanical spring in heavy vehicles to support the body of the vehicle on its running gear. In an air spring, the energy-storage element is an air-filled container with an internal elastomeric bellows or diaphragm.

air, standard sea level Dry air at 15°C (59°F) and at a pressure of 101.3 kPa (29.92 in Hg) absolute. [ARP171]

airstart envelope The region (of altitude and airspeed) in which airstarting is performed. The concept of an airstart envelope is the same as that of a flight envelope. [ARP906A]

air starting Air starting is engine starting while in flight and is accomplished by using various starting methods, such as spooldown, windmilling, or starter-assist. [ARP906A]

air supply (**AS**) *1.* The supply of air used in pneumatic instrumentation for a power. *2.* Plant air supply (PA). *3.* Instrument air (IA).

air taxi *1.* A term used to describe helicopter/ VTOL aircraft movement conducted above the surface, but normally not above 100 ft above ground level. [ARP4107] *2.* In air commerce, the carriage of persons or property for compensation or hire, as a commercial operator, in aircraft having a maximum seating capacity of less than 20 passengers or a maximum payload of less than 6000 pounds. [ARP4107]

air thermometer A device for measuring temperature in a confined space by detecting variations in pressure or volume of air in a bulb inside the space.

airtight *1.* Sealed to prevent passage of air or other gas. *2.* Impervious to leakage of gases across a boundary.

air traffic clearance (**ATC clearance**) An authorization by air traffic control for an aircraft to proceed within controlled airspace under traffic conditions specified to prevent collisions with known traffic. *See also* ATC instructions. [ARP4107]

air traffic control (**ATC**) A service operated by an appropriate authority, military or civil, to promote the safe, orderly, and expeditious flow of air traffic. [ARP4107]

air traffic control radar beacon system (**ATCRBS**) A radar system in which the object to be detected is fitted with cooperative equipment in the form of a radio receiver/transmitter (transponder). Radar pulses transmitted from the searching transmitter/ receiver(interrogator) site trigger a distinctive transmission from the transponder. This reply transmission, rather than a reflected signal, is then received back at the interrogator site for processing and display at an air traffic control facility. In this way, the responding object (usually an aircraft) can be identified as well as detected. [ARP4107]

air traffic control specialist (controller) A person authorized to provide air traffic control service. [ARP4107]

air turbine starter An engine-starting device that incorporates a turbine wheel driven by cold or heated compressed air or other gas. The starter also generally includes an inlet scroll, reduction gears, an engaging mechanism or clutch, and a switch or speed signal to signal cutoff speed. It may also include an emergency disconnect. [ARP906A]

air vehicle Any object that, by virtue of lift obtained as a result of its own buoyancy in air, or by the dynamic reaction of air over and around its surfaces, or by its reaction to a jet stream, goes through the air as a carrier of something. [AIR4073]

air vent Small outlet to prevent entrapment of gases in a molding or tooling fixture. [AIR4844]

air vessel An enclosed chamber, partly filled with pressurized air, that is connected to a piping system to counteract water hammer or to promote uniform flow of liquid.

airway A control area or portion thereof established in the form of a corridor through airspace with specified width and height. The centerline of the airway is defined by radio navigational aids.

airway beacon A lighted visual device used to mark airway segments in remote mountain areas. The light flashes Morse code to identify the beacon site. [ARP4107]

airworthiness The condition of an item (an aircraft, aircraft system, or part) in which that item operates in a safe manner to accomplish its intended function. [ARP4761]

airworthiness certificate A certificate issued by the FAA, a designee, or the airworthiness authority of another nation certifying that an aircraft met, at the time of inspection, current airworthiness standards. [ARP4107]

airworthy In a condition suitable for safe flight. [ARP4107]

airy disk The central bright spot produced by a theoretically perfect circular lens or mirror. The spot is surrounded by a series of dark and light rings, produced by diffraction effects.

AIT *See* autoignition temperature.

Aitken nuclei Microscopic particles in the atmosphere which serve as condensation nuclei for droplet growth during the rapid adiabatic expansion produced by an Aitken dust counter.

alarm *1.* An abnormal process condition. *2.* The sequence state when an abnormal process condition occurs. *3.* An instrument, such as a bell, light, printer, or buzzer, that indicates when the value of a variable is out of limits.

alarm point The level at which an environmental condition or process variable exceeds some predetermined value.

alarm severity A selection of levels of priority for the alarming of each input, output, or rate of change.

alarm system An integrated combination of detecting instruments and visible or audible warning devices that actuates when an environmental condition or process variable exceeds some predetermined value.

alarm valve A device that detects water flow and sounds an alarm when an automatic sprinkler system is activated.

albedo The ratio of the amount of electromagnetic radiation reflected by a body to the amount incident upon it, often expressed as a percentage; for example, the albedo of the earth is 34%.

aldehydes Carbonyl groups to which a hydrogen atom is attached (–CHO).

alert Indicator (visual, auditory, or tactile) that provides information to the crew in a timely manner about a non-normal situation. [ARP4153] *See also* process condition and sequence state.

alert box In data processing, a window that appears on a computer screen to alert the user of an error condition.

alert height The minimum height above ground level at which the pilot is supposed to make the go-around if a system failure is detected. [ARP4107]

AlGaAs *See* aluminum gallium arsenides.

algae Any plants of a group of unicellular and multicellular primitive organisms that include the Chlorella, Scenedesmus, and other genera.

algal bloom *See* algae.

algebraic adder An electronic or mechanical device that can automatically find the algebraic sum of two quantities.

ALGOL *See* Algorithmic-Oriented Language.

algorithm A step-by-step procedure for solving a problem or accomplishing some end. [ARP1587] .

algorithmic language A language designed for expressing algorithms.

Algorithmic-Oriented Language (ALGOL) An international procedure-oriented language.

alias For sampled signals, a frequency that cannot be distinguished from another frequency by analysis of their equally spaced values.

aliasing *1.* A phenomenon of sampled signals. Myquist's theorem states that a signal that is sampled N times per second can be reconstructed completely from the samples if it does not contain significant components at frequencies equal to or greater than H/2 Hz. *2.* An error introduced by a sampling measurement system if the sampling rate is not at least double the highest frequency component of the signal being measured. Aliasing prevents determination of the frequency of the signal, and can result in a measurement offset. [AIR1872] *3.* A peculiar problem in data sampling, whereby data are not sampled enough times per cycle, and the sampled data cannot be constructed.

aliasing error 1. An inherent error in time-shared telemetry systems in which improper filtering is employed prior to sampling. 2. See folding error.

alidade *1.* An instrument used in the plane-table method of topographic surveying and mapping. *2.* Any sighting device for making angular measurements.

alighting gear A general term that includes all equipment or components connected with alighting on or landing on water, land, shipboard etc. Alighting gear includes wheels, shock struts, floats, and skis. *See also* landing gear. [AIR1489]

aligning torque The footprint torque which, in a rolling tire, resists rolling in a curvilinear path, i.e., tends to align the roll into a straight path. Also called self-aligning torque. [AIR1489]

alignment Performing the adjustments that are necessary to return an item to specified operation. [ARP4386]

aliphatic Organic compounds in which carbon atoms are arranged in an open or straight chain. [AIR4844]

aliquot A representative sample of the particle suspension in the calibration liquid. Multiple aliquots have identical characteristics. [ARP1192]

alkali Substance that neutralizes acids to form a salt and water. [AIR4844]

alkali metals Metals in group IA of the periodic table; namely, lithium, sodium, potassium, rubidium, cesium, and francium.

alkaline cleaner An alkali-based aqueous solution for removing soil from metal surfaces.

alkalinity *1.* The state of being alkaline. *2.* Represents the amount of carbonates, bicarbonates, hydroxides, and silicates or phosphates in the water and is reported as grains per gallon, or ppm as calcium carbonate.

alkali vapor lamps Lamps in which light is produced by an electric discharge between electrodes in an alkali vapor at low or high pressure.

alkalosis, respiratory A condition of the blood in which there is a decrease of the carbonic acid fraction of blood relative to the bicarbonate fraction, with resultant increase in pH. This occurs as a result of excessive elimination of carbon dioxide, as in hyperventilation. Also called hypocapnia. [ARP171]

alkyd plastic Thermoset plastic based on resins composed principally of polymeric esters, in which the recurring ester groups are an integral part of the main polymer chain, and in which ester groups occur in most cross-links between chains. [AIR4844]

alkylation A gasoline refining process. [AIR4783]

Allen screw A screw or bolt that has a hexagonal socket in its head, and is turned by inserting a straight or bent hexagonal rod into the socket.

all fire The minimum stimulus that must be applied to an EED for initiation under specified environmental conditions and reliability require-ments. [AIR913]

alligatoring *1.* Cracking in a film of paint or varnish characterized by broad, deep cracks extending through one or more coats. Also known as crocodiling. *2.* Surface roughening of very coarse-grained sheet metal during form-ing. *3.* Longitudinal splitting of flat slabs in a plane parallel to the rolled surface; occurs dur-ing hot rolling. Also called fishmouthing.

allobar A form of an element that has a distri-bution of isotopes different from the distribu-tion in the naturally occurring form. An allobar thus has a different apparent atomic weight than the naturally occurring form of the element.

allotropy The existence of a substance—espe-cially an element—in two or more forms. [AIR4844]

allowable free play Maximum allowable uncontrolled rotation of the wheels about the shock strut centerline due to tolerances and/or wear. [AIR1752]

allowable interface load The maximum capa-bility of a structure (airplane or galley) to react to a load without structural failure. Defined as a force in pounds at the structure attachment in interface-point airplane coordinates. [AS1426]

allowance Specified difference in limiting sizes, either minimum clearance or maximum inter-ference between mating parts. Computed math-ematically from the specified dimensions and tolerances of both parts.

alloy *1.* In plastics, a blend of polymers or copolymers with other polymers or elastomers under selected conditions. [AIR4844] *2.* Sub-stance having metallic properties and being composed of two or more chemical elements, of which at least one is an elemental metal. [ARP1931]

alloy-depleted surface layer A loss of alloy-ing elements at the surface, caused by using a suitable etchant. [AMS5914]

alloy family Refers to variations within a single alloy designation. [MAM2771]

alloy steel An alloy of iron and carbon that also contains one or more additional elements

intentionally added to increase hardenability or to enhance other properties.

all-pass network A network designed to intro-duce phase shift or delay into an electronic sig-nal without appreciably reducing amplitude at any frequency.

allyl plastic A thermoset plastic based on res-ins made by addition polymerization of mono-mers containing allyl groups. [AIR4844]

Alnico Any of a series of commercial iron-based permanent-magnet alloys containing varying amounts of aluminum, nickel, and cobalt as the chief alloying elements. The Alnicos are char-acterized by their ability to produce a strong magnetic field for a relatively small magnet mass, and to retain their magnetism with rela-tively insignificant loss in field strength when the magnetizing field is removed.

ALNOT Stands for alert notice; a message sent by a flight service station (FSS) or air-route traffic-control center (ARTCC) requesting an extensive communication search for over-due, unreported, or missing aircraft. [ARP4107]

aloha system A multiple random-access com-munications scheme in which there is a nonfixed allocation of channel capacity, so that the channel is available to any terminal when-ever it has a packet ready for transmission.

ALPA Acronym for Airline Pilots Association.

alpha The allotrope of titanium with a hexago-nal, close-packed crystal structure. [AS1814]

alpha 2 structure In titanium, a structure con-sisting of an ordered alpha phase found in highly stabilized alpha. Defined by x-ray dif-fraction, not optical metallography. [AS1814]

alpha-beta structure In titanium, a microstruc-ture that contains both alpha and beta as the principal phases at a specific temperature. It is composed of alpha, transformed beta, and retained beta. [AS1814]

alpha case The oxygen-, nitrogen-, or carbon-enriched, alpha-stabilized surface that results from exposure at elevated temperature to environments containing these elements. Alpha case is normally and hard, brittle, and is con-sidered detrimental. [AS1814]

alpha counter *1.* A system for detecting and counting energetic alpha particles. It consists of an alpha counter tube, amplifier, pulse-height discriminator, scaler, and recording or indicating mechanism. *2.* A term sometimes loosely used to describe the alpha counter tube or chamber.

alpha decay The radioactive transformation of a nuclide by alpha-particle emission.

alpha double prime A supersaturated nonequilibrium orthorhombic phase formed by a diffusionless transformation of the beta phase in certain alloys. Alpha double prime occurs when cooling rates are too high to permit transformation by nucleation and growth. It may be strain-induced during working operations and may be avoided by appropriate in-process annealing treatments. Also referred to as orthorhombic martensite. [AS1814]

alpha emitter A radionuclide that disintegrates by emitting an alpha particle from its nucleus.

alphanumeric display Letters and numerals used to show identification, altitude, beacon code, and other information concerning a target on a radar display, as used in automated radar terminal systems (ARTS). Also referred to as a data block. [ARP4107]

alpha particle A positively charged particle emitted from the nuclei of certain atoms during radioactive disintegration.

alpha prime In titanium, a supersaturated, acicular, nonequilibrium hexagonal phase formed by a diffusionless transformation of the beta phase. Alpha prime occurs when cooling rates are too high to permit transformation by nucleation and growth. It exhibits an aspect ratio of 10:1 or greater. Also known as martensite or martensite alpha. [AS1814]

alpha radiation *See* alpha particles.

alpha-ray spectrometer An instrument used to determine the energy distribution in a beam of alpha particles.

alpha stabilizer For titanium, an alloying element that dissolves preferentially in the alpha phase and raises the alpha-beta transformation temperature. Aluminum is the most commonly used alpha stabilizer. Interstitial elements such as oxygen and nitrogen are also potent alpha-stabilizing elements. [AS1814]

alpha-transus In titanium, the temperature that designates the phase boundary between the alpha and alpha-plus-beta fields. [AS1814]

alpine meteorology Wind, precipitation, atmospheric physics, and other climatological phenomena peculiar to the Alps and/or other similar mountainous areas.

ALS *See* approach light system.

altazimuth A sighting instrument having both horizontal and vertical graduated circles so that both azimuth and declination can be determined from a single reading. Also known as an astronomical theodolite or universal instrument.

alteration switch A manual switch on the computer console, or a program-simulated switch, which can be set "on" or "off" to control coded machine instructions. Also called sense switch.

alternate An item that fully meets required functional and structural specifications, but differs either in overall external dimensions, connections, installations and/or mounting provisions, and required additional parts, rework, or modification to install in a specific application. [AIR4896]

alternate code complement In a frame synchronization scheme, complementing of a frame synchronization pattern on alternate frames to give better synchronization.

alternate immersion test A type of accelerated corrosion test in which a test specimen is repeatedly immersed in a corrosive medium, then withdrawn and allowed to drain and dry.

alternate power sources Power sources such as auxiliary power units, emergency power units, and batteries. [AS1831]

alternating current (AC or a-c) An electrical current that reverses direction at regular intervals. The rate of this reversal is expressed as hertz (cycles per second). [ARP1931]

alternating-current bridge A bridge circuit that utilizes an a-c signal source and a-c null detector. In an alternating-current bridge, generally, both in-phase (resistive) and quadrature (reactive) balance conditions must be established to balance the bridge; however, some bridges require only one balance (resistive or reactive) and use a phase-sensitive detector.

alternating current generators *See* AC generators.

alternating stress *1.* A stress that varies between two maximum values that are equal but of opposite signs, according to a law determined in terms of time. *2.* Changes between stresses of different levels. [AIR4844]

alternating stress amplitude A test parameter of a dynamic fatigue test, equal to one-half the algebraic difference between the maximum and minimum stress in one cycle. [AIR4844]

alternator rating The full-load value expressed in kVA at a power factor of 0.8. [ARP1148A]

alternators *See* AC generators.

alternator speed The nominal speed at which the alternator operates to produce 400 Hz. [ARP1148A]

altigraph A recording pressure altimeter.

altimeter An instrument for determining height of an object above a fixed level or reference plane (e.g., sea level). The aneroid altimeter and the radio altimeter are the most common types.

altimeter setting The barometric pressure reading used to adjust a pressure altimeter for variations in the local mean sea-level reference atmospheric pressure, or to adjust to the standard altimeter setting (29.92 in. Hg). [ARP4107]

altitude *1.* The maximum height in feet above sea level at which a unit must operate and maintain characteristics within recommended limits. [ARP1148A] *2.* In astronomy, angular displacement above the horizon. *3.* Height, especially radial distance as measured above a given datum, for example, average sea level.

altitude acclimatization A physiological adaptation to reduced atmospheric and oxygen pressure.

altitude control A function, or mode of operation that provides constant altitude by controlling the pitch attitude of the aircraft in response to signals from an altitude sensor. [ARP419]

altitude, density The altitude corresponding to a given density in a standard atmosphere. [ARP171]

altitude, equivalent or cabin The standard altitude at which atmospheric pressure is equal to the cabin pressure. [ARP147C]

altitude, pressure *See* pressure altitude.

altitude restriction An altitude or altitudes that are to be maintained until reaching a specific point or time, stated in the order in which they are to be flown. Such restrictions are issued by ATC to ensure proper altitude separation of traffic or clearances from other known hazards. [ARP4107]

altitude sensor A barometric device that provides output signals proportional to the deviation from a reference pressure altitude. [ARP419]

altitude sickness In general, any sickness brought on by exposure to reduced oxygen tension and barometric pressure.

altitude signals Reflected radio signals returned to an airborne electronic device from the land or sea surface directly underneath the vehicle.

altitude, standard The altitude corresponding to the temperature and pressure tabulated in an accepted standard atmosphere table. [ARP147C]

ALU *See* arithmetic and logical unit.

alum General name for a class of double sulfates containing aluminum plus another cation, such as potassium, ammonium, or iron.

alumina The oxide of aluminum—Al_2O_3.

aluminides Intermetallic compounds of aluminum and a transition metal.

aluminizing Forming a protective coating on metal by depositing aluminum on the surface; or reacting surface material with an aluminum compound, and diffusing the aluminum into the surface layer at elevated temperature.

aluminum A soft, white metal which, in pure form, exhibits excellent electrical conductivity and oxidation resistance. Aluminum is the base metal for an extensive series of lightweight structural alloys used in aircraft frames and skin panels.

aluminum arsenides Binary compounds of aluminum with negative, trivalent arsenic.

aluminum boron composites Structural materials composed of aluminum alloys reinforced with boron fibers (filaments).

aluminum casting, premium grade Aluminum castings that have guaranteed mechanical properties at drawing-designated areas. [ARP1917]

aluminum conductor An aluminum wire or group of wires, not insulated from one other, which carry electrical current. [ARP1931]

aluminum gallium arsenides (AlGaAs) Compounds exhibiting characteristics suitable for use in laser devices, light-emitting diodes, solar cells, etc.

aluminum graphite composites Structural materials composed of aluminum alloys reinforced with graphite.

alveolar duct A portion of the terminal air passages of the lung from which air sacs (alveoli) arise. [ARP171]

alveolar membrane Thin layer of tissue that partitions the air in the alveoli from capillary blood, and through which gas exchange occurs between blood and alveolar air. [ARP171]

alveolar ventilation The air that enters the alveoli. In principle, an amount of air equal to the tidal volume minus the physiological dead space times the respiratory rate. [ARP171]

AM *See* amplitude modulation.

AMAD *See* airframe mounted accessory drive.

AMAD disconnect *See* disconnect, AMAD.

AMAG *See* airframe mounted accessory drive.

amalgamation *1.* Forming an alloy of any metal with mercury. *2.* A process for separating a metal from its ore by extracting it with mercury in the form of an amalgam. The process was formerly used to recover gold and silver, which are now extracted chiefly through the cyanide process.

ambient A surrounding or prevailing condition, especially one that is not affected by a body or process contained in it.

ambient air The air that surrounds the equipment. The standard ambient air for performance calculations is air at 59°F, 60% relative humidity, and a barometric pressure of 29.921 in. Hg. This translates to a specific humidity of 0.013 lb of water vapor per lb of air.

ambient background Background light on a display face other than that generated by the display. [ARP1782]

ambient conditions The environment of an enclosure (room, cabinet, etc.) surrounding a given device or piece of equipment.

ambient illumination Light that illuminates the face of a display and its surroundings from sources such as the sky or cockpit and cabin lighting. [ARP1782]

ambient orientation *1.* A means of maintaining gross orientation in space without "thinking" about it. Ambient orientation is the result of the preconscious level of awareness. *2.* Keeping track of various sensory inputs, including visual, tactile, kinesthetic, and auditory modes, in order to keep a person oriented with respect to the horizon. [ARP4107]

ambient pressure The static pressure surrounding a component. [ARP4386]

ambient temperature *1.* Temperature of the environment in which an apparatus is working. [ARP4386] *2.* The temperature of the medium, such as air, water, or earth, into which the heat of a device is dissipated. [ARP1199]

ambiguous information Information that can be understood in a different sense than that intended. Also referred to as erroneus information. [AS8034]

American standard pipe thread A series of specified sizes for tapered, straight, and dryseal pipe threads, established as a standard in the United States. Also known as Briggs pipe thread.

American standard screw thread A series of specified sizes for threaded fasteners, such as bolts, nuts, and machine screws, established as a standard in the United States.

American Wire Gauge (AWG) A standard system for identifying the physical size of wire. [ARP1931]

AMI *See* airspeed mach indicator.

Amici prism Also known as a roof prism. A right-angle prism in which the hypotenuse has been replaced by a roof, in which two flat faces meet at a 90° angle. The prism performs image erection, while deflecting the light by 90°. This is the same as rotating the image by 180° (reversing it left to right) and at the same time inverting it top to bottom.

amine resin A synthetic resin derived from the reaction of urea, thiourea, melamine or similar compounds with aldehydes, particularly formaldehyde. [AIR4844]

amino Indicates the presence of an NH_2 or NH group. [AIR4844]

amino plastics Plastics based on resins made by the condensations of amines, such as urea and melamine, with aldehydes. [AIR4844]

amino-silane finish Finish applied to glass fibers to give a good bond to epoxide, phenolic, and melamine resins. [AIR4844]

ammeter An instrument for determining the magnitude of an electric current.

ammonia (NH₃) A pungent, colorless, gaseous compound of hydrogen and nitrogen. Ammonia is readily soluble in water, where it reacts to form the base ammonium hydroxide.

AMOOS *See* aeromaneuvering orbit to orbit shuttle.

amorphous film A film of material deposited on a substrate for corrosion protection, insulation, conductive properties, or a variety of other purposes. An amorphous film is noncrystalline and can be deposited by evaporation, chemical deposition, or by condensation. The method employed for deposition is dictated by the composition and ultimate use of the film.

amorphous plastic A plastic that has no crystalline component. [AIR4844]

amount of unbalance Quantitative measure of unbalance in a rotor (with reference to a plane), without referring to its angular position. It is obtained by multiplying the unbalance mass by the distance of its center of gravity from the shaft axis. [ARP587]

amp (ampere) (A or a) *1.* Metric unit for electric current produced by one volt acting through a resistance of one ohm. *2.* The current that will deposit silver at the rate of 0.001118 grams per second, with the current flowing at 1 coulomb per second.

ampere-hour A quantity of electricity equal to the amount of electrical energy passing a given point when a current of one ampere flows for one hour.

amp-hour meter An integrating meter that measures electric current flowing in a circuit and indicates the integral of current with respect to time.

ampere per meter The SI unit of magnetic field strength. One ampere per meter is the field strength developed in the interior of an elongated, uniformly wound coil excited with a linear current density in the winding of one ampere per meter of axial distance.

amphiboles A group of dark, rock-forming, ferromagnesian silicate minerals closely related in crystal form and composition.

amphoteric Having the property of behaving as an acid or a base according to the condition of the reaction. [AIR4844]

amplification *1.* Increasing the amplitude of a signal by using a signal input to control the amplitude of a second signal supplied from another source. *2.* The ratio of the output-signal amplitude from an amplifier circuit to the input-signal amplitude from the control network, both expressed in the same units.

amplification factor The μ factor for plate and control electrodes of an electron tube when the plate current is held constant.

amplifier *1.* An active fluidic component that provides a variation in output signal greater than the impressured control-signal variation. The polarity of the output signal may be either positive or negative relative to the control signal. The level of the control signal may be greater or less than the output level. [ARP993A] *2.* Any device that can increase the magnitude of a physical quantity, such as mechanical force or electric current, without significant distortion of the wave shape of any variation with time associated with the quantity.

amplifier, laser A device that amplifies the light produced by an external laser, but lacks the mirrors needed to sustain oscillation and independently produce a laser beam.

amplitude *1.* The extent of vibratory movement measured from the mean position to an extreme. *2.* The maximum departure of alternating voltage or current from the average value; indicated by vertical height on an A-scan presentation. [ARP5089] *3.* The maximum value of the displacement of a wave or other periodic phenomenon from a reference position. *4.* The angular distance north or south of the prime vertical; the arc of the horizon, or the angle at the zenith between the prime vertical and a vertical circle, measured north or south from the prime vertical to the vertical circle.

amplitude distortion A condition in an amplifier or other device in which the amplitude of the output signal is not an exact linear function of the input (control) signal.

amplitude-frequency response *See* frequency response.

amplitude modulation (AM) In general, modulation in which the amplitude of a wave is the feature subject to variation.

amplitude noise Random fluctuations in the output of a light source or signal caused by other generating or detecting means.

amplitude ratio (AR) *1.* The ratio of the sinusoidal-flow amplitude to the sinusoidal current amplitude at a particular frequency divided by the same ratio at zero frequency or some specified low frequency. [ARP4386] *2.* The ratio of the control-flow amplitude to the input-current amplitude at a particular frequency divided by the same ratio at the same input-current amplitude at a specified low frequency (usually 5 or 10 Hz). Amplitude ratio (AR) may be expressed in decibels, where dB = $20 \log_{10}(AR)$. [ARP490]

amplitude response A measure of the time taken for a defined change of amplitude.

AM rejection The removal of unwanted amplitude modulation of a signal. Usually performed by using signal clipping or limiting circuitry.

AMS spec Abbreviation for Aerospace Materials Specification, published by SAE, Aerospace Division. [AIR4069]

A/M station (automatic/manual station) In control systems, a device that enables the process operator to manually position one or more valves. A single-loop station enables manual positioning of a single valve; a shared station enables control of multiple valves; and a cascade station provides control of paired loops.

AMTV *See* automated mixed traffic vehicles.

anaerobic Free of air or uncombined oxygen.

anaerobic adhesive An adhesive that cures only in the absence of air after being confined between assembled parts. [AIR4844]

anaerobic sealant A faying-surface sealant in which cure is inhibited by exposure to air. The sealant can be applied to the faying surface and will not cure until the surfaces are brought together. [AIR4069]

analog *1.* A variable quantity, or instrument for measuring it, which varies continuously over some defined operating range. [AS5116] *2.* The general class of devices or circuits whose output is a continuous function of its input. [ARP993] *3.* The representation of numerical quantities by means of physical variables, such as translation, rotation, voltage, or resistance. *4.* Contrasted with digital, a waveform is analog if it is continuous and varies over an arbitrary range.

analog back-up An alternate method of process control by conventional analog instrumentation in the event of a failure in the computer system.

analog channel A channel on which the information transmitted can take any value between the limits defined by the channel. Voice channels are analog channels.

analog computer A computer that works on the principle of measuring (rather than counting) in which the input data is analogous to a measurement continuum, such as linear lengths, voltages, or resistances which can be manipulated by the computer.

analog control Implementation of automatic control loops with analog (pneumatic or electronic) equipment. Contrast with direct digital control.

analog control system Classically, a system that consists of electronic or pneumatic single-loop analog controllers, in which each loop is controlled by a single, manually adjusted device.

analog data Data represented in a continuous form, as contrasted with digital data represented in a discrete, discontinuous form. Analog data are usually represented by means of physical variables, such as voltage, resistance, and rotation.

analog device A mechanism that represents numbers by physical quantities, e.g., by lengths, as in a slide rule; or by voltage or currents, as in a differential analyzer or an analog computer.

analog electronic controller Any of several adaptations of analog computers to perform control functions. An analog electronic controller may produce an output signal directly related to the difference between a measured value and a predetermined setpoint, or it may produce an output signal modified by rate-of-change or other feedback signals.

analog hardware description language (AHDL) A modeling language capable of representing both the structural and behavioral properties of analog circuits. NOTE: "Structural" refers to the connectivity or net-list properties of a circuit; "behavioral" refers to the mathematical equations for individual components.

analog information Information generated in analog form from a source. [AIR4911]

analog input *1.* A continuously variable input. *2.* A termination panel used to connect field wiring from the input device.

analog input module I/O module that converts a process voltage or current signal into a multiple-bit form for use in the PC. The signal is the analog of some process variable.

analog millivoltmeter Galvanometer-type instrument with pointers and scales that may read in millivolts or directly in degrees. Since the deflection of a galvanometer is a function of the current flowing through it, the reading of a particular millivoltmeter-thermocouple system depends on the total resistance of the circuit that includes the instrument. [AIR46]

analog output A continuously variable output (generally 4-20 mA or 3-15 psi).

analog output module I/O module that converts a multiple-bit number calculated in the PC to a voltage or current output signal for use in control.

analog signal A continuously variable representation of a physical quantity, property, or condition, such as pressure, flow, temperatures. An analog signal may be transmitted as pneumatic, mechanical, or electronical energy.

analog simulation The calculation of the time or frequency domain response of electrical circuits to input stimulus. Through analog simulation, a set of simultaneous equations associated with circuit topology is assembled and solved.

analog system *1.* A continuous time system. *2.* A system in which all signals can change continuously with time. [ARP89]

analog-to-digital (A/D) A device or subsystem that changes real-world analog data (as from transducers) to a form compatible with binary (digital) processing (as done in a microprocessor).

analog-to-digital converter (ADC) Any unit or device used to convert analog information to approximate corresponding digital information.

analogue readout A readout on a continuous scale whose readability depends on the length of the scale. [ARP587]

analysis *1.* An evaluation based on decomposition into simple elements. [ARP4754] *2.* That part of the field of mathematics which arises from calculus, and which deals primarily with functions. *3.* Quantitative determination of the constituent parts.

analysis criticality A procedure by which each potential failure mode is ranked according to the combined influence of severity and probability of occurrence. [ARD50010]

analysis, damage tolerance Application of engineering principles to determine periods of safe, unrepaired service usage in the presence of assumed structural defects. [AIR4896]

analysis, maintenance Process of identifying required maintenance functions through analysis of a fixed or assumed design, and determining the most effective means of accomplishing these functions. [AIR4896]

analysis, performance Assessment of the performance level of an item, aircraft, or fleet from performance data, or measurements, or statistical information. [AIR4896]

analysis program That part of EASY which performs the ECS transient analysis. [AIR1823]

analysis, reliability The assessment, by means of statistical studies, of probabilities used to determine satisfactory performance of an item under specified conditions of use over a given service period. [AIR4896]

analysis, ultimate Chemical analysis of solid, liquid, or gaseous fuels. In the case of coal or coke, ultimate analysis is the determination of carbon, hydrogen, sulfur, nitrogen, oxygen, and ash.

analyte The substance that is being identified in an analysis.

analytical balance Any weighing device having a sensitivity of at least 0.1 mg.

analytical curve In spectrographic analysis, graphical representation of some function of relative intensity plotted against some function of concentration.

analytical gap In spectroscopic analysis, the separation between the source electrodes in a spectrograph.

analytical line In spectrographic analysis, the spectral line of an element used to determine its concentration.

analytical maintenance program A program that implements the reliability-centered maintenance (RCM) philosophy; for example, a program that holds that additional maintenance cannot improve the reliability inherent in the design of hardware. [AIR4896]

analytical redundancy A concept wherein an estimate of a physical parameter is computed by combining and filtering information from other sensed data which have a known physical relationship to the signal of interest. [ARP4386]

analytical scale In spectroscopic analysis, the scale that results when an analytical curve is projected onto the intensity axis. The analytical scale is often used in lieu of an analytical curve to permit direct reading of spectral intensity as element concentration.

analyzer *1.* Any of several types of test instruments; usually one that can measure several different variables either simultaneously or sequentially. *2.* In an absorption refrigeration system, the component that allows the mixture of water and ammonia vapors leaving the generator to come in contact with the relatively cool ammonia solution entering the generator.

anchor bond That portion of the beam lead which adheres to the device dielectric layer and has the required beam-lead thickness. [AS1346]

AND A logic operator having the property that if P is an expression, Q is an expression, R is an expression..., then the AND of P, Q, R...is true if all expressions are true, false if any expression is false.

Anderson bridge A type of a-c bridge especially suited to measuring the characteristics of extremely low-Q coils.

andesite Volcanic rock composed essentially of andesine and one or more mafic constituents.

AND gate A basic electronic circuit used in microprocessor systems. With an AND gate, a logical 1 value on output is produced only if all of the inputs have logical 1 values.

anechoic chamber A test room having all surfaces lined with a sound-absorbing material. Also known as dead room.

anelasticity A characteristic exhibited by certain materials in which strain is a function of both stress and time. No permanent deformations are involved, but a finite time is required to establish equilibrium between stress and strain in both the loading and unloading directions. [AIR4844]

anemia A quantitative or qualitative deficiency of the blood. This may be a reduction in the total number of red blood cells, and/or a reduction of the hemoglobin content within each cell (hemoglobinemia). [ARP171]

anemobiagraph A recording pressure-tube anemometer, such as a Dines anemometer, in which springs are used to make the output from the float manometer linear with wind speed.

anemoclinometer An instrument for determining the inclination of the wind to a horizontal plane.

anemometer A device for measuring wind speed. If the device produces a recorded output, it is known as an anemograph.

anemoscope A device for indicating wind direction.

aneroid *1.* A sealed, flexible, evacuated chamber that expands when exposed to a reduced ambient pressure and contracts when subjected to increased ambient pressure. *See* control, barometric. [ARP171] 2. Describing a device or system that does not contain or use liquid.

angel In radar meteorology, an echo caused by physical phenomena not discernible to the eye. Such phenomena have been observed when abnormally strong temperature and/or moisture gradients are known to have existed. They are also sometimes attributed to insects or birds flying in the radar beam. [ARP4107]

angle The inclination to each other of two intersecting lines. Measured by the arc of a circle intercepted between the two lines forming the angle, the center of the circle being the point of intersection.

angle beam In ultrasonic testing, a longitudinal wave from an ultrasonic search unit that enters the test surface at an acute angle.

angle modulation A type of modulation involving the variation of carrier-wave angle in accordance with some characteristic of a modulating wave. Angle modulation can take the form of either phase modulation or frequency modulation.

angle of attack The angle between the wing chord plane and the relative wind.

angle of attack indicator An instrument that indicates the angle between the wing chord plane and the relative wind. [ARP4107]

angle of cord (tires) The angle at which the cords in adjacent layers of fabric are set in the body or carcass of a tire. [AIR1489]

angle of elevation The angle between a horizontal plane and the line of sight to an object lying above the plane of the observer.

angle of extinction The phase angle of the stopping instant of anode current flow in a gas tube with respect to the starting instant of the corresponding half-cycle of anode voltage.

angle of ignition The phase angle of the starting instant of anode current flow in a gas tube with respect to the starting instant of the corresponding positive half-cycle of anode voltage.

angle of incidence *1.* The acute angle between a fixed reference, usually the longitudinal axis of the aircraft, and the chord of a wing or other airfoil. [ARP4107] *2.* The angle formed between the direction of propagation of a ray of incident radiation and a normal to the surface it strikes. For a reflected wave, the angle of reflection and the angle of incidence are equal.

angle of repose A characteristic of bulk solids equal to the maximum angle with the horizontal at which an object on an inclined plane will retain its position without tending to slide. The tangent of the angle of repose equals the coefficient of static friction.

angle of unbalance Given a polar coordinate system fixed in a plane perpendicular to the shaft axis, and rotating with the rotor, the polar angle at which an unbalance mass is located with reference to the given coordinate system. [ARP587]

angle-ply laminate *1.* A laminate having fibers of adjacent plies oriented at alternating angles. *2.* Any filamentary laminate that is not uniaxial. *3.* A laminate built from orthotropic plies that is symmetrical with respect to the laminate mid-plane, and in which the ply orientations of adjacent plies are alternating. [AIR4844]

angle of yaw *See* yawed angle.

angle valve A valve design in which one port is colinear with the valve stem or actuator, and the other port is at a right angle to the valve stem.

angle wrap Refers to tape fabric wrapped on a starter dam mandrel at an angle to the centerline. [AIR4844]

Angstrom Abbreviated A. A unit of length defined as 1/6438.4696 of the wavelength of the red line in the Cd spectrum. One angstrom equals almost exactly 10^{-10} meter. This unit was once used almost exclusively for expressing wavelengths of light and x-rays, but it has now been largely replaced by the SI unit nanometer, or 10^{-9} meter.

angstrom unit The unit used to define the short wavelengths of the electromagnetic spectrum, such as visible light, ultraviolet light, and x-rays. [AIR4844]

angular acceleration The rate of change of angular velocity.

angular accelerometer A device for measuring the rate of change of angular velocity between two objects.

angular frequency A frequency expressed in radians per second. Angular frequency equals two times the frequency in Hz.

angular misalignment The maximum included angle between the two axes of the coupling halves. [AIR4728]

angular momentum The product of a body's moment of inertia and its angular velocity.

angular momentum flowmeter A device for determining mass flow rate in which an impeller turning at constant speed imparts angular momentum to a stream of fluid passing through the meter. A restrained turbine located just downstream of the impeller removes the angular momentum, and the reaction torque is taken as the meter output. Under proper calibration conditions, the reaction torque is directly proportional to mass flow rate. Also called an axial flowmeter.

angular resolution Specifically, the ability of a radar to distinguish between two targets solely by the measurement of angles.

angular velocity The change of angle per unit time; specifically, the change in angle of the radius vector per unit time. [AIR1489]

anhydride 1. A mixture from which water has been extracted. 2. An oxide of a metal or nonmetal that forms an acid or base when mixed with water.

anhydrous Describes a chemical or other solid substance whose water of crystallization has been removed.

aniline A substance produced from coal tar or the indigo plant which is used to make inks, dyes, and plastics. Also called phenyamine.

anion The negatively charged ion which, during electrolysis, travels toward the positive electrode or anode. [AIR4844]

ANIP *See* Army-Navy Instrument Program.

anisotropic Not isotropic; exhibiting different properties along axes in different directions. [AIR4844]

anisotropic laminate A laminate in which the properties are different in different directions along the laminate plane. [AIR4844]

anisotropy Exhibiting different properties or other characteristics (e.g., strength, coefficient of thermal expansion) in different directions with respect to a given reference, such as a specific lattice direction in a crystalline substance. Also known as nonisotropy.

anneal To heat and then gradually cool in order to relieve mechanical and thermal stress. [ARP1931]

annealing Treating metals, alloys, or glass by heating and controlled, slow cooling—primarily to soften the materials and remove residual internal stress, but sometimes to simultaneously produce desired changes in other properties or in microstructure.

annotate To add explanatory text to computer programming or any other instructions.

annular suspension and pointing system In the shuttle era, high accuracy pointing and stabilization of an experiment payload.

annulus Any ring-shaped cavity or opening.

annunciator An electromagnetic, electronic, or pneumatic signaling device that either displays or removes a signal light, metal flag, or similar indicator, and/or sounds an alarm, when occurrence of a specific event is detected. In most cases, the display or alarm is single-acting, i.e., after being tripped, it must be reset before it can indicate another occurrence of the event.

anode 1. An electrode at which an oxidation reaction (loss of electrons) occurs. In secondary cells either electrode may become the anode depending on the direction of current flow. [ARP4386] 2. The positive pole or electrode of an electron emitter, such as an electron tube or an electric cell. 3. The metal plate or surface that acts as an electron donor in an electrochemical circuit. In an electrolyte (e.g., during electroplating or electrochemical corrosion) metal ions go into solution at the anode 4. The positive electrode in an x-ray tube or vacuum tube, at which electrons leave the interelectrode space.

anode circuit A circuit that includes the anode-cathode path of an electron tube connected in series with other circuit elements.

anode supply voltage The voltage across the terminals of an electric power source connected in series in the anode circuit.

anodic coating An oxide film produced on a metal by treating in an electrolytic cell with the metal as the cell anode.

anodic protection Reducing the corrosion rate of a metal that exhibits active-passive behavior by imposing an external electrical potential on the metal part.

anodic stripping The removal of metallic coatings.

anodize To form a protective passive film (conversion coating) on a metal part, such as a film of Al_2O_3 on aluminum, by making the part an anode in an electrolytic cell and passing a controlled electric current through the cell.

anodizing A method of producing a film on a metal surface which is particularly well suited for aluminum.

anomalies In general, deviations from the norm.

anomalous dispersion Inversion of the derivative of refractive index with respect to wavelength in the vicinity of an absorption band.

anorthosite A group of essentially monomineralic plutonic igneous rocks composed almost entirely of plagioclase feldspar.

anoxemia A condition of absolute lack of oxygen, as opposed to hypoxemia or hypoxia, which refer to oxygen deficiency or lowered oxygen content. [ARP171]

anoxia A complete lack of oxygen available for physiological use within the body.

ANSI Acronym for American National Standards Institute

ANSI screen control An ANSI standard that specifies a specific set of character sequences which instruct the computer to perform certain actions on the computer screen.

antecedent events (mishap) *See* mishap, antecedent events.

antenna A device for sending or receiving radio waves; does not include the means of connecting the device to a transmitter or receiver. *See also* dipole antenna and horn antenna.

antenna array A single mounting containing two or more individual antennas coupled together to give specific directional characteristics.

antennas Conductors or systems of conductors for radiating or receiving radio waves.

anthropometrics Measurements of the height, weight, build, and other physical dimensions of a person.

anticipation A means of sensing that some modification of controller action is necessary

before the main sensing element has specified such a need. [ARP89]

anticipator A sensitive element in a control system designed to respond to a change (or rate of change) in pressure or temperature and to reset the pressure- or temperature-controlling instrument to counteract the tendency of the controlling system to hunt. [ARP147C]

anticoincidence circuit A circuit with two inputs and one output, which produces an output pulse only if one of the input terminal receives a pulse within a specified time interval, but does not produce a pulse if both input terminals receive a pulse within that interval.

anticorrosive Describes a substance, such as paint or grease, that contains a chemical that counteracts corrosion or produces a corrosion-resistant film by reacting with the underlying surface.

anti-drive end The end of a starter or accessory opposite the end which mates with the engine or drive device. [ARP906A]

antifouling Refers to measures taken to prevent corrosion or the accumulation of organic or other residues or growths on operating mechanisms, especially in underwater environments.

antifriction Describes a device, such as a bearing or other mechanism, that employs rolling contact with another part, rather than sliding contact.

antigravity A hypothetical effect that would arise from cancellation by some energy field of the effect of the gravitational field of the earth or other body.

anti-icing The prevention of ice buildup on a protected surface, either by evaporating impinging water, or by allowing it to run back and freeze on noncritical areas. [AIR1168/9]

antimagnetic Describes a device that is made of nonmagnetic materials or employs magnetic shielding to avoid being influenced by magnetic fields during operation.

antimisting fuels Fuels that include an additive to reduce misting and thus improve safety.

antinodes *1*. Either of the two points on an orbit where a line in the orbit plane, perpendicular to the line of nodes and passing through the focus intersects the orbit. *2.* A point, line, or surface in a standing wave at which some characteristic of the wave field has maximum amplitude. *3.* The points, lines, or surfaces in a medium containing a standing wave at which some characteristic of the wave field is at maximum amplitude. Also known as loops.

antioxidant A substance used to prevent or retard the degradation of material that can occur by exposure to oxygen or to an oxygen-containing environment. [ARP1931]

antiozonant A substance used to prevent or retard the degradation of material that can occur by exposure to ozone. [ARP1931]

antiparticles Particles with a charge of opposite sign to the same particles in normal matter.

antipodes Anything exactly opposite to something else. Particularly, that point on the earth 180° from a given place.

anti-prop end The end of the engine opposite the prop end. Also known as the rear of the engine. [ARP169]

antiradiation missiles Missiles that attack radiating targets, such as radar transmitters.

antireflective coating A coating designed to suppress reflections from an optical surface.

antiresonance A condition existing between an externally excited system and the external sinusoidal excitation wherein any small increase or decrease in the frequency of the excitation signal causes the peak-to-peak amplitude of a specified response to increase.

antiresonant Describing an electric, acoustic, or other dynamic system whose impedance is very high—approaching infinity.

antiresonant frequency A frequency at which antiresonance exists between a system and its external sinusoidal excitation.

anti-rotation connector Connector design that provides keying or locking provisions to maintain positive orientation for accessory hardware. [ARP914A]

antiskid Describes a material, surface, or coating that has been roughened, or that contains abrasive particles to increase the coefficient of friction and prevent sliding or slipping. Also known as antislip.

antiskid brake A wheel brake system that avoids skidding by automatic reduction of the hydraulic system pressure when initial skidding is sensed. [ARP4386]

antiskid system *See* skid control system.

antislip *See* antiskid.

antistatic agents Agents that, when added to a molding material or applied to the surface of a molded object, hinder the fixation of dust or the buildup of electrical charge. [AIR4844]

antisurge control Control by which the unstable operating mode of compressors known as "surge" is avoided.

antisymmetric laminate *1.* A special type of laminate that is balanced but unsymmetric. *2.* A laminate in which, for a given ply configuration in the lower half, there is an identical ply configuration in the upper half, but with an alternating ply angle. [AIR4844]

anvil A heavy iron or steel block with a smooth, flat top on which metals are shaped by hammering. *See also* resizing tool.

AOG *See* aircraft on ground.

AOPA Acronym for Aircraft Owners and Pilots Association.

APD *See* avalanche photodiode.

aperiodic Varying in a manner that is not periodically repeated.

aperiodically damped Reaching a constant value or steady state of change without introducing oscillation.

aperture A hole in a surface through which light is transmitted. When placed in the Fourier (focal) plane, apertures are sometimes called spatial filters.

aperture time In a sample-and-hold circuit, the time required for the switch to open after the "hold" command has been given.

API adapter An outlet of a hydrant valve or bottom loading adapter that is in accordance with the American Petroleum Institute (API). [AIR4783]

APL (A programming language) A high-level, interactive computer language primarily designed for mathematical applications. APL was developed by Kenneth Iverson in 1962. It is characterized by extensive operators and array-handling capability. NASA Goddard was one of the first users of APL and was instrumental in introducing it to the computer community.

apogee The point at which a missile trajectory or satellite orbit is farthest from the center of the gravitational field of the controlling body or bodies. [ARP4386]

Apollo asteroids Earth-grazing asteroids in orbits between Mars and Jupiter, which cross the earth's orbit. This group contains nine known asteroids.

apparent bond width On a bonded device, the maximum width of the beam lead in the bonded area. Also called the squash dimension. [AS1346]

apparent candlepower For an extended source measured at a specific distance, apparent candlepower is the candlepower of a point source that would produce the same illumination at the same distance. Also called luminous intensity. [ARP798]

apparent density The density of loose or compacted particulate matter determined by dividing actual weight by volume occupied. Apparent density is always less than true density of the material comprising the particulate matter because the volume occupied takes into account pores or cavities between particles.

apparent viscosity The viscosity or resistance to continuous deformation in a non-Newtonian fluid subjected to shear stress.

appearance potential The minimum electron-beam energy required to produce ions of a particular type in the ion source of a mass spectrometer.

application *1.* Capability provided by a system that is specific to the satisfaction of a set of user requirements. [AS4893] *2.* The system or problem to which a computer is applied. Reference is often made to computation, data processing, and control as the three application categories.

application platform The set of computer hardware on which a software application will run. [AS4893]

application program A program that performs a task specific to a particular end-user's needs. Generally, an application program is any program written on a program development operating system that is not part of a basic operating system.

application program interface The interface between the application software and the application platform, across which all services are provided. [AS4893]

application tasks Those processes that are passing data to one another across a data communications network [ARD50008]

application time The time available for sealant application after a sealant has been mixed or after a premixed and frozen cartridge of sealant has been thawed. Acceptability limits established for Class A brushable sealants are expressed as the time required for the viscosity to increase to a specified level at 77°F (25°C) and 50% relative humidity. Acceptability limits for Class B extrudable sealants are expressed in terms of the extrusion rate of the sealant from a 6 fl oz (177 mL) cartridge, through a nozzle with a 0.125 in. (3.18 mm) diameter orifice, using air pressure of 90 psi ±5 (621 kPa ±34) in a pneumatic sealant gun. The extrusion rate is expressed in grams/minute or (per some specifications) in cubic centimeters (cm^3)/minute. A minimum extrusion rate after the stated application time is given as the acceptable limit. [ARP4069]

applied load *1.* Weight carried or force sustained by a structural member in service. In most cases, the load includes the weight of the member itself. *2.* Material carried by the load-receiving member of a weighing scale, not including any load necessary to bring the scale into initial balance.

approach *1.* To maneuver an aircraft into position relative to the landing area. *2.* Airspace over a designated region of the terminal area, controlled by an approach control unit.

approach channel The passage(s) through which gas must flow from a vessel to reach the operating parts of a safety relief device. [ARP4386]

approach chart A graphical depiction of the details of an approved IFR landing approach path. Also known as an approach plate.

approach clearance Authorization by ATC for a pilot to conduct an instrument approach. [ARP4107]

approach control A service established to provide air traffic control when an aircraft is arriving at, departing from, or operating in the vicinity of, the terminal area.

approach gate A point on the final approach track representing the minimum distance from the threshold at which vectored or instructed aircraft should intercept the final course. [AIR4102/9]

approach light system (ALS) An airport lighting facility that provides visual guidance to landing aircraft by radiating light beams in a directional pattern. On final approach for landing, the pilot uses the ALS to align the aircraft with the extended centerline of the runway. *See also* runway edge light system. [ARP4107]

approach phase *See* mishap, maneuver.

approach plate *See* approach chart.

approach sequence The order in which aircraft are positioned while on approach or while awaiting approach clearance. [ARP4107]

approach speed The velocity of an aircraft during an approach to landing. Often refers to recommended speeds presented in the aircraft flight manual.

approval The act of formal sanction of an implementation by a certification authority. [ARP4754]

approved Accepted by the certification authority as suitable for a particular purpose. [ARP4754]

APT *See* automatically programmed tools.

APU (auxiliary power unit) A propellant-powered device used to generate electrical, gas, or fluid power independent of the main propulsion unit. [AIR913]

aquaplane *See* hydroplane.

aqueous Water-containing or water-based. [AIR4844]

aqueous contaminant Water-borne contaminant. [AIR4783]

aqueous vapor *See* water vapor.

aramid A type of highly oriented organic material derived from polyamide, but incorporating aromatic ring structure. [AIR4844]

arbitration The process by which a single bus master is selected from competing potential bus masters. [AS4710]

arbitration bar A test bar cast from molten metal at the same time as a lot of castings. The arbitration bar is used to determine mechanical properties in a standard tensile test, which are then evaluated to determine the acceptability of the lot of castings.

arc *1.* The track over the ground of an aircraft flying at a constant distance from a navigational aid by reference to distance measuring equipment (DME). [ARP4107]. *2.* A discharge of electricity across a gap between electrical conductors.

architecture framework A skeleton that serves to organize the architecture of a system without dictating the detailed implementation. [AS4893]

architecture *1.* In computers, the design of system and logic organization, and information flow relationships in a computer, rather than the circuit and component features. *2.* The structure, functional, and performance characteristics of a system, specified in an implementation-independent way.

archival file In data processing, storage of seldom-used data that must be retained for several years.

arcing time *1.* In regard to fuses, the time measured from that point at which element melt time ends to that point at which current is interrupted and permanently becomes zero. NOTE: If a mechanical indicator (not presently recommended) is utilized which incorporates a secondary element parallel to the fusible element, arcing time is measured from that point at which indicator melt time ends. *2.* In regard to breakers, the time measured from that point at which contacts first separate to that point at which the current is interrupted and permanently becomes zero. [ARP1199A]

arc lamp A high-intensity lamp in which a direct-current electric discharge produces light that is continuous, as opposed to a flashlamp, which produces pulsed light.

arc line A spectral line in spectroscopy.

arc melting Raising the temperature of a metal to its melting point using heat generated by an electric arc. Usually refers to melting in a specially designed furnace to refine a metal, produce an alloy, or prepare a metallic material for casting.

arc resistance 1. Ability to withstand exposure to an electric voltage. 2. The total time in seconds that an intermittent arc may play across a plastic surface without rendering the surface conductive. [AIR4844]

arc strike *See* strike.

arc welding A group of welding processes in which metals are made to coalesce by heating them with an arc, with or without pressure or the use of filler metal.

area classification The classification of hazardous (classified) locations by Class I, II, or III—depending on the presence of flammable gases or vapors, flammable liquids, combustible dust, or ignitable fibers or flyings; by Division 1 or 2, depending upon the ability of these materials to exist in an ignitable concentration under normal or abnormal conditions; and by Groups A through F, depending on the ease of ignition and explosive characteristics of air mixtures of the chemicals and compounds.

area classification, class *Class I locations*–those in which flammable gases or vapors are or may be present in the air in quantities sufficient to produce explosive or ignitable mixtures. *Class II locations*–those that are hazardous because of the presence of combustible dust. *Class III locations*–those that are hazardous because of the presence of easily ignitable fibers or flyings, but in which such fibers or flyings are not likely to be in suspension in the air in quantities sufficient to produce ignitable mixtures.

area classification, division *Division 1 (hazardous)*–locations in which concentrations of flammable gases or vapors exist: (a) continuously or periodically during normal operations; (b) frequently during repair or maintenance or because of leakage; or (c) due to equipment breakdown or faulty operation which could cause simultaneous failure of electrical equipment. *Division 2 (normally nonhazardous)*–locations in which the atmosphere is normally non-hazardous and may become hazardous only through the failure of the ventilating system, opening of pipe lines, or other unusual situations.

area classification, group *Group A*–atmospheres containing acetylene. *Group B*–atmosphere containing butadiene, ethylene oxide, propylene oxide, acrolein, or hydrogen (or gases or vapors equivalent in hazard to hydrogen). *Group C*–atmospheres such as cyclopropene, ethyl ether, ethylene, or gases or vapors of equivalent hazard. *Group D*–atmospheres such as acetone, alcohol, ammonia, benzene, benzol, butane, gasoline, hexane, lacquer solvent vapors, naphtha, natural gas, propane, or gases or vapors of equivalent hazard. *Group E*–atmospheres containing metal dusts. *Group F*–atmospheres containing combustible dusts having resistivity of $<10^5$ ohm-cm^2.

area, light VTOL landing A special category landing area that will normally require no construction effort other than the clearing of vegetation. [AIR1489]

areal weight The weight of fiber per unit area of tape or fabric. [AIR4844]

area meter A device for measuring the flow of fluid through a passage of fixed cross-sectional area, usually through the use of a weighted piston or float supported by the flowing fluid.

area navigation (RNAV) A method of navigation that permits aircraft operations on any desired course within the coverage of station-referenced navigation signals, or within the limits of self-contained system capabilities. [ARP4107]

area navigation high route A route of flight using area navigation within the airspace that extends upward from, and including, 18,000 ft MSL to flight level 45. [ARP4107]

area navigation low route A route of flight using area navigation within the airspace that extends upward from 1200 ft above the surface of the earth to, but not including, 18,000 ft MSL. [ARP4107]

area navigation, random routes Direct routes, based on area navigation capability, between waypoints defined in terms of degree/distance fixes or offset from published or established routes/airways at specified distance and direction. Also referred to as random RNAV routes. [ARP4107]

area navigation, RNAV waypoint (W/P) A predetermined geographical position used for route or instrument-approach definition, or for progress-reporting purposes, defined relative to a VORTAC station position. [ARP4107]

area, negative pressure Any region in which the static pressure is less than that of the static pressure of the undisturbed air stream. [ARP147C]

area, positive pressure Any region in which the static pressure is greater than that of the static pressure of the undisturbed air stream. [ARP147C]

area, sheltered A reinforced concrete structure offering protection against conventional attack.

argentometer A hydrometer used to determine the concentration of a silver salt in water solution.

argument *1.* An independent variable. In looking up a quantity in a table, the argument is the number or any of the numbers that identifies the location of the desired value. In a mathematical function, the argument is the variable that, when substituted by a certain value, determines the value of the function. *2.* An operand in an operation on one or more variables. *See* also parameter and independent variable.

ARINC Acronym for Aeronautical Radio, Inc., an FAA contractor providing international communication support and meteorological services to aircraft.

ARIP (impact prediction) *See* computerized simulation.

arithmetic and logical unit (ALU) A compon-ent of the central processing unit in a computer in which data items are compared, arithmetic operations performed, and logical operations executed.

arithmetic check *See* mathematical check.

arithmetic element The portion of a mechanical calculator or electronic computer that performs arithmetic operations.

arithmetic expression An expression containing any combination of data names, numeric literals, and named constants, joined by one or more arithmetic operators in such a way that the expression as a whole can be reduced to a single numeric value.

arithmetic operation A computer operation in which the ordinary, elementary arithmetic operations are performed on numerical quantities. Contrast with logical operation.

arithmetic operator Any of the operators, + and -, or the infix operators, +, -, *, /, and **.

arithmetic unit The unit of a computing system that contains the circuits that perform arithmetic operations.

arm *1.* A rigid member that extends to support or provide contact beyond the perimeter of the basic item. [ARP480A] *2.* Allows a hardware interrupt to be recognized and remembered. Contrast with disarm. *See* enable.

armature *1.* The core and windings of the rotor in an electric motor or generator. *2.* The portion of the moving element of an instrument which is acted upon by magnetic flux to produce torque.

arming Pre-setting a skid-control function, such as touchdown protection, by some specific operating sequence of the aircraft (e.g., gear retraction/extension, squat switch activation). May also be applied to systems other than skid control. [AIR1489]

armor Mechanical protection—usually a metallic layer of tape, braid, or wires. [ARP1931]

armored cable A wire or cable covered with armor. [ARP1931]

armored meter tube Variable-area meter tube (rotometer) of all-metal construction, utilizing magnetic coupling between the float and an external follower.

Army-Navy Instrument Program (ANIP) A display research and development program initiated in 1952, which advanced the "contact analog" pictorial display concept, including the "highway in the sky" feature. [ARP4107]

aromatic Unsaturated hydrocarbon with one or more benzene ring structures in the molecule. [AIR4844]

array 1. An arrangement of elements in one or more dimensions. *See* also matrix and vector. 2. In a computer program, a numbered, ordered collection of elements, all of which have identical data attributes. 3. A group of detecting elements, usually arranged in a straight line (linear array), or in two-dimensional matrix (imaging array). 4. A series of data samples, all from the same measurement point. Typically, an array is assembled at the telemetry ground station for frequency analysis.

array dimension The number of subscripts needed to identify an element in the array.

array process A hardware device that processes data arrays. Fast Fourier transforms (FFT) and power-spectral density (PSD) are typical processes.

array processor The capability of a computer to operate at a variety of data locations at the same time.

arrester A device that impedes the flow of large dust particles or sparks from a stack; usually consists of screening at the top of the stack.

arresting cable A wire rope that is stretched across a deck or runway, and which is engaged by the aircraft arresting hook to decelerate the aircraft. Also called an arresting wire. [AIR1489]

arresting gear Any gear or apparatus designed to arrest something in its motion—either all or part of such gear being external to the object being arrested. Specifically, any such apparatus used in carrier landings to arrest airplanes in the landing roll. [AIR1489]

arresting hook The hook assembly, usually mounted on the aircraft, that engages arresting gear mounted on the deck or runway; used for deceleration (arrestment) of the aircraft. [AIR1489]

arresting system *See* system, arresting.

arresting wire *See* arresting cable.

arrival time The time an aircraft touches down on arrival. [ARP4107]

arrow keys Keys on a computer keyboard that will move the cursor.

arrow wings Aircraft wings of V-shaped planform, either tapering or of constant chord, suggesting a stylized arrowhead.

ARTCC Acronym for air route traffic control center.

article Components, assemblies, subassemblies, and parts connected or associated together to perform an operational function. [AIR4896]

articulated On a landing gear, an arrangement in which the gear (e.g., an axle) is hinged to permit folding as desired for retraction or operation. [AIR1489]

articulated arms (waveguides) A beam-direction arrangement in which light passes through a series of jointed pipes containing optics.

articulated structure A structure, either stationary or movable (e.g., a motor vehicle or train), that is permanently or semipermanently connected in such a way that different sections of the structure can move relative to the others. Usually involves pinned or sliding joints.

artificial aging Heat treating a metal at a moderately elevated temperature to hasten age hardening.

artificial gravity A simulated gravity established within a space vehicle by rotation or acceleration.

artificial ice Real ice, but formed by artificial means, such as a spray rig or tanker. [AIR1667]

artificial intelligence (AI) 1. A characteristic of a knowledge-based concept or process that involves decision making and reasoning based on stored knowledge. A system possessing such a characteristic is sometimes referred to as an "expert system," because it uses knowledge and inference procedures to solve problems, or diagnose conditions by their symptoms. [ARP4386] 2. A subfield of computer science concerned with the concepts and methods of symbolic inference by a computer and the symbolic representation of the knowledge to be used in making inferences. 3. The use of computers to simulate the way the human mind operates, as in learning or adaptation.

artificial language A language that has been specifically designed for ease of communication in a particular area of endeavor, but that is not yet natural to that area. By contrast, a natural language is one that has evolved through long usage.

artificial radioactivity Radioactivity induced by bombarding a material with a beam of energetic particles or with electromagnetic radiation.

artificial satellites Man-made satellites.

artificial weathering Producing controlled changes (e.g., in surface appearance) in materials under laboratory conditions that simulate outdoor exposure.

ARTS See automated radar terminal systems.

AS See air supply.

asbestos A fibrous variety of the mineral hornblend, used extensively for its fire-resistant qualities to make insulation and fire barriers. Asbestos has been stitched, bonded, or woven into blankets; mixed with portland cement and water to make sheet roofing, wall cladding, drainage tiles and corrosion-resistant pipe; and combined with binders, such as asphalt or bentonite, to make asbestos felt or plaster.

ASC See Accredited Standard Committee.

A-scan 1. A method of data presentation on a CRT utilizing a horizontal baseline, which indicates distance or time; and a vertical deflection from the baseline, which indicates amplitude. 2. A nondestructive inspection technique for finding voids, delaminations, and defects in laminates. [AIR4844]

ASCII (American Standard Code for Information Interchange) A widely used code in which alphanumerics, punctuation marks, and certain special machine characters are represented by unique, 7-bit, binary numbers. 128 (2^7) different binary combinations are possible, thus 128 characters may be represented.

ASCII file A text file that uses the ASCII character set.

as-fabricated Describes the condition of a structure or material after assembly, and without any conditioning treatment, such as a stress-relieving heat treatment. Specific terms such as as-welded, as-brazed, or as-polished are used to designate the nature of the final step in fabrication.

as-fired fuel Fuel in the condition as fed to the fuel burning equipment.

ash content The incombustible residue remaining after completely burning a combustible material.

ash-free basis The method of reporting fuel analysis whereby ash is deducted and other constituents are recalculated to total 100%.

aspect ratio 1. In aircraft design, the ratio of the wingspan to the chord of the wing. [AIR4888] 2. In an essentially two-dimensional rectangular structure, the ratio of the long dimension to the short dimension. 3. In compression loading, sometimes considered to be the ratio of the load-direction dimension to the transverse dimension. 4. For a fiber, the ratio of length to diameter. 5. See tire aspcet ration.

aspheric For optical elements, surfaces that are not spherical or flat. Lenses with aspheric surfaces are sometimes called aspheres.

asphyxia Suffocation from lack of available or usable oxygen, plus retention of carbon dioxide in vital tissues. The retention of carbon dioxide is what distinguishes asphyxia from hypoxia, merely a diminished oxygen supply. [ARP171]

aspirate Removal of gases or liquids by means of suction.

aspirated thermocouple assembly Thermocouple assembly in which gas flows over the measuring junction at a rate higher than the free-stream velocity, thereby increasing the rate of heat transfer at the junction. [ARP485]

aspirating burner A burner in which the fuel, in a gaseous or finely divided form, is burned in suspension. The air for combustion is drawn through one or more openings by the lower static pressure created by the velocity of the fuel stream, and is brought into contact with the fuel.

aspiration 1. The use of suction to draw a sample of ambient air for ice detection with low forward velocity. [AIR4367] 2. Removal of accumulated mucus and foreign bodies from the airway by suction. [ARP171] 3. Using a vacuum to draw up gas or granular material, often by passing a stream of water across the end of an open tube, or through the run of a tee joint, with the open tube or branch pipe extending into a reservoir containing the gas or granular material. *See* vacuum.

aspiscrubber An aircraft underwing refueling system component used with on-board inert gas generating systems or liquid nitrogen inerting systems to remove dissolved oxygen from the fuel during pressure refueling. [AIR4783]

ASR *1. See* airport surveillance radar. *2. See* automatic send/receive.

ASR approach Surveillance approach.

as-received fuel Fuel in the condition as received at the plant.

assembled coupling A completed, coupled connection, formed by using a coupling assembly to connect two ferrules that are attached to tube ends. [MA2241]

assembly A number of parts, subassemblies, or any combination thereof, which are joined together to perform a specific function, and which can be disassembled without destroying their designed use. [ARP4754]

assembly (fitting) Assembled and torque-tightened fittings, nuts, sleeves, and tubing. [MA2005]

assembly language A computer programming language, similar to computer language, in which the instructions usually have a one-to-one correspondence with computer instructions in machine language, and which utilizes mnemonics for representing instructions.

assembly list A printed list that is the by-product of an assembly procedure; lists in logical instruction sequence all details of a routine, showing the coded and symbolic notation next to the actual notations established by the assembly procedure. The assembly list is highly useful in the debugging of a routine.

assembly load *See* load assembly.

assembly program *See* assembly system.

assembly system A system comprising two elements: (a) a symbolic language; and (b) an assembly program that translates source programs written in the symbolic language into machine language.

assembly time Used in reference to faying-surface sealants. Refers to the amount of time available, after a two-part sealant has been mixed, for closing or "squeezing out" the faying surfaces to which sealant has been applied. If the assembly time is exceeded, the sealant will become too cured (too firm) and will therefore not be able to be squeezed out for the desired surface-to-surface contact. Also called work life and open time. [AIR4069]

assert (signal) The action of changing the state of a bus signal line from released, logic 0, to asserted, logic 1; or of ensuring that the line remains in the asserted state. [AS4710]

assessment An evaluation based on engineering judgment. [ARP4754]

assignment statement A program statement that calculates the value of an expression and assigns it a name.

associated electrical apparatus Electrical apparatus in which the circuits are not all intrinsically safe, but which contains circuits that can affect the safety of intrinsically safe circuits connected to it.

association The combining of ions into larger ion clusters in concentrated solutions.

association reactions Gas-phase chemical processes in which two molecular species, A and B, react to form a larger molecule, AB. In astrophysics, these processes are involved in the "condensation" of small gaseous molecules into larger species.

associative processing (computers) Byte-variable computer processing with multifield search, arithmetic, and logic capability.

associative storage A storage device in which storage locations are identified by their contents, not by names or positions. Synonymous with content-addressed storage. Contrast with parallel search storage.

assumptions Statements, principles, and/or premises offered without proof. [ARP4754]

assurance The planned and systematic actions necessary to provide adequate confidence that a product or process satisfies given requirements. [ARP4754]

A-stage The initial state of a resin as produced by the manufacturer. [AIR4844]

astatic *1.* Without polarity. *2.* Independent of the earth's magnetic field.

asteroid Small celestial bodies revolving around the sun, most having orbits between those of Mars and Jupiter.

asteroid belts The location of the orbits of most of the minor planets (asteroids, estimated at a half million) between Mars and Jupiter. About 2000 asteroids have been assigned numbers and names.

asteroid capture The transfer of an asteroid or comet from the influence of one planet into that of another planet or neutral satellite.

asteroid missions Space missions for the study of asteroids and related celestial bodies.

astigmatism A defect in an optical element that causes rays from a single point in the outer portion of a field of view to fall on different points in the focused image.

astrakanite *See* bloedite.

astrionics The science of adapting electronics to aerospace flight. [ARP4386]

astrochanite *See* bloedite.

astrodynamics The practical application of celestial mechanics, astroballistics, propulsion theory, and allied fields to the problem of planning and directing the trajectories of space vehicles.

astrolabe *1.* Instrument designed to observe the positions and measure the altitudes of celestial bodies. *2.* An instrument formerly used to determine the altitudes of celestial bodies; a predecessor of the sextant.

astronaut Any person who travels in a spacecraft. Specifically, someone chosen by NASA for space flight.

astronautic centrifuge *See* centrifuge.

astronautics The art or science of designing, building, or operating space vehicles. [ARP4386]

astronomical coordinates Coordinates defining a point on the surface of the earth, or of the geoid, in which the local direction of gravity is used as a reference.

astronomical theodolite *See* altazimuth.

astronomy The science that treats the location, magnitudes, motions, and constitution of celestial bodies and structures.

astrophysics A branch of astronomy that treats the physical properties of celestial bodies, such as luminosity, size, mass, density, temperature, and chemical composition.

asymmetric rotor A rotating machine element whose axis of rotation is not the same as its axis of symmetry.

asymmetry potential The difference in potential between the inside and outside pH-sensitive glass layers when they are both in contact with solutions of pH 7.

asymptotic properties Properties of any mathematical relation or corresponding physical system characterized by an approach to a given value as a expression, containing a variable, tends to infinity.

asynchronous *1.* A mode in which an operation is started by a signal before the one on which it depends is completed. *2.* For hardware devices, the method whereby each character is sent with its own synchronizing information and hardware operations are scheduled by "ready" and "done" signals, rather than by time intervals. This implies that a second operation can begin before the first operation is completed. *3.* For machinery, not synchronous with the line frequency (e.g., as applied to rotating machinery). *4.* The use of an independent clock source in each terminal for message transmission. Decoding is achieved in receiving terminals using clock information derived from the message. [AS15531]

asynchronous communication Often called start/stop transmission, a method of transmitting data in which each character is preceded by a start bit and followed by a stop bit.

asynchronous transmission *1.* Transmission in which each information character, or sometimes each word or small block, is individually synchronized, usually by the use of start and stop elements. The gap between each character (or word) is not of a necessarily fixed length. Synonymous with start-stop transmission. Compare with synchronous transmission. *2.* Data transmission mode in which the timing is self-determined and not controlled by an external clock.

ATA Acronym for Airline Transport Association.

atactic A molecular chain in which the methyl groups are more or less in random order. [AIR4844]

ATARS *See* automatic traffic advisory and resolution.

ATC *See* air traffic control.

ATCA *See* ATC assigned airspace.

ATC advises A phrase used as a prefix to a message of noncontrol information when it is relayed to an aircraft by someone other than an air traffic controller. [ARP4107]

ATC assigned airspace (ATCA) Airspace of defined vertical/lateral limits, assigned by ATC, for the purpose of providing air traffic segregation between the specified activities being conducted within the assigned airspace and other IFR air traffic. [ARP4107]

ATC clearance *See* air traffic clearance.

ATC instructions Directives issued by air traffic control that require a pilot to take specific actions. [ARP4107]

ATCRBS *See* air traffic control radar beacon system.

atelectasis Partial or complete collapse, or imperfect expansion, of the air sacs of the lungs; may include entire lobes or entire lung fields. [ARP171]

athletic *See* physical condition.

athodyds *See* ramjet engines.

ATIS *See* automatic terminal information service.

atmidometer *See* atmometer.

atmometer A generic name for any instrument that measures evaporation rates. Also known as an atmidometer; an evaporimeter; or an evaporation gage.

atmosphere Gaseous envelope surrounding the earth, other planets, and celestial bodies.

atmosphere, standard The atmosphere as defined by the latest US Standard Atmosphere issued by the Government Printing Office. [ARP147C]

atmospheric air Air under the prevailing atmospheric conditions.

atmospheric and oceanographic information system A data system designed primarily for the interactive manipulation of meteorological satellite images. Capabilities include displaying, analyzing, storing, and manipulating digital data in the field of meteorology and earth resources.

atmospheric braking The action of atmospheric drag in decelerating a body that is approaching a planet. Where sufficient atmosphere exists, can be deliberately used to lose much of the vehicle velocity before landing. [ARP4386]

atmospheric chemistry Study of the production, transport, modification, and removal of atmospheric constituents in the troposphere and stratosphere.

atmospheric circulation Global or hemispheric air movements that can be treated by equations of motion. Contrasted with atmospheric diffusion, which is small random movement not amenable to treatment by these equations.

atmospheric cloud physics lab (APCL–Spacelab) A NASA Spacelab mission involving cloud physics experiments in zero-gravity environment.

atmospheric communication Sending signals in the form of modulated light through the atmosphere, without the use of fiber optics to contain and direct the beam.

atmospheric corrosion Corrosion that occurs naturally due to exposure to climatic conditions. Atmospheric corrosion rates vary depending on specific global location because of variations in average temperature; humidity; rainfall; airborne substances, such as sea spray, dust, and pollen; and airborne pollutants, such as sulfur dioxide, chlorine compounds, fly ash, and other combustion products.

atmospheric electricity *1.* Electrical phenomena, regarded collectively, which occur in the earth's atmosphere. *2.* The study of electrical processes occurring within the atmosphere.

atmospheric entry The penetration of any planetary atmosphere by any object from outer space. Specifically, the penetration of the earth's atmosphere by a manned or unmanned capsule or spacecraft.

atmospheric general circulation experiment Model experiment of the earth's atmospheric circulation as proposed for a Spacelab flight. In this experiment, a liquid is contained between two concentric spheres and subjected to rotation. The thermal driving force is a stable radial temperature gradient and an unstable latitudinal gradient.

atmospheric lasers Refers to the theoretical phenomenon whereby the upper atmosphere is used as the lasing medium.

atmospheric monochromator A monochromator in which the optical path is through air. This is the standard type used for visible and infrared wavelengths transmitted by air.

atmospheric noise See atmospherics.

atmospheric optics The study of the topical characteristics of the atmosphere and of the optical phenomena produced by suspensoids and hydrometeors in the atmosphere. Atmospheric optics embraces the study of refraction, reflection, diffraction, scattering, and polarization of light, but is not commonly regarded as including the study of any other kinds of radiation.

atmospheric pressure *1.* Absolute pressure of the atmosphere at a given location and time. [ARP4386] *2.* The barometric reading of pressure exerted by the atmosphere. At sea level, this is 14.7 lb per sq in. or 29.92 in. of mercury.

atmospheric radiation Infrared radiation emitted by or being propagated through the atmosphere.

atmospheric refraction Refraction resulting when a ray of radiant energy passes obliquely through an atmosphere.

atmospherics The radio-frequency electromagnetic radiation originating, principally, in the irregular surges of charge in thunderstorm lightning discharges. Atmospherics are heard as a quasi-steady background of crackling noise (static) in ordinary amplitude-modulated radio receivers.

atmospheric sounding Measurement of atmospheric phenomena, generally with instruments carried aloft by spacecraft, rockets, etc.

atmospheric stratification The presence of strata or layers in the earth's atmosphere.

atmospheric tides Defined in analogy to the oceanic tide as an atmospheric motion on a worldwide scale. In analyzing atmospheric tides, vertical accelerations are neglected, but compressibility is taken into account.

ATO (Aborted Takeoff) *See* abort and rejected takeoff.

ATO (Assisted Takeoff) A procedure in which an assistant is utilized for takeoff operation, such as rocket assist (RATO) or jet assist (JATO). Takeoff assistance may also be provided by a catapult (shipboard or shorebased), a launching device that can be reused repeatedly and rapidly. [AIR1489]

atom The smallest particle or unit of matter. Consists of a dense, positively charged nucleus surrounded by a system of electrons.

atomic clocks Timekeeping devices controlled by the frequency of the natural vibrations of certain atoms.

atomic mass unit (amu) A unit for expressing atomic weights and other small masses; exactly equal to one-twelfth the mass of the carbon-12 nuclide.

atomic number An integer that designates the position of an element in the periodic table of the elements; equal to the number of protons in the nucleus, as well as the number of electrons in the electrically neutral atom.

atomic weight The weight of an atom in atomic mass units (amu), valued as one-twelfth the mass of the carbon-12 nuclide.

atomizer Device to reduce a liquid or solid to small droplets, in the form of a spray. *See* also nebulizer. [ARP171]

atom probe An instrument that consists of a field-ion microscope with a probe hole in its screen that opens into a mass spectrometer. Used to identify a single atom or molecule on a metal surface.

ATP *See* airline transport pilot.

ATPD Acronym for ambient temperature and pressure, dry. [ARP171]

ATPS Acronym for ambient temperature and pressure, saturated with water vapor. [ARP171]

ATS message A clearance or flight-plan message. Includes strategic messages associated with the user flight plan; revisions to the initial clearance; tactical messages, such as horizontal, vertical or speed/time/delay instructions; procedure-based instructions; and traffic and urgent advisories. [ARP4791]

attachment hardware Any hardware other than retractors designed for terminating the webbing of a torso-restraint system. [AS8043]

attemperation Regulating the temperature of a substance; for example, passing super-heated steam through a heat exchanger, or injecting water mist into steam to regulate its final temperature.

attention The active selection of, and emphasis on, one component of a complex experience; and the narrowing of the range of objects to which the organism is responding. [ARP4107]

attention, anomalies of boredom A psychological state resulting from any activity that lacks motivation, or from enforced continuance in an uninteresting situation. [ARP4107]

attention, anomalies of channelized attention The focusing of conscious attention on a limited number of environmental cues, which may lead to the exclusion of others of an objectively higher or more immediate priority. Channelized attention is an active anomaly of attention, sometimes referred to as fixation. [ARP4107]

attention, anomalies of cognitive task saturation That state in which the cognitive task demands exceed the individual's capacity to respond appropriately to all of them. Under such a situation, the individual often focuses attention on a subset of the cognitive demands present. [ARP4107]

attention, anomalies of complacency *1*. A state of adjustment, or a dynamic balance between organism and environment, typified by established habits and responses that are in a quiescent stage. *2*. Self-satisfaction accompanied by unawareness of actual dangers or deficiencies. [ARP4107]

attention, anomalies of distraction *1*. An undesired redirection of the focus of attention by an environmental cue or mental process. *External Distraction*–interruption of attention by a nontask-related environmental cue. *Internal Distraction*–interruption of attention by a nontask-related mental process. [ARP4107] *2*. A stimulus that causes an undesired shift in attention. The distracting stimulus may be either an environmental cue (external distraction) or a mental process (internal distraction). [ARP4107]

attention, anomalies of fascination An anomaly of attention in which a person monitors the relevant environmental cues around him/her, but fails to respond to them because of a sense of unreality or detachment from events, as if he/she were viewing them from the outside. Fascination is usually associated with a high-stress or crisis situation. [ARP4107]

attention, anomalies of habit-pattern interference Reverting to previously learned response modes which are objectively inappropriate to the task at hand. Habit-pattern interference usually occurs at the preconscious level of awareness. [ARP4107]

attention, anomalies of habituation Adaptation and subsequent inattention to an environmental cue after prolonged or repeated exposure to it. [ARP4107]

attention, anomalies of inattention A state of self-reduced conscious attention due to a sense of security, self-confidence, or a perceived absence of threat from the environment. *General Inattention*–nonselective inattention typically due to boredom or complacency. *Selective Inattention*–insufficient attending to relevant environmental cues due to lack of knowledge or an inappropriate perceptual or attitudinal set. [ARP4107]

attention, focus of The part of the span of attention directed toward conscious information processing. [ARP4107]

attention, level of The relative proportions of the span, focus, and margin of attention afforded to information processing. [ARP4107]

attention, margin of The span of attention minus the focus of attention—or a person's remaining capacity to focus conscious attention. [ARP4107]

attenuate To weaken or make thinner; for example, to reduce the intensity of sound or ultrasonic waves by passing them through an absorbing medium.

attenuation *1*. Loss of energy caused by scattering of the sound beam within a material or at an interface or an electronic device in or attached to the instrument. [ARP5089] *2*. Power loss in an electrical system. In cables the loss is expressed in decibels per unit length of cable, at a given frequency. [ARP1931] *3*. Reducing in intensity. *4*. The loss of amplitude in a signal as it is transmitted through a conductor. *See* also gain.

attenuation coefficients A measure of the space rate of attenuation of any transmitted electromagnetic radiation.

attenuator *1*. An optical device that reduces the intensity of a beam of light passing through it. *2*. An electrical component that reduces the amplitude of a signal in a controlled manner.

attitude *1*. The orientation of the three major axes of an aircraft (longitudinal, lateral, and vertical) with respect to a fixed reference, such as the horizon, the relative wind, or direction of flight. [ARP4017] *2*. An enduring, learned predisposition to behave in a consistent way toward a given class of objects. *3*. A persistent mental and/or neural state of readiness to react to a certain object or class of objects, not as they are but as they are conceived to be. [ARP4107] *4*. The position of an object in space determined by the angles between its axes and a selected set of planes.

attitude control *1*. A control mode of the flight control system which directs and maintains the desired attitude of the vehicle. [ARP4386] *2*. The regulation of the attitude of an aircraft, spacecraft, etc. *3*. A device or system that automatically regulates and corrects attitude, especially of a pilotless vehicle.

attitude director indicator (**ADI**) An integrated flight display in aircraft cockpits which combines the attitude indicator with steering bars (or command bars) that aid the pilot in navigating with VOR or ILS signals. [ARP4107]

attitude gyro A gyroscope that provides an output signal which is a measure of the aircraft's attitude with respect to a gravitational reference. Also called a vertical gyro. [ARP419]

attitude hold An automatic flight control mode that maintains a desired attitude for the vehicle. [ARP4386]

attitude indicator An instrument that shows the pitch and roll attitudes of the aircraft with respect to the horizon. [ARP4107]

attitudinal set A predisposition, rooted in attitude(s), which may cause an individual to respond in a particular manner to a set of stimuli. [ARP4107]

attribute A characteristic or property that a product either does or does not have, e.g., shorts and opens in electronic parts; leaks in hydraulic lines; "stiction" in bearings. [AIR4896]

attribute sampling A type of sampling inspection in which an entire production lot is accepted or rejected depending on the number of items in a statistical sample that have at least one characteristic (attribute) that does not meet specifications.

audio Pertaining to audible sound—usually taken as sound frequencies in the range or 20 to 20,000 Hz.

audio data Useful information at audio signal frequency.

audio frequencies Frequencies corresponding to normally audible sound waves.

audio frequency The band of frequency that is audible to the human ear. Usually 20 to 20,000 Hz. [ARP1931]

audiometer An instrument used to measure the ability of people to hear sounds. Consists of an oscillator, amplifier, and attenuator, and may be adapted to generate pure tones, speech, or bone-conducted vibrations.

audio signals Signals with a bandwidth of less than 20 kilohertz.

auditory sensation areas In acoustics, the frequency region enclosed by the curves defining the threshold of pain and the threshold of audibility.

aufeis (ice) Icing of ground or river water in Arctic areas with continuous permafrost on which the water has continued to flow.

auger conveyor *See* screw conveyor.

augmented support Support provided by the system contractor during the initial introduction of new development or production equipment. Purpose is to furnish technical assistance, spares, and repair parts; and to perform actual maintenance as required during the introductory period. [AIR4896]

AUI *See* access unit interface.

auroral activity *See* auroras.

auroral zones Roughly circular bands around either geomagnetic pole, above which there is a maximum of auroral activity. Auroral zones lie about 10° to 15° of geometric latitude from the geomagnetic poles.

auroras Sporadic radiant emissions from the upper atmosphere over middle and high latitudes.

austenite A solid solution of carbon in gamma-iron.

austenitic stainless steel *1*. Steels that, at room temperature, have a microstructure consisting predominantly of austenite. Their austenitic microstructure is attained mostly through alloying conditions, e.g., manganese and nickel. *2*. An alloy of iron containing at least 12% Cr, plus sufficient Ni (or in some specialty stainless steels, Mn), to stabilize the face-centered cubic crystal structure of iron at room temperature.

austere airfields Those airfields that: (a) are within navigation aids; and (b) in most cases, have short landing areas within paved landing surfaces, or other facilities necessary for operation of typical medium/large-size transport aircraft. [ARP4107]

autocatalytic degradation The phenomenon whereby the breakdown products of the initial phase of degradation act to accelerate the rate at which subsequent degradation proceeds. [ARP1931]

autoclave A closed vessel for providing an environment of fluid pressure, with or without heat, to an enclosed object while it undergoes a chemical reaction or other process. [AIR4888]

autoclave molding A molding process in which an entire assembly is placed in a heated autoclave, after lay-up, winding, or wrapping; usually at 340 to 1380 kPa. [AIR4844]

autocollimator A telescopic sight including a light source and a partially reflecting mirror, focused to infinity. Used for measuring small angular motion and checking alignment. *See* also collimators.

autocorrelation In statistics, the simple, linear internal correlation of members of a time series (ordered in time or other domains).

AUTOEXEC The name of the file in MS-DOS that has commands to be executed when the computer is booted. Usually named AUTOEXEC.BAT.

autoignition temperature (AIT) Temperature at which a fluid flashes into flame without an external ignition source, and continues burning. The AIT must be determined by one of several approved test methods. [ARP4386]

autokinesis *See* illusions, visual.

automated cone penetrometer A truck-mounted, automated system for making trafficability penetrometer measurements of soil strength. [AIR1780]

automated en route ATC An air traffic control technology that allows computers to make decisions about conflict resolution, the generation of clearances, and their automatic transmission, with the operator standing by to take over in an emergency.

automated guideway transit (AGT) vehicles A system of a large number of captive vehicles traveling at relatively close headways on an exclusive guideway controlled by a computer.

automated mixed traffic vehicles (AMTV) Low-speed surface vehicles automatically operated and controlled in a pedestrian environment by following a buried wire in the roadways, sensing obstacles, and stopping at predetermined spots for passenger exit and entry.

automated pilot advisory system An airport advisory system and an air traffic advisory system designed to improve airport and air traffic advisories at high-density uncontrolled airports.

automated radar terminal systems (ARTS) A highly automatic radar system that displays, for terminal aircraft controllers, information about the aircraft they are controlling. ARTS gives identification, flight plan data, and other flight-associated information (for example altitude and speed). *ARTS II*–a programmable, nontracking, computer-aided display subsystem capable of modular expansion. ARTS II systems provide a level of automated air traffic control capability at terminals having low-to-medium activity. Flight identification and altitude may be associated with the display of secondary radar targets. *ARTS III*–the Beacon Tracking Level (BTL) of the modular programmable ARTS in use at medium- to high-activity terminals. ARTS III detects, tracks, and predicts secondary radar-derived aircraft targets. These are displayed by computer-generated symbols and alphanumeric characters depicting flight identification, aircraft altitude, ground speed, and flight plan data. *ARTS IIIA*–the Radar Tracking and Beacon Tracking Level (RT&BTL) of the modular, programmable ARTS. ARTS IIIA detects, tracks, and predicts primary as well as secondary radar-derived aircraft targets. ARTS IIIA is a more sophisticated computer-driven system. It upgrades the ARTS III system by providing improved tracking, continuous data recording, and fail-soft operations. [ARP4107]

automatic Self-acting; operating by its own mechanism when actuated by some impersonal influence, for example, a change in current strength, pressure, temperature, or mechanical configuration. [ARP1199A]

automatically programmed tools (APT) A numerical language.

automatic altitude reporting That function of a transponder which responds to Mode C interrogations by transmitting the altitude of the aircraft in 100-ft increments. [ARP4107]

automatic carrier landing system (ACLS) US Navy final approach equipment, consisting of precision tracking radar coupled to a computer data link. Provides continuous information to the aircraft, monitoring capability to the pilot, and a backup approach system. [ARP4107]

automatic control *1.* Control of devices and equipment, including aerospace vehicles, by automatic means. *2.* The type of control in which there is no direct action of man on the controlling device.

automatic control engineering The branch of science and technology that deals with the design and use of automatic control devices and systems

automatic controller Any device that measures the value of a process variable, and generates a signal or some controlling action to maintain the value in correspondence with a reference value or setpoint.

automatic control panel A panel of indicator lights and switches on which are displayed an indication of process conditions, and from which an operator can control the operation of the process.

automatic direction finder (ADF) A radio device that is used to locate the direction (bearing) toward a signal on a selected radio frequency. Used in navigation, for instrument approaches to airfields, and for locating lost aircraft or persons who might be transmitting a particular radio frequency. [ARP4107]

automatic error correction A technique usually requiring the use of special codes or automatic retransmission, which detects and corrects errors occurring in transmission. The degree of correction depends on coding and equipment configuration.

automatic flight control systems (AFCS) Systems consisting of electrical, mechanical, and hydraulic components that generate and transmit automatic control commands. These systems provide pilot assistance or relief through automatic or semiautomatic flight path control or automatically control airframe response to disturbances. AFCSs include automatic pilots, stick or wheel steering, autothrottles, and similar control mechanizations. [ARP4386]

automatic frequency control (AFC) *1.* A device or circuit designed to maintain the frequency of an oscillator within a preselected band of frequencies. *2.* In an FM radio receiver, the circuitry that senses frequency drift and automatically controls an internal oscillator to compensate for the drift.

automatic gain control (AGC) *1.* A process by which gain is automatically adjusted as a function of input or other specified parameter. *2.* An auxiliary circuit that adjusts gain of the main circuit in a predetermined manner when the value of a selected input signal varies.

automatic lighter A means for starting ignition of fuel without manual intervention. Usually applied to liquid, gaseous, or pulverized fuel.

automatic mold A mold for injection or compression molding that repeatedly goes through the entire cycle, including ejection, without human assistance. [AIR4844]

automatic pilot Also known as autopilot. An aircraft subsystem that automatically flies the airplane, maintaining a specified course or heading and altitude. When the automatic pilot is used, it substitutes for the pilot as an active component in the flight control loop, and the pilot takes the role of monitoring the automatic pilot. [ARP4107]

automatic press *1.* A hydraulic press for compression molding. *2.* An injection machine that operates continuously, and is controlled mechanically, electrically, or hydraulically; or by a combination of any of these methods. [AIR4844]

automatic reset *See* reset.

automatic reset circuit breakers Thermal-type circuit breakers that have no provisions for manual operation—either in opening or closing. [ARP4404]

automatic rocket impact predictors *See* computerized simulation.

automatic self test Self-test to that degree of fault detection and isolation which can be achieved entirely under computer control, without human intervention. [ARD50010]

automatic send/receive (ASR) A teletypewriter unit with a keyboard, printer, paper tape, reader/transmitter, and paper tape punch. The ASR may be based on-line or off-line; in some cases, on-line and off-line simultaneously.

automatic terminal information service (ATIS) The continuous broadcast of recorded noncontrol information in selected terminal areas. Purpose is to improve controller effectiveness and to relieve radio-frequency congestion by automating the repetitive transmission of essential but routine information. [ARP4107]

automatic test Performance assessment, fault detection, diagnosis, isolation, and prognosis that is performed with a minimum of reliance on human intervention. This may include BIT. [ARD50010]

automatic test equipment *See* equipment, automatic test.

automatic traffic advisory and resolution (ATARS) Ground-based collision-avoidance system using the surveillance and data link capabilities of a discrete address beacon system (DABS).

automatic valve A valve in which operation is controlled entirely by action of fluid which passes through it. [ARP4386]

automatic weather stations Weather stations at which the services of observers are not required; usually equipped with telemetric apparatus.

automatic zero- and full-scale calibration Zero- and sensitivity stabilization by servos for comparison of demodulated zero- and full-scale signals with zero- and full-scale references.

automix A valve operated by barometric pressure which regulates the mixture of oxygen with ambient air to attain a percentage of oxygen according to altitude. [ARP171]

autonomous space clocks Standard time-scale instruments aboard spacecraft with provisions for synchronization with an existing satellite-based system (e.g., a global positioning system).

autopilot *See* automatic pilot.

autoradiography A technique for producing a radiographic image using ionizing radiation produced by radioactive decay of atoms within the test object itself.

autorotation A rotorcraft flight condition in which the lifting rotor is driven entirely by the action of the air when the rotorcraft is in motion. [ARP4107]

auto throttle(s) A control system that positions and adjusts the throttle(s) of an aircraft to maintain a constant airspeed. The auto throttle is set (and can be changed) by the pilot. [ARP4107]

autothrottle system Automatic control of engine throttle setting to produce desired engine power change. [ARP4102/5]

autothrust system Automatic control of engine power change with or without throttle-lever movement. [ARP4102/5]

auto-tracking antenna A receiving antenna that always points to the transmitting site (a vehicle being telemetered), automatically tracking all movements of the vehicle.

autotransformer A type of transformer in which certain portions of the windings are shared by the primary and secondary circuits.

autotrophs Organisms capable of synthesizing organic nutrients directly from simple inorganic substances, such as carbon dioxide and inorganic nitrogen.

auto-zero logic module A component of a digital controller whose function is primarily to establish an arbitrary zero-reference value for each individual measurement.

AUTRAN Acronym for automatic utility translator.

auxiliary actuator A mechanism added to modify a switch-operating characteristic, or to translate direction and/or amplitude of motion to direction and/or levels suitable for a particular switch. [AIR4077]

auxiliary airfoil A secondary airfoil, such as a slat, flap, or tab, which supplements or aids flight in some manner, for example by creating an additional force, or by providing a smooth airflow. [ARP4107]

auxiliary device *1.* Generally, any device that is separate from a main device, but which is necessary or desirable for effective operation of the system. 2. Specifically, any device used in conjunction with an instrument to extend its range, increase its accuracy, otherwise assist in making a measurement, or perform a function not directly involved in making the measurement.

auxiliary means A device or subsystem, usually placed ahead of the primary detector, which alters the magnitude of the measured quantity to make it more suitable for the primary detector; the nature of the measured quantity is unchanged.

auxiliary output A secondary output.

auxiliary panel A panel that is not in the main control room. The front of an auxiliary panel is normally accessible to an operator, but the rear is normally accessible only to maintenance personnel.

auxiliary power Those elements of secondary power related to main engine bleed air and shaft power extraction, or power generation separate from the main engines. Included are engine bleed air systems, remote engine-driven gearboxes, engine-starting systems, auxiliary power units, and emergency power systems. *See* secondary power. [ARP906A]

auxiliary power unit (APU) An internal combustion (usually gas turbine) engine used as the prime mover for equipment; supplies a starter with a desired medium of energy. Also frequently used as a source of bleed air for environmental-control systems and as a source of shaft power for electric- or hydraulic-power generation. [ARP906A]

auxiliary storage A storage device in addition to the main storage of a computer, e.g., magnetic tape, disk, magnetic drum, or core. Auxiliary storage usually holds much larger amounts of information than the main storage, and the information is accessible less rapidly. Contrast with main storage.

auxiliary tires Tires that support the remaining weight of the aircraft not supported by the main tires. [AS4833]

auxiliary unit A separate galley unit for food or beverage and liquor service. [AS1426]

availability *1.* Probability that an item is in a functioning state at a given point in time. [ARP4754] 2. A complex function of equipment mean time between failures (MTBF), mean corrective maintenance time (Mct), duty cycle (d), failure delectability (k) of checkout and test equipment, mission time (tm), and the functional performance threshold (Gmin) at which the system can be classified as operationally ready for a particular mission assignment. [AIR4896] 3. The number of hours in the reporting period, less the total downtime for the reporting period, divided by the number of hours in the reporting period (expressed in percent).

availability factor The fraction of the time during which a unit is in operable condition.

available draft The draft that may be utilized to cause the flow of air for combustion, or the flow of products of combustion.

available energy Energy that theoretically can be converted to mechanical power.

available heat In a thermodynamic working fluid, the amount of heat that could be transformed into mechanical work under ideal conditions by reducing the temperature of the working fluid to the lowest temperature available for heat discard.

available power An attribute of a linear source of electric power, defined as $V_{rms}/4R$, where V_{rms} is the open-circuit rms voltage of the power source, and R is the resistive component of the internal impedance of the power source.

available power gain An attribute of a linear transducer, defined as the ratio of power available from the output terminals of the transducer to the power available from the input circuit under specified conditions of input termination.

available work The capacity of a fluid or body to do work if applied to an ideal engine.

avalanche Production of a large number of ions by cascade action. In this process, a single charged particle, accelerated by a strong electric field, collides with neutral gas molecules and ionizes them.

avalanche photodiode (APD) A photodiode designed to take advantage of avalanche multiplication of photocurrent. As the reverse-bias voltage approaches the breakdown voltage, hole-electron pairs created by absorbed photons acquire sufficient energy to create additional hole-electron pairs when they collide with substrate atoms, producing a multiplication effect.

average The sum of a group of test values divided by the number of test values summed. [AIR4844

average life The mean value for a normal distribution of lives. The term is generally applied to mechanical failures resulting from "wearout." [ARD50010]

average outgoing quality limit The average percentage of defective units that remain undetected in all lots that pass final inspection. This is a measure of the ability of sampling inspection to limit the probability of shipping defective product. Here, a defective unit is considered to be one containing at least one attribute that does not meet specifications.

average-position action A type of control-system action in which the final control element is positioned in either of two fixed positions, the average time at each position being determined from some function of the measured value of the controlled variable.

average rms value The average root mean square (rms) value of phase quantities is the arithmetic sum of the phase rms values divided by the number of phases. [AS1212]

averaging pitot tube An adaptation of the pitot tube in which a multiple-ported pitot tube spans the process tube. Total pressure is measured as a composite of the pressures on several ports facing upstream, while static pressure is measured using one or more ports facing downstream. The device works best for clean liquids, vapors, and gases; but can be used for streams containing suspended solids or viscous contaminants, if the purging flow is supplied to the measuring tube.

averaging system Type of fault-tolerant system using two or more active channels, in which the individual channel outputs are summed to provide an average output. All channels are normally operative, so performance degradation may occur after a failure. An example of an averaging system is the use of multiple control surfaces on an airplane, each individually actuated. [ARP4386]

aviation meteorology Weather conditions and meteorological studies pertaining to aeronautics.

avionics Electrical and electronic equipment used in aviation, principally for navigation and communication. [ARP4107]

avoidance procedure A procedure that attempts avoidance of an area of known or predicted windshear (not yet penetrated). [ARP4109]

awareness, conscious level of A theoretical level of mental awareness at which active information processing or "thinking" takes place. Only one operation at a time can take place at the conscious level. [ARP4107]

awareness, level of The theoretical sources of the mental activity which operates in our behavior. The extent to which an organism is conscious of something; the act of "taking account" of an object or state of affairs. [ARP4107]

awareness, preconscious level A theoretical level of mental awareness which is the repository of short-term and long-term memory and overlearned response modes and habit patterns. Actions controlled by the preconscious allow us to do more than one thing at a time. [ARP4107]

awareness, subconscious level The theoretical repository of information and response modes not available at the conscious level. Reflexes and psychological defense mechanisms operate at the subconscious level. [ARP4107]

AWG *See* American Wire Gage.

axes (coordinates) *See* coordinate.

axial displacement The incremental difference between an initial position and a final position resulting from a force applied along the axis of a component. [ARP914A]

axial fan Consists of a propeller or disc-type of wheel within a cylinder, discharging the air parallel to the axis of the wheel.

axial-flow Describing a machine, such as a pump or compressor, in which the general direction of fluid flow is parallel to the axis of its rotating shaft.

axial flowmeter *See* angular momentum flowmeter.

axial hydraulic thrust In single-stage and multiple-stage pumps, the axial component of the summation of all unbalanced impeller forces.

axial modes Regimes of vibration along a given axis.

axial piston motor A motor utilizing multiple pistons which are pressurized in sequence as the output shaft rotates. [ARP4386]

axial piston pump A pump utilizing multiple pistons arranged such that the pistons stroke in sequence as the pump drive shaft rotates. [ARP4386]

axial runout For a rotating member, the total amount that a specific surface deviates from a plane perpendicular to the axis of rotation in one complete revolution. Usually expressed in 0.001 in., or some other suitable unit of measure, taken at a specific radial distance from the axis of rotation.

axial strain A linear strain in a plane parallel to the longitudinal axis of the specimen.

axial winding In filament-wound reinforced plastics, a winding with the filaments parallel or at a small angle to the axis. [AIR4844]

axisymmetric focused-jet amplifier An amplifier that utilizes control of the attachment of an annular jet to an axisymmetric flow separator (i.e, control of the focus of the jet) to modulate the output. [ARP993A]

axle *1.* A supporting shaft upon which a wheel turns, or which provides support for a rotating member. [AIR1489] *2.* A rod, shaft, or other supporting member that carries wheels, and either transmits rotating motion to the wheels, or allows the wheels to rotate freely about it.

axle base The dimension between axle centerlines on a multiple-axle landing gear. [AIR1489]

axle beam The structural member that connects and supports axles on a multiple-axle gear arrangement. [AIR1489]

axle sleeve A tubular sleeve that fits over the axle and supports the wheel bearings. The axle sleeve: (a) protects the axle from galling or damage from bearing races; and (b) permits wheel tire assembly buildup and torquing adjustment in the shop, and thereby protects against sand, dust, and foreign object damage in installation on the aircraft. [AIR1489]

azeotrope A mixture whose evolved vapor composition is the same as the liquid it evolves from. This phenomenon occurs at one fixed composition for a given system. At either side of the azeotropic point, the vapor will have a different composition from that of the liquid it evolves from. Such mixtures act as pure substances in distillation, and thus are inseparable by standard distillation methods. Azeotropic distillation is necessary to separate such a mixture.

azeotropic distillation A distillation technique in which one of the product streams is an azeotrope. Sometimes used to separate two components by adding a third, which forms an azeotrope with one of the original two components.

azimuth Horizontal direction or bearing.

azimuth angle An angular measurement in a horizontal plane about some arbitrary center point, using true North or some other arbitrary direction as a reference direction (0°).

azimuth circle A ring scale graduated from 0° to 360°, and used with a compass, radar plan position indicator, direction finder, or other device to indicate compass direction, relative bearing, or azimuth angle.

azole A compounds that contains a five-membered heterocyclic ring containing one or more nitrogen atoms.

B

babbitt Any of the white alloys, composed principally of lead or tin, which are used extensively to make linings for sliding bearings.

babbitt metal Any of the white alloys composed primarily of tin or lead—and of lesser amounts of antimony, copper, and other metals—which are used for bearings.

babble The composite signal resulting from crosstalk among a large number of interfering channels.

backbone The trunk media of a multi-media LAN that is separated into sections by bridges, routers, or gateways.

backcoating Coating material that is deposited on the side opposite that which is being coated in the evaporated coating process. Generally results in an objectionable-appearing film or haze. [ARP924]

back course (ILS) *See* course.

back draft A reverse taper on the sidewall of a casting mold or forging die that prevents a casting pattern or forged part from being removed from the cavity of the mold or die.

backfigure antennas Antennas consisting of radiating feeds, reflector elements, and reflecting surfaces. Such antennas function as open resonators, with radiation from the open end of the resonator.

back flow preventer *See* vacuum breaker.

background *1.* In radiation counting, a low-level signal caused by radiation from sources other than the source of radiation being measured. *2.* The surface of the part upon which the indication is viewed. May be either the natural surface of the part or the developer coating on the surface. [AMS2647A] *3.* In multirate, real-time digital machines, refers to the "left-over" processor time between frames in which non-time-critical tasks run. This includes any output monitor port updates, engine trend monitoring algorithms, system data logging, etc. [AIR4548]

background discontinuities Small, scattered, subsurface discontinuities, such as shrink cavities, porosity, more-dense or less-dense foreign material, and gas holes, appearing over extended areas of a casting. [AMS4991]

background discrimination The ability of a measuring instrument or detection circuit to distinguish an input signal from electronic noise or other background signals.

background fluorescence Fluorescent residues observed over the general surface of a test part. [AMS2647A]

background luminance The luminance of an area on the face of a display that is not illuminated by the graphics presented upon the face of the display. [ARP1782]

background noise *1.* In recording and reproducing, the total system noise independent of whether a signal is present. The signal is not included as part of the noise. *2.* In receivers, the noise in the absence of signal modulation on the carrier.

background program A program that is of the lowest urgency with regard to time, and which may be preempted by a program of higher urgency and priority. Contrast with foreground program.

backhand welding Laying down a weld bead with the back of the welder's principal hand (the one holding the torch or welding electrode) facing the direction of welding. In torch welding, this directs the flame backward against the weld bead to provide postheating.

backing pump In a vacuum system using two pumps, the pump discharging directly to the atmosphere which reduces system pressure to an intermediate value, usually 10^{-2} to 10^{-5} psia. Also known as the fore pump.

backing ring When joining tubes or pipes by welding, a ring of steel or other material placed behind the welding groove to confine the weld metal.

backing strip A piece of metal, asbestos, or other nonflammable material placed behind a joint prior to welding to enhance weld quality.

backlash *1.* In an aircraft control system, a looseness or freeplay in the linkage between crew station controls (input) and the device being controlled (output). *2.* In a landing gear system, any looseness, freeplay, or slop in the system, e.g., nose wheel steering backlash/freeplay. [AIR1489] *3.* The uncontrolled load motion, due to clearance in actuation elements including the load attach point. Usually expressed in terms of absolute load motion. [ARP4386] *4.* In a mechanical linkage or gear train, the amount by which the driving shaft must rotate, when reversing direction, in order to merely take up looseness in the linkage or gear train before it begins to transmit motion in the reverse direction. *5.* The difference in actual values of a controlled variable when a control dial is brought to the same indicated position from opposite rotational directions.

backlobes Radiation lobes whose axes make angles of approximately 180° with respect to the axes of the major lobes of the antennas; and, by extension, radiation lobes in the half-space opposed to the direction of peak activity.

backmounted A connector installed with its mounting flange positioned behind the mounting surface, when looking at the mating face or front side of the connector. [ARP914A]

backplane A motherboard comprising wiring for the bus and connectors to the modules attached to the bus. [AS4710]

back pressure *1.* Pressure applied to specified fuel connections of a component during normal operation, or during tests for the purpose of evaluating possible effects on performance, etc. [AIR1749] *2.* The pressure level measured in the exhaust collector of the starter or auxiliary power unit. [ARP906A]

back reflection *1.* Indication of an echo from the far boundary of a product under test.

[AMS2633B] *2.* Signal from the far boundary of a test part. [ARP5089]

backscattering The scattering of light in the direction opposite to the direction in which it was originally traveling.

backshell A connector accessory or component, which may or may not be supplied with the connector, to provide for strain relief, tighter harness routing in restricted space, shielding from electrical interferences, or positive moisture. [ARP914A]

backstep sequence A method of laying down a weld bead in which a segment is welded in one direction, then the torch is moved in the opposite direction a distance approximately twice the length of the first segment, and another segment is welded back toward the first. When using this method, the general direction of progress along the joint is opposite to the direction of welding individual segments, with the end point of each segment coinciding with the starting point of the preceding segment.

backstreaming The "upstream" motion of some oil vapor molecules, for example, in the top portions of a hot diffusion pump. In backstreaming, pump oil vapors move in the reverse direction, into the furnace. [AMS2769]

back-to-back A test performed with the same engine before and after a modification to the facility or associated equipment. [ARP741]

backtracking A technique used to synchronize mixed-signal simulation systems in which an analog simulator is required to back up to a previous time point in order to process a signal originating in the digital simulator.

backup *1.* An item kept available to replace an item which fails to perform satisfactorily. *2.* An item under development intended to perform the same general function of another item also under development. *3.* An item or system kept available to replace an item or system, e.g., a backup gear extension system. [AIR1489]. *4.* Equipment that is available to complete an operation in the event that the primary equipment fails. *5.* A copy of a computer diskette which protects against destruction or loss of the original.

back-up block A piece of metal, usually steel, shaped to support the end of a multi-contact electric connector. The mass of the back-up block is used to resist the impact of driving taper pins into the taper pin receptacles of the connector. [ARP592]

backup bus controller A terminal on a bus that has the capability to become primary bus controller. [AS4115]

back-up oxygen An oxygen supply located on the air-frame to accommodate failure of OBOGS (on-board oxygen generation system). The back-up oxygen supply may be activated manually or automatically. [ARP171]

backup seal A seal on the dry side of the integral fuel tank. It is considered a secondary or redundant seal; never a primary seal. [AIR4069]

back-up stiffness The mechanical stiffness of the actuator attach point(s) through which load reaction forces are transmitted. [ARP4386]

backup system A mode of control that is engaged upon failure of the primary operational system. Usually used to refer to a system that is as independent as possible from the primary system. Sometimes used as protection against multiple generic failures. [ARP4386]

backward differencing A method of solving a parabolic problem for approximating a time derivative in terms of a previous time step.

backward facing steps A step structure that faces away from an oncoming flow. Also referred to as rearward facing steps.

backward waves In traveling wave tubes, waves whose group velocity is opposite to the direction of electron-stream motion.

baffle *1.* Plate that regulates the flow of a fluid, e.g., a heat exchanger, boiler flue, or automotive muffler. *2.* A plate or vane, plain or perforated, used to regulate or direct the flow of fluid. *3.* A cabinet or partition used with a loudspeaker.

baffle-nozzle amplifier A device for converting mechanical motion to a pneumatic signal. Consists of a supply tube ending in a small nozzle, and a movable baffle plate attached to a mechanical arm. The supply tube has a restriction a short distance before the nozzle, so that as the baffle plate moves closer to the nozzle opening the pressure rises in the section of the supply tube between the restriction and the nozzle. In this device, arm motion and nozzle clearance are small—on the order of 0.2 mm or less. A baffle-nozzle amplifier serves as the primary detector in almost all pneumatic transmitters and controllers. Often referred to as a flapper-nozzle amplifier because the baffle plate is mounted on a pivoting arm.

baffle plate A tray or partition, solid or perforated, positioned in the flowpath through a process vessel so as to cause the process stream to flow in a certain direction, to reverse its direction of flow, or to slow its velocity.

baffle-type collector In gas paths utilizing baffles, a device so arranged as to deflect dust particles out of the gas stream.

bag A deep bulge in the shell of a furnace or fire-tube boiler.

bag, economizer A bag (connected to a mask) to which oxygen is admitted continuously at a fixed rate of flow; and which, during expiration, is isolated from the mask by a check valve so that oxygen delivered to the bag during expiration is available for the next inspiration, thereby reducing the peak demand on the oxygen system. [ARP171]

bag filter A device containing one or more cloth bags for recovering particles from dust-laden gas or air which is blown through it.

bagging Applying an impermeable layer of film over an uncured part and sealing edges so that a vacuum can be drawn. [AIR4844]

bag molding A process in which the consolidation of the material in a mold is effected by the application of fluid or gas pressure through a flexible membrane. [AIR4844]

bag, rebreather A bag connected through an open passage to a mask, so that oxygen delivered to the bag at a continuous, fixed rate of flow becomes mixed with a portion of the expired gas, thereby providing a large volume of oxygen-enriched gas for the next inspiration. This reduces the peak demand on the oxygen system and reduces the oxygen consumption by providing for rebreathing of a portion of each expiration. [ARP171]

bag side The side of a part that is cured against the vacuum bag. [AIR4844]

bagside surface The side of a composite part that was cured against the vacuum bag. The bagside surface usually has a slightly rough texture, unlike the smooth tool-die surface of the part. [ARP5089]

bag-type collector A filter in which the cloth filtering medium is made in the form of cylindrical bags.

bakeout Heating the surfaces of a vacuum system during evacuation to degas them and aid in the process of reaching a stable final vacuum level. *See also* degassing.

baking Heating parts in order to remove hydrogen from them. [AMS2759B]

balance *1.* Generically, a state of equilibrium. May be static, as when forces on a body exactly counteract each other; or dynamic, as when material flowing into and out of a pipeline or process has reached steady state and there is no discernible rate of change in process variables. *2.* An instrument for making precise measurements of mass or weight.

balanced construction Equal parts of warp and fill in fiber fabric. [AIR4844]

balanced design In filament-wound reinforced plastics, a winding pattern so designed that the stresses in all filaments are equal. [AIR4844]

balanced field length Designates a condition in which the actual field length is equal to the critical field length. *See* critical field length. [AIR1489]

balanced-in-plane contour In a filament-wound part, a head contour in which the filaments are oriented within a plane, and the radii of curvature are adjusted to balance the stresses along the filaments with the pressure loading. [AIR4844]

balanced interconnecting cable A cable of two or more wires having impedance characteristics that provide similar impedance paths for each of two paired conductors. [AIR4258]

balanced laminate A composite laminate in which all of the laminae at angles other than 0° and 90° occur only in plus and minus pairs, and are symmetrical around the centerline. [AIR4844]

balanced (to ground) *See* unbalanced (to ground).

balanced transmission Transmission of discrete information that takes place over two wires, such that the common mode voltage ideally remains constant, while the differential voltage takes two or more states to represent the discrete signal. [AIR4258]

balanced twist An arrangement of twists in a yarn of two or more strands that does not cause the yarn to kink or twist on itself when held in the form of an open loop. [AIR4844]

balance weight A mass positioned on the balance arms of a weighing device so that the arms can be brought to a predetermined position (null position) for all conditions of use.

balancing equipment Refers to the balance machine, proving rotor, tooling, and accessories necessary to accomplish a balancing operation. [ARP533A]

balancing machine A machine that provides a measure of the unbalance in a rotor. Can be used for adjusting the mass distribution of a rotor mounted on it so that once per revolution vibratory motion of the journals or force on the bearings can be reduced if necessary. [ARP533A]

ballast *1.* Relatively dense material placed in the keel of a ship or the gondola of a lighter-than-air craft to increase stability and/or control buoyancy. *2.* Weight added to a flight vehicle to move its center of gravity or change its moment of inertia.

ball bearing A type of antifriction bearing in which the load is borne on a series of hard, spherical elements (balls) confined between inner and outer retaining rings (races).

ball burnishing *1.* Producing a smooth, dimensionally precise hole by forcing a slightly over-sized tungsten-carbide ball at high speed through a slightly undersized hole. *2.* A method of producing a lustrous finish on small parts by tumbling them in a wood-lined barrel with burnishing soap, water, and hardened steel balls.

ball bushing A variation of ball bearing that permits axial motion of a shaft instead of rotating motion.

ball check valve A valve that permits flow in one direction only by lifting a spring-loaded ball off its seat when a pressure differential acts in that direction, and by forcing the ball more tightly against the seat when a pressure differential acts in the opposite flow direction.

ball-float liquid-level meter A device consisting of a hollow or low-density float attached by means of a linkage to a pointer. In operation, the float rises and falls with the level of liquid in a tank, while the pointer indicates the position of the float on a scale outside the tank.

ball, full A closure component that has a complete, spherical surface with a flow passage through it.

ballistic cameras Ground-based cameras using multiple exposures on the same plate to record the trajectories of rockets.

ballistic damage Damage resulting from armament projectile strikes. [AIR4844]

ballistic missile Any missile that does not rely on aerodynamic surfaces to produce lift. [ARP4386]

ballistics *1.* The science that deals with the motion, behavior, and effects of projectiles—especially bullets, aerial bombs, rockets, or the like. 2. The science or art of designing and hurling projectiles so as to achieve a desired performance.

ballistic trajectories Trajectories followed by a body that is acted on only by gravitational forces and the resistance of the medium through which it passes.

ball lightning A relatively rare form of lightning consisting of a reddish, luminous ball, on the order of one foot in diameter, which may move rapidly along solid objects or remain floating in midair. Hissing noises emanate from such balls, and they sometimes explode noisily, but may also appear noiselessly.

balloon *1.* A non-power-driven, lighter-than-air aircraft. 2. To rise slightly—either just before or after—touching down (landing). [ARP4107] *3.* The circular symbol used to denote and identify the purpose of an instrument or function; may contain a tag number. *4.* Synonym for bubble.

ball screw actuator A linear mechanical actuator consisting of a ball race, threaded shaft, and ball race threaded nut, coupled together through spherical balls. [ARP4386]

ball, segmented A closure component that is a segment of a spherical surface, one edge of which may be contoured to yield a desired flow characteristic.

ball-type viscometer An apparatus for determining viscosity, especially of high-viscosity oils and other fluids, in which the time required for a ball to fall through the test liquid confined in a tube is measured.

ball valve A type of shutoff valve consisting of a solid ball with a diametral hole through it, which can be rotated within a spherical seat about an axis perpendicular to the axis of the hole. To permit flow, the ball is rotated so that the hole lines up with inlet and outlet ports of the valve; to shut off flow, the ball is rotated so that the hole does not line up with the ports.

Banbury mixer A heavy-duty batch mixer with two counterrotating rotors. Designed for blending doughy materials, such as uncured rubber and plastics.

band 1.The gamut or range of frequencies. 2. A group of channels. *See* channel. 3. A group of recording tracks on a computer magnetic disk or drum.

band brake A device for stopping or slowing rotational motion by increasing the tension in a flexible band, thereby tightening it around a drum that is attached to the rotating member.

band density In filament winding, the quantity of fiberglass reinforcement per inch of bandwidth, expressed in strands per inch or per centimeter. [AIR4844]

band-elimination filter A wave filter having a single attenuation band whose critical and cut-off frequencies are finite, nonzero values.

bandgap *See* energy gaps (solid state).

band marking A circular band applied at regular intervals to the insulation of a conductor for the purpose of size designations or circuit identification. [ARP1931]

bandpass *See* bandwidth.

bandpass filter A process or device in which all signals outside a selected band are strongly attenuated, while the signal components lying within the band are passed with a minimum of change.

B and S gage Abbreviation for Brown and Sharp gage.

band spectrum A spectral distribution of light or other complex wave in which the wave components can be separated into a series of discrete bands of wavelengths. *See also* continuous spectrum.

band thickness In filament winding, the thickness of the reinforcement as it is applied to the mandrel. [AIR4844]

bandwidth The difference, expressed in hertz, between the two boundaries of a frequency range.

bandwidth (bandpass) The frequency range over which an actuation system has acceptable dynamic response.

bang-bang system *See* on-off system.

bank A group of cylinders arranged in a common plane parallel to the axis of the crankshaft. [ARP169]

bank switching A method of equipping a computer with greater memory by giving the same address to added memory chips.

bar A solid, elongated piece of metal, usually having a simple cross section and usually produced by hot rolling or extrusion, which may or may not be followed by cold drawing.

BAR One atmosphere.

Barany chair A type of chair in which a person is revolved to test his susceptibility to vertigo. It is named after the Swedish physician Robert Barany who lived from 1876 to 1936.

bar code A pattern of narrow and wide bars that can be scanned and interpreted into alpha and numeric characters.

bar-code scanner A type of optical scanner developed to read the 12-character Universal Product Code used to identify groceries and other products.

barcol hardness A hardness value obtained by measuring the resistance to penetration of a sharp steel point under a spring load. [AIR4844]

bare glass Glass (e.g., yarns, rovings, and fabrics) from which the sizing or finish has been removed. [AIR4844]

bare junction (exposed junction) A thermocouple assembly that has no shielding or support-tube covering of any type over the measuring junction. [ARP485]

bark A decarburized layer on steel, just beneath oxide scale formed by heating the steel in air.

Barkometer scale A specific gravity scale used primarily in the tanning industry. On this scale, the specific gravity of a water solution is determined from the formula: sp gr = $1.000 \pm 0.001n$, where n is degrees Barkometer. On this scale, water has a specific gravity of zero Barkometer.

barn A unit of nuclear cross section in which the probability of a specific nuclear interaction, such as neutron capture, is expressed as an apparent area. In this context, one barn equals 10^{-28} m^2.

baroclinic instability Hydrodynamic instability arising from the existence of a meridional temperature gradient (and hence a thermal wind) in an atmosphere that is in quasigeostrophic equilibrium and that possesses static stability.

baroclinity The state of stratification in a fluid in which surfaces of constant pressure (isobaric) intersect surfaces of constant density (isosteric). The number, per unit area, of isobaric-isosteric solenoids intersecting a given surface is a measure of baroclinity.

barometer An absolute pressure gage for determining atmospheric pressure. If the gage is a recording instrument, it is known as a barograph.

barometric control *See* control, barometric. [ARP171]

barometric hypsometry Determining elevation above some arbitrary reference plane (usually sea level) through the use of mercury or aneroid barometers.

barometric pressure Atmospheric pressure as determined by a barometer. Usually expressed in inches of mercury. *See also* atmospheric pressure.

barometry The study of atmospheric pressure measurement, in particular, determining errors in barometric instrument readings and correcting them.

barostat A device for maintaining constant pressure within a chamber.

barothermograph An instrument for automatically recording both atmospheric temperature and pressure.

barothermohygrograph An instrument for automatically recording atmospheric pressure, temperature, and humidity on the same chart.

barotropism The state of a fluid in which surfaces of constant density (or temperature) are coincident with surfaces of constant pressure. Barotropism is the state of zero baroclinity. *See* baroclinity.

barred galaxies Spiral galaxies whose nuclei are in the shape of bars, at the ends of which the spiral arms begin. About one-fifth of all spiral galaxies are barred spirals.

barrel *1.* A cylindrical component of an actuating cylinder, accumulator, etc. in which a piston or sealed separator moves. [ARP4386] *2.* A unit of volume. For petroleum, one barrel equals 9702 in^3; for fruits, vegetables, other dry commodities, and some liquids, a different standard barrel is used.

barrel chamfer The flared entrance or internal bevel at the wire-entry end of a termination device, intended to facilitate entry of the conductor. [ARP914A]

barrel, conductor The section of a contact, splice, or terminal that accommodates the stripped cable conductor. In the case of insulation-displacing barrels, the cable need not be stripped. [ARP914A]

barrel finishing Producing a lustrous surface finish on metal parts by tumbling them in bulk in a barrel partly filled with an abrasive slurry. Similar processes are used for cleaning and electroplating; however, these use detergent solutions or electrolytes instead of an abrasive slurry.

barrel, insulation The portion of a contact, terminal, or splice that accommodates the wire insulation. [ARP914A]

barricade Used for deceleration and recovery of an aircraft with a known landing emergency. Generally a strap or webbing arrangement connected to an energy-absorbing system; similar to a barrier. Whereas a barrier often is designed to engage an aircraft landing gear, a barricade primarily is designed to engage the wings or other airframe structure, but may also engage the landing gear. [AIR1489]

barrier *1.* A film or material applied to a tank inner liner to resist fluid permeation. Used on nonself-sealing and self-sealing fluid tanks containing aviation fuels. [AIR1664] *2.* Any material through which passage of solids, liquids, semisolids, gases, or forms of energy such as ultraviolet light is limited. *3.* Dielectric material used to insulate electrical circuits from each other or from ground. [ARP914A]

barrier coat An exterior coating applied to a composite wound structure to provide protection. [AIR4844]

barrier film The layer of film used to permit removal of air and volatiles from a composite lay-up during cure, while minimizing resin loss. [AIR4844]

barrier injection transit time diodes *See* Barritt diodes.

barrier, mid-field A type of arresting gear located at the mid-point of the runway to decelerate and stop an aircraft. May be utilized as the aircraft lands in either direction on the runway. [AIR1489]

barrier, overrun A type of arresting gear located near the end of a runway to decelerate and stop an aircraft in case of brake system or other failure. Generally an overrun area is provided off the end of the runway for runout. [AIR1489]

barrier plastics A general term applied to a group of lightweight, transparent, impact-resistant plastics, usually rigid copolymers of high acrylonitrile content. [AIR4844]

barrier, pop-up A type of arresting gear designed so that aircraft can roll over it during normal operations. As an emergency requires and a command signal is given, the barrier pops up into a position for engagement as the aircraft passes. [AIR1489]

barrier strip A continuous section of dielectric material that insulates electrical circuits from one another or from ground. [ARP914A]

Barritt diode Barrier injection transit time diode that operates similarly to the IMPATT diode. The operating frequencies are determined by the transit times across the drift.

baryon resonance An anomaly found in scattering cross sections indicating the existence of an unstable, excited-state baryon.

base *1.* A structural foundation upon which an item is to be permanently assembled. The base is an integral part of the item for which it is a foundation. [ARP480A] *2.* The foundation or support upon which a machine or instrument rests. *3.* The fundamental number of characters available for use in each digital position in a numbering system. *See* radix number. *4.* A chemical substance that hydrolyzes to yield OH⁻ ions. *5.* A reference value. *6.* A number that is multiplied by itself as many times as indicated by an exponent.

base address *1.* A number that appears as an address in a computer instruction, but serves as the base, index, initial, or starting point for subsequent addresses to be modified. Synonymous with presumptive address and reference address. *2.* A number used in symbolic coding in conjunction with a relative address. *3.* An address used as the basis for computing the value of some other relative address.

baseband *1.* A single-channel signaling technique in which the digital signal is encoded and impressed on a physical medium. *2.* The frequencies starting at or near d-c.

base compound The major component of a two-part curing-type sealant. Contains the prepolymer. [AIR4069]

base flow Fluid flow at the base or extreme aft end of a body.

base leg *See* traffic pattern.

baseline *1.* The horizontal trace across the A-scan CRT display. Represents distance or time. [ARP5089] *2.* A quantifiable physical condition or level of performance from which changes are measured. [ARP1587] *3.* Generally, a reference set of data against which operating data or test results are compared to determine such characteristics as operating efficiency or system degradation with time. *4.* In navigation, the geodesic line between two stations operating in conjunction with each other.

baseline facility A facility designated as the standard for certification of an engine. [ARP741]

base load The term applied to that portion of a station or boiler load that is practically constant for long periods.

base metal *1.* The metallic element present in greatest proportion in an alloy. *2.* The type of metal to be welded, brazed, cut, or soldered. *3.* In a welded joint, the metal that was not melted during welding. *4.* Any metal that will oxidize in air, or that will form metallic ions in an aqueous solution. *5.* Metal to which a plated, sprayed, or conversion coating is applied. Also known as a basis metal.

base number *See* radix and radix number.

base point *See* radix point.

base pressure In aerodynamics, the pressure exerted on the base, or extreme aft end, of a body (e.g., cylindrical or boattailed body or a blunt-trailing-edge wing) in a fluid flow.

BASIC *See* Beginner's All-Purpose Symbolic Instruction Code.

basic cause *See* cause, basic.

basic element A single component or subsystem that performs one necessary and distinct function in a measurement sequence. In order to be considered a basic element, a component must perform one and only one of the smallest steps into which the measuring sequence can be conveniently divided.

basic empty weight The empty weight of an airplane with fixed ballast, hydraulic fluid, and other items required by regulatory standards. *See* empty weight. [ARP4107]

basic failure rate For a product, a failure rate that is derived from the catastrophic failure rate of the parts of the product, but does not take into account use and tolerance factors. [AIR4896]

basic frequency In a waveform made up of several sinusoidal components of different frequencies, the single component having the largest amplitude, or having some other characteristic that makes it the principal component of the composite wave.

basic input output system (BIOS) That part of a computer operating system that handles input and output.

basic mission The basic intended function or capability of an aircraft, such as bomber, fighter, patrol, observation, or utility. [AIR4896]

basic rack A section of the toothed surface of a cylindrical gear of infinitely large diameter on a plane at right angles to the tooth surfaces, the profile of which is used as the basis for defining the standard tooth dimensions of a system of involute gears. [AS1560]

basic recipe A generic, transportable recipe consisting of header information, equipment requirements, formula, and procedure.

basic reliability The duration or probability of failure-free performance under stated conditions. [AIR4896]

basic research Research work in the design concept stages of a system project prior to project definition. Includes theoretical studies and new technology development. [ARP4293]

basic T A standardized arrangement of four basic flight instruments in a T pattern. Across the top are the airspeed indicator, attitude indicator, and altimeter; below the attitude indicator is the heading indicator to form the T. [ARP4107]

basis metal The metal from which a connector, contact, terminal, or splice is made. [ARP914A]

basis weight For paper and certain other sheet products, the weight per unit area.

basketweave Alpha platelets, with or without interweaved beta platelets, that occur in colonies. Also known as Widmanstatten structure. Forms during cooling through the beta transus at intermediate cooling rates. [AS1814]

batch *1.* A mass of base compound or catalyst manufactured at one time. The identity of a batch is maintained by an assigned batch number. [AS7200/1] *2.* The quantity of material required for or produced by a production operation at a single time. *3.* A group of similar computer transactions joined together for processing as a single unit.

batch distillation A distillation process in which a fixed amount of a mixture is charged, followed by an increase in temperature to boil off the volatile components. This process differs from continuous distillation, in which the feed is charged continuously.

batch mixer A type of mixer in which starting ingredients are fed all at once and the mixture is removed all at once at some later time. Contrast with continuous mixer.

batch monitor *See* monitor software.

batch process A process by which a finite quantity of material is manufactured by subjecting measured quantities of raw materials to a time-sequential order of processing actions using one or more pieces of equipment.

batch processing Pertaining to the technique of executing a set of programs such that each program is completed before the next program of the set is started.

bat file A file name ending in .bat which contains a list of commands most often used to initiate a computer program.

bathochrome An agent or chemical group that causes the absorption band of a solution to shift to lower frequencies.

bathometer An instrument for measuring depth in the ocean or other body of water.

battery A complete cased assembly of interconnected electrochemical cells, ready for installation in an aircraft. [AS8033]

battery power The maximum power discharge current, declared as a minimum by the manufacturer and expressed in amperes, which a fully charged cell or battery at 23°C is capable of delivering immediately prior to the conclusion of a 15 s maximum power discharge, controlled so as to maintain a constant terminal voltage at one-half of the rated value. [ARP906A]

battery setting Describes a setting of two or more boilers with common division walls.

baud Same as bits per second when each signal event represents only one bit condition.

Baudot code A three-part teletype code consisting of a start pulse (always a space), five data pulses, and a stop pulse (1.42 times the length of the other pulses) for each character transmitted. In this code, various combinations of data pulses are used to designate letters of the alphabet, numerals 0 to 9, and certain standard symbols.

baud rate Any of the standard transmission rates for sending or receiving binary coded data. Standard rates are generally between 50 and 19,200 baud.

Baumé scale Either of two specific gravity scales devised by French chemist Antoine Baumé in 1768 and often used to express the specific gravity of acids, syrups, and other liquids. For light liquids, the scale is determined from the formula: $°Bé = (140/sp \, gr) - 130$; for heavy liquids, it is determined from: $°Bé = 145 - (145/sp \, gr)$. $60°F$ is the standard temperature used.

Bauschinger effect The phenomenon wherein plastic deformation of a metal raises its tensile yield strength, but decreases its compressive yield strength.

B-A-W devices *See* bulk acoustic wave devices.

Bayard-Alpert ionization gage Ionization vacuum gage that includes a tube with an electrode structure designed to minimize x-ray-induced electron emission from the ion collector.

bayonet coupling, cylindrical *See* coupling, bayonet, cylindrical.

bayonet flange The three-lug or slot portion of either an aircraft refueling adapter or a bottom loading adapter. [AIR4783]

B-basis The B-basis or "B" mechanical property value is the value above which at least 90% of the population of values is expected to fall, with a confidence of 95%. [AIR4844]

BCD *See* binary coded decimal.

BCOMP *See* buffer complete.

BDC *See* buffered data channel.

BDHI *See* bearing distance heading indicator.

beaching gear Any wheeled device attached to a seaplane or flying boat to permit it to be moved onto the beach or shore or moved about ashore. May also be gear normally separate from the aircraft, but attached temporarily for the same purpose. [AIR1489]

beacon Light, group of lights, electronic apparatus, or other device that guides, orients, or warns aircraft, spacecraft, etc. in flight.

bead *1.* Layers of steel (usually) wire imbedded in rubber and wrapped with fabric. Gives a base around which the plies are anchored to provide a firm fit on the wheel. To provide increased stiffness, "flippers" are sometimes wrapped around the enclosed beads. Beads restrict air pressure expansion of the tire and prevent jumping off the bead seat of the wheel. [AIR1489] *2.* A rolled or folded seam along the edge of metal sheet. *3.* A projecting band or rim. *4.* A drop of precious metal produced during cupellation in fire assaying. *5.* An elongated seam produced by welding in a single pass.

bead bundle Bead wires wrapped together to form a tire bead. [AIR1489]

beaded tube end The rounded exposed end of a rolled tube when the tube metal is formed over against the sheet in which the tube is rolled.

bead heel The outer bead edge that fits against the wheel flange. [AIR1489]

bead toe The inner bead edge closest to the tire centerline. [AIR1489]

bead wire Steel wire from which tire beads are wound. [AIR1489]

beam *1.* An elongated structural member that carries lateral loads or bending moments. *2.* A confined or unidirectional ray of light, sound, electromagnetic radiation, or vibrational energy, usually of relatively small cross section.

beam center A point located equidistant from the 50% luminance points of the primary beam intensity distribution. [ARP1782]

beam divergence For a laser, the increase in beam diameter with increase in distance from the exit aperture of the laser. Divergence, expressed in milliradians, is measured at specified points across the diameter of the beam.

beam expander An optical system in which a narrow beam is expanded to a larger diameter, ideally without changing the divergence of the beam.

beam injection The introduction of a particle radiation beam into a plasma or ionized gas for the purpose of diagnostics, plasma control, or the study of beam/plasma interactions.

beam integrator A device that integrates the energy in a light beam to make it uniform across the beam cross section.

beam lead as-plated surface The surface of a beam lead that is plated last, and is bonded to the conductor film metallization during the bonding operation. [AS1346]

beam lead bond misalignment On a bonded device, that portion of the unbonded beam lead width not over the conductor film metallization. [AS1346]

beam lead length On an unbonded device, the distance from the edge of the dielectric layer to the outer edge of the beam lead. [AS1346]

beam lead thickness The thickness of an unbonded (undeformed) beam lead. [AS1346]

beam lead width The width of an unbonded (undeformed) beam lead. [AS1346]

beam neutralization Neutralization that takes place by means of charge exchange with a neutral gas.

beam penetration CRT A tube with multiple phosphors in close association whose color and/or persistence can be altered by modulation of the screen voltage. *Layer type*–beam penetration CRT in which the phosphors are deposited separately with an electron-attenuating layer between them. *Onionskin type*–beam penetration CRT in which there is a single deposit of coated granules using several phosphors and attenuating layers between them. [ARP1782]

beam rider guidance System for guiding aircraft, spacecraft, or missiles along a desired path by means of a radar beam, light beam, etc. The center of the beam axis forms a line along which the vehicle senses its location and corrects its course relative to the beam axis.

beam splitter *1.* Partially reflecting mirror that permits some incident light to pass through, and reflects the remainder. *2.* A device that separates a light beam into two beams. Some types affect polarization of the beam.

beam spread Divergence of a sound beam as it travels through a material. [ARP5089]

bearing *1.* The angle, usually expressed in deg (0–360, clockwise), between the direction from an observer (for example, on an aircraft) to an object or point and a reference line. The reference line may be the fore-and-aft axis of an aircraft (relative bearing), true north (true bearing), or magnetic north (compass bearing or magnetic bearing). [ARP4107] *2.* The angle, in the horizontal plane, between the longitudinal axis of the own aircraft and the relative location of an intruder aircraft, measured in the clockwise direction when viewed from above. [ARP4153] *3.* A part in which a journal, gudgeon, pivot, pin, shaft, or similar revolving part is supported to reduce friction. [ARP480A] *4.* A machine part that supports another machine part while the latter undergoes rotating, sliding, or oscillating motion. *5.* That portion of a beam, truss, or other structural member which rests on the supports.

bearing area *1.* The diameter of a hole times the thickness of the material containing the hole. *2.* The cross-sectional area of a bearing load member on a sample. [AIR4844]

bearing circle A ring-shaped device that fits over a compass or compass repeater to facilitate taking compass bearings.

bearing distance heading indicator (BDHI) A flight instrument that displays bearing to a station as well as the distance to the station (DME). [ARP4107]

bearing load A compressive load on an interface. [AIR4844]

bearing mounted actuator An actuator mounted between a support structure and load with spherical bearings at each end. [ARP4386]

bearing strain *1.* The ratio of the deformation of a bearing hole, in the direction of the applied force, to the pin diameter. *2.* The stretch or deformation strain for a sample under bearing load. [AIR4844]

bearing strength The maximum bearing stress that can be sustained. [AIR4844]

bearing stress The applied load divided by the bearing area. Maximum bearing stress is the maximum load sustained by a specimen

69

during a test, divided by the original bearing area. [AIR4844]

bearing yield strength The bearing stress at which a material exhibits a specified limiting deviation from the proportionality of bearing stress to bearing strain. [AIR4844]

beat *See* synchronism.

beat frequencies The frequencies obtained when two simple harmonic quantities of different frequencies 11 and 12 are superimposed. The beat frequency equals 11–12.

beat-frequency oscillator An electrical oscillator that generates a frequency that, in turn, is beat against another frequency to generate a third, usually audible, frequency. Generally used in communications receivers to provide an audible signal for CW reception or to reinsert a carrier for reception of single side band signals.

beating A resultant pulsating waveform sometimes produced when two or more periodic quantities of different frequencies combine.

beat note The wave of different frequency resulting when two sinusoidal waves whose frequencies differ from each other are supplied to a nonlinear device.

beats Periodic pulsations in amplitude that are created when a wave of one frequency is combined with a wave of a different frequency.

beauty defects Those imperfections of components and elements of an optical system that do not affect the optical characteristics. Beauty defects are undesirable, but may be accepted if they do not cause a significant degradation of image quality or environmental stability. [ARP924]

bed *1.* The part of a machine having precisely machined ways or bearing surfaces for supporting and aligning other parts, such as toolholders or dies. *2.* A perforated floor, lining, or support structure, often covered with a layer of granular material, in a furnace, chemical processing tank, or filtration tank.

Beer's law The law relating absorption coefficient to molar density.

Beginner's All-Purpose Symbolic Instruction Code (BASIC) A widely used computer language for personal computers.

behavior The way in which an organism, organ, body, or substance acts in an environment or responds to excitation; for example, the behavior of steel under stress, or the behavior of an animal in a test.

behavioral modeling Modeling a device or component directly in terms of its underlying mathematical equations.

behind the panel *1.* Applies to a location that is within an area that: (a) contains the instrument panel; (b) contains the rack-mounted hardware associated with the instrument panel; or (c) is enclosed within the instrument panel. *2.* Refers to devices that are not accessible for the operator's normal use. *3.* Refers to devices that are not designated as local or front-of-panel-mounted.

bel A dimensionless unit for expressing the ratio of two power levels. The ratio in bels is equal to log (P_2/P_1), where P_1 and P_2 are the two power levels.

Belleville washer *See* disk spring.

bellows *1.* A unit in which one or more convolutions are used to provide flexibility. [ARP699D] *2.* An enclosed chamber whose walls are pleated or corrugated so that the interior volume may be varied. Used either to alternately draw in and expel a gas or other fluid, or to expand and contract in response to variations in internal pressure.

bellows, braided A convoluted unit surrounded by a woven wire sleeve attached to the ends. The sleeve restricts the movements of the unit. [ARP699D]

bellows expansion joint A type of coupling between two pieces of pipe in which a flexible metal bellows is used to prevent leakage, while allowing limited linear movement (e.g., accommodation of thermal expansion and contraction).

bellows, free A convoluted unit that does not incorporate any device or part for the purpose of restricting movement. [ARP699D]

bellows gage A pressure-measuring device in which variations in internal pressure within a flexible bellows causes movement of an end plate against spring force. The position of the end plate is directly related to bellows internal pressure.

bellows meter A differential-pressure-measuring instrument whose a measuring element is a set of opposed metal bellows, the motion of which positions the output actuator.

bellows, restrained A convoluted unit incorporating a means of restraint, other than braid, to prevent axial movement. [ARP699D]

bellows seal *1.* A multi-convolution-type element used as a protective barrier between an instrument and the process fluid. *2.* A seal in the shape of a bellows used to prevent air or gas leakage.

bellows sealed valve A valve in which a bellows replaces the conventional packing gland. One end of the bellows is welded to the rising stem; the other is sealed against the valve body.

bellows stem seal A thin-walled, convoluted, flexible component that makes a seal between the stem and bonnet or body, and thus allows stem motion while maintaining a hermetic seal.

bell-type manometer A gage for measuring differential pressure which consists essentially of a cup inverted in a container of liquid. In this gage, pressure from one source is fed to the inside of the cup, while pressure from a second source is applied to the exterior of the cup. Pressure difference is indicated by the position of the cup in relation to the liquid level.

below minimums Weather conditions below the minimums prescribed by regulation for the particular action involved (for example, landing minimums and takeoff minimums). [ARP4107]

belt drive Power transmitted by a continuous loop belt and pulleys. [ARP4386]

bench (optical) A mounting surface for optical components.

bench check A physical inspection or functional test of a part or item removed for an alleged malfunction. Used to determine if the part or item is serviceable/repairable. A bench check also includes a determination of the extent of maintenance, repair, or possible overhaul required to return it to serviceable status. [ARD50010]

bench mark A natural or artificial object having a specific point marked to identify a reference location, such as a reference elevation.

benchmark program A routine used to determine the performance of a computer or piece of software.

bench test The subjection of aircraft, engines, accessories, equipment, and equipage to prescribed conditions and specifications with the use of shop test equipment to ensure proper functioning. [AIR4896]

bender A device for use in precision and non-precision bending of material [ARP480A]

bending Applying mechanical force or pressure to form a metal part by plastic deformation around an axis lying parallel to the metal surface. Commonly used to produce angular, curved, or flanged parts from sheet metal, rod, or wire.

bending stiffness The sandwich property which resists bending deflections. [AIR4844]

bending-twisting coupling A property of certain classes of laminates that exhibit twisting curvatures when subjected to bending moments. [AIR4844]

bend loss Attenuation caused by high-order modes radiating from the side of a fiber. The two common types of bend losses are: (a) those occurring when the fiber is curved around a restrictive radius of curvature; and (b) microbends caused by small distortions of the fiber imposed by externally induced perturbations, such as poor cabling techniques.

bends A form of decompression sickness produced when gaseous emboli (bubbles; primarily nitrogen) evolve from solution in and around the bending joints. [ARP171]

bend test Ductility test in which a metal specimen is bent through a specified arc around a support of known radius. Used primarily to evaluate the inherent formability of metal sheet, rod, or wire; or to evaluate the weld quality produced with specific materials, joint design, and welding technique.

bent tube boiler A water-tube boiler consisting of two or more drums connected by tubes, practically all of which are bent near the ends to permit attachment to the drum shell on radial lines.

BER *See* bit error rate.

Bernoulli coefficient (K$_b$) Dimensionless coefficient used for the purpose of calculating the change in velocity and corresponding change in static pressure, or "head," when the area of a stream is changed (e.g., by a reducer). This pressure change is measured in units of velocity head.

Bernoulli equation *See* Bernoulli theorem.

Bernoulli theorem Also known as Bernoulli equation. In aeronautics, a law or theorem stating that in a flow of incompressible fluid the sum of the static pressure and the dynamic pressure along a streamline is constant if gravity and frictional effects are disregarded. Named for Daniel Bernoulli, a Swiss scientist who lived from 1700 to 1782.

beryllium A metal that is lighter than aluminum, non-magnetic, and characterized by good electrical conductivity and high thermal conductivity. Used in alloys, especially beryllium copper alloy. [ARP1931]

Bessel A filter characteristic in which phase-linearity across the pass band, rather than amplitude linearity, is emphasized. Also known as constant-delay.

best angle of climb speed *See* V$_x$.

best angle of climb speed with one engine inoperative *See* V$_{xse}$.

best rate of climb speed *See* V$_y$.

best rate of climb speed with one engine inoperative *See* V$_{yse}$.

best straight line A line drawn through the points of a calibration curve such that the distances from the line to a minimum of three of the most distant points are equal. [ARP794]

best-straight-line linearity An average of the deviation of all calibration points. Also called independent linearity.

beta The allotrope of titanium with a body-centered cubic crystal structure occurring at temperatures between the solidification of molten titanium and the beta transus. [AS1814]

beta emitter A radioactive nuclide that disintegrates by emitting a beta particle.

beta eutectoid stabilizer An alloying element that dissolves preferentially in the beta phase; lowers the alpha-beta to beta transformation temperature, under equilibrium conditions; and results in the beta decomposition to alpha plus a compound. The latter is a eutectoid reaction and can be very sluggish for some alloys. Commonly used beta eutectoid stabilizers are iron, chromium, and manganese. [AS1814]

beta factor In plasma physics, the ratio of the plasma kinetic pressure to the magnetic pressure.

beta fleck Transformed alpha-lean and/or beta-rich region in the alpha-beta microstructure, with a beta transus measurably below that of the matrix. Beta flecks have reduced amounts of primary alpha, which may exhibit a morphology different from the primary alpha in the surrounding alpha/beta matrix. [AS1814]

beta interactions *See* weak interactions (field theory).

beta isomorphous stabilizer An alloying element that is soluble in beta titanium in all proportions; lowers the alpha-beta to beta transformation temperature without a eutectoid reaction; and forms a continuous series of solid solutions with beta titanium. Commonly used beta isomorphous stabilizers are vanadium and molybdenum. [AS1814]

beta particle An electron or positron emitted from the nucleus of a radioactive nuclide.

beta ratio The ratio of the diameter of the constriction to the pipe diameter, $B = D_{const}/D_{pipe}$.

beta ray A stream of beta particles.

beta-ray spectrometer An instrument used to measure the energy distribution in a stream of beta particles or secondary electrons.

beta test The second stage of testing a new software program.

beta transus The temperature that designates the phase boundary between the alpha plus beta and beta fields. Commercially pure grades of titanium transform in a range of 1630 to 1760°F (890 to 960°C), depending on oxygen and iron content. In general, aircraft alloys vary in transformation temperature from 1380 to 1900°F (750 to 1040°C). [AS1814]

betatron Particle accelerator in which magnetic induction is used to accelerate electrons.

bevel gear One of a pair of gears whose teeth run parallel to a conical surface so that they can transmit power and motion between two shafts whose axes intersect.

bezel A ring-shaped member surrounding a cover glass, window, cathode-ray tube face, or similar area to protect its edges and often to also provide a decorative appearance.

B-H meter An instrument used to determine the intrinsic hysteresis loop of a magnetic material.

bias *1.* The difference between the average of a measurement population and the true value. [AIR1678] *2.* A constant or systematic error as opposed to a random error. A bias manifests itself as a persistent positive or negative deviation of the method average from the accepted reference value. *3.* The departure from a reference value of the average of a set of values. Bias is thus a measurement of the amount of unbalance of a set of measurements or conditions; that is, error having an average value that is non-zero. *4.* The average d-c voltage or current maintained between a control electrode and the common electrode in a transistor or vacuum tube.

bias (tape) The sine wave, typically 10 times the amplitude and 3.5 times the top frequency, that is applied to tape recording heads along with a signal in order to eliminate most signal distortion.

bias limit The expected upper limit of the true bias error. [AIR1678]

bias load A steady-state load that is unidirectional and constant over full load travel. [ARP4386]

bias/preload A steady-state load that is unidirectional and constant over full load travel. [ARP4386]

bias tire A pneumatic tire in which the ply cords extend to the beads, and are laid at alternate angles substantially less than 90° to the centerline of the tread, the carcass being stabilized by an essentially inextensible circumferential belt. [AS4833]

biaxial load A loading condition in which a laminate is stressed in two perpendicular directions. [AIR4844]

biaxial winding In filament winding, a type of winding in which the helical band is laid in sequence, side by side, with crossover of the fibers eliminated. [AIR4844]

BICEPS A General Electric process-oriented language.

bicycle gear configuration *See* gear configuration, bicycle.

bidirectional Capable of free movement in two opposing directions.

bidirectional laminate A reinforced plastic laminate with the fibers oriented in two directions in its plane. [AIR4844]

bidirectional load cell A column-type strain-gage load cell with female or male fittings at both ends for attaching load hardware. Can be used to measure either tension or compression loading. Also known as a universal load cell.

bidirectional pulse A wave pulse in which intended deviations from the normally constant values occur in two opposing directions.

Bielby layer An amorphous layer at the surface of mechanically polished metal.

bifurcated contact A spring-type contact with a lengthwise slot to provide two segments which apply contact force in the same direction. [ARP914A]

bilateral *1.* Of or having two sides. *2.* Affecting two sides equally; reciprocal. *3.* Fabric used in composite laminates which has substantially equal strength in both the warp and fill directions.

bilateral tolerance The amount of allowable variation about a given dimension. Usually expressed as plus-or-minus a specific fraction or decimal.

bilateral transducer A transducer that can transmit signals simultaneously in both directions between two or more terminations.

billet *1.* A semifinished primary mill product ordinarily produced by hot-rolling metal ingot to a cylinder or prism of simple cross-sectional shape and limited cross-sectional area. *2.* A general term for the starting stock used to make forgings and extrusions.

bimetal A bonded laminate consisting of two strips of dissimilar metals. The bond is usually a stable metallic bond produced by corolling or diffusion bonding. The composite material or bimetal is used most often as an element for detecting temperature changes by means of differential thermal expansion in the two layers.

bimetallic corrosion A type of accelerated corrosion induced by differences in galvanic potential between dissimilar metals immersed in the same liquid medium (electrolyte) and also in electrical contact with each other.

bimetallic element A temperature-sensitive device composed of two materials having different thermal coefficients of expansion. This results in a proportional movement of the free segment of the device with changes in temperature. [AIR1900]

bimetallic thermometer element A temperature-sensitive strip of metal (or other configuration) made by bonding or mechanically joining two dissimilar strips of metal together in such a manner that small changes in temperature will cause the composite assembly to distort elastically, and produce a predictable deflection. The element is designed to take advantage of the fact that different metals have different coefficients of thermal expansion.

bimetric theories Theories of gravitation.

bin Storage hopper to promote steady discharge of dry, granular material.

BIN *See* blockend interrupt.

bin activator A vibratory device sometimes installed in the discharge path of a mass-flow bin or storage hopper to promote steady discharge of dry granular material.

binary *1.* A computer numbering system that uses two as its base, rather than ten. The binary system uses only 0 and 1 in its written form. *2.* A device that uses only two states or levels to perform its functions, such as a computer.

binary alloy A metallic material composed of only two chemical elements (neglecting minor impurities), at least one of which is a metal.

binary cell An information-storage element that can assume either of two stable conditions, and no others.

binary code *1.* Code composed of a combination of entities, each of which can assume one or two possible states. Each entity must be identifiable in time or space. *2.* A code that uses two distinct characters, usually 0 and 1.

binary coded decimal (BCD) Describes a decimal notation in which the individual decimal digits are represented by a group of binary bits. For example, in the 8-4-2-1 coded decimal notation, each decimal digit is represented by a group of four binary bits. In this notation, the number twelve is represented as 0001 0010 for 1 and 2, respectively, whereas in binary notation it is represented as 1100. *See* binary.

binary-coded decimal system A system of number representation in which each digit in a decimal number is expressed as a binary number.

binary counter A counter that counts according to the binary number system.

binary digit *1.* In binary notation, either of the characters 0 or 1; same as bit. *2. See* equivalent binary digits.

binary distillation A distillation process that separates only two components.

binary file In electronics, refers to a file that is not a text file.

binary notation A numbering system that uses the digits 0 and 1, with a base of 2.

binary number A number composed of the characters 0 and 1, in which each character represents a power of two. The number 2 is 10; the number 12 is 1100; the number 31 is 11111, etc.

binary point The radix point in a binary number system.

binary scaler A signal-modifying device (scaler) with a scaling factor of 2.

binary stars Systems of two stars revolving about a barycenter.

binary synchronous A procedure for connecting many terminals that share a single link.

binary synchronous communications (BISYNC) A communications procedure that uses special characters for control of synchronized transmission

binary unit *1.* A binary digit; same as bit. *See also* check bit and parity bit. *2.* A unit of information content, equal to one binary decision, or the designation of one of two possible and equally likely values or states of anything used to store or convey information.

binary word A group of binary digits with place values in increasing powers of two.

binder *1.* A spirally served tape or thread wrap used for holding in place assembled cable components that are awaiting further manufacturing operations. [ARP1931] *2.* In metal founding, a material other than water added to foundry sand to make the particles stick together. *3.* In powder metallurgy, a substance added to the powder to increase green strength of the compact, or a material (usually of relatively low melting point) added to a powder mixture to bond particles together during sintering that otherwise would not bond into a strong sintered body. *4.* A composition that holds together a charge of finely divided particles and increases the mechanical strength of the resulting propellant grain when it is consolidated under pressure. Binders are usually resins, plastics, or asphaltics used dry or in solution. [ARP4386]

binding post A fixed support, generally screw-type, to which conductors are connected. [ARP914A]

Bingham body A non-Newtonian substance that exhibits true plastic behavior; that is, it flows when subjected to a continually increasing shear stress only after a definite yield point has been exceeded.

Bingham viscometer A time-of-discharge device for measuring fluid viscosity. In this device, the fluid is discharged through a capillary tube instead of an orifice or nozzle.

binomial random variable The number of successes in independent trials at which the probability of success is the same for each trial. [AIR4844]

bioastronautics The study of biological, behavioral, and medical problems pertaining to astronautics. This includes systems functioning in the environments expected to be found in space, vehicles designed to travel in space, and the conditions on celestial bodies earth.

biochemical oxygen demand (BOD) The amount of oxygen necessary for the oxidative decomposition of a material by microorganisms. The amount of oxygen consumed in mg/L of water (or waste water) over a period of 5 days at 20°C under laboratory conditions.

bioconversion The transformation of algae and/or other biomass materials in successive stages to aliphatic organic acids, to aliphatic hydrocarbons, to diesel and/or other liquid fuels.

biodynamics The study of the effects of dynamic processes (motion, acceleration, weightlessness, etc.) on living organisms.

biofeedback Originally confined to the presenting of a subject with sensory information about his ongoing physiological activities. Now includes the controlling of specific physiological activities through trained mental effort.

bioinstrumentation Instruments that can be attached to humans or animals to record biological parameters, such as pulse rate, breathing rate, or body temperature.

biological corrosion Deterioration of metal surfaces due to the presence of plant or animal life. May be caused by chemicals excreted by the life form; or by concentration cells, such as those under a barnacle; or by other interactions.

biomagnetism *1.* Magnetic fields surrounding all or parts of a living biological system. *2.* The effects of magnetism on all or parts of a biological entity.

biomass The dry weight of living matter in a given area; expressed in terms of mass or weight per unit of volume or area.

bionics The study of systems, particularly electronic systems, which function after the manner characteristic of, or resembling living systems.

bioreactors Biological processors to remove or produce certain chemicals or a particular chemical.

BIOS *See* basic input output system.

biosatellites Artificial satellites that are specifically designed to contain and support man, animals, or other living material in a reasonably normal manner for an adequate period of time and that, particularly for man and animals, possesses the proper means for safe return to the earth.

biosphere That transition zone between earth and atmosphere within which most forms of terrestrial life are commonly found. Refers to the outer portion of the geosphere and inner or lower portion of the atmosphere.

biotechnology The application of engineering and technological principles to the life sciences.

biotelemetry The remote sensing and evaluation of life functions, e.g., as in spacecraft and artificial satellites.

biotite A widely distributed and important rock-forming mineral of the mica group.

Biot number A standard heat-transfer dimensionless number.

bi-phase A method of bit encoding for serial data transmission or recording whereby there is a signal transition every bit period.

bipolar A digital signaling format in which one logic level is defined as a condition in which energy of one form is present and the other logic level is defined as a condition in which energy of a different form is present. [AIR4288]

bipolarity Capability of assuming negative or positive values.

bipolar technology Technology that uses two different polarity electrical signals to represent logic states of 1 and 0.

bipolar transistor A transistor created by placing a layer of p- or n-type semiconductors between two regions of an opposite type of semiconductor.

bipropellant A propellant, usually a liquid, consisting of two unmixed or uncombined chemicals (a fuel and an oxidizer) fed separately to the gas generator. [ARP4386]

biquinary code A method of coding decimal digits in which each numeral is coded in two parts—the first being either 0 or 5, and the second any value from 0 to 4. The digit equals the sum of the two parts.

birdcage A defect in stranded wire in which the strands have separated from the normal lay. [ARP1931]

birefringence *1.* The difference between the two principal refractive indices. *2.* The ratio between the retardation and thickness of a material at a given point. [AIR4844]

birefringent element A device that has a refractive index which is different for lightwaves of different orthogonal polarizations. Because of this difference, light of the two orthogonal polarizations travels at different speeds and is refracted slightly differently.

Birmingham wire gage (Bwg) A system of standard sizes used in the United States for brass wire, and for strip, bands, hoops and wire made of ferrous and nonferrous metals. The decimal equivalent of standard Bwg sizes is generally larger than for the same gage number in both the American wire gage (AWG) and US steel wire gage systems.

biscuit *1.* A piece of pottery that has been fired but not glazed. *2.* An upset blank for drop forging. *3.* A small cake of primary metal, generally one produced by bomb reduction or a similar process.

bismaleimide A type of polyimide that cures by an addition rather than a condensation reaction, thus avoiding problems with volatiles formation. The bismaleimide is produced by a vinyl-type polymerization of a prepolymer terminated with two maleimide groups. [AIR4844]

bisphenol A A condensation product formed by the reaction of two molecules of phenol with acetone. [AIR4844]

bistable and tristable control A control system in which the power to control the load is fully "on" in either polarity, or fully "on" in one polarity (bistable); "off," or fully "on" in the other polarity (tristable). These systems are sometimes called "on-off" systems or "bang-bang" systems. In such systems, when the time duration of the application of power is modulated by the input, the system is called pulse width modulated (PWM). [ARP4386]

bistable Capable of assuming either of two stable states, hence of storing one bit of information.

bistatic reflectivity The characteristic of a reflector that reflects energy along a line or lines different from, or in addition to, that of the incident ray.

BISYNC *See* binary synchronous communications.

bit *1.* A cutting tool for drilling or boring. *2.* A removable tooth of a saw, milling cutter, or carbide-tipped cutting tool. *3.* The heated tip of a soldering iron. *4.* An abbreviation of the term "binary digit." *5.* A single pulse in a group of pulses. *6.* The smallest unit of information that can be recognized by a computer.

BIT *See* built in test.

bit density A measure of the number of bits recorded per unit of length or area.

BITE *See* built in test equipment.

bit error rate (BER) The number of erroneous bits or characters received from some fixed number of bits transmitted.

bit error rate tester A system that measures the fraction of bits transmitted incorrectly by a digital communication system.

bit map A table that describes the state of each member of a related set. Most often used to describe the allocation of storage space; each bit in the table indicates whether a particular block in the storage medium is occupied or free.

bit pattern A combination of n binary digits to represent 2^n possible choices, e.g., a 3-bit pattern represents 8 possible combinations.

bit rate The rate at which binary digits, or pulses representing them, pass a given point on a communications line or channel. *See also* baud.

bits per second (BPS) In a serial transmission, the instantaneous bit speed within one character, as transmitted by a machine or a channel. *See* baud.

bit stream A binary signal transmitted without regard to grouping by character.

bit string A string of binary digits in which each bit position is considered as an independent unit.

bit synchronizer A hardware device that establishes a series of clock pulses in synchronism with an incoming bit stream and identifies each bit.

bituminous Describes a substance that contains organic matter, mostly in the form of tarry hydrocarbons (bitumen).

blab A type of station failure characterized by continuous generation of a signal on the physical media. [AIR4288]

black body A physical object that absorbs incident radiation, regardless of the spectral character or directional preference of the incident radiation. A perfect black body is most closely approximated by a hollow sphere with a small hole in its wall—the plane of the hole being the black body. A perfect black body is used as an ideal reference concept in the study of radiant energy.

black body radiation The electromagnetic radiation emitted by an ideal black body; the theoretical maximum amount of radiant energy of all wavelengths that can be emitted by a body at a given temperature.

blackbody temperature The true temperature of a blackbody source.

black box A generic term used to describe an unspecified device that performs a special function or in which known inputs produce known outputs in a fixed relationship.

black-bulb thermometer A thermometer whose sensitive element is covered with lampblack to make it approximate a black body.

black data Data that do not require safeguards. [AIR4271]

black light Electromagnetic radiation in the near ultraviolet range of wavelength 320 to 400 nm. [AMS2647A]

blacklight meter A meter containing photosensitive cells for reading of blacklight intensity in microwatts per square centimeter. [AMS2647A]

blackout A temporary loss of vision—sometimes even consciousness—resulting from stagnant hypoxia. Commonly induced by positive G forces of severe intensity and/or duration. *See also* grayout. [ARP4107]

bladder An elastomeric lining for the containment of hydroproof or hydroburst pressurization medium in filament-wound structures. [AIR4844]

blade *1.* A fuse or limiter terminal having a substantially rectangular cross-section. [ARP1199A] *2.* Elements of a turbine rotor that convert gas energy to mechanical energy. [AIR1639] *3.* Arms of propellers or of rotating wings. Specifically (and restrictively), those parts of propellers or rotating wings from the shank outward, i.e., those parts that have efficient airfoil shapes, and that cleave the air. *4.* Vanes, such as rotating vanes or stationary vanes in rotary air compressors, or vanes of turbine wheels.

blade angle The acute angle between the chord of a section of a propeller, or of a rotary wing system, and a plane perpendicular to the axis of rotation. [ARP4107]

blade element theory Analysis of fluid-blade interaction based on one or more stream tubes passing through the blade row. [AIR4548]

blade slap noise Impulsive noise (short, high pressure sound waves) of rotating blades, primarily helicopter blades.

blade-type consistency sensor A pneumatic device for determining changes in consistency of a flowing non-Newtonian substance such as a slurry. Senses the force required for a shaped blade to shear through the flowing stock, and transmits a pneumatic output signal proportional to changes in consistency. Normal operating range is 1.75 to 6.0% suspended solids, with a sensitivity of ±0.02% in many applications.

blank alarm point *See* alarm point.

blank common *See* global common.

blankets (fission reactors) Damper materials for fusion reactors.

blanking *1*. Inserting a solid disc at a pipe joint or union to close off flow during maintenance, repair, or testing. *2*. Using a punch and die to cut a shaped piece from sheet metal or plastic for use in a subsequent forming operation. *3*. Using a punch and die to make a semifinished powder-metal compact.

blast deflectors Devices used to divert the exhaust of a rocket fired from a vertical position.

blast furnace gas Lean, combustible by-product gas resulting from burning coke with a deficiency of air in a blast furnace.

blasting *1*. Detonating an explosive. *2*. Using abrasive grit, sand, or shot carried in a strong stream of air or other medium to remove soil or scale from a surface.

blast vane *See* jet vanes.

blazars Strongly optical polarized active galactic nuclei objects exhibiting BL lacertae-like and quasar-like characteristics.

bleed *1*. A port for removal of air, water, etc., in servicing. [ARP4386] *2*. To evacuate excess resin from an assembly during the curing cycle. [AIR4844] *3*. To give up color when brought into contact with water or solvents. *4*. An undesired movement of certain materials in adhesives to the surface of the bonded article or into adjacent materials.

bleed air Air extracted from the compressor of a gas turbine engine or propulsion unit. [ARP699D]

bleed air regulator Regulator incorporated between the heat exchanger and the OEAS (oxygen enriched air system) concentrator in some OBOGS (on-board oxygen generation system) installations to prohibit excessive concentrator supply pressure, and to improve heat exchanger performance through flow reduction. [ARP171]

bleed air shutoff valve A valve incorporated in some MSOGS (molecular sieve oxygen generation system) applications. Located in the bleed air line between the aircraft engine outlet port and the concentrator; may be automatically and/or manually activated. Purpose is to shut off bleed air to prevent conditions that may cause concentrator damage or malfunction. [ARP171]

bleed and burn Describes a process whereby air is extracted from a compressor or air accumulator, and to which fuel is added and is burned to increase the air temperature. [ARP906A]

bleeder cloth A nonstructural layer of material used in the manufacture or repair of composite parts to allow the escape of excess gas and resin during the cure. [AIR4844]

bleeding *1*. Allowing a fluid to drain or escape to the atmosphere through a small valve or cock. Used to provide controlled, slow reduction of slight overpressure; to withdraw a sample for analysis; to drain condensation from compressed air lines; or to reduce the airspace above the liquid level in a pressurized tank. *2*. Withdrawing steam from an intermediate stage of a turbine to heat a process fluid or boiler feedwater. *3*. Natural separation of liquid from a semisolid mixture; for example, oil separation from a lubricating grease; or water separation from freshly poured concrete. *4*. The removal of excess resin from a laminate during cure. *5*. The diffusion of color out of a plastic part into the surrounding surface or part. [AIR4844]

bleeding cycle A type of steam cycle in which steam is withdrawn from the turbine at one or more intermediate stages and used to heat feedwater before it enters the boiler.

bleedout *1*. The excess liquid resin that migrates to the surface of a winding. Primarily occurs in filament winding. *2*. The spread of adhesive away from the bond area. [AIR4844]

bleed port An outlet in a gas turbine compressor casing through which air is extracted or bled from the compressor. [ARP699D]

blend *1*. To mix ingredients so that they are indistinguishable from each other in the mixture. *2*. To produce a smooth transition between two intersecting surfaces, such as at the edges of a radiused fillet between a shaft and an integral flange or collar. *3*. Powder from different lots that have been thoroughly blended by means such as V-cone blenders or other demonstrated means that ensure adequate particle size distributions. [AMS4994]

blind fastener A fastener that is installed with access from one side only. [AIR4844]

blind hole A hole in a piece of material that does not completely penetrate to the back surface.

blinding glare Glare that is so intense that, for an appreciable length of time, no object can be seen. [AIR1151]

blind nipple A short piece of pipe or tubing with one end closed and sealed.

blind pressure transmitter A pressure transmitter that does not have an integral readout device.

blind speed The rate of departure or closing of a target relative to the radar antenna at which cancellation of the primary radar target by moving target indicator (MTI) circuits (in the radar equipment) causes a reduction or complete loss of signal from that target. [ARP4107]

blinker A type of oxygen-flow indicator that has a shutter which opens and closes during breathing. [ARP171]

blip *1*. On radar screens, a streak of light caused by an object, vehicle, or some electronic disturbance passing through the path of the radar beam. [ARP4107] *2*. Any erratic signal on a computer screen.

blister *1*. A small area on the surface of metal or plastic where a thin layer of the material has

been separated from underlying material and is raised due to gas trapped between the layers, yet remains attached around the edges of the raised area. *2*. An enclosed, macroscopic cavity in a glaze or other fired ceramic coating. *3*. A raised area where a paint, electroplate, or other coating has become detached from the substrate due to accumulation of gas or moisture at the coating-substrate interface.

BL lacertae objects One of a class of astronomical objects exhibiting: (a) rapid variations in intensity at radio, infrared, and optical wavelengths; (b) energy distributions largely at infrared wavelengths; (c) absence of discrete features in low dispersion spectra; and (d) strong and rapidly varying polarization at visual and radio wavelengths.

block *1*. A piece of material, such as wood, stone, or metal, usually with one or more plane or approximately plane faces, used to strengthen or sustain. [ARP480A] *2*. A set of things, such as words, characters, or digits, handled as a unit. *3*. A collection of contiguous records recorded as a unit. Blocks are separated by interblock gaps, and each block may contain one or more records. *See* cell. *4*. A more or less integral group of cylinders constituting an entire bank or a part of a bank of cylinders. [ARP169] *5. See* cylinder block.

blockage seal A type of isolation seal used with a faying-surface seal; a prepack or injection seal that must join the primary seal. [AIR4069]

block and fall *See* block and tackle.

block-and-tackle A hoisting gear consisting of a rope or cable and one or more independently rotating frictionless pulleys. Also known as block and fall.

block, data A set of associated characters or words handled as a unit.

block diagram *1*. A graphical representation of the hardware in a computer system. Primary purpose is to indicate the paths along which information or control flows between the various parts of the computer system. Should not be confused with the term "flow chart." *2*. A coarser and less symbolic representation than a flow chart.

blocked impedance Of an electro-mechanical transducer, the electrical impedance at the input terminals when the mechanical system is "blocked," or is prevented from moving.

blocked load pressure *See* control pressure.

blockend interrupt (BIN) A signal in TELEVENT that indicates that a buffer is completely filled with data.

blocker-type forging A shape forging designed for easy forging and extraction from the die through the use of generous radii, large draft angles, smooth contours, and generous machining allowances. Used as a preliminary stage in multiple-die forging or when machining to final shape is less costly than forging to final shape.

block flow The distance that freshly mixed, uncured sealant will sag while hanging from a vertical surface for approximately one-half hour at standard conditions. [AS7200/1]

blocking *1.* Producing a semifinished forging of approximate shape suitable for further forging or machining to final size and shape. *2.* Reducing the oxygen content of the bath in an open-hearth furnace. *3.* Undesired adhesion between plastics surfaces during storage or use.

block maintenance program *See* program, block maintenance.

block out device A device utilized to prevent a hose-end pressure-control valve from closing or tending to close. [AIR4783]

block sequence A welding sequence in which separated lengths of a continuous multiple-pass weld are built up to full cross section before gaps between the segments are filled in. Compare with cascade sequence.

block switching A two-level multiplexing technique used in data transmission, whereby one level selects the input channel to be transmitted and the second level selects the group of first-level input channels to be addressed. The chief advantage of block switching is reduction of leakage currents from "off" channels, which interfere with data signals being transmitted. Also known as submultiplexing.

blocky alpha Alpha phase that is considerably larger and more polygonal in appearance than the primary alpha present. Blocky alpha is induced by unidirectional metal working and has an aspect ratio of 3:1 or higher. It may result from extended exposure high in the alpha-beta phase field following rapid cooling through the beta transus during forging or heat treating operations. May accompany grain boundary alpha. Blocky alpha may be removed by beta recrystallization or by all-beta working followed by further alpha-beta work. The microhardness of blocky alpha is not significantly different from that of the surrounding normal alpha-beta matrix. [AS1814]

bloedite A mineral consisting of colorless hydrous sodium magnesium sulfate. Also known as astrakanite or astrochanite.

blood-brain barrier A mechanism that maintains the constancy of the neurons in the central nervous system by preventing certain substances from leaving the bloodstream and entering the neural tissue.

bloom *1.* A section of a strand after it has been cut from the strand; as-cast material that is no longer part of the whole strand. [MAM2304] *2.* A visible, local exudation or finish change on the surface of a plastic. [AIR4844] *3.* A semifinished metal bar of large cross section (usually a square or rectangle exceeding 36 sq in.) that has been hot rolled or forged from ingot. *4.* Visible fluorescence on the surface of lubricating oil or an electroplating bath. *5.* A bluish fluorescent cast on a painted surface caused by a thin film of smoke, dust, or oil. *6.* A loose, flower-like corrosion product formed when certain nonferrous metals are exposed to a moist environment. *7.* To apply an antireflection coating to glass. *8.* To hammer or roll metal to brighten its surface.

blowback The difference between the pressure at which a safety valve opens and the pressure at which it closes, usually about 3% of the pressure at which the valve opens.

blowby *1.* Leakage of fluid through the clearance between a piston and its cylinder during operation. *2.* A condition in which fluid leaks at a high rate across a sealing surface, permitting high-pressure fluid to pass from one cavity to one of much lower pressure.

Usually occurs during rapid pressure reversals. [ARP4386]

blowdown *1.* In a safety valve, the difference between opening and closing pressures. *2.* In a steam boiler, the practice of periodically opening valves attached to the bottom of steam drums and water drums, during boiler operation, to drain off accumulations of sediment.

blow down accumulator Similar to a direct blow down system except that the stored, cold gas is applied to a vessel containing hydraulic fluid with provision for fluid separation. The incompressible fluid then powers the actuation components. *See* direct blowdown system. [ARP4386]

blow down valve A valve generally used to continuously regulate the concentration of solids in a boiler (but not a drain valve).

blower A fan used to force air under pressure.

blowhole A pocket of air or gas trapped during solidification of a cast metal.

blowing agent A substance used to cause expansion in the manufacture of hollow or cellular articles. [AIR4844]

blow-off valve A specially designed, manually operated valve connected to a boiler for the purpose of reducing the concentration of solids in the boiler or for draining purposes.

blowout Term used to indicate that a combuster has become unlit. [ARP906A]

blue brittleness In some steels, loss of ductility associated with tempering or service temperatures in the blue heat range, 400 to 600°F.

blueing *See* bluing.

blue stars Stars of spectral type B, A, or F according to the Draper catalog.

blue vitriol A solution of copper sulfate sometimes applied to metal surfaces to make scribed layout lines more visible.

bluff bodies Bodies having a broad, flattened front, as in some reentry vehicles.

bluing Also spelled blueing. *1.* Forming a bluish oxide film on steel by exposing it to steam, air, or other agents at a suitable temperature. Provides a scale-free surface, an attractive appearance, and improved corrosion resistance. *2.* Heating formed springs after fabrication to improve their properties and reduce residual stress.

blunt leading edges The obtuse cross sections of certain front edges of airfoils or wings.

blunt trailing edges The rounded or obtuse-angled trailing edges of wings and/or control surfaces designed to enhance aerodynamic characteristics.

BNI *See* Bureau d'Orientation de la Normalisatin en Informatique.

board In computers, a flat sheet on which integrated circuits are mounted. *See also* panel.

boattails The rear portions of elongated bodies, such as in rockets, having decreasing cross-sectional area toward the rear.

BOD *1. See* biochemical oxygen demand. *2. See* end of descent.

bode diagram A graphical depiction of servovalve frequency response, wherein phase lag and logarithmic amplitude ratio (usually expressed in dB) are plotted versus input signal frequency. [ARP490]

Bode plot *1.* Presentation of system frequency response data plotted in rectangular coordinate form. The magnitude ratio (output-to-input) and phase angle (time relationship of output-to-input) are plotted against frequency of a change in input. [AIR1823] *2.* A graph of transfer function versus frequency wherein the gain (often in decibels) and phase (in degrees) are plotted against the frequency on log scale. Also called a bode diagram.

bodies of revolution Symmetrical bodies having the form described by rotating a plane curve about an axis in its plane.

body axes A set of orthogonal axes rigidly fixed in the airplane, with origin located at the center of gravity, and extending forward, downward, and to the right. [ARP4386]

body, connector The main portion of a connector consisting of the housing and insert assembly, to which contacts and accessories are attached. [ARP914A]

body, encapsulated A body with all surfaces covered by a continuous surface layer of a different material, usually an elastomeric or polymeric material.

body pressure *See* case drain (body pressure).

body putty A paste-like mixture of plastic resin and talc used in repair of metal surfaces. [AIR4844]

body, split A valve-body design in which trim is secured between two segments of a valve body.

body temperature (non-biological) *See* temperature.

body temperature regulation *See* thermoregulation.

body, wafer A thin, annular section body whose end surfaces are located and clamped between the piping flanges by bolts extending from flange to flange.

body, wafer, lugged A thin, annular section body whose end surfaces mount between the pipeline flanges, or may be attached to the end of a pipeline without any additional flange or retaining parts, using either through-bolting and/or tapped holes.

body, weir type A body having a raised contour contacted by a diaphragm to shut off fluid flow.

bogey *See* bogie.

bogie Also spelled bogey or bogy. *1*. A low, strongly built cart. In landing gear usage, refers to two or more wheels or sets of wheels on axles connected by means of an axle beam (or bogie beam), and pivoted at the lower end of a shock strut assembly or structure to permit pitching motion as it rides over the runway or other supporting surface. *See also* truck. [AIR1489] *2*. A type of aircraft landing gear consisting of two sets of wheels in tandem with a central strut. *3*. A supporting and aligning idler wheel or roller on the inside of an endless track.

bogie gear A type of landing gear configuration utilizing a shock strut or structural support assembly; and, at the lower end, a pivoted bogie assembly. [AIR1489]

bogy *See* bogie.

Bohr magneton A constant equivalent to the magnetic moment of an electron.

boiler That part of an air-conditioning system or heat exchanger in which a source of heat is utilized to vaporize a liquid heat transfer medium. [ARP147C]

boiler horsepower The evaporation of 34-1/2 lbs of water per hour from a temperature of 212°F into dry, saturated steam at the same temperature. Equivalent to 33,475 Btu.

boiler water A term construed to mean a representative sample of the circulating boiler water, after the generated steam has been separated, and before the incoming feed water or added chemical becomes mixed with it so that its composition is affected.

boiling out The boiling of a highly alkaline water in boiler pressure parts for the removal of oils, greases, etc.

boiling point The temperature at which a liquid vaporizes at standard atmospheric pressure. [AIR1116]

boilup Vapors that are generated in a column reboiler.

bolides Brilliant meteors—especially ones that explode, detonating fireballs.

bolometer *1*. Instrument that measures the intensity of radiant energy by employing thermally sensitive electrical resistors; a type of actinometer. Also called thermal detector. *2*. A sensitive infrared detector whose operation is based on a change in temperature induced by absorbing infrared radiation. Made of two thin, blackened gratings of platinum, one illuminated and the other kept in the dark. The absorption of heat changes the electrical resistance, which is detected by comparing the resistances of the two gratings in an electrical circuit.

bolster *1*. A steel block or plate used to support dies and attach them to a press bed. *2*. In drop forging, a device used to attach dies to the ram.

bolt A threaded fastener consisting of a rod, usually made of metal, having threads at one end and an integral round, square, or hexagonal head at the other end. Short bolts usually have threads running the entire length below the head, and longer bolts often have an unthreaded shank between the head and threaded end.

bolter An attempted arrested landing wherein: (a) the arresting hook fails to engage the arresting wire due to hook skip; (b) the aircraft touches down beyond the wire area;

(c) the arresting hook fails to retain the wire; or (d) the arresting hook or wire fails after engagement. The aircraft then becomes airborne and resumes normal flight. [AIR1489]

bomb calorimeter An apparatus for measuring the quantity of heat released by a chemical reaction. Consists of a strong-walled metal container (bomb) immersed in about 2.5 liters of water in an insulated container. A sample is sealed in the bomb, the bomb immersed, the sample is ignited (or a reaction started) by remote control, and the heat released is measured by observing the rise in temperature of the water bath.

bombs (ordnance) Explosive devices designed to be detonated under specified conditions.

bond *1.* An electrical connection between conductive parts which provides the required electrical conductivity. [ARP1870] *2.* A wire rope that attaches a load to a crane hook. *3.* Adhesion between cement or mortar and masonry. *4.* In an adhesive bonded or diffusion bonded joint, the junction between faying surfaces. *5.* In welding, brazing, or soldering, the junction between assembled parts. Where filler metal is used, it is the junction between fused metal and heat-affected base metal. *6.* In grinding wheels and other rigid abrasives, the material that holds abrasive grains together. *7.* Material added to molding sand to hold the grains together. *8.* The junction between base metal and cladding in a clad metal product.

bonded Describes conductive parts that are mechanically interconnected to maintain a common electrical potential. [ARP1870]

bonded liner In a butterfly valve body, a liner vulcanized or cemented to the body bore.

bonded strain gage A device for measuring strain which consists of a fine-wire resistance element, usually in zigzag form, embedded in nonconductive backing material, such as impregnated paper or plastic. The backing material is cemented to the test surface or sensing element.

bonded structure The structure resulting when a combination of parts is assembled and the parts are intimately attached to one another by applying a structural adhesive to the faying

surfaces, followed by curing of the adhesives by pressure and/or heat. [AIR4844]

bonded transducer A pressure sensor that uses a bonded strain gage to generate the output signal.

bond face The part or surface of an adherend that serves as a substrate for an adhesive. [AIR4844]

bonding *1.* Specifically, a system of connections between all metal parts of an aircraft or other structure forming a continuous electrical unit and preventing jumping or arcing of static electricity. *2.* Gluing or cementing together for structural strength. *3.* A general term applied to the process of electrically connecting two or more conducting objects. [ARP4404]

bonding cable An electrically conductive wire that is used to equalize differences in potential due to accumulated static electricity, both before and during fuel-transfer operations, such as aircraft fueling and vehicle loading or discharge. *See also* grounding cable. [AIR4783]

bonding connector A device used to connect exposed metal to ground. Normally carries no current, but is used as a current path to eliminate shock or spark hazards, and insures the operation of circuit protective devices in cases of insulation breakdown. [ARP914A]

bonding reel A device onto which a bonding cable is wound and stored. [AIR4783]

bond length On a bonded device, the distance from the outer edge of the bonded beam lead to the apparent inner edge of the bonded area. [AS1346]

bond line The layer of adhesive that attaches two adherends. [AIR4844]

bondline cracks Cracking in an adhesive layer as a result of strain. [AIR4844]

bond ply The ply of a prepreg material that is placed against the fluted core of a radome. [AIR4844]

bond strength The unit load applied in tension, compression, flexure, peel, impact, or shear required to break an adhesively bonded assembly, with the failure occurring either within the adhesive or at the adhesive/adherent interface. [AIR4844]

bondtester An electronic device used for testing the integrity of an adhesive bond. [ARP5089]

Bonne projection A type of conical map projection in which meridians are plotted as curves and the parallels are spaced along them at true distances.

bonnet, seal-welded A bonnet welded to a body to provide a zero-leakage joint.

Boolean Pertaining to logic quantities.

Boolean algebra *1*. The study of the manipulation of symbols representing operations according to the rules of logic. Boolean algebra corresponds to an algebra using only the number 0 and 1, and therefore can be used in programming digital computers which operate on the binary principle. *2*. A process of reasoning or a deductive system of theorems using a symbolic logic, and dealing with classes, propositions, or on-off circuit elements. Boolean algebra employs symbols to represent operators, such as AND, OR, NOT, EXCEPT, IF, THEN, to permit mathematical calculation. Named after George Boole, famous English mathematician.

Boolean expression A quantity expressed as the result of Boolean operations, such as AND, OR, and NOT, upon Boolean variables.

Boolean functions A system of mathematical logic often executed in circuits to provide digital computations, such as OR, AND, NOR, and NOT.

Boolean operator A logic operator, each of whose operands, and whose result have one of two values.

Boolean variable *See* logical variable.

boost control valve *See* control valve.

booster *1*. An intermediate explosive charge to augment the initiating component of an explosive train and cause detonation or deflagration of the main explosive charge. [AIR913] *2*. Propulsion system employed in the launching phase of a vehicle flight. [AIR913]

booster charge The ignition component, ignited by the primer charge, that either pressurizes the generator to operating pressure or generates a higher-energy shock pulse. [ARP4386]

booster fan A device for increasing the pressure or flow of a gas.

booster relay A volume- or pressure-amplifying pneumatic relay that is used to reduce the time lag in pneumatic circuits by reproducing pneumatic signals with high-volume and/or high-pressure outputs.

boostglide vehicles Vehicles designed to glide in the atmosphere following a rocket-powered phase. Portions of the flights may be ballistic, out of the atmosphere.

boot *1*. A connector accessory, usually made from a flexible or semi-rigid insulating material, designed to house wire/cable terminations as a protective device; provide harness direction; and provide a moisture seal when bonded or used as a potting form. [ARP914A] *2*. A computer routine in which a few instructions are loaded, which then cause the rest of the system to be loaded.

bootstrap A technique for loading the first few instructions of a routine into storage, then using these instructions to bring in the rest of the routine. This usually involves either the entering of a few instructions manually or the use of a special key on the console.

bootstrap loader A routine whose first instruction is sufficient to load the remainder of itself into memory from an input device. Normally used to start a complete system of programs.

bootstrap reservoir *1*. An airless reservoir with two pistons of unequal area. *2*. A large-area piston connected to the pump inlet and a small-area piston connected to the pump outlet. [ARP4386]

bore *1*. The inner cavity in a pipe or tube. *2*. The diameter of the cylinder of a piston-cylinder device, such as a reciprocating compressor, engine, or pump; or a hydraulic or pneumatic power cylinder. *3*. To penetrate or pierce a workpiece with a rotating cutting tool. *4*. To increase the size of an existing hole, generally with a single-point cutting tool, while either the work or the cutting tool rotates about the central axis of the hole. *5*. The inner surface of a gun tube. *6*. The central hole in a laser or other type of tube (e.g., a capillary, a waveguide, or a hole in a micro-channel plate).

boredom *See* attention, anomalies of boredom.

boreholes Holes made by drilling into the ground to study stratification, to search for or to obtain natural resources, or to release underground pressures.

bore Reynolds number Calculated Reynolds number including R_d using V_{bore}, P_{bore}, μ_{bore}, d_{bore}; also $R_d = R_D/\beta$.

borescope A viewing device used to visually inspect items (e.g., cannon bores, internal combustion engine cylinder walls, propellant perforations of rocket motors or similar items) for defects of manufacture, corrosion, scars, erosion, etc., without the necessity of disassembling the item(s). A borescope is basically a straight-tube telescope using a mirror or prism. It may or may not have extension tubes to cover the various ranges of inspection. [ARP480A]

boresight error Linear displacement between two parallel lines of sight.

boresighting Initially aligning a gun, directional antenna, or other device by optical means, or by observing a return signal from a fixed target at a known location. The term is derived from an early military practice of looking down the bore of an artillery piece to obtain an initial line of sight to a target.

boron counter tube A type of radiation counter tube used to detect slow neutrons. Contains electrodes coated with a boron compound, and also may be filled with BF_3. A slow neutron is easily absorbed by a B^{10} nucleus, with subsequent emission of an alpha particle.

boron fiber A fiber produced by vapor deposition of elemental boron, usually onto a tungsten filament core, to impart strength and stiffness. [AIR4844]

borosilicate glass A type of heat-resisting glass that contains at least 5% boric acid.

Borsic (tradename) Trademark of United Aircraft Products, Inc. for its boron aluminum composite materials.

bort Industrial diamonds or diamond fragments.

boss *1.* A raised portion of metal or small area and limited thickness on flat or curved metal surfaces. *2.* A short, projecting section of a casting, forging, or molded plastics part, often cylindrical in shape, used to add strength or to provide for alignment or fastening of assembled parts.

bottle A hollow vessel, usually having a neck smaller than the body, and a narrow mouth for a stopper or other type closure. [ARP480A]

bottom contraction The vertical distance from the crest to the floor of the weir box or channel bed.

bottom dead center The position of a piston and its connecting rod when at the extreme downstroke position.

bottoming dies *See* butting dies.

bottoming drill A flat-end twist drill that converts the conical bottom of a blind hole into a cylinder.

bottoming tap A tap designed for cutting full threads all the way to the bottom of a blind hole. Usually used to finish the bottom of a hole tapped with a regular, tapered-end tap.

bottom loading Method of loading a tank utilizing a closed system through the bottom of the tank. [AIR4783]

bottom loading adapter A quick disconnect, dry-break connection that mates either a nozzle or coupler for the purposes of loading a tank vehicle. [AIR4783]

bottom plate A steel plate fixed to the lower section of a mold, often used to join the lower section of the mold to the platen of the press. [AIR4844]

bottoms The higher-boiling product streams usually taken from the bottom of a distillation column—but sometimes taken from the reboiler, and sometimes from a separate surge vessel.

Bouguer law A relationship describing the rate of decrease of flux density of a plane-parallel beam of monochromatic radiation as it penetrates a medium which both scatters and absorbs at that wavelength. Also known as Lambert law.

boundary echo A reflection of an ultrasonic wave from an interface. [ARP5089]

boundary element method Technique for solving two- and three-dimensional boundary value problems in thermodynamics, mechanics, etc.

boundary integral method Technique related to the boundary element method, and used for laminar and turbulent flow problems.

boundary layer *1.* The layer of air immediately adjacent to a moving surface, such as an airfoil. The movement of air in the boundary layer relative to that at the surface is relatively small. [ARP4107] *2.* In a flowing fluid, a low-velocity region along a tube wall or other boundary surface.

boundary layer amplifier An amplifier that utilizes the control of the separation point of a power stream from a curved or plane surface to modulate the output. [ARP993A]

boundary layer plasmas Plasmas resulting from the frictional heat of hypersonic spacecraft entering the earth's atmosphere.

boundary lubrication A condition in sliding contact in which contact pressures are high enough and sliding velocities low enough that hydrodynamic lubrication is completely absent. Mating surfaces slide across each other on a multimolecular layer of lubricant, often with some solid-to-solid surface contact; for liquid lubricants, a bearing-characteristic (Sommerfield) number of 0.01 is considered to be the upper limit of boundary lubrication.

boundary structure The fuel-tight primary structure of an integral fuel tank, including skin panels, bulkheads, and spars (all elements that form the tank boundaries). [AIR4069]

boundary value problem Physical problem completely specified by a differential equation in an unknown and certain information (boundary condition) about the unknown. The differential equation is valid in a certain region of space. The information required to determine the solution depends completely and uniquely on the particular problem.

bound water In a moist solid to be dried, that portion of the water content which is chemically combined with the solid matter.

Bourdon element *See* Bourdon tube.

Bourdon pressure gage *See* Bourdon tube.

Bourdon tube A flattened tube, twisted or curved, and closed at one end, which is used as the pressure-sensing element in a mechanical pressure gage or recorder. A process stream pressure is routed to the open end of the tube, and the tube flexes or untwists in relation to the internal pressure, with the change in shape of the tube being used to operate a mechanical

pointer or pen positioner. Also known as a Bourdon element or Bourdon pressure gage.

Boussinesq approximation The assumption (frequently used in the theory of convection) that the fluid is incompressible except insofar as the thermal expansion produces a buoyancy.

bow shock waves *See* bow waves and shock waves.

bow waves Shock waves in front of a body, such as an airfoil, or apparently attached to the forward tip of the body. Also called bow shock waves.

box A flow-chart symbol.

box beam *See* box girder.

boxcar averager A signal-processing instrument that averages selected portions of repetitive signals to improve signal quality. Sometimes called a gated integrator because it passes or gates portions of the signal, then integrates them.

box girder A hollow girder or beam, usually having a spare or rectangular cross section. Also known as box beam.

box header boiler A horizontal boiler of the longitudinal or cross-drum type, consisting of front and rear inclined rectangular headers connected by tubes.

Boyle's Law Gas law stating that the volume of a gas is inversely proportional to pressure, temperature remaining constant. [ARP171] *See* gas laws.

B power supply An electrical power supply connected in the plate circuit of a vacuum tube electronic device.

BPS *See* bits per second and baud.

bracket A support or structural part that attaches and holds a duct or other component to the vehicle structure.

bracket, adjustable A bracket that allows some freedom in the location of the duct, which it supports, during installation. [ARP699D]

bracket, anchor A main duct support at a point on the duct system that remains fixed with respect to duct system expansions, constrictions, or deflections. [ARP699D]

bracket, guide A support designed to provide for movement of a duct in a predetermined direction, while supporting it in all other directions. [ARP699D]

bracket, sliding A bracket that allows a duct to slide in a controlled direction while supporting it in all other directions. [ARP699D]

bracket, swing A linkage that supports a duct and allows movement along the arc of the swinging bracket. [ARP699D]

Bragg curve A curve showing the average specific ionization of an ionizing particle of a particular kind as a function of its kinetic energy, velocity, or residual range.

Bragg law A principle describing the apparent reflection of x-rays (and DeBroglie waves associated with certain particulate beams) from atomic planes in crystals. Maximum reflected intensity occurs along the family of directions defined by $\theta = \arcsin \lambda n 1/2d$, where θ is the Bragg angle (angle of reflection and of incidence), n is an integer, λ is the wavelength of monochromatic radiation reflected from the crystal, and d is the interplanar spacing of the reflecting parallel planes in the crystal.

braid *1.* Woven wire sleeving used to limit movement of a bellows. [ARP699D] *2.* A woven or braided sheath made from conductive or nonconductive material. [ARP914A] *3.* An assembly of fibrous or metallic filaments woven to form a protective and/or conductive covering over one or more wires; or as a flexible, metallic, conductive cable, such as a grounding strap. [ARP1931]

braid angle The angle of the braided filaments or fibers in relation to the longitudinal axis of the wire or cable. [ARP1931]

braid carrier The yarn or strand, or group of yarns or strands, laid parallel in the braid by a single bobbin of the braider. [ARP1931]

braid end An individual yarn or strand in a braid carrier. [ARP1931]

braiding Weaving of fibers into a tubular shape instead of a flat fabric, as for graphite-fiber-reinforced golf club shafts. [AIR4844]

brake A machine used to bend sheet metal.

brake, annular piston A type of brake utilizing for operation an annular or ring-shaped piston/cylinder arrangement and powered by hydraulic or air pressure to apply force. When force is applied, a set of alternate rotating/stationary discs is squeezed together, producing torque and a decelerating force. [AIR1489]

brake back plate The structural end plate on a multiple disc brake—opposite the actuators. It's purpose is to restrain the discs and resist the force exerted by the actuators. [AIR1489]

brake carrier *See* brake housing.

brake chatter A self-induced brake vibration of less than 100 cycles per second, excited by the friction characteristics of the rubbing surfaces. [AIR1489]

brake control system The total system for control of a brake or brakes. May be a mechanical system, hydraulic system, electrical/electronic system, etc., but generally includes a combination of types; for example, mechanical pilot input to brake pedals, cable system to a metering valve and to the brakes actuators via an electronically/mechanically controlled antiskid system. [AIR1489]

brake disc A single disc of a brake. May be a rotating disc (rotor) in a single- or multiple-disc brake or stationary disc (stator) in a multiple-disc brake. Functions in a heat-absorbing capacity and to produce a frictional retarding force. May be a solid disc or an assembly of segments. [AIR1489]

brake, disc A type of brake utilizing one or more rotating discs with a clamping force applied to produce a frictional retarding force. [AIR1489]

brake displacement The volumetric displacement required in a brake actuation system to actuate from full off to full on. [AIR1489]

brake drag force Any resistance to motion produced by a brake assembly. May apply to an on or off condition but generally considered as the decelerating force produced by the brake when powered to on condition. [AIR1489]

braked roll The portion of an air-vehicle landing roll during which the brakes are applied for deceleration of the vehicle. [AIR1489]

brake drum Cylindrical, rotating part fastened to the wheel. Acts as a friction surface, heat sink, or transmits torque. [AIR1489]

brake, drum *See* drum, brake.

brake energy (kinetic energy) The energy due to the motion of the vehicle mass which must be converted by the brake assembly to some other form of energy and either dissipated or absorbed. [AIR1489]

brake, expander tube A drum-type brake using a rubberlike tube to expand and press the brake lining segments against the friction surface of the drum. [AIR1489]

brake horsepower The power of an engine or other motor as calculated from the force exerted on a friction brake or absorption dynamometer applied to the flywheel or the shaft. *See also* shaft horsepower. [ARP4107]

brake housing The brake part that is fixed to the axle and is used to contain hydraulic or pneumatic fluid passages and the actuating pistons and cylinders. Also called brake carrier, power plate, or actuator housing. [AIR1489]

brake lining *See* lining, brake.

brake linkage *1.* Links, levers, or structural members used to connect an axle-mounted, free-rotating brake assembly to the fixed structure of the gear. *2.* A linkage system for the purpose of actuating a brake assembly. [AIR1489]

brake master cylinder In a fluid-operated brake system, the cylinder assembly that receives an input command force and, in turn, transmits a fluid (hydraulic/pneumatic) force to one or more actuating cylinders, usually located at or on the individual brake assemblies. [AIR1489]

brake, multiple disc A brake assembly utilizing a multiplicity of alternate rotating and stationary discs (rotors and stators). [AIR1489]

brake, multiple piston A brake configuration utilizing two or more actuating cylinders and pistons, interconnected and circumferentially spaced around the brake housing. [AIR1489]

brake overheat indicator A device or system for detecting and indicating to the pilot that an overheat condition, i.e., a caution and/or warning, exists in the brakes. [AIR1489]

brake port An outlet port carrying modulated brake actuating pressure. [ARP4386]

brake pressure metering valve A fluid-metering valve in the brake control system. Upon receiving an input command (mechanical or electrical), the valve meters fluid pressure and flow (in proportion to the input command) to the brakes. [AIR1489]

brake pressure plate A stationary disc with a prescribed cross-section used to distribute the applied piston-actuating loads or forces to the brake discs. Also called a primary disc or retractor plate. [AIR1489]

brake rotor Rotating disc. A part normally keyed to the wheel and rotating with it relative to the fixed axle. [AIR1489]

brake shoe A structural part of the brake assembly which is actuated by the system force input, and which presses the brake lining against the rotating brake drum. *See also* shoe. [AIR1489]

brake, single disc A brake configuration utilizing a single rotating disc. [AIR1489]

brake, spot type A brake configuration utilizing a single cylinder and piston arrangement with a "spot" type lining carrier to rub against the rotating disc. [AIR1489]

brake squeal A self-induced brake vibration mode with a frequency greater than 100 cycles per second. [AIR1489]

brake stator Stationary disc. A part normally keyed to a stationary part of the brake assembly. [AIR1489]

brake torque The torque developed by the friction elements of the brake assembly. This torque is transmitted through the wheel and tire where it is reacted by a drag force at the tire/runway interface. Other resultant forces are reacted by the air vehicle structure. *See also* brake torque compensating linkage. [AIR1489]

brake torque compensating linkage A mechanical linkage for the purpose of transmitting brake torque reaction loads to the fixed structure of a landing gear structure. Especially in a bogie-type landing gear configuration where brake drag loads reacted at the tire/runway interface would cause excessive pitching of the bogie assembly if uncompensated with the linkage to fixed-gear structure. The linkage system may also contain hydromechanical and/or automatic electronic control elements. Also known as brake torque reaction linkage. [AIR1489]

brake torque reaction linkage *See* brake torque compensating linkage.

brake torque tube A tubular-shaped part of the brake assembly which fits around the axle, and to which the stators (stationary discs) are keyed. The torque tube transmits the stator torque to the brake housing and/or other fixed brake structure, and to the axle, compensating linkage, or gear structure. [AIR1489]

brake valve A valve for control of wheel brake actuating pressure. [ARP4386]

braking action Refers to a report of conditions on the airport surface area which provides the pilot with an estimate of the degree/quality of braking that might be expected. Braking action is reported as good, fair, poor, or nil. *See also* runway condition reading (RCR). [ARP4107]

braking, antiskid *See* antiskid brake.

braking, automatic Control of air vehicle braking with an automatic system and the pilot out of the control loop. An input of the desired rate of deceleration is made prior to landing. The system generally includes an antiskid system and manual selection of a deceleration rate. Eliminates pilot technique during the braked roll. [AIR1489]

Brale A 120° conical diamond indenter used in Rockwell hardness testing of relatively hard metals.

branch The selection, based on some criterion, of one of two or more possible paths in the control of flow. The instructions that mechanize this concept are sometimes called branch instructions; however, the terms transfer of control and jump are more widely used. *See also* conditional transfer.

branch circuit That portion of a wiring system extending beyond the final overcurrent device protecting the circuit. [ARP1199A]

branched polymer In molecular structure of polymers, a main chain with attached side chains, in contrast to a linear polymer. [AIR4844]

branch instruction An instruction that performs a branch.

branchpoint A point in a routine at which one of two or more choices is selected under control of the routine. *See also* conditional transfer.

brass Any of the many alloys based on the binary system copper-zinc. Most brasses contain no more than 40 wt% zinc.

Brayton cycle A thermodynamic cycle consisting of two constant-pressure processes inter-spersed with two constant-antropy cycles. Named after George B. Brayton, American engineer.

braze welding A joining process similar to brazing, but in which the filler metal is not distributed in the joint by capillary action.

Brazilian space program The space program of Brazil, which is under the jurisdiction of the Instituto de Pesquisas Espaciais (INPE).

brazing A joining process in which a filler material that melts at a temperature in excess of 800°F, but less than the melting point of the base metal, combines with the base metal. [ARP1931]

breadboard Describes nonproduction engine or system control that can be easily modified. [AIR4548]

breadboard model *1.* Assembly of preliminary circuits or parts used to prove the feasibility of a device, circuit, system, or principle without regard to the final configuration or packaging of the parts. *2.* A prototype or uncased assembly of an instrument or electronic device whose parts are laid out on a flat surface and connected together to demonstrate or check its operation.

break To interrupt the sending end and take control of the circuit of the receiving end.

breakaway torque *See* torque, breakaway.

breakdown *1.* Initial hot working of ingot-cast or slab-cast metal to reduce its size prior to final working to finished size. *2.* A preliminary press-forging operation.

breakdown voltage The electrical potential necessary to cause the passage of a specified electric current through an insulator or insulating material. [ARP1931]

break-in An initial run procedure required to produce desired surface conditions on component parts (e.g., gears, piston and cylinder) operating relative to each other. Also referred to as run-in. [ARP906A]

breaking extension *1.* The elongation necessary to cause rupture of a test specimen. *2.* The tensile strain at the moment of rupture. [AIR4844]

breaking factor The breaking load divided by the original width of a test specimen, expressed in lb/in. [AIR4844]

breaking length A measure of the breaking strength of yarn; the length of a specimen whose weight is equal to the breaking load. [AIR4844]

break loose torque Torque required to produce nut rotation from the seated condition. [AS1895]

breakout *1.* Fiber separation or break on surface plies at drilled or machined edges. [AIR4844] *2.* The point along the length of a harness or other multiconductor configuration (other than at the end), at which a wire or group of wires leaves the configuration. [ARP1931]

breakout load The minimum load under which the system starts moving while the controller is gradually moved toward saturation. [ARP4386]

breakout pressure The pressure required to overcome static friction in a component. [ARP4386]

break point *1.* A location at which program operation is suspended so that partial results can be examined. *2.* A preset point in a program at which control passes to a debugging routine.

breakpoint instruction *1.* An instruction that will cause a computer to stop or to transfer control, in some standard fashion, to a supervisory routine which can monitor the progress of the interrupted program. *2.* An instruction that, if some specified switch is set, will cause the computer to stop or take other special action.

break rate The percent of time that an aircraft will return from an assigned mission with one or more previously working system/subsystems on the mission essential subsystem list (MESL) inoperable (code 3 including ground and air aborts). [ARD50010]

breaks Creases or ridges, usually appearing in aged sheet or strip, where the yield point has been locally exceeded. Depending on its origin, a break may be termed a coil break, cross break, edge break, or sticker break.

breath The air inhaled and exhaled during respiration. [ARP171]

breathing *1.* The opening and closing of a mold to allow gas to escape early in the molding

cycle. *2.* Permeability to air of plastic sheeting. *3.* The removal of air or gases from an assembly during autoclave molding by use of a breather. [AIR4844]

breech *1.* A reloadable pressure vessel used to contain a propellant cartridge. [ARP4386] *2.* The container provided on cartridge starters to receive and fire the cartridge. [ARP906A]

breech cap The removable portion of the breech which contains the igniting mechanism. [ARP906A]

breech chamber The fixed portion of the breech which directs the cartridge gas to the starter. [ARP906A]

breeching A duct for the transport of the products of combustion between parts of a steam generating unit or to the stack.

Bremsstrahlung *1.* Electromagnetic radiation produced by the rapid change in the velocity of an electron or another fast, charged particle as it approaches an atomic nucleus and is deflected by it. In German it means braking radiation. *2.* X-rays having a broad spectrum of wavelengths; formed due to deceleration of a beam of energetic electrons as they penetrate a target. Also known as white radiation.

Brewster-angle window A window inserted into an optical path at Brewster's angle—the angle at which unpolarized light must be incident upon a nonmetallic surface for the reflected radiation to acquire maximum plane polarization. At Brewster's angle, the reflected, plane-polarized beam and the refracted beam through the window are at 90° to one another.

bridge *1.* A device that interfaces and transfers data between two or more similar communication systems. [AIR4271] Also called a gateway. *2.* A network device that interconnects two local area networks that use the same LLC but may use different MACs. A bridge requires only OSI Level 1 and 2 protocols. *See also* router. *3.* The strain-to-voltage converter in many measurement systems (actually, a Wheatstone bridge).

bridge amplifier A type of amplifier circuit used extensively in instrumentation to provide gains up to 1000 at bandwidths up to 50 kHz. Generally configured as a direct-coupled amplifier constructed of four subamplifiers and suitable fixed resistances.

bridge circuit An electronic network in which an input voltage is applied across two parallel elements and the output voltage—to an indicating device or load—is taken across two intermediate points on the parallel elements.

bridged-T network A T-network having a fourth branch connected in parallel with the two series branches of the T, the fourth branch termination at one input and one output terminal.

bridgewall In a furnace, a wall over which the products of combustion pass.

bridgewire A resistance wire that thermally or explosively initiates the charge in an EED. [AIR913]

bridging *1.* Premature solidification of metal across a mold section before adjacent metal solidifies. 2. Welding or mechanical jamming of the charge in a downfeed furnace. 3. Forming an arched cavity in a powder metal compact. *4.* Forming an unintended solder connection between two or more conductors— either a secure connection, or merely an undesired electrical path without mechanical strength. Also known as crossed joint or solder short.

bridle, catapult A wire rope or synthetic fiber assembly designed to connect an aircraft to a catapult for the purpose of accelerating the aircraft for takeoff. The bridle connects to two catapult hooks on the aircraft and passes around the catapult shuttle spreader. The bridle falls off the aircraft (sheds) as it overtakes the shuttle, and is generally restrained by a bridle-arrester system on the deck. Also called catapult pendant. [AIR1489]

Briggs pipe thread *See* American standard pipe thread.

bright dipping Producing a bright surface on metal, such as by immersion in an acid bath.

brightness *1.* Perception of the intensity of light reflected by an object or emitted by a source of light. [ARP4032] 2. Term used in non-quantitative statements to refer to sensations and perceptions of light; in quantified statements to refer to the description of brightness by photometric units.

brightness contrast The brightness difference been a viewed object and its adjacent background. [AIR1093]

brightness distribution The statistical distribution based on brightness, or the distribution of brightness over the surface of an object.

brightness ratio The ratio of the brightnesses of any two surfaces. When the two surfaces are adjacent, the brightness ratio is called the brightness contrast. [ARP798]

brightness temperature *1.* In astrophysics, the temperature of a black body radiating the same amount of energy per unit area as the observed body at the wavelengths under consideration. *2.* The temperature of a nonblack body determined by measurement with an optical pyrometer.

bright plating Electroplating to yield a highly reflective coated surface.

bright switch A solid-state switch consisting of two bipolar transistors connected in an inverted configuration to achieve a low offset voltage. Used in only limited applications today.

Brinell test A standard bulk hardness test in which a 10-mm-diameter ball is pressed into the surface of a test piece and a hardness number determined by dividing applied load in kg by area of the circular impression in sq mm.

briquetting Producing relatively small lumps or block of compressed granular material, often incorporating a binder to help hold the particles together.

brisance The shattering effect of a high explosive. [AIR913]

bristle A relatively thick, short section cut from a monofilament. [AIR4844]

British thermal unit (Btu) The mean British thermal unit is 1/180 of the heat required to raise the temperature of 1 lb of water from 32°F to 212°F at a constant atmospheric pressure. It is about equal to the quantity of heat required to raise 1 lb of water 1°F. A Btu is essentially 252 calories.

brittle fracture Separation of solid material with little or no evidence of macroscopic plastic deformation, usually by rapid crack propagation involving less energy than for ductile fracture of a similar structure.

brittleness The tendency of a material to fracture without apparent plastic deformation.

Brix scale A specific gravity scale used almost exclusively in sugar refining. The degrees Brix represent the weight percent pure sucrose in water solution at 17.5°C.

broaching Cutting a finished hole or contour in solid material by axially pulling or pushing a bar-shaped, toothed, tapered cutting tool across a workpiece surface or through a pilot hole.

broadband A medium based on CATV technology in which multiple signals are frequency division multiplexed. Because broadband cables use CATV technology, they are unidirectional (within any given block of frequencies); thus, two types of broadband systems are in common use, single cable and dual cable. In a single cable system, stations transmit and receive on the same cable but at different frequencies. The station transmits on one frequency, the signal travels down the network to the head end, gets translated into a different frequency, and is sent back down the network where it is received by all stations. In a dual cable system, stations transmit and receive at the same frequency but on different cables. The end of the transmit cable is connected to the beginning of the receive cable, forming a double loop through the plant.

broad-banded Having a relatively large bandwidth. Used to describe instruments having an initial pulse with a relatively wide bandwidth and an amplifier with response to a relatively wide range of frequencies; opposite of narrow-banded or tuned. [ARP5089]

broadband filter A filter that has fixed low-pass and high-pass corner frequency break points. [AS8054]

broadband pyrometer *See* wideband radiation thermometer.

broadband transmission (fiber optic) Transmission of signals with a large bandwidth, such as video transmission.

broadcast The simultaneous dissemination of information to one or more stations—one-way, with no acknowledgment of receipt.

broadgoods *1.* Fiber woven to form fabric up to 1270 mm (50 in) wide. *2.* A term loosely applied to prepreg material greater than about 12 inches in width, usually furnished by suppliers in continuous rolls. [AIR4844]

broken symmetry Phenomena in which a loss of symmetry is present, such as in piezoelectricity.

broken wires On hose with a wire-braided jacket or cover, a strand of wire, or a plait formed from multiple strands of wire, that is severed or broken in two due to abrasion, flexing, or stress fatigue. [ARP1658A]

bronze *1.* A copper-rich alloy of copper and tin, with or without small amounts of additional alloying elements. *2.* By extension, certain copper-base alloys containing less tin than other elements, such as manganese bronze and leaded tin bronze; and certain other copper-base alloys that do not contain tin, such as aluminum bronze, beryllium bronze, and silicon bronze. *3.* Trade name for certain copper-zinc alloys (brasses), such as architectural bronze and commercial bronze.

bruise resistance The capability of a tire to resist bruise damage, i.e., from rocks and foreign objects. [AIR1489]

Brunt-Vaisala frequency The frequency at which an air parcel will oscillate when subjected to an infinitesimal perturbation in a stably stratified atmosphere.

brushcoat The thin layer of a curing-type sealant usually applied over fasteners, seams, and various parts and small openings prior to the application of the primary Class B fillet sealant. [AIR4069]

brushes (electrical contacts) Conductive metal or carbon blocks used to make sliding electrical contact with a moving part, as in an electric motor.

brushless permanent magnet (PM) motor A d-c motor in which d-c current supplied to conductors in the stator reacts with the PM field developed by permanent magnets located in the rotor. This reaction results in motor torque. [ARP4386]

brush permanent magnet (PM) motor A d-c motor in which d-c current supplied to conductors in the rotor reacts with a magnetic field developed by permanent magnets located in the stator. This reaction results in motor torque. [ARP4386]

brush plating An electroplating process in which the surface to be plated is not immersed, but rather rubbed with an electrode containing an absorbent pad or brush which holds (or is fed) a concentrated electrolyte solution or gel.

brush wound field motor A d-c motor similar to a brush PM motor, except that a wound coil supplied with continuous and unidirectional d-c current is used in place of a PM field in the stator. [ARP4386]

B-scan A data presentation method generally applied to pulse echo techniques. Yields a two dimensional view of a cross-sectional plane through the test piece. [ARP5089]

B-stage An intermediate stage in the reaction of certain thermosetting resins in which the material softens when heated and is plastic and fusible, but may not entirely dissolve or fuse. [AIR4844]

BTPD Abbreviation for body temperature (37°C or 98.6°F), ambient pressure, dry. [ARP171]

BTPS Abbreviation for body temperature (37°C or 98.6°F), ambient pressure, saturated with water vapor ($pH_2O = 47$ mm Hg = 6.3 kPa). [ARP171]

Btu *See* British thermal unit.

bubble *See* balloon.

bubblegas Any gas selected to bubble from the end of a tube immersed in liquid for level measurement. The basis for such a measurement is the hydrostatic back pressure created in the tube.

bubble tube A length of pipe or tubing placed in a vessel at a specified depth to transport a gas injected into the liquid; used to measure level from hydrostatic back pressure in the tube.

bubble memory *See* magnetic bubble memory.

bubble point *1*. The value of pressure at which a constant stream of bubbles first appears from a filter mesh that is immersed a given depth below the surface of a given liquid with known temperature, density, and surface tension. [ARP4386] *2*. The temperature at which a liquid mixture begins to boil and evolve vapors.

bubble sort In data processing, a method of arranging a group of numbers in some order.

bubble tight A nonstandard term used to refer to control valve seat leakage. Refer to ANSI/FC1 70-2 for specification of seat-leakage classifications.

bubble-type viscometer A device similar to a ball-type viscometer; however, in this device, viscosity is determined from the timed rise of a standard-size bubble through the sample liquid, rather than the timed fall of a ball.

buckle *1*. Localized waviness in a metal bar or sheet, usually transverse to the direction of rolling. *2*. An indentation in a casting due to expansion of molding sand into the cavity. *3*. A quick-release connector in a torso-restraint system. [AS8043]

buckle line A line of collapsed honeycomb cells, two to three cells wide, with undistorted cells on either side. [AIR4844]

Buckley gage A device that measures very low gas pressure by sensing the amount of ionization produced by a prescribed electric current.

buckling *1*. Crimping of fibers in a composite material, often occurring in glass-reinforced thermoset due to resin shrinkage during cure. *2*. A mode of failure generally characterized by an unstable lateral material deflection due to compressive action on the structural element involved. [AIR4844] *3*. Producing a lateral bulge, bend, bow, kink, or wavy condition in a beam, bar, column, plate, or sheet by applying compressive loading.

buckstay A structural member placed against a furnace or boiler wall to restrain the motion of the wall.

buffer *1*. An internal portion of a data-processing system serving as intermediate storage between two storage or data-handling systems with different access times or formats. Usually used to connect an input or output device with the main or internal high-speed storage. *See also* storage buffer. *2*. An isolating component designed to eliminate the reaction of a driven circuit on the circuits driving it, e.g., a buffer amplifier.

buffer, catapult In nose gear catapulting, a device or system for decelerating the aircraft to zero velocity at hookup condition, by the engagement of the holdback bar with the slider of the catapult deck hardware. [AIR1489]

buffer complete (BCOMP) In TELEVENT, the signal that indicates when the computer buffer is complete.

buffered computer A computing system with a storage device that permits input and output data to be stored temporarily in order to match the slow speeds of input and output devices. In such a system, simultaneous input-output and computer operations are therefore possible. NOTE: A data transmission trap is essential for effective use of buffering since it obviates frequent testing for the availability of a data channel.

buffered data channel (BDC) A device that provides high-speed parallel data interfaces into and out of the computer memory.

buffered I/O channel A computer I/O channel that controls the movement of data between an external device and memory, under the control of self-contained registers (i.e., independent of the operating program). *See* buffered data channel (BDC).

buffer storage In computer operations, storage used to compensate for a difference in rate of flow or time of occurrence when transferring information from one device to another.

buffeting *1.* A forced vibration of aircraft structural components caused by unsteady aerodynamic forces due to wake effects, mach effect, or pre-stall conditions. [ARP4386] *2.* The beating of an aerodynamic structure or surfaces by unsteady flow, gusts, etc. *3.* The irregular shaking or oscillation of a vehicle component owing to turbulent air or separated flow.

buffing Producing a very smooth and bright surface by rubbing it with a soft wheel, belt, or cloth impregnated with fine abrasive, such as jeweler's rouge.

bug An error, defect, or malfunction in a computer program.

bugging distance *See* device lift.

building form A male shape duplicating the shape of the finished tank. [AIR1664]

buildup *1.* An area within a laminate made thicker by addition of more plies or layers of material. [AIR4844] *2.* Excessive electrodeposition on areas of high current density, such as at corners and edges. *3.* Small amounts of work metal that adhere to the cutting edge of a tool and reduce its cutting efficiency. *4.* Deposition of metal by electrodeposition or spraying to restore required dimensions of worn or undersize machine parts.

built in test (BIT) An operational status checkout or test system which is integrated into a control system or function. Usually used to verify operational status of as many aspects of a control function as possible within the limits of integrated test capability. May be initiated automatically or on command. [ARP4386]

built in test equipment (BITE) Systems and avionics in new generation aircraft that, when prompted, perform internal automated test procedures and store or provide systems status and test results. [AIR4548]

bulb *See* envelope.

bulb, thermostatic A liquid-vapor-filled bulb wherein pressure changes are proportional to temperature changes. Used as a means of control. [ARP147C]

bulge A local distortion or outward swelling caused by internal pressure on a tube wall or boiler shell when overheated. Also applies to a similar distortion of a cylindrical furnace due to external pressure when overheated—provided the distortion is of a degree that can be driven back.

bulk acoustic wave (B-A-W) devices Acoustooptic devices utilizing bulk sound waves at megahertz frequencies in thin film transducers.

bulk density Mass per unit volume of a bulk material, averaged over a relatively large number of samples.

bulk factor The ratio of the volume of a raw molding compound or powdered plastic to the volume of the finished solid piece produced therefrom. [AIR4844]

bulkhead *1.* A partition in the host vehicle through which a cable must pass. [AIR4288] *2.* A liquid-tight, transverse closure between compartments of a tank. [AIR4783]

bulkhead seal A fitting allowing the passage of a duct through a wall or bulkhead; used to prevent the leakage through the bulkhead around the outer periphery of the duct. [ARP699D]

bulk hose Hose without couplings or fittings. [AS1933]

bulk modulus *1.* A measure of the degree of compressibility of a fluid; usually expressed in pounds per square inch. Bulk modulus is the reciprocal of compressibility. The higher the bulk modulus, the stiffer is the fluid. [AIR1116] *2.* An elastic modulus determined by dividing hydrostatic stress by the associated volumetric strain (usually computed as the fractional change in volume).

bulk molding compound Thermosetting resin mixed with strand reinforcement, fillers, etc., into a viscous compound for compression or injection molding. [AIR4844]

bulk storage In a computer system, a hardware device that supplements computer memory. Typically, a magnetic tape or disk.

bulk storage memory Any non-programmed large memory; for example, discs, drums, or magnetic tape units.

bull block A machine with a power-driven rotating drum for pulling wire through a drawing die.

bull gear A bull wheel with gear teeth around its periphery.

bullion *1.* A semirefined alloy containing enough precious metal to make its recovery economically feasible. *2.* Refined gold or silver, ready for coining.

bull wheel *1.* The main wheel or gear of a machine, usually the largest and strongest. *2.* A cylinder with a rope wound around it for lifting or hauling.

bump A raised or flattened portion of a boiler drum head or shell formed by fabrication. Generally used for nozzle or pipe attachments.

bumpless transfer Change from manual mode to automatic mode of control, or vice versa, without change in control signal to the process.

bumpy toruses The doughnut shapes of certain plasmas.

Buna-N A nitrile synthetic rubber known for resistance to oils and solvents.

bunched lay Refers to a conductor or cable in which the strands or wires are twisted together in the same direction without regard to geometrical arrangement, with the same lay length. [AS1198]

bunched stranding Any number of conductor strands of the same diameter twisted together in the same direction without regard to geometric arrangement of the individual strands. [ARP1931]

bundle A group of wires fastened or held together by auxiliary means, such as straps, ties, clamps, lacing tape/twine, or flexible wrappings (jackets) or sheaths. Also called cable. [ARP914A]

bundle (fiber optic) A group of fibers packaged together which collectively transmit light in a coherent bundle. The end fibers of the bundle are in a fixed relationship to each other and can transmit an image.

bungee A spring, elastic cord, or other tension (or compression) device used in a system to balance an opposing force; or in a landing gear system to assist in retracting or extending the gear; or to absorb shock. Examples include the downlock bungee, uplock bungee, and overcenter bungee, etc. [AIR1489]

bungee cord *See* shock cord.

bunker C oil Residual fuel oil of high viscosity, commonly used in marine and stationary steam power plants (No. 6 fuel oil).

buoyancy The tendency of a fluid to lift any object submerged within it. The buoyancy force applied to the body equals the product of fluid density and volume of fluid displaced.

buoyancy displacement The technique of measuring liquid level by measuring the buoyant force on a partially immersed volumetric displacing device.

buoyancy-type liquid-level detector Any of several designs of level gage that depend for their operation on the buoyant force acting on a float or similar device located inside the tank or vessel.

burden *1.* The amount of power consumed in the measuring circuit of an instrument, usually given as the volt-amperes consumed under normal operating conditions. *2.* The property of a circuit connected to the secondary winding of an instrument transformer which determines active and reactive power at the transformer output terminals.

Bureau d'Orientation de la Normalisatin en Informatique (BNI) The French national standards body for computer-related standards.

burner *1.* Any device for producing a flame using liquid or gaseous fuel. *2.* A device in the firebox of a fossil-fuel-fired boiler that mixes and directs the flow of fuel and air to give rapid and complete combustion. *3.* A worker who cuts metal using an oxyfuel-gas torch.

burner windbox A plenum chamber around a burner in which a certain air pressure is maintained to insure proper distribution and discharge of secondary air.

burner windbox pressure The air pressure maintained in the windbox or plenum chamber; measured above atmospheric pressure.

burn-in *1.* The operation of an item to stabilize its characteristics. Basically, a reliability-conditioning procedure—a method of aging an item by operating it under specified environmental and test conditions in accordance with an established procedure in order to eliminate early failures and age or stabilize the item prior to final test and shipment. [ARD50010] *2.* The operation of an item under stress to stabilize its characteristics. [ARP4386] *3. See* predicted preconditioning.

burning *See* combustion.

burning process *See* combustion.

burning rate The rate at which a solid propellant burns normal to its surface at a specified chamber pressure and propellant temperature. [AIR913]

burnish *1.* To polish or make shiny. *2.* Specifically, to produce a smooth, lustrous surface finish on metal parts by tumbling them with hardened metal balls or rubbing them with a hard metal pad.

burnout The termination of combustion in a rocket engine because of exhaustion of the propellant.

burnout velocity The velocity attained by a missile at the point of burnout. [ARP4386]

burnout weight Weight of a missile at the time of engine burnout, but including any unusable fuel. [ARP4386]

burr *1.* Metal piece outside geometrical surface defined by the drawing. [ARP4784] *2.* A thin, turned-over edge or fin produced by a grinding wheel, cutting tool, or punch. *3.* A rotary tool having teeth similar to those on a hand file.

burst *1.* Surface fissures or ruptures caused by metal movement during rolling or forging. [ARP4784] *2.* A period in which data are present on the network, which is preceded and succeeded by periods in which data are not present. [AIR4288]

bursting In data processing, the act of separating continuous forms into single sheets.

burst pressure *1.* Pressure higher than proof pressure which is applied to the fuel connections of a component for a specified duration during qualification burst tests. [AIR1749] *2.* The pressure required to rupture or burst. Generally a design includes allowance for maximum expected pressure in service—multiplied by a reasonable safety factor. [AIR1489]

burst pressure, actual Pressure at which a component bursts or exhibits massive leakage due to permanent or non-permanent structural failure, or structural deflection. [ARP4386]

burst pressure, minimum (ultimate pressure) Pressure during burst pressure testing up to which no externally visible bursting and no significant external leakage occurs. Deformation and permanent set are permitted. Function may be impaired. [ARP4386]

burst strength Measure of the ability of a material to withstand internal hydrostatic or gas dynamic pressure without rupture. [AIR4844]

bus *1.* A generic term describing a channel along which signals travel from one or several sources (stations) to one or more destinations (stations). [AIR4271] *2.* A group of wires or conductors, considered as a single entity, which interconnects parts of a system. *3.* In a computer, signal paths, such as the address bus or the data bus. *4.* A circuit over which data or power is transmitted; often one that acts as a common connection among a number of locations. Synonymous with trunk. *5.* A communications path between two switching points. *6.* A common connector circuit, usually multiwire, for transfer of power, data, timing, etc., between the several modules or units on the bus.

bus acquisition latency The time from the request for bus mastership by the highest-priority module to the time at which that module becomes bus master. [AS4710]

bus activity timer A timer in each station that measures the period of inactivity between consecutive bursts. [AIR4288]

bus connection Connection designated for attachment of the coupler to the bus cable. [AS4117]

bus controller The terminal assigned the task of initiating information transfers on the data bus. [AS1773]

bus cycle The transfer of one word or byte between two devices.

bushing *1.* A replaceable part, cylindrical in shape, hollow, and designed primarily to be inserted in a hole to reduce the effective inside diameter of the hole, and to protect the body structure about the hole from damage resulting from stress, strain, and vibration. [ARP480A] *2.* A removable piece of soft metal or impregnated, sintered-metal sleeve used as a bearing or guide. *3.* A ring-shaped device made of ceramic or other nonconductive material used to support an electrical conductor while preventing it from becoming grounded to the support structure.

busing The joining of two or more circuits to provide a common electrical connection. [ARP914A]

bus interface *See* active bus interface. [AS4710]

bus interface unit Electronics within each station that implement the network protocol. [AIR4288]

bus master The module currently in control of the bus. [AS4710]

bus monitor The terminal assigned the task of receiving bus traffic and extracting selec-

ted information to be used at a later time. [AS15531]

bus request The DEC PDP-11 priority system for determining which external device will obtain control of the UNIBUS to interrupt the CPU for service. There are seven bus requests and one bus grant.

bus tenure The period between the time a bus master gains control of the bus and the time at which control is released. [AS4710]

butt braze Joining of two conductors end-to-end, with no overlap and with axis in line, using the process of brazing. [ARP1931]

butterfly valve A valve consisting of a disc inside a valve body which operates by rotating about an axis in the plane of the disc to shut off or regulate flow in a piping system. A similar device used in heating or ventilating ductwork is called a butterfly damper.

butterfly type damper *See* damper.

buttering Coating the faces of a weld joint prior to welding to preclude cross-contamination of a weld metal and base metal.

Butterworth The filter characteristic in which constant amplitude across the pass band is the objective. Also known as constant amplitude (CA).

butting dies Crimping dies so designed that the opposing die faces touch at the closed condition of the crimping cycle. Also called bottoming dies. [ARP914A]

butt joint *1.* A joint between two members lying approximately in the same plane. In welded joints, the edges may be machined or otherwise prepared to create any of several types of grooves prior to welding. *2.* A type of edge joint in which the edge faces of the two adherents are at right angles to the other faces of the adherents. [AIR4844]

buttock line A vertical fore and aft plane used for identifying inboard/outboard locations within the airplane. [AS1426]

button samples The small amount of sealant extruded from each mixed-sealant cartridge onto an indexed card before the cartridge is frozen. Used as an indicator of the condition of the sealant in each tube, i.e., uniformity of mix, presence of air, cure rate, and hardness on cure. [AIR4069]

97

butt splice A device for joining conductors end-to-end with their axes in line and not overlapping. *See* splice. [ARP914A]

buttstrap A narrow strip of boiler plate overlapping the joint of two butted plates, used for connecting by riveting.

butt weld A weld that joins the edges or ends of two pieces of metal having similar cross sections, without overlap or offset along the joint line.

butt wrap A spiral wrapping of tape over a cable core in which the trailing edge of one wrap just meets the leading edge of the preceding wrap, with neither overlap nor spacing. [ARP1931]

buzz Undesirable vibration of a flight control surface induced and maintained by aerodynamic forces. [ARP4386]

bypass *1.* A duct that conveys air around a system or system components. [ARP699D] *2.* A passage for a fluid, permitting all or a portion of the fluid to flow around its normal pass flow channel.

bypass air In a turbofan engine, air that the engine draws in, but that is not used in the combustion process. Bypass air is compressed slightly and discharged together with the core engine exhaust. [AS5116]

bypass capacitor A capacitor connected in parallel with a circuit element to provide an alternative a-c current path of relatively low impedance.

bypass ratio In a turbofan engine, the mass flowrate of bypass air divided by the mass flowrate of combustion air. [AIR5026]

bypass valve A valve utilized on a refueler to control pressure. [AIR4783]

by-product Incidental or secondary output of a chemical production or manufacturing process that is obtained in addition to the principal product, with little or no additional investment or allocation of resources.

byte Eight contiguous bits starting on an addressable byte boundary. In a byte, bits are numbered from the right, 0 through 7, with 0 being the low-order bit. A byte can be used to store one ASCII character.

C

C_1 The numerical value of the minimum expected capacity in units of ampere hours, at a discharge rate of C_1 amperes (the one-hour rate), and a battery temperature of 23°C. The C_1 is the rating that indicates the relative amount of available emergency electrical energy capability of a particular cell/battery design. [AS8033]

CA *See* constant amplitude.

C-8A augmentor wing aircraft NASA's research, short haul, jet aircraft.

cabin, nonpressurized An airplane cabin that is not designed or equipped for pressurizing and which will, therefore, have a cabin pressure equal to that of the surrounding atmosphere. [ARP171]

cabin pressurization The process of producing pressures in an aircraft that are higher than ambient pressures outside the aircraft. [ARP4107]

cabin, pressurized An airplane cabin that is constructed, sealed, and equipped with an auxiliary system to maintain a pressure within the cabin greater than that of the surrounding atmosphere. [ARP147C]

cable Two or more wires contained in a common jacket or covering; or two or more wires twisted or molded together without a common covering; or one or more wires contained in a shield or in a shield and jacket. [AR1931] Unless otherwise specified, a cable or wire is one on which all manufacturing operations have been completed. *See* wire and bundle. [AS1198]

cable assembly A completed cable and its associated hardware. [AR1931]

cable attenuation Cable power loss per unit length at a specified frequency. [AS4117]

cable clamp A connector accessory or portion of a component that is designed to grip a wire or cable to provide strain relief and absorb mechanical stress that would otherwise be transmitted to the termination. [ARP914A]

cable core The portion of an insulated cable that lies under the protective covering or coverings. [AR1931]

cable core binder A wrapping of tapes or cords around the several conductors of a multiple-conductor cable; used to hold them together. Cable core binder is usually supplemented by an outer covering of braid, jacket, or sheath. [AR1931]

cable, electric Two or more individually insulated conductors, solid or stranded, contained in a common covering; or two or more individually insulated conductors twisted or molded together without a common covering; or one or more insulated conductors with a metallic covering or sheath. [ARP4404]

cable, fiber-optic *See* fiber-optic cable.

cable filler Material used in multiple conductor cables to occupy the interstices formed by the assembly of the insulated conductors, thus forming a cable core of the desired shape. [AR1931]

cable shielding clamp A connector accessory device consisting of a sealing member and cable support, designed to terminate the shield of the electrical cable at the connector. [ARP914A]

cabling The act of twisting together two or more insulated conductors to form a cable. [AR1931]

cache memory A small, high-speed memory placed between the slower main memory and the processor. A cache increases effective memory transfer rates and processor speed. It contains copies of data recently used by the processor and fetches several bytes of data from memory in anticipation that the processor will access the next sequential series of bytes.

CAD *See* computer aided design.

cadmium mercury tellurides *See* mercury-cadmium tellurides.

cadmium plating An electroplated coating of cadmium on a steel surface which resists atmospheric corrosion. Applications include nuts, bolts, screws, enclosures, and many hardware items.

cage A circular frame for maintaining uniform separation between balls or rollers in a rolling-element bearing. Also known as a separator.

caking Producing a solid mass from a slurry or mass of loose particles by any of several methods involving filtration, evaporation, heating, pressure, or a combination of these.

calcine *1.* To heat a material such as coke, limestone, or clay without fusing. Purpose is to decompose compounds such as carbonates and drive off volatiles, such as moisture, trapped gases, and water of hydration. *2.* To heat a material under oxidizing conditions. *3.* The product of a calcining or roasting process.

calculate To compute mathematically.

calculating action A type of control system action in which one or more feedback signals are combined with one or more actuating signals to provide an output signal that is some function of the combination.

calculation A group of numbers and mathematical symbols that is executed according to a series of instructions.

calculation, V/L Involves determining the amount of air that will come out of solution between two partial pressures ($P_{initial}$ - P_{TVP}) and (P_{final} - P_{TVP}), while flowing in a closed system. [AIR1326]

calculus of variations The theory of maxima and minima of definite integrals whose integrand is a function of the dependent variables, the independent variables, and their derivatives.

calderas Large, basin-shaped volcanic depressions, more or less circular in form. The diameter of a caldera is many times greater than that of the included vent or vents.

calefaction *1.* A warming process. *2.* The resulting warmed condition.

calender *1.* To pass a material, such as rubber or paper, between rollers or plates to make it into sheets, or to make it smooth and glossy. *2.* A machine for performing such an operation.

calibrate To check or adjust the graduations of a quantitative measuring instrument.

calibrated airspeed (CAS) The indicated airspeed of an aircraft corrected for position and instrument error. Calibrated airspeed is equal to true airspeed in standard atmosphere at sea level. [ARP4107]

calibrated flow In a unit that controls or limits rate or quantity of flow, that rate or quantity of flow for which the unit is calibrated or adjusted. [ARP4386]

calibrating tank A liquid vessel of known capacity that is used to check the volumetric accuracy of positive-displacement meters. Also known as a meter-proving tank.

calibration *1.* The comparison of a particular instrument or system with a standard of known accuracy. [ARP741] *2.* The comparison of a measurement system or device of unverified accuracy to a measurement system or device of known and greater accuracy. Purpose is to detect and correct any variation from required performance specifications of the measurement system or device. [ARD50010]

calibration angle *1.* A preselected shaft position in the positive quadrant. *2.* A straight line drawn through the calibration angle and electrical zero, terminating at the positive and negative rated excursion angles; represents the ideal output voltage of the linear resolver. [ARP826]

calibration curve A plot of indicated value versus true value used to adjust instrument readings for inherent error. A calibration curve is usually determined for each calibrated instrument in a standard procedure, and its validity confirmed or a new calibration curve determined by periodically repeating the procedure.

calibration cycle *1.* The frequency that a device is due for calibration. This cycle could be dependent on calendar, cycles, or hours. *2. See* calibration interval, period, or cycle.

calibration gas A mixture of gases of specified and known composition used as the basis for interpreting analyzer response in terms of the concentration of the gas to which the analyzer is responding. [ARP1256]

calibration hierarchy The chain of calibrations that links or traces a measuring instrument to a primary standard. [AIR1678]

calibration interval, period, or cycle The time during which the instrument should remain within specific performance levels, with a specified probability, under normal conditions of handling and use. [AIR4896]

calibration period *See* calibration interval, period, or cycle.

calibration procedure The specific steps and operations to be followed by activity personnel in the performance of an instrument calibration

calibration record A record of the measured relationship of the transducer output to the applied measurand over the transducer range.

calibration speed A specified speed for purposes of output voltage standardization and reference. [ARP667]

california bearing ratio A widely used index to express soil strength. Used in empirical procedures for the design of unsurfaced, expediently surfaced, and flexible airfield surfaces. [AIR1780]

caliper A gaging device with at least one adjustable jaw; used to measure linear dimensions, such as lengths, diameters, and thicknesses.

caliper, brake A C-shaped structural part of a brake assembly which fits around the edge and adjacent to both sides of a rotating disc. The caliper usually contains an actuator(s) on both sides to provide opposing forces against the rotating disc, to produce the decelerating force. [AIR1489]

calk *See* caulk.

call *1*. To transfer control to a specified closed subroutine. *2*. In communications, the action performed by the calling party; or the operations necessary to making a call; or the effective use made of a connection between two stations.

calling sequence A specified arrangement of instructions and data necessary to set up and call a given subroutine.

Callisto A satellite of Jupiter orbiting at a mean distance of 1,884,000 kilometers. Also called Jupiter IV.

call-up Initial voice contact between a facility and an aircraft, using the identification of the unit being called and the unit initiating the call. [ARP4107]

calorie *1*. 1/100 of the heat required to raise the temperature of 1 gram of water from 0°C to 100°C at a constant atmospheric pressure. About equal to the quantity of heat required to raise one gram of water 1°C. *2*. 3600/860 joules, a joule being the amount of heat produced by a watt in one second (a more recent definition).

calorific value The number of heat units liberated per unit of fuel burned in a calorimeter under prescribed conditions.

calorimeter *1*. An instrument designed to measure heat evolved or absorbed. *2*. An instrument or detector that measures the amount of heat in a light beam. Can be used to measure incident radiation if the percentage of absorbed radiation is known.

calorimetric detector A detector that operates by measuring the amount of heat absorbed. With such a detector, incident radiation must be absorbed as heat to be detected.

calorize To produce a protective coating of aluminum and aluminum-iron alloys on iron or steel (or, less commonly, on brass, copper, or nickel). A calorized coating is protective at temperatures up to about 1800°F.

CALS *See* computer-aided acquisition and logistic support.

cam An irregularly shaped device that revolves around an axis or slides within limits. Actuates other parts by contact to change the direction, speed, and/or timing of a part in motion. [ARP480A]

CAM *See* computer aided manufacturing.

camber *1*. The angle that the plane of a wheel or tire makes with a plane normal to the ground. Usually considered positive when the angle vector moves up and outboard relative to the centerline of the vehicle. [AIR1489] *2*. The maximum distance between the mean line and the chord of an airfoil. The convexity or rise of the curve of an airfoil from its chord, and the ratio of the maximum departure of that curve from the chord to the length of the chord. [ARP4107]

camera tube An electron-beam tube in which an optical image is converted to an electron-current or charge-density image. This image is scanned in a predetermined pattern to provide an electrical output signal whose magnitude corresponds to the intensity of the scanned image.

cam follower The output link of a cam mechanism.

Campbell bridge A type of a-c bridge used to measure mutual inductance of a coil or other inductor in terms of a mutual inductance standard.

camshaft The rotating member that drives a cam.

cam-type timer Any of several designs of timing device using a single contoured cam to continually adjust a process parameter, such as a setpoint; or employing several cams mounted on a single timer shaft to provide interlocked sequence control of a complex operation without using relays.

canard *1.* An aerodynamic vehicle in which trim surfaces used for longitudinal (or pitch) control are forward of the main lifting surface. *2.* Such a forward-placed control surface itself. [ARP4107]

canard configurations *1.* Pertaining to an aerodynamic vehicle in which horizontal surfaces used for trim and control are forward of the main lifting surface. *2.* The horizontal trim and control surfaces in such an arrangement.

candela (cd) *1.* Metric unit for luminous intensity. Luminous intensity is used to measure light coming from a point source in a given direction, such as the light coming from a landing light or anticollision light on an aircraft. It is roughly equal to the obsolete units of candles and candlepower. [ARP1782] *2.* Used to express the intensity of light visible to the human eye. Corresponds to the emission from 1/60 of a square centimeter of a black body operating at the solidification temperature of platinum, and emitting one lumen per steradian.

candela/m² *1.* Luminous intensity per unit area of a radiating source with an extended area, rather than a point source. (The unit area is normal to the direction of observation.) *2.* Metric unit for luminance (photometric brightness). Equal to 0.2919 fL (1 fL = 3.426 cd/m²). [ARP1782]

candle The unit of luminous intensity. Defined as 1/60 of the intensity of one square centimeter of a black body radiator at the temperature of solidification of platinum (2046 K). *See also* candela. [ARP798]

candlepower *1.* Luminous intensity expressed in candles. [ARP798] *2.* An obsolete unit of measure for luminous intensity. *See also* candela.

canned *1.* Describes a pump or motor enclosed within a watertight casing. A "canned" motor is usually enclosed within the same casing as the driven element (such as a pump), and designed so that its bearings are lubricated by the pumped liquid. *2.* Describes a composite billet or slab consisting of a reactive metal core encased in metal that is relatively inert. This enables the reactive metal to be hot worked in air by rolling, forging, or extrusion without excessive oxidation.

cannibalize *1.* To remove serviceable parts from one aircraft for installation on another. [ARD50010] *2.* To disassemble or remove parts from one assembly and use the parts to repair other, like assemblies.

cannot duplicate A fault indicated by BIT or other monitoring circuitry which cannot be confirmed at the next level of maintenance. [ARD50010]

canopies The topmost layers of leaves and branches of forest trees or other plants.

cantilever A beam or other structural member fixed at one end and hanging free at the other end.

capability A measure of the ability of an item to achieve mission objectives given the conditions during the mission. [AIR4896]

capacitance That property of a system of conductors and dielectrics which permits the storage of electricity when potential differences exist between the conductors. Its value is expressed as the ratio of the electrostatic charge on a conductor to the potential difference between the conductors required to maintain that charge. Capacitance is measured in farads. [AR1931]

capacitance meter An instrument for determining electrical capacitance of a circuit or circuit element. *See also* microfaradmeter.

capacitance-voltage characteristics Those characteristics of a metal semiconductor contact or a semiconductor junction that are manifested as a measured capacitance as a function of a d-c bias voltage when a small, superimposed a-c voltage is applied to that junction or contact.

capacitive coupling Electrical interaction between two conductors caused by the capacitance between them. [AR1931]

capacitive instrument A measuring device whose output signal is developed by varying the capacitive reactance of a sensitive element.

capacitor *1.* A passive fluidic element that, because of fluid compressibility, produces a pressure across the device which lags net flow into it by essentially 90°. [ARP993A] *2.* A device used for storing an electrical charge.

capacitor discharge system An ignition system in which the spark energy is primarily the result of a capacitor discharge. [AIR784]

capacity The dischargeable ampere hours available from a fully charged cell/battery at any specified discharge rate/temperature condition. [AS8033]

capacity factor The ratio of the average load carried to the maximum design capacity.

capacity lag In any process, the amount of time it takes to supply energy or material to a storage element at one point in the process from a storage point elsewhere in the process. Also known as transfer lag.

capillary action *1.* The tendency of certain liquids to travel, climb, or draw into tight crack-like interface areas due to such properties as surface tension, wetting, cohesion, and adhesion. *2.* Spontaneous elevation or depression of a liquid level in a fine, hair-like tube when it is dipped into the body of a liquid. Capillary action is induced by differences in surface energy between the liquid and the tube material.

capillary drying Progressive removal of moisture from a porous solid by evaporation at an exposed surface, followed by movement of liquid from the interior to the surface by capillary action, until the surface and core reach the same stable moisture concentration.

capillary tube *1.* A small-diameter tube connecting a sensor with a bellows or diaphragm in a control device. *2.* A small-diameter tube used as an expansion orifice in small refrigeration systems. [ARP147C] *3.* A tube sufficiently fine that capillary action is significant.

cap, pressure A part that can be used to cover tightly the end of a duct to sustain the internal pressure. [ARP699D]

cap, protective A part used to cover an end or opening in a duct for the purpose of excluding foreign matter. [ARP699D]

cap screw A threaded fastener, similar to a bolt, but generally used without a nut by threading it into a tapped hole in one part of an assembly.

capstan A vertical-axis drum used for pulling or hauling. May be power driven or may be turned manually by means of a bar extending radially from a hole in the drum.

capstan clutch A coaxial spring and shaft with a preloaded fit wherein power can be transmitted through the shaft-spring interface. [ARP4386]

capstan coil Coil that operates on the principle that a mechanical coil, under a torque in an unwind direction, radially expands and engages the housing, thus grounding the input torque above a predetermined level. [ARP4386]

capstan coil (wrap spring) no back A coil spring placed in a housing with a slight interference fit. Motion in either direction (to unwind the spring), initiated at the output end, will prevent backdriving motion. Motion in either direction (to wind the spring), initiated at the input end, will result in coupling and drive. [ARP4386]

capsules *See* space capsules.

captive device A multi-part fastener, usually screw-type, whose components are retained without separation when loosened from the base assembly. [ARP914A]

captive device-fastener A fastener, usually screw-type, whose components are retained without separation when loosened from the base assembly. [ARP914A]

captive tests Holddown tests of a propulsive subsystem, rocket engine, or motor—as distinguished from a flight test.

captured sample (grab sample) A sample that has been taken into a container, or adsorbed onto a substrate from a flowing stream, and that will be analyzed subsequently by an off-line measurement technique. [ARP4418]

capture effect An effect in frequency-modulation (FM) reception in which the stronger signal of two stations on the same frequency completely suppresses the weaker signal.

carbene An organic radical containing divalent carbon.

carbide Compound of carbon with one or more metallic elements.

carbide tool A cutting tool whose working edges and faces are made of tungsten, titanium, or tantalum carbide particles, compacted and sintered into a hard, heat-resistant, and wear-resistant solid by powder metallurgy. The heat-resistant properties of the material are derived in part from a matrix alloy, usually cobalt, used to cements the carbide particles together.

carbon The element that provides the backbone for all organic polymers. Graphite is an ordered form of carbon. Diamond is the densest crystalline form of carbon. [AIR4844]

carbonaceous materials Substances composed of or containing carbon or carbon compounds.

carbon-carbon A composite material consisting of carbon or graphite fibers in a carbon or graphite matrix. [AIR4844]

carbon cycle The path of carbon in living beings. In the carbon cycle, carbon dioxide is fixed by photosynthesis to form organic nutrients and is ultimately restored to the inorganic state by respiration and protoplasmic decay.

carbon dioxide (CO$_2$) An odorless, colorless gas that neither burns nor supports combustion. CO$_2$ is one of the chief products of combustion of carbon-containing substances; it is also an end product of the living metabolic processes in the body; excreted in exhaled air. [ARP171]

carbon equivalent An empirical relationship that is used to estimate the ability to produce gray cast iron. Also used to rate the weldability of alloy steels. For production of grey cast iron, the formula is CE = TC + 1/3(Si + P), where CE is the carbon equivalent, TC is the total carbon content, Si is the silicon content, and P is the phosphorus content, all in wt%. For weldability of alloy steels, the formula is CE = C + Mn/6 + (Cr + Mo + V)/5 + (Ni + Cr)/15, where each symbol stands for the concentration of the indicated element in wt%.

carbon fiber Fiber produced by the pyrolysis of organic precursor fibers, such as rayon, polyacrylonitrile, and pitch, in an inert environment. [AIR4844]

carbonitriding A surface-hardening process in which a suitable ferrous material is heated at a temperature above the lower transformation temperature in an atmosphere that causes simultaneous absorption of carbon and nitrogen at the surface; and, by diffusion, creates a concentration gradient. Final properties are achieved by controlled cooling, and sometimes by subsequent tempering.

carbonization The process of converting coal to carbon by removing other ingredients.

carbon loss Refers to the unliberated thermal energy resulting from failure to oxidize some of the carbon in the fuel.

carbon monoxide (CO) A colorless, odorless, toxic gas usually resulting from combustion of carbonaceous compounds in an insufficient supply of oxygen. [ARP171]

carbon-pile pressure transducer A resistive-type pressure transducer that depends for its operation on the change in resistance that occurs when irregular carbon granules or smooth carbon disks are pressed together. Because of their low resistance, carbon-pile transducers can often provide sufficient output current to actuate electrical instruments without amplification.

carbon potential A measure of the ability of an environment to alter or maintain the surface-carbon content of ferrous alloys. The effect on surface-carbon content also depends on temperature, time, and steel composition.

carbon steel An alloy of carbon and iron that contains no more than 2% carbon, and no alloying elements other than a small amount of manganese.

carbon suboxides Colorless lacrimatory gases having unpleasant odors and boiling points of approximately -7°C.

carburetor A component of a spark-ignition internal combustion engine that mixes fuel with air, in proper proportions, and delivers a controlled quantity of the mixture to the cylinders.

carburizing A surface-hardening process in which a suitable ferrous material is heated at a temperature above the transformation range in the presence of carbon-rich environment. The carbon-rich environment may be produced from solid carbon, vaporized liquid hydrocarbons, or gaseous hydrocarbons. Following production of a carbon concentration gradient in the alloy, it is either quenched from the carburizing temperature and tempered; or reheated, quenched, and tempered, to achieve desired properties in both the carbon-rich outer case and the carbon-lean inner core.

carcass The structural body of a tire. Generally includes all tire components except the tread. [AIR1489]

carcass saturation Condition in which fuel has permeated the materials of a hose carcass. [AIR4783]

card A circuit board within a computer or other electronic instrument or system.

cardinal altitudes (cardinal flight levels) "Odd" or "even" thousand-foot altitudes or flight levels; for example, 1000 ft, 12,000 ft, or FL 350, as distinct from 1500 ft, 12,276 ft, or FL 355. *See* altitude and flight level. [ARP4107]

cardinal heading A heading toward one of the cardinal points of the compass (that is, north, south, east, or west). [ARP4107]

card reader (Hollerith cards) A hardware device for reading computer-standard punched cards for computer entry.

cargo tank A container having a liquid capacity in excess of 100 gal, used for carrying aviation fuels, and mounted permanently or otherwise secured on the vehicle. [AIR4783]

Carnot cycle An idealized reversible thermodynamic cycle. The Carnot cycle consists of four stages: (a) an isothermal expansion of the gas at temperature T1; (b) an adiabatic expansion to temperature T2; (c) an isothermal compression at temperature T2; and (d) an adiabatic compression to the original state of the gas to complete the cycle.

carriage A mechanism that moves along a predetermined path in a machine to carry and position another component.

carriage bolt A threaded fastener with a plain (unslotted) head and a square shoulder below the head, which keeps the bolt from turning as the nut is tightened. This type of bolt is designed primarily for bolting wood members, but can be used with metal members if the member next to the bolt head has a square bolt hole to accommodate the shoulder of the bolt.

carriage stop A device attached to the outer way of a lathe bed. Permits accurate and repeatable positioning of the tool carriage for cutting grooves, turning multiple diameters and lengths, and cutting off pieces of specific lengths.

carrier *1.* The element or combination of several elements laid parallel in a braid by a single bobbin of the braider. [AR1931] *2.* A continuous-frequency signal capable of being modulated to carry information.

carrier band A single-channel signaling technique in which the digital signal is modulated on a carrier and transmitted.

carrier density The charge carrier concentration of holes and/or electrons in a semiconductor. Determines the electronic characteristics and function of the semiconductor.

carrier frequency The basic frequency or pulse repetition rate of a transmitted signal, bearing no intrinsic intelligence until it is modulated by another signal that bears intelligence.

carrier modulation *See* modulation.

carrier sense multiple access with collision detect (CSMA/CD) An access procedure in which a device with data to transmit first listens to the medium; when the medium is not busy, the device starts transmitting. While the device is transmitting, it listens for collisions (simultaneous transmission by another station); if a collision occurs, the node stops transmitting, waits, and tries again.

carrier-to-noise ratio *1.* RF signal power input to the receiver divided by the noise power input. *2.* Carrier amplitude divided by noise amplitude. *3.* Carrier power divided by noise power.

carrier transport The mobility of conduction electrons or holes in semiconductors.

carrier waves Waves generated at a point in the transmitting system and modulated by the signal.

carryover The chemical solids and liquid entrained with the steam from a boiler.

Cartesian coordinates A coordinate system in which the locations of points in space are expressed in reference to three planes, called coordinate planes, no two of which are parallel. Also called rectangular coordinates.

cartridge *1.* A gas-generation device, packaged to include ignition, propellant and other required items. *Monopropellant*–a cartridge using a single propellant. *Bi-propellant*–a cartridge using a fuel and an oxidizer, generally in liquid form. *Solid propellant*–a cartridge incorporating a fuel and oxidizer in a solid form. [AIR906A] *2.* A small unit used for storing computer programs or data values. The amount of information stored in a cartridge tends to be small and access times large compared to discs; however, they are widely used in hobbyist applications.

cartridge disk A relatively low-capacity data- or program-storage medium; generally removable.

cartridge heater Cylindrical-bodied, electrical heater used to provide heat for injection, compression, and transfer molds; injection nozzles; runnerless mold system; hot stamping dies; sealing; etc. [AIR4844]

cartridge tape Small magnetic tape for digital program storage; stores discrete records.

CAS *1. See* calibrated airspeed. *2. See* collision avoidance system.

CASA/SME *See* Computer and Automated Systems Association of the Society of Manufacturing Engineers.

cascade circuit A circuit in which more than one protector is connected in series between the power source and the load. *See also* coordination. [ARP1199A]

cascade control *1.* A control system composed of two loops in which the setpoint of one loop (the inner loop) is the output of the controller of the other loop (the outer loop). *2.* A control technique that incorporates a master and a slave loop. The master loop controls the primary control parameters and establishes the slave-loop setpoint. The purpose of the slave loop is to reduce the effect of disturbances on the primary control parameter, and to improve the dynamic performance of the loop.

cascade control action Control action in which the output of one controller is the setpoint for another controller.

cascaded Describes a series of machines, devices, or machine elements, so arranged that the output from one feeds directly into the next.

cascade sequence A welding sequence in which a continuous, multiple-pass weld is built up by depositing weld beads in overlapping layers. Usually, weld beads are laid in a backstep sequence, starting with the root bead, which extends only part way along the joint length, then starting successive beads a short distance farther along the joint from the start of the previous bead. Compare with block sequence.

cascading failure A failure whose probability of occurrence is substantially increased by the existence of a previous failure. [ARP4754]

case *1.* A pressure vessel designed to contain a propellant charge before and during burning. [ARP4386] *2.* A hardened outer layer on a ferrous alloy produced by suitable heat treatment. Sometimes involves altering the chemical composition of the outer layer before hardening.

CASE *See* Common Applications Service Elements.

case-bonded grain A solid propellant grain that is cast in place in the surrounding rocket case. [AIR913]

cased glass Glass composed of two or more layers of different glasses; usually a clear, transparent layer to which is added a layer of white or colored glass. Cased glass is sometimes referred to as flashed, multilayer, or polycased glass. [ARP798]

case drain (body pressure) Pressure in the case or body passages of a component that causes internal fuel leakage to be returned independently to the reservoir, return circuit, overboard, etc. [AIR1749]

case drain flow Cooling flow and internal leakage of a hydraulic pump or motor to the housing, and out of the case drain port; not provided in all models. [ARP4386]

case drain patch An indication of contaminant level or internal wear rate of a unit other than that indicated by the outlet patch. [ARP575A]

case hardening Producing a hardened outer layer on a ferrous alloy by any of several surface-hardening processes, including carburizing, carbonitriding, nitriding, flame hardening, and induction hardening. Also known as surface hardening.

casing *1.* The structural part of a tire. [ARP4834] *2.* A covering of sheets of metal or other material, such as fire-resistant composition board, used to enclose all or a portion of a steam-generating unit.

cast *1.* To produce a solid shape from liquid or semisolid bulk material by allowing it to harden in a mold. *2.* A tinge of a specific color. *3.* A slight overtint of a color different from the main color; for instance, white with a bluish cast.

castellated nut A hexagonal nut with a slotted, cylindrical projection above one of the hexagonal sides. Used in conjunction with a cotter pin or safety wire. The cotter pin or safety wire passes through a lateral hole in the bolt or stud, which is aligned with two of the slots in the nut. The cotter pin or safety wire keeps the nut from turning so the joint stays tight.

caster *1.* Also castor. To swivel about an eccentric vertical axis. *2.* The dimension that defines this characteristic, i.e., the distance from the point where the swivel-axis intersects the ground and the centroid of the tire contact area. Generally considered positive when the

tire centroid point falls aft of the swivel-axis intersect point. [AIR1489]

castering gear A gear configuration that has a designed-in capability to caster, such as a nose gear or tail gear. *See also* cross wind gear. [AIR1489]

castering wheel(s) Landing wheel(s) mounted in a frame that can rotate about a vertical axis. Castering wheels are of two types: (a) free-castering wheels; and (b) those whose vertical alignment is controlled from the cockpit. [ARP4107]

casting *1.* The process of making a solid shape by pouring molten metal into a cavity, or mold, and allowing it to cool and solidify. *2.* A near-net-shape object produced by this process. A rough casting, cylindrical, square, or rectangular in cross section, and intended for subsequent hot working or remelting is called an ingot.

casting alloy An alloy having suitable fluidity when molten and suitable solidification characteristics for use in producing shape castings. Most casting alloys are not suitable for rolling or forging and can only be shaped by casting.

casting resin A resin in liquid form that can be poured or otherwise introduced into a mold and shaped without pressure into solid articles. [AIR4844]

castings A designation for low-quality drill diamonds.

casting shrinkage *1.* Total reduction in volume due to the three stages of shrinkage: (a) shrinkage during cooling from casting temperature to the liquidus; (b) shrinkage during solidification; and (c) shrinkage during cooling from the solidus to room temperature. *2.* Reduction in volume at each stage in the solidification of a casting.

casting slip A slurry of clay and additives suitable for casting into molds to make unfired ceramic products.

casting wheel A large turntable with molds positioned around its periphery so that each can be moved, in turn, into position for receiving molten metal.

cast iron Any iron-carbon alloy containing at least 1.8% carbon and suitable for casting to shape.

castor *See* caster.

cast tape (film) A material that is formed directly into a tape (or film) by means of flowing or casting a solution or dispersion of the film-forming material onto a suitable carrier, then removing the solvents. [AR1931]

CAT *See* clear air turbulence.

CAT II (Category II) *See* instrument landing system (ILS) categories.

catalyst *1.* A substance that, by its mere presence, changes the rate of a reaction (decomposition), and may be recovered unaltered in nature or amount at the end of the reaction. [ARP4386] *2.* The component of a two-part, curing-type sealant that causes the prepolymer to polymerize. [AIR4069]

cataphoresis Movement of suspended solid particles in a liquid medium due to the influence of electromotive force.

catapult *1.* A device or machine used to accelerate an airplane to takeoff speed. The catapult has the capability for rapid and repeated re-use, and this distinguishes it from rocket or JATO-assisted takeoff devices. *2.* A device (usually explosive) for ejecting a person from an aircraft. [AIR1489]

catapult bridle *See* bridle, catapult.

catapult buffer *See* buffer, catapult.

catapult dolly *See* dolly, catapult.

catapult hook A hook or hooks built into an aircraft for the purpose of engaging the catapult, pendant, or bridle for the purpose of catapulting the aircraft. [AIR1489]

catapult shuttle The deck-mounted or field-mounted arrangement or device used to tow an aircraft for catapulting. Usually includes an inverted hook (spreader) for engagement with the catapult launching bridle or pendant. [AIR1489]

catastrophic failure Failure of a mechanism or component that renders an entire machine or system inoperable.

category *See* instrument landing system (ILS categories).

category "A" utilization equipment Utilization equipments whose installation in aircraft is controlled so that line drops are limited to 2 volts a-c line drop and/or 1 volt d-c. The line drop is the voltage difference between the point of voltage regulation and the power input terminals of the equipment. Use of this category should be held to a minimum, and its use is subject to approval by the procuring activity. [AS1212]

category "B" utilization equipment Utilization equipments destined for aircraft for which the line drops are less than 4 volts a-c and/or 2 volts d-c. When a detail equipment specification does not designate a category, the equipment is considered category "B" equipment. This category includes the majority of aircraft electric equipments and is the preferred category. [AS1212]

category "C" utilization equipment Equipments that are operated intermittently. During operation, voltage limits include allowance for 8 volts a-c line drop and/or 3 volts d-c line drop. [AS1212]

catenary The shape produced by holding a rope or cable at its ends and allowing the center section to sag under its own weight.

cathetometer An optical instrument for measuring small differences in height; for instance, the difference in height between two columns of mercury.

cathode *1.* An electrode at which a reduction reaction (gain of electrons) occurs. In secondary cells, either electrode may become the cathode depending on direction of current flow. [ARP4386] *2.* In electron tubes, electrodes through which a primary stream of electrons enters the interelectrode space. *3.* The metal plate or surface that acts as an electron acceptor in an electrochemical circuit. Metal ions in an electrolytic solution plate on the cathode during electroplating, and hydrogen may be formed at the cathode during electroplating or electrochemical corrosion. *4.* The positive electrode in a storage battery, or the negative electrode in an electrolytic cell. *5.* The negative electrode in an x-ray tube or vacuum tube, at which electrons enter the interelectrode space.

cathode burn Burning sensation of the eyes resulting from prolonged staring at cathode ray tubes (electronic displays). [ARP4107]

cathode corrosion *1.* Corrosion of the cathode in an electrochemical circuit, usually involving the production of alkaline corrosion products. *2.* Corrosion of the cathodic member of a galvanic couple.

cathode follower A type of electronic circuit in which the output load is connected in the cathode circuit of an electron tube or equivalent transistor, and the input signal is impressed across a terminal pair, at which one is connected directly to the control grid and the other to the remote end of the output load.

cathode ray In an electron tube or similar device, a stream of electrons emitted by the cathode.

cathode-ray oscillograph An instrument that produces a record of a waveform by photographing the graph of the waveform produced on a cathode-ray tube, or by otherwise recording such an image.

cathode ray tube (CRT) An electron beam tube in which the beam, or beams, can be focused to a desired cross section on a surface, and varied in position and intensity to produce a visible or otherwise detectable pattern. Unless otherwise stated, the term cathode-ray tube is reserved for devices in which the screen is cathodoluminescent, and in which the output information is presented in the form of a pattern of light. [ARP1782]

cathodic coating Material forming a continuous film on a base metal; applied by mechanical coating or by electroplating.

cathodic protection Preventing electrochemical corrosion of a metal object by making it the cathode of a cell using either a galvanic or impressed current.

cathodoluminescence *1.* Luminescence produced when high-velocity electrons bombard a metal in a vacuum, thus vaporizing small amounts of the metal. The vaporized metal is in an excited state and emits characteristic radiation. *2.* Luminescence induced by exposure of a suitable material to cathode rays.

CATT devices Controlled avalanche transit time (CATT) triodes that use avalanche multiplication in the collector-depletion region of a silicon, bipolar, transistor-like structure to increase the gain, and thereby achieve a higher frequency operation.

CATV Acronym for community antenna television. *See* broadband.

caulk *1.* A heavy paste, such as a mixture of a synthetic or rubber compound and a curing agent; or a natural product, such as oakum; used to seal cracks or seams, and make them airtight, steamtight, or watertight. Also known as caulking compound or calk. *2.* To seal a crack or seam with caulk.

caulking compound *See* caulk.

caul plates Smooth metal plates, free of surface defects, the same size and shape as a composite lay-up, used immediately in contact with the lay-up during the curing process to transmit normal pressure and temperature, and to provide a smooth surface on the finished laminate. [AIR4844]

cause, basic The cause of a defect, failure, or damage which results in malfunctioning of an item when: (a) the item was being operated and maintained in a manner for which it was designed; and (b) the cause was not externally induced.

caustic cracking *See* caustic embrittlement.

caustic dip A strongly alkaline solution for immersing metal parts to etch them; to neutralize an acid residue; or to remove organic material, such as grease or paint.

caustic embrittlement Intergranular cracking of carbon steel or Fe-Cr-Ni alloy exposed to an aqueous caustic solution at a temperature of at least 150°F while stressed in tension; a form of stress-corrosion cracking. Also known as caustic cracking.

caustic lines The locations of wave-front interactions induced by the maneuvers of supersonic aircraft in changing direction and/or attitude.

caustics The envelope of rays diffracted by surface defects in materials.

caustic soda The most important of the commercial caustic materials. Consists of sodium hydroxide that contains 76 to 78% sodium oxide.

caution alert Abnormal operational or aircraft system condition that requires immediate crew awareness and subsequent corrective or compensatory crew action. [ARP4153]

caution signal A signal that alerts the operator to an impending dangerous condition requiring attention, but not necessarily immediate action. [AIR4896]

Cavendish balance A torsional instrument for determining the gravitational constant by measuring the displacement of two spheres of known small mass, mounted on opposite ends of a thin rod suspended on a fine wire, when two spheres of known large mass are brought near the small spheres.

cavitation Formation of cavities, either gaseous or vapor within a liquid stream, occurring where the pressure is locally reduced to the vapor pressure of the liquid. Cavitation may include gas coming out of solution in the liquid as pressure is reduced (soft cavitation). [ARP4386] *See also* cavitation flow.

cavitation damage *See* cavitation erosion.

cavitation erosion Progressive removal of surface material due to localized hydrodynamic impact forces associated with the formation and subsequent collapse of bubbles in a liquid in contact with the damaged surface. Also known as cavitation damage or liquid-erosion failure.

cavitation flow The formation of bubbles in a liquid, occurring whenever the static pressure at any point in the fluid flow becomes less than the fluid vapor pressure.

cavitons Density cavities created by localized oscillating electric fields.

cavity *1.* Gaseous or vapor space within a liquid stream. Occurs where the pressure is locally reduced to the vapor pressure of the liquid. *2.* The space inside a mold in which a resin or molding compound is poured or injected. *3.* The female portion of mold. [AIR4844]

cavity resonator A space normally enclosed by an electrically conducting surface; used to store electromagnetic energy. The resonant frequency of a cavity resonator is determined by the shape of the enclosure.

cavity-type wavemeter An instrument used to determine frequency in a waveguide system. Typically, the position of a piston inside a cylindrical cavity is tuned to resonance, which is determined by a drop in transmitted power; the meter is then detuned for normal operation.

CBL *See* control-by-light.

CBR *See* California Bearing Ratio.

CBW *See* constant bandwidth.

CCD *See* charge coupled devices.

CCD star tracker Navigation instrument designed for the NASA space transportation system.

CCF *See* charge flow device.

CCR *See* control complexity ratio.

CCV *See* control-configured vehicle (aircraft).

cd *See* candela.

CD Abbreviation for constant delay. *See* Bessel.

CDF *See* confined detonating fuse.

CDI *See* course deviation indicator.

CDMA *See* code division multiple access.

CDU *See* control and display unit.

ceiling *1.* The height above the earth's surface of the lowest layer of clouds or other obscuring phenomena that are reported as "broken," "overcast," or "obscuration," and not classified as "thin" or "partial." *2.* The upper operating limit(s) of an aircraft. *See also* absolute ceiling and service ceiling. [ARP4107]

ceilometer A device or apparatus for measuring the height of a cloud ceiling or determining the vertical visibility to an obscuration. [ARP4107]

celestial geodesy The determination of the form of the earth, of the earth's gravitational field, and of relative positions of satellite trajectories.

celestial mechanics The study of the theory of motions of celestial bodies under the influence of gravitational fields.

celestial navigation The process of directing a craft from one point to another by reference to celestial bodies of known constants.

celestial sphere An imaginary sphere of infinite radius, concentric with the earth, on which all celestial bodies except the earth are assumed to be projected.

cell *1.* An electrochemical device composed of positive and negative plates, a separator, and an electrolyte, which is capable of storing electrical energy. When encased in a container and fitted with terminals, a cell is the basic building block of a battery. [ARP4386] *2.* The basic battery building block. An electrochemical storage device consisting of positive and negative plates, separator, gas barrier, and electrolyte contained in a cell case (jar), with suitable electrical terminals for connection to additional cells. [AS8033] *3.* A location specified by whole or part of the address and possessed of the faculty to store. Specific terms such as column, field, location, and block, are preferable when appropriate. *See also* storage cell.

cell reversal The reversal of polarity of the terminals of a cell in a multicell battery due to overdischarge. [ARP4386]

cell size The diameter of an inscribed circle within a cell of honeycomb core. [AIR4844]

cellular plastic A plastic whose density is reduced by the presence of numerous small cavities dispersed throughout the mass. [AIR4844]

Celsius A scale for temperature measurement based on the definition of 0°C and 100°C as the freezing point and boiling point, respectively, of pure water at standard pressure. Formerly referred to as centigrade.

cementation *1.* High-temperature impregnation of a metal surface with another material. *2.* Conversion of wrought iron into steel by packing it in charcoal and heating it at about 1800°F for 7 to 10 days.

cemented carbide A powder-metallurgy product consisting of granular tungsten, titanium, or tantalum carbides in a temperature-resistant matrix, usually cobalt. (The proportion of matrix material is small compared to the amount of carbide.) Used for high-performance cutting tools, punches, and dies.

cementite An intermetallic compound containing iron and carbon.

censoring *Right censoring*–data is right censored at M if, whenever an observation is less than or equal to M, the actual value of the observation is recorded. *Left censoring*–data is left censored at M if, whenever an observation is greater than or equal to M, the actual value of the observation is recorded. [AIR4844]

cent The interval between two sound frequencies, where the ratio of the two frequencies is the twelve-hundredth root of two. Also equal to one-hundredth of a semi-tone.

center gage A gage used to check angles, such as the angle of a cutting-tool point or screw thread.

centerline groove An injection-sealing groove machined along the fastener line. [AIR4069]

center of gravity The center of mass of a system of masses; for example, the barycenter of the earth-moon system.

center of lift *1.* The location of the resultant lift vector. *2.* The mean of all airfoil centers of pressure on a wing.

center of mass A point on a material body or system of bodies that moves as though the total mass of the system existed at that point and all external forces were applied at the point.

center of pressure The point on the chord of an airfoil section that is at the intersection of the chord and the line of action of the combined air forces. [ARP4386]

center tapped winding A winding having an internal tap at the center. [ARP667]

centigrade A nonpreferred term, formerly used to designate the scale now referred to as the Celsius scale. *See* Celsius.

centimeter waves Electromagnetic radiation in the 3,000 to 30,000 MHz range.

centimeter-gram-second (CGS) A standard metric system of units used largely for scientific work prior to adoption of the international SI system currently preferred for both scientific and engineering work.

central flow control function (CFCF) Air traffic control command center function that is responsible for coordination and approval of all major intercenter flow-control restrictions on a system basis in order to obtain maximum utilization of the airspace. *See also* fuel advisory departure and quota flow control. [ARP4107]

central integrated test system An on-line test system that processes, records, or displays at a central location information gathered by test-point datasensors at more than one remotely located equipment or system under test. [AIR4896]

centralized control In data communications systems, a method of media access control in which a designated station exercises positive control over all message traffic. [AIR4271]

centralized maintenance shops One maintenance shop that has responsibility to maintain all equipment in the facility. Usually several crafts work out of this one centralized maintenance shop.

central 80% of useful display area Area within the locus of points that are 80% of the distance from display center to the edge of the useful display area. [ARP1782]

central processing unit (CPU) 1. The unit of a computing system that includes the circuits controlling the interpretation of instructions and their execution. 2. *See* mainframe.

centrifugal casting A production technique for fabricating cylindrical composites, such as pipes, in which composite material is positioned inside a hollow mandrel designed to be heated and rotated as the resin is cured. [AIR4844]

centrifugal fan A fan rotor or wheel within a housing which discharges the air at right angle to the axis of the wheel.

centrifugal force The apparent force in a rotating system, deflecting masses radically outward from the axis of rotation. The magnitude of this force per unit mass is $\omega^2 R$ where ω is the angular speed of rotation and R is the radius of curvature of the path. This magnitude may also be written V^2/R, in terms of the linear speed, V. Centrifugal force (per unit mass) is equal and opposite to the centripetal acceleration. [AIR1489]

centrifugal growth The growth in diameter of a rotating tire due to centrifugal force. [AIR1489]

centrifugal pump A pump that creates high fluid velocity through centrifugal action. Fluid momentum is recovered as a pressure head at the pump outlet. [ARP4386]

centrifugal separator A device that utilizes centrifugal force to separate materials of differing densities, such as water droplets or impurities from air. [ARP147C]

centrifugal tachometer An instrument that measures the instantaneous angular speed of a rotating member, such as a shaft, by measuring the centrifugal force on a mass that rotates with it.

centrifuge *1.* Specifically in aerospace, large motor-driven apparatus with long arms at the end. Human and animal subjects or equipment can be secured to the arms. With the subjects or equipment in place, the centrifuge is revolved and rotated at various speeds to simulate (very closely) the (prolonged) acceleration in high performance aircraft, rockets, and spacecraft. Sometimes called astronautic centrifuges. *2.* A rotating device that separates suspended, fine or colloidal particles from a liquid; or separates two liquids of different specific gravities by means of centrifugal force.

cepstra The Fourier transformation of the logarithm of a power spectrum.

cepstral analysis The application of cepstral methods to wave or signal phenomena in seismology, speech analysis, echoes, underwater acoustics, etc.

ceramic A heat-resistant natural or synthetic inorganic product made by firing a nonmetallic mineral.

ceramic fibers Fibers composed of ceramic materials. Usually used for reinforcement.

ceramic matrix composites Composite materials consisting of a reinforced ceramic matrix.

ceramics Inorganic compounds or mixtures requiring heat treatment to fuse them into homogeneous masses; usually possess high temperature strength, but low ductility. Uses range from china for dishes to refractory liners for nozzles.

ceramic tool A cutting tool made from fused, sintered, or cemented metallic oxides.

ceramic transducer *See* electrostriction transducer.

Cerenkov effect *See* Cerenkov radiation.

Cerenkov radiation *1.* The radiation from a charged particle whose velocity is greater than the phase velocity that an electromagnetic wave would have if it were propagating in the medium. A particle emitting Cerenkov radiation will continue to lose energy by radiation until its velocity is less than this phase velocity. *2.* Visible light produced when charged particles pass through a transparent medium at a speed exceeding the speed of light in the medium.

cermet A body consisting of ceramic particles bonded with a metal. Used in aircraft, rockets, and spacecraft for high-strength, high-temperature applications. The name is derived from a combination of ceramic and metal.

certified takeoff weight The maximum takeoff weight stated on the type certificate data sheet of the aircraft; or, for aircraft manufactured prior to 1957, on the type specification sheet of the aircraft. [ARP4107]

CFCF *See* central flow control function.

CFD *See* charge flow devices.

CFR engine Abbreviation for Cooperative Fuel Research engine—a standard test engine for determining the octane number of motor fuels.

C-glass A glass with a soda-lime-borosilicate composition; used for its chemical stability in corrosive environments. [AIR4844]

CGM Acronym for computer graphics metafile.

CGS *See* centimeter-gram-second.

chafer strip The strips of rubber-coated fabric that reinforce the bead area and protect the carcass plies against wheel chafing and against damage when mounting or demounting a tire. [AIR1489]

chafing Repeated relative motion between wires, cables, groups, harnesses, or bundles; or between these wiring system components and structure or equipment. Results in deleterious wear. [AR1931]

chain-balanced density meter A submerged-float meter, containing an iron-core float that moves up and down within a pickup coil. A slack chain attached to the bottom of the float applies more weight as the float rises, and establishes a definite equilibrium position for any given fluid density within the range of the instrument.

chain drive *1.* A series of links used to transmit power from one rotating member to another. [ARP4386] *2.* A device for transmitting power and motion without slipping. Consists of an endless chain that meshes with driving and driven sprockets. Chain drives are used on bicycles and motorcycles to provide the motive power; on conveyors to drive the belts; and in hoisting mechanisms to provide the lifting power.

chain-float liquid-level gage A device for indicating liquid level in a tank. Consists of a float connected to a counterweight by a chain running over a sprocket. As the float rises and falls with liquid level in the tank, the chain rotates the sprocket, which, in turn, positions a pointer to indicate liquid level.

chain growth polymerization A chemical reaction in which polymer formation is initiated by a reactive species, R*, produced from some compound, I, termed an initiator. [AIR4844]

chain-link fence illusion *See* illusions, visual.

chain length The length of the stretched linear macromolecule, most often expressed by the number of identical links. [AIR4844]

chalking The formation of a powdery surface condition due to disintegration of firesleeving or hose cover material by weathering or other destructive environments. [ARP1658A]

chamber pressure The pressure within the case during combustion. [ARP4386]

chamber temperature The temperature of combustion products in the combustion chamber, generally considered to be the adiabatic isobaric flame temperature. [ARP4386]

chamfer *1.* A beveled edge that relieves an otherwise sharp corner. *2.* A relieved, angular cutting edge at a tooth corner on a milling cutter or similar tool.

chance failure Any failure whose occurrence is unpredictable in an absolute sense, but which is predictable in a probabilistic or statistical sense. [AIR4896]

Chandler motion *See* polar wandering.

change control (configuration control) The process of recording, evaluating, approving or disapproving, and coordinating changes to configuration items after formal establishment of their configuration identification or to baselines after their establishment.

change detection A process of examining imagery to detect changes on a planetary surface or astronomical body.

changeover effect In ultrasonic inspection, refers to the reduction in amplitude and complete disappearance of back reflection as the delamination signal rises. [ARP5089]

channel *1.* One signal or control path of a redundant set. A channel is an entity within itself and contains elements individual to that channel. [ARP4386] *2.* A passage formed by a structural discontinuity, such as the opening under a joggle or other void. [AIR4069] *3.* A groove machined in a faying surface to accept a uniform bead or section of sealant. [AIR4069] *4.* A path along which information, particularly a series of digits or characters, may flow. *5.* One or more parallel tracks treated as a unit. *6.* In a circulating storage, one recirculating path containing a fixed number of words stored serially by word. *7.* A path for electrical communication. *8.* A band of frequencies used for communication.

channel buffering Technique used to minimize the possibility of a failure in one channel inducing a failure in another channel. [ARP4386]

channeling leak A leak that develops at a source located some distance from the leak exit point. [AIR4069]

channelized attention *See* attention, anomalies of channelized attention.

channel noise In communications, bursts of interruptive pulses caused mainly by contact closures in electromagnetic equipment, or by transient voltages in electric cables during transmission of signals or data. Impulsive noise is the frequent cause of transmission errors.

channel priority The order of authority of the various channels in a redundant system in which the channels are not equivalent. [ARP4386]

channel sampling rate The number of times a given data input is sampled during a specified time interval.

channel selector In an FM discriminator, the plug-in module that causes the device to select one of the channels and demodulate the subcarrier to recover data.

channel summing The combining of multiple channels to provide a control function. Examples of channel summing techniques are: flow summing, flux summing, force summing,

position summing, torque summing, and velocity summing. [ARP4386]

CHAR *See* character.

character A single computer symbol. Abbreviated CHAR.

character codes The binary code patterns used to create characters in a computer.

characteristic *1.* Distinguishing quality, property, feature, or capability of an entity. *2.* The integral part of a common logarithm, e.g., in the logarithm 2.5, the characteristic is 2 and the mantissa is 0.5. *3.* That portion of a floating point number indicating the exponent. *4.* A distinctive property of an individual, document, item, etc.

characteristic curve *1.* A curve expressing a relation between two variable properties of a luminous source; for example, candlepower and volts or candlepower and rate of fuel consumption. [ARP798] *2.* Of a photographic or radiographic film, the graph of relative transmittance of the emulsion versus exposure, or a graph of functions of these two quantities. Also known as characteristic emulsion curve.

characteristic emulsion curve *See* characteristic curve.

characteristic impedance (Zo) Of a uniform line, the ratio of an applied potential difference to the resultant current at the point where the potential difference is applied, when the line is of infinite length. [AR1931]

charge *1.* A given quantity of explosive. [AIR913] *2.* The starting stock loaded into a batch process. *3.* A measure of the accumulation or depletion of electrons at any given point. *4.* The amount of substance loaded into a closed system, such as refrigerant into a refrigeration system.

chargeable Within the responsibility of a given organizational entity. Applied to terms such as failures and maintenance time. [ARD50010]

chargeable failure A relevant, independent failure of equipment under test, and any dependent failures caused thereby, which are classified as one failure and used to determine contractual compliance with acceptance and rejection criteria. [AIR4896]

charge coupled device (CCD) Semiconductor devices arrayed so that the electric charge at the output of one provides the input stimulus to the next.

charge efficiency The efficiency of electric cell recharging.

charge exchange The collisional transfer of an electron from a neutral atom or molecule to an ion.

charge flow devices (CFD) Metal oxide semiconductor (MOS) devices used for fire detectors and humidity sensors.

charge oil The amount of refrigerant oil added to a vapor-cycle system along with the refrigerant. Purpose is to lubricate the compressor and valves. [ARP147C]

charge, refrigerant The amount and type of refrigerant contained in a system. [ARP147C]

charge retention The tendency of a charged cell to resist self-discharge. [ARP4386]

charging 1. The process of filling a vapor-cycle system with the refrigerant and oil charge. [ARP147C] 2. The return of electrical energy to a battery and its storage in electrochemical form. [AS8033]

charm A quantum number that has been proposed to account for an apparent lack of symmetry in the behavior of hadrons relative to that of leptons; to explain why certain reactions of elementary particles do not occur; and to account for the longevity of the J particle.

Charles' Law *See* Gay-Lussac's or Charles' law. *See also* gas laws.

Charon Natural satellite of the planet Pluto; discovered and named by Dr. James W. Christy.

charpy impact test A test for shock loading in which a centrally notched sample bar is held at both ends and broken by striking the back face in the same plane as the notch. [AIR4844]

charring The heating of a composite in air to reduce the polymer matrix to ash, allowing the fiber content to be determined by weight. [AIR4844]

chart A sheet or plate giving printed information in tabular and/or graphic form. Is not intended for, and does it have provisions for, recording additional information thereon. Excludes maps. [ARP481A]

chart datum *See* datum plane.

chart recorder A device for automatically plotting a dependent variable against an independent variable. The dependent variable is proportional to the input signal from a transducer. The independent variable may be proportional to a transducer signal, too, but is most often time or a time-dependent variable that can be produced by controlling the rate of advance of the rolled chart paper.

chase 1. The main body of a mold that contains one or more mold cavities. 2. To make a series of cuts, each following the path of a preceding cut, such as is done to produce a thread in lathe turning using a single-point tool. 3. To straighten and clean damaged or debris-filled threads on a screw or pipe end.

chassis 1. A frame or box-like sheet-metal support for mounting the components of an electronic device. 2. A frame for a wheeled vehicle that provides most of the stiffness and strength of the vehicle body; and supports the body, engine, and passenger or load compartment on the running gear.

chatter 1. In machining or grinding, a vibration of the tool, wheel, or workpiece, producing a wavy surface on the work. 2. The finish produced by such vibration. [ARP4784] 3. In a servoactuator, a low-amplitude (usually) high-frequency oscillation of the output. [ARP1281A]

check 1. A process of partial or complete testing of the correctness of machine operations. 2. The existence of certain prescribed conditions within the computer, or the correctness of the results produced by a program. A check of any of these conditions may be made automatically by the equipment or may be programmed. *See also* marginal check.

check analysis Analysis made by purchaser of parts and materials to verify the composition of a heat or lot, or to determine variations in composition within a heat or lot. [AMS2269E]

check, bench *See* bench check.

check bit A binary check digit; often a parity bit. *See also* parity check.

check, C A heavy maintenance check. *See also* letter check. [ARP4386]

check digit In data transmission, one or more redundant digits appended to a machine word, and used in relation to the other digits in the word to detect errors in data transmission.

check, functional A quantitative check to determine if one or more functions of an item performs within specified limits. [AIR4896]

checking *1.* Short, shallow cracks in the surface of an elastomeric hose cover material resulting from damaging action of environmental conditions. [ARP1658A] *2.* A network of fine cracks in a coating or at the surface of a metal part. May appear during processing, but are more often associated with service, especially when the service involves thermal cycling.

check, operational A task to determine if an item is fulfilling its intended purpose. The task does not require quantitative tolerances. [AIR4896]

checkout *1.* A sequence of actions taken to test or examine a thing as to its readiness for incorporation into a new phase of use, or for the performance of its intended function. *2.* The sequence of steps taken to familiarize a person with the operation of an airplane or other piece of equipment. *3.* Man/machine task to determine that the equipment is operating satisfactorily and ready for return to service.

checkout time The time required to check out equipment, to complete a maintenance action, or to otherwise verify that a system or equipment is in satisfactorily operating condition. [AIR4896]

check, parity *See* parity check.

checkpoint *1.* A point in time in a machine run at which processing is temporarily halted to make a record of the condition of all the variables of the machine run, such as the status of input and output devices; or to make a copy of working storage. Checkpoints are used in conjunction with a restart routine to minimize reprocessing time occasioned by functional failures. *2.* A particular point in a program at which processing is halted for checking.

check problem A problem used to test the operation of a computer or to test a computer program. If the result given by the computer for the check problem does not match the known result, it indicates an error in programming or operation.

check, scheduled maintenance Any of the maintenance opportunities that are prepackaged and are accomplished on a regular basis. [AIR4896]

checksum *1.* A routine for checking the accuracy of data transmission by dividing the data into small segments, such as disk sectors, and computing a sum for each segment. *2.* Entry at the end of a block of data corresponding to the binary sum of all information in the block. Used in error-checking procedures.

check, validity *See* validity check.

check valve A valve that allows flow in a single direction only. [ARP986]

chemical affinity *1.* The relative ease with which two elements or compounds react with each other to form one or more specific compounds. *2.* The ability of two chemical elements to react to form a stable valence compound.

chemical analysis Determination of the principal chemical constituents.

chemical attack Damage to a resin matrix by accidental contact with, or unauthorized use of, chemical products. [AIR4844]

chemical cleaning Processing using nitric, hydrofluoric, or nitric/hydrofluoric acid solutions in which material removal exceeds 0.0004 inch. [AMS4985B]

chemical clouds Artificial clouds of chemical compounds released in the ionosphere for observation of dispersion and other characteristics.

chemical conversion coating A decorative or protective surface coating produced by inducing a chemical reaction between surface layers of a part and a specific chemical environment, such as in chromate treatment or phosphating.

chemical core An object, composed of sodium chlorate or an analogous alkali metal chlorate or perchlorate, formulated with fuels, catalysts, and other modifiers and additives as required by the particular design. Evolves oxygen by a controlled chemical decomposition reaction when actuated. [AS1303]

chemical defense All actions and counteractions designed for the protection of personnel and material against offensive chemical agents.

chemical energy Energy produced or absorbed in the process of a chemical reaction. In any such reaction, energy losses or gains usually involve only the outermost electrons of the atoms or ions of the system undergoing change. A chemical bond of some type is established or broken without disrupting the original atomic or ionic identities of the constituents.

chemical feed pipe A pipe inside a boiler drum through which chemicals for treating the boiler water are introduced.

chemical fuels Fuels, such as liquid or solid rocket fuel or internal combustion engine fuel, that depend on an oxidizer for combustion or for development of thrust. Distinguished from nuclear fuel.

chemically foamed plastic A cellular plastic in which the cells are formed by gases generated from thermal decomposition or chemical reaction of the constituents. [AIR4844]

chemical oxygen generator A device containing a compound with chemically bonded oxygen which, when properly activated, will produce a supply of gaseous oxygen at a purity, rate, and quantity suitable for breathing. [ARP171]

chemical release modules Shuttle-launched, free-flying spacecraft that contain canisters for injecting chemicals into the upper atmosphere and measuring the resultant reactions.

chemical vapor deposited carbon Carbon deposited on a substrate by pyrolysis of a hydrocarbon, such as methane. [AIR4844]

chemical vapor deposition Process used in the manufacture of several composite reinforcements, especially boron and silicon carbide, in which desired reinforcement material is deposited from the vapor phase onto a continuous core. [AIR4844]

chemiluminescence Any luminescence produced by chemical action.

chemisorption The binding—through chemical bonds or forces—of a liquid or gas on the surface or the interior of a solid.

chemosphere The vaguely defined region of the upper atmosphere in which photochemical reactions take place. Generally considered to include the stratosphere (or the top thereof) and the mesosphere; sometimes the lower part of the thermosphere.

cherry picker Any of several types of small, traveling cranes, especially one consisting of an open passenger compartment at the free end of a jointed boom.

chiller A liquid transport system whose heat sink is a vapor-cycle system. [ARP147C]

chimney core The inner cylindrical section of a double-wall chimney, which is separated from the outer section by an air space.

chimney lining The material that forms the inner surface of the chimney.

chip Single, large-scale integrated circuit.

chip breaker An attachment or a relieving channel behind the cutting edge of a lathe tool, which causes removed stock to break up into pieces, rather than to come off as long, unbroken curls.

chip (clamshell type) Mark or flaw made by the breaking off of a small piece of material. The resulting damage looks like a clam shell, with curved periphery and many fine hairlines or ridges that follow the outline of the outer edge. [AIR5122]

chipping *1.* Using a manual or pneumatic chisel to remove seams, surface defects, or excess metal from semifinished mill products. *2.* Using a hand-tool or pneumatic hammer with chisel-shaped or pointed faces to remove rust, scale, or other deposits from metal surfaces.

chips (electronics) Integrated microcircuits mounted on substrates and performing significant numbers of functions.

chips (memory devices) Integrated microcircuit devices used collectively to perform the functions of data storage: accepting, retaining, and emitting bits of data.

Chiron Minor planet 2060, a solar system asteroid discovered by Charles T. Kowal of Hale Observatories.

chirp *1.* An all-encompassing term for the various techniques of pulse expansion-pulse compression applied to pulse radar. *2.* A technique to expand narrow pulses to wide pulses for transmission, and compress wide received pulses to the original narrow pulse width and wave shape. Used to gain improvement in signal-to-noise ratio without degradation of range resolution and range discrimination.

chlorate candle *1.* An alternate term sometimes applied to a chemical core, possibly because the progress of the organized decomposition reaction along the length of the core can be likened to the burning of a candle. [AS1303] *2.* "Candles" that can store oxygen in a compact volume approaching that attainable with liquid oxygen. The liberated oxygen contains a small amount of fine sodium chloride which can be readily removed by filters. Chlorate candles have desirable weight, cost, and storage-life characteristics. [AIR1246]

Chlorella A genus of unicellular green algae that can be adapted to convert carbon dioxide into oxygen in a closed ecological system.

chlorocarbons Compounds that contain chlorine and carbon, with or without other elements.

choke *1.* A flow restriction in which the length of the reduced area passage is of sufficient magnitude that fluid viscosity becomes of major importance in determining pressure drop. [ARP4386] *2.* A valve that increases suction in an internal combustion engine so that an excess proportion of fuel is drawn in to facilitate starting a cold engine.

choke coil An inductor that allows direct current to pass, but presents relatively large impedance to alternating current.

choked flow *1.* Flow of a compressible fluid (gas), limited by the speed of sound in the fluid at the throat of a control section. [ARP4386] *2.* The condition that exists when, with the upstream conditions remaining constant, the flow through a valve cannot be further increased by lowering the downstream pressure.

chokes *True chokes*–a form of decompression sickness in which a gas (usually nitrogen) evolves from solution in the lung area, causing respiratory distress; a dry, non-productive cough; sharp pain in the chest; a sense of suffocation; and/or severe substernal pain. *False chokes*–dryness of the throat and coughing resulting from prolonged breathing of aviator's oxygen. This is not decompression sickness and is distinguished from true chokes by the lack of pain in the chest. [ARP171]

chondrites Meteoritic stones characterized by small, rounded grains or spherules.

chop (fuel) Shutoff of fuel when the engine is running. [AIR906A]

chop (throttle) Rapid throttle movement from any power setting to idle or cutoff. [AIR906A]

chopped strand mat A mat formed of strands cut to a short length, randomly distributed, without intentional orientation, and held together by a binder. [AIR4844]

chopped strands Short strands cut from continuous filament strands, not held together by any means. [AIR4844]

chopper Any device for periodically interrupting a continuous current or flux.

chord (chord line) In aeronautics, a straight line, parallel to the plane of symmetry, connecting the leading and trailing edges of an airfoil. The ordinates and angles of the airfoil are measured from the chord line. [ARP4107]

chord line *See* chord.

chords *1.* Straight lines intersecting circles or other curves, or straight lines connecting the ends of arcs. *2.* In aeronautics, straight lines intersecting or touching airfoil profiles at two points; specifically, those parts of lines between two points of intersections.

Christiansen filter A device for admitting monochromatic radiation to a lens system. Consists of coarse powder of a transparent solid confined between parallel windows, with the spaces between particles being filled with a liquid whose refractive index is the same as that of the powder for a certain wavelength; only that wavelength is transmitted by the filter without deviation.

chroma *See* Munsell chroma.

chromadizing Improving paint adhesion on aluminum and its alloys by treating the surface with chromic acid.

chromate treatment Applying a solution of hexavalent chromic acid to produce a protective conversion coating of trivalent and hexavalent chromium compounds.

chromatic aberration In a laser, the focusing of light rays of different wavelengths at different distances from the lens. This is not a significant effect with a single-wavelength laser source, but can be when working at different or multiple wavelengths.

chromatic contrast The difference between two areas such as a symbol and its background. [ARP4032]

chromaticity *1.* The expression of color in terms if CIE codents. [ARP798] *2.* The color quality of a color stimulus definable by its chromaticity coordinates. [ARP1782]

chromaticity coordinate The ratio of each of a set of three tristimulus values to their sum. Given in ordered pairs, for example, ((x, y), u^1 7_1) etc. [ARP1782]

chromaticity diagram Graphic representation of all possible colors on a two-dimensional diagram. Colors of similar dominant wavelength and excitation purity will plot close to one another. [ARP4032]

chromaticity difference (CD) Distance between two color points. [ARP4067]

chromatogram In chromatography, a plot of detector response against peak volume of solution emerging from the system for each of the constituents that have been separated. [AIR4844]

chromatography The separation of chemical substances by making use of differences in the rates at which the substances travel through or along a stationary medium.

chrome finish Applied to glass fibers to give good bonding to polyester and epoxy resins. [AIR4844]

chrome plating *See* chromium plating.

chromic dispersion Dispersion in a fiber caused by different wavelengths of the signal traveling at different speeds down the fiber. [AIR4288]

chromium coating *See* chromium plating.

chromium plating Electrodeposition of either a bright, reflective coating, or a hard, less-reflective coating of chromium on a metal surface. Also known as chrome plating and chromium coating.

chromium steels Steels containing chromium as the main alloying element.

chromizing Producing an alloyed layer on the surface of a metal by deposition and subsequent diffusion of metallic chromium.

chromophore The group of atoms within a molecule that contributes most heavily to its light-absorption qualities.

chromosphere A thin layer of relatively transparent gases above the photosphere of the sun.

chronograph An instrument used to record the time at which an event occurs or the time interval between two events.

chronotron A device for measuring elapsed time between two events; the time is determined by measuring the position of the superimposed loci of a pair of pulses initiated by the events. *See also* time lag.

chuffing Intermittent or irregular burning in a solid propellant rocket motor with corresponding low-frequency pressure oscillations. [AIR913]

chugging An irregular combustion of liquid fuels in a rocket engine, with corresponding low-frequency pressure oscillations. [AIR913]

chute, drag *See* drag chute.

chute, drogue A parachute, deployed for the purpose of providing the necessary force for extracting and deploying a main drag chute or chutes. [AIR1489]

chute, pilot A parachute deployed to initiate a sequence of events in a drag chute system: The pilot chute is mechanically (or otherwise) ejected into the airstream. It provides enough drag to pull the drogue chute, which, in turn, provides adequate power to extract or deploy the main chute(s). [AIR1489]

Ci *See* curie.

CID *See* computer interface device.

CIE *See* Commission Internationale de l'Eclairage.

CIM *See* computer integrated manufacturing.

cinder A particle of gas-borne, partially burned fuel larger than 100 micrometers in diameter.

CIP *See* component improvement program.

Cipolletti weir An open-channel flow-measurement device, similar to a rectangular weir but having sloping sides, which results in a simplified discharge equation.

circadian desynchronization That state in which the body's "normal" 24-hour rhythmic biological cycle (circadian rhythm) is disturbed. Typically caused by movement across several time zones, and generally having an adverse effect on pilot performance. Colloquially referred to as "jet lag." [ARP4107]

circadian rhythm The tendency for some biological processes to occur at approximately the same time in each 24-hour period.

circle of confusion A circular image in the focal plane of an optical system; the image formed by the optical system of a distant point object.

circle-to-land maneuver (circling maneuver) A maneuver initiated by the pilot to align the aircraft with a runway for landing when a straight-in landing from an instrument approach is not possible or is not desirable. This maneuver is made only after ATC authorization has been obtained and the pilot has established the required visual reference to the airport. [ARP4107]

circling minimums *See* landing minimums/ IFR landing minimums.

circuit *1.* A complete path over which electrons can flow from the negative terminals of a voltage source, through parts and wires, to the positive terminals of the same voltage source. [AR1931] *2.* Network providing one or more closed paths. *3.* Any group of related electronic paths and components that electronic signals pass through to perform a specific function.

circuit analyzer *1.* A multipurpose assembly of several instruments or instrument circuits in one housing which are to be used in measuring two or more operating characteristics of an electronic circuit. *2. See* volt-ohm-milliammeter.

circuit board A sheet of insulating material that is laminated to foil. The foil is etched to produce a circuit pattern on one or both sides. [AIR4844]

circuit breaker *1.* Automatic circuit-interrupting device that is capable of repeatedly performing its function should a prescribed set of conditions exist. [ARP4404] *2.* A device designed to open and close a circuit by non-automatic means. Also designed to open the circuit automatically on a prede-

termined overload of current without injury to itself when properly applied within its rating. [ARP1199A] *3.* A resettable circuit-protective device. Circuit breakers can be divided by function into three classes: control circuit breakers, power circuit breakers, and remote circuit breakers. *Control circuit breaker*–a circuit breaker whose function is to protect the wiring used to operate control devices, such as relays. *Power circuit breaker*–a circuit breaker, as distinguished from a control circuit breaker, whose function is to protect the wiring carrying the power to using equipment. *Remote circuit breaker*–a circuit breaker that is not accessible to the crew during flight. [AS486B]

circuit breaker, companion-trip multiple-pole *See* companion trip multiple pole circuit breaker.

circuit breaker, nontrip-free A circuit breaker so designed that the circuit can be maintained closed when carrying overload current that would automatically trip the breaker to the open position. [ARP1199A]

circuit breaker, trip-free multiple-pole *See* trip-free circuit breaker.

circuit diagram A line drawing of an electronic/ electrical system which identifies components and diagrams how they are connected.

circuit, fail-safe A circuit that has characteristics such that any probable malfunction will not adversely affect the safe operation of the aircraft or the safety of the passengers. [ARP4404]

circuit-noise meter An instrument that uses frequency-weighting networks and other components to measure electronic noise in a circuit, giving approximately equal readings for noises that produce equal levels of interference.

circuit protection Automatic protection of a consequence-limiting nature used to minimize the danger of fire and/or smoke, as well as the disturbance to the rest of the system, that may result from electrical faults or prolonged electrical overloads. [ARP4404]

circuits, emergency Essential circuits, the failure of which may result in the inability of the aircraft to maintain controlled flight and effect a safe landing. [ARP4404]

circuits, essential Those circuits necessary to accomplish the mission of the aircraft under the most adverse environmental conditions for which the aircraft was designed. [ARP4404]

circular-chart recorder A type of recording instrument in which the input signal from a temperature, pressure, flow, or other transducer moves a pivoted pen over a circular piece of chart paper that rotates about its center at a fixed rate with time.

circularity In data processing, a warning message that the commands for two separate but interdependent cells in a program cannot proceed until a value for one of the cells is determined.

circularity (roundness) One-half the variation in pitch diameter of a screw thread around the circumference of the pitch cylinder. [AS8879]

circularly polarized light *1.* Light in which the polarization vector rotates periodically, but does not change magnitude, describing a circle. *2.* Light wave formed by the superposition of two plane-polarized (or linearly polarized) light waves, of equal magnitude, one 90° in phase behind the other.

circular mil A wire-gage measurement equal to the cross-sectional area of a wire that is one mil (0.001 in.) in diameter; actual area is 7.8540 $\times$ 10^{-7} in^2.

circular mil area (CMA) The cross-sectional area of the current-carrying portion of a conductor, expressed in circular mils. [AR1931]

circular polarized wave An electromagnetic wave in which the electric field vector, magnetic field vector, or both, describe a circle.

circularvection *See* illusion, vection.

circular velocity Critical velocity at which a satellite will move in a circular orbit around its primary. [ARP4386]

circular waveguides Small, hollow tubes that are designed to transmit a specific wavelength along the length of the tube.

circulating memory In an electronic memory device, a means of delaying information, combined with a means for regenerating the information and reinserting it into the delaying means.

circulation *1.* The flow or motion of a fluid in or through a given area or volume. *2.* A precise measure of the average flow of a fluid along a given closed curve. *3.* The movement of water and steam within a steam-generating unit.

circulation control airfoils Airfoils in which a high lift capability is produced by supercirculation. Control of the stagnation points by the jet sheet produces high lift coefficients.

circulation control rotors Rotors that provide STOL capability on high-performance aircraft by means of tangential blowing over a rounded trailing edge, and mass flow characteristic of turbine engine bleed.

circulation distribution The line integral of the velocity component around a curve along a closed contour.

circulation ratio The ratio of the water entering a circuit to the steam generated within that circuit in a unit of time.

circulator *1.* A pipe or tube used to pass steam or water between upper boiler drums; usually located where the best absorption is low. *2.* Tubes connecting headers of horizontal water tube boilers with drums.

circumferential crimp A type of crimp in which the crimping dies completely surround a barrel, resulting in a symmetrical reshaping of the barrel. Circumferential crimps may be oval, hexagonal, or circular, to name a few. [ARP914A]

circumferential winding In filament-wound, reinforced plastics, a winding in which the filaments are essentially perpendicular to the axis. [AIR4844]

circumsolar radiation Radiation from small-angle scattering of direct sunlight by atmospheric aerosols with dimensions on the order of or greater than the wavelength of light.

circumsolar telescopes Optical instruments for measuring circumsolar radiation for application to solar-energy systems. Mirrors and lenses are utilized for incident sunlight concentration.

cirrus clouds A high-level, principal, stratiform type of cloud, composed of detached cirriform elements in the form of white patches or narrow bands. Composed mostly of ice crystals; usually very thin, and appear transparent to the extent that a halo phenomenon may be observed. [AIR4367]

cislunar space Of or pertaining to phenomena, projects, or activity in the space between the earth and the moon, or between the earth and the moon's orbit.

cladding *1.* A method of applying a layer of metal over another metal whereby the junction of the two metals is continuously bonded. [AR1931] *2.* The low-refractive index material that surrounds the core of a fiber and protects against surface-contaminant scattering.

cladding strippers Chemicals or devices that remove the cladding from an optical fiber to expose the light-carrying core. The term is sometimes misapplied to chemicals or devices that remove the protective coating applied over the cladding to protect the fiber from the environmental stress.

claim token process The process by which network initialization occurs. [AIR4288]

clamp *1.* A normally circular device that adjusts the circumferential length of a band. Used to bind two members together by the exertion of radial pressure. [ARP699D] *2.* A device that, by rigid compression, holds a piece or part in position; or retains units in close proximity or parts in alignment. Compression quality depends on an integral-screw mechanism or screws, bolts, and similar mechanical fasteners. [ARP481A]

clamping area Largest molding area that an injection-molding machine can hold closed under full pressure. [AIR4844]

clamping circuit *1.* Circuit that maintains either extremity of a waveform at a prescribed potential. *2.* Networks for adjusting the absolute voltage level of waveforms.

clamping force Residual load or preload as effected by relaxation. [ARP700]

clamping plate A plate for attaching a mold to a plastics-molding or die-casting machine.

clamping pressure *1.* The pressure that is applied to a mold to keep it closed, in opposition to the fluid pressure of the compressed molding material. *2.* The pressure on two plates

in a bolted joint after the bolts are tightened, each to a specific torque loading . *3.* The pressure applied to a composite or bonded joint or repair during curing—when clamps are used instead of a press or vacuum. [AIR4844]

clamp tonnage Rated clamping capacity of an injection- or transfer-molding machine. [AIR4844]

clamshell chip *See* chip (clamshell type).

class With respect to the certification, ratings, privileges, and limitations of airmen, a classification of aircraft within a category having similar operating characteristics, such as single-engine, land, water, and free-balloon.

Class I location A location in which flammable gases or vapors are or may be present in the air in quantities sufficient to produce ignitable mixtures.

Class II location A location that is hazardous because of the presence of combustible dust.

Class III location A location in which easily ignitable fibers or materials producing combustible flyings are handled, manufactured, or used.

Class A amplifier An amplifier in which the grid bias and alternating grid voltages are such that plate current always flows in a specified tube.

Class AB amplifier An amplifier in which the grid bias and alternating grid voltages are such that plate current in a specified tube flows for considerably more than one-half of the electrical cycle, but less than the entire electrical cycle.

Class A steering system *See* classification of nosewheel steering systems.

Class B amplifier An amplifier in which the grid bias is approximately equal to the cutoff value, therefore making the plate current in a specified tube approximately zero when the grid voltage is zero.

Class B steering system *See* classification of nosewheel steering systems.

Class C amplifier An amplifier in which the grid bias is considerably more negative than the zero plate current value.

Class C steering system *See* classification of nosewheel steering systems.

classification of nosewheel steering systems Classification of nosewheel steering (NWS) systems by their importance and operational usage as follows: Primary (Class A), Secondary (Class B), and Tertiary (Class C). *Primary (Class A) steering system*–one that is essential to safe ground operation of the aircraft. Redundancy (fail operative) is an implied requisite. *Secondary (Class B) steering system*–one that is normally in full-time use during ground operation, but is not essential for safe ground operation of the aircraft. Fail-safety (fail passive) is an implied requisite. *Tertiary (Class C) steering system*–one that is used primarily for taxi-parking -catapult spotting, and is not normally required or used for takeoff or landing operations. A totally passive disengaged mode is an implied requisite. [ARP1595]

class of hydraulic system A pressure standard for military aircraft hydraulic systems based on the nominal pump output pressure or other supply pressure, defined in ISO 6771. [ARP4386]

clay atmometer A simple device for determining evaporation rate to the atmosphere. Consists of a porous porcelain dish connected to a calibrated reservoir filled with distilled water.

clay filter Describes a system for treating fuel to remove surface-active agents. [AIR4783]

clean aircraft *1.* An aircraft in flight configuration versus landing configuration. *2.* An aircraft whose external stores are removed. [ARP4386]

cleanliness Term used to describe type and amount of nonmetallic inclusions in steel. [ARP1917]

cleanliness level The amount of particles by size range and number in a given volume of fluid as determined by a particle count. [ARP4386]

clean room A special room where composite or bonded-metal components are assembled prior to bonding in a clinical atmosphere. [AIR4844]

cleanup *1.* Removing small amounts of stock by an imprecise machining operation, prima-

rily to improve surface smoothness, flatness, or appearance. *2.* The time required for an electronic leak-testing instrument to reduce its output signal to 37% of the initial signal transmitted when tracer gas is first detected. *3.* The gradual disappearance of internal gases during operation of a discharge tube.

clear air Air in which no visible liquid water droplets, snow, ice crystals, etc. are present. [AIR4367]

clear-air turbulence (CAT) Turbulence encountered in air where no clouds are present. *See also* wind shear and jet stream. [ARP4107]

clearance *1.* Authorization by a traffic-control facility for an aircraft to proceed within controlled airspace, taking into account the location of other known aircraft. [ARP4107] *2.* The lineal distance between two adjacent parts that do not touch. *3.* Unobstructed space for insertion of tools or removal of parts during maintenance or repair.

clearance, brake running The clearance provided between the actuators and the pressure plate in a brake assembly in the "full off" condition, i.e., the actuator travel required from "full off" to initial application of force to the friction surfaces. [AIR1489]

clearance fit A type of mechanical fit in which the tolerance envelopes for mating parts always results in clearance when the parts are assembled.

clearance, lateral The clearance provided at the side of a wheel or tire to ensure against interference during rotation and/or variations in a predetermined location, for example, lateral runout or looseness in the system. [AIR1489]

clearance limit The fix, point, or location to which an aircraft is cleared when issued an air traffic clearance. [ARP4107]

clearance, radial The clearance provided radially to ensure against interference during rotation and/or variations in a predetermined location. Tire radial clearance should include allowances for new tire growth due to pressure, service relaxation, and centrifugal growth. [AIR1489]

clear, bright, and dry test The simplest test to determine if fuel is free of visible water or particulates. [AIR4783]

cleared for approach Phrase denoting ATC authorization for an aircraft to execute any standard or special instrument approach procedure for that airport. [ARP4107]

cleared for the option Phrase denoting ATC authorization for an aircraft to make a touch-and-go, low approach, missed approach, stop-and-go, or full-stop landing at the discretion of the pilot. [ARP4107]

cleared to land Phrase denoting ATC authorization for an aircraft to land. [ARP4107]

clear icing/clear ice *See* ice, glaze or clear.

clear status The status word may have the busy and/or service request bit set. All other status code bits in the status word must be zero and the associated message must have the proper word count. [AS15531]

clearway An area beyond the takeoff runway that is under the control of airport authorities and within which terrain or fixed obstacles may not extend above certain limits. [ARP4107]

cleaver A device used to cut or break optical fibers in a precise way so the ends can be connected with low loss.

clevis A U-shaped metal fitting with holes at the open ends of the legs. Pins or bolts are inserted into the holes to make a closed link for attaching or suspending a load.

climb Any gain in height by an aircraft.

climb intercept point The geographical position at which a designated climb altitude is attained. [AIR4102/9]

climbout phase *See* mishap, maneuver.

clinometer A divided-circle instrument for determining the angle between mutually inclined surfaces.

clipping circuit *1.* A circuit that prevents the peak amplitude of a signal from exceeding some specific level. *2.* A circuit that eliminates the tail of a signal pulse after some specific time. *3.* A circuit element in a pulse amplifier that reduces the pulse amplitude at frequencies less than some specific value.

clock frequency The master frequency of periodic pulses, used to schedule the operation of a computer.

clock mode A system circuit that is synchronized with a clock pulse, that changes states only when the pulse occurs, and will change state no more than once for each clock pulse.

clock pulse A synchronization signal provided by a clock.

clock rate The time rate at which pulses are emitted from the clock. In a synchronous computer, the clock rate determines the rate at which logical arithmetic gating is performed.

clock recovery unit Circuitry within a receiver that regenerates the clock-signal component from the modulation waveform. [AIR4288]

clock skew A phase shift between the clock inputs of devices in a single clock system; the result of variations in gate delays and stray capacitance in a circuit.

clockwise Comparable to the direction of rotation of the hands of a clock as viewed from the anti-prop end. [ARP169]

closed bomb A fixed-volume chamber used for testing the pressure-time and chemical reaction characteristics of combustible materials. [AIR913]

closed cell A cell totally enclosed by its walls, and hence noninterconnecting with other cells. [AIR4844]

closed-cell cellular plastic A cellular plastic in which almost all the cells are noninterconnecting. [AIR4844]

closed center Refers to a hydraulic system in which no service is actuated and the system is closed to flow, as distinguished from an open center system. [ARP4386]

closed center valve Valve in which no position allows flow from the pressure port to the return port, and the pressure port is blocked in off position. [ARP4386]

closed circuit Any device or operation in which all or part of the output is returned to the inlet for further processing.

closed die A forming or forging operation in which metal flow takes place only within the die cavity.

closed ecological systems Systems that provide for the maintenance of life in an isolated living chamber through complete reutilization of the material available; in particular, by means of a cycle wherein exhaled carbon dioxide, urine, and other waste matter are converted chemically or by photosynthesis into oxygen, water, and food.

closed end splice A splice, open at one end only, designed to terminate two or more conductors. *See also* splice. [ARP914A]

closed entry A socket contact or insert cavity design feature that prevents the entry of oversize pin contacts, test probes, or other insertable components. [ARP914A]

closed-fireroom system A forced-draft system in which combustion air is supplied by elevating the air pressure in the fireroom.

closed ladder circuit A parallel circuit in which the bus bars are electrically continuous. [ARP485]

closed loop *1.* A combination of control units in which the process variable is measured and compared with the desired value (or set point). If the measured value differs from the desired value, a corrective signal is sent to the final control element to bring the controlled variable to the proper value. *2.* A hydraulic or pneumatic system in which flow is recirculated following the power cycle. The system contains a limited amount of fluid, which is continually reused.

closed loop control A system in which the control action is dependent on the output. The controller acts on a process in such a way to correct an error detected by direct measurement. [ARP89]

closed-loop control system A control system in which the command is compared with a measurement of system output, and the resulting error signal is used to drive the load towards the desired output. [ARP4386]

closed loop failure reporting system A controlled system assuring that all failures and faults are reported and analyzed; that positive corrective actions are identified to prevent recurrence; and that the adequacy of implemented corrective actions is verified by test. [AIR4896]

closed-loop frequency response The frequency response between command input and control system output, with the feedback signal summed with command. Actuation system response for a closed-loop system is usually specified as closed-loop frequency response. [ARP4386]

closed loop numerical control A type of numerical-control system in which position feedback, and often velocity feedback as well, is used to control the dynamic behavior and successive positions of machine slides or equivalent machine members.

closed-loop system *1.* A system in which the output is used to control the input. *See* feedback control loop. [AIR1489] *2.* A control system that includes feedback, a reference mechanism, a capability to detect error, and a means of correcting error so that the output of the system can be modified in progress. For example, the system of a pilot manually flying an aircraft and adjusting its altitude involves the controller's (pilot's) reception of feedback of his/her performance (altimeter, vertical velocity indicator, and visual perception of height), comparison with desired or commanded altitude, detection of any difference (error), and output to the flight-control system to adjust the altitude of the aircraft so that the result will be zero error. In this situation the pilot is said to be "in-the-loop." *See also* in-the-loop. [ARP4107]

closed neutral Refers to the condition in a directional control valve in which a cylinder or load ports are blocked in neutral or off position. [ARP4386]

closed pass A metal rolling arrangement in which a collar or flange on one roll fits into a groove on the opposing roll, thus permitting production of a flash-free shape.

closed position A position that is zero percent closed.

closed runway A runway that is unusable for aircraft operations. [ARP4107]

closed traffic Refers to successive operations involving takeoffs and landings or low approaches in which the aircraft does not depart from the traffic pattern. [ARP4107]

close-grained Consisting of fine, closely spaced particles or crystals.

close-tolerance forging Hot forging in which draft angles, forging tolerances, and cleanup allowances are considerably smaller than those used for commercial-grade forgings.

closing plate A plate used to cover or close openings in non-pressure parts.

closing pressure In a safety relief valve, the static inlet pressure at the point at which the disc has zero lift off the seat.

closing time Time for a valve or system to close from full-rated flow or any intermediate flow rate. [AIR4783]

closure The complete coverage of a mandrel with one layer of fiber. [AIR4844]

closure component, characterized A closure component with a contoured surface, such as the "vee plug," to provide various flow characteristics.

closure component, cylindrical A cylindrical closure component with a flow passage (or a partial cylinder) through it.

closure component, eccentric A closure component face that is not concentric with the shaft centerline and moves into the seat when closing.

closure component, eccentric spherical disk A disk that is a spherical segment, not concentric with the disk shaft.

closure component, linear A closure component that moves in a line perpendicular to the seating plane.

closure component, rotary A closure component that is rotated into or away from a seat to modulate flow.

closure component, tapered A closure component that is tapered and may be lifted from the seating surface before rotating to close or open.

cloud A large agglomeration of liquid droplets or ice crystals suspended in the atmosphere.

cloud base The lower surface of a cloud. [ARP4107]

cloud chamber An enclosure filled with super-saturated vapor; used to indicate the paths of energetic particles when vapor condenses along the trail of ionized molecules created as the particles passes through the enclosure.

cloud deck The upper surface of a cloud. [ARP4107]

clusec A unit of power used to express the pumping power of a vacuum pump. Equals about 1.333×10^{-6} watt, or the power associated with a leak rate of 10 mL/s at a pressure of 1 mtorr.

cluster Two or more discrete discontinuities separated by less than three times the length of the largest adjacent discontinuity. [AMS4991]

cluster analysis Analysis of data with the object of finding natural groupings within the data—either by hand or with the aid of a computer.

clutch A machine element that allows a shaft in an equipment drive to be connected and disconnected from the power train, especially while the shaft is running. *See also* engaging mechanism.

clutch couple turbomechanical actuator A turbomechanical actuator in which a constant speed turbine is coupled to the load through a double-acting servocontrolled clutch. [ARP4386]

clutch coupling drive A motion transmission that allows interruption of the load path for control purposes. [ARP4386]

clutch, face jaw A type of engaging mechanism that makes use of a pair of face teeth to transmit torque from the starter to the engine in one direction. These teeth are generally moved axially into engagement with one other by the starter engaging mechanism during the start cycle. [AS943A]

clutch, overrun A type of clutch mechanism that employs sprags, rollers or pawls, etc., with appropriate driving and driven members. Such a clutch drives in one direction only, overrunning in the other direction. [AS943A]

clutter Atmospheric noise, extraneous signals, etc. that tend to obscure the reception of a desired signal in a radio receiver, radarscope, etc.

CM *See* condition monitoring.

CMA *See* circular mil area.

C-M diagram *See* color-magnitude diagram.

CMOS The combination of a PMOS (p-type channel metal oxide semiconductor) with an NMOS (n-type channel metal oxide semiconductor). *See* complementary metal oxide semi-conductor.

CMRR *See* common-mode rejection ratio.

CMV *See* voltage, common mode.

CN (cyanide) emission Radio waves emitted from incandescent gaseous cyanide (CN) in space under low pressures at wavelengths characteristic of the elements comprising the gas.

cnoidal waves Finite-amplitude progressive waves in shallow water having a wave profile represented by the Jacobian elliptic function 'CN.'

CO *See* carbon monoxide.

CO$_2$ *See* carbon dioxide.

CO$_2$ welding *See* gas metal-arc welding.

coal chemicals A group of chemicals used to make antiseptics, dyes, drugs, and solvents; obtained initially as byproducts of the conversion of coal to metallurgical coke.

coal derived gases The gases that are derived from various coal gasification processes.

coal derived liquids Fluid hydrocarbons derived from the liquefaction of coal.

coalesce To unite into a whole; to grow together.

coalescence A term used to describe the bonding of materials into one continuous body, with or without melting along the bond line, as in welding or diffusion bonding.

coalescer A device used in a water separator to cause the small-diameter water particles entering the separator from the expansion turbine to be combined into particles large enough to be affected by centrifugal forces. [AIR4073]

coalescer element A cylindrical filter cartridge that removes particulate contamination and also causes free water in the fuel to coalesce or gather together into droplets large enough to separate from the continuous phases by gravity. [AIR4783]

coal gas Gas formed by the destructive distillation of coal.

Coanda effect A phenomenon of fluid attachment to one wall in the presence of two walls.

coarse aggregate Crushed stone or gravel used in making concrete which will not pass through a sieve with 1/4-in. (6 mm) holes.

coarse grained *1.* Having a coarse texture. *2.* In metals, having a grain size larger than about ASTM No. 5.

coarse vacuum An absolute pressure between about 1 and 760 torr.

coastal fix The navigation aid or intersection by which an aircraft transitions between the domestic route structure and the oceanic route structure. [ARP4107]

coasting flight The flight of a rocket between burnout or thrust cutoff of one stage and ignition of another; or between burnout and summit altitude or maximum horizontal range.

coating A continuous film of some material on a surface.

coating (fiber optic) A layer of plastic or other material applied over the cladding of an optical fiber to prevent environmental degradation and to simplify handling.

coating (optics) A thin layer or layers applied to the surface of an optical component to enhance or suppress reflection of light, and/or to filter out certain wavelengths.

coating holes (voids) Areas devoid of coating. Coating holes or voids are caused by dust, dirt, lint, or improperly cleaned surfaces beneath the film. [ARP924]

coaxial *1.* A construction of two (usually cylindrical) entities sharing a common axis. [ARP1931] *2.* Having coincident axes. A coaxial cable, for example, is one in which a central insulated conductor is surrounded by one or more metallic sheaths that act as ground leads or secondary conductors.

coaxial cable *1.* A cable in which one conductor completely surrounds the other, the two being coaxial, and separated by a continuous (usually solid) insulating material. [ARP1931] *2.* Waveguide consisting of two concentric conductors insulated from each other. *3.* Cable with a center conductor surrounded by a dielectric sheath and an external conductor. Has controlled impedance characteristics that make it valuable for data transmission.

coaxial nozzles Class of nozzle configurations for reducing noise in jet aircraft.

coaxial propellers (shafts) The concentric arrangement in which each component of the propeller rotates in an opposite direction wherein the speed of each component can vary independently because no fixed gear ratio is incorporated between the shafts. [ARP355]

COBOL *See* Common Business Oriented Language.

cobonding The curing together of two or more elements, of which at least one has already been fully cured and at least one is uncured. [AIR4844]

cobra dane (radar) Radar installation for monitoring Soviet missiles.

cock A valve or other mechanism that starts, stops, or regulates the flow of liquid, especially into or out of a tank or other large-volume container.

cockpit indicator The readout device for any system or component, placed in the cockpit of an aircraft to provide information directly to the flight crew.

cockpit voice recorder (CVR) An approved device for recording electronically detected voice communications within the aircraft cockpit and between the aircraft and others. The recorder must operate continuously from the use of the checklist before the flight to completion of the final check at the end of the flight. The device may have erasure features, but the most recent 30 minutes of recording must be retained. [ARP4107]

co-consolidation A processing step in which two or more thermoplastic, preformed parts are joined by properly locating in a fixture or tool and reheating to melt under pressure. [AIR4844]

co-curing *1.* The act of curing a composite laminate and simultaneously bonding it to some other prepared surface. *2.* The act of curing together an inner and outer tube of similar or dissimilar fiber-resin combination after each has been wound or wrapped separately. *3.* The simultaneous curing together of two or more uncured elements. [AIR4844]

COD *See* crack opening displacement.

COD (carrier onboard delivery) Operation involving aircraft delivery of personnel or material on board an aircraft carrier. [AIR1489]

CODAB *See* configuration data block.

code *1.* A system of symbols for meaningful communication. *See also* instruction. *2.* To translate the program for the solution of a problem on a given computer into a sequence of machine language, assembly language, or pseudo instructions and addresses acceptable to that computer. *See also* encode.

code division multiple access (CDMA) Multiple-access system in which users are segregated by means of pseudorandom signal coding and bandwidth spreading so that the complete time and frequency axes are occupied and only the power is shared.

code division multiplexing The separation of two or more simultaneous radio transmissions over a common path by signal coding and bandwidth spreading.

codes In PCM telemetry, the manner in which ones and zeros in each binary number are denoted.

codes (transponder codes) The number assigned for a particular electronic multiple-pulse reply signal transmitted by the transponder in an aircraft. This code makes it easier for the air traffic controller to identify that particular aircraft on his/her radar display screen. [ARP4107]

CODIL *See* control diagram language.

coding In computer code or pseudo code, the ordered list of the successive computer instructions representing successive computer operations for solving a specific problem.

coding sheet A fill-in form on which computer programming instructions are written.

coefficient of discharge The ratio of actual flow to theoretical flow. Includes the effects of jet contraction and turbulence.

coefficient of elasticity The reciprocal of Young's modulus in a tension test. [AIR4844]

coefficient of expansion The fractional change in dimension of a material with a unit change in temperature. [ARP1931]

coefficient of friction A measure of the resistance to sliding of one surface in contact with another surface. [AIR4844]

coefficient of friction, maximum instantaneous A peak value for coefficient of friction; usually cannot be sustained over a period of time. [AIR1489]

coefficient of performance (COP) For a refrigeration cycle, the ratio of refrigeration produced to work supplied, where refrigeration produced and work supplied are expressed in consistent units. [ARP147C]

coefficient of thermal expansion The change in length or volume per unit length or volume produced by a 1° rise in temperature. [AIR4844]

coefficient of variation (CV) The difference in size (diameter) between the largest and smallest referee particles in a given size sample, expressed as a percent deviation from the nominal size. Can also be expressed as the ratio of the standard deviation to the mean. [ARP1192]

coefficient, side friction For a tire, coefficient of friction in a lateral direction. May be affected by tread pattern. [AIR1489]

coercimeter An instrument for measuring the magnetic intensity of a magnet or electromagnet.

coesite A polymorph of silicon dioxide.

coextrusion A process for bonding two metal or plastic materials by forcing them simultaneously through the same extrusion die.

cofferdam A raised projection surrounding a hatch or trapdoor to keep water out of the opening.

Coffin-Manson law A relationship that is used to estimate fatigue life from cyclic plastic strain range.

cogeneration The generation of electricity or shaft power by an energy-conversion system and the concurrent use of the rejected thermal energy from the conversion system as an auxiliary energy source.

cognitive disorientation A situation in which a person has lost proper perspective within his/her environment, and which results in confusion as to the sequence or priority of tasks to perform. This is referred to colloquially as "getting behind the power curve" or "losing situational awareness." [ARP4107]

cognitive engineering The application of knowledge from cognitive psychology (psychology of information processing) to the engineering design of systems. [ARP4107]

cognitive flexibility An individual's ability to shift from one mental task to another or to effectively time-share between several tasks,

while maintaining situational awareness. [ARP4107]

cognitive psychology The study of the acquisition, storage, and retrieval of information; the processing of that information; and the consequent decisionmaking processes. [ARP4107]

cognitive task saturation *See* attention, anomalies of cognitive task saturation.

cognizant The engineering organization responsible for the design of the parts, its allied quality assurance organization, or a designee of that organization. [AMS2432B]

cogwheel A wheel with radial teeth on its rim.

coherence A property of electromagnetic waves that are all the same wavelength and precisely in phase with one another.

coherence length The distance over which light from a laser retains its coherence after it emerges from the laser.

coherent fiber bundle A bundle of optical fibers with input and output ends in the same spatial relationship to each other, allowing them to transmit an image.

coherent radar A type of radar that employs circuitry that permits comparison of the phase of successive received target signals.

coherent scattering Scattering of electromagnetic or particulate rays in which definite phase relationships exist between the incident and scattered waves; coherent waves scattered from two or more scattering centers are capable of interfering with each other.

cohesion (cohesive strength) The strength of internal forces holding a sealant together. [AS7200/1]

cohesive failure Failure of an adhesive joint occurring primarily in an adhesive layer. [AIR4844]

cohesive strength *See* cohesion (cohesive strength).

coil breaks Creases or ridges in metal sheet or strip that appear as parallel lines across the direction of rolling, generally extending the full width of the material.

coil impedance The complex ratio of coil voltage to coil current. [ARP4386]

coil resistance The DC resistance of each torque motor coil, expressed in ohms. [ARP490]

coil spring A flexible, elastic member in a helical or spiral shape which stores mechanical energy or provides a pulling or restraining force directly related to the amount of elastic deflection.

coincidence circuits Circuits that produce a usable output only when each of two or more input circuits receive pulses simultaneously or within an assignable time interval.

coincidence error In particle counting, a statistical function of particle concentration and the sensing zone volume. A coincidence error of 10% is the maximum that is considered acceptable in using APCs. Maximum particle concentration introduced into the particle counter should be limited to values resulting in coincidence error below 10%. NOTE: Coincidence is defined as the probability of more than one particle being present in the optical sensing zone at any one time. When coincidence occurs, the APC cannot distinguish individual particles and will record one or more particles larger than those present in the sensing zone. [ARP1192]

coining *1.* Squeezing a metal blank between closed dies to form well-defined imprints on both front and back surfaces. *2.* Compressing a sintered powder-metal part to final shape. The process is usually done cold, and involves relatively small amounts of plastic deformation.

coke When heating a carbonaceous material, such as coal, pitch, or petroleum residues, the solid residue remaining after most of the volatile constituents have been driven out. Consists chiefly of coherent, cellular carbon, with some minerals and a small amount of undistilled volatiles.

coke oven gas Gas produced by destructive distillation of bituminous coal in closed chambers. Heating value is 500–550 Btu/ft^3.

cold bend A test used to determine the effect of low temperatures on the insulation system of wire or cable when the wire or cable is flexed. Failure is characterized by the appearance of cracks or other defects in the insulation system. [ARP1931]

cold cathode fluorescent lamps Lamps that are similar in design and construction to hot cath-ode lamps, except that no filaments are provided in the electrode ends of the lamps. The electrode in each end is shaped with a cavity and is coated with a special barium oxide compound to optimize conduction and minimize impedance. To start conduction or "strike the arc" in a cold cathode lamp, a minimum voltage is required that is 2 to 4 times greater than that required for starting a hot cathode lamp. [AIR512B]

cold cathodes A cathode whose operation does not depend on its temperature being above the ambient temperature.

cold cathode tubes Electron tubes containing cold cathodes.

cold drawing Pulling rod, tubing, or wire through one or more dies to reduce its cross section, without applying heat either before or during this process.

cold extrusion Striking a cold metal slug in a punch-and-die operation so that metal is forced back around the die. Also known as cold forging; cold pressing; extrusion pressing; and impact extrusion.

cold-finished Refers to a primary-mill metal product, such as strip, bar, tubing, or wire, whose final shaping operation was performed cold. Cold-finished material has more precise dimensions, and usually higher tensile and yield strength, than a comparable shape whose final shaping operation was performed hot.

cold finishing Process by which final dimension and surface characteristics are produced below the recrystallization temperature. [ARP1917]

cold flow *1.* The distortion that takes place in a material under continuous load, at temperatures within the working range of the material, without a phase or chemical change. [AIR4844] *2.* Permanent deformation of wire insulation due to mechanical forces, without the aid of heat softening. [ARP1931]

cold flow tests Tests of liquid rockets without firing, to check or verify the efficiency of a propulsion subsystem providing for the conditioning and flow of propellants (including tank pressurization, propellant loading, and propellant feeding).

cold forging *See* cold extrusion.

cold forming Shaping sheet metal, rod, or wire by bending, drawing, stretching, or other stamping operations without the application of heat. *See also* cold working.

cold galvanizing Painting a metal with a suspension of zinc particles in a solvent, so that a thin zinc coating remains after the organic solvent evaporates.

cold gas Gas at essentially room temperature, or at a temperature that is generally available from a pressure source without burning, decomposition, or external heating. [ARP4386]

cold heading Cold working a metal by application of axial compressive forces that upset metal and increase the cross sectional area over at least a portion of the length of the starting stock. Also known as upsetting.

cold joint In soldering, making a soldered connection without adequate heating, so that the solder does not flow to fill the spaces, but merely makes a mechanical bond. A cold joint typically exhibits poor to nonexistent electrical conduction across the joint, is not leak-tight, and may break loose under vibration or other mechanical forces.

cold junction *See* reference junction.

cold neutrons Neutrons of less velocity than thermal neutrons; at 152°C their energy is below 0.01 eV.

cold plate A mounting plate for electronic components which has tubing or internal passages through which liquid is circulated to remove heat generated by the electronic components during operation. Also known as liquid-cooled dissipator.

cold pressing *See* cold extrusion.

cold rolling *1*. Forming material below the recrystallization temperature. [AS7481] *2*. Rolling metal at about room temperature. Cold rolling reduces thickness; increases tensile and yield strengths; improves fatigue resistance; and produces a smooth, lustrous, or semi-lustrous finish.

cold-setting adhesive A synthetic resin adhesive capable of hardening at normal room temperature in the presence of a hardener. [AIR4844]

cold shut A portion of a part that is partially separated from the main body of metal by oxide, or by the failure of two streams of metal to unite. [AS3071A]

cold solder joint A solder connection exhibiting poor wetting and a grayish, porous appearance. Can be caused by insufficient heat, inadequate cleaning prior to soldering, or excessive impurities in the solder solution. [ARP1931]

cold start temperature The temperature at which a hydraulic system will start to operate, but need not necessarily meet full performance. [ARP4386]

cold trap A length of tubing between a vacuum system and a diffusion pump or instrument, which is cooled by liquid nitrogen, and used to help remove condensable vapors.

cold treatment Subzero treatment of a metal part—usually at -65°F, -100°F, or liquid-nitrogen temperature—to induce metallurgical changes that either stabilize dimensions, complete a phase transformation, or condition the metal and prepare it for further processing.

cold working Deforming a metal plastically at a temperature lower than its recrystallization temperature. Examples of cold working include: fillet rolling, thread rolling, shot peening, cold upsetting (below the recrystallization temperature of the metal) hot upsetting, hot-cold upsetting, wire drawing, and extruding. [ARP700]

collar A rigid, ring-shaped machine element that is forced onto or clamped around a shaft or similar member to restrict axial motion, provide a locating surface, or cover an opening.

collating sequence In data processing, the order of the ASCII numeric codes for the characters.

collector *1*. Any of a class of instruments for determining electrical potential at a point in the atmosphere, and ultimately the atmospheric electric field. All collectors consist of a device for bringing a conductor rapidly to the potential of the surrounding air, and an electrometer for measuring its potential with respect to the earth. *2*. A device used for removing gas-borne solids from flue gas. *3*. One of the functional regions in a transistor.

collet *1.* A rigid, lateral container for mold-forming material. *2.* The drive wheel that pulls glass fibers from the bushing. *3.* A metal band, ferrule, collar, or flange often used to hold a tool or workpiece. [AIR4844]

collimate *1.* To make parallel. *2.* To project to infinity.

collimated roving Roving made by using a special process that causes the strands to be more parallel than in standard roving. [AIR4844]

collimating optics Optical components of head-up displays used to collimate (project to infinity) the display image. [ARP4102/8]

collimation Producing a beam of light or other electromagnetic radiation whose rays are essentially parallel.

collimator An optical system that focuses a beam of light so that all of the rays form a parallel beam.

collision *1.* A close approach of two or more bodies (including energetic particles) that results in an interchange of energy, momentum, or charge. *See also* elastic collision and inelastic collision. *2.* The network state in which two or more stations transmit concurrently, resulting in their transmissions being superimposed on the network. [AIR4288]

collision parameters *1.* In orbit computation, the distances between centers of attraction of central force fields and the extension of velocity vectors of moving objects at great distances from the centers. *2.* In gas dynamics and atomic physics, any of several parameters, such as cross section, collision rate, mean free path, etc., that provide a measure of the probability of collision.

collision rates Ratios defined by the average number of collisions per second suffered by a molecule or other particle moving through a gas.

colloid *1.* A dispersion of particles of one phase in a second phase, in which the particles are so small that surface phenomena play a dominant role in their chemical behavior. Typical colloids include mists or aerosols (liquid dispersed phase in gaseous dispersion medium); smoke (solid in gas); foam (gas in liquid); emulsions (liquid in liquid); suspensions (solid in liquid); solid foam such as pumice (gas in solid); and solid solution, such as colloidal gold in glass (solid in solid). *2.* A finely divided organic substance that tends to inhibit the formation of dense scale and results in the deposition of sludge, or causes it to remain in suspension, so that it may be blown from the boiler.

colloidal A state of suspension in a liquid medium in which extremely small particles are suspended and dispersed but not dissolved. [AIR4844]

Colmonoy A series of high nickel alloys (manufactured by Wall-Colmonoy Corp.) used for hard facing of surfaces subject to erosion.

colonies Regions within prior beta grains in which alpha platelets have nearly identical orientations. Colonies arise as transformation products during cooling from the beta field at cooling rates slow enough to allow platelet nucleation and growth. In commercially pure titanium, colonies often have serrated boundaries. [AS1814]

color That characteristic of mental percepts variously stimulated by and corresponding to the quantitative aspects of visible light energy as sensed through the shaped response of the eyes as sensors of vision. Color is that aspect of visual perception by which an observer may distinguish differences between two structure-free fields of view of the same size and shape, such as may be caused by differences in the spectral composition of the radiant energy concerned in the observation. Color may be stated in terms of luminance, dominant wavelength, and purity. Neutral color qualities, such as black, white and grey, that possess a zero saturation (or chroma) are called achromatic colors. Colors having a finite saturation or chroma are chromatic or colored. [ARP1782]

color (particle physics) *See* quantum chromodynamics.

colorants Substances that are used to produce the colors of objects. Includes dyes, pigments, inks, paints, and decorative coatings. [ARP798]

color code Any system of colors used to differentiate a specific type or class of objects from other, similar objects; for example, to differentiate steel bars of different grades in a warehouse.

color coding A system for the identification of components, materials, tools, and related devices by means of color.

color-color diagram A two-axis coordinate graph showing the distribution of stars or other objects with reference to difference color indices.

colored federal airways L/MF airways depicted in brown on aeronautical charts, and identified by color name and number (for example, Amber One). Green and red airways are plotted east and west. Amber and blue airways are plotted north and south. NOTE: The term colored airways is no longer used in the US. [ARP4107]

color filter A filter containing a colored dye, which absorbs some of the incident light and transmits the remainder.

colorimetry Any analytical process that uses absorption of selected bands of visible light (or sometimes ultraviolet radiation), to determine a chemical property, such as the end point of a reaction or the concentration of a substance whose color is indicative of product purity or uniformity.

color infrared photography A representation of temperature differences using false colors.

color-magnitude diagram The plot of the absolute or apparent magnitude against the color index for a group of stars. Also known as a C-M diagram.

color of objects The capacity of the object to modify the color of the light incident upon it. [ARP798]

color shift A change in the dominant wavelength or the color purity of light which has been reflected from, or has passed through, the glass. [ARP924]

color stimulus Radiant power of given magnitude and spectral composition entering the eye and producing a sensation of color. [ARP1782]

color temperature The temperature at which a black body must be operated to give a color matching to that of the source in question. [ARP798]

color vision deficiencies A decreased ability, relative to the normal observer, to discriminate and identify certain colors. [ARP4032]

column *1*. A vertical structural member of substantial length designed to bear axial compressive loads. 2. *See* cell.

column loading A factor that takes into consideration the quantity of liquid descending in the column and the quantity of vapor ascending in the column. If either the liquid or the vapor flow rate becomes too high, column flooding will occur.

coma A lens aberration in which light rays from an off-axis source that pass through the center of a lens arrive at the image plane at different distances from the axis than do rays from the same source that pass through the edges of the lens.

combination automatic controller A type of control system arrangement in which more than one closed control loop are coupled through primary feedback or through any of the controller elements.

combination die A forging, forming, or casting die with more than one cavity.

combined cycle power generation Power generation that combines an open-cycle gas turbine and a closed-cycle steam turbine.

combined flight control/utility system A system that supplies a portion of the power required to operate the flight control system and also supplies power to the utility system. Also called a combined system. [ARP578]

combined system *See* combined flight control/utility system.

combiner Component located in the pilot's forward field of view which provides superimposition of the symbology on the external field of view. [ARP4102/8]

combustibility *See* flammability.

combustible The heat-producing constituents of a fuel.

combustible loss The unliberated thermal energy resulting from failure to oxidize completely some of the combustible matter in the fuel.

combustion An exothermic chemical reaction that liberates heat and usually produces high-temperature gases and light. Also known as burning or burning process. [AIR913]

combustion chamber Any chamber or enclosure designed to confine and control the generation of heat and power from burning fuels.

combustion control Control of factors (temperature, preheating, draft, excess or deficient air, etc.) that affect combustion efficiency.

combustion efficiency The percentage ratio of the energy actually released by the combustion process to the energy that would be realized if all the carbon in the fuel were oxidized to CO_2 and the hydrogen to water vapor. [AIR1533]

combustion engine An energy-conversion machine that operates by converting heat from the burning of a fuel to motion.

combustion (flame) safeguard A system for sensing the presence or absence of flame and indicating, alarming, or initiating control action.

combustion pressure Increase in pressure within a fuel tank due to combustion of a flammable mixture. [AIR4170]

combustion rate The quantity of fuel fired per unit of time, expressed as pounds of coal per hour, or cubic feet of gas per minute.

combustion safety control, programming type A combustion-safety control that provides for various operations at definite periods of time in predetermined sequences.

combustion volume Percent of ullage space that is ignited by a projectile or some other ignition source. [AIR4170]

cometary atmospheres The region of the coma of a comet, as well as the gaseous part surrounding the coma, often a hydrogen atmosphere containing particulate matter.

comets Luminous members of the solar system composed of a head, or coma, and often with a spectacular gaseous tail extending a great distance from the head.

COM file A computer file name ending in .COM which most often contains a machine code program. It is short for "command" file.

co-mingled tank farm Fuel tankage system in which various fuel suppliers may deposit a specific type of fuel into a common storage tank to be used by various users. [AIR4783]

command *1.* An input that represents the desired output of the control system. [ARP4386] *2.* A signal that causes a computer to start, stop, or to continue a specific operation.

command bars Needles on the attitude director indicator that assist the pilot in intercepting and maintaining a glide slope and/or course. The needles present "command" information in that the pilot must fly to the bars to bring them to a neutral position (zero displacement). Keeping the bars in the neutral position, the system will cause the pilot to intercept and maintain the course (and the glide slope, if on an ILS approach). Also known as steering bars. [ARP4107]

command/control The orderly distribution of authority and responsibility designed to systematically accomplish a mission; the continuous-feedback loop communications network connecting all levels of command so that decisions can be made, efforts coordinated, and discipline maintained. [ARP4107]

command file *See* COM file.

command guidance The guidance of a spacecraft or rocket by means of electronic signals sent to receiving devices in the vehicle.

command input An input that represents the desired output of the control system. [ARP4386]

command language Vocabulary to interactively execute activities such as computer retrieval or input.

command message segment A command word and associated data words. [AS4113]

command post A place at which the commander of a unit receives orders from his/her superiors and from which command is exercised over a unit. [ARP4107]

command resolution The maximum change in the value of a command signal that can be made without inducing a change in the controlled variable.

command response (data communication systems) A method of media access control in which the remote controllers receive and transmit data only when commanded by the bus controller. [AIR4271]

command systems *See* command guidance.

comment An expression that explains or identifies a particular step in a routine, but which has no effect on the operation of the computer in performing the instructions for the routine.

commercial aircraft Any civilian aircraft being used in the transportation of persons or property for compensation or hire. [ARP4107]

commercial spacecraft Commercial satellites and other spacecraft operated by the private sector.

Commission Internationale de l'Eclairage (CIE) The principal international body in the field of illumination. Purpose is to establish agreeable standards for industry, such as the Standard Colorimetric Observer (or standard observer). The standard observer is the basis for most trichromatic systems and has not changed since 1931 when it was adopted. The standard observer was expanded in 1964 to fields of view larger than 4 degrees. [ARP1782]

common *1.* A reference within a system having the same electrical potential throughout. Usually connected to ground at one point. Often different commons are used throughout a system, such as power common, signal common, etc., depending on the accuracy to which the reference is held. *2. See* power common.

commonality The factors that are common in equipment or systems.

Common Applications Service Elements (CASE) One of the application protocols specified by MAP.

common area A section in memory that is set aside for common use by many separate programs or modules.

Common Business Oriented Language (COBOL) A specific language by which business data-processing procedures may be precisely described in a standard form. The language is intended not only as a means for directly presenting any business program to any suitable computer for which a compiler exists, but also as a means of communicating such procedures among individuals.

common cause analysis Generic term encompassing zonal analysis, particular risk analysis, and common mode analysis. [ARP4754]

common field A field that can be accessed by two or more independent routines.

common junction circuit A circuit in which each temperature-measuring junction is paralleled by means of individual leads to a common point. [ARP485]

common machine language In data processing, coded information that is in a form common to a related group of data-processing machines.

common mode In analog data, an interfering voltage from both sides of a differential input pair (in common) to ground.

common mode failure An event that simultaneously affects a number of elements otherwise considered to be independent. [ARP4754]

common mode interference *See* interference, common mode.

common-mode rejection ratio (CMRR) A measure of the ability of a detector to damp out the effect of a common-mode-generated interference voltage; usually expressed in decibels.

common mode voltage (CMV) *1.* In-phase, equal-amplitude signals that are applied to both inputs of a differential amplifier, usually referred to as a guard shield or chassis ground. *2.* The mean of the voltages measured between each signal conductor and a common 0 V reference point. [AIR4258]

common trip multipole circuit breaker A circuit breaker in which an overload on any pole will cause all poles to open simultaneously. [ARP1199A]

communication Transmission of intelligence between points of origin and reception without alteration of sequence or structure of the information content.

communication functions Those functions that provide the necessary services to the application tasks so that they can use the data communications network. [ARD50008]

communication link The physical means of connecting one location to another for the purpose of transmitting and receiving information.

communication networks Organization of facilities for the rapid reception of, transmission of, and/or relaying of electrical impulses for reproduction as printed messages, pictures, or other data.

communication satellites Satellites designed to reflect or relay electromagnetic signals used for communication.

communications media The communications system that is used by the data link system to effect up- and downlink communications, e.g., VHF, UHF, HF, Mode S, SATCOMM, MLS. [ARP4102/13]

communication transaction The cycle of ground- and air-initiated messages required for the full handshake that constitutes an information exchange between ground and airborne ATS system elements. [ARP4791]

commutation Sequential sampling, on a repetitive timesharing basis, of multiple data sources for transmitting and/or recording, on a single channel.

commutation duty cycle A channel dwell period, expressed as a percentage of a channel interval.

commutation frame period The time required for sequential sampling of all input signals. For a simple multicontact rotary switch, this would correspond to one revolution.

commutation rate The number of commutator inputs sampled per specified time interval.

commutator *1.* Device used to accomplish time division multiplexing by repetitive sequential switching. *2.* A segmented ring, usually constructed of hard-drawn copper segments separated by an insulator such as mica. Used to energize only the correct windings of a d-c generator or motor at any given instant.

compact A powder-metallurgy part made by pressing metal powder, with or without a binder or other additives. Prior to sintering, known as a green compact; after sintering, as a sintered compact or simply a compact.

compaction The application of a temporary vacuum bag and vacuum to remove trapped air and compact a lay-up. Sometimes called de-bulking. [AIR4844]

companding A process in which compression is followed by expansion, as in noise reduction systems.

companion body A nose cone, last stage rocket, or other body that orbits along with an earth satellite. [ARP4386]

companion-trip multiple-pole circuit breaker A circuit breaker in which the unfaulted poles may be maintained closed while a tripping condition persists on one pole or combination of poles; however, the faulted pole(s) is trip-free. [ARP4404]

comparative tracking index (CTI) The numerical value of the maximum voltage in volts at which a material withstands 50 drops without tracking.

comparator *1.* In computer operations, a device or circuit for comparing information from two sources. *2.* A device for inspecting a part to determine any deviation from a specific dimension. The device may operate by electrical, optical, pneumatic, or mechanical means.

comparison monitoring The technique of comparing a set of computed variables with a corresponding set from an independent source. [ARP1834]

compartment A liquid-tight division in a cargo tank. [AIR4783]

compass bearing *See* bearing.

compass locator A low-power, low- or medium-frequency (L/MF) radio beacon installed at the site of the outer or middle marker of an instrument landing system (ILS). The compass locator can be used for navigation at distances of approximately 15 miles or as otherwise authorized in the approach procedure. [ARP4107]

compass locator, middle (LLM) A compass locator installed at the site of the middle marker of an instrument landing system. [ARP4107]

compass locator, outer (LOM) A compass locator installed at the site of the outer marker of an instrument landing system. [ARP4107]

compass rose A circle, graduated in degrees, printed on some charts or marked on the ground at an airport. Used as a reference to either true or magnetic direction. [ARP4107]

compatibility interface A point at which hardware, logic, and signal levels are defined to allow the interconnection of independently designed and manufactured components.

compatible Refers to the ability of different kinds of computers or equipment to use the same programs or data.

compatible casing A casing capable of passing all acceptance criteria of the retreader for the given size, ply rating, and speed rating. [ARP4834]

compensated pendulum A pendulum made of two materials having different coefficients of linear expansion, and so constructed that the distance between the center of oscillation and the point of suspension remains the same over the normal range of ambient temperatures.

compensated reference junction A junction maintained at a known signal output level by circuitry designed to compensate for changes in ambient temperatures. [AIR1900]

compensating accumulator *See* accumulator, compensating.

compensating provision Actions that are available or can be taken by an operator to negate or mitigate the effect of a failure on a system. [AIR4896]

compensation signals In telemetry, a set of reference signals recorded on tape along with the data, and used during playback to automatically compensate for any nonuniformity in tape speed.

compensator A device that compensates, makes up for, or offsets. Device may be of mechanical, electrical, or hydraulic design. For example, a hydraulic pressure compensator (nose wheel steering system, etc.) maintains a positive pressure on the unit in case of loss of system pressure. [AIR1489]

compensator winding An additional winding that provides a feedback voltage to adjust transformation ratio, phase shift, or loading effects. [ARP826]

compensatory leads An arrangement of connecting elements between an instrument and a transducer or other observation device that permits compensation for variations in the properties of any of the connecting elements, such as temperature effects that induce changes in resistance, so that they do not affect instrument accuracy.

compile *1.* A computer function that translates symbolic language into machine language. *2.* To prepare a machine-language program from a computer program written in another programming language by making use of the overall logic and structure of the program and/or generating more than one machine instruction for each symbolic statement, as well as performing the function of an assembler.

compiler *1.* A program that translates a high-level source language (such as FORTRAN IV or BASIC) into a machine language suitable for a particular machine. *2.* A computer program more powerful than an assembler. In addition to its translating function, which is generally the same process as that used in an assembler, a compiler is able to replace certain items of input with a series of instructions, usually called subroutines. Thus, an assembler translates item for item, and produces as output the same number of instructions or constants that were put into it; but a compiler will do more than this. The program that results from compiling is a translated and expanded version of the original. Synonymous with compiling routine. *See also* assembler.

compile time In general, the time during which a source program is translated into an object program.

compiling routine *See* compiler.

complacency *See* attention, anomalies of complacency.

complaint A known or suspected malfunction or defect found by flight crew or maintenance personnel which is documented and requires maintenance action. [AIR4896]

complaints, pilot Suspected or known malfunctions or unsatisfactory conditions that are entered by the flight crew into the aircraft log and that require maintenance action. [AIR4896]

complement *1.* An angle equal to 90° minus a given angle. *2.* A quantity expressed to the base n, which is derived from a given quantity by a particular rule. Frequently used to represent the negative of the given quantity.

Complementary Metal Oxide Semiconductor (CMOS) A type of computer semiconductor memory. The main feature of CMOS memory is its low power consumption.

complementary operator The logic operator that is the NOT of a given logic operator.

complementary wavelength As applied to colorimetry, the monochromatic wavelength of light that matches a standard reference light when combined with the sample color in suitable proportions.

complete combustion The complete oxidation of all the combustible constituents of a fuel.

complete contraction A combination of both end and bottom contractions in a weir.

completed item *See* item, completed.

completion network In a strain gage signal conditioner, the one to three resistors that must be added to make a four-arm bridge (the transducer being the active arm or arms).

complex compounds Chemical compounds in which part of the molecular bonding is of the coordinate type.

complex curvature A surface that curves in more than one direction, such as a saddle or spherical shape. [AIR4844]

complex dielectric constant The vectorial sum of the dielectric constant and the loss factor. [AIR4844]

complex frequency A complex number used to characterize exponential or damped sinusoidal waves in the same way as an ordinary frequency is used to characterize a simple harmonic wave.

complex lens A lens system consisting of more than one optical element.

complex shear modulus The vectorial sum of the shear modulus and the loss modulus. [AIR4844]

complex tone A sound wave produced by combining simple sinusoidal component waves of different frequencies.

complex Young's modulus The vectorial sum of the Young's modulus and the loss modulus. [AIR4844]

compliance Successful performance of all mandatory activities; agreement between the expected or specified result and the actual result. [ARP4754]

component *1.* An element of a higher-level unit or system. [AS8054] *2.* An article that is a self-contained element of a complete operating unit, and performs a function necessary to the operation of that unit.

component improvement program (CIP) cost Total system CIP cost over the life of the aircraft system. [ARP4293]

component void Cutout in the foam which provides clearance between the foam and an internal element of the fuel tank. [AIR4170]

composite A material or structure made up of physically distinct components that are mechanically, adhesively, or metallurgically bonded together; examples include filled plastics, laminates, filament-wound structures, cermets, and adhesive-bonded honeycomb-sandwich structures. *See also* composite material.

composite class A major subdivision of fibrous composite materials, defined by the geometric characteristic of the fiber arrangement. [AIR4844]

composite flight plan A flight plan that specifies VFR operation for one portion of flight and IFR for another portion. [ARP4107]

composite joint A connection between two parts that involves both mechanical joining and welding or brazing, and where both contribute to total joint strength.

composite materials Structural materials of metals, ceramics, or plastics with built-in strengthening agents. The strengthening agents may be in the form of filaments, foils, powders, or flakes of a different compatible material.

composite part An individual part that is either of the following: (a) an inseparable assembly of composite materials cured, consolidated, cobonded, or secondary bonded together, alone or in combination with other composite or noncomposite parts; or (b) an uncured assembly of composite materials that have been stacked up and compacted together, alone or in combination with other composite or noncomposite parts. [AIR4844]

composite propellants Solid rocket propellants consisting of a fuel and an oxidizer, neither of which would burn without the presence of the other.

composite route system An organized oceanic route structure, incorporating reduced lateral spacing between routes, in which composite separation (of aircraft) is authorized. [ARP4107]

composite separation A method of managing route and altitude assignments separating aircraft so that a combination of half the lateral and half the vertical minimums specified for the area concerned is applied. *See also* separation minima. [ARP4107]

composite solid propellants Those propellants in which a granular, inorganic oxidizer is suspended in an organic fuel binder, neither of which would burn without the presence of the other. [ARP4386]

composite subcarrier Two or more subcarriers that are combined in a frequency-division multiplexing (FDM) scheme.

composite wave filter A selective transducer made up of two or more filters—the filters being any combination of high-pass, low-pass, band-pass, or band-elimination types.

composition controller A device incorporated in some MSOGS (molecular sieve oxygen generation systems) to limit the oxygen concentration in the breathing gas supply to an upper bound. The composition controller is in operation at ground level and above. Its purpose is to prevent or minimize the occurrence of acceleration-induced lung collapse (atelectasis) at an aircraft pressure altitude range where high-G maneuvers are possible and routinely performed. [ARP171]

compound *1.* The intimate admixture of a polymer with other ingredients, such as fillers, softeners, plasticizers, reinforcement, catalysts, pigments, or dyes. [AIR4844] *2.* A homogeneous substance, composed of two or more essentially different chemicals that are present in definite proportions, having properties that are different from those of its constituent elements. [ARP1931]

compound angle The surface contour formed by two intersecting mitered angles.

compound die Any die so constructed that it performs more than one operation on a given part with a single stroke of the punch.

compound document A digital representation of a paper document consisting of multiple object types such as text, graphics, and photo images. [AS4159]

compound engine A multicylinder engine in which the working fluid—steam, air, or hot gas—expands successively as it passes from one cylinder to another through the engine.

compound layer Layer on the surface of steel that may be converted to one or more compounds of essentially pure nitrides of metal, primarily iron. [AMS2759/8]

compound lever A device consisting of two or more levers, in which force or motion is transferred from the arm of one lever to the next lever in the train.

compound screw A screw having threads of different pitches or opposite helixes on opposite ends of the shank.

compound semiconductor A semiconductor, such as gallium arsenide, that is made up of two or more materials, in contrast to simple single-element semiconductors, such as silicon and germanium.

compound wound motor A wound-field d-c motor that incorporates both shunt and series windings in the stator. [ARP4386]

compressed gas in solution A nonliquefied gas that is dissolved in a solvent. [ARP4386]

compressible Capable of being compressed. Gas and vapor are compressible fluids.

compressible flow *1.* Flow of a fluid wherein the conditions of pressure, density, and temperature may vary between cross sections of the stream in the direction of flow. [ARP4386] *2.* In aerodynamics, flow at speeds sufficiently high that density changes in the fluid cannot be neglected. *3.* Fluid flow under conditions that cause significant changes in density.

compressibility *1.* The property whereby the density of a substance (e.g., air) increases with increase in pressure. *2.* Volumetric strain per unit change in hydrostatic pressure.

compressibility factor (Z) A factor used to compensate for deviation from the laws of perfect gases. If the gas laws are used to compute the specific weight of a gas, the computed value must be adjusted by the compressibility factor, Z, to obtain the true specific weight.

compression, adiabatic Compression of a gas or mixture of gases without transmission of heat to or from it. [ARP171]

compression after impact strength *1*. The compression strength of a laminate when tested after impact damage has been cause by a controlled impact. *2*. The actual compression strength of a component in service after impact of unknown severity has occurred. [AIR4844]

compressional wave *1*. Waves in which the particle motion or vibration is in the same direction as the propagated wave. Same as longitudinal wave. [ARP5089] *2*. A wave in an elastic medium which causes an element of the medium to undergo changes in volume without rotating.

compression failure Buckling, collapse, or fracture of a structural member that is loaded in compression.

compression, isothermal Compression of a gas or mixture of gases with subtraction of sufficient heat to maintain a constant temperature. [ARP171]

compression machine *See* compressor.

compression member A beam, column, or other structural component that is loaded in such a way as to be under predominantly compressive stress.

compression mold A type of plastics mold that is opened to introduce starting material, closed to shape the part, and reopened to remove the part and restart the cycle. Pressure to shape the part is supplied by closing the mold.

compression molding pressure The unit pressure applied to the molding material in the mold. [AIR4844]

compression ratio *1*. In internal combustion engines, the ratio between the volume displaced by the piston plus the clearance space, to the volume of the clearance space. *2*. In powder metallurgy, the ratio of the volume of loose powder used to make a part to the volume of the pressed compact.

compression ring A separate ring within the backshell assembly that is chamfered to provide an environmental seal by compressing the rear grommet. [ARP914A]

compression set The amount of compression an elastomer retains. Expressed as a percentage of original dimension. [ARP1931]

compression spring An elastic member, usually made by bending metal wire into a helical coil, that resists a force tending to compress it.

compression system A duct system wherein the fluid column loads due to internal pressure are reacted by the support structure. [ARP699D]

compression test A destructive test for determining fracture strength, yield strength, ductility, and elastic modulus by progressively loading a short-column specimen in compression.

compression waves *1*. In acoustics, waves in an elastic medium which cause an element of the medium to change volume without undergoing rotation. *2*. Mathematically, a wave whose velocity wave has zero curl.

compressive modulus Ratio of compressive stress to compressive strain below the proportional limit. [AIR4844]

compressive strength The ability of a material to resist a force that tends to crush or buckle. [AIR4844]

compressive stress The normal stress caused by forces directed toward the plane on which they act. [AIR4844]

compressor *1*. A device in which work is done on a fluid to raise its total pressure and temperature. [ARP147C] *2*. Machine for compressing air or other fluids. *3*. A machine—usually a reciprocating-piston, centrifugal, or axial-flow design—that is used to increase pressure in a gas or vapor. Also known as compression machine. *4*. A hardware or software process for removing redundant or otherwise uninteresting words from a stream, thereby "compressing" the data quantity.

compressor, axial A compressor that inducts and delivers a fluid axially by one or more rotating elements, compressing the fluid. [ARP147C]

compressor blades Blades that are either rotor blades or stator blades in axial-flow compressors; sometimes used restrictively (and ambiguously) for compressor rotor blades.

compressor bleed valve A valve used to control air flow from a driven compressor to provide: (a) compressor stall margin during starting and/or acceleration; and (b) compressed air for starting and/or other purposes. [ARP906A]

compressor, cabin A compressor that compresses and delivers air to a pressurized cabin. [ARP147C]

compressor, centrifugal A compressor that inducts a fluid (liquid or gas) axially, delivers it radially outward relative to the rotating impeller, and compresses the fluid. [ARP171]

compressor, lysholm type A positive-displacement, lobe-type compressor with internal compression. [ARP147C]

compressor, positive displacement A compressor that compresses the fluid by mechanical displacement. [ARP147C]

compressor, reciprocating A positive-displacement, piston-type compressor. [ARP147C]

compressor, roots-type A positive-displacement, lobe-type compressor without internal compression. [ARP147C]

compressor stall A condition in which one or more blades in the turbocompressor cannot maintain the pressure differential across them. [AS5116]

compressor, vane A positive displacement compressor of the vane type. [ARP147C]

Compton effect The decrease in frequency and increase in wavelength of x-rays or gamma-rays when scattered by free electrons.

Compton scattering A form of interaction between x-rays and loosely-bound electrons in which a collision between them results in deflection of the radiation from its previous path, accompanied by random phase shift and slight increase in wavelength.

compulsators Compensated pulsed alternators, i.e., single-phased alternators designed for pulsed-power duty with air-gap armature windings and air-gap compensating windings.

compulsory reporting points Reporting points that must be reported to ATC as each is reached. Designated on aeronautical charts by solid triangles or filed in a flight plan as fixes selected to define direct routes. [ARP4107]

computation The numerical solution of complex equations.

computational chemistry A complementary method for determining properties of gases and solids, and their interactions, from first principle calculations. Computational chemistry extends testing capabilities to realms that are too dangerous or too costly to test experimentally.

computational fluid dynamics The application of large computer systems for the numerical solutions of complex fluid dynamics equations.

computational process An instance of execution of a segment by a processor using a data area.

computational stability The degree to which a computational process remains valid when subjected to effects such as errors, mistakes, or malfunctions.

computer 1. A data processor that can perform substantial computation, including numerous arithmetic or logic operations, without intervention by a human operator during the run. 2. A device capable of solving problems by accepting data, performing described operations on the data, and supplying the results of those operations. Calculators, digital computers, and analog computers are some of the types of computer. *See also* analog computer, digital computer, and hybrid computer.

computer-aided acquisition and logistic support (CALS) The strategy aimed at using digital techniques to integrate technical information flowing from digital systems to facilitate the design, development, manufacturing, and support of aircraft systems. [ARP4293]

computer aided design (CAD) The use of the computer in design work.

computer aided engineering *See* computer aided design.

computer aided manufacturing (CAM) Interactive computing in support of manufacturing.

computer aided mapping Creating databases of topographic and man-made features for the production of traditional maps and digital maps. Resultant digital maps have great flexibility and can be easily updated. The user can select the appropriate scale, view selected features, and view any desired area.

Computer and Automated Systems Association of the Society of Manufacturing Engineers (CASA/SME) A professional engineering association dedicated to the advancement of engineering technology. CASA/SME supports the administrative functions of the MAP/TOP users group.

computer code *See* machine-language code.

computer control Device in which control and/or display actions are generated for use by other system devices. When used with other control devices on the communication link, the computer normally performs or functions in a hierarchical relationship to the other control devices.

computer-dependent language A relative term for a programming language that can be translated only by a specific model (or models) of computer.

computer graphics The representation of a picture in terms of computer graphic primitives, such as lines, markers, polygons, text, and cell arrays. [AS4159]

computer-independent language A language in which computer programs can be created without regard for the actual computers that will be used to process them. *See also* transportability.

computer information security Protective measures to prevent destruction, larceny, and/or unauthorized use of information in computerized files.

computer instruction A machine instruction for a specific computer.

Computer Integrated Manufacturing (CIM) Manufacturing process in which a central computer gathers all types of data, provides information stored in the database for decisions, and controls production input and output.

computer interface The interface device between the host computer and other devices

on the data highway. Converts data from the protocol of the computer to that of the highway, and vice versa.

computer interface device (CID) Hardware that allows a general-purpose computer to share data with the rest of the distributed control system.

computerized simulation Computer-calculated representation of a process, device, or concept in mathematical form. Used for ARIP (impact prediction), automatic rocket impact predictors, computer simulation, and IP (impact prediction).

computer-limited Pertaining to a situation in which the time required for computation exceeds the time available.

computer network The interconnection of two or more computers for the mutual or individual processing of data to and from a multitude of terminals or stations. Accomplished by utilizing appropriate switching techniques, transmission systems, or miniprocessors.

computer networking Interconnection of two or more geographically separated computers so that information can be exchanged between them, usually under the direction of individual, autonomous control programs. *See also* distributed processing.

computer operator A person who performs standard system operations, such as adjusting system operation parameters at the system console, loading a tape transport, placing cards in a card reader, and removing listings from the line printer.

computer part-programming In numerical control, the preparation of a part program to obtain a machine program. The computer and an appropriate processor and post processor are used for this process.

computer program A series of instructions or statements in a form acceptable to a computer prepared in order to achieve a certain result.

computer program integrity The completeness of a program to execute its intended function.

computer security *See* computer information security.

computer simulation A logical-mathematical representation of a simulation concept, system, or operation programmed for solution on an analog or digital computer. *See also* computerized simulation.

computer systems performance The efficiency and reliability that characterize the real operation of the system.

computer systems simulation Forecasting of computer requirements by the use of predictive modeling and estimating computer workloads.

computer vision Capability of computers to analyze and act on visual input.

computer word A sequence of bits or characters treated as a unit and capable of being stored in one computer location. Synonymous with machine word.

COMSEC Abbreviation for communication security.

ComStar satellites Series of domestic Comsat communication satellites.

CON *See* control field.

concatenate To combine several files into one file, or several strings of characters into one string, by appending one file or string after another.

concatenated codes Two or more codes that are encoded and decoded in series.

concave A term describing a surface whose central region is depressed with respect to a flat plane approximately passing through its periphery.

concentrate *1.* To separate metal-bearing minerals from the gangue in an ore. *2.* The enriched product resulting from an ore-separation process. *3.* An enriched substance that must be diluted, usually with water, before it is used.

concentration *1.* The amount of a given constituent present in a unit volume. May be expressed as a ratio, as a percentage, in parts per million, or in milligrams per liter (mg/L). [ARP171] *2.* The volume fraction of the component of interest in a gas mixture—expressed as volume percentage or as parts per million. [ARP1256] *3.* The weight of solids contained in a unit weight of boiler or feed water. *4.* The

number of times that the dissolved solids have increased from the original amount in the feedwater to that in the boiler water due to evaporation in generating steam.

concentrator A network component that multiplexes a number of nodes to/from a trunk bus. [AIR4288]

concentricity *1.* Measurement expressing the location of the center of the conductor with respect to the geometric center of the circular insulation. [ARP1931] *2.* The quality of two or more geometric shapes having the same center; usually applied to plane shapes or cross sections of solid shapes that are approximately circular.

concentric lay Describes a concentric lay conductor or cable that is composed of a central core surrounded by one or more layers of helically wound strands or wires. Unless otherwise specified, it is optional for the successive layers to be alternately reversed in direction of lay (true concentric lay) or to be in the same direction (unidirectional lay). *See* unidirectional lay. [AS1198]

concentric lay conductor A conductor with one or more layers of helically wound strands in a fixed, round geometric arrangement. In a concentric lay conductor, it is optional for the direction of lay for successive layers to be alternately reversed or in the same direction. If the direction of the lay for successive layers is the same, the lay length should increase with each successive layer. The standard direction of the lay of the outer layer is left hand. [ARP1931]

concentric orifice plate A fluid-meter orifice plate having a circular opening whose center coincides with the axis of the center of the pipe in which it is installed.

concentric spheres Structures in which the space between the spheres is utilized for experiments involving fluid flow, etc.

concept development Development of a system concept from a system detailed design. [ARP4033]

concept, fleet leader Refers to the performance of inspections on specific aircraft selected from those that have been the highest operating age/

usage in order to identify the first evidence of deterioration in their condition caused by fatigue damage. [AIR4896]

conceptual phase The identification and exploration of alternative solutions or solution concepts to satisfy a validated need. [ARD50013]

conceptual studies cost The sum of all contractor- and government-funded costs required to complete the feasibility studies and subsequent definition of any system or equipment conceived in response to meet a mission requirement. [ARP4293]

concurrent fault simulation Fault-list-type simulations during which fault lists, associated with each primitive block, are propagated by using the same elemental evaluator routines used by the failure-free simulation. [AIR4896]

concurrent processing Two (or more) computer operations that appear to be processed simultaneously when, in fact, the CPU is rapidly switching between them.

cond *See* conductivity.

condensate *1.* The liquid product of a condensing cycle. Also known as condensate liquid. *2.* A light hydrocarbon mixture formed by expanding and cooling gas in a gas-recycling plant to produce a liquid output.

condensate liquid *See* condensate.

condensate pot A section of pipe (4 in. diameter) installed horizontally at the orifice flange union to provide a large-area surge surface for movement of the impulse line fluid with instrument element position change. Used to reduce measurement error from hydrostatic head difference in the impulse lines.

condensate trap A device used to trap and retain condensate in a measurement impulse line to prevent hot vapors from reaching the instrument.

condensation *1.* The physical process by which a vapor becomes a liquid or solid; the opposite of evaporation. *2.* Specifically, in meteorology, the transformation from vapor to liquid.

condensation nuclei Liquid or solid particles upon which condensation of water begins in the atmosphere.

condensation polymerization A chemical reaction in which two or more molecules combine, with the separation of water or some other simple substance. [AIR4844]

condensation reaction A chemical reaction in which two different molecules react to form a new compound of greater complexity, with the formation of water, alcohol, ammonia etc., as a byproduct. [AIR4844]

condensation test Test used to evaluate the rust-preventive properties of greases covering steel objects under alternate low-temperature and moderate-temperature, high-humidity conditions.

condensation-type hygrometer Any of several designs of dew-point instruments that operate by detecting the equilibrium temperature at which dew or frost forms on a thermoelectrically, mechanically, or chemically cooled surface. In a condensation-type hygrometer, the surface condensation may be detected by optical, electrical, or nuclear techniques.

condenser *1.* A heat exchanger in which the state of a fluid is changed from a gas or vapor to a liquid. [ARP147C] *2.* The heat exchanger, located at the top of the column, that condenses overhead vapors. For distillation, the common condenser cooling media are water, air, and refrigerants such as propane. The condenser may be partial or total. In a partial condenser only part of the vapors are condensed, with the remainder usually withdrawn as a vapor product.

condenser boiler A boiler in which steam is generated by the condensation of a vapor.

conditional branch *See* conditional transfer.

conditional jump *See* conditional transfer.

conditional stability *1.* Refers to a linear system that is stable for a certain interval of values of the open-loop gain, and unstable for certain lower and higher values. *2.* The property of a controlled process whereby it can function in either a stable or unstable mode, depending on conditions imposed.

conditional transfer An instruction that, if a specified condition or set of conditions is satisfied, is interpreted as an unconditional transfer. If the conditions are not satisfied, the instruction causes the computer to proceed in its normal sequence of control. A conditional transfer also includes the testing of the condition. Synonymous with conditional jump and conditional branch. *See also* branch.

conditioner A device that warms, purifies, humidifies, or medicates inspired air. [ARP171]

conditioning *1.* Subjecting a material to a prescribed environmental and/or stress history before testing. [AIR4844] *2.* A maintenance procedure consisting of deep discharge, short, and constant-current charge; used to correct cell imbalance that may have been acquired during battery use. [ARP4386]

conditioning, air The simultaneous control of all, or at least the first three, of the following factors affecting both the physical and chemical conditions of the atmosphere within a structure: temperature, humidity, motion, distribution, pressure, dust, and bacteria. [ARP171]

conditioning, preflight air The process of air conditioning aircraft compartment(s) while the aircraft is on the ground. [ARP147C]

condition monitoring (CM) *1.* A data gathering process (not a preventative maintenance process). Condition monitoring allows failures to occur, and relies on analysis of operating experience information to indicate the need for appropriate action. If the removal rate is deemed to be excessive, the component may temporarily be assigned a hard time removal limit until corrective action can restore its removal performance to an acceptable level. *2.* Refers to an item that has neither hard-time removal nor scheduled test to determine its continued serviceability. All removals are for unscheduled cause. All units removed must, upon arrival in the shop, be given a check in accordance with the procedures in the appropriate shop manual. If the unit "passes," it is returned to service. Failure to pass the check requires the unit to be repaired to a state that it will pass and be serviceable. [ARP4386]

condition monitoring system A system designed to monitor the condition of a machine or process.

conductance A measure of the ability of any material to conduct an electric charge. Conductance is a ratio of the current flow to the potential difference causing the current flow. [ARP1931] *See also* resistance.

conductance, air space The heat transfer coefficient of an air space. Includes the combined influence of conduction, convection, and radiation for a specified air space width. [ARP147C]

conducted susceptibility The tendency for the performance of a piece of equipment to degrade in response to interference on its connecting wires. [ARP1931]

conducting *See* conduction.

conducting media *See* conductors.

conducting polymer A plastics material having electrical conductivity approaching that of metals.

conduction *1.* The transfer of energy within and through a conductor by means of internal particle or molecular activity, and without any net external motion. *2.* Flow of heat through or across a conductor.

conduction band(s) *1.* A range of states in the energy spectrum of a solid in which electrons can move freely. *2.* A partially filled or empty energy band in which electrons are free to move easily.

conduction error The difference between the gas temperature and the sensed temperature, caused by conduction of heat along the thermocouple assembly support tubes and leads. [AIR1900]

conduction pump A device for pumping a conductive liquid, such as a liquid metal, by passing an electric current across the stream of liquid and applying a magnetic field at right angles to the electrical current.

conductive Capable of conducting electricity.

conductive elastomer An elastomeric material that conducts electricity; usually made by mixing powdered metal into a silicone before it is cured.

conductivity *1.* The capability of a material to carry an electric charge. The conductivity of a metal is usually expressed as a percentage of copper conductivity—copper being one hundred percent (100%). [ARP1931] *2.* The amount of heat (Btu) transmitted in one hour through one square foot of a homogeneous material, 1 in. thick, for a difference in temperature of 1°F between the two surfaces of the material.

conductivity bridge A simple, four-arm a-c bridge circuit in which a conductivity cell is the unknown circuit element. Electrically, the cell is equivalent to a resistance and a capacitance in series. In such a device, higher a-c frequencies lead to lower cell-polarization errors, but introduce greater errors due to capacitance impedance. The latter can be reduced by using a phase-sensitive detector.

conductivity, thermal *See* thermal conductivity.

conductivity-type moisture sensor An instrument for measuring moisture content of fibrous organic materials, such as wood, paper, textiles, and grain, at moisture contents up to saturation.

conductometer An instrument that measures thermal conductivity, especially one that does so by comparing the rates at which different rods conduct heat.

conductor Any material through which electrical current can flow.

conductor, card edge A rectangular connector into which the edge of a circuit board is inserted so as to make electrical contact with conductive traces located on the circuit board. [ARP914A]

conductor stop A device or design feature on a terminal, splice, contact, or tool which correctly positions the conductor in the conductor barrel. [ARP914A]

conduit *1.* A rigid or semi-rigid tube within which the signal-carrying medium may be placed. Also serves the functions of strength member and scuff jacket. [ARD50020] *2.* Any channel, duct, pipe, or tube for transmitting fluid along a defined flow path. *3.* A thin-walled pipe used to enclose wiring.

cone Stabilizing and drag member incorporating pressure and static heads beneath aircraft under test in undisturbed air conditions.

cone bearing A tapered sleeve bearing in the shape of a truncated cone; runs in a correspondingly tapered bearing block.

cone index An index of soil-shearing resistance obtained using a trafficability cone penetrometer. [AIR1780]

cone of ambiguity An inverted cone extending upward from the site of a VOR/TACAN facility in which navigational signals tend to be unreliable. [ARP4107]

cone penetrometer A probe-type instrument designed to measure soil strength. Consists of a shaft attached to the base of a right circular cone. [AIR1780]

cone-plate viscometer An instrument for routinely determining the absolute viscosity of fluids in small sample volumes by sensing the resistance to rotation of a moving cone. This resistance is caused by the presence of the test fluid in a space between the cone and a stationary flat plate.

cones Geometric configurations having a circular bottom and sides, tapering off to an apex (as in nose cone).

confidence An attitudinal set in which a person is predisposed to think that he/she can perform a task.

confidence factor The percentage figure that expresses confidence level. [ARP4386]

confidence level *1.* The probability that a given statement is correct; usually associated with statistical predictions. [ARP4386] *2.* In acceptance sampling, the probability that accepted lots will be better than a specific value known as the rejectable quality level (RQL). A confidence level of 90% indicates that 90 out of every 100 lots accepted will have a quality better than the RQL. *3.* In statistical work, the degree of assurance that a particular probability applies to a specific circumstance.

confidence limits *See* limits, confidence.

confidence test *1.* A brief ATE self-test performed during ATE initialization. [ARP4904] *2.* A test primarily performed to increase the confidence that the unit under test is operating acceptably. [ARP4386]

CONFIG.SYS A basic computer file that outlines how that particular device is designed to operate.

configuration *1.* The functional and/or physical characteristics of hardware/software as achieved and set forth in technical documentation. [ARD50013] *2.* The arrangement of the parts or elements of something. *3.* A low-level, fill-in-the-blank form of programming a process control device.

configuration control The systematic evaluation, coordination, approval or disapproval, and implementation of all approved changes in the configuration of a configuration item—after formal establishment of its configuration identification. [AIR4896]

configuration data block (CODAB) In TELEVENT, the data section that identifies the "personality" of a hardware-software combination.

configuration deviation list *See* list, configuration deviation.

configuration interaction In physical chemistry, the interaction between two different possible arrangements of the electrons in an atom or molecule.

configuration management The process that identifies functional and physical characteristics of an item during its life cycle, controls changes to those characteristics, provides information on status of change action, and audits the conformance of configuration items to approved configurations. [ARP4293]

configure The installation procedure that sets up software to operate on a particular computer and printer.

confined detonating fuse (CDF) *1.* A mild detonating fuse, completely contained within a shock-absorbing sheath to prevent damage to the surroundings when the fuse is detonated. [ARP4386] *2.* A detonating cord with a flexible outer sheath that retains the products of detonation. [AIR913]

confined flow Flow of a continuous stream of fluid within a process vessel or conduit.

conflict alert A feature of certain automated air traffic control systems designed to alert radar controllers to existing or pending conflict situations recognized by the computer program parameters. [ARP4107]

conflict situation A state of affairs that exists when aircraft come within certain vertical, lateral, or longitudinal distances from one another.

These distances may vary depending on the density of the airspace and the type of aircraft. [ARP4107]

conformance *1.* Established as correct with reference to a standard, specification, or drawing. [ARP4754] *2.* Conformance of a device to the manufacturer's specifications. *See also* accuracy.

confusion Loss of situational awareness that is recognized by the individual concerned. A state characterized by bewilderment, emotional disturbance, lack of clear thinking, and (sometimes) perceptual disorientation. [ARP4107]

conical orifice An orifice having a 45° bevel on the inlet edge to yield more constant and predictable discharge coefficient at low flow velocity (Reynolds number less than 10,000).

conical scan antenna An automatic-tracking antenna system in which the beam is driven in a circular path such that it forms a cone. The antenna is steered automatically so that the telemetry source is kept at the center of the cone.

conical scanning Scanning in which the direction of maximum radiation generates a cone whose vertex angle is on the order of the beam width. Such scanning may be either rotating or nutating, depending on whether the direction of polarization rotates or remains unchanged.

coniscope *See* koniscope.

conjugate bridge An arrangement of electrical or electronic components in which the supply circuit and detector circuit are interchanged as compared with the normal arrangement for that type of bridge.

conjugated circuits Branches of an electrical network configured so that a change in the electromotive force in either branch does not result in a current change in the other.

conjugate gradient method An interactive method for solving a system of linear equations of dimension N, which terminates in, at most, N steps if no rounding errors are encountered. Each iteration will bring one closer to the solution.

conjugate impedances An impedance pair in which the magnitudes of resistance and reactive components of one are equal to the corresponding values of the other; but whose reactive components are of opposite signs.

conjugate profiles Refers to the situation in which the tooth profiles of an involute gear pair determine each other mutually; that is, the profile of one gear determines the profile of the other gear. [AS1560]

connect To establish linkage between an interrupt and a designated interrupt servicing program. *See also* disconnect.

connecting rod Any straight link that transmits power or motion from one part of a mechanism to another, especially one that links a rotating member to a reciprocating member; for example, the link that attaches a piston to the crankshaft in a reciprocating internal-combustion engine.

connector(s) *1.* Used to describe provision for an electrical connection; not recommended for any other uses. *See also* fitting. [ARP4386] *2.* Any detachable device for providing electrical continuity between two conductors. *3.* In fiber optics, a device that joins the ends of two optical fibers together temporarily. *4.* Where multiple cells are used, molded strips of lead that connect the posts over the top of the cover. [AIR1898]

connector cutout *See* cutout, connector.

connector, electrical A conductor-terminating device that allows for the separation of one or more electrical circuits. Used to facilitate servicing. [ARP914A]

connector, hermaphroditic A connector that has features enabling it to be mated with an identical connector. [ARP914A]

connector, plug An electrical connector, intended to be attached to the free end of a conductor, wire, cable, or bundle, which couples or mates to a receptacle connector. [ARP914A]

connector, receptacle An electrical connector, intended to be mounted or installed into a fixed structure, such as a panel, electrical case, or chassis, which couples or mates to a plug connector. [ARP914A]

connector, right angle A rectangular connector that is generally mounted onto a printed circuit board, whose contacts are inserted into a matching pattern or plated through holes in the circuit board and soldered in place. [ARP914A]

connector set, electrical Two or more separate plug and receptacle connectors designed to be mated together. The set may include mixed connectors mated together, such as one plug connector and one dummy receptacle connector, or one receptacle connector and one dummy plug connector. [ARP914A]

connector, umbilical An electrical connector used to connect a cable to a vehicle (e.g., an aircraft or rocket). This connector is unmated prior to or during initial movement or launching of the vehicle. [ARP914A]

conscious level *See* awareness, level of conscious.

consecutive access A method of data access that is characterized by the sequential nature of the I/O device involved, for example, a card reader. In a card reader, each card must be read one after another, and no distinction is made between logical sets of data in or among the cards in the input hopper.

consecutive starts Start cycles that follow each other at close intervals. The engine may or may not have come to rest before the second or successive starts. [ARP906A]

consistency A qualitative means of classifying substances, especially semisolids, according to their resistance to dynamic changes in shape.

console An array of controls and indicators for the monitoring and control for a particular sequence of actions, such as the checkout of a rocket, a countdown action, or a launch procedure.

consolidation *1.* In metal matrix or thermoplastic composites, a processing step in which fiber and matrix are compressed by one of several methods to reduce voids and achieve the desired density. *2.* The joining together under heat and pressure of multiple plies of thermoplastic composite materials, which may be resoftened without undergoing a chemical change when heated. *3.* A process that fuses each ply together by tacking and flowing the matrix between the plies. [AIR4844]

constant *1.* A value that remains the same throughout the distinct operation; opposite of variable. *2.* A data item that takes as its value its name (hence, its value is fixed during program execution).

constant amplitude Amplitude that is invariant in specified circumstances

constant-amplitude (CA) (filter) A reference to the characteristic of a Butterworth filter. *See* Butterworth.

constantan An alloy of 55% copper and 45% nickel; used with copper in thermocouples in the temperature range 169°C to 386°C. Normally, the copper is the positive wire, and the constantan is the negative wire. The temperature coefficient of resistivity for constantan is 0.0002/°C. [ARP1931]

constant-bandwidth (CBW) The spacing of FM subcarriers equally with relation to one another. *See also* proportional bandwidth.

constant-current potentiometer A type of null-balance instrument for determining an unknown d-c voltage, usually less than 10 V, under conditions that maintain constant current in the detector circuit. Resolution up to one part in 10^3 can be achieved with a single potentiometer slidewire; up to one part in 10^7 with a multidecade device.

constant-current transformer A type of transformer that automatically adjusts the output of its secondary circuit to maintain a constant current under varying load impedances when its primary windings are connected to a constant-voltage power supply.

constant-delay (CD) (filter) *See* Bessel.

constant failure period The period during which failures of some items occur at an approximately uniform rate. [AIR4896]

constant-head meter A flow-measurement device that maintains a constant pressure differential by varying the cross section of a flowpath through the meter, such as in a piston meter or rotameter.

constant horsepower line The locus of points on a curve at which the product of torque and speed is constant. [ARP906A]

constant-load balance A single-pan weighing device having a constant load. When using a constant-load balance, the sample weight is determined by hanging precision weights from a counterpoised beam.

constant-resistance potentiometer A type of null-balance instrument for determining an unknown d-c voltage, usually less than 10 V. Employs a constant-scaling resistor in parallel with the potentiometer circuit.

constant-speed propeller A propeller designed to maintain engine speed at a constant RPM, automatically increasing or decreasing pitch as engine speed tends to increase or decrease. [ARP4107]

constant-volume gas thermometer A device for detecting and indicating temperature based on Charles' Law, which states that the pressure of a confined gas varies directly with absolute temperature. In practical instruments, a bulb immersed in the thermal medium is connected to a Bourdon tube by means of a capillary. Changes in temperature are indicated directly by movement of the Bourdon tube, due to changes in bulb pressure.

constellation *1.* Conspicuous configuration of stars (original definition). *2.* Regions of the celestial sphere marked by arbitrary boundary lines (modern definition).

constituent An element of a larger grouping. In advanced composites, the principal constituents are the fibers and the matrix. [AIR4844]

constrained mechanism A mechanical device in which all members move only along predetermined paths.

constraint *1.* A speed or altitude or time restriction usually related to some downstream waypoint. [ARP1570] *2.* The limit of normal operating range. *3.* Anything that keeps a member under longitudinal tension from contracting laterally; this sets up a condition of biaxial tension in the member. The term is used most often in connection with welded joints that cannot shrink laterally as the weld solidifies and cools.

consumable electrode An arc-welding electrode that melts during welding to provide the filler metal.

consumable insert A piece of metal that is placed in the root of a weld prior to welding, and which melts during welding to supply part of the filler metal.

consumable item *See* item, consumable.

consumables Those materials, such as grease, safety wire, packing, etc., that are not covered by a part number, but are required in the assembly of an engine. [ARP4294]

contact *1*. To establish communications with a facility; and, if appropriate, to specify the frequency that should be used. Some examples of the use of this term are: "Contact departure control." or "Contact tower on one nineteen point niner." *2*. A flight condition wherein the pilot ascertains the attitudes of his/her aircraft and navigates by maintaining visual reference (contact) with the surface. [ARP4107] *3*. In hardware, a set of conductors that can be brought into contact by electromechanical action and thereby produce switching. *4*. In software, a symbolic set of points whose open or closed condition depends on the logic status assigned to them by internal or external conditions.

contact adhesive An adhesive that is apparently dry to the touch and that will adhere to itself simultaneously upon contact. [AIR4844]

contact analog A flight display that takes the essential visual cues from the contact view that the pilot uses in landing an aircraft and in other ground-referenced maneuvers, and incorporates them into a vertical situational display. This vertical situational display presents a (sometimes highly stylized) contact view in which the dynamic responses of the pictured elements are analogous to those of their visual-world counterparts in contact flight. In a true contact analog display, all elements obey the same laws of motion perspective as their visual-world counterparts. [ARP4107]

contact angle Refers to the angle formed with the surface when a drop of liquid is placed on a surface and either retracts into a ball, like a drop of mercury, or spread out like water on a clean, high-energy surface. [AIR4844]

contact approach An approach in which an aircraft on an IFR flight plan, operating clear of clouds, with at least 1 mile flight visibility, having received an air traffic control authorization, may deviate from the prescribed instrument approach procedure and proceed to the airport of destination by visual reference to the surface. [ARP4107]

contact area *1*. The area of a loaded tire that is actually in contact with the supporting surface; generally elliptical in shape. *See also* tire footprint, gross. [AIR1489] *2*. The designated surface required to contact a mating surface within the limits specified. The contact area should be distributed uniformly over the surface. [AS291D] *3*. The area that is in contact between two conductive elements, through which current flow can take place. [ARP914A]

contact arrangement The position and layout pattern of contacts in a connector. [ARP914A]

contact back wipe On an actuated contact surface, refers to the travel of a contact on the surface of its mating contact during the actuation cycle, and its movement back to a clean wiped surface at the completion of the actuation or engagement cycle. [ARP914A]

contact corrosion Corrosion of a metal at an area in which contact is made with another material, usually nonmetallic. [AIR4844]

contact, electrical The conductive element in a connector or other device which mates with a corresponding element to provide an electrical path or circuit. [ARP914A]

contact element A temperature-measuring element that is in intimate contact with the solid state body to be measured. [ARP485]

contact engaging force *See* force, contact engaging.

contact float The allowable, free axial, lateral or angular movement of a contact in a connector. [ARP914A]

contact, hermaphroditic An electrical contact with features that enable it to be mated with an identical contact. [ARP914A]

contact input *See* input, contact.

contact inspection In ultrasonic testing, a method of scanning a test piece. Involves placing a search unit directly on the test piece surface, which is covered with a thin film of couplant.

contact loads Dynamic loading by contact between two bodies.

contact method An inspection method in which the search unit face makes direct contact with the test part and ultrasonic energy is transmitted through a thin film of couplant. [ARP5089]

contact molding A process for molding reinforced plastics in which reinforcement and resin are placed on a mold. [AIR4844]

contactor A mechanical or electromechanical device for repeatedly making and breaking electrical continuity between two branches of a power circuit, thereby establishing or interrupting current flow.

contact, pin A contact intended to make electrical contact on its outer surface upon engagement with a socket contact. [ARP914A]

contact pressure resins Liquid resins that thicken or polymerize upon heating. When used for bonding laminates, contact pressure resins require little or no pressure. [AIR4844]

contact rectifier A device for converting a-c electrical power to d-c power; constructed of two different solids in contact with one another. Contact rectifiers operate on the principle that the selected combination of solids yields greater electrical conductivity in one direction across the interface between them than in the other direction.

contact resistance The electrical resistance through a pair of engaged contacts or terminals. Resistance may be measured in ohms or millivolt drop at a specified test current. A contact resistance measurement does not include the resistance of the terminating conductor joint or the conductor. [ARP914A]

contact retainer A device either on the contact or in the connector insert whose purpose is to retain the contact in the insert. [ARP914A]

contact retention *1.* In an electrical connector, the provision or means by which the contacts are retained. *2.* The ability of a connector to retain contacts. *See also* force, contact retention. [ARP914A]

contacts The electrically conducting parts in a contactor that repeatedly come in contact, or separate, to make or break electrical continuity.

contact sense module A device that monitors and converts program-specified groups of field-switch contacts into digital codes for input to the computer.

contact separation force *See* force, contact separation.

contact size Either a single number designator based on the AWG size number most closely corresponding in circular mil area (CMA) to the CMA of the pin contact of a given contact set; or a double number designator, similarly based, in which the first number corresponds to the CMA of the pin contact, and the second number corresponds to the maximum wire size accommodated by the conductor barrel of the contact. [ARP914A]

contact, socket A contact having an engagement end that will accept entry of a pin contact; has a point of electrical contact on the inside diameter of the contacting surface. [ARP914A]

contact symbology Representation of logic schemes in contact or ladder diagram form.

contact testing Testing in which the transducer assembly, through a thin layer of couplant, is in direct contact with the test material. [ARP5089]

contact thermography A method of measuring surface temperature in which the surface of an object is covered with a thin layer of luminescent material and then viewed under ultraviolet light in a darkened room. The brightness viewed indicates surface temperature.

contact time The total time between application and removal of the penetrant or emulsifier and/or remover. [AMS2647A]

contact transducer A transducer that is coupled to a test surface either directly or through a thin film of couplant. [ARP5089]

contact tube In gas metal arc welding and flux cored arc welding, a metal part with a hole in it. The wire electrode is fed continuously through the hole, thus providing electrical contact with the welding machine.

contact-type membrane switch A disk-shaped momentary-contact switch of multilayer construction, the active element of which consists of two conductive buttons separated by an insulating washer. Finger pressure on one face of the disk brings the buttons into contact, completing the electrical circuit. When the pressure is released, the contacts separate, breaking the electrical circuit.

contact-wear allowance The thickness that may be lost due to wear from either of a pair of mating electrical contacts before they cease to adequately perform their intended function.

contact wipe The distance a contact travels on the surface of its mating contact during engagement or separation. *See also* wiping action. [ARP914A]

containment *1.* The ability of the starter of APU to retain within its envelope all energy-laden fragments of the turbine(s) or other rotating components if these parts are caused to fail. [ARP906A] *2.* The restriction of damage within a specified physical envelope following failure of an item. [AIR4896]

contaminant Any material, usually a solid, that is suspended or dissolved in a fluid system and is not there by design. [ARP4386]

contamination *1.* Foreign matter of solid, gelatinous, liquid, or gaseous form that is contained in, or potentially capable of being contained in, a system fluid (liquid or gas); capable of moving with the system fluid and of being deposited upon or forming upon system filters. [AIR787] *2.* Presence of an unwanted substance, usually a substance that causes an undesired effect or interferes with a desired effect.

contender A potential bus master module that is actively vying for bus mastership. [AS4710]

content-addressed storage *See* associative storage.

contention *1.* A condition on a multidrop communication channel in which two or more locations try to transmit at the same time. *2.* Unregulated bidding for a line or other device by multiple users.

context *1.* The composition, structure, or manner in which something is put together. *2.* The situation or environment of an event.

contiguous file On a mass-storage device, a file consisting of physically adjacent blocks.

continental control area *See* controlled airspace, continental control area.

contingency One or more unplanned occurrences that make the performance of a task more difficult, but are not inherently hazardous

(for example, unforecast bad weather or ATC delays). [ARP4107]

continuity tester *1.* A device for testing fiber optic communication systems; used to determine if there is a continuous optical path between two points. *2.* A device used to check the electrical continuity of a part or set of parts. [AIR4783]

continuous blowdown The uninterrupted removal of concentrated boiler water from a boiler for the purpose of controlling total solids concentration in the remaining water.

continuous dilution A technique for continuously supplying a protective gas flow to an enclosure housing electrical circuitry and containing an internal potential source of flammable gas or vapor. Purpose is to dilute any flammable gas or vapor that is present to a level well below the lower-explosion-limit.

continuous duty Describes a service that demands operation at a substantially constant load current for an indefinitely long time. [ARP1199A]

continuous-duty rating The maximum power or other operating characteristic that a specific device can sustain indefinitely without significant degradation of its functions.

continuous filament *1.* An individual flexible rod of small diameter, with great or indefinite length. *2.* A yarn or strand in which the individual filaments are substantially the same length as the strand. [AIR4844]

continuous-filament yarn Yarn formed by twisting two or more continuous filaments into a single, continuous strand. [AIR4844]

continuous furnace A type of reheating furnace in which the charge is loaded at one end, moves through the furnace to accomplish the intended treatment, and is discharged at the other end.

continuous maintenance program *See* program, continuous maintenance.

continuous mixer A type of mixer in which starting ingredients are fed continuously and the final mixture is withdrawn continuously, without stopping or interrupting the mixing process. Generally, unmixed ingredients are fed at one end of the machine and blended progressively as they move towards the other end, where the mixture is discharged. Contrast with batch mixer.

continuous operation A process that operates on the basis of continuous flow, as opposed to batch, intermittent, or sequenced operations.

continuous-path numerical control A type of numerical-control system involving not only specification of successive end positions of machine slides or equivalent machine members, but also automatic generation of the linear, circular, or parabolic path to be followed in moving from one end position to the next. Also known as contouring numerical control.

continuous phase The basic product flowing from a filter separator or monitor system which continues on through the system after being subjected to solids and/or water removal. [AIR4783]

continuous power The maximum power that can be delivered by an actuation system for an indefinite period without sustaining damage or reducing life. [ARP4386]

continuous rating A defined power input or set of operating variables that represent the maximum values for operating a device continuously for an indefinite time without reducing its normal service life.

continuous roving Parallel filaments coated with a sizing, gathered together into single or multiple strands. [AIR4844]

continuous sampling The presentation of a flowing sample to an analytical analyzer so as to obtain continuous measurement of concentrations of the components of interest. [ARP1256]

continuous spectra Spectra in which wavelengths, wave numbers, and frequencies are represented by the continuum of real numbers or a portion thereof, rather than by a discrete sequence of numbers. For electromagnetic radiation, this corresponds to spectra that exhibit no detailed structure and represent a gradual variation of intensity with wavelength from one end to the other, such as the spectra of incandescent solids. For particles, this corresponds to spectra that exhibit a continuous variation of the momentum or energy.

continuous spectrum A distribution of wavelengths in a beam of electromagnetic radiation in which the intensity varies continuously with wavelength. A continuous spectrum exhibits no characteristic structure, such as a series of bands, in which the intensity does not abruptly change at discrete wavelengths. *See also* band spectrum.

continuous weld A welded joint in which the fusion zone is continuous along the entire length of the joint.

continuum Things that are continuous; that have no discrete parts; for example, the continuum of real numbers as opposed to the sequence of discrete integers; or the background continuum of a spectrogram due to thermal radiation.

contour control system A system of control in which two or more controlled motions occur in relation to one another so that a desired angular path or contour is generated.

contouring numerical control *See* continuous-path numerical control.

contour sensors The sensing of image coincidences by means of optical processing techniques.

contraction The narrowing of a stream of liquid passing through a notch of a weir.

contrahelical A method of applying two or more layers of spirally twisted, served, or wrapped materials, in which each successive layer is wrapped in the opposite direction of the preceding layer. [ARP1931]

contraorbit missile A missile that is sent backward along the calculated orbit of an approaching spacecraft, satellite, or aerospace weapon for the purpose of destroying it in a head-on collision with an explosive warhead or missile. [ARP4386]

contrarotating propellers Two propellers mounted on concentric shafts having a common drive and rotating in opposite directions.

contrast *1.* Ratio of display symbology brightness to external visual cue brightness, as viewed through the head-up display. [ARP4102/8] *2.* In general, the degree of differentiation between different tones in an image. *3.* In a photographic or radiographic image, the ability to record small differences in light or x-ray intensity as discernible differences in photographic density.

contrast as modulation *See* contrast.

contrast factor For a given photographic or radiographic emulsion, the slope of the central portion of a graph of photographic density versus exposure.

contrast, radiographic The difference in density between two adjacent image areas on a radiograph. [ARP5089]

contrast ratio (CR) *1.* For a symbol in the presence of incident light, the ratio of symbol luminance (Bs) to background luminance (Bb). *2. See* contrast. [ARP4067]

control *1.* A product of known characteristics that is included in a series of similar service or bench tests to provide a basis for evaluation of one or more unknown products. [AIR4844] *2.* (Mechanical) A device used to govern a machine or mechanism in operation. [ARP480A] *3.* (Electrical) A component that governs the operation of another component or grouping of components. [ARP480A]

control accuracy The degree to which a controlled process variable corresponds to the desired value or setpoint.

control action The act of changing some variable (position, power, etc.) to effect a correction of the controlled temperature. [ARP89C]

control action, derivative *See* derivative control action.

control action, integral *See* integral control action.

control agent The energy or material comprising the process element that is controlled by manipulating one or more of its attributes. Such attributes are commonly termed the controlled variables.

control algorithm A mathematical representation of the control action to be performed.

control amplifier The device that mixes the signals from the input and feedback devices, detects an error if there is one, and causes a correction to be made. [ARP89C]

control and display unit (CDU) A device that provides the pilot interface with the flight management system for inserting data and controlling and selecting parameters and modes. Includes an alphanumeric display that permits pilot verification of data. [AIR4102/9]

control apparatus An assembly containing one or more control devices which acts to manipulate a controlled variable.

control area *See* controlled airspace, control area.

control augmentation system (CAS) *1.* A vehicle flight control system in which the control system responds to the error between the commanded vehicle motion and the actual vehicle motion. [ARP4386] *2.* A function of the flight control system, including command shaping, sensors, actuators, etc., that perform in such a manner so as to augment the static and dynamic stability and maneuver response of the aircraft. When considered as an entity, it is essentially a closed-loop tracking control system, responding to pilot commands. [ARP4386]

control authority The amount of control surface or force effector deflection that can be produced by AFCS signals relative to the total available control deflection. This phrase is often preceded by the word "electrical," or by abbreviations such as "CAS" or "SAS," so as to be more explicit. [ARP4386]

control, automatic *See* automatic control.

control, barometric A method of control that depends on the barometric pressure of the atmosphere. [ARP147C]

control bench Facility for testing an engine control using a model or other artificial input in lieu of a hardware engine. [AIR4548]

control block A storage area through which a particular type of information required for control of the operating system is communicated among its parts.

control board A panel that contains control devices, instrument indicators, and (sometimes) recorders that display the status of a system or subsystem; and from which switches, dials, and controllers can be manipulated to alter system operating variables. Also known as a control panel or panel board.

control bus The data highway used for carrying control signals.

control-by-light (CBL) (fly-by-light) system A control system in which control information is transmitted by light through a fiberoptic cable. A true CBL system has neither CBW or mechanical backup, nor CBW or mechanical override. [ARP4386]

control-by-wire (fly-by-wire/CAS-by-wire) system A control system in which control information is transmitted completely by electrical means. [ARP4386]

control calculations Installation-dependent calculations that determine output signals from the computer to operate the process plant. These may or may not use generalized equation forms, such as PID forms.

control card A card that contains input data or parameters for a specific application of a general routine.

control character A character whose purpose is to control an action, rather than to pass data to a program. ASCII control characters have an octal code between 0 and 37; normally typed by holding down the CTRL key on a terminal keyboard while striking a character key.

control chart A plot of some measured quantity, such as a dimension, versus sample number, time, or quantity of goods produced. Can be used to determine a quality trend or to make adjustments in process controls as necessary to keep the measured quantity within prescribed limits.

control circuit *1*. A circuit in a control apparatus which carries the electrical signal used to determine the magnitude or duration of control action. The control circuit does not carry the main power used to energize instrumentation, controllers, motors, or other control devices. *2*. A circuit in a digital computer which performs any of the following functions: (a) directing the sequencing of program commands; (b) interpreting program commands; or (c) controlling operation of the arithmetic element and other computer circuits in accordance with the interpretation. *3*. The hydraulic system that controls the four-way valve opening rates affecting the rate-of-rise of the impulse curve. [AIR1228]

control circuit breaker Circuit breaker whose function is to protect the wiring used to operate control devices, such as relays. *See* circuit breaker. [ARP4101/5]

control column A lever or post having a wheel, half-wheel, or other device that is manipulated in controlling the attitude of the aircraft. [ARP4107]

control complexity ratio (CCR) A measure of the complexity of the logic configuration of a particular control system.

control computer A process computer that directly controls all or part of the elements in the process. *See also* process computer and industrial computer.

control configured vehicle (CCV) *1*. A vehicle in which performance benefits are attained through a design that exploits the capabilities of the flight control system to augment the inherent aerodynamic design characteristics of the basic aircraft. [ARP4386] *2*. A fixed-wing aircraft designed to allow modulation of positive aerodynamic and/or reactive forces along and about all three axes, thereby providing limited independent maneuverability in all six degrees of freedom. A control-configured vehicle can be moved vertically without changing pitch, and can be moved laterally without changing bank or heading. [ARP4107]

control counter A physical or logical device in a computer that records the storage locations of one or more instruction words that are to be used in sequence, unless a transfer or special instruction is encountered.

control device Any device, such as a heater, valve, electron tube, contactor, pump, or actuator, that is used to directly effect a change in some process attribute.

Control Diagram Language (CODIL) A process-oriented language and system offered by Leeds and Northrup Company.

control, differential pressure A method of control that limits the maximum pressure differential between cabin pressure and atmospheric pressure and maintains this differential at all altitudes above those of the isobaric control. When operating, the differential control always overrides the isobaric control. [ARP171]

control electrode In an electron tube or similar device, an electrode whose potential can be varied to induce variations in the current flowing between two other electrodes.

control element A component of a control system that reacts to manipulate a process attribute when stimulated by an actuating signal.

control feel The "feel," or reaction, that a pilot perceives through the cockpit controls, either from the aerodynamic forces acting on the control surfaces, or from artificial input simulating these aerodynamic forces. [ARP4107]

control field (CON) Field that contains ring access control information, including token status, priority, priority reservation, and short message count. [AIR4075]

control flow The flow through the valve control ports, expressed in cis, gpm, or mL/s. Control flow is referred to as "no-load flow" when there is zero load-pressure drop; and as "loaded flow" when there is load-pressure drop. [ARP490]

control-force reversal A reversal or disappearance of the conventional forces acting on the aircraft control system; an example would be when, owing to abnormal conditions, a forward movement of the stick or yoke causes the nose of the plane to rise instead of drop, with dropping being the the normal response. [ARP4107]

control grid An element of an electron tube ordinarily positioned between the anode and cathode to act as a control electrode.

control impedance The ratio of control pressure change to flow change measured at a control port. [ARP993]

control initiation The signal introduced into a measurement sequence to regulate any subsequent control action as a function of the measured quantity.

control instruction A computer instruction that directs the sequence of operations.

control, isobaric A method of control that maintains essentially constant cabin air pressure. [ARP171]

control key A computer control key that, when pressed with another key, gives that key a different meaning.

controllability *1.* The capability of an aircraft rocket or other vehicle to respond to control, especially in direction or attitude. *2.* An attribute of equipment design that defines or describes the extent to which signals of interest may be controlled. [AIR4896]

controllable check valve A two-position, manually operated valve that functions as a conventional valve in one position, but permits free flow in either direction when actuated to the other operating position. [ARP4386]

controllable-pitch propeller/variable pitch propeller A propeller whose blade angle may be changed from the cockpit while the propeller is rotating. [ARP4107]

controlled airspace Airspace designated as continental control area, control area, control zone, terminal control area, transition area, or positive control area, within which some or all aircraft may be subject to air traffic control. [ARP4107]

controlled airspace, continental control area The airspace of the 48 contiguous states, the District of Columbia, and Alaska (excluding the Alaska peninsula west of Longitude 160W) at and above 14,500 ft MSL, but not including the airspace less than 1500 ft above the surface of the earth. [ARP4107]

controlled airspace, control area Airspace designated as colored federal airways, VOR Federal Airways, control areas associated with jet routes outside the continental control area, additional control areas, control area extensions, and area low routes. [ARP4107]

controlled airspace, control zone Controlled airspace that extends upward from the surface and terminates at the base of the continental control area. [ARP4107]

controlled airspace, terminal control area (TCA) Controlled airspace extending upward from the surface or higher to specified altitudes, within which all aircraft are subject to operating rules and pilot and equipment requirements. [ARP4107]

controlled airspace, transition area Controlled airspace extending upward from 700 ft or more above the surface of the earth—when designated in conjunction with an airport for which an approved instrument approach procedure has been prescribed; or from 1200 ft or more above the surface of the earth—when designated in conjunction with airway route structures or segments. [ARP4107]

controlled avalanche transit time (CATT) devices *See* CATT devices.

controlled cooling Cooling a part from elevated temperature in a specific medium to produce desired properties or microstructure, or to avoid cracking, distortion, or high residual stress. For a controlled cooling process, the usual cooling mediums, in descending order of severity, are: brine, water, soluble oil, fused salt, oil, fan-blown air, and still air.

controlled device The final element, such as a valve, damper, programmer, or resistance heater, that is under the direct control of the control amplifier. [ARP89C]

controlled medium The process fluid or other substance containing the controlled variable.

controlled test A test designed to control or balance out the effects of environmental differences and to minimize the chance of bias in the selection, treatment, and analysis of test samples. [AIR4896]

controlled variable *1.* The variable that the control system attempts to keep at the setpoint value. The setpoint may be constant or variable. *2.* The part of a process that one desires to control (flow, level, temperature, pressure, etc.). *3.* A process variable that is to be controlled at some desired value by means of manipulating another process variable.

controller *1.* The component of the actuation system that controls the power modulator as a function of the command or error signal. Many electronic controllers have dynamic compensation as well as significant gain, so they must be recognized as separate components in the actuator loop. [ARP4386] *2. See* air traffic control specialist. *3.* A device that permits the human pilot to initiate turn, pitch, and bank signals in the automatic pilot. [ARP419] *4.* A device for interfacing a peripheral unit or subsystem in a computer; for example, a tape controller or a disk controller. *5.* Device that

contains all of the circuitry needed for receiving data from external devices, both analog and digital; processes the data according to preselected algorithms; then provides the results to external devices.

control limit An automatic safety control for limiting the operation of the controlled equipment; responsive to changes in liquid level, pressure or temperature, or position.

controlling extensions A controller that derives its input from the motion of the float. Can be installed within the extension housing.

controlling means The components of an automatic controller that are directly involved in producing an output control signal or other controlling action.

control lock A securing device that prevents movement of control surfaces. [ARP4107]

control logic The sequence of steps or events necessary to perform a particular function. Each step or event is defined to be either a single arithmetic or a single Boolean expression.

control loop A combination of two or more instruments or control functions arranged so that signals pass from one to another for the purpose of measurement and/or control of a process variable. *See also* closed loop and open loop.

control, manual *See* manual control.

control, modulating A continuous, automatic regulating-type of control. [ARP147C]

control motor A reversible motor designed to be operated from two independent voltage sources of the same frequency in such a way that variations in the voltage source will determine the direction, speed, and torque of the motor. Inherent characteristics of this type of motor include a high torque-to-inertia ratio and a straight line speed torque curve. [ARP667]

control operation An action, such as the starting or stopping of a particular process, performed by a single device. Conventionally, carriage return, fault change, rewind, end of transmission, etc., are control operations, whereas the actual reading and transmission of data are not.

control output module A device that stores commands from the computer and translates them into signals that can be used for control purposes. A control output module can generate digital outputs to control on-off devices or to pulse setpoint stations; or it can generate analog output (voltage or current) to operate valves and other process control devices.

control panel *1.* A part of a computer console that contains manual controls. *See also* pinboard and plugboard. *2.* A means of providing readout of information on pressures and their control. [AIR1228] *3. See* control board.

control point *1.* A design attribute that enhances testability by enabling, disabling, blocking, resetting, etc., functions within a system to effect efficient test control. [AIR4896] *2.* The actual value of the controlled temperature at which the system is controlling. [ARP89C] *3.* The setpoint or other reference value that an automatic controller acts to maintain as the measured value of a process variable under a given set of conditions.

control precision The degree to which a given value of a controlled variable can be reproduced for several independent control initiations, using the same control point and the same system operating conditions.

control pressure The pressure at the control port of a three-way valve; or the differential pressure between the control ports of a four-way valve, expressed in psi or kPa. Control pressure is referred to as blocked load pressure when there is zero steady flow to the load. [ARP490]

control pressure (pilot pressure) The pressure required to control or influence any motion or change in motion. [ARP4386]

control pressure curve The graphical representation of control pressure versus input current; usually a continuous plot throughout a complete cycle between limits of plus and minus rated current values for four-way valves, and between limits of zero and rated current for three-way valves. [ARP490]

control, pressure ratio *1.* A method of control that limits the maximum pressure ratio between cabin pressure and atmospheric pressure, and maintains this ratio at all altitudes above those

of the isobaric and differential controls. *2.* A control that operates to maintain a specific pressure ratio between two points in a system. [ARP147C]

control program *1.* A group of programs that provides such functions as the handling of input/output operations, error detection and recovery, and program loading and communication between the program and the operator. In the disk and tape operating systems, IPL, supervisor, and job control make up the control program. *2.* Specific programs that control an industrial process.

control programming Writing a user program for a computer that will control a process in the sense of reacting to random disturbances in time to prevent impairment of yield or dangerous conditions.

control resolution The smallest increment of change that can be induced in the controlled process variable as a result of control-system action.

control reversal The change in the direction of the lift increment from that normally produced by a movable control surface. [ARP4386]

control rockets Vernier engines, retrorockets, or other such rockets used to change the attitude of, guide, or make small changes in the speed of a rocket, spacecraft, or the like.

control rod A long piece of neutron-absorbing material that controls the number of neutrons available to trigger nuclear fission and/or absorbs sufficient neutrons to stop fission in case of an emergency.

control room area *See* area, control room.

control, safety Controls that are intended to prevent unsafe operation of the controlled equipment; includes relays, switches, and other auxiliary equipment used in conjunction therewith to form a safety control system.

control, safety, combustion *See* combustion (flame) safeguard.

control sector An airspace area of defined horizontal and vertical dimensions for which a controller, or group of controllers, has air traffic control responsibility. Normally, such an area is within an air route traffic control center or an approach control facility. [ARP4107]

control spring A spring designed to produce a torque equal and opposite to the torque produced by the moving element of an instrument, for any position of the moving element within the limits of its operating range.

control/status function The logical function provided by the sensor/video interconnect subsystem which establishes and maintains information transfer paths through the transmission network. [AIR4911]

control/status information Digital information that is generated in the form of standard, discrete digital messages. [ARD50012]

control stick A cockpit control lever of a particular type used to produce longitudinal and lateral control inputs to the aircraft. [ARP4386]

control system A system in which deliberate guidance or manipulation is used to achieve a prescribed value of a variable. A control system has at least one input and one output. [ARP4386]

control system model A mathematical representation of the control system; may include sensors, controllers, and actuators. [ARP4148]

control unit *1.* The portion of a computer that directs the sequence of operations, interprets the coded instructions, and initiates the proper commands to the computer circuits preparatory to execution. *2.* A device designed to regulate the fuel, air, water, or electrical supply to the controlled equipment; may be automatic, semi-automatic, or manual.

control valve A valve used to control the fluid flow of a hydraulic type servo. Operates in response to signals from a controller, selector, or sensor. Also called a boost control valve or power boost control valve. [ARP419]

control variable The variable that the control system attempts to keep at the setpoint value.

control voltage winding The motor winding that is energized by a varying voltage, at a time phase difference from the voltage applied to the fixed voltage winding. [ARP667]

control zone *1.* A portion of the working zone of a piece of thermal processing equipment having a separate sensor/instrument/heat input or output mechanism to control its temperature. [AMS2750] *2. See* controlled airspace.

CONUS Acronym for the continental United States.

convection *1.* In general, mass motion within a fluid resulting in transport and mixing of the properties of that fluid. *2.* Specifically, in meteorology, atmospheric motions that are predominately vertical. *3.* The transmission of heat by the circulation of a liquid or a gas, such as air. Convection may be natural or forced.

convection cooling Removing heat from a body by means of heat transfer, using a moving fluid as the transfer medium. Usually involves only the motion caused by differences in heat content between fluid near the hot surface and fluid at some distance from the surface.

convection-type superheater *See* superheater.

convective SIGMET (convective significant meteorological information) A weather advisory concerning convective weather significant to the safety of all aircraft. Convective SIGMETs are issued for tornadoes; lines of thunderstorms; embedded thunderstorms of any intensity level; areas of thunderstorms greater than or equal to radar weather echo intensity level 4, with an area coverage of 4/10 (40 percent) or more; and hail 3/4 inches or more in diameter. *See also* SIGMET. [ARP4107]

conventional (refractive/reflective) collimating optics Refers to optics using lenses and mirrors for collimation/superimposition. [ARP4102/8]

convergence *1.* Approach to a limit, e.g., by an infinite sequence. *2.* The condition in which all the electron beams of a multibeam (color) cathode ray tube intersect at a specific point. *See also* misconvergence.

convergence, dynamic Describes the incorporation of a means (electrostatic or magnetic) for automatically converging the multiple beam spots of a color CRT over the entire screen as a function of scanning. [ARP1782]

conversational mode Communication between a terminal and a computer in which each entry from the terminal elicits a response from the computer, and vice versa.

conversion The process of changing from one type of control or operational state to another; e.g. from an active to a standby control, or from a primary to a secondary system. [ARP4386]

conversion (to engineering units) Scaling signals from their raw input form to the form used internally—usually into floating point engineering units.

conversion coating A protective surface layer on a metal, created by chemical reaction between the metal and a chemical solution.

conversion system Part of the electric system; includes the transformers, rectifiers, converters, etc. [ARP4404]

conversion time The time required for a complete measurement by an analog-to-digital converter.

conversion transducer Any transducer whose output-signal frequency is different from its input-signal frequency.

converter *1.* A type of refining furnace in which impurities are oxidized and removed by blowing air or oxygen through the molten metal. *2. A/D*–analog to digital converter; *D/A*–digital to analog converter; *I/P*–current to pneumatic pressure converter; *P/I*–pneumatic pressure to current converter; *P/V*–pneumatic pressure to voltage converter; *V/P*–voltage to pneumatic pressure converter.

convex Describes a surface whose central region is raised with respect to a flat plane approximately passing through its periphery.

convex programming In operations research, a particular case of nonlinear programming in which the function to be maximized or minimized and the constraints are appropriately convex or concave functions of the controllable variables. Contrast with dynamic programming, integer programming, linear programming, mathematical programming, and quadratic programming.

convolution Generally, a recess projecting into the tank surface to clear structural ribs, formers, longerons, etc. Convolutions permit utilization of the maximum volume of the structure for fluid storage. [AIR1664]

cook-off The detonation or deflagration of an explosive device caused by externally applied heat. [AIR913]

coolant *1.* Liquids or gases used to cool something, such as a rocket combustion chamber. *2.* Any fluid used primarily to remove heat from an object and carry it away. *3.* In a machining operation, any cutting fluid whose chief function is to keep the tool and workpiece cool.

Coolidge-type x-ray tube A high-vacuum tube in which electrons emitted from a high-voltage cathode impinge on a water-cooled metal target inclined with respect to the tube axis. X-rays emitted from the focal spot on the target are directed through a side window in the metal tube enclosure, where a material that is relatively transparent to x-rays (e.g., beryllium foil, mica, aluminum, or special low-absorption glass) allows them to escape.

cooling effect The ability of a fluid stream to carry away heat. For a given fluid, cooling effect is a function of temperature and mass flow. [ARP89C]

cooling effect detector (sensor) A device used to determine the cooling capability of the fluid in which it is immersed. [ARP147C]

cooling fixture A fixture used to maintain the shape or dimensional accuracy of a molding or casting after it is removed from the mold, and until the molded/cast material is cool enough to hold its own shape. [AIR4844]

cooling load, cabin The heat transfer rate to the cabin air as measured by the difference between the total enthalpy of the air discharged from the cabin and the total enthalpy of the air entering the cabin during a specified period of time. [ARP147C]

cooling load, total The heat transfer rate from the compartment coolant supply as measured by the difference between the total enthalpy of the coolant entering the cooling device or devices and the total enthalpy of the coolant leaving the cooling devices during a specified period of time. [ARP147C]

cooling system, bootstrap An air-cycle refrigeration system in which air from a pressure source flows successively through a compressor, a heat exchanger, and a turbine. [ARP147C]

cooling system, reduced ambient An air-cycle refrigeration system incorporating two turbines in which ambient air passes first through a turbine; then through a heat exchanger where it is used to cool air from a pressure source; and then through a compressor that raises its pressure to ambient pressure. After passing through the heat exchanger, the air from the pressure source is further cooled by the second turbine; it then passes into the compartment to be cooled. [ARP147C]

cooling system, regenerative *See* regenerative cooling.

cooling system, shoestring Refers to a cooling system in which air from a pressure source flows, successively, through a heat exchanger and a turbine, or section of a turbine. [ARP147C]

cooling system, simple/bootstrap A cooling system that combines both the bootstrap cooling system and the simple cooling system. Air from a pressure source flows, successively, through a heat exchanger, a compressor, another heat exchanger, and a turbine. [ARP147C]

cooling system, three wheel recirculating A recirculating or shoestring cooling system that utilizes a three-wheel air cycle machine. The third wheel is a fan that consumes a fraction of the shaft power and provides heat exchanger cooling air. [ARP147C]

coordinate A set of measures defining points in space.

coordinate system *See* coordinate.

coordination Defines the ability of the protector with the lowest rating (in a cascade arrangement) to open before protectors with higher ratings when a fault occurs downstream from the lowest rated protector. *See also* cascade circuit. [ARP1199A]

COP *See* coefficient of performance.

copilot A licensed pilot serving in any piloting capacity other than as pilot-in-command, but excluding a pilot who is on board the aircraft for the sole purpose of receiving flight instruction.

copilot syndrome Describes an attitude that results in ineffective crew coordination; based on the comforting premise that one or more

of the other crewmembers have the situation under control and are looking out for one's best interest. Implicit in the term "other crewmembers" are nonflight members, such as personnel in the ARTCC, the command post, or a RAPCON facility. [ARP4107]

copper alloy An alloy in which copper is the predominant element. Generally, the addition of sulfur, lead, or tellurium improves machinability. Cadmium improves tensile strength and wearing qualities. Chromium gives very good mechanical properties at temperatures well above 200°C. Zirconium provides hardness, ductility, strength, and relatively high electrical conductivity at temperatures at which copper and common high-conductivity copper alloys tend to weaken. Nickel improves corrosion resistance, while silicon offers much higher mechanical properties. Beryllium, when present in an approximate 2% content, permits maximum strength, while about 0.5% content offers high conductivity. [ARP1931]

copper, ETP (electrolytic tough pitch copper; ETPC) Copper with a minimum copper content of 99.9%. Annealed conductivity averages 101% with a 100% minimum. [ARP1931]

copper, OFHC (oxygen-free high conductivity copper; OFHC) Copper with a 99.95% minimum copper content, with an average annealed conductivity of 101%. Suitable for apparatus that is welded or exposed to reducing gases at high temperatures. OFHC copper has no residual deoxidant. [ARP1931]

copper, silver bearing Copper with a 99.9% copper content that provides nearly the same electrical conductivity as ETP copper, but offers a higher softening point, greater resistance to creep, and higher strength at elevated temperatures. Silver-bearing copper also offers higher resistance to wear and oxidation, and improved machinability. [ARP1931]

coprocessor A device added to a CPU which performs special functions more efficiently than the CPU alone.

corbinotron A device consisting of a corbino disc made of high-mobility semiconductor material, and a coil that produces a magnetic field perpendicular to the plane of the disc.

cord body The main body of the tire; includes layer(s) of cord fabric (plies). [AIR1489]

cord fabric The fabric from which the plies, breakers, and chafer strips are made. May be rayon, nylon, fiberglass, etc. [AIR1489]

cordite *See* double base propellant.

core *1.* A component or assembly of components over which another component or components, such as a shield, jacket, sheath, or armor are applied in order to form a cable. *See* cable core. [ARP1931] *2.* A strongly ferromagnetic material used to concentrate and direct lines of flux produced by an electromagnetic coil. *3.* The inner layer in a composite material or structure. *4.* The central portion of a case-hardened part, which supports the hard outer case and gives the part its toughness and shock resistance. *5.* An insert placed in a casting mold to form a cavity, recess, or hole in the finished part. *6.* A rod or closed tube inserted in a tube to reduce the flow area. *7. See* magnetic core. *8.* Magnetic memory elements; typically the main memory in a computer system. *9.* The central volume of an optical fiber through which light is propagated. [AIR4288]

core corrosion Oxidation or other chemical or electrolytic attack that adversely affects the core. [AIR4844]

core crush A collapse, distortion, or compression of the core. [ARP5089]

core depression A localized indentation or gouge in the core. [AIR4844]

cored electrode A tubular welding electrode that contains flux or some other material in the central cavity.

cored mold A mold that incorporates passages for electrical heating elements. [AIR4844]

cored solder Wire solder having a flux-filled central cavity.

core iron A grade of soft steel suitable for making cores used in electromagnetic devices, such as chokes, relays and transformers.

core memory *1.* The most common form of main memory storage used by a central processing unit, in which binary data are represented by switching the polarity of magnetic cores. *2. See* main memory.

core nodes The points at which honeycomb cells are bonded to each other. [AIR4844]

core node separation A partial or complete breaking of the core node bond. [AIR4844]

core resident Pertains to programs or data permanently stored in core memory for fast access.

core splice A joint between two sections of honeycomb core, formed by bonding with foaming adhesive or crushing the core together to interlock the cells. [ARP5089]

core splice gap That gap between core segment edges and other core segments or solid members that is filled with core splice adhesive or equivalent. [AMS3920A]

core splice void or disbond Any unbonded area existing in the plane of a bond between two core segments. [AMS3920A]

core splicing The joining of segments of a core by bonding or by overlapping each segment, and then driving them together. [AIR4844]

core stabilization A process to rigidize honeycomb core materials to prevent distortion during machining. [AIR4844]

core wire Copper wire having a steel core; often used to make antennas.

coring A metallurgical condition in which individual grains or dendrites vary in composition from center to grain boundary due to nonequilibrium cooling during solidification in an alloy that solidifies over a range of temperatures.

Coriolis effect *1.* The physiological effect felt by a person moving radially in a rotating system, such as a rotating space station; results in nausea, vertigo, dizziness, etc. *2.* An accelerating force acting on any body moving freely above the earth's surface. Arises from the rotation of the earth with respect to a given axis through its center. The Coriolis effect causes, for instance, a level bubble carried in an airplane to be deflected perpendicular to the direction of flight; and a river in the Northern Hemisphere to scour its right bank more than its left bank, whereas a river in the Southern Hemisphere scours its left bank more than its right. The Coriolis effect is also the basis for mass-flow meters. *See also* illusions, vestibular.

Coriolis illusion *See* illusions, vestibular.

Coriolis force Force resulting from Coriolis acceleration acting on a mass moving with a velocity radially outward in a rotating plane.

Coriolis-type mass flowmeter An instrument for measuring mass flowrate by determining the torque from radial acceleration of the fluid.

Corliss valve A type of valve used to admit steam to, or exhaust steam from, a reciprocating engine cylinder.

corner-cube prism A prism in which three flat surfaces meet at right angles, as the faces do at the corner of a cube. Incident light through a planar face of such a prism is reflected back to the source.

cornering force Lateral or cornering force of a tire (perpendicular to the direction of motion). [AIR1489]

corner taps The differential pressure signal location in an orifice flange union defined by the corner formed between the orifice plate and the internal diameter of the flange.

corona An electrically detectable (usually luminous) field-intensified ionization that occurs in an insulating system; caused by a potential gradient that exceeds a certain critical level. [ARP1931]

corona discharge *See* electric corona.

coronal holes Solar areas where extreme UV and x-ray coronal emission is abnormally low or absent; apparently associated with diverging magnetic fields.

coronal loops Loop like structures revealed in soft x-ray images of the solar limb, and believed to evolve from the introduction of energy and density perturbations at the top of an arched, cylindrical magnetic flux tube initially in equilibrium in the coronal plasma.

corona onset point The critical value of electrical potential where corona is first detected. Also known as ignition voltage. [ARP1931]

corona resistance Resistance to an ionizing process. [AIR4844]

corona voltmeter A type of voltmeter that uses the inception of corona to determine the crest value of voltage in an a-c electric current.

corotating *See* wheels, corotating.

corpuscular radiation Nonelectromagnetic radiation consisting of energetic charged or neutral particles.

corrected flow *See* air flow parameter.

corrected speed The speed of a starter or APU that is utilized to present performance corrected for temperature. Usually expressed as $N / \sqrt{T}$ or $N / \sqrt{\theta}$, where T or θ are defined at the inlet of the starter or APU. [ARD50013]

correction A quantity, equal in absolute magnitude to the error, added to a calculated or observed value to obtain a true value.

correction factor or deviation *1.* Number of degrees, determined from the most recent calibration, that must be added to or subtracted from the temperature reading of a sensor, instrument, or combination thereof to obtain true temperature. [AMS2750] *2. See* correlation factor.

correction time *See* time, settling.

corrective action *1.* A documented design, process, procedure, or materials change implemented and validated to correct the cause of failure or design deficiency. [ARD50010] *2.* Controller output that results in a change in controlled temperature in the direction of the control point. [ARP89C] *3.* The change produced in a controlled variable in response to a control signal.

corrective action affectivity The date or item serial number indicating when corrective action will be or has been incorporated into the item. [AIR4896]

corrective downtime rate Corrective maintenance downtime per hour of operation. [AIR4896]

corrective maintenance Maintenance specifically intended to eliminate an existing fault. Synonymous with emergency maintenance. Contrast with preventive maintenance.

corrective network An electronic network incorporated into a circuit to improve its transmission and/or impedance properties.

correlation *1.* The comparison of engine performance parameters measured on a common engine that has been tested in two test facilities, one of which serves as the reference facility. [AIR4755] *2.* In statistics, a relationship between two occurrences that is expressed as a number between minus one (-1) and plus one (+1). *3.* Measurement of the degree of similarity of two images as a function of detail and relative position of the images. Obtained by multiplying the Fourier transforms of the two images, then taking the Fourier transform of the product.

correlation coefficients Dimensionless numbers that describe the relative trend in size of one measurement as compared to another or to several others. [AIR5145]

correlation detection A method of detection in which a signal is compared, point-to-point, with an internally generated reference.

correlation engine An engine of known and repeatable performance used for test cell correlation. [AIR4755]

correlation factor A multiplier or delta used where appropriate to adjust for the difference in performance between the customer facility and the reference facility. Also known as a facility modifier or correction factor. [AIR4755]

correlation function *See* correlation.

correlator *1.* Device that detects weak signals in noise by performing an electronic operation. *2.* A logic device that compares a series of bits in a data stream with a known bit sequence and puts out a signal when correlation is achieved. One use of the correlator is as a PCM frame synchronizer.

corresponding states Refers to the principle that two substances should have similar properties at corresponding conditions with reference to some basic properties, e.g., critical pressure and critical temperature.

Corrodekote test An accelerated corrosion test for electrodeposits in which a specimen is coated with a slurry of clay in a salt solution, and then is exposed for a specified time in a high-humidity environment.

corrosion The chemical deterioration of a portion of a part by oxidation. Can be caused by oxygen, moisture, or attack by other oxidizing chemicals. [AIR4069]

corrosion coating Refers to a material applied to the surfaces of integral tanks to protect the metal, thereby preventing corrosion. NOTE: Although the term "corrosion coating" is used widely, what is actually meant is "corrosion-preventive coating." [AIR4069]

corrosion fatigue A synergistic interaction between the failure mechanisms corrosion and fatigue, such that cracking occurs much more rapidly than would be predicted by simply adding the separate effects of the two failure mechanisms. Failure by corrosion fatigue requires the simultaneous presence of a cyclic stress and a corrosive environment.

corrosion preventive coating *See* corrosion coating.

corrosion protection Preventing corrosion or reducing the rate of corrosive attack by any of several means, including coating a metal surface with a paint, electroplate, rust-preventive oil, anodized coating, or conversion coating; adding a corrosion-inhibiting chemical to the environment; using a sacrificial anode; or using an impressed electric current.

corrosion resistance The ability of a material to withstand contact with ambient natural factors, or those of a particular artificially created atmosphere, without degradation or change in properties. [AIR4844]

corrosive Any substance or environment that causes corrosion.

corrosive flux A soldering flux that removes oxides from the base metal when the joint is heated to apply solder. Usually composed of inorganic salts and acids that are corrosive; thus to ensure maximum service life, must be removed before placing the soldered components in service.

corrosiveness The degree to which a substance causes corrosion.

corrugated fastener A thin, corrugated strip of steel used to fasten two pieces of wood together by hammering it into the wood at approximately right angles to the joint line.

corrugating Forming sheet metal into a series of alternating parallel ridges and grooves. May be done by rolling the metal between matched grooved rolls, or by forming it in a press brake equipped with a special-shaped punch and die.

corrupt In data processing, the inclusion of errors in programs or data.

cosmetic defect A variation from the conventional appearance of an item, such as a slight deviation from its usual color, which is not detrimental to the performance of the item. [ARP1931]

cosmetic irregularities Surface conditions that do not affect the integrity of the component and require no further action. [ARP5089]

Cosmic Background Explorer satellite A NASA satellite designed to measure background radiation in order to confirm or deny the big bang theory.

cosmic dust Finely divided solid matter with particle sizes smaller than a micrometeorite (i.e., with diameters much smaller than a millimeter) moving in interplanetary space.

cosmic gamma ray bursts *See* gamma ray bursts.

cosmic noise Interference caused by cosmic radio waves.

cosmic radiation *See* cosmic rays.

cosmic rays *1.* The aggregate of extremely high-energy subatomic particles that travel the solar system and bombard the earth from all directions. Cosmic ray primaries seem to be mostly protons (hydrogen nuclei), but also contain heavier nuclei. On colliding with atmospheric particles, cosmic rays produce many different kinds of lower-energy secondary cosmic radiation. *2.* Penetrating ionizing radiation whose ultimate origin is outside the earth's atmosphere; also known as cosmic radiation. Some of the constituents of cosmic rays can penetrate many feet of material, such as rock.

cosmochemistry The branch of chemistry that deals with the chemical composition and changes in the universe.

COSPAS The USSR satellite of the COSPAS-SarSat project, a satellite-aided project for the search and rescue of distressed vehicles, administered by USSR, US, French, and Canadian agencies.

cottered joint A joint across which power is transferred via shear force transverse to the longitudinal axis of a bar known as a cotter. The cotter holds the joint together; it is usually tapered along one side to ensure a tight fit.

cotter key *See* cotter pin.

cotter pin A split pin, usually formed by folding a length of half-round wire back on itself. A cotter pin is inserted into a hole and then is bent; it can be used to keep a castle nut from

turning on a bolt, to hold a cotter securely in place, to hold hinge plates together, or to pin various other machine parts together. Also known as a cotter key.

Cottrell precipitator A device for removing dust or mist from a gas by passing the gas through a vertical, electrically grounded pipe where the particulates become ionized by corona discharge from an axial wire maintained at a high negative voltage. The ionized particles migrate to the inner wall of the pipe where they collect for later removal by mechanical means.

coulomb Metric unit for quantity of electricity.

coulomb collisions The collisions of sets of two particles, both of which are charged.

coulomb damping The dissipation of energy that occurs when a particle in a vibrating system is resisted by a force whose magnitude is a constant, independent of displacement and velocity, and whose direction is opposite to the direction of the velocity of the particle. Also called dry friction damping. [AIR1489]

coulomb friction load A constant friction load opposing motion. [ARP4386]

coulombmeter An instrument for measuring the quantity of electricity (in coulombs) by integrating a stored charge in a circuit that has a high impedance.

coulometer An electrolytic cell constructed and operated to measure a quantity of electricity in terms of the electrochemical action it produces.

coulometric titration A method of wet chemical analysis in which the amount of an unknown substance taking part in a chemical reaction is determined by measuring the number of coulombs required to reach the end point in electrolysis.

count In computer programming, the total number of times a given instruction is performed.

countdown A step-by-step process that culminates in a climatic event, each step being performed in accordance with a schedule marked by a count in inverse numerical order. Specifically, a countdown is used in leading up to the launch of a large or complicated rocket vehicle; or in leading up to a captive test, a readiness firing, a mock firing, or other firing test.

counter *1*. A device or register in a digital processor for determining and displaying the total number of occurrences of a specific event. *2*. In the opposite direction. *3*. Device or PC program element that can total binary events and perform ON/OFF actions based on the value of the total. *4*. A device, register, or location in storage for storing numbers or number representations in a manner that permits these numbers to be increased or decreased by the value of another number; or to be changed or reset to zero, or to an arbitrary value.

counter balance *See* counterweight.

counter balance valve *See* lock valve.

counterbore *1*. An end cutting tool having: (a) two or more cutting edges; (b) flutes or grooves adjacent to the cutting edges, for the passage of cuttings; usually helical to give a positive rake at the cutting edge; and (c) an integral or inserted pilot, or centrally located hole for insertion of a pilot, for guiding the tool. A counterbore may have a straight shank, tapered shank, or special shank requiring a special holder, for holding in a power- or hand-operated machine. It is used to form a flat-bottomed enlargement of the mouth of a previously formed cylindrical hole. [ARP480A] *2*. A drilled or bored flat-bottomed hole, often concentric with another, smaller hole.

counter-clockwise The opposite direction of clockwise, as viewed from the anti-prop end. [ARP169]

countercurrent flow Flow of two fluids in opposite directions within the same device, such as a tube-in-shell heat exchanger. Contrast with counterflow.

counterflow Flow of a single fluid in opposite directions in adjacent portions of the same device, such as a U-bend tube. Contrast with countercurrent flow.

counterglow *See* gegenschein.

counterpoise *See* counterweight.

counter rotating propellers (shafts) The airplane arrangement in which shafts and propellers rotate in opposite directions, but in which the shafts are not concentric. [ARP355]

counter rotation Movement of sets of bodies or fluids around a common axis such that movement in one rotational direction is opposed by movement in the opposite direction.

countershaft A secondary shaft, driven by the main shaft of a machine, and used to supply power to one or more machine parts.

countersink *1*. An angular cutting tool, usually made with angular relief, having two or more flutes with specific size angle cutting edges. Used for chamfering and countersinking holes. A countersink may have a straight shank, tapered shank, bit stock shank, or special shank requiring a special holder, for holding in a power- or hand-operated machine. It may be designed to be driven or turned by means of a wrench or carpenter's brace. [ARP480A] *2*. A chamfer around the edge of a circular hole, which removes burrs, provides a seat for a flat-head screw or other fastener, or provides a tapered surface in which a machine center may rest.

counterweight *1*. A mass that counterbalances the weight of the lifting device or load platform of an elevator or hoist so that the engine must only work against the payload, friction, and any remaining unbalanced machine loads. *2*. Any mass incorporated into a mechanism to compensate for an out-of-balance condition and maintain static equilibrium. Also known as counterbalance or counterpoise.

counting rate The average number of ionizing events that occur per unit time, as determined by a counting tube or similar device.

counting-rate meter An instrument whose indicated output is related to the average rate of occurrence of ionizing events.

counting scale Any of several designs of weighing device in which the total weight of a large number of identical parts is compared with the weight of one part, or the weight of a small, easily counted number of parts; and the number of parts in the unknown quantity determined by automatic indication, readout, or calculation.

counts *1*. An alternate form of representing raw data corresponding to the numerical representation of a signal received from or applied to external hardware. 2. The accumulated total of a series of discrete inputs to a counter. 3. The discrete inputs to an accumulating counter. 4. The units associated with the digital numbers that go into or out of D/A and A/D converters. [AIR4548]

couplant 1. A substance (usually liquid) used between a search unit and test part to permit or improve transmission of ultrasonic energy in the test part. [ARP5089] 2. A substance used to transmit sound waves from an ultrasonic search unit to the surface of a test piece, thus reducing losses and improving test accuracy. Usual couplants include water, oil, grease, paste, and other liquid or semisolid substances.

coupled *See* wheels, co-rotating.

coupled control-element action A type of control system action in which two or more actuating signals or control element actions are used in concert to operate one control device.

coupled mode Mode of vibration that is not independent, but in which one mode influences the other.

coupler 1. In data processing, a device that joins similar items. 2. In fiber optics, a device that joins together three or more fiber ends—splitting the signal from one fiber so it can be transmitted to two or more other fibers. Directional, star, and tee couplers are the most common.

couple unbalance Condition of unbalance in which the central principal axis intersects the shaft axis at the center of gravity. NOTES: The quantitative measure of couple unbalance can be given by the vector sum of the movements of the two dynamic unbalance vectors about a certain reference point in the plane containing the center of gravity and the shaft axis. If static unbalance in a rotor is corrected in any plane other than that containing the reference point, the couple unbalance will be changed. [ARP588A]

coupling 1. A fitting or clamping device that serves to join the mating ends of adjacent ducts or other components. [ARP699D] 2. Interdependence in a computer system.

coupling agent A double-ended molecule, one end of which bonds to the substrate and the

other to the resin or adhesive. Purpose is to assist both the strength and (especially) durability of bonding. [AIR4844]

coupling assembly A connecting device that constrains the ferrules and provides a pressure seal and electrical continuity across the ferrules. [MA2241]

coupling, automatic outlet An oxygen connection that, when disconnected, automatically closes a valve to prevent loss of oxygen, and reopens the valve when reconnected. [ARP171]

coupling, bayonet, cylindrical A coupling mechanism utilizing spiral ramps in one half of the cylindrical connector to engage projections in the mating half, so as to provide jacking and locking together of the mating halves through limited rotation of the coupling ring. [ARP914A]

coupling, breech A coupling mechanism that distributes the coupling load over large, solid metal engaging and locking lands for positive coupling alignment and complete connector mating with a limited rotation of the coupling ring. [ARP914A]

coupling, quick disconnect A design feature that permits relatively rapid joining and separation of mating parts. [ARP914A]

coupling ring The portion of a cylindrical plug connector housing that, by rotation, aids in the mating, captivation, or unmating of the plug to the receptacle connector. *See also* coupling, bayonet, cylindrical; coupling, quick disconnect; and coupling, threaded. [ARP914A]

coupling, threaded A coupling mechanism utilizing matching screw threads for mating and unmating of cylindrical connectors or other devices. [ARP914A]

coupling, threaded self locking A coupling mechanism utilizing matching screw threads for mating and unmating of cylindrical connectors or devices; incorporates an automatically actuated locking mechanism to prevent the coupling ring from disengaging under vibration conditions. [ARP914A]

coupling torque The force required to rotate a coupling ring or jackscrew when engaging a mating pair of connectors. [ARP914A]

coupling triple start, self locking A coupling mechanism using a triple-start thread for quick connector mating with one full turn of the coupling ring which is also designed with an anti-coupling device. [ARP914A]

course *1.* The intended direction or path of flight in the horizontal plane measured in degrees from north. The actual course or path of an aircraft should be distinguished from its heading. (They may coincide, but usually do not, the difference being a function of heading, side-slip, and drift.) *See also* heading. *2.* The ILS localizer signal pattern usually specified as front course or back course. *Front course (ILS)*–the approach course of the localizer, which is used with other functional parts (for example, glide slope, marker beacons, etc.). The signal of the front course (ILS) is transmitted from the far end of the runway (opposite from the runway approach threshold), and is adjusted so that the distance between full-scale deflections (left to right) of the course deviation indicator needle equates to approximately 700 ft of linear width at the runway threshold. *Back course (ILS)*–the course line along the extended centerline of a runway, in the opposite direction to the front course (ILS). [ARP4107]

course deviation indicator (CDI) A display device for presenting the magnitude (deg, ft, miles) of displacement of a vehicle from a selected course. [ARP4107]

course selector A manually operated navigation instrument used in conjunction with a VHF omnidirectional radio range to ascertain station direction. [ARP4107]

coverage *1.* The frequency with which an interval estimate of a parameter may be expected to contain the true value. [AIR1678] *2.* The calculated percentage that defines the completeness with which a metal braid covers the underlying surface. The higher percentage of coverage, the greater the protection against external interference. [ARP1931] *3.* The load repetition factor. For an air vehicle, the number of passes of loaded tires in adjacent tire paths sufficient to just cover a given width of pavement one time. Determining factors are:

width of traffic lane, number of wheels, width of tire contact area, and traffic distribution. [AIR1489]

cover blisters Raised areas or spots on an elastomeric hose cover surface, usually forming a void or air-filled space in the material. [ARP1658A]

covering power The ability of an electroplating solution to give a satisfactory plate at low current densities (such as occur in recesses), but not necessarily to build up a uniform coating. Contrast with throwing power.

cover plate *1.* Any flat metal or glass plate used to cover an opening. *2.* Specifically, a piece of glass used to protect the tinted glass in a welder's helmet or goggles from being damaged by weld spatter.

cover, protective An accessory used to cover the mating portion of a connector for mechanical, environmental, and/or electrical protection. [ARP914A]

cover the six A request to protect the vulnerable area(s) of a person or a vehicle, i.e., those areas that cannot be monitored by the operator of the system. Generally used by pilots of combat aircraft when requesting their wingman to guard the area to their rear (6 o'clock position). [ARP4107]

covert lighting A light source with a relative absence of energy in the visible range; usually designed for use in night-vision imaging systems with a high concentration of energy in the near infrared region. [ARP4392]

covolume The space occupied by a gas when compressed to its limit. [AIR913]

cowling A cover that provides a streamlined enclosure for an engine.

CPA Closest point of approach of two aircraft in conflict. [ARP4153]

CP/M An operating system for microcomputers.

C power supply An electrical power supply connected between the cathode and grid of a vacuum tube to provide a grid-bias voltage.

CPU-bound A state of program execution in which all operations are dependent on the activity of the central processor; for example, when a large number of calculations are being performed. Compare to I/O-bound.

CR *See* constant ratio.

crab To turn an aircraft partly into the wind to compensate for drift. [ARP4386]

crack *1.* A discontinuity that has a relatively large cross section in one direction, and a small or negligible cross section when viewed in a direction perpendicular to the first. [ARP5089] *2.* A clean (crystalline) fracture passing through or across the grain boundaries; may possibly follow inclusions of foreign elements. Cracks are normally caused by overstressing the metal during forging or other forming operations, or during heat treatments. [MA1568] Where parts are subjected to significant reheating, cracks usually are discolored by scale. [MA2005] *3.* To incompletely sever a solid material, usually by overstressing it. *4.* To open a valve, hatch, door, or other similar device a very slight amount.

crack closure Phenomenon that occurs when the cyclic plasticity of a material gives rise to the development of residual plastic deformations in the vicinity of a crack tip, causing the fatigue crack to close at positive load.

cracked residue The fuel residue obtained by cracking crude oils.

cracked shot One that exhibits a linear discontinuity with length greater than three times its width, and length greater than 20% of the particle diameter. [AMS2431/2A]

cracked sleeve A firesleeve that evidences surface cracks or fissures caused by strain and environmental conditions. [ARP1658A]

crack geometry The shape and size of partial fractures or flaws in materials.

crack growth Rate of propagation of a crack through a material due to a static or dynamic applied load. [AIR4844]

cracking Formation of a fissure of visible width; can occur at any angle to the surface. A crack typically propagates from a stress riser, such as a scratch, craze, or improperly machined edge. [AIR5122]

cracking (chemical engineering) A process used to reduce the molecular weight of hydrocarbons by breaking molecular bonds by thermal, catalytic, or hydrocracking methods.

cracking, coating Cracking that extends through the surface coating. Occurs where the elongation of the coating is less than the substrate, and where pressurization or thermal expansion causes strains greater than the tolerance of the coating. [AIR5122]

cracking, edge Short, linear cracks that originate along the polished transition edge of outer panes and have propagated radially inward. [AIR5122]

cracking pressure The differential pressure at which a closed valve allows a specified flow rate to occur. [AIR1749]

cracking process A method of manufacturing gasoline and other hydrocarbon products by heating crude petroleum distillation fractions or residues in the presence of a catalyst so that they are broken down into lighter hydrocarbon products, some of which can be distilled off.

cracking star A group of several roughly linear surface micro-cracks that originate at a point and propagate outward to form a star-like pattern. [AIR5122]

crack opening displacement (COD) The displacement at the mouth of a crack in a material.

crack tip The boundary between cracked and uncracked material.

cradle A device to cradle a part by engaging the contour of the part. *See also* saddle. [ARP480A]

crane A hoisting machine with a power-driven horizontal or inclined boom and lifting tackle.

crane hoist A mobile hoisting machine used principally for lifting loads by means of cables. Consists of a mobile undercarriage and support structure; a power unit and winch enclosed in a cab or house (often one that swivels on the undercarriage); a movable boom; and various lifting, boom positioning, and support cables.

crane scale A type of lifting device with an integral crane hook (or attached to a crane hook), which has an internal load cell that automatically weighs a load as it is lifted. When a strain-gage load cell is used, weight can be indicated or recorded remotely.

crank *1.* To rotate the main shaft of an internal combustion or gas turbine engine by means of a starter. [ARP906A] *2.* A mechanical link that can revolve about a center of rotation.

cranking torque Torque required to crank the engine until it reaches self-sustaining rotation. [AIR4152]

Crank-Nicholson method A method for solving parabolic partial differential equations, whose main feature is an implicit method that avoids the need for using very small time steps.

crankpin A cylindrical projection on a crank; used for attaching a connecting rod.

crankshaft *1.* A straight shaft to which one or more cranks are attached. *2.* A cast, forged, or machined shaft with integral cranks, such as is used in a reciprocating automobile engine.

crank throw *1.* The web or arm of a crank. *2.* The radial displacement of the crankpin from the crankshaft axis. *3. See* crank web.

crank web The portion of a crank that connects a crankpin to the crankshaft, or to another adjacent crankpin. Also known as a crank throw.

crash A computer hardware or software malfunction that causes the system to be reset or restarted.

crash engagement *See* engagement, crash.

crash re-engagement *See* engagement, crash.

crashworthiness The ability of a vehicle to withstand a crash.

C-rate The discharge or charge rate in amperes; numerically equal to the rated capacity of a cell in ampere-hours. [ARP4386]

crater *1.* A spot on the face of a cutting tool where it has been worn by contact with chips. *2.* A depression at the finishing end of a weld bead.

Cray computer Supercomputer built by Cray Research Inc. that requires the supporting services of another front-end general-purpose computer for operation. The Cray computer incorporates very fast scalar and vector hardware. It is used primarily for the simulation of physical phenomena, and is programmed in FORTRAN.

crazing *1.* Minute cracks on or near the surface of materials. [ARP1931] *2.* Development of such a network of minute cracks.

CRC *See* cyclic redundancy check.

create To open, write data to, and close a file for the first time.

creel A device for holding the required number of roving balls or supply packages in the desired position for unwinding in the next processing step, that is, weaving, braiding, or filament winding. [AIR4844]

creep *1.* The time- and temperature-dependent deformation that occurs in a material as a result of the application of a constant load. Although creep can occur at any temperature in some materials, it is of concern in metals at higher temperature, generally above one-half of the melting temperature (absolute). The calculation of life usage due to creep requires accurate time and temperature information since a change of 15°C, at high temperatures, can affect the life of a part by a factor of two. [AIR1872] *2.* Time-dependent plastic strain occurring in a metal or other material under stress, usually at elevated temperature.

creep (leakage) distance On the surface of an insulator, the shortest distance separating two electrically conductive surfaces. [ARP914A]

creep rate The rate at which creep deformation occurs; always measured at a specified temperature and stress. Creep rate is classified in four stages, and can further be expressed as the slope of the creep/time curve. [ARP700]

creep strength The constant nominal stress that will cause a specified quantity of creep in a given time at constant temperature.

crest *1.* The top of a screw thread. *2.* The bottom edge of a weir notch, sometimes referred to as the sill.

crest factor The ratio of the peak voltage to the RMS voltage. For a true sine wave, this ratio is 1.414. [ARP1148A]

crest voltmeter An instrument whose indicated value is the average positive peak amplitude of a sinusoidal a-c electric voltage.

crest width The distance along the crest between the sides.

crevice corrosion A type of concentration-cell corrosion associated with the stagnant conditions in crevices, fissures, pockets, and recesses away from the flow of a principal fluid stream. Under such conditions, concentration or depletion of dissolved salts, ions, or gases (such as oxygen) leads to deep pitting.

crew Flight members assigned tasks during flight times.

crew coordination The systematic division of subtasks between or among crew or flight members so as to accomplish a larger task more efficiently. Crew coordination is the most basic level of command/control. [ARP4107]

crewmember A person assigned to perform duty in an aircraft during flight time. The term "flight crewmember" refers to the pilot, copilot, navigator, or (where applicable) flight engineer. [ARP4107]

crew procedure (inflight) Operations performed by crews aboard aircraft or spacecraft during flight. Includes flight operations as well as spaceborne experiment procedures.

crew procedure (preflight) Operations performed by crews aboard aircraft or spacecraft, and by ground support crews before flight or launching.

crew rest facility An area designed to provide sleeping, stowage, and clothes-changing facilities for off-duty flight crew. [ARP4101/3]

crew size The number of people in a crew.

crew systems Those portions of aircraft systems/subsystems that are affected by the aircrew; for example, the flight control system, flight display system, radar systems, and environmental control systems. [ARP4107]

CRI *See* criterion referenced instruction.

crimp The physical compression or reshaping of a conductor barrel or ferrule around a conductor in order to obtain a mechanical and an electrical connection. [ARP914A]

crimp-and-cleave Refers to the connectorization approach in which the end of a fiber is prepared using a precision fracture (cleave) after having been captivated by being crimped to the connector. [AIR4288]

crimp anvil (nest) The portion of a crimping die that supports a barrel or ferrule during crimping. [ARP914A]

crimper A device to flute, corrugate, compress, or otherwise deform a part to change its initial shape. [ARP480A]

crimp indentor The portion of the crimping die that indents or reshapes the barrel or ferrule. [ARP914A]

crimping *1.* Forming small corrugations in order to set down and lock a seam, create an arc in a metal strip, or reduce the radius of an existing arc or circle. *See also* swaging (swedging). *2.* Causing something to become wavy, crinkled, or warped.

crimping dies The portion of a crimping tool that compresses and reshapes the conductor barrel or ferrule to form the crimp. [ARP914A]

crimping tool *1.* A mechanical device having specially shaped dies for reforming the wire barrel of a taper pin around a wire. [ARP592] *2.* Any device used to perform a crimp.

crimp pot adapter A sleeve that fits around the stripped conductor and allows for a small wire to fit into a large-gage crimp pot. [ARP914A]

crimp tensile strength The axial force required to separate a wire from the crimped conductor barrel. The wire may pull out of or break in the crimped area of the conductor barrel. [ARP914A]

crimp termination A termination accomplished by the controlled reforming of a wire barrel portion of a terminating device, through physical compression exerted by appropriate tooling. [ARP1931]

criterion referenced instruction (CRI) A training system methodology in which an individual's achievement in the instructional program is measured in terms of a predetermined set of absolute criteria, rather than relative to the performance of other individuals.

critical altitude The maximum altitude in standard atmospheric conditions at which it is possible to maintain a specified engine power or a specified manifold pressure at a specified rotational speed (engine RPM). [ARP4107]

critical angle of attack *1.* The angle of attack of an airfoil at which the flow of air about the airfoil changes abruptly in such a manner that lift is sharply reduced and drag is sharply increased. [ARP4386] *2.* The minimum angle of attack of a given airfoil or airfoil section at which extensive flow separation occurs, with consequent loss of lift and increase of drag; generally results in stalling of the airfoil. [ARP4107]

critical cooling rate The minimum cooling rate that will suppress undesired transformations during a hardening heat treatment.

critical damping *1*. Exponential convergence to stable, steady-state operation. A system is considered to have critical damping when two roots of the system characteristic equation are real and identical. [ARP89] *2*. The minimum damping that will allow a displaced system to return to its initial position without oscillation. [AIR1489] *See also* damping

critical defect *1*. A defect that judgment and experience indicate is likely to result in hazardous or unsafe conditions for individuals using, maintaining, or depending upon the product; or a defect that judgment and experience indicate is likely to prevent performance of the tactical function of a major end item, such as an aircraft or missile. Synonymous with failure. [ARD50010] *2. See* defect.

critical dimension *1*. Generally, any physical measurement whose value or accuracy is considered vital to the function of the involved component or assembly. *2*. In a waveguide, the cross-sectional dimension that determines the critical frequency of the waveguide.

critical engine The engine whose failure would most adversely affect the performance or handling qualities of an aircraft. [ARP4107]

critical engine failure speed *1*. A minimum speed at which it is considered safe to attempt to complete the takeoff with one engine inoperative. [AIR1489] *2. See* V_1.

critical field length The length of field (runway) required to accelerate a multiengined aircraft to the critical engine failure speed, experience failure of the critical engine, and either: (a) continue takeoff and have just enough runway left to execute a safe takeoff; or (b) elect to reject the takeoff and be able to stop the aircraft exactly at the end of the runway. [AIR1489]

critical flow Somewhat ambiguous term that signifies a point at which the characteristics of flow suffer a finite change. In the case of a liquid, critical flow could mean the point at which the flow regime changes from laminar to transitional. More often, however, "critical

flow" is used to mean "choked flow." In the case of a gas, critical flow may mean the point at which the velocity at the vena contracta attains the velocity of sound; or it may mean the point at which the flow is fully choked.

critical frequency The frequency below which a traveling wave of a given mode cannot be maintained in a given waveguide.

critical hazard A hazard that could result in serious injury to personnel and/or damage to flight or ground equipment that would cause mission abort or a significant program delay (one or more days). [AIR4728]

critical item *1*. Any item that: (a) if it failed, would critically affect system safety, cause the system to become unavailable or unable to achieve mission objectives, or cause extensive/expensive maintenance and repair; (b) has stringent performance requirement(s) in its intended application relative to state-of-the-art techniques for the item; (c) has a known operating life, shelf life, or environmental exposure, such as vibration, thermal, propellant; or a limitation that warrants controlled surveillance under specified conditions; (d) has exhibited an unsatisfactory operating history; (e) does not have sufficient history of its own, or similarity to other items having demonstrated high reliability, to provide confidence in its reliability. [ARD50010] *2*. An item that has a limitation to warrant controlled surveillance under specified conditions. [ARP4386]

criticality *1*. Indication of the hazard level associated with a function, hardware, software, etc., considering abnormal behavior alone, or in combination with external events. [ARP4754] *2*. A measure of the impact of a failure mode on the mission objective. Criticality combines the frequency of the occurrence and the level of severity of a failure mode. [ARP926A]

critical length The minimum fiber length required for shear loading to its ultimate strength by the matrix. [AIR4844]

critical longitudinal stress Applied to fibers, the longitudinal stress necessary to cause internal slippage and separation of a spun yarn. [AIR4844]

critical Mach number The flight Mach number at which the highest local velocity on a vehicle, usually on the wing, becomes equal to 1.0, resulting in a significant drag increase.

critical mass The amount of concentrated fissionable material that can just support a self-sustaining fission reaction.

critical material Any raw or partially-processed material essential in a national emergency, and not expected to be available in quantity, quality, or time to meet requirements unless stockpiling action is taken. [AIR4073]

critical part A life-limited part, whose failure would likely impair the ability of the aircraft to continue safe flight—either through damage to the structure or controls, or injury to the crew. [AIR1872]

critical point The point at which the liquid and the vapor have identical properties. Critical temperature, critical pressure, and critical volume are the terms given to the temperature, pressure, and volume at the critical point. NOTES: Above the critical temperature, gas cannot be liquefied by pressure alone. Critical pressure is saturation pressure corresponding to critical temperature. [ARP147C]

critical pressure *1.* In rocketry, the pressure in the nozzle throat for which the isentropic weight flow rate is maximum. *2.* The pressure of a gas at the critical point, which is the highest pressure under which a liquid can exist in equilibrium with its vapor. *3.* The equilibrium pressure of a fluid that is at its critical temperature.

critical-pressure ratio The ratio of downstream pressure to upstream pressure that corresponds to the onset of turbulent flow in a moving stream of fluid.

critical size *1.* The established flaw size deemed to be detrimental to the serviceability criteria of a product. *2.* The acceptance/rejection levels established by design-engineering-required limits to meet design performance. [ARP5089]

critical speed *1.* A speed of a rotating system that corresponds to a resonance frequency of the system. [AIR1489] *2.* The speed of angular rotation at which a shaft becomes dynamically unstable due to lateral resonant vibration. *See also* critical velocity.

critical strain The amount of prior plastic strain that is just sufficient to trigger recrystallization when a deformed metal is heated.

critical temperature *1.* The temperature above which a substance cannot exist in the liquid state regardless of the pressure. *2.* As applied to reactor overheat or afterheat, the temperature at which the least-resistant component of the reactor core begins to melt down. *3.* As applied to materials, the temperature at which a change in phase takes place, causing an appreciable change in the properties of the material.

critical velocity *1.* In rocketry, the speed of sound at the conditions prevailing at the nozzle throat. *2.* For a given fluid, the average linear velocity marking the upper limit of streamline flow and the lower limit of turbulent flow, at a given temperature and pressure, in a given confined flowpath.

critical viewing sector Those geometric sectors of an instrument that include the normal viewing angles of the various crew members for the particular instrument. [ARP1161]

critical volume The volume occupied by a unit mass of a fluid at its critical temperature and pressure. [AIR1116]

cross-assembler An assembler program run on a larger host computer, and used for producing machine code to be executed on another, usually smaller, computer.

cross-axis sensitivity *See* transverse sensitivity.

crossbar micrometer An instrument for determining differences in right ascension and declination of celestial objects. Consists of two bars mounted perpendicular to each other in the focal plane of a telescope and inclined at 45° to the east-west path of the stars.

cross-bleed Refers to air ducted from the compressor section of an operating engine to another engine, in such a way that it can be used for starting or other purposes. [ARP906A]

cross bleed starting The starting of one engine utilizing cross bleed air from another operating engine. [ARP906A]

cross-cell The comparison of engine performance parameters measured on a common engine (though not necessarily a correlation engine) in at least two previously correlated test cells. Purpose is to check facility correlation of a third test cell. [ARP741]

cross-channel monitoring *See* monitor.

cross-compiler A computer program run on a larger host computer and used for translating a high-level language program into the machine code to be executed on another computer.

cross controls To position aircraft controls in an uncoordinated fashion; for example, to deflect the right aileron downward while holding right rudder. [ARP4107]

cross drum boiler A section header or box boiler in which the axis of the horizontal drum is at right angles to the center lines of the tubes in the main bank.

crossed joint *See* bridging.

cross feed On a multiengine aircraft, the feeding or transfer of fuel or oil from engine to engine, or from tank to tank. [ARP4107]

cross flow A flow going across another flow, such as a spanwise flow over a wing.

crosshair An inscribed line or a thin hair, wire, or thread used in the optical path of a telescope, microscope, or other optical device to obtain accurate sightings or measurements. Originally, a pair of hairs at right angles was used for this purpose; this is the original source of the term.

crosshead *1.* A sliding block that moves back and forth between guides and that contains a wrist pin for converting reciprocating motion to rotary motion. *2.* A device designed to extrude material at an angle; used most extensively at the discharge end of an extruder in a wire-coating operation.

cross laminated Describes a laminated material in which some of the layers are oriented at various angles to the other layers, with respect to the laminate reference axis. [AIR4844]

cross-linking In polymers, the establishing of chemical links between the molecular chains through the process of chemical reaction, electron bombardment, or vulcanization. [ARP1931]

crosslinking agent A substance that promotes or regulates intermolecular covalent or ionic bonding between polymer chains. [AIR4844]

cross-modulation Carrier and signal harmonics of one or more channels appearing in other channels of a system. When there is a large number of cross-modulation products, the resultant cross-talk noise approaches the characteristics of fluctuation noise (AM).

crossover frequency *1.* The frequency at which a dividing network delivers equal power to upper-band and lower-band channels. *2.* For an acoustic recording system, the frequency at which the asymptotes to the constant-amplitude and constant-velocity portions of the frequency-response curve intersect. Also known as the transition frequency or the turnover frequency.

crossover network A selective network that divides the audio-frequency output of an amplifier into two or more bands of frequencies to supply two or more loudspeakers. Also known as a dividing network or a loudspeaker dividing network.

cross ply Describes a laminate in which the laminae are at right angles to each other. [AIR4844]

cross-ply laminate A laminate with plies usually oriented at 0° and 90° only. [AIR4844]

cross-pointer indicator An aircraft instrument having two crossing needles that indicate the position of the aircraft with respect to an instrument landing system localizer and glide slope. [ARP4107]

cross polarization The component of the electric field vector normal to the desired polarization component.

cross section *1.* A measure of the effectiveness of a particular process expressed either as the area that would produce the observed results (geometric cross section), or as ratio. *2.* For a given confined flowpath or a given elongated structural member, the dimensions, shape, or area determined by its intersection with a plane perpendicular to its longitudinal axis. *3.* In characterizing interactions between moving atomic particles, the probability per unit flux and per unit time that a given interaction will occur.

cross-sectional area (of a conductor) The sum of the cross-sectional areas of the component wires of the conductor, that of each wire being measured perpendicular to its axis. [ARP1931]

crosstalk *1.* Term used in multiple channel units to describe the voltage produced in the secondary of one channel by the primary excitation of another channel. [ARP4386] *2.* Electrical disturbances in a communication channel as a result of coupling with other communication channels.

cross track distance The shortest distance between the present position of an aircraft and the desired track. [AIR4102/9]

crosswind *1.* When used concerning the traffic pattern, the word means "crosswind leg." *See* traffic pattern. *2.* When used concerning wind conditions, the word means a wind not parallel to the runway or the path of an aircraft. *See* crosswind component. [ARP4107]

crosswind component The wind component measured in knots at 90° to the longitudinal axis of the runway. [ARP4107]

cross wind gear A landing gear configuration that permits alignment of the wheels to compensate for crab angle of the air vehicle (due to crosswinds) at touchdown and for rollout. A fixed angle may be set into the gear or it may be allowed the freedom to caster and align with the direction of aircraft travel. [AIR1489]

cross-wire weld A resistance weld made by passing a controlled electric current through the junction of a pair of crossed wires or bars; used extensively to make mesh or screening.

crosswise direction Refers to the cutting of specimens and to the application of load. [AIR4844]

crown glass An optical glass of alkali-lime-silica composition with index of refraction usually 1.5 to 1.6.

crown sheet In a firebox boiler, the plate forming the top of the furnace.

CRS *See* desired course.

CRT *See* cathode ray tube.

crucible A pot or vessel made of a high-melting-point material, such as a ceramic or refractory metal, used for melting metals and other materials.

crude oil Unrefined petroleum.

cruise control A control mode whereby the aircraft is controlled to obtain the greatest practical efficiency for a given flight or mission. [ARP4386]

cruise missile A guided missile, the major portion of whose flight path to its target is conducted at approximately constant velocity. A cruise missile depends on the dynamic reaction of air for lift, and on propulsion forces to balance drag. [ARP4386]

cruise phase *See* mishap, maneuver.

cruising altitude An altitude maintained during the cruise portion of flight, as shown by a constant altimeter indication in relation to a fixed and defined datum. [ARP4386]

crush *1.* A casting defect caused by displacement of sand as the mold is closed. *2.* Buckling or breaking of a section of a casting mold caused by incorrect register as the mold is closed.

crush splicing The joining of segments of core by overlapping each segment two to four cells and then driving them together. [AIR4844]

cryochemistry The study of chemical phenomena in very low-temperature environments.

cryogenic Any process carried out at very low temperature, usually considered to be -60°F (-50°C) or lower.

cryogenic cooling Use of cryogenic fluids to reach temperatures near absolute zero.

cryogenic fluid A liquid that boils below -123 K (-238°F; -150°C) at one atmosphere absolute pressure.

cryogenic liquid Liquefied gas at very low temperature, such as liquid oxygen, nitrogen, or argon. [ARP4386]

cryogenic rocket propellants Rocket fuels, oxidizers, or propulsion fluids that are liquid only at very low temperatures.

cryogenics *1.* The subject of physical phenomena in the temperature range below about -150°C (-238°F). [ARP4386] *2.* The study of the methods of producing very low temperatures. *3.* The study of the behavior of materials and processes at cryogenic temperatures.

cryogenic wind tunnel Wind tunnel employing a cryogenic environment and utilizing independent control over Mach number, Reynolds number, aeroelastic effects, and model-tunnel interactions.

cryometer A thermometer for measuring very low temperatures.

cryoscope A device for determining the freezing point of a liquid.

cryostat An apparatus for establishing the very low-temperature environment needed for carrying out a cryogenic operation.

cryotron Device based upon the principle that superconductivity established at temperatures near absolute zero is destroyed by the application of a magnetic field.

cryptography The science of preparing messages in a form that cannot be read by those not privy to the secrets of the form.

crystal dislocations Types of lattice imperfections whose existence in metals is postulated in order to account for the phenomenon of crystal growth and of slip, particularly for the low value of shear stress required to initiate slip.

crystal lattices Three-dimensional, recurring patterns in which the atoms of crystals are arranged.

crystalline fracture A type of fracture surface appearance characterized by numerous brightly reflecting facets resulting from cleavage fracture of a polycrystalline material.

crystalline plastic A polymeric material having an internal structure in which the atoms are arranged in an orderly, three-dimensional configuration. [AIR4844]

crystallinity In polymers, a microstructure in which the linear molecular chains are arranged in an orderly fashion. [AIR4844]

crystal oscillator A device for generating an a-c signal whose frequency is determined by the properties of a piezoelectric crystal.

crystal spectrometer An instrument that uses diffraction from a crystal to determine the component wavelengths in a beam of x-rays or gamma rays.

C-scan *1.* The search unit that is moved over the surface of a test piece in a search pattern. *2.* A nondestructive inspection technique that uses ultrasonics for finding voids, delaminations, defects in fiber distribution, etc. [AIR4844]

CSMA/CD *See* carrier sense multiple access with collision detect.

C-stage In the reaction of certain thermosetting resins, the final stage in which the material is practically insoluble and infusible. [AIR4844]

C-star A longitudinal handling qualities criterion based on time response envelopes of a linear combination of normal acceleration, pitch rate, and pitch acceleration per unit stick force. [ARP4386]

CTI *See* comparative tracking index.

cull Material remaining in a transfer chamber after the mold has been filled. [AIR4844]

cultural stereotypes Information, interpretations, or expectations that are characteristic of a particular culture, discipline, or user community. [ARP4155]

cumulative/chronic fatigue *See* fatigue.

cumulative dose The total amount of penetrating radiation absorbed by the whole body, or by a specific region of the body, during repeated exposures.

cumuliform clouds Clouds that are like cumulus clouds, the principal characteristic of which is vertical development in the form of rising mounds, domes, or towers. [AIR4367]

cumulus A principal cloud type in the form of individual detailed elements that are dense and possess sharp nonfibrous outlines; develops vertically with tops often resembling a cauliflower. [AIR4367]

cup-and-cone fracture *See* cup fracture.

cup fracture A mixed mode fracture in ductile metals, usually observed in round tensile specimens, in which part of the fracture occurs under plane-strain conditions and the remainder under plane-stress conditions, such that in a round tensile bar one of the mating fracture surfaces looks like a miniature cup and the other like a truncated cone. Also known as cup-and-cone fracture.

cupping *1.* The first step in deep drawing. *2.* The fracture of severely worked rod or wire in which one of the fracture surfaces is roughly conical and the other cup-shaped.

cure To change the physical properties of a material by chemical reaction, through the action of heat and catalysts alone or in combination with pressure. [ARP1931]

cure cycle The time/temperature/pressure cycle used to cure a thermosetting resin system or prepreg. [AIR4844]

cured angle The angle at which the cords are set in a tire after cure. In practice, this angle may be measured from centerline of axle or from the plane of symmetry. [AIR1489]

cure-date The date a compounded, uncured elastomer is vulcanized to produce an elastomeric product. [AS1933]

cured end count Cords per inch in the tire ply fabric—after cure. [AIR1489]

cure monitoring, electrical Use of electrical techniques to detect changes in the electrical properties and/or mobility of the resin molecules during cure. [AIR4844]

cure rate A measure of the rate of polymerization based on the increasing hardness of the sealant with time. Testing is performed at standard conditions. Cure time varies widely with the type of sealant, and with the temperature and relative humidity. [AS7200/1]

cure stress In composite structures, a residual internal stress produced during the curing cycle. [AIR4844]

cure time A measure of the time required for the polymerization to advance to a given hardness at standard conditions.

Curie Abbreviated Ci. The standard unit of measure for radioactivity of a substance. Defined as the quantity of a radioactive nuclide that is disintegrating at the rate of 3.7×10^{-10} disintegrations per second.

Curie temperature For a ferromagnetic material, the temperature above which the material becomes substantially nonmagnetic.

curing *1.* Allowing a substance, such as a polymeric adhesive or poured concrete, to rest under controlled conditions; may include clamping, heating, or providing residual moisture, until the material undergoes a slow chemical reaction to reach final bond strength or hardness. *2.* In thermoplastics molding, stopping all movement for an interval prior to releasing mold pressure, so that the molded part has sufficient time to stabilize.

curing agent A catalytic or reactive agent that, when added to a resin, causes polymerization. [AIR4844]

curing temperature Temperature at which a cast, molded, or extruded product, a resin-impregnated reinforcement, an adhesive, etc. is subjected to curing. [AIR4844]

curing time The length of time that a part must be subjected to heat or pressure in order to fully cure the resin. [AIR4844]

curing type sealants Sealants that polymerize to form a cured, nonreversible, polymeric elastomer either by mixing two parts together or by exposure to moisture from the air. [AIR4069]

curl (vectors) A vector operation on a vector field that represents the rotation of the field, related to the circulation of the field at each point.

current *1.* Rate of transfer of electricity expressed in amperes. [ARP1931] *2.* The rate of flow of an electrical charge in an electric circuit; analogous to the rate of flow of water in a pipe.

current amplification For a given amplifier, the ratio of current delivered to the output circuit to the corresponding current supplied to the input circuit.

current carrying capacity The maximum current that a wire or cable with a given circular mil area is capable of carrying without exceeding its temperature limit. [ARP1931]

current limitation The ability of a protective device to reduce the short-circuit peak current to a value less than that which would be available if no protective device were in the circuit. [ARP1199A]

current limiting control A motor control for restricting the maximum motor torque or force by limiting the current supplied to the motor. [ARP4386]

current loop (20 mA) A serial transmission standard widely used for VDUs and teletypes. 0 and 1 are represented by the absence or presence, respectively, of a current (20 mA).

current meter *1.* Any of a wide variety of devices for measuring a-c or d-c electric current, including moving-coil, moving iron, electronic, and electrodynamic instruments. *2. See* velocity-type flowmeter.

current rating *1.* For a protective device, the nominal direct current or alternating current, in amperes, at rated frequency, that the device will carry continuously under defined conditions. [ARP1199A] *2.* The maximum continuous electrical flow of current recommended for a given wire in a given configuration, expressed in amperes. [ARP1931] *3.* The maximum current that a device is designed to conduct for a specified time at a specified operating temperature. [ARP914A]

current-responsive element (fusible element) The part of a fuse or limiter that carries current and melts when the current exceeds a predetermined value. [ARP1199A]

current return path The part of a power circuit to an electric device that is between the device and basic structure. [ARP4404]

current-to-pressure transducer (I/P) A device that receives an analog electrical signal and converts it to a corresponding air pressure.

current transformer An instrument transformer designed to have its primary winding connected in series with a circuit carrying the current being measured or controlled.

curtain type damper *See* damper.

curvature of field A defect in an optical lens or system that causes the focused image of a plane field to lie along a curved surface, rather than a flat plane.

curve data Data in tabular, spline, or polynomial form. [AIR4548]

curve-fit The process of determining the coefficients in a curve by mathematically fitting a given set of data to that curve class; for example, a linear curve-fit, or an nth-order polynomial curve-fit.

customer facility The test facility that is to be correlated against the reference facility. [ARP741]

custom LSI A large-scale integrated circuit that is designed for a specific purpose and thus has a dedicated function.

cutoff *1.* An act or instance of shutting something off. Specifically, in rocketry, an act or instance of shutting off the propellant flow in a rocket or stopping the combustion of the propellant. *2.* The parting line on a compression-molded plastics part. Also known as a flash groove or pinch-off. *3.* The point in the stroke of an engine at which admission of the working fluid to the cylinder is shut off.

cutoff tool A lathe tool with a narrow cutting edge used to sever a finished piece from remaining bar stock. Also known as a parting tool.

cutoff valve A quick-acting valve used to stop the flow of working fluid into an engine cylinder.

cutoff velocity The velocity attained by a missile at the point of cutoff. [ARP4386]

cutoff voltage The voltage at which a discharge or charge is terminated. [ARP4386]

cutoff wheel A thin abrasive wheel used to cut stock or to make slots in a part.

cutout, connector A hole or group of holes cut in a panel, case, or chassis for the purpose of mounting a connector. [ARP914A]

cut-out pressure The pressure at which the sequence of reduced flow of a component or system begins. [ARP4386]

cutout speed *See* speed, starter cutoff.

cutout switch A speed-sensing device, located on either the starter or the engine side of the engaging mechanism, used to terminate starter operation at a predetermined cutout speed. [ARP906A]

cutter *1.* A cutting tool, especially a rotary, toothed cutting wheel. *2. See* cutting tool.

cutter bar A supporting member for the cutting tool in a lathe or other machine tool.

cut-through resistance The ability of a material to withstand mechanical pressure, usually a sharp edge, without penetration of the impinging item through the material. [ARP1931]

cutting angle The angle between the face of a cutting tool and the uncut stock surface.

cutting edge *1.* In a diamond or ceramic tool, the point or edge of the insert material that actually cuts the work. *2.* Generally, the sharpened edge of any cutting tool that contacts the work during machining.

cutting fluid In a metal-cutting operation, any liquid that is introduced into the area where the tool contacts the work, especially a liquid used to provide lubrication at the cutting edge, to carry away the heat generated during machining, and to flush out chips or other machining debris. Some cutting fluids contain chemical compounds that react with the tool and material being cut to enhance cutting action.

cutting speed The relative velocity between cutting tool and workpiece along the main direction of cutting. Also known as peripheral speed.

cutting tool A sharp-edged single-point or toothed tool that comes in contact with the workpiece and removes stock in a machining operation. Also known as a cutter.

cutting torch A device for producing a controlled flame which has an additional supply line for introducing a jet of oxygen into the flame. Cuts metal and other materials by first heating a small area, then rapidly oxidizing and melting the material along a thin line when the jet of oxygen is turned on. Usually a special plasma torch is needed for stainless steel because of its oxidation resistance.

CV *See* coefficient of variation.

CVR *See* cockpit voice recorder.

cyanide emission *See* CN emission.

cyaniding A surface-hardening process similar to carbonitriding. Produces a carbon- and nitrogen-rich surface layer on steel by immersing parts in a bath of molten cyanide salts. Can also be done in the gas phase.

cyanosis A bluish-gray tinge in the color of mucous membranes and skin (usually first noticeable in the lips, ear lobes, and nail beds) associated with blood-oxygen deficiency; usually caused by the presence of excessive amounts of reduced hemoglobin in the capillaries. [ARP171]

cyanotic Showing signs of cyanosis. [ARP171]

cybernetics *1.* The study of methods of control and communication that are common to living organisms and machines. *2.* The branch of learning that brings together theories and studies on communication and control in living organisms and machines.

cycle *1.* One discharge and one charge of a battery. [AIR1898] *2.* The complete sequence, including reversal of the flow, of an alternating current. [ARP1931] *3.* The complete sequence of values of a periodic quantity that occur during a period. *4.* An interval of space or time in which one set of events or phenomena is completed. *5.* Any set of operations that is repeated regularly in the same sequence. The operations may be subject to variations in each repetition. *6.* In any repetitive variable process, variation of a given variable through one complete range of values. *7.* To run a machine through a complete set of operating steps. *8.* The fundamental time interval for operations inside a computer. *9.* Describes a condition in a sequential circuit in which the circuit passes from an initial, unstable state through more unstable states before reaching a stable state.

cycle, aircraft operating A completed takeoff and landing sequence. NOTE: Touch-and-go landings are counted as aircraft operating cycles. [ARD50010]

cycle, closed A cycle in which the working substance is returned regularly to a particular state or condition at each point in the cycle during steady operation. [ARP147C]

cycle, engine operating A completed engine thermal cycle, including the application of takeoff power. [ARD50010]

cycle, gear extension In a retractable landing gear system, the transition cycle from the up and locked (stowed) position to the down and locked (extended) position. [AIR1489]

cycle, gear retraction In a retractable landing gear system, the transition cycle from the gear down and locked (extended) position to the up and locked (stowed) position. [AIR1489]

cycle-index The number of times a cycle has been executed; or the difference, or the negative of the difference, between the number or repetitions that have been executed and the number of repetitions desired.

cycle mode A computer simulation of an engine cycle normally consisting of a specific set of component and process representations. [AIR4548]

cycle redundancy check (CRC) An error-detection scheme, usually hardware implemented, in which a check character is generated by taking the remainder after dividing all the serialized bits in a block of data by a predetermined binary number. This remainder is then appended to the transmitted date and recalculated and compared at the receiving point to verify data accuracy.

cycle stealing Describes the process by which data is transferred over the data bus during a direct memory access while little disruption occurs to the normal operation of the microprocessor.

cycle, supersonic A completed supersonic flight sequence comprising acceleration through Mach 1 and deceleration to subsonic flight. [AIR4896]

cycle time The time required by a computer to read from or write into the system memory. If system memory is core, the read cycle time includes a write-after-read (restore) subcycle.

cyclic *1.* Helicopter control mechanism for periodically varying the blade angle of each rotor, producing a tilt in the tip-path plane, and effecting motion in a desired direction. [ARP4107] *2.* A condition of either steady-state or transient oscillation of a signal about the nominal value.

cyclic adenosine monophosphate *See* cyclic AMP.

cyclic AMP (cyclic adenosine monophosphate) A nucleotide that is implicated as an intracellular messenger in a wide variety of cellular processes. Prototypically, cyclic AMP acts, through a series of reactions, as a molecular transducer of nonsteroid signals from outside the cell to relevant cellular enzymes.

cyclic code A form of gray code, used for expressing numbers. In cyclic code, when coded values are arranged in the numeric order of real values, each digit of the coded value assumes its entire range of values alternately in ascending and descending order. *See* gray code. *See also* shaft encoder.

cyclic compounds In organic chemistry, compounds containing a ring of atoms.

cyclic redundancy check (CRC) An error-checking technique in which a checking number is generated by taking the remainder after dividing all the bits in a block (in serial form) by a predetermined binary number. Can easily be achieved by shift operations.

cyclic redundancy check character (CRC) A character used in a modified cyclic code for error detection and correction.

cyclic shift A shift in which the data moved out of one end of the storing register are re-entered into the other end, as in a closed loop.

cycling Periodic repeated variation in a controlled variable or process action. *See also* cycle.

cyclograph A device for electromagnetically sorting or testing metal parts by means of the pattern produced on a cathode-ray tube when a sample part is placed in an electromagnetic sensing coil. When using a cyclograph, the CRT pattern is different in shape for different values of carbon content, case depth, core hardness, or other metallurgical properties.

cyclones (equipment) *See* centrifuges.

cyclotron A device that utilizes an alternating electric field between electrodes positioned in a constant magnetic field to accelerate ions or charged subatomic particles to high energies.

cyclotron frequency Frequency at which a charged particle orbits in a uniform magnetic field. This frequency depends on the charge-to-mass ratio of the particle times the magnetic field. While the frequency is independent of the particle energy, Lamor orbit increases with energy.

cyclotron radiation The electromagnetic radiation emitted by charged particles as they orbit in a magnetic field. Arises from the centripetal acceleration of the particles as they move in circular orbits.

cyclotron resonance Energy transfer to charged particles in a magnetic field from an alternating-current electric field whose frequency is equal to the cyclotron frequency.

cyclotron resonance device Microwave amplifier based on the interaction between electromagnetic waves and transverse electron streams moving along helical trajectories.

cylinder *1.* A port selectively carrying operating pressure, to be numbered when more than one. [ARP4386] 2. Falex ring and mandrel assembly. [AIR1794] 3. A domed, closed storage tank for hot water. Also known as a storage calorifier. 4. A strong, thick-walled container for storing and transporting compressed gases. 5. A round, straight-walled cavity, closed at one or both ends, in which a piston rides. Used to convert the potential energy in pressurized gas to linear mechanical motion and power; or for applying mechanical power to compress a gas.

cylinder, actuating Sometimes called a jack, ram, or strut. A linear motion device in which the thrust or force is proportional to the effective cross-sectional area and the pressure differential. [ARP4386]

cylinder, balanced An actuating cylinder in which the effective thrust-producing area is equal in both directions. [ARP4386]

cylinder block A massive piece of metal, usually made by casting, that contains the piston chambers of a multicylinder engine or compressor. Also known as block or engine block.

cylinder bore The inside diameter of a piston chamber.

cylinder, disk All like-numbered tracks on a disk pack; a portion of the disk that can be recorded or reproduced without moving the heads.

cylinder, double acting A cylinder with provisions for applying fluid pressure at each end, and thus capable of exerting a force in either direction. [ARP4386]

cylinder, fixed end A cylinder that is held in a rigid position. [ARP4386]

cylinder, flight control An actuating cylinder designed for in-flight actuation of aerodynamic surfaces. [ARP4386]

cylinder, gas In aviation, common name for a container of pressurized gas; may be portable or fixed to the aircraft, low- or high-pressure. [ARP171]

cylinder head The cap used to close the end of a piston chamber in a reciprocating engine, pump, or compressor; usually has a specially shaped recess and usually provides valve openings, spark-plug taps, and other penetrations necessary for machine operation.

cylinder, landing gear A tubular structural member that houses the piston, orifice, oil, gas, and other elements of an air-oil shock absorber, or the mechanical elements of a mechanical system. Carries lugs and/or provisions for various attachments on the exterior. [AIR1489]

cylinder liner A separate cylindrical sleeve that is inserted into a piston chamber to provide a cylinder wall with properties different from those of the cylinder block. Normally used to furnish a better-wearing material for piston rings than the block, e.g., a cast iron liner in an aluminum block.

cylinder, rotating end A cylinder mounted to permit limited rotary movement about a fixed point. [ARP4386]

cylinder, single acting A cylinder in which fluid pressure is introduced in one end so that fluid force is exerted in one direction only. Gravity, spring forces, or other means are used to accomplish the return stroke. [ARP4386]

cylinder, swivel end A cylinder in which one or both ends is provided with a joint that not only allows oscillation of the cylinder, but that also incorporates stationary fluid connections. [ARP4386]

cylinder, transfer A device for transmitting fluid pressure from one circuit to another without intermixture of fluid between the circuits. [ARP4386]

cylindrical cam A mechanism consisting of a cylinder that rotates on its longitudinal axis and causes linear motion parallel to the axis in a cam follower that rolls in a groove cut in the cylindrical surface.

cylindrical lens A lens that is cylindrical in cross section, and thus is curved in one direction, but not in the perpendicular direction; used to expand a laser beam into a plane of light.

cylindrical plasmas Magnetic self-attraction of parallel electric currents causing constriction of a conducting plasma through which a large current is flowing.

cylindrical waves Waves in which the wave fronts are coaxial cylinders.

D

4D *See* speed management-time control.

D-1 nozzle An underwing nozzle designed for military use. Has a characteristic 45° elbow as a part of its inlet body to allow for vertical hose drape when refueling on fuselage-mounted refueling adapters. [AIR4783]

D-1R nozzle Same as the D-1 nozzle except a nominal 55 psi hose end control valve is fitted to the unit. [AIR4783]

D-2 nozzle Same as the D-1 nozzle except the inlet body is straight. [AIR4783]

D-2R nozzle Same as the D-2 nozzle except a nominal 55 psi hose end control valve is fitted to the unit. [AIR4783]

DA *See* destination address.

D/A *See* converter and digital-to-analog converter.

DAC *1. See* digital-to-analog converter *2. See* distance amplitude correction.

daily inspection *See* inspection, aircraft engine.

daisy chain *1.* A serial interconnection of devices. Signals are passed from one device to another, generally in the order of high priority to low priority. *2.* A method of propagating signals along a bus, often used in applications in which devices are connected in series.

Dalton's law Empirical generalization stating that, for many so-called perfect gases, a mixture of gases will have a pressure equal to the sum of the partial pressures that each gas would have as a sole component with the same volume and temperature, provided there is no chemical interaction. *See also* gas laws.

dam Boundary support or ridge used to prevent excessive edge bleeding or resin runout of a laminate, and to prevent crowning of the bag during cure. [AIR4844]

damage, accidental Physical deterioration of an item caused by contact or impact with an object or influence, which may/may not be a part of the aircraft; or by improper manufacturing or maintenance practices. [ARD50010]

damage assessment Estimate of injury and loss to components, subsystems, or entire systems, as well as the cost of repairs or replacement to restore serviceability.

damage, environmental *See* environmental damage/deterioration.

damage factor A relative number assigned to indicate a defined amount or unit of engine component or piece part life usage; e.g., LOF counts, hot section factors. [ARP1587]

damage, ingestion Internal damage to an engine caused by outside objects, such as birds, stones, or other foreign objects. [AIR4896]

damage tolerance The ability of the airframe and engine structure to resist failure due to the presence of flaws, cracks, or other damage for a specific period of unrepaired usage. [ARD50010]

damage tolerance analysis *See* analysis, damage tolerance.

damage tolerance critical part A part whose structural failure could cause loss of aircraft. [AIR4896]

damp To suppress oscillations or disturbances. [AIR1489]

damped frequency *See* frequency, damped.

damped wave A wave in which the source amplitude diminishes with each succeeding cycle.

dampener A device for progressively reducing the amplitude of spring oscillations after abrupt application or removal of a load.

damper *1.* A device or system used to suppress oscillations or disturbances. May be hydraulic or mechanical. [AIR1489] *2.* A device for introducing a variable resistance for regulating the volumetric flow of gas or air: *Butterfly type damper*–a single-blade damper pivoted about its center. *Curtain type damper*–a damper, composed of flexible material, moving in a vertical plane as it is rolled. *Flat type damper*–a damper consisting of one or more blades, each pivoted about one edge. *Louver type damper*–a damper consisting of several blades, each pivoted about its center and linked together for simultaneous operation. *Slide type damper*–a damper consisting of a single blade that moves substantially normal to the flow.

damper, drag A damper for the purpose of damping drag forces. Sometimes built into a landing gear drag strut. [AIR1489]

damper, hop A damper designed into a landing gear system for the purpose of suppressing pitching oscillations of the bogie assembly in the bogie-type gear. [AIR1489]

damper loss The reduction in the static pressure of a gas flowing across a damper.

damper, shimmy A damper designed into a landing gear system for the purpose of suppressing shimmy or oscillation of the wheel system. Used especially on gear configurations with castering provisions, such as nose or tail wheel. [AIR1489]

damping *1.* Limiting the duration and/or decreasing the amplitude of vibrations or oscillations in the motion of a body or in an electrical system subjected to influences that are capable of causing vibration or oscillation. [ARP5089] *2.* Reducing or eliminating vibrations—especially reducing noise or reverberations by using sound-absorbing materials.

damping behavior In a tire, the characteristic that tends to damp out periodically applied forces. [AIR1489]

damping coefficient Hydraulic damping torque divided by the square of the angular velocity of the shock strut piston about its longitudinal centerline. [AIR1752]

damping constant The slope of the speed torque curve. The value of the damping constant is approximately equal to the stall torque divided by the no-load speed. [ARP667]

damping factor In any damped oscillation, the ratio of the amplitude of any given half-cycle to the amplitude of the succeeding half-cycle. *See also* damping.

damping, friction *See* coulomb damping.

damping, hydraulic A damping system in which a hydraulic fluid is utilized. External applied forces cause the oil to be moved through an orifice or restriction. Resisting force is proportional to the square of the velocity. [AIR1489]

damping magnet A permanent magnet used in conjunction with a moving conductor to produce an opposing torque when there is relative motion between the magnet and the conductor; a secondary function is to dissipate kinetic energy resulting from eddy currents that may be induced in the moving conductor.

damping ratio The ratio of the deviations of the indicator following an abrupt change in the measurand in two consecutive swings from the position of equilibrium, the greater deviation being divided by the lesser. The deviations are expressed in angular measure.

damping, structural Damping due to internal friction within the material itself. Also known as solid damping. Structural damping is independent of frequency and proportional to the maximum stress of the vibration cycle. [AIR1489]

damping tachometer generator Generator characterized by relatively low output-to-null ratios and therefore employed in low-gain loop servo-mechanism applications for damping purposes. [ARP667]

damping, viscous *See* viscous damping.

dark adaptation The process by which the iris and retina of the eye adjust to allow maximum vision in dim illumination, following exposure of the eye to a relatively brighter illumination.

dark ambient Any ambient light level that contributes less than 1% of the display luminance value measured by a luminance measurement device focused on the display surface. This ambient contribution can result from reflections off the display surface or any other means. [ARP4067]

dark current The current that flows in photosensitive detectors when there is no incident radiant flux (i.e., when there is total darkness).

d'Arsonval galvanometer A galvanometer made by suspending a light coil of wire on thin gold or copper ribbons in the field of a permanent magnet. When current is carried to the coil via the suspending ribbons, the coil rotates, and the amount of rotation is indicated by reflecting a beam of light from a small mirror carried on the coil onto a fixed linear scale. Also known as a light-beam galvanometer.

d'Arsonval movement The mechanism of a permanent-magnet moving-coil instrument, such as a d'Arsonval galvanometer.

DAS *See* data acquisition system.

dashpot *1.* A type of shock absorber that utilizes the resistance of oil being forced through an orifice or restriction to oppose an impact load. Differs from an air-oil shock absorber in that no air charge system is utilized. [AIR1489] *2.* A fluid-filled cylinder containing a loose-fitting piston; used to damp vibratory motion, or to change the effect of a sharp change in load from an instantaneous change in position to a more gradual change.

dashpot relay A timing device in which the delay is provided by the restrictive action of an orifice on a fluid. When the relay coil is energized, the armature piston moves against a reservoir of fluid, forcing it through a restriction, which slows down the action. Timing is achieved by variations in orifice size.

DAST program Stands for drones for aerodynamic and structural test program, a NASA program that uses the Firebee 2 target drone aircraft as a test bed for obtaining flight data on research wings. The drone is launched from the wing of a B52 and recovered by parachute. The purpose of the program is to study flight loads and load control.

data *1.* Information of any type. *2.* A common term used to indicate the basic elements that can be processed or produced by a computer.

data acquisition The function of obtaining data from sources external to a microprocess or a computer system, converting it to binary form, and processing it.

data acquisition system (DAS) A system used for acquiring data from sensors via amplifiers, multiplexers, and any necessary analog-to-digital converters.

data bank A comprehensive collection of data; for example, several automated files, a library,

or a set of loaded disks. Synonymous with database.

database *1.* Any body of information. *2.* A specific set of information available to a computer. *3. See* data bank.

data base management systems Software products that control data structures containing interrelated data. In a database management system, data is stored so as to optimize accessibility and control, minimize redundancy, and offer multiple views of the data to various applications programs.

data block *See* alphanumeric display.

data bus The hardware in the signal distribution network, including the harness assembly of fiber optic cables, couplers, connectors, etc., required to provide a path between all terminals. [AS1773]

data bus cable Two twisted, shielded, and jacketed conductors. [AS4117]

data bus coupler The electronic module in which the connections between the stubs and the data bus are made. [AS4117]

data capture (logging) The systematic collection of data for use in a particular data processing routine; for example, monitoring and recording temperature changes over a period of time.

data file In a computer, a portion of memory allocated to a specific set of organized data, including codes that identify the file name and sometimes the file type. Also referred to as a data set.

data graph An exhibit of relationship between sets of numeric data, either as points with coordinates or lines. Used to extract data values by interpolating or extrapolating; or to simply illustrate a trend or formula. [AS4159]

data input/output unit (DI/OU) A device that interfaces to a process for the sole purpose of acquiring or sending data.

data integration Refers to the process of taking data from multiple sources and merging it into a single data file.

data link *1.* Digital communication of data; may be initiated automatically or manually from the aircraft or ground. [ARP4102/13] *2.* Communication channel or circuit used to transmit data from a sensor to a computer, readout device, or storage device.

data link system Digital telecommunications capability that supports communication between airborne and ground-based computers and their operators. [ARP4791]

data logger A system or subsystem with a primary function of acquiring and storing data in a form (e.g., computer-language tape) that is suitable for later reduction and analysis.

data management A general term that collectively describes those functions of the control program that provide access to data sets, enforce data storage conventions, and regulate the use of input/output devices.

data processing Application of procedures—mechanical, electrical, computational, or other—whereby data are changed from one form to another. Also referred to as information processing.

data reduction The process of transforming masses of raw test data or experimentally obtained data, usually gathered by automatic recording equipment, into useful, condensed, or simplified intelligence.

data set *See* data file.

data signaling rate In communications, the data transmission capacity of a set of parallel channels. The data signaling rate is expressed in bits per second.

data smoothing The mathematical process of fitting a smooth curve to dispersed data points.

data structures *1.* Refers to the organization of computer memory used to represent information in a computer program or database. *2.* Refers to the storage of related data in computer memory by use of arrays, records, or data lists.

data transmission The sending of data from one part of a system to another part.

data type Any one of several different types of data, such as integer, real, double precision, complex, logical, and Hollerith. Each data type has a different mathematical significance and may have different internal representation.

datum *1.* A point, direction, or level used as a convenient reference for measuring angles, distances, heights, speeds, or similar attributes.

2. Any value that serves as a reference for measuring other values of the same quantity.

datum level *See* datum plane.

datum plane A permanently established reference level, usually average sea level, used for determining the value of a specific altitude, depth sounding, ground elevation, or water-surface elevation. Also known as chart datum, datum level, reference level, or reference plane.

daughter A nuclide formed as a result of nuclear fission or radioactive decay.

dawsonite A mineral consisting of aluminum sodium carbonate.

dB *See* decibel.

DBS (satellites) *See* direct broadcast satellites.

DC or d-c *See* direct current.

DCA Stands for dual, central active: A network topology characterized as having two redundant paths (D), centralized coupling (C), and embedded gain components (A). [AIR4288]

d controller *See* controller, derivative (D).

DDA Stands for dual, distributed active: A network topology characterized as having two redundant paths (D), distributed coupling (D), and embedded gain components (A). [AIR4288]

DDC *See* direct digital control.

DDP Dual, distributed passive: A network topology characterized as having two redundant paths (D), distributed coupling (D), and no gain components (P). [AIR4288]

dead Defines the network state in which signaling symbols are not present. [AIR4288]

dead band *1.* For some three-way valves, the region around zero input current (expressed in mA) in which there is no change of control pressure corresponding to input current changes. For these valves, null shift is the change in deadband at a specified supply pressure. [ARP490] *2.* Free motion at the wheel(s) under control (output) over which the controlling system (input) has no authority. *See also* backlash. May also apply to other mechanical systems. May also be defined as movement of the input control which produces no response in the output. [AIR1489] *3.* The complete range of values of the controlled temperature in which no corrective action will be taken by the controller. [ARP89C] *4. See* dead zone.

dead center Either of two positions of a crank at which the turning force between the crank and its connecting rod are zero. Dead center occurs when the centerline of the crank and the centerline of the connecting rod lie in the same plane.

dead-end shutoff A nonstandard term used to refer to control valve leakage. Refer to ANSI/FCI 70-2 for specifications of leakage classifications.

dead-end tube A tube with a closed end, for example, a tube in a porcupine boiler.

dead-front switchboard A switching panel constructed so that all of the live terminations are made on the rear of the panel.

deadhead pressure The pump outlet/system pressure at no flow. [AIR4783]

dead length, actuator Nominal length between attach points of an actuator when in the fully retracted condition, less the available stroke. [AIR1489]

deadman A control design in which continuous, deliberate pressure on the control is necessary for activation and continuous operation; and in which relief of that pressure will cause control deactivation. [AIR1375]

deadman control The manual capability of open-close control of flow through the hydrant system and hydrant servicer by an operator. [AIR4782]

deadman's brake A safety device that automatically stops a vehicle when the driver does not have his foot on the pedal. Also used on other operator-controlled mechanisms, such as cranes and lift trucks.

deadman's handle A hand grip or handle that an operator must squeeze or press on continuously to keep a machine running.

dead man timer (DMT) Device in which a circuit monitors operation of the processor cards and signals if a failure occurs.

deadman valve An on-off valve used to stop the flow of fuel through the system unless the operator is physically acting upon a control valve or switch in an active manner. [AIR4783]

dead reckoning (DR) Navigational method for determining the location of an aircraft based on time, heading, and estimated true airspeed

(with allowances made for winds and compass errors) flown since the last observed position of the aircraft. [ARP4107]

dead room *See* anechoic chamber.

dead space, anatomical Volume of gas in connecting passages to the lungs, including throat, mouth, and nasal airway, in which no respiratory exchange takes place. Average value is about 100 to 150 cc in the 70 kg male and 70 to 90 cc in the 55 kg female during quiet breathing. Enlarges with deep breathing. [ARP171]

dead space, mechanical Space in breathing apparatus, outside the body of the subject, in which the expired air is trapped and then re-inhaled. [ARP171]

dead space, physiological The anatomical dead space plus the volume of inspired gas-ventilating alveoli that have no pulmonary capillary perfusion. [ARP171]

dead-stick Without power, as in "dead-stick landing," or "into land dead-stick." [ARP4107]

dead time *1.* The interval of time between initiation of an input change or stimulus and the start of the resulting response. *2.* Any definite delay deliberately placed between two related actions in order to avoid overlap that might cause confusion; or to permit a particular different event, such as a control decision, switching event, or similar action, to take place. Also known as insensitive time.

dead-time correction A correction applied to an instrument reading to account for events or stimuli actually occurring during the dead time of the instrument.

deadweight gage A device used to generate accurate pressures for the purpose of calibrating pressure gages. In this device, freely balanced weights (dead weights) are loaded on a calibrated piston to give a static hydraulic pressure output.

dead zone *1.* Zone in the test part directly underneath the sound entry surface where discontinuities cannot be detected; caused by the finite length of the initial pulse, ringing time of the transducer element, and/or electronic characteristics of the instrument. [ARP5089] *2.* Also called dead band. A range of values around the setpoint; when the controlled variable is within this range, no control action takes place.

deaeration Removing a gas, such as air, oxygen or carbon dioxide, from a liquid or semisolid substance, such as boiler feedwater or food.

debond A deliberate separation of a bonded joint or interface, usually for repair or rework purposes. [AIR4844]

debug *1*. To detect and correct malfunctions in the computer itself. *See also* diagnostic routine. *2*. To submit a newly designed process, mechanism, or computer program to simulated or actual operating conditions for the purpose of detecting and eliminating flaws or inefficiencies. *See also* trouble-shoot.

debuggers System programs that enable computer programs to be debugged.

debugging The process of detecting, diagnosing, and then correcting program faults.

debugging aid routine A routine to aid programmers in the debugging of their routines. Some typical debugging aid routines are storage printout, tape print-out, and drum printout routines.

debulking *1*. Compacting of a thick laminate under moderate heat and pressure, i.e., noncuring conditions and/or vacuum, to remove most of the air, to ensure seating on the tool, and to prevent wrinkles. *2*. *See* compaction. [AIR4844]

deburr To remove burrs, fins, sharp edges, and the like from corners and edges of parts, or from around holes, by any of several methods; often involves the use of abrasives.

Debye length A theoretical length that describes the maximum separation at which a given electron will be influenced by the electric field of a given positive ion.

decade A group or assembly of ten units; for example, a counter that counts to ten in one column, or a resistor box that inserts resistance quantities in multiples of powers of ten.

decade scaler A scaling device that produces one output pulse for each ten input pulses.

decalescence Darkening of a metal surface upon undergoing a phase transformation on heating. This phenomenon is caused by isothermal absorption of the latent heat of transformation.

decant *1*. To pour off without disturbing the sediment. *2*. To drain dregs of fuel from the lowest point of integral or other tank.

decanting Boiling or pouring off liquid near the top of a vessel that contains two immiscible liquids, or a liquid-solid mixture that has separated by sedimentation, without disturbing the heavier liquid or settled solid.

decarburization The loss of carbon from the surface of a ferrous alloy as a result of heating in a medium that reacts with the carbon at the surface. [ARP700]

decarburizing Removing carbon from the surface layer of a steel or other ferrous alloy by heating it in an atmosphere that reacts selectively with carbon.

decay The spontaneous transformation of a nuclide into one or more other nuclides either by emitting one or more subatomic particles or gamma rays from the nucleus or by nuclear fission. Radioactive decay of a specific nuclide is characterized by a quantity known as half life—the time it takes for one-half of the original mass to spontaneously transform.

decay time In a circuit, the time in which a voltage or current pulse will decrease to one-tenth of its maximum value. Decay time is proportional to the time constant of the circuit.

Decca navigation A long-range, ambiguous, two-dimensional navigation system using continuous wave transmission to provide hyperbolic lines of position; uses the radio-frequency phase comparison techniques from four transmitters.

decelerate To reduce in speed.

decelerating electrode An intermediate electrode in an electron tube which is maintained at a potential that induces decelerating forces on a beam of electrons.

deceleration The act or process of moving, or of causing to move, with decreasing speed. Sometimes called negative acceleration. [AIR1489] *See also* impact deceleration.

decelerometer An instrument for measuring the rate at which speed decreases.

decentralized Refers to the distribution of functions among several authorities; for example, decentralized maintenance distributes maintenance functions among areas of responsibilities or areas of the physical plant.

decibel (dB) The unit used to express differences of power level; equal to ten times the common logarithm of the power ratio. The decibel is used to express power loss in cables; a 3dB loss approximates a 50% decrease; a 2dB loss approximates a 27% decrease. [ARP1931]

decibel meter An instrument calibrated in logarithmic steps and used for measuring power levels, in decibel units, of audio or communication circuits.

decimal balance A type of balance having one arm ten times as long as the other, so that heavy objects can be balanced with light weights.

decimal coded digit A digit or character defined by a set of decimal digits; for example, a letter or special character specified by a pair of decimal digits.

decimal digit In decimal notation, one of the characters 0 through 9.

decimal notation A fixed radix notation, in which the radix is ten; for example, in decimal notation, the numeral 576.2 represents the number 5×10 squared plus 7×10 to the first power, plus 6×10 to the zero power, plus 2×10 to the minus 1 power.

decimal numbering system A system of reckoning by ten or powers of ten using the digits 0–9 to express numerical quantities.

decimal point *See* radix point.

decimal-to-binary conversion The process of converting a number written to the base ten, or decimal, into the equivalent number written to the base two, or binary.

decision The selection of a response designed to achieve a desired goal after having made a judgment as to the significance and priority of available information. [ARP4107]

decision delay Failure to select a response in a timely manner due to an anomaly of attention or motivation. [ARP4107]

decision height (DH) The height at which a decision must be made during an ILS or PAR instrument approach—either to continue the approach or to execute a missed approach. [ARP4107]

decision instruction An instruction that effects the selection of a branch in a program, for example, a conditional jump instruction.

decision, poor Selection of an inappropriate response (assuming adequate information and time to decide) due to an anomaly of attention or motivation. [ARP4107]

decision table A table of all contingencies that are to be considered in the description of a problem, together with the actions to be taken. Decision tables are sometimes used in place of flow charts for problem description and documentation.

deck A moveable platform used to elevate the refueling operator to be able to easily reach taller aircraft. [AIR4783]

deck hoses Relatively short output hoses used on a moveable deck of a refueling vehicle; for refueling aircraft with refueling adapters requiring such access. [AIR4783]

deck run, catapult The distance from the end of the catapult power stroke to the end of the deck. [AIR1489]

deck scale A low-profile weighing device used for moderate to heavy loads—up to 20,000 lb. Because the load platform of a deck scale is 2–10 in. above floor level, loads must be lifted onto the scale; or ramps must be provided to enable wheeled vehicles to move onto the platform and off again. The frame of a deck scale rests directly on the existing floor, rather than in a pit, and most models can be moved to different locations as needed.

declaration As used in many programming languages, a statement that is not to be executed, but usually is used for descriptive purposes.

declination Angular distance north or south of the celestial equator; the arc of an hour circle between the celestial equator and a point on the celestial sphere, measured northward or southward from the celestial equator through 90°, and labeled N or S to indicate the direction of measurement.

declinometer An instrument similar to a surveyor's compass, used for determining the variation of magnetic directions from true directions. In this instrument, the horizontal circle is constructed so that the line of sight can be aligned with the magnetic needle or with any other desired setting.

decode To determine the meaning of instructions from the status of bits that describe the instruction, command, or operation to be performed.

decoder *1.* Device for translating electrical signals into predetermined functions. *2.* In computer operations, networks or devices in which one or two or more possible outputs results from a prescribed combination of inputs.

decollate The separation of multi-part computer forms.

decommissioning Disposal or deactivation of equipment or sites whose usefulness has diminished to a point where it is no longer required for its original purpose.

decommutation A reversal of the commutation process; separation of information in a commutated data stream into as many independent information channels as were originally commutated.

decommutator Equipment for separation, demodulation, or demultiplexing commutated signals.

decomposition A chemical and physical material breakdown due to an excess exposure to heat and oxidation, or due to the effects of bacterial contamination of the adhesive. [AIR4844]

decomposition/disassociation The chemical separation of a substance into two or more substances, which may differ from each other and from the original substances. [ARP4386]

decompression *1.* Commonly refers to the loss of pressurization of an aircraft cabin or cockpit. When the decompression occurs in 1 s or less, it is termed an explosive decompression. [ARP171] *2.* Any method for relieving pressure.

decontamination Removing or neutralizing an unwanted chemical, biological, or radiological substance.

decoration aging treatment A relatively low-temperature, short-duration aging treatment used to determine the degree of recrystallization in cold- or hot-worked metastable beta tita-

nium alloys that have been solution heat treated. [AMS4897]

decoupling The technique of reducing process interaction through coordination of control loops.

decoupling control A technique in which interacting control loops are automatically compensated when any one control loop takes a control action.

decrement *1.* The quantity by which a variable is decreased. *2.* In some binary computers, a specific part of an instruction word; thus, a set of digits.

decremeter An instrument for measuring the damping of a train of waves by determining its logarithmic decrement.

decryption Translating computer data from an unreadable format to a readable format.

dedicated In data processing, refers to a device that performs only one function.

dedicated combiner Dedicated to a head-up display role. [ARP4102/8]

deductive Describes those analytical approaches involving the reasoning from a defined unwanted event or premise to the causative factors of that event or premise by means of a logical methodology. [ARP926A]

deep discharge Removal of the rated energy capacity of a battery. [AIR1898]

deep drawing A press operation for forming cup-shaped or deeply recessed parts from sheet metal by forcing the metal to undergo plastic deformation between dies without substantial thinning.

deep-draw mold A mold having a core that is long in relation to the wall thickness. [AIR4844]

deep stall A stabilized high angle of attack assumed by an aircraft after it reaches the stall angle. [ARP4107]

deep well injection (wastes) Storage of liquid wastes, particularly chlorohydrocarbons, by injection into subsurface geologic strata for long-term isolation from the environment.

default The value of an argument, operand, or field assumed by a program if a specific assignment is not supplied by the user.

default directory In MS-DOS, the directory in which the computer looks for files if no directory is specified.

default drive In MS-DOS, the disk drive that the computer will use to search for files if no disk drive is specified.

defect *1.* A discontinuity that interferes with the usefulness of a part. *2.* A fault in any material or part that is detrimental to its serviceability. [ARP4784] *3.* Any nonconformance of a unit or product that has specified requirements. Defects are normally grouped into one or more of the following classes, but may be grouped into other classes or subclasses with these classes: (a) critical defects; (b) major defects; and (c) minor defects. *Critical defect*–a defect that constitutes a hazardous or unsafe condition; or, as determined by experience and judgment, could conceivably become so, relative to its deleterious effect on the prime intended function, or mission capability of the aircraft or its operating personnel. *Major defect*–a defect, other than critical, that could result in failure, or materially reduce the usability of the unit or part for its intended purpose. *Minor defect*–a defect that does not materially reduce the usability of the unit or part for its intended purpose; or is a departure from standard, but has no significant bearing on the effective use or operation of the unit or part. [ARD50010]

defect density Average number of latent defects per item. [AIR4896]

defective A unit of product that contains one or more defects. [AS7481]

defect resolution *See* resolution, defect.

Defense Meteorological Satellite Program (DMSP) *See* DMSP satellites.

defibrillate To divide longitudinally into fibers of smaller diameter. [AIR4844]

definition *1.* The resolution and sharpness of an image, or the extent to which an image is brought into sharp relief. *2.* The degree with which a communication system reproduces sound images or messages.

deflagration The chemical decomposition (burning) of a material in which the reaction front advances into the unreacted material at less than sonic velocity. [AIR913]

deflashing Removing fins or protrusions from the parting line of a die casting or molded plastics part.

deflecting electrode An intermediate electrode in an electron tube whose surrounding electric field induces constant or variable deflecting forces on an electron beam.

deflecting force In a direct-acting recording instrument, the force produced at the marking device for any position of the scale; produced by the positioning mechanism acting in response to the electrical quantity being measured.

deflecting yoke An assembly of one or more coils that induces a magnetic field to deflect an electron beam in a manner related to the oscillating frequency and magnitude of the current flowing through the coils.

deflection *1.* A displacement of a duct or joint due to operating conditions. [ARP699D] *2.* The radial compression or deflection of a tire under load; may be expressed by a finite value or as a ratio of the percentage of the actual deflection measurement to the total available (undeflected outside diameter less the wheel rim flange diameter, divided by two). [AIR1489] *3.* Movement of a pointer away from its zero or null position. *4.* Elastic movement of a structural member under load. *5.* Shape change or change in diameter of a tubular member without fracturing the material.

deflection factor The reciprocal of the instrument sensitivity.

deflection polarity In an oscilloscope, the relationship between the direction of electron-beam displacement and the polarity of the applied signal voltage.

deflection temperature under load The temperature at which a simple cantilever beam deflects a given amount under load. [AIR4844]

deflectometer An instrument for determining minute elastic movements that occur when a structure is loaded.

deflector A device for changing the direction of a stream of air or a mixture of pulverized fuel and air.

defocus To cause a beam of electrons, light, x-rays, or other type of radiation to depart from accurate focus at a specific point in space, ordinarily the surface of a workpiece or test object.

deformation under load The dimensional change of a material under load for a specified time following the instantaneous elastic deformation caused by the initial application of the load. [AIR4844]

deformed beam lead thickness On a bonded device, the mean thickness of the beam lead in the bonded area. [AS1346]

defrost To remove ice from a surface, usually by melting or sublimation.

defueling Off-loading fuel from an aircraft. [AIR4783]

deg (or °) *See* degree.

degas To remove dissolved, entrained, or adsorbed gas from a solid or liquid.

degasification *1.* Removal of gases from samples of steam taken for purity tests. *2.* Removal of CO_2 from water, as in the ion-exchange method of softening.

degasifier An element or compound added to molten metal to remove dissolved gases.

degassing A technique of accelerating outgassing, usually by the application of heat.

degenerate matter A state of matter found in white dwarf stars and other ultrahigh-density objects in which the electrons follow Fermi-Dirac statistics. In such matter, the density becomes so high that the pressure increases to the point where it becomes independent of the temperature and is a function of the density only, thereby departing from the classical laws of physics.

degenerate waveguide modes A set of waveguide modes having the same propagation constant for all frequencies of interest.

degeneration *1.* A gradual impairment in ability to perform. [ARD50010] *2.* Negative feedback.

degradation The condition or status indicating impaired or deteriorating condition, function, or physical state. [ARP1587]

degradation factor A factor that, when multiplied by the predicted mean time between failures, yields a reasonable estimate of the operational mean time between failures. [AIR4896]

degradation failure Gradual shift of an attribute or operating characteristic to a point where the device no longer can fulfill its intended purpose.

degrease To remove oil and grease from adherend surfaces. [AIR4844]

degreasing An industrial process for removing grease, oil, or other fatty substances from the surfaces of metal parts, usually by exposing the parts to condensing vapors of a polyhalogenated hydrocarbon solvent.

degree *1.* A unit of measure of a temperature scale. Abbreviated °. *2.* A unit of angular measure equal to 1/360 of a circle. Abbreviated deg or °. *3.* Relative intensity. *4.* One of a series of stages or steps.

degree of freedom A mode of motion—either angular or linear—with respect to a coordinate system, independent of any other mode. A body in motion has six possible degrees of freedom (three linear and three angular).

degree of polymerization Number of structural units, or mers, in the average polymer molecule. [AIR4844]

degree rise The increase in temperature caused by the flow of electrical current through a wire. [ARP1931]

dehumidification Reducing the moisture content of air, which increases its cooling power.

dehumidify To decrease the moisture content (humidity) of.

dehydration Removal of water as such from a substance, or after formation from a hydrogen and hydroxyl group in a compound, by heat or a dehydrating substance. [AIR4844]

deice Removal of ice accretion by thermal, mechanical, or chemical means.

deicing *1.* Using heat, chemicals, or mechanical rupture to remove ice deposits, especially those that form on motor vehicles and aircraft at low temperatures or high altitudes. *2.* The periodic shedding, either by mechanical or thermal means, of small ice buildups by destroying the bond between the ice and the protected surface. [AIR1168/9]

Deimos A satellite of Mars orbiting at a mean distance of 23,500 kilometers.

deionization time The time it takes for the grid in a gas tube to regain control of tube output after the anode current has been interrupted.

deionize *1.* To remove ions from. *2.* To restore gas that has become ionized to its former condition.

delamination The separation of layers in a laminate through failure of the adhesive bond. [ARP1931]

delay A pyrotechnic device that introduces a controlled time-delay between initiation and functioning of an explosive device. [AIR913]

delay distortion A form of distortion in a transmitted radio wave that occurs when the rate of change of phase shift with frequency is not constant over the transmission-frequency range.

delayed combustion A continuation of combustion beyond the furnace. *See also* secondary combustion.

delayed response *See* response, delayed.

delayed sweep An A-scan or B-scan presentation in which an initial part of the time scale is not displayed. [ARP5089]

delay-interval timer A timing device that is electrically reset to delay energization or deenergization of a circuit for an interval of time up to 10 min following a specific event, such as restoration of power after a power failure or turning a manual switch off.

delay line *1.* Material (liquid or solid) placed in front of the search unit to cause a time delay between the initial pulse and front surface signal. [ARP5089] *2.* A cable constructed so as to provide very low velocity of propagation with a specific electrical delay for transmitted signals. [ARP1931] *3.* In electronic computers, devices for producing a time delay of a signal. *4.* A transmission medium that delays a signal passing through it by a known amount of time; typically used in timing events.

delay-line memory A type of circulating memory having a delay circuit as the chief element in the path of circulation.

delay-line register An acoustic or electric delay line, one or more words long, combined with appropriate input, output, and circulation circuits.

delay modulation A method of data-encoding for serial data-transmission or recording. According to this method, a logic ONE (or ZERO) is represented by a signal transition at midbit time and a logic ZERO (or ONE) followed by a logic ZERO (or ONE) is represented by a transition at the end of the first ZERO (or ONE) bit.

delay-on-make timer A timing device in which the main contacts are held open for a preset period of time after the device receives an initiating signal, then the contacts are closed and current is allowed to flow in the main circuit. When the timer receives a stopping signal, the contacts open; and, after a short interval, the timer automatically resets so it can repeat the cycle.

delay (technical) Delay occurring when the malfunctioning of an item, the checking of the item, or necessary corrective action, causes the final departure to be delayed by more than a specified time after the programmed departure time. [AIR4896]

delay time That part of downtime during which no maintenance is being accomplished on the item because of either supply or administrative delay. [AIR4896]

delimiter A symbol or sequence of symbols used to mark the boundary of a field. [AIR4289]

deliquescence The absorption of atmospheric water vapor by a crystalline solid until the crystal eventually dissolves into a saturated solution. [AIR4844]

delta network A set of three circuit branches connected in series, end-to-end, to form a mesh having three nodes.

delta wing(s) A symmetrical, triangular wing having a low aspect ratio, tapered leading edge, and straight trailing edge. Also referred to as triangular wing(s). [ARP4107]

demand assignment multiple access A technique of assigning communication resources on an "as needed basis," such as in satellite communications.

demand meter Any of several types of instruments used to determine the amount of electricity used over a fixed period of time.

demand system An oxygen system using demand regulators. *See* regulator, demand. [ARP171]

demodulation The process of retrieving intelligence (data) from a modulated carrier wave; the reverse of modulation.

demodulator *1.* Electronic device that operates on an input of a modulated carrier to recover the modulating wave as an output. *2.* A device that recovers information from a carrier or subcarrier. A telemetry receiver contains a demodulator; an FM discriminator is a demodulator.

demography Statistical study of human populations, especially with reference to size, density, distribution, and vital data.

demonstrate To display, operate, or explain by reasoning or evidence.

demonstrated MTBF interval (THETA D) The probable range of true MTBF under test conditions; that is, an interval estimate of MTBF at a stated confidence level. [ARD50010]

demonstration and validation (valid) phase The period during which selected candidate solutions are refined through extensive study and analysis; hardware development, if appropriate; test; and evaluation. [AIR4896]

demultiplexer *1.* The device that enables the telemetry operator to observe individual measurements from within a multiplexer; has the opposite function to a "multiplexer." *2.* A device that separates two or more signals that have been multiplexed together for transmission through a single optical fiber. *3.* A reverse multiplexer that allows the transfer of data from one microprocessor port to a number of output devices, such as actuators.

denaturant A substance that renders alcohol unfit for use as a beverage. [AIR4844]

dendrite arm spacing The spacing between the secondary arms of a dendrite structure. [ARP1947]

denier A term that describes the weight of a yarn, which in turn determines the physical size of the yarn. [ARP1931]

denitrogenation The reduction of the nitrogen concentration in the body by respiring 100% oxygen over a period of time; an attempt to promote the diffusion of nitrogen from the blood to the lungs, thereby eliminating much

of the nitrogen dissolved in the body tissues. [ARP171]

densification process Consolidation of a loose or bulky material. [AIR4844]

densimeter See densitometer.

densitometer *1.* Instrument for measuring the density or specific gravity of liquids, gases, or solids. Also known as densimeter or gravitometer. *2.* Instruments for measuring the optical density (photographic transmission, photographic reflection, visual transmission, etc.) of a material, generally of a photographic image.

density *1.* The mass of a unit volume of a fluid; unless otherwise stated, in grams per milliliter at 77°F. [AIR1116] *2.* Weight per unit volume of a substance. [ARP1931] *3.* Closeness of texture or consistency. *4.* Degree of opacity, often referred to as optical density.

density bottle *See* specific gravity bottle.

density correction Any correction made to an instrument reading to compensate for the deviation of density from a fixed reference value; may be applied because the fluid being measured is not at standard temperature and pressure, because ambient temperature affects density of the fluid in a fluid-filled instrument, or because of other similar effects.

density, mass The mass of any substance per unit volume. [ARP147C]

density (rate/area) *See* flux density.

density transmitter An instrument used to determine liquid density by measuring the buoyant force on an air-filled float immersed in a flowing liquid stream.

dent A concave depression that does not rupture plies or debond the composite structure. [AIR4844]

DEP *See* design eye position.

Department of Defense Flight Information Publications (DoD FLIP) Publications used for flight planning, en route, and terminal operations. FLIPs are produced by the Defense Mapping Agency for worldwide use. En route charts and instrument approach procedure charts are incorporated into DoD FLIPs for use in the National Airspace System (NAS). [ARP4107]

departure Movement of an aircraft from the blocks for the purposes of intended flight. [AIR4896]

departure control A function of an approach control facility providing air traffic control service for departing IFR (and, under certain conditions, VFR) aircraft. [ARP4107]

dependability *1.* A measure of the degree to which an item is operable and capable of performing its required function at any (random) time during a specified mission profile, given item availability at the start of the mission. (Item state during a mission includes the combined effects of the mission-related system R&M parameters, but excludes non-mission time.) *See* availability. [ARD50010] *2.* The quality of service that a network is capable of providing. [AIR4288]

dependent variables Variables considered as a function of other variables, the latter being called independent.

dependent warning/caution A signal indicating the specific cause of activation of the master warning or caution signals, respectively. [AIR1161]

depolarizers Optical components that scramble the polarization of light passing through them, effectively turning a polarized beam into an unpolarized beam.

depolymerization Separation of a more complex molecule into two or more simpler molecules that are chemically similar to and have the same empirical composition as the original. [AIR4844]

deposit *1.* Any substance intentionally laid down on a surface by chemical, electrical, electrochemical, mechanical, vacuum, or vapor transfer methods. *2.* Solid or semisolid material accumulated by corrosion or sedimentation on the interior of a tube or pipe.

deposited metal In a weldment, filler metal added to the joint during welding.

deposit gage Any instrument used for assessing atmospheric quality by measuring the amount of particulate matter that settles out on a specific area during a defined period of time.

deposition The process of applying a material to a base by means of vacuum, electrical, chemical, screening, or vapor methods, often with the assistance of a temperature and pressure container. [AIR4844]

deposition rate *1.* The amount of filler metal deposited per unit time by a specific welding procedure, usually expressed in pounds per hour. *2.* The rate at which a coating material is deposited on a surface, usually expressed as weight per unit area per unit time, or as thickness per unit time.

deposition sequence The order in which increments of a weld deposit are laid down.

depot maintenance *See* maintenance levels.

depth filter *1.* A filter medium that retains contaminant, primarily within tortuous passages, at different levels within the filtration media. In a depth filter, other materials are frequently used in combination with wire to improve filtration. [AIR888] *2.* A filter that removes particles from a fluid system by the combined principles of direct interception and entrapment. [ARP4386]

depth gage An instrument or micrometer device capable of measuring distance below a reference surface to the nearest 0.001 in. Most often used to measure the depth of a blind hole, slot, or recess below the normal part surface surrounding it; or to measure the height of a shoulder or projection above the adjacent part surface.

depth of crimp The distance a crimp die indentor indents the conductor barrel or ferrule. [ARP914A]

depth of discharge The capacity removed from a battery during a discharge relative to the available capacity. Depth of discharge may be expressed as a percent. [ARP4386]

depth of engagement The radial contact distance between mating threads.

depth of fusion The distance that the molten zone extends from the original surface into the base metal during welding.

depth of thread For a screw thread, the radial distance from crest to root.

depth perception *See* space perception.

derandomizer The circuit that removes the effect of data-randomizing; used to recover data that has been randomized for tape storage. *See also* randomizer.

derate To operate at a lower-than-normal value.

derating *1*. Using an item in such a way that applied stresses are below rated values. Derating is an intentional reduction of the stress/strength ratio in the application of an item, usually for the purpose of achieving a "reliability margin" in design, which should reduce the occurrence of stress-related failures. [ARD50010] 2. The lowering of the rating of an item in one stress field to allow an increase in another stress field. [ARP4386]

derating factor *1*. The ratio of operating to design stress. 2. A factor used to determine the acceptable reduced current-carrying capacity of a wire when that wire is used in an environment or application other than that for which its original current-carrying capacity was determined. [ARP1931]

derivative *1*. Mathematically, the reciprocal of rate. 2. Control action that will cause the output signal to change according to the rate at which input signal variations occur during a certain time interval.

derivative action A type of control-system action in which a predetermined relation exists between the position of the final control element and the derivative of the controlled variable with respect to time.

derivative control Refers to a type of control in which change in the output is proportional to the rate of change of the input. Also called rate control.

derivative control action (rate control action) Control action in which the output is proportional to the rate of change of the input. *See* control action.

derivative control mode A controller mode in which controller output is directly proportional to the rate of change of controlled variable error.

derivative time The time interval by which rate action advances the effect of proportional action on the final control element.

derived requirements Additional requirements resulting from design or implementation decisions during the development process. [ARP4754]

descaling Removing adherent deposits from a metal surface, such as thick oxide from hot-rolled or forged steel, or inorganic compounds from the interior of boiler tubes. Descaling may be accomplished by chemical attack, mechanical action, electrolytic dissolution, or other means, alone or in combination.

descent intercept point The geographical position at which designated descent altitude is attained. [AIR4102/9]

descent phase *See* mishap, maneuver.

descent profile A curve relating altitude to elapsed time following a decompression. Can be utilized in determining the flow rate of oxygen that should be supplied by a chemical oxygen generator at each point in time following initiation of an emergency descent. [AS1304]

describing function For a nonlinear element in sinusoidal steady state, the frequency response obtained by taking only the fundamental component of the output signal. The describing function may depend on the frequency and on the amplitude of the input signal, or only on the amplitude of the input signal.

design A process, usually iterative, by which the details of a system are selected, analyzed, and documented in order to produce a system that meets a specified set of operation criteria.

design acceptance tests Tests performed to demonstrate conformance to specified requirements. [AIR4896]

design adequacy The probability that the system will successfully accomplish its mission, given that the system is operating within design specifications. [AIR4896]

design allowables Material-property-allowable strengths used for design purposes; usually refer to stress or strain. Design allowables are based on a sufficient number of tests to be statistically significant and to give values with specified levels of confidence. [AIR4844]

design approval test *See* reliability qualification test.

design cruising speed *See* V_c.

design development test Test conducted to establish or verify design concepts for items which have not been proved by previous use. [AS1426]

design diving speed *See* V_d.

design drawings The design drawing prepared by and available from the procuring activity. [ARP1493]

design error A mistake in the design process resulting from incorrect methods or incorrect application of methods or knowledge. [ARP4754]

design evaluation tests Tests performed to evaluate the design under environmental conditions and to verify compatibility of interfaces, adequacy of tools and test equipment, etc., for the established maintenance concept. [AIR4896]

design eye position (DEP) A point fixed in relation to the aircraft structure (neutral seat reference point) at which the midpoint of the pilot's eyes should be located at the normal position. The DEP is the principal dimensional reference point for the location of flight deck panels, controls, displays, and external vision. [ARP4067]

design for testability A design process or characteristic thereof such that deliberate effort is expended to assure that a product may be thoroughly tested with minimum effort, and that high confidence may be ascribed to test results. [AIR4896]

design landing ground line L Ground line related to aircraft centerline, at the aircraft landing attitude for the design landing configuration, with the shock struts and tires deflected to the corresponding landing loads. [ARP1538]

design load *1.* A specified load that a structural member or part should withstand without failing. Determined by multiplying some particular load by an appropriate factor, usually the limit load multiplied by a factor of safety. [ARP4107] *2.* The load for which a steam-generating unit is designed; considered the maximum load to be carried.

design maneuvering speed *See* V_a.

design mission scenario(s) Those portions of the total mission scenario(s) selected for use in designing a particular system. In the design mission scenario, segments of the total mission scenario are eliminated because they are contained within other segments, determined to be noncritical, determined to be redundant, or for other similar reasons. The design mission scenario may be described in the same variety of ways as the total mission scenario, i.e., by summary-of-mission narrative, mission narrative, ribbon-in-the-sky, altitude/timeline curves, or design scenario timeline; typically, it is described in all of these ways during the process. [ARP4107]

design pressure The maximum allowable working pressure permitted under the rules of the ASME Construction Code.

design process The process of creating a system or an item from a set of requirements. [ARP4754]

design steam temperature The temperature of steam for which a boiler is designed.

design stress The maximum permissible load per unit area that a given structure can withstand in service, including all allowances for such things as unexpected or impact loads, corrosion, dimensional variations during fabrication, and possible underestimation of service loading.

design support tests Tests performed to determine the need for parts; materials; and component evaluation or qualification to satisfy characteristic stability, interchangeability, failure rate, tolerances, design margins and other reliability design criteria. [AIR4896]

design thickness The sum of thickness required to support service loads. Design thickness is specified particularly when designing boilers, chemical process equipment, and metal structures that will be exposed to atmospheric environments, soils, or seawater.

design to cost A process whereby cost factors are determined and calculated for the life cycle of a product as an integral part of its design.

design ultimate load 1.5 times the limit load. [ARP1493]

design verification tests Tests performed to verify the functional adequacy of the design. [AIR4896]

design volts The voltage at which a lamp is designed for the tabulated ampere, candlepower, and rated laboratory life characteristics. [ARP881]

desired course (CRS) The intended horizontal direction of travel, expressed as an angular distance from a reference direction (usually true or magnetic north). [ARP1570]

desired track The imaginary line on the earth's surface connecting successive points, over which flight is desired; describes the great circle course between successive waypoints and is further defined by the intersection of a plane and the earth's surface when the plane passes through two successive waypoints and the center of the earth. *See also* ground track. [ARP1570]

desired track angle (DTK or DSRTK) The clockwise angle from true north to an imaginary line or path on the earth's surface connecting successive points over which flight is desired. [ARP1570]

desizing The process of eliminating sizing (generally starch), from gray goods before applying special finishes or bleaches. [AIR4844]

desorption The process of removing sorbed gas.

destination address (DA) The field of a frame that contains instructions showing which station is to receive the information. [AIR4288]

destructive testing Testing in which the preparation of the test specimen or the test itself may adversely affect the life expectancy of the unit under-test (UUT) or render the sample unfit for its intended use. [ARD50013]

destruct system A system that, when operated by external command or preset internal means, destroys the missile or similar vehicle. [ARP4386]

detachment A particular state of isolation in which man is separated or detached from his accustomed behavioral environment by inordinate physical and psychological distances. This condition may compromise his performance.

detail specification A military or aircraft manufacturer-prepared document used to define the applicable design and performance parameters required for a particular starter application. [AS943A]

detectability The quality of a measured variable in a specific environment that is determined by relative freedom from interfering energy or other characteristics of the same general nature as the measured variable.

detection-correction system Type of fault-tolerant system in which a failure or out-of-operating tolerance condition is detected and corrective action is taken automatically. The corrective action may involve switching to a standby system; or, if two or more systems are normally operating, it may involve switching-out the failed channel. A finite time for detection and correction are inherent in this system. With detection-correction systems, it is possible to use a model of an active system as a reference in order to extend the failure-correction capability of the total system. [ARP4386]

detection mechanism The means or methods by which a failure can be discovered by an operator under normal system operation, or can be discovered by the maintenance crew by some diagnostic action. [AIR4896]

detector *1.* A transduction device that transforms optical to electrical energy according to a defined mathematical law; a photodetector. [ARD50024] *2.* A sensor, or an instrument employing a sensor, to detect the presence of something in the surrounding environment, e.g., a wheel speed sensor or a proximity sensor. [AIR1489]

detector-amplifier A device in which an optical detector is packaged together with electronic amplification circuitry.

detent A catch or lever that initiates or prevents movement in a mechanism, especially an escapement.

deterioration Decline in the quality of a device, mechanism, or structure over time due to environmental effects, corrosion, wear, or gradual changes in material properties. If allowed to continue unchecked, deterioration often leads to degradation failure.

deterministic Refers to a property whose future behavior can be predicted precisely. [AIR4075]

detonate To explode by use of mechanical or electrical action.

detonating cord Flexible tube containing a core of high explosive. [AIR913]

detonation The extremely rapid chemical decomposition (explosion) of a material in which the reaction front advances into the unreacted material at greater than sonic velocity. [AIR913]

detonation waves Shock waves that accompany detonation and have a shock front followed by a region of decreasing pressure in which the reaction occurs.

detonator *1.* An explosive train component capable of initiating high-order detonation in a subsequent high-explosive component. [AIR913] 2. In an ignition train, the component that, when detonated by the primer, in turn detonates a less sensitive but larger explosive (usually the booster); or a component that contains its own primer and initiates the detonation. [ARP4386]

deuterium A heavy isotope of hydrogen having one proton and one neutron in the nucleus; hydrogen-2.

deuterium fluoride lasers *See* DF lasers.

deuterium fluorides (DF) Fluorides of deuterium, a heavy isotope of hydrogen.

deuterium oxide *See* heavy water.

deuteron detector A type of specialized radiation detector used in some nuclear reactors to detect the concentration of deuterium nuclei present.

deuterons The nuclei of deuterium atoms.

developed boiler horsepower The boiler horsepower generated by a steam-generating unit.

developer A material that is applied to the test part surface after the excess penetrant has been removed; designed to enhance the penetrant bleedout to form indications. The developer may be a fine, dry powder, or a suspension (in solvent or water) that dries, leaving an adsorptive film on the test part surface. [AMS2647A]

developer, dry A dry, fine powder that is applied as a dust to the test part after the excess penetrant is removed and the surface dried. [AMS2647A]

developer, nonaqueous wet A developer consisting of fine particles suspended in a volatile solvent; applied by spraying onto the test part surface after the excess penetrant is removed and the surface dried. [AMS2647A]

developer, soluble A material completely soluble in its carrier (usually water) which dries to an adsorptive coating; applied to the part after removal of the excess penetrant and prior to drying. [AMS2647A]

developing time The elapsed time between the application of the developer and the examination of the part for indications. [AMS2647A]

development and validation cost The cost of all engineering effort, including theoretical studies and design, hardware, tooling, rig testing, management, and other supporting activities, including LCC and ILS considerations; used to transform the results of conceptual studies into design proposals suitable for full-scale development. [ARP4293]

development assurance All of those planned and systematic actions used to substantiate, at an adequate level of confidence, that development errors have been identified and corrected such that the system satisfies the applicable certification basis. [ARP4754]

development error A mistake in requirements determination or design. [ARP4754]

development system A system used to develop both the hardware and software for a microcomputer system. The development system may contain an editor, assembler and/or high-level language, compiler, and debugging and in-circuit emulation facilities.

development test A test performed by the developing agency to verify the operation or performance of a system or component design, or to produce data that will permit improving the design of the item under test. [ARD50013]

deviation *1.* Variation from a specified dimension or requirement, usually defining the upper and lower limits. [AIR4844]. *2. See* droop.

deviation alarm *1.* Alarm that is set whenever the deviation exceeds the preset limits. 2. An alarm caused by a variable departing from its desired value by a specified amount.

deviation controller A type of automatic control device that acts in response to any difference between the value of a process variable and the instrument setpoint, independent of their actual values.

deviation ratio The ratio given by $M = f/f_{max}$, where f is the maximum frequency difference between the modulated carrier and the unmodulated carrier, and f_{max} is the maximum modulation frequency.

device *1.* The portion of a module, excluding the bus interface, that does the application-dependent function of the module. [AS4710] *2.* A component or assembly designed to perform a specific function by harnessing mechanical, electrical, magnetic, thermal, or chemical energy. *3.* Any piece of machinery or computer hardware that can perform a specific task. *4.* A component in a control system; for example, the primary element, transmitter, controller, recorder, or final control element.

device control character One of a class of control characters intended for the control of peripheral devices associated with a data processing or telecommunication system, usually for switching devices "on" or "off."

device controller A hardware unit that electronically supervises one or more of the same types of devices; acts as the link between the CPU and I/O devices.

device driver A program or routine that controls the physical hardware activities on a peripheral device. A device driver is generally the device-dependent software interface between a device and the common, device-independent I/O code in an operating system.

device flags One-bit registers that record the current status of a device.

device handler A program or routine that drives or services an I/O device. A device handler is similar to a device driver, but provides more control and interfacing functions than a device driver.

device independence The ability to request input/output operations without regard to the characteristics of the input/output devices.

device lift On a bonded device, the vertical distance from the conductor film metallization to the dielectric surface of the semiconductor device. Also called the "bugging distance." [AS1346]

devitrification The formation of crystals in a glass melt, usually occurring when the melt is too cold. [AIR4844]

devitrify To deprive of vitreous quality; to make glass or vitreous rock opaque and crystalline. [AIR4844]

dewars Insulated thermos-like containers for cryogenic liquids. Can be designed to house detectors or lasers requiring cooling.

dewatering Removing water from solid or semisolid material; for example, by centrifuging, filtering, settling, or evaporation.

dew cell An instrument consisting of two bare electrical wires wound spirally around an electrical insulator and covered by wicking wetted with an aqueous solution containing an excess of LiCl. The dew point of the surrounding atmosphere is determined by passing an electric current between the two wires, which raises the temperature of the LiCl solution until its vapor pressure is the same as that of the ambient atmosphere.

dewetting *1.* Generally, loss of surface attraction between a solid and a liquid. *2.* Specifically, flow of solder away from a soldered joint upon reheating.

dew point The temperature at which the air would become saturated (with respect to water) if cooled at constant pressure and without the addition or removal of water vapor. Expressed in degrees Fahrenheit. [AIR1335]

dew-point recorder An instrument that determines dew-point temperature by alternately heating and cooling a metal plate and using a photocell to automatically detect and record the temperature at which condensed moisture appears and disappears on the target. Also known as a mechanized dew-point meter.

dew-point temperature The temperature at which condensation of moisture from the vapor phase begins.

DF See deuterium fluorides.

DF lasers (deuterium fluoride lasers) Gas lasers in which the active material is deuterium fluoride.

DFO *See* dual-fail operative.

DFT *See* diagnostic function test.

DG *See* directional gyroscope.

D-glass A high-boron-content glass made especially for laminates requiring a precisely controlled dielectric constant. [AIR4844]

DH *See* decision height.

DHC 2 aircraft De Havilland Canada STOL utility aircraft. Also known as DHC Beaver aircraft.

DHC Beaver aircraft *See* DHC 2 aircraft.

diagnosis The functions performed and the techniques used in determining and isolating the cause of malfunctions. [AIR4896]

diagnostic Program or other system feature designed to help identify malfunctions in a system; an aid to debugging.

diagnostic accuracy The percentage of failures correctly diagnosed, based on the possible failure population.

diagnostic alarm Alarm that is set whenever the diagnostic program reports a malfunction.

diagnostic capability All capabilities associated with the detection and isolation of faults, including built-in test, automatic test systems, and manual test. [AIR4896]

diagnostic flow chart A test-oriented logical description of branching routines used in a test sequence to describe the steps taken to successfully diagnose a failure. [AIR4896]

diagnostic function test (DFT) A program to test overall system reliability.

diagnostic message An error message in a programming routine to help the programmer identify the error.

diagnostic programs Computer programs that isolate equipment malfunctions or programming errors.

diagnostic routine A sequence of tests or fault tree logic designed to use data inputs and predetermined standards or operational limits to establish condition status and locate a malfunction or discrepancy. Also called malfunction routine. *See also* debug. [ARP1587]

diagnostic(s) *1.* A system of self-generated routines that detect and isolate faults. [ARD50024] *2.* An analysis result pertaining to the detection and isolation of a malfunction or discrepancy. [ARP1587]

diagnostic sensitivity A measure of the threshold level at which a change of condition or functional status yields symptomatic indications with a given diagnostic routine or technique. The threshold level is an accumulation of all error contributions that input into the diagnostic routine or technique. [ARP1587]

diagonal stay A brace used in fire-tube boilers between a flat head or tube sheet and the shell.

dial The graduated scale adjacent to a control knob that is used to indicate the value or relative position of the control setting.

dial indicator A type of measuring gage used to determine fine linear measurements, such as radial or lateral runout of a rotating member. Operates by resting a feeler against a surface and noting the change in position of a pivoted pointer relative to the calibrated gage face as the part is rotated. A dial indicator gage can also be adapted to other setups where precise relative position is to be determined.

diamagnetic material A substance whose specific permeability is less than 1.00 and is therefore weakly repelled by a magnetic field.

diameter For wire strand and cable, refers to the diameter of the circumscribing circle, or across diametrically opposite wires. [AS4536]

diamond-pyramid hardness A material hardness determined by indenting a specimen with a diamond-pyramid indenter having a 136° angle between opposite faces. The hardness number is calculated by dividing the indenting load by the pyramidal area of the impression. Also known as Vickers hardness.

diamond-turned mirror A mirror in which the surface has been formed by machining away material with a diamond tool.

diamond wheel A grinding wheel for cutting very hard materials; uses synthetic diamond dust as the bonded abrasive material.

diaphragm *1.* A thin, flexible disc that is supported around the edges and whose center is allowed to move in a direction perpendicular to the plane of the disc. Used for a wide variety of purposes, such as detecting or reproducing sound waves; keeping two fluids separate, while transmitting pressure or motion between them; or producing a mechanical or electrical signal proportional to the deflection produced by differential pressure across the diaphragm.

2. A partition of metal or other material placed in a header, duct, or pipe to separate portions thereof.

diaphragm motor A diaphragm mechanism used to position a pneumatically operated control element in response to the action of a pneumatic controller or pneumatic positioning relay.

diaphragm seal A thin, flexible sheet of material clamped between two body halves to form a physical barrier between the instrument and process fluid.

diaphragm valve A valve in which the difference in pressure across a diaphragm assembly or flexible membrane is used to vary the position of the assembly, thus enabling it to control flow rate, pressure, or other parameters. [AIR4783]

dichroic filter A filter that selectively transmits some wavelengths of light and reflects others. Typically, such filters are based on multilayer interference coatings.

dichromate treatment A technique for producing a corrosion-resistant conversion coating on magnesium parts by boiling them in a sodium dichromate solution.

didymium A mixture of rare earth elements that is free of cerium. Once regarded as an element, but now know to contain chiefly neodymium and praseodymium, usually associated with lanthanum. Didymium is used in coloring glass for optical filters.

die The movable block of characteristic shape by which solid material is formed or shaped in a forming operation. [ARP480A]

die block A heavy block, usually of tool steel, into which the desired impressions are sunk, formed, or machined. The die block is bolted to the bed of a press.

die casting *1.* A casting process in which molten metal is forced under pressure into the cavity of a metal mold. *2.* A part made by this process.

die chaser One of the cutting parts of a threading die.

die equipment/cavity/pattern identification Numbers, letters, or symbols that identify the die equipment or die cavity used to produce the forging when more than one such set of tooling is used. [AMS2808C]

die forging *1.* The process of forming shaped metal parts by pressure or impact between two dies. *2.* A part formed in this way.

die holder A plate or block mounted between the die block and press bed.

dielectric *1.* An insulator or nonconductor. *2.* An insulating medium that intervenes between two conductors. *3.* A material in which the energy required to establish an electric field may be stored, and later recovered in whole or in part as electrical energy. [ARP1931]

dielectric breakdown The voltage required to cause an electrical failure or breakthrough of the insulation. *See also* breakdown voltage. [ARP1931]

dielectric coating An optical coating made up of one or more layers of dielectric (nonconductive) materials. The layer structure determines what fractions of incident light at various wavelengths are transmitted and reflected.

dielectric constant A material characteristic expressed as the capacitance between two plates when the intervening space is filled with a given insulating material divided by the capacitance of the same plate arrangement when the space is filled with air or is evacuated.

dielectric curing The curing of a synthetic thermosetting resin by the passage of an electric charge through it. [AIR4844]

dielectric heating The heating of materials by dielectric loss in a high-frequency electrostatic field. [AIR4844]

dielectric loss The time rate at which electric energy is transformed into heat in a dielectric when it is subjected to a changing electric field. [ARP1931]

dielectric loss angle The difference between 90° and the dielectric phase angle. [AIR4844]

dielectric loss factor For a given material, the product of the dielectric constant and the tangent of the dielectric loss angle. [AIR4844]

dielectric materials *See* dielectrics.

dielectric monitoring A means of tracking the cure of thermosets by monitoring changes in their electrical properties during material processing. [AIR4844]

dielectric phase angle The angular difference in phase between the sinusoidal alternating potential difference applied to the dielectric and the component of the resulting alternating current having the same period as the potential difference. [AIR4844]

dielectric power factor The cosine of the dielectric phase angle. [AIR4844]

dielectrics Substances that contain few or no free charges and that can support electrostatic stresses.

dielectric strength *1.* The average voltage gradient at which electrical failure or breakdown occurs under prescribed conditions. [AIR1116] 2. The voltage that an insulating material can withstand before breakdown occurs; usually expressed as a voltage gradient, such as "volts per mil." [ARP1931] *See also* breakdown voltage and insulation resistance.

dielectric test *1.* The high voltages impressed between a component and the frame of the alternator to check insulation characteristics. [ARP1148A] 2. Test that consists of the application of a voltage higher than the rated voltage for a specified time for the purpose of determining the adequacy against breakdown of the insulation under normal conditions. [ARP1931]

dielectrometry During cure of the resin in a laminate, use of electrical techniques to measure the changes in loss factor and in capacitance. [AIR4844]

die scalping Improving the surface quality of bar stock, rod tubing, or wire by drawing it through a sharp-edged die to remove a thin surface layer containing minor defects.

die set A tool or tool holder consisting of a die base and punch plate, used for attaching matched upper and lower dies. Can be inserted into a press and removed from it as a single unit.

diesinking Making a shaped recess in a working face of a die, usually by mechanical, electrochemical, or spark-discharge machining.

die welding Forge welding using shaped dies.

difference limen The increment in a stimulus that is barely noticed in a specified fraction of independent observations in which the same increment is imposed.

differential *1.* As applied to two-position control action, the difference between the value of the controlled temperature at which the controller operates to one position and that value of controlled temperature at which it operates to the other position. As applied to a control with a deadband, the difference between the value of the controlled temperature at which the controller action in a given direction is started and the value at which it is stopped. The differential is not necessarily the same on both sides of null. [ARP89C] 2. Any arrangement of epicyclic gears that allows two driven shafts to revolve at different speeds, with the speed of the main driving shaft being the algebraic mean of the speeds of the driven shafts. Also known as differential gear.

differential amplifier A device that compares two input signals and amplifies the difference between them.

differential analyzer An analog computer designed and used primarily for solving differential equations.

differential delay The difference between the maximum and the minimum frequency delays occurring across a band.

differential fault current The difference between the positive and return currents in a two-wire distribution system. [AS1831]

differential gap The smallest increment of change in a controlled variable required to cause the final control element in a two-position control system to move from one position to its alternative position.

differential gear *See* differential.

differential input The difference between the instantaneous values of two voltages, both being biased by a common mode voltage.

differential input (to a signal conditioner) An input in which both sides are isolated from the chassis and power supply ground. The differential input signal is applied as a differential voltage across the two sides.

differential instrument Any instrument that has an output signal or indication proportional to the algebraic difference between two input signals.

differential linearity *See* linearity, differential.

differential motion A mechanism in which the net motion of a single driven element is the difference between motions that would be imparted by each of two driving elements acting alone.

differential (of a control) The difference between cut-in and cut-out points.

differential pressure The difference in value between two functionally related pressures occurring simultaneously at different points, such as at opposite sides of an actuator piston. [ARP4386]

differential-pressure gage Any of several instruments designed to measure the difference in pressure between two enclosed spaces, independent of their absolute pressures.

differential-pressure transmitter Any of several transducers designed to measure the pressure difference between two points in a process and transmit a signal proportional to this difference, without regard to the absolute pressure at either point.

differential-pressure-type liquid-level meter Any of several devices designed to measure the head of liquid in a tank above some minimum level and produce an indication proportional to this value; or to measure the head below some maximum level and display it similarly.

differential propagation delay The difference in propagation delay between the high-frequency components and the low-frequency components of a waveform. [AIR4288]

differential pulse code modulation (DPCM) An efficient signal-encoding method for reducing the transmission rate of digital signals. The basic principle of DPCM is to quantize code and transmit the difference between the actual sample and prediction value.

differential quantum efficiency Used in describing quantum efficiency in devices having nonlinear output/input characteristics; the slope of the characteristic curve is the differential quantum efficiency.

differential scanning calorimetry Technique that provides a measurement of the energy absorbed or produced as a resin system is cured. [AIR4844]

differential screw A type of compound screw that produces a motion equal to the difference in motion between the two components of the screw.

differential thermal analysis (DTA) An experimental analysis technique in which a specimen and a control are heated simultaneously and the difference in their temperatures is monitored. [AIR4844]

differential travel The distance from the operating point to the release point. [AIR4077]

differential voltage The voltage difference observed between the two signal conductors in a balanced transmission system. [AIR4258]

differential windlass A windlass that has a barrel with two sections of different diameter. In the differential windlass, the pulling rope passes around one section, then through a pulley and around the other section; the pulley is attached to the load.

differentiator A device whose output function is proportional to the derivative, i.e., the rate of change, of its input function with respect to one or more variables (usually with respect to time).

diffracted beam In x-ray crystallography, a beam of radiation composed of a large number of scattered rays mutually reinforcing one another.

diffracted wave The wave component existing in the primary propagation medium after an interaction occurs between the wave and a discontinuity or a second medium occurs. The diffracted wave coexists in the primary medium with incident waves, and with waves reflected from suitable plane boundaries.

diffraction A phenomenon associated with the scattering of waves when they encounter obstacles whose size is about the same order of magnitude as the wavelength; in effect, each scattering point produces a secondary wave superimposed on the unscattered portion of the incident wave, the intensity of the scattered wave varying with direction from the scattering point. Diffraction effects form the basis for x-ray crystallography; they also tend to produce aberrations that must be dealt with in the design and construction of high-quality acoustical and optical systems.

diffraction grating An array of fine, parallel, equally spaced reflecting or transmitting lines that diffract light in a direction characteristic of the spacing of the lines and the wavelength of the diffracted light.

diffraction-limited beam A beam with a far-field spot size dependent only on the theoretical diffraction limit, the function of output wavelength divided by output aperture diameter.

diffraction propagation Wave propagation around objects, or over the horizon, by diffraction.

diffraction radiation Electromagnetic radiation excited by an electron flux passing near a diffractive, periodic structure, such as a wiggler magnet in a free electron laser.

diffraction x-ray machine An apparatus consisting of an x-ray tube, power supply, controls, and auxiliary equipment, used in the study of crystals, semiconductors, and polymeric materials.

diffused-semiconductor strain gage A component used in manufacturing transducers, principally diaphragm-type pressure transducers; consists of a slice of silicon about 2.5 to 22 mm in diameter into which an impurity element, such as boron, has been diffused.

diffuser *1.* A device for converting the velocity pressure of a fluid stream into pressure head, usually accomplished by efficiently reducing the velocity of air. [ARP147C] *2.* Specially designed duct, chamber, or section, sometimes equipped with guide vanes, that decreases the velocity of a fluid, such as air, and increases its pressure; for example, as used in jet engines or wind tunnels. *3.* As applied to oil or gas burners, a metal plate with openings so placed as to protect the fuel spray from high-velocity air, while admitting sufficient air to promote the ignition and combustion of fuel. Sometimes called an impeller.

diffuse radiation Radiant energy propagating in many different directions through a given small volume of space. Contrast with parallel radiation.

diffuse reflection Reflection in which the light is reflected in all directions. [ARP798]

diffuse reflection factor The ratio of the diffusely reflected light to the incident light. [ARP798]

diffuse-specular Surfaces that are essentially diffuse, but contain an outer layer of glazed material that reflects specularly. Porcelain-enamel is a common example. [ARP798]

diffuse transmission Transmission in which the transmitted light is emitted in all directions from the transmitting body. [ARP798]

diffuse transmission factor The ratio of the diffusely transmitted light to the diffuse incident light. [ARP798]

diffusing surfaces and media Surfaces that break up the incident light and distribute it more or less in accordance with Lambert's cosine law of emission; for example, rough plaster and white glass. [ARP798]

diffusion *1.* In an atmosphere, or in any gaseous system, the exchange of fluid parcels between regions, in apparently random motions of a scale too small to be treated by the equations of motion. *2.* In materials, the movement of atoms of one material into the crystal lattice of an adjoining material, e.g., penetration of the atoms in a ceramic coating into the lattice of the protected metal. *3.* In ion engines, the migration of neutral atoms through a porous structure incident to ionization at the emitting surface. *4.* Conversion of gas-flow velocity into static pressure, as in the diffuser casing of a centrifugal fan. *5.* The movement of ions from a point of high concentration to low concentration.

diffusion coefficient *1.* For a material, the rate of moisture or other fluid absorption with time at a given temperature. [AIR4844] *2.* For a gas diffusing through a gas or a porous medium, the absolute value of the ratio of the molecular flux per unit area to the concentration gradient, where the molecular flux is evaluated across a surface perpendicular to the direction of the concentration gradient.

diffusion effect *See* diffusion.

diffusion, law of gaseous Refers to the movement of a gas of higher partial pressure to a volume of gas of lower partial pressure; for example, the movement of gases, such as oxygen and carbon dioxide, through the membranes of the lung, into and out of the lung. [ARP171]

diffusion pump A vacuum pump in which a stream of heavy particles, such as oil or mercury vapors, carries gas molecules out of the vacuum chamber.

diffusion, random gaseous In solution, the spontaneous movement of molecules or other particles due to their random thermal motion, such that the molecules or particles reach a uniform concentration throughout. [ARP171]

diffusion zone Refers to the hardened zone in steel produced by nitrogen diffusion according to classical principles, leading to precipitation and solid solution hardening. [AMS2759/8]

diffusivity A measure of the rate of diffusion of a substance, expressed as a diffusivity coefficient K.

digital Describes a variable quantity, or instrument for measuring it, that is represented by discrete, usually numerical, values. [AIR5026]

digital back-up In a computer system, an alternate method of digital process control initiated by use of special-purpose digital logic in the event of a failure in the system.

digital circuits See digital electronics.

digital computer Computers that operate with information—numerical or otherwise—represented in a digital form.

digital controller A control device consisting of a microprocessor plus associated A/D input converters and D/A output converters. The digital controller receives one or more analog inputs related to current process variables; uses a predetermined control algorithm to compute an output signal from the digitized information; and converts the result to an analog signal, which operates the final control element. A digital controller may also be adapted to furnish additional outputs, such as alarms, totalizer signals, and displays.

digital data Data represented in discrete, discontinuous form, as contrasted with analog data represented in continuous form. Digital data is usually represented by means of coded characters, for example, numbers, signs, or symbols.

digital delay generator An electronic instrument that can be programmed digitally to delay a signal by a specific interval; a time-delay generator.

digital differential analyzer *1.* An incremental computer in which the principal type of computing unit is a digital integrator whose operation is similar to the operation of an integrating mechanism. *2.* A differential analyzer that uses digital representation for the analog quantities.

digital electronics The use of circuits in which there are usually only two states possible at any point. The two states can represent any of a variety of binary digits (bits) of information.

digital filter *1.* Computational means of attenuating undesired frequencies in sets of time-dependent data. *2.* An algorithm that reduces undesirable frequencies in the signal.

digital indicator A device that displays the value of a measured variable in digitized form. In most instances, the measurement range is not displayed simultaneously, which is considered an inherent disadvantage.

digital information Information that is generated in digital form from a source. [ARD50012]

digital input A number value input.

digital logic Describes a system in which signal level is represented as a number value with a most significant and least significant bit. Binary digital logic uses numbers consisting of strings of ones and zeros.

digitally-directed analog (DDA) control system See supervisory control.

digital manometer A manometer equipped with a sonar device that measures column height and produces a digitized display.

digital millivoltmeter, high impedance Device used by many modern aircraft. The high impedance eliminates the need for controlling input resistance. [AIR46]

digital multiplexer A data selection device that permits sharing a common information path between multiple groups of digital devices, for example, the path from a computer CPU to any of several groups of digital output devices.

digital readout An electrically powered device that interprets a continuously variable signal and displays the amplitude, or another signal attribute, as a series of numerals or other characters that correspond to the measured value and can be read directly. With a digital readout, the accuracy of measurement is limited by the decimal position of the rightmost character in the display, rather than by characteristics of the measurement circuit alone.

digital resolution The value of the least significant digit in a digitally coded representation.

digital signal A discrete or discontinuous signal—one whose various states are discrete intervals apart.

digital system A system in which the inputs and outputs are discrete data time. [ARP89]

digital tachometer Any of several instruments that are designed to determine rotational speed and display the indication in digital form.

digital television Television in which picture redundancy is reduced or eliminated by transmitting only the data needed to define motion in the picture, as represented by changes in the areas of continuous white or black.

digital-to-analog converter (D/A or DAC) A device or sub-system that converts binary (digital) data into continuous analog data, for example, to drive actuators of various types or motor-sped controllers.

digital valve A single valve casing containing multiple solenoid valves whose flow capacities vary in binary sequence (1, 2, 4, 8, 16, ...); to regulate flow, the control device sends operating signals to various combinations of the solenoids. Applications of digital valves are limited to very clean fluids at moderate temperatures and pressures; but, within these limitations, precise flow control and rapid response are possible. An eight-element valve, for example, yields flow resolution of 0.39% (1 part in 256).

digitize To convert an analog measurement of a physical variable into a numerical value, thereby expressing the quantity in digital form. *See also* analog-to-digital converter.

digitized signal Signal in which information is represented by a set of discrete values, in accordance with a prescribed law. Every discrete value represents a definite range of the original undigitized signal. *See also* analog-to-digital converter.

digitizer A device that converts an analog measurement into digital form. *See also* analog to digital converter.

dihedral The spanwise inclination of wing or other surface relative to horizontal. [ARP4107]

dilatant substance A material that flows under low shear stress, but whose rate of flow decreases with increasing shear stress.

dilatometer An apparatus for accurately measuring thermal expansion of materials. *See also* extensometer.

diluent A substance that reduces the viscosity and pour point of the lubrication oil of an engine. [S-4, 6]

diluter A device for mixing atmospheric air with oxygen. [ARP171]

diluter-demand *See* regulator, diluter-demand. [ARP171]

dilution *1.* Addition of solvent to a solution to lower its concentration. *2.* Melting low-alloy base metal or previously deposited weld metal into high-alloy filler metal to produce a weld deposit of intermediate composition.

dimensional stability The ability of a material to retain its size and shape over an extended period of time, under a defined set of environmental conditions, especially temperature.

dimetcote An inorganic zinc coating composed of two materials that are mixed together: (a) a reactive liquid; and (b) a finely divided powder. The mixture reacts in place with a steel surface to form an insoluble coating.

diminished radix complement A number obtained by subtracting each digit of the given number from one less than the radix; typical examples are the nine's-complement in decimal notation and the one's-complement in binary notation.

diode A two-electrode electronic component containing merely an anode and a cathode.

diode laser A laser in which stimulated emission is produced at a p-n junction in a semiconductor material. Only certain materials are suited for diode-laser operation; among them are gallium arsenide, indium phosphide, and certain lead salts.

diode laser array A device in which the outputs of several diode lasers are brought together in one beam. The lasers may be integrated on the same substrate, or discrete devices may be coupled optically and electronically.

diopter A measurement of the refractive power of a lens, equal to the reciprocal of the focal length in meters. A lens with a 20-centimeter focal length has power of five diopters, while one with a 2-meter focal length has a power of 0.5 diopter.

DI/OU *See* data input/output unit.

DIP *See* dual in-line package.

dip brazing Producing a brazed joint by immersing the assembly in a bath of hot molten chemicals or hot metal. In dip brazing, a chemical bath may provide the brazing flux; molten metal, the brazing alloy.

dip coating Covering the surface of a part by immersing it in a bath containing the coating material.

diplexer *See* duplexer.

dip needle A device for indicating the angle, in a vertical plane, between a magnetic field and the horizontal plane.

dipole System composed of two separated, equal, electric or magnetic charges of opposite sign.

dipole antenna *1.* A straight radiator, usually fed in the center, and producing a maximum of radiation in the plane normal to its axis. *2.* A center-fed antenna that is approximately half as long as the wavelength of the radio waves it is primarily intended to transmit or receive.

dip soldering A process similar to dip brazing, but using a lower-melting filler metal.

dipstick A calibrated stick for manually determining the quantity of fuel contained in a fuel tank. [AIR4783]

direct access *1.* The retrieval or storage of data by a reference to its location on a volume, rather than relative to the previously retrieved or stored data. *2. See* random access.

direct acting controller A controller in which the value of the output signal increases as the

value of the input (measured variable or controlled variable) increases.

direct acting recorder A recorder in which the pen or other writing device is directly connected to, or directly operated by, the primary sensor.

direct action A controller in which the value of the output signal increases as the value of the input (measured variable or controlled variable) increases.

direct address An address that indicates the location at which the referenced operand is to be found or stored, with no reference to an index register. Synonymous with first-level address.

direct blow down system A simple power-conversion system in which stored cold gas is used directly by a pneumatic actuator. [ARP4386]

direct broadcast satellite (DBS) Domestic satellite used for direct TV transmission to home receivers.

direct code A code that specifies the use of actual computer command and address configurations.

direct-connected An arrangement in which a meter or other driving mechanism is connected to a driven mechanism without intervening gears, pulleys, or other speed-changing devices.

direct coupling The association of two circuits, accomplished by capacitance, resistance, or self-inductance common to both circuits.

direct coupling drive A motion transmission with a continuous load path. [ARP4386]

direct current (DC) An electrical current that travels uniformly in one direction. [ARP1931]

direct-current amplifier An amplifier designed to amplify signals of infinitesimally small frequency.

direct current resistance The resistance offered by any circuit or circuit component to the flow of direct current. [ARP1931]

direct digital control (DDC) A computer control technique in which the position of the final control element is set directly by the computer output.

direct drive Any powered mechanism in which the driven portion is on the same shaft as the driving portion, or is coupled directly to the driving portion.

direct drive valve A servovalve in which the power control valve is driven directly by the electric control signal. [ARP4386]

direct entry In data processing, the input of data directly to computer memory and disk, in contrast to earlier methods of keying to punched cards, which were then read into a computer.

direct extensions Refers to a device that provides flow rate indication by means of viewing the position of the extension of the metering float within a glass extension tube.

direct glare The sensation produced by brightnesses within the visual field that are sufficiently greater than the luminance to which the eyes are adapted; causes annoyance, discomfort, or loss in visual performance and visibility. [AIR1151]

direct interface *1.* An interface that defines a service/consumer relationship between adjacent GOA layers in the GOA model. *2.* The connection between an entity sending data and another entity receiving data; used for transmission of that data along with routing path. [AS4893]

directional antennas Antennas that radiate or receive radio signals more efficiently in some directions than in others.

directional axis The vertical (Z) axis of an aircraft about which the aircraft revolves in yawing. [ARP4386]

directional control Control of the directional motion of an aircraft. Directional control is primarily achieved through control of yaw and sideslip. [ARP4386]

directional control valve *1.* A selector valve having four working ports—pressure, return, and two cylinder or load ports—and a reversible flow pattern. [ARP4386] *2.* A valve whose chief function is to control the direction of flow within a fluid system.

directional coupler *1.* A device for separately sampling either the forward or backward oscillations in a transmission line. *2.* Describes a fiber optic coupler that preferentially transmits light in one direction.

directional gyroscope (DG) A navigational instrument for indicating direction. Contains a free gyroscope that holds its position in azimuth, thus allowing the instrument scale to indicate deviation from the reference direction.

directional motion Rotational motion of a vehicle about the vertical axis. [ARP4386]

directional property For a material, any mechanical or physical property whose value varies with orientation of the test axis within the test specimen.

directional solidification (crystals) Controlled solidification (crystal growth) of molten metal in a casting so as to provide feed metal to the solidifying front of the casting.

directional stability Refers to the properties of an aircraft, rocket, tire, etc. that enable the vehicle to restore itself from a yawing or sideslipping condition. Also called weathercock stability. [AIR1489]

direction finder (radio) *See* radio direction finder.

direction finding A procedure or process for locating or localizing the origin of radar, acoustical, or optical emissions.

direction of lay In a conductor or cable, the lateral direction, either right-hand or left-hand, in which a strand or wire passes over the top as it recedes from an observer looking along the axis of the conductor or cable. [AS1198]

direction of polarization The direction of the electric field vector of an electromagnetic wave.

direction of propagation The direction of average energy flow with respect to time at any point in a homogeneous, isotropic medium.

directive An operator command that is recognized by computer software.

directivity The ability of an antenna to radiate or receive more energy in some directions.

direct lift control Employment of lifting surfaces independent of the elevators or elevons, such as flaps, symmetrical ailerons, spoilers, canards, to exert vertical flight path control by means of applied lift forces. [ARP4386]

direct maintenance man hours *See* manhours, direct.

direct memory access (DMA) A method of fast data transfer between the peripherals and the computer memory. The transfer does not involve the CPU.

direct numerical control (DNC) A distributed numerical control system in which the supervisory computer controls several CNC or NC machines.

directory device A mass-storage retrieval device, such as disk or DECtape, that contains a directory of the files stored on the device.

directory service The network management function that provides all addressing information required to access an application process.

direct power generation Any method of producing electric power directly from thermal or chemical energy without first converting it to mechanical energy; examples include thermopiles, primary batteries, and fuel cells.

direct process Any method for producing a commercial metal directly from metal ore, without an intervening step, such as roasting or smelting, to produce a semirefined metal or another intermediate product.

direct-reading gage Any instrument that indicates a measured value directly, rather than by inference; for example, a gage that indicates liquid level by means of a sight glass partly filled with liquid from the tank, or by means of a pointer directly connected to a float in the tank.

direct record In instrumentation tape, the mode in which tape magnetization is directly related to data voltage level.

direct wave A wave that is propagated through space without relying on the properties of any gas or other substance occupying the space.

direct-writing recorder A pen-and-ink recorder in which the position of the pen on the chart is controlled directly by a mechanical link to the coil of a galvanometer, or indirectly by a motor controlled by the galvanometer.

disability glare Glare that reduces visual performance and visibility, and often is accompanied by discomfort. [AIR1151]

disable To disallow the processing of an established interrupt until interrupts are enabled. Contrast with enable. *See also* disarm.

disarm To cause an interrupt to be completely ignored. Contrast with arm. *See also* disable.

disassemble To reduce an assembly to its component parts by loosening or removing threaded fasteners, pins, clips, snap rings, or other mechanical devices—in most instances, for some purpose, such as cleaning, inspection, maintenance, or repair, followed by reassembly.

disaster Large-scale drought, glacier movement, flood, fire, storm, etc.

disbonding Separation of the adhesive bondline between two adhesively bonded parts or subassemblies. [ARP5089]

discard task The removal from service of an item at specified life limit. [AIR4896]

discernible Term intended to separate symbol movement associated with jitter from that associated with drift or dither; based on the time constants of jitter, drift, and dither. The drift and dither time constants are long enough that symbol movement would not be discernible. [AS8034]

discharge channel The passage(s) beyond the operating parts through which gas must flow to reach the atmosphere—exclusive of any piping attached to the outlet of the safety relief device. [ARP4386]

discharge head The pressure at which a pump discharges freely to the atmosphere, usually measured as feet of water above the intake level.

discharging The removal of electrical energy from a battery. [AS8033]

DISCO Abbreviation for distributed star coupled—a family of distributed topologies that includes DDP, MDP, DDA, and MDA. [AIR4288]

discomfort glare Glare that produces discomfort; does not necessarily interfere with visual performance or visibility. [AIR1151]

disconnect To disengage the apparatus used in a connection and to restore it to its ready condition when not in use.

disconnect, AMAD A device used to decouple the AMAD from the engine. [ARP906A]

disconnect switch An electrical switch for interrupting power supplied to a machine. Usually separate from the machine controls (often mounted nearby on the wall), and serves mainly to deenergize the equipment for safety during setup or maintenance.

discontinuity An interruption in the normal physical structure or configuration of a part. A discontinuity may or may not affect the usefulness of a part. [ARP5089]

discontinuity depth The extent of a discontinuity normal to the plane established by the circle defining the discontinuity length; not the location of the discontinuity below a surface. [AMS4991]

discontinuous phase The contaminated product, containing water or filtered contaminants, that is separated from the continuous phase of a filter system. [AIR4783]

Discos (satellite attitude control) A satellite orbit "disturbance compensation system" designed to maintain an object (proof object) in correct orbit by detecting forces and compensating for them by using thrusters.

discrepancy For a material or item, any difference or inconsistency between a requirement specified in a contract, drawing specification standard, test procedures, or other document and the actual characteristic of the material or item. [ARD50010]

discrete address beacon system Radar beacon system with discretely addressable transponders and a ground-air-ground data link for automated air traffic control (FAA).

discrete component circuit A circuit implemented through the use of individual transistors, resistors, diodes, capacitors, etc. Contrast with integrated circuit.

discrete continuity A well-defined, individual recess, cavity, or inclusion. [AMS4991]

discrete frequency In direct pilot-controller ATC communications, a particular radio frequency that is selected to reduce radio-frequency congestion by controlling the number of aircraft operating on a particular frequency at one time. [ARP4107]

discrete instrument Pertaining to distinct elements or to representation by means of distinct elements.

discrete programming *See* integer programming.

discrete signal A signal that takes on different states to represent a function or piece of information that has two or more states or values. [AIR4258]

discrete signal interface input The input to the transmitter. [AIR4258]

discrete signal interface output The output from the transmitter. [AIR4258]

discriminant analysis (statistics) A linear combination of a set of N variables that will classify (into two different classes) the events or items for which the measurements of the N variables are available, with the smallest proportion of misclassifications.

discrimination ratio One of the standard test plan parameters; the ratio of the upper test MTBF (THETA 0) to the lower MTBF (THETA 1). [ARD50010]

discriminator 1. In general, a circuit in which output depends on the difference between an input signal and a reference signal. 2. A hardware device used to demodulate a frequency-modulated carrier or subcarrier to produce analog data.

discriminator type no back A device that contains a discriminator and a brake. When the discriminator detects that motion was initiated from the output end of the device, it activates the brake. [ARP4386]

disdrometer An apparatus capable of measuring and recording the size distribution of raindrops in the atmosphere.

disengage To intentionally pull apart two normally meshing or interlocking parts, such as gears or splines, especially for the purpose of interrupting the transmission of mechanical power.

disengagement The uncoupling of the starter from the engine by means of the starter clutch or engaging device. Normally occurs at starter cutoff speed or starter maximum speed. [ARP906A]

disengagement force Force required to demate the couplings. [AIR4728]

disengaging surface The surface of the boiler water from which steam is released.

dish *See* parabolic reflector.

dish antenna An antenna in which a parabola-shaped "dish" serves as the reflector to increase antenna gain.

dishing A metalforming operation in which a shallow concave surface is formed.

disinfectant A chemical agent that destroys microorganisms, bacteria, and viruses—or renders them inactive.

disintegrated Separated or decomposed into fragments; having lost its original form. [ARP4107]

disk A high-speed, rotating magnetic platter for storing computer data.

disk area The area of the circle described by the blade tips of a rotating propeller or rotor. [ARP4107]

disk brake A mechanical brake in which the friction elements, normally called pads, press against opposite sides of a spinning disk attached to the rotating element to slow or stop its motion.

disk cam A flat cam with a contoured edge that rotates about an axis perpendicular to the plane of the cam, communicating radial linear motion to a follower that rides on the edge of the cam.

disk clutch A device for engaging or disengaging a connection between two shafts. The chief clutch element is a pair of disks, one coupled to each shaft, which transmit power when engaged by means of disk-face linings made of friction materials.

disk coupling A flexible coupling in which power and motion is transmitted by means of a disk made of elastomeric or other flexible material.

disk directory A table for storing the location of files held on a disk.

disk drive A device that reads and writes computer data on disks.

diskette *1.* A round, flat, flexible platter coated with magnetic material and used for storage of software or data. *2. See* floppy disk.

disk galaxies Galaxies with a central bulge consisting of a spheroidal aggregation of stars, and a surrounding disk of stars fanning outward in a thin layer.

disk meter A flow-measurement device that contains a nutating disk mounted in such a way that each time the disk nutates, a known volume of fluid passes through the meter.

disk operating system (DOS) 1. The program with which a computer performs such mundane but useful tasks as storing, locating, and retrieving files on disk; reading the keyboard; and issuing display and print information. 2. A collection of system programs for operating a microcomputer system.

disk spring A mechanical spring consisting of a dished circular plate and washer supported in such a way that one opposing force is distributed uniformly around the periphery and the second acts at the center. Washer-type disk springs are sometimes known as Belleville washers.

dispatch To supervise the departure of aircraft on schedule.

dispatching priority For a central processing unit in a multitask situation, a number assigned to tasks, and used to determine precedence.

disperse To disseminate or distribute. Can be applied, for example, to light waves; or to globs of oil in water when mixing them to make an emulsion.

disperser *See* emulsifier.

dispersing agent *See* emulsifier.

dispersing prism A prism designed to spread out the wavelengths of light to form a spectrum.

dispersion *1.* Finely divided particles in suspension in another substance. [ARP1931] *2.* Any process that breaks up an inhomogeneous, lumpy mixture, and converts it to a smooth paste or suspension in which the particles of the solid component are more uniform and small in size. *3.* The process by which an electromagnetic signal becomes distorted due to the different propagation characteristics of the various frequency components of the signal. *4.* In wave mechanics, the rate of change of distance along a spectrum with frequency (linear dispersion); or the rate of change of frequency with distance along a spectrum (reciprocal linear dispersion). *5.* The broadening effect of an optical pulse that is being propagated through a component of limited bandwidth. [AIR4288]

dispersion forces Forces responsible for most of the adhesion forces involved in adhesive bonding. Often called van der Waals forces. [AIR4844]

dispersion limited operation Denotes operation in which the dispersion of the pulse, rather than its amplitude, limits the distance between repeaters. In this regime of operation, waveguide and material dispersion preclude an intelligent decision on the presence or absence of a pulse.

displaced threshold A threshold that is located somewhere other than at the designated beginning of the runway. [ARP4107]

displacement *1.* A vector quantity that specifies the change of position of a body or particle, usually measured from the mean position or position of rest. *2.* The volume swept out by a piston as it moves inside a cylinder from one extreme of its stroke to the other extreme.

displacement angle In filament winding, the advancement distance of the winding ribbon on the equator after one complete circuit. [AIR4844]

displacement antiresonance A condition of antiresonance in which the external sinusoidal excitation is a force and the specified response is displacement at the point at which the force is applied.

displacement meter A meter that measures the amount of a material flowing through a system by recording the number of times a vessel or cavity of known volume is filled and emptied.

displacement resonance A condition of resonance in which the external sinusoidal excitation is a force and the specified response is displacement at the point at which the force is applied.

displacement-type density meter A device that measures liquid density by means of a float and balance beam used in conjunction with a pneumatic sensing system. The float is confined within a small chamber through which the test liquid continually flows, so that density variations with time can be determined.

displacer-type liquid-level detector A device for determining liquid level by means of force measurements on a cylindrical element partly submerged in the liquid in a vessel. As the level in the vessel rises and falls, the displacement (buoyant) force on the cylinder varies and is measured by a lever system, torque tube, or other force-measurement device.

displacer-type meter An apparatus for detecting liquid level or for determining gas density by measuring the effect of the fluid on the buoyancy of a displacer unit immersed in it.

display *1.* A component that converts a liquid signal into an equivalent visual output. [ARP993A] *2.* The cockpit or crew station presentation of information regarding condition of a component or system, such as landing gear position (up and locked, down and locked, in transit) or break overheat warning. [AIR1489]

display center The center of the useful display area. [ARP1782]

display element The smallest addressable entity of the display. [ARP4256]

display media The various basic technologies through which information can be communicated visually; includes printed paper, cathode ray tube (CRT) display, and liquid crystal display (LCD). [ARP4155]

display surface The outermost surface of the display unit through which the display image is projected. The display surface includes the CRT imaging surface, envelope glass, and all contrast and anti-reflective filter surfaces. [ARP4067]

display technology The technical means by which symbology is presented to an operator. [ARP4155]

display tube A cathode ray tube used to display information.

display unit A device that provides a temporary visual representation of data. *See also* cathode ray tube.

dissector tube A camera tube that produces an output signal by moving the electron-optical image formed by photoelectric emission on a continuous photocathode surface past an aperture.

dissipation factor *1.* A measure of the AC power loss. Dissipation factor is proportional to the power loss per cycle (f) per potential gradient (E^2) per unit volume, as follows: Dissipation Factor = power loss/($E^2 \times f \times$ volume $\times$ constant). [ARP1931] *2.* The ratio of the loss index to its relative permittivity. *3.* The ratio of the energy dissipated to the energy stored in the dielectric per cycle. *4.* The tangent of the loss angle.

dissociation The separation of a complex molecule into constituents by collision with a second body or by absorption of a photon. The product of dissociation of a molecule is two ions, one positively charged and one negatively charged.

dissolved gases Gases in solution.

dissolved solids In water, those solids that are in solution.

dissymmetrical transducer A transducer in which interchanging at least one pair of specified terminals will change the output signal delivered, when the input signal remains the same.

dissymmetry 1. Absence of symmetry. 2. Symmetry in opposite directions, as in left and right from centerline.

distal At the greatest distance from a central point; peripheral. [ARP171]

distance amplitude correction (DAC) Electronic change of amplification to provide equal amplitude from equal reflectors at different depths. [AMS2633B]

distance (audio communications) *See* DME.

distance measuring equipment (DME) 1. Equipment (airborne and ground) used to measure, in nautical miles, the slant-range distance of an aircraft from the position at altitude to the DME navigational aid on the ground. *See also* TACAN, VORTAC. [ARP4107] 2. A radio aid to navigation which provides distance information by measuring total round trip time of transmission from an integrator to a transponder and return.

distance perception See space perception.

distillate 1. The overhead product from a distillation column. When a partial condenser is used, there may be both a liquid and a vapor distillate stream. 2. In the oil and gas industry, refers to a specific product withdrawn from the column, usually near the bottom.

distillate fuel Any of the fuel hydrocarbons obtained during the distillation of petroleum that have boiling points higher than that of gasoline.

distillation 1. A unit operation used to separate a mixture into its individual chemical components. 2. Vaporization of a substance with subsequent recovery of the vapor by condensation.

distortion 1. Refers to the condition in which objects appear misshapen, tilted, or oversized and undersized when viewed through a distorted window. [AIR5122] 2. Inhomogeneities in the glass or irregularities in the surface of the element causing displacement of images. [ARP924] 3. An undesired change in waveform. In a system used for transmission or reproduction of sound, a failure by the system to transit or reproduce a received waveform with exactness. An undesired change in the dimensions or shape of a structure; for example, the distortion of a fuel tank due to abnormal stresses or extreme temperature gradients. 4. A lens defect that causes the images of straight lines to appear geometrically other than straight lines.

distortion descriptor A non-dimensional numerical representation of the measured inlet pressure distribution. [ARP1420]

distortion extent The circumferential arc size of a distorted region. [ARP1420]

distortion intensity The amplitude of a distortion pattern. [ARP1420]

distortion meter An instrument that visually indicates the harmonic content of an audio-frequency signal.

distortion sensitivity Loss in compressor surge pressure ratio per unit of numerical distortion descriptor. [ARP1420]

distraction *See* attention, anomalies of distraction.

distributed In a control system, refers to control achieved by intelligence that is distributed about the process to be controlled, rather than by a centrally located single unit.

distributed control A method of media access control in which responsibility for control is distributed among all stations. [AIR4271]

distributed control system 1. Describes a system in which: (a) processors and consoles are distributed physically in different areas of the plant or building; or (b) data processing is distributed, such as a system in which several processors are running in parallel (concurrent), each with a different function. 2. A system of dividing plant or process control into several areas of responsibility, each managed by its own controller (processor), with the whole interconnected to form a single entity, usually by communication buses of various kinds.

distributed feedback lasers Lasers containing a periodic medium that provides the necessary feedback for laser action.

distributed processing Processing with multiple small computers that are capable of operating independently, but can communicate over a network with one other and/or a central computer.

distribution functions The density functions or number of particles per unit volume of phase space. The distribution functions are a function of the three space coordinates and the three velocity coordinates.

distribution system *1.* The combination of ducts, cabin inlet openings, valves, and orifices to distribute fluids to satisfy cabin and equipment cooling and heating demands. [ARP147C] *2.* The part of the electrical system used for conveying energy to the point of utilization from the 270-VDC source. [AS1831]

distributor *1.* A block-type unit with an inlet and several outlets to provide force to multiple points, generally hydraulic or air. *See also* manifold. [ARP480A] *2.* Any device for apportioning current or flow among various output paths. *3.* A formula that gives the probability that a value will fall within prescribed limits. [AIR4844]

disturbance resolution The minimum change caused by a disturbance in a measured variable that will induce a net change of the ultimately controlled variable.

disturbance variable A measured variable that is uncontrolled, and that affects the operations of the process.

dither *1.* A low-amplitude, relatively high-frequency periodic electrical signal, sometimes superimposed on the servovalve input to reduce threshold. Dither is expressed by the dither frequency (Hz) and the peak-to-peak dither current amplitude (mA). [ARP490] *2.* The deliberate, slow, periodic motion of symbology on the face of a CRT, intended to prevent degradation of phosphors. [ARP1782]

dive brakes/speed brakes Movable aerodynamic devices on aircraft which reduce airspeed during descent and landing. [ARP4107]

divergence *1.* The expansion or spreading out of a vector field; or a precise measure thereof. *2.* A static instability of a lifting surface or of a body on a vehicle in which the aerodynamic loads tending to deform the surface or body are greater than the elastic restoring forces. *3.* The spreading out of a laser beam with distance, measured as an angle.

divergence loss The portion of energy in a radiated beam that is lost due to nonparallel transmission or spreading.

diversion (technical) The landing of an aircraft at an airport other than the airport of origin or destination, as a result of the malfunction or suspected malfunction of any item on the aircraft. [AIR4896]

diversion valve A type of fluidic control device that uses the Coanda effect to either switch flow from one outlet port to another or proportion flow between two divergent outlet ports.

diversity The fact or quality of being different; of having variety; multiformity.

diversity combiner The device that accepts two radio signals from a single source that have been received with polarization, frequency, or space diversity, and combines them to yield an output that is better than either original signal.

diversity reception The use of two or more radio receivers, each being connected to different antennas, to improve the signal level. The antennas have diversity in space, phasing, and polarity.

dividing network *See* crossover network.

divisions *See* zones.

dizziness *See* vertigo.

DMA *See* direct memory access.

DME *1. See* distance measuring equipment. *2.* An abbreviation used in audio communications to mean simply distance.

DME separation Spacing of aircraft in terms of distances (nautical miles) determined by distance measuring equipment (DME). [ARP4107]

DMSP satellites Satellites of the defense meteorological satellite program, a program sponsored by the Space Division of the United States Air Force System Command, which provides timely global imagery and specialized meteorological data for supporting a variety of Department of Defense operations.

DMT *See* dead man timer.

DNC *See* direct numerical control.

docking *1.* The act of coupling two or more orbiting objects. *2.* The operation of mechanically connecting together, or in some manner bringing together, orbital payloads. [AIR1489]

docking speed The average speed of a vehicle during the last 18 in. of travel before contact with the aircraft. [AIR1558]

doctor blade A straight piece of material used to spread resin, as in application of a thin film of resin for use in hot melt prepregging or for use as an adhesive film. [AIR4844]

doctor roll In an adhesive spreader, a roller mechanism that revolves at a different surface speed, or in an opposite direction, resulting in a wiping action that regulates the adhesive supplied to the spreader roll. [AIR4844]

DoD Acronym for Department of Defense.

Dod FLIP *See* Department of Defense Flight Information Publications.

Dodge-Romig tables A set of standard tables with known statistical characteristics that are used in lot-tolerance and AOQL acceptance sampling.

DODISS Acronym for Department of Defense Index of Standards and Specifications. [AIR4779]

dog A machine tool accessory used as a clamp for gripping a piece of work and conveying motion to it.

dog clutch Two interfacing, rotating elements that transmit mechanical power when in contact by means of pins, teeth, dowels, or other load-carrying elements. [ARP4386]

doghouses (electronics) Small enclosures placed at the base of transmitting antenna towers to house antenna-tuning equipment.

dolly *1.* In filament winding, the planar reinforcement applied to a local area between windings to provide extra strength in an area where a cut-out is to be made, for example, port openings. [AIR4844]

dolly, catapult *1.* A low, mobile platform that rolls on casters. *2.* A wheeled apparatus used to support an aircraft for launching from a catapult. *3.* A low, wheeled platform used to support the nose wheel of an airplane for launching from an SATS catapult, and to which the aircraft is connected for launching. The airplane lifts off from the dolly at the end of the power stroke, and the dolly remains on the ground and is braked to a halt. [AIR1489]

DO loop A FORTRAN statement that directs the computer to perform that sequence to which it is keyed.

dome In filament winding, the portion of a cylindrical container that forms the spherical or elliptical shell ends of the container. [AIR4844]

dome nuts Plate nuts with a mechanical seal at the base and a cap over the top, which together provide a fuel-tight seal. Commonly used to permit screw or bolt attachments to inaccessible areas and/or access doors. [AIR4069]

dominant wavelength The wavelength of monochromatic light that matches a given color when combined in suitable proportions with a standard reference light.

door, strut fairing A door in the landing gear installation that allows fairing the gear (strut) well to the contours of the air vehicle after the strut is retracted into the vehicle. [AIR1489]

door, wheel fairing A door in the landing gear installation that allows fairing the wheel well to the contours of the air vehicle after the wheels are retracted inside the vehicle. [AIR1489]

dope A cellulose ester lacquer used as an adhesive or coating.

doped germanium A type of detector in which impurities are added to germanium to make the material respond to infrared radiation at wavelengths much longer than those detectable by pure germanium.

doping *1.* Adding a small amount of a substance to a material or mixture to achieve a special effect. *See also* additive. *2.* Coating a mold or mandrel to prevent a molded part from sticking to it.

Doppler effect The change in frequency with which energy reaches a receiver when the receiver and the energy source are in motion relative to each other.

Doppler-effect flowmeter A device that uses ultrasonic techniques to determine flow rate. In this device, a continuous ultrasonic beam is projected across fluid flowing through the pipe; the difference between incident-beam and transmitted-beam frequencies provides a measure of fluid flow rate.

Doppler-Fizeau effect The Doppler effect as applied to a source of light. When the distance between the observer and the source of light is diminishing, the lines of the spectrum are displaced toward the violet; when the distance is increasing, they are displaced toward the red; in each case, the displacement being proportional to the relative velocity of approach or recession.

Doppler navigation Dead reckoning performed automatically by a device that gives a continuous indication of position. This indication is produced by integrating the speed derived from measurement of the Doppler effect of echoes from directed beams of radiant energy transmitted from the craft.

Doppler radar Radar that utilizes the Doppler effect to determine the radial components of relative radar target velocities, or to select targets having particular radial velocities.

Doppler shift A phenomenon that causes electromagnetic or compression waves emanating from an object to have a longer wavelength if the object moves away from an observer than would be the case if the object were stationary with respect to the observer; and to have a shorter wavelength if the object moves toward the observer. Doppler shift is the physical phenomenon that forms the basis for analyzing certain sonar data and certain astronomical observations.

DOS *See* disk operating system.

dose The amount of radiation received at a specific location per unit area or unit volume; or the amount received by the whole body.

dose meter Any of several instruments for directly indicating radiation dose.

dose rate Radiation dose per unit time.

dose-rate meter Any of several instruments for directly indicating radiation dose rate.

dosimeter *1.* An instrument for measuring the ultraviolet in solar and sky radiation. *2.* Device worn by persons working around radioactive material, which indicates the dose of radiation to which they have been exposed.

DoT Acronym for Department of Transportation.

double acting Acting in two directions. Examples of double acting devices include a reciprocating compressor in which each piston has a working chamber at both ends of the cylinder; a pawl that drives in both directions; and a forging hammer that is raised and driven down by air or steam pressure.

double-action forming A metalforming process in which one stroke of the press performs two die operations.

double amplitude The peak-to-peak amplitude. *See* peak-to-peak.

double base powder (propellant) A powder or propellant containing nitrocellulose and another principle explosive ingredient, usually nitroglycerin. [AIR913]

double base propellants Solid rocket propellants using two unstable compounds, such as nitrocellulose and nitroglycerin. The unstable compounds used in a double based propellant do not require a separate oxidizer.

double-buffered I/O An input or output operation that uses two buffers to transfer data; while one buffer is being used by the program, the other buffer is being read from or written to by an I/O device.

double compression A weave in which the shute wires are compressed tightly by the comb of the loom so that they are deformed in the machine direction; the shute wires are in contact with one another. [AIR888]

double ended actuator An actuator with a single cylinder and a piston that has two rods extending to atmosphere. [ARP4386]

double groove weld A weldment in which the joint is beveled or grooved from both sides to prepare the joint for welding.

double pole A type of device, such as a switch, relay, or circuit breaker, that is capable of either closing or opening two electrical paths.

double precision *1.* Pertaining to the use of two computer words to represent a number. *2.* In floating-point arithmetic, the use of additional bytes or words to represent a number, in order to double the number of bits in the mantissa.

doubler *1.* Refers to a localized area of extra layers of reinforcement, usually to provide stiffness or strength for fastening or other abrupt load transfers. *2.* An extra piece of facing attached to strength or stiffen the panel or to distribute the load more widely into the core. [AIR4844]

double sampling A type of sampling inspection in which the lot can be accepted or rejected based on results from a single sample, or the decision can be deferred until the results from a second sample are known.

double-sided A computer diskette that stores data on both sides.

double spread The application of adhesive to both adherends in a joint. [AIR4844]

double stars Stars that appear as single points of light to the eye, but which can be resolved into two points by a telescope. A double star is not necessarily a binary (a two-star system revolving about a common center), but may be an optical double (two unconnected stars in the same line of sight).

doublet lens A lens with two components of different refractive index, generally designed to be achromatic.

double turbine turbomechanical actuator A turbomechanical actuator utilizing two turbines of opposite directional sense mechanically coupled together. [ARP4386]

double-welded joint A weldment in which the joint is welded from both sides.

double window fibers Optical fibers that are designed for transmission at two wavelength regions, 0.8 to 0.9 micrometer and around 1.3 micrometers.

double word An ordered set of 32 bits operated on as a pair of words or as a single unit. The most significant bit of a double word is labeled bit 31, and the least significant is labeled bit 0. [AS4710]

doughnut shape wheel *See* toroidal wheel.

downhand welding *See* flat-position welding.

downing event The event that causes an item to become unavailable to initiate its mission; the transition from up-time to down-time. [ARD50010]

downlink Message or data transmitted by an aircraft to the ground network. [ARP4102/13]

downlinking The transmission of signals (data, information, etc.) from satellites to ground terminals.

download Data or program transfer, usually from a larger computer to a PC.

downlock A mechanism or device for locking the landing gear in the down or extended position preparatory to landing. The downlock locks the structure in the proper position for taking loads imposed by ground operations. Some types used are: (a) an "on center" or "past center" side or drag brace held in position hydraulically and/or mechanically; (b) a secondary "jury brace," held on or past center hydraulically and/or mechanically; (c) a lock pin engagement; and (d) a hook and roller engagement. [AIR1489]

downrange The airspace extending downstream on a given rocket test range.

downtime That element of time during which an item is not in condition to perform its intended function. [ARD50010]

downtime, active maintenance The maintenance downtime during which work is being done on the item or aircraft. [AIR4896]

downtime, ground mean (GMDT) The average elapsed time between loss of function and restoration to performance-capable status. [AIR4896]

downtime, logistics That portion of downtime during which repair is delayed solely because of the necessity to wait for a replacement part or other subdivision of the system. [ARD50010]

downtime, maintenance The interval between the time an item or aircraft is made available for preventive or corrective maintenance until the item or aircraft is returned to, or considered available for, service. [AIR4896]

downtime, mean The average elapsed time between loss of mission-capable status and restoration of the system to mission-capable status. [ARD50010]

downtime, nonactive maintenance The maintenance downtime during which no work is done on a component or aircraft. [AIR4896]

downwash A flow of air deflected or forced downward, as by the passage of a wing, or by the action of a rotor or a rotor blade. Also referred to as rotor wash. *See also* wake turbulence. [ARP4107]

dowtherm A constant-boiling mixture of phenyl oxide and diphenyl oxide used in high-temperature heat transfer systems (boiling point 494°F, 257°C).

DP *See* draft proposal.

DP cell A pressure transducer that responds to the difference in pressure between two sources. Most often used to measure flow from the pressure difference across a restriction in a flow line.

DPCM (modulation) *See* differential pulse code modulation.

DR *See* dead reckoning.

draft Also spelled draught. 1. The side taper on molds and dies that makes it easier to remove finished parts from the cavity. 2. The small, positive pressure that propels exhaust gas out of a furnace and up the stack. 3. The difference between atmospheric pressure and some lower pressure existing in the furnace or gas passages of a steam-generating unit. 4. A preliminary document.

draft angle 1. The angle of a taper on a mandrel or mold that facilitates removal of the finished part. 2. The angle between the tangent to the surface at that point and the direction of ejection. [AIR4844]

draft differential The difference in static pressure between two points in a system.

draft gage A type of manometer used to measure small gas heads, such as the draft pressure in a furnace.

draft loss A decrease in the static pressure in a boiler or furnace due to flow resistance.

draft proposal (DP) The first stage of the ISO standard process.

drag *1.* The aerodynamic force in a direction opposite to that of flight; due to the resistance to movement brought to bear on an aerospace vehicle by the atmosphere through which it passes. [ARP4386] *2.* Resistance of a vehicle body to motion through the air due to total force acting parallel to and opposite to the direction of motion. *3.* In data processing, the movement of an object on a screen by using a mouse.

drag-body flowmeter A device that measures the net force on a submerged solid body in a direction parallel to the direction of flow, and converts this value to an indication of flow or flow rate.

drag brace A structural brace of a landing gear system whose primary function is to react drag loads imposed on the system. The drag brace may be a folding member and often serves a secondary function of locking the gear system in the extended and/or retracted position. [AIR1489]

drag brake A device that provides a retarding force on a body in motion through a fluid, parallel to the direction of motion of the body. *See also* ground spoilers and speed brakes. [AIR1489]

drag chute Any of various types of parachutes attached to high-performance aircraft that can be deployed, usually during landings, to decrease speed; and also, under certain flight conditions, to control and stabilize the aircraft. [AIR1489]

drag coefficients The ratios of drag to the products of dynamic pressures and reference areas.

drag effect *See* drag.

drag efficiency A measure of efficiency of a skid-control system, expressed as the ratio (percentage) of the actual drag performance produced over a given period of time (or distance) relative to the maximum (100%) drag performance obtainable for the given set of conditions. [AIR1489]

drag force anemometer Instrument for measuring both the static and dynamic velocity head and flow in high-frequency, unsteady flow.

drain A port carrying internal leakage flow for independent return to a reservoir, or for connection of a valved drainage connection. [ARP4386]

drain time That portion of the contact or dwell time during which the excess penetrant, emulsifier, or remover drains off the test piece. [AMS2647A]

draught *See* draft.

draw *1*. To shape plastic by stretching or deforming through dies. [AIR4844] *2*. To pull a load. *3*. To form cup-shaped parts from sheet metal. *4*. To reduce the size of wire or bar stock by pulling it through a die. *5*. To remove a pattern from a sand-mold cavity. *6*. A fissure or pocket in a casting caused by inadequate feeding of molten metal during solidification. *7*. A shop term for temper.

draw bead A contoured rib or projection on a draw ring or holddown to control metal flow in deep drawing.

drawbench The stand that holds the die and draw head used for reducing the size of wire, rod, bar stock, or tubing.

drawdown The curvature of the liquid surface upstream of the weir plate.

drawdown ratio The ratio of die opening to product thickness in a deep drawing operation.

drawhead *1*. The die holder on a drawbench. *2*. A group of rollers through which strip, tubing, or solid stock is pulled to form angle stock.

drawing A process used in the manufacture of wire. Consists of pulling the metal through a die, or series of dies, for the purpose of reducing the diameter. [ARP1931]

drawing back *1*. A shop term for tempering. *2*. Reheating hardened steel below the critical temperature to reduce its hardness.

drawing compound A lubricating substance, such as soap or oil, applied to prevent draw marks, scoring, or other defects caused by metal-to-metal contact during a stamping, wire-drawing, or similar metalforming operation.

drawing of samples The drawing at random of one or more units of product from a lot. [AIR4896]

drawing tower Equipment for making optical fibers, in which optical fibers are drawn from heated glass preforms.

draw mark Any surface flaw or blemish that occurs during drawing, including scoring, galling, pickup, or die lines.

drawn fiber Fiber with a certain amount of orientation imparted by the drawing process by which it was formed. [AIR4844]

draw radius The curvature at the edge of the cavity in a deep-drawing die.

draw ring A ring-shaped die part over which the punch pulls the draw blank during a drawing operation.

dredged material Sand, mud, silt, gravel, etc. recovered from the bottom of harbors, canals, etc. during dredging operations.

dredging Mechanical or hydraulic excavation of underwater material. Used in maintaining and building of channels and ports, as well as underwater mining of sand, gravel, and minerals.

dress To restore a tool to its original contour and sharpness.

drift *1*. A lateral divergence of an aircraft from the projected line of its heading, arising from movement due to wind. [ARP4386] *2*. A bar tube or shaft-type tool used for the specific purpose of assembling or separating two or more parts by axially applying impact or pressure by means of an external force. [ARP480A] *3*. When used in connection with analog transducers, analyzers, etc., usually expressed as the change in output over a specified time with fixed input and operation conditions.

drift angle The angle between the aircraft centerline and ground track; or the angular difference between true heading and ground track angle. Drift angle is "right" when ground track angle is greater than true heading; and "left" when ground track angle is less than true heading. [ARP1570]

driftpin A round, tapered metal rod that is driven into matching holes in mating parts to stretch them and bring them into alignment, such as for riveting or bolting.

drift plug A tapered rod that can be driven into a pipe to straighten it or flare its end.

drift rate The amount of drift, in any of its several senses, per unit time. Drift rate has many specific meanings in different fields; the type of drift rate should always be specified.

drill drift A flat, tapered piece of steel used to remove taper shank drills and other tools from their tool holders.

drill gage A flat, thin steel plate with numerous holes of accurate sizes that can be used to check the size of drills.

drill jig A tool constructed to guide a drill during repeated drilling of the same size holes—either at many locations in a given piece, or at the same location in many identical pieces—especially where exceptional straightness or accuracy of location is desired.

dripless dripstick A calibrated device installed and operated from the bottom of a fuel tank; used to manually determine the quantity of fuel contained in a fuel tank without leakage or loss of fuel from the tank. [AIR4783]

dripstick A calibrated device installed and operated from the bottom of a fuel tank, used to manually check a fuel tank to determine the quantity of fuel contained in the fuel tank. [AIR4783]

drive fit A type of interference fit requiring light to moderate force to assemble.

driven end The threaded end on a stud that is assembled into the aluminum or magnesium component with an interference fit and remains fast. [MAP1670]

driven gear The member(s) of a gear train that receive power and motion from another gear.

driver *1.* A hand-tool for assembling or removing components. *See also* extractor and pusher. [ARP480A] *2.* A software element that converts operator instructions into suitable language to drive a hardware device; for example, unit or stream drivers.

drive shaft A shaft that transmits power and motion from a motor or engine to the other elements of a machine.

drive stiffness (actuator stiffness) The stiffness of the actuator between the mounting and output motion attach points. Drive stiffness is the resultant of the mechanical stiffness of the load-carrying elements and of the actuation system stiffness. [ARP4386]

driving pinion In a gear train, the gear that receives power and motion by means of a shaft connected to the source of power, and transmits the power and motion via its teeth to the next gear in the train.

driving-point impedance The complex ratio of applied sinusoidal voltage, force, or pressure at the driving point of a transducer to resulting current, velocity, or volume velocity, respectively, at the same point, all inputs and outputs being terminated in some specified manner.

driving-point reactance The imaginary component of driving-point impedance.

driving-point resistance The real component of driving-point impedance.

drizzle Fairly uniform precipitation, composed exclusively of fine drops of water (diameter less than 1/50 of an inch) very close to one another. Drizzle drops are too small to cause appreciable ripples on the surface of still water.

drone A remotely controlled, self-powered aircraft or missile.

drone aircraft Remotely controlled aircraft.

drones for aerodynamic and struct test (DAST) *See* DAST program.

droop The difference between the control point and the setpoint due to some inherent control characteristic. Also called offset or deviation. [ARP89C]

drooped airfoils A baseline airfoil with an abrupt change in cross-section at about midspan from the fuselage. The outboard portion of the wing has a cross section with a nearly flat bottom and a drooped (downward) leading edge in relation to the inboard baseline wing.

drop *1.* Adjustment by airborne artillery observer to indicate decrease in range along spotting line. *2.* Dropping of airborne troops and supplies in a designated drop zone.

drop leg The section of measurement piping below the process tap location to the instrument.

drop number Sequential identification number of aluminum ingots poured simultaneously from the same cast. [ARP1917]

dropout Discrete variation in signal levels during the reproduction of recorded data, which results in data reduction error.

dropsondes Radiosondes equipped with a parachute, dropped from an aircraft to transmit measurement of atmospheric conditions as it descends.

drop test A type of test to determine the dynamic capabilities of a landing-gear shock-absorber system to accept the loads and kinetic energies specified and imposed on it. The shock-absorber system may be mounted in a test rig and dropped at specified loads and velocities, or a complete air vehicle may be free-dropped. The purpose of a drop test is to verify the energy-absorption capability and to optimize the energy control system (hydraulic, pneumatic, mechanical, etc.). [AIR1489]

drop tester May refer to either: (a) a hammer assembly that drops upon a test item; or (b) a device that is used to drop the test item in a prescribed manner onto a suitable surface. [ARP476A]

drop tower Large device for low-gravity processing of molten material. Consists of a capsule that is dropped, and/or a drop tube in which containerless low-gravity studies are conducted.

drosometer An instrument for measuring the amount of dew that condenses on a given surface.

drum *1.* Any machine element consisting essentially of a thin-walled, hollow cylinder. *2.* A thin-walled, cylindrical container, especially a flat-ended shipping container holding liquids or bulk solids and having a capacity of 12 to 110 gallons (50 to 400 liters). *3.* The cylindrical member around which a hoisting rope is wound. *4.* A high-capacity computer storage device.

drum baffle A plate or series of plates or screens placed within a drum to divert or change the direction of the flow of water or water and steam.

drum brake A mechanical brake in which the friction elements, normally called shoes, press against the inside surface of a cylindrical member (the drum) attached to the rotating element to slow or stop its motion.

drum course A cylindrical section of a drum.

drum head A plate closing the end of a boiler drum or shell.

drum internals All apparatus within a drum.

drum operating pressure The pressure of the steam maintained in the steam drum or steam-and-water drum of a boiler in operation.

dry To change the physical state of an adhesive on an adherent through the loss of solvent constituents by evaporation or absorption. [AIR4844]

dry air Air with which no water vapor is mixed. This term is used comparatively, since in nature there is always some water vapor included in air—and such water vapor, being a gas, is dry.

dry ash Refuse in the solid state, usually in granular or dust form.

dry assay Determining the amount of a metal or compound in an alloy, ore, or metallurgical residue by means that do not involve the use of liquid to separate or analyze for constituents.

dry back The baffle in a firetube boiler that joins the furnace to the second pass to direct the products of combustion; separate from the pressure vessel and constructed of heat-resistant material (generally refractory and insulating material).

dry-back boiler The baffle in a firetube boiler that joins the furnace to the second-pass to direct the products of combustion. The dry-back boiler is designed to be separate from the pressure vessel and is constructed of heat-resistant material (generally refractory and insulating material).

dry basis A method of expressing moisture content in which the amount of moisture present is calculated as a percentage of the weight of bone-dry material; used extensively in the textile industry. Also called regain moisture.

dry bay Space (cavity) within the mold line of an aircraft that is normally dry; but, in the event of combustible fluid leakage due to combat damage or natural causes, may contain an explosive gas mixture. [AIR4170]

dry blast cleaning Using a dry abrasive medium, such as grit, sand, or shot, to clean metal surfaces. The abrasive medium is driven against the surface with a blast of air or by centrifugal force.

dry-break disconnect A device normally mated to the inlet of a nozzle for ease of checking a strainer, or interchanging an underwing nozzle to an overwing nozzle, or vice versa, without the draining or spillage of the upstream hose or system. [AIR4783]

dry-bulb temperature The temperature of the air indicated by a thermometer that is not affected by the water vapor content of the air.

dry corrosion Atmospheric corrosion taking place at temperatures above the dew point.

dry-coupled bondtester A bondtester that operates on the pitch-catch surface wave principle. Couplant is not required. [ARP5089]

dryer A device used to remove water or water vapor from a refrigerant or other fluid. [ARP147C]

dry fiber area Area of fiber not totally encapsulated by resin. [AIR4844]

dry fuel rocket A rocket that uses a mixture of fast-burning powder. [ARP4386]

dry friction damping *See* coulomb damping.

dry gas Gas containing no water vapor.

dry-gas loss The loss representing the difference between the heat content of the dry exhaust gases and the heat content of the gases at the temperature of ambient air.

drying oil One of the many natural, usually vegetable oils (glyceryl esters of unsaturated fatty acids) that harden in air by oxidation to a resinous skin. [AIR4844]

drying oven A closed chamber for driving moisture from surfaces or bulk materials by heating them at relatively low temperatures.

drying temperature The temperature to which an adhesive on an adherent or in any assembly—or the assembly itself—is subjected to dry the adhesive. [AIR4844]

drying time *1.* The time required to dry a component with absorbed moisture to a specified level that can be measured. *2.* The time specified for drying a component, considered sufficient to achieve an adequate condition for bonding when measurement methods are not available. [AIR4844]

drying tower In prepregging via a solvent process, the drying section in which heated air is used to remove excess solvent from the prepreg; a conveyor belt carries the prepreg through the drying tower. [AIR4844]

dry joints Lack of adhesion due to insufficient adhesive or poor contact of mating surfaces. [AIR4844]

dry laminate A laminate containing insufficient resin for complete bonding of the reinforcement. [AIR4844]

dry lay-up Construction of a laminate by the layering of preimpregnated reinforcement in a female mold or on a male mold, usually followed by bag molding or autoclave molding. [AIR4844]

dry pipe A perforated pipe in the steam space above the water level in a boiler; helps keep entrained liquid from entering steam outlet lines.

dry spot On laminated plastics, an area of incomplete surface film. [AIR4844]

dry steam *1.* Steam containing no moisture. *2.* Commercially dry steam containing not more than one half of one percent moisture.

dry steam drum A pressure chamber, usually serving as the steam offtake drum, located above and in communication with the steam space of a boiler steam-and-water drum.

dry strength The strength of an adhesive joint determined immediately after drying under specified conditions or after a period of conditioning in a standard laboratory atmosphere. [AIR4844]

dry test meter A type of meter used extensively to determine gas flow rates for billing purposes and to calibrate other flow-measuring instruments. Has two chambers separated by a flexible diaphragm connected to a dial by means of a gear train. In operation, the chambers are filled alternately, with a flow-control valve switching from one chamber to the other as the first becomes completely filled; flow rate is indicated indirectly from movement of the diaphragm.

dry winding Describes filament winding using preimpregnated roving, as contrasted with wet winding, in which unimpregnated roving is pulled through a resin bath just before being wound on to a mandrel. [AIR4844]

DSRTK *See* desired track angle.

DTA (analysis) *See* thermal analysis.

DTK *See* desired track angle.

dual A level-control valve/system that has redundant, parallel/series elements that provide mode control independently. [AIR1660A]

dual-axis tracking antenna A tracking antenna that is steered automatically in both azimuth and elevation.

dual-beam analyzer *1.* A type of radiation-absorption analyzer that compares the intensity of a transmitted beam with the intensity of a reference beam of the same wavelength. *2. See* split-beam ultraviolet analyzer.

dual, central active *See* DCA.

dual control A double set of cockpit controls to permit (a) pilot or copilot or (b) instructor or student to fly the aircraft. [ARP4386]

dual, distributed active *See* DDA.

dual, distributed passive. *See* DDP.

dual-fail operative (DFO) A condition or requirement wherein an active control device or system can sustain any two failures within the system and remain operative. It is implicit with DFO that the system be able to accept identical but non-simultaneous failures in two of its channels and continue to operate with no nominal loss of performance. Unless specifically stated, it is understood that no nominal loss of performance occurs after one or two failures. [ARP4386]

dual harness reserve parachute assembly A certificated parachute assembly that is used for a premeditated jump by two people (a parachutist in command and a passenger); contains one main parachute assembly and one reserve parachute assembly. [AS8015]

dual in-line package (DIP) A common way of packaging semiconductor components.

dual input servovalve A servovalve that receives two inputs (electrical and mechanical or two independent electrical). [ARP4386]

duality principle Principle holding that for any theorem in electric circuit analysis, there is a dual theorem in which quantities are replaced with dual quantities. Examples are current and voltage or impedance and admittance.

duality theorem Theorem stating that if either of two dual linear programming problems has a solution, then so does the other.

dual load path A type of mechanical paralleling in which two separate load-carrying paths exist from the control system input to the system output. Each load path is capable of carrying sufficient load such that failure of any one member will not jeopardize system performance. [ARP4386]

dual ramp ADC Technique for converting analog data into digital format. The unknown voltage is input to a ramp generator and integrated for a specified time. At the end of this time, a counter is started and a reference voltage applied to cause a controlled ramp down. The counter is stopped when the voltage becomes zero. The count gives the digital number output.

dual ratio Dual-purpose gear box that permits engine starting at one speed and pumping or generating capability from the starter at another speed. The two gear systems are isolated from one another by overrunning clutches. [ARP906A]

dual rotation propellers (shafts) The concentric arrangement in which each component of the propeller rotates in an opposite direction at the same speed because of a fixed gear ratio between shafts. [ARP355]

dual search unit A single search unit containing two transducer elements, one used as a transmitter of ultrasonic energy, the other used as a receiver of ultrasonic energy. [ARP5089]

dual-slope converter An integrating analog-to-digital converter in which the unknown signal is converted to a proportional time interval.

dual system Special configurations that use two computers to receive identical input and execute the same routines, with the results of such parallel processing subject to comparison. Exceptional high-reliability requirements are usually involved.

dual-tandem valve A tandem valve having two separate control sections. [ARP4386]

dual thrust A rocket thrust derived from two propellant grains using the same propulsion section of a missile. [ARP4386]

dual twin gear configuration *See* gear configuration, dual twin.

dual twin tandem gear configuration *See* gear configuration, dual twin tandem.

dual wing configuration A configuration of two wings of nearly the same planform and area, one behind the other.

duct *1.* A single duct designed to be attached to the rear of a turbine engine, providing an uninterrupted discharge for the exhaust gases; not to be used when exhaust gases are to be controlled for proper velocity. [ARP480A] *2.* Specifically, a tube or passage that confines and conducts fluids; for example, a passage for the flow of air to the compressor of a gas turbine engine, or a pipe leading air to a supercharger.

duct assembly Detail parts fitted and joined together to form an integral part. [ARP699D]

ducted fan Fan enclosed in a duct.

ducted fan engine Aircraft engine incorporating a fan or propeller enclosed in a duct—especially jet engines in which an enclosed fan or propeller is used to ingest ambient air to augment the gases of combustion in the jetstream.

duct geometry The shape and dimension of ports or other openings designed for passage of fluid (gas, liquid, or mixtures) in or external to engines.

ductile fracture *See* fibrous structure.

ductile iron The term preferred in the United States for cast iron containing spheroidal nodules of graphite in the as-cast condition. Also known as nodular cast iron; nodular iron; spherulitic-graphite cast iron.

ductility The property of a metal that indicates its relative ability to deform without fracturing; usually measured as percent elongation or reduction of area in a uniaxial tensile test.

dud An explosive device that has failed to initiate as intended. [AIR913]

Dumet wire Wire made of Fe-42Ni, covered with a layer of copper. Used to replace expensive platinum as the seal-in wire in incandescent lamps and vacuum tubes; the copper coating prevents gassing at the seal.

dump *1.* A printout of computer memory or a file in hexadecimal and character form. *2.* The transfer of data without regard for its significance. Same as storage dump.

dump combustor Combustor having a means of reducing flow velocity and forming recirculation zones through the sudden enlargement area between the inlet duct and the combustion chamber.

dump valve *1.* A two-position control valve having two or more ports with all ports blocked in "off" position and all ports in "on" position. [ARP4386] *2.* A large valve in the bottom of a tank or container that can quickly empty the tank in an emergency.

duodecimal number A number consisting of successive characters and representing a sum, in which the individual quantity represented by each character is based on a radix of twelve. The characters used are 0, 1, 2, 3, 4, 5, 6, 7, 8, 9, T (for ten), and E (for eleven). *See also* number system.

duplex *1.* Pertaining to a twin, pair, or a two-in-one situation, e.g., a channel providing simultaneous transmission in both directions, or a second set of equipment to be used in event of the failure of the primary device. *2.* Referring to any item or process consisting of two parts working in connection with each other.

duplex cable A cable that contains two optical fibers in a single cable structure. Light is not coupled between the two fibers; typically one is used to transmit signals in one direction and the other to transmit in the opposite direction.

duplexer Device that permits a single antenna system to be used for both transmitting and receiving. Duplexers should not be confused with diplexers, devices permitting an antenna system to be used simultaneously or separately by two transmitters.

duplex film An adhesive film in which two different adhesives separated by a scrim cloth are manufactured into one film. [AIR4844]

duplex operation The operation of associated transmitting and receiving apparatus in which the processes of transmission and reception are concurrent.

duplex pump A reciprocating or diaphragm pump having two parallel flow paths through the same housing, with a common inlet and a common outlet.

durability *1.* The ability of an item of material to resist cracking, corrosion, thermal shock and degradation, delamination, wear, and the effects of foreign object damage for a specified period of time. *2.* A measure of the resistance of a material to wear. *3.* Physico-chemical change specified conditions of use and/or storage. [AIR4896]   *4.* A measure of useful life (a special case of reliability). [ARD50010]

durability analysis Application of engineering principles to determine the period of service usage that would degrade the structure below functional or economic limits. [AIR4896]

durability and damage tolerance analysis An analysis performed to determine the growth of flaws, cracks, and other damage in a structure versus time. [AS8879]

durability critical part Structure that is not safety-of-flight, or is not sized by damage tolerance requirements, but is sized by durability requirements and meets the criteria for critical parts selection. [AIR4896]

durability proof test *1.* A missionized engine duty schedule performed at the engine maximum limiting specification design requirements *2.* A stair-step schedule of engine operation at speed increments to maximum rotor speed, including thrust transients. [AIR4896]

durability structure Structure that is not practical to design or does not qualify as damage-tolerant. [AIR4896]

durometer *1.* A measurement of the compressibility of rubber. The lower the durometer number, the more flexible/compressible. [AIR1558]   *2.* A mechanical device used to measure the hardness of a sealant. A Shore A or REX A reading is produced. Some coatings are measured on a Shore D scale, which encompasses a higher hardness range. [AS7200/1]

dust Particles of gas-borne solid matter larger than one micrometer in diameter.

dust counter A photoelectric instrument that measures the number and size of dust particles in a known volume of air. Also known as Kern counter.

dust loading The amount of dust in a gas, usually expressed in grains per cu ft or lb per thousand lb of gas.

dustproof So constructed or protected that the accumulation of dust will not interfere with successful operation. [ARP4404]

dust-tight So constructed that dust will not enter the enclosing case. [ARP4404]

dutch oven A furnace that extends forward of the wall of a boiler setting. A dutch oven is usually of all refractory construction, although in some cases it is water-cooled.

dutch roll Oscillating motion of an aircraft combining rolling and yawing (roll-induced yaw); so named for the resemblance to the characteristic rhythm of an ice skater. [ARP4107]

Dutch weave A weave in which the warp wires are straight and heavier than the shute wires. [AIR888]

duty The specification of service conditions defining the type, duration, and constancy of applied load or driving power.

duty cycle *1.* The operating cycle required of the ignition system; expressed as a function of time ON and time OFF or continuous, as applicable. Generally associated with the ignition exciter specification. [AIR784]   *2.* The period during which a component is performing its designed function. The duty cycle may consist of multiple cycles. [ARP906A]   *3.* A description of the load throughout the total mission time with sufficient detail to determine load velocity requirements, frequency of occurrence, and dynamic load characteristics. A complete duty cycle description defines the actuation energy required for the total mission. [ARP4386]   *4.* For a device that operates repeatedly, but not continuously, the time intervals involved in starting, running, and stopping, plus any idling or warm-up time. *5.* For a device that operates intermittently, the ratio of working time to total time, usually expressed as a percent. Also known as the duty factor.   *6.* In an electric resistance welding machine, the percent of total operating time that current flows.

duty cyclometer A meter for directly indicating duty cycle.

duty factor *See* duty cycle.

dwarf galaxy Galaxy with low luminosity.

dwarf novae Short-period binary systems in which a red quasi-main sequence star fills its Roche lobe and transfers matter, via an accretion disk, onto a white dwarf.

dwell *1.* A pause in the application of pressure or temperature to a mold, made just before it is completely closed, to allow the escape of gas from the molding material. *2.* In filament winding, the time that the traverse mechanism is stationary while the mandrel continues to rotate to the appropriate point for the traverse to begin a new pass. *3.* In a standard autoclave cure cycle, an intermediate step in which the resin matrix is held at a temperature below the cure temperature for a specified period of time sufficient to produce a desired degree of staging; used primarily to control resin flow. [AIR4844] *4.* A contour on a cam that causes the follower to remain at maximum lift for an extended portion of the cycle. *5.* In a hydraulic or pneumatic operating cycle, a pause during which pressure is neither increased nor decreased.

dwell time In any variable cycle, the portion of the cycle during which all controlled variables are held constant; for example, to allow a parameter, such as temperature or pressure, to stabilize; or to allow a chemical reaction to go to completion.

dye penetrant A low-viscosity liquid containing a dye used in nondestructive examination to detect surface discontinuities, such as cracks and laps, in both magnetic and nonmagnetic materials.

dynamic In motion; active; moving. [AIR4783]

dynamically relocatable coding Coding for a computer that has special hardware to perform the derelativization. With an appropriately designed computer system, coding can be loaded into various sections of core, the appropriate addresses can be changed, and the program executed.

dynamic braking load The load applied to a nose tire during deceleration of an aircraft during a landing or rejected takeoff. [AS4833]

dynamic bulk modulus For a fluid, the product of the mass fluid density and the square of the speed of sound through the fluid. [AIR1116]

dynamic bus control The operation of a data bus system in which designated terminals are offered control of the data bus. [AS15531]

dynamic bus control acceptance bit The status word bit at bit time 18. [AS4113]

dynamic calibration A calibration procedure in which the quantity of liquid is measured while liquid is flowing into or out of the measuring vessel.

dynamic compensation A technique used in control to compensate for dynamic response differences to different input streams to a process. A combination of lead and lag algorithms will handle most situations.

dynamic convergence *See* convergence, dynamic.

dynamic impedance (stiffness) The impedance associated with the output deflections of an active closed-loop actuation system caused by externally applied dynamic forces (usually sinusoidal) over a specified frequency range. Dynamic impedance includes the load mass or inertia, load friction, system stiffness, and any other load-related compliance effects. Since it is a complex quantity, the term dynamic impedance is preferred to the term dynamic stiffness. [ARP4386]

dynamic information Information that changes during the time of relevant task performance. [ARP4155]

dynamic load A load imposed by dynamic action, as distinguished from a static load. Specifically, with respect to aircraft, rockets, or spacecraft, a load due to an acceleration, as imposed by gusts, maneuvering, landing, firing rockets, braking, etc. [AIR1489]

dynamic luminance (shades of gray) The difference between adjacent luminance values that is readily distinguishable by the human eye. This follows a power law relationship to luminance and is limited to colors that are not highly saturated. A minimum ratio between two adjacent luminance values of 1.414 (square root of 2) is normally used. [ARP1782]

dynamic memory *See* dynamic storage.

dynamic model Model of aircraft or other objects in which linear dimensions, weight, and moments of inertia are reproduced in scale, in proportion to the original.

dynamic modulus The ratio of stress to strain under vibratory conditions. [AIR4844]

dynamic optimization A type of control, frequently multivariable and adaptive in nature, that optimizes some criterion function in bringing the system to the setpoints of the controlled variables. The sum of the weighted, time-absolute errors is an example of a typical criterion function to be minimized. Contrast with steady-state optimization.

dynamic pressure *1.* The pressure of the air through which an aircraft is flying, determined by the velocity of the aircraft and the density of the air. [ARP4386] *2.* The portion of the local pressure in a fluid that is recovered when the fluid is brought to rest. [ARP4386]

dynamic programming In operations research, a procedure for optimization of a multi-stage problem in which a number of decisions are available at each stage of the process. Contrast with convex programming, integer programming, linear programming, mathematical programming, nonlinear programming, and quadratic programming.

dynamic range *1.* For an APC, the ratio of smallest to the largest particle that can be processed by the instrument; for example, a size range of 2.0-120 m would yield a ratio or dynamic range of 60:1. [ARP1192] *2.* The range of signals that is accepted by a device without manual adjustment.

dynamic response The behavior of an output in response to a changing input.

dynamics Study of the motion of a system of material particles under the influence of forces, especially those that originate outside the system under consideration.

dynamic sensitivity In leak testing, the minimum leak rate that a particular device is capable of detecting.

Dynamics Explorer satellites Two satellites that have been designed to occupy different orbits and supply comparative data for studying the boundary region between earth and space. Of the 24 goals of the program, one half require data from both satellites; one fourth, data from one satellite; and one fourth, data from the other satellite. The satellites were launched together in August of 1981.

dynamic stability For an aircraft or other body, the characteristics that cause the body, when disturbed from an original state of steady flight or motion, to damp the oscillations set up by restoring moments and gradually return to the original state; specifically, the aerodynamic characteristics. [AIR1489]

dynamic stiffness *1.* The stiffness associated with the output deflections of an active actuation system caused by externally applied dynamic forces, usually sinusoidal, over a specified frequency range. *2. See* dynamic impedance. [ARP4386]

dynamic storage The storage of data on a device or in a manner that permits the data to move or vary with time, thus causing the data to be not always available instantly for recovery; for example, the use of an acoustic delay line, storage on a magnetic drum, or circulation or recirculation of information in a medium. Synonymous with dynamic memory.

dynamic storage allocation A storage allocation technique in which the location of programs and data is determined by criteria applied at the moment of need.

dynamic subroutine A subroutine that involves parameters such as decimal point position or item size, from which a relatively coded subroutine is derived. The computer is expected to adjust or generate the subroutine according to the parametric values chosen.

dynamic suction lift In a pump, differential height measured between the fuel surface level and flow inlet of the pump during operation. [AIR4783]

dynamic test A test of a device or mechanism conducted under variable loading or stimulation.

dynamic torque The torque that a brake will produce for deceleration dynamically, as opposed to static torque. Dynamic torque is affected by velocity, temperature, lining coefficient, and other factors. [AIR1489]

dynamic unbalance *1.* Condition in which the central principal axis is not coincident with the shaft axis. NOTE: The quantitative measure of dynamic unbalance can be given by two complementary unbalance vectors in two specified planes (perpendicular to the shaft axis), which completely represent the total unbalance of the rotor. [ARP588A] *2.* A condition in rotating equipment in which the axle of rotation does not exactly coincide with one of the principal axes of inertia for the mechanism. Dynamic unbalance produces additional forces and vibrations that, if severe, can lead to failure or malfunction.

dynamic variable Process variables that can change from moment to moment due to unspecified or unknown sources.

dynamometer Instrument for measuring power or force; specifically, an instrument for measuring the power, torque, or thrust of aircraft engines or rockets.

dynamometer, brake test A device for measuring (testing) forces and power or capability of brakes. Generally, the system includes a large, rotating steel wheel (mass), which is brought up to speed by external power, and represents an inertia equivalent to which the brake is to be tested. The test brake/wheel/tire assembly is loaded against the dynamometer wheel and the power of the brake is utilized to stop the rotating mass. *See also* dynamometer, tire test. [AIR1489]

dynamometer, tire test A device for measuring (testing) capabilities of tires. Differs from a brake test dynamometer in that an inertia equivalent is generally not required. The dynamometer wheel is powered to rotate and the tire is loaded radially against it. Tire test parameters of interest are usually load, speed, time, and distance. *See also* dynamometer, brake test. [AIR1489]

dyna-soar A boost-glide, manned spacecraft system for demonstrating the feasibility of operating at near and orbital speeds, reentering the atmosphere at hypersonic speeds, and maneuvering to a controlled landing. [ARP4386]

Dzus Fastener A trade name for a quick-release type fastener, designed to permit rapid removal of inspection plates or aircraft cowling. [ARP4107]

E

e The base of natural logarithms.

EADI *See* electronic attitude direction indicator.

earing Forming a scalloped edge around a deep-drawn sheet-metal part due to directional properties in the blank material.

early failure period The period of life, after final assembly, in which failures occur at an initially high rate because of the presence of defective parts and workmanship errors. [AIR4896]

earphones Electroacoustic transducers operating from an electrical system to an acoustical system and intended to be closely coupled acoustically to the ear. Synonymous with headset.

earth axis Any one of a set of mutually perpendicular reference axes established with the upright axis (the Z axis) pointing to the center of the earth, used in describing the position or performance of an aircraft or other body in flight. The earth axes may remain fixed or may move with the aircraft or other object.

earth currents *See* telluric currents.

earth hydrosphere The part of the earth that consists of the oceans, seas, lakes, and rivers.

earth mantle The zone of the earth below the crust and above the core (to a depth of 3480 km). The earth mantle is divided into the upper mantle and the lower mantle, with a transition zone between.

earth observations (from space) The acquisition of earth surface data from aircraft or spacecraft.

earth observing system (EOS) NASA orbital multisensor observatory system for the long-term acquisition of earth sciences data—operated in conjunction with an integrated ground-based science information system.

earth reference or inertial (I) Referenced with respect to the earth, as opposed to being referenced with respect to air mass. [ARP1570]

earth shape *See* geodesy.

earth terminal measurement system NBS system for measuring electromagnetic parameters of communication satellites and ground stations relative to antenna gain, ratio of carrier power to operating noise temperature, and satellite effective isotropic power.

earth terminals Portable or stationary ground-based equipment used to transmit and receive signals and other data via satellites in communications networks.

EAS *See* equivalent airspeed.

EBCDIC *See* extended binary coded decimal interchange code.

ebullition The act of boiling or bubbling.

EBW *See* exploding bridge wire.

eccentric Describes a rotating mechanism whose center of rotation does not coincide with the geometric center of the rotating member.

eccentricity A measure of the center of location of a conductor with respect to the circular cross section of the insulation. Expressed as a percentage of center displacement of one circle within another. [ARP1931]

eccentric orifice An orifice whose center does not coincide with the centerline of the pipe or tube. Usually, the eccentricity is toward the bottom of a pipe carrying flowing gas and toward the top of a pipe carrying liquid, which tends to promote the passage of entrained water or gas, rather than allowing entrained water or gas to build up in front of the orifice.

echo check A check of accuracy of transmission in which the information that was transmitted to an output device is returned to the information source and is compared with the original information to insure accuracy of output.

echoes *1.* Waves that have been reflected or otherwise returned with sufficient magnitude and delay to be detected as a wave distinct from that directly transmitted. *2.* In radar, a pulse of reflected radio-frequency energy; the appearance on a radar indicator of the energy returned from a target.

echo ranging A form of active sonar in which the sonar equipment generates pulses of sound, then determines the distance to underwater objects. In echo ranging, distance to an object is measured by precisely measuring the time it takes for a pulse to reach the object, be reflected, and return to a known location, usually one adjacent to the transmitter; by using a narrow, focused sound beam, direction to the object can also be determined.

eclipse *1.* The reduction in visibility or disappearance of a nonluminous body when it passes into the shadows cast by another nonluminous body. *2.* The apparent cutting off, wholly or partially, of the light from a luminous body by a dark body coming between it and the observer.

ecliptic The apparent annual path of the sun among the stars; the intersection of the plane of the earth's orbit with the celestial sphere. The ecliptic is a great circle of the celestial sphere inclined at an angle of about 23 degrees 27 minutes to the celestial equator.

ECM *See* electrochemical machining.

econometrics The application of mathematics and statistical techniques to the testing and quantifying of economic theories and the solution of economic problems.

economic analysis A systematic approach to the problem of choosing how to employ scarce resources; and an investigation of the full implications of achieving a given objective in the most efficient and effective manner. [ARP4293]

economizer bag *See* bag, economizer. [ARP171]

economizers Heat exchangers used to recover excess thermal energy from process streams. Economizers are used for feed preheat and as column reboilers; in some systems, the reboiler for one column is the condenser for another.

eddy A whirlpool of fluid. *See also* vortices.

eddy current An electric current set up in the near-surface region of a metal part by induction resulting from the electromagnetic field of an external coil carrying an alternating current. Eddy currents are used to generate heat or electromagnetic fields for use in such applications as induction heating, electromagnetic sorting and testing of materials, vibration damping in spacecraft, and various types of instrumentation.

eddy-current tachometer A device for measuring rotational speed; consists of a permanent magnet revolving in close proximity to an aluminum disk, which is pivoted to turn against a spring. As the magnet revolves, it induces eddy currents in the disk, setting up torque that acts against the spring. The amount of disk deflection is indicated by a moving pointer directly coupled to the disk. Eddy-current tachometers have been used extensively in automotive speedometers.

eddy-sonic Refers to a process in which sonic or ultrasonic energy is produced in a test part by a coil on or near the surface of the test part; the coil is used to produce eddy currents in the test part. Vibrations in the test part result from the interaction of the magnetic field from the eddy currents in the test part with the magnetic field of the coil. [ARP5089]

eddy-sonic bondtester A dry-coupled bondtester operating on the eddy-sonic principle. [ARP5089]

eddy viscosity The turbulent transfer of momentum by eddies, giving rise to an internal fluid friction. Occurs in a manner analogous to the action of molecular viscosity in laminar flow, but taking place on a much larger scale.

edge bleed Removal of volatiles and excess resin through the edge of the laminate, as in matched die molding of a laminate. [AIR4844]

edge close-outs Members placed around the sides of a panel to protect the sandwich from damage or to attach the panel to a support or to another panel. [AIR4844]

edge delamination A separation of the detail parts along an edge after the assembly has been cured. [AIR4844]

edge distance ratio The distance from the center of the bearing hole to the edge of the specimen, in the direction of the principal stress, divided by the diameter of the hole. [AIR4844]

edge filter An interference filter that abruptly shifts from transmitting to reflecting over a narrow range of wavelengths.

edge joint A joint made by bonding the edge faces of two adherents. [AIR4844]

EDM *See* electrical discharge machining.

EDP *See* electronic data processing.

eductor A device that withdraws a fluid by aspiration and mixes it with another fluid. *See also* injector.

EED *See* electro-explosive device.

effective bandwidth An operating characteristic of a specific transmission system equal to the bandwidth of an ideal system whose uniform pass-band transmission equals maximum transmission of the real system, and whose transmitted power is the same as the real system for equal input signals having a uniform distribution of energy at all frequencies.

effective case depth The maximum depth below the surface at which a hardness of 50 HRC, or equivalent, is achieved. [AMS2759/8]

effective cutoff frequency For a transducer, the frequency at which the insertion loss between two terminating impedances exceeds the loss at some reference frequency in the transmission band by a specified value.

effective internal resistance The apparent opposition to current within a battery; manifests itself as a drop in battery voltage proportional to the discharge current. [ARP4386]

effectiveness The probability that a material will operate successfully when required. [ARP4386]

effectiveness, built in test A measure of the automated capability of a system to: (a) correctly ascertain the operating condition of a subsystem/function; and (b) isolate defective item(s) at a designated ambiguity level, without the use of equipment external to the system. The built-in test (BIT) function may be contained within the item being evaluated; it may be a separate piece of system equipment; or it may incorporate both of these characteristics. [ARD50010]

effectiveness, heat exchanger Effectiveness as defined for either the hot side or the cold side.

Hot side–the temperature drop of the hotter fluid divided by the maximum temperature drop theoretically obtainable through the use of infinite heat transfer surface. *Cold side*–the temperature rise of the cooler fluid divided by the maximum temperature rise theoretically obtainable through the use of infinite heat transfer surface. [ARP147C]

effective resistance The equivalent pure d-c resistance that, when substituted for the winding being checked, will draw the same power. [ARP667]

effective tread depth T_2 Effective dimension from one stair nosing to the adjacent stair vertical riser. [ARP836A]

effectivity An indication of the applicability of items, materials, and/or technical data to a type, series, model, or individual item. [AIR4896]

efficiency *1.* The ratio of the output to the input; for example, the efficiency of a steam-generating unit is the ratio of the heat absorbed by water and steam to the heat in the fuel fired. *2.* In manufacturing, the average output of a process or production line expressed as a percent of its expected output under ideal conditions. *3.* The ratio of useful energy supplied by a dynamic system to the energy supplied to it over a given period of time.

efficiency, adiabatic *1.* For turbines, the ratio of the actual dry air enthalpy drop through the turbine to the enthalpy drop for a reversible adiabatic expansion. *2.* For compressors, the ratio of the enthalpy rise of dry air in the compressor for a reversible adiabatic compression to the actual air enthalpy rise. [ARP147C]

efficiency, volumetric For a specific item, the percentage ratio of actual flow to the exact flow established by design parameters of record (e.g., speed, displacement). [ARP906A]

effluent Liquid waste discharged from an industrial processing facility or waste-treatment plant.

effluvium Waste byproducts of food or chemical processing.

E format In FORTRAN, an exponential type of data conversion denoted by Ew.d, where w is the number of characters to be converted as a floating point number, with d spaces reserved for the digits to the right of the decimal point; for example, E11.4 yields 000.5432E03 as input, 543.2 internally, and 0.5432E+03 as output.

E glass A low-alkali lime borosilicate glass made into glass fiber filaments; used in composite materials.

EGT Literally, exhaust gas temperature; however, as used, EGT refers to the temperature measurement defined by the engine manufacturer as a limiting parameter (for example, MGT, TIT, or TOT). [AIR1873]

eigenvalues The roots (or solution parameters) of the characteristic equation of a system. The eigenvalues are complex numbers, each with a real and an imaginary part. [AIR1823]

eigenvalue sensitivity A quantative measure of the change in system eigenvalues for a given change in the magnitude of a design parameter. [AIR1823]

EIS *See* extended instruction set.

EISCAT radar system (Europe) The European incoherent scatter radar system.

ejecta Matter ejected during impact cratering processes, usually meteoritic.

ejection *1.* Physical removal of an object from a specific site, such as removal of a cast or molded product from a die cavity by hand, compressed air, or mechanical means. *2.* Emergency expulsion of a passenger compartment from an aircraft or spacecraft. *3.* Withdrawal of fluid from a chamber by the action of a jet pump or eductor.

ejector *1.* A device used in conjunction with the mechanism of a machine, fixture, or the like, which automatically sorts or throws out completed, accepted, or rejected items from the working station. [ARP480A] *2.* Device consisting of a nozzle, mixing tube, and diffuser; utilizes the kinetic energy of a fluid from a low-pressure region by direct mixing and ejecting both streams.

ejector condenser A direct-contact condenser in which vacuum is maintained by a jet of high-velocity injection water, which simultaneously condenses the steam and discharges water, condensate, and noncondensible gases to the atmosphere or to the next stage ejector.

Ekman layer The layer of transition between the surface boundary layer of the atmosphere, where the shearing stress is constant, and the free atmosphere. The Ekman layer is treated as an ideal fluid in approximate geostrophic equilibrium.

elapsed time The duration for which an engine operates at a particular mission element. [ARP1352]

elastance A measure of the increase of airway pressure for a given increase of lung volume. [ARP171]

elastic chamber The portion of a pressure-measuring system that is filled with the medium whose pressure is being measured, and that expands and collapses elastically with changes in pressure; examples include the Bourdon tube, bellows, flat or corrugated diaphragm, spring-loaded piston, or a combination of two or more single elements, which may be the same or different types.

elastic collision A collision between two or more bodies in which the internal energy of the participating bodies remains constant, and in which the kinetic energy of translation for the combination of bodies is conserved.

elastic deformation The changes in dimensions of an item in response to stress, such that the item returns to its original dimensions when the stress is removed. [ARP700]

elasticity Refers to the property of a material that enables it to regain its original dimension when stress is relieved.

elastic limit The greatest stress that a material is capable of sustaining without permanent strain remaining after the complete release of the stress. [AIR4844]

elastic modulus *See* modulus of elasticity.

elastic recovery The fraction of a given deformation that behaves elastically. A perfectly elastic material has an elastic recovery of 1; a perfectly plastic material has an elastic recovery of 0. [AIR4844]

elastic scattering A collision between two particles, or between a particle and a photon, in which total kinetic energy and momentum are conserved.

elastomer A material that possesses elastic properties similar to those of natural rubber in the vulcanized state. At room temperature, an elastomer can be stretched repeatedly to at least twice its original length and will, upon release of stress, return to its approximate original length. [AS1933]

elastomeric tooling A tooling system that uses the pressure from the thermal expansion of rubber materials to form composite parts during cure. [AIR4844]

Elber equation In fatigue crack propagation studies, an equation describing the effective stress range ratio as a function of stress ratio: the effective stress range ratio $U = 0.5 + 0.4R$, where R is the stress ratio.

elbow *1*. A fitting that connects two pipes at an angle, usually 90°, but may be any other angle less than 100°. *2*. Any sharp bend in a pipe.

elbow meter A pipe elbow that is used as a flow-measurement device by placing a pressure tap at both the inner and outer radius and measuring the pressure differential caused by differences in flow velocity between the two flow paths.

electric Containing, producing, arising from, actuated by, or carrying electricity; or designed to carry electricity and capable of doing so. [ARP4404]

electrical Related to, pertaining to, or associated with electricity, but not having its properties or characteristics. [ARP4404]

electrical actuation systems (EAS) All electrical, mechanical, optical, or fluidic components necessary to convert a command signal and electrical power from the vehicle into a controlled linear or rotary force while using an electromagnetic prime mover. [ARP4255]

electrical computing resolver An electromagnetic device having primary and secondary windings so oriented that the voltages on the secondaries are sine and cosine functions of the angular position of the rotor with respect to the stator, multiplied by linear functions of the voltages applied to the primaries. [ARP826]

electrical conductivity A characteristic that indicates the relative ease with which electrons flow through a material; usual units are %IACS, which relate the conductivity of the material to that of annealed pure copper. Electrical conductivity is the reciprocal of electrical resistivity.

electrical contact *See* contact, electrical.

electrical control power The power dissipation required for control of a valve. Control power is a maximum with full input signal, and is zero with zero-input signal. It is independent of the coil connection (series, parallel, or differential) for any conventional two-coil operation. NOTE: For differential operation, the control power is the power consumed in excess of the electrical quiescent power. This power increase is a result of the differential current change. [ARP490]

electrical cure monitoring *See* cure monitoring, electrical.

electrical discharge machining (EDM) Often referred to by its abbreviation, EDM. A machining method in which stock is removed by melting and vaporization under the action of rapid, repetitive spark discharges between a shaped electrode and the workpiece through a dielectric fluid flowing in the intervening space. Process variations include electrical discharge grinding and electrical discharge drilling. Also known as electroerosive machining; electron discharge machining; electrospark machining.

electrical dissipation factor For a dielectric material, the ratio of the power loss in the material to the total power transmitted through it. [AIR4844]

electrical-electronic system *See* system, electrical-electronic.

electrical engineering A branch of engineering that deals with practical applications of electricity, especially the generation, transmission, and utilization of electric power by means of current flow in conductors.

electrical grounding The connecting of the aircraft or other vehicle to be refueled or defueled, or the vehicle accomplishing the refueling operation, to a grounding rod suitably inserted into the earth. [AIR4783]

electrical load management center Equipment that contains localized power buses and utilizes power controllers for protection and remote control of branch and load circuits. [AS1831]

electrical logic Logic for mode switching or failure detection and correction, performed with electronic or electrical components. [ARP181A]

electrically actuated valves Hose valves that rely only on electrical power to position the valve. [ARP986]

electrically operated extensions Usually a highly sensitive induction-type device for signaling high or low flows or deviations from any set flow. Consists of a sensing coil positioned around the extension tube of the rotameter. Movement of the metering float into the field of the coil causes a low-level signal change, which is usually amplified to a level suitable for performing annunciator or control functions.

electrical pumping Deposition of energy into a laser medium by passing an electrical current or discharge, or a beam of electrons, through the material.

electrical quiescent power The dissipation required for differential operation when the current through each coil is equal and opposite in polarity. [ARP490]

electrical resistivity A characteristic that indicates the relative resistance of a material to the flow of electrons; usual units are ohm-m (SI) or ohms per circular-mil foot (US customary). Electrical resistivity is the reciprocal of electrical conductivity.

electrical steel Low-carbon steel that contains 0.5 to 5% Si or other material. Produced specifically to have enhanced electromagnetic properties suitable for making the cores of transformers, alternators, motors, and other iron-core electric machines. Contrast with electric steel.

electrical stroke *1.* The displacement range over which the output signal and all electrical parameters are achieved. [ARP4386] *2. See* measurement range.

electric contact Either of two opposing, electrically conductive buttons or other shapes that allow current to flow in a circuit when they touch each other. The electric contacts are usu-

ally attached to a spring-loaded mechanism that is mechanically or electromagnetically operated to control whether or not the contacts touch.

electric controller An assembly of devices and circuits that turns electric current to an electrically driven system "off" and "on" in response to a stimulus. In most instances, the assembly also monitors and regulates one or more characteristics of the electric supply—voltage or amperage, for example.

electric corona A luminous, and often audible, electric discharge that is intermediate in nature between a spark discharge (which usually has a single discharge channel) and a non-point discharge (which has a diffuse, quiescent, nonluminous character). Also known as corona discharge.

electric discharge The flowing of electricity through a gas, resulting in the emission of radiation that is characteristic of the gas and the intensity of the current.

electric/electronic extensions A system that converts float position to a proportional electric signal (either a-c or d-c); or to a proportional shift or unbalance in impedance, which is balanced by a corresponding shift in impedance in the receiving instrument.

electric field Within a medium or evacuated space, a condition that imposes forces on stationary or moving electrified bodies in direct relation to their electric charges.

electric field strength The magnitude of an electric field vector.

electric-furnace steel *See* electric steel.

electric hybrid vehicle Surface vehicle whose propulsion systems consist of both electric motors and conventional internal combustion engines.

electric hygrometer An instrument that uses an electrically powered sensing means to determine ambient atmospheric humidity.

electric instrument An indicating device for measuring electrical attributes of a system or circuit. Contrast with electric meter.

electricity meter A device for indicating the time integral of an electrical quantity.

electric meter A recording or totalizing instrument that measures the amount of electric power generated or used as a function of time. Contrast with electric instrument.

electric motor Motors powered by direct current (DC) or alternating current (AC). In pumps, electric motors are wound to operate on the specified voltage and frequency of the input current to provide the speed and torque conditions required by the pump. [ARP4386]

electric potential In electrostatics, the work done in moving unit positive charge from infinity to the point whose potential is being specified.

electric propulsion A general term encompassing all of the various types of propulsion in which the propellant consists of charged electrical particles accelerated by electrical and/or magnetic fields; for example, electrostatic propulsion, electromagnetic propulsion, and electrothermal propulsion.

electric steel Any steel melted in an electric furnace, thereby allowing close control of composition. Also known as electric-furnace steel. Contrast with electrical steel.

electric strength In a dielectric, the property that opposes a disruptive discharge. [AIR4844]

electric stroboscope A device in which an electric oscillator or similar element is used to produce precisely timed pulses of light. In an electric stroboscope, oscillator frequency can be controlled over a wide range so that the device can be used to determine the frequency of a mechanical oscillation (e.g., rpm of a rotating shaft or frequency of a mechanical vibration); this is done by determining light-pulse frequency at which the object appears motionless.

electric telemeter An apparatus for remotely detecting and measuring a quantity—including the detector intermediate means, transmitter, receiver, and indicating device. In an electric telemeter, the transmitted signal is conducted electrically to the remote indicating or recording station.

electric wire A single metallic conductor of solid, stranded, or tinsel construction, designed to carry current in an electric circuit. [ARP4404]

electroacoustic transducer A transducer for receiving waves from an electric system and delivering waves to an acoustic system, or vice versa. Microphones and earphones are examples of electroacoustic transducers.

electrochemical cleaning Removing soil by the chemical action induced by passing an electric current through an electrolyte. Also known as electrolytic cleaning.

electrochemical coating A coating formed on the surface of a part due to chemical action induced by passing an electric current through an electrolyte.

electrochemical corrosion Corrosion of a metal due to chemical action induced by electric current flowing in an electrolyte. Also known as electrolytic corrosion.

electrochemical machining (ECM) A machining method in which stock is removed by electrolytic dissolution. Accomplished through the action of a flow of electric current between a tool cathode and the workpiece, through an electrolyte flowing in the intervening space. Also known as electrochemical milling or electrolytic machining.

electrochemical milling *See* electrochemical machining.

electrochemical recording A type of recording system in which a signal-controlled electric current passes through a sensitized recording medium, usually in sheet form, inducing a chemical reaction in the medium.

electrochemical transducer A device in which a chemical change is used to measure an input parameter, and in which the output electrical signal is proportional to the input parameter.

electrochromism A phenomenon whereby a select number of solid materials will change color when an electric field is applied.

electrode *1.* Terminal at which electricity passes from one medium into another. The positive electrode is called the anode; the negative electrode is called the cathode. *2.* In a semiconductor device, an element that performs one or more of the functions of emitting or collecting electrons or holes; or of controlling their movements by an electric field. *3.* In an electron tube, a conducting element that performs one or more of the functions of emitting, collecting, or controlling, by an electromagnetic field, the movements of electrons or ions.

electrode characteristic For a given electrode in a system, the relation between electrode voltage and electrode amperage, with the voltages of all other electrodes in the system held constant; usually shown as a graph.

electrode force In electric-resistance spot, seam, or projection welding, the force that tends to compress the electrodes against the workpiece. Also known as welding force.

electrode holder The material that is joined to the top of the electrode or electrode stub to provide the connection between the electrode and the VAR furnace electrical equipment. [AMS2380C]

electrodeposition Any electrolytic process that results in deposition of a metal from a solution of its ions; includes processes such as electroplating and electroforming. Also known as electrolytic deposition.

electrodes or plates A conducting body containing active materials, by which the electrochemical reaction occurs. [ARP4386]

electrode stub The material that may be joined to the top of an electrode to provide the connection between the electrode and the electrode holder. [AMS2380C]

electrode voltage The electric potential difference between a given electrode and the system cathode or a specific point on the cathode; the latter is especially applicable when the cathode is a long wire or filament.

electrodynamic instrument An electrical instrument having both a fixed and a moving coil, both of which carry all or part of the current to be measured. If the coils are connected in series, interaction of the fields induced by the coils produces a torque proportional to the square of the current (as in an a-c voltmeter); if connected in parallel, the torque is proportional to the product of the two coil currents (as in an a-c ammeter). In both cases, the indication is an effective (rms) value.

electrodynamics The science dealing with forces and energy transformations or electric currents and the magnetic fields associated with them.

electroepitaxy Crystal growth process achieved by passing an electric current through the substrate solution.

electroerosive machining *See* electrical discharge machining.

electro-explosive device (EED) Any cartridge, squib, ignitor, detonator, etc. that is initiated electrically. [AIR913]

electroformed molds A mold made by electroplating metal on the reverse pattern of the cavity. [AIR4844]

electroforming Shaping a component by electrodeposition of a thick metal plate on a conductive pattern. In electroforming, the part may be used as formed, or it may be sprayed on the back with molten metal or other material to increase its strength.

electrogalvanizing Coating a metal with electrodeposited zinc.

electrograph *1.* A tracing produced on prepared, sensitized paper or other material by passing an electric current or electric spark through the paper. *2.* A plot or graph produced by an electrically controlled stylus or pen.

electrohydraulic actuator A self-contained EAS in which an electric motor drives a hydraulic pump to operate a hydraulic actuation system. [ARP4255]

electrohydraulic servovalve A servovalve in which the input is electrical and the output is hydraulic fluid; the electric control signal is sometimes amplified hydraulically in the first stage to drive the second stage power-control valve spool. [ARP4386]

electrohydrostatic actuator An actuator configuration in which a variable-speed-reversible electric motor is used to drive a hydraulic pump coupled to a piston or vane hydraulic motor. [ARP4386]

electrojets Laterally limited, relatively intense electric currents located in the ionosphere.

electroless deposition Controlled autocatalytic reduction method of depositing coatings.

electroless plating Deposition of a metal from a solution of its ions, without the use of impressed electric current; occurs through a chemical reduction induced when the basis metal is immersed in the solution.

electroluminescence Emission of light caused by the application of electric fields to solids or gases. In gas electroluminescence, light is emitted when the kinetic energy of electrons or ions accelerated in an electric field is transferred to the atoms or molecules of the gas in which the discharge takes place.

electrolysis The production of chemical changes by passage of an electric current through an electrolyte. [ARP1931]

electrolyte *1.* Any substance that, when dissolved in water or other suitable solvent, forms a solution that conducts electricity, the conductivity being due to ionic dissociation of the dissolved substance. [AIR4844] *2.* The ionically conductive alkaline solution used in the nickel cadmium cell. [AS8033]

electrolytic cleaning *See* electrochemical cleaning.

electrolytic corrosion *1.* Corrosion caused by electrochemical reactions. [ARP1931] *2. See* electrochemical corrosion.

electrolytic deposition *See* electrodeposition.

electrolytic etching Engraving a pattern on a metal surface by electrolytic dissolution.

electrolytic grinding A combined grinding and electrochemical machining operation in which an electrically conductive grinding wheel is made the cathode and the workpiece the anode, and an electric current is impressed between them in the presence of a chemical electrolyte.

electrolytic hygrometer An apparatus for determining water-vapor content of a gas by directing it at known flow rate through a teflon or glass tube coated on the inside with a thin film of P_2O_5 (phosphorus pentoxide), which absorbs water from the flowing gas. Once absorbed, the water is dissociated by a d-c voltage impressed on a winding embedded in the hygroscopic film; the resulting current represents the number of molecules dissociated. A calculation based on flow rate, current, and temperature then yields water concentration in ppm.

electrolytic machining *See* electrochemical machining.

electrolytic pickling Removal of scale and surface deposits by electrolytic action in a chemically active solution.

electrolytic polishing *See* electropolishing.

electrolytic powder Metal powder that is produced directly or indirectly by electrodeposition.

electromagnet Any magnet assembly whose magnetic field strength is determined by the magnitude of an electric current passing through some portion of the assembly.

electromagnetic *See* electromagnetism.

electromagnetic acceleration The use of perpendicular components of electric and magnetic fields to accelerate a current carrier.

electromagnetic compatibility The capability of systems and associated equipment to perform within required levels in the specified electromagnetic environment. [AS1831]

electromagnetic environment experiment Shuttleborne radio-frequency experiment.

electromagnetic instrument Any instrument in which the indicating means or recording means is positioned by mechanical motion controlled by the strength of an induced electromagnetic field.

electromagnetic interference (EMI) *1.* Any electrical or electronic disturbance phenomenon, signal, or emission (man-made or natural)—except deliberately generated interference—that causes undesirable responses; unacceptable responses; malfunctions; degradation of performance; or premature and undesired location, detection, or discovery by enemy forces. [ARP1281A] *2.* Generally refers to interference at frequencies that are generated inside the system, as contrasted with interference coming from sources outside the system. [AIR4728]

electromagnetic pulse (EMP) A type of disturbance that leads to noise in radio-frequency electric or electronic circuits.

electromagnetic radiation *1.* Energy propagated through space or through material media in the form of an advancing disturbance in electric and magnetic fields. The term "radiation" alone is commonly used for this type of energy, although it actually has a broader meaning. *2.* An all-inclusive term for any wave having both an electric and a magnetic component. The spectrum of electromagnetic waves includes (in order of increasing photon energy, increasing frequency, and decreasing wavelength): radio waves, infrared, visible light, ultraviolet, x-rays, gamma rays, and cosmic rays.

electromagnetic spectra Spectra of known electromagnetic radiations, extending from the shortest cosmic rays, through gamma rays, x-rays, ultraviolet radiation, visible radiation, and including microwave and all other wavelengths of radio energy.

electromagnetic wave A wave in which both the electric and magnetic fields vary periodically, usually at the same frequency. *See also* electromagnetic radiation.

electromagnetism *1.* Magnetism produced by an electric current. *2.* The science dealing with the physical relations between electricity and magnetism.

electromechanical actuator An EAS that utilizes an electric motor mechanically coupled to the load. [ARP4255]

electromechanics The technology associated with mechanical devices and systems that are electromagnetically or electrostatically actuated or controlled.

electrometallurgy The technology associated with the recovery and processing of metals using electrolytic and electrical methods.

electrometer *1.* An instrument for measuring differences in electric potential. *2.* An instrument for measuring electric charge, usually by means of the forces exerted on one or more charged electrodes in an electric field.

electrometer tube A high-vacuum tube with an exceptionally low control-electrode conductance, used to measure extremely small d-c voltages or amperages.

electromotive force Force capable of maintaining a potential difference—and thus a current—within a circuit. Electromotive forces can be established by chemical action or by mechanical work.

electron An elementary subatomic particle having a rest mass of 9.107×10^{-28} g and a negative charge of 4.802×10^{-10} statcoulomb. Also known as a negatron. A subatomic particle of identical weight and positive charge is termed a positron.

electron acceleration The acceleration of electrons by action of solar cosmic rays.

electron avalanche In a gas subjected to a strong electric field, the process whereby a relatively small number of free electrons accelerate, ionize gas atoms by collision, and thus form new free electrons, which undergo the same process in cumulative fashion.

electron beam *1.* A focused stream of electrons used to neutralize the positively charged ion beam in an ion engine. *2.* A device used to melt or weld materials with externally high melting points.

electron cyclotron heating A type of radio-frequency plasma heating in which high-power microwave energy is introduced into the plasma region.

electron device Any device whose operation depends on conduction by the flow of electrons through a vacuum, gas-filled space, or semiconductor material.

electron discharge machining *See* electrical discharge machining.

electron emission Ejection of free electrons from the surface of an electrode into the adjacent space.

electron flux *See* flux (rate).

electron gun An electron-tube subassembly that generates a beam of electrons, and may also accelerate control, focus, or deflect the beam.

electron-hole drops Exciton condensations exhibiting the properties of electrically conducting plasmas. Electron-hole drops form in germanium and silicon crystals at sufficiently low cryogenic temperatures.

electronic *1.* Of, pertaining to, or involving electrons or electronics. *2.* The science and technology of electronic devices and systems.

electronic aircraft Designation for tactical electronic warfare aircraft.

electronic amplifier *See* amplifier.

electronic data processing (EDP) Data processing performed largely by electronic equipment.

electronic engineering A branch of engineering that deals chiefly with the design, fabrication, and operation of electron-tube or transistorized equipment, used to generate, transmit, analyze, and control radio-frequency electromagnetic waves or similar electrical signals.

electronic heating Producing heat by the use of radio-frequency current generated and controlled by an electron-tube oscillator or similar power source. Also known as high-frequency heating or radio-frequency heating.

electronic photometer *See* photoelectric photometer.

electronics That branch of physics that treats the emission, transmission, behavior, and effects of electrons.

electronic switch A circuit element causing a start and stop action or a switching action electronically, usually at high speeds.

electronic transition A transition in which an electron in an atom or molecule moves from one energy level to another.

electronic unit An item that can be removed and replaced within the end item, such as a weapon replaceable assembly or a line replaceable unit. [AIR4896]

electronic warfare Military action involving the use of electromagnetic energy to determine, exploit, reduce, or prevent hostile use of the electromagnetic spectrum; and action that retains friendly use of the electromagnetic spectrum.

electron metallography Using an electron microscope to study the structure of metals and alloys.

electron microprobe analysis A technique for determining concentration and distribution of chemical elements over a microscopic area of a specimen. The specimen is bombarded with high-energy electrons in an evacuated chamber, then x-ray fluorescent analysis is performed on the secondary x-radiation emitted by the specimen.

electron microscope Any of several designs of apparatus in which diffracted electron beams are used to make enlarged images of tiny objects.

electron multiplier tube A type of electron tube that uses cascaded secondary emission to amplify small amperages.

electron spectroscopy The study and interpretation of atomic, molecular, and solid-state structure of substances based on x-ray induced electron emission from the substances.

electron tube *See* tube.

electron volt Abbreviated eV (preferred) or EV. A unit of energy equal to the work done in accelerating one electron through an electric potential difference of one volt.

electro-optic effect A change in the refractive index of a material under the influence of an electric field. Kerr and Pockels effects are, respectively, quadratic and linear in electric field strength.

electropainting Electrodeposition of a thin layer of paint on metal parts that have been made anodic. Also known as electrophoretic painting.

electrophoretic painting *See* electropainting.

electroplate The application of a metallic coating to a surface by means of electrolytic action. [ARP1931]

electroplating Electrodeposition of a thin layer of metal on a surface of a part that is in contact with a solution (electrolyte) containing ions of the deposited metal. In most electroplating processes, the part to be plated is the cathode, and the concentration of metal ions in the solution is maintained by placing a sacrificial anode of the deposited metal in the electrolyte.

electropneumatic controller An electrically powered controller in which some or all of the basic functions are performed by pneumatic devices.

electropolishing Smoothing and polishing a metal surface by closely controlled electrochemical action; similar to electrochemical machining or electrolytic pickling. Also known as electrolytic polishing.

electroscope An instrument for detecting an electric charge by observing the effects of mechanical force exerted between two or more electrically charged bodies.

electrospark machining *See* electrical discharge machining.

electrostatic *1.* Of, pertaining to, or produced by static electric charges. *2.* The scientific study of static electricity.

electrostatic bonding Use of the particle-attracting property of electrostatic charges to bond particles of one charge to those of the opposite charge.

electrostatic discharge A large electrical potential moving from one surface or substance to another. [AIR4844]

electrostatic instrument Any instrument whose operation depends on forces of electrostatic attraction or repulsion between charged bodies.

electrostatic lens A set of electrodes arranged so that their composite electric field acts to focus a beam of electrons or other charged particles.

electrostatic memory A memory device that retains information by means of electrostatic charge, usually involving a special type of cathode-ray tube and associated circuits.

electrostatic microphone An electroacoustic transducer for converting sound into an electrical signal by means of variation in electrostatic capacitance of the active transducer element.

electrostatic painting A spray-painting process in which the paint particles are charged by spraying them through a grid of wires that is held at a d-c potential of about 100 kV. The parts being painted are connected to the opposite terminal of the high-voltage circuit so that they attract the charged paint particles. Electrostatic painting yields more uniform coverage than conventional spray painting, especially at corners, edges, recesses, and oblique surfaces.

electrostatic precipitator A device for removing dust and other finely divided matter from a flowing gas stream by electrostatically charging the particles, then passing the gas stream over charged collector plates, which attract and hold the particles.

electrostatic voltmeter An instrument for measuring electrical potential by means of electrostatic forces between elements in the instrument.

electrostriction The phenomenon wherein some dielectric materials experience an elastic strain when subjected to an electric field, this strain being independent of the polarity of the field.

electrostriction transducer A device that consists of a crystalline material that produces elastic strain when subjected to an electric field; or that produces an electric field when strained elastically. Also known as a ceramic transducer or piezoelectric transducer.

electrothermal process Any process that produces heat by means of an electric current; may be an electric arc, induction, or resistance method. Electrothermal processes are used especially when temperatures higher than those obtained by burning a fuel are required.

elemental error The bias and/or precision error associated with a single error source. [AIR1678]

element melt time The time elapsed from the moment a fusing current begins to flow to the moment the current sharply drops in value and arcing commences. [ARP1199A]

elevated temperature repair A repair using a resin system that will be cured at temperatures above 70°C (160°F). [AIR4844]

elevation Vertical distance above a reference level, or datum, such as sea level.

elevation error A type of error in temperature-measuring or pressure-measuring systems that incorporate capillary tubes partly filled with liquid; introduced when the liquid-filled portion of the system is at a different level than the instrument case, the amount of error varying with distance of elevation or depression.

elevator A movable horizontal airfoil, usually attached to the horizontal stabilizer, that is used to control pitch. [ARP4107]

elevator angle The acute angle between the chord of the aircraft elevator moved from its neutral position and its chord in neutral position. The angle is positive when the trailing edge of the elevator is below the neutral position. [ARP4386]

elevator control Control of the pitch motion of an aircraft as effected by deflections of the ailerons, spoilers, and rudder. [ARP4386]

elevator illusion See illusions, vestibular.

elevons Wing control surfaces combining functions of ailerons and elevators.

elinvar An iron-nickel-chromium alloy that also contains varying amounts of manganese and tungsten; has low thermal expansion and almost invariable modulus of elasticity. Chief uses are for chronometer balances, watch balance springs, instrument springs, and other gage parts.

ellipse Plane curve constituting the locus of all points the sum of whose distances from two fixed points (called focuses or foci) is constant; an elongated circle.

ellipsoid Surface whose plane sections (cross sections) are all ellipses or circles, or the solid enclosed by such a surface.

ellipsometer An optical instrument that measures the constants of elliptically polarized light; most often used in thin-film measurements.

elliptically polarized light *1.* Light in which the polarization vector rotates periodically, changing in magnitude with a period of 360°, so it describes an ellipse. *2.* The result of two plane-polarized beams of light (each approximately a sine wave) perpendicular to each other and having a constant phase difference; the resultant plane-polarized wave in the direction of the common beams will describe an ellipse. A special case called circular polarization occurs when the amplitudes of the two plane-polarized waves are equal and the phase difference is an odd multiple of $\pi/2$.

elliptically polarized wave Any electromagnetic wave whose electric and/or magnetic field vector at a given point describes an ellipse.

elliptical polarization The polarization of a wave radiated by an electric vector rotating in a plane and simultaneously varying in amplitude so as to describe an ellipse.

ellipticity The amount by which a spheroid differs from a circle; calculated by dividing the difference in the length of the axes by the length of the major axis.

elongated alpha In titanium, the hexagonal crystal phase appearing as stringer-like arrays, considerably larger in appearance than the primary alpha. Commonly exhibits an aspect ratio of 3:1 or higher. [AS1814]

elongation *1.* The permanent extension in the gage length of a test specimen, measured after rupture. (The extension measured at the moment of rupture is termed "ultimate elongation.") Unless otherwise specified, elongation is expressed as a percentage of the original gage length; for example, if a 1-inch gage is marked on an unstretched specimen and the specimen is stretched until the gage marks are 7 inches apart, the elongation is 6 inches or 600 percent. [AS1198] *2.* Change in dimensions per unit of dimension; expressed as a percentage. [ARP700]

elongation at break Elongation recorded at the moment of rupture of the specimen, often expressed as a percentage of the original length. [AIR4844]

elongation rating The maximum elongation/distortion measured in percent; determined by pulling the fender until failure. [AIR1558]

ELT *See* emergency locator transmitter.

eluent The mobile phase used to sweep or elute the sample (solute) components into, through, and out of the column. [AIR4844]

elutriation Separation of fine, light particles from coarser, heavier particles by passing a slow stream of fluid upward through a mixture so that the finer particles are carried along with it.

embedded model An engine model included as part of a larger digital control application. [AIR4548]

embolism The occlusion of a blood vessel by an embolus. *See* embolus. [ARP171]

embolus Undissolved material, such as a clot, plug, fat globule, or gas bubble, carried by the blood from one vessel and forced into a smaller one so as to occlude or obstruct the circulation. [ARP171]

embrittlement cracking A form of metal failure that occurs in steam boilers at riveted joints and at tube ends, the cracking being predominantly intercrystalline.

emergency break away/emergency dry break A device mounted strategically in a fuel system that, when acted upon by a specific force, will separate and break the hose or system connection with a minimum of spillage. [AIR4783]

emergency electric-system operation Condition of the electric-system during flight in which the primary electric-system becomes unable to supply sufficient or proper electric power, thus requiring the use of a limited independent source(s) of emergency power for essential utilization equipments. [AS1212]

emergency equipment The class of equipment intended to assist the flight crew in coping with emergency operating conditions; includes portable and fixed protective breathing equipment, smoke protective devices, fire extinguishers, crash axes, first aid kits, and evacuation devices. [ARP917A]

emergency exit An opening that may be used for emergency evacuation of the aircraft, divided into three classes: Class A, Class B, and Class C. *Class A*–openings primarily intended for personnel use. *Class B*–openings primarily intended for servicing or personnel use. *Class C*–openings primarily intended for emergency use. [ARP4101/7]

emergency locator transmitter (ELT) A radio transmitter, attached to the aircraft structure, that operates from its own power source on 121.5 MHz and 243.0 MHz. Aids in locating downed aircraft by radiating a downward-sweeping audio tone between two and four times per second. Designed to function independently after an accident. [ARP4107]

emergency locking retractor (inertia reel) A retractor incorporating adjustment hardware by means of a locking mechanism that is activated by aircraft acceleration, webbing movement relative to the aircraft, or other automatic action during an emergency; when locked, capable of withstanding restraint forces. [AS8043]

emergency maintenance *1.* An urgent need for repair or upkeep that was unpredicted or not previously planned work. *2. See* corrective maintenance.

emergency mode Refers to a condition of the electric system in which the emergency system is used to power a selected, reduced complement of distribution and utilization equipment. [AS1831]

emergency oxygen Oxygen normally used upon ejection from the aircraft; may also be used as an emergency source of oxygen in the event of failure of the bleed-air supply or on-board oxygen generation system. [ARP171]

emergency parachute assembly A certificated parachute assembly worn for emergency; for unpremeditated use only. [AS8015]

emergency placarding Any durable visual intelligence, fixed in place, that provides instructions for locating and/or operating airplane exits and emergency equipment. [ARP577]

emergency power unit (EPU) A device that can provide short-term emergency electric, pneumatic, or hydraulic power for engine starting, flight controls, etc. in the even of primary source failure. [ARP906A]

emergency safe altitude *See* minimum safe altitude.

emergency valve A valve positioned at the outlet of a vehicle fuel tank so that it can be closed to prevent spillage in an emergency. [AIR4783]

emery An abrasive material composed of pulverized, impure corundum; used in various forms, including cloth or paper with an adhesive-bonded layer of emery grains, and compacted emery-binder mixtures shaped into cakes, sticks, stones, grinding wheels, and other implements.

EMI *See* electromagnetic interference. [ARP1161]

emission characteristic The relation between rate of electron emission and some controlling factor (e.g., temperature, voltage, or current of a filament or heater) for a specific element of a system—all other factors being held constant.

emission index The mass of emissions of a given constituent per unit mass of fuel, multiplied by 1000. [AIR1533]

emission spectra The spectra of wavelengths and relative intensities of electromagnetic radiation emitted by a given radiator. Each radiating substance has a unique, characteristic emission spectrum, just as every medium of transmission has its individual absorption spectrum.

emissivity Also known as emittance. *1.* For a given material, a property measured as the emittance of a specimen that is thick enough to be completely opaque and has an optically smooth surface. *2.* For a given material, a characteristic determined as the ratio of radiant-energy emission rate due solely to temperature for an opaque, polished surface of the material, divided by the emission rate for an equal area of a blackbody at the same temperature. *3.* The rate at which electrons are emitted from a solid or liquid surface when additional energy is imparted to the system by radiant energy, such as heat or light; or by energetic particles, such as a beam of electrons.

emittance *See* emissivity.

emitter A transduction device that transforms electrical to optical energy and launches optical radiation, according to a defined mathematical law; an optical source. [ARD50024]

emotion *See* affective states, emotions.

EMP *See* electromagnetic pulse.

empennage The fuselage assembly located at the aft end of an aircraft, comprised of the horizontal and vertical stabilizers and their associated control surfaces. [ARP4386]

empty field myopia *See* illusions, visual.

empty weight The weight of an aircraft with fixed ballast, unusable fuel, and full operating fluids, including oil, hydraulic fluid, and other fluids required for normal operation of the systems of the airplane. *See also* basic empty weight. [ARP4107]

EMR The earlier name of the Data Systems Division within Fairchild-Weston, a Schlumberger corporation, supplier of telemetry/computer systems.

EMS *See* engine monitoring system.

E-MSAW *See* en route minimum safe altitude warning.

EMT *See* maintenance time, elapsed.

emulate To imitate one system with another such that the imitating system accepts the same data, executes the same programs, and achieves the same results as the imitated system.

emulsification time The total time that an emulsifier is permitted to combine with the penetrant prior to removal by water. [AMS2647A]

emulsifier A substance that can be mixed with two immiscible liquids to form an emulsion. Also known as a disperser or dispersing agent.

emulsion A suspension of fine particles or globules of a liquid within a liquid. [AIR4844]

emulsion characteristic curve A graph of relative transmittance of a developed photographic or radiographic emulsion versus exposure; alternatively, a graph of a function of transmittance versus a function of exposure.

emulsion corrosion test A test used to evaluate the rust-preventive property of grease-water emulsions when coated steel objects are exposed to the severely corrosive atmosphere of the salt spray test. [S-5C,40-55]

enable *1.* To restore a computer system to ordinary operating conditions. *2.* To "arm" a software or hardware element to receive and respond to a stimulus. *3.* To allow the processing of an established interrupt. *4.* To remove a blocking device, e.g., switch, to permit operation. Contrast with disable. *See* arm.

enamel *1.* Thin ceramic coating, usually of high glass content, applied to a substrate, generally a metal. *2.* A type of oil paint that contains a finely ground resin, and that dries to a harder, smoother, glossier finish than other types of paint. *3.* Any relatively glossy coating, but especially a vitreous coating on metal or ceramic obtained by covering it with a slurry of glass frit and firing the object in a kiln to fuse the coating. Also known as porcelain enamel and vitreous enamel.

encapsulated microcircuit Microelectronic circuit enclosed in plastic.

encapsulation The enclosure of an item in plastic or other material. [AIR4844]

Enceladus A satellite of Saturn orbiting at a mean distance of 238,000 kilometers.

enclosed Surrounded by a case that will prevent accidental contact of a person with internal parts carrying electric current. [ARP4404]

enclosed, totally So enclosed as to prevent circulation of air between the inside and the outside of the case, but not necessarily sufficiently to be termed airtight. [ARP4404]

encode To substitute letters, numbers, or characters for other numbers, letters, or characters, usually to intentionally hide the meaning of the message—except from certain individuals who know the enciphering scheme.

encoder A device capable of translating from one method of expression to another method of expression; for example, translating a message, "add the contents of A to the contents of B," into a series of binary digits. Contrast with decoder.

encrustation The buildup of slag, corrosion products, biological organisms (e.g., barnacles) or other solids on a structure or exposed surface.

encryption Converting data into codes that cannot be read without a key or password.

end *1.* An individual yarn or strand in a braid or shield. [AS1198] *2.* In computer programming, a word indicating the completion of a program structure.

end around carry A carry from the most significant digit place to the least significant digit place.

end burning Describes a solid propellant grain that is inhibited and burns from one end only, such that burning progresses in the direction of the longitudinal axis. [AIR913]

end device The last device in a chain of devices that performs a measurement function; the device that performs the final conversion of a measured value into an indication, record, or control-system input signal.

end of descent (EOD) The metering fix, clearance, or other scheduled end point for the main descent from en route cruise. If a level flight deceleration segment is programmed at the end flight level, the EOD or BOD (bottom of descent) will be the downstream end of that segment. The EOD or BOD includes a geographical position, altitude, and speed. It may also include a required or estimated time of arrival. [ARP1570]

endothermic A chemical reaction that absorbs heat energy. [AIR4844]

endothermic reaction A reaction that occurs with the absorption of heat.

end play *1.* The maximum difference between the high and low readings of a dial indicator suitably arranged to measure the total axial

travel of the shaft in its own bearings upon the reversal of a specified axial load. [ARP667] *2.* Axial movement in a shaft-bearing assembly due to clearances within the assembly.

end point In titration, an experimentally determined point close to the equivalence point, which is used as the signal to terminate titration. End point is used instead of equivalence point in most calculations, and corrections for the error between end point and equivalence point usually are not applied.

endpoint control The exact balancing of process inputs required to satisfy stoichiometric demands.

endpoint linearity The linearity of the object taken between the end points of calibration.

ends See braid end.

end scale value On a given instrument, the value of an actuating electrical quantity that corresponds to the high end of the indicating or recording scale.

end-system System that originates a message or is designated as the recipient of a message. [ARP4791]

end-to-end data system Comprehensive data system that demonstrates the processing of sensor data to the user, thus reducing data fragmentation.

endurance *1.* The time an aircraft can continue flying under given conditions without refueling. *2.* The time a system continues to perform without deviating from the norm. [ARP4386]

endurance limit The maximum stress below which a material can presumably withstand an infinite number of stress cycles; if the stress is not completely reversed, the minimum stress also should be given. *See also* fatigue strength.

endurance testing The process of subjecting material to stress levels within design limits until failure occurs or until the desired life has been demonstrated. [ARP4386]

energetic particles Charged particles having energies equaling or exceeding 100 MeV.

energize To apply rated voltage to a circuit or device in order to activate it. [ARP1931]

energizing frequency sensitivity The changes in phase speed sensitive output voltages, phase shift, and zero speed voltages due only to changes in energizing frequency within a specified frequency range. [ARP667]

energizing voltage winding The tachometer generator winding that is energized by a fixed voltage. [ARP667]

energy *1.* The output capacity of a cell or battery expressed as capacity times voltage or watthours. [ARP4386] *2.* Any quantity with dimension that can be represented as mass times length squared divided by time squared. *3.* The capacity of a body for doing work or its equivalent. May be classified as potential or kinetic, depending on whether it is associated with bodies at rest or bodies in motion; or it may be classified as chemical, electrical, electromagnetic, electrochemical, mechanical, radiant, thermal, vibrational—or any other type, depending on its source or nature.

energy balance The balance relating the energy in and energy out of a column. In control applications, the energy balance manipulative variables are reflux and boilup.

energy beam An intense ray of electromagnetic radiation, such as a laser beam, or of nuclear particles, such as electrons, that can be used to test materials or to process them by cutting, drilling, forming, welding, or heat treating.

energy, brake The portion of the total energy of an air vehicle that, in the deceleration process, goes to the brake. (Other portions go to aerodynamic drag, rolling resistance, etc.) [AIR1489]

energy density *1.* A figure of merit for batteries, expressed by the stored energy per unit of battery weight or volume. Energy density is dependent on the discharge rate. [ARP4386] *2.* Light energy per unit area, expressed in joules per square meter—equivalent to the radiometric term "irradiance." *See also* flux density.

energy, drag chute The portion of the total energy of an air vehicle that, in the deceleration process, goes into the drag chute (in the form of aerodynamic drag). [AIR1489]

energy efficiency transport program *See* ACEE program.

energy exchanger A generic term for any of several devices whose primary function is to transfer energy from one medium to another; examples include heat exchangers, boilers, and electrical transformers.

energy gap (solid state) A range of forbidden energies in the band theory of solids.

energy level Any one of different values of energy that a particle, atom, or molecule may adopt under the given conditions; the possible values are restricted by quantizing conditions.

energy limited Describes a starter that can provide only a fixed amount of energy on a given start. Examples of energy-limited starters include the cartridge starter, fuel-air starter, hydraulic starter operating from an accumulator, and battery-operated electric starter. [ARP906A]

energy power brake A dynamic brake that brings to a stop a shaft that is in motion. [ARP4386]

energy, shock absorber The portion of the total air vehicle energy that, in the landing process, goes into the shock absorber(s). Generally, this energy is the part due to the vertical sink speed of the vehicle at touchdown, although some of this energy is damped out in the air vehicle structure and tires. [AIR1489]

energy, tire The portion of the total air vehicle energy that, in the landing process, goes into the tires. The tires also absorb energy during the braking process. [AIR1489]

engagement The connecting of a starter shaft to the engine. Sometimes called re-engagement if the engine is not at rest. [ARP906A]

engagement, crash The connecting of the starter shaft to the engine while the engine is decelerating and the starter is running at a no-load speed greater than that of the engine. The differential speed that exists at engagement can cause a destructive crash engagement. Sometimes called crash re-engagement. [ARP906A]

engagement force Force required to mate the couplings to permit fluid flow. [AIR4728]

engagement, running The connecting of the starter shaft to the engine while the engine is running (usually decelerating or steady windmilling) at some speed below starter cutoff speed. Except for the effects of backlash in the drive train, the engagement takes place when the starter speed is the same as that of the engine. Sometimes called running re-engagement. [ARP906A]

engaging mechanism A device for connecting the starter shaft to the engine. [ARP906A]

engaging mechanism, jaw type An engaging mechanism that uses matching elements with jaw teeth formed on their faces. One of the elements is typically attached to the engine-side accessory shaft; the other is attached to the starter and is caused to move axially into engagement with the engine-side jaw. Separation is automatic and the mechanism permits the starter elements to come to rest between start cycles. [ARP906A]

engaging mechanism, overrunning clutch A type of clutch mechanism that employs sprags, rollers, pawls, etc., with appropriate driving and driven members. This type of clutch drives in one direction only, overrunning in the other direction. [ARP906A]

engine airframe integration Physics of the interface between the engine and the airframe.

engine, basic Those units and components that are used to induce and convert a fuel/air mixture into thrust/power. [AIR4896]

engine bleed systems Engine compressor bleed systems used for engine starting, auxiliary power, and wing anti-icing. Bleed gas is normally defined as "high-pressure" air that is extracted from the compressor section of turbine engine. Pressures may range to 250 psig and 1000°F in present-day aircraft. [AIR744A]

engine block *See* cylinder block.

engine condition monitoring *See* engine health monitoring.

engine dress kit Typically consists of an engine-mounted nacelle, aerodynamic hardware, accessories, and test instrumentation required to permit operation of the engine in the test cell. [ARP741]

engine duty cycle A composite cycle (or cycles) derived from the mission profiles and mission mix. The engine duty cycle is usually expressed in terms of power lever position versus time. [ARD50010]

engineering The application of scientific principles to practical purposes, such as the design, construction, and operation of efficient and economical structures, equipment, and systems.

engineering development Refers to testing to determine tactical suitability of an item for military use in the environments (real or simulated) for which the item was designed. [AIR4896]

engineering plastics Plastics materials that are suitable for making into structural members and machine elements.

engineering units Terms of data measurement, such as degrees Celsius, pounds, and grams.

engine exhaust gas to air heat exchanger A heater in which the heat of the exhaust gases (either undiluted or diluted) from an aircraft engine (either main or auxiliary) may be utilized for the purpose of heating the air that is supplied to the airplane. [ARP86]

engine generated contaminants Those substances produced directly by the engine, including those formed by thermal degradation within the engine. [ARP4418]

engine health monitoring The general discipline or technique for indication of status of the mechanical or functional condition of an engine or engine components; sometimes referred to as engine condition monitoring. [ARP1587]

engine lathe A manually operated lathe whose headstock is driven by a gear train, by a stepped pulley mechanism, or by a combination of gears and pulleys.

engine maintainability That quality of an engine and its parts whereby maintenance is permitted within a specified period of time without excessive expenditure of maintenance manpower, personnel skill levels, test equipment, and maintenance support facilities. [AIR4896]

engine, maximum neutral An engine plus those parts making it peculiar to an aircraft type, but not to any particular position on the aircraft. [AIR4896]

engine model A mathematical representation of a gas turbine engine that accepts as its inputs the controlled variables of the system. [ARP4148]

engine monitoring system (EMS) A complete system approach to define engine, engine component, and sub-system health status through the use of sensor inputs, data collection, data processing, data analysis, and the human decision process. This system approach can consist of an integrated set of hardware and software and several separate engine-monitoring system elements; and can be manual, computer aided, or automated. [AIR1873]

engine operating hours The total number of operating hours accumulated by all engines in the sample. [AIR4896]

engine pod On an airplane, a streamlined structure or nacelle, usually mounted beneath the wing or attached to the wing tip, housing one or more jet engines. [ARP4107]

engine repair interval The interval at which parts must be removed from the engine for repair or replacement. [AIR4896]

engine rotor component A rotating part whose life is governed by low-cycle fatigue limits or other mechanical property limits. [AMS2647A]

engine start valve A valve whose function is to initiate or terminate hydraulic starter operation by opening or closing the fluid-inlet line in response to an external signal, and that usually provides automatic shutoff at the end of an engine-starting cycle. [ARP4386]

engine structural integrity program A time-phased set of required actions performed at the optimum time during the life cycle (design through phase-out) of an aircraft engine to ensure the structural integrity (strength, rigidity, damage tolerance, durability, and service life capability) of the engine. [ARD50010]

Engler viscosity A standard time-based viscosity scale used primarily in Europe.

enhanced vision system System that generates and displays visual imagery, based on sensor information, of the out-of-cockpit scene in front of an aircraft. [ARD50019]

enhancement, serial data A method in which a continuous string of logical ONEs or ZEROs is modified to introduce bit transitions; this enables bit synchronization for recording purposes and preserves bandwidth. In an incoming serial data stream, for example, a number of words are all logical ZEROs, and therefore a DC level that the bit synchronizer cannot synchronize on; the data are enhanced by making the LSB of the words a logical ONE.

en route descent Descent from the en route cruising altitude which takes place along the route of flight. [ARP4107]

en route flight advisory service (flight watch) A service specifically designed to provide, upon pilot request, timely weather information pertinent to the pilot's type of flight, intended route of flight, and altitude. [ARP4107]

en route minimum safe altitude warning (E-MSAW) A function of the NAS State A en route computer that aids the controller by alerting him/her when a tracked aircraft is below (or is predicted by the computer to go below) a predetermined minimum IFR altitude (MIA). [ARP4107]

enthalpy (h) A mathematically defined thermodynamic function of state; a property of a substance in an energy term defined as: $h = u + PV$ Btu per lb (J/kg), where u = internal energy; P = pressure; and V = volume. NOTE: If the fluid can be regarded as a perfect gas, its enthalpy can alternately be expressed as the product of its constant pressure specific heat and its absolute temperature. [ARP147C]

entity *1.* An abstract element that represents an object in the real world, its data attributes, and essential services along with the respective performance and quality characteristics. [AS4893] *2.* An active element within an OSI layer (e.g., Token Bus MAC is an entity in the Layer 2).

entrained air Air forced into liquid systems by the action of applicator mechanisms working in the mass. [AIR4844]

entrained moisture Water that is carried along in the air as liquid water. [AIR4073]

entrained water Discrete water droplets carried by a continuous hydrocarbon phase. [AIR4783]

entrainment The conveying of particles of water or solids from the boiler water by the steam.

entropy A measure of the extent to which the energy of a system is unavailable.

entropy (statistics) A factor or quantity that is a function of a mechanical system and is equal to the logarithm of the probability of the particular arrangement in that state.

entry guidance (STS) The precise steering commands for trajectory from initial penetration of the earth's atmosphere until the terminal area guidance is activated at an earth-relative speed (about 2500 fps).

entry name The alphanumeric name given to an entry point. *See* entry point.

entry point In a routine, any place to which control can be passed.

envelope *1.* Generally, the boundaries of an enclosed system or mechanism. *2.* Specifically, the glass or metal housing of an electron tube, or the glass enclosure of an incandescent lamp. Also known as the bulb.

environment The aggregate of all external and internal conditions (such as temperature, humidity, radiation, magnetic and electric fields, shock, and vibration)—natural, man-made, or self-induced—that influences the form, performance, reliability, or survival of an item. [ARD50013]

environmental chemistry Collective term comprising the complex chemical relationships involving the atmosphere, climatology, air and water pollution, fuels, pesticides, energy, biochemistry, geochemistry, etc.

environmental damage/deterioration Physical deterioration of the strength of an item or the resistance of the item to failure as a result of interaction with climate or environment. [ARD50010]

environmentally sealed Describes a device that is protected against the entry of moisture, fluids, and foreign, particulate contaminants that could otherwise affect the performance of the device. [ARP914A]

environmental stress cracking The susceptibility of a thermoplastic resin to crack or craze when in the presence of surface-active agents or other environments. [AIR4844]

environmental stress screening A series of tests conducted under environmental stresses to disclose weak parts and workmanship defects for correction. [ARD50010]

environmental temperature *See* ambient temperature.

environmental test Any laboratory test conducted under conditions that simulate the expected operating environment in order to determine the effect of the environment on component operation or service life.

EOD *See* end of descent.

EOS *See* earth observing system.

eosinophils A type of white blood cell or leukocyte that stains a red color with resin stain; normally about 2 or 3 percent of white cells in the blood, but tending to decrease during stressful situations and thus usable as an index for stress.

ephemeride Periodical publication tabulating the predicted positions of celestial bodies at regular intervals, such as daily, and containing other data of interest to astronomers. A publication giving similar information useful to a navigator is called an almanac.

ephemeris time The uniform measure of time defined by the laws of dynamics and determined in principle from the orbital motions of the planets, specifically the orbital motion of the earth as represented by Newcomb's Tables of the Sun.

EP lubricant Extreme-pressure lubricant; an oil or grease containing additives that enhance the ability of the lubricant to adhere to a surface and reduce friction under high bearing loads.

epoxy Any of a large number of carbon-chain polymers that contain an epoxide group (two carbon atoms, one oxygen, and three hydrogen).

epoxy adhesive An adhesive made of epoxy resin.

epoxy matrix composite A composite material consisting of an epoxy resin matrix reinforced by imbedded fibers such as glass, boron, or graphite.

Eppley pyrheliometer A thermoelectric device for measuring direct and diffuse solar radiation. In this device, radiation is directed onto two concentric silver rings, the outer covered with MgO and the inner covered with lampblack, and a thermopile is used to determine the difference in temperature between the two rings.

EPU *See* emergency power unit.

equal area actuator An actuator having equal effective piston areas and equal force output for both directions of motion. [ARP4386]

equalization (channel balancing) The use of feedback to achieve close coincidence between the outputs of two or more elements or channels in a fault-tolerant control system. Equalization may be necessary to reduce the transient that could occur while shutting off a failed channel, or it may be necessary to minimize the adverse effects of normal tolerances. [ARP181A]

equalized maintenance program *See* program, equalized maintenance.

equalizer *1.* A device that connects parts of a boiler to equalize pressures. *2.* The electronic circuit in a tape reproducer whose gain across the spectrum of interest compensates for the unequal gain characteristic of the record/reproduce heads, thereby providing "equalized" gain across the band.

equal resistance branch circuit A circuit in which the temperature-measuring junctions are paired in parallel and symmetrically arranged about junction points, so that equal resistance paths are maintained around each loop. [ARP485]

equations of motion A set of equations that give information regarding the motion of a body or a point in space as a function of time when initial position and initial velocity are known.

equations of state Equations relating temperature, pressure, and volume of a system in thermodynamic equilibrium.

equator In filament winding, the line in a pressure vessel described by the junction of the cylindrical portion and the end dome. [AIR4844]

equatorial atmosphere The composition and characteristics of the earth's atmosphere at and/or near the equator.

equiaxed structure A polygonal or spheroidal microstructural feature having approximately equal dimensions in all directions. In alpha-beta titanium alloys, such a term commonly refers to a microstructure in which most of the alpha phase appears spheroidal, primarily in the transverse direction. [AS1814]

equilay stranding Stranding composed of more than one layer of helically laid strands, with a reversed direction of lay, and the same length of lay for each successive layer. [ARP1931]

equilibrium flow Gas flow in which energy is constant along streamlines and the composition of the gas at any point is not time dependent.

equilibrium glide A flight trajectory along which the total vertical acceleration of the vehicle is zero. [ARP4386]

equilibrium, mass A state of balance; a condition in which the materials taken into a body or system are balanced by other materials given off. [ARP171]

equilibrium state Any set of conditions that results in perfect stability; for example, mechanical forces that completely balance each other and do not produce acceleration, or a reversible chemical reaction in which there is no net increase or decrease in the concentration of reactants or reaction products.

equilibrium water uptake Point at which the rate of increase of water uptake is virtually nil. [AIR4844]

equinox One of two points of intersection of the ecliptic and the celestial equator occupied by the sun when its declination is zero degrees.

equipment, automatic test Equipment that is designed to automatically conduct analysis of functional or static parameters and to evaluate the degree of unit under test (UUT) performance degradation; and may be used to perform fault isolation of UUT malfunction. The decision-making, control, or evaluative functions are conducted with minimum reliance on human intervention, usually under computer control. [ARD50010]

equipment compatibility As applied to computers, the characteristic by which one computer may accept and process data prepared by another computer without conversion or code modification.

equipment depreciation For equipment dedicated to a system, the loss of monetary value ascribed to the equipment as a result of its age. [ARP4293]

equipment maintenance cost For equipment dedicated to a system, labor and material costs incurred in order to maintain the equipment in a serviceable condition. [ARP4293]

equipment replaceable unit In consonance with the maintenance concept, the lowest assembly or individual part that can be fault-detected, isolated, removed, replaced, and verified functional at the organization level without disassembly of the equipment to which it is attached. [AIR4896]

equipotential For conducting item(s), an identical state of electrical potential (practical definition). [ARP1870]

equivalence point Point on the titration curve at which the acid ion concentration equals the base ion concentration.

equivalent airspeed (EAS) The calibrated airspeed of an aircraft corrected for adiabatic compressible flow for the particular altitude. Equivalent airspeed is equal to calibrated airspeed in standard atmosphere at sea level. [ARP4107]

equivalent binary digits The number of binary digits required to express a number in another base to the same precision, e.g., approximately 3-1/3 binary digits are required to express in binary form each digit of a decimal number.

equivalent evaporation Evaporation expressed in pounds of water evaporated from a temperature of 212°F to dry saturated steam at 212°F.

equivalent network A network that can perform the functions of another network under certain conditions; the two networks may be of different forms, for example, one mechanical and one electrical.

equivalent optical diameter The size perceived by the APC; based on calibration data for the particular instrument. Normally, the diameter reported is that of a sphere with an equal projected area, and specific optical properties. [ARP1192]

equivalent single wheel load A theoretical calculated load that, if applied to a single tire, with a contact area equal to that of one tire of the assembly, would produce the same effect on the airfield as does a multiple wheel assembly. [AIR1489]

equivalent static wheel load A static wheel or tire load rating based on a dynamic load reduced by a specified factor. May be higher than the actual required static load rating for a specific application, in which case the tire/wheel static rating is dictated by this consideration rather than the actual static rating requirement. [AIR1489]

equivalent step function A mathematical function that is used to evaluate actual surges found to exist in power systems. [AS1212]

erg The unit of energy in the CGS system; the amount of energy consumed (work) when a force of one dyne is applied through a distance of one centimeter.

ergonometrics Human factors engineering that deals with machine design and workspace environment to make them compatible with human capacities and limitations. [ARP4107]

Erichsen test A cupping test for determining the suitability of metal sheet for use in a deep drawing operation. The result is expressed as the depth in millimeters of a cup-shaped impression in a sheet of metal, supported on a ring. The sheet of metal is deformed at the center by a spherical tool until it breaks or tears.

erosion *1.* In a solid rocket, burning of the propellant at a rate greater than the rate normally associated with the existing motor pressure and propellant temperature, usually due to high gas velocity parallel to the burning surface. [AIR913] *2.* The wearing away of refractory or of metal parts by gas-borne dust particles.

erosion-corrosion Progressive destruction of a structural member by the combined effects of corrosion and erosion acting simultaneously.

erosive burning Combustion of solid propellants accompanied by nonsteady, high-velocity flows of product gases across burning propellant surfaces.

erroneous information *See* ambiguous information.

error *1.* A mistake in specification, design, production, maintenance, or operation that causes an undesired performance of a function. [AIR1916] *2.* The occurrence of a difference between the information transmitted and the information received. [AIR4271] *3.* In mathematics, the difference between the true value and a calculated or observed value.

error, adjustment Operating a control too slowly or too rapidly, moving a control/switch to the wrong position, or following the wrong sequence in operating several controls/switches. [ARP4107]

error, built Those faulty actions that occur and are not rectified during the assembly and/or subsequent test of an item during manufacture, overhaul, or repair. [AIR4896]

error burst In data transmission, a sequence of signals containing one or more errors, but counted as only one unit in accordance with some specific criterion or measure; such a criterion might be that if three consecutive correct bits follow an erroneous bit, an error burst is terminated.

error checking Data quality assurance usually attempted by calculating some property of the data block before transmission. The resulting property or check character is also sent to the receiver, where it may be inspected and compared with a recalculated value based on the received data.

error correcting code A code in which each acceptable expression conforms to specific rules of construction that also define one or more equivalent nonacceptable expressions. In such a code, if certain errors occur in an acceptable expression, the result will be one of its equivalents, and thus the error can be corrected.

error detecting code A code in which each expression conforms to specific rules of construction. In such a code, if certain errors occur in an expression, the resulting expression will not conform to the rules of construction, and thus the presence of the errors is detected. Synonymous with self-checking code.

error index The total error incurred in a system over the present time (t) and future time (T) at which the control system operates. Sym-

bolically, it is e(t), where e is a function of t, and t $\leq$ T. [ARP4386]

error message An audible or visual indication of a software or hardware malfunction, or a non-acceptable data entry attempt.

error range *1.* The range of all possible values of the error of a particular quantity. *2.* The difference between the highest and the lowest of these values.

error ratio The ratio of the number of data units in error to the total number of data units.

error signal *1.* The output of the summing point at which algebraic summation of two or more control loop variables is performed. [AIR1916] *2.* The output of a comparing element.

error, zero Conceptual goal of perfection; ideal performance or product.

ES Symbol for the eddy-sonic method of nondestructive testing inspection. [ARP5089]

ESA spacecraft Spacecraft of the European Space Agency.

escape For a particle of larger body, to achieve an escape velocity and a flightpath outward from a primary body so as neither to fall back to the body nor to orbit it.

escape maneuver A computed maneuver to prevent a potential collision; can be any single maneuver or combination of maneuvers that resolves a conflict. [ARP4153]

escapement A ratchet device that permits motion only in one direction, such as the device that controls motion in the works of a mechanical watch or clock.

escape orbit One of the various paths that a body or particle escaping from a central force field must follow in order to escape. [ARP4386]

escape rocket Small rocket engine attached to the leading end of an escape tower; may be used to provide additional thrust to the capsule to obtain separation of the capsule from the booster vehicle in an emergency.

escapes A proportion of incoming defect density that is not detected by a screen and test, and that is passed on to the next level. [AIR4896]

escape velocity The radial speed that a particle of larger body must attain in order to escape from the gravitation of a planet, satellite, or star.

essential flight information A message that contains information relevant to the safety of the flight. [ARP4102/13]

established reliability A quantitative maximum failure rate demonstrated under controlled test conditions specified in a military specification, and usually expressed as percent failures per thousand hours of test. [AIR4896]

estimate A predicted value of performance. [AS1607]

ET Symbol for the electromagnetic (or eddy current) method of nondestructive testing inspection. [ARP5089]

Etalon A type of Fabry-Perot interferometer in which the distance between two highly reflecting mirrors is fixed. Used to separate light into different wavelengths when the wavelengths are closely spaced.

etalons Two adjustable, parallel mirrors mounted so that either one may serve as one of the mirrors in a Michelson interferometer; used to measure distance in terms of wavelengths of spectral lines.

etch cleaning Removing soil by electrolytic or chemical action. Etch cleaning also removes some of the underlying metal.

etch cracks Shallow cracks in the surface of hardened steel due to hydrogen embrittlement; sometimes occurs when the metal comes in contact with an acidic environment.

etched wire insulation Refers to a process in which a fluoroplastic insulated wire is passed through a sodium bath to create a rough surface, thereby allowing a material to bond to the surface of the fluoroplastic. [ARP1931]

etching *1.* Controlled corrosion of a metal surface to reveal its metallurgical structure. *2.* Controlled corrosion of a metal part to create a design. The design may consist of alternating raised and depressed areas, or it may consist of alternating polished and roughened areas, depending on the conditions and corrodent used.

ETP copper *See* copper, ETP.

Euler equations The equations of motion for a fluid without friction.

Europa A satellite of Jupiter orbiting at a mean distance of 761,000 kilometers. Also called Jupiter II.

European Incoherent Scatter Radar *See* EISCAT radar system (Europe).

eutectic *1.* A process by which a liquid solution undergoes isothermal decomposition to form two homogeneous solids—one richer in solute than the original liquid, and one leaner. *2.* The composition of the liquid that undergoes eutectic decomposition; possesses the lowest coherent melting point of any composition in the range over which the liquid remains single-phase. *3.* The solid resulting from eutectic decomposition, consists of an intimate mixture of two phases.

eutectic composites Composite materials in which the matrix is of a mixture of metallic solids including eutectoids.

eutectic mixture A mixture of two or more substances in such a ratio that it has the lowest melting point of any combination. [AIR4844]

eutectoid A decomposition process having the same general characteristics as a eutectic, but taking place entirely within the solid state.

eV (or EV) *See* electron volt.

evacuating (transportation) The organized withdrawal or removal of people from a place or area as a protective measure.

evaluation The process of deciding as to the severity of a condition after the indication has been interpreted. Evaluation leads to the decision as to whether the part must be rejected or salvaged, or may be accepted for use. [ARP5089]

evaluation, operational The test and analysis of a specific end item or system, in so far as practical under service operating conditions, to determine if quantity production is warranted. [AIR4896]

evaluation program cost The cost, if it occurs, of any system evaluation program subsequent to development. [ARP4293]

evaluation, service An evaluation of an item while it is performing its intended function during normal operation of the aircraft. [AIR4896]

evaluation, structural (airframe) The assessment of any data that relates to the structural integrity of the airframe. [AIR4896]

evaluation, technical Studies and investigations by a developing agency to determine the technical suitability of material, equipment, or systems for use in the military services. [AIR4896]

evaporated make-up Distilled water used to supplement returned condensate for boiler feed water.

evaporation The physical process by which a liquid or solid is transformed into the gaseous state; the opposite of condensation.

evaporation gage *See* atmometer.

evaporation rate *1.* The mass of material evaporated per unit time from unit surface of a liquid or solid. *2.* The number of molecules of a given substance evaporated per second per square centimeter from the free surface of the condensed phase.

evaporative cooling *1.* Lowering the temperature of a mass of liquid by evaporating part of it, using the latent heat of vaporization to dissipate a significant amount of heat. *2. See* vaporization cooling.

evaporator *1.* As pertains to a refrigeration system, that part of the system in which heat is transferred to the refrigerant, resulting in its change of phase from a liquid to a gas. [ARP147C] *2.* Any of several devices in which a liquid undergoes a change of state to a gas under relatively low temperature and low pressure.

evaporimeter *See* atmometer.

event An occurrence that causes a change of state.

even tension In a ball of roving, the process whereby each end of roving is kept in the same degree of tension as the other ends making up that ball. [AIR4844]

event recorder An instrument that detects and records the occurrence of specific events, often by recording on-off information against time to show when an event starts and stops, and how often it recurs.

exactness *See* precision.

examination An element of inspection consisting of investigation, without the use of special laboratory appliances or procedures, of sup-

plies and services to determine conformance to those specified requirements that can be determined by such investigations. [AIR4896]

exceedance Surpassing or exceeding a life limit. [AIR4896]

exception reporting An information system that reports on situations only when actual results differ from planned results. When results occur within a normal range they are not reported.

excess air Air supplied for combustion in excess of that theoretically required for complete oxidation.

excess flow control A function of some hydrant valves or hydrant couplers, which will automatically stop flow through the system if the flow rate exceeds a predetermined value. [AIR4783]

excessive intergranular corrosion Degree of intergranular corrosion greater than what is usually experienced from test samples of the same alloy, temper, thickness, heat treated to the applicable specification. [ARP1917]

excimer laser A laser in which the active medium is an "excimer" molecule—a diatomic molecule that can exist only in its excited state. The internal physics of an excimer laser are conducive to high powers in short pulses, with wavelengths in the ultraviolet.

excimers Molecules characterized by repulsive or very weakly bound ground electronic states.

excitation *1.* Addition of energy to a nuclear, atomic, or molecular system, thereby transferring it to another energy state. *2.* Voltage supplied by a signal conditioner to certain types of physical measurement transducers (bridges, for example).

excitation purity Represents the relative purity of a color. Colors made up of only a single wavelength (monochromatic) have the highest excitation purity, while the achromatic white point has the least. Excitation purity is defined by the ratio of two lengths on a chromaticity diagram. [ARP4032]

excitation voltage *1.* The electrical potential used to excite the transformer. The frequency and waveform of this voltage must be stated along with magnitude. [ARP4386] *2.* A precision voltage applied to transducers; when pressure, strain, or the like are sensed by the transducer, a small portion of this voltage appears on the signal lines. The value of this signal voltage is proportional to the stimulus applied.

excited state *See* excitation.

exciter/regulator The devices necessary to provide voltage regulation and to provide or control the excitation of the alternator. [AS8011A]

execution time The period of time required for a particular machine instruction.

EXE file In MS-DOS, refers to a compiled, executable program; indicated by the file extension .EXE.

exercise To operate an equipment in such a manner that it performs all of its intended functions, such that observation, testing, measurement and diagnosis of operational condition can be performed. [AIR4896]

exerciser A machine that simulates the strains and vibrations to which a missile is subjected; used to test the missile for structural integrity. [ARP4386]

exfoliation A surface defect on composite parts where the resin appears scaled or flaky. [AIR4844]

exfoliation corrosion A type of corrosion that proceeds parallel to the surface of a material, causing thin outer layers to be undermined and lifted by corrosion products.

exhaust *1.* Discharge of working fluid from an engine cylinder or from turbine vanes after it has expanded to perform work on the piston or rotor. *2.* A duct for conducting waste gases, fumes, or odors from an enclosed space, especially the discharge duct from a steam turbine, gas turbine, internal combustion engine, or similar prime mover; gas movement may be assisted by fans.

exhaust clouds Clouds formed from the exhaust aerosols of launch vehicle engines and boosters at liftoff.

exhaust emission The movement of gaseous or other particles and radiation from the nozzle of a rocket or other reaction engine.

exhaust-gas analyzer An instrument that measures the concentrations of various combustion products in waste gases to determine the effectiveness of combustion.

exhaust steam Steam discharged from a prime mover.

exhaust stroke The portion of the cycle in an engine, pump, or compressor during which working fluid is expelled from the cylinder.

exhaust system As relates to compartment ventilation, that combination of air discharge ducts, vents, and outlet grills utilized for the discharge of air from the compartment to the outside. [ARP147C]

exhaust valve A valve in the headspace of a cylinder that opens during the exhaust stroke to allow working fluid to pass out of the cylinder.

exhaust velocity The velocity of gases or particles (exhaust stream) that exhaust through the nozzle or a reaction engine, relative to the nozzle.

exit closure The door, window, or other device used to close or otherwise fill or occupy the exit opening. [ARP4101/7]

exit plane pressure The static pressure of a gas stream existing at the nozzle exit plane. [ARP4386]

exobiology The field of biology that deals with the effects of extraterrestrial environments on living organisms, and with the search for extraterrestrial life.

exosphere The outermost, or topmost, portion of the atmosphere; lower boundary is the critical level of escape, variously estimated at 500 to 1000 kilometers above the earth's surface.

exotherm The liberation or evolution of heat during the curing of a plastic product, or during any chemical reaction. [AIR4844]

exothermic reaction A reaction that occurs with the evolution of heat.

exotic fuels High-energy fuels, especially the hydroborons, which have higher calorific values than the corresponding hydrocarbons. At one time, these fuels were proposed for use in high-performance aircraft and missiles.

expandable tire A tire that in the free, uninflated state assumes a shape that is smaller in diameter and cross section than when in the inflated state; used to conserve stowage volume in the air vehicle. Requires deflation after takeoff and prior to retraction, and reinflation after gear extension and prior to landing. [AIR1489]

expandable tooling Use of a hollow rubber mandrel that can be pressurized to form composite hardware during cure. [AIR4844]

expanded EMS An EMS in which the instrumentation is improved and supplemented such that the engine components and modules can be individually monitored. Data processing and analysis are more extensive and may require integration with the control of individual engine components and modules. [AIR1873]

expanded joint The pressure tight joint formed by enlarging a tube end in a tube seat.

expanded metal A form of coarse screening made by lancing sheet metal in alternating rows of short slits, each offset from the adjacent rows; then stretching the sheet in a direction transverse to the rows of slits so that each slit expands to give a roughly diamond-shaped opening.

expanded plastic A light, spongy plastics material made by introducing air or gas into solidifying plastic to make it foamy. Also known as foamed plastic or plastic foam.

expander *1.* A device that has a main function of temporarily increasing an expansible type part to a desired dimension. *See also* spreader. [ARP480A] *2.* The tool used to expand tubes.

expansion *1.* Increasing the volume of a working fluid, with a corresponding decrease in pressure, and usually with an accompanying decrease in temperature, as in an engine, turbine or other prime mover. *2.* Generally, any increase in volume or dimension that causes a body to occupy more physical space.

expansion factor (Y) Correction for the change in density between two pressure-measurement stations in a constricted flow.

expansion joint A joint that permits movement due to expansion without undue stress.

expansion ratio In rocketry, the ratio of nozzle exit area to nozzle throat area. [AIR913]

expansion, thermal *See* thermal expansion.

expectancy *1.* A mental set in which environmental conditions are assumed prior to their occurrence. This may lead to a perceptual response, or attitudinal set. *2.* An acquired disposition whereby a response to a certain sign, object, or cue stimulus is expected to bring about a certain other situation. [ARP4107]

expedite A command used by ATC when prompt compliance is required to avoid the development of a dangerous situation. [ARP4107]

expendable item *See* item, expendable.

expendables Those components (e.g., standard parts found in bins) that are normally not tracked, but do have part numbers. [ARP4294]

expendable thermocouples Thermocouple made of fabric- or plastic-covered wire. [AMS2750]

expert system Computer program that manipulates symbolic information to produce the same results as human experts would. Expert systems deal with uncertain data and make decisions on that data; input and design of such systems, however, relies on human experts.

expiratory reserve *See* reserve, expiratory.

expiratory resistance The dynamic pressure differential related to a unit expiratory flow change. [ARP171]

exploding bridge wire (EBW) A bridge wire designed to be exploded by a high-energy electrical discharge. [AIR913]

exploding bridgewire detonator or initiator A detonator or initiator that is initiated by capacitor discharge that explodes (rather than merely heats) the bridgewire. [ARP4386]

exploratory development An item used for experimentation or tests to investigate or evaluate the feasibility and practicality of a concept, device, circuits, or system in breadboard or rough experimental form, without regard to the eventual overall fit or final form. [AIR4896]

explosion *1.* A rapid chemical reaction accompanied by the generation of a high temperature and, usually, a large quantity of gas. [AIR913] *2.* The sudden production of large quantities of gases, usually hot, from much smaller amounts of gases, liquids, or solids. *3.* Combustion that proceeds so rapidly that a high pressure is generated suddenly.

explosion door A door in a furnace or boiler setting designed to be opened by a predetermined gas pressure.

explosion proof apparatus Apparatus enclosed in a case that is capable of withstanding an explosion of a specified gas or vapor that may occur within it; and of preventing the ignition of a specified gas or vapor surrounding the enclosure by sparks, flashes, or explosion of the gas or vapor within; and that operates at such an external temperature that a surrounding flammable atmosphere will not be ignited thereby.

explosion suppression Any method used to confine or suppress an explosion.

explosion welding A solid-state process for creating a metallurgical bond by driving one piece of metal rapidly against another with the force of a controlled explosive detonation.

explosive cladding Producing a bimetallic material by explosion welding a thin layer of one metal on a substrate. Used most advantageously to yield a material with one surface having a unique property, such as resistance to corrosion by certain strong chemicals, while the bulk of the material possesses good fabrication and structural properties.

explosive decompression Rapid reduction of air pressure inside an aircraft, coming to a condition of balance with the external pressure; generally, any decompression that occurs in less than 0.5 seconds. [ARP4107]

explosive forming Shaping parts in dies through the use of explosives to generate the forming pressure. Most often, a sheet metal part is placed over an open die and covered with a sheet of explosive, which is then detonated to drive the metal into the die.

explosive train A series of explosive elements. [AIR913]

exponential case The reliability characteristics of those products known to exhibit a constant failure rate. [AIR4896]

EXPOS (Spacelab payload) X-ray spectropolarimetry payload for Spacelab.

exposed junction *See* bare junction.

exposure *1.* For x-rays, the product of the x-ray intensity (measured by filament current in milliamperes) and time (in seconds); for gamma rays, the product of source strength (in curies) and time (in seconds or minutes).

[ARP5089] *2.* For a photographic or radiographic emulsion, the product of incident radiation intensity and interval of time it is allowed to impinge on the emulsion. *3.* A term loosely used to indicate time of exposure in photography.

exposure time *1.* The period (in hours or cycles) during which a system, subsystem, unit, or part is exposed to failure; measured from when it was last verified as functioning to when its proper performance is or may be required. [AIR1916] *2.* The elapsed time during which radiant energy is allowed to impinge on a photographic or radiographic emulsion.

expression A combination of operands and operators that can be evaluated to a distinct result by a computing system.

extend *1.* To add fillers or low-cost materials in an economy-producing endeavor. *2.* To add inert materials to improve void-filling characteristics and reduce crazing. [AIR4844] *3.* Port to which pressure is applied for extension of an actuator, such as a cylinder. [ARP4386]

Extended Binary Coded Decimal Interchange Code (EBCDIC) An 8-bit code that represents an extension of a 6-bit "BCD" code, which has been widely used in computers of the first and second generations. EBCDIC can represent up to 256 distinct characters and is the principal code used in many of the current computers.

extended duration space flight *See* long duration space flight.

extended instruction set (EIS) In the DEC system, software that provides hardware with fixed-point arithmetic and direct implementation of multiply, divide, and multiple shifting.

extended position *1.* The configuration of a retractable landing gear when it is down and locked and ready for landing, and supporting the air vehicle. *2.* The state of a shock absorber when it is in the unloaded or uncompressed position. [AIR1489]

extenders Low-cost materials used to dilute or extend high-cost resins without extensive lessening of properties. [AIR4844]

extensibility The ability of material to extend or elongate upon application of sufficient force, expressed as percent of the original length. [AIR4844]

extensible flap A flap that can be extended and rotated downward, effectively increasing both the area and the camber of the wing. [ARP4107]

extension *1.* An attachment for extending the length of a boring bit, socket wrench or handle, tow bar, and like items. The body of the item may be either a coil spring or a solid bar, as applicable. Extensions permit the connection of ends that are designed for direct mating to each other. [ARP480A] *2.* A multiple character set that follows a computer filename, which further clarifies the filename.

extensional-bending coupling A property of certain classes of laminates that exhibit bending curvatures when subjected to extensional loading. [AIR4844]

extensional-shear coupling A property of certain classes of laminates that exhibit shear strains when subjected to extensional loading. [AIR4844]

extension spring A tightly coiled helical spring designed to resist a tensile force.

extensometer *1.* An apparatus for studying seismic displacements by measuring the change in distance between two reference points that are separated by 20 to 30 meters or more. *2.* An instrument for measuring minute elastic and plastic strains in small objects under stress, especially the strains prior to fracture in standard tensile-test specimens. *See also* dilatometer.

external *1.* Of, on, or for the outside of an outer part. *2.* Acting or coming from the outside.

external environment A set of external entities with which the application software or application platform exchanges information. [AS4893]

external environment interface The interface between the application software or application platform and the external environment interface across which information is exchanged. [AS4893]

external event Those events over which the designer of an item has no authority.

external gas leakage Leakage of precharge gas from within the vessel as evidenced by free gas bubbles when the accumulator is immersed in a test fluid. [ARP4379]

external leakage Leakage from the interior of a device to the exterior, other than out of the fluid ports. [AIR1916]

external loads Aerodynamic loads imposed on the actuator output. NOTE: External loads vary generally as a function of surface displacement, rate of displacement, and operating conditions such as Mach number and altitude. [ARP1281A]

externally fired boiler A boiler in which the furnace is essentially surrounded by refractory or water-cooled tubes.

externally quenched counter tube A radiation counter tube equipped with an external circuit that inhibits reignition of the counting cycle by internal ionizing events.

external memory *See* external storage.

external-mix oil burner A burner having an atomizer in which the liquid fuel is struck, after it has left an orifice, by a jet of high-velocity steam or air.

external multiplexors Scanivalves, switching temperature indicators, and other devices that permit input of several signals on one computer input channel.

external storage A facility or device, not an integral part of a computer, on which data usable by a computer is stored, such as off-line magnetic tape units, or punch-card devices. Synonymous with external memory. Contrast with internal storage.

external stores Mission-oriented items carried externally to the normal configuration of the aircraft. [ARP4386]

external tank A fuel tank of pressure-vessel construction with aerodynamic fairing; may be hung on either the aircraft fuselage, under the wing, or on the wing tip. External tanks are generally hung on the wing and fuselage by pylons and may be jettisonable. [AIR4783]

external treatment Treatment of boiler feed water prior to its introduction into the boiler.

extinction ratio Ratio between the signal high and signal low powers. [ARP4290]

extract instruction An instruction that requests the formation of a new expression from selected parts of given expressions.

extractive distillation A distillation technique (employing the addition of a solvent) used when the boiling points of the components being separated are very close—within 3°C (5°F)—or the components are constant-boiling mixtures; this technique is a combination of fractionation and solvent extraction. In extractive distillation, the solvent is generally added to the top of the column and recovered from the bottom product by means of subsequent distillation. *See also* azeotrope.

extractor A device that has a main function of removing threaded or sleeved part from blind holes; requires hand gripping. *See also* driver and pusher. [ARP480A]

extra hard temper In nonferrous alloys and some ferrous alloys, a level of hardness and strength corresponding approximately to a cold-worked state, one-third of the way from full hard to extra spring temper.

extra spring temper For nonferrous alloys and some ferrous alloys, a level of hardness and strength corresponding to a cold-worked state above full hard, beyond which hardness and strength cannot be measurably increased by further cold work.

extraterrestrial intelligence Intelligent life existing elsewhere than on earth.

extraterrestrial life Life forms evolved and existing outside the terrestrial biosphere.

extraterrestrial radiation In general, solar radiation received just outside the earth's atmosphere.

extremely improbable For airworthiness purposes, refers to the condition in which the likelihood of a failure is less than once in a billion flight hours (10^{-9}). [ARP4107]

extreme ultraviolet Explorer satellite An explorer satellite carrying scientific instruments for scanning the sky in the 100 nanometer region of the spectrum; purpose is to study the very hot celestial bodies (white dwarfs, for example).

extreme ultraviolet radiation Ultraviolet emission in the 10–100 nanometer range.

extremum values In statistics, the upper or lower bound of the random variable, not expected to be exceeded by a specified percentage of the population within a given confidence interval.

extrinsic Refers to a processs wherein modulation by the measurand of the light beam within the sensor occurs externally to the light-guiding regions of the fiber. [ARD50024]

extrusion *1*. A process for forming elongated metal or plastic shapes of simple to moderately complex cross section by forcing ductile, semi-soft solid material through a die orifice. *2*. A length of product made by this process.

extrusion billet In extrusion processing, a slug of metal, usually heated into the forging temperature range, that is forced through a die by a ram.

extrusion pressing *See* cold extrusion.

eye A device whose main function is lifting with a hoist. [ARP480A]

eyebar A metal bar having an enlarged section at each end, with a hole through each.

eyebolt A bolt with a loop formed at one end in place of a head.

eyelet *1*. A reinforced, conductive device or hole into which conductors are passed/routed or terminated. [ARP914A] *2*. A small ring or barrel-shaped piece of metal used to reinforce a hole, especially in fabric.

eyeleting Forming a lip around the rim of a hole in sheet metal.

F

F *See* farad and Fahrenheit.

FAA Acronym for Federal Aviation Administration.

fabricating The manufacture of products from molded parts, rods, tubes, sheeting, extrusions, or other forms by appropriate operations, such as punching, cutting, drilling, and tapping. [AIR4844]

fabrication *1.* A general term for parts manufacture, especially structural or mechanical parts. *2.* Assembly of components into a completed structure.

fabry-perot A pair of highly reflecting mirrors, whose separation can be adjusted to select light of particular wavelengths. When used as a laser resonator, this type of cavity can narrow the range of wavelengths emitted by the laser.

face *1.* The portion of the glass envelope through which the luminous pattern is viewed. [ARP1782] *2.* An exposed structural surface. *3.* In a weldment, the exposed surface of the fusion zone. *4. See* facings.

face clutch Two interfacing, rotating elements that transmit mechanical power when in contact, by means of friction surfaces. [ARP4386]

face curtain A sheet of heavy fabric designed to be pulled in front of the face for protection against wind blast during ejection from an aircraft. [ARP4107]

face dimpling Buckling of a compressive facing into a honeycomb cell. [AIR4844]

faceplate *1.* A circular plate attached to the spindle of a lathe; the plane of the plate is perpendicular to the spindle axis. Used to attach and align certain types of workpieces. *2.* A protective cover for holes in an equipment enclosure. *3.* A glass or plastic window in personal protective gear such as welding helmets, respirator masks, or diving masks. *4.* A two-dimensional array of separate optical fibers, fused together, serving to strongly direct light forward.

facesheet *See* facings.

facet The plane surface of a crystal or fracture surface.

face-to-core voids and disbonds Voids are areas in which adhesive is not present; disbonds are any unbonded areas occurring between the facing and the honeycomb core or between doublers and core. [AMS3920A]

face wrinkling Buckling of the compressive facing into or away from the core. [AIR4844]

facility modifier *See* correlation factor.

facing *1.* Machining a flat, planar surface in lathe turning by positioning a single-point tool against the workpiece at the axis of rotation and moving the tool radially outward so that it cuts a spiral path in a plane perpendicular to the axis of rotation. *2.* Fine molding sand applied to the surface of the mold cavity.

facings *1.* Skins and doublers in any layup. *2.* The outermost layer or composite component of a sandwich construction, generally thin and of high density; resists most of the edgewise loads and flatwise bending moments. Synonymous with face, skin, and facesheet. [AIR4844]

factor, acceleration The ratio between the times necessary to obtain a stated proportion of failures for two different sets of stress conditions involving the same failure modes and/or mechanisms. [AIR4896]

factorization Process or instance of factoring.

factor, operating The ratio of the operating hours of the in-service equipment under consideration to the number of flying hours incurred by the equipment. [AIR4896]

faculae Large patches of bright material forming a veined network in the vicinity of sunspots. Faculae appear to be more permanent than sunspots and are probably due to elevated clouds of luminous gas.

FAD *See* fuel advisory departure.

fadeometer An apparatus for determining the resistance of resins and other materials to fading. This device accelerates the fading by subjecting the article to high-intensity ultraviolet rays of approximately the same wavelength as sunlight. [AIR4844]

fading A drop in signal intensity, or a slow undulation, caused by changes in the properties of the transmission medium.

FAF *See* final approach fix.

Fahrenheit A temperature scale in which the freezing point of pure water occurs at 32°F and the span between the freezing point and boiling point of pure water at standard pressure is defined to be 180 scale divisions (180 degrees).

fail all simulator Device in which all faults are simulated one at a time in serial fashion. Also known as a sequential simulator. [AIR4896]

fail closed A condition in which the valve-closure component moves to a closed position when the actuating energy source fails.

failed-off A display element, row, or column that is failed permanently or sporadically in the "dark" or nonemitting state. [ARP4256]

failed-on A display element, row, or column that is failed permanently or sporadically in the "bright" or emitting state. [ARP4256]

fail-functional A more limited case of fail-operative in which performance is degraded following a failure. [AIR1916]

fail-hardover The type of failure in which the output of the failed element is at an extreme condition (e.g., a position or force). In cases where there is a polarity of output, a hardover failure may be of either polarity. [AIR1916]

fail-neutral A failure mode in which the control device or system fails to a passive null or locked-at-null condition. [AIR1916]

fail-open The type of failure in which the failed element disconnects the normal control path within a device. Such a failure either prevents the signal from passing or seriously alters the signal that passes through the system. [AIR1916]

fail operational A design feature that enables a system to continue to operate despite the malfunction or failure of one or more components. *See also* fail safe and redundant design. [ARP4107]

fail-operative Refers to the quality whereby a control device or system can continue operation after a failure or failures. A more explicit description is given by single fail operative (SFO) or dual fail operative (DFO). In a true fail-operative situation, a failure will cause no nominal loss of performance. [ARP1181A]

fail passive *1.* Refers to the quality whereby failure in the steering system will cause the nose gear to revert to a free castoring mode. [AIR1752] *2.* Refers to the quality whereby the failed device or system ceases to create any active output. In the purest sense, a fail passive device is one that would simply remove its presence from the control system; however, a device is still considered fail-passive if it remains a part of the system but acts only as an additional load. Sometimes referred to as fail-soft. [AIR1916]

fail-safe (FS) *1.* Refers to the quality whereby the control device or system ceases to function, but the conditions or consequences resulting from the failure are not hazardous and do not preclude continued safe flight. The condition following failure may be completely passive, or it may involve driving to a predetermined nonactive condition. [ARP1181A] *2.* A design or systems used to minimize risk in case of a malfunction; for example, a steering control system designed to fail in a free caster mode rather than a wheel hard over condition. *See also* fail operational and redundant design. [AIR1489]

fail safe design Design in which a failure will not adversely affect the safe operation of the system, equipment, or facility. [AIR4896]

fail soft *1.* A non-specific condition of a system that has manifested a number of failures, but still provides most of its functional capability. [AIR4896] *2.* With reference to electronic equipment, a design feature that enables the equipment to compensate automatically when a partial failure occurs. [ARP4107] *3. See* fail passive.

failure *1.* Any condition that would contribute to joint leakage or would promote premature fatigue breakage, unless otherwise determined to be due to a tubing defect. [ARP899] *2.* Describes the state of having failed. In dealing with fault-tolerant flight control systems, a failure occurs when a device within the system fails to function within prescribed limits without regard to the cause of the failure. Thus a failure may be: (a) any loss of function of any element within the control system; (b) loss of supply power to the system; (c) erroneous hardover conditions or loss of control intelligence at the signal input; or (d) any out-of-tolerance condition that exceeds normal operating limits. [ARP1181A] *3.* A functional status or physical condition characterized by the inability of an engine, engine component, or sub-assembly to fulfill its design purpose; the most severe degree of malfunction. [ARP1587] *4. See* critical defect.

failure analysis The logical, systematic examination of an item or its diagram(s) to identify and analyze the probability causes and consequences of potential and real failures. [ARD50010]

failure, basic A defect, failure, or damage as a result of malfunctioning of a system, unit, or part while being used in the manner for which it was designed; not externally induced. [AIR4896]

failure, catastrophic *See* catastrophic failure.

failure condition, major (failure condition, hazardous) Failure conditions that would reduce the capability of the aircraft or the ability of the crew to cope with adverse operating conditions to the extent that there would be: (a) a significant reduction in the safety margin; (b) a significant increase in crew workload, or in conditions that impair crew efficiency; or (c) in more severe cases, adverse effects on occupants. [AIR1916]

failure condition, minor Failure conditions that would not significantly reduce aircraft safety, and that involve crew actions that are well within their capabilities. Minor failure conditions may include: (a) a slight reduction in safety margins; (b) a slight increase in workload, such as routine flight plan changes; and (c) some possible discomfort to occupants. [AIR1916]

failure coverage The ratio of failures detected to failure population, expressed as a percentage. [AIR4896]

failure criteria Rules for failure relevancy, such as specified limits for the acceptability of an item. [AIR4896]

failure criticality A relative measure of a consequence of a failure mode and its frequency of occurrence. [AIR4896]

failure degradation A failure that occurs as a result of a gradual or partial change in the characteristics of parts or materials. [AIR4896]

failure, detectable A failure that can be detected with 100% test detection efficiency. [AIR4896]

failure effect The consequence(s) that a failure mode has on the operation, function, or status of an item. Failure effects are classified as local effects, next higher levels, and end effects. [ARD50010]

failure effects, nonoperational Failure effects that do not prevent aircraft operation, but are economically undesirable due to added labor and material cost for aircraft or shop repair. [AIR4896]

failure effects, operational Failure effects that interfere with the completion of the aircraft mission. [AIR4896]

failure, equipment The cessation of the ability to meet the minimum performance requirements of the equipment specifications. [AIR4896]

failure, equipment design Any failure that can be traced directly to the design of the equipment; that is, any failure in which the design of the equipment caused the part to degrade or fail. [AIR4896]

failure, equipment manufacturing A failure cased by poor workmanship or inadequate manufacturing process control during equipment construction, testing, or repair prior to the start of testing. [AIR4896]

failure free period A contiguous period of time during which an item is to operate without the occurrence of a failure while under environmental stress. [AIR4896]

failure free warranty A procurement requirement in which the manufacturer or the design control agent is intended to continuously upgrade the field reliability of designated equipment. [ARP4386]

failure, functional Refers to how an item failed to perform its designed function. [AIR4896]

failure, independent A failure that occurs without being related to the failure of associated items, distinguished from dependent failure

failure, intermittent A failure that occurs randomly in time. [AIR4896]

failure mechanism *1.* The process involved in the cause of failure. *2.* The physical, chemical, electrical, thermal, or other process that results in failure. [ARD50013]

failure, mission Failure to complete the intended mission as a consequence of equipment failure. [AIR1916]

failure mode A manner in which a device can or did fail. Simple devices may have only one failure mode; whereas, more complex devices can have several failure modes. [ARP1181A]

failure, nonchargeable *1.* A non-relevant failure. *2.* A relevant failure caused by a condition previously specified as not within the responsibility of a given organizational entity. (All relevant failures are chargeable to one organizational entity or another.) [ARD50010]

failure, noncritical Any failure that results in degraded operation requiring special operating techniques or alternative modes of operation, which could be tolerated through a mission, but should be corrected immediately upon completion of the mission. [AIR4896]

failure, partial Failure resulting from deviations in characteristics beyond specified limits, but not sufficient to cause a complete lack of function. [AIR4896]

failure path The chain of events or set of circumstances that result in an engine, engine component, or sub-system failure because of interrelationships between components and sub-systems. [ARP1587]

failure population Those failures that are used as a basis for the design and evaluation of tests. [AIR4896]

failure, random Any failure whose occurrence is unpredictable in an absolute sense, but which is predictable only in a probabilistic or statistical sense. [ARD50013]

failure rate *1.* The total number of failures within an item population, divided by the total number of life units expended by that population during a particular measurement interval under stated conditions. [ARD50013] The symbol lambda is used to represent failure rate. *2.* A reliability measure related to MTBF. [ARD50010]

failure rate, initial An initial estimate of the expected failure frequency of an item; may later be adjusted as actual experience is gained. [AIR1916]

failure rate, instantaneous The conditional probability of failure in a small time interval given the item has survived to the beginning of that interval. [AIR4896]

failure rate, smooth Failure rate determined when sufficient failures and exposures per age intervals are available—enough to eliminate peaks and valleys from the data. [AIR1916]

failure resistance The ability of a system or an item to withstand stresses imposed on it by its operating environment. [AIR4896]

failures, excluded *1.* Failures resulting from transportation, storage, inspection, maintenance, repair, installation, overhaul, or replacement improperly performed by using service personnel contrary to currently applicable instruction or reasonable standards of aircraft quality workmanship. This exclusion does not apply to improper actions by contractor personnel. [ARD50010] *2.* Failures in which the primary failure cause was not directly attributable to the design or quality of the engine; for example, failures attributed to foreign object damage (FOD) in excess of engine specification requirements. [ARD50010] *3.* Any reported malfunctions that cannot be verified by subsequent investigation and do not occur in subsequent operation. [ARD50010]

failure, soft Failure resulting from deviations in characteristics beyond specified limits, but not sufficient to cause a complete lack of function. [AIR4896]

failure, subsidiary A failure found after removal, which is not related to the reason for removal. [AIR4896]

failure symptom Any circumstances, event, or condition associated with a failure and that indicates the existence or occurrence of the failure. [AIR4896]

failure, tolerance A system or equipment failure resulting from multiple drift and instability problems—even though part failures may not have occurred. [AIR4896]

failure, transient A temporary failure induced by a momentary or temporary external factor, such as input power fluctuation, excessive ambient temperature excursion, electromagnetic interference; or by factors internal to the system. [AIR4896]

failure, undetectable A postulated failure mode for which there is no failure detection method by which the operator is made aware of the failure. [AIR4896]

failure universe The totality of failures being considered; if all of these failures are detected, then 100% fault coverage has been achieved. [AIR4896]

faint object camera One of the five components of the first scientific payload of the Hubble Space Telescope. The faint object camera is used to observe extremely faint astronomical objects with wavelengths between 120 and 700 nm.

fairing (feathering) *1.* A shape that produces a smooth transition from one angular direction to another; or the act of producing this smooth contour. In tooling sealants, the purpose of fairing or feathering is to ensure good contact with the surfaces and to minimize air entrapment. [AIR4069] *2.* A stationary member or structure, whose primary function is to produce a smooth contour; serves to cover projecting parts that would offer resistance to air flow. [ARP480A]

fall time The time required for the output voltage of a digital circuit to change from a logical high level (1) to a logical low level (0).

false advisory A resolution advisory or traffic advisory given when the design criteria for issuing such an advisory do not actually exist. [AIR4102/10]

false alarm *1.* A fault indicated by BIT or other monitoring circuitry where no fault exists. [AIR4896] *2.* In general, the unwanted detection of input noise. In radar, an indication of a

detected target even though one does not exist, due to noise or interference levels exceeding the set threshold of detection.

false alarm rate The number of false alarms per unit time, or number of false alarms per BIT alarms, expressed as a percentage. [AIR4896]

false alert *1.* An alert, caused by a false track or a system malfunction, that is given when no threat exists in the TCAS operational envelope. [ARP4153] *2.* An alert that occurs when the design windshear threshold conditions do not exist. [ARP4109]

false brinelling Fretting between the rolling elements and races of ball or roller bearings.

false horizon illusion *See* illusions, visual.

false indication A penetrant indication that might be erroneously interpreted as a defect or discontinuity in the test part. [AMS2647A]

false removal The removal of an item from its normal location, which, after testing, is found to be operating properly. [AIR4896]

false set Rapid hardening of freshly mixed cement, mortar, or concrete with a minimum evolution of heat; plasticity can be restored by mixing without adding more water.

family The complete series of compatible materials from one manufacturer, designed to perform a specific process of penetrant inspection. [AMS2647A]

fan *1.* In glass-fiber forming, the fan shape that is made by the filaments between the bushing and the shoe. [AIR4844] *2.* A rotating mechanism used to induce movement in air or other gases. Usually, either a paddle-wheel type "squirrel cage" or a multiple-blade propeller, often encased in a duct or housing.

fan blade One or more revolving vanes attached to a rotary hub and operated by a motor.

fan inlet area The inside area of the fan outlet.

fan jet A jet engine having a ducted fan in its forward end that draws in extra air, the compression and expulsion of which provides additional thrust. [ARP4107]

fanout The number of digital elements that can be controlled from the output of a single identical element operating at a common power nozzle supply pressure. [ARP993]

fan performance A measure of fan operation in terms of volume, total pressures, static pressures, speed, power input, and mechanical and static efficiency, at a stated air density.

FAR *1.* Acronym for Federal Aviation Regulations. *2. See* fuel-air ratio. [AIR4548]

farad (F) Unit of electrical capacitance; the capacitance of a capacitor that, when charged with one coulomb, gives a difference of potential of one volt. [ARP1931]

Faraday rotation A rotation of the plane of polarization of light caused by the application of a magnetic field to the material transmitting the light.

Faraday rotator A device that relies on the Faraday effect to rotate the plane of polarization of a beam of light passing through it. Faraday rotator glass is a type of glass with composition designed to display the Faraday effect.

far field *1.* Sound beam zone in which equal reflectors give signals of exponentially decreasing amplitude with increasing distance. [ARP5089] *2.* Distant from the source of light. This qualification is often used in measuring beam quality, to indicate that the measurement is made far enough away from the laser that local aberrations in the vicinity of the laser have been averaged out.

far-infrared laser Generically, may be taken to mean any laser emitting in the far infrared, a vaguely defined region of wavelengths from around 10 micrometers to 1 millimeter. The far-infrared family of lasers requires optical pumping by an external laser, usually carbon dioxide.

FAS *See* flight deck alerting system.

fascination *See* attention, anomalies of fascination.

fast break In magnetic particle testing of ferromagnetic materials, refers to interrupting the current in the magnetizing coil to induce eddy currents and strong magnetization as the magnetizing field collapses.

fast charge battery A battery that can be charged at a fast charging rate, and that gives a signal that can be used to terminate the fast charge without damage to the battery. [ARP4386]

fast charging The rapid return of energy to a battery at the C rate or greater. [ARP4386]

fast curve read routines Curve read routines in which the pointer location used for interpolation is remembered from the previous evaluation, and can be moved only one location per time step. [AIR4548]

fastener Any of several types of devices used to hold parts firmly together in an assembly. Some fasteners hold parts firmly in position, but allow free or limited relative rotation.

Fast-Fourier Transform (FFT) A type of frequency analysis that can be done on data by computer using special software, or by an array processor, or by a special-purpose hardware device.

fatal injury Any injury that results in death within 30 days of occurrence. [ARP4107]

fatigue The progressive failure of materials (especially metals) as a result of repeated cyclic loading, which causes cracks to form and grow, eventually resulting in complete failure if loading is continued. [AIR1872]

fatigue, acute/transient The type of fatigue associated with physical or mental activity between two regular sleep periods. Acute/transient fatigue is eliminated after a regular sleep period. [ARP4107]

fatigue life The number of applied stress reversals, at a particular stress level, at which a material fails. [AIR1872]

fatigue limit (fatigue endurance limit) 1. The level of applied stress at which an infinite number of reversals can be endured without the material failing. 2. The number of cycles that can be accumulated by a part before being retired from service. [AIR1872]

fatigue notch factor The ratio of the fatigue strength of an unnotched specimen to the fatigue strength of a notched specimen of the same material and condition, the notch being of a specified size and contour. In determining this factor, the strengths are compared at the same number of stress cycles.

fatigue notch sensitivity An estimate of the effect of a notch or hole on the fatigue properties of a material; expressed as $q = (K_f - 1)/(K_t - 1)$, where q is the fatigue notch sensitivity, K_f is the fatigue notch factor, and K_t is the stress concentration factor for a specimen of the material containing a notch of a specific size and shape.

fatigue ratio The ratio of fatigue strength to tensile strength. [AIR4844]

fatigue strength The maximum stress that ordinarily leads to fatigue fracture in a specified number of stress cycles; if the stress is not completely reversed during each stress cycle, the minimum stress should also be given. *See also* endurance limit.

fatigue, subjective The type of fatigue associated with the wearing effects of such psychosocial problems as unresolved conflicts, prolonged frustration, or constant worrying. Subjective fatigue is not eliminated by any number of sleep periods without first resolving the conflict or removing the frustrations. [ARP4107]

fault A physical condition that causes a device, component, or element to fail to perform in a required manner; for example, a short circuit or a broken wire. [ARD50010]

fault, catastrophic A defect or malfunction in a component, assembly, or system, causing a sudden change in its operating characteristics, and resulting in a substantial lack of useful performance of the system. [AIR4896]

fault class Refers to the grouping of equivalent faults. [AIR4896]

fault current *See* short-circuit current.

fault delay A fault in a digital device such that switching occurs to a proper level, but does so outside of a specified time interval. [AIR4896]

fault, dependent A fault that is caused by the failure of an associated element. [AIR4896]

fault, design A fault due to inadequate hardware or software design. [AIR4896]

fault detection The process, technique, or capability of identifying a discrepancy. [ARP1587]

fault detection time The extent or duration of time during which the existence of a fault is not known. [AIR4896]

fault dictionary A list containing each fault signature and the associated failed item causing the fault signature, to be developed by the test program and displayed by the ATE. [AIR4896]

fault, functional A fault that can be described by a change in the operation of some portion of a system. [AIR4896]

fault, hard A physical condition that causes a device, component, or element to fail to perform in a required manner. [AIR4896]

fault indicator A device that presents a visual display, audible alarm, or other indication, when a failure or marginal condition exists. [AIR4896]

fault, input A fault at the input terminals or components within the unit under test. [AIR4896]

fault insertion 1. A transformation that maps the original network into a new network. 2. The process of inserting actual or simulated faults in a unit under test for the purpose of demonstrating BIT or test program set (TPS) performance.

fault, internal A fault internal to an integrated component or device such as an integrated circuit. [AIR4896]

fault isolated replaceable units The replaceable subsystems, assembly, subassembly, or components identified through diagnostic testing of the unit under test. [AIR4896]

fault isolation The process of determining the location of a fault to the extent necessary to effect repair. [ARD50010]

fault isolation analysis (FIA) A systematic evaluation of failure causes, indications, and probabilities to determine the level of fault isolability provided for in the system design. [AIR1266]

fault isolation time A component of mean time to repair; the time between detection and isolation of a fault. [AIR4896]

fault latency time The extent or duration of time between fault occurrence and fault indication. [AIR4896]

fault list analysis An analysis of faults prior to fault simulation. [AIR4896]

fault localization The process of determining the approximate location of a fault. [AIR1916]

fault location time The time spent arriving at a decision as to which items caused the system to malfunction. [AIR4896]

fault management Those aspects of the system design that cover fault monitoring (detection), fault response, fault storage, and fault annunciation, for both operational and maintenance purposes. [ARP1834]

fault masking Refers to an equipment design that prevents complete unique fault isolation. [AIR4896]

fault, nondetectable A fault that results in nonrelevant failure. [AIR4896]

fault, open A fault caused by an electrical separation of normal electronically connected points. [AIR4896]

fault, out of tolerance A defect or malfunction in a component, assembly, or system in which a performance parameter approximates, but fails outside the prescribed upper or lower limit for the parameter. [AIR4896]

fault, probable A hard or soft fault that is most likely to occur—relative to all possible faults—within the unit under test. [AIR4896]

fault resolution Describes how well the test program can pinpoint the failed item from among other items in the unit under test. [AIR4896]

fault sensor Device used to detect ground faults in the HVDC (high voltage DC) generation and distribution system. [AS1831]

faults, equivalent Two or more faults that create the same response for all possible tests. [AIR4896]

fault signature Data developed by the test program and used by the ATE (automated test equipment) to indicate the ambiguity group. [AIR4896]

fault simulator A computer program that inserts and studies simulated faults at the nodes of a represented digital circuit being exercised by test stimulus patterns. [AIR4896]

fault, soft A fault causing a degraded performance of the unit under test. A condition manifested only under certain conditions of unit under test operation; when those conditions change the fault disappears. [AIR4896]

fault, struck A failure in which the digital signal is permanently held in one of its binary states. [AIR4896]

fault symptom A measurable or visible abnormality in an equipment parameter. [AIR4896]

fault time The built-in delay in a fault-detection device before it initiates corrective action. [AS1831]

fault tolerance The ability of a system to perform a designated set of functions after having suffered one of a specified set of faults. [AIR4271]

fault tree *1.* A graphic representation of the various parallel and series combinations of subsystem and component failures that can result in a specified system fault. The fault tree, when fully developed, may be mathematically evaluated to establish the probability of the ultimate undesired event occurring as a function of the estimated probabilities of identifiable contributory events. *2.* An expression for a logic path used to establish engine, engine component, or subsystem functional status and condition. [ARP1587] *3.* Acyclic directed graphs used in the analysis or prediction of faults and defects.

faying surface Either of two surfaces in contact with each other in a welded, fastened, or bonded joint; or in one about to be welded, fastened, or bonded.

faying-surface seal A preassembly seal installed between two mating (overlapping) surfaces. Faying-surface sealants are used to prevent corrosion; and, in conjunction with fillet seals, to prevent a leak path from extending through a faying surface to another area. When modified by a groove, a faying-surface seal has been used as a primary seal. [AIR4069]

FCAW *See* flux-cored arc welding.

FCC *1.* Acronym for Federal Communications Commission. *2. See* frame code complement.

F-center laser A solid-state laser in which optical pumping by light from a visible-wavelength laser produces tunable, near-infrared emission from defects—called "color centers" or "F centers"—in certain crystals.

FCS *See* flight control system.

FDM *See* frequency-division multiplex.

FDR (flight data recorder) Device that records a variety of parameters detailing the operation of an aircraft. Referred to commonly in accident investigations as one of the two black boxes. *See also* cockpit voice recorder.

feather angle The blade angle setting at which the nonrotating propeller produces the least drag. [ARP4107]

feathering *See* fairing.

feed *1.* The act of supplying material to a process or to a specific processing unit. *2.* The material supplied to a process or to a specific processing unit. Also known as feedstock. *3.* In a machining operation, forward motion tending to advance a tool or cutter into the stock.

feedback *1.* The return of a portion of the output of a device to the input. Positive feedback adds to the input; negative feedback subtracts from the input. [AIR1489] *2.* Information, as to progress, results, etc., returned to an originating source. [AIR1489] *3.* Part of a closed-loop system that provides information about a given condition for comparison with the desired condition.

feedback control An error-driven control system in which the control signal to the actuators is proportional to the difference between a command signal and a feedback signal from the process variable being controlled.

feedback control unit *See* follow-up control unit.

feedback elements Those elements in a closed loop control system that relate the feedback signal to the controlled variable. [ARP4386]

feedback loop The components and processes involved in correcting or controlling a system by using part of the output as input. *See also* closed loop.

feedback signal *1.* The signal used to cancel further action when the command is carried out. [AIR1916] *2.* A signal derived from some attribute of the controlled variable, or from a control-system output, which is combined with one or more input or reference signals to produce a composite actuating signal.

feeder *1.* A circuit conductor originating at the power source bus from which the branch circuit loads are served. [ARP1199A] *2.* A conveyor adapted to control the rate of delivery of bulk materials, packages, or objects to a specific point or operation

feeder drop The total voltage drop in the feeders, including both the positive and return

feeder drop between the power source terminals and point of regulation. [AS1831]

feed-forward *1.* An industry-standard process control program in which mathematically predicted errors are corrected before they occur; used mainly for process loops with long lags or response times. *2.* Open loop control.

feedforward control action Control action in which information concerning one or more external conditions that can disturb the controlled variable is converted into corrective action to minimize deviations of the controlled variable. Feedforward control is usually combined with other types of control to anticipate and minimize deviations of the controlled variable. *See also* open loop.

feedhead A reservoir of molten metal that extends above a casting to supply additional molten metal and compensate for solidification shrinkage. Also known as riser or sinkhead.

feed pipe A pipe through which water is conducted into a boiler.

feed rate In a machining operation, the relative velocity between tool holder and workpiece along the main direction of cutting.

feedscrew An externally threaded rod used to control the advance of a tool or tool slide on a lathe, diamond drilling rig, percussion drill, or other equipment.

feedstock 1. Material delivered to a process or processing unit, especially raw material delivered to a chemical process or reaction vessel. 2. *See* feed.

feed-thru A connector, terminal block, or terminal device having conductive elements accessible from opposite sides of an insulator or a partition, for termination or connection with mating devices. [ARP914A]

feedwater Process water supplied to a vessel, such as a boiler or still, as opposed to circulating water or cooling water.

feed-water treatment The treatment of boiler feed water by the addition of chemicals to prevent the formation of scale or eliminate other objectionable characteristics.

feel The control stick force felt by the pilot per g in the pitch axis or per degree/second of yaw. [ARP4386]

feet of head Measurement of pressure in feet of the liquid being pumped. [AIR4783]

female fitting An element of a connection in pipe, tubing, electrical conductors, or mechanical assemblies that surrounds or receives the mating (male) element; for example, the internally threaded end of a pipe fitting is termed female.

Fermat principle *1.* Principle stating that the path along which electromagnetic radiation travels between any two points will be that path for which the elapsed time for the travel is in minimum. *2.* Also called the principle of least time; states that a ray of light traveling from one point to another, including reflections and refractions which may occur, follows that path that requires the least time. Stated another way, the optical path is an extreme path—in the terminology of the calculus of variations.

Fermi-Dirac statistics The statistics of an assembly of identical half-integer spin particles. Such particles have wave functions antisymmetrical with respect to particle interchange and satisfy the Pauli exclusion principle.

ferroalloy An alloy, usually a binary allow, of iron and another chemical element, which contains enough of the second element to make it suitable for introduction into molten steel, to produce alloy steel; or, in the case of ferrosilicon or ferroaluminum, to produce controlled deoxidation.

ferrodynamic instrument An electrodynamic instrument in which the presence of ferromagnetic material (such as an iron core in an electromagnetic coil) enhances the forces ordinarily developed in the instrument.

ferrography Wear analysis conducted by withdrawing lubricating oil from an oil reservoir and using a ferrograph analyzer to determine the size distribution of wear particles picked up as the oil circulates between moving mechanical parts. Ferrography may also be used to assess deterioration of human joints or joint-replacement prostheses by analyzing for the presence of bone, cartilage, and prosthetic-material fragments in human synovial fluid.

ferromagnetic material Any material that, like the chemical element iron, exhibits the phenomena of magnetic hysteresis and saturation, and whose permeability depends on the magnetizing force.

ferrometer An instrument for measuring magnetic permeability and hysteresis in iron, steel, and other ferromagnetic materials.

ferrous alloy Any alloy containing at least 50% of the element iron by weight.

ferrule *1.* A short tube or sheath of conductive material used to make connections to shielded or coaxial cable. *2.* A specially formed metal ring used in connector accessories to reduce the transmission of torque to the connector grommet. [ARP914A] *3.* A fuse or limiter terminal of a cylindrical shape which encloses the end of a fuse or limiter. [ARP1199A] *4.* A metal ring or cap that is fitted onto the end of a tool handle, post, or other similar member to strengthen and protect it. *5.* A bushing inserted in the end of a boiler flue to spread and tighten it. *6.* A tapered bushing used in compression-type tubing fittings to provide the wedging action that creates a mechanical seal. *7.* An element of a fiber optic connector, typically used to house or align fibers.

ferrule/end fitting Metal sleeve used for crimping onto the cable to maintain tension in the cable. [AS4536]

FFD *See* film focal distance.

FFT *See* Fast-Fourier Transform.

FIA *See* fault isolation analysis.

fiber *1.* Any particle of at least 100 μm in length that has a length-to-width ratio of 10 or more. [ARP4386] *2.* The characteristic of wrought metal that indicates directionality, and can be revealed by etching or fractography. *3.* The pattern of preferred orientation in a polycrystalline metal after directional plastic deformation, such as by rolling or wiredrawing. *4.* A filament or filamentary fragment of natural or synthetic materials used to make thread, rope, matting, or fabric. *5.* In stress analysis, a theoretical element representing a filamentary section of solid material aligned with the direction of stress; usually used to characterize nonuniform stress distributions, as in a beam subjected to a bending load.

fiber composite Structural material consisting of combinations of metals, alloys, or plastics reinforced with one or more types of fibers.

fiberglass An individual filament made by drawing molten glass. [AIR4844]

fiberglass reinforcement Material used to reinforce a resin matrix using continuous or discontinuous glass fibers. [AIR4844]

fiber-lock fitting Numerous individual fibers fabricated radially into the tank wall as means of retaining the metal fitting in the tank. [AIR1664]

fiber metal A material composed of metal fibers that have been pressed or sintered together, and that may also have been impregnated with resin, molten metal, or other material that subsequently hardened.

fiber migration Carry-over of fibers from filter or separator media material into the effluent. [AIR4783]

fiber-optic-based sensor Any sensor employing a fiber-optic signal path. [ARD50024]

fiber optic cable A glass fiber conduit handled in a similar manner as electrical wiring and capable of transmission of light signals. [AIR4783]

fiber optic gyroscope A device used to measure rotation speed; measurement is based on changes in the wavelength of light going in different directions through a long length of optical fiber wound many times around a ring.

fiber optics The technique of transmitting light through long thin, flexible fibers of glass, plastic, or other transparent materials.

fiber optic sensor/probe A device that is used to sense the existence of a liquid in a tank and transmit the change by light signal to a control device. [AIR4783]

fiber release The release of carbon or graphite when graphite-reinforced composites are burned, especially in aircraft crashes or fires.

fiber sensor A sensing device in which the active element is an optical fiber or an element attached directly to an optical fiber. The quantity being measured changes the optical properties of the fiber in a way that can be detected and measured.

fibrillation, ventricular Heart condition in which the ventricular beat is rapid, irregular, and ineffective. The spontaneous contraction of individual muscle fibers (fibrils) leads to irregular and ineffective beats. [ARP171]

fibrous composite A material consisting of natural, synthetic, or metallic fibers embedded in a matrix, usually a matrix of molded plastics material or hardenable resin.

fibrous fracture A type of fracture-surface appearance characterized by a smooth, dull gray surface.

fibrous structure *1*. In fractography, a ropy fracture-surface appearance; generally synonymous with silky or ductile fracture. *2*. In forgings, a characteristic macrostructure indicative of metal flow during the forging process; revealed as a ropy appearance on a fracture surface or as a laminar appearance on a macroetched section. NOTE: A ropy appearance on the fracture surface of a forging does not carry the same implication as a ropy fracture of other wrought metals, and should not be considered the same as a silky or ductile fracture. *3*. In wrought iron, a microscopic structure consisting of elongated slag fibers embedded in a matrix of ferrite.

fidelity The degree to which a system, subsystem or component accurately reproduces the essential characteristics of an input signal in its output signal. *See also* accuracy.

field The part of a computer record containing a specific portion of information. *See also* cell.

field aligned current Ionospheric and magnetospheric current aligned along the electric field of a planet.

field bus Under development in ISA SP50, a standard for a bus to interconnect process control sensors, actuators, and control devices.

field coil A stationary or rotating electromagnetic coil.

field curvature Formation of an image that lies on a curved surface rather than a flat plane. For single and double element lenses, curvature is always inward; but for other types, the curvature can be in either direction.

field emission Induced electron emission from an unheated metal surface resulting from application of a strong electric field.

field excitation Controlling the speed of a series-wound electric motor or a diesel-electric locomotive engine by changing the relationship between armature current and field strength, either through the use of shunts to reduce field current or through the use of field taps.

field-free emission current The electron current flowing from a cathode when the electric gradient at the cathode surface is zero.

field installable Describes a fiber optic splice or cable that can be mounted by technicians working in the field, without a lab full of equipment at hand.

field level repairable A low-cost repairable, capable of being restored to serviceable condition only at the intermediate maintenance activity, as indicated by the source, maintenance, and recoverability code. [AIR4896]

field of view (FOV) Described as total FOV or instantaneous FOV; instantaneous FOV is further subdivided as monocular, ambinocular, and binocular. *Total FOV*–Spatial angle in which the symbology can be displayed, measured vertically and laterally from within the total viewing cone/wedge. *Instantaneous FOV, monocular*–spatial angle in which the symbology can be viewed from any single eye position. *Instantaneous FOV, ambinocular*–envelope of both left and right eye monocular instantaneous FOV. *Instantaneous FOV, binocular*–envelope within the ambinocular FOV that is common to both monocular FOV, and in which the symbology can be viewed by both eyes simultaneously. [ARP4102/8] *2.* The area or solid angle that can be viewed through or scanned by an optical instrument.

field-replaceable unit Computer hardware modules that are easily replaced.

field strength For any physical field, the flux density, intensity, or gradient of the field at the point in question.

field weld A weld made at a construction or installation site, as opposed to a weld made in a fabrication shop.

FIFO *See* first-in, first-out.

filament *1.* A very fine single strand of metal wire, extruded plastic, or other material. *2. See* solar prominence.

filament winding Fabricating a composite structure by winding a continuous fiber rein-forcement on a rotating core under tension. The reinforcement usually consists of glass, boron, or silicon carbide thread, either previously impregnated with resin or impregnated during winding.

filament-wound structure A composite structure made by fabricating one or more structural elements by filament winding, then curing them and assembling them; or by assembling them first and curing the entire structure.

filar micrometer An attachment for a microscope or telescope consisting of two parallel fine wires or knife edges in an eyepiece, one of them in a fixed position and the other capable of being moved in a direction perpendicular to its length by means of a very accurate micrometer screw. The filar micrometer is used to make accurate measurements of linear distances in the optical field of view; actual distances are determined by dividing the micrometer reading by the magnification of a microscope, although in some cases the micrometer scale is calibrated for direct reading at a specific magnification.

filed A term normally used in conjunction with flight plans; means that a flight plan has been submitted to air traffic controller (ATC). [ARP4107]

fill *1.* A port provided for filling purposes in servicing of a unit. [ARP4386] *2. See* pick.

filled composite A plastics material made of short-strand fibers or a granular solid mixed into thermoplastic or thermosetting resin prior to molding.

filled-system thermometer Any of several devices consisting of a temperature-sensitive element (bulb), an element sensitive to changes in pressure or volume (Bourdon tube, bellows, or diaphragm), capillary tubing, and an indicating or recording device. The bulb, capillary tube, and pressure- or volume-sensitive element are partly or completely filled with a fluid that changes its volume or pressure in a predictable manner with changes in temperature.

filler *1.* The material used to effect a bond between fitting and tubing. [ARP899] *2.* A material used in cable to fill large interstices. [ARP1931] *3.* A substance, often inert, added to a compound to improve properties and/or reduce cost. [ARP1931] *4.* A metal or alloy deposited in a joint during welding, brazing, or soldering; usually referred to as filler metal.

filler metal *See* filler.

filler wires *See* shute.

fillet *1.* A rounded filling of adhesive that fills the corner or angle where two adherents are joined. [AIR4844] *2.* A concave transition surface between two surfaces that meet at an angle. *3.* A molding or corner piece placed at the junction of two perpendicular surfaces to lessen the likelihood of cracking.

fillet seal A primary seal (post assembly) applied at the juncture of two adjoining parts or surfaces and along the edges of faying surfaces as a continuous bead of sealing material. Can be applied over, along the edges of, and between installed parts. [AIR4069]

fillet weld A roughly triangular weld that joins two members along the intersection of two surfaces that are approximately perpendicular to each other.

fill pressure The nominal pressure to which the vessel is charged. [ARP4386]

fill time The time required to completely fill the combuster fuel manifold after fuel flow is initiated. [ARP906A]

film Thin sheeting; the finished form of a material that is: (a) processed by casting a fluid material on a large surface (usually a rotating drum) and exposing it to a curing process, or (b) melt-extruded directly into the sheets. *See also* cast tape. [ARP1931]

film adhesive A synthetic resin adhesive, usually of the thermosetting type, in the form of a thin, dry film of resin with or without a paper, glass, or other carrier. [AIR4844]

film cooling The cooling of a body or surface, such as the inner surface of a rocket combustion chamber, by maintaining a thin fluid layer over the affected area.

film edge-bond spacing For a bonded device, the distance from the end of the substrate conductor film metallization to the closest edge of the bonded area. [AS1346]

film focal distance (FFD) Distance between film and tube target. [ARP5089]

filmogen The material or binder in paint that imparts continuity to the coating.

film strength *1.* Generally, the resistance of a film to disruption. *2.* In lubricants, a measure of the ability to maintain an unbroken film over surfaces under varying conditions of load and speed.

filter *1.* A device serving to remove solid particles from a flowing fluid by passing it through a porous element. [ARP4386] *2.* Electrical circuits designed to eliminate various frequencies from a circuit output or input. May be low pass, high pass, or band pass. [ARP5089].

filter aid An inert, powdery or granular material, such as diatomaceous earth, fly ash, or sand, that is added to a liquid about to be filtered; function is to form a porous bed on the filter surface, thereby increasing the rate and effectiveness of the filtering process.

filter, air *See* air filter.

filter, bandpass *See* bandpass filter.

filter capacitor A capacitor used as an element of an electronic filter circuit.

filter inductor An inductor used as an element of an electronic filter circuit.

filter medium The portion of a filter or filtration system that actually performs the function of separating out the solid material; may consist of metal or nonmetal screening, closely woven fabric, paper, matted fibers, a granular bed, a porous ceramic cup or plate, or other porous component.

filter micron rating Unit of measure associated with filtration quality; one micron equals 1.0×10^{-6} m (3.9379×10^{-5} in). [ARP906A]

filter monitor Same as filter separator, a device designed to filter contaminants and water from fuel, except the stripped water is retained within a portion of the elements of the unit. [AIR4783]

filter, optical Optical coating, glass, or plastic panel placed on or near the display face to provide selective spectral transmission of light or to reduce the amount of light transmitted. Used to improve contrast ratio, for color selection, and for anti-reflection and/or absorption. A neutral density filter, for example, improves contrast ratio by doubly attenuating the ambient light (compared to the single attenuation of the display luminance). [ARP1782]

filter patch A membrane on which pump- or motor-generated contaminants have been collected. This patch, when compared to a "standard patch," is the basis for rejection or acceptance of a pump or motor prior to delivery. [ARP575A]

filter separator A device designed to filter contaminants and water from fuel. [AIR4783]

filter, sintered A filter made by sintering together minute globules of metal (or ceramic), forming tortuous passages through which gas can flow, but particulates cannot. [ARP171]

filter vessel The outer container of a filter separator or monitor which contains the filtration devices and other associated apparatus. [AIR4783]

fin *1.* Fixed or adjustable airfoil or vane attached longitudinally to an aircraft, rocket, or a similar body to provide a stabilizing effect. *2.* A flat plate of structure, such as a cooling fin. *3.* A thin, flat, or curved projecting plate, typically used to stabilize a structure surrounded by flowing fluid, or to provide an extended surface to improve convective or radiative heat transfer. *4.* A defect consisting of a very thin projection of excess material at a corner, edge, or hole in a cast, forged molded, or upset part, which must be removed before the part can be used.

final approach The last leg of a landing pattern, or the last heading flown by an aircraft before touchdown, during which the aircraft is lined up with the runway and is held to nearly constant speed and rate of descent. [ARP4386]

final approach fix (FAF) The designated fix from or over which the final approach (IFR) to an airport is executed. The FAF identifies the beginning of the final approach segment of the instrument approach. *See also* segments of an instrument approach procedure. [ARP4107]

final approach segment *See* segments of an instrument approach procedure. [ARP4107]

final control element *1.* An instrument that takes action to adjust the manipulated variable in a process. This action moves the value of the controlled variable back toward the setpoint. *2.* The last system element that responds quantitatively to a control signal and performs the actual control action. Examples include valves, solenoids, and servometers.

final controller The controller providing information and final approach guidance during precision approach radar (PAR) and surveillance approach radar (SAR) approaches utilizing radar equipment. [ARP4107]

final controlling element The element in a control system that directly changes the value of the manipulated variable.

final level The liquid level after final closure of the shutoff valve; a function of the liquid surface area vs. the liquid height above/below the sensing level. [AIR1660A]

fine grinding *1.* Mechanical reduction of a powdery material to a final size of at least -100 mesh, usually in a ball mill or similar grinding apparatus. *2.* In metallography or abrasive finishing, producing a surface finish of fine scratches by use of an abrasive having a particle size of 320 grit or smaller.

fineness Purity of gold or silver expressed in parts per thousand; for instance, gold having a fineness of 999.8 has only 0.02%, or 200 parts per million, of impurities by weight.

fineness ratio *1.* The ratio of the length of a body to its maximum diameter (or, sometimes, to some equivalent dimension); used especially in reference to a body such as an airship hull or rocket. *2.* *See* slenderness ratio.

fines In a granular substance having mixed particle sizes, those particles smaller than the average particle size.

finish *1.* A chemical or other substance applied to the surface of virtually any solid material to protect it, alter its appearance, or modify its physical properties. *2.* The degree of reflectivity of a lustrous material, especially metal. Usually described by one of the following imprecise terms, listed in order of increasing luster and freedom from scratches: (a) machined; (b) ground brushed; (c) matte; (d) dull lustrous; (e) bright; (f) polished; and (g) mirror.

finished inserts An insert ready for use, inclusive of any possible treatments and/or surface coatings, as specified in the dimensional standard or definition document. [AS3506]

finished thickness The thickness after final finishing operations, measured as the shortest distance between the two opposite surfaces of the feature being considered. [AMS4991]

finish grinding The final step in a grinding operation, which imparts the desired surface appearance, contour, and dimensions.

finishing temperature In a rolling or forging operation, the metal temperature during the last reduction and sizing step, or the temperature at which hot working is completed.

finite element method *See* panel method (fluid dynamics).

finite impulse response filter *See* FIR filter.

finite volume method A moving mesh method for analyzing transonic flow over airfoils.

fin tube In a boiler, a tube having water on the outside and carrying the products of combustion on the inside.

fiord Arm of the sea having steep sides, deep bottom, and shallow sills separating it from the sea.

fire assay Determining the metal content of an ore or other substance through the use of techniques involving high temperatures.

fire crack A crack starting on the heated side of a tube, shell, or header, resulting from excessive temperature stresses.

fire cuff Flame- and fire-retardant element, normally rubber, molded over the hose and hose fittings. [AS2078]

fire detection, aircraft cargo compartment Divided into three types: Type I, Carbon Monoxide; Type II, Smoke Detector, Electronic; and Type III, Smoke Detector, Visual. *Type I, Carbon Monoxide*–an instrument that will actuate an alarm signal when the concentration of carbon monoxide in air exceeds a specified value. *Type II, Smoke Detector, Electronic*–an instrument operating on the principle that smoke particles modify the relationship between a light beam and electronic light sensor, which will actuate an alarm signal when the concentration of smoke in air exceeds a specified value. *Type III, Smoke Detector, Visual*–an instrument that, by visual means, will show in a positive manner the presence of smoke when the concentration of smoke in air exceeds a specified value. [AS446]

fired pressure vessel A vessel containing a fluid under pressure exposed to heat from the combustion of fuel.

fired torque The torque produced by an engine during the start cycle after engine light-off has occurred. [ARP906A]

fire point The lowest temperature at which, under specified conditions, fuel oil gives off enough vapor to burn continuously when ignited.

fire polish A specular reflective finish that is produced by flame or moulding; or an equivalent surface that may be produced by mechanical means. [ARP924]

fireproof Resistant to combustion or to damage by fire under all but the most severe conditions.

fire-resistant *1*. With respect to sheet or structural members, the capacity to withstand heat at least as well as aluminum alloy, in dimensions appropriate for the purpose for which they are used; not readily ignited, requiring considerable heat input for ignition and flame formation. *2*. With respect to fluid-carrying lines, other flammable-fluid system parts, wiring, air ducts, fittings, and powerplant controls, the capacity to withstand heat at least as well as aluminum alloy, in dimensions appropriate for the purpose for which they are used, under the heat and other conditions likely to occur at the place concerned. [ARP171] *3*. Resistant to combustion and to heat of standard intensity for a specified time without catching fire or failing structurally.

fire-resistant electric equipment Equipment and components, as installed in the aircraft, that are capable of withstanding a 2000°F oxidizing flame impinging on their surfaces for at least 5 min without adverse effect on their circuit function. [ARP4404]

fire retardant Describes a normally combustible material (e.g., wood, paper, or textile) that has been treated by coating or impregnation so that it catches fire less readily and burns more slowly than untreated material.

fire tube In a boiler, a tube having water on the outside and carrying the products of combustion on the inside.

fire tube boiler A boiler with straight tubes, which are surrounded by water and steam, and through which the products of combustion pass.

firewall Also fire wall. *1.* A fireproof or fire-resistant wall or bulkhead separating an engine from the aircraft structure; designed to prevent the spread of any fire originating at the engine. 2. To move the throttle forward toward the firewall; this movement produces an increase in power from the engine. [ARP4107]

FIR filter Stands for finite impulse response filter—a physically unrealizable, nonrecursive digital filter.

firing rate control A pressure, temperature, or flow controller that controls the firing rate of a burner according to the deviation from pressure or temperature setpoint. The system may be arranged to operate the burner on-off, high-low, or in proportion to load demand.

firmware *1.* Hard-wired software that often encompasses microcodes. 2. Programs or instructions that are permanently stored in hardware memory devices (usually read-only memories), which control hardware at a primitive level.

first air In air-monitoring of environmentally-controlled environments, a term that refers to the quality of the air as it first enters the controlled environment. [ARP4386]

first degree repair The repair of gas turbine engines to a depth that includes and goes beyond that repair authorized for second- and third-degree intermediate maintenance activities. [AIR4896]

first-in, first-out (FIFO) An ordered queue; a discipline wherein the first transaction to enter a queue is also the first to leave it. Contrast with last-in, first-out.

first-level address *See* direct address.

first-order system A system definable by a first-order differential equation.

first-order transition A change of state associated with crystallization or melting of a polymer. [AIR4844]

first word address (FWA) A program/routine.

Fisher loop test One of several Wheatstone bridge test arrangements commonly used to determine the distance to a fault (grounded or crossed wires) in a communications cable.

fisheye *1.* An area on a fracture surface having a characteristic white crystalline appearance, usually caused by internal hydrogen cracking. 2. A small globular mass in a blended material, such as plastic or glass, that is not completely homogeneous with the surrounding material; particularly noticeable in transparent or translucent materials. 3. *See* flake.

fishing tool An elongated or telescopic tool with a magnet, hook, or grapple at one end; used to retrieve objects from inaccessible places.

fishmouthing *See* alligatoring.

fish plate Either of the two plates bolted or riveted to the webs of abutting rails or beams, on opposite sides, to secure a mechanical joint.

fishtail Excess metal at the trailing end of an extrusion or a rolled billet or bar, which is generally cropped and either discarded or recycled into a melting operation.

fissure A small, cracklike surface discontinuity, often one whose sides are slightly opened or displaced with respect to each other.

fit The closeness of mating parts in an assembly, as determined by their respective dimensions and tolerances. Fits may be classified as running (sliding), locational, transition or force (shrink) fits, depending on the size and direction (positive for running or negative for force fits) of the dimensional allowance. Fits may also be termed clearance or interference depending on whether there is always a gap between mating parts or always interference, as long as the parts are within specified tolerances.

fit check fixture A fixture accurately simulating the size and shape of the aircraft flexible tank cavity. Used as a quality-assurance acceptance test tool. [AIR1664]

fitting A connector that permanently joints the appropriate unit(s) of tubing ready for service. [ARP899]

fitting flange Plies of rubber-coated fabric or molded rubber that are bonded to the metal rings of the fitting and to the tank wall, and become an integral part of the tank assembly during vulcanization. [AIR1664]

fittings, tank mounted Subassemblies fabricated into the tank wall for the purpose of: (a) attaching the tank to structure; (b) providing a sealing surface for a means of attaching equipment items to the tank; or (c) providing for the attachment and sealing of vent lines and fluid flow lines to the tank. [AIR1664]

fitting, tube A self-contained detachable device, including a fluid passage, used for attaching or connecting fluid-carrying lines. [ARP243B]

fitting-tube junction Area that includes one tube-outside diameter in length beyond the fitting envelope. The envelope is the area from free parent-tube material to free parent-tube material. [ARP899]

fix A geographical position determined by visual reference to the surface, by reference to one or more radio NAVAIDs, by celestial plotting, or by another navigational device. [ARP4107]

fixed body actuator An actuator with mounting provisions such that the body is rigidly attached to the load-bearing support. [ARP4386]

fixed carbon In making the proximate analysis of a solid fuel, the carbonaceous residue less the ash remaining in the test container after the volatile matter has been driven off.

fixed cavity A seal cavity that is fixed or defined at the time of installation. [ARP4968]

fixed displacement pump A pumping unit in which the displaced volume is fixed by design; pump outlet is essentially constant for any given pump speed, regardless of system pressure level. [ARP4386]

fixed point *1.* In mathematics, a positional notation in which corresponding places in different quantities are occupied by coefficients of the same power of the base; notation in which the base point is assumed to remain fixed with respect to one end of the numeric expressions. *2.* A reproducible standard value, usually derived from a physical property of a pure substance, which can be used to standardize a measurement or check an instrument calibration.

fixed-point arithmetic A type of arithmetic in which the operands and results of all arithmetic operations must be properly scaled so as to have a magnitude between certain fixed values.

fixed-point data In data processing, the representation of information by means of the set of positive and negative integers. Fixed-point data is faster than floating point data and requires fewer circuits to implement.

fixed program computer *See* wired program computer.

fixed restrictor A fixed physical restriction to fluid flow.

fixed-riser stairway A stairway in which the riser to tread ratio (R/T) is a constant, which will result in a fixed angle of inclination. [ARP836A]

fixed storage A storage device that stores data not alterable by computer instructions; for example, magnetic core storage with a lockout feature, or photographic disk.

fixed surface An airfoil or fin that is rigidly fixed to the body of an aircraft. [ARP4386]

fixed voltage winding A motor winding that is energized by a fixed voltage. [ARP667]

fixed wing A wing (permanently fixed, foldable, or adjustable), that is fixed to the airplane fuselage and outspread in flight; that is, a nonrotating wing. [ARP4107]

fixed word length Describes the property whereby a machine word always contains the same number of characters or bits.

fix phase The portion of a scheduled inspection that involves the correction of discrepancies found during the look phase. [AIR4896]

fix rate The percent of aircraft that return "code 3," that must be repaired in a specified number of clock hours. [AIR4896]

fixture In a machining operation, a special holder that positions the work, but does not guide the tool.

fL *See* footlambert.

FL See flight level.

flag/flag alarm A warning device incorporated in certain airborne navigation and flight instruments indicating that: (a) instruments are inoperative or otherwise not operating satisfactorily; or (b) the signal strength or quality of the received signal falls below acceptable values. [ARP4107]

flag terminal A terminal having a tongue protruding from the side of its barrel. [ARP914A]

flake *1.* Dry, unplasticized cellulosic plastics-base material. *2.* An internal hydrogen crack such as may be formed in steel during cooling from high temperature. Also known as fisheye, shattercrack, or snowflake. 3. Metal powder in the form of fish-scale particles. Also known as flaked powder.

flaked powder *See* flake.

flame A luminous body of burning gas or vapor.

flame cutting Using an oxyfuel-gas flame and an auxiliary oxygen jet to sever thick metal sections or blanks.

flame deflector *1.* In a vertical launch, any of variously designed obstructions that intercept hot gases of rocket engines so as to deflect them away from the ground or from a structure. 2. In captive tests, elbows in the exhaust conduits or flame buckets that deflect the flame into the open.

flame detector A device that indicates if fuel (liquid, gaseous, or pulverized) is burning, or if ignition has been lost. The indication may be transmitted to a signal or to a control system.

flame hardening A form of surface hardening in which the inherent hardenability of a steel or other hardenable alloy is used to produce a hardened surface layer; accomplished by spot-heating the metal with a fuel-gas flame to a shallow depth and then rapidly cooling the heated metal.

flame ionization detector *1.* An analyzer that quantifies organic hydrocarbon species concentrations in gaseous samples or standard gas mixtures. [ARP4418] 2. A hydrogen-air diffusion flame detector that produces a signal nominally proportional to the mass-flow rate of hydrocarbons entering the flame per unit of time; generally assumed responsive to the number of carbon atoms entering the flame. [ARP1256]

flameout Unintended loss of combustion in turbine engines resulting in the loss of engine power. [ARP4107]

flame photometer An instrument for determining compositions of solutions by spectral analysis of the light emitted when the solution is sprayed into a flame.

flame plate A baffle of metal or other material for directing gases of combustion.

flameproof apparatus Apparatus so treated that it will not maintain a flame or will not be injured readily when subjected to flame. [ARP4404]

flame propagation rate Speed of travel of ignition through a combustible mixture.

flame resistance Describes a characteristic of a material whereby the tendency of the material, when burning, is to self-extinguish once the ignition source is removed. [ARP1931]

flame retardants Certain chemicals that are used to reduce or eliminate the tendency of a resin to burn. [AIR4844]

flame retarded resin A resin compounded with certain chemicals to reduce or eliminate its tendency to burn. [AIR4844]

flame shield The metal shield adjacent to the case insulation; prevents erosion of the insulation and objectionable insulation pyrolysis products from entering the gas stream. [ARP4386]

flame spraying *1.* Applying a plastic coating on a surface by projecting finely powdered plastic material mixed with suitable fluxes through a cone of flame toward the target surface. 2. Thermal spraying by feeding an alloy or ceramic coating material into an oxyfuel-gas flame. Compressed gas may or may not be used to atomize the molten material and propel it onto the target surface.

flame-spray strain gage A fine-wire strain gage element attached to a substrate by flame spraying a ceramic encapsulation over the element. This attaches the element without damaging either the gage or the substrate; and produces a bond suitable for operating over the temperature range -270–820°C (-450–1500°F).

flame treating Bathing inert thermoplastics parts in open flames to promote surface oxidation, thereby making the parts receptive to inks, lacquers, paints, or adhesives.

flammability Susceptibility to combustion. Also called combustibility.

flammable Describes a characteristic of a material whereby the material has a tendency to ignite and burn when an ignition source is brought sufficiently close. [ARP1931]

flammable liquid A liquid, usually a liquid hydrocarbon, that gives off combustible vapors.

flammable vapor area *See* vapor area, flammable.

flange Projecting rim of a mechanical part. [ARP4784]

flange, connector A projection extending from or around the periphery of a connector for the purpose of attaching the connector to a rigid surface. [ARP914A]

flanged spade tongue terminal A slotted-tongue terminal in which the ends of the tongue are formed up or down to the tongue plane; provides a degree of protection against the terminal slipping out from under its captive hardware. [ARP914A]

flank *1.* On a cutting tool, the end surface adjacent to the cutting edge. *2.* On a screw thread, the side of the thread.

flap *1.* A control surface used primarily to alter the lift and drag of an airplane, usually by effectively changing the camber of an airfoil. [ARP4386] *2.* A hinged, pivoted, or sliding airfoil or plate, or a combination of the former, regarded as a single surface; normally located at the trailing edge of the wing, and designed to add camber well aft on the chord and/or increase wing area. *See also* Fowler flap and leading edge flaps. [ARP4107]

flaperons Aircraft control surfaces that serve the function of both aileron and flap.

flapper nozzle amplifier *See* baffle-nozzle amplifier.

flapper/nozzle servovalve An electrohydraulic sevovalve in which the electric control signal moves a flapper to modulate flow between two nozzles in the first stage spool. [ARP4386]

flap valve A valve with a hinged flap or disc that swings in only one direction.

flare The increase in pitch angle of an aircraft just before touchdown. This change in pitch attitude on final approach to touchdown attitude allows the airspeed to dissipate and the aircraft to settle down to the runway in the proper attitude. [ARP4107]

flareback A burst of flame from a furnace in a direction opposed to the normal flow, usually caused by the ignition of an accumulation of combustible gases.

flared tube-end The projecting end of a rolled tube that is expanded or rolled to a conical shape.

flare stars Members of a class of dwarf stars that show sudden intensive outbursts of energy.

flaring Increasing the diameter at the end of a pipe or tube to form a conical section.

flash In plastics molding, elastomer molding, or metal die casting, a portion of the molded material that overflows the cavity at the mold-parting line.

flashback Backward burning of a flame into the lip of a burner or torch.

flashbreaker tape A tape used around the edge of a bonded joint so that any adhesive that spews out of the joint during curing can be easily removed. [AIR4844]

flash converter A converter in which all bit choices are made at the same time.

flashed glass *See* cased glass.

flash groove *See* cutoff.

flashing *1.* A desired, and usually controlled, variation in the luminance of a symbol or group of symbols. [AS8034] *2.* The evaporation of a heated liquid as a consequence of rapid pressure reduction. *3.* Steam produced by discharging water at saturation temperature into a region of lower pressure.

flashlamp A gas-filled lamp that is excited by an electrical pulse passing through it to emit a short, bright flash of light. A broad range of wavelengths are produced, with their precise nature depending on the gas or gases used.

flash line A raised line on the surface of a molded or die-cast part that corresponds to the parting line between mold faces.

flash mold A mold designed to permit the escape of excess molding material; such a mold relies on back pressure to seal the mold and put the piece under pressure. [AIR4844]

flash plating Electrodeposition of a very thin film of metal, usually just barely enough to completely cover the surface.

flash point Temperature at which a fluid gives off sufficient vapor to cause it to ignite when a small flame is applied, under controlled conditions. [AIR1916]

flash-resistant Not susceptible to burning violently or rapidly when ignited. [ARP171]

flash welded ring Ring produced by roll-forming of extruded or rolled stock, the ends being joined by flash butt welding. This ring contains a heat-affected zone at the weld area. [AMS2355F]

flash welding A resistance welding process commonly applied to wide, thin members, irregularly shaped parts, and tube-to-tube joints. In this process, the faying surfaces are brought into close proximity; electric current is passed between them to partly melt the surfaces (by combined arcing and resistance heating); and the surfaces are then upset-forged together to complete the bond.

flask In foundry work, a wood or metal frame for holding a sand mold. The flask is open ended, and usually consists of two halves— the cope (upper half) and the drag (lower half)—although three or more flask sections are occasionally used.

flat cable A cable whose geometric configuration is flat or essentially flat, rather than round. In a multiconductor flat cable, the conductors are laid side by side in the same plane. *See also* ribbon cable. [ARP1931]

flat conductor A conductor with a rectangular cross section. [ARP1931]

flat conductor cable A cable constructed using flat conductors. [ARP1931]

flat lay *1.* In laminating adhesives, the property of nonwarping. *2.* Describes an adhesive material with good noncurling and nondistension characteristics. [AIR4844]

flat-position welding Welding from above the work, with the face of the weld in the horizontal plane. Also known as downhand welding.

flat spin A spin in which the longitudinal axis of the aircraft inclines downward at an angle less than 45°. *See* spin. [ARP4107]

flatspotting The wearing of a flat spot on a tire by means of skidding, due to overbraking or faulty skid control. [AIR1489]

flattening Straightening metal sheet by passing it through a set of staggered and opposing rollers, which bend the sheet slightly to flatten it without reducing its thickness.

flattening test A test that evaluates the ductility, formability, and weld quality of metal tubing by flattening it between parallel plates to a specified height.

flat type damper *See* damper.

flavor (particle physics) The specific identifiers of quarks which distinguish various combinations of electric charge and mass.

flaw A discontinuity or other physical attribute in a material which exceeds acceptable limits. NOTE: The term flaw is nonspecific, and more specific terms, such as defect, discontinuity, or imperfection, are often preferred.

fleet leader concept *See* concept, fleet leader.

FLEETSATCOM *See* fleet satellite communication system.

fleet satellite communication system (FLEETSATCOM or FLTSATCOM) Global communication system utilizing satellites.

flettner tab *See* servo tab.

flexibility A 4D term representing the FMCS total range of control capability over arrival time at a designated downstream waypoint or waypoints. May be expressed as a TRA Max and TRA Min in respective GMT hrs:min:sec.[ARP1570]

flexibilizer An additive that makes a finished plastic more flexible. [AIR4844]

flexible coupling A fluid-coupling assembly designed to accommodate relative movement between the mating tube-end components. [ARP4968]

flexible manufacturing systems (FMS) A manufacturing system under computer control, with automatic material handling; primarily designed for batch manufacturing.

flexible molds Molds made of rubber or elastomeric plastics, used for casting plastics. [AIR4844]

flexible pavement *See* pavement, flexible.

flexible shafting A mechanical connection that is composed of an inner core formed from a series of helical-wrapped wire layers, running inside of a flexible tube that allows torque to be transmitted from end to end of the inner core, through one or more curved bends. [ARP4386]

flexible spacecraft Space vehicles (usually space structures or rotating satellites) whose surfaces and/or appendages may be subject to elastic flexural deformations (vibrations).

flexible thermocouple harness An assembly in which the lead supporting and protective structure is easily bent, and adapts itself readily to change of shape. [ARP485]

flexivity Temperature rate of flexure for a bimetal strip of given dimensions and material composition.

flex life The number of flexes that can be achieved when flexing an item under given conditions (temperature, radius of bend, load, arc, etc.) before the failure point is reached. [ARP1931]

flex-section Device incorporated in a duct system that permits relative motion in one or more planes. [ARP699D]

flexural modulus For a test specimen in flexure, the ratio (within the elastic limit) of the applied stress on the specimen to the corresponding strain in the outermost fibers of the specimen. [AIR4844]

flexural strength The maximum stress that can be borne by the surface fibers in a beam in bending. [AIR4844]

flicker Undesired temporal variation of luminance in a portion or the total display, at a rate detectable by the eye. [ARP1782]

flicker vertigo *See* illusions, visual.

flight *1.* The movement of an object through the atmosphere or through space, sustained by aerodynamic, aerostatic, or reaction forces, or by orbital speed; especially, the movement of a man-operated or man-controlled device, such as a rocket, space probe, space vehicle, or aircraft. *2.* Refers to the period beginning when the aircraft first moves forward on its takeoff run, or takes off vertically from rest at any point of support, and ending after airborne flight when the aircraft is on the surface. [AIR4896]

flight characteristic Characteristic exhibited by an aircraft, rocket, or the like in flight, such as a tendency to stall or to yaw, or an ability to remain stable at certain speeds.

flight check *1.* In-flight investigation and evaluation of a navigational aid (NAVAID) to determine whether it meets established tolerances. *2.* An in-flight evaluation of a flight crew-

member's ability to perform assigned duties. *3.* A call-sign prefix used by Federal Aviation Administration (FAA) aircraft engaged in flight inspection/certification of navigational aids and flight procedures. [ARP4107]

flight control system (FCS) A system that includes all aircraft subsystems and components used by the pilot or other sources to control one or more of the following: aircraft flight path, attitude, airspeed, aerodynamic configuration, ride, and structural modes. [ARP1181A]

flight crew member A pilot, copilot, flight engineer, or flight navigator assigned to an aircraft during flight time. [ARP4107]

flight deck alerting system (FAS) Alerting system that consists of the following levels of alerts: warning (level 3); caution (level 2); advisory (level 1); and information (level 0). *Warning (level 3)*–corresponds to an emergency situation. *Caution (level 2)*–corresponds to an abnormal situation. *Advisory (level 1)*–corresponds to a recognition situation. *Information (level 0)*–corresponds to an information situation. [ARP4102/4]

flight deck door The door that connects the flight deck area with any cabin area or intervening area, whether passenger or cargo; does not include any door leading from the flight deck directly to the outside of the aircraft. [ARP4101/8]

flight deck interior door The door that connects the flight deck area with any cabin area or area leading to any cabin area, whether passenger or cargo; does not include any door leading from the flight deck directly to the outside of the aircraft. [ARP807]

flight discipline Adherence to established procedures throughout the course of a sortie. This includes not pursuing irrational or impulsive courses of action, actions that are inconsistent with established procedure, or actions not prebriefed. [ARP4107]

flight envelope The bounds within which a certain flight system can operate, especially a graphic representation of these bounds showing interrelationships of operational parameters.

flight envelope awareness A function that incorporates modification of control characteristics or control laws outside the normal flight envelope as a means to alert the pilot flying of a specific exceedance of aerodynamic, structural, or other boundaries. [ARP4101/1]

flight envelope protection A function that incorporates a modification of control laws outside the normal flight envelope in order to prevent or minimize the risk of exceedance of aerodynamic, structural, or other boundaries. [ARP4101/1]

flight information message An informational message that does not imply any change in operating behavior on the part of the pilot or the controller. Includes routine weather observations and forecasts. [ARP4791]

flight leg Any of the sequential aircraft-operating cycles that together constitute a flight. [AIR4896]

flight level (FL) A level of constant atmospheric pressure related to a reference datum of 29.92 in. of mercury; stated in digits representing hundreds of feet. For example, flight level 250 represents a barometric altimeter indication of 25,000 ft; flight level 255 indicates 25,500 feet. [ARP4107]

flight monitor package A special software system that enables several operators to monitor data from an aircraft or other source in real time.

flight operation Collective term for ground support operations by flight crew or supported personnel preparatory to space flight, or tasks performed by crew during flight.

flight path Path made or followed in the air or in space by an aircraft or rocket; the continuous series of positions occupied by a flying body; more strictly, the path of the center of gravity of the flying body, referred to the earth or other fixed reference.

flight-path angle The angle between the flight path of the aircraft and the horizontal. [ARP4107]

flight path control The process of guiding or directing a vehicle so that its position vector describes a desired locus in space. [ARP4386]

flight plan Specified information relating to the intended flight of an aircraft which is filed orally or in writing with a flight service station (FSS) or an air traffic control (ATC) facility. [ARP4107]

flight plan–progress display The pilot's primary navigation display. May include, but is not limited to, the following: aircraft relative position and orientation; altitude intercept range; course, desired track or localizer; course or localizer deviation; drift angle and/or track angle error; ETAs and RTAs; heading; predictive track; track; time control speed reference; waypoints; and weather. [ARP1570]

flight profile A graphic, vertical-plane portrayal of the flight path of an aircraft. [ARP4107]

flight recorder An instrument or device that records information about the performance of an aircraft in flight, or about conditions encountered during flight. [ARP4107]

flight service station (FSS) Air traffic facilities that: (a) provide pilot briefing, en route communications, and VFR search and rescue services; (b) assist lost aircraft and aircraft in emergency situations; (c) relay ATC clearances; (d) originate Notices to Airmen (NOTAM); (e) broadcast aviation weather and NAS information; (f) receive and process IFR flight plans; and (g) monitor NAVAIDs. Selected flight service stations provide en route flight advisory service (flight watch), take weather observations, issue airport advisories, and advise Customs and Immigration of transborder flights. [ARP4107]

flight simulator Training device or apparatus that simulates certain conditions of flight or of flight operations.

flight test *1.* Test by means of actual or attempted flight to see how an aircraft, spacecraft, space-air vehicle, or missile flies. *2.* To test a component part of a flying vehicle (or an object carried in such a vehicle) by making it endure actual flight; used to determine the suitability or reliability of the part or object in terms of its intended function.

flight vehicle power *See* secondary power.

flight visibility For an aircraft in flight, the average forward horizontal distance from the cockpit of at which prominent, unlighted objects may be seen and identified by day, and prominent, lighted objects may be seen and identified by night. [ARP4107]

flight watch Refers to air-ground contacts on frequency 122.0 MHz, used to identify the flight service station providing en route flight advisory service. *See* en route flight advisory service. [ARP4107]

flint glass An optical glass that contains lead or other elements, which raise its refractive index between 1.6 and 1.9—higher than other types of optical glass.

FLIP *See* Department of Defense Flight Information Publications.

flip-flop *1.* A digital component or circuit with two stable states and sufficient hysteresis so that it has "memory." The state of a flip-flop is changed with a control pulse; a continuous control signal is not necessary for it to remain in a given state. [ARP993A] *2.* A control device for opening or closing gates, i.e., a toggle.

FLIR detector Stands for forward-looking infrared detector, a detector for sensing all emissions of heat or light.

float A device or component part of an air vehicle that permits the vehicle to remain suspended upon the surface of a body of water without sinking; also to maneuver upon, take off, and land upon a body of water. [AIR1489]

float chamber A vessel in which a float regulates the liquid level.

float gage In a tank or vessel, any of several types of devices that use pulleys, levers, or other mechanisms to transmit the position of a float to a scale that indicates liquid level.

floating Describes the condition of a line in a logic circuit that is not grounded or tied to any established potential.

floating charge Describes the use condition of a storage battery wherein charge is maintained by continuous, long-term constant-potential charge. [ARP4386]

floating plug A short-nosed mandrel attached to a rod, inserted into pipe or tubing during reduction by drawing. Also known as plug die.

floating point *1.* An arithmetic notation in which the decimal point can be manipulated; values are sign, magnitude, and exponent (e.g., $+0.833 \times 10^2$). *2.* A form of number representation in which quantities are represented by a bounded number (mantissa) and a scale factor (characteristic or exponent) consisting of a power of the number base, e.g.,

$127.6 = 0.1276*10**3$, where the bounds are 0 and 1 for the mantissa and the base is ten.

floating point arithmetic A method of calculation that automatically accounts for the location of the radix point. This usually is accomplished by handling the number as a signed mantissa times the radix raised to an integral exponent; for example, the decimal number $+88.3$ might be written as $+.883*10**2$; the binary number $-.0011$, as $-.11*2**-2$. Contrast with fixed-point arithmetic.

floatless level control Any device for measuring or controlling liquid level in a tank or vessel without the use of a float; methods include manometers, electrical probes, capacitance devices, radiation instruments, and sonic or ultrasonic instruments.

float switch An on-off switch that is activated by the position of a float.

float valve An on-off-type valve whose action is triggered by the rise or fall of a float.

FLOLS Acronym for Fresnel lens optical landing system.

flood lighting Similar to normal home lighting, lighting in which a light source floods a general area or is directed by reflectors to light a particular area. [ARP1161]

flood lit systems Integrally lit displays in which the light source or a light-distributing material directs light to the front of the display, where it is reflected to the observer. Examples of flood lit systems are wedge lighting, ring lighting, and parallel plate lighting. [ARP1161]

floppy disk Any flexible platter with a magnetic coating that can accept computer data. Also called a diskette.

flospinning Forming cylindrical, conical, or curvilinear parts from light plate by power-spinning the metal over a rotating mandrel.

flotation A process for separating particulate matter in which differences in surface chemical properties are used to make one group of particles float on water while other particles do not. Used primarily to separate minerals from gangue, but also used in some chemical and biological processes. In mining engineering, also known as froth flotation.

flotation, ground *See* ground flotation.

flow *1.* Rate of fluid movement, usually expressed in GPM. [ARP243B] *2.* Quantity of a fluid (volume or mass) crossing the transverse plane of a flow path per unit of time. Gas volume flow may be expressed at standard reference conditions of sea level atmospheric pressure and ambient temperature. [AIR1916] *3.* The order of events in the computer solution to a problem. *4.* The movement of a resin under pressure, allowing it to fill all parts of a mold. *5.* The gradual but continuous distortion of a material under continued load, usually at high temperatures. *6.* A qualitative description of the fluidity of an adhesive material during the process of bonding, before the adhesive is set. [AIR4844]

flowability A general term describing the ability of a slurry, plasticized material, or semi-solid to behave like a fluid.

flow amplifier A component designed specifically for amplifying flow signals. [ARP993A]

flow brazing A brazing process in which the joint is heated by pouring hot molten nonferrous filler metal over the assembled parts until brazing temperature is attained.

flow, calibrated *See* calibrated flow.

flow chart *1.* Graphical representation of sequences of operations using symbols to represent the operations. *2.* A system-analysis tool that provides a graphical presentation of a procedure. Includes block diagrams, routine sequence diagrams, general flow symbols, and so forth. *3.* A chart to represent the flow of data, procedures, growth, equipment, methods, documents, machine instructions, etc. for a specific problem. *4.* A graphical representation of a sequence of operations by using symbols to represent the operations such as COMPUTE, SUBSTITUTE, COMPARE, JUMP, COPY, READ, WRITE, etc.

flow coat To apply a coating by pouring liquid over an object and allowing the excess to drain off.

flow compensation Using secondary signals to correct flow values for changes in density or viscosity.

flow control *1.* Measures used by ATC to adjust the flow of traffic into a given airspace, along a given route, or bound for a given airport, to ensure the most effective utilization of the airspace. [ARP4107] *2.* Any method for controlling the flow of a material through piping, ductwork, or channels.

flow control servovalve A servovalve that modulates output flow with respect to an input. [ARP4386]

flow curve The graphical representation of control flow versus input current; usually a continuous plot of a complete cycle between plus and minus rated current values of no-load flow. [ARP490]

flow divider A flow proportioner that operates only with dividing flow. [ARP243B]

flow equalizer A flow proportioner in which the portions are equal. [ARP243B]

flow gage A device that indicates the flow rate of liquid in a system. May be mechanically driven or respond to electronic pulses. [AIR4783]

flow gain The slope of the control flow versus input current curve in any specific operating region, expressed in cis/mA, gpm/mA, or mL/s/mA. Three operating regions are usually significant with flow-control servovalves: (a) the null region; (b) the region of normal flow control; and (c) the region where flow saturation effects may occur. NOTE: Where this term is used without qualification, it is assumed to mean normal flow gain. [ARP490]

flow indicator A device for indicating that oxygen is flowing through a regulator, or to a mask. [ARP171]

flow, laminar *See* laminar flow.

flow limit The condition in which control flow no longer increases with increasing input current. [ARP4386]

flow line(s) *1.* The texture revealed by etching a metal surface or section, showing direction of metal flow. [ARP700] *2.* The connecting line or arrow between symbols on a flow chart. *3.* A mark on a molded plastic part where two flow fronts met during molding. Also known as a weld mark.

flow marks Wavy surface marks on a molded thermoplastic part resulting from improper flow of resin during molding.

flow, mass *See* mass flow.

flowmeter An instrument used to measure linear, non-linear, or volumetric flow rate or discharge rate of a fluid flowing in a pipe. Also known as fluid meter.

flowmeter secondary device The device that responds to the signal from the primary device and converts it to a display or to an output signal that can be translated relative to flow rate or quantity.

flow mixer A device for mixing two solids, liquids, or gases together in which the mixing action occurs as the materials pass through the device. Also known as a line mixer.

flow nozzle A type of differential pressure-producing element having a contoured entrance. Characterized by its ability to be mounted between flanges, with lower permanent pressure loss than an orifice plate.

flow pattern The paths of fluid flow connecting various ports in a given valve position. [ARP4386]

flow proportioner A device that automatically maintains a relatively constant ratio between the portions of dividing or combining flow passing through it, regardless of differences in pressure between the portions. A flow proportioner may operate only with combining flow and/or dividing flow. [ARP243B]

flow rate The quantity of fluid that moves through a pipe or channel within a given period of time.

flow, rated *See* rated flow.

flow-rate range Range of flow rates bounded by the minimum and maximum flow rates.

flow regulating valve A valve that limits flow in a line to a predetermined value, irrespective of variation in pressure differential caused by back pressure or working against load. [ARP4386]

flow saturation region The region in which flow gain decreases with increasing input current. [ARP4386]

flow soldering *See* wave soldering.

flow steady A continuous flow of constant quantity under the prevailing condition. [ARP171]

flow surge (surge) Temporary rise and fall of flow. [AIR1916]

flow symmetry The degree of equality between flow outputs of one direction and those of the reversed direction produced by changing only the polarity of the electrical input. [ARP4386]

flow transmitter A device that senses the flow of liquids in a pipe and converts the sensor output into electric signals proportional to flow rate, which can be transmitted to a remote indicator or controller.

flow, turbulent *See* turbulent flow.

flow, volumetric The volume rate of fluid flow at a specified temperature and pressure, usually expressed in units of ft^3/min or m^3/s. [ARP147C]

flow, weight The weight rate of fluid flow, usually expressed in units of lbs/min or kg/s. [ARP147C]

FLTSATCOM *See* fleet satellite communication system.

flue-gas analyzer An instrument that monitors the composition of flue gas as it passes out of a boiler or heating unit. The readout from this analyzer is used to guide adjustment of combustion controls to achieve maximum combustion efficiency or heat output.

flueric Describes fluidic devices and systems performing sensing, logic, amplification, and control functions with no use whatsoever of moving mechanical elements. [ARP993A]

fluid *1.* A substance that has such a low cohesive force that it does not remain in the solid state under normal temperature and pressure conditions. [ARP4386] *2.* May refer to a gas or a liquid, both of which have the property of undergoing continuous deformation when subjected to any finite shear stress as long as the shear stress is maintained.

fluid bulk modulus For a trapped fluid volume, the quantity equal to incremental pressure/(incremental volume/total volume). [ARP4386]

fluid capacitance The ratio of the volume flow to the rate of change of pressure. [ARP993]

fluid coupling *1.* A device for transmitting rotational motion and power between shafts by means of the acceleration or deceleration of oil or another suitable liquid. Also known as hydraulic coupling. *2.* A quick-disconnect fluid connector, consisting of a tanker half and spacecraft half; when mated by some external means, permits fluid flow across the interface. [AIR4728]

fluid current The volume of fluid per unit time passing an arbitrary cross section of a fluid conductor. [ARP993]

fluid filled shell Shell of revolution containing a gas or liquid.

fluidics A control technology that relies on the interaction of fluid streams to provide sensing, amplification, filtering, signal conditioning, and other control functions. [AIR1245]

fluid inertance The ratio of pressure difference to the rate of change of volume flow. [ARP993]

fluidity The degree to which a substance flows freely.

fluidized bed A dynamic mixture of a gas and/or vapor and minute solid particles of such a size that the mixture resembles a fluid in motion.

fluid management The isolation and separation of liquids from gas in a storage vessel that operates in a reduced or zero-gravity environment, using liquid-acquisition devices such as those used in the Space Shuttle RCS (Reacting Control System) tankage.

fluid meter *See* flowmeter.

fluid, operating The medium or fluid to be used in a unit or system. [ARP243B]

fluid port The fluid pressure port of a hydropneumatic accumulator; sometimes called the oil port. [ARP4386]

fluid resistance The ratio of pressure drop to volume flow rate entering the resistor. [ARP993]

fluid-solid interaction The interaction of a rigid or elastic structure with an incompressible or compressible fluid. Airblast loading and response, acoustic interaction, aeroelasticity, and hydroelasticity comprise the major divisions of this area of inquiry.

fluid stiffness Refers to the stiffness exerted by the volume of fluid trapped in the pressurized fluid chamber on each side of the piston in an actuator when the control valve is at null. This stiffness opposes the change in fluid volume when an external load is applied to the piston.

fluid temperature Temperature of the fluid measured at a specified point in the system. [AIR1916]

fluid transpiration *See* transpiration.

flume An adaptation of the venturi concept of flow constriction applied to open-channel flow measurement.

fluorescence *1.* Emission of light or other radiant energy as a result of and only during absorption of radiation of a different wavelength from some other source. *2.* Characteristic x-rays produced due to absorption of higher-energy x-rays.

fluorescence spectroscopy The study of materials by analyzing the light they emit when irradiated by other light. Many materials emit visible light after they have been illuminated by ultraviolet light. The intensity and wavelengths of the emitted light can be used to identify the material and its concentration.

fluorescent contamination Contamination of parts or developer with fluorescent penetrant. [AMS2647A]

fluorescent emission *See* fluorescence.

fluorescent lamp An electric-discharge light source in which light is predominantly produced by fluorescent powders activated by ultraviolet energy generated by a mercury arc. There are two basic types of fluorescent lamps, hot cathode and cold cathode. In general, fluorescent lamps are used for area lighting, including cabin, galley and lavatory, and decorative lighting. [AIR512B]

fluorimeter *See* fluorometer.

fluorocarbon A polymer or gas containing only carbon and fluorine. [ARP1931]

fluorometer An instrument for measuring the fluorescent radiation emitted by a material when excited by monochromatic incident radiation, usually filtered radiation from a mercury-arc lamp or from a tungsten or molybdenum x-ray tube. Also spelled fluorimeter.

fluoroplastics A family of plastics resins based on fluorine substitution of hydrogen atoms in certain hydrocarbon molecules. *See also* fluoropolymers.

fluoropolymers A family of polymers based on fluorine replacement of hydrogen atoms in hydrocarbon molecules. Compounds are characterized by chemical inertness, thermal stability, and low coefficient of friction. *See also* fluoroplastics.

fluoroscopy X-ray examination similar to radiography except the image is produced on a fluorescent screen instead of on radiographic film.

fluorosilicone sealant A sealant based on a fluorosilicone polymer. There are two types used in integral fuel tanks: (a) a one-part, moisture-cured, polymerizing elastomer used in brushcoat and fillet (extrusion type) applications inside the fuel tank; and (b) a one-part, noncuring channel or groove sealant, which can be prepacked into the grooves in the faying surface in assembly or injection through external ports. [AIR4069]

flushing Removing debris, deposits, wear particles, or used lubricating oil from a piping system, chamber, or mechanism by circulating a liquid, such as a solvent oil or water, then draining the system to carry off unwanted substances.

flushing connection A connection on an instrument, manifold, or piping to permit periodic back-flow of an external fluid for clearing purposes.

flushing pressure Pressure required to flush a system at defined conditions (for instance at defined flow). [AIR1916]

flute *1.* In a drill, reamer, or tap, a channel or groove in the body of the tool which exposes the cutting edge and provides a passage for cutting fluid and chips. 2. In a milling cutter or hob, the chip space between the back of one cutting tooth and the face of the following tooth.

flutter *1.* An aeroelastic self-excited vibration in which the external source of energy is the airstream; this vibration depends on the elastic, inertial, and dissipative forces of the system in addition to the aerodynamic forces. 2. Irregular, alternating motion of a control surface, often due to turbulence in a fluid flowing past the surface.

flux *1.* The rate of flow of some quantity, often used in reference to the flow of some form of energy. 2. In nuclear physics, generally, the number of radioactive particles per unit volume times their mean velocity. *3.* In metal refining, a substance added to the melt to remove undesirable substances, such as sand, dirt, or ash, and sometimes to absorb undesirable elements or compounds, such as sulfur in steelmaking or iron oxide in copper refining. *4.* In welding, brazing, and soldering, a substance preplaced in the joint or fed into the molten zone to prevent formation of oxides or other undesirable compounds, or to dissolve and make them easy to remove. *5.* In magnetic or electromagnetic applications, the integral of magnetic field strength over the cross-sectional area of the field.

flux-cored arc welding (FCAW) A form of electric-arc welding in which the electrode is a continuous tubular wire of filler metal whose central cavity contains welding flux. Welding may be performed with or without a shielding gas such as CO_2 or argon.

flux density The flux (rate of flow) of any quantity, usually a form of energy, through a unit area of specified surface. NOTE: This is not a volumetric density like radiant density.

flux gate A detector that produces an electric signal whose magnitude and phase are proportional to the magnitude and direction of an external magnetic field aligned with the axis of the detector.

flux guide A shaped piece of metal used in magnetic or electromagnetic applications to direct magnetic flux along preferred paths or to prevent it from spreading beyond specific boundaries.

fluxmeter An instrument for measuring the intensity of magnetic flux.

flux pinning In superconductors, the interaction between the magnetic and the metallurgical microstructures. Flux pinning controls the critical current density in a given superconducting material.

flux pump Cryogenic d-c generator.

flux (rate) The total emanation of energy, material, or particles from a single source per unit time. Used for electron flux, neutron flux, and particle flux.

flux (rate per unit area) *See* flux density.

flux vector splitting The splitting of the non-linear flux vectors of the conservation law form of the inviscid gasdynamic equations into subvectors by similarity transformations so that each subvector has associated with it a specified eigenvalue spectrum.

fly A fan with two or more blades that is used in timepieces or light machinery to control rotational speed by means of air resistance.

fly ash Fine particulate, essentially noncombustible refuse, carried in a gas stream from a furnace.

fly-by-light A flight-control system in which vehicle control input is transmitted by light through a fiber-optic cable. [ARP4386]

flyby mission Interplanetary mission in which the vehicle passes close to the target planet but does not impact it or go into orbit around it.

fly by tube control In aircraft, a fluidic light control in which a hydraulic control-signal link connects the pilot's controls to the control surface actuators.

fly-by-wire A flight-control system in which vehicle control input is transmitted completely by electrical means. [ARP4386]

fly cutting Machining using a rotating, single-point tool or a milling cutter having only one tooth.

fly heading (degrees) A controller command that informs the pilot of the heading that he/she should fly. [ARP4107]

flying *See* flight.

flying tab *See* servo tab.

flywheel A balanced, rotating element attached to a shaft which utilizes inertial forces to maintain uniform rotational speed and damp out small variations in power generated by the driving elements.

FM (frequency modulation) The process (or the result of the process) in which the frequency deviates from the unmodulated carrier in proportion to the instantaneous value of the modulating signal.

FM (tape record/reproduce) The tape-record/reproduce process whereby data modulate an FM oscillator for recording, and are demodulated by an FM discriminator.

FMC *See* full mission capable.

FM discriminator A device that converts frequency variations to proportional variations in voltage or current.

FM/FM Frequency modulation of a carrier by subcarriers that are frequency modulated by information.

FM/PM (modulation) Phase modulation of a carrier by subcarriers that are frequency modulated by information.

FMS *See* flexible manufacturing system.

F number For a lens, the ratio of the principle length to the diameter.

foamed plastics *1.* Resins in sponge form, flexible or rigid, with cells closed or interconnected and density over a range from that of the solid parent resin to 0.030 g/cm^3. [AIR4844] *2. See* expanded plastic.

foaming agent Chemicals added to plastics and rubbers which generate inert gases on heating, causing the resin to assume a cellular structure. [AIR4844]

foaming film adhesive An adhesive film used to join honeycomb core in bonded assemblies. [AIR4844]

foam-in-place A method widely used to apply foamed insulation to industrial equipment; in this method, two or more reactive substances are deposited onto a surface to be covered, where the foaming reaction takes place.

foam porosity Numerous, small openings in the foam structure; classified as coarse or fine, based on the size of the openings and on air pressure drop tests. [AIR4170]

focal length The distance from the focal point of a lens or lens system to a reference plane at the lens location, measured along the focal axis of the lens system.

focal plane device Radiation-sensitive device positioned at the focal area of electromagnetic detectors.

focal point *1.* The location on the opposite side of a lens or lens system at which rays of light from a distant object meet at a point. Also known as a focus. *2.* The point in space at which a beam of electromagnetic energy (such as light, x-rays, or laser energy) or of particles (such as electrons) has its greatest concentration of energy; corresponds to the point at which a converging beam of energy undergoes a transition to become a diverging beam.

focal spot In a target in an x-ray tube, the area where the stream of electrons from the cathode strikes the target.

focus *1.* To adjust the position of a lens with respect to an imaging surface so that sharp features of the object appear sharp in the image. *2. See* focal point.

focused transducer A transducer with a concave face which converges the acoustic beam to a focal point or line at a definite distance from the face. [ARP5089]

focusing coil An assembly containing one or more electromagnetic coils which is used to focus an electron beam.

focusing electrode An electrode configured so that its electric field acts to control the cross-sectional area of an electron beam.

focusing magnet An assembly containing one or more permanent magnets or electromagnets which is used to focus an electron beam.

focus of attention *See* attention, focus of.

FOD Stands for foreign object damage; damage to a jet engine caused by the ingestion of debris (foreign objects); also, the debris that damages a jet engine when it is ingested. [ARP4107]

fog *1.* A suspension of very small water droplets in the air. [AIR1335] *2.* A defect in developed radiographic, photographic, or spectrographic emulsions consisting of uniform blackening due to unintentional exposure to low-intensity light or penetrating radiation.

fogged metal A metal surface whose luster has been greatly reduced by the creation of a film of oxide or other reaction products.

fog quenching Rapidly cooling an item by subjecting it to a fine mist, usually of water.

fog time During a ground or airstart attempt, the period between a fuel manifold full condition and lightoff; typified by a light fuel mist exiting the exhaust nozzle. [ARP906A]

foil Very thin metal sheet, usually less than 0.006 in. (0.15 mm) thick.

foil strain gage A type of metallic strain gage usually made in the form of a back-and-forth grid by photoetching a precise pattern on foil

of a special alloy having high resistivity and low temperature coefficient of resistivity.

FOL *See* fuel, oil, and lubrication (one entry).

fold A doubling over of metal that may occur during the forging operation. Folds may occur at or near the intersection of diameter changes, and are especially prevalent with noncircular necks, shoulders, and heads. [MA2005]

folding error An error in sampling an electronic signal arising from failure to sample at a high enough rate. (Sampling rate should be at least double the maximum signal frequency.). When folding error occurs, the sampling device perceives high-frequency components of the signal as low-frequency components. Also known as aliasing error.

follow-on operational test Tests conducted on a continuing basis to ensure that the established reliability and accuracy factors are preserved during the life of the vehicle, aircraft, etc. [ARP4386]

follow-up control unit A device that provides an output signal proportional to the displacement, or rate of movement, of the device being driven by the servo. Also called a feedback or repeatback control unit. [ARP419]

follow-up signal A signal produced by a follow-up control unit. [ARP419]

footcandle (fc) A measure of the intensity or level of illumination. One footcandle is the intensity of illumination at a point on a surface one foot from a uniform point source of one standard candle. [ARP1161]

footcandle/lux conversion One footcandle equals 10.76 lux. [AIR512B]

footlambert (fL) A unit of brightness equal to the uniform brightness of a perfectly diffusing surface emitting or reflecting light at the rate of one lumen per square foot. NOTE: The average brightness of any reflecting surface in footlamberts is the product of the illumination in footcandles by the reflection factor of the surface. [ARP798]

foot-pound A force of one pound applied to a lever one foot long.

footprint *1.* The tire area in contact with the ground surface. [AIR1780] *2.* The area of airplane floor space reserved for locating the galleys; also, the lower surface of galley envelope areas shown in plan view. [AS1426] *3*. Ground pattern or contour of an acoustical or microwave nature that is predictable and measurable.

Forbush decrease The observed decrease in cosmic ray activity in the earth's atmosphere about a day after a solar flare.

Forbush effect *See* Forbush decrease.

force *1*. The force used in testing during the use of a gasket, expressed in kg/cm of length of gasket or seam. [ARP1173] *2*. The male half of a mold, which enters the cavity, exerting pressure on the resin and causing it to flow. Also called punch. [AIR4844] *3*. The cause of the acceleration of material bodies, measured by the rate of change of momentum produced on a free body.

force-balance transmitter A transmitter-design technique utilizing feedback of the output signal to balance the primary input signal from the measuring element. The balanced output signal is proportional to the measured variable.

force, contact engaging The force required to fully engage a pair of mating contacts. [ARP914A]

force contact retention The maximum allowable force that, if applied axially in either direction on a contact, does not displace the contact permanently from its normal position in the connector or jeopardize or damage the contact or retention provision. *See also* pull-out force. [ARP914A]

force, contact separation The force required to separate a pair of fully mated contacts. [ARP914A]

forced circulation Using a pump or fan to move fluid through a conduit or process vessel; for instance, air or gases through a furnace or combustion chamber (often referred to as forced draft), ambient or conditioned air through ductwork (often referred to as forced ventilation), or a mixture of water and steam through tubes in a boiler.

forced draft fan A fan supplying air under pressure to the fuel-burning equipment.

forced oscillation Oscillation of a system attribute in which the period of oscillation is determined by an external periodic force. *See also* forced vibration.

forced vibration An oscillation of a system in which the response is imposed by the excitation. If the excitation is periodic and continuing, the oscillation is steady state. *See also* forced oscillation.

force factor The complex ratio of the force required to block the mechanical system of an electromechanical transducer to the corresponding current in the electrical system.

force fit A class of interference fit involving relatively large amounts of negative allowance; requires large amounts of force to assemble and results in relatively large induced stresses in the assembled parts. *See also* shrink fit.

force flight Refers to forces produced internally in a redundant actuator caused by unsynchronization of the independent channels. [ARP4386]

force, insert retention The maximum allowable force that, if applied to the mating face of a connector insert, does not displace the insert permanently from its normal position in the connector housing or jeopardize or damage the insert or connector housing retention provision. [ARP914A]

force structural maintenance plan A plan that identifies the inspection and modification requirements and the estimated economic life of the airframe structure. [AIR4896]

force vector recorder Instrument for recording force displacements in a variety of disciplines.

forcing Applying control signals greater than those warranted by a given deviation in the controlled variable in order to induce a more rapid rate of adjustment in the controlled variable.

Ford cup viscometer A time-to-discharge apparatus used primarily for determining the viscosity of paints and varnishes.

foreground *1*. The area in memory designated for use by high-priority programs; the program, set of programs, or functions that gain the use of machine facilities immediately upon request. *2*. Describes control tasks that are time-critical; all other tasks are called background tasks. [AIR4548]

foreground/background A control system that uses two computers, one performing the control functions and the other used for data logging, off-line evaluation of performance, financial operations, and so on. Either computer is able to perform the control functions.

foreground/background processing A computer system organized so that primary tasks dominate computer processing time when required, and secondary tasks fill the remaining time.

foreground program A time-dependent program initiated via request, whose urgency preempts operation of a background program. Contrast with background program.

forehand welding Welding in which the palm of the welder's torch- or electrode-hand faces the direction of weld travel; has special significance in oxyfuel-gas welding, where the welding flame is directed ahead of the weld puddle and provides preheating. Contrast with backhand welding.

foreign object Any object that is considered to be an undesirable constituent of the incoming airstream to a gas turbine engine. [AIR4096]

foreign object damage (FOD) In a narrow sense, any physical damage to the gas turbine engine or its internal flow path resulting from the ingestion of foreign objects. [AIR4096]

fore pump A vacuum pump operated in series with another vacuum pump to produce vacuum at the discharge of the latter; the second pump is not capable of discharging gases at atmospheric pressure.

fore vacuum A space on the exhaust side of a vapor jet or pump where the ambient static pressure is below atmospheric pressure.

forged ring Ring produced from cast stock or pre-formed biscuit that has been pierced and subsequently forged into the final ring size. [AMS2355F]

forging Using compressive force to plastically deform and shape metal; usually done hot, in dies or between rolls.

forging billet *See* forging stock.

forging cracks In metal forging, rupture of the metal that develops during the forging operation due to overstressing the metal by forging at too low a temperature. [ARP4784]

forging/part number A numeric or alphanumeric designation that identifies a particular forging configuration. [AMS2808C]

forging range In hot forging, the optimum temperature range for shaping the metal.

forging stock A piece of semifinished metal used to make a forging. Also known as forging billet.

fork The structural member of a shock-strut assembly which supports the axle and extends alongside the wheel and tire and connects to the basic shock strut. [AIR1489]

fork, full A configuration of landing gear fork that supports the axle on both ends, extends adjacent to the wheel on both sides, and joins around the tire to attach to the basic shock strut. [AIR1489]

fork, half A configuration of landing gear fork that supports the axle on one end only, lies adjacent to the wheel on one side only, and attaches to the shock strut assembly. [AIR1489]

formal demonstration phase A period of time during which maintainability demonstration tests are performed and data are acquired and analyzed, to determine conformance to specified requirements. [AIR4896]

formation flight Describes the situation in which more than one aircraft that, by prior arrangement between the pilots, operate as a single aircraft with regard to navigation and position reporting. Separation between aircraft within the formation is the responsibility of the flight leader and the pilots of the other aircraft in the flight. This responsibility extends to transition periods when aircraft within the formation are maneuvering to attain separation from each other to effect individual control, and during join-up and breakaway. [ARP4107]

form grinding Producing a contoured surface on a part by grinding it with an abrasive wheel whose face has been shaped to the reverse of the desired contour.

forming Applying pressure to shape a material by plastic deformation without intentionally altering its thickness.

forming die A die for producing a contoured shape in sheet metal or other material.

form tool A single-edged, nonrotating cutting tool that produces its inverse or reverse form on a workpiece.

formula translating system (FORTRAN) A procedure-oriented language for solution of arithmetic and logical programs.

FORTRAN *See* formula translating system.

forward-loop control elements Elements situated between the error signal and the output of the control system. [AIR1916]

forward scattering The scattering of radiant energy into the hemisphere of space bounded by a plane normal to the direction of the incident radiation and lying on the side toward which the incident radiation was advancing; the opposite of backward scatter.

forward slip A slip in which the direction of motion of the aircraft continues the same as before the slip was begun. The primary purpose of the forward slip is to dissipate altitude without increasing the airspeed of the aircraft, particularly in aircraft not equipped with flaps. *See* slip. [ARP4107]

forward thrust *See* thrust.

fossil fuel Coal or petroleum hydrocarbon fuel as distinguished from nuclear fuel.

fouling *1.* Growth of adherent plant or animal life on submerged structures; often leads to biological corrosion or degradation of performance, such as reduction of heat transfer or increase of fluid friction. *2.* The accumulation of refuse in gas passages or on heat-absorbing surfaces, which results in undesirable restrictions to the flow of gas or heat.

foundry A commercial enterprise, plant, or portion of a factory where metal or glass is melted and cast.

four-ball tester An apparatus for determining lubrication efficiency by driving one ball against three stationary balls clamped together in a cup filled with the test lubricant; effectiveness of lubrication is expressed relatively in terms of wear-scar diameters on the stationary balls.

four-bar linkage *See* quadrilateral mechanism.

Fourier analysis The representation of physical or mathematical data by the use of the Fourier series or Fourier integral.

Fourier optics A prism or grating monochromator that essentially performs a Fourier transform on the light incident on the entrance slit.

four-way servoactuator A servoactuator having two control ports connecting to a four-way servovalve. A four-way linear servoactuator is usually, but not necessarily, a double-ended actuator. [ARP4386]

four-way valve A multi-orifice flow-control valve with a supply port, return port, and two control ports (No. 1 and No. 2) arranged so that the valve action in one direction opens the supply port to control port No. 1 and opens control port No. 2 to the return port. [ARP4386]

FOV *See* field of view.

fovea The central part of the retina, which contains a high concentration of the color-sensitive receptors known as cones.

foveal visual cues Visual stimuli occurring within an approximately 60-degree cone from a person's normal sight line. Visual cues in this region are typically detected phototropically (with cones). Foveal vision is mostly used for discerning fine detail, estimating depth and distance, and differentiating colors. [ARP4107]

Fowler flap Named after Harlan D. Fowler, American aeronautical engineer; a type of extendible trailing-edge flap that effectively increases both the camber and wing area. [ARP4107]

fractal Highly irregular geometrical figure, such as snowflakes or the boundary of a cloud, whose capacity dimension is not an integer. Capacity dimension is determined by measuring of the number of different-sized, superimposed squares needed to cover the geometric shape.

fraction *1.* In classification of powdered or granular solids, the proportion of the sample (by weight) that lies between two stated particle sizes. *2.* In chemical distillation, the proportion of a solution of two liquids consisting of a specific chemical substance.

fractional distillation A thermal process whereby a mixture of liquids that boil at different temperatures is heated at a series of increasing temperatures, and the distillates boiled off at each temperature are collected separately.

fractionating column An apparatus for fractional distillation in which rising vapor and falling liquid are brought into intimate contact.

fraction defective In quality control, the average number of units of product containing one or more defects for each 100 units of product in a given lot.

fractography The study of fracture surfaces, especially for the purpose of determining the causes of failure and relating these causes to macrostructural and microstructural characteristics of parts and materials.

fracture *1.* The breaking and subsequent separation of parts of a body because of external or internal forces. *2.* The rupture of a surface without complete separation of the laminate. [AIR4844]

fracture ductility The true plastic strain at fracture. [AIR4844]

fracture mechanics A solid-mechanics analytical method in which the state of stress in a cracked, scratched, or otherwise flawed material is used to describe the condition under which the crack, scratch, or flaw may grow and/or to predict the growth rate of the crack, scratch or flaw. [AIR1872]

fracture stress The true, normal stress on the minimum cross-sectional area at the beginning of fracture. [AIR4844]

fracture test A method for determining composition, grain size, case depth, or material soundness by breaking a test specimen and examining the fracture surface for certain characteristic features.

fracture toughness A measure of the damage tolerance of a material containing initial flaws or cracks. [AIR4844]

fragmentation In data processing, refers to an effect of often-used files whereby they grow in size to become non-contiguous when stored on a soft or hard disk.

frame *1.* Structured collection of bits by which information is transferred. [ARP4290] *2.* The image in a computer display terminal. *3.* In time-division multiplexing, one complete commutator revolution that includes a single synchronizing signal or code.

frame code complement (FCC) Refers to the subframe synchronization method whereby the frame synchronization code is complemented to signal the beginning of each subframe.

frame rate (FRATE) The rate, or the pulses that clock the rate, of rotation of a data multiplexer "wheel."

frame synchronization pattern A unique code, coded pulse, or interval to mark the start of a commutation frame period.

frame synchronizer (FSY) Telemetry hardware that recognizes the unique signal that indicates the beginning of a frame of data. A typical frame synchronizer "searches" for the code, "checks" the recurrence of the code in the same position for several frame periods, and then "locks" on the code.

framework The load-carrying members of an assembled structure.

framing error An error resulting from transmitting or receiving data at the wrong speed. When framing error occurs, the character of data will appear to have an incorrect number of bits.

FRATE *See* frame rate.

Fraunhofer lines Dark lines in the absorption spectrum of solar radiation due to absorption by gases in the outer portions of the sun and in the earth's atmosphere.

fraying The unraveling of a material, usually of a woven, fibrous braid. [ARP1931]

free air or free gas Air or gas measured at sea level, Standard Day conditions, pressure of 14.7 lb/sq in. (absolute) and a temperature of 59°F (15°C). [ARP4386]

free ash Ash that is not included in the fixed ash.

free atmosphere That portion of the earth's atmosphere, above the planetary boundary layer, in which the effect of the earth's surface friction on the air is negligible, and in which the air is usually treated (dynamically) as an ideal fluid. The base of the free atmosphere is usually taken as the geostrophic wind level.

free-cutting *See* free-machining.

free-electron laser Multifrequency laser utilizing optical radiation amplification by a beam of free electrons passing through a vacuum in a transverse periodic magnetic field, as opposed to a conventional laser in which the oscillating electrons are bound to atoms and molecules and have a specific wavelength.

free electrons Electrons that are not bound to an atom.

free fall *1.* Describes a retractable landing-gear configuration capable of moving from the stowed (up and locked) position to the extended (down and locked) position by means of gravitational forces. [AIR1489] *2.* The fall or drop of a body, such as a rocket, not guided, not under thrust, and not retarded by a parachute or a braking device. [ARP4107]

free field Ideally, a wave field or potential-energy field in a homogeneous, unbounded medium; practically, a field where boundary effects are negligible over the useful portion of the medium.

free field test An engine test performed on an outdoor test stand to determine the baseline performance of the engine. [ARP741]

freefit A type of clearance fit having a relatively large allowance; used when accuracy of assembly is not essential and/or when large temperature variations may occur. Also known as free-running fit.

free flight Unconstrained or unassisted flight, as in the flight of a rocket after consumption of its propellant or after motor shutoff; the flight of an unguided projectile; and the flight—in certain kinds of wind tunnels—of unmounted models.

free flow A condition in which the liquid surface downstream of the weir plate is far enough below the crest so that air has free access beneath the nappe.

free gyroscope A gyro wheel mounted in two or more gimbal rings so that its spin axis can maintain a fixed position in space.

free impedance A transducer characteristic equal to the input impedance when the load impedance is zero.

free jet Fluid jet without solid boundaries, such as a jet discharging into the open.

free-machining Describes a material in which chemical composition has been altered somehow to substantially improve machinability. Examples would be steel to which sulfur, phos-

phorus or lead have been added; or nonferrous metals to which lead has been added. Also termed free-cutting.

free oscillation *See* free vibration.

free-piston engine Engine in which the pistons are not connected to the crank.

free piston pump A reciprocating piston device that converts pneumatic power to hydraulic power. Usually acts as an intensifier with the pneumatic pressure increased to a higher hydraulic pressure. [ARP4386]

free radicals Atoms or groups of atoms broken away from stable compounds by application of external energy. Free radicals contain unpaired electrons, but remain free for transitory or longer periods.

free rolling Describes a wheel/tire that is neither powered (driven) nor braked. The peripheral velocity in the contact zone then is the same as ground velocity. [AIR1489]

free-running fit *See* freefit.

free space An area designated for the use of the crew, for changing and stowing clothing. [ARP4101/3]

free vibration Oscillation of a system in the absence of external forces. *See also* free oscillation.

free water The amount of water released when a wet solid is dried to its equilibrium moisture content.

freeze-up Stoppage of rotational motion due to radial expansion or adhesive welding between a bearing and its journal. Also called seizure.

freezing drizzle Drizzle, the drops of which freeze on impact with the ground or with other objects at or near the earth's surface. [AIR1335]

freezing fog Cloud of supercooled water droplets that forms a deposit of ice on objects in cold weather conditions. [AIR4737]

freezing point The temperature at which equilibrium is attained between liquid and solid phases of a pure substance. Also applied to compounds and alloys that undergo isothermal liquid-solid phase transformation.

freezing precipitation Snow, sleet, freezing rain, drizzle, or hail that adheres to aircraft surfaces. [AIR4737]

freezing rain *See* ice, freezing rain.

french coupling A coupling with a right- and left-hand thread.

frequency *1.* The number of revolutions or cycles completed in unit time. [ARP5089] *2.* The number of times an alternating current repeats its cycle in one second. [ARP1931] *3.* Rate of signal oscillation in Hertz.

frequency (fundamental) In resonance testing, the frequency at which the wavelength is twice the thickness of the examined material. [ARP5089]

frequency assignment The specific frequency or frequencies authorized by competent authority; expressed for each channel by: (a) the authorized carrier frequency, the frequency tolerance, and the authorized emission bandwidth; (b) the authorized emission bandwidth in reference to a specific assigned frequency (when a carrier does not exist); or (c) the authorized frequency band (when a carrier does not exist).

frequency band The continuum between two specified limiting frequencies. *See also* frequency.

frequency characteristic of voltage modulation The component frequencies that make up the modulation envelope wave form. [ARP1148A]

frequency departure The amount that a carrier frequency or center frequency varies from its assigned value.

frequency deviation The peak difference between the instantaneous frequency of a modulated wave and the frequency of the unmodulated carrier wave.

frequency discriminator Electronic circuit that delivers output voltages proportional to the deviations of signals from predetermined frequency values.

frequency distortion A form of distortion in which the relative magnitudes of the components of a complex wave are changed during transmission.

frequency divider An electronic circuit or device whose output-signal frequency is a proper fraction of its input-signal frequency.

frequency-divisional multiplexing The combination of two or more signals at different frequencies so they can be transmitted as one signal. This can be done electronically, or it can be done optically by using two or more light sources of different wavelengths.

frequency division multiple access Multiple access communication system in which the user has a specific frequency allocation and uses all of the time axis while sharing the available power.

frequency-division multiplex (FDM) A system for transmitting information about two or more quantities (measurands) over a common channel by dividing the available frequency bands; subcarriers may be amplitude, frequency, or phase modulated.

frequency drift The slow and random variation of the controlled frequency level within the steady-state limits; may occur, for example, as a result of environmental effects and wear on the electric power-drive system. [AS1212]

frequency drift rate The rate of change of frequency owing to frequency drift when plotted against time. [AS1212]

frequency hopping In transmission, random changing of frequencies to mislead or prevent interception by unauthorized equipment.

frequency-independent dynamic stiffness *See* infinite frequency stiffness.

frequency meter *1.* An instrument for determining the frequency of a cyclic signal, such as an alternating current or radio wave. *2. See* wave meter.

frequency modulation *1.* The cyclic and/or random variation of instantaneous frequency about a mean frequency during steady-state electric-system operation. This variation is normally within narrow frequency limits and occurs as a result of speed variations in a generator rotor owing to the dynamic operation of the rotor-coupling and drive-speed regulation; it is frequently nonsinusoidal. [AS1212] *2.* Angle modulation of a sine-wave carrier in which the instantaneous frequency of the modulated wave differs from the carrier frequency by an amount proportional to the instantaneous value of the modulating wave. *3.* A type of electronic circuit that produces an output signal whose frequency has been modified by one or more input signals. *See also* modulated wave. *4.* In telemetry, modulation of the frequency of an oscillator to indicate data magnitude.

frequency modulation rate The rate of change of frequency due to frequency modulation when plotted against time. [ARP1148A]

frequency modulation repetition rate The reciprocal of the period of the modulation waveform. [AS1212]

frequency monitor An instrument that determines the amount that a frequency deviates from its assigned value.

frequency multiplication The generation of harmonics of the frequency of a lightwave by nonlinear interactions of the lightwave with certain materials. High power beams are needed for the nonlinear interaction to occur.

frequency multiplier An electronic circuit or device whose output-signal frequency is an exact multiple of its input-signal frequency.

frequency regulation The band within which the output frequency stays—except during transients. [ARP1148A]

frequency regulation steady state The band within which the output frequency stays during any fixed load. [ARP1148A]

frequency response *1.* The complex ratio of control pressure to input current as the current is varied sinusoidally over a range of frequencies. Frequency response is normally measured with constant input current amplitude and expressed by the amplitude ratio and phase angle. [ARP490] *2.* For a servoactuator, defined as the steady-state relationship of the output amplitude to the input amplitude and the output to input phase difference, when the input is subjected to constant amplitude sinusoidal signals of various frequencies. [ARP1281A] *3.* The portion of the frequency spectrum that can be sensed by a device within specified limits of amplitude error. *4.* Response of a system as a function of the frequency of excitation. *5.* A measure of the effectiveness with which a circuit or device transmits signals of different frequencies, usually expressed as a graph of magnitude or phase of an output signal as a function of frequency. Also known as amplitude-frequency response or sine-wave response.

frequency response (Bode plot) The complex ratio of the actuation system output to the command input, while the input is cycled sinusoidally and the frequency is varied. Frequency response is usually presented as a combined plot of normalized amplitude ratio in dB, and of the input to output phase angle, in degrees, versus the logarithm of frequency (called a Bode plot). [AIR1916]

frequency response method A method of tuning a process-control loop for optimum operation by proper selection of controller settings. This method is based on a study of the frequency response of the open process-control loop.

frequency reuse A digital satellite communication technique that features the reuse of frequency bands in a downlink transmission to provide high-power utilization and flexible accommodations of dynamic-source-destination traffic patterns.

frequency shift keying (FSK) The form of frequency modulation in which the modulating wave shifts the output frequency between predetermined values, and the output wave is coherent with no phase discontinuity.

frequency spectrum A plot of sound pressure level or sound power level against frequency; determines the quality of sound. [AIR1826]

frequency stability *1.* A measurement of how well the output frequency (or equivalently emitted wavelength) of a laser stays constant. (In some types of lasers, the emitted wavelength tends to drift because of factors such as changing temperature of the laser itself.) *2.* For an electronic oscillator, a statement of deviation with time, temperature, or supply voltages when compared to a standard.

frequency swing A characteristic of a frequency-modulation system equal to the difference between the maximum and minimum design values of instantaneous frequency in the modulated wave.

frequency telemetering Refers to a system for transmitting measurements in which the information values are represented by frequencies within a specific band; the specific frequency is determined by the percent of full-scale equivalent to the current value of the measured variable.

frequency transient recovery The time required for the output frequency to recover to and remain within the prescribed limits after full load application or removal. [ARP1148A]

frequency transients The maximum instantaneous deviation of the output frequency from the frequency regulation band. [ARP1148A]

Fresnel lens(es) Thin lenses constructed with stepped setbacks, such that they have the optical properties of much thicker lenses.

Fresnel reflector Device characterized by a set of mirrors with varying orientation arranged so as to have the optical properties of a smooth reflector, e.g., parabolic reflector.

Fresnel region The region between the antenna and the Fraunhofer region.

Fresnel zone *See* near field.

fretting Wear caused by oscillatory slip of small amplitude between two surfaces. The amplitude below which wear is to be considered fretting is considered to be that which is large enough to maintain a lubricant film of finite thickness between the surfaces at all times. Fretting is divided into three classes, Class 1, Class 2, and Class 3. *Class 1–*fretting that occurs in parts of a mechanism that were never intended to have relative motion. *Class 2–*fretting that occurs in parts of a mechanism intended to undergo relative motion some or all of the time, but not through the action of antifriction bearings. *Class 3–*fretting that occurs in anti-friction bearings. [S-2,47]

fretting corrosion Corrosion resulting from fretting. [S-2,47]

fretting, molecular attrition theory Theory holding that fretting is a form of molecular attrition (or transfer) caused when the molecules of the two mating surfaces alternately approach closely and subsequently separate. On close approach, the fields of force of the molecules overlap, and on separation the molecules are torn out of both surfaces as a result of breaking the cohesional bonds. [S-2,47]

fretting, weld-breaking theory Theory based on the modern concepts of friction, and holding that when two solids come in contact, the force bringing them together is concentrated on a small number of high points in both surfaces. These high points flow plastically under the load, and cold-weld, equilibrium being finally reached when the cross-sectional area of the welded junctions is sufficiently large to support the load. Friction and wear result from breaking of the welds and from plowing of the asperities of one material through the other, the weld-breaking process being the dominant one for most practical purposes. [S-2,47]

friable Capable of being easily crumbled, pulverized, or otherwise reduced to power.

friction Force generated between solids, liquids, or gases opposing relative motion.

friction head Pressure loss, measured in height, caused by product moving through pipe and fittings. [AIR4783]

friction tape A type of cotton tape impregnated with a sticky, moisture-resistant compound, which is used to cover and insulate exposed electrical connections or terminations. Friction tape has been largely replaced by electrical tape made of polyvinyl chloride resin backed with a sticky adhesive.

friction-tube viscometer A device for measuring viscosity by determining the pressure drop across a friction tube as the test fluid is pumped through it.

friction unit force A measure of the shearing force on the brake friction material. Equals brake torque divided by the product of brake radius and lining area. [AIR1489]

frigorimeter A thermometer for measuring low temperatures.

fringe multiplication The duplicating effect of a family of curves superimposed on another family of curves so that the curves intersect at angles less than 45°. A new family of curves appears which pass through intersections of the original curves.

frise aileron A type of aileron having its leading edge projection well ahead of the hinge axis. [ARP4386]

frit A powdered ceramic prepared by fusing a physical mixture of oxides into a uniform melt, and then quenched and milled this into a fine, homogeneous powder.

frit seal A hermetical seal for enclosing integrated circuits and other electronic components; made by fusing a mixture of metallic powder and glass binder.

from-to tester A type of electronic test equipment for checking continuity between two points in a circuit.

front course (ILS) *See* course.

front-end processor *1.* A device that receives computer data from other input devices, organizes such data as specified, and then transmits it to another computer for processing. *2.* The computer equipment used to receive plant signals, including analog-to-digital converters and the associated controls.

front mounted A connector mounted with its mounting flange positioned in front of the mounting surface, when looking at the mating face or front side of the connector. [ARP914A]

front of the engine *See* prop end.

frost A crystallized deposit, formed from water vapor on surfaces that are at or below 0°C (32°F). [AIR4737]

frost plug A device for determining liquid level when the contents of a tank are at a temperature below 0°C. Consists of a side tube resembling a sight glass, but having a series of closed tubes (plugs) at different levels instead of the glass. The tubes below liquid level are cooled so that moisture from the atmosphere forms frost on them, while the tubes above liquid level remain frost free.

froth flotation *See* flotation.

Froude number The nondimensional ratio of the inertial force to the force of gravity for a given fluid flow; the reciprocal of the Reech number.

FS *See* fail-safe.

FSK *See* frequency shift keying.

FSS *See* flight service station.

FSY *See* frame synchronizer.

fuel Any material that will burn or otherwise react to release heat energy. Common fuels include coal, charcoal, wood, and petroleum products (fossil fuels), which burn; and uranium, which undergoes nuclear fission.

fuel advisory departure (FAD) Procedures (for example, postponement of takeoff) to minimize engine running time for aircraft destined for an airport that is experiencing prolonged arrival delays. [ARP4107]

fuel-air mixture Mixture of fuel and air.

fuel-air ratio (FAR) The mass rate of fuel flow to the engine divided by the mass rate of dry airflow through the engine. [AIR1533]

fuel cell Device that converts chemical energy directly into electrical energy; differs from a storage battery in that the reacting chemicals are supplied continuously as needed to meet output requirements.

fuel cell power plant Power-generating device that directly produces electrical energy from chemical energy. Consists of fuel processors, stacked fuel cells, and d-c to a-c converters. The main types, distinguished by electrolytes that are heated to different temperatures, are base, phosphoric acid, molten carbonate, and solid oxide.

fuel consumption *1.* The using of fuel by an engine or power plant. *2.* The rate of this consumption, measured, for example, in gallons or pounds per minute.

fuel-defuel valve Within a ground refueling system, a multi-position valve that is manually positioned to change a mobile refueling system from a fueling mode to a defueling mode. [AIR4783]

fuel gas A combustible, gaseous substance that is used as a fuel.

fuel-no-air A level-control valve/system that closes to the passage of gas and opens to permit the passage of liquid. [AIR1660A]

fuel oil Any oily hydrocarbon liquid, having a flash point of at least 100°F (38°C), that can be burned to generate heat.

fuel, oil, and lubrication (FOL) All ground running, testing and mission fuel, oil, and lubricants consumed by the fleet during the total O&S phase of a program. [ARP4293]

fuel production Producing of conventional and/or alternative fuels by various technologies.

fuel quantity measurement system On-board aircraft system for measuring and indication of fuel quantity per tank and total fuel on-board. [AIR4783]

fuel sense pressure A pressure signal obtained from a point remote from, and downstream of, a pressure-control valve on the refueling vehicle. [AIR4783]

fuel servicing cabinet A fixed, above-ground structure containing equipment connected to an airport fueling system to enable fuel to be dispensed into the aircraft. [AIR4783]

fuel shutoff valve Aircraft control valve used to enable fuel to enter a tank during refueling, and shut off on command (or automatically, if part of the level-control system). [AIR4783]

fuel storage facilities Tanks and associated facilities for the storage of aviation fuel at an airport. [AIR4783]

fuel temperature (t) Final equilibrium test temperature in °F of the liquid and vapor phase. [AIR1326]

fulchronograph An instrument for recording lightning strikes electromagnetically.

fulgurator An atomizer used in flame analysis to spray the salt solution to be analyzed into the flame.

full adder A computer logic device that accepts two addends and a carry input, and produces a sum and a carry output.

full annealing An imprecise term that implies heating to a suitable temperature followed by controlled cooling to produce a condition of minimum strength and hardness.

full coupling A flexible coupling joint with comparable end fittings on each mating tube end. [ARP4968]

full cycling control Control placed on the crimping cycle of a crimping tool, forcing the tool to be closed to its fullest extent, requiring completion of the crimping cycle before the tool can be opened. [ARP914A]

full duplex Describes communications that appear to have information transfer in both directions (transmit and receive) at the same time.

full hard temper For nonferrous alloys and some ferrous alloys, a level of hardness and strength corresponding to a cold-worked state, beyond which the material can no longer be formed by bending.

full mission capable (FMC) Refers to systems and equipment that are safe and have all mis-sion-essential subsystems installed and operating as designated by the Military Service. [ARD50010]

full overtravel point That position of the plunger beyond which further overtravel would cause damage to the switch or actuator. [AIR4077]

full-scale development phase The period during which a system and the principal items necessary for its support are designed, fabricated, tested, and evaluated. [ARD50013]

full-scale value *1.* The largest value of a measured quantity that can be indicated on an instrument scale. *2.* For an instrument whose zero is between the ends of the scale, the sum of the absolute values of the measured quantity corresponding to the two ends in the scale.

full-wave rectifier An electronic circuit that converts an a-c input signal to a d-c output signal, with current flowing in the output circuit during both halves of each cycle in the input signal.

fully-processed Denotes those materials that have been magnetically annealed prior to shipment by the vendor. [AMS7716C]

fume-resistant So constructed that it will not be injured readily by specific fumes. [ARP4404]

functional design The specification of the working relations between the parts of a system in terms of their characteristic actions.

functional design specifications Those levels of design in which all subtasks are specified and their relationships defined so that the total collection of subsystems will perform the intended task of the entire system.

functional diagram A diagram that represents the functional relationships among the parts of a system.

functional hazard assessment A systematic, comprehensive examination of aircraft functions to identify and classify failure conditions of those functions according to their severity. [ARP4754]

functional isolation The property of a system whereby effective separation of functions is provided to minimize adverse interaction. [ARP1834]

functional modularity A design feature in which individual units that perform portions of a function being tested are packaged on a single replaceable unit for the maintenance level doing the test. [AIR4896]

functional program A routine or group of routines that, when considered as a whole, completes some task with a minimum of interaction of other functional programs other than to obtain data and signal completion of its task. For example, a group of routines that take data from an analog scanner and store it on a bulk storage device might be considered to be a functional program.

functional requirements A specification of required functional behavior, operation, performance, or purpose.

functional significant item Those items other than structures judged to be relatively important from a safety, reliability, or economic standpoint. [AIR4896]

functional specification A document that tells exactly what the system should do, what will be supplied to the system, and what is expected to come out of it.

functional test See test, functional.

function analysis for maintainability Analytical basis for allocating tasks to personnel and equipment to achieve optimum system maintainability. [AIR4896]

function, hidden See hidden function.

functioning time In an EED, the lapsed time between application of initiating energy and some later function, such as bridge wire beak, case opening or start of pressure rise, or peak pressure. [AIR913]

function scan A feature of the EASY program that provides a graphic output of algebraic functional relationships or of tabulated data input by a user. [AIR1823]

function selector A device that permits the human pilot to select automatic pilot functions or modes of operation. [ARP419]

function, structural See structural function.

function subprogram An independently written program that is treated as such by the compiler. May consist of any number of statements that are executed when it is called.

function switch A circuit having a fixed number of inputs and outputs, designed so that the output information is a function of the input information, each expressed in a certain code, signal configuration, or pattern.

function table 1. Two or more sets of information so arranged that an entry in one set selects one or more entries in the remaining sets. 2. A dictionary. 3. A device constructed of hardware, or a subroutine, that can either decode multiple inputs into a single output or encode a single output into multiple outputs. 4. A tabulation of the values of a function for a set of values of the variable.

fundamental frequency The frequency of a sinusoidal function having the same period as a complex periodic quantity.

fundamental mode 1. The mode of a waveguide having the lowest critical frequency. 2. A type of sequential circuit in which there is only one input change at a time and no further change occurs until all states are stabilized.

fundamental natural frequency The lowest frequency in a set of natural frequencies.

fundamental null Obtained at the minimum voltage positions; is the fundamental component of the residual voltage when the in-phase voltage is zero. This residual voltage consists entirely of quadrature voltage. [ARP826]

furnace An apparatus for liberating heat and using it to produce a physical or chemical change in a solid or liquid mass. Most often, the heat is produced by burning a fossil fuel, passing electric current through a heavy-duty resistance element, generating and sustaining an electric arc, or electromagnetically inducing large eddy currents in the charge.

furnace draft The draft in a furnace, measured at a point immediately in front of the highest point at which the combustion cases leave the furnace.

furnace volume The contents of the furnace or combustion chamber, in cubical units.

fuse 1. A device that automatically shuts off flow in a line in event of downstream system rupture. [ARP243B] 2. A replaceable circuit-protecting device that depends on the melting of a conductor for circuit interruption.

[AS486B] *3.* Any of several devices for detonating an explosive, for example, by elapsed time, command, impact, proximity, or thermal effects.

fuse clips The contacts of the fuseholder, which support the fuse or limiter and connect their terminals with the circuit. [ARP1199A]

fused fiber optics A number of separate fibers that are melted together to form a rigid, fused bundle to transmit light. Fused fiber optics may be used for transmitting images or simply illumination; they are not necessarily coherent bundles of fibers.

fused silica The term usually applied to synthetic fused silica, formed by the chemical combination of silicon and oxygen to produce a high-purity silica.

fused slag Slag that has coalesced into a homogeneous solid mass by fusing.

fuse, hydraulic A hydraulic device, designed to shut off flow of fluid if the flow exceeds a specified value. Used in brake systems to stop depletion of the fluid in the system if a line rupture or similar failure occurs. [AIR1489]

fuse or limiter holder A mounting device with contacts and terminals for the purpose of accepting fuses or limiters for easy connection within a circuit. [ARP1199A]

fuse pull-out A removable fuseholder that can be removed to replace fuses or to open an electrical circuit.

fuse, quantity measuring A fuse that closes when more than a predetermined quantity of fluid has passed through it. [ARP243B]

fuse, return flow A fuse that closes both pressure and return lines when the ratio between the flows deviates beyond a predetermined value. [ARP243B]

fuse tube A tube of insulting material that sur-rounds the current-responsive element. [ARP1199A]

fusibility Describes the property of slag whereby it fuses and coalesces into a homogeneous mass.

fusible alloy An alloy with a very low melting point, in some instances approaching 150°F (65°C), usually based on Bi, Cd, Sn or Pb. Fusible alloys have varied uses, the most widely known being solders and fusible links for automatic sprinklers, fire alarms, and other safety devices.

fusible plug *1.* A thermal-sensitive pressure-release device used to prevent tubeless tire or wheel assembly failure when an overheat condition occurs which would cause degradation of wheel and/or tire structural properties. [AIR1489] *2.* A hollowed, threaded plug having the hollowed portion filled with a low-melting point material, usually located at the lowest permissible water level.

fusion The combining of atoms and consequent release of energy.

fusion welding Any welding process that involves melting of a portion of the base metal.

fusion zone In a weldment, the area of base metal melted, as determined on a cross section through the weld.

fuze A device designed to initiate a munition. [AIR913]

fuzzy sets Mathematical models coupled with a provision for the effect of human factors and construction process and experience.

fuzzy systems Systems that involve fuzzy sets.

FWA *See* first word address.

FW-SIFR Acronym for fixed-wing special IRF.

G

g *See* gram.

G 1. The basic unit of acceleration. 1G = 32.2 ft/s² (9.81 m/s²). [ARP4107] *2. See* specific gravity.

G-adaptation illusion *See* illusions, kinesthetic.

gage Also spelled gauge. *1*. Device to ascertain a specific or comparative dimension. [ARP480A] *2*. The thickness of metal sheet, or the diameter of rod or wire.

gage block A rectangular chromium steel block having two flat parallel surfaces, with flatness and parallelism guaranteed within a few millionths of an inch. Gage blocks are usually manufactured and sold in sets for use as standards in linear measurement. Also known as Johanssen block, Jo block, precision block, or size block.

gage cock A valve attached to a water column or drum for checking water level.

gage factor *See* strain sensitivity.

gage glass A glass or plastic tube for measuring liquid level in a tank or pressure vessel, usually by direct sight; usually connected directly to the vessel through suitable fittings and shutoff valves.

gage length In materials testing, the original length of an elongated specimen over which measurements of strain, thermal expansion, or other properties are taken.

gage point A specific location used to position a part in a jig, fixture, or qualifying gage.

gage, pressure *1*. An instrument that shows the pressure at a given point in a system. May be calibrated to allow either gage pressure or absolute pressure. [ARP171] *2*. Pressure measured relative to ambient pressure. *3*. The difference between the local absolute pressure of the system and the atmospheric pressure at the place of the measurement.

gage, quantity An instrument similar to a pressure gage, except that it is calibrated to read the quantity of gas or liquid remaining in the storage container. [ARP171]

gain *1*. The relative degree of amplification in an electronic circuit. *2*. The ratio of the change in output to the change in input that caused the change. *3*. In a controller, a quantity related to the proportional band setting; for example, if the proportional band is set at 25%, the controller gain is 0.25.

gain, antenna By common definition, the difference in signal strengths between a given antenna and an isotropic antenna.

gain margin *1*. A measure of system stability defined as the reciprocal of the magnitude of the open loop gain at 180 deg of phase lag crossover frequency, usually expressed in dB. [ARP4386] *2*. The factor by which the loop gain must be multiplied for a system to reach the stability boundary. [AIR1823]

gal Abbreviation for gallon.

Gal A unit of acceleration equal to 1 cm/s². The milligal is frequently used because it is about 0.001 times the earth's gravity.

galactic cosmic rays Energetic particles that come from outside the solar systems; generally come from within our galaxy.

galactic mass The total amount of matter contained in a galaxy.

galactic radio waves Radio waves emanating from our galaxy.

galaxies Vast assemblages of stars or nebulae, composing island universes separated from other such assemblages by great distances.

galley(s) All galley hardware items furnished for installation on the airplane, exclusive of airplane provisions. This includes galley complex structure, decorative panels, ceilings, furnishings, lighting, galley units, modules, food, beverages, utensils, ovens, coffee makers, floor tie-down fittings, refrigeration equipment, electrical, water, and drain system connections from the galley equipment to the airplane interface connectors. [AS1426]

galley envelope The three-dimensional outside surface limits of the volume(s) within the airplane reserved for locating the galleys. [AS1426]

galling Localized adhesive welding with subsequent spalling and roughening of rubbing metal surfaces; occurs as a result of excessive friction and metal-to-metal contact at high spots.

galvanic cell A cell made up of two dissimilar conductors in contact with an electrolyte or two similar conductors in contact with dissimilar electrolytes. [AIR4844]

galvanic corrosion Electrochemical corrosion associated with current in a galvanic cell; set up when two dissimilar metals (or the same metal in two different metallurgical conditions) are in electrical contact and are immersed in an electrolyte solution.

galvanic series A list of metals and one nonmetal, graphite, starting with the least noble (magnesium) and ending with the most noble (platinum). [AIR4844]

galvanizing Coating a metal with zinc, using any of several processes, the most common being hot dipping and electroplating.

galvanometer An instrument for measuring small electric currents using electromagnetic or electrodynamic forces to create mechanical motion, such as changing the position of a suspended moving coil.

GAMA Abbreviation for General Aviation Manufacturers Association.

gamma A measure of the contrast properties of a photographic or radiographic emulsion; equal to the slope of the straight-line portion of its H-D curve.

gamma correction Correction to compensate for the fact that the relationship between video amplifier drive level (voltage) and luminance output of CRT phosphors is nonlinear and different for each primary. [ARP4032]

gamma counter An instrument for detecting gamma radiation, either by measuring integrated intensity over a period of time, or by detecting each photon separately.

gamma ray *1.* Quantum of electromagnetic radiation emitted by nuclei, each such photon being emitted as the result of a quantum transition between two energy levels of the nucleus. Gamma rays have energies usually between 10 thousand electron volts and 10 million electron volts, with correspondingly short wavelengths and high frequencies. *2.* A term sometimes used to describe any high-energy electromagnetic radiation, such as x-rays exceeding about 1 MeV or photons of annihilation radiation.

gamma-ray astronomy Astronomy based on the detection of gamma-ray emission and interactions from supernova remnants, neutron stars, flare stars, galactic core and disc, black holes, etc.

gamma-ray bursts Short (about 0.1–4 sec.), intense, low-energy (about 0.1–1.2 MeV) bursts recorded by the Vela satellite system in 1967. The isotropic distribution of these bursts suggests an extragalactic origin, but a galactic disk origin cannot be ruled out.

gamma-ray lasers Stimulated emission devices producing coherent gamma radiation.

gamma-ray spectra The energy distribution of gamma rays emitted by nuclei.

gamma-ray telescope Special telescope for the observation (and recording) of astronomical phenomena in the gamma-ray spectrum.

gamma structure An ordered structure of a titanium-aluminum compound with a stoichiometric ratio of TiAl and a face-centered tetragonal crystal structure. [AS1814]

gang disconnect A connector that permits the simultaneous connection or disconnection of two or more electrical circuits. [ARP914A]

gantry cranes Large cranes mounted on platforms, which usually run back and forth on parallel tracks astride the work area.

Gantt chart A style of bar chart used in production planning and control to display both work planned and work done, in relation to time.

gap *1.* In filament winding, the space between successive windings that are usually intended to lay next to each other. [AIR4844] *2.* An interval of space or time that is used as an automatic sentinel to indicate the end of a word, record, or file of data on a tape; for example, a word gap at the end of a word, a record or item gap at the end of a group of words, or a file gap at the end of a group of records or items. *3.* The absence of information for a specified length of time, or space on a recording medium, as contrasted with marks and sentinels that indicate the presence of specific information to achieve a similar purpose. *4.* The space between the reading or recording head and the recording medium, such as tape, drum, or disk. *5.* In a weldment, the space between members, prior to welding, at the point of closest approach for opposing faces.

gap-filling adhesive An adhesive subject to low shrinkage in setting, used as sealant. [AIR4844]

gap scanning In ultrasonic examination, projecting the sound beam through a short column of fluid produced by pumping couplant through a nozzle in the ultrasonic search unit.

garter spring A closed ring made by welding together the ends of a closely wound helical spring.

gas *1.* The gaseous product(s) resulting from the decomposition, dissociation, or combustion of liquid or solid mono- or bi-propellants. [ARP777] *2.* A substance possessing perfect molecular mobility and the property of indefinite expansion, as opposed to a solid or liquid. [ARP171]

gas amplification For a counter-tube or ionization-chamber, a characteristic equal to the charge collected divided by the charge produced in the active volume by a given ionizing event.

gas analysis The determination of the constituents of a gaseous mixture.

gas atomization Atomization of fluids by high-velocity gas jets.

gas barrier The ionically conductive film between the cell plates that forces all plate-generated gases to the top of the electrolyte and prevents them from reaching plates of opposite polarity. [AS8033]

gas bearing A journal or thrust bearing in which a film of gas is used to lubricate the running surfaces. Also known as gas-lubricated bearing.

gas burner A burner for use with gaseous fuel.

gas carburizing A surface-hardening process in which steel or an alloy of suitable alternative composition is exposed at elevated temperature to a gaseous atmosphere with a high carbon potential; the resulting carbon-rich surface layers are hardened by quenching the part from the carburizing temperature or by reheating and quenching.

gas concentration The volume fraction of the component of interest in the sample or calibration gas mixture, expressed either as a volume percentage or as parts per million by volume. [ARP4418]

gas counter A type of counter tube into which a gaseous sample is directly introduced when measuring its radiation.

gas current A current of positive ions flowing to a negatively biased electrode, the positive ions being produced when electrons flowing between two other electrodes collide with residual gas molecules.

gasdynamic pumping The production of a population inversion by a gasdynamic process. In this process, a hot, dense gas is expanded into a near vacuum, causing the gas to cool rapidly; when the gas cools faster than energy can be redistributed, a population inversion is generated.

gaseous emission Substance emitted in the form of gas downstream of the combustion chamber; limited to carbon monoxide, carbon dioxide, nitric oxide, nitrogen dioxide, and hydrocarbons. [AIR1533]

gaseous oxygen system *See* GOX.

gas etching Removing material from a semiconductor material by reacting it with a gas to form a volatile compound.

gas generator A device in which a propellant is burned to produce a sustained flow of pressurized gas. [AIR913]

gasification The process of converting a solid or liquid fuel into a gaseous fuel, for example, the gasification of coal.

gasket The flexible sealing element in a stationary or static fluid seal. Also called a static seal. [ARP243B]

gas laws Laws governing the behavior of gases; includes the following: Boyle's Law,

Gay-Lussac's or Charles' Law, Dalton's Law, Henry's Law, and Graham's Law. *Boyle's Law*– at a constant temperature, the volume of a gas is inversely proportional to the pressure to which it is subjected. *Gay-Lussac's or Charles' Law*–with constant pressure, the volume of a gas will vary directly with the temperature. *Dalton's Law*–the pressure exerted by each gas in a gaseous mixture is independent of other gases in the mixture, and the total pressure of the mixture of gases is equal to the sum of the separate pressures that each gas would exert if it alone occupied the whole volume. *Henry's Law*–the weight of a gas absorbed by a given liquid, with which it does not combine chemically, is directly proportional to the pressure of the gas above the liquid. *Graham's Law*–the rate of diffusion of a gas is directly proportional to the pressure and temperature and inversely proportional to the square root of the density (molecular weight) of the gas. [ARP171]

gas lift Refers to the technique of raising a liquid in a vertical flow line by injecting a gas below a portion of the liquid column, causing upward flow.

gas-lubricating bearing *See* gas bearing.

gas metal-arc welding (GMAW) A form of electric-arc welding in which the electrode is a continuous filler metal wire, and the welding arc is shielded by supplying a gas such as argon, helium or CO_2 through a nozzle in the torch or welding head. The term GMAW includes the methods known as MIG welding.

gas meter An instrument for measuring and recording the volume or mass of a gaseous fluid that flows past a given point in a piping system.

gasohol Synthetic fuel consisting of a mixture of gasoline and grain alcohol (ethanol).

gasometer A piece of apparatus typically used in analytical chemistry to hold and measure the quantity of gas evolved in a reaction; similar equipment is used in some industrial applications.

gas path analysis Mathematical process of determining overall engine performance, individual module performances, and sensor performances from any specific set of engine-related measurements.

gasper air outlet A supplementary conditioned-air outlet, adjustable in flow quantity and direction; usually provided at each passenger seat and crew station to create additional air motion as required by the individual; tends to alleviate motion sickness. Also known as a velocity outlet. [ARP147C]

gas plasma display A data-display screen used on some laptop computers. Characters are easier to read than those on liquid-crystal display screens, but the unit is more expensive.

gas pliers A pinchers-type tool for grasping round objects, such as pipes, tubes and rods.

gas pocket A cavity within a solid or liquid body that is filled with gas.

gas power In a gas, the amount of energy theoretically available to do work. [ARP4386]

gasproof So constructed or protected that the specified gas will not enter the enclosing case under specified conditions of pressure. [ARP4404]

gas release The capability of pressurizing a specified downstream volume to a specified pressure level within a defined time. [ARP4386]

gas seal A type of shaft seal that prevents gas from leaking axially along a shaft where it penetrates a machine casing.

gas-shielded arc welding An all-inclusive term for any arc-welding process that utilizes a gas stream to prevent direct contact between the ambient atmosphere and the welding arc and weld puddle.

gassing *1.* Absorption of gas by a material. *2.* Formation of gas pockets in a material. *3.* Evolution of gas during a process; for example, evolution of hydrogen at the cathode during electroplating, gas evolution from a metal during melting or solidification, or desorption of gas from internal surfaces during evacuation of a vacuum system; the last is sometimes referred to as outgassing.

gas-solid interaction Effect of the impingement of gases (particles) on solid surfaces in various environments.

gas specific gravity balance A weighing device consisting of a tall gas column with a floating bottom; a pointer mechanically linked to the floating bottom indicates density or specific gravity directly, depending on scale calibration.

gas thermometer For gas in a closed system, a temperature transducer that converts temperature to pressure. The relation between temperature and pressure is based on the gas laws at constant volume.

gastight So constructed that the specified gas will not enter the enclosing case under specified conditions of pressure. [ARP4404]

gas to fluid conversion Describes a process whereby a gas source is driving a component whose output is pressurized hydraulic fluid. [ARP4386]

gas tube An electron tube whose operating characteristics are substantially affected by the presence of gas or vapor within the tube envelope.

gas tungsten-arc welding (GTAW) A form of metal-arc welding in which the electrode is a nonconsumable, pointed tungsten rod. Shielding is provided by a stream of inert gas, usually helium or argon; filler metal wire may or may not be fed into the weld puddle; and pressure may or may not be applied to the joint. The term GTAW includes the method known as Heliarc or tungsten inert-gas (TIG) welding; however, the latter is a non-preferred term.

gas turbine Turbine rotated by expanding gases, as in a turbojet engine or in a turbosupercharger.

gas turbine starter *See* jet fuel starter.

gate *1.* Electronic device to monitor signals in a selected segment of the distance trace on an A-scan display. [ARP5089] *2.* The area from which cargo and/or persons are put on or taken off an aircraft. At some airports, fueling also takes place in the gate area; at others, there is a separate area for fueling. *3.* A particular point in the air-route structure from which aircraft proceed to enter the terminal control area for approach to an airfield; a limiting point of the en route phase of a flight. [ARP4107] *4.* A device, such as a valve or door, that controls the rate at which materials are admitted to a conduit, pipe, or conveyor. *5.* The passage in a casting mold that connects the sprue to the mold cavity. Also known as the in-gate. *6.* An electronic component that allows only signals

of predetermined amplitudes, frequencies, or phases to pass.

gated integrator *See* boxcar averager.

gate-hold procedures At selected airports, procedures to hold aircraft at the gate or other ground location whenever departure delays exceed or are anticipated to exceed 5 minutes. [ARP4107]

gate valve A type of valve whose flow-control element is a disc or plate that undergoes translational motion in a plane transverse to the flow passage through the valve body.

gateway *See* bridge.

gauge A term used to denote the physical diameter of a wire or conductor. [ARP1931] *See also* gage.

gauge length Length over which deformation is measured for a tensile or compressive test specimen. [AIR4844]

gauge pressure Absolute pressure minus atmospheric pressure. [AIR1916]

gauge theory A field theory in which symmetries of the theory are implemented locally in space and time. This leads to theories in which forces are generally carried by vector bosons. Examples of gauge theories are electrodynamics, quantum chromodynamics, and Yang Mills theory.

gauss The CGS unit of magnetic flux density or magnetic induction. NOTE: The SI unit, the tesla, is preferred.

Gaussian beam A laser beam in which the intensity has its peak at the center of the beam, then drops off gradually toward the edges; the intensity profile measured across the center of the beam is a classical Gaussian curve.

Gaussian elimination A technique for solving linear equations by progressive differencing.

Gaussian noise *See* random noise.

Gaussian or normal distribution For a population, a density function that is bell-shaped and symmetrical; and that is completely defined by two independent parameters, the mean and the standard deviation. [ARP4386]

gaussmeter A magnetometer for measuring only the intensity, not the direction, of a magnetic field; scale is graduated in gauss or kilogauss. *See* magnetometer.

Gay-Lussac's or Charles' Law Gas law stating that, with constant pressure, the volume of a gas will vary directly with the temperature. [ARP171] *See also* gas laws.

GCA See ground-controlled approach-landing.

G-differential illusion *See* illusions, kinesthetic.

GDOP *See* geometric dilution of precision.

gear *1.* A toothed machine element for transmitting power and motion between rotating shafts whose axes are relatively close to each other or are intersecting. *2.* A collective term for equipment that performs a specific functions—lifting gear, for example.

gear configuration, bicycle Configuration in which there are two gear assemblies, arranged in tandem design, one behind the other. [AIR1489]

gear configuration, dual twin Configuration in which the axle arrangement is such that two wheel/tire assemblies are located on either side of the shock strut or supporting member. [AIR1489]

gear configuration, dual twin tandem Configuration in which there are two axles, one behind the other, connected by an axle beam, and each of which is a dual twin wheel/tire arrangement. [AIR1489]

gear configuration, quadracycle Configuration in which four main landing gear assemblies are located, one in each aircraft quadrant. [AIR1489]

gear configuration, single wheel Configuration utilizing only one wheel/tire assembly on the gear assembly. [AIR1489]

gear configuration, tandem (also single tandem) Configuration on which the two wheel/tire assemblies are mounted on separate axles one behind the other or "in tandem." [AIR1489]

gear configuration, tricycle General descriptive term for an arrangement with three gear assemblies, one in front and two main gears located aft, in a tricycle arrangement. Each gear may have its own configuration depending on the number of wheel/tire assemblies and the individual arrangement. [AIR1489]

gear configuration, triple wheel Configuration in which there is one axle on which are located three wheel/tire assemblies. [AIR1489]

gear configuration, tri-tandem Configuration in which three wheel/tire assemblies are arranged one behind the other or "in tandem" on a single gear assembly. [AIR1489]

gear configuration, tri-twin tandem Configuration in which there are three axles, one behind the other, connected by means of an axle beam, and each of which carries two wheel/tires assemblies. [AIR1489]

gear configuration, twin tandem Configuration in which there are two axles, one behind the other, connected by an axle beam, and each axle having two wheel/tire assemblies. [AIR1489]

gear configuration, twin tricycle Configuration of an individual landing gear assembly in which there are three axles, each mounting two wheel/tire assemblies. The axle arrangement is one in front, and two aft in a tricycle arrangement. [AIR1489]

gear configuration, twin wheel Configuration in which there is one axle on which are mounted two wheel/tire assemblies. [AIR1489]

gear down To arrange a gear train so that the driven shaft rotates at a slower speed than the driving shaft.

gear drive A mechanism for transmitting power (torque) and motion from one shaft to another by means of direct contact between toothed wheels.

geared rotary actuator A mechanical device utilized for amplifying force through a series of gears converting high-speed, low-torque input into high-torque, low-speed output. [ARP4386]

gear level To arrange a gear train so that the driving and driven shafts rotate at the same speed.

gear, lever suspension A gear arrangement in which the axle is mounted at the aft end of a lever (structural member), which is pivoted at the forward end to permit freedom for the wheels to ride up and over bumps and/or obstructions. The lever motion is controlled by a shock absorber connected between the lever and the strut or other fixed structure. [AIR1489]

gear, main The principle support of the aircraft on land or water. May include any combination of wheels, tires, floats, skis, shock-absorbing mechanisms, brakes, retracting mechanisms, and structural provisions necessary for attachment and operation. The combination may be in varied configurations. [AIR1489]

gear meter A positive-displacement fluid meter in which two meshing gear wheels provide the metering action.

gear motor *1.* A motor utilizing two or more counter-rotating gears which provide a fixed volumetric displacement per revolution of the output shaft. [ARP4386] *2.* A device consisting of an electric motor and a direct-coupled gear train. This arrangement allows the motor to run at optimum speed—usually 1800 or 3600 rpm—while delivering rotational motion at a substantially lower speed.

gear, neutral A gear assembly that may be utilized on either right-hand or left-hand positions on the aircraft. [AIR1489]

gear, nose An auxiliary landing gear unit. In addition to supporting some weight of the aircraft, the nose gear provides balance, controls the aircraft attitude, and gives ground stability; in many instances, it includes provisions for steering the aircraft. The nose gear is located forward on the aircraft or at the nose. [AIR1489]

gear, outrigger An auxiliary gear that supports some weight of the aircraft under some conditions. Located at the outer extremities (e.g., wing tip) of the airplane, and designed to alleviate overturning and to provide stability during ground operations. [AIR1489]

gear positioning system *See* system, gear positioning.

gear pump *1.* A pump in which two or more counter-rotating gears are used to provide a positive volume displacement. Gear pumps are usually fixed displacement; and at a given speed, output flow is essentially constant at any system pressure. [ARP4386] *2.* A pump in which fluid is fed to one side of a set of meshing gears, which entrain the fluid and discharge it on the other side. *3.* A pump supplied with pressurized fluid, which converts fluid flow to rotary motion.

gear ratio The ratio of the revolutions per unit time of one shaft to the revolutions in the same unit time of another shaft connected to it by a gear device. Starter gear ratio is usually expressed as the ratio of starter turbine speed to starter output speed. [ARP906A]

gear, scooter A gear arrangement in which there are usually two axles located one behind the other and connected by an axle beam. The main support for the bogie assembly is a pin-ended shock absorber attached between the axles and connected to the gear structure at the upper end. The forward end of the axle beam is pin-jointed to a post, which is free to slide within a structural cylinder assembly. In the side view, the forward-located post with one axle under the post and the other aft, gives the appearance of a "scooter." In a scooter gear, the post assembly carries drag, side, and torque loads; vertical load is carried basically by the shock absorber. [AIR1489]

gear, tail An auxiliary gear that supports a small portion of the aircraft weight, and usually consists of a shock strut, wheel/tire assembly (or skid), and controlling devices; sometimes steerable. The tail gear is located in the aft or tail portion of the aircraft. [AIR1489]

gear train A combination of two or more gears, arranged to transmit power and motion between two rotating shafts or between a rotating shaft and a member that moves linearly.

gear up To arrange a gear train so that the driven shaft rotates at a higher speed than the driving shaft.

gear walk Cyclic fore and aft motion of the landing gear strut assembly about a normally static vertical strut centerline. Caused by drag loads applied at the tire/ground interface and the natural spring rate of the gear structure. Sometimes aggravated by anti-skid braking action (cycling) and resonant frequency of vibration of the strut. [AIR1489]

gear wheel A wheel with integral gear teeth that mesh with another gear, a rack, or a worm.

gegenschein A round or elongated spot of light in the sky at a point 180 degrees from the sun. Also called counterglow.

Geiger threshold The lowest voltage applied to a counter tube that results in output pulses of essentially equal amplitude, regardless of the magnitude of the ionizing event.

gel The initial, jellylike solid phase that develops during the formation of a resin from a liquid. [AIR4844]

gelation In a resin cure, describes the condition wherein the resin viscosity has increased to the point that it barely moves when probed with a sharp instrument. [AIR4844]

gelation time In connection with the use of synthetic thermosetting resins, that interval of time extending from the introduction of a catalyst into a liquid adhesive system until the start of gel formation. [AIR4844]

gel coat Refers to the resin gelled on the internal surface of a plastics mold prior to filling it with a molding material; when the finished part is a two-layer laminate, the gel coat provides improved surface quality.

gel permeation chromatography A form of liquid chromatography in which the polymer molecules are separated by their ability or inability to penetrate the material in the separation column. [AIR4844]

gel point The stage at which a liquid begins to exhibit pseudoelastic properties. [AIR4844]

general aviation The portion of civil aviation that encompasses all facets of aviation, with the exception of air carriers holding a certificate of public convenience and necessity, and large commercial aircraft operators. [ARP4107]

generally orthotropic ply An orthotropic ply in which the loads are not in the direction of the principal plane of elastic symmetry. The reference coordinate system does not coincide with the ply axis of an orthotropic ply. [AIR4844]

general purpose electronic test equipment Test equipment containing the capability, without modification, to generate, modify, or measure a range of parameters of electronic functions; required to test two or more prime equipments or systems of basically different design. [AIR4896]

generated background Internally generated imagery upon which symbology may be superimposed. [AS8034]

generating electric field meter An instrument for measuring electric field strength in which a flat conductor is alternately exposed to the field and shielded from it; the potential gradient of the field is determined by measuring the rectified current through the conductor.

generating magnetometer An instrument for measuring magnetic field strength by means of the electromotive force generated in a rotating coil immersed in the field being measured.

generation system Part of the electrical system including that portion from the generators up to and including the bus contactors. [ARP4404]

generator Any mechanically driven device that supplies DC power, including, but not limited to, those with commutators, conductors, homopolar and flux switch types; alternators with rectified output with or without slip rings; and generators used as starters. [AS8020]

generator cartridge An assembly that is made up of a chemical core and its surrounding mechanical enclosure, insulating material, initiation means for the chemical core, pressure-relief means, and any necessary filtering means; designed as a replaceable component, and intended for insertion into a reusable, portable chemical oxygen device, as a means of recharging the device. [AS1303]

geodesic The shortest distance between two points on a surface. Also called geodetic. [AIR4844]

geodesic isotensoid Constant stress level in any given filament at all points in its path. [AIR4844]

geodesic-isotensoid contour In filament-wound, reinforced-plastic pressure vessels, a dome contour in which the filaments are placed on geodesic paths so that the filaments will exhibit uniform tensions throughout their length under pressure loading. [AIR4844]

geodesic ovaloid A contour of end domes, the fibers forming a geodesic line—the shortest distance between two points on a surface of revolution. The forces exerted by the filaments are proportioned to meet hoop and meridional stresses at any point. [AIR4844]

geodesy The science that deals mathematically with the size and shape of the earth, and the earth's external gravity field; and with surveys of such precision that the overall size and shape of the earth must be taken into consideration.

geodetic *See* geodesic.

geodetic accuracy The degree to which point positions or boundaries indicated on maps or imagery correspond with true geodetic positions.

geodetic coordinates Quantities that define the position of a point on the spheroid of reference, with respect to the planes of the geodetic equator and of a reference meridian.

geodetic survey Survey that takes into account the size and shape of the earth.

Geodynamic Experimental Ocean Satellite *See* GEOS-D satellite.

geographic disorientation The type of spatial disorientation in which a person is correctly oriented with reference to the horizon, but not oriented in relation to known ground references or navigational fixes; simply stated, the person is lost. [ARP4107]

geographic information system Computer-assisted system that acquires, stores, manipulates, and displays geographic data. Some systems are not automated.

geographic north *See* true north.

geomagnetic latitude Angular distances from the geomagnetic equator, measured northward or southward through 90 degrees, and labeled N or S to indicate the direction of measurement.

geomagnetism *1.* Collectively considered, the magnetic phenomena exhibited by the earth and its atmosphere; and, by extension, the magnetic phenomena in interplanetary space. *2.* The study of the magnetic field of the earth. Also called terrestrial magnetism.

geometric accuracy The internal geometric fidelity of an imaging system.

geometrical acoustics The study of the behavior of sound under the assumption that sound traversing a homogeneous medium travels along straight line or rays. Also called ray acoustics.

geometrical optics The geometry of paths of light rays and their imagery through optical systems.

geometrical theory of diffraction A ray theory of diffraction process.

geometric dilution of precision (GDOP) A navigation and positioning system performance index expressing the dilution of range measurement precision due to the geometric relationship between user and satellites. Formulated as the square root of the sum of the variances of position estimates in the three orthogonal directions; can be employed to determine the optimal locations for network satellites, and in the selection of optimal satellite signals sources.

geometric distortion An aberration that causes a displayed pattern to be contorted from the desired pattern; includes the linearity errors in a display. Linearity errors are defined as on-axis horizontal and vertical deflection non-uniformities, whereas geometrical distortion is a general term applying to deflection non-uniformities over the entire phosphor screen. [ARP1782]

geometric-perspective illusion *See* illusions, visual.

geometric pitch The tangent of the acute angle between the extended chord plane of a propeller airfoil and its plane of rotation. *See also* standard pitch. [ARP4107]

geometric rectification (imagery) The correction of image distortions due to sensor view angle, platform attitude, or target surface features.

geometric trail *See* trail, geometric.

geomorphology A science that deals with the land and submarine relief features of the earth's surface and genetic interpretation of them by use of the principles of physiography in its descriptive aspects, and of dynamic and structural geology in its explanatory phases.

geophysical fluid General term for the liquids and gases on or in the earth—from water in all forms, to petroleum and hydrocarbons in liquid and gaseous form, to molten rock material within the earth.

geophysical fluid flow cell Apparatus used in model experiments for deep solar convection and Jovian atmospheric circulation, for Spacelab 1 and Spacelab 3.

geopotential height The height of a given point in the atmosphere in units proportional to the potential energy of unit mass (geopotential) at this height, relative to sea level.

geopotential research mission A NASA gravity-field-mapping mission utilizing the low-low satellite tracking concept to measure the Doppler shift between two coorbiting polar satellites.

geopressure Pressure that exceeds the normal hydrostatic pressure of about 0.465 psi per foot of depth.

GEOS-D satellite Stands for geodynamic experimental ocean satellite; another in a series of the European Space Agency's geostationary scientific satellites launched by NASA for long-term cosmic radiation studies.

geostationary platform *See* synchronous platform.

geostationary satellite *See* synchronous satellite.

geostrophic wind The horizontal wind velocity for which the coriolis acceleration exactly balances the horizontal pressure force.

gerber cutter A computer-controlled reciprocating knife process for cutting, kitting, and labelling prepreg fabric and tape plies. [AIR4844]

getter A material exposed to the interior of a vacuum system in order to reduce the concentration of residual gas by absorption or adsorption.

getter-ion pump A type of vacuum pump that produces and maintains high vacuum by continuously or intermittently depositing chemically active metal layers on the wall of the pump, where they trap and hold inert gas atoms that have been ionized by an electric discharge and drawn to the activated pump wall. Also known as sputter-ion pump.

ghost An indication that has no direct relation to pulses reflected from discontinuities in the materials being tested. [ARP5089]

GHz Abbreviation for gigahertz.

giant-hand illusion *See* illusions, vestibular.

gib A removable plate that holds other parts or that acts as a bearing or wear surface.

gilbert The CGS unit of magnetomotive force; the SI unit, the ampere (or ampere-turn), is preferred.

gimbal Device with two mutually perpendicular and intersecting axes of rotation (and thus free angular movement in two directions) on which engines or other objects may be mounted in gyros, supports that provide the spin axes with degrees of freedom.

gimbal lock In a gyro having two degrees of freedom, a position in which the spin axis becomes aligned with an axis of freedom, thus depriving it of a degree of freedom—and of its useful properties.

gimbal mount An optical mount that allows the position of a component to be adjusted by rotating it independently around two orthogonal axes.

Giotto mission The European Space Agency's mission to fly through the head of Halley's Comet in order to make in situ measurements of the composition and physical state as well as the structure of the head.

girt An assembly that connects the evacuation device of an aircraft to a floor-level exit; typically constructed of coated fabric materials, and may have more than one component. [ARP4277]

gland *1.* The cavity or space provided for the accommodation and operation of an elastic packing or gasket for sealing of a fluid vessel or compartment. [ARP243B] *2.* A movable part that compresses the packing in a stuffing box.

glare The sensation produced by luminances from extraneous sources within the visual field that are sufficiently greater than the emitted luminance; the eyes may adapt to this sensation, or it may cause loss in visual performance. [ARP1782]

glare source Any light-emitting source or reflective surface of similar brightness; normally associated with integral flood-lit systems. [ARP1161]

glass bead rating For a porous medium, a size designation in micrometers that represents the largest pore through which a glass sphere of a designated size can pass. [ARP4386]

glass filament A form of glass that has been drawn to a smaller diameter and extreme length. [AIR4844]

glass filament bushing The unit through which molten glass is drawn in making glass filaments. [AIR4844]

glass finish A material applied to the surface of a glass reinforcement to improve the bond between the glass and the plastic resin matrix. [AIR4844]

glass flake Thin, irregularly shaped flakes of glass, typically made by shattering a thin-walled tube of glass. [AIR4844]

glass former An oxide that easily forms a glass. [AIR4844]

glass laser High-power laser used in laser-fusion technology research.

glass, percent by volume The product of the specific gravity of a laminate and the percent glass by weight, divided by the specific gravity of the glass. [AIR4844]

glass stress In a filament-wound part, usually a pressure vessel, the stress calculated using the load and the cross-sectional area of the reinforcement only. [AIR4844]

glass transition In an amorphous polymer or in amorphous regions of a partially crystalline polymer, the reversible change from (or to) a viscous or rubbery condition to (or from) a hard and relatively brittle condition. [AIR4844]

glass transition temperature *1.* The approximate midpoint of the temperature range over which the glass transition takes place. *2.* The temperature at which increased molecular mobility results in significant changes in the properties of a cured-matrix resin system or plastic fiber; the point at which rigid behavior changes to a rubbery behavior. [AIR4844]

glassy alloy *See* metallic glass.

glaze *1. See* ice, glaze or clear. *2.* A glossy, highly reflective, glasslike, inorganic fused coating. *See also* enamel.

glazed surface An undesirable, hard, glossy surface developed where rubbing action (friction) breaks down parts of the surface or the lubricant. [ARP4107]

glazing *1.* Cutting and fitting glass panes into frames. *2.* Smoothing the exposed solder of a wiped pipe joint with a hot iron.

glazing compound A caulking compound, such as putty, used to seal the edges of a pane of glass where it fits into its frame.

glide Controlled descent of airplane under little or no engine thrust, in which forward motion is maintained by gravity and vertical descent is controlled by lift forces.

glide angle *See* glide slope.

glide path *1.* The line to be followed by an aircraft as it descends from horizontal flight to land on the surface. *2.* An automatic flight-control mode in which the glide path radio beam is used by the flight-control system to continually maintain the aircraft on the glide path without pilot intervention. [ARP4386] *3.* Flight path of an aeronautical vehicle in a glide, seen from the side. *4.* The path used by aircraft or spacecraft in approach procedure, generated by instrument landing facilities.

glide path, on/above/below Terms used by ATC to inform an aircraft making a precision approach radar (PAR approach) of its vertical position (elevation) relative to the desired descent profile to the runway. [ARP4107]

glide rocket A rocket vehicle kept within or near the sensible atmosphere so as to assume a flat gliding altitude within the atmosphere after power shutoff. [ARP4386]

glide slope (GS) A means of providing vertical guidance for aircraft during approach and landing. The glide slope consists of the following: (a) electronic components emitting signals, which provide vertical guidance by reference to airborne instruments during instrument approaches such as ILS; or (b) visual ground aids, such as VASI, which provide vertical guidance for VFR approach or for the visual portion of an instrument approach and landing. [ARP4107]

glide slope intercept altitude The minimum altitude of the intermediate approach segment prescribed for a precision approach that assures required obstacle clearance; depicted on instrument approach procedure charts. *See* segments of an instrument approach procedure and instrument landing system. [ARP4107]

glime *See* ice, glime.

Glimm method Numerical technique for solving gas dynamics problems involving hyperbolic systems of conservation laws.

glitch Undesirable electronic pulses that cause processing errors.

glitter Decorative flaked powder having a particle size large enough so that the individual flakes produce a visible reflection or sparkle; used in certain decorative paints and in some compounded plastics stock.

global *1*. Describes a name whose scope is the entire system in which it resides. *2*. A computer instruction that causes the computer to locate all occurrences of specific data. *3*. A value defined in one program module and used in others; globals are often referred to as entry points in the module in which they are defined, and externals in the other modules that use them.

global array A set of data listings that can be referenced by other parts of the software.

global common An unnamed data area that is accessible by all programs in the system. Sometimes referred to as blank common.

global positioning system A satellite navigation system that will display many (up to 24) satellites in three sets of orbits by means of a precise time standard and three-dimensional information on position and velocity.

global variable Any variable available to all programs in the system. Contrast with reserved variable.

globe An enclosing device made of clear or diffusing material, designed to protect a lamp, to diffuse or redirect its light, or to modify its color. [ARP798]

globe valve A type of valve in which the flow direction through the controlling portion is generally at 90° to the inlet and outlet ports. [AIR4783]

globular alpha In titanium, a spheroidal form of equiaxed alpha. [AS1814]

gloss The shine, sheen, or luster of a dried film. [AIR4844]

glossimeter An instrument for measuring the "glossiness" of a surface—that is, the ratio of light reflected in a specific direction to light reflected in all directions—usually by means of a photoelectric device. Also known as glossmeter.

glossmeter *See* glossimeter.

glow discharge *1*. Electrical discharge that produces luminosity. *2*. A discharge of electrical energy through a gas, in which the space potential near the cathode is substantially higher than the ionization potential of the gas.

glue A crude, impure form of commercial gelatin that softens to a gel consistency when wetted with water and dries to form a strong adhesive layer.

glued Describes a mixed-signal simulation system that combines existing analog and digital simulation software into a hybrid analog or digital simulation system.

gluons The carriers of the strong force that holds atomic nuclei together (holding together groups of quarks making up stable particles, which in turn are bound together in the atomic nuclei).

GMAW *See* gas metal-arc welding.

GMT *1*. Stands for Greenwich mean time, the mean solar time of the meridian of Greenwich, England, used as the prime basis of standard time throughout the world. Expressed in hs GMT or as Z (Zulu phonetically). [ARP4107] *2*. *See* universal time.

gnomonic projection A projection on a plane tangent to the surface of a sphere having the point of project at the center of the sphere.

gnotobiotics The study of germ-free animals.

goal A long-term requirement implied by specification or contract and used primarily for guidance. [AIR4896]

go around Instructions for a pilot to abandon his/her approach to landing. A pilot on an instrument approach should execute the published missed approach procedure or follow instructions provided by ATC.

"Go" functional diameter The pitch diameter of an enveloping thread at the minimum material condition, with perfect form but reduced thread height. [AS8879]

Golay cell An infrared detector in which the incident radiation is absorbed in a gas cell, thereby heating the gas; the temperature-induced expansion of the gas deflects a diaphragm, and a measurement of this deflection indicates the amount of incident radiation.

goniometer *1*. Instrument for measuring angles. *2*. Specifically, an instrument used in crystallography to determine angles between crystal planes, using x-ray diffraction or other means. *3*. An instrument used to measure refractive index and other optical properties of transparent optical materials or optical scattering in materials at UV, visible, or IR wavelengths.

go/no-go gage A composite gaging device that enables an inspector to quickly judge whether specific dimensions or contours are within specified tolerances. In many instances, the device is so constructed that the part being inspected will fit one part of the gage easily and will not fit another part if it is within tolerance, and will pass both parts or pass neither if it is not within tolerance.

go/no go test A test designed to yield a "test pass" or "go" indication in the absence of faults in a unit under test, and a "test fail" or "no go" indication when faults have been detected.

goodness of fit In an experiment, the degree to which the observed frequencies of occurrence of events correspond to the probabilities in a model of the experiment.

gouges Elongated grooves or cavities. [ARP4784]

gouging Forming a groove in an object by electrically, mechanically, thermomechanically, or manually removing material. Gouging is typically used to remove shallow defects prior to repair welding.

governor A device for automatically regulating the speed or power of a prime mover—especially a device that relies on centrifugal force in whirling weights opposed by springs or gravity to actuate the controlling element.

GOX (gaseous oxygen systems) Refers to the provision of facilities to store gaseous oxygen in cylinders at either high or low pressure and to deliver it to the aircrewman at a reduced pressure for breathing. [ARP171]

grab sample *See* captured sample. [ARP4418]

graceful degradation A system attribute whereby, when a piece of equipment fails, the system falls back to a degraded mode of operation, rather than failing catastrophically and giving no response to its users.

grade *1*. A classification of materials, alloys, ores, units of product, or characteristic according to some attribute or level of quality. *2*. To sort and classify according to attributes or quality levels. *3*. Oil classification according to quality.

graded index fiber (GRIN) An optical fiber in which the refractive index changes gradually between the core and cladding, in a way designed to refract light so it stays in the fiber core. Such fibers have lower dispersion and broader bandwidth than step-index fibers.

graded refractive index lens A lens in which the refractive index of the glass is not uniform. Typically the refractive index will differ with distance from the center of the lens.

gradient (scale factor) The approximate slope of the curve of speed-sensitive output voltage vs speed. Expressed as volts per 1000 rpm. [ARP4418]

gradient index optics Optical systems with components whose refractive indexes vary continuously within the material used for the optical elements.

graduation Any of the major or minor index marks on an instrument scale.

Graham's Law Gas law stating that the rate of diffusion of a gas is directly proportional to the pressure and temperature and inversely proportional to the square root of the density (molecular weight) of the gas. [ARP171] *See also* gas laws.

grain *1*. A single piece of solid propellant, regardless of size or shape. [ARP913] *2*. In metals and other crystalline substances, an individual crystallite in a polycrystalline mass. *3*. In crumbled or pulverized solids, a single particle too large to be called powder.

grain boundary The plane of mismatch between adjacent crystallites in a polycrystalline mass, as revealed on a polished and etched cross section through the material.

grain boundary alpha In titanium, primary or transformed alpha outlining prior beta grain boundaries. May be continuous unless broken up by subsequent work; also may accompany blocky alpha. Occurs by slow cooling from the beta field into alpha-beta field. [AS1814]

grain flow Fibrous appearance on a polished and etched section through a forging, caused by orientation of impurities and inhomogeneities along the direction of working during the forging process.

grain growth In a metal, an increase in the average grain size, usually as a result of exposure to high temperature.

graininess Visible coarseness in a photographic or radiographic emulsion, caused by countless small grains of silver clumping together into relatively large masses visible to the naked eye or with slight magnification.

graining Working a translucent stain while still wet to simulate the grain in wood or marble; tools such as special brushes, combs, and rags are used by hand to create the desired irregular patterns.

grain size For metals, the size of crystallites in a polycrystalline solid, which may be expressed as a diameter, number of grains per unit area, or standard grain size number determined by comparison with a chart, such as those published by ASTM. In most instances the grain size is given as an average, unless there are substantial proportions that can be given as two distinct sizes; if two or more phases are present, grain size of the matrix is given.

grains per cu ft The term for expressing dust loading in weight per unit of gas volume (7000 grains equals one pound).

gram Abbreviated g. The CGS unit of mass, equal to 0.001 kilogram.

grandfathered Certification of individuals who are considered qualified under a prior qualification program. [AIR4844]

grand mean The average of the averages. [AIR4979]

grand unified theory (GUT) A theory describing the unification of gravity with the other elementary forces in physics, i.e., the weak force, the strong force, and the electromagnetic force.

granular fracture A rough, irregular fracture surface, which can be either transcrystalline or intercrystalline, and which often indicates that fracture took place in a relatively brittle mode, even though the material involved is inherently ductile.

granular structure Nonuniform appearance of molded or compressed material due to the presence of particles of varying composition.

graphic panel A master control panel that pictorially (and usually colorfully) traces the relationship of control equipment and the process operation. Permits an operator, at a glance, to check on the operation of a far-flung control system by noting dials, valves, scales, and lights.

graphite *1.* A crystalline allotropic form of carbon. [AIR4844] *2.* Soft, naturally occurring allotropic form of carbon, also produced artificially in the form of strong fibers with perfect hexagonal crystalline structure.

graphite-epoxy composites Structural materials composed of epoxy resins reinforced with graphite.

graphite fiber A fiber made from a precursor by an oxidation, carbonization, and graphitization process. [AIR4844]

graphite flake A form of graphite present in gray cast iron which appears in the microstructure as an elongated, curved inclusion.

graphite-polyimide composite Composite material utilizing graphite reinforcing fibers in a resin matrix.

graphite rosette A form of graphite present in gray cast iron which appears in the microstructure as graphite flakes extending radially outward from a center of crystallization.

graphitic carbon Free carbon present in the microstructure of steel or cast iron; an essential feature of most cast irons, but almost always undesirable in steel.

graphitic corrosion Corrosion of gray cast iron in which the iron matrix is slowly leached away, leaving a porous structure behind that is largely graphite, but that may also be held together by corrosion products. Graphitic corrosion occurs in relatively mild aqueous solutions and on buried pipe and fittings.

graphitic steel Alloy steel in which some of the carbon is present in the form of graphite.

graphitization Formation of graphite in iron or steel. Termed "primary graphitization" if it forms during solidification; and "secondary graphitization if it forms during subsequent heat treatment or extended service at high temperature.

graphitizing Annealing a ferrous alloy in such a way that at least some of the carbon present is converted to graphite.

graphoepitaxy The use of artificial surface-relief structures to induce crystallographic orientation in thin films.

Grashof number A nondimensional parameter used in the theory of heat transfer. The Grashof number is associated with the Reynolds number and the Prandtl number in the study of convection.

graveyard spin A sequence of repeated spins occurring because the pilot's motion-sensing system, in proper recovery from a spin, tends to create an illusion of spinning in the opposite direction. Responding to this somatogyral illusion, the pilot returns the plane to its original spin. [ARP4107]

graveyard spiral A progressively steepening spiral resulting from the somatological illusion during a coordinated constant rate turn that has ceased to stimulate the motion-sensing system of the pilot. When, in this situation, the pilot observes a loss of altitude, the tendency is to pull back on the controls, thus tightening the spiral and increasing the loss of altitude. [ARP4107]

gravimeter A device for measuring the relative force of gravity by detecting small differences in weight of a constant mass at different points on the earth's surface. Also known as gravity meter.

gravireceptors Highly specialized nerve endings and receptor organs located in skeletal muscles, tendons, joints, and in the inner ear, which furnish information to the brain with respect to body position, equilibrium, and the direction of gravitational forces.

gravitation The acceleration produced by the mutual attraction of two masses; the magnitude of this acceleration is inversely proportional to the square of the distance between the two centers of mass.

gravitational constant *1.* The coefficient of proportionality in Newton's law of gravitation. *2.* A dimensionless conversion factor in English units that arises from Newton's sec-

ond law ($F = ma$), where mass is expressed in pounds-mass (lb_m).

gravitational wave antennas Devices for receiving propagating gravitational fields produced by some change in the distribution of matter.

gravitometer *See* densitometer.

gravitons The hypothetical elementary units of gravitation.

gravity *1.* Weight index of fuels. That of liquid petroleum products is expressed either as specific, Baumé, or API. (American Petroleum Institute) gravity; that of gaseous fuels, as specific gravity related to air under specified conditions; and that of solid fuels, as specific gravity related to water under specified conditions. *2. See* gravitation.

gravity fill point or port An opening in the fuel tank of an aircraft which allows for filling by use of an overwing nozzle. [AIR4783]

gravity meter *1.* A device in which a U-tube manometer is used to determine specific gravities of solutions by direct reading. *2.* An electrical device for measuring variations in gravitational forces through different geological formations. *3. See* gravimeter.

gravity separation Separation of immiscible phases resulting from a difference in specific gravity. [AIR4783]

gravity wave Wave in an interface between fluids of different density in which the restoring force is gravity.

gray Metric unit for absorbed dose.

gray body An object having the same spectral emissivity at every wavelength, or one whose spectral emissivity equals its total emissivity.

gray code A generic name for a family of binary codes that have the property whereby a change from one number to the next sequential number can be accomplished by changing only one bit in a code for the original number. This type of code is commonly used in rotary shaft encoders to avoid ambiguous readings when moving from one position to the next. *See* shaft encoder. *See also* cyclic code.

gray goods *See* greige goods.

gray iron Cast iron containing free graphite in flake form; so named because a freshly broken bar of the alloy appears gray.

grayout A temporary condition in which vision is hazy, restricted, or otherwise impaired, owing to insufficient oxygen supply to the brain; commonly induced by positive G forces of moderate intensity and duration. *See also* blackout. [ARP4107]

gray scale *1.* The incremental levels of display-element light transmission that exist between fully off and fully on. [ARP4256] *2.* Images that are not colored or multispectral.

grazing incidence Incidence at a small glancing angle.

grazing incidence solar telescope *See* GRIST (telescope).

great circle A circle on the surface of a sphere, especially the earth, whose plane passes through the center of the sphere. [ARP4107]

great circle course *1.* The direction of the great circle through the point of departure and the destination expressed as an angular distance from a reference direction (usually north) to the great circle. The angle varies from point to point along the great circle. *2.* An imaginary line of intersection on the earth's surface of any plane through the earth's center. [ARP1570]

greaves *1.* Armor to protect the legs. *2.* In landing gear, armor to protect the struts or equipment mounted thereon from arresting cables, barrier straps, etc. [AIR1489]

green run Test stand time interrupted by a malfunction, resulting in engine disassembly. Time accumulated during a green run is not recorded as accumulated time. [AIR4896]

greenstick fracture Fracture in which the crack does not go right through the material, but is deflected part way through, allowing the remainder of the cross section to deform elastically. [AIR4844]

Greenwich mean time *See* universal time.

greige goods Any fabric before finishing, as well as any yarn or fiber before bleaching or dyeing, i.e., goods with no finish or size. Also known as gray goods. [AIR4844]

grid *1.* A network of lines, typically forming squares, used in layout work or in creating charts and graphs. *2.* A criss-cross network of conductors used for shielding or controlling a beam of electrons.

grid circuit An electronic circuit that includes the grid-cathode path of an electron tube in series with other circuit elements.

grid control A method of controlling anode current in an electron tube by varying the potential of the grid electrode with respect to the cathode.

grid emission Emission of electrons or ions from the grid electrode of an electron tube.

grid nephoscope A device for determining the direction of cloud motion by sighting through a gridwork of bars and adjusting the angular position of the grid until some feature of the cloud in the field of view appears to move along the major axis of the grid.

grid spaced contacts Contacts in a multiple contact connector spaced in a geometric pattern. [ARP914A]

GRIN *See* graded index fiber.

grinding *1.* Removing material from the surface of a workpiece using an abrasive wheel or belt. *2.* Reducing the particle size of a powder or granular solid.

grinding burn Localized overheating of a workpiece surface due to excessive grinding pressures and/or inadequate supply of coolant.

grinding check Fine thermal cracks that develop from overheating of the area being ground. Such cracks are generally at right angles to the direction of grinding, but may appear as a complete network. [AS3071A]

grinding cracks Shallow cracks in the surface of a ground workpiece. Appear most often in relatively hard materials due to excessive grinding friction or high material sensitivity.

GRIST (telescope) An ESA Spacelab payload designed for grazing incidence solar phenomena.

grit blasting Abrasively cleaning metal surfaces by blowing steel grit, sand, or other hard particulate against the surfaces to remove soil, rust, and scale. Also known as sandblasting.

grommet *1.* An elastomeric or plastic sealing device that supports and protects terminations and wires/cables from adverse mechanical and environmental conditions. [ARP914A] *2.* A rubber or soft-plastic eyelet inserted in a hole through sheet metal, such as an electronic equipment chassis or enclosure, to prevent a wire from chafing against the side of the hole, damaging its insulation, or shorting out to the chassis. *3.* A circular piece of fibrous packing material used under a bolt head or nut to seal the bolthole.

grommet nut A blind nut with a round head that is sometimes used with a screw to attach a hinge to a door.

groove *1.* A long, narrow channel or furrow in a solid surface. *2.* In a weldment, a straight-sided, angled, or curved gap between joint members prior to welding, which helps confine the weld puddle and ensure full joint penetration to produce a sound weld.

grooved drum A windlass drum whose face has been grooved, usually in a helical fashion, to support and guide the rope or cable wound on it.

grooved tube-seat A tube seat having one or more shallow grooves into which the tube may be forced by the expander.

groove seal A postassembly seal formed by injecting a noncuring sealant material into a groove machined in one faying surface of the mating or overlapping structure. [AIR4069]

gross porosity In weld metal or castings, large or numerous gas holes, pores, or voids that are indicative of substandard quality or poor technique.

ground *1.* For power generation and power utilization systems, refers to the primary aircraft structure, the referenced ground for the negative of the d-c and neutral of the a-c. [AS1212] *2.* A conducting connection, either intentional or accidental, by which an electric current or equipment is connected to the earth, or to a conducting structure that serves a function similar to that of an earth ground (i.e., a structure such as a frame of an air, space, or land vehicle that is not conductively connected to earth). [ARP1870] *3.* A (neutral) reference level for electrical potential, equivalent to the level of electrical potential of the earth's crust. *4.* A secure connection to earth that is used to reference an entire system. Usually the connection is in the form of a rod driven or buried in the soil or a series of rods connected into a grid buried in the soil.

ground clutter A pattern produced on the radar scope by reflections from local ground features. Ground clutter may make it difficult to detect other radar returns in the affected area. [ARP4107]

ground-controlled approach /landing (GCA) A radar approach system operated from the ground by air-traffic-control personnel transmitting instructions to the pilot by radio. [ARP4107]

ground effect The apparent increase in aerodynamic lift experienced by an aircraft when flying near the ground and observed up to a distance above the ground approximately equal to the wingspan of the aircraft. [ARP4107]

ground effect (communications) The effect of ground conditions on radio communications.

ground fault An abnormally low impedance between the positive bus, the negative bus, and/ or the ground return. [AS1831]

ground flotation Characteristic definition used in describing the capability of an aircraft to operate satisfactorily on or from a specific airfield with respect to bearing and support parameters of that airfield. [AIR1489]

ground icing conditions Weather conditions, on the ground, that are conducive to formation of frost or ice, or the accumulation of snow or slush on airplane surfaces and components. [AS5116]

grounding A particular form of bonding, usually the process of electrically connecting a conducting object to a basic metal structure for the purpose of safely completing either a normal or fault circuit at power frequency. [ARP4404]

grounding cable As commonly used in the aircraft fuel industry, a misnomer for a bonding cable. [AIR4783]

grounding conductor A conductor that provides a current-return path from an electrical device to ground. [ARP914A]

ground interruption *See* interruption, ground.

ground lead *See* work lead.

ground loads Any load imposed on the landing-gear assembly by contact with the ground during any ground operation, i.e., drag, side, vertical, torsion, spin up, springback, towing, turning, etc. [AIR1489]

ground loop *1.* An uncontrolled, violent turn of an airplane while taxiing, or during the landing or takeoff run. [ARP4107] *2.* The generation of undesirable current flow within a ground conductor, owing to the circulation currents that originate from a second source of voltage—frequently as a result of connecting two separate grounds to a single circuit. [ARP1931]

ground plane A surface, all points of which are assumed to be at the same potential, usually the zero reference potential for the system. NOTE: A true, equipotential ground plane does not exist in practice. The deviations from the ideal increase with the frequency of the signals appearing on the ground plane conductor and can become a very important consideration in system design. [ARP1870]

ground power A power source that remains on the ground and is not airborne. [ARP906A]

ground power unit A gas turbine engine, similar in function to an auxiliary power unit, but mounted on a cart or truck and brought to the aircraft that it is to serve on the ground. [AS5116]

ground resonance Refers to the phenomenon during the process of landing, hovering, or take-off whereby a landing gear component touches and causes a sudden displacement of the rotor, inducing vibrations in the plane of rotation of the rotor or in the pylon system.

ground roll The distance between touchdown and final stop when an aircraft lands. [ARP4386]

ground safety lock A device designed to prevent inadvertent retraction of the gear when the aircraft is on the ground and supported by the gear. Generally installed manually after landing and removed prior to flight. Also used on landing gear doors and arresting hooks to provide safety for ground maintenance personnel. [AIR1489]

ground service Airplane servicing operations during which rotatable galley equipment, such as containers, carts, modules, and similar inserts, are unloaded from galleys on the aircraft, transported to ground kitchens, washed and cleaned, and stored or recycled with food and supplies for installation in galleys ready for flight. [AS1426]

ground speed The speed of an aircraft relative to the surface of the earth, typically expressed in knots or statute miles per hour. [ARP4107]

ground spoilers A device or system designed to "spoil" or reduce the aerodynamic lift on the aircraft quickly after touchdown, thus applying weight to the gears and enabling braking action for deceleration of the aircraft. *See also* speed brakes and drag brake. [AIR1489]

ground support equipment That equipment on the ground, including all implements, tools, and devices (mobile or fixed), required to inspect, test, adjust, calibrate, appraise, gage, measure, repair, overhaul, assemble, transport, safeguard, record, store, or otherwise function in support of a rocket, space vehicle, or the like, either in the research and development phase or in the operational phase, or in support of the guidance system used with the missile, vehicle, or the like. *See also* aircraft ground equipment.

ground support equipment recommendation A document that reflects a contractor's recommendations for major items of ground support equipment for a specific end item. [AIR4896]

ground track (track, actual track) The imaginary line or path on the earth's surface connecting successive points over which the aircraft has flown. *See also* desired track. [ARP1570]

ground track angle (track angle, actual track angle) The clockwise angle from true north to an imaginary line on the earth's surface connecting successive points over which the aircraft has flown (ground track). The normal usage implies present track angle, i.e., using the most recent successive points over a relatively short period of time. [ARP1570]

ground truth Data obtained on the ground concerning the significance of anomalies observed in remote sensing, to help interpretation.

ground visibility The prevailing horizontal visibility near the earth's surface as reported in the United States by the National Weather Service or an accredited observer. [ARP4107]

group A collection of units, assemblies, or subassemblies that is not capable of performing a complete operational function. [AIR4896]

group velocity *1.* The velocity of a wave disturbance as a whole, i.e., of an entire group of component simple harmonic waves. *2.* The velocity corresponding to the rate of change of average position of a wave packet as it travels through a medium.

grub screw A headless screw that is slotted at one end to receive a screw driver.

GS See glide slope.

GTAW *See* gas tungsten-arc welding.

guard ring In a counter tube or ionization chamber, an auxiliary ring-shaped electrode whose chief functions are to control potential gradients, reduce insulation leakage, or define the active region of the tube.

guard vacuum An enclosed, evacuated space between a primary vacuum system and the atmosphere whose primary purpose is to reduce seal leakage into the primary system.

guidance (motion) The process of directing the movements of an aeronautical vehicle or space vehicle, with particular reference to the selection of a flight path.

guidance, celestial The guidance of a vehicle by reference to celestial bodies. [ARP4386]

guidance, stellar A system in which a guided missile may follow a predetermined course with reference primarily to the relative position of the missile and certain bodies. [ARP4386]

guidance, terminal *1.* The guidance applied to a guided missile between midcourse guidance and arrival in the vicinity of the target. *2.* Electronic, mechanical, visual, or other assistance given an aircraft pilot to facilitate arrival at, operation within or over, landing upon, or departure from an air landing or air drop facility. [ARP4386]

guide *1.* A device whose prime function is attaching to engine parts for the purposes of locating two or more parts for assembly or disassembly. [ARP480A] *2.* A pulley, idler roll, or channel member that keeps a rope, cable, or belt traveling in a predetermined path. *3.* A stationary machine element, such as a beam, bushing, rod, or pin, whose primary function is to maintain one or more moving elements confined to a specific path of travel.

guide bearing A plain bushing used to prevent lateral movement of a machine element while allowing free axial translation, with or without (usually without) simultaneous rotation. Also known as guide bushing.

guide bushing *See* guide bearing.

guided aircraft missile A type of self-propelled missile, normally carried by a parent aircraft, which, after launching, can be guided to surface targets. [ARP4386]

guided bend test A bend test in which the specimen is bent to a predetermined shape in a jig or around a grooved mandrel.

guided missile An unmanned airborne vehicle whose flight path or trajectory can be altered by some mechanism within or attached to the vehicle, in response to either a preprogrammed control sequence or a control sequence transmitted to the vehicle while in flight.

guided wave A wave whose energy is confined by one or more extended boundary surfaces, and whose direction of propagation is effectively parallel to the boundary.

guide pin A pin or rod extending beyond the mating force or body of a connector, designed to position and guide connectors during mating so as to ensure proper engagement of the contacts. [ARP914A]

guide pin bushing The bushing through which the guide pin moves when the mold is closed. [AIR4844]

guide socket A socket or hole in a connector designed to accept a guide pin of a mating connector and thereby position and guide the connectors during mating so as to ensure proper engagement of the contacts. [ARP914A]

guide vanes Control surfaces that may be moved into or against a rocket's jetstream, used to change the direction of the jet flow for thrust-vector control.

gun launcher Ordnance device for firing missiles and rockets with initial attitude control.

gusset *1.* A piece used to give added size or strength in a particular location of an object. *2.* The folded-in portion of a flattened tubular film. [AIR4844] *3.* The transition between the terminal tongue and conductor barrel. [ARP914A]

gust A sudden, brief increase in the wind. According to United States Weather Bureau practice, gusts are reported when the variation in wind speed between peaks and lulls is at least 10 knots. [ARP4107]

GUT *See* grand unified theory.

guy A wire, rope, or rod used to secure a pole, derrick, truss, or temporary structure in an upright position; or to hold it securely against the wind.

guyed-steel stack A steel stack of insufficient strength to be self-supporting, and laterally stayed by guys.

gyratory screen A sieving machine having a series of nested screens whose mesh sizes are progressively smaller from top to bottom of the stack. A mechanism shakes the stacked screens in a nearly circular fashion, which causes fines to sift through each screen until an entire sample or batch has been classified.

gyro *See* gyroscope.

gyrodamper Single-gimbal control-moment gyro actively controlled to extract the structural vibratory energy through the local rotational deformations of a structure; used in large space structures.

gyrofrequency The natural period of revolution of a free electron in the earth's magnetic field.

gyro horizons Artificial horizons or attitude gyroscopes.

gyromagnetic ratio The magnetic moment of a system divided by its angular momentum.

gyroscope (gyro) *1.* Device that utilizes the angular momentum of a spinning mass (rotor) to sense angular motion of its base about one or two axes orthogonal to the spin axis. *2.* An instrument that maintains a stable, angular reference direction by virtue of the application of Newton's second law of motion to a mechanism whose chief component is a rapidly spinning heavy mass. Also known as a gyrostat.

gyroscopic couple The turning moment generated by a gyroscope to oppose any change in the position of its axis of rotation.

gyroscopic horizon A gyroscopic instrument that simulates the position of the natural horizon and indicates the attitude of an aircraft with respect to this horizon.

gyrostat *See* gyroscope.

gyro wheel The heavy, rotating element of a gyroscope, consisting of a wheel whose rather large mass is distributed uniformly around its rim. In precision gyroscopes, the gyro wheel is specially constructed to have nearly perfect balance.

H

h *See* enthalpy.

H *See* henry.

HAA *See* height above airport.

habit pattern interference *See* attention, anomalies of habit pattern interference.

habituation *See* attention, anomalies of habituation.

hackles Raised strips or striations on a fracture surface caused by an array of small cracks produced during a shear failure. [AIR4844]

HAD *See* high aluminum defect.

hair-line cracks Fine, random cracks in a coating (e.g., paint) or on any rigid surface.

HAL *See* height above landing.

half-adjust To round a number so that the least significant digit(s) determines whether a "one" is to be added to the digit next higher in significance than the digit(s) used as criterion for the determination. After the adjustment is made, if required, the digit(s) used as criterion will be dropped, e.g., 432.784 using the terminal 4 as criterion yields 432.78 as the half-adjusted value. The number 432.785 half-adjusts to 432.79, since the terminal digit is "one half, or more."

half-and-half solder A lead-tin alloy (50Pb-50Sn) used primarily to join copper tubing and fittings.

half coupling A simple joint using only one seal with selected components of a full coupling to join a tube end with a special machined fitting. [ARP4968]

half cycle In alternating circuits, the time to complete one half of a full cycle at the operating frequency.

half duplex Communications in both directions (transmit and receive), but in only one direction at a given instant in time. *See also* full duplex.

half-life The average time required for one half of the atoms in a sample of radioactive element to decay.

half shell Die-formed duct halves. [ARP699D]

half-thickness The thickness of an absorbing medium that will depreciate the intensity of the radiation beam by one-half.

half-wave plate A polarization retarder that causes light of one linear polarization to be retarded by a half wavelength 180° relative to the phase of the orthogonal polarization.

half-wave rectification Modification of an ultrasonic signal so that only one half (either positive or negative-going) of the RF waveform is displayed. [ARP5089]

half-wave rectifier An electronic circuit that converts an a-c input signal to a d-c output signal, with current flowing in the output circuit during only one half of each cycle of the input signal.

Hall current *See* Hall effect.

Hall effect The electrical polarization of a horizontal conducting sheet of limited extent, when that sheet moves laterally through a magnetic field having a component vertical to the sheet. The Hall effect is important in determining the behavior of the electrical currents generated by winds in the lower atmosphere.

halogen occultation experiment Shuttle experiment to provide global stratospheric vertical concentration profiles of key chemical species involved in the catalytic destruction of ozone due to chlorine compounds.

hammerhead stall An abrupt maneuver in which the aircraft zooms into a vertical climb, stalls and yaws simultaneously, and goes into a dive from which recovery is made opposite to the direction of entry. [ARP4107]

H and D curve A measurement of photographic emulsion; shown as a curve in which density

320

is expressed as a function of the logarithm of exposure.

handhole An opening in a pressure part for access, usually not exceeding 6 in. in longest dimension.

handhole cover A handhole closure.

hand lance A manually manipulated length of pipe carrying air, steam, or water, used for blowing ash and slag accumulations from heat-absorbing surfaces.

hand lay-up The process of placing successive plies of reinforcing material or resin-impregnated reinforcement in position on a mold by hand. [AIR4844]

handling life The out-of-refrigeration time over which a material retains its handleability. [AIR4844]

handling strength A low level of strength initially obtained by an adhesive; allows specimens to be handled, moved, or unclamped without causing disruption of the curing process or affecting bond strength. [AIR4844]

handoff An action taken to transfer the radar identification of an aircraft from one controller to another if the aircraft will enter the receiving controller's airspace and radio communications with the aircraft will be transferred to that controller. [ARP4107]

handrail height - H The distance to the center of the handrail as measured at the nose of the step and perpendicular to the tread surface. [ARP836A]

handshake The recognition between two computers that they are able to communicate.

hanger fittings Fittings that provide a means of externally attaching or supporting the flexible tank to the aircraft structure. May also apply to fittings used on the inside of the tank to support items of equipment and fluid flow lines. [AIR1664]

hang fire An undesired delay in the functioning of an explosive device after initiating energy is applied. [AIR913]

hang glider Ultralight, unpowered aircraft in which the pilot controls the flight attitude and glide path by shifting his position on a suspended seat (swing seat).

hang pick A pick (fill yarn) caught in a warp yarn row, producing a triangular shaped hole in the fabric. [AIR4844]

hard clad silica fibers Silica optical fibers that are coated with hard plastic material, not with the soft materials typically used in plastic-clad silica.

hard detect Describes failures that can be positively detected. [AIR4896]

hard disk A computer storage medium with a large storage capacity as compared to floppy disks.

hard drawn copper wire Copper wire that has been drawn to size and not annealed. [ARP1931]

hard-drawn wire Cold-drawn metal wire of relatively high tensile strength and low ductility.

hardener A substance or mixture added to a plastic composition to promote or control the curing action; does this by taking part in the curing reaction. [AIR4844]

hardening Heating to a suitable temperature above the austenite transformation temperature (austenitizing), followed by quenching to develop a martensitic structure suitable for tempering to desired properties. [AMS2759B]

hardening (system) Technique for decreasing the susceptibility or vulnerability of weapon systems and components.

hard landing An impact landing of a spacecraft on the surface of a planet or natural satellite, destroying all equipment—except possibly a very rugged package.

hard lead Any of a series of lead-antimony alloys of low ductility. Hard lead typically contains 1 to 12% Sb.

hardness *1.* Generally, resistance of metal to plastic deformation, usually by indentation; however, may also refer to stiffness or temper, or resistance to scratching, abrasion, or cutting. *2.* A measure of the amount of calcium and magnesium salts in boiler water. Usually expressed as grains per gallon or ppm as $CaCO_3$.

hardover *1.* The displacement of a control member from its normal or neutral position to maximum (extreme) travel either in direction or rotation. [ARP4386] *2.* Response or movement of an aircraft control system at the maximum rate toward its maximum deflection. [ARP4107]

hard points *See* insert.

hardware *1.* An item that has physical being. Generally refers to such items as line replaceable units or modules, circuit cards, and power supplies. [ARP4754] *2.* Any part of a torsorestraint system, other than webbing. [AS8043] *3.* Physical equipment, as contrasted to ideas or design that may exist only on paper. *4.* In data processing, refers to the physical equipment associated with the computer.

hardware system A composite—at any level of complexity—of equipment that is designated to perform a specific function or mission. [ARP926A]

hard water Water that contains calcium or magnesium in amounts that require an excessive amount of soap to form a lather.

harmonic Having a frequency that is a multiple of the basic cyclical quantity to which it is related.

harmonic analysis A statistical method for determining the amplitude and period of certain harmonic or wave components in a set of data with the aid of Fourier series.

harmonic analyzer An instrument for measuring from a graph the magnitude and phase of harmonic segments of a cyclical function.

harmonic conversion transducer A transducer in which the output frequency is a multiple of the input frequency.

harmonic distortion Distortion caused by the presence of harmonics of a desired signal.

harmonic function Any solution of the Laplace equations.

harmonic generation The multiplication of the frequency of a lightwave by nonlinear interactions of the lightwave with certain materials. Generating the second harmonic is equivalent to dividing the wavelength in half.

harmonics *1.* Those vibrations that are integral multiples of the fundamental frequency; used in resonance testing. [ARP5089] *2.* Eigenfrequency oscillations excited in a vibrating system.

harness *1.* A multi-conductor cable with leads spaced along its length. [ARP480A] *2.* A group of wires or cables routed together, with or without attached components, and secured in a manner to provide a pre-shaped electrical wire or cable assembly. [ARP914A]

harness–high density A harness designed to save weight and space. [ARP1931]

harness satin Weaving pattern producing a satin appearance. [AIR4844]

Hartley information unit In information theory, a unit of measurement of the decision content of a set of 10 mutually exclusive events, expressed as the logarithm to the base 10; for example, the decision content of an eight-character set equals log 8, or 0.903 Hartley.

Hartree-Fock-Slater method A refined approximation method using wave functions to calculate for chemical elements quantities such as electron total energies and kinetic energies.

hash Numerous, small indications appearing on the viewing screen of an ultrasonic instrument, corresponding to many small inhomogeneities in the material and/or background noise. [ARP5089]

hashing From a group of records, the generation of a meaningless number that can be used as a location address.

Hastelloy B An International Nickel Co. alloy having a nominal composition of nickel (Ni) 66.7%; iron (Fe) 5%; molybdenum (Mo) 28%; and vanadium (V) 0.3%.

Hastelloy C An International Nickel Co. alloy having a nominal composition of nickel (Ni) 59%; iron (Fe) 5%; molybdenum (Mo) 16%; tungsten (W) 4%; and chromium (Cr) 16%.

HAT *See* height above touchdown.

Hay bridge A general-purpose a-c bridge circuit in which two opposing sides of the bridge are fixed resistances; the unknown leg is a combination of resistance and inductance, and the remaining side consists of a variable resistor and a variable capacitor.

HAZ *See* heat affected zone.

hazard A potentially unsafe condition resulting from failures, malfunctions, external events, errors, or combinations thereof. [ARP4754]

hazardous area An area in which explosive gas/air mixtures are, or may be expected to be, present in quantities that require special precautions for the construction and use of electrical apparatus.

hazardous atmosphere A combustible mixture of gases and/or vapors.

hazardous material Any substance that requires special handling to avoid endangering human life, health, or well being. Such substances include poisons; corrosives; and flammable, explosive, or radioactive chemicals.

hazardous material disposal (in space) The disposal in space of hazardous material. When radioactive materials are involved, the expected lifetime of orbit exceeds the lifetime of the radioactivity.

hazard rate The instantaneous failure rate. [AIR4896]

haze Describes the smoky or foggy appearance in a window; also describes the cloudy halo effect around a point light source. [AIR5122]

hazemeter *See* transmissometer.

HCL argon laser Gas laser in which the active material is gaseous hydrogen chloride and argon.

HCL laser Gas laser in which the active material is gaseous hydrogen chloride.

HDI *See* high density inclusion.

HDLC *See* high-level data link control.

header *1.* A conduit or chamber that receives fluid flow from a series of smaller conduits connected to it, or that distributes fluid flow among a series of smaller conduits. *2.* In data processing, data placed at the beginning of a file for identification. *3.* The portion of a batch recipe that contains information about the purpose, source, and version of the recipe, such as recipe and product identification, originator, issue date, and so on.

heading *1.* The upsetting of wire, rod, or bar stock in dies to form parts having some of the cross-sectional area larger than the original. [ARP700] *2.* The direction, usually expressed in deg relative to true or magnetic north, in which the longitudinal axis of an aircraft points; for example, "Fly heading 270 magnetic." *See also* course. [ARP4107]

heading hold An automatic flight-control mode in which a desired aircraft heading is maintained continuously without pilot intervention. [ARP4386]

head loss Pressure loss in terms of a length parameter, such as inches of water or millimeters of mercury.

head pressure Static pressure exerted on the fuel as a result of elevation (gravity effect). [AIR1749]

headset *See* earphones.

head up display *See* HUD.

health monitoring The acquisition of data from specialized sensors that measure various parameters related to structural integrity, readiness, ability to perform a mission, or other similar attributes of a major subsystem. [ARD50024]

heat *1.* In batch melting, material that was cast at the same time from the same furnace and was identified with the same heat number. In continuous melting, material that was poured without interruption. [ARP1917] *2.* Energy that flows between bodies because of a difference in temperature; same as thermal energy.

heat absorbing filter A glass filter that transmits most visible light, but strongly absorbs infrared light.

heat-activated adhesive A dry adhesive that is rendered tacky or fluid by application of heat, or heat and pressure, to the assembly. [AIR4844]

heat-affected area The area adjacent to the repair which may be affected by the heat of the heat blanket. [ARP5089]

heat affected zone (HAZ) That portion of the base metal in which the structure or properties have been altered by the heat of a welding or gas-cutting operation.

heat aging A test used to determine the ability of a wire or material to withstand a specific temperature for a specific length of time. The test temperature is usually higher than the related temperature of the test specimen. Sometimes referred to as "life cycle." [ARP1931]

heat available The thermal energy above a fixed datum that is capable of being absorbed for useful work. In boiler practice, the heat available in a furnace is usually taken to be the higher heating value of the fuel corrected by subtracting radiation losses, unburned combustible, latent heat of the water in the fuel formed by the burning of hydrogen, and adding the sensible heat in the air for combustion, all above ambient temperatures.

heat balance *1.* The equilibrium that exists on average between the radiation received from the sun by a planet and its atmosphere, and that emitted by the planet and its atmosphere. *2.* The equilibrium that is known to exist when all sources of heat-gain and -loss for a given region of body are accounted for. In general, this balance includes advective or evaporative terms as well as a radiation term. *3.* An accounting of the distribution of the heat input and output.

heat blanket A fabric blanket interwoven with a heating element; used in thermographic inspection to preheat the surface of a component prior to inspection. [ARP5089]

heat buildup In a part, the rise in temperature resulting from the dissipation of applied strain energy as heat or from applied mold cure heat. [AIR4844]

heat cleaned A condition in which glass or other fibers are exposed to elevated temperatures to remove preliminary sizings or binders not compatible with the resin system to be applied [AIR4844]

heat content The amount of heat per unit mass that can be released when a substance undergoes a drop in temperature, a change in state, or a chemical reaction. *See also* enthalpy.

heat-convertible resin A thermosetting resin convertible by heat into an infusible and insoluble mass. [AIR4844]

heat damage Charring, chalking, or hardening and cracking of hose cover caused by excessive or prolonged exposure to heat. Other evidence may include fluid leakage from a socket where a fitting or coupling is attached to the hose. [AIR1658A]

heat-deflection temperature The temperature at which a standard test bar deflects a specified amount under a stated load. [AIR4844]

heat distortion point The temperature at which a standard test bar deflects a specified amount under a stated load. [AIR4844]

heater A device that provides controlled heat to a specific area. [ARP480A]

heater, cabin/compartment A heat exchanger, usually employing combustion gases, compressed air, or electrical energy, from which heat is transferred to the cabin/compartment air supply. [ARP147C]

heater, exhaust hot air type A type of heater employed in aircraft in which the heat of the exhaust gases from the engine is used to directly heat, by means of a heat exchanger, the air supplied to the aircraft. [ARP147C]

heater, internal combustion type A heater in which the heat produced by combustion of a fuel is utilized within the heater itself. [ARP147C]

heater, muff type A heater designed to surround a duct or pipe carrying hot gases. When a muff-type heater is used, heat will be transferred to air passed through the annular space between the hot pipe and the muff. Also known as a shroud heater. [ARP147C]

heater, radiant A device that accomplishes heating by means of direct radiation. [ARP147C]

heater, shroud *See* muff heater. [ARP147C]

heat exchanger, cabin A heat exchanger designed to reduce the temperature of air discharged by a cabin air compressor. [ARP147C]

heat exchanger, MSOGS Located between the pressurized air source and the OEAS (oxygen enriched air system) oxygen concentrator; conditions the air supplied to the concentrator. [ARP171]

heat exchanger, precooler A heat exchanger used to cool engine-bleed air before it enters the air-conditioning compartment or unit. Usually serves the function of a primary heat exchanger in a bootstrap cooling system. [ARP147C]

heat exchanger, primary In a bootstrap cooling system, the heat exchanger that is used to cool the air from the pressure source before it enters the compressor. [ARP147C]

heat exchanger, regenerative A heat exchanger used to recover energy from a fluid whose cooling (or heating) potential was developed in another part of the system. [ARP147C]

heat exchanger, rotating A heat exchanger in which one fluid is passed inside hollow blades of a blower, while the other fluid is drawn through the heat exchanger and over the exterior of the blades. [ARP147C]

heat exchangers *1.* Part of the air-cycle air-conditioning system (precoolers, primary, secondary, intercoolers, regenerative, and evaporative); device that transfers heat from one medium to another without mixing of the media. [AIR4073] *2.* Device for transferring heat from one fluid to another without inter-mixing the fluids; for example, a regenerator or an apparatus for cooling or heating the air in a wind tunnel.

heat exchanger, secondary In a bootstrap cooling system, the heat exchanger that is used to cool the air that passes from the compressor to the cooling turbine. [ARP147C]

heat-fail temperature The temperature at which delamination occurs under static loading in shear. [AIR4844]

heating load, cabin The rate of heat transfer to the cabin, as measured by the difference between the total enthalpy of air entering the cabin and the total enthalpy of the air leaving the cabin during a specified period of time. [ARP147C]

heating load, total The rate of heat transfer, as measured by the difference between the total enthalpy of the fluid leaving the heating system and the total enthalpy of the fluid entering the heating system during a specified period of time. [ARP147C]

heating surface Includes those surfaces that are exposed to products of combustion on one side and water on the other. Measured on the side receiving the heat (as provided in the ASME Power Test Code); expressed in square feet.

heat load A source of heat generation in an aircraft compartment; or a source of heat transfer into an aircraft compartment. [ARP147C]

heat load, equipment Heat load associated with the heat generation of equipment within an aircraft compartment; usually results from the conversion of electrical energy into heat. [ARP147C]

heat load, infiltration Heat load associated with a fluid that infiltrates into an aircraft compartment at a different temperature than the compartment. [ARP147C]

heat load, latent Heat load associated with the complete change of state of a substance at a constant temperature. [ARP147C]

heat load, sensible Heat load associated with the change in temperature of a substance. [ARP147C]

heat load, solar Heat load associated with solar radiation through transparent areas, directly upon flight personnel and equipment, and upon the cabin interior surfaces. [ARP147C]

heat of fusion The increase in enthalpy accompanying the conversion of one mole, or a unit mass, of a solid to a liquid at its melting point at constant pressure and temperature. Also called latent heat of fusion. *See* latent heat.

heat of sublimation *See* latent heat.

heat of vaporization *See* latent heat.

heat rate For a thermal power plant, the ratio of heat input to work output; a measure of power plant efficiency.

heat release The total quantity of thermal energy above a fixed datum introduced into a furnace by the fuel; considered to be the product of the hourly fuel rate and its high heat value, expressed in Btu per hour per cubic foot of furnace volume.

heat resistance The property or ability of plastics and elastomers to resist the deteriorating effects of elevated temperatures. [AIR4844]

heat resistant alloys Alloys developed for very high-temperature service where relatively high stresses (tensile, thermal, vibratory, and shock) are encountered, and where oxidation resistance is frequently required. *See also* high-temperature alloy.

heat ring A ring or flange used between the duct and support to reduce the transfer of heat from the ducts to the supporting structure. [ARP699D]

heat sealing A method of joining plastic films by simultaneous application of heat and pressure to areas in contact. [AIR4844]

heat sealing adhesive A thermoplastic film adhesive that is melted between the adherent surfaces by heat application to one or both of the adjacent adherent surfaces. [AIR4844]

heat shielding *1.* The use of devices for protection from heat. *2.* Specifically, the protective structure necessary to protect a reentry body from aerodynamic heating.

heat shock Refers to a test used to determine the stability of a material by sudden exposure to a high temperature for a short period of time. [ARP1931]

heat sink *1.* The portion of the brake that provides the primary storage medium for the heat generated in deceleration of the aircraft (i.e., aircraft kinetic energy converted to heat energy). Generally includes the rotors, stators, and sometimes portions of the back plate pressure plate. [AIR1489] *2.* Any device, usually a static device, used primarily to absorb heat and thereby protect another component from damage due to excessive heat. *3.* A contrivance for the absorption or transfer of heat away from a critical element or part. [AIR4844]

heat sink loading A measure of the energy absorbed per unit weight by the brake heat sink during a single stop. Equals the total kinetic energy divided by the heat sink weight (foot pounds per pound or joules per kilogram.) [AIR1489]

heat soak Refers to the transfer of energy between the gas path and the engine hardware. [AIR4548]

heat tracing The technique of adding heat to a process or instrument measurement line by placing a steam line or electric heating element adjacent to the line. *See also* steam trace.

heat transfer The transfer or exchange of heat by radiation, conduction, or convection with a substance and between the substance and its surroundings.

heat transfer, coefficient of Heat flow per unit time across a unit area of a specified surface, under the driving force of a unit temperature difference between two specified points along the direction of heat flow. Also known as overall coefficient of heat transfer.

heat transfer coefficient, overall *See* heat transfer, coefficient of.

heat transfer coefficient, surface or film The conductance of the thin layer of fluid immediately adjacent to the surface. [ARP147C]

heat transmission Heat transmitted from one substance to another.

heat treating Term used to describe annealing, hardening, tempering, and similar processes. [AIR4844]

heat treat lot All product of the same mill form, alloy, temper, section, and size traceable to one heat treat furnace load; or, if heat treated in a continuous furnace, charged consecutively during an eight-hour period. [AMS2355F]

heat treatment Heating and cooling a solid metal or alloy in such a way as to obtain desired conditions or properties.

heavier-than-heavy key The remaining components in the bottoms stream, other than the heavy key.

Heaviside bridge A type of a-c bridge for making mutual inductance measurements when the inductance of the primary winding is already known.

heavy duty cable Generally a type of fiber-optic or electrical cable designed to withstand unfriendly conditions, such as those encountered outdoors; some varieties are armored to withstand hostile conditions.

heavy ends The fraction of a petroleum mixture having the highest boiling point.

heavy fraction The final products retrieved by distilling crude oil.

heavy key In multicomponent distillation, the component that is removable in the bottoms stream, and that has the highest vapor pressure of the components at the bottoms. If more reboiler heat is added, the heavy key component is the first component to be put in the overhead product.

heavy lift airships Airships designed to lift heavy materials.

heavy oil A viscous fraction of petroleum or coal-tar oil having a high boiling point.

heavy water Water in which the hydrogen of the water molecule consists entirely of the heavy hydrogen isotope of mass 2 (deuterium). Also known as deuterium oxide.

heel block A block or plate attached to a die which keeps the punch from deflecting too much.

HEI Acronym for high explosive incendiary projectile. [AIR4170]

height above airport (HAA) The height of the minimum descent altitude above the published airport elevation. *See* minimum descent altitude. [ARP4107]

height above landing (HAL) The height above a designated helicopter landing area used for helicopter instrument approach procedures. [ARP4107]

height above touchdown (HAT) The height of the decision height or minimum descent altitude above the highest runway elevation in the touchdown zone (the first 3000 ft of the runway). *See* decision height and minimum descent altitude. [ARP4107]

height gage A mechanical device, usually having a vernier or micrometer scale, used for measuring precise distances above a reference plane.

helical antennas Antennas used where circular polarization is required. The driven element consists of a helix supported above a ground plane.

helical stripe A continuous, colored spiral stripe applied over the outer perimeter of an insulated conductor for circuit identification purposes. [ARP1931]

helical winding In filament-wound items, a winding in which a filament band advances along a helical path, not necessarily at a constant angle, except in the case of a cylinder. [AIR4844]

helicopter A rotorcraft that depends principally on its engine-driven rotors for movement as well as lift. [ARP4107]

helicopter dynamic component The part or series of parts that transmit power from the aircraft power plant to the rotary wing and rotary rudder. [AIR4896]

helicopter impulsive noise *See* blade slap noise.

heliosphere The region around the sun in which plasma processes are dominated by solar wind.

heliostat Instrument for reflecting the sun's rays in a fixed direction; consists of mirrors moved by clockwork.

helipad A particular location on an airfield, building, or other area specifically designated as the place of helicopter landings and take-offs. [ARP4107]

heliport An area of land or water, or a structure, used or intended to be used for the landing and takeoff of helicopters. Includes the helipad and other associated structures and facilities, such as fences, terminal buildings, and fire-suppression equipment. [ARP4107]

helix A spiral winding. [ARP1931]

helix tube *See* traveling wave tube.

Hellige turbidimeter A variable-depth instrument for visually determining the cloudiness of a liquid caused by the presence of finely divided suspended matter.

Helmholtz resonator An enclosure having a small opening consisting of a straight tube of such dimension that the enclosure resonates at a single frequency determined by the geometry of the resonator.

hemeostasis The state of a body in equilibrium; the complex interaction of all systemic mechanisms whereby the body is kept in overall balance so that normal body function is carried out. [ARP171]

hemoperfusion Type of poison treatment in which the patient's blood is passed over a bed of absorbent material (e.g., activated carbon or resin) to remove the toxin from the bloodstream.

henry *1.* Unit of inductance; equal to the inductance of a circuit in which an electromotive force (emf) of one volt is produced when the current in the circuit changes uniformly at the rate of one ampere per second. [ARP1931] *2.* Metric unit for inductance. Abbreviated H.

Henry's law Gas law stating that the amount of gas dissolved in a solution is directly proportional to the pressure of the gas over the solution—a principle of physical chemistry that relates equilibrium partial pressure of a substance in the atmosphere above a liquid solution to the concentration of the same substance in the liquid. The ratio of concentration to equilibrium partial pressure equals the Henry's-law constant, a temperature-sensitive characteristic. Henry's law generally applies only at low liquid concentrations of a volatile component. *See also* gas laws.

Herbig-Haro objects Celestial objects having many of the characteristics of a T Tauri star (e.g., their spectra show a weak continuum with strong emission lines), believed to be stars in the very early stages of development. All known Herbig-Haro objects have been found within the boundaries of dark clouds.

Hering-Breuer reflex Pulmonary nerve impulses that help to regulate the depth and rate of respiration. Also called pulmonary stretch reflex. [ARP171]

hermaphroditic contact *See* contact, hermaphroditic.

hermetically sealed Describes an enclosure in which one of the walls are glass, ceramic, or metal; and the closure is a fused joint of the appropriate material, which so seals the enclosure that it does not breathe under any combination of environmental conditions. [ARP4404]

hertz (Hz) Unit of frequency equal to one cycle per second. [ARP1931]

Hessian matrices Given a real value function of N variables, an N by N symmetric of all second order partial derivatives.

heterodyne A combination of a-c signals of two different frequencies, coupled so as to produce beats whose frequency is the sum or difference of the frequencies of the original signals.

heterodyne conversion transducer A transducer in which output frequency is the sum and difference of the input frequency and the local oscillator frequency.

heterodyning Mixing two radio signals of different frequencies to produce a third signal of lower frequency, i.e., to produce beating.

heterogeneous Describes a material consisting of dissimilar constituents separately identifiable. [AIR4844]

heterogeneous radiation A beam of radiation containing rays of several different wavelengths or particles of different energies or different types.

heterojunction A junction between semiconductors that differ in doping levels and in atomic compositions; for example, a junction between layers of GaAs and GaAlAs. A double heterojunction laser contains two heterojunctions, a single heterojunction laser contains only one.

heterojunction devices Electron devices whose mode of operation is based on junctions between different semiconducting materials. The characteristics and performance of such devices are dependent on the relative lineup of the energy bands at the junctions.

heterosphere The upper portion of a two-part division of the atmosphere according to the general homogeneity of atmospheric composition; the layer above the homosphere. The heterosphere is characterized by variation in composition and mean molecular weight of constituent gases. This region starts at 80 to 100 kilometers above the earth, and therefore closely coincides with the ionosphere and the thermosphere.

heuristic *1.* Refers to a method of problem-solving in which solutions are discovered by evaluation of the progress made toward the final solution, such as a controlled trial-and error-method. *2.* An exploratory method of tackling a problem, or sequencing of investigation, experimentation, and trial solution in closed loops, gradually closing in on the solution. A heuristic approach usually implies or encourages further investigation, and makes use of intuitive decisions and inductive logic in the absence of direct proof known to the user. Thus, heuristic methods lead to solutions of problems or inventions through continuous analysis of results obtained thus far, permitting a determination of the next step. Contrast with stochastic.

heuristic program A program that monitors its performance with the objective of improved performance.

hex A number-representation system of base 16. The hex number system is very useful in cases where computer words are composed of multiples of four bits (that is, 4-bit words, 8-bit words, 16-bit words, and so on). *See* hexadecimal.

hexadecimal Number system using base 16 and the digit symbols from 0 to 9 and A to F. *See* hex.

hexagonal-head bolt A standard threaded fastener with an integral hexagon-shaped head.

hexagonal nut A hexagon-shaped fastener with internal threads used with a mating, externally threaded bolt, stud, or machine screw.

hex code A low-level code in which the machine code is represented by numbers using a base of 16.

Heydweiller bridge A type of a-c bridge circuit suitable for determining the mutual inductance between two interacting windings, both having unknown inductances.

HF *See* high frequency.

HID *See* high interstitial defect.

hidden function An item whose function is normally active and whose failure is not evident to the operating crew during performance of normal duties. [ARD50010]

hierarchical distributed control A hierarchy of computer systems in which one computer acting as supervisor controls several lower-level computers.

hierarchy Specified rank or order of items, thus, a series of items classified by rank or order.

high-alloy steel An iron-carbon alloy containing at least 5% by weight of additional elements.

high altitude flight *See* flight.

high aluminum defect (HAD) In titanium, an aluminum-rich alpha-stabilized region containing an abnormally large amount of aluminum, which may extend across a large number of beta grains. Contains an inordinate fraction of primary alpha, but has a microhardness only slightly higher than that of the adjacent matrix. [AS1814]

high-ambient lighting conditions In a transport aircraft flight deck, refers to a maximum ambient illumination of at least 86,000 lux (8000 foot candles), while in a bubble-canopy aircraft the value is at least 107,500 lux (10,000 ft candles). [ARP4032]

high brass A commercial wrought brass containing 65% copper and 35% zinc.

high-carbon steel A plain carbon steel with a carbon content of at least 0.6%.

high cycle fatigue Failure caused by cyclic loading to levels less than the material elastic limit, resulting in complete failure in more than 50,000 loading cycles. High cycle fatigue life usage is difficult to calculate as it tends to be very localized on engine components on which measurements are not made during service. [AIR1872]

high density inclusion (HDI) A region having a concentration of elements, usually tungsten or columbium, with a higher density than the matrix. Such regions are readily detectable by x-ray analysis and will appear brighter than the matrix. [AS1814]

high electron mobility transistors A recently developed field-effect transistor whose mode of operation is based on the technique of modulation doping of GaAs/Al(x)Ga(1-x) as heterojunctions. This technique achieves high mobility in part by introducing carriers into high purity GaAs from donor ions in an adjacent AlGaAs layer, the electrons and ions being separated by the built-in heterojunction potential.

high energy fragments Any fragment that has sufficient energy to pass through a shield of soft aluminum with a thickness of 0.040 in. maximum during the test demonstration. [AIR1639]

highest phase voltage limiting A means of limiting the highest-phase voltage of the alternator during any unbalanced load condition. [ARP1148A]

high frequency (HF) The radio frequency band between 3 and 30 MHz. [ARP4107]

high-frequency bias A sinusoidal signal that is mixed with the data signal during a magnetic tape direct recording process; purpose is to increase the linearity and dynamic range of the recording medium. Bias frequency is usually three to four times the highest data frequency to be recorded.

high frequency bondtester Wet-coupled bondtester generally operating at frequencies above 100 kHz. [ARP5089]

high-frequency heating The heating of materials by dielectric loss in a high-frequency electrostatic field. [AIR4844]

high gas pressure switch A switch to stop the burner if the gas pressure is too high.

high-intensity runway light system (HIRL)
See runway edge light system.

high interstitial defect (HID) Interstitially stabilized alpha phase region of substantially higher hardness than surrounding material. Arise from very high local nitrogen, oxygen, or carbon concentrations, which increase the beta transus and produce the high-hardness, often brittle, alpha phase. The boundaries of HIDs are always diffused. [AS1814]

high-level data link control (HDLC) A type of data link protocol.

high-level human interference (HLHI) A device that allows a human to interact with the total distributed control system over the shared communications facility.

high-level language(s) *1.* Computer languages in which instructions or statements each correspond to several machine language instructions. *2.* A programming language in which statements are translated into more than one machine-language instruction. Examples of high-level languages are BASIC, FORTRAN, COBOL, and TELEVENT.

high-level operator interface (HLOI) A type of HLHI designed for use by a process operator.

high-lift device Any device, such as a flap, slat, or boundary-layer-control device, used to increase the maximum lifting capacity (or maximum lift coefficient) of a wing. [ARP4107]

high-low bias test *See* marginal check.

high order Pertaining to the weight or significance assigned to the digits of a number, e.g., in the number 123456, the highest-order digit is 1, the lowest-order digit is 6. The term "higher-order" may also be used to refer to the bits of a binary word.

highpass filter *1.* Wave filter having a single transmission band extending from some critical or cutoff frequency—not zero—up to infinite frequency. *2.* A filter that passes high frequencies above the cutoff frequency, with little attenuation.

high pitch Refers to the setting of a propeller blade at a high angle (other than the feather angle) relative to the plane of rotation. The specific angle considered to be "high pitch"

may vary from one model of propeller to another. High pitch settings are used to conserve fuel by reducing engine rpm when power requirements are low to moderate. [ARP4107]

high power laser Stimulated emission device having high-energy flux-density outputs.

high-pressure boiler A boiler furnishing steam at pressure in excess of 15 pounds per square inch; or hot water at temperatures in excess of 250° F or at pressures in excess of 160 pounds per square inch.

high-pressure laminates Laminates molded and cured at pressures not lower than 6.9 MPa (1.0 ksi), and more commonly in the range of 8.3 to 13.8 MPa (1.2 to 2.0 ksi). [AIR4844]

high pressure laminating A term usually reserved for matched-die molding, typically done under high pressures in a press. [AIR4844]

high pressure molding A molding process in which the pressure used is greater than 6.9 MPa (1.0 ksi) [AIR4844]

high pressure stored gas system System consisting of four elements: (a) the fluid medium or gas; (b) a container; (c) an actuator to open the container upon command; and (d) a pressure reducer, which reduces the container gas pressure down to the required system discharge level. Simpler systems may operate without the use of a pressure reducer, thereby providing unregulated output.

high-rate discharge The withdrawal of large currents for short intervals of time, usually at a rate that would completely discharge the cell or battery in less than a given period of time. [ARP4386]

high resolution graphics A finely defined graphical display on a computer monitor screen.

high Reynolds number A Reynolds number above the critical Reynolds number of a sphere.

high speed flight *See* flight.

high-speed taxiway/exit/turnoff A long-radius taxiway provided with lighting or marking to define the path of aircraft, traveling at speeds up to 60 knots, from the runway center to the point on the center of a nearby taxiway that is clear of the active runway. The high-speed taxiway is designed to expedite aircraft turning off the runway after landing, thus reducing runway occupancy time. [ARP4107]

high-strength alloy A metallic material having a strength considerably above that of most other alloys of the same type or classification.

high strength alloy conductor A conductor that shows a maximum of 20% increase in resistance and a minimum of a 70% increase in breaking strength over the equivalent construction in pure copper, while exhibiting a minimum elongation of 5% in 10 inches. As required, the alloy should be capable of sustaining continuous exposure to temperatures as high as 300°C without suffering an appreciable permanent change in properties. [ARP1931]

high-temperature alloy A metallic material suitable for use at 500°C (930°F) or above. This classification includes iron-base, nickel-base and cobalt-base superalloys, and the refractory metals and their alloys. These materials retain enough strength at elevated temperature to be structurally useful and generally do not undergo metallurgical changes that weaken or embrittle them. *See also* heat resistant alloy.

high-temperature hot-water boiler A water-heating boiler operating at pressure exceeding 160 psig or temperatures exceeding 250°F.

high temperature, pressure sensitive tape Tape used for a variety of applications in composite fabrication processes. [AIR4844]

high temperature superconductor Superconducting material consisting of mixed metal oxide ceramics which maintain their superconductivity at higher temperature ranges (above 24 K) than the more traditional superconductors.

high tension system Ignition system capable of delivering voltages in excess of 5 kV to the firing tip of the spark igniter. [AIR784]

highway in the sky A pictorial representation of the path that the pilot is to fly, generally presented on a forward-looking vertical situation display. [ARP4107]

hinge line A line corresponding to the axis about which a control surface is deflected. [ARP4386]

hinge moment The product of the aerodynamic force acting upon the center of pressure of the control surface and perpendicular distance from the center of pressure to the hinge line. [ARP4386]

HIP (process) *See* hot isostatic pressing.

HIRL *See* runway edge light system.

hiss Random noise in the audiofrequency range, having subjective characteristics analogous to prolonged sibilant sounds.

HLHI *See* high-level human interference.

HLOI *See* high-level operator interface.

hoarfrost *See* ice, hoarfrost.

hohlraums In radiation thermodynamics, cavities whose walls are in radiative equilibrium with the radiant energy within the cavity.

hoist A mechanical device consisting of a supporting frame and integral mechanism specifically designed to raise or lower a load by tensile force. [ARP480A]

holdback A device or assembly designed for restraining an aircraft during engine runup. [AIR1489]

holdback and release assembly A device or assembly designed to restrain an aircraft against engine thrust, ship motion, and tensioning forces prior to catapulting. Once the catapult is fired, the holdback and release assembly is designed to release at a specific load, usually by means of a frangible link (tension bar), thus permitting the aircraft to be accelerated by the catapult. [AIR1489]

holdback bar A rigid holdback and release assembly used to restrain the aircraft prior to launch, specifically from a nose-gear launch catapult system. [AIR1489]

holdback, engine run up A device or assembly designed to restrain an aircraft during engine trim procedures and full-power checkouts. [AIR1489]

holdback fitting, aircraft The attachment point or device on an aircraft to which the holdback is coupled for engine runup or for catapulting. [AIR1489]

holdback, repeatable A device or assembly designed for catapult holdback and release, which may be used repeatedly without dependence on or replacement of a frangible link. [AIR1489]

hold down A device or assembly that is part of the arresting-hook installation of an aircraft. Designed to hold down the hook after deployment and contact with the deck (runway) to prevent hook bounce and insure engagement with the arresting gear cable. *See also* dashpot and snubber. [AIR1489]

hold/holding procedure A predetermined maneuver that keeps aircraft within a specified airspace while awaiting further clearance from air traffic control. [ARP4107]

holding beam An electron beam for reactivating the charge on the surface of an electronic device.

holding fix A specified fix, identifiable to a pilot by NAVAIDs or visual reference to the ground, that is used as a reference point in establishing and maintaining the position of an aircraft while holding. [ARP4107]

holding pump A small mechanical pump automatically valved to the foreline of the diffusion pump at all times when the main mechanical pumps are serving elsewhere. [AMS2769]

holdover time The estimated time that anti-icing fluid will prevent the formation of frost or ice and the accumulation of snow or slush on the protected surfaces of an aircraft. [AIR4737]

hold time In any process cycle, an interval during which no changes are imposed on the system. Hold time is usually used to allow a chemical or metallurgical reaction to reach completion, or to allow a physical or chemical condition to stabilize before proceeding to the next step.

hole burning A laser process that depletes, spatially or spectrally, the electron/hole pair density in a region of space (in the case of spatial hole burning) or frequency of high coherent light (in the case of spectral hole burning).

hollow A center void with a diameter that is greater than 10% of the diameter of the shot particle. [AMS2431/2A]

hollow cathode An effect created by gaps, holes, or cavities in parts, resulting in an intense localized glow discharge. [AMS2759/8]

hologram Three-dimensional photograph or image produced by interference between two sets of coherent light waves. [AIR4844]

holographic (diffractive) collimating optics Refers to the use of one or several holograms for collimation/superimposition. [ARP4102/8]

holographic diffraction grating Diffraction grating in which the pattern of light-diffracting lines has been recorded holographically rather than mechanically ruled into the surface.

holographic optical elements Holograms that have been made to diffract light in the same pattern as other optical components; for example, through computer synthesis, a hologram can be produced that mimics the function of the lens. In some applications, such holographic optical elements are less costly than conventional optics.

holographic subtraction A holographic technique by which two dissimilar optical fields can be subtracted to yield only their difference.

homing Following a path of energy waves to or toward their source or point of reflection.

homogeneous Describes a material that is of uniform composition throughout. [AIR4844]

homogeneous glass Glass of essentially uniform composition throughout its structure. [ARP798]

homogeneous radiation A beam of radiation containing rays whose wavelengths all fall within a narrow band of wavelengths; or containing particles of a single type having about the same energy.

homojunction 1. Solar cell in which both sides are made of the same material. 2. A junction between semiconductors that differ in doping levels but not in atomic composition; for example a junction between n-type and p-type GaAs.

homologous pair In optical spectroscopy, two lines so chosen that the ratio of their radiant powers changes minimally with variations in the input conditions.

homopolar generator Rotating electric machine for converting mechanical power into pure direct current; this is accomplished by utilizing poles that have the same polarity as the armature.

homopolymerized A condition whereby a monomeric material is polymerized only with itself. [AIR4844]

homosphere The lower portion of a two-part division of the atmosphere according to the general homogeneity of atmospheric composition; below the heterosphere. The homosphere exhibits no gross change in atmospheric composition. It includes all of the atmosphere from the earth's surface to about 90 kilometers. The homosphere is about equivalent to the neutrosphere, and includes the troposphere, stratosphere, and mesosphere; it also includes the ozonosphere and at least part of the chemosphere.

honeycomb Manufactured product of resin-impregnated sheet material or metal foil, formed into hexagonal-shaped cells. [AIR4844]

honeycomb core Lightweight strengthening material of structures resembling honeycomb meshes.

honeycomb sandwich assembly A structural composition consisting of relatively dense, high-strength facings bonded to a lightweight, cellular honeycomb core. [AIR4844]

hook, downlock A hook designed to engage a roller or restraining device and lock the gear assembly in the down position, ready for touchdown and ready to resist applied ground loads. [AIR1489]

Hookean behavior In liquid expansion, a condition in which the fractional change in volume is proportional to the hydrostatic stress, if under such stress the liquid evidences ideal elastic behavior.

hook gage An instrument consisting of a pointed metal hook mounted on a micrometer slide, used to measure the level of a liquid in an evaporation pan. The level with respect to a reference height is determined when the point of the hook just breaks the liquid surface.

hook point *1.* The point of a hook as designed for proper engagement or operation in a mechanism. *2.* In an arresting hook installation, the portion of the arresting hook assembly that engages the arresting gear cable. Usually detachable from the arresting hook shank for wear-replacement purposes. [AIR1489]

hook tongue terminal A terminal with a hook-shaped tongue. [ARP914A]

hookup *1.* The process or operation of preparing an aircraft for launch from a catapult. Includes engagement of the holdback and attachment of the catapult bridle or launch bar. *2.* The condition in which the aircraft is spotted on the catapult ready for tensioning. [AIR1489]

hook, uplock A hook designed to engage a roller or restraining device and lock the gear and/or door assembly in the up or stowed position, in a safe condition for all modes of flight maneuvering of the aircraft. [AIR1489]

hoop stress Circumferential stress in a material of cylindrical form subjected to internal or external pressure. [AIR4844]

horizon The great circle of the celestial sphere midway between the zenith and nadir, or a line resembling or approximating such a circle.

horizon misplacement *See* illusions, visual.

horizontal boiler A water-tube boiler in which the main bank of tubes are straight and on a slope of 5 to 15 degrees from the horizontal.

horizontal branch stars Horizontal strips of stars on the Hertzsprung-Russell diagram of globular clusters, to the left of the red giant branch.

horizontal flight A flight in which the aircraft maintains constant altitude above sea level. [ARP4386]

horizontal orientation The attitude of an object in reference to the plane perpendicular to the direction of gravity.

horizontal return-tubular boiler A fire-tube boiler consisting of a cylindrical shell, with tubes inside the shell attached to both end closures. In such a boiler, the products of combustion pass under the bottom half of the shell and return through the tubes.

horn antenna *1.* Antenna shaped like a horn. *2.* The flared end of a radar waveguide, the dimensions of which are chosen to give efficient radiation of electromagnetic energy into the surrounding environment.

hose Tubing—either flexible by construction or made of flexible material—that is attached to the ends of adjacent ducts, tubes, or fittings. [ARP699D]

hose assembly A length of hose with a coupling attached to one or each end. [AS1933]

hose assembly date The date the hose end fittings are applied to the hose to form a hose assembly. [AS1933]

hose end ball valve A ball valve mounted to the inlet of the nozzle to allow the strainer to be checked without disconnection of the unit, and with minimum or no spillage of fuel when fitted with a strainer in the ball bore. [AIR4783]

hose end control valve or regulator A direct-acting pressure-control device that is mounted as an integral part of the underwing nozzle. [AIR4782]

hose end pressure control valve A direct-acting pressure regulator that is mounted on the inlet of the nozzle to limit pressure at its outlet and control surge pressure limits in the downstream aircraft manifold system. [AIR4783]

hose reel A device used to store hose and to allow the hose to be used easily in either a full- or partial-length condition. [AIR4783]

host (computer) *1.* The primary computer in a multielement system; the system that issues commands, has access to the most important data, and is the most versatile processing element in a system. Compare with target computer. *2.* Condition-monitoring system software that performs all equipment-independent functions. [ARD50002]

host equipment Individual life preservers, life rafts, and slide/rafts and other survival equipment to which a survivor-locator light may be fitted. [AS4492]

hot atoms Atoms with high internal or kinetic energy as a result of a nuclear process such as beta decay or neutron capture.

hot cathode Cathode that functions primarily by the process of thermionic emission.

hot cathode fluorescent lamps Lamps that employ coiled tungsten filaments as electrodes. The filaments are coated with one or more of the alkaline earth oxides. This electron-emissive coating provides an abundance of free electrons when hot. By suitable circuit arrangements, these cathodes can be heated to a satis-

factory electron-emitting temperature before the arc is struck (preheat, trigger or rapid start); or they may be required to act momentarily as cold cathodes until heated by the electron stream after starting (instant start or slimline). [AIR512B]

hot corrosion Corrosion at high temperatures as a result of the reduction of protective oxide coatings and scales and the subsequent accelerated oxidation.

hot dip A method of coating in which the item to be coated is immersed in a molten bath of the coating material. [ARP1931]

hot dip galvanizing A process for rust-proofing iron and steel products by the application of a coating of metallic zinc.

hot dipping A process for coating parts by briefly immersing them in a molten metal bath, then withdrawing them and allowing the metal to solidify and cool.

hot finishing Final finishing of metal above the recrystallization temperature and resulting in no strain hardening. [ARP1917]

hot gas Gas at a temperature over 2500°F (1400°C); typical of gases used in rocket motors for propulsion. [ARP4386]

hot head tape layer A computer-controlled, automated tape-placement process for thermoplastic prepregs. In this process a gantry-mounted hot shoe is used to partially consolidate each consecutive ply in a programmed orientation and laminate size. [AIR4844]

hot hydrant valve A hydrant valve whose inlet shutoff portion has been left open or whose inlet shutoff valve seals are leaking sufficiently to prevent reduction of the pressure in the cavity between the inlet shutoff valve and the outlet adapter. (Pressure reduction is normally accomplished by compressing the pressure equalization valve in the outlet adapter poppet.) [AIR4783]

hot isostatic pressing (HIP) A process for fabricating certain metal-matrix composites. [AIR4844]

hot-melt adhesive An adhesive that is applied in a molten state and forms a bond after cooling to a solid state. [AIR4844]

HOTOL launch vehicle A British, unmanned (or later, manned), horizontal-takeoff and landing, single-stage-to-orbit launch vehicle.

hot parts Engine parts exposed to a hot gas stream, such as the combustion chamber, turbine components, augmentor, and exhaust nozzle. [AIR4896]

hot parts life The interval over which hot parts operate without repair or replacement.

hours, APU The auxiliary power unit-operating time from startup to shutdown. [AIR4896]

hot-setting adhesive An adhesive that requires a temperature of 100°C (212°F) or above to set. [AIR4844]

hot start An engine start during which allowable engine turbine gas temperature limits are exceeded. [ARP906A]

hot tear A fracture formed in a metal during solidification because of hindered contraction; usually forms on the surface of a part. [ARP4784]

hot well *See* well.

hot/wet properties The mechanical properties required of a composite or bonded-metal assembly under prescribed conditions of temperature, time, and relative humidity or water immersion. [AIR4844]

hot-wire ammeter *See* thermoammeter.

hot-wire instrument A measuring device that depends on the heating reaction of a wire carrying a current for its operation.

hot working Any form of mechanical deformation processing carried out on a metal or alloy above its recrystallization temperature but below its melting point. [AIR4844]

hours, block The number of hours incurred by an aircraft from the moment it first moves for a flight until it comes to rest at its intended blocks at the next point of landing, or returns to its departure point prior to takeoff. [AIR4896]

hours, flying The accumulated time intervals between wheels-off and wheels-on. [AIR4896]

hours, out of service The number of elapsed hours that an aircraft is not available for operation when scheduled to be available. [AIR4896]

hours, unit flying The accumulated flying hours of all like units installed in aircraft during a specified reporting period. [AIR4896]

housekeeping Administrative or overhead operations or functions that are necessary in order to maintain control of a situation. For a computer program, for example, housekeeping involves the setting up of constants and variables to be used in the program. Synonymous with "red tape."

housing The outer component of an assembly designed to enclose, support, and/or protect the internal mechanism. [ARP480A]

housing, electrical connector The portion of a connector into which the insert is assembled. Also called the shell. [ARP914A]

hover Maintenance of a relatively stable position over a point on the surface, at a particular height above the surface; a capability of helicopters, hovercraft, hummingbirds, and many insects. [ARP4107]

hover check Indicates that a helicopter/VTOL aircraft requires a stabilized hover to conduct a performance/power check prior to hover taxi, air taxi, or takeoff. The altitude of the hover will vary based on the purpose of the check. [ARP4107]

hover taxi Describes the movement of a helicopter/VTOL aircraft conducted above the surface and in ground effect at airspeeds less than approximately 20 knots. The actual height may vary, and some helicopters may require hover taxi above 25 ft AGL to reduce ground-effect turbulence or provide clearance for cargo slingloads. [ARP4107]

HPLC Acronym for high-performance liquid chromatography.

HSI Acronym for horizontal situation indicator.

HTPB propellant Solid rocket propellant containing hydroxyl-terminated polybutadiene (HTPB) as bonding material.

hub The inner portion of the rotor. [AIR1639]

Hubble constant The rate at which the velocity of recession of the galaxies increases with distance.

HUD Stands for head up display; a method of presenting images to the pilot of an aircraft while he/she is looking forward through the windscreen. These images are generated from a device out of the pilot's field of view and reflected from a transparent surface in front of the pilot. Thus, the pilot looks through this transparent surface and sees the information superimposed on his/her vision of the real world. [ARP4107]

Huggenberger tensometer A magnifying extensometer that employs a compound lever system to intensify by about 1200 times the changes taking place in a 10 to 20 mm gage length.

hum Electrical disturbance at the power-supply frequency, or harmonics thereof.

human-computer interface The physical and logical interface between a user and a computer, including keyboards, mouse devices, monitors, display formals, and communication/command logic.

human error reliability criteria Criteria used in the design of a complex system to adapt its physical features to the response characteristics of the person who is ultimately charged with its operation. [AIR4896]

humidistat An instrument for measuring and controlling relative humidity.

humidity A measure of the water vapor content of the air. [AIR1335]

humidity, absolute *See* absolute humidity.

humidity, relative *See* relative humidity.

humidity, specific *See* specific humidity.

humidity test A corrosion test for comparing relative resistance of specimens to a high-humidity environment at constant temperature.

hung start The operation of starting a gas-turbine engine wherein the combined torque output of the engine and starter are insufficient to provide positive acceleration during an attempt to start. [ARP906A]

hunting A term applied to the undesirable oscillation of a control device resulting in a poor degree of control. [ARP147C]

Huygens principle A very general principle applying to all forms of wave motion. States that every point on the instantaneous position of an advancing phase front (wave front) may be regarded as a source of secondary spherical wavelets; the position of the phase front a moment later is then determined as the envelope of all the secondary wavelets (ad infinitum).

hybrid *1*. A composite laminate consisting of laminae of two or more composite material systems. [AIR4844] *2*. A combination of two

or more integrated circuits (ICs) in one package. *3*. A combination of an analog and digital computer. [ARP89]

hybrid cable A multiconductor cable containing two or more types of conductors. Such a cable could include optical-fiber and electrical conductors. [ARP1931]

hybrid computer A computer for data processing using both analog representation and discrete representation of data.

hybrid power controller A device utilizing solid-state components in conjunction with electromechanical switching contractors to perform a power-control function. [AS1831]

hybrid structure An assembly constructed of interconnected, rigid and flexible structural shapes; designed to sustain dynamic, static, and other loads.

hybrid T Series T and shunt T junctions located at the same point in a waveguide, and designed to restrict energy flow to specified channels.

hydrant cart A vehicle, usually limited to a towable unit that is not self-propelled, which contains the necessary equipment to allow for connection to the hydrant valve and the aircraft refueling adapter.

hydrant coupler The mechanical quick connection between the hydrant servicer and the hydrant valve. [AIR4782]

hydrant pit A covered, small chamber containing the hydrant system pipe terminus, hydrant valve, and associated equipment. The hydrant pit is strategically located to provide access to the aircraft underwing refueling adapter. [AIR4782]

hydrant pit lid The cover or lid to a hydrant pit. [AIR4783]

hydrant servicer A vehicle, self-propelled or towable, which contains the necessary equipment to allow for connection to the hydrant valve and the aircraft refueling adapter. [AIR4783]

hydrant system System that provides a consistent source of fuel from a remotely located storage facility or tank farm to the refueling apron or parking positions of the various aircraft to be refueled. [AIR4783]

hydrant valve The terminus of the hydrant system piping system, which provide (as a minimum) quick-coupling capability with a hydrant coupler, and may provide controls. [AIR4782]

hydraulic Describes any device, operation, or effect that uses pressure or flow of oil, water, or any other liquid of low viscosity.

hydraulic actuator *See* actuator, hydraulic.

hydraulically actuated valves Valves in which an incompressible fluid, such as hydraulic oil or fuel, is utilized for operation. Includes valves that are solenoid-controlled, but hydraulically actuated. [ARP986]

hydraulic amplifier A fluid-valving device that acts as a power amplifier; for example, a sliding spool, a nozzle/flapper, or a jet pipe with receivers. [ARP4386]

hydraulic booster The use of hydraulic-power actuation to reduce the pilot effort needed for control of a vehicle. With a hydraulic booster, the actuator output force is in direct proportion to the manual input force. [ARP4386]

hydraulic circuit A fluid-flow circuit that operates somewhat like an electric circuit.

hydraulic coupling *See* fluid coupling.

hydraulic engineering A branch of civil engineering that deals with the design and construction of such structures as dams and other flood-control devices; sewers and sewage-disposal plants; water-driven electric power stations; and water treatment and distribution systems.

hydraulic fluid A light oil or other low-viscosity liquid used in a hydraulic circuit.

hydraulic gage A gage designed for service at extremely high pressure.

hydraulic power A source of mechanical power, usually rotary, driven hydraulically.

hydraulic power transfer unit A device that transfers power by mechanical interconnection, but does not transfer fluid from one system to another. The power may be either unilateral or bilateral. [ARP578]

hydraulic press A press in which the molding force is created by the pressure exerted by a fluid. [AIR4844]

hydraulic starter A device for converting fluid energy into rotary mechanical energy, intended to provide continuous torque for engine-starting purposes, and usually incorporating a suitable mechanism to connect the starter to the engine during starting cycles only. [ARP4386]

hydraulic starter, fixed displacement A starter whose displacement remains constant under all operating conditions and which theoretically requires a constant volume of fluid-per-revolution. [ARP4386]

hydraulic starter/pump A rotary device capable of converting fluid energy into mechanical energy when operating as a starter, and capable of converting mechanical energy into fluid energy when operating as a pump. [ARP4386]

hydraulic starter, variable displacement A starter whose displacement during the starting cycle is automatically controlled, usually to limit the flow rate to a predetermined maximum. [ARP4386]

hydraulic system A closed system using the same fluid throughout the entire circuit; and composed of components and elements to generate, transmit, and control hydraulic power, as well as convert fluid energy to mechanical power. [ARP578]

hydraulic utility system An aircraft hydraulic system whose basic purpose is to provide hydraulic power for operation of equipment not part of primary subsystems used for flight control. [ARP4386]

hydride phase The phase formed in titanium when the hydrogen content exceeds the solubility limit. Hydrogen and, therefore, hydrides tend to accumulate at areas of high stress. [AS1814]

hydroblast Refers to a test in which a fluid is used to load a pressure vessel to failure. [AIR4844]

hydroburst Refers to a procedure used for burst-testing of tires in which water is used as the pressurizing medium rather than air. (This avoids the explosive decompression effect of air.) [AIR1489]

hydroclave An autoclave that uses water as its pressurizing medium. [AIR4844]

hydrocracker A chemical reactor in which large hydrocarbon molecules are fractured in the presence of hydrogen.

hydrocracking Technique for the catalytic conversion of coal into liquid fuels.

hydrodynamic coefficients The factors producing motions of objects floating in liquids.

hydrodynamic ram *1.* The pressure loads that develop in a fluid-filled tank as a result of shock loading. *2.* The penetration and travel of a projectile through a fluid. [AIR1664]

hydrodynamic ram effect The physical effect (force) transmitted to the walls of a liquid-filled container by the action of a projectile penetrating the container and transferring its energy (as kinetic energy) to the liquid; the fluid, in turn, transfers this kinetic energy to the walls of the container, causing excessive structural damage.

hydrodynamics Science of fluid motion.

hydroelectricity Electric power produced by water power, using water wheels, turbo-generators, or other conversion equipment.

hydrogen bonding A mechanism for intermolecular attraction and therefore adhesion. [AIR4844]

hydrogen chloride laser *See* HCL laser.

hydrogen damage Any of several forms of metal failure caused by dissolved hydrogen, including blistering, internal void formation, and hydrogen-induced delayed cracking.

hydrogen embrittlement A decrease in fracture strength of metals due to the incorporation of hydrogen in the metal lattice.

hydrogen engine Internal combustion engine utilizing gaseous hydrogen as the fuel.

hydrogen masers A stimulated emission device in which hydrogen gas provides an output signal with a high degree of stability and spectral purity.

hydrogen metabolism The physical and chemical processes by which an organism transforms the complex hydrogen components of foodstuffs into simple hydrogen compounds by dissimilation and catabolism in the production of energy.

hydrogen oxygen engine Engine using liquid hydrogen as fuel and liquid oxygen as oxidizer.

hydrogen production Production of hydrogen for fuel purposes by photosynthetic, chemical, electrical, thermal, electrochemical, or other means.

hydrokineter A device for recirculating or causing flow of water by the use of a jet of steam or water at higher pressure than the water caused to flow.

hydrology model Mathematical or physical representation by which the circulation, distribution, and properties of the waters of the earth can be studied.

hydrolysis Chemical decomposition of a substance involving the addition of water. [AIR4844]

hydrolytic stability Refers to the tendency of fluids to be chemically affected by the presence of water. Instability effects are generally accelerated by elevated temperature. [AIR1116]

hydromagnetics *See* magnetohydrodynamics.

hydromatic propeller A constant-speed, hydraulically operated propeller. [ARP4107]

hydromechanical logic Logic for mode switching or failure detection and correction, performed only with mechanical elements, using hydraulic information in the form of hydraulic pressures or flows. [ARP4386]

hydromechanical press A press in which the molding forces are created partly by a mechanical system and partly by a hydraulic system. [AIR4844]

hydrometer An instrument for directly indicating the density or specific gravity of a liquid.

hydrophilic Having an attraction for water; capable of adsorbing or absorbing water. [AIR4844]

hydrophilic emulsifier A water-soluble detergent concentrate used with the post-emulsifiable penetrants. [AMS2647A]

hydrophobic Capable of repelling water; poorly wetted by water. The opposite of hydrophilic. [AIR4844]

hydrophone *1.* Microphone suitable for use in water or other liquid. *2.* A transducer that reacts to water-borne sound waves.

hydroplaning For a pneumatic tire, refers to an operating condition in which the water on a wet runway or surface is not displaced from the nominal tire-ground contact area by the rolling tire or by a moving but non-rotating (full skidding) tire at a rate fast enough to allow the tire to make contact with the ground surface over its complete nominal footprint area, as would be the case on a dry ground surface. When hydroplaning occurs, the tire rides on a wedge or film of water over a part or all of its footprint area, depending upon conditions. [AIR1489]

hydroplaning, dynamic The process of hydroplaning associated with movement of a body over water; to skim over water at high speed. *See* hydroplaning. [AIR1489]

hydroplaning speed The speed at which hydroplaning occurs. Tire hydroplaning develops progressively from low speed through an intermediate range, over which the tire progressively loses contact with the runway; finally, at sufficient forward speed, total hydroplaning develops. [AIR1489]

hydroplaning, total A hydroplaning situation in which the total tire footprint (contact area) is supported by a film of water. [AIR1489] *See also* hydroplaning, viscous.

hydroplaning, viscous Another term for total hydroplaning, i.e., the tire contact footprint is supported by a film of viscous fluid (water). *See* hydroplaning, total. [AIR1489]

hydropneumatic Describes a device that is operated by both liquid and gas power.

hydroponics Growing of plants in a nutrient with the mechanical support of an inert medium such as sand.

hydroproof Refers to a test in which a fluid is used to carry out a proof loading requirement for a pressure vessel. [AIR4844]

hydropyrolysis A coal-to-liquid process in which bituminous coal, lignite, tars, sand, and related materials are rapidly heated to 1000–1100 K in pressurized hydrogen gasification reactors to generate pure methane.

hydroscopic Describes any material that easily absorbs and retains moisture.

hydro ski A type of landing gear designed to enable an aircraft to take off and land on water. Also called a water ski. [AIR1489]

hydrosphere (earth) *See* earth hydrosphere.

hydrostatic head The pressure created by a height of liquid above a given point.

hydrostatic-head gage A pressure gage that is unique from others in that the graduation of the pressure scale is usually in feet.

hydrostatics Science dealing with the characteristics of fluids at rest.

hydrostatic test Determining the burst resistance or leak tightness of a fluid component or system by imposing internal pressure.

hydrothermal stress analysis The evaluation of the combined effects of temperature-humidity cycling.

hydrothermal system An energy system utilizing hot water from geysers, hot springs, solar heating, and other sources.

hydrox engine *See* hydrogen oxygen engine.

hygral properties The affinity of a substance or material for moisture.

hygrometer An instrument for directly indicating humidity.

hygrometry Any process for determining the amount of moisture present in air or another gas.

hygroscopic Capable of attracting, absorbing, and retaining atmospheric moisture. [AIR4844]

hygrothermal effect Refers to a change in properties due to moisture adsorption and temperature change. [AIR4844]

hygrothermograph An instrument that records both temperature and humidity on the same chart.

hyperbolas Open curves with two branches, all points of which are at a constant difference in distance from two fixed points called focuses.

hyperbolic navigation Radio navigation in which a hyperbolic line of position is established by signals received from two stations at a constant time difference.

hypercapnia Excess carbon dioxide in the blood and body fluids, usually causing increased respiration. [ARP171]

hypercube multiprocessor A distributed-memory, message-passing multiprocessor designed to reduce the number of interconnections compared to the number of processors. Other simple geometries such as rings, meshes, or trees of processors can be embedded in hypercubes.

hypergolic Describes a multi-phase propellant system that will spontaneously ignite upon mixing of the constituents. [AIR913]

hypergolic fuel Fuel that will spontaneously ignite with an oxidizer; for example, aniline with fuming nitric acid. [ARP4386]

hyperkinesia Excessive exercise, often accompanied by uncontrollable muscular movement.

hypermodel A behavioral model of the analog-digital interface in a mixed-mode simulator.

hyperon In the classification of subatomic particles according to mass, the heaviest of such particles. Some large and highly unstable components of cosmic rays are hyperons.

hyperoxia A condition in which the total oxygen content of the body is increased above that normally existing at sea level.

hyperpnea *See* respiration, types of. [ARP171]

hypersonic Of or pertaining to speeds equal to or in excess of five times the speed of sound. [ARP4386]

hypersonic flow In aerodynamics, flow of a fluid over a body at speeds much greater than the speed of sound, and in which the shock waves start at a finite distance from the surface of the body.

hypersonic glider Unpowered vehicle, specifically reentry vehicle, designed to flow at hypersonic speeds.

hypersonics That branch of aerodynamics that deals with hypersonic flow.

hypervelocity Extremely high velocity. Applied by physicists to speeds approaching the speed of light, but generally implies speeds on order of satellite speeds and greater.

hypocapnia *See* alkalosis, respiratory.

hypopnea *See* respiration, types of. [ARP171]

hypotension Low arterial blood pressure. [ARP171]

hypoxia A state of oxygen deficiency in the body sufficient to impair functions of the brain and other organs.

hypsometer An instrument that determines elevation above a reference plane (such as sea level) by measuring the boiling point of a liquid, and from that measurement finding atmospheric pressure.

hysteresimeter A device for measuring a lagging effect related to physical change, such as the relationship between magnetizing force and magnetic induction.

hysteresis *1.* The difference between the value of the control parameter that will switch the element in one direction and the value that will switch it in the opposite direction. [ARP993] *2.* The difference in command inputs required to produce the same actuation system output as the output is cycled through the full plus or minus range of travel. The cycling rate must be significantly below the control bandpass so that velocity error signals are not included in this parameter. [ARP4386] *3.* Any of several effects resembling a kind of internal friction, accompanied by the generation of heat within the substance affected. *4.* The delay of an indicator in registering a change in a parameter being measured. *5.* A phenomenon demonstrated by materials whose behavior is a function of the history of the environment to which they have been subjected. *6.* The tendency of an instrument to produce a different output for a given input, depending on whether the input resulted from an increase or decrease from the previous value. *7.* The lagging in the response of a unit of a system behind an increase or a decrease in the strength of signal.

Hz *See* hertz.

I

I *See* moment of inertia.

IAE *See* integral absolute error.

IAF *See* initial approach fix.

IAS *See* indicated airspeed.

IC *See* integrated circuit.

ICAO *See* International Civil Aviation Organization.

ICE *See* in-circuit emulation.

ice crystals Frozen, supercooled water droplets. May exist in the form of snow crystals or ice nodules such as sleet. [AIR1667]

ice detector A device that provides an indication that the aircraft is operating in atmospheric conditions that will or may cause ice to be accreted. [AIR4367]

ice, freezing rain Precipitation in the form of large, above-freezing water droplets, which become supercooled and freeze upon contact with a below-freezing surface, within a below-freezing air mass. [AIR1667]

ice, glaze or clear Transparent ice formed during flight in clouds by the slower freezing of supercooled water droplets; this is most likely to occur at ambient temperatures near freezing when the droplets may flow along the surface or remain liquid before freezing occurs. The ice formed during freezing rain is also an example of glaze ice. This type of ice may occur at conditions above the Ludlam Limit, reducing the apparent LWC (liquid water content). [AIR1667]

ice, glime A mixture of glaze and rime, generally with rough surfaces and runback ice. [AIR1667]

ice, hoarfrost Ice crystals deposited directly from water vapor onto surfaces that are below freezing. [AIR1667]

ice pellets Precipitation of transparent or translucent pellets of ice that are spherical or irregular, rarely conical, having a diameter of 1/5 of an inch or less. Ice pellets are subdivided into two main types: (a) frozen raindrops or snowflakes that have largely melted and then refrozen—the freezing process usually taking place near the ground; and (b) pellets of snow encased in a thin layer of ice, which has formed from the freezing either of droplets intercepted by the pellets or of water resulting from the partial melting of the pellets. [AIR1335]

ice prisms A fall from the air of unbranched ice crystals, in the form of needles, columns, or plates—often so tiny that they seem to be suspended in the air. [AIR1335]

ice, rime Opaque ice formed during flight in clouds by the rapid freezing of small, supercooled water droplets, producing a streamlined spear shape. This type of ice occurs below the Ludlam Limit. [AIR1667]

ice-warning indicator An instrument that detects the presence of ice on the aircraft or of icing conditions. [ARP4107]

icing The formation of any type of ice that adheres to a structure, especially on airfoils or other parts of an airframe. [ARP4107]

icing cloud Cloud containing supercooled water droplets in sufficient concentration to produce ice on an aircraft surface. [AIR1168/9]

icing intensity The relationship of the icing intensity terms of trace, light, moderate, and severe to the corresponding cloud liquid water content (LWC). [AIR1667]

icing, light Icing in which the rate of accumulation may create a problem if flight is prolonged in the environment (over 1 h). Occasional use of deicing/anti-icing equipment removes/prevents accumulation. Light icing does not present a problem if the deicing/anti-icing equipment is used. [AIR1667]

icing, moderate Icing in which the rate of accumulation is such that even short encounters become potentially hazardous and use of deicing/anti-icing equipment or diversion is necessary. [AIR1667]

icing, natural Icing that occurs during flight in a cloud formed by nature. [AIR1667]

icing rate indicator A device to provide an indication of the rate at which ice is accreting on the sensor. [AIR4367]

icing, severe Icing in which the rate of accumulation is such that deicing/anti-icing equipment fails to reduce or control the hazard. Immediate diversion is necessary. [AIR1667]

icing severity system A system that provides information regarding the severity of the icing encounter, either in terms of liquid water content (LWC) or in terms of light, moderate, and heavy icing. [AIR4367]

icing, trace Icing in which the ice becomes perceptible. The rate of accumulation is slightly greater than the rate of sublimation. Trace icing is not hazardous even though deicing/anti-icing equipment is not utilized, unless encountered for an extended period of time (over 1 h). [AIR1667]

ICL computers Family of British digital computers produced by International Corporation, Ltd.

icon In data processing, a picture that represents a particular command; used with a mouse.

ICP *See* integrated circuit piezoelectric.

ID (Time Code) A three-numeral identification that can be inserted manually into time code, in the place of the "day of the year" information.

ideal elastic behavior A characteristic of a material whereby, under given conditions, the strain is a unique straight-line function of stress and is independent of previous stress history.

ideal gas A hypothetical gas whose distinguishing characteristic is that it obeys precisely the equation for a perfect gas, $PV = nRT$. Also called perfect gas.

ideal transducer A hypothetical passive transducer that produces the maximum possible output for a given input.

ident Refers to a request for a pilot to activate the aircraft transponder identification feature to help the controller identify an aircraft. [ARP4107]

ident feature A special feature in air-traffic-control radar beacon system (ATCRBS) equipment used to immediately distinguish one displayed beacon target from other beacon targets. [ARP4107]

identifier A symbol used in data processing whose purpose is to identify, indicate, or name a body of data.

idle characters Control characters interchanged by a synchronized transmitter and receiver to maintain synchronization during non-data periods.

idle power The minimum power for continued safe engine operations in existing conditions. [ARP4102/5]

idle time That part of available time during which computer hardware is not being used. Contrast with operating time.

idling pressure Pressure required to maintain a system or component at the idling speed, or flow. [ARP4386]

ID synchronization A count contained in one word of a telemetry frame to indicate which subframe is being sampled at any given time.

ID synchronizer A method of PCM-telemetry subframe recognition in which a specific word in the format activates a counter that identifies the number of the subframe word being received.

IF *1. See* intermediate fix. *2. See* intermediate frequency.

IF amplifier An intermediate-frequency stage in a typical superheterodyne radio receiver.

if and only if (IFF) A conditional statement implying that an action is to be taken or that a result is true if, and only if, stated prerequisite conditions are satisfied.

I format Notation used in FORTRAN; Iw indicates that w characters are to be converted as a decimal integer.

IFR *See* instrument flight rules.

IFR aircraft/IFR flight An aircraft conducting flight in accordance with instrument flight rules. [ARP4107]

IFR conditions Weather conditions below the minimum for flight under visual flight rules. [ARP4107]

IFR departure procedure *See* IFR takeoff minimums and departure procedures.

IFR landing minimums *See* landing minimums.

IFR takeoff minimums and departure procedures Standard takeoff rules for certain civil aviation users, prescribed in FAR, Part 91. At some airports, obstructions or other factors require the establishment of nonstandard takeoff minimums, departure procedures, or both, to assist pilots in avoiding obstacles during a climb to the minimum en route altitude. [ARP4107]

igniter An explosive device specifically designed to initiate burning of a fuel mixture or a propellant. [AIR913]

ignitiator The primary stimulus component in all explosive and pyrotechnic devices. [AIR913]

ignition The initiation of combustion.

ignition capable equipment and wiring Equipment and wiring that, in its normal operating condition, releases sufficient electrical or thermal energy to cause ignition of a specific hazardous atmosphere, under normal operating conditions.

ignition delay *See* ignition lag.

ignition exciter An assembly of component parts which provides a means of changing low-voltage alternating current or low-voltage direct current to a condition suitable to provide (with or without additional devices) a spark discharge for ignition purposes. [ARP667]

ignition lag In an internal combustion engine, the time interval between spark discharge and fuel ignition. Also known as ignition delay.

ignition lead, high tension A definite length of electrical cable having at least one end terminated in a single, common fitting. Construction, materials used, etc. must be such as to conduct the discharge energy from a high-tension ignition exciter (in excess of 5 kV) to a high-tension spark igniter. [ARP667]

ignition lead, low tension A definite length of electrical cable having at least one end terminated in a single, common fitting. Construction, materials used, etc. must be such as to conduct the discharge energy from a low-tension ignition exciter (less than 5 kV) to a low-tension spark igniter. [ARP667]

ignition loss The difference in weight before and after burning. [AIR4844]

ignition period *See* trial-for-ignition.

ignition system *1.* The system associated with rocket engines which provides for igniting the propellant. [AIR913] *2.* The portion of the electrical subsystems of an internal combustion engine that produces a spark to ignite the fuel.

ignition temperature For a fuel, the lowest temperature at which combustion becomes self-sustaining.

ignition train In a pyrotechnic or propellant mixture, the step-by-step arrangement of charges by which initial fire from a primer is transmitted and intensified until it reaches and sets off the main charge. [ARP4386]

ignition voltage *See* corona onset point.

ignitor A flame or high-energy spark that is utilized to ignite the fuel at the main burner.

ignitor intermittent An electric-ignited pilot that is automatically lighted each time there is a call for heat. This pilot burns during the entire period that the main burner is firing.

ignitor interrupted An electric-ignited pilot that is automatically lighted each time there is a call for heat. The pilot fuel is cut off automatically at the end of the trial-for-ignition period of the main burner.

illegal bus traffic Traffic consisting of illegal commands and the data and responses associated with those commands. [AS4115]

illuminance *1.* At a point on a surface, the luminous flux incident on an infinitesimal element of the surface containing that point, divided by the area of that surface element. *2.* As applied to electronic displays, the metric of measurement of light from a source, such as the sun, that impinges upon the face of the display. The units of illuminance are the foot-candle (fcd) and lux or lumen per square meter (lm/m^2) [ARP1782]

illuminant C A source of illumination having an energy distribution similar to that adopted by the International Commission on Illumination as "average daylight." Normally produced by a tungsten lamp and an appropriate filter.

Similar to "North skylight" illumination. [ARP1161]

illuminants Light oil or coal compounds, such as ethylene, propylene and benzene, that readily burn with a luminous flame.

illuminated dial A transparent, semitransparent, or nontransparent circular scale that is artificially illuminated.

illumination The density of luminous flux on a surface; equal to the flux divided by the area—when the latter is uniformly illuminated. [ARP798]

illusion An erroneous perception of sensory input due to limitations of sensory receptors and/or the manner in which sensory information is presented; the incorrect perception of an object(s). Often, the laws of physics explain the erroneous perception. [ARP4107]

illusions, kinesthetic An erroneous perception of somatosensory stimuli to the ligaments, muscles, or joints of the body. Specific types of kinesthetic illusions include the G-adaptation illusion and the G-differential illusion. *G-adaptation illusion*–an erroneous perception that motion has ceased after continued exposure to a sustained velocity; for example, movement in an elevator is only perceived at the beginning and end of the ascent or descent. *G-differential illusion*–an erroneous perception of aircraft attitude based on "seat of the pants" sensations; for example, without other sensory inputs, a 30-deg-bank level turn feels the same as a 60-deg-bank turn. [ARP4107]

illusions, vection Visual illusions of motion, erroneously detected peripherally, in which a person perceives that he/she is moving when in fact an external object is moving. Specific types of vection illusions include circularvection and linearvection. *Circularvection*–an erroneous sensation of rotation due to movement detected in the visual field, especially peripherally. *Linearvection*–an erroneous perception of linear movement due to motion detected in the visual field, especially peripherally. [ARP4107]

illusions, vestibular Erroneous perceptions of orienting stimuli to the semicircular ducts or otolith organs of the vestibular apparatus. Spe-

cific types of vestibular illusions include the Coriolis illusion, the elevator illusion, the giant-hand illusion, the leans, somatogravic illusion, and somatogyral illusion. *Coriolis illusion*–an erroneous sensation of rotation due to the movement of the head into a plane of angular or linear acceleration which induces fluid movement in the semicircular ducts. *Elevator illusion*–an erroneous sensation of pitch-up after level off from a steep descent, or pitch-down after level off from a steep climb, or when in turbulence. *Giant-hand illusion*–the erroneous sensation that controls will not respond to inputs, even with seemingly great effort, when the source of resistance is in fact the operator himself/herself attempting to respond to conflicting sensory cues. *Leans*–an illusion of angular displacement (bank) due to an undetected, subthreshold angular acceleration followed by a detected, transthreshold angular acceleration. *Somatogravic illusion*–an erroneous sensation of tilt in the vertical plane due to linear acceleration. This illusion is most common during rapid acceleration or deceleration. *Somatogyral illusion*–an erroneous perception that rotation has ceased because the semicircular canal fluid has stabilized after angular acceleration. The graveyard spin and graveyard spiral are results of the somatogyral illusion. [ARP4107]

illusions, visual Erroneous perceptions of stimuli to the visual system. Specific types of visual illusions include autokinesis, the chain-link fence illusion, empty field myopia, flicker vertigo, false horizon illusion, and the geometric perspective illusion. *Autokinesis*–an erroneous perception of movement of a light when stared at for a length of time in a dark visual field. *Chain-link-fence illusion*–the blending into the foreground of nearby objects when focusing on a distant object. *Empty field myopia*–the tendency for the eyes to focus at a distance of about one meter when viewing a visually non-stimulating field. *Flicker vertigo*–the disruptive psychological effects of cyclic visual stimulation of about 10 to 15 cycles per second. *False horizon (horizon misplacement) illusion*–an illusion created by sloping cloud formations, an obscure horizon, a dark scene with ground lights and stars, or certain geometric patterns of ground light, which results in the pilot placing the aircraft in a dangerous attitude because of the perception of not being aligned properly with the actual horizon. *Geometric-perspective illusion*–an erroneous perception of being nearer to, or farther away from, an object than one actually is, due to equating retinal image size to distance or angular displacement of familiar objects. For example, an 8000-ft runway viewed from 1000 ft up may appear the same size as a 10,000-ft runway viewed from 1500 ft up; another example is the tendency to flare high on a wider-than-usual runway. [ARP4107]

ILS (landing system) *See* instrument landing system.

IM *See* inner marker.

image A representation of an object by optical or electronic means.

image analysis Technique for understanding or quantifying digital data as presented in a two-dimensional format.

image converter camera A camera that converts images from one wavelength region to another, typically from the infrared to the visible.

image digitizer A device that measures light intensity at each point in an image and generates a corresponding digital signal to indicate that intensity. Such a device converts an analog image to a digital data set.

image impedances For a transducer, the impedances that will simultaneously produce equal impedances in both directions at each of the inputs and outputs.

image intensifier A viewing system that functions as a light amplifier; takes a faint image and amplifies it so that it can be viewed more easily.

image inverter A fused fiber-optic bundle that is permanently twisted during manufacture to turn the image it transmits upside down. The same can be done with conventional optics, but a fiber-optic image inverter can do it in a distance of less than an inch.

image orthicon A camera tube whose output is generated using a low-velocity electron beam to scan the reverse side of a storage target. The storage target contains an image produced by focusing the electron image from a photoemitting surface on it.

image processing Conversion of optical images into digital data form for storage and reconstruction by computer techniques.

image reconstruction The reproduction of an original scene from data stored or transmitted after scanning with an electron beam.

image resolution In optics, a measure of the ability of an optical instrument to produce separable images of different points on an object.

image retention An undesired afterimage that persists on the display. [ARP4256]

image rotation Mechanized or digital rotation of an image.

image source Component of head-up displays which provides the optical origin of the symbology. [ARP4102/8]

IMC *1. See* instrument meteorological conditions. *2.* Acronym for Institute of Measurement and Control.

IMD *See* intermodulation distortion.

immersion length For a thermometer, the distance along the thermometer body from the boundary of the medium whose temperature is being determined to the free end of the well, bulb, or element, if unprotected.

immersion test A test used to determine the relative abilities of greases to prevent corrosion on metal surfaces. In this test, a grease-coated metal object is immersed in water for an extended period of time. [S-5C,40-55]

immiscible Describes two or more fluids, not mutually soluble; incapable of attaining homogeneity.

immunity An inherent or induced electrochemical property whereby a metal resists attack by a corrosive solution.

immunoassay An assay that utilizes antigen-antibody reactions for the determination of biochemical substances.

I_{mp} The numerical value of discharge current capability of a battery at maximum power delivery, partway through a simulated turbine engine start. The I_{mp} rating indicates relative engine-starting power capability of the battery, and is the minimum value of I_{mp} expected from a fully charged battery at 23°C. [AS8033]

impact The force transmitted by or as if by a collision.

impact acceleration The acceleration generated by very sudden starts or stops of a vehicle; usually applied in the context of physiological acceleration.

impact angle The angle at which an aircraft or object strikes the terrain relative to the slope of the impact site terrain. [ARP4107]

impact damage Damage from foreign objects (other than ballistic). [AIR4844]

impact deceleration *See* deceleration and impact acceleration.

impacted airport An airport affected by adverse weather or runway conditions, equipment failure, personnel shortages, or other phenomena that impair its ability to accommodate its normal flow of aircraft traffic. [ARP4107]

impact extrusion *See* cold extrusion.

impact fusion The conversion of the kinetic energy of a fast-moving, initially stationary, macroparticle projectile into the internal energy of fusile material using a particle accelerator. Impact fusion is generally an inertial confinement fusion concept.

impact ice Ice that forms when snow, sleet, or supercooled water droplets impinge upon aircraft surfaces that are at or below freezing temperature. [ARP4107]

impact melts Molten material resulting from hypervelocity impact.

impact modulator An amplifier in which the intensity of two directly opposed, impacting power jets is controlled, thereby controlling the position of the impact plane to modulate the output. [ARP993A]

impact pressure The pressure that a moving stream of fluid produces against a surface that brings part of the moving stream abruptly to rest. For subsonic flow in a fluid medium, impact pressure is approximately equal to the stagnation pressure.

impact strength *1.* A test for determining the resistance of an insulating material or system to damage by impacting it with a given weight dropped from a given distance in a controlled environment. [ARP1931] *2.* The amount of energy required to fracture a material. This value is affected by the type of specimen and the testing conditions, thus these should be specified. *3.* Property of a material that indicates it's ability to resist breaking under extremely rapid loading; usually expressed as energy absorbed during fracture.

impact temperature The temperature of a gas after impact with a solid body; the impact converts some of the kinetic energy of the gas to heat, and thus raises the gas temperature above ambient.

impact test A test that measures the energy necessary to fracture a standard notched bar by an impulse load. [AIR4844]

impact torque *See* torque, impact.

impact tube A small diameter tube, immersed in a fluid, and oriented so that the fluid stream impinges normally on its open end.

impact value The energy absorbed by a specimen of standard design when sheared by a single blow from a testing machine hammer. [AIR4844]

impedance (acoustic) Resistance to flow of ultrasonic energy in a medium; a product of particle velocity and material density. [ARP5089]

impedance (Z) The complex ratio of a forcelike parameter to a related velocity-like parameter; for example, force to velocity; pressure volume velocity; electric voltage to current; temperature to heat flow; or electric field strength to magnetic field strength.

impedance bridge A four-arm bridge circuit in which one or more of the arms have reactive components instead of purely resistive components. An impedance bridge must be excited by an a-c signal to yield complete analysis of the unknown bridge element.

impedance plane display A graphical representation of phase and amplitude data, resulting in a dot location on a screen. [ARP5089]

impeller *1.* Device that imparts motion to a fluid. *2.* In centrifugal compressors, rotary disks faced on one or both sides with radial vanes, which accelerate the incoming fluid outward into diffusers. *3. See* diffuser.

impingement A method of removing entrained liquid droplets from a gas stream by allowing the stream to collide with a baffle plate. Also called liquid knockout.

impingement attack A form of accelerated corrosion in which a moving corrosive liquid erodes a protective surface layer, thus exposing the underlying metal to renewed attack.

impingement limit The location farthest aft on a body at which ice impinges. The impingement limit governs the extent of protection to be provided. [AIR1168/9]

impingement starting A starting method in which nozzle(s) are built into the engine such that the starting fluid works directly on one of the engine turbine or compressor rotors. [ARP906A]

implementation The act of creating a physical reality from a specification. [ARP4754]

implosion The rapid inward collapsing of the walls of vacuum systems or devices as the result of failure of the walls to sustain the ambient pressure.

impregnate In reinforced plastics, to saturate the reinforcement with a resin. [AIR4844]

impregnated bit A diamond cutting tool made of fragmented bort or screened whole diamonds in a sintered powder-metal matrix.

impregnated fabric A fabric impregnated with a synthetic resin. [AIR4844]

impregnator A mechanical device for wetting or impregnating fabrics with resin. [AIR4844]

improbable For airworthiness purposes, the likelihood of a failure that is equal to or less than once in 1,000,000 flight hours (10^{-6}). [ARP4107]

improvement maintenance Efforts to reduce or eliminate the need for maintenance. Includes modification, retrofit, redesign, or change-order.

impulse *1.* For a rocket engine, may refer to total impulse or specific impulse. *Total impulse—* the product of the engine thrust in pounds and the burning time in seconds. *Specific impulse—* the thrust in pounds of a specified fuel with its oxidizer in one second. *2.* A pulsed signal or force whose duration is small compared to the response time of the dynamical system in which the impulse is applied. [ARP4386] *3.* The product of the force and the time during which the force is applied.

impulse excitation A method of producing oscillations in which the duration of stimulus is relatively short in relation to the duration of oscillation.

impulse line The conduit that transfers the pressure signal from the process to the measuring instrument.

impulse pressure A rapidly occurring pressure rise, peaking at a prescribed multiple of the nominal or operating pressure. After the impulse peak, the pressure trace follows a prescribed curve, with a hold at nominal and zero pressure during one impulse-pressure cycle. [MA2005]

impulse response optimization The selection of the composite impulse response of both the dynamic process and the controller to minimize the error index. [ARP4386]

impulse, specific *See* impulse.

impulse strength The voltage breakdown of insulation under voltage surges on the order of microseconds in duration. [ARP1931]

impulse, total *1.* The integral of the thrust with respect to time over the entire burning period. [AIR913] *2. See* impulse.

impulse-type telemetering Employing intermittent electrical impulses to transmit instrument readings to remote locations.

IMS Acronym for International Magnetospheric Study.

in Fluid-entrance port, not otherwise designated, at which direction of flow is critical. [ARP4386]

inaccessible area An area to which entry may be dangerous without special controls.

inaccuracy *See* error.

inattention *See* attention, anomalies of inattention.

incandescence Emission of light due to high temperature of the emitting material. (Any other emission of light is called luminescence.)

incandescent filament lamp As used in aircraft cabin lighting, lamps that comprise a range from 0.3 watt, 0.03 candela to 36 watt, 50 candelas; used in every facet of cabin lighting, including indications, signs, area, reading, and decorative. The efficacy of light produced by incandescent filament lamps ranges from approximately 1.2 lumens per watt to 15 lumens per watt, depending on size, wattage and design life. [AIR512B]

incandescent lamp A light source consisting of a glass bulb containing a filament electrically maintained at incandescence. [ARP798]

inches water gage ("w.g.) Usual term for expressing a measurement of relatively low pressures or differentials by means of a U-tube. One inch w.g. equals 5.2 lb per sq ft or 0.036 lb per sq in.

incidence *1.* Partial coincidence, as a circle and a tangent line. *2.* The impingement of a ray on a surface. *3.* The angle between the chord line of a wing and the reference line of the fuselage to which it is attached.

incident An occurrence other than an aircraft accident, associated with the operation of an aircraft, which adversely affects or could affect the safety of operations. [ARP4107]

incident related part A part associated with an abnormal operational circumstance. [AS7104]

incident, technical Any event of a technical nature which may be considered to significantly affect the potential airworthiness of an aircraft. [AIR4896]

incident wave A wave in a given medium that impinges on a discontinuity or a medium of different propagation characteristics.

incipient failure A functional status or condition that exists at the beginning of a failure of engine, engine component, or subsystem. [ARP1587]

incipient skid The point of wheel instability at which brake torque exceeds resisting tire-runway friction torque, and the wheel therefore begins a deceleration condition, which if continued, would result in abrupt wheel lock-up. [AIR1489]

in-circuit emulation (ICE) A development aid for testing the software in computer hardware. Involves an umbilical link between a development system and the target hardware being plugged into the microprocessor socket.

inclination The angle between the plane of an orbit and the reference plane. The equator is the reference plane for geocentric orbits and the elliptic is the reference plane for heliocentric orbits.

inclined-tube manometer A glass-tube manometer having one leg inclined from the vertical to give more precise readings.

inclinometer An instrument for determining the angle of the earth's magnetic field vector from the horizontal.

inclusions *1.* Foreign materials, such as backing paper or peel ply, that are unintentionally incorporated in a composite part. [ARP5089] *2.* Nonmetallic impurities, such as slag, oxide, and sulfides, that were present in the original ingot. [AS3071A]

inclusive Applies to both upper and lower limits of a specified range. [AMS2243G]

incoherent fiber optics A bundle of fibers in which the fibers are randomly arranged at each end. The pattern may be truly random to achieve uniform illumination, or the manufacturer may have failed to align the individual fibers. In either case, the fiber bundle cannot transmit an image along its length.

incoherent scatter radar Radar used in the study of the ionosphere, thermosphere, etc.

incomplete combustion The partial oxidation of the combustible constituents of a fuel.

incompressible For liquids, refers to the property whereby the change in volume due to pressure is negligible.

incompressible flow Fluid flow under conditions of constant density.

Inconel A series of International Nickel Co. high-nickel, chromium, and iron alloys characterized by inertness to certain corrosive fluids.

increaser A pipe fitting identical to a reducer, except specifically referred to for enlargements in the direction of flow. *See* reducer.

incremental encoder An electronic or electromechanical device that produces a coded digital output based on the amount of movement from an arbitrary starting position; the output for any given position with respect to a fixed point of reference is not unique.

incremental feedback In numerical control, assignment of a value for any given position of machine slide or actuating member based on its last previous stationary position.

incremental plotter A discrete X-Y plotter.

incremental representation A method of representing a variable in which changes in the value of the variables are represented, rather than the values themselves.

indentation hardness Hardness evaluated from measurements of area or indentation depth caused by pressing a specified indentor into the surface of the material with a specified force. [AIR4844]

indenture levels The item levels that identify or describe relative complexity of assembly or function. The levels progress from the more complex to the simpler divisions. [AIR4896]

independence *1.* Refers to a design concept that ensures that the failure of one item does not cause a failure of another item. *2.* Separation of responsibilities assuring the accomplishment of objective evaluation. [ARP4754]

independent linearity *See* best-straight-line linearity.

independent variable *1.* Any of the variables of a problem, chosen according to convenience, which may arbitrarily be specified, and which then determine the other or dependent variables of the problem. The independent variable is often called the coordinate, particularly in problems involving motion in space. *2.* A process or control-system parameter that can change only due to external stimulus.

independent warning A signal indicating a condition requiring immediate action. The specific condition is defined by the location of the signal or the legend associated with the signal. [AIR1161]

index *1.* An ordered reference list of the contents of a computer file or document, together with keys or reference notations for identification or location of those contents. *2.* To prepare a list as in definition 1. *3.* A symbol or number used to identify a particular quantity

in an array of similar quantities. *4.* To move a machine part to a predetermined position, or by a predetermined amount, on a quantized scale.

index address modification (indexing) *See* address modification.

indexed address In a computer instruction, an address that indicates a location at which the address of the reference operand is to be found. In some computers, the machine address indicated can itself be indirect. Such multiple levels of addressing are terminated either by prior control or by a termination symbol. Synonymous with second-level address.

index graduations The heaviest or longest division marks on a graduated scale, opposite the scale numerals.

index matching fluid A liquid with refractive index that matches that of the core or cladding of an optical fiber. Used in coupling light into or out of optical fibers, and can help in suppressing reflections at glass surfaces.

indicated airspeed (IAS) The speed of an aircraft as shown on its airspeed indicator. This is the speed used in pilot/controller communications under the general term "airspeed." [ARP4107]

indicating gage Any measuring device whose output can be read visually, but is not automatically transcribed on a chart or other permanent record.

indicating lead That portion of the thermocouple indicating system from the engine-airframe disconnect to the reference junction. [ARP485]

indicating scale On a recording instrument, a scale that allows a recorded quantity to be simultaneously observed.

indication In nondestructive testing, any visible sign or instrument reading that must be interpreted to determine whether a flaw exists.

indicator A mechanism for amplifying and measuring the displacement of a movable contact point that is to measure a determination or variation from a standard determination. The mechanism consists essentially of a case with means for mounting the indicator, a spindle carrying the contact point, an amplifying mechanism, a point, and a graduated dial. May include accessories, and/or attachments. [ARP480A]

indicator card A chart for recording an indicator diagram.

indicator diagram A graphic representation of work done by or on the working fluid in a positive-displacement device such as a reciprocating engine.

indicator, gear position An instrument or indicator whose purpose is to provide information to the pilot or crew regarding position of the landing gear. May be located at the pilot or crew station or remotely (e.g., a pop-up indicator projecting above upper surface of wing above the gear). Green lights are used for safe configurations and red lights for unsafe or in-transition conditions; and "up" and "down" are used to indicate up and locked or down and locked. [AIR1489]

indicator melt time The elapsed time from the moment element-melt time ends and arcing commences. [ARP1199A]

indicator, oxygen flow A device that gives a visual signal when oxygen flows. [ARP171]

indicator tube An electron-beam tube used to convey useful information, through variations in beam cross section at a luminescent target.

indirect-acting recording instrument An instrument in which the output level of the primary detector is increased through intermediate mechanical, electric, electronic, or photoelectric means to actuate a writing or marking device.

indirectly heated cathode A cathode in a thermionic tube that is heated by an independent heating element.

indirect method *See* Liapunov's second method.

indium-tin-oxide semiconductors *See* ITO semiconductors.

individual harmonic The RMS value of any individual harmonic voltage when measured with a harmonic analyzer. This value is expressed as a percentage of the fundamental. [ARP1148A]

individual wheel control A type of control used in antiskid systems in which each wheel uti-

lizes its own valve and control circuit. [AIR1489]

indoor test cell A facility for the testing of gas turbine engines in a restricted environment. [ARP741]

induced draft Airflow through a device, such as a firebox or drying unit, which is produced by placing a fan or suction jets in the exit duct.

induced draft fan A fan exhausting hot gases from heat-absorbing equipment.

induced drag In subsonic flow over a finite airfoil or other body, that part of the drag induced by lift. Induced drag is inversely proportional to the airspeed; it is the direct result of the aerodynamic force resulting from the downward velocity imparted to the air as the airfoil moves through the air. [ARP4107]

inductance Symbolized by L; the property of a circuit or circuit element that opposes a change in current flow. Inductance causes current changes to lag behind voltage changes. Inductance is measured in henrys. [ARP1931]

inductance-type pressure transducer Any of several designs of pressure sensor in which motion of the primary sensor element, such as a Bourdon tube or diaphragm, is detected and measured by a variable-inductance element and measuring circuit.

induction heating Raising the temperature of an electrically conductive material by electromagnetically inducing eddy currents in the material.

induction instrument A type of meter whose indicated output is determined by the reaction between magnetic flux in fixed windings and flux in a moving coil, the two fluxes being induced by electric currents from different sources.

induction motor An a-c motor in which alternating current supplied to the stator produces a rotating magnetic field. [ARP4386]

induction motor meter A type of meter resembling an induction motor, in which the rotor moves in direct relation to the reaction force between a magnetic field and currents induced in the rotor.

induction system That combination of scoops and ducts that introduce outside air to the air-distribution equipment of the airplane. [ARP147C]

inductive Describes an analytical approach that involves the systematic evaluation of the defined parts or elements of a given system or subsystem to determine specific characteristics of interest. [ARP926A]

inductive bridge position transducer A device for measuring linear position by means of induction between a fixed member slightly longer than the limits of motion and a movable member approximately half as long. With this device, position is determined by: (a) selecting appropriate taps from the longer member, which are connected in a successive decade to the external inductors to form a bridge circuit; and (b) relating the configuration that balances the bridge to the actual position of the movable member. The chief advantage of this device is the relatively high output voltage developed for a relatively small change in position.

inductive coupling *1.* Crosstalk resulting from the action of the electromagnetic field of one conductor on the other. [ARP1931] *2.* Using common or mutual inductance to cause signals in one circuit to vary in accordance with signals in another.

inductive plate position transducer A device for measuring rotary position by means of induction between a stationary and rotary plate, each having an etched winding projected onto a nonconductive surface; or for measuring linear motion by means of induction between a stator plate and a sliding member, each also having etched windings. Advantages of this device include eliminating wear and backlash, as well as providing good resolution—often within 0.001 in. or less.

inductive system An ignition system in which the spark energy is primarily the result of a rapid variation in magnetic flux in an induction coil. [ARP667]

inductive transducer A device to provide an indication of a change in resonance resulting from a change in self-inductance. [AIR4367]

inductor A wire coil that will store energy in the form of a magnetic field.

industrial adapter A dry-break disconnect with a three-slotted (inverted bayonet) portion to allow connection to the mating industrial coupling. [AIR4783]

industrial computer A computer used on-line in various areas of manufacturing, including process industries (e.g., chemical or petroleum), numerical control, and production lines. *See* on-line. *See also* control computer and process computer.

industrial computer language *1.* A computer language for industrial computers. *2.* A language used for programming computer control applications and system development, e.g., assembly language, FORTRAN, RTL, PROSPRO, BICEPS, and AUTRAN.

industrial controls A collective term for control instrumentation used in industry.

industrial coupling The mating coupling for an industrial adapter. [AIR4783]

industrial engine A turboshaft engine that is designed for nonaircraft applications, such as electrical generation and driving petroleum pumps. [AS5116]

industrial engineering A branch of engineering that deals with the design and operation of integrated systems of personnel, equipment, materials, and facilities.

inelastic collision Collision between two particles in which changes occur both in the internal energy of one or both of the particles and in the sums, before and after collision, of their kinetic energies.

inelastic stress A force acting on a solid and producing a deformation such that the original shape and the size of the solid are not restored after the force is removed.

inert Describes a rocket system or component thereof that contains no explosive, pyrotechnic, or other reactive material. [AIR913]

inertance A passive fluidic element that has a pressure drop across it because of fluid inertia. The pressure drop leads flow through the element by essentially 90°. [ARP4386]

inert atmosphere A gaseous medium that is used to enclose tests or equipment because of its lack of chemical reaction.

inert filler A material added to a plastic to alter the end-item properties through physical rather than chemical means. [AIR4844]

inert gas *See* rare gas.

inert gaseous constituents Incombustible gases such as nitrogen which may be present in a fuel.

inertia Inherent resistance of a body to changes in its state of motion.

inertia bonding The joining of materials with friction and pressure.

inertia equivalent In brake testing, the equivalent amount of the total air-vehicle energy that goes to a single brake assembly. This equivalent energy is translated into dynamometer energy, which is absorbed by the brake in the test process. [AIR1489]

inertial confinement fusion The process of using intense beams of heavy ions to convey the energy needed to compress and heat small pellets containing deuterium-tritium fuels in order to achieve ignition of the pellets.

inertial force *See* inertia.

inertial fusion (reactor) Reactors in which pellet fusion is initiated by high-energy sources, including lasers.

inertial guidance Guidance by means of the measurement and integration of acceleration from within the craft.

inertial navigation Dead reckoning performed automatically by a device that gives a continuous indicational position by integration of accelerations since leaving a starting point.

inertia load A load proportional to load acceleration or deceleration. The load is usually opposing for acceleration and aiding during deceleration. [ARP4386]

inertial space An assumed stationary frame of reference. A nonrotating set of coordinates in space, relative to which the trajectory of an aerospace vehicle is calculated. [ARP4386]

inertia, moment of *See* moment of inertia.

inertia-type timer Any of several types of relay devices that incorporate extra weights or flywheels to achieve brief time delay in normal relay action. The delay is achieved by providing additional inertia to be overcome. In such a device, delays are usually on the order of 80 to 120 milliseconds.

infant mortality *1.* Refers to the initial phase of the lifetime of a population of a particular component when failures occur as a result of manufacturing errors, etc. Infant mortalities are screened out by burn-in. [ARD50010] *2.* Premature catastrophic-type failures occurring at a rate substantially greater than that observed during subsequent life prior to wearout. Infant mortality is usually reduced by stringent quality control. [AIR4896]

infiltration Casting molten metal to be drawn into void spaces in a powder-metal compact, formed-metal shape, or fiber-metal layup.

infinite frequency stiffness (frequency-independent dynamic stiffness) The stiffness associated with the output deflections of an active actuation system, caused by externally applied loads when the frequency of the load disturbance is significantly above the bandpass of the actuation system. Since the system cannot actively react against high-frequency load disturbances, this stiffness is identical to stiffness to ground at frequencies well beyond the system bandwidth. [ARP4386]

infinite loop In data processing, a routine that can be ended only by terminating the program.

infinity A point, line, or region beyond measurable limits.

inflation pressure The pressure (gage) to which a tire is to be inflated for a specific load and deflection, usually the rated pressure for a given service application. [AIR1489]

in-flight engine status In-flight indications (real time or near real time) of potential failures and warnings of a cautionary or advisory nature, e.g., high vibration. NOTE: Warnings to the cockpit should be only those to which the flight crew can react. Event detection and exceedance documentation should be provided. [ARP1587]

in-flight shutdown Cessation of engine operation during flight for any reason other than training procedure. [ARP4107]

influence The change in the indicated value of an instrument caused solely by a difference in value of a specified variable or condition from its reference value or condition when all other variables are held constant.

influence coefficient Describes the effect that a small change in an input parameter has on the output. [AIR1678]

influent Stream of fluid at the inlet of a filter or filter separator. [AIR4783]

information Any facts or data that can be used, transferred, or communicated.

information adaptive system The spaceborne portion of the NASA End-to-End Data System.

information message An informational message that does not imply any change in operating behavior on the part of the pilot or the controller. [ARP4791]

information processing *1.* The mental process of receiving incoming information from the environment, assessing its meaning, and deciding on an appropriate response. *2.* A general term for the presumed operations whereby the raw sense-data are refashioned into items of knowledge and utilized for decision making that may lead to action(s). Among these operations are perceptual organization, comparison with items stored in memory, and the making of decisions as to the response to be made. *3.* The mental processes from sensory input to evoked response. [ARP4107] *4.* The organization and manipulation of data, usually by a computer. *See also* data processing.

information processing (biology) An approach to the study of perception, memory, language, and/or thought that considers organisms to be complex systems that receive, transform, store, and transmit information.

information theory The mathematical theory concerned with information rate, channels, channel width, noise, and other factors that affect information transmission. Initially developed for electrical communications, information theory is now applied to business systems and other phenomena that deal with information units and flow of information in networks.

information transfer The physical transfer of information from one physical entity to another; the information may be digital, analog, or control/status information. [ARD50012]

infrared Part of the electromagnetic spectrum between the visible light range and the radar range.

infrared absorption The taking up of energy from infrared radiation by a medium through which the radiation is passing.

infrared absorption moisture detector An instrument for determining moisture content of a material such as sheet paper. In this instrument, moisture content can be read directly by determining the ratio of two beam intensities, one at a wavelength within the resonant-absorption band for water and the other at a wavelength just outside the band.

infrared astronomy satellite A joint NASA-Netherlands-Great Britain spacecraft designed to perform astronomical observations in the infrared spectral region. Launched on January 25, 1983.

infrared imaging device Any device that receives infrared rays from an object and displays a visible image of the object.

infrared photometry Photometry in the infrared region.

infrared radar Radar covering a range from the limit of the visible spectrum to the shortest microwaves.

infrared radiation Electromagnetic radiation lying in the wavelength interval from 75 micrometers to an indefinite upper boundary, sometimes arbitrarily set at 1000 micrometers (0.01 centimeter).

infrared signatures The infrared spectral characteristics of an object or uniform land surface that uniquely defines it.

infrared source (astronomy) Celestial bodies or astronomical regions emitting a large amount of radiation in the infrared portion of the electromagnetic spectrum.

Infrared Space Observatory (ISO) An astronomical satellite observatory funded by ESA, operating at wavelengths from 3 to 2000 micrometers. The observatory is composed of a 60-cm Cassegrain telescope, a CCD infrared camera, two Michelson interferometers, and a photopolarimeter.

infrared spectroscopy A technique for determining the molecular species present in a material, and measuring their concentrations. Operates by detecting the characteristic wavelengths at which the material absorbs infrared energy and measuring the relative drop in intensity associated with each absorption band.

infrared suppression The shielding and/or protection of aircraft engines and exhausts from heat-seeking missiles and/or detecting devices.

infrared telescope Special optical instrument for astronomical observations in the range from one micrometer to one millimeter.

infrared windows A frequency region in the infrared in which there is good transmission of electromagnetic radiation through the atmosphere.

infrasonic Generating or using waves or vibrations with frequencies below that of audible sound.

infrasonic frequencies A sound-wave frequency lower than the audio-frequency range.

in-gate See gate.

ingot *1.* That which is solidified in an ingot mold. *2. See* casting. [MAM2304]

inhalator A device from which gaseous oxygen is inhaled with or without medicaments, for therapeutic purposes. [ARP171]

inherent Achievable under ideal conditions, generally derived by analysis, and potentially present in the design. [ARD50010]

inherent damping Using mechanical hysteresis of materials, such as cork or rubber, to reduce vibrational amplitude.

inherent error The error in quantities that serve as initial conditions at the beginning of a step in a step-by-step set of operations; thus, the error carried over from the previous operation from whatever source or cause.

inherent failure causes Failures due to integral operating factors of engine usage, such as creep and low cycle, high cycle, and thermal fatigue. [AIR1872]

inherent r and m value Any measure of reliability or maintainability that includes only the effects of item design and installation, and assumes an ideal operating and support environment. [AIR4896]

inherent reliability A measure of reliability that includes only the effect of an item design and

its application, and assumes an ideal operation and support environment. [ARD50010]

inherent stability In-flight stability achieved as a result of the basic shape of the aircraft.

inhibit Refers to a physical barrier or device that constrains fluid flow, such as a valve or a seal. [AIR4728]

inhibitor *1.* A substance that retards a chemical reaction. *2.* A material added to a resin to slow down curing. [AIR4844] *3.* Substances that inhibit; specifically, substances bonded, taped, or dip-dried onto a solid propellant to restrict the burning surface and to give direction to the burning process.

inhomogeneous Consisting of more than one phase, i.e., having discrete regions of different materials. [AIR4844]

initial approach For an aircraft, the portion of the flight immediately prior to arrival over the airport of destination or over the reporting point from which the final approach to the airport is commenced. [ARP4386]

initial approach fix (IAF) The fixes depicted in instrument approach procedure charts which identify the beginning of the initial approach segment(s). [ARP4107]

initial approach segment *See* segments of an instrument approach procedure.

initial delay time The time between the moment the equipment becomes available for maintenance and the moment work is commenced. [AIR4896]

initial indenture level The level of the total, overall item that is the subject of the FMEA.

initial instantaneous pressure rise rate *See* rise rate, initial instantaneous pressure.

initial isolation Isolation to the equipment/system subunit that must be replaced on line to return the equipment/system to operation. [AIR4896]

initial isolation level of ambiguity The number of possible equipment/system subunits that might contain the failed component, defined by the built-in-test, external tests equipment, or manual test procedure. It is possible that a combination of built-in-test, external special-purpose test equipment, and manual procedures may be necessary to effect isolation. For example, if an equipment-test subsystem (built-in, external, manual) isolates a fault to one or two subunits, the level of ambiguity is equal to two; if it isolates it to one of three subunits, the level of ambiguity is equal to three. [ARD50010]

initial modulus The slope of the initial straight portion of a stress-strain or load-elongation curve. [AIR4844]

initial pulse Electrical pulse generated by the ultrasonic instrument; used to excite a search unit in order to produce ultrasonic energy. [ARP5089]

initial set For a powdery material such as plaster or Portland cement, the start of a hardening reaction following water addition to the material.

initial spares provision Refers to the spares purchased to enable start up of the system, covering up to two years of base operation for military and six months for commercial operations. [ARP4293]

initial strain The strain produced in a specimen by given loading conditions before creep occurs. [AIR4844]

initial stress The stress produced by force in a specimen before stress relaxation occurs. [AIR4844]

initial support cost The cost incurred by having a system supplier provide maintenance coverage of a new system for a defined period after it has become operational. [ARP4293]

initial tooling The initial package of tooling (including software, where appropriate) to establish and maintain production of a system as required. [ARP4293]

initial training The initial training of operators and maintainers of a new system, provided by the contractor. [ARP4293]

initial yield The point at which 0.2% permanent deformation has occurred. [AS1607]

initiation The first action in the first element of an explosive train. [AIR913]

initiator A source of free radicals; a peroxide, for example. [AIR4844]

injection Insertion of a satellite into orbit, or a spacecraft onto the desired trajectory.

injection laser diode A semiconductor device in which lasing takes place within the p-n junction. Light is emitted from the diode edge.

injection molding A forming process in which a heat-softened or plasticized material is forced from a cylinder into a relatively cool cavity, which gives the product a desired shape. A similar process is used for forming solid propellants from quick-cure ingredients.

injection seal A seal accomplished by injecting sealant into holes, joggles, channels, grooves, and other voids caused by build-up of structure in the fuel-tank boundaries. An injection seal is used to provide continuity where fillet seals are interrupted by the structure and also to fill cavities completely. [AIR4069]

injector Any nozzle or nozzle-like device through which a fluid is forced into a chamber or passage.

inlet A passage or opening at which fluid enters a conduit or chamber.

inlet, air An opening or valve through which air is admitted to dilute the oxygen (as in a diluter-demand regulator or constant-flow mask). [ARP171]

inlet airframe configuration Optimum location of engine inlet for various purposes.

inlet box An enclosure at or near the entrance to a chamber or duct system for attaching a fan to the system.

inlet filter A fluid-contamination filter located upstream of the skid-control valve (or other hydraulic unit). [AIR1489]

inlet flow distortion Spatial variations in the total pressure at the inlet-engine interface plane. [ARP1420]

inlet patch An indication of contaminant level or cleanliness of the fluid, provided in the system for testing the unit. [ARP575A]

inlet pressure For pump performance data, when not otherwise specified, the total static pressure measured in a standard testing chamber by a vacuum gage located near the inlet port.

inlet pressure (supply pressure) Pressure at the inlet of a component. [ARP4386]

inlet, temperature A location for measuring the temperature of fluids, particles, etc. entering a heat system, an engine, or other machine.

inlet valve A valve for admitting the working fluid to the cylinder of a positive-displacement device, such as a reciprocating pump or engine.

in line *1.* Centered on an axis. *2.* Having several features, components, or units aligned with one another. *3.* In a motor-driven device, having the motor shaft parallel to the driven shaft, and approximately centered on each other.

in-line monitoring *See* monitor.

in-line valve A valve that is mounted in the fueling-vehicle pipe work to automatically control flow, pressure, or other parameters. [AIR4783]

inner marker (IM)/inner marker beacon A marker beacon used with an ILS (CAT II) precision approach, located between the middle marker and the approach end of the ILS runway, transmitting a radiation pattern keyed at six dots per second, and indicating to the pilot, both aurally and visually, that he/she is at the designated decision height (DH) (normally, 100 ft above touchdown zone elevation on the ILS CAT II approach). [ARP4107]

inner tube A rubber tube located within a pneumatic tire which seals in the air. [AIR1489]

inorganic Designating or pertaining to the chemistry of all elements and compounds not classified as organic. [AIR4844]

inorganic pigments Natural or synthetic metallic oxides, sulfides, and other salts that impart heat and light stability, weathering resistance, color, and migration resistance to plastics. [AIR4844]

in-phase voltage A voltage that is of the same time phase as the resolver output of fundamental frequency at maximum coupling. [ARP826]

inplane loads Loads that are parallel to the facings. [AIR4844]

input *1.* An independent variable supplied to the control system. [ARP4386] *2.* Signals taken in by an input interface as indicators of the condition of the process being controlled. *3.* Data keyed into a computer or computer peripherals.

input area An area of computer storage reserved for input. Synonymous with input block.

input block *See* input area.

input channel A channel for impressing a state on a device or logic element.

input current The current, expressed in milliamperes, to the valve that commands control flow. [ARP4386]

input device In data processing, a device or collective set of devices used for conveying data into another device.

input impedance The electrical load that the primary coil presents to the excitation source. [ARP4386]

input interface Any device that connects computer hardware or other equipment for the input of data.

input/output (I/O) The interface to a unit which provides data or signals used or generated by the unit. [AS8034]

input/output curve The graphic representation of actuation system output versus command input; usually a continuous plot throughout a complete cycle, between plus and minus rated commands. The cycling rate must be significantly below control bandpass so that velocity error signal is not included in this parameter. [ARP4386]

input ports In computer hardware, terminals for connection in external devices which input data to the computer.

input power The real power in watts, or the apparent power in volt-amperes, used for excitation. [ARP4386]

input signal A signal applied to a device, element, or system.

input strobe (INSTRB) A signal that enters set-up data into registers.

input work queue A list of summary information of job-control statements maintained by the job scheduler, from which the scheduler selects the jobs and job steps to be processed.

insensitive time *See* dead time.

insensitivity *See* sensitivity.

insert *1.* An integral part of a plastic molding, consisting of metal or other material that may be molded or pressed into position after the molding is completed. *2.* Apparatus placed into the sandwich for attaching items. Synonymous with hard points. *3.* Part used to ensure that the required loads can be transmitted, in the case of floor panels and others where high shear loads have to be carried through the panel to surrounding structure. [AIR4844]

insert, electrical connector The insulating element of a connector which supports and positions the contacts. [ARP914A]

insertion gain The ratio of the power delivered to a portion of a transmission system following a transducer, to the power delivered to the same portion without the transducer in place.

insertion loss The loss in load power resulting from the insertion of a cable. Expressed in decibels as the ratio of power received at the load before insertion to the power received at the load after insertion. [ARP1931]

insertion point Usually indicated by a computer cursor, the place where characters will appear when an operator starts typing.

insertion tool A tool used to insert contacts into their retaining device. [ARP914A]

insert pin A pin that keeps an inserted part inside the mold by screwing or friction. The insert pin is removed when the object is being withdrawn from the mold. [AIR4844]

insert retention force *See* force, insert retention.

inside caliper A caliper having outward-turned feet on each leg for measuring inside dimensions.

inside diameter The maximum dimension across a cylindrical or spherical cavity. Ideally, this is a line passing through the exact center of the cavity and perpendicular to the inner surface of the cavity.

inside gage *1.* A fixed-dimension device for checking inside diameters. *2.* The inside diameter of a bit, measured between opposing cutting points.

inside micrometer A micrometer caliper designed for measuring inside diameters and similar inside dimensions between opposing surfaces.

inside strand A strand of a multiple-strand heat that flows from an inside location of the molten reservoir. [MAM2304]

in situ *1.* In natural order. *2.* Refers to a foam that has a conductive additive included in the chemical foam formulation prior to blowing and curing. [AIR4170]

insolation *1.* In general, solar radiation received at the earth's surface. *2.* The rate at which direct solar radiation is incident upon a unit horizontal surface at any point on or above the surface of earth. Contracted from incoming solar radiation.

inspect To compare the characteristics of an item with established standards. [AIR4896]

inspection, aircraft engine Inspection classified as preflight, postflight, turnaround, daily, special, and phase. *Preflight*–an inspection conducted prior to each flight to ensure the aircraft is safe for flight and to verify proper servicing. *Postflight*–an inspection conducted after each flight to detect degradation or damage that may have occurred during the flight and to determine the need for servicing. *Turnaround*–an inspection conducted between flights to ensure the integrity of the aircraft for flight, verify proper servicing, and detect degradation that may have occurred during the previous flight. *Daily*–an inspection conducted to detect defects, to a greater depth than the turnaround or preflight inspections. *Special*–a scheduled inspection with a prescribed interval other than daily, calendar/phase, major engine, or standard depot-level maintenance. *Phase*–a series of related inspections that are performed sequentially at specific intervals. [AIR4896]

inspection, detailed An intensive visual check of a specified detail, assembly, or installation. [AIR4896]

inspection, directed A collective term that includes the detailed inspection and the special detailed inspection. [AIR4896]

inspection door A small door in the outer enclosure allowing certain parts of the interior of the apparatus to be observed.

inspection, external surveillance A visual check that is used to detect obvious unsatisfactory conditions/discrepancies in externally visible structure and components. [AIR4896]

inspection, final Quality-control inspector certification performed after the last in-plant fabrication or assembly operation, and prior to preparation for shipment. [AS7200/1]

inspection, general visual A collective term that includes the external surveillance inspection, internal surveillance inspection, and the walk-around check. [AIR4896]

inspection hatch A covered opening in a tank which allows for access. [AIR4783]

inspection hole A hole located in the conductor barrel which permits inspection to determine that the conductor is properly located before crimping, and that the conductor is properly located after crimping, thus ensuring a proper termination. [ARP914A]

inspection, internal surveillance A visual check that is used to detect obvious unsatisfactory conditions/discrepancies in internal structure and components. [AIR4896]

inspection, nondestructive (NDT) *See* nondestructive inspection.

inspection, periodic A thorough and close check of the overall aircraft repeated at regular intervals of calendar time or hours of operation. Each periodic inspection comprises all intermediate pre-flight, thru-flight and post-flight inspections. It is repeated at regular intervals of calender time or hours of operation. [ARP4386]

inspection, quality control *See* quality control.

inspection, sample *See* sample inspection.

inspection, special detailed An intensive check of a specific location, similar to the detailed inspection except that some special technique such as nondestructive test techniques, dye penetrant, or high-powered magnification is requested, and disassembly procedures may be required. *See* inspection, detailed. [AIR4896]

inspection tests Those tests performed on a production article to determine acceptability prior to shipment. Also referred to as production acceptance tests. [ARP906A]

inspection, walk around check A visual check conducted from ground level to detect obvious discrepancies. [AIR4896]

inspiratory reserve *See* reserve, inspiratory.

instability *See* stability.

installation *1.* A completed set of duct assemblies and duct supports incorporated in a vehicle. [ARP699D] *2.* Putting equipment or software in place prior to commencing operation.

installation drawings Drawings that define the exterior size of the unit and include data on installation interfaces. [ARP906A]

installation torque The required nut torque to properly install a coupling. [AS1895]

installed Refers to hose assemblies joined to other parts to form assemblies, or components offered for acceptance against procurement contracts. [AS1933]

installer The party responsible for the installation of the equipment in the aircraft, and for obtaining certification for the complete system as installed. [AS8054]

instantaneous Any scenerio in which there is zero lag between stimulus and response.

instantaneous frequency In an angle-modulated wave, the derivative of the angle with respect to time.

instantaneous sampling Taking a series of readings of the instantaneous values of one or more wave parameters.

instantaneous trip (opening) Indicates that delay is not purposely introduced into the action of the device. [ARP1199A]

instantons Field configurations of Yang-Mills theory which are localized in space and time. These configurations are solutions of the Yang-Mills field equations in Euclidean space time which allow the transitions (tunneling) from one vacuum state to another.

INSTRB See input strobe.

instruction A written document describing the specifics of "how-to" accomplish a given task. [AS7102]

instruction counter *See* location counter.

instrument A device for measuring the value of an observable attribute; may merely indicate the observed value, or may also record or control the value.

instrumental analysis Any analytical procedure in which an instrument is used to measure a value, detect the presence or absence of an attribute, or signal a change or end point in a process.

instrumentation amplifiers High-precision amplifiers with high noise-rejection capabilities.

instrumentation tape Analog magnetic tape, ungapped, for continuous data (as in PCM or FM telemetry).

instrument correction A quantity added to, subtracted from, or multiplied into an instrument reading to compensate for inherent inaccuracy or degradation of instrument function.

instrument flight rules (IFR) *1.* Rules governing the procedures for conducting flight by reference to instruments. *2.* A term used by pilots and controllers to indicate a type of flight plan. [ARP4107]

instrument landing system (ILS) *1.* A precision-instrument approach system that normally consists of the following electronic components and visual: (a) localizer, (b) glide slope, (c) outer marker, (d) middle marker, and (e) approach lights. *See* localizer, glide slope, outer marker, middle marker, and approach lights. [ARP4107] *2.* In an aircraft, a system that provides, in the aircraft, a display of the lateral, longitudinal, and vertical references necessary for a landing.

instrument landing system (ILS) categories Landing systems (ILS) procedures that are classified as follows: ILS Category I, ILS Category II, and ILS Category III. ILS Category III is further divided into IIIA, IIIB, and IIIC. *ILS Category I*–an ILS approach procedure that provides for approach to a height above touchdown of not less than 200 ft and with runway visual range of not less than 1800 feet. *ILS Category II*–an ILS approach procedure that provides for approach to a height above touchdown of not less than 100 ft and with runway visual range of not less than 1200 feet. *ILS Category IIIA*–an ILS approach procedure that provides for approach without a decision height minimum and with runway visual range of not less than 700 feet. *ILS Category IIIB*–an ILS approach procedure that provides for approach without a decision height minimum and with runway visual range of not less than 150 feet. *ILS Category IIIC*–an ILS approach procedure that provides for approach without a decision height minimum and without runway visual range minimum. [ARP4107]

instrument loop diagram A diagram containing the information needed to understand the operation of the loop, and also showing all connections; purpose is to facilitate instrument startup and maintenance of the instruments. The loop diagram must show the components and accessories of the instrument loop, highlighting special safety requirements and other requirements.

instrument meteorological conditions (IMC) Meteorological (weather) conditions expressed in terms of visibility, distance from cloud, and ceilings that are less than the minima specified for visual meteorological conditions. [ARP4107]

instrument, millivoltmeter type An instrument that is operated and actuated by varying EMF output of a thermocouple. The varying EMF input to the instrument is obtained by temperature changes of the temperature-sensing thermocouple. [AS413B]

instrument oil A special grade of lubricating oil for instruments and other delicate mechanisms; formulated to resist oxidation and gumming, to be compatible with electric insulation, and to inhibit metals from tarnishing.

instrument, ratiometer type An instrument that is actuated by changes in electrical resistance of a temperature-sensing electrical resistance element. The resistance changes are obtained by temperature changes of the temperature-sensing resistance element. [AS413B]

instrument reading time On a continuous-reading instrument, the time lag between an actual change in an attribute and stable indication of that change .

instrument specification A detailed and exact statement of particulars, especially a statement prescribing performance, dimensions, construction, tolerances, bill of materials, features, and operating conditions.

instrument system A device that consists of more than one assembly or device (e.g., an electronic display system). [AIR818C]

instrument torque The turning moment on the moving element of an instrument, produced directly or indirectly by the quantity being measured.

instrument transformer A precision transformer capable of reproducing a signal in a secondary circuit; suitable for use in measuring, control, or protective devices.

insulated terminal A terminal having its conductor barrel and insulation support, if any, covered with a dielectric material. [ARP914A]

insulation *1.* Material having a high resistance to the flow of electric current, used to prevent leakage of current from a conductor. [ARP1931] *2.* Nonconducting material used to separate a conductor or shield from other conductors and shields and from ground. [AS1198] *3.* A material of low thermal conductivity used to reduce heat loss. *4.* A material of specific electrical properties used to cover wire and electrical cable.

insulation crimp Refers to the physical reshaping of an insulation sleeve to close or compress around the wire insulation. [ARP914A]

insulation grip The portion of an insulation barrel that, when closed or compressed around the conductor insulation, makes contact with the insulation on the cable. *See* barrel, insulation. [ARP914A]

insulation piercing terminal A terminal having a barrel with a design that displays the wire insulation and makes contact with the conductor. [ARP914A]

insulation resistance *1.* The ratio of applied voltage to the total current between two electrodes in contact with a specific insulation, usually expressed in megohms for 1000 feet. [ARP1931] *2.* For a wire or cable, the electrical resistance offered by its insulation to an impressed direct-current potential tending to produce a leakage of current through the insulation. [AS1198]

insulation shroud *See* shroud, insulation.

insulation support The portion of an insulation barrel that extends around the cable insulation, but is not necessarily compressed or closed to the point of making contact with the cable insulation. [ARP914A]

insulator A prefabricated item specifically designed to prevent any undesirable flow of energy between a conductor and/or other objects. [ARP480A]

insurance item *See* item, insurance.

intake *1.* An opening through which a fluid enters a chamber or conduit; an inlet. *2.* The amount of fluid entering through an opening.

intake hose The hose connection between the hydrant coupler and the plumbing of the hydrant dispenser. [AIR4783]

integer programming *1.* In operations research, a class of procedures for locating the maximum or minimum of a function subject to constraints, where some or all variables must have integer values. Contrast with convex programming, dynamic programming, linear programming, mathematical programming. *2.* Loosely refers to discrete programming.

integral *1.* A control action that will cause the output signal to change according to the summation of the input signal values sampled at regular intervals up to the present time. *2.* Mathematically, the reciprocal of reset.

integral absolute error (IAE) A measure of controller error defined by the integral of the absolute value of a time-dependent error function. Used in tuning automatic controllers to respond properly to process transients. *See also* integral time absolute error.

integral action A type of controller function in which the output (control) signal or action is a time integral of the input (sensor) signal.

integral blower A blower built as an integral part of a device to supply air thereto.

integral-blower burner A burner of which the blower is an integral part.

integral composite structure Composite structure in which several structural elements that would conventionally be assembled together by bonding or mechanical fasteners after separate fabrication are instead laid up and cured as a single, complex, continuous structure; for example, spars, ribs, and one stiffened cover of a wing box fabricated as a single integral part. [AIR4844]

integral control *1.* Describes a system that uses integration in the control-loop elements to provide an output in response to an error signal; referred to as a Type N system, where N is the order of the integration. [ARP4386] *2.* A form of control action that returns the value of the controlled variable to the set point whereas sustained offset would occur without this action. Also called reset control.

integral control action (reset) Control action in which the output is proportional to the time integral of the error input, i.e., the rate of change of output is proportional to the error input.

integral control mode A controller mode in which the controller output increases at a rate proportional to the controlled variable error; thus, the controller output is the integral of the error overtime, with a gain factor (called the integral gain).

integral flange A flange on a length of pipe, a nozzle or a pressure vessel which is cast or forged with the item itself, or is permanently attached to it by welding.

integral fuel tank On an aircraft, a load-carrying structure that is absolutely sealed to provide for fuel containment. The integral fuel tank exists as a cavity in a wing and/or the fuselage. [AIR4069]

integrally-cast specimen An attached specimen that is cast in the mold. [AMS5402B]

integrally heated Refers to tooling that is self-heating—through use of electrical heaters, such as cal rods. [AIR4844]

integrally lighted instrument An instrument in which the lighting system is designed such that the light sources are contained within the instrument enclosure, thus not relying on any outside source of illumination. An integrally lighted instrument may employ reflected or transilluminated light. [ARP924]

integral orifice A differential-pressure-measuring technique for small flow rates in which the fluid flows through a miniature orifice plate integral with a special flow fitting.

integral rocket ramjet A combination of a solid-propellant rocket and a ramjet which uses the empty booster case as a ramjet combustor.

integral skin foam Urethane foam with a cellular core structure and a relatively nonporous skin. [AIR4844]

integral time absolute error (ITAE) A measure of controller error defined by the integral of the product of time and the absolute value of a time-dependent error function; whereas the

absolute value prevents opposite excursions in the process variable from canceling each other, the multiplication by time places a more severe penalty on sustained transients. *See also* integral absolute error.

integrated actuator package A self-contained hydraulic servoactuator designed as a packaged unit, with an electric motor driving a hydraulic pump to pressurize the hydraulic system in the package. The package includes a reservoir, filter, check valve, control valve, and other elements that are necessary to provide an output. The output controls the position of a device, usually an airfoil. [ARP4386]

integrated circuit (IC) A complete electronic circuit containing active and passive elements fabricated and assembled as a single unit, usually as a single piece of semiconducting material, resulting in an assembly that cannot be disassembled without destroying it.

integrated circuit piezoelectric (ICP) A type of pressure-sensitive sensor that combines a piezoelectric element with isolation amplifier and signal conditioning microelectronics inside the sensor housing. Such a sensor allows the output signal to be transmitted over ordinary two-wire cable instead of special low-noise cable.

integrated energy systems Community systems for energy generation and distribution.

integrated flight propulsion control Control designed to manage both the engine and aircraft to the benefit of both, as opposed to conventional controls in which the aircraft and engine controls are designed separately. [AIR4548]

integrated logistic support (ILS) The management and technical process through which supportability and logistic support considerations of systems/equipment are integrated during the early phases of, and throughout the life cycle of, the program; and through which all elements of logistic support are planned, acquired, tested, and provided in a timely and cost-effective manner. [ARP4293]

integrated optics Thin-film devices containing tiny lenses, prisms, and switches to transmit very thin laser beams, which serve the same purposes as the manipulation of electrons in thin-film devices of integrated electronics.

integrated software A computer program that combines several functions for ease of use.

Integrated Systems Digital Network (ISDN) A suite of protocols being defined by CCITT to provide voice and data services over wide-area networks (WANs).

integrating accelerometer A device that measures the acceleration of an object, and converts the measurement to an output signal proportional to speed or distance traveled.

integrating ADC A type of analog-to-digital converter in which the analog input is integrated over a specific time. The integrating ADC has the advantages of high resolution, noise rejection, and linearity.

integrating control A control system in which rate of change of valve position is proportional to temperature error; or valve position is the integral of temperature error. [ARP89C]

integrating extensions An integrator that derives its input from the motion of the float.

integrating frequency meter A master-frequency meter for an electric power system which measures the actual number of cycles of alternating voltage for comparison with the theoretical number of cycles for the same time at the prescribed frequency.

integrating meter *1.* A totalizing meter, such as for electric energy consumed. *2.* An instrument whose output is proportional to the single (or higher order) integral of the quantity measured.

integrating network A transducer circuit whose output waveform is a time integral of its input waveform.

integrating sphere A sphere used in optical measurements which is intended to integrate the input light over the output aperture to provide uniform illumination.

integrating temperature compensated tachometer generator A generator used in computing applications requiring integration of a variable with respect to time. This type of tachometer generator is characterized by very small deviations of the output voltage as a function of temperature and a minimum warm-up

time. Temperature-control and compensation networks are usually integral parts of integrating generators. [ARP667]

integration *1.* The act of causing elements of an item to function together. *2.* The act of gathering a number of separate functions within a single implementation. [ARP4754]

integrator(s) *1.* A device whose output is proportional to the integral of an input signal. *2.* In digital computers, a device for accomplishing a numeric approximation of the mathematical process of integration. [AIR1489] *3.* A device that continually totalizes or adds up the value of a quantity for a given time. *4.* A device whose output is proportional to the integral of the input variable with respect to time.

integrity *1.* Attribute of a system or an item indicating that it can be relied upon to work correctly on demand. [ARP4754] *2.* In data processing, a word to describe data that has not been corrupted.

intelligence In data processing, the processing capability of a computer.

intelligent terminal A computer terminal with some local processing capability.

intensifier A device that converts an input pressure to a higher output pressure, proportional to its input pressure. [ARP4386]

intensifier tube A tube that passes within the ducts or pipes carrying engine exhaust gases; designed to transfer heat from the exhaust gas to the air flowing within the tube. [ARP147C]

intensifying screen A sheet of material placed in contact with radiographic film, which undergoes secondary fluorescence when struck with x-rays or gamma rays, thereby increasing image density for a given exposure.

intensity *1.* In general, the degree or amount, usually expressed by the elemental time rate or spatial distribution of some condition or physical quantity, such as electric field, sound, or magnetism. *2.* With respect to electromagnetic radiation, a measure of the radiant flux per unit solid angle emanating from some source. Frequently, it is desirable to specify this as radiant intensity in order to distinguish it clearly from luminous intensity. *3.* The amount of light incident per unit area. For human viewing of visible light, the usual term is illuminance; for electromagnetic radiation in general, the term is radiant flux.

intensity level The amplitude of a sound wave, commonly measured in decibels.

interaction *1.* A phenomenon, characteristic of a multivariable process, in which the effect of a manipulative variable change in one control loop not only affects its own controlled variable, but also the controlled variable in another loop. In distillation, the primary consideration is interaction between the overhead composition control loop and the bottoms composition control loop. *2.* The influence of one subsystem or subassembly on the performance, maintainability, and reliability of another. [AIR4896]

interaction analysis A technique used in determining the pairing of manipulative and controlled variables in a control loop.

interactive In data processing, refers to a technique of user/system communication in which the operating system immediately acknowledges and sets upon a request entered by the user at a terminal. Compare with batch.

interactive control The sending of multiple commands that are selected on the basis of data received from an experiment in real time.

interaxis error The angular deviation of the null positions for all rotor-stator winding combinations from space quadrature. [ARP826]

interblock gap Blank space on a computer storage medium between two adjacent blocks of data.

intercalation Production of layer-type semiconducting materials as well as other conducting materials.

intercept method A method for estimating the quantity of particles or number of grains within a unit area of a microscopic image by counting the number intercepted by a series of straight lines through the image. This is one of the standard methods of determining grain size of a polycrystalline metal.

interchange Removing the item that is to be replaced, and installing the replacement item. [ARD50010]

interchangeability The ability to interchange, without restriction, like equipments or portions thereof in manufacture, maintenance, or operation. [ARD50010]

interchangeable Refers to a part, subassembly, assembly, or unit that meets or exceeds required functional and structural specifications for a given application. [AIR4896]

interchangeable item An item that: (a) possesses such functional and physical characteristics as to be equivalent in performance, reliability, and maintainability to another item of similar or identical purposes; and (b) is capable of being exchanged for the other item, without selection for fit or performance; and without alteration of the items themselves or of adjoining items, except for adjustment. [ARD50010]

interchangeable, one way Refers to the introduction of a new item and the placing of restrictions on the use of the old item. The new item may be used in place of either the old or the new, but the old item can only be used in an application where it has previously been installed. The new item is considered to supersede the old. For example, considering items A and B, describes the case where B may freely be used in applications where A is specified, but A must not be used in applications where B is specified. [ARD50010]

intercom A voice-communication system among different stations in an aircraft. [ARP4107]

interconnect function The physical function that establishes and maintains information-transfer paths through the transmission network. [ARD50012]

interconnect wire A type of wire used in general-purpose application in aerospace vehicles for interconnecting individual electrical or electronic devices. [ARP1931]

intercooler A heat exchanger in the path of fluid flow between stages of a compressor; purpose is to cool the fluid and allow it to be further compressed at lower power demand.

intercooler, cabin pressurizing A heat exchanger designed to reduce the temperature between two stages of air compression. [ARP147C]

interelectrode capacitance *1.* The capacitance between electrodes of a vacuum tube. *2.* A capacitance determined by measuring the short-circuit transfer admittance between two electrodes.

interface *1.* A shared boundary. [ARD50020] *2.* The part of a sensor system that includes the receiver, plus circuits to reformat the signal (electrically) for the end user, plus an emitter to generate a transducer-driving signal. [ARD50020] *3.* Those parts of a data-bus system in which optical/electrical conversion takes place. [ARD50020] *4.* The physical, functional, or procedural relationships established as a basis for division of responsibilities between two or more independent design, manufacturing, or test activities, as related to the hardware for which they are jointly responsible. [AS1426]

interface-functional The relationship between functions at the hardware physical interface, such as electrical voltage and current, fluid pressure and flow, temperature, acceleration, and acoustic and other environments. [AS1426]

interface, optical An arbitrary point at which signals are in optical form. [ARD50024]

interface-physical The hardware physical relationship at the juncture between items designed, manufactured, or tested by different activities jointly responsible for the hardware. Includes dimensional geometric relationship, tolerances, materials, and finishes at the juncture. [AS1426]

interface-procedural For hardware that has physical or function interfaces, the matters related to successful conduct of the program, such as design reviews, tests, inspections, approvals, government certification, data preparation, schedules, and related procedural activities. [AS1426]

interfaces The systems, external to the system being analyzed, that provide a common boundary or service, and are necessary for the system to perform its mission in an undegraded mode, e.g., systems that supply power, cooling, heating, air services, or input signals. [ARD50013]

interface, signal An arbitrary point at which the effect of the measurand is available in a form directly usable for control or measurement purposes. [ARD50024]

interference *1.* Instrument response due to presence of components other than the gas (or vapor) to be measured. [AIR1533] *2.* Any undesirable electromagnetic emission or any electrical or electromagnetic disturbance, phenomenon, signal, or emission—man-made or natural—which causes an undesirable response, malfunction, or degradation of the performance of electrical or electronic equipment. [ARP1931]

interference (electrical) Any spurious voltage or current arising from external sources and appearing in the circuits of a device. *See also* noise.

interference, common mode A form of interference that appears between measuring-circuit terminals and ground.

interference, electromagnetic *See* electromagnetic interference.

interference filter An optical filter that selectively transmits specific wavelengths of light as a result of interference due to dielectric coatings on the surface of the material. Multilayer interference coatings may include metallic layers.

interference fits A joint or mating of two parts in which the male part has an external dimension larger than the internal dimension of the mating female part. [AIR4844]

interference pattern The pattern of some characteristic of a stationary wave produced by superimposing one wave train on another. The interference pattern may be the distribution in space of energy density, energy flux, particle velocity, pressure, or some other characteristic.

interference seal A seal produced between a fastener and its hole when a fastener of a given diameter is driven into a hole of smaller diameter. (The hole diameter must be approximately 0.003 in. (0.08 mm) smaller to be effective.) An interference seal is also produced when a fastener shank is expanded by the installation process. [AIR4069]

interferometer An instrument so designed that the variance of wavelengths and light-path lengths within the mechanism allows very accurate measurement of distances.

interferometric pressure transducer A type of pressure sensor developed to read pressure differentials on the order of 200 Pa (0.030 psi), with a resolution of 1 Pa (0.00015 psi), by detecting very small deflections of a fragile diaphragm through optical interferometry.

interferon A protein (lymphokine) released by cells in response to virus infection. When taken up by other cells, interferon inhibits the replication of viruses within them.

interflow Flow between ports of a valve at intermediate positions, not in accordance with nominal or intended flow pattern. [ARP4386]

intergranular beta Beta phase situated between alpha grains. May be at grain corners, as in the case of equiaxed alpha-type of microstructures in alloys having low beta-stabilizer content. [AS1814]

interior ballistics The branch of ballistics that deals with the propulsion of projectiles, e.g., the motion and behavior of projectiles in a gun barrel, or the temperatures and pressures developed inside a gun barrel or rocket.

interlaminar Describes an object, event, or potential field referenced as existing or occurring between two or more adjacent laminae. [AIR4844]

interlaminar shear In a laminate, a shearing force tending to produce a relative displacement between two laminae along the plane of their interface. [AIR4844]

interlaminar shear strength The maximum shear stress between layers that a laminated material can resist. [AIR4844]

interleaving The alternating of two or more operations or functions through the overlapped use of a computer facility.

interlock *1.* Instrument that will not allow one part of a process to function unless another part is functioning. *2.* A device such as a switch that prevents a piece of equipment from operating when a hazard exists. *3.* To join two parts together in such a way that they remain rigidly attached to each other solely by physical interference. *4.* A device to prove the physical state of a required condition, and to furnish that proof to the primary safety-control circuit.

interlock, motor start A connection made through contacts on the motor controller which is wired in series with the safety circuit, so that the motor must be energized before the system is allowed to proceed.

interlock switch A switch used in the stowage bucket, or at the bottom-loading adapter, to actuate the interlock system when the nozzle or coupler is not stowed, or when the bottom-loading system is being utilized. [AIR4783]

interlock system A system incorporated into the stowage of some or all of the movable equipment that must be put into a stowage position prior to allowing movement of the vehicle. [AIR4783]

interlock valve A valve used in the stowage bucket or at the bottom-loading adapter to actuate the interlock system when the nozzle or coupler is stowed, or when the bottom-loading system is being utilized. [AIR4783]

intermediate addressing Method of addressing data stored in a computer memory. In this method, the instruction operand is the data to be used with the instruction.

intermediate approach segment *See* segments of an instrument approach procedure.

intermediate band A mode of recording and playback in which the frequency response at a given tape speed is "intermediate."

intermediate bearing stress The bearing stress at the point on the bearing-load-deformation curve where the tangent is equal to the bearing stress divided by a designated percentage of the original hole diameter. [AIR4844]

intermediate fix (IF) For an instrument approach procedure, the fix that identifies the beginning of the intermediate approach segment. This fix is not normally identified on the instrument approach chart as an intermediate fix (IF). *See also* segments of an instrument approach procedure. [ARP4107]

intermediate frequency (IF) The beat frequency used in a heterodyne receiver, usually the difference between the received radio-frequency signal and a locally generated signal.

intermediate level maintenance Maintenance performed in direct support of using organizations. Tasks normally consist of calibration, repair, or replacement of damaged or unserviceable parts, components, or assemblies; emergency manufacture of nonavailable parts; and provision of technical assistance to using organizations. Referred to as Level 2 maintenance. Synonymous with intermediate maintenance. [ARD50010] *See also* maintenance levels.

intermediate maintenance *See* intermediate level maintenance.

intermediate means In an instrumentation or control system, all system elements between the primary detector and the end device which transmit or modify the output of the former to make it compatible with input requirements of the latter.

intermediate phase In an alloy system, a distinct compound or solid solution whose composition limits do not extend to any of the pure constituents.

intermediate temperature setting adhesive An adhesive that sets in the temperature range from 30 to 100°C (87 to 121°F) [AIR4844]

intermetallic compound In an alloy system, a phase that usually occurs at a definite atomic ratio and exhibits a narrow solubility range. Nearly all such phases are brittle. [AS1814]

intermittency effect In photography or radiography, refers to a departure from the reciprocity law when the emulsion is exposed in a series of discrete increments, compared to the response when it is exposed continuously to the same total energy level.

intermittent duty Refers to an operating cycle that consists of alternating periods of use and idle time; for example, on and off; load and no-load; load and rest; or load, no-load and rest. In most instances, successive periods of use or idle time vary widely in length, although some intermittent-duty cycles follow well-defined patterns.

intermittent firing A method of firing in which fuel and air are introduced into and burned in a furnace for a short period, after which the flow is stopped, this succession occurring in a sequence of frequent cycles.

intermittent precipitation Describes precipitation that has stopped and recommenced at least once during the hour (60 minutes) preceding the actual time of observation. [AIR1335]

intermodulation The modulation of the components of a complex wave by one another, producing new waves whose frequencies are equal to the sums and differences of integral multiples of the component frequencies of the original complex wave.

intermodulation distortion (IMD) Defined as 20 log (rms sum of the sum and difference distortion products)/(rms amplitude of the fundamental).

internal adhesive stress Stress created within an adhesive layer by the movement of the adherents at differential rates or by the contraction or expansion of the adhesive layer. [AIR4844]

internal combustion engine A mechanical prime mover that uses as the thermodynamic working fluid exhaust gases resulting from the burning of fuel within the engine.

internal combustion heat exchanger A heat exchanger that utilizes the heat produced by combustion of a fuel within the heater to heat the air supplied to various aircraft systems. [AS8040]

internal energy A mathematically defined thermodynamic function of state, interpretable through statistical mechanics as a measure of the molecular activity of the system.

internal furnace Within a boiler, a furnace consisting of a straight or corrugated flue, surrounded with water.

internal gear Any ring-type or annular gear whose teeth are on the inner surface of the rim.

internalized unit values A system of values, motives, and prioritized goals held by a unit and adopted by a member of that unit. A person who has internalized unit values is referred to colloquially as a "team player." [ARP4107]

internal leakage Leakage between internal cavities of a device. [ARP4386]

internally fired boiler A fire-tube boiler having an internal furnace, such as a scotch, locomotive fire-box, vertical tubular, or other type having a water-cooled plate-type furnace.

internal-mix oil burner A burner having a mixing chamber in which high-velocity steam or air impinges on jets of incoming liquid fuel, which is then discharged in a completely atomized form.

internal oxidation In a material, a form of degradation involving absorption of oxygen at the surface and diffusion of oxygen to the interior, where it forms subsurface scale or oxide inclusions.

internal pressure *1.* The pressure inside a system or component. [ARP4386] *2.* The pressure inside a portion of matter due to the attraction between molecules.

internal respiration *See* respiration, types of. [ARP171]

internal standard In chemical analysis, especially instrumental analysis, a material present in or added to a sample in known amounts to serve as a reference in determining composition.

internal storage Addressable storage directly controlled by the central processing unit of a digital computer.

internal stress *See* residual stress.

internal system load A load that is dynamically generated by the system, and that may exceed the externally applied loads. [ARP4386]

internal treatment The treatment of boiler water by introducing chemicals directly into the boiler.

internal wave In fluid mechanics, wave motions of stably stratified fluids in which the maximal vertical motions occur below the surface of the fluids.

International Civil Aviation Organization (ICAO) A specialized agency of the United Nations whose objective is to develop the principles and techniques of international air navigation and to foster planning and development of international civil air transport. [ARP4107]

international rubber hardness degree A measure of hardness, the magnitude of which is derived from the depth of penetration of a specified indentor into a test piece under specified conditions. [AIR4844]

International Standard (IS) The third (and highest) stage of the ISO standard process. Prospective ISO standards are balloted three times. The first stage is a Draft Proposal (DP). After a Draft Proposal has been in use a period of time (typically six months to a year) the standard, frequently with corrections and changes, is re-balloted as a Draft International Standard (DIS). After the Draft International Standard has been in use for a period of time (typically one to two years) it is reballoted as an International Standard (IS).

International System of Units (SI Units) The metric system of units based on the meter, kilogram, second, ampere, Kelvin degree, and candela. Other SI units are hertz, radian, newton, joule, watt, coulomb, volt, ohm, farad, weber, and tesla.

interpass temperature The lowest temperature reached by weld metal before the next pass is deposited in a multiple-pass weld.

interphase The boundary region between a bulk resin or polymer and an adherent; in this region, the polymer has a high degree of molecular orientation. [AIR4844]

interply cracks Through-cracking of individual layers within a CFRP layup, perpendicular to the ply interfaces. [ARP5089]

interply hybrid A composite in which adjacent laminae are composed of different materials. [AIR4844]

interpolation Estimation of a value between two known values to account for the uncertainty, applicable to the temperature read. Interpolation is dependent on the scale-division spacing, the width of the indicator point, and the operator's skill in discerning the ratio of the widths of the spaces between the indicator and the two nearest scale markings. [AS7102]

interpretation *1.* The determination of the cause of an indication. *2.* The evaluation of the significance of discontinuities as to whether they are detrimental defects or superficial blemishes. [ARP5089]

interprocessor communication Communication between two or more processors in a computer system.

Inter-Range Instrumentation Group (IRIG) A working group that is responsible for specifying the industrywide standards and practices of telemetry.

interrogator *1.* The ground-based surveillance radar beacon transmitter-receiver that normally scans in synchronism with a primary radar system. The interrogator transmits signals that cause properly functioning transponders located in aircraft to reply with a unique signal. The reply signals from the interrogated aircraft are mixed with the primary radar returns and displayed on the radar scope of the controller. 2. The airborne element of the TACAN/DME system. [ARP4107]

interrupting capacity A rated value that represents the maximum short-circuit current at rated voltage which a protective device is required to interrupt under the operating duty specified, with a normal frequency-recovery voltage not less than rated voltage. *See also* rupture capacity. [ARP1199A]

interruption, air A change from original flight plan due to a known or suspected malfunction and/or defect during flight. [AIR4896]

interruption, ground Can refer to either of the following occurrences: (a) when an aircraft leaves the block and returns for a technical reason before becoming airborne; or (b) when, after landing, a technical problem is experienced prior to reaching the block. [AIR4896]

intersection In Boolean algebra, the operation whereby concepts are described by stating that they have all the characteristics of the classes involved. Intersection is expressed as AND.

intersection departure/intersection takeoff A takeoff or proposed takeoff on a runway commencing from the intersection of a taxiway with a runway or the intersection of two runways. [ARP4107]

interstellar chemistry Molecular formation/dissociation in interstellar space due to radiation, collision, and other forces.

interstice The space or void left between or around the cabled or stranded components. [ARP1931]

interstitial Situated between the cellular components of an organ or structure. [ARP171]

interstitial element An element with relatively small atomic diameter which can assume position in the interstices of the titanium crystal lattice. Common examples are oxygen, nitrogen, hydrogen, and carbon. [AS1814]

interval The number of word times that occur between successive repetitive samples of the same channel. Synonymous with super-commutation and strapping interval.

interval timer A device that provides an interrupt signal upon completion of a predetermined or programmed interval of time. *See* timer.

in-the-loop Describes a component that is a necessary part of the closed control loop, and that if removed would cause inaction, failure, or malfunction, thus interrupting the control-action-feedback loop of the system. For example, when an aircraft is flying on autopilot, the autopilot subsystem is in-the-loop, and the pilot is no longer in-the-loop; the pilot is merely monitoring the closed-loop flight control system. If the autopilot fails, the pilot must get back in-the-loop, that is, resort to manual reversion. *See also* closed-loop system. [ARP4107]

intralaminar Descriptive term pertaining to an object event or potential field existing entirely within a single lamina, without reference to any adjacent laminae. [AIR4844]

intraorbit transfer vehicle Small, scooter-type tugs proposed to move men and materials within an orbit.

intrapleural pressure *See* pressure, intrapleural. [ARP171]

intraply hybrid A composite in which different materials are used within a specific layer or band. [AIR4844]

intrapulmonary pressure *See* pressure, intrapulmonary. [ARP171]

intrinsic Refers to the situation in which modulation by the measurand of the light beam within the sensor occurs internally to the light-guiding regions of the fiber. [ARD50024]

intrinsic availability The probability that the system is operating satisfactorily at any point in time when used under stated conditions, and where the time considered is operating time and active repair time. [AIR4896]

intrinsic joint loss A loss intrinsic to the fiber, caused by fiber-parameter mismatches when joining two nonidentical fibers.

intrinsic safety A method to provide safe operation of electric process-control instrumentation where hazardous atmospheres exist. The method keeps the available electrical energy so low that ignition of the hazardous atmosphere cannot occur.

intrinsic safety barrier A device inserted in wire, between process-control instrumentation and the point where the wire passes into the hazardous area. This device limits the voltage and current on the wire to safe levels.

intruder Any aircraft (including all threat aircraft) in the airspace of another aircraft which is tracked by the collision-avoidance system. [ARP4153]

intrusive Describes the situation when the sensor protrudes into the flow stream and disturbs the flow field. [AIR4367]

invalid status message segment A condition that occurs within a bus controller when the bus controller determines that a status-message segment is not a valid status-message segment. [AS4113]

inventory, active The group of items assigned to an operational status. [AIR4896]

inventory, inactive The group of items being held in reserve for possible future assignments to an operational status. [AIR4896]

inverse response For a process, a dynamic characteristic whereby the output of the process responds to an input change by moving initially in one direction, but finally in the other.

inverse scattering Method of analyzing some classic wave scattering.

inverse-time *See* time-inverse.

inversion temperature In a thermocouple, the temperature of the "hot" junction when the thermoelectric emf of the circuit is equal to zero.

inverted spin A spin throughout which the airplane is upside down. *See* spin. [ARP4107]

inverter A NOT element. With an inverter, the output signal is the reverse of the input signal.

involute splines, wear limit The amount of tooth thickness decrease for an external involute spline, or the amount of space width increase for an internal involute spline which may be considered allowable without expectation of failure. [ARP1200]

I/O *See* input/output.

I/O hardware Computer hardware used to carry signals into and out of processing hardware.

ion A charged atom or radical which may be positive or negative.

ion chamber *See* ionization chamber.

ion density (concentration) In atmospheric electricity, the number of ions per unit volume of a given sample of air; more specifically, the number of ions of a given type (e.g., positive small ion, negative small ion, positive large ion, or negative large ion) per unit volume of air.

ion engine A type of engine in which the thrust to propel the missile or spacecraft is obtained from a stream of ionized atomic particles, generated by atomic fusion, fission, or solar energy. [ARP4386]

ion exchange A chemical process for removing unwanted dissolved ions from water by inducing an ion-exchange reaction (either cationic or anionic) as the water passes through a bed of special resin containing the substitute ion.

ion-exchange resin A synthetic organic compound (resin) that can remove unwanted ions from a dilute solution by combining with the ions or by exchanging them for ions that produce desirable or neutral effects.

ion gage *See* ionization gage.

ionic mobility In gaseous electric conduction, the average velocity with which a given ion drifts through a specified gas under the influence of an electric field of unit strength. Mobilities are commonly expressed in units of centimeters per second per volt per centimeter.

ionic propellant *See* ion engine.

ionic strength Effective strength of all ions in a solution; for dilute solutions, equal to the sum of one-half of the product of the individual ion concentration and their ion valence or charge squared.

ion implantation A process for enhancing surface properties of a solid by bombarding it with a beam of high-energy ions, which are absorbed into the surface layer of the material.

ionization *1.* Generally, the disassociation of an atom or molecule into positive or negative ions or electrons. *2.* A property of an insulator whereby it facilitates the passage of current due to the presence of charged particles, usually induced artificially. [ARP1931]

ionization chamber Apparatus used to study the production of small ions in the atmosphere by cosmic-ray and radioactive bombardment of air molecules. Also known as an ion chamber.

ionization constant A measure of the degree of dissociation of a polar compound in dilute solution at equilibrium. Equal to the product of the concentrations of the dissociated compound (ions) divided by the concentration of the undissociated compound.

ionization gage A pressure transducer whose mode of operation is based on conduction of electric current through ionized gas of the system whose pressure is to be measured. Useful only for very low pressures (for example, below 10^{-3} atm). Also known as an ion gage.

ionization potential The energy required to ionize an atom or molecule; usually given in terms of electron volts.

ionization time In a gas tube, the interval between the time when conduction conditions are established and the time when conduction actually begins at some stated value of tube potential.

ionization vacuum gage An instrument for measuring very low pressures (high vacuums) by means of a current of positive ions produced in the gas. The positive ions are produced by electrons emitted from a hot cathode and accelerated across a portion of the evacuated space toward another electrode.

ionization voltage (corona level) The minimum value of falling rms voltage that sustains electrical discharge within the vacuous or gas-filled spaces in the cable construction or insulation. [ARP1931]

ionizer In ion engines or other devices, a filament, grid, or porous body that strips electrons from the outer shells of neutral atoms to form positively charged ions.

ionizing event Any interaction between an atom or molecule and an energy beam, particle, atom, or molecule which causes one or more ions to be generated.

ionizing radiation Any electromagnetic or particulate radiation capable of producing ions—directly or indirectly—in its passage through matter.

ion laser A laser in which the active medium is an ionized gas, typically one of the rare gases, argon or krypton, or a mixture of the two.

ion nitriding A vacuum process for producing a nitrided case on ferrous materials using a glow discharge. [AMS2759/8]

ionopause The upper boundary of the ionosphere of certain planets (excluding the earth) and comets, at which electrons decline sharply.

ionosphere The portion of the earth's atmosphere where ionization takes place due to ultraviolet radiation of the sun, or from bombardment by hydrogen bursts from sunspots. The various layers, identified as B, C, D, E and F, have characteristics that cause them to reflect and refract radio waves according to their frequency, time of day, sunspot cycle, and earth weather.

ionospheric storms Disturbances of the ionosphere, resulting in anomalous variations in its characteristics and effects on radio communication.

ionospheric tilts Ionospheric conditions in which the number of the electrons as a function of altitude is variable. Ionospheric tilts are sometimes created by traveling ionospheric disturbances (TIDs). They deflect radio waves in unexpected directions, adversely affecting radio reception.

ion pair The combination of a positive ion and a negative ion having the same magnitude of charge, and formed from a neutral molecule due to absorption of the energy in radiation.

ions 1. Charged atoms or molecularly bound groups of atoms; or, sometimes, free electrons or other charged subatomic particles. 2. In atmospheric electricity, any of several types of electrically charged submicroscopic particles normally found in the atmosphere. Atmospheric ions are of two principal types, small ions and large ions, although a class of intermediate ions has occasionally been reported. 3. In chemistry, atoms or specific groupings of atoms that have gained or lost one or more electrons; for example, the chloride ion or ammonium ion. Such ions exist in aqueous solutions and in certain crystal structures.

ion storage The placing of ions within an electromagnetic trap, cooled to sub-Kelvin temperatures with lasers. Potential uses are for frequency standards.

ion stripping In a reactor to be used for particle-beam pellet fusion, the procedure following the focusing of ion beams in the target chamber.

I/P converter A device that linearly converts electric current into gas pressure; for example, 4–20 mA into 3–15 psi. *See* converter.

IR camera A camera used to view or record temperature differentials during thermographic testing. [ARP5089]

IRIG *See* inter-range instrumentation group.

iron bird model Model used in developing an aircraft control system whereby higher fidelity and nonstandard conditions are expected. [AIR4548]

irradiance The power per unit area incident upon a surface. Also called radiant flux density.

irradiation *1.* As applied to plastics, the bombardment with a variety of subatomic particles, usually alpha-, beta-, or gamma-rays. [AIR4844] *2.* The exposure of an insulating or jacketing material to high-energy emissions for the purpose of favorably altering the molecular structure by providing cross-linking of molecular chains. [ARP1931] *3.* Exposing an object or person to ultraviolet, visible, or infrared energy.

irregular galaxies Galaxies with amorphous structure and relatively low mass (10^8 to 10^{10} solar masses). Fewer than 10% of all galaxies are classified as irregular.

irreversible Not capable of redissolving or remelting. [AIR4844]

irreversible component Any component of the system that has a forward-driven mechanical efficiency of less than 50%. An irreversible component can fulfill the function of a no back. [ARP4386]

irreversible drive A drive that, without input torque, cannot be back-driven from the output by a torque less than the maximum design load at the output. [ARP4386]

irreversible surface A control surface whereby the control mechanism prevents aerodynamic forces from moving the surface. [ARP4386]

IS *See* International Standard.

ISA Acronym for Instrument Society of America.

ISDN *See* integrated systems digital network.

isentrope A line of equal or constant pressure, with respect to either space or time.

isentropic Proceeding at constant entropy.

isentropic exponent A ratio defined by the specific heat at constant pressure divided by the specific heat at constant volume.

ISO *See* infrared space observatory.

isobaric Proceeding at constant pressure.

isobaric flame temperature The temperature of a propellant flame under constant-pressure conditions. [AIR913]

isobaric process Of equal or constant pressure, with respect to either space or time. [ARP4386]

isobars (pressure) Lines of equal or constant pressure, specifically such lines on a weather map.

isochoric flame temperature The temperature of a propellant flame under constant-volume conditions. [AIR913]

isochoric process Of equal or constant volume; usually applied to a thermodynamic process during which the volume of the system remains unchanged. [ARP4386]

isochronous governor A device that maintains constant engine rotational speed, regardless of load.

isocyanate plastics Plastics based on resins made by the condensation of organic isocyanates with other compounds. [AIR4844]

isolated Describes a metallographic feature that occurs in 3% or less of the microstructure. [AS1814]

isolation A technique used in fault-tolerant systems whereby the effects of a failure are removed or a failure is prevented from propagating or affecting the continued operation of the system. [ARP4386]

isolation level The functional level to which a failure can be isolated using accessory test equipment at designated test points. [ARD50010]

isolation seal A secondary seal used to isolate potential fuel-leakage paths. Purpose is to prevent channeling of fuel along a leak path between structural members. Can be a repair seal installed to reestablish seal continuity in areas of direct contact with the fuel. [AIR4069]

isomerization Process of converting hydrocarbon or other organic compound to an isomer.

isomers *1.* Nuclides having the same mass number A and atomic number Z, but existing for measurable times in different quantum states with different energies and radioactive properties. *2.* Molecules having the same atomic composition and molecular weight, but differing in geometrical configuration.

isoparametric finite elements The basis for the calculation of physical properties of structural shapes, including stress analyses.

isopotential point Point on the millivolt versus pH plot at which a change in temperature has no effect. This is at 7 pH and zero millivolts, unless shifted by the standardization and meter-zero adjustments, or an electrode-asymmetry potential.

isopycnic process Of equal or constant density with respect to either space or time. Equivalent to an isosteric process. [ARP4386]

isostasy A supposed equality existing in vertical sections of the earth, whereby the weight of any column from the surface of the earth to a constant depth is approximately the same as that of any other column of equal area, the equilibrium being maintained by plastic flow from one part of the earth to another.

isostatic pressing Pressing powder under a gas or liquid so that pressure is transmitted equally in all directions; for example, as in sintering. [AIR4844]

isosteric process Of equal or constant specific volume with respect to either space or time. Equivalent to an isopycnic process. [ARP4386]

isotensoid structure Filamentary structure in which the filaments are uniformly stressed throughout as a result of the design loading conditions.

isothermal Proceeding at constant temperature.

isothermal expansion An expansion process in which the gas temperature remains constant. [ARP4386]

isothermal process For a system, a thermodynamic change of state that takes place at constant temperature.

isothermal region A region of constant temperature. [ARP147C]

isotherms Lines connecting points of equal temperature.

isotope(s) Any of two or more nuclides that have the same number of protons in their nuclei but different numbers of neutrons. Isotopes are of the same element, and thus cannot be separated from each other by chemical means; however, because they have different masses, they can be separated by physical means.

isotope effect The effect of nuclear properties, other than the number of protons, on the non-nuclear physical and chemical behavior of nuclides.

isotopic enrichment Process by which the relative abundance of the isotopes of a given element are altered in a batch to produce a form of the element enriched in a particular isotope.

isotropic Invariant with respect to direction.

isotropic laminate A laminate in which the strength properties are the same in all directions. [AIR4844]

isotropic ply A ply with similar properties in all directions. [AIR4844]

isotropic turbulence Turbulence in which the products and squares of the velocity components and their derivatives are independent of direction; or more precisely, invariant with respect to rotation and reflection of the coordinate axes in a coordinate system moving with the mean motion of the fluid.

ITAE *See* integral time absolute error.

item One or more hardware and/or software elements treated as a unit. [ARP4761]

item, completed Items of equipment (including basic or end items, components, and assemblies) that have been overhauled, modified, renovated, and completed in accordance with terms of contracts, project orders, or other work directives and authorizations, and are ready for their intended use after receiving final mechanical acceptance inspection. [ARP4386]

item, consumable An item that is used only once. [ARD50010]

item, critical *See* critical item.

item, expendable Items for which no authorized repair procedure exists, and for which cost of repair would normally exceed that of replacement. [ARD50010]

item, insurance An item held by either the manufacturer or an airline, purely as a precaution, in order to preclude undue scheduling problems and/or economic hardship which might otherwise occur, should the part be out of stock when a requirement occurs. [AIR4896]

item levels From the simplest to the more complex, designated as follows: part, subassembly, assembly, unit, group, set, subsystem, system. [ARP4386]

item, life limited An item that has a limited and predictable useful life and could be considered for replacement on a preplanned basis for reliability, safety, or economic reasons. [ARD50010]

item, line maintenance Any item that can be readily changed on an aircraft during the maintenance operations. [ARD50010]

item, maintenance significant (MSI) Item identified by the manufacturer whose failure: (a) could affect safety (ground or flight), and/or (b) is undetectable during operations, and/or (c) could have significant operational economic impact, and/or (d) could have significant nonoperational economic impact. [ARD50010]

item, mandatory replacement An item that, if disturbed or removed during the course of maintenance or overhaul, must be replaced to comply with specifications and procedures. [ARD50010]

item obtainment time The time required for the technician to obtain replacement parts, assemblies, or units, depending on the maintenance concept and the location and method of storing spare items. [AIR4896]

item, repairable An item comprising or including replaceable parts, commonly economical to repair, and subject to being rehabilitated to a fully serviceable condition over a period less than the life of the flight equipment to which it is related. [ARD50010]

item, rotable An item that can be economically restored to a serviceable condition; and, in the normal course of operations, can be repeatedly rehabilitated to a fully serviceable condition over a period approximating the life of the flight equipment to which it is related. [AIR4896]

item, serviceable An item that can be returned to service with appropriate airworthiness documentation. [ARP4386]

item, structural significant *See* structural significant item.

ITO semiconductor Semiconductor device consisting of a layer of tin sandwiched between an indium layer and an oxide layer.

izod impact test A test for shock-loading in which a notched specimen bar is held at one end, the bar is broken by striking, and the energy absorbed is measured. [AIR4844]

J

J *See* joule.

jack *1.* A device, usually hydraulic (but may be pneumatic or mechanical), for the purpose of lifting or raising heavy weights such as aircraft. Various types and configurations are used, such as axle jacks, and fuselage jacks. [AIR1489] *2.* A connecting device to which a wire or wires of a circuit may be attached and which is arranged for the insertion of a plug. *3. See* cylinder, actuating.

jacket *1.* The outermost separable layer of insulating material on a wire or cable. [ARP1931] *2.* Covering or casing of some kind; specifically, a shell around the combustion chamber of a liquid-fuel rocket, through which the propellant is circulated in regenerative cooling. *3.* Coatings of one material over another to prevent damage such as oxidation or micrometeroid penetration. *4.* A plastic layer applied over the coating of an optical fiber, or sometimes over the bare fiber. Used for color coding in optical cables, to make handling easier, or for protection of the fiber against mechanical stress and strain. *5.* Stiff plastic protective material that encases a floppy disk, with slots for a disk drive to access the data.

jack pad A pad or surface built on a landing gear for the purpose of accepting jacking loads or a separate attachable jack fitting. [AIR1489]

jack point A raised point, usually spherical, on a landing gear for the purpose of accepting the lifting forces imposed by a jack. [AIR1489]

jackscrew *1.* A captive screw-and-nut assembly, attached to connectors, for use in mating and unmating connectors; may also facilitate connector mounting and polarization. [ARP914A] *2.* A type of puller that separates two parts, utilizing threaded action by one of its details. Used to lift or exert pressure. [ARP480A] *3.* A portable device for lifting a heavy load a short distance by means of a screw mechanism. *4.* The screw of such a device.

jackshaft A countershaft, especially an auxiliary shaft used between two other shafts.

Jahn-Teller effect The effect whereby degenerate orbital states in molecules (other than linear molecules) are unstable.

jamming Electronic or mechanical interference that may disrupt the display of an aircraft on radar or the transmission/reception of radio communications/navigation. [ARP4107]

JATO See jet assisted takeoff.

jet assisted takeoff (JATO) *1.* A method of enabling aircraft to take off in a shorter distance than otherwise possible by the use of boosters (rockets) temporarily attached, usually to the underside of the wings. [ARP4107] *2.* A rocket motor used to assist the takeoff of an aircraft. Also known as rocket assisted takeoff (RATO). [AIR913]

jetavator A control surface that may be moved into or against the jet stream of an engine, used to change the direction of the jet flow for thrust-vector control. [ARP4386]

jet blast The equivalent of a wind gust produced by the jet engines of an aircraft. [ARP1328]

jet damping *See* damping.

jet engine(s) *1.* An aircraft engine that produces thrust by the rearward expulsion of combustion products from the burning of fuel with air. [ARP4107] *2.* Broadly, engines that eject jets or streams of gas or fluids, obtaining all or most of their thrust by reaction to the ejection. Specifically, aircraft engines that derive all or most of their thrust by reaction to their ejection of combustion products (or heated air) in a jet. Jet engines obtain oxygen from the atmosphere for fuel combustion (or outside air for heating, as in the case of the nuclear jet engine), distinguished in this sense from a rocket engine. Jet engines of this kind may have compressors, commonly turbine-driven to take in and compress air (turbojets); or they may be compressorless, taking in and compressing air by other means (pulsejets, ramjets).

jet fuel starter A small gas-turbine engine mounted on the engine gearbox or AMAD to provide cranking power for starting. Also referred to as a gas turbine starter. [ARP906A]

jet interaction amplifier An amplifier that utilizes control jets to deflect a power jet and modulate the output. Usually employed as an analog amplifier. [ARP993A]

jet level sensor Used in a bottom loading system, a fluidic control device that discriminates between liquid and gas to signal a bottom loading valve to open and close when the liquid level reaches a preset level. [AIR4783]

jet membrane process Method for separating or enriching isotopes of the same element by using a condensable vapor as the carrier fluid. A process gas containing the isotopes enters a chamber into which a heavy condensable gas (the jet) flows. The lighter of the two isotopes is enriched relative to the heavier species, and is collected by a probe downstream for further enrichment or analysis.

jet pipe servovalve An electrohydraulic servovalve in which the electric control signal moves a pipe discharging a high-velocity jet of hydraulic fluid; this modulates flow between two ports in the first stage of the valve. [ARP4386]

jet pump A type of pump in which a jet of fluid is used to induce flow in another fluid.

jet lag *See* circadian desynchronization.

jet reference fluid (JRF) A combination of aliphatic and aromatic liquids, also containing mercaptans, used as a harsh environment to test the resistance of integral fuel tank sealants to jet fuels. [AS7200/1]

jet route A route designed to serve aircraft operations from 18,000 ft MSL on up to and including flight level 450. The routes are referred to as "J" routes, with numbering to identify the designated route (for example, J 105). [ARP4107]

jet stream A narrow, shallow, meandering river of strong winds that usually extends around the temperate zone of the earth. A jet stream is considered to exist whenever winds of 50 knots or stronger, embedded in the high tropospheric or lower stratospheric general wind flow, are concentrated in a band at least 300 miles long. Jet streams are generally found in segments of 1000 to 3000 miles in length, 100 to 400 miles in width, and 3000 to 7000 ft in depth, with winds generally between 100 to 150 knots. [ARP4107]

jet thrust The thrust of a fluid, especially as distinguished from the thrust of a propeller.

jet vanes Vanes either fixed or movable, used in a jetstream, especially in the jetstream of a rocket, for purposes of stability or control under conditions in which external aerodynamic controls are ineffective. Also called blast vanes.

jet wash *See* wake turbulence.

jewels Recessed bearings of glass, sapphire, or diamond that support the ends of a pivot pin in an instrument or a fine, mechanical watch or clock.

JFET Stands for junction field effect transistors, devices in which semiconductor channels of low conductivity join the source and drain. In these devices, these channels are reduced and cut off by the junction-depletion regions. This reduces the conductivity and causes a voltage to be applied between the gate electrodes.

J integral A contour energy integral formulated by Rice and used for evaluating fracture toughness of elastoplastic materials.

jitter Undesired, rapid movement of symbology on a display face, discernible to a human eye located at the design eye position. [ARP1782]

Jo block *See* gage block.

joggle A displacement machined or formed in a structural member to accommodate the base of an adjacent member. Although joggles are sealed by prepacking during preassembly whenever possible, in some cases they must be sealed by injection during postassembly operations. [AIR4069]

Johanssen block *See* gage block.

Johansson curved crystal spectrometer A type of spectrometer having a reflecting crystal whose face is concave, thus causing x-rays that diverge slightly after passing through the primary slit to be refocused at the detector slit.

joint, ball A joint that permits relative angular movement and rotation of two adjacent ducts. [ARP699D]

joint, bonded A part of a structure at which two adherents are held together with a layer of adhesive. [AIR4844]

joint, compensating A pressure-compensated assembly allowing axial motion, used to maintain duct walls in tension. [ARP699D]

joint, expansion *See* expansion joint.

joint, flexible A non-rigid joint, convoluted tubing, hose, or ball-joint assembly that joins two ducts and permits relative motion of the ducts in one or more planes. [ARP699D]

joint, restrained A flex-section assembly in which an angular deflection can occur with a tension load being transmitted by an external or internal device. [ARP699D]

joint, rotary A joint that permits relative rotation of two adjacent ducts. [ARP699D]

joint, slip *See* slip joint.

joule (J) Metric unit for energy, quantity of heat, or work.

Joule-Thomson effect A change of temperature in a gas undergoing Joule-Thomson expansion.

journal bearings/bushings A cylindrical shell installed in a housing in conjunction with a mating pin or bolt. [AIR1594]

joy stick A device by which an individual can communicate with an electronic information system through a cathode ray tube.

JRF *See* jet reference fluid.

J-route *See* jet route.

judder A mode of vibration in which the oscillation of the landing gear leg is due to externally applied forces and the inertia forces are due to the combined inertia of the wheel, brake, axle, and the leg. [AIR1489]

judgment delay Failure, due to an anomaly of attention or motivation, to assess the sig-

nificance and priority of information from the environment in a timely manner, assuming adequate quality and quantity of information. [ARP4107]

judgment, poor Failure, due to an anomaly of attention or an anomaly of motivation, to realistically assess the significance and priority of information from the environment, assuming adequate quality and quantity of information. [ARP4107]

jump 1. An instruction that causes a new address to be entered into a computer program counter; the program continues execution from the new program counter address. 2. *See* transfer.

jumper(s) *1.* Short lengths of conductors used to complete electrical circuits, usually temporarily, between terminals, or bypassing an existing circuit. *2.* A temporary wire used to bypass a portion of an electrical circuit or to attach an instrument or other device during testing or troubleshooting.

jumper tube A short tube connection for bypassing, routing, or directing the flow of fluid as desired.

junction The portion of a transistor where opposite types of semiconductor material meet.

junction box A protective enclosure around connections between electric wires or cables.

junction field effect transistors *See* JFET.

junctions In semiconductor devices, regions of transition between semiconducting regions of different electrical properties.

Jupiter II *See* Europa.

Jupiter IV *See* Callisto.

jury strut (or brace) *1.* A secondary toggle kinematic arrangement to provide a lock for the gear assembly in a specific configuration. *2.* A secondary brace that locks the primary brace in position, both of which are "on center" or near "on center" condition. *3.* A piece of ground-support equipment used to provide extra bracing for movable or folded assemblies against abnormally high loads, e.g., to brace folded wings on carrier-based aircraft against heavy weather conditions. [AIR1489]

K

K *1. See* kilobyte. *2. See* Kelvin. *3.* See Kurtosis number.

Kalman filter *1.* An optimal estimator or filtering technique for estimating the state of a linear system. [AIR1823] *2.* A technique for calculating the optimum estimates of process variables in the presence of noise. This technique, which generates recursion formulas suitable for computer solutions, also can be used to design an optimal controller.

kaolinite A hydrous silicate of aluminum. Constitutes the principle mineral in kaolin.

Karl Fischer technique A titration method for accurately determining moisture content of solid, liquid, or gas samples using Karl Fischer reagent (a solution of iodine, sulfur dioxide, and pyridine in methanol or methyl Cellusolve). The titration is highly suitable for automation, has high sensitivity (5 ppm), and good accuracy (±1%) over a wide range of moisture content (10 ppm to 100%).

Karman vortex street A double trail of vortices formed alternately on both sides of a cylinder or similar body moving at right angles to its axis through a fluid, the vortices in one row rotating in a direction opposite to that of the other row.

Karnaugh map A tubular arrangement that facilitates the combination and elimination of logical functions by listing similar logical expressions, thereby taking advantage of the human brain's ability to recognize visual patterns to perform the minimization.

K_b *See* Bernoulli coefficient.

KByte 1024 (2^{10}) bytes.

Kelvin (K) *1.* Metric unit for thermodynamic temperature. NOTE: In the Kelvin system, the scale divisions are not referred to as degrees as they are in other temperature-measurement systems. *2.* An absolute temperature scale in which the zero point is defined as absolute zero (the point where all spontaneous molecular motion ceases) and the scale divisions are equal to the scale divisions in the Celsius system. 0°C equals approximately 273.16 K.

Kelvin bridge A type of d-c bridge circuit similar to a Wheatstone bridge, but incorporating two extra resistances in parallel with two of the known resistances; this minimizes the inaccuracies introduced because of finite lead and contact resistances in the circuit.

Kelvin-Varney voltage divider (KVVD) A resistive-type voltage divider used in some d-c bridge circuits to provide greater sensitivity at low values of the unknown resistance.

Kennison nozzle A specially shaped nozzle designed for measuring flow through partially filled pipes. Because of its self-scouring, nonclogging design, the Kennison nozzle is especially useful for measuring flow of raw sewage, raw and digested sludge, final effluent, trade wastes, and other liquids containing suspended solids or debris; it also functions well at low flow rates or when flow rates vary widely.

Kepler laws The three empirical laws governing the motions of the planets in their orbits, discovered by Johannes Kepler (1571-1630). These are: (a) the orbits of the planets are ellipses, with the sun at a common focus; (b) as a planet moves in its orbit, the line joining the planet and the sun sweeps over equal areas in equal intervals of time (also called law of equal areas); and (c) the squares of the periods of revolution of any two planets are proportional to the cubes of their mean distances from the sun.

kerf The width of a cut made by a saw blade, torch, water jet, laser beam, etc. [AIR4844]

Kern counter *See* dust counter.

Kerr cell A device in which the Kerr effect is used to modulate light passing through the material. The modulation depends on rotation of beam polarization caused by the application of an electric field to the material. The degree of rotation determines how much of the beam can pass through a polarizing filter.

Kevlar A Dupont synthetic textile material, lightweight and nonflammable, and with high impact resistance.

key *1.* A projection on a connector which engages a keyway in a mating connector so as to guide the connectors halves during mating; may be positioned and/or used in multiples to provide a polarization feature. [ARP914A] *2.* A device that moves or pivots to secure or tighten components in an assembly. *3.* A component, usually having notches or grooves in its working face, that is inserted into a lock to engage or disengage the locking mechanism.

key characteristics For a material, those characteristics that have a significant influence on product fit, performance, service life, or manufacturability. [ARD9000]

keyway A slot or groove into which a key engages. [ARP914A]

keyword One of the significant and informative words in a title or document which describe the content of that document.

K factor The coefficient of thermal conductivity; the amount of heat that passes through a unit cube of material in a given time, when the difference in temperature of two opposite faces is 1°. [AIR4844]

kg *See* kilogram.

kiln An oven or similar heated chamber for drying, curing, or firing materials or parts.

kilo A decimal prefix denoting 1000.

kilobyte Symbolized by K, a measure of a computer memory. One kilobyte (one K) of memory can store 1024 characters.

kilogram (kg) Metric unit of mass.

kilohm (kohm) Unit of electrical power equivalent to 1000 ohms.

kilometric waves Electromagnetic waves with wavelengths between 1000 and 10,000 meters.

kilovolt ampere (kVA) 1000 volt × amperes. [ARP1931]

kilovolt (kV) 1000 volts. [ARP1931]

kilowatt (kW) A unit of power equal to one thousand watts. [ARP1931]

kilowatt-hour capacity A unit of measurement of the rated energy capacity of a battery. [AIR1898]

kinematics The branch of mechanics dealing with the description of the motion of bodies or fluids without reference to the forces producing the motion.

kinematic viscosity Absolute viscosity of a fluid divided by its density.

kinesthesia The sensation of motion, position, and tension of the body perceived through nerve-end organs in muscles, tendons, and joints.

kinesthetic Relating to kinesthesia.

kinesthetic illusion *See* illusions, kinesthetic.

kinetic energy *1.* The energy that a body possesses as a consequence of its motion. *2.* Energy related to the fluid of dynamic pressure, $1/2 \, p \, V^2$.

kinetic theory The derivation of the bulk properties of fluids from the properties of their constituent molecules, their motions, and interactions.

kinetic vacuum system A vacuum system capable of attaining and sustaining limiting pressures of 5×10^{-5} to 5×10^{-7} torr despite a relatively high outgassing load or the presence of small leaks.

kink A tight loop in wire or wire rope that results in permanent damage due to deformation.

kinking A temporary or permanent distortion of a hose, induced by winding or doubling upon itself and consequently exceeding the recommended minimum bend radius established for that hose. [ARP1658A]

Kirchhoff law of radiation A radiation law that states that, at a given temperature, the ratio of the emissivity to the absorbtivity for a given wavelength is the same for all bodies, and is equal to the emissivity of an ideal black body at that temperature and wavelength.

Kirchhoff's Law Law stating that the sum of the voltage across a device in a circuit series is equal to the total voltage applied to the circuit.

kit *1*. A container or containers in which a one-part or a two-part sealant is packaged. Two component kits consist of the base compound in one container and the catalyst in the other. [AS7200/1] *2*. A group of related but non-homogeneous items used for service or modification purposes. [ARP480A]

klystrons Electron tubes for converting direct-current energy into radio frequency energy by alternately speeding up and slowing down the electrons.

K Modulus The total load in pounds divided by the total volume displaced in cubic inches. [AIR1489]

kneeling gear A landing gear assembly or installation designed to "kneel" (lower) the aircraft for ground handling purposes, such as parking, loading, maintenance, and catapulting. The gear may retract to an intermediate position or fold in a manner different from normal retraction. [AIR1489]

knoop hardness Hardness that is measured by calibrated machines which force a rhomb shape, pyramidal diamond indentor, having specified edge angles, under specified conditions, into the surface of the test material. [AIR4844]

knot A velocity of one nautical mile per hour, 1.150779 statute miles per hour; not a measure of distance. NOTE: Although the term "knot" is defined in some dictionaries (based on lay usage) as "1 nautical mile," this usage (as in "knots per h") is not acceptable in professional aviation or nautical contexts. [ARP4107]

known defect test standard A test standard containing known defects, used to perform system performance checks and classify penetrants. [AMS2647A]

knuckle area In a filament-wound part, the area of transition between sections of different geometry; for example, where the skirt joins the cylinder of the pressure vessel. [AIR4844]

Knudsen flow Gas flow in a long tube at pressures such that the mean free path of a gas molecule is significantly greater than the tube radius.

Knudsen gages Gages that measure pressure in terms of the net rate of transfer of momentum by molecules between two surfaces maintained at different temperatures and separated by a distance smaller than the mean free path of the gas molecules.

kohm *See* kilohm.

kondo effect Refers to the change in superconductivity characteristics resulting from magnetic impurities in the compounds involved.

konimeter A device for determining dust concentration by drawing in a measured volume of air, directing the air jet against a coated glass surface, and thereby depositing dust particles for subsequent counting under a microscope.

koniscope An indicating instrument for detecting dust in the air.

Korteweg-Devries equation The mathematical representation describing the propagation of long waves of small but finite amplitude.

Kramers-Kronig formula The relationship between the attenuation coefficient and the dispersion (frequency dependent phase velocity) for viscoelastic waves.

kriging A method of providing unbiased estimates of variables in regions where the available data exhibit spatial autocorrelation, these estimates being obtained in such a way that they have minimum variance.

kruger flap A movable flap running underneath the leading edge of the wing, with the hinge line on the upstream side of the flap. Rotation of the flap into the airstream modifies the leading edge profile to increase the stalling angle of the wing. [ARP4386]

krypton fluoride lasers Rare-gas halide ultraviolet stimulated emission devices in which krypton fluoride is the active lasing medium.

K-sample data A collection of data consisting of values observed when sampling from k-batches. [AIR4844]

kurtosis In statistics, the extent to which a frequency distribution is peaked or concentrated about the mean. Sometimes defined as the ratio of the fourth moment of the distribution to the square of the second moment.

Kurtosis number Figure of merit, K, used for monitoring impulsive type vibrations of ball bearings.

kV *See* kilovolt.

kVA *See* kilovolt ampere.

KVVD *See* Kelvin-Varney voltage divider.

kW *See* kilowatt.

Kynar Tradename for polyvinylidene fluoride, manufactured by Pennwalt Corp.

L

L *1. See* liter. *2. See* inductance. *3. See* lambert.

La Chemical symbol for lanthanum.

labeled molecule A molecule of a specific chemical substance in which one or more component atoms is an abnormal nuclide; that is, a nuclide that is radioactive when the molecules normally are composed of stable isotopes, or vice versa.

laboratory ambient conditions 18–24°C (65–75°F) and 45–55 percent relative humidity. [AS8043]

laboratory standard An instrument that is calibrated periodically against a primary standard. [AIR1678]

labyrinth seals Minimum leakage seals that offer resistance to fluid flow while providing radial or axial clearance.

lace line A high-strength nylon cord used to support a flexible tank to the structure. [AIR1664]

lack of fill out For a laminated plastic, describes an area, usually occurring at the edge, where the reinforcement has not been wetted with resin. [AIR4844]

lacquer A liquid material applied to fibrous braid to prevent fraying, wicking, or moisture absorption in the braid; a saturant. [ARP1931]

ladder circuit Temperature-measuring junctions geometrically and electrically paralleled to two bus bars. [ARP485]

ladder diagram Symbolic representation of a control scheme in which the power lines form the two sides of the ladder-like structure, and the program elements are arranged to form the rungs. The basic program elements are contacts and coils, as in electromechanical logic systems.

lag *1.* The delay between change of conditions and the indication of the change on an instrument. [AIR1489] *2.* Delay in human reaction. *3.* The amount that one cyclic motion is behind another, expressed in degrees; the opposite of lead. [AIR1489] *4.* In control theory, a transfer function term in the form, $1/(Ts + 1)$.

lag coefficient *See* time constant.

lagging *1.* In an a-c circuit, refers to a condition in which peak current occurs at a later time in each cycle than does peak voltage. *2.* A thermal insulation, usually made of rock wood and magnesia plaster; used to prevent heat transfer through the walls of process equipment, pressure vessels, or piping systems.

Lagrange coordinates Coordinate system whereby fluid parcels can be identified for all times. In this system, the parcels are assigned coordinates that do not vary with time. Examples of such coordinates are: (a) the values of any properties of the fluid conserved in the motion; or (b) more generally, the positions in space of the parcels at some arbitrarily selected moment. Subsequent positions in space of the parcels are then the dependent variables, functions of time and of the Lagrange coordinates.

lambda *See* failure rate.

lambert Abbreviated L. A unit of luminance equal to the uniform luminance of a perfectly diffusing surface emitting or reflecting light at one lumen per square centimeter.

Lambert law *See* Bouguer law.

Lambert's cosine law Law stating that the radiance of certain surfaces (known as Lambertian reflectors, Lambertian radiators, or Lambertian sources) is independent of the angle from which the surface is viewed.

lamb wave A type of wave that propagates within the thickness of a thin plate; can only be generated at particular values of angle of incidence, frequency, and plate thickness; and can exist only when the plate is thicker than one wavelength. The velocity of a lamb wave is dependent on the product of plate thickness and frequency. [ARP5089]

lamella (metallurgy) Describes crystalline materials whose grains are in the form of thin sheets.

lamina A single ply or layer in a laminate made up of a series of layers. [AIR4844]

laminar boundary layer In fluid flow, refers to a layer of fluid immediately adjacent to a solid boundary which is characterized by laminar rather than turbulent flow. *See* laminar flow.

laminar flow Fluid flow characterized by the gliding of fluid layers (laminae) past one another in orderly fashion. *See also* Poiseuille flow, streamline flow, and viscous flow. [ARP4386]

laminar flow amplifier A jet amplifier in which the nozzle Reynolds number is maintained at such a level that laminar flow conditions exist in the device. [ARP4386]

laminate *1.* A product made by bonding together two or more layers of material. *2.* To unite laminae with a bonding material, usually with pressure and heat. [AIR4844]

laminate coordinates A reference coordinate system generally in the direction of principal axes, when they exist. [AIR4844]

laminated materials *See* laminate.

laminated molding A molded plastic article produced by bonding together, under heat and pressure, in a mold, layers of resin-impregnated laminating reinforcement. [AIR4844]

laminated tape A tape consisting of two or more layers of different materials bonded together. [ARP1931]

laminate orientation The configuration of a cross-plied composite laminate with regard to the angles of cross-plying, the number of laminae at each angle, and the exact sequence of the lamina lay-up. [AIR4844]

laminate ply Refers to a single fabric-resin or fiber-resin layer of a product, which is bonded to adjacent layers in the curing process. [AIR4844]

laminates Structures created by stacking several thin layers of material; laminated materials.

laminations Thin, flat discontinuities seen only at the edge or end of a plate; found only in plate steel. [AS3071A]

lamination sequence Refers to the composite lay-up sequence, which usually begins with the ply nearest the tool surface. [AIR4844]

LAN (computer networks) *See* local area network.

Landau damping The damping of a space charge wave by electrons that move at the phase velocity of the wave and gain energy transferred from the wave.

landing The complete action of bringing an airborne airplane down to the surface and bringing it to a stop, including the approach, the touchdown, and the landing roll. [ARP4386]

landing approach line A A theoretical line determined by increasing the angle of line L by vectorially adding to the design landing speed the design sinking speed less the maximum design headwind. [ARP1538]

landing gear The apparatus comprising those components of an aircraft or spacecraft that support and provide mobility for the craft on land, water, or other surfaces. The landing gear consists of wheels, floats, skids, bogies, and treads, or other devices, together with all associated struts, bracing, shock absorbers, etc. *See also* undercarriage. NOTE: Landing gear includes all supporting components, such as the tail wheel or tail skid and outrigger wheels or pontoons, but the term is often conceived to apply only to the principal components, i.e., to the main wheels, floats, etc., and the nose gear if any. For military aircraft, catapulting provisions and the arresting hook installation are often considered an ancillary component of the landing gear. [AIR1489]

landing gear bearing A component that, when installed between two or more landing gear structural members having relative motion, can perform the following: (a) transmit loads between those members; (b) control the friction coefficient of the joint; and (c) control the wear of the joint. [AIR1594]

landing-gear-extended speed The maximum speed at which an aircraft can be safely flown with the landing gear extended. [ARP4107]

landing-gear operating speed The maximum speed at which the landing gear can be safely extended or retracted. [ARP4107]

landing gear warning *See* warning, landing gear.

landing hot Landing an aircraft at a speed substantially greater than its stalling speed. [ARP4107]

landing impact system *See* system, landing impact.

landing mat A ground cover laid down for the purpose of supporting aircraft operations, i.e., parking, taxi, takeoff, and landing. Several types are used, including those made of pierced metal strips or aluminum planks. [AIR1489]

landing minimums/IFR landing minimums The minimum ceiling and visibility prescribed for landing civil aircraft while using an instrument approach; classified as straight-in landing minimums and circling minimums. These minimums apply along with other limitations set forth in FAR Part 91 with respect to the minimum descent altitude (MDA) or decision height (DH) prescribed in the instrument approach procedures. *Straight-in landing minimums*–the MDA and visibility, or the DH and visibility, required for straight-in landing on a specified runway. *Circling minimums*– the MDA and visibility required for the circle-to-land maneuver. Descent below the established MDA or DH is not authorized during an approach unless the aircraft is in a position from which a normal approach to the runway of intended landing can be made, and adequate visual reference to required visual cues is maintained. [ARP4107]

landing phase *See* mishap, maneuver.

landing roll/rollout The distance from point of touchdown to the point where the aircraft can be brought to a stop or exit the runway. [ARP4107]

language translator A general term for any assembler, compiler, or other routine that accepts statements in one language and produces equivalent statements in another language.

lap *1.* In a sliding spool valve, the relative axial position relationship between the fixed and movable flow-metering edges with the spool at null. *2.* For a servovalve, the total separation at zero flow of straight line extensions of the nearly straight portions of the normal flow curve, drawn separately for each polarity, expressed as percent of rated current. [ARP490] *3.* In metalworking, a feature similar to a seam, which may result from improper rolling practices; or, in the case of forging, the metal being folded over but failing to weld into a single piece. Rolling laps generally run lengthwise with the bar but can extend into the bar at an angle. [AS3071A] *4.* A controlled surface, used to remove insignificant amounts of metal. [ARP480A]

lap joint A connection between two parts made by overlapping members at the junction and welding, riveting, or bolting them together.

Laplace transform *1.* A mathematical relationship that is used to transform a function from the time domain to a transformed domain. [AIR1823] *2.* In control theory, a mathematical method for solution of differential equations.

lapping Smoothing or polishing a surface by rubbing it with a tool made of cloth, leather, plastic, wood, or metal in the presence of a fine abrasive.

lapping in A process of mating contact surfaces by grinding and/or polishing.

lapse rate The decrease of an atmospheric variable (temperature, unless otherwise specified) with height. Used ambiguously to refer to the environmental lapse rate and the process lapse rate; the meaning must often be ascertained from the context.

laptop A small, portable computer, usually with a flip-up screen.

lap weld A lap joint made by welding.

large aircraft Aircraft having a maximum certified takeoff weight of more than 12,500 pounds. [ARP4107]

large core fiber An optical fiber with a comparatively large core, usually with a diameter of 400 micrometers or more, and usually a step index type.

large scale integration (LSI) *1.* Describes a computer chip containing a large number of digital circuits in a small area. *2.* Describes integrated circuits with more than 100 logic gates.

larmor radius For a charged particle moving transversely in a uniform magnetic field, the radius of curvature of the projection of its path on a plane perpendicular to the field.

laser From light amplification by stimulated emission of radiation; a device for producing light by emission of energy stored in a molecular or atomic system when stimulated by an input signal.

laser anemometer Wind-measuring instrument in which the wind passes through two perpendicular light beams and the resulting change in velocity of one or both beams is measured.

laser annealing Rapid heating of metals and/or alloys by means of lasers.

laser cutting The cutting of material by means of lasers.

laser diode array A device in which the output of several diode lasers is brought together in one beam. The lasers may be integrated on the same substrate, or may be discrete devices coupled optically and electronically.

laser doppler flowmeter An apparatus for determining flow velocity and velocity profile of a moving fluid stream by measuring the Doppler shift in laser radiation scattered from particles in the stream; contaminants such as smoke may have to be introduced into a gas stream to provide scattering centers. A laser doppler flowmeter can be used to measure velocities of 0.01 to 5000 in./s (0.25 mm/s to 125 m/s).

laser glass An optical glass doped with a small concentration of a laser material (referred to as an impurity). When the impurity atoms are excited by light, they are stimulated to emit laser light.

laser guidance Guidance system for rockets or projectiles, utilizing a laser beam for a precise trajectory to a designated target.

laser gyroscope Ring-laser angular rotation sensor for stabilizing and controlling large space structures; used, for example, for space vehicle guidance.

laser induced fluorescence (LIF) Emission of electromagnetic radiation that is caused by the flow of laser radiation into the emitting body, and that ceases abruptly with the excitation.

laser interferometer A type of optical interferometer in which a laser is used as the source of monochromatic light; accuracies of better than 20 microinch ($1.25 \mu m$) are achieved when measuring lengths up to 200 in. (5.08 m).

laser interferometry The design and use of interferometers in which a laser is the light source. The monochromaticity and brilliance of the laser light enables the differentiation between interfering beams of hundreds of meters, in contrast to a maximum of 20 centimeters for the classical interferometers.

laser line filter A filter that transmits light in a narrow range of wavelengths centered on the wavelength of a laser; light at other wavelengths is reflected. Such filters are used to remove light from non-laser sources, i.e., light that could interfere with the operation of the laser system.

laser microscope A microscope having a ceramic tube in which a metal vapor is formed at 1600°C. Copper (or other metal atoms) are excited and amplify light, thus when used with a projection microscope, the object to be magnified is illuminated. The power of the emitted beam on the screen remains constant.

laser plasma interactions The results of the actions of laser beams on electrically ducting fluids, such as plasmas or ionized gases.

laser propulsion The use of high-power lasers for aircraft, rocket, or spacecraft propulsion by: (a) indirect conversion of laser-heated propellants or working fluids to produce thrust; (b) direct thrust generation with laser-light pressure on the vehicle; or (c) direct conversion of laser energy into electricity for propulsion.

laser pumping The application of a laser beam of appropriate frequency to a laser medium whereby absorption of the radiation increases the population of atoms or molecules in higher energy states.

laser simulator A light source that simulates the output of a laser. In practice, the light source is a 1.06 micrometer LED, which simulates the output of a neodymium laser at much lower power levels.

laser spectrometer Spectrometer using a laser.

laser spectroscopy The use of lasers for spectroscopic analysis; particularly in Raman spectroscopy.

laser stability Characteristic of a laser beam free from oscillations.

laser target designator Laser equipment aboard spacecraft for identifying satellites, missiles, and objects in space.

laser target interactions Interactions in which lasers are used to produce heating, fusion, or damage in targets.

laser targets Objects subjected to laser radiation, especially for laser fusion applications.

laser weapons Refers to the military applications of high-power lasers (mainly gasdynamic and chemical mixing lasers).

laser welding Microspot welding with a laser beam.

lasing Generation of visible or IR light waves having very nearly a single frequency as a result of pumping or exciting electrons into high-energy states in a stimulated emission device (laser).

latches Devices that fasten one thing to another (e.g., a rocket to a launcher), but are subject to ready release so that the things may be separated.

latching digital output A contact closure output that holds its condition (set or reset) until changed by later execution of a computer program. *See also* momentary digital output.

latching relay Real device or program element that retains a changed state without power. In a computer program element, the power removal is only in terms of the logic power expressed in the diagram; real power removal will affect the PC outputs according to some scheme provided by the manufacturer.

latching solenoid valve An electrically pulsed solenoid valve that remains at its last operating position when electrical power is removed. [ARP4386]

latch switch A control to prevent fuel-valve opening if the burner is not secured in the firing position.

latch-up A pnpn self-sustaining low-impedance state; a type of electronic malfunction.

late arrival (technical) An aircraft arrival after scheduled arrival time, caused by a known or suspected equipment malfunction and/or defect. [AIR4896]

latency *1.* A measure of time delay. [AIR4271] *2.* In data processing, the time between the completion of the interpretation of an address and the start of the actual transfer from the addressed location. Latency includes the delay associated with access to storage devices, such as drums and delay lines.

latent curing agent A curing agent that produces long-term stability at room temperature, but rapid cure at elevated temperatures. [AIR4844]

latent defect An inherent or induced weakness, not detectable by ordinary means, which will either be precipitated to early failure under environmental stress screening conditions or eventually fail in the intended use environment. [ARD50010]

latent failure A failure that is not inherently revealed at the time it occurs. [ARP926A]

latent heat *1.* The unit quantity of heat required for isothermal change in a state of a unit mass of matter. Latent heat is termed heat of fusion, heat of sublimation, heat of vaporization, depending on the change of state involved. *2.* Heat that does not cause a temperature change.

latent heat of fusion *See* heat of fusion.

lateral Of, at, from; or toward the side; sideways; for example, lateral movement.

lateral axis *1.* The horizontal (Y) axis of an aircraft about which the aircraft revolves in pitching. The positive direction is to the right when looking forward. [ARP4386] *2.* An imaginary lateral line at right angles to the longitudinal axis, passing through the center of gravity of the airplane, and lying within a plane normal to the plane of symmetry. Angular movement about this axis is called pitching. [ARP4107]

lateral control Control of the lateral motion of an aircraft as effected by deflections of the ailerons, spoilers, and rudder. [ARP4386]

lateral motion Rolling motion (of a vehicle). [ARP4386]

lateral navigation (LNAV) Those functions that provide navigation guidance in the horizontal plane and that provide command signals to the roll channel of the auto flight system and displays. Currently used interchangeably with RNAV (random navigation), a term created by ARINC. [ARP1570]

lateral offset misalignment The maximum distance between the two centerlines of coupling halves. [AIR4728]

lateral separation The lateral spacing of aircraft at the same altitudes, achieved by requiring operation on different routes or in different geographical locations. [ARP4107]

lateral spring constant Also called lateral spring rate. The number of unit loads required to produce a unit deflection in a lateral direction (normal to plane of the wheel). [AIR1489]

lateral spring rate *See* lateral spring constant.

lateral stability *See* stability, lateral.

latex A generic term describing a stable dispersion of insoluble resin particles in a water system. [AIR4844]

Latin square method In mathematics, the use of an n × n square array of n different symbols, each symbol appearing once in each row and once in each column.

latitude *1.* Angular distance from a primary great circle or plane. *2.* For a photographic emulsion, the ratio of the exposure limits between which the film density curve (Hurter and Driffield curve, or H & D curve) is essentially linear.

lattice network An electronic network composed of four branches connected end-to-end to form a mesh, and having two nonadjacent junctions as the input terminals and the two remaining nonadjacent junctions as the output terminals.

lattice parameter In crystallography, the length of any side of the unit cell in a given space lattice; if the sides are unequal, all unequal lengths must be specified.

lattice pattern A pattern of filament winding with a fixed arrangement of open voids. [AIR4844]

launch bar In a nose gear launch system, the link that provides the means of steering the nose gear during tracking and couples the aircraft to the catapult during the power run; part of the nose gear assembly. [AIR1489]

launch clouds *See* exhaust clouds.

launcher A structure or device often incorporating a tube, a group of tubes, or a set of tracks, from which self-propelled missiles or aircraft

are sent forth; and by means of which the missiles or aircraft are aimed or imparted inertial guidance (distinguished in this specific sense from a catapult). [AIR1489]

launching bases Areas, such as Cape Kennedy or Vandenburg Air Force Base, that have several launch sites.

launching pad The load-bearing base or platform from which a rocket vehicle is launched.

launching run During catapulting, the extent of aircraft travel from "release" to the end of "deck run." The launching run is the sum of the power run or power stroke and the deck run. [AIR1489]

launching site Defined area from which a rocket vehicle is launched, either operationally or for test purposes; specifically, at Cape Kennedy or Vandenburg, any of the several areas equipped to launch a rocket.

launch time *See* launch windows.

launch windows The postulated openings in the continuum of time or of space, through which a spacecraft or missile must be launched in order to achieve a desired encounter, rendezvous, impact, or the like.

Lauritsen electroscope An electroscope in which the sensitive element is a metallized quartz fiber.

law of exception In the investigation of aircraft accidents, the principle holding that, if all other possible causes have been ruled out, it is concluded that the operative cause was pilot behavior. [ARP4107]

lay *1.* In glass fiber, the spacing of the roving bands indicated on the roving package, expressed as the number of bands per inch. *2.* In filament winding, the orientation of the ribbon with some reference, usually the axis of rotation. [AIR4844] *3.* The direction of the predominant surface pattern produced by tool marks. [AS291D]

layer *1.* In reference to sky cover, clouds or other obscuring phenomena whose bases are approximately at the same level. A layer may be continuous or composed of detached elements. NOTE: The term layer does not imply that a clear space exists between the layers or that the clouds or obscuring phenomena composing them are of the same type. [ARP4107] 2. A subdivision of the OSI architecture.

layered laminate Describes a laminate in which two or more plies, either of the same or different materials, are bonded and stacked one on top of the other to act as a single structural layered element. [AIR4844]

lay-up *1.* Reinforcing material placed in position in a mold. 2. The process of placing reinforcing material in position in a mold. 3. Resin-impregnated reinforcement. 4. A description of the component materials, geometry, etc., of a laminate. 5. A stackup of composite materials that forms either a cured or uncured part. 6. A process of fabrication involving the assembly of successive layers of resin-impregnated material. [AIR4844]

L-band In telemetry, the radio spectrum that is available for manned vehicles; 1435-1540 MHz.

LCD *See* liquid crystal display.

LCF *See* low cycle fatigue.

LCU *See* local control unit.

LCVASI See low cost visual approach slope indicator.

L/D (reflux-to-distillate ratio) *1.* A quantity used in analyzing column operations. *See also* reflux ratio. 2. Lift to drag ratio of flight vehicles.

LDA *See* localizer-type directional aid.

L-direction The ribbon direction; that is, the direction of the continuous sheets of honeycomb. [AIR4844]

lead *1.* The amount that one cyclic motion is ahead of another, expressed in degrees; the opposite of lag. [AIR1489] 2. A definite length of one-conductor electrical wire, wire braid, or other conductive material (except cable or cord), of which one or both ends are processed or terminated. A lead may be of any size or shape; insulated or uninsulated. [ARP480A] 3. A column of high explosive used as one component of an explosive train. [AIR913] 4. In control theory, a transfer function term in the form, (Ts+1). 5. The distance a screw mechanism will advance along its axis in a single rotation.

lead acid batteries Common automobile batteries whose electrodes are grids of metallic

lead containing lead oxides, which change in composition during charging and discharging. The electrolyte is generally dilute sulfuric acid.

lead angle *1.* In welding, the angle between the axis of the electrode and the axis of the weld. 2. For a helix, the angle between the tangent to the helix and a plane perpendicular to the axis of the helix.

lead equivalent The radiation-absorption rating of a specific material, expressed in terms of the thickness of lead that reduces radiation dose an equal amount under given conditions.

leading In an a-c circuit, a condition in which peak current occurs earlier in each cycle than does peak voltage.

leading edge *1.* The first transition of a pulse going in either a positive (high) or a negative (low) direction. 2. The front or upstream edge of a wing or other lifting surface.

leading edge flaps *1.* Control surfaces at the leading edges of airfoils; hinged panels deflected downward to induce and control separation of the air flow. 2. Movable surfaces on the leading edge of wings which increase either the chamber or the chord or both, usually for the purpose of increasing lift.

leading edge thrust The increase in lift produced by highly-swept, low-aspect ratio wings, which develop a strong separation vortex; this, however, produces an even larger increase in drag.

lead-lag compensator A dynamic compensator combining lead action (the inverse of lag) with lag.

lead wire Any wire connecting two points in an electrical circuit, but especially a wire connecting an electric device to a source of power, or connecting an indicating or controlling instrument to a sensor.

lead zirconate titanates Dense ceramics with high piezoelectric coefficients and a high relative permittivity.

leakage *1.* Flow through a passage that is in a nominally closed position or at a location that normally should permit no flow, usually of relatively small magnitude. [ARP4386] 2. A type of streamline flow most often observed in viscous fluids near solid boundaries, and characterized by the tendency for fluid to remain in thin, parallel layers to maintain uniform velocity. *3*. Undesirable loss or entry across the boundary of a system; usually applied to slow passage of a fluid through a crack or fissure, but may also be used to describe passage of small quantities of particles, radiation, electricity, or magnetic lines of force beyond desired boundaries.

leakage flow Flow through a passage that is in a nominally closed position or at a location that normally should permit no flow, usually of relatively small magnitude. [AIR4728]

leakage rate The amount of leakage across a defined boundary per unit time.

leak detector An instrument, such as a helium mass spectrometer, used for detecting small cracks or fissures in a vessel wall.

leaker *1*. A crack or hole in a tube which allows fluids to escape. [ARP1658A] *2*. A hose assembly that allows fluids to escape at the fittings or couplings. [ARP1658A]

leak exit The point outside a tank where a leak appears. [AIR4069]

leak path The path that a fuel leak follows— from the leak source to its external exit. [AIR4069]

leak rate A general indication of vacuum integrity of a system. [AMS2769]

leak source The point inside a fuel tank where a leak starts. [AIR4069]

lean mixture Refers to an insufficient amount of fuel in the fuel-to-air mixture (weight-ratio basis) required by an internal combustion engine; a relatively large amount of air compared to fuel, generally greater than 16:1. *See also* rich mixture. [ARP4107]

leans *See* illusions, vestibular.

least significant bit (LSB) The smallest bit in a string of bits, usually at the extreme right.

least significant digit (LSD) The right-most digit of a number.

least squares method Any statistical procedure that involves minimizing the sum of squared differences.

least time, principle of *See* Fermat principle.

LED Stands for light-emitting diode. A semiconductor diode, the junction of which emits light when passing a current in the forward (junction on) direction. Light from an LED is incoherent spontaneous emission, as distinct from the coherent stimulated emission produced by diode lasers and other types of lasers. The indicator lights on most I/O modules are LEDs.

Ledoux bell meter A type of manometer whose reading is directly proportional to the flow rate sensed by a head-producing measuring device, such as a pitot tube.

left justified A field of numbers (decimal, binary, etc.) existing in a memory cell, location, or register, and possessing no zeroes to its left.

leg *1*. A single landing gear structure or assembly. [AIR1489] *2*. One of the members of a branched object or system. *3*. The distance between the root of a fillet weld and the toe. *4*. Any structural member that supports an object above the horizontal.

length For a discontinuity, weld-deposited material, or blended area, refers to the diameter of a circle encompassing the limits of the discontinuity, weld deposited material, or blended area. [AMS4991]

length of lay For any helically wound strand of wire or cable, the axial length of one complete turn of the helix, usually expressed in inches. [AS1198]

length of life Denotes the length of time it takes for a unit of product to fail after being placed on life test. [AIR4896]

leptons In the classification of subatomic particles according to mass, the lightest of all particles; examples of leptons are the electron and positron.

letdown The descent of an aircraft from cruising altitude in preparation for an approach or landing. [ARP4107]

letdown procedure Procedure used for descending from cruising altitude to the airport. [ARP4107]

letter check In the airline industry, the alphabetic designation given to a scheduled-maintenance package. A maintenance package is a group of maintenance tasks scheduled for accomplishment at the same time. *See also* check, C. [ARP4386]

let-through current The current that actually passes through a protective device after initiation of a fault. [ARP1199A]

level Any bubble-tube device used to establish a horizontal line or plane.

level (logic) A signal that remains at the "0" or "1" level for long amounts of time.

level, confidence *See* confidence level.

level control valve A device mounted within an aircraft fuel system which controls the flow of fuel into a tank based on the level of the fuel within the tank. [AIR4783]

level indicator *1.* An indicating instrument for determining the position of a liquid surface within a vessel. *2.* An instrument containing a meter, neon lamp, or cathode-ray tube which shows audio voltage level in an operating sound-recording system.

leveling saddle A pipe clamp anchoring a swivel joint 2 in. threaded socket, thereby allowing a 2 in. pipestand to be properly positioned.

level of attention *See* attention, level of.

level of awareness *See* awareness, level of.

level-off To make the flight path of an airplane horizontal after a climb, glide, or dive. [ARP4107]

level of repair analysis A technique that establishes: (a) whether an item should be repaired or discarded; and (b) if repaired, at what maintenance level, i.e., organizational, intermediate, or depot. [ARD50010]

levels/decibels Describes the magnitude of sound. The level is a measure of the ratio of sound power, or sound pressure to a reference value, and is expressed in decibels. [AIR1826]

levered suspension A type of landing gear arrangement in which an axle is supported on a trailing arm from a structural post or support. The trailing arm rotates about the pivot as a lever to provide the vertical wheel motion for shock absorption. [AIR1489]

lever suspension gear *See* gear, lever suspension.

levitation melting A metallurgical process in which a piece of metal placed above a coil carrying a high-frequency current can be supported against gravity by the Lorentz force caused by the induced surface currents in the metal; at the same time, the heat produced by Joule dissipation melts the metal.

lexical analysis In data processing, a stage in the compilation of a program during which statements such as IF, AND, and END are replaced by codes.

LF *See* low frequency.

L/F (reflux-to-feed ratio) A quantity used in analyzing column operations.

LFR *See* low frequency range.

Liapunov's second method A method analogous to the rate-of-change-of energy method for mechanical systems whereby stability or instability or a process control system can be determined. Also referred to as the indirect method.

libration A real or apparent oscillatory motion, particularly the apparent oscillation of the moon. Because of libration, more than half of the moon's surface is revealed to an observer on the earth—even though the same side of the moon is always toward the earth as a result of the moon's periods of rotation and revolution being the same. Other motions regarded as librations are long period orbital motions and periodic perturbations in orbital elements.

Lichtenberg figure camera A device for indicating the polarity and approximate crest value of a voltage surge. Consists of a photographic film or plate backed by an extended plane electrode. The emulsion on the photographic film or plate contacts a small electrode, and this electrode is connected to the circuit in which the surge occurs.

LIF (fluorescence) *See* laser induced fluorescence.

life A period of time that is related to the usability of an item. [AIR4896]

life, achieved overhaul The life achieved by an item when an overhaul becomes necessary. [AIR4896]

life cycle *1.* The total life span of an aeronautical system, beginning with the concept formulation phase and extending through the operational phase up to retirement from the inventory. Synonymous with life cycle profile. [ARD50010] *2. See* "heat aging."

life cycle costs The sum of the acquisition costs and maintenance costs for the life of a system.

life cycle profile *1.* A thorough time-life description of the events and conditions associated with an item of equipment from the time of final factory acceptance until its ultimate disposition. [AIR4896] *2. See* life cycle.

life limited item *See* item, life limited.

life, maximum permitted The time specified by an appropriate authority after which a particular item must be removed from service. [AIR4896]

life profile A time-phased description of the events and environments that an item experiences from manufacture to final expenditures or removal from the operational inventory, to include one or more mission profiles. [AIR4896]

life, service The life of an item at which it is no longer physically or economically feasible to repair or overhaul the items to acceptable standards. [ARD50010]

life, storage *See* storage life.

life test *1.* The process of placing a unit of product under a specified set of test conditions and measuring the time it takes until failure. [ARD50010] *2.* A destructive test in which a device is operated under conditions that simulate a lifetime of use.

life test sampling plan A procedure that specifies the number of units of product from a lot which are to be tested, and the criterion for determining acceptability of the lot. [AIR4896]

life units A measure of use duration applicable to an item; for example, operating hours, cycles, distance, rounds fired, or attempts to operate. [ARD50010]

life usage For a life-limited part in a gas turbine engine, refers to a method of tracking either the useful life remaining or life consumed on the part. [AIR1873]

life, useful *1.* The length of time that a population of items is expected to operate with a constant failure rate. This excludes any infant mortality and wearout periods. [ARD50010] *2.* The second phase of the lifetime of a population of a particular component when only random failures occur. [AIR4896]

lift *1.* The force component acting perpendicular to the line of flight (and in general usage, parallel to the plane of symmetry of the aircraft), and produced primarily by the pressure forces acting on the surfaces of the aircraft. [ARP4107] *2.* To lift off, take off in vertical ascent.

lift coefficients *See* aerodynamic coefficient.

lift drag ratio The ratio of lift to drag, obtained by dividing the lift by the drag or the lift coefficient by the drag coefficient.

lift forces *See* lift.

lifting surface divergence The static aeroelastic instability of a lifting surface which usually occurs in torsion when the torsional rigidity of the structure is exceeded by the aerodynamic twisting moments. [ARP4386]

liftoff The initial separation of an aircraft or other vehicle from the ground, especially the vertical takeoff of a rocket or VTOL aircraft. [ARP4107]

liftoff speed *See* V_{lof}.

lift ratio The ratio of the tire outside diameter to the rim diameter of the wheel (bead seat diameters). [AIR1489]

ligament The minimum cross section of solid metal in a header, shell, or tube sheet between two adjacent holes.

light Visible radiation (about 400 to 700 nm in wavelength), considered in terms of its luminous efficiency, i.e., evaluated in proportion to its ability to stimulate the sense of sight.

light amplification by stimulated emission of radiation *See* laser.

light-beam galvanometer A type of sensitive galvanometer whose null-balance point is indicated by the position of a beam of light reflected from a mirror carried in the moving coil of the instrument. Also known as d'Arsonval galvanometer.

light-beam instrument A measurement device that indicates measured values by means of the position of a beam of light on a scale.

light-coupled switch A switch in which the switching signal is transmitted to the activating device by means of a light beam.

light crude Crude oil rich in low-viscosity hydrocarbons of low molecular weight.

light curtain An arrangement whereby a wide, thin beam of invisible modulated light is used to detect passage of objects through a plane up to about 8 by 78 in. (200 mm by 2 m).

light duty cable Generally, a type of fiber-optic cable designed to withstand conditions encountered in a building—not in outdoor conditions.

lighted dial A dial or indicating scale and pointer that has a small lamp within the assembly so that the scale and pointer are illuminated for viewing in darkness. Compare with luminous dial.

light emitting diode See LED.

light ends The fraction of a petroleum mixture having the lowest boiling point.

lighter-than-air aircraft Aircraft that can rise and remain suspended through the use of contained gas, such that the total weight of the aircraft including the gas is less than the weight of the total volume of air it displaces at the given altitude. [ARP4107]

lighter-than-light key The remaining components in the overhead stream other than the light key.

light freezing rain Precipitation of liquid water particles that freeze upon impact with exposed objects; may be in the form of widely separated drops or drizzle. [ARP4737]

light icing See icing, light.

lighting off torch A torch used for igniting fuel from a burner. May consist of asbestos wrapped around an iron rod and saturated with oil, or may be a small oil or gas burner.

light intensity See luminous intensity.

light ions Ions of helium, boron, and other elements used in implantation experiments.

light key The component in multicomponent distillation that is removed in the overhead stream and that has the lowest vapor pressure of the components in the overhead. If the reboiler head is decreased or the reflux flow increased, the light key component is the first component to fall into the bottoms product.

light leaks Light emitted by a surface intended to be opaque; normally associated with integral transilluminated systems. [ARP1161]

light meter A small, hand-held instrument for measuring intensity of illumination.

light-modulating display device A display device made of material whose reflectance or transmittance changes as a function of the application of an electric field. [ARP4032]

light modulator An apparatus that produces a sound track by means of a source of light, an appropriate optical system, and a device for inducing controlled variations in light-beam characteristics.

light-off The initiation of combustion in the combustor of a gas turbine. [ARP906A]

light oil Any oil whose boiling point is in the temperature range 110–210°C, especially a coal tar fraction obtained by distillation.

light pen A device that enables an individual to communicate with an information system through a cathode ray tube.

light source As utilized for exterior compatible lighting, may be incandescent, electroluminescent, or infrared light-emitting diodes. [ARP4392]

light transport aircraft A classification of multiengine airplanes having a maximum passenger capacity of 30 seats and a gross weight of about 35,000 pounds.

light valve(s) 1. Optical shutters that, when activated by light, become either transparent or opaque. 2. A device whose ability to transmit light can be made to vary by applying an external electrical quantity, such as a current, voltage, electric field, electron beam, or magnetic field.

light water Water in which both hydrogen atoms in each molecule are of the isotope protium.

light water reactors Nuclear reactors using ordinary (rather than heavy) water as the moderator.

likelihood ratio The probability of a random drawing of a specified sample from a population, assuring a given hypothesis about the parameters of the population, divided by the probability of a random drawing of the same sample, assuring that the parameters of the population are such that this probability is maximized.

limb brightening The increase in the intensity of radio or x-ray brightness of the sun or other stars from the center of the object to its limbs.

limb darkening A condition, sometimes observed on celestial bodies, in which the brightness of the object decreases as the edges or limbs of the object are approached. The sun and Jupiter exhibit limb darkening.

limit The extreme of the designated range through which a measured value or characteristics may vary and still be considered acceptable. [AIR4896]

limit-check The comparison of data from a specific source with pre-established allowable limits for that source.

limit checking Internal program checks for high, low, rate-of-change, and deviation from a reference. These checks are to detect signals indicating undesirable or unsafe plant operation.

limit control A sensing device that shuts down an operation or terminates a process step when a prescribed limiting condition is reached.

limit cycle A sustained oscillation of finite amplitude.

limit cycling A repetitive on-off cycling of brake pressure by the locked wheel circuit. [AIR1489]

limited EMS An EMS in which the measurement parameters are usually limited to those that are provided as part of the standard aircraft instrumentation. [AIR1873]

limiter(s) *1.* A fuse designed specifically with a high-temperature melting point to provide protection of electric power distribution systems against fault short-circuit current. A limiter is relatively insensitive to ambient temperature. [ARP1199A] *2.* In fusion power reactors, material aperture that collect particles from the outer surfaces of the plasmas to control their transport to regions of low density. *3.* A device that applies limits to a signal.

limit exceedances Parameter excursions beyond pre-established values. [ARP1587]

limiting Describes those actions that causes a transducer output to become constant even though its input continues to rise above a certain value.

limiting zone temperature The maximum permissible zone operating temperature. [ARP485]

limit load The highest load factors that can be expected in normal operation under various operational situations. FAR requires that aircraft structures be capable of supporting 1-1/2 times the limit load factor without failure. [ARP4107]

limit of detection In any instrument or measurement system, the smallest value of the measured quantity that produces discernible movement of the indicator.

limit of error In an instrument or control device, the maximum error over the entire scale or range of use under specific conditions.

limit of measurement In an instrument or measurement system, the smallest value of the measured quantity that can be accurately indicated or recorded.

limit priority A priority specification associated with every task in a multitask operation, representing the highest dispatching priority that the task may assign to itself or to any of its subtasks.

limits The prescribed maximum and minimum values of a dimension or other attribute.

limits, confidence The values, upper and lower, between which a true value can be expected to fall, with a pre-established level of confidence. [ARD50010]

limit switch An electromechanical device that is operated by some moving part of a power-driven machine; functions to alter an electrical circuit associated with the machine.

lin *See* linear.

Lindemann electrometer An electrometer in which a metallized quartz fiber mounted on a quartz torsion fiber perpendicular to the axis of the fiber is positioned within a system of electrodes to produce a visual indication of electric potential.

line *1.* A tube, pipe, or hose that acts as a conductor of fluid. [ARP4386] *2.* In word processing, a string of characters that terminates with a vertical tab, form-feed, line-feed, or carriage return. *3.* In communications, a device that provides a data transmission link. *4.* In process plants, a collection of one or more associated units and equipment modules, arranged in serial and/or parallel paths, used to make a complete batch of material or finished product. Also called production line or train.

lineal scale length The distance from one end of an instrument scale to the other, measured along the arc if the scale is curved or circular.

linear The type of relationship that exists between two variables when the ratio of the value of one variable to the corresponding value of the other is constant over the entire range of possible values.

linear accelerator Device for accelerating charged particles; employs alternate electrodes and gaps arranged in a straight line, so proportioned that when their potentials are varied in the proper amplitudes and frequency, particles passing through them receive successive increments of energy.

linear actuator *1.* Any gas-powered device, generally consisting of a piston and cylinder, which in its energy-converting stage changes gas energy to a directional (push- and/or pull-type) mechanical power. [ARP777] *2.* An actuator that develops rectilinear motion and force as outputs. [ARP4386] *3.* A device for converting power into linear motion.

linear (or nonlinear) analysis Method for analyzing nonlinear systems; a linear approximation of the nonlinear equations is made about a specific point. [AIR1823]

linear arrays Antenna arrays whose elements are equally spaced along a straight line.

linear bus A bus with a single shared medium segment. [AS4710]

linear control system A control system in which the transfer function between the controlled condition and the command signal is independent of the amplitude of the command signal.

linear dc motor A d-c motor in which direct current supplied to the armature winding reacts with the stator field to produce a rectilinear force. [ARP4386]

linear discontinuity A discontinuity whose ratio of length to average width is greater than three. [AMS4991]

linear dispersion *See* dispersion.

linear evolution equations Denotes a large class of differential or integral differential equations that are used to describe the evolution in time of some physical systems from an initial state. An equation is said to be linear if the unknown functions and their derivatives appear linearly.

linear expansion The increase of a given dimension, measured by the expansion or contraction of a specimen or component subject to a thermal gradient or changing temperature. [AIR4844]

linear induction motor A linear motor in which polyphase a-c supplied to the primary winding produces a translating magnetic field. [ARP4386]

linearity *1.* The degree to which the normal control pressure curve conforms to the normal pressure gain line, with other operational variables held constant. Linearity is measured as the maximum deviation of the normal control pressure curve from the normal pressure gain line, expressed as percent of rated current. [ARP490] *2.* A measure of the linear relationship between input/output over the entire operating range. [ARP1281A] *3.* The degree to which the normal output curve conforms to a straight line under specified load conditions, usually expressed as percentage of rated command. [ARP4386]

linearity, differential Describes the condition in which any two adjacent digital codes should result in measured output values that are exactly 1 LSB apart. Any deviation of the measured "step" from the ideal difference is called differential nonlinearity expressed in multiples of 1 LSB.

linearity of LVDT The maximum deviation of any calibration point from a straight line. [ARP4386]

linearization The process of converting a nonlinear (nonstraight-line) response into a linear response.

linear meter An instrument whose indicated output is proportional to the quantity measured.

linear offset Distance between the duct centerlines of two joining ducts. [ARP699D]

linear polarization *1.* Polarization of an electromagnetic wave in which the electric vector of a fixed point in space remains pointing in a fixed direction although varying in magnitude. *2.* Light in which the electric field vector points in only a single direction.

linear position sensing detector An optical detector that can measure the position of a light spot along its length.

linear potentiometer A variable-resistance device whose effective resistance is a linear function of the position of a control arm or other adjustment. Most often, the device is constructed so that a single length of straight or coiled wire, whose resistance varies uniformly along its length, is in contact with a shoe or similar sliding member; effective resistance is varied by connecting the circuit to one end of the wire and to the shoe, and then varying the position of the shoe. Use of a wire-wound resistor, thin film, or printed circuit element allows greater voltage drop per unit length along the potentiometer, and therefore stronger and more useful output signals.

linear programming (LP) A method for solving problems in which a linear function of a number of variables is subject to a number of constraints in the form of linear inequalities. Contrast with convex programming, dynamic programming, integer programming, mathematical programming, nonlinear programming, and quadratic programming.

linear quadratic Gaussian control A type of optimal-state feedback control that is designed to consider noise; primarily used to control aircraft and spacecraft systems.

linear quadratic regulator A type of optimal-state feedback controller that does not consider noise; primarily used to control aircraft and spacecraft. *See also* linear regulator.

linear regulator Refers to a special case used for minimization of the performance index to obtain optimal performance. [AIR1823] *See also* linear quadratic regulator.

linear resolver *1.* That position of the rotor for which the output windings experience minimum coupling. [ARP826] *2.* A device that converts a mechanical input, rotor position, into an electrical output, which is a linear function of rotor position over a specified angular travel. [ARP826]

linear resolver functional error The difference between the in-phase component of the output of the secondary winding and the theoretical output voltage; expressed as percent linearity. [ARP826]

linear transducer A type of transducer for which a plot of input signal level versus output signal level is a straight line.

linear variable differential transformer (LVDT) A type of position sensor consisting of a central primary coil and two secondary coils wound on the same core. In this sensor, a moving-iron element linked to a mechanical member induces changes in self induction that are directly proportional to movement of the member.

linear variable reluctance transducer (LVRT) A type of position sensor consisting of a center-tapped coil and an opposing moving coil attached to a linear probe. In this sensor, the winding is continuous over the length of the core, instead of being segmented as in an LVDT. The chief disadvantage of an LVRT is that the overall length must be at least double the stroke, whereas the chief advantage is its excellent linearity over an effective stroke up to 24 in. (610 mm).

linearvection *See* illusions, vection.

linear velocity A vector quantity whose magnitude is expressed in units of length per unit time, and whose direction is invariant. (If the direction varies in circular fashion with time, the quantity is known as angular or rotational velocity; and if it varies along a fluctuating or noncircular path the quantity is known as curvilinear velocity.)

line, bleed A line, selectively open to overboard, which serves only for removing foreign substances from a system or unit, e.g., for removal of entrapped air from a hydraulic circuit. [ARP4386]

line-class valve A valve qualified by its design characteristics to be used as the first valve off the process line. *See also* process block valve.

394

line combustor A fuel-burning device in the air line of a pneumatic starting system, utilized to heat the air directed to the starter. [ARP906A]

line, drain A line returning leakage fluid independently to a reservoir or return circuit. Also, a line selectively open to overboard for removing fluid from the system. [ARP4386]

line drop compensation A system whereby alternator voltage is increased in proportion to the current and power factor in the output cables, such that the voltage is held constant at the aircraft receptacle. [ARP1148A]

line fill A cache-loading mechanism. [ARD50022]

line maintenance item *See* item, line maintenance.

line mixer *See* flow mixer.

line mounting Refers to a component mounted on and supported by a duct, instead of being directly attached to a support or bracket. [ARP699D]

line of sight An aim or observation taken with mechanical or optical aid to establish a direct path to an objective, target, etc.

line of sight communication Electromagnetic wave propagation, usually microwaves, in a straight line between the transmitter and receiver. The useful transmission distance of line of sight communication is generally limited to the horizon as sighted from the elevation of the transmitter.

line-of-sight-transmissions Transmissions that follow a straight line rather than the curvature of the earth. UHF and VHF transmissions tend to be more line-of-sight than LF or MF transmissions. [ARP4107]

line oriented flight training *See* LOFT.

line, pilot *See* pilot line.

line pressure A line that conducts fluid from the pressure source to a control unit or units. [ARP4386]

liner *1.* In a filament-wound pressure vessel, the continuous, usually flexible coating on the inside surface of the vessel, used to protect the laminate from chemical attack or to prevent leakage under stress. [AIR4844] *2.* The innermost ply of a flexible fluid tank. Can be

either a rubber layer, a rubber-coated fabric, or a spray-on polyurethane. [AIR1664]

line replaceable unit *1.* A unit that can be readily changed on an aircraft during flight line maintenance operations. [ARP4386] *2.* A propulsion component or assembly that may be replaced at the lowest level of maintenance; sometimes called a weapon replacement assembly (WRA). [ARP1587]

line, return A line that conducts working fluid back to the reservoir. [ARP4386]

line spectra In an atom, the spontaneous emission of electromagnetic radiation from the bound electrons as they jump from high to low energy levels.

line spectrum The spectrum of a complex wave consisting of several components having discrete frequencies.

line, supply A line that conducts a fluid supply, for example, from a reservoir to a pump. [ARP4386]

line, vent A line that is continuously open to atmosphere. [ARP4386]

line width *1.* Width at 50 percent of peak luminance of the line luminance distribution. [AS8034] *2.* On a CRT, the width of a luminance distribution at 50% of the maximum amplitude of the luminance distribution. [ARP1782]

lining The material used on the furnace side of a furnace wall; usually of high-grade refractory tile or brick, or plastic refractory material.

lining, brake In a brake assembly, a specially compounded material applied to the rotor and/or stator to create a predictable coefficient of friction and act as a wear surface. [AIR1489]

lining loading A measurement of the total amount of energy absorbed through the interface of each square unit of lining and its mating surface over a short period of time or a single stop. [AIR1489]

lining power A measure of the average amount of energy absorbed by a square unit of lining area and its mating surface during each second. [AIR1489]

link *1.* A logical path between a source and one or more sinks. [ARD50012] *2.* An intermediate rod or piece of material that transmits force or motion. [ARP480A]

linkage *1.* A technique for providing interconnections between routines. *2.* A mechanism consisting of bars, slides, pivots, and rotating members, which transfers motion from one part of a machine to another.

link time The amount of time required to establish a link. [ARD50012]

lipophilic emulsifier A ready-to-use oil base emulsifier that is miscible with penetrant. [AMS2647A]

liquefied compressed gas A gas that, under the charging pressure, is partially liquid at a temperature of 70°F (21°C). [ARP4386]

liquid barometer A simple device for measuring atmospheric pressure. Can be constructed by filling a glass tube having one closed end with a liquid such as mercury, then temporarily plugging the open end, inverting the tube into a container partly filled with the liquid, and unplugging the open end. If the liquid used in such a device is mercury, the tube must be at least 30 in. (76.2 mm) long; liquids of different densities require tubes of different lengths.

liquid cooled dissipator *See* cold plate.

liquid couplant The liquid interface between a transducer and the subject of a nondestructive inspection. [AIR4844]

liquid crystal display (LCD) A type of digital display device.

liquid crystal light valve A device used in optical processing to convert an incoherent light image into a coherent light image.

liquid crystal polymer A newer thermoplastic polymer that is melt-processable and develops high orientation in molding, with resultant tensile strength and high-temperature capability that is notably improved over thermoplastic polymers developed earlier. [AIR4844]

liquid-erosion failure *See* cavitation erosion.

liquid-filled thermometer Any of several designs of temperature-measurement devices whose principle of operation is the predictable change in volume with temperature of a liquid medium confined in a closed system.

liquid knockout *See* impingement.

liquid level control A device for sensing and regulating the position of a liquid surface within a vessel.

liquid-level manometer A differential pressure gage in which the reading is obtained by viewing the change in level of one or both of the free surfaces of a liquid column spanning both gage legs.

liquid limit The moisture content of fine-grained soil when it passes from the liquid state to the plastic state. [AIR1780]

liquid-metal embrittlement A decrease in strength or ductility of a solid metal caused by contact with a liquid metal.

liquid metal infiltration Process for immersion of fibers in a molten metal bath to achieve a metal matrix composite. [AIR4844]

liquid oxygen A light blue, magnetic, transparent, or water-like fluid that is produced by the fractional distillation of purified liquid air. When cooled to -182.9°C (at 14.7 psia) oxygen passes from the gaseous to the liquid state. [ARP171]

liquid oxygen converter assembly A self-powered system for the storage of liquid oxygen and for its conversion to gaseous oxygen as and when required. [ARP171]

liquid phase epitaxy A liquid-phase transformation during crystal growth.

liquid plus solid zones *See* mushy zones.

liquid propellant A propellant in liquid form, consisting of a single liquid (monopropellant) or two liquids (a bipropellant). [ARP4386]

liquid propellant auxiliary power system In its simplest form, consists of propellant, propellant tankage, propellant expulsion system, flow-control devices, and the combustion/decomposition chamber. [AIR744A]

liquid propellant rocket engines Rocket engines using a propellant or propellants in liquid form.

liquid receiver, vapor cycle A vessel permanently connected to a system, used for the storage of liquid refrigerant. [ARP147C]

liquid resin An organic, polymeric liquid that becomes a solid when converted to its final state for use. [AIR4844]

liquid rocket propellants Specifically, rocket propellants in liquid form. Examples of liquid propellants include fuels such as alcohol, gasoline, aniline, liquid ammonia, and liquid hydrogen; oxidants such as liquid oxygen, hydrogen peroxide (also applicable as a mono-propellant), and nitric acid; additives such as water; and monopropellants such as nitromethane.

liquids Substances in a state in which the individual particles move freely with relation to one another and take the shape of the container, but do not expand to fill the container.

liquid shim Material used to position components in an assembly where dimensional alignment is critical. [AIR4844]

liquid sloshing The back-and forth-movement of a liquid fuel in its tank, creating problems of stability and control in the vehicle.

liquid spring A type of shock absorber that is completely filled with fluid; the compressibility of the fluid is utilized for the shock absorption function. A liquid spring provides a relatively high spring rate compared to an air-oil type shock absorber. [AIR1489]

liquid transport system A system for extracting heat at one location and rejecting it at another location. Such a system usually consists of a coolant, a coolant reservoir, a recirculation pump, and heat exchangers. [ARP147C]

liquidus The maximum temperature at which equilibrium exists between a molten glass and its primary crystalline phase. [AIR4844]

liquid wastes The liquid counterpart of solid wastes from industrial, chemical, metabolic, and/or mineral sources.

liquid water content The amount of super-cooled water, in grams of water per m^3 of air, contained in an icing cloud. [AIR1168/9]

LIRL *See* runway edge light system.

Lissajous figures Figures describing the path of a particle moving in a plane when the components of its position along two perpendicular axes each undergo simple harmonic motions and the ratio of their frequencies is a rational number.

list, configuration deviation Those items, such as access panels, caps, or fairings, which normally form part of the exterior profile of the aircraft, but whose absence does not prevent dispatch. [AIR4896]

liter Abbreviated L. The SI unit of volume; equals 0.001 m^3 or 1.057 quarts.

lithium sulfur batteries Primary cells for producing electrical energy using lithium metal for one electrode and sulfur for the other.

lithology Description of the physical character of a rock as determined by eye or with a low-power magnifier, and based on color, structure, mineralogic components, and grain size.

litmus A blue, water-soluble powder derived from lichens and used as an acid-base indicator. Litmus is blue at pH 8.3 and above, and red at pH 4.5 and below.

live center A lathe center that is held in the headstock and rotates with the headstock and part being turned.

live front An assembly arrangement in which all moving or energized parts are exposed on the front of the panel, framework, or cabinet.

live length The convoluted length of a flex-section assembly. [ARP699D]

live part A part that is considered capable of rendering an electric shock.

live room An enclosed space characterized by unusually small capacity for absorbing sound.

live steam Steam that has not performed any of the work for which it was generated.

lixiscopes Portable, lightweight, battery-operated low intensity x-ray imaging systems with medical, industrial, and scientific applications.

LLC *See* logical link control.

LLHI *See* low-level human interface.

LLM *See* compass locator, middle.

LLWSAS *See* low-level, wind-shear alert system.

lm *See* lumen.

L/MF *See* low to middle frequency.

L/MF airway(s) Airways whose NAVAIDs utilize low or medium frequencies. [ARP4107]

LMM *See* compass locator, middle.

LNAV *See* lateral navigation.

L network An electronic network composed of two branches in series, with the junction and the free end of one branch being connected to one pair of terminals, and the free ends of both branches being connected to another pair of terminals.

load *1.* The resultant force exerted upon restraining brackets (and structure) due to internal fluid static and dynamic pressures, thermal expansion or contraction, structural deflection, and component weight. [ARP699D] *2.* A port that carries output or outlet pressure to actuated load. [ARP4386] *3.* The quantity that can be carried at one time; for example, cargo load or passenger load. 4. The total weight of a vehicle; for example, as in the calculation of wing loading, the load supported by the wings. *5.* The stress or forces placed upon a structure either by external weights or air pressures. [ARP4107] *6.* To store a computer program or data into memory. *7.* In an electric power circuit, the resistive and reactive components that comprise the device being powered by the circuit.

load-and-go In data processing, an automatic coding procedure that not only compiles the program, thereby creating machine language, but also proceeds to execute the created program. Load-and-go procedures are usually part of a monitor.

load, assembly The total load imposed on a landing gear assembly or leg. May be further broken down to individual wheel loads. [AIR1489]

load cell A transducer for the measurement of force or weight; action is based on strain gages mounted within the cell on a force beam.

load circuit A circuit or a branch of a network that carries the main portion of current flow.

load classification number A rating number that indicates the capability of an aircraft with specific tire/wheel equipment to operate from a given type of airbase. For intercontinental express, load classification number is 100; for light service, it is less than 14. [AIR1489]

load-deflection curve A curve resulting when the increasing tension, compression, or flexural loads are plotted on the ordinate axis and the deflections caused by those loads are plotted on the abscissa axis. [AIR4844]

load, dynamic *See* dynamic load.

loaded flow *See* control flow.

loaded galley weight The weight of all galleys, including maximum allowable weight of food, carts, beverages, inserts, and galley service items. [AS1426]

loaded inflation pressure The recommended pressure determined by multiplying an unloaded calculated tire inflation pressure by the factor 1.04. [AS4833]

load, equivalent single wheel *See* equivalent single wheel load.

load factor *1.* A number that yields the inertial load (g load) when multiplied by the weight of the object. [AIR1489] *2.* The ratio of two loads (basic load is the denominator). The load factor for a landing gear assembly is obtained by dividing the sum of the external forces acting upon it by the static load. NOTE: The force of gravity does not appear in the sum of external forces because on each particle of mass, the gravity force is cancelled by the inertial force of free-fall acceleration. [AIR1489] *3.* The ratio of a specified load to the total weight of the aircraft. The specified load is expressed in terms of any of the following: aerodynamic forces, inertia forces, or ground or water reactions. [ARP4107]

load, holdback The load applied to the aircraft holdback fitting. This load results from the buffing force during deceleration of the aircraft from taxi-in velocity, or from the simultaneous application of catapult tensioning force, aircraft full takeoff power, and ship motion before firing the catapult. [AIR1489]

loading Buildup of material along the cutting edge of a bit or other tool; or, similarly, buildup of grinding debris on the working face of a grinding wheel or abrasive disc.

loading density The weight of explosive per unit volume. [AIR913]

loading sensitivity Refers to the changes of in-phase speed-sensitive output voltages, phase shift, and zero speed voltages due only to changes in tachometer load within a specified impedance range. [ARP667]

load, limit *See* limit load.

load module A program prepared in a format that is ready for loading and executing.

load natural frequency The undamped resonant frequency of the load mass coupled with the load-related stiffness. [ARP4386]

load on LVDT The impedance of the circuit to which the secondary winding of the LVDT is connected. [ARP4386]

load, pneumatic The force determined by the product of the internal differential pressure and the duct cross-sectional area. [ARP699D]

load point (mag tape) The point, near the beginning of the tape, at which the computer can start to record data.

load pressure The pressure reacting to a static or dynamic load. [ARP4386]

load-pressure drop The differential pressure between the control ports, expressed in psi (or kPa). In conventional three-way servovalves, load-pressure drop may be expressed mathematically as the supply pressure, less return pressure, and less the pressure drop across the single active control orifice. [ARP490]

load range The range of load over which a tire is capable of operating. For a tire, variation of loads is usually accompanied by a proportional increase or decrease of inflation pressure within design limitations. [AIR1489]

load, rated *See* rated load.

load rating The maximum permissible load at a specific tire inflation pressure. [AS4833]

load, release The maximum load applied to the aircraft holdback fitting by application of the catapult firing force, in addition to the existing holdback load. The release load of predetermined magnitude ruptures a release element or disengages a repeatable release fitting, thereby allowing the aircraft to be released. [AIR1489]

load repetition factor The number of passes of load-carrying tires in adjacent tire paths required to just cover a given width of pavement one time; identical with "coverage" as used in pavement design. *See* coverage. [AIR1489]

load, speed, time curves Curves in which the abscissa is time, showing the relationships of load and speed vs. time. [AIR1489]

load thermocouple A thermocouple attached to and in direct contact with the heaviest section of a part or representative sample. [AMS2728]

load, ultimate The load that should cause destructive failure according to stress analy-

sis; or the load that causes failure during a test of strength. [ARP4107]

load weighing torque limiter A limiter in which the magnitude of the torque is detected by a suitable device. When the torque exceeds the predetermined value, the weighing device activates the brake. [ARP4386]

LOC *See* localizer only approach.

local area network (LAN) *1.* A network, generally microcomputer based, that enables users in the same location to use the same programs and equipment such as printers. *2.* A communications mechanism by which computers and peripherals in a limited geographical area can be connected. LANs provide a physical channel of moderate to high data rate (1 to 20 Mbit), which has a consistently low error rate (typically 10^{-9}). *See also* network.

local catch efficiency The local catch divided by the catch that would be observed for 1 sq. ft. of surface area with a catch efficiency of unity. [AIR1168/9]

local control unit (LCU) A control device that performs closed-loop control and interfaces directly with the process.

local effect The impact of the failure mode on the item that is being analyzed. [ARP926A]

localization level The functional level to which a failure can be traced or located without using accessory test equipment. [AIR4896]

localizer *1.* A radio facility that provides signals for use in lateral guidance of aircraft with respect to a runway centerline. [ARP419] *2.* The component of an ILS that provides course guidance to the runway. [ARP4107]

localizer only approach (LOC) A nonprecision instrument approach procedure utilizing the localizer signal for course guidance to the runway, the glide slope not being available. [ARP4107]

localizer-type directional aid (LDA) A NAVAID that is used for nonprecision instrument approaches with utility and accuracy comparable to a localizer, but is not part of a complete ILS, and is not aligned with the runway. [ARP4107]

local oscillator An oscillator whose output is combined with another frequency to generate a sum or difference frequency, either of which may be easier to amplify and use; for example, the oscillator in a superhetrodyne receiver.

local processing unit A field station with input/output circuitry and the main processor. Local processing units measure analog and discrete inputs; convert these inputs to engineering units; perform analog and logical calculations (including control calculations) on these inputs; and provide both analog and discrete (digital) outputs.

local reference A copper bar mounted on the cabinet of a subsystem which becomes the signal reference point for the entire subsystem. All power commons and signal commons of a subsystem are tied to the local reference. Each local reference is tied to the master reference, by a separate wire.

local traffic Aircraft that are: (a) operating in the local traffic pattern or within sight of the tower; or (b) known to be departing for or arriving from flight in local practice areas; or, (c) executing simulated instrument approaches at the airport. [ARP4107]

local water catch The point-by-point distribution of water (or ice) over the impingement area. [AIR1168/9]

location In computer storage or memory, an address at which a unit of data or an instruction can be stored. *See also* cell.

location counter *1.* In data processing, the control-section register that contains the address of the instruction currently being executed. *2.* A register in which the address of the current instruction is recorded. Synonymous with instruction counter and program address counter.

locator A device for positioning a contact, terminal, or splice in the crimping dies. [ARP914A]

locator beacon, personnel A portable, lightweight, manually operated beacon, designed to be carried on the person, in the cockpit of an aircraft, or attached to a parachute, and capable of operation by unskilled persons. Operates from its own power source on 121.5 MHz and/ or 243 MHz, (preferably on both emergency

frequencies), transmitting a distinctive downward-swept audio tone for homing purposes. May or may not have voice capability. [ARP4107]

lock *1.* Describes a forging that has a flash line in more than one plane. *2.* To prevent a movable part from moving; to seize.

locked-in liner In a butterfly valve body, a liner retained in the body bore by a key ring or other means.

locked wheel protection Refers to the temporary removal of brake pressure by a secondary control circuit when the controlled wheel(s) reaches a low predetermined wheel velocity, when the aircraft is at a significantly higher velocity. [AIR1489]

lock-in amplifier An amplifier that selects signals at one prespecified frequency and amplifies them, while discriminating against signals at other frequencies.

locking Pertaining to code extension characters that change the interpretation of an unspecified number of following characters. Contrast with nonlocking.

lockout Any condition that prevents any or all senders or receivers from communicating.

lock-step A method to synchronize a mixed-signal simulation system whereby each simulator progresses one time step and passes all interacting signals to the other simulator.

lockup pressure The maximum permissible output pressure when the regulator flow demand is zero. [ARP4386]

lock valve A pressure-operated controllable check valve, sometimes known as a counter balance valve or ratchet valve. [ARP4386]

locomotive boiler A horizontal fire-tube boiler with an internal furnace, the rear of which is a tube sheet. The tube sheet is directly attached to a shell containing tubes through which the products of combustion leave the furnace.

locus of control An attitudinal set whereby a person believes either that he/she is in control of his/her destiny (internal locus of control), or that outside influences control his/her destiny (external locus of control). [ARP4107]

locus of failure Site of failure. [AIR4844]

LOFT Stands for line oriented flight training; a technique of conducting aircrew flight training in simulators by having the crew fly full mission scenarios together in the simulator as they would in an operational aircraft. This type of flight training emphasizes the whole mission concept and is aimed at developing effective communications among crewmembers, effective human resource utilization, and good crew coordination. [ARP4107]

lofting Transcribing the lines from the basic geometry drawing of the aircraft onto metal in order to make a master layout for use by tooling departments. [AIR4844]

log *1.* A record of everything pertinent to a machine run, including identification of the machine run, a record of alteration switch settings, identification of input and output tapes, a copy of manual key-ins, identification of all stops, and a record of action taken on all stops. *2.* To record occurrences in a chronological sequence.

logarithm The power to which a fixed number, called the base, usually 10 or e (2.7182818), must be raised to produce the value to which the logarithm corresponds.

logarithmic amplifier An amplifier whose output is a logarithmic function of its input.

logarithmic decrement In an exponentially damped oscillation, the natural logarithm of the ratio of one peak value to the next successive peak value in the same direction.

logger A device that automatically records physical processes and events, usually chronologically.

logic *1.* A means of solving complex problems through the repeated use of simple functions that define basic concepts. Basic logic functions include AND, OR, and NOT. *2.* The science dealing with the criteria or formal principles of reasoning and thought. *3.* The systematic scheme that defines the interactions of signals in the design of an automatic data-processing system. *4.* The basic principles and the application of truth tables and interconnection between logical elements required for arithmetic computation in an automatic data processing system.

logical addressing A mode of addressing in which message content or type is labeled. [AIR4271]

logical block An arbitrarily defined, fixed number of contiguous bytes used as the standard I/O transfer unit throughout a computer operating system; for example, the commonly used logical block in PDP-11 systems is 512 bytes long. An I/O device is treated as if its block length is 512 bytes, although the actual (physical) block length of the device may be different. Logical blocks on a device are numbered from block 0 consecutively up to the last block on the volume.

logical connectives The computer operators or words, such as AND, OR, ELSE, IF, THEN, NEITHER, NOR, and EXCEPT, which make new expressions from given expressions, and which have the property that the truth or falsity of the new expressions can be calculated from the truth or falsity of the given expressions and the logical meaning of the operator.

logical decision *1.* The choice or ability to choose between alternatives; basically, this amounts to an ability to answer yes or no with respect to certain fundamental questions involving equality and relative magnitude. *2.* The utilization of a logic instruction.

logical device name An alphanumeric name assigned by a user to represent a physical device; the name can be used synonymously with the physical device name in all references to the device. Logical device names are used in device-independent systems to enable a program to refer to a logical device name that can be assigned to a physical device at run time.

logical difference Corresponds to all elements belonging to Class A but not to Class B, when two classes of elements, Class A and Class B, are given.

logical element(s) In computers or data-processing systems, the smallest building blocks that can be represented by operators in an appropriate system of symbolic logic. Typical logical elements are the AND gate and flip-flop, which can be represented as operators in a suitable symbolic logic.

logical expression An expression consisting of logical constants, variables, array elements, function references, and combinations of those operands, separated by logical operators and parentheses.

logical interface An interface that defines a peer-to-peer relationship between entities within the same GOA layer of the GOA model. [AS4893]

logical link control (LLC) The upper sublayer of the Data Link Layer (Layer 2) used by all types of IEEE 802 Local Area Networks. LLC provides a common set of services and interfaces to higher layer protocols. Three types of services are specified: Type 1, Connectionless; Type 2, Connection Oriented; and Type 3, Acknowledged Connectionless. *Type 1, Connectionless*–a set of services that permit peer entities to transmit data to each other without the establishment of connections. Type 1 service is used by both MAP and TOP. *Type 2, Connection Oriented*–a set of services that permit peer entities to establish, use, and terminate connections with each other in order to transmit data. *Type 3, Acknowledged Connectionless*–a set of services that permit a peer entity to send messages requiring immediate response to another peer entity. This class of service can also be used for polled (master-slave) operation.

logical operation Logical shifting, masking, and other nonarithmetic operations of a computer. Contrast with arithmetic operation.

logical operator *See* logical connectives.

logical product Same as AND.

logical record Within a file, a logical unit of data whose length is defined by the user and whose contents have significance to the user; a group of related fields treated as a unit.

logical sum A result obtained in the process of ordinary addition which is similar to an arithmetic sum, except that the rules are such that a result of one is obtained when either one or both input variables is a one, and an output of zero is obtained when the input variables are both zero. The logical sum is the name given to the result produced by the inclusive OR operator.

logical unit number A number associated with a physical device unit during the I/O operations of a task; each task in the system can establish its own correspondence between logical unit numbers and physical device units.

logical variable A variable that may have only the value true or false. Also called a Boolean variable.

logic analyzer A device used to analyze the logical operation of a microcomputer; a test device used for debugging systems.

logic design The specification of the working relations between the parts of a computer system, in terms of symbolic logic, and without primary regard for hardware implementation.

logic device Refers to the general category of digital fluidic components that perform logic functions; for example AND, NOT, OR, NOR, and NAND. Logic devices can gate or inhibit signal transmission with the application, removal, or other combinations of control signals. [ARP993A]

logic diagram Graphic method of representing a logic operation or set of operations.

logic gate A device that takes binary bits as input and produces an output bit to some specification.

logic instruction A computer instruction that executes an operation that is defined in symbolic logic, such as AND, OR, NOR.

logic levels Electrical convention for representing logic states. For TTL systems, the logic levels are nominally 5 V for logic 1, and 0 V for logic 0.

logic network In data processing, an arrangement of logic gates designed to achieve specific outputs.

logistics The science of planning and carrying out the movement and maintenance of forces; in its most comprehensive sense, those aspects of military operations that deal with: (a) design and development, acquisition, storage, movement, distribution, maintenance, evaluation, and hospital inspection of personnel; (b) acquisition of construction, maintenance, operation, and disposition of facilities; and (c) acquisition or furnishing of services. [ARD50010]

logistics critical component An item that is selected for multiple source procurement. [AIR4073]

logistics support The materials and services required to enable the operating forces to operate, maintain, and repair the end item within the maintenance concept defined for that end item. Logistics support encompasses the identification, selection, procurement, scheduling, stocking, and distribution of spares, repair parts, facilities, ground support equipment, trainers, technical publications, contractor engineering and technical services; and personnel training as necessary to provide the operating forces with the capability needed to keep the end item in a functioning status. [ARD50010]

logistic support analysis A structured process that includes actions to define, analyze, and quantify logistic support requirements, and to influence design for supportability, through system development. [ARP4293]

logistic time All replacement procurement time, except that time when the maintenance man is engaged in the procurement activity. [ARD50010]

lognormal distribution A probability distribution in which the probability that an observation selected at random from the population falls between a and b is given by the area under the normal distribution between log a and log b. [AIR4844]

LOM *See* compass locator, outer.

long duration space flight Space flight involving interplanetary and/or interstellar travel. Also known as extended duration space flight.

longevity Length of useful life of a product, to its ultimate wearout requiring complete rehabilitation. [AIR4896]

long flame burner A burner in which the fuel and air do not readily mix, resulting in a comparatively long flame. This may be due to the way the fuel emerges or the way the air for combustion is admitted into the burner.

longitude Angular distance, along a primary great circle, from the adopted reference point; the angle between a reference plane through the polar axis and a second plane through that axis.

longitudinal *1.* Direction of maximum extension of a metal during forging. [AMS6474] *2.* Along the length of a material. *3.* In the 0° direction. [AIR4844]

longitudinal axis *1.* The tail-to-nose (X) axis of an aircraft about which the aircraft revolves in rolling. [ARP4386] *2.* An imaginary straight line lying in the plane of symmetry extending from the nose to the tail of an aircraft, and passing through the center of gravity. Movement about this axis is called roll. [ARP4107]

longitudinal drum boiler A sectional header or box header boiler in which the axis on the horizontal drum or drums is parallel to the tubes in a vertical plane.

longitudinal redundancy check (LRC) A system of error control based on the formulation of a block check following preset rules. The check formation rule is applied in the same manner to each character.

longitudinal separation The longitudinal spacing of aircraft at the same altitude by a minimum distance expressed in units of time or miles. The amount of separation required is a function of the availability of radar coverage and the distance the aircraft is from the radar antenna site. [ARP4107]

longitudinal stability *See* stability, longitudinal.

longitudinal wave(s) *1.* Waves in which the direction of displacement at each point of the medium is normal to the wave front. *2. See* compressional wave.

longos Low-angle helical or longitudinal windings. [AIR4844]

long range navigation *See* LORAN.

long range predictive alerting system A system that senses and identifies windshear far enough in advance of a possible encounter to allow the crew to consider maneuvering away from the hazard. [AIR4102/11]

long term trending Tracking of engine, engine component, or subsystem degradation on a periodic basis, often by flight. This type of tracking indicates a deviation of monitoring data from an established trend. [ARP1587]

look angles (electronics) The solid angle in which an instrument operates effectively;

generally used to describe radars, optical instruments, and space radiation detectors.

look angles (tracking) The elevation and azimuth at which a particular satellite is predicted to be found at a specified time.

look phase The portion of an inspection that includes the basic requirements outlined by the periodic maintenance information cards, excluding repair of discrepancies, which cannot be completed within the time alloted on maintenance requirement cards (MRCs). [AIR4896]

loom state Describes reinforcement fibers as supplied by the manufacturer; the fibers usually have a small amount of size on their surface, designed to reduce filament damage and facilitate handling during weaving. [AIR4844]

loop *1.* In aeronautics, a movement whereby an aircraft describes a closed curve or circle in the vertical plane. Referred to as an inside loop when the top of the aircraft is toward the center of the circle, and an outside loop when it faces away from the center. *2.* The signal path in a closed-loop control system beginning with the error signal and ending with the resultant feedback signal that functionally interconnects the forward-loop and feedback-loop elements. [ARP4386] *3.* A sequence of instructions that is executed repeatedly until a terminal condition prevails. *4.* A complete hydraulic, electric, magnetic, or pneumatic circuit.

loop (computing) Instructions that actually perform the primary function of a loop, as distinguished from loop initialization, modification, and testing.

loop (initialization) Instructions that immediately precede the loop proper, for setting addresses, counters, or data to initial values.

loop, closed *See* closed loop.

loop diagram A schematic representation of a complete hydraulic, electric, magnetic or pneumatic circuit.

loop, feedback *See* feedback loop. *See also* closed loop.

loop gain The product of the gains of all the elements in a loop.

loop identification Identification consisting of a first letter and the number of the instrument loop. Each instrument within a loop is assigned the same loop number and, in the case of parallel numbering, the same first letter.

loop load The algebraic sum of the applied loads at the anchorages of a torso-restraint system segment. A balanced loop load is achieved when the reaction loads at each lap belt anchorage are equal. [AS8043]

loop-modification In a loop, instructions that alter instruction addresses, counters, or data.

loop tenacity The tenacity or strength value obtained by pulling two loops, as two links in a chain, against each other in order to demonstrate the susceptibility of a fibrous material to cutting or crushing itself; loop strength. [AIR4844]

loop-testing In a loop, instructions that determine whether the loop is complete.

loose cover A separation of the cover from the carcass or reinforcements. [ARP1658A]

loose pick A filling yarn that is not flush with the surrounding fabric, usually caused by insufficient tension. [AIR4844]

LORAN Contraction of long range navigation. An electronic navigational system whereby hyperbolic lines of position are determined by measuring the difference in the time of reception of synchronized pulse signals from two fixed transmitters. The position of the vehicle is determined by the intersection of these two hyperbolic lines of position generated by the two transmitters. Loran A operates in the 1750 to 1950 kHz frequency band; Loran C and D operate in the 100 to 110 kHz frequency band. [ARP4107]

Lorentz force The force affecting a charged particle due to the motion of the particle in a magnetic field.

loss *1.* Energy dissipated without accomplishing useful work. [ARP1931] *2.* Dissipation of power which reduces the efficiency of a machine or system. *3.* Dissipation of material or energy due to leakage.

loss factor For an insulating material, a factor that is equal to the product of the dissipation factor and the dielectric constant of the material. [ARP1931]

loss frequency The expected failure rate of a particular item or function in its operational mode and environment. [ARP926A]

loss in surge pressure ratio The percent change in surge pressure ratio, at constant corrected airflow, between the undistorted surge line and the distorted surge point. [ARP1420]

lossless materials Dielectric materials that do not dissipate energy or that do not dampen oscillations.

loss modulus A damping term describing the dissipation of energy into heat when a material is deformed. [AIR4844]

loss of back reflection Absence of an indication of the far surface of the article being inspected. [ARP5089]

loss on ignition Weight loss, usually expressed as percent of total, after burning off an organic sizing from glass fibers, or an organic resin from a glass fiber laminate. [AIR4844]

lossy line A cable having large attenuation per unit of length. [ARP1931]

lossy media A material that dissipates electromagnetic or acoustic energy passing through it.

lost cluster A group of one or more disk sectors that are not available for storage use.

lot Described as an inspection lot or a preproduction lot. *Inspection lot*–a collection of units of product, manufactured under essentially the same conditions, from which a sample is drawn and tested to determine compliance with the acceptability criterion. *Preproduction lot*–one or more units of product submitted prior to initiation of production for test to determine compliance with the acceptability criterion. [AIR4896]

loudness The intensive attribute of an auditory sensation, in terms of which sounds may be ordered on a scale extending from soft to loud; measured in sones. Loudness depends primarily on the sound pressure of the stimulus, but it also depends on the frequency and waveform of the stimulus.

loudness level A measurement of sound intensity numerically equal to the sound pressure, in decibels, relative to 0.0002 microbar, of a simple tone whose frequency is 1000 Hz and that is judged by the listeners to be equivalent in loudness; the units of measure determined in this way are called phons.

loudspeaker An electroacoustic transducer usually constructed to effectively radiate sound of varying frequencies into the air.

loudspeaker dividing network *See* crossover network.

louver type damper *See* damper.

low-alloy steel An iron-carbon alloy that contains up to about 1% C, and less than 5% by weight of additional elements.

low altitude airway structure/federal airways The network of airways serving aircraft operation up to but not including 18,000 ft MSL. [ARP4107]

low approach An approach over an airport or runway following an instrument (IFR) approach, or a visual flight rules (VFR) approach including the go-around maneuver, in which the pilot intentionally does not make contact with the runway. Low approaches are often performed during training flights in which a landing is not required at the completion of each approach; or under those circumstances when the pilot feels that it would be unsafe to continue the approach for a landing. [ARP4107]

low carbon steel *1.* An iron-carbon alloy containing about 0.05 to 0.25% C, and up to about 0.7% Mn. *2.* An iron alloy containing carbon in low percentages, and displaying temper and malleability characteristics not found in ordinary carbon steels.

low cost visual approach slope indicator (LCVASI) A visual approach slope indicator system consisting of painted plywood panels (normally black and white or fluorescent orange) whose alignment as seen by the pilot indicates the relative position of the aircraft with regard to a fixed glide path to a touchdown point on the runway. [ARP4107]

low cycle fatigue (LCF) *1.* A type of fatigue that results from high stress, which can be below or above the proportional limit of the material, but is higher than its endurance strength. [AIR1639] *2.* Failure caused by cyclic loading to levels resulting in non-elastic deformation, which results in complete failure of the component in less than 50,000 loading cycles. Most rotating gas turbine components are prone to low-cycle fatigue failure as a result of centrifugal and thermally induced loads, and their life usage can be measured with reasonable accuracy. [AIR784] *3.* Component material life usage incurred by cyclic stress excursions. [ARP1587]

low draft switch A control to prevent burner operation if the draft is too low. Used primarily with mechanical draft.

lower atmosphere Generally, and quite loosely, that part of the atmosphere in which most weather phenomena occur (i.e., the troposphere and lower stratosphere); hence, used in contrast to the common meaning for the upper atmosphere.

lower body negative pressure Application and/or measurement of reduced pressure in the portion of the body below the iliac crests. Used as a simulator of orthostatic stress or as an indicator of cardiovascular deconditioning in weightless environment.

lower limit The signal corresponding to the minimum value of the transmitted input.

lower test MTBF (THETA 1) The MTBF value that is unacceptable. Standard test plans will be reject, with high probability, equipment with a true MTBF that approaches THETA 1. [ARD50010]

low-fire start The firing of a burner with controls in a low-fire position to provide safe operating condition during light-off.

low frequency (LF) The frequency band between 30 and 300 kHz. [ARP4107]

low frequency bondtester A dry-coupled bondtester generally operating at frequencies below 100 kHz. [ARP5089]

low frequency range (LFR) A directional NAVAID, in the frequency band between 30 and 300 kHz, no longer used in the United States. [ARP4107]

low frequency stiffness (frequency dependent dynamic stiffness) The stiffness associated with the output deflection of an active actuation system, caused by externally applied loads when the frequency of the load disturbance is within the bandpass (bandwidth) of the actuation system. [ARP4386]

low gas pressure switch A control to stop the burner if gas pressure is too low.

low gravity *See* reduced gravity.

low head boiler A bent-tube boiler having three drums with relatively short tubes in a vertical plane.

low-heat value The high heating value minus the latent heat of vaporization of the water formed by burning the hydrogen in the fuel.

low-hydrogen electrode A covered welding electrode that provides an atmosphere around a welding arc which is low in hydrogen.

low intensity readability The visual clarity of a lighted display when compared with other lighted displays, both energized at a selected low level of excitation. [ARP1161]

low intensity runway light system (LIRL) *See* runway edge light system.

low intensity x-ray imaging scopes *See* lixiscopes.

low-level human interface (LLHI) A device that allows a human to interact with a local control unit.

low-level, wind-shear alert system (LLWSAS) A system of five or six anemometers around the periphery of an airport, the readouts of which are automatically compared with that of the center-field anemometer. A wind vector difference of 15 knots or more between the center-field anemometer and any peripheral anemometer is indicative of potential wind shear, and the tower will advise pilots of the potential for wind shear. [ARP4107]

low loss Term applied to a dielectric material or cable that has a small amount of power loss over long lengths, making it suitable for transmission of radio-frequency energy. [ARP1931]

low noise cable A cable configuration specially constructed to eliminate spurious electrical disturbances caused by capacitance changes or self-generated noise. [ARP1931]

low oil temperature switch (cold oil switch) A control to prevent burner operation if the temperature of the oil is too low.

low order Pertaining to the weight or significance assigned to the digits of a number, e.g., in the number, 123456, the low-order digit is six. As another example, one may refer to the three low-order bits of a binary word.

lowpass filter(s) 1. Wave filters having a single transmission band extending from zero frequency up to some critical or bounding frequency, not infinite. 2. A filter that passes frequencies below its cutoff frequency with little attenuation.

low-pass output filter (LPOF) In a subcarrier discriminator, the filter that rejects subcarrier components and all extraneous noise while passing the frequencies that are known to contain data.

low pitch A propeller setting whereby the chord of the blade is at a relatively acute angle to the plane of rotation, resulting in a high propeller speed. The specific angle considered to be low pitch will vary from one model propeller to another. Low pitch settings are used when very high or maximum power is required at low airspeeds, as in takeoff or go-around. [ARP4107]

low point The lowest point in a pipe line or tank where drains may be installed. [AIR4783]

low-pressure hot-water and low-pressure steam boiler A boiler furnishing hot water at pressures not exceeding 160 pounds per square inch or at temperatures not more than 250°F, or steam at pressures not more than 15 pounds per square inch.

low pressure laminates Laminates molded and cured in the range of pressures from 2760 kPa (400 psi) down to and including pressure obtained by the mere contact of the plies. [AIR4844]

low pressure molding The distribution of relatively uniform low pressure over a resin-bearing fibrous assembly of cellulose, glass, asbestos, or other material, with or without application of heat from an external source, to form a structure possessing definite physical properties. [AIR4844]

low rate discharge The withdrawal of small currents for long periods of time, usually longer than 1 h. [ARP4386]

low resolution graphics In data processing, the ability of a dot-matrix printer to reproduce simple forms or pictures.

low Reynolds number A Reynolds number below the critical Reynolds number of a sphere.

low stress grinding Grinding/polishing under controlled conditions to minimize and produce required compressive surface stresses. [AS7101]

low-temperature hygrometry The measurement of water vapor at low temperatures. Low-temperature hygrometry requires special techniques because of the small amounts of moisture typically present and because of unusual instrument operating characteristics at such temperatures.

low temperature repair (room temperature repair) A repair using a resin system that can be cured at temperature not to exceed 70°C (160°F). [AIR4844]

low tension Describes an item incorporating an electrode(s) across which an electric spark is discharged to ignite a combustible mixture in a continuous-burning cycle engine; categorized by "shunted surface gap." This type requires less than 5 kV potential to create a spark between the electrodes. General practice dictates that a "new" spark igniter shall spark when 1000 V is applied. [AIR784]

low tension system Ignition systems capable of delivering voltages up to 5 kV inclusive to the firing tip of the spark igniter. [AIR784]

low to middle frequency (L/MF) The frequency band of nondirectional radio beacons between 200 and 1750 kHz. [ARP4107]

low vacuum The condition in a gas-filled space at pressures less than 760 torr corresponding approximately to the vapor pressure of water at 25°C and to 1 inch of mercury.

LOX (liquid oxygen) systems Systems providing facilities to store and convert liquid oxygen to gaseous oxygen at a breathable temperature and pressure for the aircrewman. Generally used in aircraft where space and weight and mission considerations are paramount. [ARP171]

L_{po} Sound pressure level at a point four feet downstream of a valve and three feet from the surface of the pipe.

LPOF *See* low-pass output filter.

LQG control *See* linear quadratic Gaussian control.

LQR *See* linear quadratic regulator.

LRC *See* longitudinal redundancy check.

L-Sat A communications satellite designed by European Space Agency member states to meet future communications satellite market needs, such as European broadcast services, global telecommunications trunk services, and mobile services.

LSB *See* least significant bit.

LSD *See* least significant digit.

LSI *See* large-scale integration.

lubricant A material added to most sizings to improve the handling and processing properties of textile strands, especially during weaving. [AIR4844]

lubrication and servicing The internal or external application of fluids, oils, or grease to an item for the purpose of maintaining its inherent design operating capabilities. [AIR4896]

lubricator A device for automatically applying lubricant.

lubricity Property of a fluid, measured by the wear scar in millimeters, produced on a stationary ball from contact with the fuel-wetted rotating cylinder operating under closely controlled conditions. [AIR1794]

ludlam limit The point at which some supercooled water droplets no longer freeze within their catchment area and the forward growth of ice is diminished. [AIR1667]

lug Any projection, like an ear, used for supporting or grasping. *See also* terminal.

lugged body *See* body, wafer, lugged.

lumen(s) *1.* The space in the interior of a tubular structure. [ARP171] *2.* The unit of luminous flux equal to the flux in a unit solid angle (one steradian) from a uniform point source of one candela. Abbreviated lm. [ARP1782]

lumens per square meter (lm/m²) The lux (light flux); the metric unit for illuminance (commonly called illumination). Illuminance is used to measure light falling on a surface, such as a display surface. A lux is approximately equal to 0.0929 foot candles (fc); 1 fc is approximately equal to 10.8 lux. [ARP1782]

luminaire A complete lighting unit consisting of a light source together with its direct appurtenances, such as the globe, reflector, housing, and such support as is integral with the housing. [ARP798]

luminance *1.* A measure of what the human eye perceives as brightness of a display. Luminance is defined as the luminous intensity per unit area that is emitted by a surface in a given direction. Luminance is measured in units of footlamberts or Nits. [ARP1782] *2.* In photometry, a measure of the intrinsic luminous intensity emitted by a source in a given direction; the illuminance produced by light from the source upon a unit surface area oriented normal to the line of sight at any distance from the source, divided by the solid angle subtended by the source at the receiving surface. *3.* The luminous intensity of any surface in a given direction per unit of projected area in a plane perpendicular to that direction.

luminance (photometric brightness) *1.* The quotient consisting of the luminous flux leaving, passing through, or arriving at an element of the surface surrounding the point, and propagated in directions defined by the elementary cone containing the given direction, divided by the product of the solid angle of the cone and the area of the orthogonal projection of the element of the surface on a plane perpendicular to the given direction. *2.* Luminous intensity of any surface in a given direction per unit of projected area of the surface as viewed from that direction. [AIR512B]

luminance contrast *1.* The relative luminance difference between two areas. [ARP4032] *2. See* contrast.

luminance intensity The luminous flux emitted into a given solid angle by a source; measured in units of candelas. Also called light intensity. [ARP1782]

luminance modulation *See* contrast.

luminance uniformity For a display, a measure of the variation of luminance across the display surface. Defined in terms of a ratio or percent of luminance measured on the display. [ARP1782]

luminescence Light emission by a process whereby kinetic heat energy is not essential for the mechanism of excitation.

luminosity *1.* The quality or state of being luminous. [AIR1093] *2.* Emissive power with respect to visible radiation.

luminosity coefficients The constant multipliers for the respective tristimulus values of any color, such that the sum of the three products is the luminance of the color.

luminous *1.* Emitting or seeming to emit a steady, diffused light that is reflected or produced from within. [AIR1093] *2.* Emitting radiation in the form of visible light.

luminous clouds *See* noctilucent clouds.

luminous dial A dial or indicating scale and pointer whose scale divisions, numerals, and pointer are made of or coated with a light-emitting substance, such as luminous paint, so that they can be seen in the dark. Compare with lighted dial.

luminous efficiency Luminous flux divided by radiant flux.

luminous flux *1.* Luminous power (i.e., radiant power that has been adjusted for human eye response) that is emitted by a source. The process of correcting radiant power for the human eye response has been defined by the CIE. The unit of luminous flux is the lumen. [ARP1782] *2.* The amount of light passing a given point per unit time.

luminous intensity Luminous energy per unit time per unit solid angle; the intensity (flux per unit solid angle) of visible radiation weighted to take into account the variable response of the human eye as a function of the wavelength of light; usually expressed in candles.

lumped-constant wavemeter A device for determining frequency using a tunable reso-nant LC circuit coupled to a crystal detector; the circuit generally utilizes plug-in coils of various inductances and a continuously variable capacitor with a dial calibrated in frequency.

lumped parameter systems Systems in which the parameters may be considered to represent, for purposes of analysis, a single inductance, capacitance, resistance, etc., throughout the frequency range of interest.

lunar crater A depression, usually circular, on the surface of the moon, usually with a raised rim called a ringwall.

lunar eclipse The phenomenon observed when the moon enters the shadow of the earth.

lunar probes Probes for exploring and reporting on conditions on or about the moon.

lunik lunar probes Russian term for a space probe launched to the moon's vicinity or to impact on the moon.

lux Abbreviated lx. Metric unit of illuminance.

LVDT *See* linear variable differential transformer. [ARP4386]

LVRT *See* linear variable reluctance transducer.

lx *See* lux.

Lyman alpha radiation The radiation emitted by hydrogen at 121.6 nanometers, first observed in the solar spectrum by rocket-borne spectrographs. Lyman alpha is very important in the heating of the upper atmosphere, thus affecting other atmospheric phenomena.

lysimeters Instruments for measuring the water percolating through soils, and for determining the materials dissolved by the water.

M

m *See* meter.

M *See* Mach number.

MAA *See* maximum authorized IFR altitude.

MAC *1. See* media access control. *2.* Acronym for mean aerodynamic chord.

Mach angle The angle between the path of a body moving with supersonic velocity and a corresponding Mach line; the speed of sound divided by the body's velocity equals the sine of the Mach angle.

Mach cones *1.* The cone-shaped shock waves theoretically emanating from an infinitesimally small particle moving at supersonic speed through a fluid medium; the locus of the Mach lines. *2.* The cone-shaped shock waves generated by a sharp, pointed body, as at the nose of a high-speed aircraft.

machine language *1.* A language that is used directly by a machine. *2.* In software, the language that a computer understands, i.e., ones and zeros.

machine-language programming Basically, refers to programming using machine language.

machine-oriented language A language designed for interpretation and use by a machine without translation.

machine-readable medium A medium that can convey data to a given sensing device.

machine word *See* computer word.

machining center A versatile CNC machine tool with multi-axis control and, usually, automatic tool loading. Machining centers are designed to carry out a range of operations.

machining tear Describes a pattern of short, jagged, individual cracks, generally at right angles to the direction of machining; the result of improperly set cutting tools or dull cutting tools. [AIR1667]

mach meter An instrument that indicates the ratio of aircraft speed to the speed of sound at a particular altitude and temperature. [ARP4107]

Mach number Symbolized by M. *1.* The ratio of true airspeed to the speed of sound. *2.* A number expressing the ratio of the speed of a body or of a point on a body with respect to the surrounding air or other fluid; or the speed of a flow to the speed of sound in the medium. *3.* The speed represented by the Mach number. Named after Ernst Mach (1838–1916), Austrian scientist. [ARP4107]

Mach reflection A reflection of a shock wave from a rigid wall in which the shock strength of the reflected wave and the angle of reflection both have the smaller of the two values theoretically possible.

macro *1.* Directions for expanding abbreviated text. *2.* A boilerplate that generates a known set of instructions, data, or symbols. A macro is used to eliminate the need to write a set of instructions that are used repeatedly; for example, an assembly-language macro instruction enables the programmer to request the assembler to generate a predefined set of machine instructions.

macro instruction The more powerful instructions that allow a programmer to refer to several instructions as though they were a single instruction. When a programmer uses the name of a macro instruction, all of the instructions are inserted at that point in the coding by the macroprocessor.

macro modeling The representation of a component or device in terms of a net-list description of an equivalent circuit. Standard components, such as resistors or capacitors, are typically employed.

macroprocessor A program that translates a single symbolic statement into one or more assembly language statements.

macroprogramming Programming with macro instructions.

macroscopic stress Load per unit area distributed over an entire structure, or over a visible region of the structure.

macrostructure The features of a polycrystalline metal revealed by etching and visible at magnifications of 10 diameters or less.

magic tees Compound waveguides or coaxial tees with four arms which exhibit directional characteristics, when properly matched. A signal entering one arm of such a tee will be split between two of the other arms but not the third. A signal entering another arm is likewise split, with half the energy entering one of the arms common to the other input, but not its second arm; and the other half of the energy entering the arm not used by the other input. Magic tees are used in radar as transmitter receiver duplexers.

magma Naturally occurring mobile rock materials, generated within the earth and capable of intrusion and extrusion. Igneous rocks are thought to have been derived by solidification and related processes in magma.

magnet *1.* Anything that attracts. *2.* Any piece of iron or steel that has the property of attracting iron.

magnetically actuated extensions A device attached to the meter body which contains an electrical switch and which is magnetically actuated by the metering float extension to signal a high or low flow. The switch is adjustable with respect to the float position over a range equal to the travel of the metering float. Standard switch ratings are usually 0.3 amperes for 110 volt, 60 cycle a-c supply (and five amperes or more if relays are used).

magnetic amplifier An electronic amplification or control device that functions through the use of saturable reactors, either alone or in combination with other circuit elements.

magnetic bearing(s) *1.* Describes any application in which something capable of rotation and translation is held by the use of electromagnetic force, without being touched. *2.* The angle between the line of sight to an object and the direction from the observer to magnetic North, measured in a plane parallel to the earth's surface. *See* bearing.

magnetic biasing Simultaneous conditioning of a magnetic recording medium by superimposing a second magnetic field on the magnetic signal being recorded.

magnetic blowout switch A special type of switch designed to switch high d-c loads; a small permanent magnet contained in the switch housing deflects the arc to quench it when the contacts open.

magnetic bubble memory A high-density information-storage device composed of a magnetic film (only a few micrometers thick) deposited on a garnet substrate. In such a device, information is stored in small magnetized regions (bubbles) of magnetic polarity opposite to that of the surrounding region. Also known as bubble memory.

magnetic card A card with a magnetic surface on which data can be stored by selective magnetization of portions of the flat surface.

magnetic circuit breaker Circuit breakers utilizing a trip mechanism that functions in response to the magnetic effect, rather than the heating effect of the current carried by the thermal circuit breaker. [ARP4404]

magnetic compass Any of several devices for indicating the direction of the horizontal component of a magnetic field, but especially for indicating magnetic North in the earth's magnetic field.

magnetic compression The force exerted by a magnetic field on an electrically conducting fluid or on a plasma.

magnetic contactor A device for opening and closing one or more sets of electrical contacts; actuated by either energizing or deenergizing an electromagnet within the device.

magnetic cooling Keeping a substance cooled to about 0.2 K through the use of a working substance (paramagnetic salt) in a cycle of processes between a high-temperature reservoir (liquid helium) at 1.2 K and a low-temperature reservoir containing the substance to be cooled.

magnetic core *1.* A configuration of magnetic material that is, or is intended to be, placed in a spatial relationship to current-carrying conductors, and whose magnetic properties are essential to its use. A magnetic core may be used to concentrate an induced magnetic field, as in transformer, induction coil, or armature; to retain a magnetic polarization for the purpose of storing data; or for its non-linear properties, as in a logic element. A magnetic core may be made of such material as iron, iron oxide, or ferrite; and in such shapes as wires, tapes, toroids, or thin film. 2. A storage device in which binary data is represented by the direction of magnetization in each unit of an array of magnetic material, usually in the shape of toroidal rings, but also forms such as wraps on bobbins. Synonymous with core.

magnetic damping Progressive reduction of oscillation amplitude by means of current induced in electrical conductors due to changes in magnetic flux.

magnetic disk A flat, circular plate with a magnetic surface on which data can be stored by selective magnetization of portions of the flat surface.

magnetic drums Memory devices used in computers; rotating cylinders on which information may be stored as magnetically polarized areas, usually along several parallel tracks around the periphery.

magnetic equator That line on the surface of the earth connecting all points at which the magnetic dip is zero.

magnetic extensions A device that provides flow-rate indication by means of a magnetic coupling between the extension of the metering float and an external indicator follower surrounding the extension tube.

magnetic field interference A form of interference induced in the circuits of a device due to the presence of a magnetic field; may appear as common mode or normal mode interference in the measuring circuits. *See also* electromagnetic interference.

magnetic field reconnection A change in topology of the magnetic field configuration resulting from a localized breakdown of the requirement for "connection" of fluid elements at one time on a common magnetic field line. May also occur when an electric field exists

with a component parallel to a locally two-dimensional X-type magnetic neutral line, which is equivalent to a breakdown in connection.

magnetic fields Regions of space in which magnetic dipoles experience a magnetic force of torque; often represented as a geometric array of the imaginary magnetic lines of force that exist in relation to magnetic poles.

magnetic float gage Any of several designs of liquid-level indicator that use a magnetic float to position a pointer or to change the orientation of bicolor wafers.

magnetic float switch A device for operating a mercury switch by repositioning a magnetic piston with respect to a small permanent magnet attached to the pivoting mercury switch capsule; in the usual configuration, a float attached to the piston positions it near the small magnet when liquid level is high, and drops the piston out of proximity when the level is low, allowing a light spring to retract the magnet and pivot the mercury capsule.

magnetic flux The magnetic force exerted on an imaginary unit magnetic pole placed at any specified point of space; a vector quantity. The direction of magnetic flux is taken as the direction toward which a north magnetic pole would tend to move under the influence of the field. If the force is measured in dynes and the unit pole is a cgs unit pole, the field intensity is given in oersteds.

magnetic focusing Causing an electron beam to become diverging or converging in order to position an image or beam on an object (usually a CRT screen); this is done by interacting the beam with a magnetic or electromagnetic field.

magnetic hardness comparator A device for determining hardness of a steel part by comparing its response to electromagnetic induction with the response of a similar part of known hardness.

magnetic lens Electric-coil electromagnets or permanent magnets assembled into such a configuration that they can accomplish magnetic focusing.

magnetic mirrors Magnetic fields so arranged that they will theoretically confine a hot plasma.

magnetic moments *1.* The quantities obtained by multiplying the distances between two magnetic poles by the average strength of the poles. 2. Measures of the magnetic flux set up by the gyration of an electric field in a magnetic field. Moments are negative, indicating they are diagrammatic, and equal to the energy of rotation divided by the magnetic field. *3.* In atomic and nuclear physics, quantities measured in Bohr magnetrons, and associated with the intrinsic spin of the particle and with the orbital motion of the particle in a system.

magnetic particle clutch Two rotating interfaces that are coupled by a magnetically oriented material when current is applied to create a magnetic field. The coupling is both mechanical and viscous in nature. [ARP4386]

magnetic proximity sensor Any of several devices that are activated when a magnetized or ferromagnetic object passes within a defined distance of their active element. There are four types of magnetic proximity sensors: variable-reluctance sensors, hermetically sealed dry-reed switches, Hall-effect switches, and Wiegand-effect sensors.

magnetic resistance *See* reluctance.

magnetic separator A machine in which strong magnetic fields are used to remove pieces of magnetic material from a mixture of magnetic material and nonmagnetic or less strongly magnetic material.

magnetic shield A metal shield that insulates the contents from external magnetic fields. Such shields are often used with photomultiplier tubes.

magnetic storms Worldwide disturbances of the earth's magnetic field.

magnetic tape encoder An electronic device that will accept data from a keyboard and write it to magnetic tape.

magnetic test coil A coil used in conjunction with a suitable indicating or recording instrument to measure variations or changes in magnetic flux when the coil is linked with a magnetic field.

magnetic variation The angular difference between true north and magnetic north. [ARP4107]

magnetic variometer An instrument for measuring variations in magnetic field strength with respect to space or time.

magnet meter An instrument for measuring the magnetic flux of a permanent magnet under specified conditions. Usually incorporates a torque coil or a moving-magnet magnetometer with a unique arrangement of pole pieces.

magneto A type of electric generator in which permanent magnets are used to supply an electric current for engine ignition. [ARP4107]

magnetohydrodynamics Science of interaction between magnetic fields and electrically conductive fluids, especially plasmas. Also known as hydromagnetics.

magnetohydrodynamic waves *1.* Low-frequency waves in an electrically, highly conducting fluid (such as a plasma) permeated by a static magnetic field. The restoring forces of the waves are, in general, the combination of a magnetic tensile stress along the magnetic field lines and the comprehensive stress between the field lines and the fluid pressure. 2. Transverse waves in a magnetohydrodynamic field; the driving force is the tension introduced by the magnetic field along the lines of force.

magnetomechanics (physics) Study of the effects that the magnetization of a material and its strain have on each other.

magnetometer *1.* An instrument used in the study of geomagnetism for measuring a magnetic element. 2. An instrument for measuring the magnitude of a magnetic field, and sometimes for also determining its direction. *See also* gaussmeter.

magnetoplasmadynamics The study of the dynamics of generating electricity by passing a beam of ionized gas through a magnetic field.

magnetostriction *1.* The phenomenon whereby ferromagnetic materials experience an elastic strain when subjected to an external magnetic field; or the converse, whereby mechanical stresses cause a change in the magnetic induction of a ferromagnetic material. 2. A characteristic of some ferromagnetic materials whereby their physical dimensions vary with the intensity of an applied magnetic or electromagnetic field.

magnetostrictive The property of a ferromagnetic material whereby it changes dimensions and vibrates when subjected to a high-frequency magnetic field. [AIR4367]

magnetostrictive effect An inherent property of some ferromagnetic materials whereby they deform elastically, and thus generate mechanical force, when subjected to a magnetic field.

magnetostrictive resonator A ferromagnetic rod so constructed that an alternating magnetic field can excite it into resonance at one or more frequencies.

magnetron An electron tube in which electrons interact with the electric field of a circuit element in crossed steady electric and magnetic fields to produce alternating-current power output.

magnetron sputtering A deposition method whereby a microwave tube is utilized to confine a plasma magnetically to produce high deposition rates and a low working-gas partial pressure.

magnets Bodies that produce magnetic fields around themselves.

magnification *1.* The ratio of output to input signal magnitudes. *2.* Attaining a change in magnitude without a change in power. *3.* Producing an enlarged visual image. *4.* The ratio of a specific dimension on a virtual image to the corresponding dimension on the physical object being viewed.

main fractionator A large, multiproduct distillation column used to separate the effluent from a reacting system.

mainframe *1.* The central processor of a computer system. Contains the main storage, arithmetic unit, and special register groups. Synonymous with central processing unit. *2.* All that portion of a computer exclusive of the input, output, peripheral, and in some instances, storage units.

main gear *See* gear, main.

main memory The set of storage locations connected directly to the central processing unit. Also called (generically) core memory.

main parachute assembly A noncertificated parachute assembly that is worn in conjunction with a certificated reserve parachute assembly as the primary parachute for premeditated jumps. [AS8015]

main power generation unit The source of high-voltage DC power. [AS1831]

main program The module of a computer program that contains the instructions to begin program execution. Normally, the main program exercises primary control over the operations performed and calls subroutines or subprograms to perform specific functions.

main rotor The rotor that supplies the principal lift to a rotorcraft. [ARP4107]

main storage The fastest general-purpose storage of a computer.

maintain *1.* To keep in continuance or in a certain state, e.g., a certain state of repair. *2.* To preserve or keep in a given existing condition, e.g., a given level of efficiency or state of repair.

maintainability *1.* The ability of an item to be retained in or restored to specified condition when maintenance is performed by personnel having specified skill levels, using prescribed procedures and resources, at each prescribed level of maintenance and repair. [ARD50010-91] *2.* The relative ability of a device or system to remain in operation, requiring only routine scheduled maintenance and occasional unscheduled maintenance, without extensive periods of downtime for major repairs. *3.* The probability that a device will be restored to operating condition within a specified period of time when maintenance is done with prescribed resources and procedures. *4.* The inherent characteristic of a design or installation that determines the ease, economy, safety, and accuracy with which maintenance actions can be performed. *5.* The ability to restore a product to service or to perform preventive maintenance within required limits.

maintainability analysis The formal procedure for evaluating system and equipment design, using prediction techniques, failure modes and effects analysis procedures, and design data, to evolve a comprehensive quantitative description of maintainability design status, problem areas, and corrective action requirements. [AIR4896]

maintainability data Data resulting from the performance of maintainability tasks in direct support of an equipment or system acquisition program. [AIR4896]

maintainability demonstration test Government acceptance test, usually at the equipment or subsystem level for the major items that will comprise the integrated system, to demonstrate conformance to specified quantitative maintainability requirements. [AIR4896]

maintainability, design for Refers to design considerations directed toward achieving those combined characteristics of equipment and facilities which will enable the accomplishment of necessary maintenance with the minimum requirements and expenditures. [AIR4896]

maintainability feasibility estimation A simplified procedure for the gross prediction of maintainability based on experience data related to the few major contributing factors to system downtime. [AIR4896]

maintainability function A plot of the probability of repair within time t, versus maintenance time. [AIR4896]

maintainability index A measure of the total maintenance manhours required to maintain a product in operational status per hour of operation. [AIR4896]

maintainability, mission A measure of the ability of an item to be retained in or restored to specified condition when maintenance is performed during the course of a specified mission profile. [AIR4896]

maintainability model A quantifiable representation of a test or process, the purpose of which is to analyze results to determine specific relationships of a set of quantifiable maintainability parameters. [AIR4896]

maintainability program *See* program, maintainability.

maintainability requirement A comprehensive statement of required maintenance characteristics, to be satisfied by the design of an item. [ARP4386]

maintainability requirements analysis An evaluation of the operational requirements, performed in order to identify the specific sys-tem effectiveness parameters, logistic planning factors, cost limitations, etc. [AIR4896]

maintainability task An action required to preclude the occurrence of a malfunction or restore an equipment to satisfactory operating condition. [AIR4896]

maintained alarm An alarm that returns to normal after being acknowledged.

maintenance *1.* All actions (corrective or preventive) necessary for retaining an item in or restoring it to a specified condition. [ARD50010] *2.* All actions necessary to sustain or restore the integrity and performance of an item to a specified condition. Includes inspection, testing, servicing, classification as to serviceability, repair, rebuilding, and reclamation. [AIR1916] *3.* Any act that either prevents equipment failure or malfunction, or restores operating capability following a failure or malfunction. *4.* Any activity intended to eliminate faults or to keep computer hardware or programs in satisfactory working condition, including test, measurements, replacements, adjustments, and repairs.

maintenance action An element of a maintenance event; one or more tasks (e.g., fault localization, fault isolation, servicing, and inspection) necessary to retain an item in or restore it to a specified condition. [ARD50010]

maintenance allocation table A table that describes the function to be performed in the repair of gas turbine engines, identifying the degree of repair. [AIR4896]

maintenance analysis The process of identifying required maintenance functions by analysis of the design, to determine the most effective means to accomplish these functions. [AIR4896]

maintenance complaints Reports of discrepancies that are found by maintenance personnel and that require maintenance action. [AIR4896]

maintenance concept *1.* Describes the planned general scheme for maintenance and support of an item in the operational environment. The maintenance concept provides the practical basis for design, layout, and packaging of the system and its test equipment, and establishes the scope of maintenance responsibility for each level (echelon) of maintenance and the personnel resources (maintenance manning and skill levels) required to maintain the system. [ARD50010] *2.* A collection of ideas and philosophies in maintenance which are used as the basis for the development of a specific maintenance program, procedure, or policy. [AIR4896]

maintenance, condition monitoring *See* condition monitoring.

maintenance, corrective *See* corrective maintenance.

maintenance, deferred Maintenance that does not have any bearing on flight safety, and whose accomplishment is deferred to a convenient time and/or location. [ARD50010]

maintenance, depot Maintenance that is the responsibility of, and is performed by, designated maintenance activities, to augment stocks of serviceable material, and to support organizational maintenance and intermediate maintenance activities by the use of more extensive shop facilities, equipment, and personnel of higher technical skill than are available at the lower levels of maintenance. The phases of depot maintenance normally consist of repair, modification, alteration, modernization, overhaul, reclamation, or rebuilding of assemblies, subassemblies, units, and equipment; the emergency manufacture of nonavailable parts; and providing technical assistance to using activities and intermediate maintenance organizations. Depot maintenance is normally accomplished in fixed shops, shipyards, and shore-based facilities. *See also* maintenance levels. [ARD50010]

maintenance depth The complexity or extensiveness of aircraft maintenance functions; for example, the complexity of a test, or the extent of disassembly. [AIR4896]

maintenance, direct That effort expended by maintenance personnel in the actual performance of maintenance on aircraft, aeronautical equipment, or support equipment (SE) in accordance with the applicable technical manual; applies equally to both contractor- and government-furnished equipment (GFE). [ARD50010]

maintenance direct costs Maintenance labor and material costs directly expended in performing maintenance on an item or aircraft. [AIR1916]

maintenance downtime rate Equipment downtime per operating hour, comprised of downtime required for preventive maintenance. [AIR4896]

maintenance engineering analysis The composite analytical studies, decisions, and related documentation conducted in connection with the design of an item to determine or influence the maintainability and reliability characteristics of the item, and to determine the total support requirements resulting from the design. [AIR4896]

maintenance environment The climatic, geographical, physical, and operational conditions under which an item will be maintained. [AIR4896]

maintenance error An error on the part of maintenance personnel in performing maintenance on an item, which results in subsequent failure or malfunction; or an error in published maintenance procedures which results in subsequent failure or malfunction. [AIR4896]

maintenance evaluation A process for determining that the design of an item is compatible with the maintenance plan and maintenance concept. [AIR4896]

maintenance event One or more maintenance actions required to effect corrective and preventative maintenance due to any type of failure or malfunction, false alarm, or scheduled maintenance plan. [ARD50010]

maintenance, hard time A primary maintenance process whereby an item must be removed from service at or before a previously specified time. [AIR4896]

maintenance, individual corrective Time required to complete an individual maintenance task or an individual maintenance action. [AIR4896]

maintenance, in shop Work that requires the use of shop facilities and cannot be normally performed outside the shop. Bench test and component disassembly and repair are examples of in-shop maintenance work. [ARD50010]

maintenance levels Refers to the division of maintenance, based on difficult and requisite technical skill, in which jobs are allocated to organizations in accordance with the availability of personnel, tools, supplies, and the time within the organization. Maintenance levels include organizational, intermediate, and depot. *Organizational maintenance*–embraces the maintenance performed by a using organization on its own equipment. This includes inspection, cleaning, servicing, preservation, lubrication, adjustment, minor repair not requiring detailed disassembly, and replacement not requiring high technical skill. *Intermediate maintenance*–is performed by designated maintenance activities in direct support of using organizations. This category will normally be limited to maintenance consisting of replacement of unserviceable parts, subassemblies, or assemblies. *Depot maintenance*–refers to the maintenance required for major overhaul or complete rebuilding of parts, subassemblies, assemblies, and other end items. Such maintenance is intended to augment stocks of serviceable equipment or to support lower levels of maintenance by use of more extensive shop equipment and personnel of higher technical skill than available in organizational or field maintenance activities. [ARD50010]

maintenance, line Routine check, inspection, and malfunction rectification performed en route and at base stations during transit, turnaround, or night stop. Synonymous with maintenance, line station. [ARD50010]

maintenance, line station *See* maintenance, line.

maintenance management system A part of the management information system (MIS) that is useful for maintaining a company's equipment. Such a system accesses equipment information, spare parts availability and location, maintenance work order systems, preventive maintenance systems, maintenance personnel qualifications, equipment maintenance history, and any other information that will help the maintenance engineer, supervisor, technician, or mechanic to be more proficient in his/her job.

maintenance manhours per life unit The average manhours per life unit required to maintain a system. [AIR4896]

maintenance manning level Total authorized or assigned personnel, per system at specified levels of maintenance organization. [AIR4896]

maintenance, on condition A maintenance concept whereby an engine has no fixed time limitation on its repair or replacement, or on repair or replacement of any of its components. Under this concept, repair or replacement of the engine or any of its components is determined by the condition of the unit; and the unit is subjected to periodic diagnostic checks and inspections to insure its continued ability to perform its function within specified limits. [ARD50010]

maintenance plan A document containing technical data, tailored to a specific weapon system maintenance concept, which identifies maintenance and support resource requirements to maintain aeronautical systems, equipment, and support equipment (SE) in an operationally ready state. The maintenance plan provides the interface between maintenance engineering and supply for provisioning purposes and communicates necessary (but incomplete) inputs to enable other logistic element managers to develop their hardware support requirements. The maintenance plan is designed as a tool for the shore community for integrated logistic support (ILS) planning, and is prepared in accordance with NAVAIRINST 4790.4A (NOTAL). [ARD50010]

maintenance planning The design, method, or scheme for accomplishing an aircraft mission or reaching an aircraft maintenance objective or objectives. [AIR4896]

maintenance, preventive *See* preventive maintenance.

maintenance procedures Established methods for periodically checking and servicing items to prevent failure or to effect a repair. [AIR4896]

maintenance process, primary The process relied upon to ensure that inherent design reliability is maintained. [AIR4896]

maintenance program *See* program, maintenance.

maintenance ratio A measure of the total maintenance manpower burden required to maintain an item; expressed as the cumulative number of manhours of maintenance expended in direct labor during a given period of the life units, divided by the cumulative number of end item life units during the same period. [ARD50010]

maintenance requirement A concise and direct statement of the maintenance function, which must be performed to verify, maintain, or restore an item to its intended function. [AIR4896]

maintenance resources Facilities, ground support equipment, manpower, spares, consumables, and funds available to maintain and support an item in its operational environment. [AIR4896]

maintenance resources, indirect The time and material that, while not directly expended in active maintenance tasks, contributes to the overall maintenance mission through the support of overhead operations, administration, etc. [AIR4896]

maintenance, scheduled *See* scheduled maintenance.

maintenance significant Describes maintenance items of equipment or components that are judged to be relatively the most important for safety, reliability, or economic impact. [AS1426]

maintenance significant item *See* item, maintenance significant.

maintenance steering group (MSG) An ATA (Air Transportation Association)-sponsored study group that publishes recommended methodologies and analytical procedures for developing a maintenance plan for aircraft, engines, and systems. [ARP1587]

maintenance support index The total number of direct and maintenance manhours for preventive and corrective maintenance required to support each hour of operation. [AIR4896]

maintenance task The maintenance effort necessary for retaining an item in, or changing/restoring it to, a specified condition. [ARD50010]

maintenance time, elapsed (EMT) For the purposes of maintenance data reporting (MDR), the actual clock time, in hours and tenths, that maintenance was being performed on a job. EMT does not include the clock hours and tenths for cure time, charging time, or leak test when they are being conducted without maintenance personnel actually monitoring the work. Although the EMT is directly related to job manhours, it is not to be confused with total manhours required to complete a job. For example, if five persons complete a job in 2.0 hours of continuous work, the EMT = 2.0 hours and the manhours = 10.0. [ARD50010]

maintenance, unscheduled That maintenance performed to restore an item to a satisfactory condition by providing correction of a known or suspected malfunction and/or defect. [ARD50010]

main tires Tires that support the principal weight of the aircraft. [AS4833]

major alteration A change not listed in the aircraft, aircraft engine, or propeller specifications which might appreciably affect weight, balance, structural strength, performance, power plant operation, flight characteristics, or other characteristics affecting airworthiness. [ARP4107]

major defect A defect, other than critical, that is likely to result in failure, or to reduce materially the usability of the unit of product for its intended purpose. *See* defect. [AIR4896]

major diameter The largest diameter of a screw thread; measured at the crest of an external thread and at the root of an internal thread.

major frame With reference to telemetry formats, the time period during which all data of a multiplex are sampled at least once; includes one or more minor frames. Major frame length is determined as (N)(Z) words, where N = the number of words per minor (prime) frame and Z = the number of words in the longest sub-multiple frame.

major graduations Intermediate graduation marks on a scale which are heavier or longer than other graduation marks, but which are not index graduations.

majority A logic operator having the property whereby if P is a statement, Q is a statement, and R is a statement, then the majority of P, Q, R is true if more than half the statements are true, false if half or less are true.

majority voting system A fault-tolerant system in which the outputs of three or more signals are summed to provide a single signal representative of the majority of the individual signals, often providing detection logic for identifying a failed channel. *See also* voter. [ARP4386]

major repair A repair that, if improperly performed, might adversely affect weight, balance, structural strength, performance, power plant operation flight characteristics, or other qualities affecting airworthiness. [ARP4107]

major time In telemetry computer systems, two sixteen-bit words: minutes/seconds and hours/days.

make-up The water added to boiler feed to compensate for that lost through exhaust, blowdown, leakage, etc.

male fitting An element of a connection in pipe, tubing, electrical conductors, or mechanical assemblies that fits into the mating (female) element; for example, the externally threaded end of a pipe fitting is termed "male."

malfunction *1.* The occurrence of a condition whereby the operation is outside specified limits. [ARP4754] *2.* A general term used to denote the occurrence of failure of a product to give satisfactory performance. A malfunction need not constitute a failure if readjustment or operator's controls can restore an acceptable operating condition. [ARD50010] *3.* Abnormal condition or status of an engine, component, or sub-system. [ARP1587] *4.* Improper functioning of components, causing improper operation of a system. *5.* The effect of a fault.

malfunction administrative time All time between the beginning and end of work on a malfunction, except for logistic or active maintenance time for that malfunction. [AIR4896]

malfunction final test time The time spent confirming that the malfunction in question has been corrected, after which time no further

maintenance is performed on that malfunction. [AIR4896]

malfunction routine *See* diagnostic routine.

malfunction verification time The time spent testing a system to observe previously reported symptoms of malfunction. [AIR4896]

malleable iron A somewhat ductile form of cast iron made by heat-treating white cast iron to convert the carbon-containing phase from iron carbide to nodular graphite.

MALS Stands for medium intensity approach light system. *See* approach light system.

MALSR Stands for medium intensity approach light system with runway alignment indicator lights. *See* approach light system.

management information system (MIS) A computerized system using a large database containing information on: (a) customers, (b) equipment, (c) supplies, (d) spare parts, (e) personnel, (f) process, (g) sales forecast, (h) history, (i) costs, (j) profits, etc. Selected information is available to those persons making decisions.

man-computer interface The interface between man and the computer and its interrelationships, including ergonomic factors.

mandatory replacement item *See* item, mandatory replacement.

mandrel A cylindrical-shaped tool with a slight taper on the overall length, with or without a flexible sleeve, centered on each end with a flat milled on the ends for a holder. Used for holding work for machining operations. Also used as a round anvil to form metal. [ARP480A]

maneuver (mishap) *See* mishap, maneuver.

maneuverability The property of a vehicle that determines the rate at which the attitude and direction of movement of the vehicle can be changed. [ARP4107]

manhead The head of a boiler drum or other pressure vessel having a manhole.

manhole In a pressure vessel, an opening of sufficient size to permit a man to enter.

manhours The total number of accumulated direct labor hours (in hours and tenths) expended in performing a maintenance action. Direct maintenance manhours are manhours expended by assigned personnel to complete the work described on the source document. This includes the functions of preparation, inspection, disassembly, adjustment, fault correction, replacement or reassembly of parts, and calibration/tests required in restoring the item to a serviceable status. It also includes such associated tasks as checking out and returning tools, looking up part numbers in the illustrated parts breakdown (IPB), transmitting required information to material control, and completing documentation of the visual information display system/maintenance action form (VIDS/MAF) or support action form (SAF). [ARD50010]

manhours, direct Manhours of work spent directly on the aircraft or removed items from the aircraft. [AIR4896]

manhours per flying hours A performance figure calculated by dividing the direct manhours expended to maintain a particular aircraft fleet during a given period, by the flying hours (airborne) during that period. [ARD50010]

manifold *1.* A preformed item with two or more inlet ports and passages, and a common outlet. *See also* distributor. [ARP480A] *2.* A pipe or header for collecting a fluid from, or the distributing of a fluid to, a number of pipes or tubes.

manifold (instrumentation) Any configuration of valves that can be manipulated to create zero differential pressure at the measuring instrument.

manifold equalizing line Within a manifold, the conduit that connects the high- and low-differential pressure impulse lines.

manifold pressure *1.* Absolute pressure as measured at the appropriate point in the induction system and usually expressed in inches of mercury. [ARP4107] *2.* The fluid pressure in the intake manifold of an internal combustion engine.

manifold variable A quantity or condition that is varied so as to change the value of the controlled variable.

manipulated variable *1.* In a process designed to regulate some condition, a quantity or a condition that is altered by the control in order to initiate a change in the value of the regulated condition. *2.* The part of a process that is adjusted to close the gap between the set point and the controlled variable.

manipulative variable In a control loop, the variable that is used by the controller to regulate the controlled variable.

manipulator *1.* A device used for orientation of a transducer assembly. As generally applied to immersion techniques, provides either angular or normal sound wave path. [ARP5089] *2.* Mechanical device for the remote handling of hazardous materials; usually hand-operated, often from behind a shield; may or may not be power-assisted.

man machine system System in which the functions of the man and the machine are interrelated and necessary for the operation of the system.

manometer An instrument for measuring the pressure of liquids and gases by determining the height to which the pressure raises a fluid. [ARP171]

manometric equivalent The length of a vertical column of a given liquid at standard room temperature which indicates a pressure differential equal to that indicated by a 1-mm-long column of mercury at 0°C.

mantissa *See* floating point.

mantle (earth structure) *See* earth mantle.

manual backup An alternate method of process control by means of manual adjustment of final control elements in the event of a failure in the computer system.

manual control The operation of a process by means of manual adjustment of final control elements.

manual controller A control device whose output signals, power, or motions are all varied by hand.

manual data entry module A device that monitors a number of manual input devices from one or more operator consoles and/or remote data entry devices, and transmits information from those devices to the computer.

manual flight control system (MFCS) A system that transmits manual pilot commands directly or generates and conveys commands that augment manual pilot control commands, thereby accomplishing flight control functions. Such a system may include electrical, mechanical, and hydraulic components which provide means for transmission of manual pilot commands to the control function. The MFCS classification includes the longitudinal, lateral-directional, lift, drag, and variable geometry control systems. Associated scheduling, limiting, and control devices are included. [ARP4386]

manual input *1.* The entry of data by hand into a device at the time of processing. *2.* Data entered by hand into a device at the time of processing.

manually actuated valves Those valves that are operated by hand motion applying a force through a mechanical lever or cable system. [ARP986]

manual loading station *See* manual station.

manual operation Processing of data in a system by direct manual techniques.

manual override Refers to the capability of a flight-control system to enable the pilot to override the AFCS through a cable and/or linkage system, and exert control in excess of the AFCS authority, or in opposition to the AFCS command. [ARP4386]

manual reversion The action of reverting to manual control because of failure of the automatic or semiautomatic system. [ARP4107]

manual station *1.* A single-loop hard manual control to operate the final control devices in case of control system failure. *2.* Station that provides for bypassing normal controller operation to manually vary an analog output signal in a controller. Used primarily in an emergency, or possibly during a maintenance shutdown of the controller. Synonymous with manual loading station.

manufacturer software A complex program package that develops the user's application and organizes computer procedures to obtain efficient response to the application program. Often this software is referred to as an operating system.

Manufacturing Automation Protocol (MAP) A specification for a suite of communication standards for use in manufacturing automation, initially developed under the auspices of General Motors Corporation.

Manufacturing Message Specification *See* MMS.

Manufacturing Messaging Format Standard (MMFS) One of the application protocols specified by MAP (Manufacturing Automation Protocol).

map To establish a correspondence or relationship between the members of one set and the members of another set and perform a transformation from one set to another; for example, to form a set of truth tables from a set of Boolean expressions. Information should not be lost or added when transforming the map from one set to another. *See also* memory map.

MAP *1. See* missed approach point. *2. See* Manufacturing Automation Protocol.

MAP/EPA Part of the EPA architecture, a node that contains both the MAP protocols and the protocols required for communication to MINI-MAP. This node can communicate with both MINI-MAP nodes on the same segment and full MAP nodes anywhere in the network.

mapped system A system in which the computer hardware memory management unit is used to relocate virtual memory addresses.

MAP/TOP Users Group United States' and Canada's MAP/TOP Users Group. *See also* Computer and Automated Systems Association of the Society of Manufacturing Engineers (CASA/MSE).

Marangoni convection Convective flow induced by surface tension gradients; important in both ground and space processing where a free surface is present.

marginal check A preventive-maintenance procedure in which certain operating conditions are varied about their normal values in order to detect and locate incipient defective units, e.g., supply voltage or frequency may be varied. Synonymous with marginal test and high-low bias test. *See also* check.

marginal test *See* marginal check.

marginal testing A system-checking procedure that indicates when some portion of the system has deteriorated to the point where there is a high probability of a system failure during the next operating period. [AIR4896]

margin of attention *See* attention, margin of.

margin of safety A ratio of permissible stress to actual applied stress, with 1.0 then subtracted from the result. Margins of safety should be positive but small in order to achieve a minimum weight structure.

marker beacon An electronic navigation facility transmitting a 75-MHz vertical fan or other radiation pattern. Marker beacons are identified by their modulation frequency and keying code; when received by compatible airborne equipment, they indicate to the pilot, both aurally and visually, that he/she is passing over the facility. [ARP4107]

marking pointer An adjustable stationary pointer, usually of different color from the indicating pointer, which can be positioned opposite any location on the scale of interest to the user.

markoff A step occurring in the facing of a bonded sandwich. [AMS3920A]

Markov chain A probabilistic model of events in which the probability of an event is dependent only on the event that precedes it.

marquenching (martempering) Quenching an austenitized alloy in a salt or hot-oil bath at a temperature in the upper part of, or slightly above, the martensite range, holding until temperature uniformity throughout the part is obtained, usually followed by air cooling through the martensite range to ambient temperature. [AMS2759B]

martensite *See* alpha prime.

martensite alpha *See* alpha prime.

martensitic transformation A phase transformation occurring in some metals, resulting in formation of martensite.

Marx generators A high-voltage electrical pulse generator in which capacitors are charged in parallel, then discharged in series to generate a voltage much higher than the charging voltage.

MASER *See* microwave amplification by the stimulated emission of radiation.

mask A protective covering applied over the face to provide adequate respiratory gas to the wearer and, in some cases, to prevent inhalation of the gases in the atmosphere surrounding the wearer. [ARP171]

masking *1.* The process of extracting a nonword group or a field of characters from a word or a string of words. *2.* The process of setting internal program controls to prevent transfers that otherwise would occur upon setting of internal machine latches.

mask programmed memory Computer memory dedicated to the storage of a particular set of data. A mask containing the particular pattern of bits is used in the manufacture of the memory.

mass *1.* For a body, a quantity that relates the attraction of the body toward another body. Since the mass of a body is not fixed in magnitude, all masses are referred to the standard kilogram (a lump of platinum). *2.* Amount of matter that an object contains.

mass density *See* density, mass.

mass drivers (payload delivery) Proposed method for payload delivery into earth orbit from the moon by electromagnetic acceleration; also for deliveries to Lagrange equilibrium points.

mass flow The amount of fluid, measured in mass units, that passes a given location or reference plane per unit time.

mass-flow bin A bin with steep, smooth sides which allow its contents to flow, without stagnant regions, whenever some of the contents are withdrawn.

mass flowmeter An instrument for measuring the rate of flow in a pipe, duct, or channel in terms of mass per unit time.

mass flow rate The mass of fluid moving through a pipe or channel within a given period of time.

mass number For a specific nuclide, the sum of the number of protons and the number of neutrons in the nucleus.

mass ratio The ratio of the mass of the propellant charge of a rocket to the total mass of the rocket when charged with the propellant.

mass spectrograph A mass spectroscope that records intensity distributions on a photographic plate.

mass spectrometer A mass spectroscope that uses an electronic instrument to indicate intensity distribution in the separated ion beam.

mass spectroscope An instrument for determining the masses of atoms or molecules, or the mass distribution of an ion mixture, by deflecting them with a combination of electric and magnetic fields which act on the particles according to their relative masses.

mass spectrum In a mixture of ions, the statistical distribution by mass or by mass-to-charge ratio.

mass storage Pertains to a computer device that can store large amounts of data so that the data are readily accessible to the central processing unit; for example, disks, DEC tape, or magnetic tape.

mass to light ratio The ratio of the mass of celestial body to its luminosity.

mass velocity Mass flow per unit cross-sectional area.

master *1.* A device or mechanism that controls the operation of different mechanisms or establishes a standard. [ARP480A] *2.* A precise pattern for making replicate workpieces, as in certain types of casting processes.

master change Documentation prepared by the airframe manufacturer which defines and coordinates a negotiable change to the contract specifications after contract signing. [AS1426]

master clock A device that functions as the primary source of timing signals.

master gage A device with fixed locations for positioning parts or holes in three dimensions.

master heat Refined metal of a single furnace charge melted and cast in air; may be poured directly into castings and/or converted into ingot for remelting. [AMS5402B]

master recipe A basic recipe that has been made site-specific.

master reference A signal point that is the signal reference point for an entire system. Usually a ground rod or grid. All local references are tied back to the master reference point.

master-slave Refers to a mode of operation in which one data station (the master) controls the network access of one or more data stations (the slaves). *See also* polled access.

master/slave manipulator A remote manipulator that mechanically, hydromechanically, or electromechanically reproduces hand or arm motions of an operator.

master warning/caution A signal indicating a condition requiring immediate action. The specific condition is shown by a separate indication. [AIR1161]

mate The joining, engaging, connecting, or coupling of two connectors or devices designed to be utilized together. [ARP914A]

material balance *1.* The procedure of accounting for the mass of material going into a process versus the mass leaving the process. *2.* The balance relating the material in and material out of a distillation column. Material-balance manipulative variables are overhead flow, bottoms flow, sidestreams flow, and feed flow.

material dispersion Light pulse broadening due to differential delay of various wavelengths of light in a waveguide material. This group delay is aggravated by broad-bandwidth light sources.

material noise Extraneous signals caused by the structure of the material being tested. [ARP5089]

materials recovery The treatment of a material to reclaim one or more of its components.

materials science The study of materials used in research, construction, and manufacturing; includes the fields of metallurgy, ceramics, plastics, rubber, and composites.

mathematical check A check that uses mathematical identities or other properties, occasionally with some degree of discrepancy being acceptable, e.g., checking multiplication by verifying that $A \times B = B \times A$. Synonymous with arithmetic check.

mathematical model The general characterization of a process, object, or concept, in terms of mathematics, enabling relatively simple manipulation of variables in order to determine how the process, object, or concept would behave in different situations.

mathematical programming In operations research, a procedure for locating the maximum or minimum of a function subject to constraints. Contrast with convex programming, dynamic programming, integer programming, linear programming, nonlinear programming, and quadratic programming.

mat, landing *See* landing mat.

matrix *1.* The constituent that forms the continuous or dominant phase of a two- or more phase microstructure. [AS1814] *2.* In mathematics, an n dimensional rectangular array of quantities. Matrices are manipulated in accordance with the rules of matrix algebra. *3.* In computers, a logic network in the form of an array of input leads and output leads with logic elements connected at some of their intersections.

matrix management An organized approach to administration of a program by defining and structuring all elements to form a single system with components united by interaction.

matrix materials The ingredients used as binding agents to produce composite materials.

mat-surface glass Glass whose surface has been altered by etching, sand-blasting, grinding, etc., to increase the diffusion. Either one or both surfaces may be so treated. [ARP798]

matte *1.* A smooth but relatively nonreflective surface finish. *2.* An intermediate product in the refining of sulfide ores by smelting.

maximum allowable working pressure The highest gage pressure that can safely be applied to an internally pressurized system under normal operating conditions. The maximum allowable working pressure is usually well below the design bursting pressure and the hydrostatic test pressure for the system; it is the pressure at which relief valves are set to lift.

maximum authorized altitude (MAA) The highest altitude on a federal airway, jet route, area navigation low or high route, or other direct route for which a minimum en route altitude is designated in FAR Part 95, at which adequate reception of navigation aid signals is assured. [ARP4107]

maximum continuous load The maximum load that can be maintained for a specified period.

maximum continuous power The power certificated for maximum continuous use, unaugmented. [ARP4102/5]

maximum coupling position A position of the rotor at which the secondary voltage of fundamental frequency that is in time phase with the secondary voltage is a maximum. [ARP826]

maximum demonstrated flight diving speed *See* V_{df}.

maximum droplet size The size of a droplet that is larger than 95% of the droplets; approximately double the volume median droplet size. The maximum droplet size is used to determine impingement limits. [AIR1168/9]

maximum entropy method Procedure used in estimating high-resolution power spectra from short data lengths.

maximum flow A quantity provided in the instrument specifications. The accuracy and validity of calibration depends on maintaining a flow rate within specifications. [ARP1192]

maximum fluid temperature Highest fluid temperature at which the fluid in a system is intended to be operated. [ARP4386]

maximum free run speed The operation of a starter with no-load on the output shaft in combination with any single failure of the intended pneumatic power source or starter control system which produces the highest free run speed including the effects of altitude. [AIR1639]

maximum gross weight The maximum allowable combined weight of a container and its cargo. [AS4041]

maximum instantaneous demand A sudden load demand on a boiler, beyond which an unbalanced condition may be established in the internal flow pattern and/or surface release conditions of the boiler.

maximum limit of ultimate trip The minimum current that will cause a circuit breaker to open under a given set of ambient conditions. Also known as the rated trip current. [ARP1199A]

maximum normal operating speed The highest speed obtained by a piece of rotating equipment under the most severe operational condition without any failures of components that affect rotational speed of the equipment. [AIR4073]

maximum operating limit speed *See* V_{mo}.

maximum operating pressure The maximum pressure that the aircraft system can supply to the brake under design conditions. [ARP1493]

maximum operating speed The highest speed obtained by the unit under the most severe temperature and altitude condition. [AS1606]

maximum operating weight The total weight of all individuals or dummies and their equipment. [AS8015]

maximum overload The maximum load, expressed in kVA at 0.8 power factor, that the alternator must produce for a specified time. [ARP1148A]

maximum permissible pressure The highest pressure that is permitted for safety reasons. [ARP4386]

maximum pointer A movable pointer that is repositioned as the indicating pointer of an instrument moves upscale, but remains stationary at the highest point reached when the indicating pointer moves downscale.

maximum power discharge current The discharge rate at which the terminal voltage is equal to one-half of the fully charged voltage. [ARP4386]

maximum pressure The highest transient pressure that can occur temporarily. [ARP4386]

maximum ramp weight The maximum operational weight of an aircraft. [AS4833]

maximum self-locking torque The maximum acceptable running torque value. [AS1895]

maximum structural cruising speed See V_{no}.

maximum tail down ground line T The maximum tail down attitude possible, limited by either aircraft structure or the tail bumper compressed to maximum working stroke. [ARP1538]

maximum thermometer A thermometer that indicates maximum temperature reached during a given interval of time. A clinical thermometer used to determine a patient's body temperature is one type of maximum thermometer.

maximum time to repair The maximum time required to complete a specified percentage of all maintenance actions. [ARD50010]

maximum torque The highest torque value occurring after the initial impact torque occurs. [AS1606]

maximum transient pressure Peak pressure of a pressure-time function, usually taken from a high-frequency response oscillograph or similar instrument capable of recording rapid variations of fuel pressure. [AIR1749]

maximum turbulence penetration speed See V_b.

maximum usable frequency For a given distance from a transmitter, the highest frequency at which sky waves can be received.

Maxwell The CGS unit of magnetic flux.

Maxwell bridge A type of a-c bridge circuit in which the impedance of an unknown inductor is measured in terms of an adjustable resistor and adjustable inductor; or, since the latter may be difficult to obtain, an alternative bridge arrangement may be used, in which an adjustable resistor and capacitor are in parallel with the unknown inductor.

Maxwellian distribution The velocity distribution of the moving molecules of a gas in thermal equilibrium, as determined by applying the kinetic theory of gases.

maypole antennas A class of antennas utilizing the deployable reflector concept for large space systems applications.

MBit Million bits per second.

MBM junctions (metal-barrier-metal junctions) Diode devices using metal-barrier-metal layers.

MByte 1,048,576 (2^{20}) bytes.

MC See mission capable.

MCA See minimum crossing altitude.

McLeod vacuum gage A common type of mercury-filled pressure gage that acts as a pressure amplifer; the design enables use of a manometer-type instrument for measuring vacuum on the order of 10^{-6} torr instead of the 10^{-2} torr usually achieved with precision manometers.

MDA See minimum descent altitude.

MDF See mild detonating fuse.

MEA See minimum en route IFR altitude.

mean Halfway between extremes; average.

mean chord The average length of the chord line from the wing tip to the wing root. [ARP4107]

mean corrective maintenance time The mean time required to complete a maintenance action, i.e., total maintenance downtime divided by total maintenance actions, over a given period of time.

mean cycles between failure The average number of operating cycles between failures. [AIR4896]

mean diameter The average of two diameter measurements taken at right angles to each other in the same cross-section of tubing. [AMS2243G]

mean effective pressure In a positive displacement machine such as a compressor, engine, or pump, the average net pressure difference across a piston. Commonly used to evaluate performance of such a machine.

mean free path *1.* Of any particle, the average distance that the particle travels between successive collisions with the other particles of an ensemble. Specifically, the average distance traveled by the molecules of a perfect gas between consecutive collisions with one another. *2.* For any process, the reciprocal of the cross section per unit volume for that process.

mean horizontal candlepower For a lamp or other light source, the average candlepower in the horizontal plane passing through the luminous center of the lamp or light source. It is assumed that the lamp or other light source is mounted in the usual manner, as in the case of a filament lamp, with its axis of symmetry vertical. [ARP798]

mean life The arithmetic mean of the times to failure of a group of nominally identical items. [ARD50010]

mean line A line lying in a plane parallel to the plane of symmetry that is equidistant from the upper and the lower surfaces from leading edge to the trailing edge of an airfoil. [ARP4107]

mean maintenance time A measure of item maintainability taking into account maintenance policy; the sum of preventive and corrective maintenance times, divided by the sum of scheduled and unscheduled maintenance events, during a stated period of time. [ARD50010]

mean repair time The average on- or off-equipment corrective maintenance time in an operational environment. [AIR4896]

mean sea level (MSL) Sea level between mean high tide and low water. [ARP4107]

mean square values In statistics, values representing the average of the sum of the squares of the deviations from the mean value.

mean task time A representative task time, equal to the summation of the task times required to perform a specific task a number of different times, divided by the number of times performed. [AIR4896]

mean time between critical failure (MTBCF) The average time between failure of mission essential system functions. [ARD50010]

mean time between demands A measure of the system reliability parameter related to demand for logistic support. The total number of system life units divided by the total number of item demands on the supply system during a stated period of time, e.g., shop replaceable unit (SRU), weapon replaceable unit (WRU), line replacement unit (LRU), and shop replaceable assembly (SRA). [ARD50010]

mean time between downing events A measure of the system reliability parameter related to availability and readiness. [AIR4896]

mean time between failures (MTBF) *1.* Mathematical expectation of the time interval between two consecutive failures of a hardware item. [ARP4754] *2.* Total system operating time divided by the number of system failures that have occurred during that period; the average time one could expect a given system to operate before experiencing a system failure. [ARP4107]

mean time between maintenance A measure of the reliability taking into account maintenance policy. The total number of life units expended by a given time, divided by the total number of maintenance events (scheduled and unscheduled) due to that item. [ARD50010]

mean time between removals A measure of the system reliability parameter related to demand for logistic support. The total number of system life units divided by the total number of items removed from that system during a stated period of time. Excludes removals performed to facilitate other maintenance and removals for product improvement. [ARD50010]

426

mean time between unscheduled removals A performance figure calculated by dividing the total unit flying hours (airborne) accrued in a period by the number of unscheduled unit removals that occurred during the same period. [ARD50010]

mean-time-to-failure (MTTF) The average or mean time between initial operation and the first occurrence of a failure or malfunction, as the number of measurements of such time on many pieces of identical equipment approaches infinity.

mean time to repair The total corrective maintenance time divided by the total number of corrective maintenance actions during a given period of time. [ARD50010]

mean time to restore system A measure of the system maintainability parameter related to availability and readiness. The total corrective maintenance time, associated with downing events, divided by the total number of downing events, during a stated period of time. Excludes time for off-system maintenance and repair of detached components. [ARD50010]

mean time to unscheduled removal A performance figure calculated by dividing the summation of times to unscheduled removal for a sample of removed items by the number of removed items in the sample. NOTE: This is different from mean time between unscheduled removal (MTBUR) since no allowance is given to items that have not been removed. [ARD50010]

measurand Physical phenomenon, parameter, quantity, property, or condition that is to be measured. *See also* variable. [ARD50024]

measured ceiling In US aviation weather observations, the ceiling classification that is applied when the ceiling value has been determined by means of: (a) a ceiling light or ceilometer; or (b) the known heights of unobscured portions of objects or other natural landmarks within 1.5 nautical miles of any runway of the airport. The measured ceiling applies only to clouds and obscuring phenomena aloft, and is identified by the ceiling designator M. [ARP4107]

measured variable *1.* The physical quantity, property, or condition that is to be measured.

Common measured variables are temperature, pressure, rate of flow, thickness, and speed. *2.* The part of the process that is monitored to determine the actual condition of the controlled variable.

measurement A datapoint that is or can be converted into a suitable signal for telemetry transmission.

measurement energy The energy, usually obtained from the measurand or the primary detector, required to operate a measurement device or system.

measurement error The difference between the true value and the measured value. Includes both bias and precision errors. [AIR1678]

measurement range *1.* The full-scale displacement range of the armature for which an electrical signal is produced. Also referred to as the electrical stroke. [ARP4386] *2.* The portion of the total response range of an instrument over which specific standards of accuracy are met.

measurement system Any set of interconnected components, including one or more measurement devices, that performs a complete measuring function, from initial detection to final indication, recording or control-signal output.

measuring junction The electrical connection between the two legs of a thermocouple. The measuring junction is attached to the body, or immersed in the medium, whose temperature is to be measured.

measuring modulator In a measuring system, a component that modulates a direct-current or low-frequency alternating-current input signal to produce an alternating-current output signal whose amplitude is related to the measured value, usually as a preliminary step to producing an amplified output signal.

measuring range The extreme values of the measured variable within which measurements can be made within the specified accuracy. The difference between these extreme values is called span.

mechanical actuator A mechanical device used to amplify force, primarily through the use of reduction gearing or threaded screw jacks. [ARP4386]

mechanical adhesion Adhesion due to the physical interlocking of the adhesive with the substrate irregularities. [AIR4844]

mechanical atomizing oil burner A burner that uses the pressure of the oil for atomization.

mechanical chart drive A spring-driven clock mechanism that feeds continuous chart paper past a recorder head at a predetermined speed.

mechanical classification Any of several methods for separating mixtures of particles or aggregates according to size or density, usually involving the action of a stream of water.

mechanical compliance Displacement of a mechanical element per unit force; the mechanical equivalent of capacitance in an electrical circuit.

mechanical damping Attenuating a vibrational amplitude by absorption of mechanical energy.

mechanical draft A negative pressure created by mechanical means.

mechanical efficiency The ratio of power output to power input.

mechanical geared rotary actuators Planetary geared devices typically having a rotary shaft input and rotary outputs. [ARP4058]

mechanical hygrometer A hygrometer in which an organic material, such as a bundle of human hair, is used to sense changes in humidity. In operation, the organic material expands and contracts with changes in moisture content in the air, and the change in length alters the position of a pointer through a spring-loaded mechanical linkage.

mechanical impedance The complex quotient of alternating force applied to a system, divided by the resulting alternating linear velocity in the direction of the force at its point of application.

mechanical impedance bondtester Dry-coupled, single-tipped, dual-element transducer probe instrument that operates at 2.5 to 10 kHz. [ARP5089]

mechanical input servovalve A servovalve in which the input is mechanical and the output is hydraulic fluid. [ARP4386]

mechanical linkage Describes a set of rigid bars or links that are joined together at pivot points and used to transmit motion. Frequently they are used in a mechanism along with a crank and slide to convert rotary motion to linear motion.

mechanical power transmission Power transmitted through mechanical devices from a given power source to a power user. [ARP4386]

mechanical properties Those properties of a material that are associated with the elastic and inelastic reactions that occur when force is applied; or those properties involving the relationship between stress and strain. [AIR4844]

mechanical reactance The imaginary component of mechanical impedance.

mechanical register A mechanical or electro-mechanical recording or indicating counter.

mechanical resistance The real component of mechanical impedance.

mechanical resonance The condition that occurs when the frequency of an exciting force equals a natural frequency of a spring-mass system. [AIR1839]

mechanical reversion Refers to the capability of reverting from fly-by-wire control to a state wherein the pilot's control is mechanically coupled to the actuator control valves. [ARP4386]

mechanical scale A weighing device in which objects are balanced through a system of levers against a counterweight or counterpoise.

mechanical seal A seal produced by an elastomeric O-ring or other molded or extruded shape when it is deformed by pressure mechanically, as in access doors. [AIR4069]

mechanical stroke The physical stroke of the armature as it moves. [ARP4386]

mechanical trail Distance of the axle aft of the strut centerline, measured perpendicular to the shock strut centerline. [AIR1752]

mechanical transmission An assembly of mechanical components suitable for transmitting mechanical power and motion.

mechanical tubing Tubing that is intended primarily for structural application and is not normally used for the transmission of fluids. [ARP1917]

mechanical volume control pump A variable-delivery pump whose output per cycle is controlled by external mechanical means. [ARP4386]

mechanism *1.* Generally, an arrangement of two or more mechanical parts in which motion of one part compels the motion of the others. *2.* Specifically, in an indicating instrument, the arrangement of parts that control motion of the pointer or other indicating means, excluding those parts that form the enclosure, scale, or support structure, or that adapt the instrument to the quantity being measured.

mechanized dew-point meter *See* dew-point recorder.

mechanoreceptor Nerve ending that reacts to mechanical stimuli, such as touch, tension, and acceleration.

media *1.* The physical interconnection between devices attached to a LAN. Typical LAN media are twisted pair, baseband coax, broadband coax, and fiber optics. *2.* The plural of medium. *3.* A name for the various materials used to hold or store electronic data, such as printer paper, disks, magnetic tape, or punched cards.

media access control (MAC) The lower sublayer of the Data Link Layer (Layer 2) unique to each type of IEEE 802 Local Area Network. MAC provides a mechanism by which users access (share) the network.

media migration Carryover of fibers and particles from filter or separator media material into the effluent. [AIR4783]

median *1.* Middle; intermediate. *2.* A median number, point, line, etc.

median corrective maintenance time The downtime within which 50% of all corrective maintenance actions can be completed under the specified maintenance conditions. The median value, Mct, is often referred to as the geometric mean (MTTRG) or equipment repair time (ERT) in some maintainability documents. [ARD50010]

median preventive maintenance time The equipment downtime required to perform 50% of all scheduled preventive maintenance actions on the equipment under the specified conditions. [ARD50010]

medium-carbon steel An alloy of iron and carbon containing about 0.25 to 0.6% C, and up to about 0.7% Mn.

medium frequency *See* middle frequency.

medium intensity approach light system (MALS) *See* approach light system.

medium scale integration (MSI) *1.* Refers to a medium level of chip density, lower than for LSI circuits but more than small scale integration. *2.* Describes an integrated circuit with 10 to 100 logic gates.

MEDP *See* minimum equipment and dispatch procedures.

mega Prefix denoting 1,000,000.

megabit One million bits.

megabyte A unit of computer memory size; one million bytes. *See also* MByte.

megahertz One million hertz or cycles per second.

megger *See* resistance meter.

megohmmeter A device that measures the electrical resistance of a sealant (and other materials as well). [AS7200/1]

MEL *See* minimum equipment list.

melt All castings poured from a single furnace charge. [AMS5402B]

melt extrusion An extrusion process in which the insulation material is heated above its melting point and is forced through a die. [ARP1931]

melt flow number *See* melt index.

melt index Extrusion rate of a thermoplastic material through an orifice of specified diameter and length under specified condition of time, temperature, and pressure. Also known as melt flow number. [ARP1931]

melting point The temperature at which a solid substance becomes liquid; for pure substances, and some mixtures it is a single unique temperature; for impure substances, solutions, and most mixtures it is a temperature range.

melt lot All castings poured from a single melt. [AMS4225C]

melts (crystal growth) Molten substances from which crystals are formed during the cooling or solidifying process.

melt spinning A material process by which polymers, such as nylon and polyesters, and glass are melted to permit extrusion into fibers through spinnerets.

melt time *See* element melt time.

membrane A thin sheet of metal, rubber, or treated fabric used to line cavities or ducts, or to act as a semirigid separator between two fluid chambers.

membrane filter A flat, disc-shaped, surface filter, usually of an organic material. [ARP4386]

membrane structures Shell structures, often pressurized, that do not take wall bending or compression loads.

memory Any form of computer data storage, including main memory and mass storage, in which data can be read and written. In its strictest sense, memory refers to main memory.

memory map Graphic representation of the general functional assignments of various areas in memory. Areas are defined by ranges of addresses.

memory protection A scheme for preventing read and/or write access to certain areas of memory.

memory resident Describes a program that remains in RAM memory even when other programs are operating, and that can be called up by interrupting the currently running program.

meniscus The concave or convex surface, caused by surface tension, at the top of a liquid column, as in a manometer tube.

meniscus lens A lens with one concave surface and one convex surface.

mercury cadmium tellurides Compounds of tellurium exhibiting photovoltaic characteristics, used for photodiodes and photodetectors in the 3–12 micrometer wavelength region at cryogenic temperatures.

mercury ion engines Machines providing thrust through the expulsion of accelerated or high-velocity mercury ions, often using energy provided by nuclear reactors.

mercury meter A differential pressure-measuring device utilizing mercury as the seal between the high and low chambers.

mercury switch A type of switch consisting of two wires sealed into the end of a glass capsule containing a bead of mercury; if the capsule is tipped one way, the mercury covers the exposed ends of the wires and completes the circuit; if tipped the other way, the mercury exposes the wires and breaks the circuit.

mercury vapor lamp A type of ion-discharge lamp widely used in ultraviolet analyzers because it emits several strong monochromatic lines with characteristic wavelengths, such as 254 nm, 313 nm, 360 nm, or 405 nm. Lamp emission can be made almost completely monochromatic by using special filters.

mercury-vapor tube A gas tube in which the active gas is mercury vapor.

mercury-wetted relay A device using mercury as the relay-contact closure substance.

meridian plane Any plane that contains the optical axis.

mesh *1.* A measure of screen size equal to the number of openings per inch along the principal direction of the weave. *2.* Engagement of a gear with its mating pinion or rack. *3.* A closed path through ductwork in a ventilation survey.

mesh number Refers to the number of wires per inch of wire mesh. Usually, the first number describes the warp count and the second number describes the shute count. [AIR888]

mesons In the classification of subatomic particles by mass, the second lightest of such particles. The mass of the meson is intermediate between that of the lepton and the nucleon.

mesopause The base of the inversion at the top of the mesosphere, usually found at 80 to 85 kilometers.

mesoscale phenomena Meteorological phenomena extending approximately one to one-hundred kilometers (mesoscale cloud pattern, for example).

mesosphere The atmospheric shell, in which temperature generally decreases with heights, extending from the stratopause at about 50 to 55 kilometers to the mesopause at about 80 to 85 kilometers.

metal A chemical element that is crystalline in the solid state, exhibits relatively high thermal and electrical conductivity, and has a generally lustrous or reflective surface appearance.

metal-barrier-metal junctions *See* MBM junctions.

metal corrosion *See* corrosion.

metal foams Foamed materials formed under low gravity conditions in space from sputtered metal deposits. This experimental space processing was completed in the second NASA SPAR flight.

metal-insulator-metal diodes *See* MIM diodes.

metallic Exhibiting characteristics of a metal.

metallic coating A thin layer of metal applied to an optical surface to enhance reflectivity.

metallic glass(es) Amorphous alloys (glassy metals) produced by extremely rapid quenching of molten transition-metal alloys (e.g., iron, nickel, and/or cobalt). Metallic glasses exhibit unique mechanical magnetic, and electrical properties, superconductive behavior, and anticorrosion resistance, depending on the alloys, their formation, and quenching techniques. Also known as a glassy alloy.

metallicity The abundance index of a metal or metals for a celestial body.

metallic superoxides Most commonly, superoxides of potassium and sodium with KO_2. These metallic superoxides serve as oxygen sources, and remove carbon dioxide, water, and odors. They weigh less, cost less and occupy a smaller volume than oxygen or lithium hydroxide (for carbon dioxide removal) systems. They provide a degree of sterilization and have an excellent shelf life. [AIR1246]

metal-nitride-oxide semiconductor (MNOS) *1.* Class of semiconductors utilizing silicon nitride and silicon oxide dielectrics. *2.* One type of computer semiconductor memory used in EAROMs.

metal-semiconductor-metal semiconductors *See* MSM (semiconductors).

metal-to-metal voids or disbonds Any unbonded areas occurring between two solid, nonporous members that are joined by an adhesive bondline. [AMS3920A]

metal vapor lasers Stimulated emission devices in which the active materials are vaporized metals.

metastable beta A nonequilibrium phase composition that can be partially or completely transformed to martensite, alpha, or eutectoid decomposition products with thermal or strain energy activation during subsequent processing or service exposure. [AS1814]

meteoritic ionization *See* meteor trails.

meteorograph A recording instrument for measuring meteorological data—temperature, barometric pressure, and humidity, for example.

meteoroids Solid objects moving in interplanetary space, of a size considerably smaller than asteroids and considerably larger than atoms or molecules.

meteoroid showers Groups of meteoroids with approximately parallel trajectories.

meteorological instrumentation Equipment for measuring weather data.

meteorology The study of phenomena of the atmosphere. This includes not only the physics, chemistry, and dynamics of the atmosphere, but is extended to include many of the direct effects of the atmosphere upon the earth's surface, the oceans, and life in general. A distinction can be drawn between meteorology and climatology, the latter being primarily concerned with average, not actual, weather conditions.

meteor trails Anything, such as light or ionization, left along the trajectory of the meteor after the head of the meteor has passed.

meter *1.* A device for measuring and indicating the value of an observed quantity. *2.* Abbreviated m. An international metric standard for measuring length, equivalent to approximately 39.37 in. in the US customary system of units.

meter, bulk A device that measures the quantity of liquid passing through it.

meter, conductivity A device for measuring the electrical conductivity of a liquid. [AIR4783]

metered brake pressure The fluid pressure (usually hydraulic) that is metered to the brakes by pilot demand and control to provide the desired decelerating force for the aircraft. [AIR1489]

meter factor *1.* A constant by which the actual reading on a scale or chart is multiplied to produce the measured value in actual units.

2. A correction factor applied to the indicated value of a meter to compensate for variations in ambient conditions; for example, temperature correction applied to a pressure indication.

metering *1.* A method of time-regulating the traffic flow into a terminal area so as not to exceed a predetermined terminal acceptance rate (the maximum inbound traffic flow to the field as determined by the FAA). [ARP4107] *2.* Regulating the flow of a fluid so that only a measured amount is permitted to flow past a given point in the system. *3.* Measuring any variable (flow rate, electrical power, etc.).

metering fix (MF) *1.* The airport arrival flow control points used by air traffic control to meter spacing for landing. Usually the fix is associated with a required flight level, speed, and ETA. [ARP1570] *2.* A fix along an established route over which aircraft will be metered prior to entering terminal airspace. [ARP4107]

metering pin Within a shock absorber, a pin that operates through the fluid-metering orifice as stroking occurs. Variation of the diameter of the pin provides a preprogrammed metering of the fluid through the orifice to obtain maximum efficiency for a given landing condition and optimum balance of efficiency for all conditions. *See also* shock absorber, variable orifice. [AIR1489]

metering tube An arrangement within a shock absorber for metering fluid for control of energy absorption. Passages are drilled through the wall of the metering tube to program the metering of fluid as stroking occurs. *See also* shock absorber, variable orifice. [AIR1489]

meter-proving tank *See* calibrating tank.

meter run A flowmeter installed and calibrated in a section of pipe having adequate upstream and downstream length to satisfy standards of flowmeter installation.

meter sensitivity The accuracy with which a meter can measure a value; usually expressed as percent of the full scale reading of the meter.

methanation The conversion of various organic compounds to produce methane.

methane Colorless, odorless, flammable gas formed by the decomposition of vegetable matter, as in marshes; the simplest hydrocarbon fuel, CH_4, a gas at NTP but a cryogenic liquid.

method of moments A method of estimating the parameters of a distribution by relating the parameters to moments.

metrication Also known as metric conversion. The conversion on an industry and/or nationwide basis of English units of measurement into the International System of Units (SI units), including engineering and manufacturing standards, tools and instruments, and all affected areas in the government and private sectors.

metric conversion *See* metrication.

metric photography The recording of events by means of photography (either singly or sequentially), together with appropriate metric coordinates to form the basis for accurate measurements.

MeV Abbreviation for mega-electron-volts; a unit of energy equivalent to the kinetic energy of a single electron accelerated through an electric potential of one-million volts.

MF *1. See* middle frequency. *2. See* metering fix.

MFCS *See* manual flight control system.

MGP *See* model generation program.

MHA *See* minimum holding altitudes.

mho Unit of conductance. Equivalent to one ampere of current passing through a material under a potential difference of one volt. Reciprocal of an ohm. [ARP1931]

MIA *See* minimum IFR Altitudes.

mica An inorganic material that separates into layers and has high insulation resistance, dielectric strength, and heat resistance. Used as an insulation wrap in wires and cables to a limited degree where resistance requirements are severe, and for high-temperature work demanding good heat resistance. [ARP1931]

micro *1.* Abbreviated µ. Prefix denoting one-millionth. [ARP1931] *2.* A common term meaning very small.

microbalance A small analytical balance for weighing masses of 0.1 g or less to the nearest µg.

microballoons Also called microspheres. Small, hollow glass spheres used as fillers in epoxy and polyester compounds to reduce density. [AIR4844]

microbar A unit of pressure equal to one dyne per square centimeter.

microburst A localized but very severe weather phenomenon resulting in abrupt changes in wind direction and velocity. [ARP4107]

microchannel plate(s) *1.* An array of microchannels formed into plates and contained in a photomultiplier tube. *2.* A glass device with many tiny, parallel holes passing through it; used, with suitable biasing, as an electron amplifier. Used primarily in imaging detectors.

micro code A system of coding in which suboperations are used that are not ordinarily accessible in programming, e.g., coding that makes use of parts of multiplication or division operations.

microcomputer(s) Complete digital computers utilizing a microprocessor consisting of one or more integrated circuit chips as the central arithmetic and logic unit, and added chips to provide timing, program memory, random access memory interfaces for input and output signals, and other functions. Some microcomputers consist of a single integrated-circuit chip.

microcurie A unit of radioactivity equal to one-millionth (10^{-6}) of a curie.

microdensitometer A device for measuring the density of photographic films or plates on a microscopic scale; the small-scale version of a densitometer.

microfarad (μf) One-millionth (10^{-6}) of a farad, a unit of capacitance. [ARP1931]

microfaradmeter A capacitance meter calibrated in microfarads.

micro filter element A construction, usually cylindrical and made from a variety of materials, through which fuel is passed to remove particulate contamination. [AIR4783]

microinch One millionth (10^{-6}) part of the US standard linear inch. [AS291D]

microinstruction An instruction that controls the operations of the various primitive resources of a computer: main and local store registers (both general and special purpose), arithmetic and logic units (ALUs), data paths, and so on.

Microinstructions are stored as words in a control store, which is traditionally (but not necessarily) separate from the main storage.

micromanipulator A positioning device for making small adjustments to the position of an optical component or other device.

micromechanics A study whereby the constituent materials are looked at separately and not as one entity. [AIR4844]

micrometeorites Very small meteorites or meteoritic particles with a diameter in general less than one millimeter.

micrometer (μm) *1.* A metric measure with a value of 10^{-6} meters (0.000039 in.), previously referred to as a micron. The diameter of dust particles is often expressed in micrometers. *2.* A type of calipers that incorporates a precision screw thread, and is capable of measuring distance between two opposing surfaces to the nearest 0.001 or 0.0001 in.

micron *See* micrometer.

micron cubic foot of gas The quantity of a gas that will occupy a one cubic foot volume at a pressure of one micron (one millionth of a meter of mercury). [ARP794]

micron cubic meter of gas The quantity of a gas that will occupy a one cubic meter volume at a pressure of one micron (one millionth of a meter of mercury). [AIR818C]

microphone An electroacoustic transducer that transmits an electrical output signal that is directly related to the loudness and frequency distribution of sound waves that strike the active element.

microphonism *1.* In an electron tube, modulation of one or more electrode currents as a direct result of mechanical vibrations of a tube element. *2.* An undesirable electrical output signal in response to mechanical or acoustic vibration of an electronic or electrical device.

microprocessor A usually monolithic, large-scale-integrated (LSI) central processing unit (CPU) on a single chip of semiconductor material. Memory, input/output circuits, power supply, etc. are needed to turn a microprocessor into a microcomputer.

microprogramming A method of operating the control unit of a computer, whereby each instruction initiates or calls for the execution of a sequence of more elementary instructions. The microprogram is generally a permanently stored section of nonvolatile storage. The instruction repertory of the microprogrammed system can thus be changed by replacing the microprogrammed section of storage without otherwise affecting the construction of the computer.

microradiography Production of a magnified radiographic image.

microradiometer A device for detecting radiant power which consists of a thermopile supported on, and directly connected to, the moving coil of a galvanometer.

microscopic stress Load per unit area over a very short distance, on the order of the diameter of a metal grain, or smaller. Usually reserved for characterizing residual stress patterns.

microsecond (μs) One-millionth of a second (10^{-6} second). [ARP1931]

microspheres *See* microballoons.

microwave *1.* Although sources do not agree, generally considered to be waves in the frequency band between approximately 500 MHz and 300 GHz. [ARP4107] *2.* Of, or pertaining to, radiation in the microwave region.

microwave amplification by the stimulated emission of radiation (MASER) The microwave equivalent and predecessor of the laser. A maser device produces coherent microwaves.

microwave landing system (MLS) An instrument landing system that operates in the microwave spectrum, and provides lateral and vertical guidance to aircraft having compatible avionics equipment. Also provides precise continuous three-dimensional position information anywhere within the approach zone, potentially allowing unrestricted choice of approach paths. [ARP4107]

microwave landing system DME/P interrogator A precision distance-measuring equipment interrogator that provides the slant range distance between the aircraft and the DME antenna associated with the ground-based MLS approach facilities. [AIR4102/12]

microwave landing system receiver (MLS receiver) A radio receiver capable of sensing and decoding all basic microwave landing system ground-generated angular and identification signals and data link messages. [AIR4102/12]

microwave landing system RNAV system A computation system that uses angular position information from the MLS receiver and distance information from the DME/P interrogator to compute guidance and output signals for aircraft systems that are required for flight along desired paths within the MLS proportional signal coverage. [AIR4102/12]

microwave scanning beam landing system (MSBLS) Primary position sensor of the navigation system of the Space Shuttle Orbiter during the autoland phase of the flight.

microwave spectrum The portion of the electromagnetic spectrum of frequencies lying between infrared waves and radio waves.

microyield strength Stress at which a microstructure (single crystal, for example) exhibits a specified deviation in its stress-strain relationship.

midaltitude The average of many measurements of altitudes, as with satellite instruments, for the compiling of planetary maps.

mid-bit transition The position in the optical waveform where the optical pulse goes from a high state to a completely off or low state, or from a low state to a high state. [AS1773]

middle atmosphere The portion of the earth's atmosphere extending from the troposphere to 100 kilometers.

middle compass locator (LMM) *See* compass locator, middle.

middle frequency (MF) The frequency band between 300 kHz and 3 MHz. Also called medium frequency. [ARP4107]

middle marker (MM) A marker beacon that defines a point along the glide slope of an ILS, normally located at or near the point of decision height (ILS Category I). The middle marker is keyed to transmit alternate dots and dashes, with the alternate dots and dashes keyed at the rate of 95 dot/dash combinations per minute, on a 1300 Hz tone; it is received aurally and visually by compatible airborne equipment. *See* marker beacon. [ARP4107]

mid-field barrier *See* barrier, mid-field.

mid-value logic system A fault-tolerant system that has an odd number of active channels (usually three), and in which the system output is determined by the middle of the (three) input signals. [ARP4386]

Mie scattering Any scattering produced by spherical particles without special regard to comparative size of radiation wavelength and particle diameter.

Mie theory *See* Mie scattering.

migration The movement of ions from an area of the same charge to an area of opposite charge.

MIG welding Stands for metal inert-gas welding. *See* gas metal-arc welding.

mil *1.* One-thousandth of an inch. *2.* A unit of angular measurement commonly used in the military for setting artillery elevations.

mild detonating fuse (MDF) A flexible metal tube, usually lead, containing a much smaller core of high explosive than the normal detonating cord. [AIR913]

mile A British and US unit of length commonly used to specify distances between widely separated points on the earth's surface. A statute mile, used for distances over land, is defined as 5280 ft. A nautical mile, used for distances along the surface of the oceans, is defined as one minute of arc measured along the equator, which equals 6080.27 ft or 1.1516 statute miles.

military climb corridor A restricted area established in the vicinity of certain military bases, used by military aircraft to climb out from an airfield to their desired operating altitude. [ARP4107]

millilambert A measure of the brightness of a surface that emits or reflects light. *See* lambert. [AS264D]

millimeter Abbreviated mm. *1.* A unit of length equal to 0.001 meter. *2.* A millimeter of mercury, abbreviated mm Hg, is a unit of pressure equivalent to the pressure exerted by a column of pure liquid mercury one mm high at 0°C under a standard gravity of 980.665 cm/s^2. Roughly equivalent to 1/760th of standard atmospheric pressure.

milliradian (mr) An angular measurement equal to 0.0573 deg, defined as 0.001 of an arc whose length equals the circle radius. [ARP4067]

millisecond One-thousandth of a second; 10^{-3} second. [ARP1931]

mil-spec/military specification A performance specification detailing the requirements for specific equipment, normally associated with US military use. [AIR4783]

MIM diodes (metal-insulator-metal diodes) Junction diodes, each consisting of an insulating layer sandwiched between two metallic surface layers, and exhibiting a negative differential resistance in V-1 characteristics, conceivably because of stimulated inelastic tunneling of electrons.

miniature boiler Fired pressure vessels that do not exceed the following limits: 16 in. inside diameter of shell; 42 in. overall length to outside of heads at center; 20 sq ft water heating surface; or 100 psi maximum allowable working pressure.

minimal-cut-sets The smallest set of primary events, inhibit conditions, or undeveloped fault events, all of which must occur for the top event of a fault tree to occur. [ARP926A]

minimal surfaces Surfaces for which the first variation of the area integral vanishes.

minimum bend radius The smallest radius around which a piece of sheet metal, wire, bar stock, or tubing can be bent without fracture; or in the case of tubing, without collapse.

minimum burst pressure The greatest applied internal pressure to which a valve can be subjected before ultimate strength of the valve is exceeded. [AS1607]

minimum crossing altitude (MCA) At certain fixes, the lowest altitude at which an aircraft must cross when proceeding in the direction of a higher minimum en route IFR altitude (MEA). For example, when the minimum en route IFR altitude changes from 6000 to 8000 ft at a particular intersection when flying from west to east, the minimum crossing altitude for that intersection eastbound would be set at 8000 feet. [ARP4107]

minimum descent altitude (MDA) The lowest altitude, expressed in feet above mean sea level, to which descent is authorized on final approach or during circle-to-land maneuvering in execution of a standard instrument approach procedure when no electronic glide slope is provided. *See also* height above airport (HAA). [ARP4107]

minimum en route IFR altitude (MEA) The lowest published altitude between radio fixes that assures acceptable navigational signal coverage and meets obstacle clearance requirements between those fixes. [ARP4107]

minimum entropy method. Application of entropy in statistical mechanics.

minimum equipment and dispatch procedures (MEPD) A supplement to the minimum equipment list which also contains procedures to follow in order to dispatch aircraft with specific equipment being inoperative. [ARP4386]

minimum equipment list (MEL) An approved list of equipment that must be operative for full mission capability. Any systems that are inoperative result in partial mission capable status. *See also* minimum equipment and dispatch procedures (MEDP). [ARP4386]

minimum fusing current The smallest value of current that will melt the current-responsive element at a specified ambient temperature. [ARP1199A]

minimum holding altitude (MHA) The lowest altitude prescribed for a holding pattern which assures navigational signal coverage and communications, and which meets obstruction clearance requirements. [ARP4107]

minimum IFR altitudes (MIA) Minimum altitudes for IFR operations as prescribed in FAR Part 91. [ARP4107]

minimum limit of ultimate trip The maximum current that a circuit breaker must hold without tripping under a given set of ambient conditions. Also known as the rated hold current. [ARP1199A]

minimum obstruction clearance altitude (MOCA) The lowest published altitude in effect between radio fixes on VOR airways, off-airway routes, or route segments, that meets obstacle clearance requirements for the entire route segment and that assures acceptable navigational signal coverage only within 25 statute miles of a VOR station. [ARP4107]

minimum operating pressure The lowest pressure at which a system or component must function. [ARP4386]

minimum performance standard A standard that establishes the minimum values for the performance characteristics of the equipment or software. [AS8004]

minimum pressure The lowest transient pressure than can occur temporarily. [ARP4386]

minimum reception altitude The lowest altitude at which one can receive signals to determine specific VOR/VORTAC/TACAN fixes. [ARP4107]

minimum reflux The quantity of reflux required to perform a specified separation in a column that has an unlimited number of trays. At minimum reflux, no products are withdrawn.

minimum safe altitude (MSA) Any altitude depicted on approach charts which provides at least 1000 ft obstacle clearance for emergency use within a specified distance from the navigation facility upon which an approach procedure is predicated. MSAs are identified as minimum sector altitudes or emergency safe altitudes. *Minimum sector altitudes*–altitudes depicted on approach charts which provide at least 1000 ft of obstacle clearance within a 25-mile radius of the navigation facility upon which the approach procedure is predicated. Sectors depicted on approach charts are at least 90 deg in scope radially and extend outward from the facility for 25 miles. These altitudes are for emergency use only and do not necessarily assure acceptable navigational signal coverage. *Emergency safe altitudes*–altitudes depicted on approach charts which provide at least 1000 ft of obstacle clearance in nonmountainous areas and 2000 ft of obstacle clearance in designated mountainous areas within a 1000-mile radius of the navigation facility upon which the approach procedure is predicated. Emergency safe altitudes are normally used only in military procedures. [ARP4107]

minimum safe altitude warning (MSAW) A function of the ARTS III computer that alerts the controller when an aircraft equipped with an operating Mode C transponder is being tracked by the radar facility and is below or is predicted by the computer to go below a pre-determined minimum safe altitude. [ARP4107]

minimum safe performance standard A standard that specifies the minimum instrument design requirements as established by the operational and environmental conditions encountered during normal flight (takeoff, climb, cruise, descent, and landing). This standard also specifies the minimum instrument performance necessary for safe operation of the aircraft during normal flight. [AIR818C]

minimum sector altitude *See* minimum safe altitude.

minimums/minima Weather condition requirements established for a particular operation or type of operation. [ARP4107]

minimum thermometer A thermometer that indicates the lowest temperature reached during a given interval of time.

minimum unstick speed *See* V_{mu}.

minimum vectoring altitude (MVA) The lowest altitude, expressed in feet above mean sea level, at which aircraft will be vectored by a radar controller. This altitude assures communications and radar coverage, and meets obstruction clearance criteria. [ARP4107]

minimum voltage (null) position A position of the rotor at which the secondary voltage of fundamental frequency that is in time phase with the secondary voltage at maximum coupling is zero (0). [ARP826]

minitrack optical tracking system *See* mini-track system.

minitrack system A satellite tracking system consisting of a field of separate antennas and associated receiving equipment interconnected so as to form interferometers which track a transmitting beacon in the payload itself.

minor defect A defect that is not likely to reduce materially the usability of the unit of product for its intended purpose, or is a departure from established standards having little bearing on the effective use or operation of the unit. *See* defect. [AIR4896]

minor graduations The shortest or lightest division marks on a graduated scale, which indicate subdivisions lying between successive major graduations or between an index graduation and an adjacent major graduation.

minor time In data processing systems, one sixteen-bit word: binary milliseconds.

mips (million instructions per second) A measure of computer machine code instructions per second.

MIRL *See* runway edge light system.

mirror fusion An open-ended configuration that traps low-beta plasmas. This configuration is realized by associating two identical magnetic mirrors having the same axis.

mirror scale An instrument scale and a mirror, so arranged that the indicating pointer and its reflection are aligned when the observer's eye is in the correct position to read the instrument without parallax error.

MIS *See* management information system.

misalignment Error in alignment between the axes of two joining parts; the error may be linear and/or angular. [ARP699D]

misalignment load A force component perpendicular to the actuator output motion, which is the resultant of a load not in line with the actuation output. [ARP4386]

miscellaneous equipment The class of equipment that is intended to be used by the flight crew under normal (including non-routine) operating conditions. Includes flight kits (brief cases), log books, flashlights, personal oxygen masks, etc. [ARP917A]

miscellaneous small parts Non-serialized, expendable parts, not traceable to lot/batch when outside of the small parts manufacturer's packaging. [AS7104]

miscible Liquids that are mutually soluble. [AIR4783]

misconvergence The degree to which the midpoint of the line widths of two or three primary colors are misregistered at the phosphor surface of a CRT. [ARP1782]

misfire Failure of an explosive device to fire after initiating energy is applied. [AIR913]

mishap An unplanned, unintended event that results in damage to equipment or injury to personnel. [ARP4107]

mishap, antecedent events Those events or conditions that occurred prior to flight but that relate to the conditions making the mishap more likely (for example, fatigue or "get-home-itis"). [ARP4107]

mishap, maneuver A sub-element of the mishap phase of flight, described by the sequence of tasks required to perform the maneuver (for example, turnout of traffic, formation crossover, or egress from a weapons delivery pass). Maneuver mishaps may occur in any of the following eight phases of flight: (a) approach phase, (b) climbout phase, (c) cruise phase, (d) descent phase, (e) landing phase, (f) range phase, (g) takeoff phase, and (h) taxi phase. *Approach phase*–from the final approach fix to the missed approach point for an instrument approach; from reaching traffic-pattern altitude until crossing the runway threshold for a visual approach. A go-around is considered part of the approach phase if it occurs prior to the missed approach point for an instrument approach or prior to crossing the runway threshold for a visual approach. *Climbout phase*–from crossing the field boundary to attaining cruise altitude. *Cruise phase*–from reaching cruise altitude to arriving at the area of range activity; or from leaving the area of the range activity to beginning descent into the base of intended landing. *Descent phase*–from the initial approach fix to the final approach fix for an instrument descent; from beginning descent from cruise altitude to arriving at the final approach fix for an en route descent to an instrument approach; from beginning descent from cruise altitude until reaching traffic pattern altitude for an en route descent to a visual approach. Holding is considered part of the descent phase of flight. *Landing phase*–from the missed approach point until touchdown for an instrument approach; from crossing the runway threshold until touchdown from a visual approach. A go-around is considered part of the landing phase if it occurs after the missed approach point for an instrument approach or after crossing the runway threshold for a visual approach. After touchdown, a touch-and-go is considered a takeoff. *Range phase*–from the time the aircraft enters the area designated for practicing/conducting mission activities until completion of those activities and departure from the designated area. This range phase may be a low-level route, military operating area, gunnery range, warning area, or a refueling track. *Takeoff phase*–from runway holdline to the point when the aircraft is airborne and has passed the field boundaries. *Taxi phase*–from engine start to runway holdline, and from clearing the active runway to having parked the aircraft. [ARP4107]

mishap, point of The point in the mishap sequence of events at which no preventive or evasive action by the operator would have avoided the mishap. [ARP4107]

mishap, predisposing events Those events or conditions that are more general in nature or more longstanding than mishap antecedent events, but that are predisposing to mishap occurrence (for example, risk-taking tendencies, lax supervision, etc.) [ARP4107]

mishap, sequence of events Those events or conditions related to the mishap, which begin with demonstration of intent for flight as defined in AFR 127-4, and end when damage or injury has occurred and ceased. [ARP4107]

mishap, task A sub-element of the mishap maneuver that describes each specific action required of the operator to accomplish that maneuver (for example, switchology, target tracking, or aircraft positioning). [ARP4107]

mismatch *1.* Condition in which the impedance of a source does not match or equal the impedance of the connected load or transmission line. *2.* Lateral offset between two halves of a casting mold or forging die, which produces distortion in shape across the parting line.

misperception *See* perception.

missed advisory A resolution advisory or traffic advisory not given when the threat of a potential collision does exist. [AIR4102/10]

missed alert A system alert that is not given even though an aircraft is in the TCAS operational envelope and the threat of collision or potential collision exists. [ARP4153]

missed approach A maneuver conducted by a pilot when an instrument approach cannot be completed to a landing. The route of flight and altitude to be flown in the execution of a missed approach are shown on the instrument approach procedure charts. A pilot executing a missed approach prior to the missed approach point (MAP) must continue along the final approach to the MAP. The pilot may climb immediately to the altitude specified in the missed approach procedure but may not alter course until the missed approach point has been crossed. At locations where ATC radar service is provided, the pilot should conform to radar vectors issued by the controller in lieu of the published missed approach procedure. *See also* go-around and segments of an instrument approach procedure. [ARP4107]

missed approach point (MAP) A point prescribed in each instrument approach procedure at which a missed approach is to be executed if the required visual reference with the outside world does not exist. [ARP4107]

missed approach segment *See* segments of an instrument approach procedure.

missile(s) Any object that is designed to be thrown, dropped, projected, or propelled for the purpose of making it strike a target. [AIR913]

missing mass (astrophysics) Refers to a problem related to a cluster of galaxies whereby the dynamical mass is substantially larger than the mass estimated by the mass-to-luminosity ratio of the visible parts of the galaxies, the visible mass.

mission *1.* The objective or task, together with the purpose, which clearly indicates the action to be taken. [ARP926A] *2.* That period beginning with the start of the engine prior to flight and ending at engine shutdown at the completion of the flight. [ARD50010]

mission adaptive wing A variable geometry wing in which an actuation system is used to vary wing camber, both prior to flight and in flight, in response to pilot commands; this increases the wing efficiency for that portion of the flight mission. [ARP4386]

mission capable (MC) Refers to status data consisting of the sum of full mission capable (FMC) and partial mission capable (PMC) for purposes of reporting to Office of the Secretary of Defense (OSD). [ARD50010]

mission capable rate The percent of possessed time that a system is capable of performing at least one of its assigned missions. [ARD50010]

mission effects Failure effects that preclude the completion of the aircraft mission. These failures cause delays, cancellations, ground or flight interruptions, high drag coefficients, flight envelope restrictions, etc. [ARD50010]

mission element Each specific operating condition of the mission divided into component units. [ARP1352]

mission essential Anything that is authorized and assigned to approved combat and combat support forces, and that would be immediately employed to wage war and provide support for combat actions. [AIR4896]

mission essential equipment The interdependent equipment in a missile weapon system that is assigned a mission on the battlefield in support of forces in contact with the enemy, and without which the assigned mission cannot be successfully completed. [ARD50010]

mission essential functions Those subsystem functions required to enable an aircraft to perform its designated mission(s). [ARD50010]

mission essential subsystem Anything that is authorized and assigned to approved combat and combat support forces, and that would be immediately employed to wage war and provide support for combat actions. [ARD50010]

mission mix The relative frequency with which each mission profile is encountered during a specified time period. [ARD50010]

mission narrative A report of the planned use of a system (for example, an aircraft), told in chronological order. Presented in terms of the real sequence and describes roles, activities, relations, and events.

mission profile A time-phased description of the events and environments that an item experiences from initiation to completion of a specified mission, including the criteria of mission success or critical failures. [ARD50010]

mission reliability The ability of an item to perform its required functions for the duration of a specified mission profile. [ARD50013]

mission time The time during which a system or equipment is actually operating (in an "up" status). Operating time is usually divisible among several operating periods or conditions. These include standby time, filament-on time, pre-flight checkout time, and flight. [ARD50010]

mission time between critical failures A measure of mission reliability: The total amount of mission time, divided by the total number of critical failures, during a stated series of missions. [ARD50010]

mission time to restore functions A measure of mission maintainability: The total corrective failure maintenance time, divided by the total number of critical failures, during the course of a specified mission profile. [ARD50010]

miter valve A valve in which the disc is at an angle of approximately 45° to the axis of the valve body.

mixed icing conditions Describes the condition in which a mixture of supercooled water droplets and ice crystals exists within the same cloud environment. [AIR1667]

mixed level A simulation system combining both low-level transistor and gate circuit descriptions with high-level behavioral circuit representations.

mixed oxides Mixture of oxides, particularly of radioactive metals.

mixed radix Pertaining to a numeration system that uses more than one radix, such as the biquinary system.

mixed signal Describes a simulation system combining both analog and digital circuit representations.

mixer In sound recording or reproduction equipment, a device capable of combining two or more input signals into a single linearly-proportioned output signal, usually with the additional capability of adjusting the levels of any of the inputs.

mixing ice conditions Describes the condition in which a mixture of supercooled water droplets and ice crystals exists within the same cloud environment. [AIR1667]

mixing valve A valve having more than one inlet, but only one outlet port. Used to blend two or more fluids to give a mixture of predetermined composition.

MLA *See* multispectral linear arrays.

MLS *See* microwave landing system.

mm *See* millimeter.

MM *See* middle marker.

MMFS *See* Manufacturing Messaging Format Standard.

MMS *1.* Stands for Manufacturing Message Specification, ISO/IEC 9506, a set of international standards developed to facilitate the interconnection of information-processing systems. The first part defines the service provided by the MMS. The second part specifies the protocol that supports the MMS. *2. See* multimission modular spacecraft.

mnemonic An assembly language instruction, defined by a symbol, that has some resemblance to the operation carried out. Mnemonics are easier to remember and use than the equivalent hex code or machine code.

mnemonic operation code An operation code in which the names of operations are abbreviated and expressed mnemonically to facilitate remembering the operations they represent. A mnemonic code normally needs to be converted to an actual operation code by an assembler before execution by the computer. Examples of mnemonic codes are ADD for addition, CLR for clear storage, and SQR for square root.

MNOS *See* metal-nitride-oxide semiconductor.

mobile communication systems Any configuration of mobile or transportable voice and data communication equipment that allows for communication between combinations of mobile/fixed points, with or without the aid of satellites.

mobile telemetering Any arrangement for transmitting instrument readings from a movable data-acquisition station to a remote stationary, movable indicating, or recording station, without the use of interconnecting wire.

mobility In gases, liquids, solids, or colloids, refers to the relative ease with which atoms, molecules, or particles can move from one location to another without external stimulus.

MOCA *See* minimum obstruction clearance altitude.

mockup *1.* The reassembly of an aircraft following its breakup in an accident. This procedure may supply a clue as to accident cause. *2.* A structural scale model of an aircraft or other artifact used as part of the design/fabrication process or as part of a simulator system. *3.* A model of a piece of equipment or a system, frequently full size, used for experiments, performance testing, or training.

mode *1.* The manner in which acoustic energy is propagated through a material, as characterized by the particle motion of the wave. [ARP5089] *2.* The letter or number assigned to a specific pulse spacing of radio signals transmitted to aircraft or received by ground interrogator from airborne transponder components of the air traffic control radar beacon system (ATCRBS). [ARP4107] *3.* A single component in a computer network. *4.* Real or complex (number system). *5.* A stable condition of oscillation in a laser. A laser can operate in one mode (single-mode) or in many modes (multimode).

mode C The automatic altitude-reporting capability that converts the altitude of the aircraft in 100-ft increments to coded digital information, which is transmitted, together with the transponder code of the aircraft, to the interrogating facility. The altitude of mode C-equipped (and operating) aircraft can be automatically presented on the radar display of the ATC controller. [ARP4107]

mode changer A device for changing the characteristics of a guided wave from one mode of propagation to another.

mode code A means by which the bus controller can communicate with the multiplex bus-related hardware, in order to assist in the management of information flow. [AS15531]

mode conversion Changing from one mode of vibration to another; caused by refraction at an interface. [ARP5089]

mode coupling *See* coupled mode.

mode filter In a waveguide circuit, an arrangement of waveguide elements that pass waves being propagated in certain mode(s) and exclude waves being propagated in other modes.

model *1.* A device used in a failure detection-correction system to simulate the performance of a component or a channel used for control.

Typical models are electrically implemented. *2.* An object, usually a scaled representation of prototype, built to give some idea of what a functioning prototype would be like. *3.* A mathematical description, usually idealized, of the performance of a dynamical system. [ARP4386]

model basin A large tank of water for design experiments and performance studies of ship hulls using scale models. Also known as a model tank or towing tank.

model dispersion In a multimode fiber, the component of pulse-spreading that is caused by differential optical path lengths.

model reference adaptive control (MRAC) A type of control utilizing the following: (a) an ideal adaptive control system whose response is agreed to be optimum; (b) computer simulation in which both the model system and the actual system are subjected to the same stimulus; and (c) parameters of the actual system, which are adjusted to minimize the difference in the outputs of the model and the actual system.

model specification A specification covering the essential detail and technical requirements of a specific unit design, including a description of the procedures by which it will be demonstrated that the requirements have been met. [ARP906A]

model tank *See* model basin.

model time step A time increment used in a model, for example, in a numerical integration algorithm. [ARP4148]

modem *1.* A device that provides both combining (modulation) and separation (demodulation) of data. Typically used to connect a node to a broadband network. *See also* transceiver and transmitter receivers. *2.* An electronic device for serial transmission of digital data in the audio frequency spectrum over a voice-grade telephone line.

mode of vibration *See* vibration mode.

moderation Reducing the kinetic energy of neutrons, usually by means of successive collisions with hydrogen, carbon, or other light atoms.

moderators Materials that have a high cross section for slowing down fast neutrons with a minimum of absorption, e.g., heavy water or beryllium, used in reactor cores.

modes Refers to the modes by which PID algorithms operate. These are determined by the operator and/or by states of other instructions inside the controller.

MODFETS Heterojunction field effect transistor device structures in which only the larger (Al, GaAs) bandgap is doped with donors, while the GaAS layer is left undoped. This results in high electron mobilities due to spatially separated electrons and donors.

modification Major engineering changes to an existing equipment or system to effect improvements in designed capabilities or characteristics. [AIR4896]

modification, mandatory A modification classified as compulsory by the local civil aviation authorities. [AIR4896]

modification, optional A modification that may be incorporated at the discretion of the operator. [AIR4896]

modification time That part of downtime necessary to introduce any specific changes to an item to improve its characteristics, or to add new ones. [AIR4896]

modify To change or alter through rework and/or through the installation or removal of an item. [AIR4896]

modular engine An engine consisting of several independent assemblies called modules, which by design can be removed/replaced without major disassembly of the engine or other modules, for example, compressor, combustion, turbine, afterburner, gearbox, torquemeter, or combinations thereof. [ARD50010]

modular integrated utility system A joint NASA-HUD concept incorporating various utilities—electric power plant, water supply, heating and air conditioning, sewage treatment, and waste disposal—into a single system having increased efficiency and economy.

modularity Describes the degree to which a system of programs is developed in relatively independent components, some of which may be eliminated if a reduced version of the program is acceptable.

modularization Designing a series of components, subassemblies, or devices for interchangeability of physical location, so that different assemblies can be easily constructed on a standard frame or mounted in standard enclosures.

modular programming Programming in which tasks are programmed in distinct sections or sub-sections resulting in the ability to modify one section without reference to other sections.

modulated brake pressure Metered brake pressure modulated by a skid-control valve and control system. [AIR1489]

modulated wave A radio-frequency wave in which amplitude, phase, or frequency is varied in accordance with the waveform of a modulating signal.

modulation *1.* The process of impressing information on a carrier for transmission; AM, amplitude modulation; PM, phase modulation; FM, frequency modulation. *2.* Regulation of the fuel-air mixture to a burner in response to fluctuations of load on a boiler. *3. See* contrast.

modulation doped fets *See* MODFETS.

modulation doping The process of doping only the larger bandgap of a heterojunction device with donors, while the other layer is left undoped. Since the electrons and donors are spatially separated, ionized impurity scattering is avoided and extremely high electron mobilities are obtained.

modulation factor The ratio of peak variation actually used in a given type of modulation, to the maximum design variation possible.

modulation index In frequency modulation with a sinusoidal waveform, the ratio of the peak (not peak-to-peak) frequency deviation, to the frequency of the modulating wave.

modulation meter An instrument for measuring modulation factor of a wave train, usually expressed in percent.

modulation noise The noise in an electronic or acoustic circuit caused by the presence of a signal, but not including the waveform of the signal itself.

modulator Device to effect the process of modulation.

module *1.* A combination of assemblies, sub-assemblies, and parts, contained in one package, or so arranged as to be installed in one maintenance action. [ARD50010] *2.* A subsection of a galley unit or complex that is easily removable and transferable to ground equipment for cycling through ground kitchens for cleaning and reloading with food or service items. [AS1426] *3.* A computer program unit that is discrete and identifiable with respect to compiling, combining with other units, and loading; for example the input to, or output from, an assembler, compiler, linkage editor, or executive routine. *4.* An entity that is addressable via the bus and has a single connection to the bus. [AS4710]

modulo A mathematical operation that yields the remainder function of division; thus, 39 modulo 6 equals 3.

modulo N check A check utilizing a check number that is equal to the remainder of the desired number when divided by n, e.g., in a modulo 4 check, the check number will be 0, 1, 2, or 3 and the remainder of the desired number when divided by 4 must equal the reported check number, otherwise an equipment malfunction has occurred.

modulus of elasticity (elastic modulus) In any solid, the slope of the stress-strain curve within the elastic region. For most materials, the modulus of elasticity is nearly constant up to some limiting value of stress known as the elastic limit. The modulus of elasticity can be measured in tension, compression, torsion or shear. The tension modulus is often referred to as Young's modulus.

moire *1.* A pattern seen when two out-of-phase, spatially periodic patterns are superimposed. [ARP4256] *2.* A pattern resulting from interference or light-blocking due to beat frequencies between two or more superimposed, spatially-periodic structures. On a color CRT display, moire takes the form of light and dark areas on raster; on stroke-written lines, moire takes the form of periodic light and dark modulation of the line. [ARP1782]

moire fringes The bands that appear in the moire effect.

Moire interferometry The use of intersecting families of curves as instruments for making precise measurement; the study of indices of refraction, etc. by utilizing the interference patterns.

moisture absorption Generally, the amount of moisture, in percentage, that an insulation will absorb under specified conditions. [ARP1931]

moisture barrier A material or coating that retards the passage of moisture through a wall made of more-permeable materials.

moisture loss Represents the difference in the heat content of moisture in the exit gases and that at the temperature of the ambient air.

moisture resistance The ability of a material to resist absorbing moisture. [ARP1931]

mol *See* mole.

molar units Concentration unit defined as the number of gram-moles of the component per liter of solution.

molded edge An edge that is not physically altered after molding for use in final form; particularly one that does not have fiber ends along its length. [AIR4844]

molded in place seal A proprietary commercial sealing gasket in which a partial O-ring is molded and retained in grooves positioned on both sides of a thin metal ring. [AIR1664]

molded net A molded part that requires no additional processing to meet dimensional requirements. [AIR4844]

molding The forming of a polymer or composite into a solid mass of prescribed shape and size by the application of pressure and heat for given times. [AIR4844]

molding cycle The period of time required for the complete sequence of operations on a molding press to produce one set of moldings. [AIR4844]

molding powder or compound Plastic material in varying stages of pellets or granulation, and consisting of resin, filler, pigments, reinforcements, plasticizers, and other ingredients, ready for use in a molding operation. [AIR4844]

molding pressure The pressure applied to the ram of an injection machine or compression or transfer press to force the softened plastic to fill the mold cavities completely. [AIR4844]

mold, potting An accessory used as a form for containing the potting compound around the terminations of a connector. [ARP914A]

mold-release agent A lubricant, liquid, or powder used to prevent sticking of molded articles in the cavity. [AIR4844]

mold seam A line on a molded or laminated piece, differing in color or appearance from the general surface, caused by the parting line of the mold. [AIR4844]

mold shrinkage The immediate shrinkage that a molded part undergoes when it is removed from a mold and cooled to room temperature. [AIR4844]

mold skid depth The depth of the grooves in a tire mold; or conversely, the height of the tread on the tire. Variation in this parameter increases or decreases wear life of the tire. Tread depth may be limited by speed requirement and centrifugal force/stresses. [AIR1489]

mold surface The side of a laminate that faced the mold during cure in an autoclave or hydroclave. [AIR4844]

mold wash *See* wash.

mole Abbreviated mol. Metric unit for amount of a substance.

molecular attrition *See* fretting, molecular attrition theory.

molecular beam A unidirectional beam of neutral-charge molecules passing through a vacuum.

molecular beam epitaxy Ultrahigh vacuum technique for growing very thin epitaxial layers of semiconductor crystals.

molecular clouds Thickest and densest interstellar clouds consisting mainly of molecular hydrogen, but also a high concentration of dust grains.

molecular dissociation *See* dissociation.

molecular flow Gas flow in a tube at a pressure low enough that the mean free path of the molecules is greater than the inside diameter of the tube.

molecular shields Furlable devices used in space vacuum research to permit deployment and retrieval of instruments and the performance of experiments without contamination.

molecular sieve *1.* Describes a gas-separation concept used in MSOGS (molecular sieve oxygen generation system). Incorporates a preferential adsorption medium that retains nitrogen molecules and allows molecules of oxygen and argon to pass through, thus providing oxygen-enriched breathing gas. *2.* Crystalline zeolite material that is used as a sorbent for gases and liquids. *3.* A specific zeolite of sodium calcium aluminosilicate modified for nitrogen-oxygen separation in the MSOGS. [ARP171]

molecular weight The sum of the atomic weights of all the atoms in a molecule. [AIR4844]

molecule(s) *1.* Aggregates of two or more atoms of a substance, existing as a unit. *2.* The smallest division of a unique chemical substance which maintains its unique chemical identity.

mole fraction The volume concentration of a gas per unit volume of the gas mixture of which it is a part. [AIR1533]

moles Number of molecular weights; the weight of the component divided by its molecular weight.

Moll thermopile A type of thermopile used in some radiation-measuring instruments; consists of multiple manganan-constantan thermocouples connected in series. Alternate junctions are embedded in a shielded nonconductive plate of large heat capacity; the remaining junctions are blackened and exposed directly to the radiation. The voltage across the thermopile is directly proportional to the intensity of radiation.

molten salts High-temperature inorganic salt or mixtures of salts used for thermal-energy storage, heat exchangers, high-power electric batteries, heat-treatment of alloys, etc.

moment *1.* The effectiveness of a force in producing rotation about an axis; moment of force. Equals the product of the radius perpendicular to the axis of rotation that passes through the point of force application, and the tangential component of force perpendicular to the plane defined by the radius and axis of rotation. *2.* The resistance of a body at rest or in motion to changes in its angular velocity; moment of inertia.

momentary *1.* Describes an alarm that returns to normal before being acknowledged. *2.* Returns to normal state when pressure or signal is removed.

momentary digital output A contact closure, operated by a computer, that holds its condition (set or reset) for only a short time. *See also* latching digital output.

momentary switch A spring-loaded switch whose contacts complete a circuit only while an actuating force is applied; for a typical momentary pushbutton, electric current flows through the switch only while the operator has a finger on the button.

moment of inertia A measure of the resistance offered by a body to angular acceleration. The mass moment of inertia is usually expressed in slug ft^2; or, when multiplied by the gravity constant (g), lb ft^2. Point of application must be expressed, as it varies as the square of a gear ratio. [ARP906A] *See also* inertia.

momentum *1.* Quantity of motion. *2.* The product of a body's mass and its linear velocity.

momentum balance An analytical method used to compute the correlation factor using the difference in airflow momentum measured upstream and downstream of the engine. [ARP741]

Monel A series of International Nickel Co. high-nickel, high-copper alloys used for their corrosion resistant properties to certain conditions.

monitor *1.* A device used for sensing the operation of a component or channel such that failures may be detected. In-line and cross-channel are two forms of monitoring. *In-line monitoring*–compares output performance to the command input or a model. *Cross-channel monitoring*–compares equivalent performance features of two or more channels. [ARP4386] *2.* To measure a quantity continuously or at regular intervals so that corrections to a process or condition may be made without delay if the quantity varies outside of prescribed limits. *3.* Software or hardware that observes, supervises, controls, or verifies the operations of a system. *4.* In data processing, a high-resolution viewing screen.

monitor command An instruction issued directly to a monitor from a user.

monitor console The system control terminal.

monitoring The act or technique of establishing functional status or conditions. [ARP1587]

monitoring, inflight Monitoring by on-board equipment during the period from engine start to engine shut down. [ARP1587]

monitoring, on site *1.* Utilization of data on-site. *2.* Ground test of engines using on-board or portable ground test equipment. [ARP1587]

monitoring, remote Utilization of data at a remote (off-site) location. [ARP1587]

monitor software The portion of the operational software that controls on-line and off-line events, develops new on-line applications, and assists in their debugging. Also known as a batch monitor.

monochromatic Of a single wavelength or frequency. In reality, light cannot be purely monochromatic, and actually extends over a range of wavelengths. The breadth of this range determines how monochromatic the light is.

monochromatic radiation Any electromagnetic radiation having essentially a single wavelength; or any electromagnetic radiation in which the photons all have essentially the same energy.

monochromator An optical device in which a prism or diffraction grating is used to spread out the spectrum, then pass a narrow portion of that spectrum through a slit. Used to generate monochromatic light from a non-monochromatic source.

monofilament A single fiber or filament of indefinite length, strong enough to function as a yarn in commercial textile operations. [AIR4844]

monolayer The basic laminate unit from which cross-plied or other laminate types are constructed. [AIR4844]

monomer Any molecule that can be chemically bound as a unit of a polymer. [ARP1931]

monopropellant A liquid propellant that contains an oxidizing agent and combustible matter (fuel) in a single phase. [AIR913]

monostable Pertaining to a device that has one stable state.

monotectic alloys Metallic composite materials having a dispersed phase of solidification products distributed within a matrix.

The dispersed components can be selected to provide characteristics such as superconductivity or lubricity.

monotonic Describes a digital-to-analog converter in which the output either increases or remains constant as the digital input increases.

mono wheel An aircraft landing gear configuration in which there is a single wheel. [AIR1489]

Monte Carlo method A trial-and-error method of repeated calculations to discover the best solution to a problem. Often used when a great number of variables are present, with interrelationships so extremely complex as to forestall straightforward analytical handling.

month The period of the revolution of the moon around the earth. The month is designated as sidereal, tropical, anomalistic, dracontic, or synodical, according to whether the revolution is relative to the stars (sidereal), the vernal equinox (tropical), the perigee (anomalistic), the ascending node (dracontic), or the sun (synodical). The calendar month is a rough approximation of the synodical month.

mood *See* affective states, mood.

mooring Tie-down or securing of the aircraft by cables, ropes, straps, or chains for protection against wind, propeller or rotor wash, jet blast, ship motion, etc. Also sometimes called picketing. *See also* tie down. [AIR1489]

morphology The overall form of a polymer structure, that is, crystallinity, branching, molecular weight, etc. [AIR4844]

MOS Abbreviation for metal oxide semiconductor.

most significant bit (MSB) In a digital sequence, the bit that defines the largest value; usually at the extreme left.

most significant digit The leftmost non-zero digit.

mother board A circuit board that includes the primary components of a microcomputer.

motion The act, process, or instance of change of position. Also called movement, especially when used in connection with problems involving the motion of one craft relative to another.

motion balance instrument Refers to a type of instrument design in which the motion of the measuring element against a spring is utilized to reach a balance of forces representing the magnitude of the measured variable.

motion converter The component of the actuation system that converts the motion of the actuator into the motion of the load; for example, a ball screw converts rotary motion into linear, and a crank arm converts linear motion into rotary. [ARP4386]

motion equations *See* equations of motion.

motion simulation Replication of exact motion or replication of part of a motion to provide the sensation of the motion.

motor *1.* To rotate an engine with the starter, but without providing fuel or ignition to the engine. [ARP906A] *2.* A generic term for a solid-propellant rocket consisting of an assembled propellant, a case, ignition system, nozzle, and appurtenances. NOTE: The term "rocket engine" is usually used for liquid-fueled rockets. [AIR913]

motor, hydraulic A device for converting liquid fluid energy into mechanical energy in the form of continual rotary motion of a mechanical member. [ARP4386]

motoring run Unit initiation (time zero) and acceleration of a load from zero rpm to a sustained speed. [AS1606]

motor meter A type of integrating meter. Consists of a rotor; one or more stators; a retarding device, which makes the speed of the rotor directly proportional to the integral of the quantity measured (usually power or electric current); and a counter or set of dials that indicates the number of rotor revolutions.

motor operator The electric or hydraulic power mechanism that receives a control signal and repositions a valve or other final control element.

motors Machines supplied with external energy that is converted into force and/or motion.

mounting pad The pad upon which the starter or accessory is attached. Consists of a fixed pad with studs or suitable attaching means concentrically located with respect to a shaft which has a spline or similar power-transmitting means. [ARP906A]

movement *See* motion.

movement area The runways, taxiways, and other areas of an airport that are utilized for taxiing, takeoff, and landing of aircraft (exclusive of loading ramp and parking areas). At those airports with a tower, specific approval for entry onto the movement area must be obtained from ATC. [ARP4107]

moving body actuator An actuator in which the body is attached to the moving load and the piston rod is attached to the support structure. [ARP4386]

moving-coil instrument An instrument whose output is related to the reaction between the magnetic field set up by current flow in one or more movable coils, and the magnetic field of a fixed-position permanent magnet. Also known as permanent-magnet moving-coil instrument.

moving-dial indicator A type of indicator in which a flat, circular scale (dial) is attached to a moving element; the instrument scale is continually repositioned with respect to a fixed pointer to indicate changing values of a measured variable.

moving-drum indicator A type of indicator in which a circular member (drum) with a scale along its periphery revolves in relation to a fixed pointer to indicate changing values of a measured variable.

moving element Describes the parts of an instrument that move as a direct result of a variation in the quantity being measured.

moving-iron instrument An instrument that depends on the reaction between one or more pieces of magnetically soft material, at least one of which moves, and the magnetic field set up by electric current flowing in one or more fixed coils, to produce its output indication or signal.

moving-magnet instrument An instrument that depends on the reaction between a movable permanent magnet and the resultant field produced by another permanent magnet interacting with one or more current-carrying coils, or by two or more coils interacting with each other, to produce its output indication or signal.

moving-scale indicator Any of several designs of instrument indicator in which the scale moves in relation to a fixed pointer to indicate changing values of a measured variable.

moving target indicator (MTI) An electronic device that permits radarscope presentation only from targets that are in motion. This is done by eliminating signals from nonmoving targets (such as buildings and high terrain), signals from targets at less than a predetermined threshold velocity, and random noise. Use of the moving target indicator is a partial remedy for ground clutter. [ARP4107]

mr *See* milliradian.

MRAC (systems) *See* model reference adaptive control.

Ms The maximum temperature at which a martensite reaction begins upon cooling from the beta phase. [AS1814]

MSA *See* minimum safe altitude.

MSAW *See* minimum safe altitude warning.

MSB *See* most significant bit.

MSBLS *See* microwave scanning beam landing system.

MSG *See* maintenance steering group.

MSI *1. See* item, maintenance significant. *2. See* medium scale integration.

MSL *1. See* mean sea level. 2. Used in conjunction with an altitude, means "above mean sea level," for example, "16,000 ft MSL" means "16,000 ft above mean sea level." [ARP4107]

MSM (semiconductors) Stands for metal-semiconductor-metal. Semiconductor devices consisting of a semiconductor layer sandwiched between two layers of metal.

MSOC (molecular sieve oxygen concentrator) A type of oxygen concentrator in which air is processed through molecular sieve beds to provide an oxygen-enriched breathing gas, by a process of pressure-swing adsorption. [ARP171]

MSOGS (molecular sieve oxygen generation system) An OBOGS (on-board oxygen generation system) in which a pressure-swing molecular-sieve oxygen concentrator is used as the source of oxygen-enriched breathing gas. Pressurized air may be obtained from engine bleed air or other compressed air sources. [ARP171]

M synchronization A type of linking between a camera shutter and a flash unit that gives a 15-millisecond delay so that the metal foil flash lamp reaches peak brightness before the shutter actually trips.

MTBCF *See* mean time between critical failure.

MTBF *See* mean time between failure.

MTI radar *See* moving target indictor.

MTTF *See* mean-time-to-failure.

muff heater *See* heater, shroud.

multi-anode microchannel arrays A family of photoelectric, photon-counting array detectors being developed for use in instruments on both ground-based and spaceborne telescopes.

multibeam antennas Antennas that have the ability to form more than one beam from a single radiating aperture.

multi-bus Intel's proprietary link between single-board systems used for industrial systems.

multibypass engine cycles Cycles in which bypass flows may be separated from the core flow at two or more stations in the compression process. These flows may subsequently mix with one another and/or the core flow to provide turbine and afterburner cooling. [AIR4548]

multicast A message addressed to a group of stations connected to a LAN.

multichannel actuator An actuator that contains more than one torque- or force-generating device. [ARP4386]

multichannel spectral analyzer A measurement system that sorts signals into a number of different channels, then counts and analyzes the signals channel by channel.

multicircuit winding In filament winding, a widening that requires more than one circuit before the band repeats by laying adjacent to the first band. [AIR4844]

multiconductor A cable containing more than one component wire. [ARP1931]

multicoupler A device for coupling several receivers or transmitters to one antenna; and for enabling the proper impedance match between the receivers or transmitters.

multicraft *1.* Refers to maintenance personnel who are proficient in more than one craft, such as instrument technician, electrician, and instrument mechanic. *2.* Responsible to maintain a variety of equipment used in control systems.

multidirectional Describes a laminate that has multiple-ply orientations. [AIR4844]

multi-element control system A control system utilizing input signals derived from two or more process variables for the purpose of jointly affecting the action of the control system. Examples are input signals from pressure and temperature; or from speed and flow. *See also* multivariable control.

multifiber cable A fiber-optic cable containing many fibers, which transmit signals independently and are housed in separate substructures within the cable.

multifilament yarn A large number of fine, continuous filaments, usually with some twist in the yarn to facilitate handling. [AIR4844]

multifuel burner A burner by means of which more than one fuel (e.g., pulverized fuel, oil, or gas) can be burned, either separately or simultaneously.

multifunction multiloop controller A type of microprocessor-based controller that combines the process-control functions of a dedicated loop controller with many of the logic functions of a programmable logic controller to provide the control strategy of an entire unit operation.

multilayer A type of printed circuit board that has several layers of circuit etch or pattern, one over the other, and interconnected by electroplated holes. Since these holes can also receive component leads, a given component lead can connect to several circuit points, reducing the required dimensions of a printed circuit board.

multilayer coating Optical coating in which several layers of different thicknesses of different materials are applied to an optical surface. Interference affects light passing through the layers. Reflection and transmission are influenced differently at different angles of incidence and different wavelengths.

multilayer glass *See* cased glass.

multilayer structure *See* laminate.

multimeter *See* volt-ohm-milliammeter.

multimission modular spacecraft (MMS) Future spacecraft to be operated in conjunction with the Space Shuttle orbiter vehicle and serviced by its module-exchange mechanism.

multimode fiber An optical fiber capable of carrying more than one mode of light in its core.

multioriented ply laminate A laminate made from multioriented plies. [AIR4844]

multipath transmission The process, or condition, in which radiation travels between source and receiver via more than one path. Since there can be only one direct path, some process of reflection, refraction, or scattering must be involved.

multiphoton absorption Ionization and dissociation of a molecule under the action of powerful laser radiation. Laser-flux dependent light intensities are emitted by different excited states of the molecule to indicate the various absorption processes.

multiple access The allocation of communication system resources (output) among multiple users by means of power, bandwidth, and power assignment singly or in combination.

multiple action A control-system action that is a composite of the actions of two or more individual controllers.

multiple disc clutch A series of interfacing, rotating discs that transmit mechanical power when in contact, by means of a friction facing on their surfaces. [ARP4386]

multiple failures The simultaneous occurrence of two or more independent failures. When two or more failed parts are found during troubleshooting and failures cannot be shown to be dependent, multiple failures are presumed to have occurred. [ARD50010]

multiple grain An assembly of solid propellant grains inside an explosive device or motor. [AIR913]

multiple layer adhesive A film adhesive, usually supported, with a different adhesive composition on each side. [AIR4844]

multiple LVDTs Packaging LVDT assemblies either dual, triple, etc., in tandem or parallel arrangements in order to obtain multiple outputs for purpose of redundancy. [ARP4386]

multiple-output system A system that manipulates a plurality of variables to achieve control of a single variable.

multiple-per-revolution distortion The equivalent number of distorted regions occurring in 360°. [ARP1420]

multiple-purpose meter *See* volt-ohm-milliammeter.

multiple sampling A type of statistical quality control in which several samples, each consisting of a specified number of items, are withdrawn from a lot and inspected. The lot is accepted, rejected, or resampled, depending on the number of unacceptable items found.

multiple tire applications Applications in which the load on one landing gear strut is shared by two or more tire/wheel assemblies. [AS4833]

multiple wheel control Control of two or more braked wheels utilizing one valve and control of circuit. [AIR1489]

multiplex *1.* To combine signals. Optical multiplexers combine signals at different wavelengths. Electronic multiplexers combine signals electronically before they are converted into optical form. *2.* A device that samples input and/or output channels and interleaves signals in frequency or time. *3.* The act of combining several signals on a common transmission path. [ARD50024]

multiplex channel(s) *1.* Multiple redundant channels, such as duplex hydraulic and quadruplex electric. [ARP4386] *2.* A single path that is capable of transferring data from multiple sources or to multiple destinations through the use of time multiplexing.

multiplexing *1.* The combining of signals from several sensors. [ARD50024] *2.* The simultaneous transmission of two or more signals within a single channel. The three basic methods of multiplexing involve the separation of signals by time division, frequency division, and phase division.

multiplex transmission *See* multiplexing.

multiplier phototubes *See* multiplier tube.

multipliers Devices that have two or more inputs and whose output is a representation of the product of the quantities represented by the input signals.

multiplier tube A phototube in which secondary emission from auxiliary electrodes produces an internally amplified output signal. Also known as a multiplier phototube or photomultiplier tube.

multipole circuit breaker A circuit breaker that has two or more poles controlled by a single-actuating member. [ARP1199A]

multiport burner A burner having a number of nozzles from which fuel and air are discharged.

multiposition action A type of controller action in which the final control element is positioned in one of three or more preset configurations, each corresponding to a definite range of values for the controlled variable.

multiprocessing *1.* Pertaining to the simultaneous execution of two or more programs or sequences of instructions by a computer or computer network. *2.* Loosely, parallel processing.

multiprogramming A computer processing method in which more than one task is in an executable state at any one time.

multiradar tracking *See* radar networks.

multirange Having two or more specific ranges of values over which an instrument or control device can be used; changing from one range to another usually involves simply repositioning a switch and does not require removing or replacing any internal parts.

multirate Refers to the condition wherein some tasks may run less than once every frame in order to optimize the system performance for a given processor. This is necessary because the program of a real-time digital computer may have several distinct tasks to perform in a real-time manner. [AIR4548]

multiskilled Describes maintenance personnel who are skilled in more than one craft.

multispectral linear arrays (MLA) Large number of interconnected solid-state detectors in a pushbroom mode. The forward motion of the vehicle (spacecraft) sweeps the assembly of detectors, which are oriented perpendicular to the ground track.

multispectral resource sampler An experimental remote-sensing instrument used by satellites to measure both intensity and polarization at several wavelengths.

multistage Occurring in a sequence of separate steps. In a multistage pump or compressor, for instance, pressure is raised by passing the working fluid through a series of impellers or pistons, each of which raise the pressure above the outlet pressure from the previous stage.

multistage rocket vehicles Vehicles having two or more rocket units, each unit firing after the one in back of it has exhausted its propellant. Normally, each unit, or stage, is jettisoned after completing its firing.

multistatic radar System in which successive lobes of the antenna are sequentially engaged to provide a tracking capability without physical movement of the antenna.

multivariable control A control system involving several measured and controlled variables, where the interdependences are considered in the calculation of the output variables. *See also* multi-element control system.

multivibrators Two-stage regenerative circuits with two possible states and an abrupt transition characteristic.

multi wheel Describes any landing gear assembly that utilizes two or more wheel assemblies. More appropriate for description of an aircraft or assembly with many wheels. [AIR1489]

"Mu" Meter An instrument, device, or apparatus for measuring the coefficient of friction of a runway surface at various conditions of damp, wet, flooded, icy, etc. [AIR1489]

Munsell chroma The dimension of the Munsell system of color corresponding most closely to saturation. Also known as chroma.

muon spin rotation Particle spin depolarization caused by sensitivity of muon spin to the presence of defects in certain metals.

mush To settle or to gain little or no altitude while flying in a semistalled condition, or at a relatively high angle of attack. [ARP4107]

mushy zones Regions of liquid-plus-solid phases in alloys that solidify over a range of temperatures.

mutual capacitance Capacitance between two conductors when all other conductors, including ground, are connected together and then regarded as an ignored ground. [ARP1931]

mutual coupling The voltage transfer between the halves of a control voltage winding when either half is energized. [ARP667]

MVA *See* minimum vectoring altitude.

MX missile United States strategic intercontinental ballistic missile.

N

N *See* newton.

NACA Acronym for National Advisory Committee for Aeronautics, the predecessor of NASA.

nacelle A separate streamlined enclosure attached to an aircraft, most often that containing the engine. [ARP4107]

naked singularities Singularities in space-time that will be visible and communicable to the outside world, i.e., singularities that are not shielded by an event horizon from infinity.

nano A prefix that means one billionth.

nanosecond (ns) One billionth of a second (10^{-9} second). [ARP1931]

nap of the earth flight (NOTE flight) An operational tactic whereby helicopters are flown as close to the surface of the earth as possible, taking advantage of natural and cultural terrain cover to escape detection or minimize exposure. The aircraft becomes hidden in the surface irregularities (nap) of the earth. [ARP4107]

nap-of-the-earth navigation (NOTE navigation) Low altitude flight of helicopters during night or day utilizing electronic means for detection and recognition of landmarks and targets.

nappe A sheet of liquid passing through the notch and falling over the weir crest.

narrow-angle diffusion A type of diffusion in which light is scattered in all directions from the diffusing medium, but the intensity is notably greater over a narrow angle in the general direction that the light would take by regular reflection or transmission. [ARP798]

narrowband Describes a frequency measurement whose frequency band of energy is smaller relative to the rest of the band.

narrow-banded Having a relatively narrow bandwidth. The opposite of broad-banded. [ARP5089]

narrowband radiation thermometer A type of temperature-measuring instrument that responds accurately only over a given relatively narrow band of wavelengths, often a band chosen to meet a special requirement of the intended application.

NASA Acronym for National Aeronautics and Space Administration.

National Airspace System (NAS) The common network of US airspace; air navigation facilities, equipment and services, airports and landing areas; aeronautical charts, information and services; rules, regulations and procedures; technical information; and manpower and material. Included are system components shared jointly with the military. [ARP4107]

national airspace system stage A/NAS stage A The en route ATC system radar, computers and computer programs, controller plan view displays (PVDs/radar scopes), input/output devices, and related communications equipment that are integrated to form the basis of the automated IFR air traffic control system. [ARP4107]

natural circulation The circulation of water in a boiler caused by differences in density.

natural contour of tire The uninflated tire cross-sectional contour. [AIR1489]

natural draft Convective flow of a gas (e.g., in a boiler, stack, or cooling tower) due to differences in density. Warm gas in the chamber rises toward the outlet, drawing in colder, more dense gas through inlets near the bottom of the chamber.

natural frequency *1.* For a spring-mass system, a quantity that is proportional to the square root of the ratio of the spring constant to the mass for each degree of freedom. [AIR1839] *2.* The frequency at which any physical object will vibrate when acted upon by an unsteady or impulsive force. Stiffer objects have higher natural frequencies.

natural gas A mixture of gaseous hydrocarbons trapped in rock formations below the earth's surface. The mixture consists chiefly of methane and ethane, with smaller amounts of other low-molecular-weight combustible gases; and sometimes noncombustible gases, such as nitrogen, carbon dioxide, helium, and H_2S (sour gas).

natural icing Icing that occurs during flight in a cloud formed by nature. [AIR1667]

natural language A language, the rules of which reflect and describe current usage rather than prescribed usage. Contrast with artificial language.

natural language (computers) A computer language whose rules reflect and describe current rather than prescribed usage. The language is often loose and ambiguous in interpretation.

natural laser *See* laser.

natural radioactivity Spontaneous radioactive decay of a naturally occurring nuclide.

nautical mile A unit of distance equal to 6076.11549 ft or 1852 m. [ARP4107]

NAVAID Contracted from navigational aid. Any visual or electronic device, either airborne or on the surface, that provides point-to-point guidance information or position data to aircraft in flight. NAVAIDs include VOR, VORTAC, and TACAN aids. [ARP4107]

NAVAID classes Classes of VOR, VORTAC, and TACAN aids, based on their operational use as follows: *T*–terminal; *L*–low altitude; *H*–high altitude. These classifications assure serviceable signals within specified distances and altitudes from the particular station emitting the navigational signal. [ARP4107]

Navier-Stokes equation The equation of motion for a viscous fluid.

navigable airspace Airspace at and above the minimum flight altitudes prescribed in FAR, including airspace needed for safe takeoff and landing. [ARP4107]

navigation The practice or art of directing the movement of a craft from one point to another. Navigation usually implies the presence of a human—a navigator—aboard the craft.

navigation technology satellites (NTS) Class of navigation satellites utilizing the global positioning system as well as a precise frequency and timing system.

NC *See* normally closed.

NDB *See* nondirectional beacon.

NDIR *See* nondispersive infrared analyzer.

NDT *See* inspection, nondestructive.

near field The region of the ultrasonic beam adjacent to the search unit, having complex beam profiles. Also known as the Fresnel zone. [ARP5089]

neat resin Resin to which nothing has been added. [AIR4844]

neat resin properties Mechanical properties of the cured resin itself, without reinforcement. [AIR4844]

nebulizer A type of atomizer that produces a uniform fine mist of medicament for inhalation, removing the larger droplets, usually by baffling, permitting only a mist of uniform droplet size to emerge. *See* atomizer. [ARP171]

neck A reduced section of pipe or tubing between sections of larger diameter, or between a pipe and a chamber.

necking The localized reduction in cross section that may occur in a material under tensile stress. [AIR4844]

needled mat A mat formed of strands cut to a short length, then felted together in a needle loom, with or without a carrier. [AIR4844]

needle valve A type of metering valve used chiefly for precisely controlling flow. The essential design feature is a slender, tapered rodlike control element that fits into a circular or conoidal seat. Operating the valve causes the rod to move into or out of the seat, gradually changing the effective cross-sectional area of the gap between the rod and its seat.

negative Voice communications word meaning "no."

negative acceleration *See* deceleration.

negative alert A corrective or preventive alert that requires the pilot to not do something to resolve a conflict; for example, "don't climb." [ARP4153]

negative damping *1.* The opposite of damping, i.e., the propagation of oscillations or disturbances; a buildup of energy with time. *2.* Landing gear vibrations induced by the sensitivity of the torque developed between the rotating and nonrotating brake parts to slip velocity. [AIR1489]

negative feedback *1.* Feedback that results in decreasing the amplification. [AIR1489] *2.* Returning part of an output signal and using it to reduce the value of an input signal.

negative G The opposite of positive G. Negative G occurs when, in a gravitational field or during an acceleration, the human body is so positioned that the force of gravity and/or inertia, acts in foot-to-head direction. On an aircraft, negative G acts to decrease or reverse the wing loading. [ARP4107]

negative-going edge The edge of a pulse going from a high to a low level.

negative ions Ions, singly or in groups, that acquire negative charges by gaining one or more electrons.

negatively skewed Describes a distribution that is not symmetric and in which the longest tail is on the left. [AIR4844]

negative plate The plate that has an electrical potential below that of the other plate during normal cell operation. [ARP4386]

negative pressure area *See* area, negative pressure.

negative yaw Rotation of an aircraft anticlockwise about the z-axis, as seen from above.

negatron(s) *1.* Negative electrons. Sometimes shortened to negatons. *2.* A negatively charged beta particle.

NEMA standards Consensus standards for electrical equipment approved by the majority of the members of the National Electrical Manufacturers Association.

neopheloscope An apparatus for making clouds in the laboratory by expanding moist air, or by condensing water vapor.

neoprene A synthetic rubber made by polymerization of chloroprene (2-chlorobutadiene-1,3). Color varies from amber to silver to cream. Neoprene exhibits excellent resistance to weathering, ozone, flames, various chemicals, and oils.

neper A unit of measure obtained by taking the natural logarithm of the scalar ratio of two voltages or two currents.

nephelometer *1.* An instrument that measures, at more than one angle, the scattering function of particles suspended in a medium. *2.* An analytical instrument for measuring the light scattering properties of a suspension. *3.* A general term for instruments that measure the degree of cloudiness or turbidity.

nephelometry The application of photometry to the measurement of the concentration of very dilute suspensions.

nephoscope An instrument for determining the direction in which clouds move.

NESA Contracted from non-electrostatically activated. Describes a particular type of aircraft window designed to prevent the buildup of ice. [ARP4107]

nest *1.* To embed a subroutine or block of data into a larger routine or block of data. *2.* To evaluate as nth degree polynomial by a particular algorithm which uses (n − 1) multiply operations and (n − 1) addition operations in succession. *See* crimp anvil.

nested laminate In reinforced plastics, a laminate formed by placing plies of fabric so that the yarns of one ply lie in the valleys between the yearns of the adjacent ply. [AIR4844]

net fan requirements The calculated operating conditions for a fan, excluding tolerances.

net heat of combustion The energy released per unit mass of fuel due to the complete oxidation of the fuel at constant pressure, as measured by cooling the products to the initial temperature without condensation of the water vapor formed in the reaction. [AIR1533]

net positive suction head (NPSH) For a pump, the minimum difference between the static pressure at the inlet to the pump and the vapor pressure of the liquid being pumped. Below the NPSH, fluid is not forced far enough into the pump inlet to be acted upon by the impeller.

net thrust The gross thrust of a jet engine, minus the drag imposed by the momentum of the incoming air. [ARP4107]

netting analysis An analysis of filament-wound structures in which it is assumed that stresses induced in the structure are carried entirely by the filaments, and the strength of the resin is neglected. It is also assumed that the filaments possess no bending or shearing stiffness, and carry only the axial tensile loads. [AIR4844]

network *1*. In data processing, any system consisting of an interconnection of computers and peripherals. Information is transferred between the devices in the network. Two types of networks are the LAN and the WAN. *LAN (local area network)*–a system at one location linked by cables. *WAN (wide area network)*–a widely dispersed system usually connected by telephone lines. *3*. In an electric or hydraulic circuit, any combination of circuit elements.

network analyzer An apparatus that contains numerous electric-circuit elements that can be readily combined to form models of electric networks.

network control The management of acquisition, routing, and switching—primarily in satellite communication.

network interface unit The hardware that interfaces the host to the data communications network. [ARD50008]

network management Refers to the facility by which network communication and devices are monitored and controlled.

network structure A type of alloy microstructure in which one phase occurs predominantly at grain boundaries, enveloping grains of a second phase.

neurotransmitters Chemical substances secreted by the terminal ends of axons, which stimulate a muscle fiber contraction or an impulse in other neurons.

neutral atmosphere An atmosphere that tends neither to oxidize nor reduce immersed materials.

neutral atoms Atoms in which the number of electrons surrounding the nucleus is equal to the number of protons in the nucleus, resulting in no net electric charge.

neutral burning The burning of a propellant grain whereby the reacting surface area remains approximately constant during combustion. [ARP4386]

neutral currents Weak interaction currents that carry zero electric charge.

neutral density filter A filter that has uniform transmission throughout the part of the spectrum in which it is used.

neutral filter A light-beam filter that exhibits constant transmittance at all wavelengths within a specified range.

neutral gases In astronomy, gas clouds of some nebulae which have not been ionized by hot stars.

neutral gear *See* gear, neutral.

neutral geometry A propellant grain configured in such a manner that the exposed surface area remains constant as burning progresses. [AIR913]

neutralization number A measure of the acidity or basicity of a fluid. [AIR1116]

neutral point Point on the titration curve at which the hydrogen ion concentration equals the hydroxyl ion concentration.

neutral static stability (of an airplane) *See* stability, neutral static.

neutrino beams Organized collections of neutrinos traveling outward from the source.

neutrinos Subatomic particles of zero (or near zero) rest mass, having no electric charge. Postulated by Fermi (1934) in order to explain apparent contradictions in the law of conservation of energy in beta particle emission.

neutron A nuclear particle that has a mass number of one and exhibits zero (neutral) charge.

neutron flux *See* flux (rate).

neutron radiography *1*. Nondestructive testing and inspection utilizing neutron beams from nuclear reactors, particle accelerators, and/or radioisotopes. 2. Imagery displaying structural defects, utilizing neutron image recorders or screens.

neutrons Subatomic particles with no electric charge, and mass of 1.67482×10^{-24} g.

never-exceed speed *See* V_{ne}.

new surplus New, unused parts produced by the original equipment manufacturer, purchased from an end user as surplus to their requirements. [AS7104]

newton *1.* Abbreviated N. A unit of force in the SI system; the force that gives to a mass of 1 kg an acceleration of 1 m/s^2. *2.* Metric unit for force.

Newtonian flow Flow that adheres to the linear relation between shear stress, viscosity, and velocity distribution.

Newtonian fluids Fluids whose viscosities are shear-independent and time-independent. The shear rate of a Newtonian fluid is directly proportional to the shear stress. [AIR4737]

next higher level effect The consequences that a failure mode has on the operation, functions, or status of the item in the next higher indenture level, above the indenture level under consideration. [AIR4896]

nibbling Contour cutting of sheet metal by a rapidly reciprocating punch, which makes numerous, successive small cuts.

Nichols plot A plot of the logarithm of the output-to-input magnitude ratio against the phase angle, as the frequency of a sinusoidal input is varied. [AIR1823]

nickel A metal that offers a combination of corrosion resistance, formability, and tough physical properties; thus used for alloying purposes and as a coating for copper. [ARP1931]

nickel iron batteries Alkaline-type electric cells in which the electrolyte is potassium hydroxide, the anodes are of a steel wool substrate with active iron material, and the cathodes are of nickel plated steel wool substrate with active nickel material.

night vision goggle compatible lighting Describes a lighting condition in which the spectral wavelengths, luminance level, and uniformity of the cockpit lighting do not interfere with the operation of night vision goggles. [ARP4168]

night vision imaging system (NVIS) A system in which image-intensifier tubes are used to provide an enhanced image of a scene in visible lighting conditions too low for normal visibility. [ARP4392]

nine-light indicator A remote indicator used in conjunction with a contact anemometer and a wind vane. Consists of a center lamp surrounded by eight lamps, equally spaced and labeled to indicate compass points. Wind speed is indicated by the number of flashes the center lamp makes in a certain time interval; wind direction, by the position of an illuminated lamp in the outer ring.

nipple A short piece of pipe or tube, usually with an external thread at each end.

nipple fitting A fitting generally molded entirely of rubber, but may contain fabric. Consists of a tubular section whose inside diameter is sized to match standard aluminum tubing. [AIR1664]

NIST Acronym for National Institute of Standards and Technology. [AS7101]

nitinol alloys Shape-memory alloys of titanium and nickel.

nitrogen A colorless, odorless gas that does not sustain higher forms of life or combustion. Also available in liquid form at -195.6°C (-320°F), which vaporizes into gaseous nitrogen. [ARP171]

nitrogen lasers Stimulated emission devices in which the nitrogen molecule is the lasing medium.

NMCM *See* not mission capable maintenance.

NMCS *See* not mission capable supply.

NO *See* normally open.

no back devices Mechanical, one-way mechanisms that permit power to flow in one direction only (from the prime mover toward the load). [ARP4386]

no bearing traffic A resolution advisory or traffic advisory generated by traffic from which no directional information can be derived. [AIR4102/10]

noble gases *See* rare gases.

noble metal thermocouple A thermocouple whose elements are made of platinum (Pt) or platinum-rhodium (Pt-Rh alloys), and that resist oxidation and corrosion at temperatures up to about 1550°C (2800°F); three standard alloy pairs are in common use: Pt vs Pt-10%Rh; Pt vs Pt-13%Rh; and Pt-6%Rh vs Pt-30%Rh.

noctilucent clouds Clouds of unknown composition which occur at great heights, 75 to 90 kilometers. Noctilucent clouds resemble thin cirrus clouds, but usually with a bluish or silverish color, although sometimes orange to red, standing out against a dark night sky. Sometimes called luminous clouds.

node bonds The area of the honeycomb core where the cell walls are adhesively bonded. [AIR4844]

nodes (standing waves) Points, lines, or surfaces in standing waves where some characteristic of the wave field has essentially zero amplitude.

nodular cast iron *See* ductile iron.

nodular iron *See* ductile iron.

NOE navigation *See* nap-of-the-earth navigation.

NOESS Acronym for the National Operational Environmental Satellite System.

no-flow valve pressure gain Ratio of change in controlled pressure to the corresponding change of a controlling variable at no flow. [ARP4386]

noise *1.* Any undesired signal that tends to interfere with normal reception or processing of the desired signal. Origin may be electrical or from small reflectors in a material. [ARP5089] *2.* Random variation in analyzer output not associated with characteristics of the sample, to which the analyzer is responding, and which is distinguishable from analyzer drift characteristics. [ARP1256] *3.* Meaningless stray signals in a control system similar to radio static. Some types of noise interfere with the correctness of an output signal.

noise and vibration General term that covers such conditions as tire and wheel unbalanced vibration and tire-tread pattern resonance and noise. [AIR1489]

noise equivalent power The amount of optical power that must be incident on a detector to produce an electrical signal equal to the rms level of noise inherent in the detector. Noise equivalent power is a measure of the sensitivity of the sensor.

noise factor In an electronic circuit, the ratio of total noise in the output signal to the portion thereof in the input signal under the following conditions: (a) a selected input frequency and its corresponding output frequency; (b) an input termination whose noise temperature is a standard 290 K at all frequen-

cies; and (c) a linear system, and noise expressed as power per unit bandwidth.

noise figure A calculated or measured mathematical figure that denotes the inherent noise in a unit, system, or link.

noise immunity Refers to the ability of a device to discern valid data in the presence of noise.

noise pollution Objectionable or harmful levels of noise.

noise prediction Estimation of intensity and frequencies based on analyses of probable oscillation of vibration-producing components.

noise prediction (aircraft) Estimating or forecasting of aircraft noise.

noise quantization Inherent noise that results from the quantization process.

noise temperature At a pair of terminals and at a specified frequency, the temperature of a passive system exhibiting the same noise power per unit bandwidth as the actual terminals.

no-light Failure of an engine to ignite and run on a start attempt for whatever reason. [ARP906A]

no-load cycle The operation of a unit without load on the output shaft. [AS1606]

no-load flow *See* control flow.

no-load pressure Pressure required to maintain a system at a specified speed with no external load. [ARP4386]

no-load speed The speed of the shaft rotation with no external load applied to the shaft and rated voltage and frequency of proper phase relationship applied to both motor winding and energizing voltage winding. [ARP667]

no-load valve flow gain The increment of valve flow per change in valve input variable at no load across the output. [ARP4386]

no-load velocity The output velocity of the actuation system measured with no external load. Maximum no-load velocity is determined with a saturation input to the power modulator. [ARP4386]

nol ring A parallel filament- or tape-wound hoop test specimen developed by the Naval Ordnance Laboratory for measuring various mechanical strength properties of a material, such as tension and compression, by testing the entire ring or segments of it. [AIR4844]

nomex Aramid fiber or paper. The paper form is used to make honeycomb. [AIR4844]

nominal bandwidth For an acoustic, electric, or optical filter, the difference between the upper and lower nominal cutoff frequencies.

nominal pressure The general pressure setting of the system. [ARP4386]

nominal pressure, operating pressure The maximum steady working pressure to which a fitting assembly or component may be subjected. The basic operating pressure without regard to operating pressure variations. [MA2005]

nominal size *1.* The standard dimension closest to the central value of a toleranced dimension. *2.* A size used for general identification.

nominal specimen thickness The nominal ply thickness multiplied by the number of plies. [AIR4844]

nominal surface The imaginary true surface that would result if all surface irregularities (peaks, waves, ridges, and hollows) were leveled off to zero value; or were non-existent. It is this nominal surface or "mean line" from which the surface irregularities deviate. [AS291D]

nominal value A value assigned for the purpose of a convenient designation. [AIR4844]

nominal voltage The midpoint voltage observed across a battery during discharge at a selected rate, usually 0.2 or 0.1 C-rate. [ARP4386]

nominal voltage rating The root-mean-square line-to-neutral and line-to-line voltage at which the alternator is rated. [ARP1148A]

nomograms *See* nomographs.

nomographs On charts or graphs, lines of constant value of given quantities with respect to either space or time. Also called nomograms.

nonactive maintenance time The time during which no maintenance can be accomplished on the item because of administrative or logistic reasons. [ARD50010]

nonadiabatic conditions In thermodynamics, changes in volume, temperature, flow, etc. accompanied by a transfer of heat.

non-blackbody Describes the thermal emittance of real objects, which emit less radiation than blackbodies at the same temperature, which may reflect radiant energy from other sources, and whose emitted radiation may be modified by passing through the medium between the body and a temperature-measuring instrument.

noncondensable gas The portion of a gas mixture (such as vapor from a chemical processing unit or exhaust steam from a turbine) that is not easily condensed by cooling. Normally consists of elements or compounds that have very low, often subzero, boiling points and vapor pressures.

nonconforming material Contract material that has physical, functional, and/or dimensional variation or deviation from the applicable specification; or is incomplete. [AS7200/1]

noncontact gaging A method of determining physical dimensions without actual contact between the measuring device and the object.

noncontacting tachometer Any of several devices for measuring rotational speed without physical contact between the sensor and the rotating element; for example, stroboscopes or eddy-current tachometers.

non-containment failure Any failure that results in the escape of rotor fragments through the nacelle cowling or through panels which isolate the propulsion installation from the remainder of the aircraft structure. [AIR1537]

noncritical dimension Any dimension that can be altered without affecting the basic function of the device.

noncritical viewing sectors Those geometric sectors of an instrument excluding the critical sectors. [ARP1161]

nondestructive inspection A process or procedures, such as ultrasonic or radiographic inspection, for determining the quality or characteristics of a material, part, or assembly, without permanently altering the subject or its properties. [AIR4844]

non-destructive read out Describes a method of reading from memory whereby the stored value is left intact by the reading process. Plated wire and modern semiconductor random-access memory (RAM) are examples of NDRO memory.

nondestructive testing Any testing method that does not damage or destroy the sample.

Usually consists of stimulating the sample with electricity, magnetism, electromagnetic radiation or ultrasound, and measuring the response of the sample.

non-detectable failure A failure that, upon occurrence, is not recognized by the failure-detection scheme(s) of a fault-tolerant flight control system. Unless stated otherwise, a flight control system must maintain its fail-operative status after the occurrence of a non-detectable failure. [ARP4386]

nondirectional beacon (NDB)/radio beacon A low- to middle-frequency (L/MF) or ultra-high frequency (UHF) radio beacon transmitting nondirectional signals whereby the pilot of an aircraft equipped with direction-finding equipment can determine his/her bearing to or from the radio beacon and "home" on, or track to or from, the station. [ARP4107]

nondispersive infrared analyzer (NDIR) An analyzer that, by absorption of infrared energy, selectively measures specific components. [ARP1256]

non ESS failures Any of the following types of failures: (a) failures directly attributable to improper installation in the test facility; (b) failures of test instrumentation or monitoring equipment (other than the BIT function), except where it is part of the delivered item; (c) failures resulting from test operator error in setting up or in testing the equipment; (d) failures attributable to an error in or interpretation of the test procedures; (e) dependent failures; (f) failures occurring during repair; or (g) failures clearly attributable to the environmental generation test equipment overstress condition. [ARD50010]

non-expendable thermocouple A thermocouple that is not covered with fabric or plastic insulation. [AMS2750]

non-hazardous area *See* safe area.

nonhygroscopic Lacking the property of absorbing and retaining an appreciable quantity of moisture from the air. [AIR4844]

noninherent failure causes Failures due to external factors not associated with intrinsic design, such as foreign object damage, corrosion, material defects, and maintenance error. [AIR1872]

noninteracting control system A multi-element control system designed to avoid disturbances to other controlled variables due to the process input adjustments made for the purpose of controlling a particular process variable.

nonintrusive Describes a sensor that is flush with the aerodynamic surface, causing no disturbance to the flow field. [AIR4367]

nonisothermal process In thermodynamics, compression or expansion of substances at nonuniform temperatures.

nonisotropy *See* anisotropy.

nonlinear distortion A departure from a desired linear relationship between corresponding input and output signals of a system.

nonlinear effects Optical interactions that are proportional to the square or higher powers of electromagnetic field intensities. Nonlinear effects generate harmonics of optical frequencies, and sum and difference frequencies when two lightwaves are mixed.

nonlinearity A type of error in an FM system whereby the input to a device does not relate to the output in a linear manner. *See* linearity.

nonlinear optics Study of the interaction of radiation with matter in which certain variables describing the response of the matter are not proportional to variables describing the radiation.

nonlinear optimization *See* nonlinear programming.

nonlinear programming In operations research, a procedure for locating the maximum or minimum of a function of variables that are subject to constraints, when either the function and/or the constraints are nonlinear. Synonymous with nonlinear optimization. Contrast with convex programming and dynamic programming.

nonlinear system Any system whose operation cannot be represented by a finite set of linear differential equations.

nonliquefied compressed gas A gas that: (a) is not in solution; (b) is entirely gaseous at a temperature of 70°F (21°C), under the charging pressure; and (c) is maintained in the liquid state at absolute pressure by maintaining the gas at a temperature less than 70°F (21°C). [ARP4386]

nonlocking Pertaining to code extension characters that change the interpretation of one or a specified number of characters. Contrast with locking.

nonmetallic inclusions Insoluble impurities, such as oxides, aluminates, sulfides, and silicates, that are trapped mechanically, or are formed during solidification. [ARP1917]

nonmission effects Failure effects that do not prevent mission success or equipment operation, but whose correction may or may not be economically desirable due to added labor and material cost for repair (including loss-of-use cost due to maintenance downtime). [ARD50010]

non-modulating A level-sensing device that provides controlled changes in response to predetermined liquid levels or has built-in hysteresis such that it opens and closes at significantly different liquid levels. [AIR1660A]

non-Newtonian fluids Fluids whose viscosities are shear- and time-dependent, and whose shear rate is not directly proportional to its shear stress. [AIR4737]

non-operating Describes the condition of a galley that is loaded with equipment and inserts, and is not used, but remains installed in the airplane on the ground, with water system filled, but without air conditioning or power, when the aircraft is exposed outdoors with cabin closed for a period of time of two hours or more duration. [AS1426]

nonoperating components Those items, such as heat exchangers and ducting, that do not contain moving parts or electronics. [AIR4073]

nonperception *See* perception.

nonpoint sources Undetermined or general areas from which pollutants, contaminants, and/or other unwanted materials or wastes enter the environment.

nonprecision approach procedure/ nonprecision approach A standard instrument approach procedure in which no electronic glide slope is provided; for example, VOR, TACAN, NDB, LOC, ASR, LDA, or SDF approaches. [ARP4107]

nonprocessor request The system for accomplishing data transfers between two devices without involving the CPU.

nonradar approach control An air-traffic-controlled (ATC) facility providing approach control service without the use of radar. Under this condition, the controller uses radio communications to keep track of the aircraft on approach to the airfield and issues further clearances to aircraft to approach the field based upon his/her knowledge of the location of the other aircraft as determined from these radio communications. [ARP4107]

nonreclosing pressure relief device A device for relieving internal pressure which remains open when actuated and must be replaced or reset before it can actuate again.

nonrelevant indication Signal received during an inspection which are not caused by the flaw or other condition sought. [ARP5089]

non-return-to-zero (NRZ) Coding of digital data for serial transmission or storage whereby a logic ONE is represented by one signal level and a logic ZERO is represented by a different signal level.

nonreturn valve A valve that prevents flow of fluid in one direction and permits the free flow of fluid in the opposite direction. [AIR4783]

non-scheduled maintenance An urgent need for repair or upkeep that was unpredicted or not previously planned, and must be added to or substituted for previously planned work.

nonself-sealing tank A tank having no ballistic sealing properties; however, it may possess tear-resistant characteristics. [AIR1664]

nonseparable filter A filter that is permanently encased in its housing. This is usually done to reduce potential leak paths or to save weight. [ARP4386]

nonstandard parts, materials, and processes Those nonstandard parts, materials, and processes that are not covered by federal or military specifications and standards or referenced DOD-adopted industry standards. [AIR4896]

nontransfer cycle A bus cycle immediately following a bus cycle in which a valid wait is asserted or one of the VZ, HZ, or DZ cycles is specified in the protocol. [AS4710]

nontransferred arc In arc welding and cutting, an arc sustained between the electrode and a constricting nozzle, rather than between the electrode and the work.

nonvolatile matter (NVM) test A test for the percentage of nonvolatile matter in a sealant. Generally, volatile solvents are used to produce the desired amount of flow in brushable and sprayable sealants in particular. [AS7200/1]

nonvolatile memory Computer memory that retains data when power is removed.

nonwoven fabric A planar textile structure produced by loosely compressing together fibers, yarns, rovings, etc., with or without a scrim cloth carrier. [AIR4844]

NORDO Contracted from no radio. A term used to indicate that an aircraft is unable to communicate because of radio failure. [ARP4107]

no response A condition that occurs if a bus controller determines that there has been no response to a command. [AS4113]

normal axis *See* vertical axis.

normal control pressure curve The locus of midpoints of the complete cycle control pressure curve. This locus is the zero hysteresis control pressure curve. Usually, valve hysteresis is sufficiently low that one side of the control pressure curve can be used for a normal control pressure curve. [ARP490]

normal electric-system operation All the functional electric-system operations required for aircraft operation, aircraft mission, and electric-system controlled continuity. These operations occur at any given instant and any number of times during ground operation, flight preparation, takeoff, airborne conditions, landing, and anchoring. Examples of such operations are switching of utilization equipment loads, engine speed changes, bus switching and synchronization, and paralleling of electric power sources. [AS1212]

normal flow gain The slope of a straight line drawn from the zero flow point of the normal flow curve, throughout the range of rated current of one polarity; drawn to minimize deviations of the normal flow curve from the straight line. [ARP4386]

normal fluid temperature The stabilized fluid temperature normally reached during continuous operation. [ARP4386]

normal input/output curve The locus of the midpoints of a complete input/output curve. This locus is the zero hysteresis output curve. [ARP4386]

normal input/output gain The slope of the normal input/output curve in units of output/input. [ARP4386]

normality Concentration units defined as the number of gram-ions of replaceable hydrogen or hydroxyl groups per liter of solution. A shorter notation of gram-equivalents per liter is frequently used.

normalize *1.* In programming, to adjust the exponent and fraction of a floating-point quantity such that the fraction lies in a prescribed normal standard range. *2.* In mathematical operations, to reduce a set of symbols or numbers to a normal or standard form. Synonymous with standardize. *3.* In heat treating ferrous alloys, to heat 50–100°F above the upper transformation temperature, then cool in still air.

normalized stress Stress calculated by multiplying the raw stress value by the ratio of measured fiber volume to the nominal fiber volume. This ratio is often approximated by the ratio of the measured specimen thickness to the nominal specimen thickness. [AIR4844]

normalizing Heating to a suitable temperature above the austenite transformation temperature, then air cooling to develop a uniform structure. [AMS2759B]

normally aspirated Describes an engine that takes its combustion air directly from the atmosphere without the benefit of mechanical devices to increase the pressure of the air. *See also* supercharger. [ARP4107]

normally closed (NC) *1.* A switch position in which the usual arrangement of contacts permits the flow of electricity in the circuit. *2.* In a solenoid valve, an arrangement whereby the disk or plug is seated when the solenoid is deenergized. *3.* A field contact that is closed for a normal process condition and open when the process condition is abnormal. *4.* Describes a valve with means provided to move to and/ or hold in its closed position without actuator energy supply.

normally open (NO) *1.* A switch position in which the usual arrangement of contacts provides an open circuit (no current flowing). *2.* In a solenoid valve, an arrangement whereby the disk or plug is seated when the solenoid is energized. *3.* A field contact that is open for normal process condition and closed when the process conditions are abnormal. *4.* Describes a valve with means provided to move to and/ or hold in its wide-open position without actuator energy supply.

normal mode voltage An extraneous voltage induced across the circuit path (transverse mode voltage).

normal opening fuse (fast-acting fuse) A fuse that opens a circuit without deliberate time-delay. [ARP1199A]

normal operating pressure The pressure or force required to produce an aircraft deceleration in accordance with the procurement document. [ARP1493]

normal operation Describes intrinsically safe electrical apparatus or associated electrical apparatus that complies electrically and mechanically with the requirements of its design specification and is used within the limits specified by the manufacturer.

normal output curve The locus of the midpoints of a complete input/output curve. This locus is the zero-hysteresis output curve. [ARP4386]

normal output gain The slope of the normal output curve in units of output/input. [ARP4386]

normal parking pressure The pressure or force required to lock a wheel at a load equal to the rated static load specified for the wheel. [ARP1493]

normal pressure gain The slope of a straight line drawn to minimize deviations of the normal control pressure curve, throughout the range of the curve in which control pressure varies with input current. The minimum deviation line is defined as the line whose greatest negative deviation (delta i) is equal to its greatest positive deviation. [ARP490]

normal rated power The maximum horsepower or thrust that an engine can deliver for a protracted period of time without damage; specified by the manufacturer or other qualified authority. [ARP4107]

normal stress The stress component that is perpendicular to the plane on which the forces act. [AIR4844]

Normal Thermometric Scale The first international standard temperature scale, adopted in 1887, which was based on the fundamental interval of 100° between the ice point of pure water and the condensing point of pure water vapor.

normative Prescribing or directing a norm or standard; used in standards to indicate text that poses requirements. [AS4893]

North Polar Spur (astronomy) One of the largest sources of diffuse radio emission outside the galactic plane. The Spur, a ridge of enhanced emission, may be the remnant of the shells of supernovae that exploded over 100,000 years ago.

North skylight illumination *See* illuminant C.

nose cone *1.* The cone-shaped leading end of a rocket vehicle, consisting of: (a) chambers in which satellites, instruments, animals, plants, or auxillary equipment may be carried; and (b) outer surfaces built to withstand high temperatures generated by aerodynamic heating. *2. See* warheads.

nose gear *See* gear, nose.

nose gear launch Launch of an aircraft by means of an apparatus including a launch bar attached to the nose gear, imparting tow loads to the aircraft through the nose gear system. [AIR1489]

noseheavy For an aircraft, describes a condition in which the nose tends to sink when the longitudinal control is released in any given attitude of normal flight. [ARP4107]

nose tips The foremost sharp points of bombs, rockets, missiles, and other symmetrical bodies.

nose tire An auxiliary tire that is mounted on the front of the aircraft. [AS4833]

nose wheel A wheel(s) designed for use on nose gears. Nose wheels differ from main wheels in that they usually make no provision for brakes within the wheel. They are also designed to take the dynamic vertical loads resulting from brake drag loads applied at the main wheels. [AIR1489]

nosewheel cycle (full cycle) That nosewheel motion from the center position to one extreme position, returning past the center position to the opposite extreme, and returning again to the center position. [ARP1595]

nosewheel, electrically controlled, electrically powered A nosewheel steering system with all-electrical servos; usually limited to low-speed (taxi) control, with notable exceptions on some executive-type aircraft. [ARP1595]

nosewheel, electrically controlled, hydraulically powered A nosewheel steering system that is inherently less reliable than those with mechanically controlled servos; redundancy and/or failure detection features are normally required. Such systems are now in common use on military combat aircraft and some business jet aircraft. [ARP1595]

nosewheel, mechanically controlled, hydraulically powered A nosewheel steering system with a hydro-mechanical position servo mechanism; commonly used on transport and business jet aircraft and some smaller military aircraft. Normally very reliable, but dependent on aircraft hydraulic system power. [ARP1595]

nosewheel, mechanically controlled, manually powered A nosewheel steering system that is highly reliable, independent of aircraft power generation; normally employed on small, general-aviation aircraft. [ARP1595]

nosewheel, powered steering angle The number of degrees (radians) that the nosewheel is displaced from the center (straight ahead) position while powered by the NWS control system. [ARP1595]

nosewheel rated load The maximum load needed to fulfill the output torque requirement. [ARP1595]

nose wheel steering (NWS) The system, apparatus, and/or controls for maneuvering the aircraft on the ground by rotating the nose wheels about an axis normal to the ground. [AIR1489]

nosewheel, steering rate The rate of change of nosewheel steering angle in degrees (or radians) per second. [ARP1595]

nosewheel, steering ratio Describes the relationship between input control movement and output nosewheel angle change (e.g., degrees nosewheel angle per inch of rudder pedal travel). [ARP1595]

NOTAM Contracted from notices to airmen. Notices containing information (not known sufficiently in advance to publicize by other means) concerning the establishment of, condition of, or change in any component of the National Airspace System, the timely knowledge of which is essential to personnel concerned with flight operations. These notices may be disseminated by teletypewriter or by voice. [ARP4107]

not-at-intermediate position A position that is either above or below the specified intermediate position.

notation *1.* The act, process, or method of representing facts or quantities by a system or set of marks, signs, figures, or characters. *2.* A system of such symbols or abbreviations used to express technical facts or quantities, for example, mathematical notation. *3.* An annotation.

notch *1.* A V-shaped indentation in an edge or surface. *2.* An indentation of any shape that acts as a severe stress raiser.

notched specimen A test specimen that has been deliberately cut or notched, usually in a V-shape, to induce and locate point of failure. [AIR4844]

notch factor Ratio of the resilience determined on a plain specimen to the resilience determined on a notched specimen. [AIR4844]

notching Cutting out various shapes from the edge of a metal strip, blank, or part.

notch sensitivity A measure of the sensitivity of a material to the presence of stress concentration caused by notches in the form of threads or grooves, scratches, and other stress raisers. Materials with low notch sensitivity are selected for the high-fatigue/high-stress applications in aircraft primary structure and in aircraft engines. [ARP700]

notch width The horizontal distance between opposite sides of a weir notch.

not-closed position A position that is more than zero-percent open. A device that is not closed may or may not be open.

NOTE *See* nap-of-the-earth flight and nap-of-the-earth navigation.

noticing conditions Generally, either operating in above-freezing conditions or in clear air; for engine inlets, operating at a temperature about 10°C (50°F). [AIR4367]

not mission capable maintenance (NMCM) A material condition indicating that systems are not capable of performing any of their assigned missions because of unit-level maintenance requirements. Recording of NMCM time begins with: (a) unscheduled maintenance, when a malfunction is discovered, or at mission completion, whichever is later; and (b) scheduled maintenance, when the determination is made that a system cannot be returned to mission-capable status within two hours. Recording of NMCM time stops when maintenance has been completed or is interrupted by work stoppage due to supply shortage. The period of work stoppage due to supply is measured as NMCS. NMCM time resumes when required supply items are delivered to the maintenance activity. [ARD50010]

not mission capable supply (NMCS) A material condition indicating that systems and equipment are not capable of performing any of their assigned missions because of maintenance work stoppage due to a supply shortage. Recording of NMCS time begins when work stoppage results from lack of parts, and the NMCS requisition is not satisfied one hour after the demand is initiated and remains unsatisfied. For Army and Marine Corps ground equipment, when both NMCM time and NMCS time are encountered in the same day and the sum is more than 12 hours, the whole day is carried against the condition status with the most hours. [ARD50010]

not-open position A position that is less than 100 percent open. A device that is not open may or may not be closed.

not operating time The element of uptime during which an item is not required to operate. [ARD50010]

novelty Describes systems using new technology and systems using a conventional technology not previously used in connection with the particular function in question. [ARP4761]

nowcasting A self-contained short-period meteorological forecast for the immediate future covering a period of up to six hours.

NOx Oxides of nitrogen; specifically, the sum of nitric oxide (NO) and nitrogen dioxide (NO_2). [AIR1533]

nozzle A device used to control the flow of air, gas, or liquids. May be designed with a constricting throat section and/or a divergent sections. [ARP480A]

nozzle area, effective A theoretical area of a nozzle throat, equal to the product of the geometric area of the nozzle throat and a dimensionless nozzle discharge coefficient. [ARP147C]

nozzle area, geometric The flow area in the throat of a nozzle determined by its physical measurements. [ARP147C]

nozzle efficiency The efficiency with which a nozzle converts potential energy into kinetic energy, commonly expressed as the ratio of the actual change in kinetic energy to the ideal change at the given pressure ratio.

nozzle/flapper A fundamental part of pneumatic-signal processing and pneumatic-control operations. Basically, a displacement of the flapper is converted to a pressure signal.

nozzle pressure drop The pressure drop across the nozzle and an aircraft refueling adapter. [AIR4783]

NPSH *See* net positive suction head.

NRZ *See* non-return-to-zero.

ns *See* nanosecond.

NTPD Stands for normal temperature and pressure, dry. Conditions comprising a temperature of 21.1°C (70°F) and an absolute pressure of 101.3 kPa (760 mm of Hg) on a dry basis (with partial pressure of water vapor equal to zero). [AS1304]

NTS *See* navigation technology satellites.

NTSB Acronym for National Transportation Safety Board.

nuclear *1*. Pertaining to or relating to a nucleus. *2*. Having the character of a nucleus. *3*. Using nuclear energy.

nuclear emulsion(s) Very thick photographic emulsions used in the study of cosmic rays and other energetic particles. The paths of the particles through the thick emulsion are recorded in three dimensions.

nuclear fluorescence thickness gage A device for determining the weight of an applied coating by exciting the coated material with gamma rays and measuring low-energy fluorescent radiation that results.

nuclear power systems Power systems of which there are two basic types: those powered by radioisotopes, and those powered by a nuclear reactor. These two types of nuclear power systems differ only in the heat power source, with power-conversion subsystems and power-regulation subsystems being similar for both types. The conversion of heat power to electrical power is accomplished using thermoelectric, thermionic, or dynamic power conversion devices. Only the radioisotope thermoelectric generators have seen practical application. [AIR744A]

nuclear pumped lasers Lasers in which the excitation is supplied by a nuclear reactor as a high-flux source, or by the kinetic energy of the fission fragments only.

nuclear pumping Laser-like pumping produced by electrons generated in nuclear reactions; or, in general, by beams of charged particles.

nuclear radiation Corpuscular emissions, such as alpha and beta particles, or electromagnetic radiation, such as gamma rays, originating in the nucleus of the atom.

nuclear rocket engines Rocket engines in which nuclear reactors are used as power sources, or as sources of thermal energy.

nuclear vulnerability Refers to the resistance of structures or materials to nuclear radiation or explosions.

nuclei *See* nucleus.

nucleons In the classification of subatomic particles according to mass, the second heaviest type of particles; the mass is intermediate between that of the meson and the hyperon.

nucleus *1*. Plural is nuclei. The positively charged core of an atom, which contains almost all of the mass of the atom but occupies only a small fraction of its volume. *2*. A number of atoms or molecules bound together with interatomic forces sufficiently strong to make a small particle of a new phase stable in a mass otherwise consisting of another phase; creating a stable nucleus is the first step in phase transformation by a nucleation-and-growth process. *3*. That portion of a control program that must always be present in main storage. *4*. The main storage area used by the nucleus and other transient control program routines.

nucleus counter An instrument that measures the number of condensation nuclei or ice nuclei in a sample volume of air.

nuclide(s) *1*. Individual atoms of a given atomic number Z and mass number A. *2*. A species of atom characterized by a unique combination of charge, mass number, and quantum state of its nucleus.

nude vacuum gage A hot-filament ionization gage mounted entirely within the vacuum system whose pressure is being measured.

nuisance advisory A resolution advisory or traffic advisory given in accordance with the design criteria, but when the threat of a potential collision does not exist. [AIR4102/10]

nuisance alert *1*. An alert that is given when an aircraft is in the TCAS operational envelope and a maneuver by the TCAS aircraft is not necessary to achieve satisfactory aircraft separation. [ARP4153] *2*. An alert that occurs at too low a level of windshear to warrant a windshear alert. [ARP4109]

null *1*. For a four-way valve, the condition wherein the valve supplies zero control pressure with zero flow to the load. Three-way valves do not have an equivalent null condition; however, it is often convenient to specify an operating point from which "null bias" and "null shift" can be specified. *Null shift*– if applied to a three-way valve, means a change in the null bias. *Null bias*–is the input current required to produce the specified operating point. [ARP490] *2*. A situation of no input to the controller. *3*. A condition, such as of balance, that results in a minimum absolute value of output.

null balance indicator A cockpit indicator that provides a visual indication by means of a potentiometer circuit whose accuracy is not affected by external resistance. [ARP485]

null-balance recorder An instrument that records a measured value by means of a pen or printer attached to a motor-driven slide; the position of the slide is determined by continuously balancing current or voltage in the measuring circuit against current or voltage from a sensing element.

null bias The input current required to bring a valve to null, excluding the effects of valve hysteresis; expressed as a percent of rated current. *See also* null. [ARP490]

null detector *See* null indicator.

null indicator An indicating device, such as a galvanometer, used to determine when voltage or current in a circuit is zero. Used chiefly in balancing bridge circuits. Also known as null detector.

null leakage Internal leakage of a valve when output flow is negligible. [ARP4386]

null pressure The pressure existing at both control ports at null, expressed in psi (or kPa). [ARP4386]

null pressure gain The change of load differential pressure at zero-load flow with input current; measured at the null point. [ARP4386]

null region The region about null, where effects of lap in the output stage predominate. [ARP490]

null shift *1*. In a servovalve, a change in null bias, expressed as percent of rated current. Null shift may occur with changes in supply pressure, temperature, and other operating conditions. *See also* null. *2*. In an LVDT, a change or shift in the null position, usually due to temperature, which causes linear expansion of the materials. [ARP4386] *3*. A shift, for whatever reason, of the input/output relationship with respect to reference coordinate axes. [ARP1281A]

null spacing errors The deviation between the two minimum voltage positions of an output winding, expressed in angular units from 180 deg. [ARP826]

null voltage The minimum differential output voltage that can be obtained with the armature positioned at a balanced center or biased null position of the stroke. [ARP4386]

number *1*. A mathematical entity that may indicate quantity or amount of units. *2*. Loosely, a numeral. *See also* binary number and random numbers.

numbers, the *1*. Digits written on the approach end of a runway to identify the orientation of the runway with respect to magnetic north, to the nearest even 10 deg, with the last zero deleted. For example, 02 painted on the approach end of a runway indicates that the magnetic orientation of the runway is approximately 020 deg and that an aircraft on final approach to that runway should be heading approximately 020 deg magnetic. *2*. An expression used to indicate the receipt of the current ATIS broadcast. For example, if a pilot tells the controller that he/she has "the numbers," it signifies that he/she is aware of the current ATIS information being broadcast for the destination airfield. [ARP4107]

number system A systematic method for representing numerical quantities, in which any quantity is represented as the sequence of coefficients of the successive powers of a particular base, with an appropriate point. Each succeeding coefficient from right to left is associated with and usually multiplies the next higher power of the base. The binary (base 2), octal or octonary (base 8), decimal (base 10), and sexadecimal or hexadecimal (base 16) systems are widely used in computers.

numerical analysis The study of methods of obtaining useful quantitative solutions to mathematical problems, regardless of whether an analytic solution exists or not; and the study of the errors and bounds on errors in obtaining such solutions.

numerical aperture The sine of the half-angle over which an optical fiber or optical system can accept light rays, multiplied by the index of refraction of the medium containing the rays.

numerical control Automatic control of a process performed by a device that makes use of all or part of numerical data generally introduced as the operation is in process.

numerical differentiation Approximate estimation of a derivative of a function by numerical techniques.

numerical readout The arithmetic number indicated by an indicating device without any physical size or unit. [ARP587

numeric word A word consisting of digits and possibly space characters and special characters.

Nusselt number A number expressing the ratio of convective to conductive heat transfer between a solid boundary and a moving fluid, defined as hI/k, where h is the heat transfer coefficient, I is the characteristic length, and k is the thermal conductivity of the fluid. Named after Wilhelm Nusselt, a German engineer.

nutating-disk flowmeter A type of positive-displacement flowmeter in which the advancing volume of fluid causes a measuring disk to wobble (nutate), thereby passing a precise volume of fluid through the meter with each revolution of the disk.

nutation *1.* The oscillation of the axis of any rotating body, as a gyroscope rotor. Specifically, in astronomy, irregularities in the precessional motion of the equinoxes because of varying positions of the moon and, to a lesser extent, of other celestial bodies with respect to the ecliptic. *2.* Rocking back and forth, or periodically repeating a circular, elliptical, conical, or spiral path, usually involving relatively small degrees of motion.

nut end The threaded end of the stud on which a nut is normally assembled to retain a cover or other part of the component assembly. [MAP1670]

N-value The exponent in the power function $V(T) = KT^N$, which is the calibration function for a ratio thermometer. The N-value and mean effective wavelength can be used to express operating characteristics of a given ratio thermometer.

NVIS *See* night vision imaging system.

NVIS compatible exterior lighting Lighting compatible with night vision imaging system optics, achieved when the level of radiance incident on the NVIS from the signal source, or the reflected energy from the light source, is sufficient to produce the information desired on the NVIS screen, but not great enough to cause significant reduction in gain of the NVIS image intensifier and loss of image contrast. [ARP4392]

NVIS radiant intensity The amount of energy emitted by a night vision imaging system light source that is visible with a night vision imaging system. [ARP4392]

NVM *See* nonvolatile matter test.

NWS *See* nose wheel steering.

Nyquist frequency One-half of the sampling frequency in a sampled data system.

Nyquist plot A plot that provides magnitudes and phase relationships (in polar form) between system input and response for any excitation frequency. [AIR1823]

467

O

OAT *See* outside air temperature.

object color stimulus The reference stimulus (Y_n, u_n, v_n) by which color differences are measured in the CIE LUV system. [ARP1781]

objective A hardware requirement established for design to achieve optimum performance, minimum weight, or other technical criteria. The feasibility of meeting an objective, or modifying the technical criteria to establish a firm requirement is subject to review. Renegotiation may be accomplished during development—after analyses, test reports, and cost data are available, supporting a review to determine viability for manufacturing production, cost of development, or other program factors. [AS1426]

objective variable A quantity or condition that is not measured directly in order to control it, but rather is controlled through its relation to another variable, a controlled variable.

object-oriented Describes a software design approach in which software pieces have clearly defined inputs, outputs, and attributes that facilitate reuse. [AIR4548]

oblate spheroid Ellipsoid of revolution, the shorter axis of which is the axis of revolution.

oblique coordinates Magnitudes defining a point relative to two intersecting nonperpendicular lines, called axes.

obliqueness The state of being neither perpendicular nor horizontal.

OBOGS (on-board oxygen generation system) A system that provides facilities for generation and delivery of breathable oxygen to the aircrew on board an aircraft through utilization of the bleed air and electrical power resources of the aircraft. [ARP171]

observability (systems) A property of a system whereby observations of the output variables always is sufficient to determine the initial values of all state variables.

observation port In a groove-sealed design, the open port next to the injection port. [AIR4069]

observed cumulative failure rate The number of relevant system failures accumulated by time t, symbolized as N(t), divided by t. [AIR4896]

observed mean time between failure (THETA) The total operating time of the equipment divided by the number of relevant failures. [ARD50010]

observed reliability A point estimate of reliability equal to the probability of survival for a specified operating time, t, given that the equipment was operational at the beginning of the period. [AIR4896]

observed significance level The probability of observing a more extreme value of the test statistic when the null hypothesis is true. [AIR4844

obstacle avoidance The use of sensors utilizing laser triangulation as means of preventing collisions, especially in the operation of roving vehicles on planetary surfaces.

OC *See* on condition.

occasional Describes a metallographic feature that occurs in 10% or less of the microstructure. [AS1814]

occlusion Specifically, the trapping of undissolved gas in a solid during solidification.

occultation The disappearance of a body behind another body of larger apparent size.

occupant environment The structural area and components that comprise a transport aircraft cabin, and that the passengers and cabin crew may impact during turbulent flight and emergency conditions, such as survivable crashes. [ARP767A]

octave The interval between two frequencies with a ratio of 2:1.

octave band Band in which the frequency of the upper limit of the pass band is twice that of the lower limit. [AIR1826]

octave-band analyzer A portable sound analyzer that amplifies a microphone signal, feeds it into one of several band-pass filters selected by a switch, and indicates signal amplitude on a logarithmic scale; except for the highest and lowest band, each band spans an octave in frequency.

octave-band filter A band-pass filter in which the upper and lower cutoff frequencies are in a fixed ratio of 2:1.

octet A group of eight bits treated as a unit. *See also* byte.

OD *See* outside diameter.

odd-even check *See* parity check.

odograph An instrument mounted in a vehicle to automatically plot the course of the vehicle and distance traveled on a map.

odometer An instrument for measuring and indicating distance traveled.

OEAS (oxygen enriched air system) Equipment that is common to all MSOGS (molecular sieve oxygen generation system)-equipped aircraft. [ARP171]

Oersted The CGS unit of magnetic field strength. NOTE: The SI unit, ampere-turn per meter, is preferred.

off-axis mirror Mirror in which the mechanical center does not correspond to the axis of the optical figure.

off equipment work For the purpose of maintenance data reporting, includes all maintenance actions performed on removed, repairable components, usually at the IMA. [ARD50010]

offgassing *1.* The emanation of volatile matter of any kind from materials into habitable areas. [AIR4728] *2.* The relatively high-mass-loss characteristics of many nonmetallic materials upon initial vacuum exposure.

off-line *1.* Describes the state of a subsystem or piece of computer equipment that is operable, but currently bypassed or disconnected from the main system. *2.* Describes lateral or angular deviation from the intended axis of a drilled or bored hole.

off-line diagnostics *1.* Describes the state of a control system, subsystem, or piece of computer equipment that is operable, but currently not actively monitoring or controlling the process. *2.* Refers to a program used to check out systems and subsystems, providing error codes if an error is detected. This diagnostic program is run while the system is off line.

off-line memory Any media that is capable of being stored remotely from the computer, and that can be read by the computer when placed into a suitable reading device. *See also* external storage.

off-line system Describes a type of system in which human operations are required between the original recording functions and the ultimate data processing function. This includes conversion operations as well as the necessary loading and unloading operations incident to the use of point-to-point or data-gathering systems. Contrast with on-line system.

off-on control Flicker control, especially as applied to rockets.

off-route vector A vector by ATC which takes an aircraft off a previously assigned route. Altitudes assigned by ATC during such vectors provide required obstacle clearance. [ARP4107]

offset *1.* The steady-state deviation. *2.* The count value output from an A/D converter resulting from a zero input analog voltage. Used to convert subsequent nonzero measurements. *3.* A short distance measured perpendicular to a principal line of measurement in order to locate a point with respect to the line. *4.* A constant and steady state of deviation of the measured variable from the set point. *5. See* droop.

offset modulus The ratio of the offset yield stress to the extension at the offset point. [AIR4844]

offset null The null voltage on a biased offset winding that is not at a balanced position of the secondaries. [ARP4386]

offset tongue terminal A terminal whose tongue is forward of its conductor barrel, and

469

whose stud hole is offset from the centerline of the conductor barrel. [ARP914A]

offset yield strength The stress at which the strain exceeds by a specific amount the extension of the initial, approximately linear, proportional portion of the stress-strain curve. [AIR4844]

offset yield stress The stress at which the stress-strain curve departs from linearity by a specified percentage of strain (offset). [AIR4844]

OFHC copper *See* copper, OFHC.

ogives *1.* Bodies of revolution formed by rotating a circular arc about an axis that intersects the arc; or the shape of such bodies. *2.* Noses of projectiles or similar objects shaped like an ogive.

ohm The metric unit for electrical resistance. Corresponds to the resistance (or impedance) of a conductor in which an electrical potential of one volt exists across the ends when it carries a current of one ampere.

ohmmeter *1.* A device for measuring electrical resistance. *2. See* resistance meter.

ohms per volt A standard rating of instrument sensitivity determined by dividing the electrical resistance of the instrument by its full-scale voltage.

oil Any of various viscous organic liquids that are soluble in certain organic solvents, such as naphtha or ether, but are not soluble in water; may be of animal, vegetable, mineral, or synthetic origin.

oil bath *1.* Oil in a container or chamber, into which a part or mechanism is submerged, or into which a part or mechanism dips during operation or manufacture. *2.* Oil poured on a cutting tool; or oil in which the tool is submerged during a machining operation.

oil burner A burner for firing oil.

oil can Describes the phenomenon whereby an excess of material in a localized area of a sheet causes the sheet to buckle in that area. [AMS4909E]

oil cone The cone of finely atomized oil discharged from an oil atomizer.

oil cooled motor A motor utilizing internal fluid flow from the hydraulic pump or system for cooling. [ARP4386]

oil dilution system The mechanism by which dilutant is added to the lubrication system in order to reduce the viscosity and pour point of the lubrication oil. [S-4, 6]

oil gas A heating gas made by reacting petroleum-oil vapors and steam.

oil heating and pumping set An apparatus consisting of a heater for raising the temperature of the oil to produce the desired viscosity, and a pump for delivering the oil at the desired pressure.

oil port *See* fluid port.

olemeter *1.* A device for measuring the specific gravity of oil. *2.* A device for measuring the proportion of oil in a mixture.

oleo (oleo strut) A telescoping landing gear strut consisting essentially of a piston that travels in an air-oil filled cylinder. Upon compression of the strut, the oil is forced through an orifice to provide a shock-absorbing effect. [AIR1489]

oleo-pneumatic *See* shock absorber, air-oil. [AIR1489]

Olsen ductility test A method for determining relative formability of metal sheet. A sheet metal sample is deformed at the center by a steel ball until fracture occurs; the cup height at fracture indicates relative ease of forming deep-drawn or stamped parts.

OM *See* outer marker

ombroscope An instrument for indicating when precipitation occurs. A heated, water-sensitive surface is exposed to the weather; when it rains or snows, an electrical or mechanical output trips an alarm or records the occurrence on a time chart.

omega A nonequilibrium, submicroscopic phase that can be formed either athermally or isothermally, preceding the formation of alpha from beta. Athermal omega is believed to form without change in composition and is analogous to martensite. Isothermal omega is generally formed by aging a retained beta structure in the 392 to 932°F (200 to 500°C) temperature range. Omega occurs in metastable beta alloys, alpha-plus-beta alloys rich in beta content, and unalloyed titanium. It leads to severe embrittlement. [AS1814]

omega bend A bend with a shape resembling the Greek letter omega, used to accommodate contraction or expansion in a duct system. [ARP699D]

omnidirectional With reference to a beacon or radio aid to air navigation, transmitting a signal throughout 360 deg of azimuth. [ARP4107]

omnidirectional antenna An antenna having equal gains in all directions.

omnigraph An automatic acetylene flame-cutting device that cuts several blanks simultaneously, duplicating the pattern traced by a mechanical pointer.

onboard data processing Processing of acquired data aboard an aircraft, satellite, etc., rather than transmitting it to ground stations for processing.

onboard guidance system On missiles and unmanned spacecraft, the automatic system that sends steering signals through the flight-control system during the terminal phase of propelled flight. [ARP4386]

once-per-rev Colloquial expression meaning either: (a) a signal that is provided once per engine rotor revolution; or (b) the rotational speed of a particular engine rotor. [AIR1839]

on condition (OC) Describes maintenance that is based on the functional, structural, or other condition of the unit or part, as differentiated from time schedule maintenance. [ARP1587]

oncotic pressure The osmotic pressure of the blood protein or the lymph colloids and electrolytes in the vascular system. [ARP171]

on-course indication An indication on an instrument that provides the pilot with a visual means of determining that the aircraft is located on the centerline of a given navigational track; or an indication on a radar scope that an aircraft is on a given track. [ARP4107]

one-component adhesive An adhesive material incorporating a latent hardener or catalyst that is activated by heat. [AIR4844]

on equipment work For the purpose of maintenance data reporting, includes those maintenance actions accomplished on complete end items, for example, aircraft, drones, SE, or removed engines. [ARD50010]

one watt-one ampere EED An EED that will not fire or will be rendered inoperative when one ampere and/or one watt is passed through the bridge wire circuit for a specified period of time. [AIR913]

one way interchangeable *See* interchangeable, one way.

on-line *1.* Describes the state of a subsystem or piece of computer equipment that is operable and currently connected to the main system. *2.* Describes coincidence of the axis of a drilled or bored hole and its intended axis, without measurable lateral or angular deviation. *3.* Done in real time.

on-line data-reduction The processing of information as rapidly as it is received by the computing system, or as rapidly as it is generated by the source.

on-line help A function that is available to the operator during ATE operation, providing the operator with additional informative and instructional messages to facilitate ATE operation and maintenance. [ARP4904]

on line maintenance Maintenance performed on a system or equipment without interrupting its operation. Synonymous with reliability with repair. [ARD50010]

on-off controller A device capable of both activation and deactivation operations of a circuit or system.

on-off system A type of skid-control system containing a valve that opens or closes when input signal exceeds or reduces below specific threshold levels. Sometimes called a "bang-bang" system. [AIR1489]

on-off valve A valve that controls the flow of air by opening or closing the flow passage; used in such a manner that it is not stopped in intermediate positions. [ARP986]

on orbit mean downtime The average elapsed time between loss of mission capable status and restoration to mission capable status. [AIR4896]

Oort cloud A region of millions of comets between 30,000 and 100,000 A.U. from the sun. Comets are perturbed out of the Oort cloud by passing stars; they then fall into the inner solar system.

opacity For an optical path, the reciprocal of transmission.

opcode In an instruction, the pattern of bits that indicates the addressing mode.

open barrel terminal A terminal with an open conductor and/or insulation barrel that is designed to be crimped around a conductor or wire. [ARP914A]

open-cell foam Foamed or cellular material with cells that are generally interconnected. [AIR4844]

open center Describes a hydraulic system in which no service is actuated and the system is open to flow, thus completing the circuit through the control units back to reservoir. An open center system will normally employ a fixed displacement pump. [ARP4386]

open center valve A valve designed for use in an open center system. Such a valve has an off position; when in this position, the pressure supply is passed through the valve to "return" or to other parts of the system. [ARP4386]

open circuit *1*. An interruption in an electrical or hydraulic circuit, usually due to a failure or disconnection, which renders the circuit inoperable. *2*. A nonrecirculating (once-through) system or process.

open circuit voltage The steady state of equilibrium potential of an electrode in absence of external current flow to or from the electrode.

open-end protecting tube A tube that extends from a physical boundary into the body of a medium to surround and protect a thermocouple, at the same time allowing direct contact between the measuring junction of the thermocouple and the medium.

opening pressure The static inlet pressure that initiates a discharge.

opening time Time for a valve or system to achieve 90% of full rated flow starting from zero. [AIR4783]

open ladder circuit A parallel circuit in which the bus bars are electrically discontinuous. [ARP485]

open loop *1*. A system operating without feedback, or with only partial feedback. *See also* closed loop. [AIR1489] *2*. Pertaining to a control system in which there is no self-correcting action for misses of the desired operational condition, as there is in a closed-loop system. *See also* feed-forward control action.

open loop control *1*. A system in which the control action is independent of the output. *2*. A feedback control system that has been broken at some point in the signal path. [ARP89] *3*. A control system that does not take any account of the error between the desired and actual values of the controlled variables. *4*. An operation in which computer-evaluated control action is applied by an operator. *See* open loop and closed loop. *5*. A system in which no comparison is made between the actual value and the desired value of a process variable.

open loop frequency response The frequency response of the feedback signal to the error signal. May be specified with or without a load. [ARP4386]

open loop gain The equivalent output rate per unit of position error, or the product of the forward and feedback gains in in./sec/in. [ARP1281A]

openness of scale With respect to measuring instruments, refers to the amount of change in a measured quantity that causes the pointer to move 1 mm (or in some instances, 1 in.) on the instrument scale.

open neutral Describes the condition in a directional control valve when cylinder or load ports are interconnected and open to return in neutral or off position. [ARP4386]

open port dilution rebreathing mask A mask incorporating a rebreather bag into which exhaled gases, high in oxygen content from the first portion of the previous exhalation, are introduced to be inspired again upon the next inspiration. [AS1224]

open position A position that is 100 percent open.

OPEN Project Acronym for the Origin of Plasmas in Earth Neighborhood project.

open seal An impulse line filled with a seal fluid that is open to the process.

open system A system that implements sufficient open specifications for interfaces, services, and supporting formats to enable properly engineered applications to: (a) be ported with minimal changes across systems conforming to the same open specifications; (b) interoperate with other applications on local and remote systems; and (c) interact with users in a style that facilitates user portability. [AS4893]

open system interconnection (OSI) A connection between one communication system and another using a standard protocol.

open time *See* assembly time.

open wiring Any section of an electric wire or cable bundle that is not enclosed in conduit, troughs, wireways, junction boxes, or similar enclosures. [ARP4404]

operable The state of being able to perform the intended function. [ARD50010]

operand The address of an instruction to be executed by the processor.

operating characteristic curves A curve showing the probability that a submitted lot with given mean life would meet the acceptability criterion on the basis of the chosen sampling plan. [ARD50010]

operating components Items that contain moving parts and/or electronics. [AIR4073]

operating conditions, reference Conditions to which a device is subjected, not including the variable measured by the device.

operating control A control to start and stop the burner; must be in addition to the high limit control.

operating factor *See* factor, operating.

operating force The force that must be applied to the plunger to cause the moving contact to snap from the normal contact position to the operated contact position. [AIR4077]

operating level The nominal position or output at which a system or process operates. Typical examples are water level in a boiler, production rate of a manufacturing process, or acoustical output (volume) of a loudspeaker system.

operating life The total accumulated operating time, measured in hours, over which a system, subsystem, or component performs its intended design and operational functions when installed in an aerospace vehicle. [ARP4386]

operating parts Refers to the portion of a safety relief device that normally closes the safety

discharge channel, but can move as a result of heat and/or pressure to permit the escape of gas from the vessel. [ARP4386]

operating point *1.* The position of the plunger at which the contact snaps from the normal contact position to the operated contact position. [AIR4077] *2.* On a compressor map, the location that satisfies the matching relationships that govern the engine operation. [ARP1420]

operating pressure *1.* The pressure available to a component or system during normal operation. [ARP4386] *2.* In a pneumatic or hydraulic system, the high and low values (range) of pressure that will produce full-range operation of an output device, such as a motor operator, positioning relay, or data-transmission device.

operating system(s) *1.* Computer programs for expediting, controlling, and/or recording computer use by other programs. *2. See* manufacturer software.

operating temperature The temperature at which a device may function on a continuous basis. [ARP914A]

operating time The time during which a system or equipment is actually operating (in an "up" status). Operating time is usually divisible among several operating periods or conditions. These include standby time, filament-on time, pre-flight checkout time, and flight time. [ARD50010]

operating weight Aircraft weight, usually including the weight of the oil, fuel, crew, crew's baggage, emergency equipment, and payload. [ARP4107]

operation A set of tasks or processes, usually performed at one location.

operational Of, or pertaining to, the state of actual usage. [ARD50010]

operational availability The percent of time that a subsystem, line replaceable unit (LRU), or line replaceable module (LRM) is capable of performing satisfactorily in the operational environment. NOTE: Operational availability does not depict the ability of an item to continue to operate for a specific period of time. This characteristic is covered via the weapon system reliability term. [ARD50010]

operational check *See* check, operational.

operational limit A pre-established reference for engine, engine component, or sub-system operation. [ARP1587]

operationally ready Available and in condition for serving the functions for which it is designed. [ARP4386]

operational readiness A measure of the degree to which an item is in the operable and committable state at the start of the mission, when the mission is called for at unknown (random) point in time. [ARD50010]

operational R&M value Any measure of reliability or maintainability that includes the combined effects of item design, quality, installation, environment, operation, maintenance, and repair. [AIR4896]

operational test and evaluation Test and evaluation that focuses on the development of optimum tactics, techniques, procedures, and concepts for systems and equipment; evaluation of reliability, maintainability, and operational effectiveness; and suitability of systems and equipment under realistic operational conditions. [AIR4896]

operational test program The test program for a specific unit or functionally related group of units under test. [AIR4896]

operation analysis In industrial engineering, an evaluation process in which design, materials, equipment, tools, working conditions, methods, and inspection standards are assessed, usually for the purpose of improving production output or decreasing cost.

operation code The part of a computer instruction word that specifies, in coded form, the operation to be performed.

operations research The use of analytic methods adopted from mathematics for solving operational problems. Objective is to provide management with a more logical basis for making sound predictions and decisions. Among the common scientific techniques used in operations research are the following: linear programming, probability theory, information theory, game theory, Monte Carlo method, and queuing theory.

operator *1.* Refers to the human responsible for handling data link communications. The operator can be an aircraft flight crew member or an air traffic specialist staffing an en route, terminal, tower, or oceanic operational position. [ARP4791] *2.* Any corporate entity or person who causes or authorizes the operation of an aircraft; for example, the owner, lessee, or bailee of an aircraft. [ARP4107] *3.* In a mishap sequence, the person in control of the aircraft at the point of the mishap. Other personnel involved in the mishap sequence of events are considered part of the operator's equipment. [ARP4107] *4.* The person who initiates and monitors the operation of a computer. *5.* A mathematical symbol that represents a mathematical process to be performed on an associated operand. *6.* The portion of an instruction that tells the machine what to do.

operator command A statement to the control program, issued via a console device, causing the control program to provide requested information, alter normal operations, initiate new operations, or terminate existing operations.

operator's console A device that enables the operator to communicate with the computer. Can be used, for example, to enter information into the computer, to request and display stored data, and to actuate various preprogrammed command routines.

operator station Station that serves as the interface between the operator and other devices on the data highway.

opisometer An instrument incorporating a tracing wheel, used for measuring the length of curved lines, such as those on a map.

opposing load A force or torque on an actuator that acts in the direction resisting motion. *See also* aiding load. [ARP4386]

opposing load pressure Pressure acting to oppose operating pressure. [ARP4386]

optical activity Ability to rotate the plane of vibration of polarized light to the right or left.

optical ammeter An electrothermic instrument, commonly employing a photoelectric cell and an indicating device, used to determine the magnitude of electric current by measuring the light emitted by a lamp filament carrying the current. Calibrated by determining the amount of light emitted when known currents are carried by the same filament.

optical amplifier A type of amplifier in which an electric input signal is converted to light, amplified as light, then converted back to an electric output signal.

optical attenuation meter A device that measures the loss or attenuation of an optical fiber, fiber optic cable, or fiber- optic system. Measurements are usually made in decibels.

optical bench A rigid, horizontal bar or track for holding and supporting optical devices in fixed positions, while allowing the positions to be changed or adjusted quickly and easily.

optical bistability A property of certain materials whereby the materials exhibit a nonlinear response when under the influence of an external driving coherent light, thereby allowing the materials to behave like optical switches.

optical character reader A scanning device that can recognize some typewritten characters.

optical comparator *1*. A comparator in which movement of a measuring plunger tilts a small mirror, which in turn reflects light in an optical system. *2*. A type of comparator in which the silhouette of a part is projected on a graduated screen and the dimensions or contour are evaluated from the projected image.

optical computers Computers in which all or part of the operation is based on the use of light, rather than electricity. Optical computers perform multiple tasks in parallel, as opposed to electronic computers, which would perform those tasks sequentially. Such increased processing capability makes optical computers suited for aerospace problems involving systems that have a large number of degrees of freedom, e.g., large space structures, pattern recognition activity, and robotics.

optical countermeasures Refers to equipment for exploiting the vulnerability of laser-guided weapon systems.

optical density *1*. A measurement of transmission, equal to the base 10 logarithm of the reciprocal of transmittance. An object with optical density of zero is transparent; an object with an optical density of one possesses 10% transmission. *2. See* density.

optical depth *See* optical thickness.

optical disk A large electronic storage device in which laser beam patterns are used to store data.

optical-emission spectrometry Measurement of the wavelength(s) and intensities of visible light emitted by a substance following stimulation.

optical encoder tachometer A type of instrument that combines a sensor (optical encoder) with a microprocessor to convert sensor impulses into a measurement of rotational velocity.

optical fibers Fine glass strands that transmit data using light signals.

optical filter *See* filter, optical.

optical flat A transparent disk, usually made of fused quartz, having precisely parallel faces, one face polished for clear vision and the other face ground optically flat. When an optical flat is placed on a surface and illuminated under proper conditions, interference bands can be observed and used to either assess surface contour (relative flatness) or determine differences between a reference gage or gage block and a highly accurate part or inspection gage.

optical fluid-flow measurement Describes any method for measuring the density of a fluid in motion that depends on measuring refraction and phase shift among different rays of light as they pass through a flow field of varying density.

optical gage A gage that measures the image of an object without touching the object itself.

optical glass Glass that is free of imperfections, such as bubbles, chemical inhomogeneity, or unmelted particles, that would degrade its ability to transmit light.

optical grating *1*. Diffraction grating usually employed with other appropriate optics to fabricate a monochromator. An optical grating consists of a series of parallel grooves, carefully and uniformly shaped, in an optical surface that is either flat or concave, depending on the application. The number of grooves formed and the shape of the grooves (the profile) determines the region of the spectrum to which the grating is applicable. Commonly referred to as a Ronchi grating. *2*. A highly accurate device used in precision dimensional measurement, consisting of a polished surface (commonly aluminum coating on a glass substrate) onto which close, equidistant and parallel grooves have been ruled—the distribution of grooves ranging from several hundred to many thousands of grooves per inch. Such gratings are used in conjunction with monochromatic light to produce interference patterns sometimes referred to as Moire patterns. They are used in optical testing as well as for generating the dot matrix for picture reproduction from a photographic negative.

optical indicator An instrument employing magnification in an optical system coupled with photographic recording, used to plot pressure variations as a function of time.

optical mark reader A device that, through light sensing, reads marks made on special forms.

optical material Any material that is transparent to visible light or to x-ray, ultraviolet, or infrared radiation.

optical multiplexer A bi-symmetric device used to separate two or more optical signals arriving on a single fiber path onto separate paths; or to combine such signals arriving from separate paths to a single fiber path. [ARD50024]

optical paths *1.* Lines of sight. *2.* The paths followed by rays of light through optical systems.

optical plastic Any plastics material that is transparent to light and can be used in optical devices and instruments, taking advantage of its superior physical or mechanical properties, or lower cost, as compared to glass.

optical pressure transducer Any of several devices in which optical methods are used to accurately measure the position of the sensitive element of a pressure transducer.

optical pyrometer An instrument used to determine the temperature of an object by comparing its incandescent brightness with that of an electrically heated wire. The current through the wire is adjusted until the visual image of the wire blends into the image of the hot surface; temperature is read directly from a calibrated dial attached to the current adjustment.

optical rails *See* rails, optical.

optical rangefinder An optical instrument used to measure distance, usually from the location of the instrument to a target some distance away. Measures the angle between rays of light from the target to separate windows on the body of the instrument.

optical recording Making a record of an instrument reading by focusing a tiny beam of light on photosensitive paper, thereby creating an orthogonal plot in which the position of the light along one axis of the plot is directly related to the value of the quantity being measured.

optical relay systems Systems that utilize photocouplers, and whose output devices are light-sensitive switches that provides the same on and off operations as the contacts of a relay.

optical rotation Rotation of the plane of polarization about the axis of a beam of polarized light.

optical scanner A light source and phototube combined as a single unit, used for scanning moving strips of paper or other materials in photoelectric side-register control systems.

optical sensor arrays A distributed network of multiple sensors, using laser beams, to provide real-time data for on-board processing. [ARP4386]

optical signal range The ratio of the maximum optical signal power to the minimum optical signal power, expressed in decibels. [AS1773]

optical slant range In a homogeneous atmosphere, the horizontal distance for which the attenuation is the same as that actually encountered along the true oblique path.

optical storage disk A computer storage medium in which lasers are used to form surface patterns that represent data. The CD-ROM (compact disk read-only memory) is an optical storage disk that stores data in digital form.

optical thickness Specifically, in calculations of the transfer of radiant energy, the mass of a given absorbing or emitting material lying in a vertical column of unit cross sectional area and extending between two specific levels. Also called optical depth.

optical time domain reflectometer A device that sends a very short pulse of light down a fiber-optic communication system and measures the time history of the pulse reflection. The reflection indicates fiber dispersion and discontinuities in the fiber path, such as breaks and connectors. The time it takes for the light pulse to travel to and from the discontinuity indicates how far it is from the test set.

optimal control A type of control in which the values of the control parameters (transfer function) are chosen such that a selected performance index is maximized (or minimized). [AIR1823]

optimization *1*. Theoretical analysis of a system, including all of the characteristics of the process, such as thermal lags, capacity of tanks or towers, and length and size of pipes. This analysis is sometimes performed sometimes with the aid of frequency response curves; its purpose is to obtain the most desirable instrumentation and control. *2*. Making a design, process, or system as nearly perfect in function or effectiveness as possible.

optimize *1*. To establish control parameters so as to make control as effective as possible. *2*. To rearrange the instructions or data in storage so that the program can be run in minimum time.

optimum altitude The altitude, at a specific gross weight, that results in the maximum miles per pound of fuel. Continuous optimum altitude can only be achieved by cruise climbing. [ARP1570]

optimum control system A control system that minimizes a given error index for a dynamic process subject to given design constraints. [ARP4386]

optimum cruise altitude The level flight altitude, adjusted by step climbs, that results in the maximum miles per pound of fuel for the trip. The general procedure is to initiate cruise above optimum; then maintain that flight level until approximately an equal weight burn below optimum; then step climb above, and repeat. The steps may be 4000, 2000, 1000, or as dictated by ATC. The winds and trip distance or distance remaining would also influence the optimum cruise altitude. [ARP1570]

optimum frequency The frequency that provides the highest signal-to-noise ratio obtainable for the detection of an individual property,

such as conductivity, crack, or inclusion, of the test specimen. [ARP5089]

option approach An approach requested and conducted by a pilot that will result in either a touch-and-go, missed approach, low approach, stop-and-go, or full-stop landing. [ARP4107]

option module Any additional device that expands the capability of a computer.

optoelectronic amplifier An amplifier whose input and output signals and method of amplification may be either optical or electronic.

optogalvanic spectroscopy A method whereby the absorption spectra of atomic and molecular species in flames and electrical discharges are obtained by measuring voltage and current changes upon laser irradiation.

orange peel Mottled surface appearance causing an effect sometimes similar to haze or distortion. [AIR5122]

orbit The path described by a celestial body in its revolution around another body. [ARP4386]

orbital curve On a the surface of a primary body, one of the tracks traced by a satellite that orbits about the primary body several times a day in a direction other than due east or west, each successive track being displaced to the west by an amount equal to the degrees of rotation of the primary body between each orbit. [ARP4386]

orbital elements A set of seven parameters defining the orbit of a body attracted by a central, inverse square force.

orbital lifetime The predicted lifetime of a satellite in orbit, usually based on such criteria as solar flux density, atmospheric density, the lessening of the eccentricity of elliptical orbits, or the gravitational effects of the sun or the moon.

orbital resonances (celestial mechanics) Systems of two or more satellites (including planets) that orbit the same primary, and whose orbital mean motions are in ratios of small whole numbers.

orbital servicing The replenishing of propellants, pressurants, coolants, and the replacement of modules and experiments, during some phase of a spacecraft flight to extend the mission and lifetime, or change the payloads.

orbital transfer *See* transfer orbits.

orbital velocity *1.* The average velocity at which an earth satellite or other orbiting body travels around its primary. *2.* The velocity of an earth satellite or other orbiting at any given point in its orbit; for example, "The orbital velocity at the apogee is less than at the perigee."

orbit spectrum utilization Telecommunication techniques in spectrum conservation for reducing user costs.

orbit transfer vehicles (OTV) Refers to a concept whereby propulsive (velocity-producing) rockets or stages are used with crew-transfer modules, manned sortie modules, or other payloads.

ordered structure An orderly or periodic arrangement of solute atoms on the lattice sites of the solvent. [AS1814]

organizational maintenance Maintenance that is the responsibility of, and is performed by, a using organization and its assigned equipment. *See also* maintenance levels. [ARD50010]

orient To place an instrument, particularly one for making optical measurements, so that its physical axis is aligned with a specific direction or reference line.

orientation *1.* Alignment with a specific direction or reference line. *2.* In ultrasonic testing, position of a discontinuity, part , or surface in relation to the test surface of the article or ultrasonic beam. [ARP5089]

orientation alpha A nonuniform alpha structure that results when colonies or domains of platelets or wormy alpha lie at different angles, having no significance to crystallographic orientation, such that different areas exhibit different aspect ratios and alpha grain outlines. [AS1814]

oriented materials Materials, particularly amorphous polymers and composites, in which the molecules and/or macro constituents are aligned in a specific way. [AIR4844]

orifice *1.* An opening that is intended for the passage of a fluid, has a fixed diameter and flow coefficient, and may be calibrated to pass a desired volumetric flow at the anticipated pressure differential. [ARP171] *2.* A short fluid passage that, by virtue of its cross-sectional area, produces a substantial reduc-

tion in flow. A true orifice has zero length. [ARP4386] *3.* An opening mouth; a vent. *4.* In a shock absorber, a hole through which oil is forced in the energy-dissipation process. [AIR1489] *5.* The opening from the whirling chamber of a mechanical atomizer or the mixing chamber of a steam atomizer, through which the liquid fuel is discharged. *6.* A calibrated opening in a plate, inserted in a gas stream for measuring velocity of flow.

orifice coefficient A numerical constant defining the characteristics of a specific orifice. [AIR1489]

orifice control The method or process whereby flow of oil through an orifice is controlled, e.g., variation of orifice area by metering pin or variation of orifice shape. [AIR1489]

orifice meter Describes any recording, differential-pressure-measuring instrument.

orifice mixer A piece of equipment for mixing two or more liquids by simultaneously directing the liquids, under pressure, through a constriction, where the resulting turbulence blends them together.

orifice plate A disc or platelike member with a sharp-edged hole in it, used in a pipe to measure flow or reduce static pressure.

orifice run A differential-pressure-producing arrangement consisting of selected pipe, orifice flange union, and orifice plate. An orifice run has rigid specifications defined by the American Gas Association.

orifice-type variable-area flowmeter A flow-measurement device consisting of a tube section containing an orifice, and a guided, conically tapered float that rides within the orifice. When fluid flows through the flowmeter, it causes the float to be positioned in relation to flow rate; float position is determined magnetically or by other indirect means.

original equipment manufacturer A manufacturer that produces, in a final inspection sense, a part or assembly. [AS7104]

O-ring *1.* Material that is shaped like a ring with a circular cross section, and is used as a seal between pressure sections of a fluid actuator. [ARP4386] *2.* A toroidal sealing ring made of synthetic rubber or similar material;

the cross section through the torus is usually round or oval, but may be rectangular or some other shape.

orometer A barometer for measuring elevation above sea level.

orsat A gas-analysis apparatus in which certain gaseous constituents are measured by absorption in separate chemical solutions.

orthicon A camera tube in which a low-velocity electron beam is used to scan an image stored electrically on a photoactive mosaic panel.

orthogonality The degree of perpendicularity between the mutually perpendicular trace axes on CRT screens. [ARP1782]

orthometric correction A systematic correction that must be applied to a measured difference in elevation to compensate for the fact that level surfaces at different elevations are not exactly parallel.

orthorhombic martensite *See* alpha double prime.

orthotropic Having three mutually perpendicular planes of elastic symmetry. [AIR4844]

orthotropic ply A ply in which there are two different material properties in two mutually perpendicular directions at a point. The two mutually perpendicular directions form the planes of material property symmetry at the point. [AIR4844]

oscillating-piston flowmeter A flow-measurement device that is similar to a nutating-disk flowmeter; however, motion of the piston takes place in one plane only. In the oscillating-piston flowmeter, the rotational speed of the piston is directly related to the volume of fluid passing through the meter.

oscillation *1.* Fluctuation or vibration on each side of a mean value or position; half a vibration. One oscillation is half an oscillatory cycle, and consists of a fluctuation or vibration in one direction. *2.* The variation, usually with time, of the magnitude of a quantity with respect to a specified reference when the magnitude is alternately greater and smaller than the reference.

oscillator A nonrotating device for producing alternating current, the output frequency of which is determined by characteristics of the device. In some cases the frequency is fixed, but in others it can be varied.

oscillator crystal A piezoelectric crystal device used chiefly to determine the frequency of an oscillator.

oscillator strength A quantum-mechanical analog of the number of dispersion electrons having a given natural frequency in an atom; used in an equation for the absorption coefficient of a spectral line.

oscillatory circuit A circuit that produces a periodically reversing current when energized by a direct-current voltage. Contains R, L, and C elements, which may be varied to change the characteristics of the resultant a-c output.

oscillogram *1.* The permanent record created by an oscillograph. *2.* A permanent record of the trace on an oscilloscope, such as might be recorded photographically.

oscillograph A device for determining waveform by plotting instantaneous values of a quantity such as voltage as a function of time.

oscilloscope Instrument for producing visual representations of oscillations or changes in an electric current.

OSI *See* open system interconnection.

OSI reference model A seven-layered model of communications networks defined by ISO. The seven layers are: Layer 7, Application; Layer 6, Presentation; Layer 5, Session; Layer 4, Transport; Layer 3, Network; Layer 2, Data Link; and Layer 1, Physical. *Layer 7, Application*–provides the interface for application to access the OSI environment. *Layer 6, Presentation*–provides for data conversion to preserve the meaning of the data. *Layer 5, Session*–provides user-to-user connections. *Layer 4, Transport*–provides end-to-end reliability. *Layer 3, Network*–provides routing of data through the network. *Layer 2, Data Link*–provides link access control and reliability. *Layer 1, Physical*–provides an interface to the physical medium.

otolith organs Structures of the inner ear (utricle and saccule) which respond to linear acceleration and tilting.

OTV *See* orbit transfer vehicles.

479

out The fluid outlet port of a component, not otherwise designated, in which the direction of flow is critical. [ARP4386]

outdoor test stand An open-air facility, without any enclosure, for testing engines. [ARP741]

outer compass locator (LOM) *See* compass locator, outer.

outer fix A fix in the destination terminal area, other than the approach fix, to which aircraft are normally cleared by an air route control center or a terminal area traffic control facility; and from which aircraft are cleared to the approach fix or final approach course. [ARP4107]

outer marker (OM) A marker beacon at or near the glide slope intercept altitude of an instrument landing system (ILS) approach. This beacon is keyed to transmit a signal consisting of two dashes per second, which is received visually and/or aurally (on a 400 Hz tone) by compatible airborne equipment. The OM is normally located between four and seven miles from the runway threshold on the extended centerline of the runway. It serves as an aid to the flight personnel in that it marks a particular distance from the approach end of the runway. [ARP4107]

outer skin The side of a part that is cured against the mold. [AIR4844]

outgassing *1.* The evolution of gas from a material in a vacuum. *2.* The release of adsorbed or occluded gases and water vapor, usually during evacuation or subsequent heating of an evacuated chamber. *3. See* gassing.

outlet, air Opening through which air is removed from a space being conditioned. [ARP147C]

outlet patch A device that provides an indication of the cleanliness and degree of internal wear of the equipment by the amount of contamination generated in the displacement and motion of the unit. [ARP1302]

outlet pressure Pressure at the outlet of a component. [ARP4386]

outlet temperature Fluid temperature at the plane of the outlet port. [ARP4386]

outliers (statistics) In sets of data, values so far removed from other values in the distribution that their presence cannot be attributed to the random combination of change causes.

out life The period of time that a prepreg or film adhesive material remains in a handleable form, and with properties intact, outside of the specified storage environment and at room temperature. [AIR4844]

out-of-round Describes a dimensional condition whereby diameters taken in different directions across a nominally circular object are unequal—the difference between them being the amount of "out-of-roundness."

output *1.* The controlled variable resulting from activity of the control system. [ARP4386] *2.* The information transferred from the internal storage of a computer to secondary or external storage, or to any device outside of the computer.

output block A block of computer words considered as a unit and intended or destined to be transferred from an internal storage medium to an external destination.

output device The part of a machine that translates the electrical impulses representing data processed by the machine into permanent results, such as printed forms, punched cards, or magnetic writing on tape; or into control signals for a process.

output impedance Symbolized by Zo, the impedance that, in series with an ideal generator voltage, is equivalent to the output voltage circuit. Also called source impedance. [ARP667]

output indicator A device connected to a radio receiver to indicate variations in output signal without indicating a specific signal value; usually used for alignment or tuning.

output pressure In a pressure-control device, the pressure that is produced at the outlet port. In a pressure-modulating unit, the output pressure should be specified as maximum or the range stated. [ARP4386]

output shaft The starter shaft that connects the starter to the engine accessory gearbox. [ARP906A]

output signal A signal delivered by a device, element, or system.

output stage The final stage of hydraulic amplification used in a servovalve. [ARP4386]

output variable A variable delivered by a control algorithm, e.g., the signal going to a steam valve in a temperature control loop. *See also* controlled variable.

output voltage of LVDT A voltage output proportional to the voltage applied to the primary windings. Depending on wire connections, a two- or three-wire output can be obtained. [ARP4386]

outrigger gear *See* gear, outrigger.

outriggers Stabilizer devices that are used to improve the stability of a vehicle, and that extend outside the normal envelope of the vehicle. [ARP1328]

outside air temperature (OAT) The static temperature of the ambient free stream air (outside the aircraft). [AIR4367]

outside caliper A caliper used to measure distances across two external opposing surfaces.

outside diameter (OD) The outer dimension of a circular member, such as a rod, pipe, or tube.

outside strand A strand of a multiple-strand heat that flows from an outside location of the molten reservoir. [MAM2304]

out time The time during which a prepreg is exposed to ambient temperature; namely, the total amount of time the prepreg is out of the freezer. The primary effects of out time are a decrease in the drape and tack of the prepreg and the adsorption of moisture from the air. [AIR4844]

ovality The difference between the maximum and minimum diameters of any plat at 90 deg to the axis of the tube. [AMS2243G]

ovaloid A surface of revolution symmetrical about the polar axis; forms the end closure for a filament-wound cylinder. [AIR4844]

oval-shaped gear flowmeter A type of positive-displacement flowmeter that operates by trapping a precise volume of fluid between an oval, toothed rotor and the meter housing, as the rotor revolves in mesh with a second rotor. With this type of flowmeter, the volume flow of an incompressible fluid is indicated directly by determining rotor speed.

oven A vessel at atmospheric pressure used to provide a controlled and uniform temperature. [AIR4844]

oven dry Describes the condition of a material that has been heated under prescribed conditions of temperature and humidity until there is no further significant change in its mass. [AIR4844]

overbalance The ratio of the force tending to close a shutoff valve to the force tending to open the shutoff valve. [AIR1660A]

overcast (meteorology) In surface aviation weather observations, describes sky cover of 1.0 (95 percent or more), when at least a portion of this amount is attributable to clouds or obscuring phenomena aloft; that is, when the total sky cover is not due entirely to surface-based obscuring phenomena. Implies a predominant opaque cover. [ARP4107]

overcharging The continued passage of electrical "charging" current after the cell plates are charged, resulting in the generation of hydrogen and oxygen gases. The onset of overcharge is manifested by a transition from negligible gassing to complete conversion of all (over) charge current to gas. [AS8033]

overconfidence *See* confidence.

overcontrol (by a pilot) To displace or move the controls of an aircraft more than is necessary for the desired performance. [ARP4107]

overcuring The beginning of thermal decomposition resulting from too high a temperature or too long a molding time. [AIR4844]

overcurrent Any current exceeding the rated current of the protective device (for circuit breakers, exceeding the maximum ultimate trip current). This includes both overload and short-circuit currents. [ARP1199A]

overdamped *See* damping.

overdue aircraft (VFR) For aircraft operating in accordance with VFR, describes an aircraft whose flight plan is not closed within one half hour after the estimated time of arrival (ETA) of the aircraft at its destination. If a flight plan is not properly closed, search and rescue procedures will be started. [ARP4107]

overdue aircraft/missing aircraft (IFR) For aircraft operating in accordance with IFR, describes an aircraft with which communications or radar identification cannot be established within 30 min after it: (a) fails to report over an ATC-specified reporting point or over a compulsory reporting point along the route of flight, whichever is earlier; or (b) becomes overdue at the point of intended landing. [ARP4107]

overflow *1.* In a digital computer, the condition that arises when the result of an arithmetic operation exceeds the capacity of the storage space allotted. *2.* The digit arising from this condition, if a mechanical or programmed indicator is included (otherwise the digit may be lost).

overflow pipe In a tank, a pipe with its open end protruding above the liquid level, serving to limit the height of liquid in the tank by carrying away any liquid entering the open end, usually to a drain or sewage system.

overfractionation Operation of a distillation column to produce a purer product than required.

overhaul A process of disassembly that is sufficient to allow inspection of all the operating components and the basic end article; includes repair, replacement, or servicing as necessary, followed by reassembly and bench check/flight test. Upon completion of the overhaul process, the component/end article must be capable of performing its intended service life/service tour. [ARD50010]

overhauled certified Describes a part overhauled by an airline or an authorized repair agency to a time since overhaul of 00:00. [AS7104]

overhaul, partial The reconditioning of a subassembly. [ARD50010]

overhaul, the controlled Reconditioning in accordance with a plan, according to which the time histories of individual items are monitored. A monitoring system is used to schedule the removal of items before they exceed a specified time limit. [ARD50010]

Overhauser effect In atomic physics, the effect when a radio-frequency field is applied to a substance in an external magnetic field, with nuclei of spin 1/2, and with unpaired electrons at the electron spin resonance frequency. This results in polarization of the nuclei as great as if the nuclei had the much larger electron magnetic moment.

overhead cost The summation of overhead cost attributable to the program, together with any unattributable fixed overheads. [ARP4293]

overheat *1.* To raise the temperature above a desired or safe limit. *2.* In metal heat-treating, to reach a temperature that results in degraded mechanical or physical properties.

overlap *1.* The lap condition that results in a decreased slope of the normal flow curve in the null region. [ARP490] *2.* A condition in which one portion of an item lays on another portion of itself or on some other item, for example, a "tape overlap." [ARP1931]

overlay The technique of repeatedly using the same blocks of internal storage during different stages of a problem. When one routine is no longer needed in storage, another routine can replace all or part of it.

overlay pavement A pavement that is layed on top of an existing pavement to increase the load-carrying capability or to repair or seal the existing pavement; for example, asphaltic pavement layed on top of concrete or old asphalt. [AIR1489]

overlay sheet A nonwoven fibrous mat used as the top layer in a cloth or mat lay-up to provide a smoother finish, minimize the appearance of the fibrous pattern, or permit machining or grinding to a precise dimension. [AIR4844]

overload capacity The force weight, power, pressure, or other capacity factor, usually higher than the rated capacity, beyond which permanent damage occurs to a device or structure.

overload current An overcurrent in excess of the current rating. The overload range is considered to be greater than the rated current up to approximately ten times rated current. [ARP1199A]

overload rating The normal overload value expressed in kVA at 0.8 power factor for a specified time. [ARP1148A]

override To manually apply a control force that exceeds the authority of an automatic control system; or to manually intervene in an otherwise automatic sequence. [ARP4107]

override control *1.* Generally, two control loops connected to a common final control element—

one control loop being normally in control, with the second being switched in by some logic element when an abnormal condition occurs, so that constant control is maintained. *2.* A control technique in which more than one controller manipulates a final control element. Used when constraint control is important.

overrun *1.* Refers to the relative speeds of the engine and starter after the start cycle has been completed. Power transmission has changed from starter-to-engine to engine-to-starter. [ARP906A] *2.* An extended clear area located at the end of a runway for the purpose of run out of an aircraft in an abort or emergency situation. [AIR1489]

overrun barrier *See* barrier, overrun.

overrunning The driving of the unit output shaft by some means other than by energizing the unit turbine. [AS1606]

overshoot *1.* The volume of liquid that passes through the shutoff valve after the shutoff valve has been signaled to close. [AIR1660A] *2.* The increment by which the output exceeds the final value when responding to a step command, usually expressed as a percentage of the output. [ARP4386] *3.* Movement of an aircraft whereby it coasts past the commanded reference altitude, attitude, or flight path before settling out. [ARP419-57] *4.* A transient response to a step change in an input signal which exceeds the normal or expected steady-state response.

overshoot control Refers to the capability of controlling the volume of fuel passing through the refueling system following the initiation of a deadman closure. [AIR4782]

overspeed *1.* The speed at which a rotor exceeds its yield limit. [AIR1639] *2.* The rotation of high-speed parts at a speed that causes the yield strength of the parts to be exceeded. [AS1606]

overspeed switch A speed-sensing device, usually located on the starter side of the engaging mechanism, used to terminate starter operation at the predetermined overspeed. [ARP906A]

overtemperature sensor The sensor located in the pressurized supply air line, between the heat exchanger and the OEAS (oxygen enriched air system) oxygen concentrator. When used, this sensor will activate a warning signal in the cockpit if the temperature of the air supplied to the concentrator exceeds a preset level, and may active at the OBOGS (on-board oxygen generation system) bleed air shutoff valve. [ARP171]

overtravel The distance through which a plunger moves when traveling from the operating point to the full overtravel point. [AIR4077]

overvoltage protection A protective device that interrupts power or reduces voltage supplied to an operating device in the event that incoming voltage exceeds a preset value.

overwing nozzle A manually-operated, hand-held, on-off device for use in dispensing liquid into the top of an open tank. [AIR4783]

overwing nozzle adapter A metallic structural component located at the gravity fill ports of the aircraft. [AIR4783]

overwing refueling The refueling of an aircraft by use of an overwing nozzle, through an adapter opening near the top of the wing tank. [AIR4783]

Owen bridge A type of a-c bridge circuit in which one leg contains a fixed capacitor, the opposite leg contains an unknown inductance and resistance, the third leg contains a fixed resistor, and the fourth leg contains a variable resistor and a variable capacitor. Especially useful for measuring wide ranges of inductances using reasonable ranges of standard capacitances. Can be used to measure permeability or core loss.

oxidation *1.* Chemical combination with oxygen. *2.* In carbon/graphite fiber processing, the step whereby the precursor polymer is reacted with oxygen, resulting in stabilization of the structure for the hot stretching operation. [AIR4844]

oxidation-corrosion stability The extent to which a fluid in the presence of oxygen will tend to corrode various metals, or produce products of degradation that will corrode metals. [AIR1116]

oxidation-reduction reactions An oxidizing chemical change, whereby the positive valence of an element is increased (by electron loss), accompanied by a simultaneous reduction of an associated element (by electron gain).

oxide of nitrogen The sum of nitric oxide and nitrogen dioxide in any ratio, and reported as nitrogen dioxide. [ARP4418]

oxidizer *1.* A rocket-propellant component, such as liquid oxygen, nitric acid, or fluorine, that supports combustion when in combination with a fuel. [ARP4386] *2.* Specifically, substances (not necessarily containing oxygen) that support the combustion of a fuel or propellant.

oxidizing atmosphere An atmosphere that tends to promote the oxidation of immersed materials.

oximeter A device for electrochemically measuring the oxygen saturation by determining the ratio of reduced hemoglobin to oxyhemoglobin in the arterial or venous blood. [ARP171]

oxyfuel gas welding A mixture of oxygen and combustible fuel used for cutting or welding metal.

oxygen A colorless, tasteless, odorless gas, constituting one-fifth of the atmosphere; also available in liquid form (-183°C; -297.3°F), which vaporizes into gaseous oxygen. Oxygen supports combustion and is essential to life. [ARP171]

oxygen 17 An isotope of oxygen.

oxygenation The saturation of a substance with oxygen, either by chemical combination, chelation, or by mixture. [ARP171]

oxygen attack In a boiler, corrosion or pitting caused by oxygen.

oxygen concentrator The component of the OBOGS (on-board oxygen generation system) that processes air to provide an oxygen-enriched breathing gas. [ARP171]

oxygen deficiency *See* hypoxia.

oxygen dissociation curve A graph indicating the amount of oxygen that will be given up by the hemoglobin at different oxygen tensions. [ARP171]

oxygen meter A device for measuring either the fraction of oxygen or the partial pressure of oxygen in air or in a mixture of oxygen with other gases. [ARP171]

oxygen monitor The component of the OBOGS (on-board oxygen generation system) that monitors the partial pressure of oxygen in the breathing gas delivered to the aircrew. [ARP171]

oxygen module A structure or assembly, frequently in the form of a box or container, in which an oxygen generator is installed. [AS1304]

oxygen saturation of the blood Fraction of total hemoglobin that is in the form of oxyhemoglobin. Equal to the amount of bound oxygen divided by the maximum amount of oxygen that can be bound by the hemoglobin. [ARP171]

oxygen system, high pressure A system for delivering oxygen in which the supply is contained in one or more bottles or cylinders at 1800 to 2200 psi (12.4 to 15.2 MPa). [ARP171]

oxynitrides Base for a broad field of nitrogen ceramics, in which silicon, aluminum, and other elements are used to produce high-temperature refractory materials.

ozone layer *See* ozonosphere.

ozonosphere The general stratum of the upper atmosphere in which there is an appreciable ozone concentration, and in which ozone plays an important part in the radiation balance of the atmosphere. This region lies roughly between 10 and 50 kilometers, with maximum ozone concentration at about 20 to 25 kilometers. Also called ozone layer.

P

P *1*. Symbol for physiological partial pressure. PaO_2 = partial pressure of arterial oxygen; $PaCO_2$ = partial pressure of arterial carbon dioxide, etc. [ARP171] *2. See* poise.

Pa *See* pascal.

P$_a$ *See* partial pressure of air.

packaged boiler Refers to a packaged steam or hot water firetube boiler; a modified Scotch unit engineered, built, and fire-tested before shipment, with material, workmanship, and performance warranteed by the manufacturer as stated in the manufacturer's standard conditions of sale. Components include, but are not limited to, the burner, boiler, and controls.

packed column A distillation column filled with packing (commonly Raschig rings) such that the descending liquid will mix with the ascending vapors. Packing is often used instead of trays in columns for certain applications (such as gas adsorption), or for very-low-pressure drop systems.

packed decimal A method of representing a decimal number by storing a pair of decimal digits in one eight-bit byte, taking advantage of the fact that the numbers zero through nine can be represented by four bits.

packets (communication) Digital data messages that are almost always preceded by headers (containing address information and other control characters), and are followed by control characters signifying the end of a message.

packet switching Refers to the switching circuit system for multiple-access time-division data transmission.

packet switching system (PSS) In a wide area network, a method of sending data between computers.

packet transmission Transmission of bursts of digital data.

packing *1*. In a fluid seal, the flexible sealing element that is subject to sliding motion. Also called stuffing. [ARP4386] *2*. In data processing, the compression of data to save storage space.

pad *1*. Larger than a boss, a part that is attached to a pressure vessel to reinforce an opening. *2*. A fixed-value attenuator.

paddle-wheel level detector A device for detecting the presence or absence of bulk solids in a particular location. Consists of a motor that slowly rotates a paddle in the absence of material; and that rotates against a momentary switch when material is present at or above the location of the paddle.

paired or grouped wheel control *1*. Configuration of the skid-control system to control brake pressure to two or more brakes when any one of the braked wheels require skid control. Pairing or grouping of wheels is a function of landing-gear arrangement, performance required, and aircraft directional stability required. [AS483A] *2*. Control of two brakes/wheels utilizing one hydraulic antiskid valve and one control circuit. [AIR1489]

PAM Stands for pulse amplitude modulation. The process (or the results of the process) in which a series of pulses is generated having amplitudes proportional to the measured signal samples.

PAM/FM Frequency modulation of a carrier by pulse amplitude modulated information.

PAM/FM/FM Frequency modulation of a carrier by subcarriers that are frequency modulated by pulse amplitude modulated (PAM) information.

pan The international radio-telephony urgency signal. It is repeated three times, to indicate uncertainty or alert, followed by nature of urgency. [ARP4107]

panel *1.* The structure or surface to which a device is mounted. [ARP914A] *2.* A sheet of material held in a frame. *3.* A section of an equipment cabinet or enclosure, or a metallic or nonmetallic sheet, on which operating controls, dials, instruments or subassemblies of an electronic device, or other equipment are mounted.

panel board *See* control board.

panel method (fluid dynamics) Technique for analyzing and predicting the properties and characteristics of fluid flow. Sometimes called the finite element method.

PANT program Stands for passive nosetip technology program. An investigation by the Air Force of flow phenomena over reentry vehicle nosetips.

PAR *See* precision approach radar.

parabolas Open curves in which all points are equidistant from a fixed point (the focus), and a straight line. The limiting case occurs when the point is on the line; the parabola then becomes a straight line.

parabolic body Surface of revolution generated by revolving a section of a parabola about its major axis.

parabolic reflector Reflecting surface with a cross section along the axis in the shape of a parabola. Parallel rays striking the reflector are brought to a focus at a point; or, if the source of the rays is placed at the focus, the reflected rays are parallel. Also known as a dish.

paracone A system for recovering men and objects from great distances above the earth's surface and landing them safely onto the earth.

parallax Refers to the apparent differences in spatial relations when objects in different planes are viewed from different directions. In making instrument readings, for instance, parallax will cause an error in the observed value unless the observer's eye is directly in line with the pointer.

parallel In data transfer operations, a procedure that handles a multiple-bit code, working with all bits simultaneously, usually one word at a time.

parallel actuators Two or more actuators arranged in parallel to drive a single output or load. Usually, parallel actuators are physically separated, each with its own output connection, and are tied together by the load in a force or torque-summing fashion. Sometimes referred to as side-by-side actuators. [ARP4386]

parallel computation device Computer configured to allow pieces of a program to be executed in parallel on multiple processors, in lieu of being executed in serial on one processor. [AIR4548]

parallel computer A computer that has multiple arithmetic or logic units, used to accomplish parallel operations or parallel processing. Contrast with serial computer.

parallel elements In an electric circuit, two or more two-terminal elements connected between the same pair of nodes.

parallel laminate A laminate of woven fabric in which the plies are aligned in the same position as originally aligned in the fabric roll. [AIR4844]

parallel linkage A linkage mechanism that amplifies reciprocating motion. Depending on the geometry of the drive crank, driven crank, and connecting link, a parallel linkage can amplify and attenuate, as well as characterize, the relationship of the output driven crank to the input driven crank.

parallel offset route A parallel track to the left or right of the designated or established airway/route. Normally associated with area navigation (RNAV) operations; aircraft with RNAV equipment may ask for courses that parallel the airway routes so that they may avoid congestion on the established airway. [ARP4107]

parallel operation Simultaneous performance of several actions—particularly flow or information processing—usually of a similar nature, through provision of individual similar or identical devices for each such action. Parallel operation is performed to save time.

parallel processing *See* multiprocessing.

parallels Spacers or pressure pods used in molding equipment to regulate height and prevent mold parts from being crushed.

parallel servo In a control system, a servo located so that its output drives in parallel with the major input. This arrangement usually is used with actuators that perform an alternate function to that of the pilot. The parallel servo output will drive both the pilot controls and the flight-control system. [ARP4386]

parallel splice A device for joining two or more conductors in which the conductors lie parallel and adjacent. *See also* lap joint and splice. [ARP914A]

parameter *1.* A measurable or calculated quantity that varies over a set of values. *See also* measurand. [ARP1587] *2.* In a subroutine, a quantity whose value specifies or partly specifies the process to be performed. May be given different values when the subroutine is used in different main routines, or in different parts of one main routine, but usually remains unchanged throughout any one such use. *3.* A quantity used in a generator to specify machine configuration, designate subroutines to be included, or otherwise describe the desired routine to be generated. *4.* In mathematics, a constant, or a variable that remains constant, during some calculation. *5.* A definable characteristic of an item, device, or system.

parameter identification Estimation of the unknown parameters of models of physical plants or processes from their dynamic response.

parameterize To set up for variable execution depending on run-time parameter.

parameter optimization The selection of parameters, such as gains and time constraints, to minimize the error index. [ARP4386]

parametric analysis Analysis of the impact on circuit performance of changes in individual parameters such as component values, process parameters, or temperature.

parametric estimate An estimate made by using an expression relating cost to selected system parameters, performance requirements, or goals. [ARP4293]

parametric oscillator A nonlinear device that, when pumped by light from a laser, can generate tunable output. The beam produced by a parametric oscillator relies on oscillation within the nonlinear material.

parametric variation A change in system properties (e.g., magnification, resistance, or area) that may affect the performance of a control system that incorporates a feedback loop.

parasitic oscillations Unintended, self-sustaining oscillations or transient pulsations.

parent coil A coil of sheet that has been processed to final temper as a single unit and subsequently cut into two or more smaller coils, or into individual sheets, to provide the required width and length. [AMS2355F]

parent plate A plate that has been processed to final temper as a single unit and subsequently cut into two or more smaller plates to provide the required width and/or length. [AMS2355F]

parity *1.* A symmetry property of a wave function. *2.* A code that is used to uncover data errors by making the sum of the "1" bits in a data unit either an odd or even number.

parity bit A binary digit that is appended to a group of bits to make the sum of all the bits always odd (odd parity) or always even (even parity). Used to verify data storage and transmission.

parity check A check that tests whether the number of ones or zeroes in an array of binary digits is odd or even. Synonymous with odd-even check.

parking brake *1.* A static brake that, when actuated, continuously holds a load or prevents motion. [ARP4386] *2.* Refers to the system and/or apparatus for applying brakes for parking an aircraft for an extended length of time. This system usually utilizes the same brake assemblies as for normal braking, but with a different control and power source. [AIR1489]

parking valve The hydraulic valve that applies or blocks pressure to or from the brake assembly to maintain pressure for parking the aircraft. [AIR1489]

Parr turbidimeter A device for determining the cloudiness (turbidity) of a liquid by measuring the depth of the turbid suspension necessary to extinguish the image of a lamp filament of fixed intensity.

parse To break a command string into its elemental components for the purpose of interpretation.

Parshall flume A venturi-type device for measuring flow in an open channel at flow rates up to 1.5-billion gal/day (5.7-million m³/day). Consists of a converging upstream section, a downward sloping throat, and an upward sloping discharge section. May be made of any suitable structural material, usually concrete.

parsing algorithms Computer routines for the syntactic and/or semantic analysis and restructuring of natural language instructions or data for internal processing.

part An element of an assembly or subassembly that normally is of little use by itself and cannot be disassembled further for repair or maintenance.

part fraction defective The number of defective parts contained in a part population, divided by the total number of parts in the population; expressed in PPM. [AIR4896]

partial message A sequence starting with a header and terminating prematurely due to a suspend, abort, or other exception indication prior to a normal completion. [AS4710]

partial mission capable (PMC) Systems and equipment that are safely usable and can perform one or more, but not all, assigned missions because one or more of their mission essential subsystems are inoperative for maintenance or supply reasons. This status code is not used for equipment with a single mission, such as ground launch missile systems and Army and Marine Corps round equipment. The Military Services may further subdivide PMC into maintenance and supply categories. [ARD50010]

partial node A point, line, or plane in a standing wave field where some attribute of the wave has a nonzero minimum value.

partial obscuration A designation applied to sky cover when: (a) part of the sky (0.1 to 0.9) is completely hidden by surface-based phenomena; or (b) the sky is hidden by surfaced-based phenomena, but the vertical visibility is not otherwise restricted. [ARP4107]

partial-panel flight Instrument flight in which one or more of the usual cockpit flight instruments are inoperative or missing. [ARP4107]

partial pressure The pressure exerted by a designated component or components of a gaseous mixture.

partial pressure of air (P_a) The partial pressure of the air in equilibrium with a liquid; the absolute static pressure of the liquid and vapor phase in psia, minus the true vapor pressure of fuel at t°F. [AIR1326]

particle accelerator Any of several different types of devices for imparting motion to charged atomic particles.

particle concentration The number of individual particles per unit volume of liquid. [ARP1192]

particle count method A method by which particles in a fluid sample can be counted by number in size ranges. This can be done by membrane filtration and manual microscope counting, or by any of the automatic techniques. [ARP4386]

particle flux *See* flux (rate).

particle intercepted distance The spacing between the silicon particles that are intercepted by a straight line drawn in a random manner across the microstructure. [ARP1947]

particle laden jets Fluid jets, mainly issuing from a nozzle, that are turbulent and contain dispersed particles.

particle precipitation The precipitation of particles other than electrons and protons.

particles *1.* Elementary subatomic particles, such as protons, electrons, or neutrons. *2.* Very small pieces of matter. *3.* In celestial mechanics, hypothetical entities that respond to gravitational forces, but that exert no appreciable gravitational force on other bodies, thus simplifying orbital computations.

particle size A measure of dust size, expressed in microns or percent passing through a standard mesh screen.

particular risk Risk associated with those events or influences that are outside the systems and items concerned, but that may violate failure independence claims. [ARP4754]

particulate composite Material consisting of one or more constituents suspended in a matrix of another material. [AIR4844]

particulate contamination Any foreign solid matter contained in a fluid. [AIR1116]

parting agent Material used to prevent sealant from sticking to a surface. Also known as a release agent. [AIR4069]

parting line A mark on a molded piece where the sections of the mold have met in closing. [AIR4844]

parting tool *See* cutoff tool.

part program In numerical control, an ordered set of instructions in a language and format that causes operations to be effected under automatic control. Written in the form of a machine program on an input medium or stored as input data for processing in a computer to obtain a machine program.

parts per million (ppm) The unit volume concentration of a gas per million unit volumes of the gas mixture of which it is a part. [AIR1533]

parts per million carbon (ppmC) The mole fraction of hydrocarbon multiplied by 10^6, measured on a (C_1H_n) equivalence basis. Thus, 1 ppm of methane is indicated as 1 ppmC. To convert ppm concentration of any hydrocarbon to an equivalent ppmC value, multiply ppm concentration by the number of carbon atoms per molecule of the gas. For example, 1 ppm propane translates as 3 ppmC hydrocarbon; 1 ppm hexane as 6 ppmC hydrocarbon. [AIR1533]

parts pool An arrangement whereby participants are entitled to withdraw items from the agreed stock held by any participant. [AIR4896]

pascal Abbreviated Pa. Metric unit for pressure or stress.

PASCAL High-order computer programming language developed by Niklaus Wirth, originally as an educational tool to foster structured programming.

pascal second A measure of the specific viscosity of a fluid. [AIR4844]

pass *1.* A single circuit through a process, such as gases through a boiler, metal between forging rolls, or a welding electrode along a joint. *2.* In data processing, the single execution of a loop. *3.* The shaped open space between rolls in a metal-rolling stand. *4.* A confined passageway, containing heating surface, through which a fluid flows in essentially one direction. *5.* A single circuit of an orbiting satellite around the earth. *6.* A transit of a metal-cutting tool across the surface of a workpiece with a single tool setting.

passageway doors Those doors connecting passenger-occupiable compartments or providing access to approved emergency exits. [ARP499A]

pass, aircraft An aircraft passing a given station or a runway or taxiway; a takeoff and a landing constitute one pass for determining ground flotation capabilities. [AIR1489]

pass, gear One movement of a gear assembly past a specific point on the runway or taxiway under consideration. (One aircraft pass may include more than one gear pass.) [AIR1489]

passivating A process for treating stainless steel in which the material is subjected to the action of an oxidizing solution which augments and strengthens the normal protective oxide film, providing added resistance to corrosive attack.

passivation of metal The chemical treatment of a metal to improve its resistance to corrosion.

passive A general class of devices that operate on the signal power alone. [ARP993A]

passive AND gate An electronic or fluidic device that generates an output signal only when both of two control signals appear simultaneously.

passive failure A type of failure in which the failed device or system has no effect on the operational performance of a fault-tolerant system, even when it is commanded to function. Usually associated with standby or inactive features of a fault-tolerant system. [ARP4386]

passive metal A metal that has a natural or artificially produced surface film that makes it resistant to electrochemical corrosion.

passive nosetip technology *See* PANT program.

passive paralleling The simplest and most common type of redundancy; two parallel functional devices are utilized, thus if one fails the second is still available. This approach is limited to the more simple elements of the control system—the ones that can only fail passively, such as springs and linkages. When failure of one element occurs, there may be a change in performance or capability. [ARP4386]

passive transducer A transducer that produces output waves without any direct interaction with the source of power producing the actuating waves.

pass-through current *See* let-through current.

paste solder Finely divided solder alloy combined with a semisolid flux.

pasteurizing column A column that purges either a lighter-than-light key impurity through a purge stream at the top of the column, or heavier-than-heavy key impurity through a purge stream at the bottom of the column.

patch A section of coding inserted into a routine to correct a mistake or alter the routine. Often the patch is not inserted into the actual sequence of the routine being corrected, but is placed somewhere else, with an exit to the patch and a return to the routine provided.

patent defect An inherent or induced weakness that can be detected by inspection, functional test, or other defined means without the need for stress screens. [ARD50010]

path In MS-DOS, the instructions to the computer as to how to locate a particular file.

path loss (radio) The signal loss between transmitting and receiving antennas.

path profile (PROF) An imaginary earth-referenced line in space, connecting successive three-dimensional points through which flight is desired and/or controlled (altitude, latitude, and longitude). The flight path angle may be varying or constant, resulting in a curved or straight vertical profile. *See also* profile and V-path. [ARP1570]

patterned screen An electronic optical display that consists of discrete elements, such as phosphor dots or stripes, on a shadowmask CRT. [ARP1782]

pattern failures The occurrence of two or more failures of the same part in identical or equivalent applications, when the failures are caused by the same basic failure mechanism and they occur at a rate that is inconsistent with the parts predicted failure rate. [ARD50010]

pattern flow The paths of fluid flow connecting various ports in a given valve position. [ARP4386]

pattern recognition The identification of shapes, forms, and configurations by automatic means.

pavement, flexible Pavement of flexible construction, such as asphaltic pavement. [AIR1489]

pavement, rigid Pavement of rigid construction, such as concrete or steel-reinforced concrete. [AIR1489]

PAW *See* plasma arc welding.

payload *1.* The load that a vehicle is designed to transport under specified conditions of operation, in addition to its unladen weight. *2.* The warhead, its container, and activating devices in a military missile. *3.* The satellite or research vehicle of a space probe or research missile. [ARP4386] *4. See* warheads.

payload assist module Rocket vehicle with a spinning, solid-propellant motor to attain injection velocity; used to place a payload into intended orbits from the parking orbits of the STS (space transportation system).

payload control Execution of events involved in operating the payload and supporting systems.

payload delivery (STS) The transport of payloads via a space transportation system, including ground-to-earth orbit delivery by the Space Shuttle and orbit-to-orbit delivery via orbit transfer vehicles.

payload deployment & retrieval system System of mechanical and control devices, with associated data systems, for payload handling in space.

payload integration plan Procedures providing for compatibility of spaceborne experiments with the carrier spacecraft (e.g., shuttle orbiter).

payload transfer The in-space movement of payloads from point to point.

P band In telemetry, the portion of the radio-frequency spectrum from 215 to 260 MHz. Generally, a narrow section of that band near 225 MHz is available for telemetry application.

PBW *See* proportional bandwidth.

PC *1.* Abbreviation for personal computer. *2. See* program counter.

p chart In quality control, a type of data display in which the fraction defective in a sample, or over a production period, is charted against time or number of units of production.

PCM *1. See* phase change materials. *2. See* pulse code modulation.

PCM serial recording A technique whereby a train of bits is recorded on a single track of magnetic tape.

PD control Proportional plus derivative control, used in processes in which the controlled variable is affected by several different lag times. *See* proportional control and derivative control.

PDD *See* programmable data distributor.

PDM (modulation) *See* pulse duration modulation.

PDU *1. See* protocol data unit *2. See* pilot's display unit.

peak flow Maximum flow of gas during inspiratory or expiratory phase. [AIR1109]

peak power The maximum power that an actuation system is capable of delivering. This may be higher than rated power determined by system actuation requirements. [ARP4386]

peak pressure Maximum pressure, usually of short duration. [ARP4386]

peak-to-peak Pertains to the maximum amplitude excursion of a signal; for example, in the case of a pure sine wave, the maximum value between the 90° and 270° excursion points. *See also* double amplitude.

peak-to-peak amplitude For an oscillating or alternating function, the difference between maximum and minimum instantaneous values of the function.

peak to peak value For an oscillating quantity, the algebraic difference between the extremes of the quantity. [AIR1489]

pearlite An aggregate of ferrite and cementite, found in steel.

Peclet number A nondimensional number arising in problems of heat transfer in fluids.

peculiar ground support equipment Any system, subsystem, component, or equipment designed and used solely for maintenance task(s) performance on a specified end article of hardware. [ARD50010]

peculiar stars Stars with spectra that cannot be conveniently fitted into any of the standard spectral classifications. Denoted by a "p" after their spectral type.

peculiar support equipment Support equipment that is compatible with only one item. [ARD50010]

pedal A cockpit control device operated by the foot and used to produce directional rudder control, directional steering control, and brake control. [ARP4386]

pedestal In PAM, an arbitrary minimum signal value assigned to provide for channel synchronization and decommutation.

PEEK Stands for polyether ether ketones. A class of semicrystalline polymers used as molding compounds and as composite matrix materials.

peel ply A layer of open-weave material, usually fiberglass or heat-set nylon, applied directly to the surface of a prepreg lay-up. [AIR4844]

peel strength The adhesive bond strength obtained in the peeling mode selected. [AIR4844]

peel test A test in which sealant is cured on a selected substrate, then peeled from it using one of several types of testing machines (e.g., Scott, Instron, or Tinius Olson). In this test, two factors are of particular importance: the force required to peel the cured sealant off and the site of the failure. If the site of failure is within the sealant, it is called "cohesive failure." If the site of failure is at the bond line, it is called "adhesive failure." [AS7200/1]

peening exposure time The time required to obtain 100% coverage of a part—unless the cognizant engineering organization directs it to be the time required to reach saturation. [AMS2432B]

peening media Spherical or quasi-spherical material used in the controlled shot peening process. [AMS2431A]

peep door A small door, usually provided with a shielded glass opening through which combustion may be observed.

peep hole A small hole in a door, covered by a movable cover.

peer entities Entities within the same layer.

peer-to-peer protocol Communication protocol between peer entities.

peg count meter A meter that counts the number of trunks tested, the number of circuits passed busy, the number of tests failed, or the number of repeat tests completed.

pegging The joining of two pieces of core by crush-splicing them together with a third piece of core. [AIR4844]

PEL *See* PIXEL.

pellicle An extremely thin, tough membrane stretched over a frame. Because of its thinness, such a membrane transmits some light and reflects other light, and thus can serve as a beamsplitter. Also because of its thinness, such a membrane avoids the problem of ghost reflections, sometimes produced by other beamsplitters. Usually found as a beam splitter in an interferometer.

Peltier effect In solid-state physics, the principle whereby if two dissimilar metals are brought into electrical contact at one point, the difference in electrical potential at some other point will depend on the temperature difference between the two points. This is the principle upon which thermocouples are based.

pendant, catapult *See* bridle, catapult.

pendant, holdback A line or link assembly suspended from an aircraft to the deck fitting; used to hold back the aircraft until the specified time for release for catapult, or to restrain the aircraft for engine runup. [AIR1489]

pendulum scale A type of weighing device in which the weight of the load is counterbalanced by rotation of a bent lever with a fixed weight at the free end.

penetrameter A stepped piece of metal used to assess the density of exposed and developed radiographic film, and to determine relative ability of the radiographic technique to detect flaws in a workpiece.

penetrant A liquid that is capable of entering discontinuities or defects open to the surface, and that is adapted to the inspection process by being highly visible in small traces. [AMS2647A]

penetrant, fluorescent An inspection penetrant that fluoresces when excited by blacklight. [AMS2647A]

penetrant, post emulsifiable A penetrant that requires the application of a separate emulsifier to render it water-washable. [AMS2647A]

penetrant, water-washable A penetrant with a built-in emulsifier that makes it directly water-washable. [AMS2647A]

penetration *1.* A surface discontinuity that penetrates one skin and core or both skins and core, and whose width is the same order of magnitude as its length. [AIR4844] *2.* A qualitative term used to describe the degree to which radiation is capable of passing through a given object. [ARP5089] *3.* The portion of a published high-altitude instrument approach procedure that prescribes a descent path from the fix on which the procedure is based, to a fix or altitude from which an approach to the airport is made. [ARP4107] *4.* Distance from the original base metal surface to the point where weld fusion ends. *5.* On a casting, a surface defect formed where molten metal filled surface voids in the sand mold.

penetration number A measure of the consistency of materials such as waxes and greases, expressed as the distance that a standard needle penetrates a sample under specified ASTM test conditions.

penetration rate The distance per unit time that a drill cuts into a material, measured along the drill axis.

penetrometer An instrument for determining penetration number.

pen-motor recorder A data-versus-time strip-chart recorder whereby each trace is written by a motor-driven pen.

Penning discharge A direct-current discharge whereby electrons are forced to oscillate between two opposed cathodes and are restrained from going to the surrounding anode by the presence of a magnetic field.

Penning effect Effect whereby the effective ionization rate of a gas increases due to the presence of a small number of foreign metastable atoms.

pentode An electron tube containing five electrodes: an anode, a cathode, a control electrode, and two others, which usually are grids.

percent conductivity Conductivity of a material expressed as a percentage of that of copper. [ARP1931]

percent contrast *See* contrast.

percent of dilution The amount of diluent in a mixture, expressed in whole numbers; for example, a mixture consisting of thirty parts of diluent and seventy parts of oil by volume is a thirty percent mixture and its dilution is thirty percent. Percent dilution is *not* measured as the amount of diluent with respect to the original amount of oil. [S-4, 6]

percent slip Percentage representing the ratio of reduction in wheel rotational velocity under the influence of braking force, to the equivalent free-rolling rotational velocity of an unbraked wheel. [AIR1489]

perception *1.* The detection and interpretation of transthreshold cues from the environment by one or more of the senses. *2.* The awareness, or the process of becoming aware, of extraorganic or intraorganic objects or relations or qualities, by means of sensory processes and under the influence of perceptual set and of prior experiences. [ARP4107]

percussion A method of initiating an explosive item by striking it sharply. [AIR913]

percussion primer A percussion-actuated explosive device. [ARP4386]

perfect combustion The complete oxidation of all of the combustible constituents of a fuel, utilizing all of the oxygen supplied.

perfect diffusion Diffusion in which light is scattered uniformly in all directions by the diffusing medium. [ARP798]

perfect gas *See* ideal gas.

perfect vacuum A reference datum that is analogous to a temperature of absolute zero; used to establish scales for expressing absolute pressures.

performance analysis *See* analysis, performance.

performance and energy management (PM) Those performance functions, including thrust, that provide guidance signals to control an aircraft in climb, cruise, and descent to achieve optimum or desired performance (min cost, min fuel, select IAS/Mach, max gradient, etc.). These guidance signals may be directed to the auto flight system pitch channel and auto throttles, as well as pilot displays. [ARP1570]

performance assessment The procedure used to account for inlet distortion effects on engine performance. [ARP1420]

performance characteristic A qualitative or quantitative measurement unique to a piece of equipment or a system, and evident only during its test or operation.

performance curves Plots of the abilities of rotating equipment under various operating conditions.

performance degradation The condition or status indicating impaired or deteriorated engine gas path performance, as referenced to some established or predetermined condition. [ARP1587]

performance factors Factors computed or applied to basic performance, representing individual aircraft variance with the standard. May be expressed as a percentage drag factor, fuel factor, or combined SFC. Performance factor may also represent an assigned minimum maneuver margin (buffet boundary) or minimum climb and cruise rate of climb. [ARP1570]

performance index *1.* Ratio, in percent, of the time integral of instantaneous brake pressure, divided by the time integral of a series of straight lines connecting the peak brake pressure levels during which skidding initiates. [AIR1489] *2.* In industrial engineering, the ratio of standard hours to hours of work actually used to produce a given output. A ratio greater than 1.00 (100%) indicates standard output is being exceeded.

performance management Functions, including thrust, that provide guidance signals to control an aircraft in climb, cruise, and descent to achieve optimum or desired performance. [AIR4102/9]

performance monitoring Testing technique whereby it is verified that the unit under test is operational and performing its intended function. [AIR4896]

performance number Any of a series of numbers used to rate aviation gasolines with octane values greater than 100. The performance number generally compares fuel antiknock values with those of a standard reference fuel in terms of an index that indicates relative engine performance.

performance objectives Criteria setting forth the maximum allowable deviations from the relevant dimensions of perfect performance. [ARP4155]

perfusion The act of pouring over or through, especially the passage of a fluid through the vessels of a specific organ or body part. [ARP171] *See also* diffusion.

perigee The point at which a satellite orbit is the least distance from the center of the gravitational field of the controlling body or bodies. [ARP4386]

perihelions Those points in solar orbits that are nearest the sun.

period doubling The bifurcation of a nonlinear system to two stable periodic cycles on its route to chaotic turbulence.

periodic duty A type of intermittent duty in which the load conditions are regularly recurrent. [ARP1199A]

periodic function An oscillating quantity whose values repeatedly recur for equal increments of the independent variable.

periodic processes *See* cycle.

period, infant mortality The early period, beginning at zero unit time, during which the failure rate of a family of items can be expected to decrease. [AIR4896]

period, natural *1.* The reciprocal of a natural frequency of a landing gear. *2.* The time for one complete vibrational cycle in seconds. [AIR1489]

period, wear out failure rate The period during which the failure rate of a family of items can be expected to increase due to deterioration processes. [ARD50010]

peripheral *1.* A supplementary piece of equipment that puts data into, or accepts data from, the computer (e.g., printers, floppy disk memory devices, and videocopiers). *2.* Any device, distinct from the central processor, that can provide input or accept output from the computer.

peripheral seal *See* seal, peripheral.

peripheral speed *See* cutting speed.

peripheral visual cues Visual stimuli occurring outside of an approximately 60-degree cone from a person's normal sight line. Visual cues in this region are typically detected scotopically (with rods). Peripheral vision is mostly used for detecting gross movement and aids in maintaining ambient orientation. [ARP4107]

periscope Optical instrument that displaces the line of sight parallel to itself to permit a view that may otherwise be obstructed.

permanence The property of a plastic whereby it resists appreciable changes in characteristics with time and environment. [AIR4844]

permanent magnet A shaped piece of ferromagnetic material that retains its magnetic field strength for a prolonged period of time following removal of the initial magnetizing force.

permanent-magnet moving-coil instrument *See* moving-coil instrument.

permanent marking Marking of an item that will ensure identification of the item during its normal service life. [AS478G]

permanent pattern A secondary pattern or master having all of the necessary coordination data and reference points on its surface, from which a tool can be directly made with repeated accuracy. [AIR4844]

permanent pressure drop The unrecoverable reduction in pressure that occurs when a fluid passes through a nozzle, orifice, or other throttling device.

permanent set The deformation remaining after a specimen has been stressed a prescribed amount in tension, compression, or shear for a specified time period, and released for a specified time period. [AIR4844]

permeability *1.* For a magnetic material, the ratio of the magnetic induction to the magnetic field intensity in the same region. *2.* The ability to permit penetrations or passage. In this sense the term is applied particularly to substances that permit penetration or passage of fluids. *3.* The product of the solubility coefficient and the diffusion coefficient. [AIR4844]

permeameter *1.* A device for determining the average size or surface area of small particles. Consists of a powder bed of known dimensions and degree of packing, through which the particles are forced under pressure. Particle size is determined from flow rate and pressure drop across the bed; surface area, from pressure drop. *2.* A device for determining the coefficient of permeability by measuring the gravitational flow of fluid across a sample whose permeability is to be determined. *3.* An instrument for determining magnetic permeability of a ferromagnetic material by measuring the magnetic flux or flux density in a specimen exposed to a magnetic field of a given intensity.

permissible dose The amount of ionizing radiation that a human being can absorb over a given period of time without harmful result.

permittivity Preferred term for dielectric constant. *See* dielectric constant. [ARP1931]

perpendicularity of mounting faces Total maximum difference between high and low readings of a dial indicator suitably arranged to measure the out-of-squareness relation of the mounting surface at a specified distance from the OD of the unit, with respect to the rotational axis of the shaft when the housing is rotated about its fixed shaft. [ARP667]

persistence The continuation of luminance of a phosphor after electron excitation has been removed. [ARP1782]

personality variables Those traits of a person that characterize his/her behavior, predispose him/her to certain response patterns, and allow for some generalized predictions as to how he/she will respond in different situations; the pattern of motivation and of temperamental or emotional traits of the individual. [ARP4107]

PERT *See* program evaluation and review technique.

perturbation *1.* Any departure introduced into an assumed steady state of a system, or a small departure from a nominal path, such as a desired trajectory. *2.* A disturbance in the regular motion of a celestial body, the result of a force additional to that which causes the regular motion; specifically a gravitational force.

perturbation generator An instrument that simulates typical data-link perturbations, such as blanking, noise, bit-rate jitter, baseline offset, and wow.

petri nets Abstract, formal models of the information flow in systems with discrete sequential or parallel events. Major use has been for the modeling of hardware systems and software concepts of computers.

petroleum Naturally occurring mineral oil consisting predominately of hydrocarbons.

pF *See* picofarad.

PFD *See* primary flight display.

PFM (modulation) *See* pulse frequency modulation.

PFR *See* preliminary flight rating.

pH A measure of acidity or alkalinity. On the pH scale of 0 to 14, solutions with a pH reading of less than 7 are acid; solutions with a pH reading of more than 7 are alkaline; the midpoint of 7 is neutral.

phase *1.* The relationship between voltage and current waveforms in a-c electrical circuits. *2.* A visibly separate, but not necessarily separable, portion of a system. *3.* A microstructural constituent of an alloy that is physically distinct and homogeneous.

phase analysis An instrumentation technique that discriminates between conditions in a test part. Different conditions produce different phase angle changes in the test signal. [ARP5089]

phase angle *1.* A measure of how the output response of a system lags or leads a sinusoidal input to the system. [AIR1823] *2.* In an alternating-current signal, the difference between the phase of current and the phase of voltage, usually determined as the angle between current and voltage vectors plotted on polar coordinates. *3.* A measure of the propagation of a sinusoidal wave in time or space from some reference instant or position on the wave. *See also* phase shift.

phase angle firing For an SCR stepless controller, a method of operation whereby power is turned on for the proportion of each half cycle in the a-c power supply necessary to maintain the desired heating level.

phase angle meter *See* phase meter.

phase change materials (PCM) Materials that undergo solid/liquid phase transformations and whose latent heat of fusion properties are used to store and deliver thermal energy, usually solar energy.

phase conjugation Technique for the removal of phase distortions during propagation of laser beams through the atmosphere.

phase crossover, gain margin frequency On the Bode plot of the transfer function, the point at which the phase angle is -180 degrees. The frequency at which phase crossover occurs is called the gain margin frequency. [ARP4386]

phase detectors Devices that continuously compare the phase of two signals and provide an output proportional to their difference in phase.

phase deviation The peak difference between the instantaneous phase of the modulated wave and the carrier frequency.

phase difference The maximum allowed difference in the phase shift of two secondary voltages when using a three-wire or split secondary. [ARP4386]

phase discriminator A device that detects the phase relationship of a signal to that of a reference.

phase inspection *See* inspection, aircraft engine.

phase lag The instantaneous time separation between the input current and the corresponding control-flow variation, measured at a specified frequency and expressed in degrees (time separation in seconds × frequency in Hz × 360° per cycle). [ARP4386]

phase margin *1.* A measure of system stability, defined as the phase lag to be added to achieve 180 deg of phase lag at the open loop frequency response corresponding to the 0 dB amplitude ratio. [ARP4386] *2.* The degree in magnitude that the actual phase angle would have to be increased to make the system unstable when the gain is unity. [ARP1281A]

phase matching Alignment of a nonlinear crystal with respect to the incident laser beam in the proper way to generate a harmonic of the laser frequency in the material.

phase meter An instrument for measuring electrical phase angles. Also known as phase-angle meter.

phase modulation Angle modulation whereby the angle of a sine-wave carrier is caused to depart from the carrier angle by an amount proportional to the instantaneous value of the modulation wave. Combinations of phase modulation and frequency modulation are commonly referred to as frequency modulation.

phase reference signal A signal, usually electrical, that is provided on some engines and that indicates when a particular point on the low-speed rotor passes a reference point on the engine case. [AIR1839]

phase-sequence indicator A device that indicates the sequence in which the fundamental components of a polyphase set of voltages or currents reach some particular value—their maximum positive value, for example.

phase shift *1.* The difference in angular degrees between the primary excitation voltage and the secondary output voltage when the output is taken differentially. [ARP4386] *2.* The phase difference of two periodically recurring phenomena of the same frequency, expressed in angular measure. *See also* phase angle. *3.* The angle between the lines connecting a celestial body and the sun and a celestial body and the earth. *4.* The time difference between the input and output signal, or between any two synchronized signals, of a control unit, system, or circuit, usually expressed in degrees or radians.

phase shift circuit An electronic network whose output voltage is shifted in phase when compared to a specified reference voltage.

phase shifter An electronic device whose output voltage (or current) differs from its input voltage (or current) by some desired phase relationship. In some phase shifters, the phase is shifted a fixed amount because of an inherent design feature, but in others the phase relationship can be adjusted.

phase shift keying (PSK) *1.* A form of phase modulation in which the modulating function shifts the instantaneous phase of the modulated wave among predetermined discrete values. *2.* A form of PCM achieved by shifting the phase of the carrier, e.g., ± 90 degrees, to represent "ones" and "zeros."

phase shift variation The change in numerical value of phase shift relative to ambient temperature, input voltage level, excitation frequency, or shaft position. Should be expressed as a percentage relative to the value of phase shift at maximum coupling under specified excitation voltage, specified excitation frequency, and specified temperature. [ARP826]

phase velocity For a traveling plane wave at a single frequency, the velocity of an equiphase surface along the wave normal.

phase voltage balance with balanced load The maximum deviation of any of the three phase voltages from the average of the three phase voltages with a balanced three-phase load. [ARP1148A]

phase voltage balance with unbalanced load The maximum deviation of any of the three phase voltages from the average of the three phase voltages with a prescribed unbalanced load. [ARP1148A]

phase voltage displacement with balanced load The maximum deviation in degrees from 120 degrees between phases of the alternator voltages during balanced load conditions. [ARP1148A]

phase voltage displacement with unbalanced load The maximum deviation in degrees from 120 degrees between phases of the alternator voltages during a prescribed unbalanced load. [ARP1148A]

phasing Condition in which the phase of the differential output voltage with the primary voltage as the armature is displaced through the null position. [ARP4386]

Phelps vacuum gage A modified hot-filament ionization gage useful for measuring pressures in the range 10^{-5} to 1 torr.

phenolic A thermosetting resin produced by the condensation of an aromatic alcohol with an aldehyde, particularly of phenol with formaldehyde. [AIR4844]

phenylamine *See* aniline.

Philips gage An instrument that measures very low gas pressure (vacuum) indirectly by determining current flow from a glow-discharge device.

pH meter An instrument for electronically measuring electrode potential of an aqueous chemical solution and directly converting the reading to pH (a measure of hydrogen ion concentration, or degree of acidity).

phon A unit of loudness level equivalent to a unit pressure level in decibels of a 1000-Hz tone.

phonotelemeter A sophisticated stopwatch for estimating the distance from artillery by measuring the elapsed time from gun flash to arrival of the detonation sound.

phosphatizing Forming an adherent phosphate coating on metal by dipping or spraying with a solution to produce an insoluble, crystalline coating of iron phosphate. The phosphate coating resists corrosion and serves as a base for paint.

phosphor *1.* A material that gives off visual radiant energy when bombarded by electrons or ultraviolet light. [ARP1782] *2.* A phosphorescent material.

phosphorescence Emission of radiant energy—often in the visible-light range—following excitation due to absorption of shorter wavelength radiation. Phosphorescent emission may persist for a long time after the exciting radiation stops.

phosphoric acid fuel cells Long-life fuel cells for the low- to medium-wattage range which use phosphoric acid as an electrolyte.

phot The CGS unit of illuminance, equal to one lumen per cm^2. NOTE: The SI unit, lux, is preferred.

photoacoustic spectroscopy An optical technique for investigating solid and semisolid materials. In this technique, the sample is placed in a closed chamber filled with a gas and is illuminated with monochromatic radiation of any desired wavelength, with intensity modulated at some acoustic frequency. Absorption of radiation results in a periodic heat flow from the sample, which generates sound detectable with a sensitive microphone.

photocathodes Electrodes used for obtaining photoelectric emission.

photocell A device whose electrical resistance is altered in proportion to the amount of light that impinges on it. *See also* photoelectric cell.

photoconductive cell *1.* Photoelectric cell whose electrical resistance varies with the

497

amount of illumination falling upon the sensitive area of the cell. *2.* A transducer that converts the intensity of electromagnetic radiation, usually in the IR or visible bands, into a change of cell resistance.

photoconductor A type of conductor whose resistivity changes when illuminated by light; the changes in resistance can be measured to determine the amount of incident light.

photodarlington A detector in which a phototransistor is fabricated on the same chip with a second transistor that amplifies the signal from the phototransistor. The circuit thus formed is a Darlington circuit, a simple and inexpensive type of detector with limited performance.

photodiode A diode that detects light. In a vacuum photodiode, light detection depends on the photoelectric effect producing free electrons, which are collected by a positively charged electrode.

photodissociation The dissociation (splitting) of a molecule by the absorption of a photon. The resulting components may be ionized in the process (photoionization).

photodraft A photographic reproduction of a master layout or design on an emulsion-coated sheet of metal. Used chiefly as a master in tool- and die-making.

photoelastic stress analysis A visual full-field technique for measuring stresses in parts and structures. When a photoelastic material is subjected to forces and viewed under polarized light, the resulting stresses are seen as color fringe patterns. Interpretation of these patterns reveals the overall stress distribution, and accurate measurements can be made of the stress directions and magnitudes at any point. Three broad categories are embraced by this technique: (a) two-dimensional model analysis; (b) three-dimensional model analysis; and (c) photoelastic coating analysis. [AIR1489]

photoelectric cell A transducer that converts electromagnetic radiation in the infrared, visible, and ultraviolet regions into electrical quantities such as voltage, current, or resistance. *See also* photocell.

photoelectric control Modifying a controlled variable in accordance with a control signal whose value is related to the intensity of a light-beam input signal.

photoelectric counter A counting device actuated when a physical object passes through an incident beam of light.

photoelectric effect A physical phenomenon whereby a so-called photoelectric material emits electrons when struck by light—one bound electron being emitted for each photon of light absorbed.

photoelectric hydrometer A device for measuring specific gravity of a continuously glowing liquid. In this device, a weighted float, similar to a hand hydrometer, rises or falls with changes in liquid density; this changes the amount of light that is permitted to fall on a sensitive phototube whose output is calibrated in specific gravity units.

photoelectric photometer A device in which a photocell, phototransistor, or phototube is used to measure the intensity of light. Also known as an electronic photometer.

photoelectric pyrometer An instrument that measures temperature by measuring the photoelectric emission that occurs when a phototube is struck by light radiating from an incandescent object.

photoelectric threshold The amount of energy in a photon of light that is just sufficient to cause photoelectric emission of one bound electron from a given substance.

photoelectrochemical devices Electrochemical devices powered by light or other incident radiation to produce electricity and/or chemical fuels (e.g., hydrogen).

photoelectrons Electrons that have been ejected from their parent atoms by interaction between the parent atoms and high-energy photons.

photoemissive Emitting electrons when illuminated.

photoemissive tube photometer A device in which a tube made of photoemissive material is used to measure the intensity of light. This device is very accurate, but requires electronic amplification of the output current from the tube; it is considered chiefly a laboratory instrument.

photogrammetry *1.* The art or science of obtaining reliable measurements by means of photography. *2.* The science of making maps or accurately measuring features from aerial photographs.

photographic emulsion A light-sensitive coating, usually a silver halide compound in gelatin, used in photography or radiography to capture and store the visual image.

photographic recording Using a signal-controlled light beam or spot to record information; the position and/or the intensity of the spot is recorded.

photoionization The ionization of an atom or molecule by the collision of a high-energy photon with the atom or molecule.

photoluminescence Nonthermal emission of electromagnetic radiation that occurs when certain materials are excited by absorption of visible light.

photomasks In the production of integrated circuit devices, repeated arrays of microphotographs of the circuit patterns on glass substrates. These arrays or photomasks are used to form successive patterns on single wafers, often of submicrometer sizes.

photometer Instrument for measuring the intensity of light or the relative intensity of a pair of lights.

photometric brightness The intensity of illumination reflected or emitted by a surface as measured by a photoelectric device. [ARP1161]

photometry Any of several techniques for determining the properties of a material or for measuring a variable quantity by analyzing the spectrum and/or intensity of visible light.

photomultiplier A type of electron tube in which photons incident on a photocathode produce electrons by photoemission. These electrons are then amplified (their numbers are increased) by passing them through an electron multiplier. (Electrons passing through the multiplier are accelerated by high voltages and hit metal screens, from which they free more electrons.)

photomultiplier tube A phototube with one or more dynodes between its photocathode and output electrode. Also called multiplier phototube. *See also* multiplier tube.

photon *1.* According to the quantum theory of radiation, the elementary quantities of radiant energy. Photons are regarded as discrete quantities having a momentum equal to $h\nu/c$, where h is the Planck constant, n is the frequency of the radiation, and c is the speed of light in a vacuum. Photons are never at rest, have no electric charges, and no magnetic moments; but they have spin moments. The energy of a photon (the unit quantum of energy) is equal to $h\nu$. *2.* A quantum of electromagnetic radiation.

photon counting A measurement technique used for measuring low levels of radiation, in which individual photons generate signals that can be counted.

photophoresis Production of unidirectional motion in a collection of very fine particles, suspended in a gas or falling in a vacuum, by a powerful beam of light.

photosphere The intensely bright portion of the sun visible to the unaided eye.

phototheodolite A telescopic instrument or device incorporating one or more cameras (sometimes a motion-picture camera), used for taking and recording horizontal and vertical angular measurements. In aeronautics, the phototheodolite (sometimes in conjunction with radar equipment) is used to track airborne craft and to measure and record attitude, altitude, azimuth, and elevation angles. [ARP4107]

photothermal conversion Conversion of optical radiation into thermal energy, by a photoabsorptive or photoselective material.

phototransistor A transistor in which one of the two junctions is illuminated by light, and electrons are released. The transistor treats this release of electrons (current) as an input, and amplifies it; thus the phototransistor is a simple detector-amplifier.

phototube An electron tube containing at least two electrodes, one of which functions as a photoelectric emitter.

photovoltaic cell A transducer that converts the intensity of electromagnetic radiation, usually in the IR or visible bands, into a voltage.

phugoid Refers to long-period oscillation of airplane pitch axis, perpetually hunting about level attitude and trimmed speed.

phugoid mode A longitudinal mode consisting of a long-period, lightly damped oscillation in pitch and velocity, with nearly constant angle of attack. [ARP4386]

phugoid oscillation In a flight path, a long-period longitudinal oscillation consisting of shallow climbing and diving motions about a median flight path, and involving little or no change in angle of attack. [ARP4107] *See* oscillations.

physical catalyst Radiant energy capable of promoting or modifying a chemical reaction. [AIR4844]

physical entity Equipment that produces or receives information. [ARD50012]

physical properties Inherent characteristics of a substance (e.g., electrical conductivity, magnetic permeability, density, or melting point) that can be determined without applying mechanical force.

physical resources Those functions that are based in hardware in the application platform. [AS4893]

physiological acceleration The acceleration experienced by a human or an animal test subject in an accelerating vehicle.

P/I A pressure-to-current converter; linearly converts a signal pressure range into a signal current range; for example, 3–15 psi into 4–20 mA. *See* converter.

piano wire Carbon steel wire (0.75 to 0.85% C), cold drawn to high tensile strength and uniform diameter.

pi-bus *1.* A linear, multidrop communications medium that transfers datum serial, bit parallel information among up to 32 modules residing on a single backplane. *2.* Those modules that implement the slave only or master and slave portions of the pi-bus protocol. [AS4710]

PIC *See* position independent code.

pick *1.* An individual filling yarn, running the width of a woven fabric at right angles to the warp. Also called fill, woof, and weft. *2.* To experience tack. *3.* To transfer unevenly from an adhesive applicator mechanism due to high surface tack. [AIR4844] *4.* The open area left by the crossing of any two carriers in the weave of a braid axially along its length. [ARP1931]

pick count In a woven fabric, the number of filling yarns per inch of fabric. [AIR4844]

picketing *See* mooring.

pickling (metallurgy) Preferential removal of oxide or mill scale from the surface of a metal by immersion, usually in an acidic or alkaline solution.

picks per inch In a wire or cable, the number of carriers in either direction contained in one inch of the braid, measured parallel to the axis of the wire or cable. [AS1198]

pickup *1.* A transducer or other device that converts optical, acoustical, mechanical, or thermal images or signals into electrical output signals. *See* transducer. *See also* sensor. *2.* Electrical noise or interference from a nearby device, system, or circuit. *3.* The minimum value of an input signal (e.g., voltage, current or power) needed to make a relay function as intended.

pick up roll A spreading device whereby the roll for picking up the adhesive runs in a reservoir of adhesive. [AIR4844]

picofarad (pF) A measure of capacitance equal to 10^{-12} farads. [ARP1931]

PI control Proportional plus integral control, used in combination to eliminate offset. *See* proportional control and integral control. Also called proportional plus reset control.

PID *See* proportional, integral, and derivative.

P&ID *See* piping and instrumentation drawing.

PID action A mode of controller action in which proportional, integral, and derivative action are combined.

PID control *1.* A combination of proportional, integral, and derivative control actions. Refers to a control method in which the controller output is proportional to the error, its time history, and the rate at which it is changing. The error is the difference between the observed and desired values of the variable that is under control action. Also called three-mode control. *2.* Proportional plus integral plus derivative control, used in processes in which the controlled variable is affected by long lag times.

See proportional control, integral control, and derivative control.

PID controller A three-mode controller.

piece-wise linear model Model based on a set of matrix equations that describe the variations about a base operating point. Also referred to as state-space model and state-variable model. [AIR4548]

piercing An operation in which a tool is forced through a metal part in order to cut a hole of a specific shape and size.

piezoelectric The property of a material whereby it changes dimensions and vibrates when subjected to a high-frequency electric field. [AIR4367]

piezoelectric accelerometer A device for measuring variable forces associated with acceleration, such as from an earthquake or from vibration, by means of the response of a piezoelectric crystal in physical contact with a mass that reacts to the accelerating forces.

piezoelectric ceramics Ceramic materials with piezoelectric properties similar to those of some natural crystals.

piezoelectric detector A sensing element for detecting seismic disturbances which consists of a stack of piezoelectric crystals with an inertial mass on top of the stack; metal foil between the crystals collects the charges that develop when the crystals are strained.

piezoelectric effect The generation of an electric potential when pressure is applied to certain materials; or, conversely, a (small) change in shape when a voltage is applied to such materials. Piezoelectric devices can be used to precisely control small motions of optical components.

piezoelectric gage A pressure-measuring device used to detect and measure blast pressures from explosives and internal pressure transients in guns. In this gage, a piezoelectric crystal is used to sense a pressure transient and develop an output voltage pulse in response.

piezoelectricity The property exhibited by some asymmetrical crystalline materials whereby, when subjected to strain in suitable directions, they develop polarization proportional to the strain.

piezoelectric pressure transducer Any of several sensor designs in which a force acting on the sensing element is converted to an electrical output by a piezoelectric crystal.

piezoelectric transducer *1.* Transducers utilizing piezoelectric elements. *2. See* electrostriction transducer.

piezoid A piezoelectric crystal adapted for use by attaching electrodes to its surface or by other suitable processing.

piezometer *1.* An instrument for measuring fluid pressure. *2.* An instrument for measuring compressibility of materials.

piezoresistive accelerometer A device for measuring variable forces associated with acceleration, such as from an earthquake or from vibration, by means of changes in resistance of two or four semiconductor strain gages connected in a Wheatstone bridge circuit.

pig *1.* An in-line scraper for removing scale and deposits from the inside surface of a pipeline; a holder containing brushes, blades, cutters, swabs, or a combination is forced through the pipe by fluid pressure. *2.* A crude metal casting, usually of primary refined metal, intended for remelting to make alloys.

pigtail *1.* A conductor or wire extending from an electrical or electronic device, or from a cable shield, to serve as a connection. [ARP1931] *2.* A 270° or 360° loop in pipe or tubing, forming a trap for vapor condensate; prevents high-temperature vapors from reaching the instrument. Used almost exclusively in static pressure measurement.

pilot *1.* A mechanical control system, such as may be used to guide an aircraft in flight. *2.* A bar extending in front of a reamer to guide the reamer and force it to cut concentric with the original borehole. *3.* A flame that is utilized to ignite the fuel at the main burner or burners. *See also* ignitor.

pilotage Navigation by visual reference to landmarks. [ARP4107]

pilot balloon (meteorology) A small, free balloon tracked with a theodolite to determine the direction and speed of the wind at various altitudes. [ARP4107]

pilot circuit The portion of a control circuit or system that carries the control signal from the signal-generating device to the control device.

pilot, constant A pilot that burns without turndown throughout the entire time the boiler is in service.

pilot, continuous *See* pilot, constant.

pilot control pump A variable delivery pump whose output is controlled by the pressure at a control port. [ARP4386]

pilot, expanding A pilot that normally burns at a low turndown throughout the entire time the burner is in service—whether the main burner is firing or not. Upon a call for heat, the pilot is automatically expanded so as to reliably ignite the main burner. This pilot may be turned down at the end of the trial-for-ignition period for the main burner.

pilot flame establishing period The length of time that fuel is permitted to be delivered to a proved pilot before the flame-sensing device is required to detect pilot flame.

pilot head The difference in vertical elevation between the elements connected by a pilot line. [AIR1660A]

pilot induced oscillation Oscillations of a flying aircraft caused by transients and system changeovers; by pilot overreaction upon such transients; or by misleading pilot cues or excessive pilot gain in modern high-gain, high-order aircraft control systems.

pilot line The conduit that connects a hydro-mechanical shutoff valve to a pilot float valve or a fluidic level-sensor. [AIR1660A]

pilot plant A test facility built to duplicate or simulate a planned process or full-scale manufacturing plant, and used to gain operating experience or evaluate design alternatives before the full-scale plant is built.

pilot pressure The pressure from the first stage or pilot stage of a valve which is directed to a subsequent stage of the valve to change its position. [ARP4386]

pilot, proved A pilot flame that has been proved by flame-failure controls.

pilots access terminal A display unit and its associated control facilities. [ARP4102/15]

pilot's display unit (PDU) Component of a head-up display, consisting of the image source, the collimating optics, and the combiner. [ARP4102/8]

pilot stabilization period A timed interval that, on most systems today, is synonymous with timed trial for pilot ignition. (Today's programmers prevent main valve operation for a specified number of seconds after commencement of trial for pilot ignition, even though the pilot is immediately proved.)

pin A cylindrical object used as a connector or as a guide in aligning mating parts. [ARP480A]

pinboard A type of control panel in which pins rather than wires are used to control the operation of a computer. On certain small computers that use pinboards, a program is changed by the operator removing one pinboard and inserting another. *See also* control panel and plugboard.

pinch effect *1.* The result of an electromechanical force that constricts, and sometimes momentarily ruptures, a molten conductor carrying current at a high density. *2.* The self-contradiction of a plasma column carrying large currents, due to the interaction of this current with its own magnetic field.

pinch-off *See* cutoff.

pin contact *See* contact, pin.

pi network A network consisting of three branches connected in series to form a closed mesh; one of the three junctions is an input terminal, one is an output terminal, and the third is a common terminal connected to both input and output circuits.

ping An audible sound given off as a result of an individual wire breaking in a wire strand. [AS4536]

pinhole camera A camera that has no lenses, but consists essentially of a darkened box with a small hole in one side, and photographic film on the opposite side. Inverted images of outside objects are projected through the hole and recorded onto the film.

pinhole detector A photoelectric device that can detect small holes or other defects in moving sheets of material.

pin holes *1.* Very small holes in a metal, sometimes found as a type of porosity because of microshrinkage or gas evolution during solidification of the molten metal. [ARP4784] *2.* A fault in a casting or coating resulting from small blisters that have burst, or from small voids that formed during plating.

pinion The smaller of two gear wheels; or the smallest gear in a gear train.

pinning Refers to sites within a superconducting material that are produced by localizing inclusions, dislocations, voids, etc., and that provide a means of resisting flux motion (flux jumps) due to Lorenz forces.

PIN photodiode A semiconductor-diode light detector in which a region of intrinsic silicon separates the p and n type materials. Offers particularly fast response and is often used in fiber-optic systems.

pipe *1.* A discontinuity in the center of a rolled bar; caused by internal cavities formed in the ingot during solidification, and elongated or stretched in the rolling operations. [AS3071A] *2.* A tubular structural member used primarily to conduct fluids, gases, or finely divided solids. May be made of metal, clay, ceramic, plastic, concrete, or other materials. *3.* A general class of tubular mill products made to standard combinations of diameter and wall thickness.

pipe coating An inert material that is bonded to the inside of a pipe or vessel to serve as corrosion protection and to provide a smooth internal lining. [AIR4783]

pipe elbow meter A variable-head meter used to measure flow around the bend in a pipe.

pipe fitting A piece with an internal cavity for connecting lengths of pipe together, or for connecting them to tanks or other process equipment. There are different types, including couplings, elbows, nipples, tees, and unions.

pipelining *1.* Processing techniques for improving the capability of computer systems by modeling, sequencing control, resource allocation, etc. *2.* The process of increasing data-processing speed by simultaneously executing a number of basic instructions.

pipe material A hollow product designated by nominal pipe size and ANSI schedule number. [ARP1917]

pipestem A type of mouthpiece. [ARP171]

pipe tap A small hole in the wall of a pipe for sampling its contents, or for connecting a control device or pressure-measuring instrument.

pipe tee A pipe fitting in the shape of the letter T; used to connect a branch line at 90° to the main run of pipe.

pipe thread A type of screw thread used chiefly to connect pipe and fittings; in the usual configuration, it is a 60° thread with flat roots and crests, and a longitudinal taper of about 3/4 in. per foot (about 6.3%).

piping A system of pipes for carrying a fluid stream or gaseous material.

piping and instrumentation drawing (P&ID) *1.* A drawing that shows the interconnection of process equipment and the instrumentation used to control a process. In the process industry, a standard set of symbols is used to prepare drawings of processes. The instrument symbols used in these drawings are generally based on Instrument Society of America (ISA) Standard S5.1. *2.* The primary schematic drawing used for laying out a process-control installation.

pipping pressure *1.* The pressure at which a safety valve opens. *2.* The pressure that a pipe cannot withstand without exceeding its design characteristics.

Pirani gage A pressure transducer used to measure very low gas pressure. Based on measurement of the resistance of a heated wire filament; resistance varies in accordance with thermal conduction of the gas, which in turn is related to gas pressure. Used primarily for pressures less than one atmosphere.

piston A cylindrical part that slides in a cylinder or barrel and serves to transfer force to or from the enclosed fluid. [ARP4386]

piston displacement The volume traversed by a piston in a single cycle, or stroke.

piston engines Engines, especially internal combustion engines, in which a piston or pistons moving back and forth work upon a crankshaft or other device to create rotational movement. Also called reciprocating engines.

piston meter A type of fluid flow meter. Consists of a variable-area, constant-head device in which the flow rate is indicated by a pointer attached to a piston, which in turn is positioned by the buoyant force of the fluid.

pistonphone A device consisting of a small chamber and reciprocating piston of measurable displacement, used to establish a known sound pressure in the chamber.

piston ring seal A seal ring installed in a groove on the piston circumference to minimize the clearance flow between the piston outer diameter and the cylinder bore.

piston rod A coaxial column or rod, attached to or integral with a piston, and serving to transmit force between the piston and another mechanical member. [ARP4386]

piston-type variable-area flowmeter Any of several flowmeter designs in which fluid passing through the meter exerts force on a piston such that the piston moves against a counterbalancing force to expose a portion of an exit orifice, the amount exposed being directly related to volume flow.

piston valve A valve in which a piston is the main control element. [AIR4783]

pit A small surface cavity in a metal part or coating, usually caused by corrosion, or formed during electroplating.

pitch *1.* Rotation or oscillation of an aircraft about its lateral axis. [ARP4107] *2.* The distance that a propeller would advance in one revolution if it were acting in a solid medium. [ARP4107] *3.* On a patterned CRT screen, the distance from the center of the most fundamental screen element (for example, red phosphor dot or line) to the center of the nearest like element (for example, the nearest adjacent red phosphor dot or line). Also known as phosphor dot pitch or triad pitch. [ARP1782] *4.* An auditory sensation of tone that is directly related to sound-wave frequency. *5.* A heavy, black or dark brown liquid or solid residue from distillation of tar or oil, occurring naturally as asphalt. *6.* The distance between similar mechanical elements in an array, such as gear teeth, screw threads, or screen wires. *7.* The distance between centerlines of tubes, rivets, staybolts, or braces. *8.* In computer printers, a measure of the number of characters printed per inch. Typically 10, 12, or 17.

pitch attitude The angle between the longitudinal axis of an aircraft and the horizontal plane. [ARP4107]

pitch axis The lateral (or Y) axis of an aircraft about which pitching occurs. [ARP4386]

pitch-catch An inspection method in which the ultrasonic energy is emitted by one transducer element and received by another on the same or adjacent surface. [ARP5089]

pitch diameter For a layer of strands or wires in a conductor or cable, the diameter of the circle passing through the centers of the strands or wires. [AS1198]

pitch diameter size The diameter of the cylinder that passes through the thread profile of either the internal or external screw thread of a product, in such a manner as to make the width of the thread ridge and thread groove equal. [AS8879]

pitch-down *See* pitchunder.

pitch moment A moment about a lateral axis of an aircraft, rocket, airfoil, etc. The pitch moment is positive when the angle of attack is increased, or the body is nosed upward. [ARP4107]

pitch setting The propeller blade setting as determined by the blade angle, and measured in a manner and at a radius specified by the instruction manual for the propeller. [ARP4107]

pitchunder An act or instance of an aircraft pitching nose downward; a tendency of an aircraft to pitch nose downward. Also called tuck down or pitch-down. [ARP4107]

pitchup An act or instance of an aircraft pitching nose upward; a tendency of an aircraft to pitch nose upward. [ARP4107]

pitot-static head *See* pitot-static tube.

pitot-static tube A combination of a pitot tube and a static port arranged coaxially or otherwise parallel to one another and mounted externally on an aircraft (generally on the wing, the nose, or the vertical stabilizer) in a position to sense the air flow and pressure undisturbed by the flow over or around other structures of the aircraft. Used principally to determine airspeed from the difference between impact and static pressures. Also called pitot-static head. *See* pitot tube and static port. [ARP4107]

pitot tube *1.* Tube that receives the total pressure, including impact pressure, created by the forward motion of the aircraft. [ARP4107] *2.* Open-ended tube or tube arrangement that, when pointed upstream, may be used to measure the stagnation pressure of the fluid for subsonic flow; or the stagnation pressure behind the normal shock wave of the tube for supersonic flow. *See also* pitot-static tube and static port.

pitot-venturi tube A combination of a venturi device and a pitot tube.

pitting *1.* Local area of coating removal or material removal caused by impact or other damage; not usually associated with visual cracks or crazing. [AIR5122] *2.* A concentrated attack by oxygen or other corrosive chemicals in a boiler, producing a localized depression in the metal surface.

pivot bearing *See* step bearing.

pivoting The procedure or technique of turning or pivoting an aircraft about a point, such as to pivot about one main gear. [AIR1489]

pivots The paths followed by a point on a diameter of a circle, as the circle rolls along in a straight line.

pixel In data processing, a portion of a CRT display screen.

PIXEL (PEL) Contracted from picture elements. Image resolution elements in vidicon-type detectors.

placard A posted notice on or in an aircraft, setting forth a requirement or limiting operational condition. [ARP4107]

placarding Any durable visual intelligence, fixed in place, that provides instructions for locating and/or operating inflatable aircraft emergency evacuation equipment. [ARP4277]

plain bearing configurations Those bearings whose interface motions are sliding actions between two parallel surfaces. [AIR1594]

plain binding A weave in which each warp wire and each shute wire passes over one and under the next adjacent wire in each direction. [AIR888]

plain weave A weaving pattern in which the warp and fill fibers alternate; that is, the repeat pattern is warp/fill/warp/fill. [AIR4844]

planar Lying essentially in a single plane. [AIR4844]

planar helix winding A winding in which the filament path on each dome lies on a plane that intersects the dome, while a helical path over

the cylindrical section is connected to the dome paths. [AIR4844]

planar network An electronic network that can be drawn or sketched on a plane surface without having any of the branches cross each other.

planar winding A winding in which the filament path lies on a plane that intersects the winding surface. [AIR4844]

plane of polarization In a plane-polarized electromagnetic wave, the plane that contains both the direction of propagation and the electric field vector.

plane of symmetry The vertical fore-and-aft plane that divides an airplane into symmetrical halves. [ARP4107]

plane polarized wave An electromagnetic wave in a homogeneous isotropic medium which has been generated, or modified by the use of filters, so that the electric field vector lies in a fixed plane also containing the direction of propagation.

plan equation Refers to an equation for determining horsepower (HP): HP = plan/33,000, where p is mean effective pressure in psi; l is piston stroke in feet; a is net piston area in in.^{2}; and n is number of strokes per minute.

plane strain A deformation of a body in which the displacement of all points in the body are parallel to a given plane, and the displacement values are not dependent on the distance perpendicular to the plane.

planetary boundary layer The layer of the atmosphere from the earth's surface to the geostrophic wind level, including the surface boundary layer and the Ekman layer.

planetary cores The centers of planets.

planetary craters Collective term for craters on any of the planetary surfaces.

planetary crusts The outermost layers of planets. A planetary crust is on top of the mantle and is modified by various processes of weathering, sedimentation, metamorphosis, volcanism, and bombardment by meteorites.

planetary limb In astronomy, the circular outer edge of a planet.

planetary systems Systems consisting of a star and the planets and other objects in orbit around it.

planetary twister A twisting machine whose payoff spools are mounted in rotating cradles, which hold the axes of the spools in a fixed direction as the spools are revolved about one another, so the wire will not kink as it is twisted. [ARP1931]

plane wave(s) Waves on uniform currents in two-dimensional nondivergent fluid systems rotating with varying angular speeds about the local vertical (beta plane). These waves represent a special case of barotropic disturbance, conserving absolute vorticity.

planigraphy *See* tomography.

planimeter A device for measuring the area of a plane surface, usually of irregular shape, by tracing its perimeter.

plan-position indicator A type of radar display that provides a maplike presentation on a circular screen, indicating the range and azimuth of any discernible object with respect to a point on the screen representing the location of the transmitter. Also called plan view display (PVD). [ARP4107]

plan view display (PVD) *See* plan-position indicator.

plasma antennas An air plasma made by ionizing the atmosphere which acts as the conducting element of an RF antenna.

plasma arc cutting Use of plasma torches for cutting hard materials at extremely high temperatures.

plasma arc welding (PAW) Welding in which metals are heated with a constricted arc between an electrode and the workpiece (transferred arc), or the electrode and the constricting nozzle (non-transferred arc). Shielding is obtained from the hot, ionized gas issuing from the orifice, which may be supplemented by an auxiliary source of shielding gas.

plasma bubbles Pockets of very low electron density in the equatorial F region of the ionosphere, in which the plasma density is lower than the ambient density.

plasma clouds Specifically, a mass of ionized gas flowing out of the sun.

plasma compression Decrease in volume and consequent increase in density of a plasma, usually by the application of an intense magnetic field.

plasma cooling Temperature control of plasmas in controlled-fusion operations.

plasma core reactors Nuclear reactors utilizing fissionable plasmas (such as uranium fluoride) for the fuel.

plasma currents Electric currents induced in plasmas by injection of fast ion beams or some other means.

plasma display devices Digital matrix, flat-panel devices in which small gas-discharge plasma cells are used as light-emitting sources.

plasma drift Movement in the ionosphere of ion and plasma concentration, caused by electric field variations in the upper atmosphere.

plasmadynamic lasers Stimulated emission devices in which the lasing gas flow has been replaced with a lasing plasma flow of atoms or ions.

plasma engines Reaction engines in which magnetically accelerated plasma is used as a propellant. Plasma engines are types of electrical engines.

plasma equilibrium Condition of plasma in which the constituent particles or fluid elements are unaccelerated or collectively at rest in steady flow.

plasma etching Removal of material by use of a focused plasma beam.

plasma focus A highly compressed plasma.

plasma generation *See* plasma generators.

plasma generators *1.* Machines, such as electric-arc chambers, that will generate very high-heat fluxes to convert neutral gases into plasmas. *2.* Devices in which the interaction of plasmas and electrical fields are used to generate currents.

plasma pumping Application of radiation of appropriate frequencies to plasma to increase the population of atoms or molecules in the higher energy states.

plasmas Electrically conductive gases that are comprised of neutral particles, ionized particles, and free electrons, but, taken as a whole, are electrically neutral. Plasmas are further characterized by relatively large intermolecular distances, large amounts of energy stored in the internal energy levels of the particles, and the presence of plasma sheaths at all boundaries of the plasma. Plasmas are sometimes referred to as a fourth state of matter.

plasma sheaths *1.* The boundary layers of charged particles between plasmas and their surrounding walls, electrodes, or other plasmas. *2.* Envelopes of ionized gases that surround bodies moving through an atmosphere at hypersonic velocities.

plasmasphere Envelope of highly ionized gases surrounding the earth or another planet.

plasma torches Burners that attain 50,000°C temperatures by the use of plasma gas injected into an electric arc. Plasma torches are used for welding, spraying molten metal, and cutting hard rock or hard metals.

plastic An imprecise term generally referring to any polymeric material, natural or synthetic. Its plural, plastics, is the preferred term for referring to the industry and its products.

plastic clad silica A step-index optical fiber in which a silica core is covered by a transparent plastic cladding of lower refractive index. The plastic cladding is usually a soft material, although hard-clad versions have been introduced.

plastic deformation The changes in dimensions of items caused by stress which are retained after the stress is removed. [ARP700]

plastic fibers Optical fibers in which both core and cladding are made of plastic material. Typically their transmission is much poorer than that of glass fibers.

plastic flow *1.* Deformation under the action of a sustained force. *2.* Flow of semi-solids in the molding of plastics. [AIR4844]

plastic foam *See* expanded plastic.

plasticity *See* plastic properties.

plasticity index The difference between liquid limit and plastic limit. [AIR1780]

plasticize To make a material moldable by softening it with heat or a plasticizer. [AIR4844]

plasticizer In compounding plastics, a chemical agent added to make the plastics softer and more flexible. [ARP1931]

plastic limit The moisture content at which a fine-grained soil passes into the plastic state from the semisolid state, or vice versa. [AIR1780]

plastic memory The tendency of a thermoplastic material that has been stretched to return to its unstretched shape upon being reheated. [AIR4844]

plasticorder A laboratory device for measuring temperature, viscosity, and shear-rate in a plastics material, properties which can be used to predict its performance.

plastic properties The tendency of a loaded body to assume a deformed state other than its original state when the load is removed.

plastic tooling Tools constructed of plastics, generally laminates or casting materials. [AIR4844]

plastometer An instrument for determining flow properties of a thermoplastic resin by forcing molten resin through a fixed orifice at a specified temperature and pressure.

plate *1.* Refers to the plate that is initially impinged upon by the teemed liquid metal and that then directs the liquid metal through feeder troughs to the bottom of the ingot molds. [MAM2304] *2.* Any tool that has the general configuration of a smooth, flat piece of material of uniform thickness. [ARP480A] *3.* A rolled, flat piece of metal. Depending on the type of metal, the minimum thickness for the product to be called plate (instead of sheet or strip) may vary; for example, plate steel is any hot-finished flat-rolled carbon or alloy steel product more than 8 in. wide and more than 0.230 in. thick; or more than 48 in. wide and more than 0.180 in. thick. *4. See* electroplating.

plateau *1.* Generally, any portion of a function where the value of the dependent variable is essentially constant over a range of values for the independent variable. *2.* Specifically, a portion of the output versus input characteristic of an instrument, electronic component, or control device where the output signal level is essentially independent of the input signal level.

plate baffle A metal baffle.

platelet alpha A relatively coarse acicular alpha, usually with low aspect ratios. This microstructure arises from cooling alpha or alpha-beta alloys at a slow rate from temperatures at which a significant fraction of beta phase exists. [AS1814]

platen(s) *1.* The mounting plates of a press, to which the entire mold assembly is bolted. [AIR4844] *2.* A plane surface receiving heat from both sides, and constructed with a width of one tube and depth of two or more tubes, bare or with extended surfaces.

plate polarized light Light polarized by means of optical plates set at Brewster's angle to the optic axis. The more plates, the greater the purity of the plane-polarized exit beam.

plate shear stress A type of honeycomb shear-strength test in which the honeycomb specimen is bonded between two thick steel plates, which are displaced relative to each other to place the specimen in shear. [AIR4844]

plating The electrolytic application of one metal over another. [ARP1931]

PLC *See* programmable logic controller.

plenum *1.* An accumulation device that may be either an integral part of the concentrator, or mounted on the airframe. This device provides reserve breathing gas for periods when breathing rates exceed the MSOGS (molecular sieve oxygen generation system) breathing-gas production rate. [ARP171] *2.* A condition whereby air pressure within an enclosure is greater than barometric pressure outside the enclosure. *3.* An enclosure through which gas or air passes at relatively low velocities.

plug *1.* A device that fits into a hole and serves as a stopper. [ARP480A] *2.* A male fitting for making electrical connections by insertion into a receptacle. [ARP480A] *3.* A rod or mandrel over which a pierced billet is drawn to form a tube or pipe; or that is inserted into a tube or pipe during cold reduction. *4.* A punch or mandrel over which a cup is drawn. *5.* A projecting portion of a die, intended to form a recess in a forged part.

plugboard A perforated board that accepts manually inserted plugs to control the operation of equipment; for example, a removable panel containing an ordered array of terminals, which may be interconnected by short electrical leads (plugged in by hand) according to a prescribed pattern, thereby designating a specific program or sequence of specified program steps. *See also* pinboard and control panel.

plug die *See* floating plug.

plug fuseholder A receptacle with female threads to accommodate a plug-type fuse.

plug gage A metal member used to check the dimension of a hole. The gaging element may be straight or tapered, plain or threaded, and of any cross-sectional shape.

plug meter A device for measuring flow rate in which a tapered rod extends through an orifice. When the rod is positioned so that the effective area of the annulus is just sufficient to handle the fluid flow, the rate of flow is read directly from a scale.

plug, sealing An accessory used to fill open, nonwired cavities in a connector grommet so as to prevent the entry of moisture, fluids, or foreign particulate contaminants. [ARP914A]

plug valve A type of shutoff valve consisting of a tapered rod with a lateral hole through it. As the rod is rotated 90° about its longitudinal axis, the hole is first aligned with the direction of flow through the valve, and then aligned crosswise, interrupting the flow.

plumb Indicating a true vertical position with respect to the earth's surface. This condition is usually determined by a plumb bob, which consists of a weight (plummet) suspended on a string and positioned entirely by gravity.

plumb-bob gage *1.* A device for determining liquid level in which a weighted plummet is lowered on a calibrated tape or cable until it just touches the liquid surface. *2.* A device for detecting solids level in a storage bin or hopper by lowering a plummet until the lowering cable slackens, which is usually detected by an electrical or mechanical triggering device.

plummet gage *See* plumb-bob gage.

plunger-type instrument A moving-iron instrument in which a pointer is attached to a long, specially shaped piece of iron that moves along the axis of a coil by variable electromagnetic attraction, depending on the current flowing in the coil.

ply *1.* In general, fabrics or felts consisting of one or more layers. *2.* The layers that make up a stack. *3.* Yarn resulting from twisting operations. *4.* A single layer of prepreg. *5.* A single pass in filament winding. *6.* A sheet or layer that is considered to be one discrete piece of manufactured material, such as fabric, tape, or adhesive film. [AIR4844]

ply grouping Uncured or unconsolidated plies that are only part of a cured part and are grouped together for drawing clarity or for manufacturing engineering purposes. [AIR4844]

ply orientation In laminated materials, the arrangement of bonded layers to obtain optimal strength or other characteristics.

ply rating A rating number that indicates the relative strength of a tire carcass. The ply rating is proportional to, but not necessarily the same as, the number of structural plies in the carcass. [AIR1489]

PMC *See* partial mission capable.

pneumatic Describes systems that employ gas, usually air, as the carrier of information and the medium to process and evaluate information.

pneumatically actuated valves Valves in which gas is the primary power source for positioning the valve. This includes valves that are solenoid-controlled, but pneumatically actuated. [ARP986]

pneumatic controller A device that is activated by air pressure to mechanically position another device, such as a valve stem. Also known as pneumatic positioner.

pneumatic control system A system that makes use of air for operating control valves and actuators.

pneumatic control valve A spring-loaded valve that regulates the area of a fluid-flow opening by changing position in response to variable pneumatic pressure opposing the spring force.

pneumatic extensions A system that converts float position to a proportional standard pneumatic signal. A magnetic coupling connects the internal float extension with an external mechanical system linked to a pneumatic transmitter.

pneumatic input servovalve A servovalve wherein the input is pneumatic. [ARP4386]

pneumatic intelligence transmission *See* pneumatic telemetering.

pneumatic-mechanical actuators A mechanical actuation system that converts pneumatic power to mechanical force usually through the use of an air motor coupled through shafting to reduction gearing that either provides rotary torque output or is converted through a screw device to linear force output. [ARP4386]

pneumatic positioner *See* pneumatic controller.

pneumatic signal line *1.* An air (pneumatic) signal; usually 3-15 psig is used as the energy medium. *2.* Applies to a signal using any gas as the signal medium. If a gas other than air is used, the gas may be identified by a note on the signal symbol or otherwise.

pneumatic system A system in which air is used for operating control valves and actuators (cylinders, motors).

pneumatic telemetering Remote transmission of a signal from a primary sensing element to an indicator or recorder by means of a pneumatic pressure impulse sent through small-bore tubing. May be used to monitor temperature, pressure, flow rate, or other variables in a process unit or system. Also known as pneumatic intelligence transmission.

pneumatic to current converter (P/I) A pressure to current converter; linearly converts a signal pressure range into a signal current range; for example, 3–15 psi into 4–20 mA. *See* converter.

Pockel's cell A device in which the Pockel's effect is used to modulate light passing through a material. This modulation relies on rotation of beam polarization caused by the application of an electric field to a crystal; the beam then has to pass through a polarizer, which transmits a fraction of the light dependent on its polarization.

pocket chamber A small ionization chamber that can be charged, then carried in a person's pocket and periodically read to determine cumulative radiation dose received since the instrument was last charged. Also known as a pocket dosimeter.

pocket dosimeter *See* pocket chamber.

pocket meter A pocket-size direct-reading instrument for measuring radiation dose rate.

poidometer An automatic weighing device used in conjunction with a belt conveyer.

point A process variable derived from an input signal or calculated in a process calculation.

pointer *1*. A needle-shaped or arrowhead-shaped element whose position over a scale indicates the value of a measured variable. *2*. In data processing: (a) a data string that tells the computer where to find a specific item; or (b) a cursor or similar object on a computer screen.

point of mishap *See* mishap, point of.

point spread functions Mathematical functions involved in image processing.

point, test A means of safe access to permit a measurement that will facilitate maintenance, repair, calibration, alignment, or monitoring. Test points may be accessible in the normally installed position (exposed test points) or may require disassembly of the equipment for accessibility (internal test points). [ARP4386]

point-to-point numerical control A simple form of numerical control in which machine elements are moved between programmed positions without particular regard to path or speed control. Also known as positioning control.

poise Abbreviated P. The CGS unit of dynamic viscosity, equal to one dyne-second per cm^2. The centipoise (cP) is more commonly used.

Poiseuille flow Laminar flow of gases in long tubes at pressures and velocities such that the flow can be described by Poiseuille's equation. *See* laminar flow.

Poisson's ratio The ratio of the change in lateral width per unit width, to change in axial length per unit length, caused by the axial stretching or stressing of a material. [AIR4844]

polar Describes an unsymmetrical molecule, such as water or sulphur dioxide, in which the mean center of all the electronic charges does not coincide with the mean electrical center of the nuclei. [AIR4844]

polar coordinates In a plane, a system of curvilinear coordinates whereby a point is located by its distance, r, from the origin (or pole) and by the angle, theta, that a line (radius vector) joining the given point and the origin makes with a fixed reference line, called the polar axis.

In three dimensions, short for space polar coordinates.

polar diagram *1*. A diagram showing the relative effectiveness of an antenna system for either transmitting or receiving. Principally shows directional characteristics. *2*. A schematic representation that relates events in the cycle of a piston engine to crankshaft position.

polarimeter An instrument for determining the degree of polarization of electromagnetic radiation, specifically the polarization of light.

polarimetry Chemical analysis in which the amount of substance present in a solution is estimated from the amount of optical rotation (polarization) that occurs when a beam of light passes through the sample.

polariscope Instrument for detecting polarized radiation and investigating its properties.

polarity *1*. Refers to the relative surface charge of a material resulting from the molecular structure of the adherent surface. [AIR4844] *2*. The relationship between the direction of control flow and the direction of input current. [ARP490] *3*. The sign of the electric discharge associated with a given object, such as an electrode or an ion.

polarization *1*. The arrangement or orientation of connector inserts, jackscrews, polarizing pins/sockets, keys/keyways, or housing configurations to prevent the mismating or cross-mating of connectors. [ARP914A] *2*. The state of electromagnetic radiation in which transverse vibrations take place in some regular manner, e.g., all in one plane, in a circle, in an ellipse, or in some other definite curve. *3*. With respect to particles in an electric field, the displacement of the charge centers within a particle in response to the electric force acting on it. *4*. The response of the molecules of a paramagnetic medium (such as iron) when subjected to a magnetic field.

polarization maintaining fiber A single-mode optical fiber that maintains the polarization of the light that enters it, normally by including some birefrigence within the fiber itself. Normal single-mode fibers, and all other types, allow polarization to be scrambled in light transmitted through them.

polarized meter A meter with its zero point at the center of the scale; the direction of pointer deflection indicates electrical polarity, and the distance of pointer deflection indicates the value of a measured voltage or current.

polarizer A filter that transmits light of only a single polarization.

polarizing coating Coatings that influence the polarization of light passing through them, typically by blocking or reflecting light of one polarization and passing light that is orthogonally polarized.

polarizing pin, socket, key, or keyway Devices incorporated in a connector to accomplish polarization. [ARP914A]

polar migration *See* polar wandering.

polarographic analysis A method of determining the amount of oxygen present in a gas by measuring the current in an oxygen-depolarized primary cell.

polarography A method of chemical analysis that involves automatically plotting the voltage-current characteristic between a large, non-polarizable electrode and a small polarizable electrode immersed in a dilute test solution. A curve containing a series of steps is produced, with the potential identifying the particular cation involved and the step height indicating cation concentration. Actual values are determined by comparing each potential and step height with plots generated from test solutions of known concentrations.

polar solvents Solvents such as alcohols and ketones that contain hydroxyl or carbonyl groups, have high dielectric constants, and show strong polarity. [AIR4844]

polar wandering (geology) Migration during geologic time of the earth's poles of rotation and magnetic poles. Also known as Chandler motion and polar migration.

polar winding A winding in which the filament path passes tangent to the polar opening at one end of the chamber, and tangent to the opposite side of the polar opening at the other end. A one-circuit pattern is inherent in the system. [AIR4844]

pole-dipole array An electrode array for making resistivity or induced-polarization surveys.

In this array, one current electrode is placed far away from the area being surveyed, while an assembly containing one current electrode and two potential electrodes is moved laterally across the area in a search pattern.

pole face On a magnetized part, the surface through which magnetic lines of flux enter or leave the part.

pole piece A shaped piece of ferromagnetic material, integral with or attached to one end of a magnet, whose function is to control the distribution of magnetic lines of flux.

pole-pole array An electrode array for making resistivity or induced-polarization surveys. In this array, one current electrode and one potential electrode, in close proximity, are moved laterally across the area being surveyed.

polestar recorder An instrument used to determine the amount of cloudiness during the night. Consists of a fixed, long-focus camera positioned so that Polaris is permanently within its field of view. The apparent motion of Polaris is recorded as a circular pattern on the film, with approximate span of cloudiness indicated by interruptions in the arc due to clouds passing between the star and the camera.

polished plate glass Glass whose surface irregularities have been removed by grinding and polishing, so that the surfaces are approximately plane and parallel. [ARP798]

polled access A media access method whereby the node that has the right to use the network medium delegates that right to other stations on a per message basis. *See also* master-slave.

polling A method of sequentially observing each channel to determine if it is ready to receive data or is requesting computer action.

pollution transport Dispersion or diffusion of atmospheric or water pollutants.

poloidal flux Refers to a plasma-confinement concept utilizing multipole magnetic fields.

polycased glass *See* cased glass.

polydispersed A suspension containing a mix of particle sizes. [ARP1192]

polyetheretherketones *See* PEEK.

polymerization A chemical reaction in which the molecules of a monomer are linked together to form large molecules whose molecular weight is a multiple of that of the original monomeric substance. [AIR4844]

polymer matrix composites Materials consisting of reinforcing fibers, filaments, and/or whiskers embedded in polymeric bonding matrices for increased mechanical and physical properties.

polymer quenchant A water solution of polymer used to minimize distortion and residual stresses. [AMS2770E]

polytropic expansion An expansion process in which changes of pressure and density are related. [ARP4386]

polyvinylchloride (PVC) A family of insulating compounds whose basic ingredient is either polyvinylchloride or the copolymer of polyvinylchloride with vinyl acetate. [ARP1931]

polyvinyl fluoride DuPont's Tedlar, unplasticized PVF film with outstanding resistance to ultraviolet radiation.

PONA analysis Determination of amounts of paraffins (P), olefins (O), naphthalenes (N) and aromatics (A) in gasoline in ASTM standard tests.

pooled Describes the combination of data from different data sources. [AIR4979]

pooled standard deviation The square root of averaged sample variances; gives an estimate of overall population standard deviation. [AIR4979]

pooled variance A weighted average of k variances, where the degrees of freedom are used as weights. [AIR4979]

Pope cell A type of relative humidity sensor in which a bifilar conductive grid on an insulating substrate is employed, whose resistance varies with relative humidity (RH) over a range of about 15 to 99% RH.

poppet A spring-loaded ball that engages a notch.

poppet valve A mushroom-shaped valve that controls the intake or exhaust of working fluid in a reciprocating engine. May be cam-operated or spring-loaded; direction of movement is at right angles to the plane of its seat.

popping pressure In compressible fluid systems, the inlet pressure at which a safety-relief valve opens.

population The set of measurements about which inferences are to be made; or the totality of possible measurements that might be obtained in a given testing situation. [AIR4844]

population mean The average of all potential measurements in a given population, weighed by their relative frequencies in the population. [AIR4844]

population median In a population, that value for which the probability of other values exceeding it is 0.5, and the probability of other values being less than it is 0.5. [AIR4844]

population variance A measure of dispersion in a population. [AIR4844]

pop-up barrier *See* barrier, pop-up.

porcelain enamel *See* enamel and vitreous enamel.

porcupine boiler A boiler consisting of a vertical shell from which project a number of dead-end tubes.

porosimeter A laboratory device for measuring the porosity of reservoir rock using compressed gas.

porosity *1.* A lack of soundness, usually in the form of gas holes or shrinkage voids that take on the character of gas holes. [AS3071A] *2.* A series of small, closely spaced discontinuities or voids. [AMS3920A]

porpoise To oscillate about the lateral axis, in the manner of a porpoise. [ARP4107]

port *1.* An opening at the surface of a component which incorporates provisions for attachment of a fluid-carrying passage, line, fitting, or removable plug. [ARP4386] *2.* The entry or exit point from a computer for connecting communications or peripheral devices. *3.* An aperture for passage of steam or other fluids.

portability Refers to the ability of test procedures to be used by more than one test-equipment configuration. [AIR4896]

portable chemical oxygen device A piece of equipment used as a portable supply of oxygen, the oxygen being produced by means of a chemical reaction. [AS1303]

portable standard meter A portable instrument used primarily as a reference standard for testing or calibrating other instruments.

positional notation A numeration system in which a number is represented by means of an ordered set of digits, such that the quantity contributed by each digit depends on its position as well as its value.

positioner A device attached to a crimping tool to position the conductor barrel between the indentors. [ARP914A]

positioner, amplifying A pneumatic positioner whereby the input control signal is amplified to a proportionately higher pressure, needed to drive the actuator, e.g., 3–15 psig input/ 6–30 psig output.

positioner, characterized A positioner in which the valve-position feedback is modified to produce a nonlinear response.

positioner, double acting A positioner with two outputs, suited to a double-acting actuator.

positioner, electro-pneumatic A positioner that converts an electronic control signal input to a pneumatic output.

positioner, reversing A positioner that converts the input control signal into an output that is directionally opposite to the input.

position error An error in the reading of an airspeed indicator owing to the difference between the pressure (especially the static pressure) at the pressure-measuring location and the free-stream pressure. [ARP4107]

positioner, single acting A positioner with one output, suited to a spring-opposed actuator.

positioner, split range A positioner that drives an actuator full stroke in proportion to only a part of the input signal range.

position independent code (PIC) A code that can execute properly wherever it is loaded in memory, without modification or relinking. Generally, this code uses addressing modes that form an effective memory address relative to the program counter (PC) of the central processor.

positioning Manipulating a workpiece in relation to working tools.

positioning action Controller action in which the final position of the control element has a predetermined relation to the value of the controlled variable.

positioning control *See* point-to-point numerical control.

positioning control system A system of control in which each controlled motion operates in accordance with instructions that only specify the next required position; the movement in the different axes of motion are not coordinated with each other and are executed, simultaneously or consecutively, at velocities that are not controlled by instructions on the input data medium.

position light Any of the three lights used on an aircraft to indicate its position and direction of flight (green on the right wing, red on the left wing, and white on the tail). [ARP4107]

position sensor Any device for measuring position and converting the measurement into an electrical, electromechanical, or other type of signal for remote indication or recording.

position summing A multichannel arrangement that sums the position outputs of multiple actuators into a single output. [ARP4386]

position symbol On a radar display, a computer-generated indication of the mode of tracking. [ARP4107]

position telemeter A remote-reading instrument for indicating linear or angular position of an object or machine component.

position, valve *See* valve position.

positive control The separation of all air traffic, within designated airspace, by air traffic control. [ARP4107]

positive displacement Describes any device that captures or confines definite volumes of fluid for purposes of measurement, compression, or transmission.

positive-displacement flowmeter Any of several flowmeter designs in which volumetric flow through the meter is broken up into discrete elements, and the flow rate is determined from the number of discrete elements that pass through the meter per unit time.

positive draft Refers to a pressure in a furnace, gas chamber, or duct that is greater than ambient atmospheric pressure.

positive feedback *1.* Feedback that results in increasing the amplification. Also called regenerative feedback. [AIR1489] *2.* A closed loop in which any change is reinforced until a limit is eventually reached.

positive G In a gravitational field or during an acceleration, the effect that occurs when the human body is normally positioned so that the force of gravity and/or inertia, acts on it in a head-to-foot direction; or the effect that occurs on an aircraft to increase the wing loading. [ARP4107]

positive-going edge The edge of a pulse going from a low to a high level.

positive ion A group of atoms that has acquired a positive electric charge by the loss of one or more electrons.

positively skewed Describes a distribution that is not symmetric and whose longest tail is on the right. [AIR4844]

positive meter Any of several devices for measuring fluid flow by alternately filling and emptying a container or chamber of known capacity. Fluid passes through such a device in a series of discrete amounts, by weight or volume.

positive motion Motion transmitted from one machine part to another without slippage.

positive plate The plate that has an electrical potential higher than that of the other plate during normal cell operation. [ARP4386]

positive pressure *1.* Pressure that is above normal atmospheric pressure, as differentiated from vacuum pressure, which is below normal atmospheric pressure. *2.* Refers to air pressure in a clean room, where lay-up takes place prior to bonding. [AIR4844]

positive pressure area *See* area, positive pressure.

positron(s) *1.* Subatomic particles that are identical to electrons in atomic mass, theoretical rest mass, and energy, but opposite in sign. *2.* A positively charged beta particle.

post A structural brace or member that supports the pivot point for the trailing arm of a levered-suspension gear system. [AIR1489]

postassembly seal A seal that is applied after the tank structure and subassemblies have been assembled/attached. [AIR4069]

postconversion bandwidth The bandwidth presented to a detector.

postcure Additional elevated-temperature cure, usually without pressure, to improve final properties and/or complete the cure, or to decrease the percentage of volatiles in the compound. [AIR4844]

post design services (PDS) cost As applied to hardware and software, the cost of the upkeep and amendments to technical documentation, defect investigations, development and design of modifications arising out of defects, production of modification kits; and where applicable, the cost of the supplier installing the modifications in the field. [ARP4293]

post-fab Describes a fabrication process whereby close-outs and inserts are attached or put into the panel after the facings are bonded to the core. [AIR4844]

post flight engine status An immediate GO/NO GO indication of engine availability for the next flight; and, if in a NO GO status situation, an indication of required maintenance action as appropriate. [ARP1587]

postflight inspection *See* inspection, aircraft engine.

postforming The forming, bending, or shaping of fully cured, C-staged thermoset laminates that have been heated to make them flexible. [AIR4844]

post guiding Describes a design using guide bushing or bushings fitted into the bonnet or body to guide the post of the plug.

post inspection cleaning The removal of penetrant material residues and developer from a test part after the penetrant inspection process is completed. [AMS2647A]

post insulate To insulate an electrical connection after assembly. [ARP914A]

postlaunch reports Memoranda issued following spacecraft launchings to report launch data, the launch vehicle performance, orbital elements (expected and measured), and current status.

postmission analysis (spacecraft) An analysis that deals with the scientific aspects of a mission; a broader term than postflight analysis.

post processor In numerical control, a computer program that adapts the output of a processor, applicable to a piece part, into a machine program for the production of that part on a particular combination of machine tool and controller.

potential energy Energy possessed by a body by virtue of its position in a gravity field, in contrast with kinetic energy, possessed by a body by virtue of its motion.

potential fault detection The process, technique, or capability of identifying a discrepancy whereby good outputs are "0" or "1" while the fault output is "X" (unknown). [AIR4896]

potential gradients In general, the local space rate of change of any potential, for example, the gravitational potential gradient or the velocity potential gradient.

potential transformer An instrument transformer that is connected in the circuit so that its primary winding is in parallel with a voltage to be measured or controlled.

potentiometer *1*. Instrument for measuring differences in electric potential by balancing the unknown voltage against a variable known voltage. If the balancing is accomplished automatically, the instrument is called a self-balancing potentiometer. 2. A variable electric resistor.

potentiometric titration A technique of automatic titration whereby the end point is determined by measuring a change in the electrochemical potential of the sample solution.

pot life The length of time, at some specific temperature, that a catalyzed thermosetting resin system retains a viscosity low enough to be used in processing. [AIR4844]

potting *1*. A resin mixture poured into honeycomb cell cavities to create a solid area. Potting is used to repair damaged core cells or to create strong points for fasteners. [ARP5089] 2. The sealing of the cable-end of a connector with a material compound to exclude entry of fluids or contaminants, and to provide strain relief to terminated wires or cable. [ARP914A]

pound of thrust In jet or rocket engines, a measurement unit of the reaction force generated and available for propulsion. [ARP4107]

pour coat A liquid honeycomb-sealant material of high solids content, used to stabilize honeycomb core after crushing, or as a seal against moisture entry when it is coated on the cell walls. The pour coat is applied by pouring it through the honeycomb cells. [AIR4844]

pour point *1*. Lowest temperature at which a fluid will flow under specified conditions. [ARP4386] *2*. Temperature at which molten metal is cast. *3*. Temperature at which a petroleum-base lubricating oil becomes too viscous to flow, as determined in a standard ASTM test.

pour point depressant A substance that, when added to an engine lubricating oil in relatively low concentrations, will reduce its pour point by retarding the formation of a rigid wax structure. [S-4, 6]

pour test Chilling a liquid under specified conditions to determine its ASTM pour point.

powder coating A painting process in which finely ground, dry plastic is applied to a part using electrostatic and compressed-air transfer mechanisms. The applied powder is heated to its melting point, flows out to form a smooth film, and cures by means of a chemical reaction.

powder pattern An x-ray diffraction pattern consisting of a series of rings on a flat film or a series of lines on a circular strip film, and resulting when a monochromatic beam of x-rays is reflected from a randomly oriented polycrystalline metal or from powdered crystalline material.

power absorption The amount of power that is absorbed by a tire in the rolling process. Power absorption results in a temperature rise in the tire carcass. [AIR1489]

power amplifier A component designed specifically for increasing signal power. [ARP993A]

power approach A landing approach during which the airplane is under power, in contrast with a power-off gliding approach. [ARP4107]

power assurance The measurement and interpretation of the engine instrumentation data to determine whether the power or thrust delivered by the engine will be achieved within operating limits. [AIR1873]

power boost control valve *See* control valve.

power-boosted flight control systems Reversible control systems in which the pilot effort, exerted through a set of mechanical linkages, is at some point in these linkages boosted by hydraulic power. [ARP578]

power-by-light Refers to the conversion of optical power into electrical power for use as the primary power source for otherwise isolated electrical circuits. [ARD50024]

power-by-wire (PBW) actuation Refers to an integrated servoactuator that incorporates an electric motor to receive power from the aircraft main electric power system in lieu of an actuator connected directly to the aircraft main hydraulic system.

power circuit The hydraulic system that provides the pressures for the impulse curve. [AIR1228]

power circuit breaker A circuit breaker whose function is to protect the wiring carrying the power to the equipment. *See* circuit breaker. [ARP4101/5]

power common The reference point for power supplies and return currents from powering equipment. Sometimes referred to simply as common. Power common or common is not to be confused with signal common.

power consumption The maximum amount of electrical power used by a device during normal steady-state operation.

power controller An electrical system component that provides circuit protection and switching control of a branch circuit. [AS1831]

power converter The component of the power system that transforms energy from the power source into a form compatible with the servoactuator requirements. [ARP4386]

power curve, getting behind *See* cognitive disorientation.

powered flight Describes the flight of an aircraft, missile, or spacecraft during the period or periods when it is propelled by a self-contained engine. [ARP4386]

powered models Models that can be tested in complete force equilibrium, including propulsion.

powered steering angle The number of degrees (radians) that the nosewheel is displaced from the center position while powered by the NWS control system. [ARP1595]

power/energy meter An instrument that measures the amount of optical power (watts) or energy (joules). May operate in the visible, infrared, or ultraviolet region, and detect pulsed or continuous beams.

power factor *1.* The ratio of the real power in watts to the apparent power in volt-amperes. [ARP4386] *2.* The ratio of actual power to apparent power delivered to an electrical power circuit. In practice, the power factor is determined by dividing the resistive load, in watts, by the product of voltage and current, in volt-amperes, and expressing the result in percent.

power factor controllers A solid-state electronic device that reduces excess energy waste in AC induction motors by providing only the amount of voltage required to satisfy a given load.

power-factor meter An instrument for directly indicating power factor in a circuit.

power failure *1.* Engine stoppage for a reason not directly attributable to the engine structure. For example, fuel exhaustion usually results in power failures, not engine failure. [ARP4107] *2.* The removal of all power accidentally or intentionally.

power gain *1.* The slope of the curve of output power versus control power in the vicinity of the operating point. [ARP993] *2.* The ratio of the power that a transducer delivers to a specified load, under specified operating conditions, to the power absorbed by its input circuit. *3.* For an antenna, in a given direction, 4π times the ratio of the radiation intensity in that direction to the total power delivered to the antenna.

power input The energy required to drive a fan, expressed in brake horsepower delivered to fan shaft. *See also* excitation.

power interrupt A momentary loss of power to the engine-monitoring equipment, resulting in possible loss of data. [ARP1587]

power landing The landing of a spacecraft on a body in space whereby the thrust of the spacecraft motors is used as a brake. [ARP4386]

power level *1.* At any given point in a system, the amount of power being delivered or used. *2.* The ratio of the power at a point to some arbitrary amount of power chosen as a reference. This ratio is usually expressed either in decibels that are based on 1 milliwatt (abbreviated dBm) or in decibels based on 1 watt (abbreviated dBW). *See* decibel.

power limited Describes a type of starter that can put out only a limited amount of power during a start cycle. Examples of power limited starters are those operating on a constant power supply, such as a pneumatic, electric, or hydraulic ground cart or airborne auxiliary power unit. [ARP906A]

power line protector A device used between a computer and a power outlet to absorb power surges or other interference that could damage the computer. *See also* surge protector.

power loading The ratio of the gross weight of a propeller-driven aircraft to its power, usually expressed as the gross weight of the aircraft divided by the rated horsepower of the powerplant corrected for air of standard density; with turboprop engines, the equivalent shaft horsepower is used. [ARP4107]

power loss *1.* The power absorbed by the tires, a loss as far as useful work is concerned. *See* power absorption. [AIR1489] *2.* In a power-transmission system or circuit, the difference between input power and output power, often expressed as a percent of input power. *3.* In a current- or voltage-measuring instrument, the active power at the terminals of the instrument when the pointer is at the upper end of the scale. In any other electrical circuit, the difference between active power and electrical load at a stated value of current or voltage.

power modulator The component of the actuation system that regulates the potential energy of the power source to the actuator as a function of controller output. In a simple hydraulic power-boost system, which has no separate controller, the power modulator controls the power to the actuator directly as a function of the command or error signal. [ARP4386]

power modules (STS) Modules that provide power for payloads for STS and mission-dependent equipment.

power off stalling speed *See* V_{so}.

power-operated flight control system An irreversible control system in which the pilot, through a set of mechanical linkages, actuates a power-control servomechanism; this mechanism actuates the primary control surface or corresponding device. A system of this type may have electrical or electronic pilot input modes. [ARP578]

power plate *See* brake housing.

power point *See* rated power.

power run That part of the launching run after release during which the applied catapult towing force is accelerating the aircraft. [AIR1489]

power setting A value used to rate the power output of an engine. The power setting of turbojet and similar engines should generally be expressed in terms of net thrust (corrected). For turboprop and similar engines, power setting should generally be in terms of shaft horsepower. [ARP1179A]

power setting parameter (PS) A parameter used by the engine manufacturer to determine engine power or thrust. Various parameters are used for this purpose; for example, EPR, torque, or N1. [AIR1873]

power source The component that supplies energy for load actuation. [ARP4386]

power spectral density (PSD) Refers to a type of frequency analysis that can be performed on data by a computer using special software, by an array processor, or by a special-purpose hardware device.

power splitter At the output of a telemetry radio transmitter, the device that splits the transmitter power between two or more antennas.

power stall A flight condition in which an airplane may be stalled despite the existence of a given amount of power. [ARP4107]

power supply Within a computer, the device that transforms external AC power to internal DC voltage.

power system The aircraft electric power system; includes four separate parts: (a) the generation system; (b) the distribution system; (c) the utilization systems; and (d) the conversion systems. [ARP4404]

power system capacity The capacity of the power sources, rated under the prescribed operating and environmental conditions in the aircraft. For parallel systems, this is the sum of the individual power source ratings with a paralleling factor applied; for nonparallel systems, this is the individual power source rating. [AS1212]

power transfer unit A device in which hydraulic power in one hydraulic system is used to supplement the hydraulic power in a second system without interchange of fluid between the systems. [ARP1280]

power transmission (lasers) Space-to-earth power transmission utilizing a laser (from solar-powered satellites).

power-up test Test conducted automatically when power is provided to the system. [ARP4102/13]

Poynting-Robertson effect The gradual decrease in orbital velocity of a small particle, such as a micrometeorite in orbit about the sun, due to the absorption and reemission of radiant energy by the particle.

ppm *See* parts per million.

PPM (modulation) *See* pulse-position modulation.

ppmC *See* parts per million carbon.

PPOS *See* present position.

practice instrument approach An instrument approach procedure conducted by visual flight rules (VFR) or instrument flight rules (IFR) for the purpose of pilot training or proficiency demonstrations. [ARP4107]

Prandtl number A dimensionless number representing the ratio of momentum transport to heat transport in a flow.

preamble Highest-density signal used for acquiring synchronization. [ARP4290]

preamplifier(s) *1.* An amplifier whose primary function is to raise the output of a low-level source to an intermediate level, so that the signal may be further processed without appreciable degradation in the signal-to-noise ratio. *2.* In radar, amplifiers separated from the remainder of the receiver, and located so as to provide the shortest possible input circuit path from the antenna so as to avoid deterioration of the signal-to-noise ratio.

preassembly seal A seal that is applied during or prior to the assembly of the fuel-tank structure; for example, faying-surface seals and prepack seals. [AIR4069]

precession Change in the direction of the axis of rotation of a spinning body, such as a gyro, when acted upon by a torque.

precharge The gas sealed within the accumulator and separated from the hydraulic fluid. [ARP4379]

precharged pressure (inflation pressure) *1.* The pressure at which a component is initially charged or inflated. [ARP4386] *2.* That gas pressure to which a liquid accumulator is charged, prior to loading with the liquid to be stored. [ARP906A]

precheck In level-control valves or systems, refers to a feature that permits external actuation of the valves or systems. This feature may be used for checking operation with a non-full tank, or for holding the valve closed for some other fuel-management consideration. [AIR1660A]

precheck system In an aircraft fuel system, a feature that provides a simulated closure operation of the level-control valves or fuel-shutoff valves. [AIR4783]

precipitation (of defects) The process of transforming a latent defect into a patent defect through the application of stress screens. [ARD50010]

precipitation hardening Heating a material to an intermediate temperature to cause precipitation of constituents within the material, thereby producing an increase in strength and hardness. [AMS2728]

precipitator An electrostatic-type fly-ash separator and collector.

precision *1.* The degree of agreement obtained within a set of observations or test results. [AIR4844] *2.* The closeness with which a measurement upon a given invariant sample can be reproduced in short-term repetitions of the measurement, with no intervening analyzer adjustment. [ARP1256] *3.* The quality of being exactly or sharply defined or stated. The precision of a representation can be measured by the number of distinguishable alternatives from which it was selected, which is sometimes indicated by the number of significant digits it contains. Also known as exactness.

precision approach procedure/precision approach A standard instrument approach procedure in which an electronic glide scope is provided; for example, ILS and PAR. [ARP4107]

precision approach radar (PAR) Radar equipment used in some ATC facilities operated by the FAA and/or the military services to detect and display azimuth, elevation, and range of aircraft on the final approach course to a runway. Primarily used to conduct a precision instrument approach to a runway. [ARP4107]

precision block *See* gage block.

precision control motor A reversible, two-phase induction-type of motor specifically designed to be operated from two independent voltage sources of the same frequency. One voltage source is maintained at a fixed voltage and phase direction, and the second source varied in voltage and phase direction to control speed, torque, and direction of rotation of the motor. A high torque-to-inertia ratio and an approximately straight line torque-speed curve are inherent characteristics of good control motors. [ARP497A]

precision error The random error based on a set of repeated measurements. Also known as repeatability error or sampling error. [AIR1678]

precision guided projectiles Missiles guided by precise laser radiation.

precision index The estimated standard deviation based on n measurements. [AIR1678]

precision measurement equipment Test and measurement equipment used to measure, calibrate, gauge, test, inspect, diagnose, or otherwise examine materials, supplies, and equipment to determine whether they comply with the established specifications. [AIR4896]

precleaning The removal of surface contaminant from a test part so that the contaminant cannot interfere with the penetrant inspection process. [AMS2647A]

precoat sealing The application of a coat of brushable sealant to serve as a base for a fillet sealing. [AIR4069]

preconscious level of awareness *See* awareness, preconscious level.

precure *1.* The full or partial setting of a synthetic resin or adhesive in a joint before the clamping operation is complete, or before pressure is applied. *2.* To cure a part prior to joining it with other parts to form a bonded part. [AIR4844]

precursor For carbon or graphite fiber, the rayon, pan, or pitch fibers from which the fibers are derived. [AIR4844]

predemonstration phase A period of time immediately prior to commencement of formal maintainability demonstration, during which the test team, facilities, and support material are assembled. [AIR4896]

predetection In instrumentation tape recorders, the process of recording a "low" intermediate frequency from the telemetry radio receiver (typically, 900 kHz center frequency) rather than the demodulated output of the receiver.

predicted Expected at some future time; postulated on analysis of past experience and tests. [AIR4896]

predicted preconditioning The operation of an item under stress to stabilize its characteristics, often referred to as "burn-in." [ARP4386]

predicted survival The survival expected at some future date, postulated on a given period of time or number of duty cycles, measured by the ratio of the number of survivors at time, t, to the population at the beginning of the period. [AIR4896]

prediction techniques Methods for estimating future behavior of material based on knowledge of its parts, functions, operating environments, and their interrelationships. [ARP4386]

predictive control *1.* A type of automatic control in which the current state of a process is evaluated in terms of a model of the process, and controller actions are modified to anticipate and avoid undesired excursions. *2.* Self-tuning. *3.* Artificial intelligence.

predictive maintenance A preventive maintenance program that anticipates failures that can be corrected before total failure.

predisposing events, mishap *See* mishap, predisposing events.

preface, mag tape The first few words of each tape record, identifying the record and documenting the status of the equipment.

pre-filter element A construction, usually cylindrical and made from a variety of materials, through which fuel is passed to remove particulate contamination, used in an application where it is followed in the flow direction by a filter monitor or filter separator. [AIR4783]

prefit A process for checking the fit of mating detail parts in an assembly prior to adhesive bonding, to ensure proper bond lines. [AIR4844]

preflight air conditioning *See* conditioning, preflight air.

pre-flight engine status Cockpit indications to aid in pre-takeoff checks, e.g., thrust check. [ARP1587]

preflight inspection *See* inspection, aircraft engine.

preform *1.* A preshaped, fibrous reinforcement formed by distribution of chopped fibers or cloth by air, water flotation, or vacuum over the surface of a perforated screen, to the approximate contour and thickness desired in the finished part. [AIR4844] *2.* A cylinder of glass that is made to have a refractive index profile desirable for an optical fiber. The cylinder is then heated and drawn out to produce a fiber. *3.* Brazing metal foil cut to the exact outline of the mating parts and inserted between the parts prior to placing in a brazing furnace.

preform binder A resin applied to the chopped strands of a preform, usually during its formation, and cured so that the preform will retain its shape and can be handled. [AIR4844]

pregel An unintentional, extra layer of cured resin on part of the surface of a reinforced plastic. [AIR4844]

preheater air Air at a temperature exceeding that of the ambient air.

preheating The heating of a compound before molding or casting, to facilitate the operation or reduce the molding cycle. [AIR4844]

preignition Spontaneous ignition of the explosive mixture in a cylinder of an internal combustion engine before the spark flashes.

preimpregnation The practice of mixing resin and reinforcement and effecting partial cure before use or shipment to the user. [AIR4844]

pre-insulate To insulate an electrical connection before assembly. [ARP914A]

prelaunch summaries Summaries prior to launch of the preparations and parameters of the mission.

preliminary flight rating (PFR) A rating that provides an engine configuration that has demonstrated sufficient flight safety for limited use in experimental flight tests. [ARD50010]

preliminary system safety assessment A systematic evaluation of a proposed system architecture and its implementation, based on the functional hazard assessment and failure condition classification, to determine safety requirements for all items in the architecture. [ARP4754]

preload *1.* The force imposed on a duct or duct system prior to, or during, installation. [ARP699D] *2.* The final installation load expressed in pounds. [ARP700] *3. See* bias/preload.

pre-main sequence stars Stars in which the nuclear reactions that normally take place in the core have not yet occurred.

premature ignition A type of malfunctioning in which a combustion device functions before the expected time. [ARP4386]

premix A molding compound prepared prior to and apart from the molding operations, and containing all components required for molding. [AIR4844]

premixing The mixing of ingredients prior to a specified action; for example, the mixing of fuel and air prior to ignition in combustion.

premodulator filter A lowpass filter at the input to a telemetry transmitter. Purpose is to limit modulation frequencies and thereby limit radiated frequencies outside the desired operating spectrum.

premolding For a laminated or chopped-fiber detail part, refers to the lay-up and partial cure of the part at an intermediate cure temperature to stabilize its configuration for handling an assembly with other parts for final cure. [AIR4844]

prepack seal A preassembly seal installed to fill voids or provide a support seal for subsequent fillet sealing. [AIR4069]

prepreg batch Prepreg containing reinforcement material from one batch, impregnated with one batch of resin in one continuous operation. [AIR4844]

prepreg lot Prepreg from one batch submitted for acceptance at one time. [AIR4844]

preprocessor *1.* A hardware device in front of a computer, capable of making certain decisions or calculations more rapidly than the computer can make them. *2.* The first of the two compiler stages. At this stage, the source program is examined for the preprocessor statements, which are then executed, resulting in the alteration of the source program text. *3.* Generally, a program that performs some operation prior to processing by a main program.

pre-retracted gear A gear system that is not utilized in a takeoff mode and therefore is pre-retracted or placed in the stowed position; it is then extended prior to landing the aircraft. [AIR1489]

pre-rotation The process of imparting rotation to the tires (wheels) prior to the landing touchdown to avoid inertia loads on the gear imposed by rapid spin up, and wear on the tires caused by initial impact skidding. [AIR1489]

present position (PPOS) The actual latitude and longitude coordinates at the point on the earth's surface directly below the airplane at any given instant. Could also be three-dimensional, including altitude. [ARP1570]

press A mechanical device deriving its power from a drive screw or hydraulic plunger, and used, for example, to assemble, disassemble, cut, or straighten parts when equipped with proper accessories. [ARP480A]

press clave A simulated autoclave made by using the platens of a press to seal the ends of an open chamber, providing both the force required to prevent loss of the pressurizing medium and the heat required to cure the laminate inside. [AIR4844]

pressed density The density of a powder-metal compact after pressing and before sintering.

pressure *1.* The force applied on a surface area, for example the force of a gas or liquid on a unit of area, expressed as pounds per square inch (psi), millimeters of mercury (mm Hg), or pascals (Pa). *See also* atmospheric pressure. [ARP171] *2.* A modifier that designates a portion of a system or unit that is normally exposed to system pressure. *3.* Refers to gage pressure, except where otherwise specified. [ARP4386]

pressure, absolute *See* absolute pressure.

pressure, allowable working *See* design pressure.

pressure altimeter A precision aneroid barometer that measures the air pressure at the altitude an aircraft is flying and converts the reading to indicated height above sea level.

pressure altitude The height of an aircraft as measured from an atmospheric pressure level of 1013.23 mb or 29.92 in Hg. Commonly used to indicate flight levels. [ARP4386]

pressure amplifier A component designed specifically for amplifying pressure signals. [ARP993A]

pressure, atmospheric *See* atmospheric pressure.

pressure bag molding A process for molding reinforced plastics in which a tailored, flexible bag is placed over the contact lay-up on the mold, sealed, and clamped in place. [AIR4844]

pressure breathing The breathing of oxygen or a suitable mixture of gases at a pressure higher than the surrounding pressure.

pressure, burst *See* burst pressure.

pressure compensation A method by which a control valve or coupler is able to control pressure within set limits at a remote and otherwise inaccessible point downstream of the actual sense point. [AIR4783]

pressure control *1.* To limit or modulate pressure at known value or range, as in aircraft ground-refueling delivery systems, which are designed to provide a safe pressure at the connection to the aircraft. [AIR4783] *2.* Describes a device or system that can raise, lower, or maintain the internal pressure in a vessel or process equipment.

pressure control servovalve A servovalve that modulates output pressure with respect to an input. [ARP4386]

pressure control valve A line-mounted-type valve that provides pressure control at a downstream location, either directly within the valve body or remotely from the valve by means of a fuel sense line. [AIR4783]

pressure curve Graphical expression of pressure variation in relation to another variable, for example, time. [ARP4386]

pressure-demand oxygen system A demand oxygen system that furnishes oxygen at a pressure higher than atmospheric pressure above a certain altitude. [ARP4107]

pressure dependence Study of how a rate constant changes with pressure.

pressure, design *See* design pressure.

pressure, differential *See* differential pressure.

pressure differential, static The difference between the static pressures at two points in a fluid system. [ARP171]

pressure differential, total The difference between the total pressures at two points in a fluid stream. [ARP147C]

pressure droop For a pressure-compensated pump or regulator, the change in pressure from a higher level to a lower level with flow. [ARP4386]

pressure drop The reduction in fluid pressure due to flow. When applied to a fluid-control unit, pressure drop is measured between given ports of the unit at a given flow, and does not include the loss in fittings installed in ports; normally, the value applicable to a complete flow pattern at rated flow, unless otherwise stated. [ARP4386]

pressure drop, total nonrecoverable The loss of total pressure between two points in a fluid stream; equal to the total pressure differential. [ARP171]

pressure, dynamic *See* dynamic pressure.

pressure efficiency The ratio of the measured brake pressure integral, to the peak-to-peak straight-line integral between the skid pressure levels from brake pressure time history instrumentation. [AIR1489]

pressure elements The portions of a pressure-measuring gage that move or are temporarily deformed by the system pressure, the amount of movement or deformation being proportional to the pressure.

pressure energized seal A seal ring retained in the body bore with raised flexible lip, which contacts an offset disk in the closed position, yet is clear of the disk in other positions.

pressure envelope The range of pressures to which a part is subjected in normal operation. [ARP699D]

pressure-expanded joint A tube joint in a drum, header, or tube sheet that has been expanded by a tool that forces the tube wall outward by driving a tapered pin into the center of a sectional die.

pressure fluctuation Variation of pressure with time, usually occurring randomly. [ARP4386]

pressure gage *See* gage, pressure.

pressure, gage Pressure as related to ambient atmospheric pressure. [ARP4386]

pressure gain *1.* The ratio of output pressure change to control pressure change required for switching to occur. *2.* The slope of the line drawn between the operating point and the origin of the pressure gain curve. [ARP993]

pressure gradient In a steady-state flow, the change in pressure value with distance. [ARP4386]

pressure head Equivalent height of a column of liquid required to produce a given pressure. [ARP4386]

pressure-impregnation-carbonization A densification process for carbon-carbon composites involving pitch impregnation and carbonization under high temperature and isostatic pressure conditions. [AIR4844]

pressure impulse A rapid rise and fall of pressure, or vice versa, of extremely short duration. [ARP4386]

pressure intensifier A layer of flexible material used to ensure the application of sufficient pressure to a location, such as a radius, in a lay-up being cured. [AIR4844]

pressure, intrapleural Pressure between the layers of the pleura, normally below atmospheric pressure. [ARP171]

pressure, intrapulmonary Pressure of the air within the lungs, normally below atmospheric pressure on inspiration, above atmospheric pressure on expiration. [ARP171]

pressure level In acoustic measurement, the pressure level in bels (P). $P = 1 \log (P_s/P_r)$, where P_s is the sound pressure, and P_r is a reference pressure, usually taken as 0.002 dyne/cm^2.

pressure loss *1.* A reduction in fuel pressure due to flow through a component or restriction. [AIR1749] *2.* A reduction in pressure caused by resistance to flow or by an extraction of energy that is not converted into useful work. [ARP4386]

pressure loss compensator Normally applies to a Venturi-shaped device when used for remote sense-pressure-control purposes. [AIR4783]

pressure, maximum operating The maximum pressure that a duct will experience under all possible operating conditions. [ARP699D]

pressure, maximum working The highest continuously applied pressure at which a system is approved for safe operation.

pressure measurement Any method of determining internal force per unit area in a process vessel, tank, or piping system due to fluid or compressed gas; this includes measurement of static or dynamic pressure, absolute (total) or gage (total minus atmospheric), in any system of units.

pressure microphone An acoustic transducer that converts instantaneous sound pressure of impinging sound waves into an electrical signal which directly corresponds in both frequency and amplitude.

pressure, minimum operating *See* minimum operating pressure.

pressure modulator radiometer A radiometer with a single optical path, into which a cell containing a known quantity of a gas is placed. The gas is subjected to cyclical pressure changes which alter the absorption lines in its infrared spectrum. A narrow band signal results from the different voltages at the detector at high and low cell pressures. A wideband signal is generated by physically chopping a percentage of the input beam with a rotating chopper blade.

pressure, negative Pressure lower than atmospheric. [ARP171]

pressure operated Describes a device that is operated by application of pressure through a

control port or ports, for example, a pressure control valve. [ARP4386]

pressure, output *See* output pressure.

pressure, partial *See* partial pressure.

pressure, positive *See* positive pressure.

pressure, proof *See* proof pressure.

pressure pulsation (pressure ripple) Periodical variation of pressure in synchronism with the operating speed of rotating equipment. [ARP4386]

pressure, (differential) ram air total (inlet) The maximum differential between the total pressure within the inlet and the static pressure of the undisturbed air stream (ambient atmosphere). [ARP147C]

pressure, rated *See* rated pressure.

pressure rating The maximum allowable internal force per unit area of a pressure vessel, tank, or piping system during normal operation.

pressure ratio *1.* For a starter or APU, expressed as the ratio of inlet pressure to exhaust discharge pressure. [ARP906A] *2.* Numerical ratio of the value of two pressures. [ARP4386] *3.* The relationship of a force to the deformation of a system whose deformation varies in some proportion to the force.

pressure recovery The percentage of dynamic pressure available within the throat of a scoop for promoting flow through the scoop. [ARP147C]

pressure reducer A device that is supplied with gas at a high pressure, and delivers gas at a controlled, lower pressure. [AS1297]

pressure reducing valve A valve that reduces an inlet pressure to a predetermined maximum outlet pressure, regardless of flow or inlet pressure. [ARP4386]

pressure-regulating valve *1.* A valve that can assume any position between fully open and fully closed. *2.* A valve that opens or remains closed against fluid pressure on a spring-loaded valve element, to release internal pressure or hold it and allow it to build up, as desired.

pressure regulator An in-line device that provides controlled venting from a high-pressure region to a lower-pressure region of a closed compressed gas system, to maintain a preset pressure value in the lower-pressure region.

pressure relief device A mechanism that vents fluid from an internally pressurized system to counteract system overpressure. May release all pressure and shut the system down (as does a rupture disc) or may merely reduce the pressure in a controlled manner to return the system to a safe operating pressure (as does a spring-loaded safety valve).

pressure, reseat *See* reseat pressure.

pressure ripple *See* pressure pulsation.

pressure rise The rise of inflation pressure in a tire due to the rolling process; and the energy absorption and temperature rise associated with the process. [AIR1489]

pressure rise, dynamic The maximum static pressure increase developed by the momentum of a fluid stream when its velocity is reduced to zero. [ARP171]

pressure, safety *See* safety pressure. [ARP171]

pressure-sensitive adhesive A viscoelastic material that, in solvent-free form, remains permanently tacky. Such material will adhere instantaneously to most solid surfaces with the application of very light pressure. [AIR4844]

pressure, service *See* rated pressure.

pressure, static *See* static pressure.

pressure suits Garments designed to provide pressure on the body so that the respiratory and circulatory functions may continue normally, or nearly so, under low-pressure conditions, such as occur at high altitudes or in space, without benefit of pressurized cabins.

pressure surge Pressure rise and fall over a period of time. [ARP4386]

pressure, surge *See* surge pressure.

pressure switch In a process vessel or piping system, a device that activates or deactivates an electrical circuit when a preselected pressure is exceeded.

pressure tap A small hole in the wall of an internally pressurized vessel or pipe, allowing attachment of the pressure element of an instrument to measure static pressure.

pressure, total The sum of the static pressure and the dynamic pressure at a given point in a fluid stream. [ARP147C]

pressure transducer An instrument component that senses fluid pressure and produces an electrical output signal that is related to the magnitude of the pressure.

pressure transient Pressure rise and/or drop, of extremely short duration, with negligible energy. [ARP4386]

pressure tubing Tubing intended primarily for the transmission of liquids or gases. [AMS2243G]

pressure under load (load pressure) The pressure reacting to a static or dynamic load. [ARP4386]

pressure-vacuum gage An instrument for measuring pressure both above and below atmospheric pressure.

pressure/vacuum valve A valve mounted in a tank which provides both positive and negative pressure relief within specific limits. [AIR4783]

pressure vessel A metal container designed to withstand a specified bursting pressure. Usually cylindrical with hemispherical end closures, but may be of some other shape, such as spherical. Usually fabricated by welding.

pressure volume control pump A variable-delivery pump whose output is controlled by its discharge pressure. [ARP4386]

pressure wave A cyclic variation of pressure with relatively low amplitude and long period. [ARP4386]

pressurization, fuel Fuel control movement from cutoff to initiate fuel flow causing the fuel manifold to become pressurized. [ARP906A]

pressurization, starter Opening the start valve to initiate starter operation. [ARP906A]

pressurizing, cabin *See* cabin pressurization.

presumptive address *See* base address.

pretakeoff check For an aircraft, a check of the wings or representative surfaces for frost, ice, snow, or slush within the holdover time of the aircraft. [AS5116]

pretakeoff contamination check For an aircraft, a check of the wings or representative surfaces for frost, ice, snow, or slush after exceeding any maximum holdover time in the certificate holder's timetable. [AS5116]

pre-tinned 1. Refers to the application of solder to a contact, conductor, or other connecting device prior to soldering. 2. Refers to the application of tin plating to the basis metal of connecting devices prior to fabrication. [ARP914A]

pretravel The distance through which the plunger moves when traveling from the free position to the operating point. [AIR4077]

prevailing torque *See* running torque.

prevailing visibility The greatest horizontal visibility equaled or exceeded throughout at least half (not necessarily continuous) of the horizon from a particular observation point. [ARP4107]

prevaporization The phase transformation of a liquid to a gas prior to some physical or chemical reaction.

preventive alert An alert that provides the crew with an escape maneuver (usually a vertical speed limit or a negative command), which because of their current flight path, they do not need to respond to; for example, a "LIMIT CLIMB 500 fpm" alert when the aircraft is in level flight. [ARP4153]

preventive maintenance The actions performed to retain an item at a specified level of performance by providing systematic inspection, detection, and prevention of impending failures. [ARD50010]

preventive maintenance time The part of the maintenance time during which preventive maintenance is performed on an item. [ARD50010]

primary air Air introduced with the fuel at the burners.

primary-air fan A fan to supply primary air for combustion of fuel.

primary alpha The allotrope of titanium with hexagonal, close-packed crystal structure, retained from the last high-temperature alpha-beta heating. [AS1814]

primary batteries Nonrechargeable batteries with self-contained, stored-potential electrochemical systems requiring only closing of an electrical circuit for operation. [AS4492]

primary billet Compacted powder metal product composed of the specified aluminum matrix material and the properly dispersed silicon carbide. [AMS4265]

primary calibration Calibration performed using referee particles of known size and physical properties. [ARP1192]

primary cell A cell designed to be used only once, then discarded; not capable of being returned to its original charged state by the application of current. [ARP4386]

primary controls Controls that are essential for the operation of the aircraft and, hence, must be readily accessible to the crew. [ARP4101]

primary control surfaces Movable surfaces, such as ailerons, elevators, rudders, or spoilers, that move the aircraft about its three axes. [ARP4107]

primary detector The system element or device that first responds quantitatively to the attribute or characteristic being measured, and performs the initial conversion or control of measurement energy.

primary device The part of a flowmeter that generates a signal in response to the flow, and from which the flow rate may be inferred.

primary disc *See* brake pressure plate.

primary element *1.* Synonym for sensor. *2.* Detector. *3.* The first system element that responds quantitatively to the measured variable and performs the initial measurement operation. A primary element performs the initial conversion of measurement energy. For transmitters not used with external primary elements, the sensing portion is the primary element. *See also* transducer.

primary events The normal terminus of a path of fault events within a fault tree. [ARP926A]

primary explosive An explosive that is very sensitive to initiation by impact or heat, and is usually one of the first elements in an explosive train. [AIR913]

primary failure *1.* Reported by the system software, a failure that is related to a fault in the ATE, which could result in damage to the ATE, UUT, or operator. [ARP4904] *2.* A failure that is not a result of another failure. [ARP1587]

primary feedback A signal that is a function of the controlled variable, and is used to modify an input signal to produce an actuating signal.

primary filter A filter located downstream of the main pump in a fluid system. Function is to protect components downstream from contaminants generated by the pump. [ARP4386]

primary flight control(s) *1.* The stability and/ or maneuvering control of aircraft using force/ moment generators. [ARP4386] *2.* The flight controls used for controlling aircraft flight paths (e.g., ailerons, rudders, and elevators). [ARP578]

primary flight display (PFD) The pilot's primary flight attitude display. This display may include attitude (pitch and roll), altitude, pitch steering commands, lateral-directional steering commands, expanded localizer, glide slope, radar height, decision height, heading reference, flight path angle, potential flight path, speed or speed reference, and angle of attack. [ARP1570]

primary instrument An instrument that can be calibrated without reference to another instrument.

primary insulation The layer or layers of nonconducting material designed to act as electrical insulation, excluding cosmetic top coatings. [ARP1931]

primary loop The outer loop in a cascade system.

primary luminous standard A standard by which the unit of light is established and from which the values of other standards are derived. A satisfactory primary standard must be reproducible from specifications. [ARP798]

primary measuring element A component of a measuring or sensing device that is in direct contact with the substance whose attributes are being measured.

primary mission For the purpose of maintenance data reporting, the primary purpose for which the aircraft is assigned to the unit. [AIR4896]

primary power system The electric system whose generators are driven by the aircraft propulsion engines. Power conversion systems (not part of utilization systems) powered by the primary generators are part of the primary power system. [AS1212]

primary pressure control Pressure resulting from a control valve or other such device located as close as possible to the aircraft refueling adapter, associated with the refueling pressure. [AIR4783]

primary seal A seal that forms a continuous, durable, and absolute seal in the sealing plane and requires no additional seals. [AS4069]

primary standard A primary physical standard, or one maintained at a National Standards Laboratory. [AIR1678]

primary steering system *See* classification of nosewheel steering systems.

primary structure A structure that is critical to flight safety. [AIR4844]

primary windings A winding that receives energizing power. [ARP826]

prime To introduce fuel directly into the induction system or cylinders of an engine as an aid in starting the engine. [ARP4107]

prime (primer mixture) An explosive mixture containing a primary explosive, usually the first element in an explosive train. [AIR913]

prime charge The level of dilution of the deodorant used to charge an aircraft toilet system. [AMS1476B]

prime mover Source of power for driving the alternator. [ARP1148A]

primer *1.* A "first" coating applied to a surface to provide a good bonding surface for the sealant or coating. The primer can be single or multifunctional; it can contain only an adhesion promoter, or only a metallic pacifier or a surface sealer, or other special agent; or it can contain several of these substances. [AIR4069] *2.* A primary initiating device to produce a hot flame. [AIR913]

primer charge A secondary component in an ignition train. The primer charge is ignited by an initiator, starts pressurization of a generator, and ignites the ignition booster charge. [ARP4386]

priming The discharge of steam containing excessive quantities of water in suspension from a boiler, due to violent ebullition.

principle of least time *See* Fermat principle.

printed circuit A system of conductors formed or deposited on a nonconducting substrate in a predetermined pattern to allow quick and repetitive construction of electronic devices.

printed wiring board A completely processed conductor pattern, usually formed on a stiff, flat base. [AIR4844]

prioritized significant events and conditions A dynamic hierarchy of environmental events and conditions that determines what tasks need to be performed, and in which order, to safely accomplish the assigned mission. [ARP4107]

priority Refers to the relative importance attached to different phenomena.

priority interrupt The temporary suspension of a program currently being executed in order to execute a program of higher priority. Priority interrupt functions usually include distinguishing the highest priority interrupt active, remembering lower priority interrupts that are active, selectively enabling or disabling priority interrupts, executing a jump instruction to a specific memory location, and storing the program counter register in a specific location. *See also* software priority interrupt.

priority subfield The part of a control field that indicates the priority level of a token. [AIR4075]

priority valve A valve whose relief pressure is independent of downstream pressure and that opens and permits free flow when both upstream and downstream pressures exceed its setting. [ARP4386]

prismatic glass Clear glass into whose surface is fabricated a series of prisms. Used to direct incident light in desired directions. [ARP798]

probability of acceptance Probability that an item under test will be accepted by that test. [AIR4896]

probability of mission success The likelihood that a weapon system will complete the scheduled mission without experiencing on-equipment failure or performance degradation that would result in an abort or mission deviation. [AIR4896]

probable malfunction Any single electrical or mechanical malfunction or failure that is considered probable on the basis of past service experience with similar components in aircraft applications. [ARP4404]

probe *1.* An immersion-type temperature-measuring element. [ARP485] *2.* A generic term for the hand-held sensing unit of a bondtester or ultrasonic instrument. [ARP5089] *3.* An instrument for remote inspection or sampling. [ARP480A] *4.* A small, movable capsule or holder that allows the sensing element of a remote-reading instrument, usually an electronic instrument, to be inserted into a system or environment and then withdrawn after a series of instrument readings has been taken.

probe-type consistency sensor A device in which forces exerted on a cylindrical body in the direction of flow are detected by a strain-gage bridge circuit. If the fluid is water, circuit output is a measure of flow rate; if the fluid is a solution or suspension flowing at a steady flow rate, the output varies with changes in viscosity (or consistency).

problem identification and correction A program for the continuous identification and correction of material problems that adversely affect the material condition of the systems and equipment. [AIR4896]

procedure *1.* A precise step-by-step method for effecting a solution to a problem. *2.* In data processing, a smaller program that is part of a large program. *3.* In batch processing, the part of a recipe that defines the generic strategy for producing a batch of material.

procedure turn (PT) The maneuver prescribed when it is necessary for an aircraft to reverse direction to establish itself on the intermediate approach segment or final approach course. The outbound course, direction of turn, distance within which the turn must be completed, and minimum altitude are specified in the procedure. However, unless otherwise restricted, the point at which the turn may be commenced, and the type and rate of the turn, are left to the discretion of the pilot. [ARP4107]

process alarm Alarm that occurs as a result of a process parameter exceeding some preset limit.

process and instrumentation diagram (P&ID) A diagram that shows the interconnection of process equipment and the instrumentation used to control the process. In the process industry, a standard set of symbols is used to prepare drawings of processes.

process block valve The first valve off the process line or vessel, used to isolate the measurement piping. *See also* line class valve.

process calculations Installation-dependent calculations providing derived data to supplement the input signals, e.g., efficiencies, flows by material balance.

process chart A graphical representation of the events in a process.

process computer A computer that, by means of inputs from and outputs to a process, directly controls or monitors the operation of elements in that process. *See* on-line. *See also* control computer and industrial computer.

process control *1.* Describes systems in which computers or controllers are used for automatic regulation of operations or processes. Typical operations are those in which control is applied continuously and adjustments to regulate the operation are directed by the computer to keep the value of a controlled variable constant. Contrasted with numerical control. *2.* An operation that regulates parameters by observation of the parameter, comparison with some desired value, and action to bring the parameter as close as possible to the desired value. *3.* Adapting automatic regulatory procedures to the more efficient manufacture of products or the processing of material. *4.* The ways and means by which continuous manufacturing and other industrial processes are monitored and maintained to create products of planned, uniform dimension and quality.

process control chart A table or graph of test results or inspection data for each unit of production, arranged in chronological sequence for the entire assembly or production lot.

process control computer *See* process computer.

process control loop A system of control devices linked together to control one phase of a process.

process control range A concentration range that is 75% of the specification range; a tool used to aid in the establishment of a frequency of process-control testing. [ARP4992]

process database Organized collection of data relating to the operation of a process.

process dynamics A set of dynamic interactions among process variables in a complex system, as in a petroleum refinery or chemical process plant.

process heat Increase in enthalpy accompanying chemical reactions or phase transformations at constant pressure. Two examples are heat of crystallization and heat of sublimation.

processing window The range of processing conditions, such as stock temperature, pressure, and shear rate, within which a particular grade of plastic can be fabricated with optimum or acceptable properties by a particular fabrication process, such as extrusion, injection molding, or sheet molding. [AIR4844]

process I/O Input and output operations directly associated with a process, as contrasted with I/O operations not associated with the process. For example, in a process control system, analog and digital inputs and outputs would be considered process I/O whereas inputs and outputs to bulk storage would not be process I/O.

process I/O bus A circuit over which data or power is transmitted; often one that acts as a common connection among a number of locations. Synonymous with trunk.

processor Abbreviated form of "central processing unit." *See* central processing unit.

processor based system A system in which a processor is used to control the timing and execution of all functions in a pre-determined relationship. [ARP1834]

process parameter A characteristic of a process that can be monitored and measured to provide information on the process.

process reaction method A method of determining optimum controller settings when tuning a process-control loop. The method is based on the reaction of the open loop to an imposed disturbance.

process reaction rate The rate at which a process reacts to a step change.

process specification Documents approved by the cognizant airworthiness authorities, containing information for performing specialized maintenance, such as retreading tires. [ARP4834]

process steam Steam used for industrial purposes, other than for producing power.

process time Elapsed time for the portion of the work cycle controlled by machines.

process variable In the treatment of material, any characteristic or measurable attribute whose value changes with changes in prevailing conditions. Common process variables are flow, level, pressure, and temperature.

process variable alarm An alarm that is set whenever a process variable exceeds the limits set for a given input.

producer gas Gaseous fuel obtained by burning solid fuel in a chamber in which a mixture of air and steam is passed through the incandescent fuel bed. This process results in a gas, almost oxygen-free, containing a large percentage of the original heating value of the solid fuel in the form of CO and H_2.

producer's reliability risk The producer's risk that a reliability acceptance test will reject a product when it is actually equal to or better than a specified value of reliability. [AIR4896]

product definition data A digital representation of product data that is required for the analysis, design, manufacture, test, and support of a product. [AS4159]

production acceptance tests *See* inspection tests.

production approval holder A company with an FAA-approved production system authorizing it to make and ship approved parts and products. [AS7104]

production line *See* line.

production lot Finished nuts fabricated by the same process from a single heat of alloy, heat treated at the same time to the same specified condition, produced as one continuous run, and submitted for vendor's inspection at the same time. [MA1568]

production reliability acceptance test A test conducted under specified conditions, by or on behalf of the government, using delivered or deliverable production items, to determine the producer's compliance with specified reliability requirements. [ARD50010]

productivity *1*. Production output per unit of input, such as number of items per labor manhour. *2*. Generically, the effectiveness with which labor, materials, and equipment are used in a production operation.

products of combustion The gases, vapors, and solids resulting from the combustion of fuel.

PROF *See* path profile.

proficiency flight A flight made by a pilot or other aircrew member or members to develop or improve proficiency in flying duties. [ARP4107]

profile *1*. Outline of an aircraft body at side elevation. *2*. The cross-section of a wing or aircraft fuselage. *3*. Flight profile, the orthogonal projection of the flightpath on a vertical surface containing nominal track, showing variation in height along a straight bottom axis representing track. *See also* path profile and V-path.

profile descent An uninterrupted descent (except where level flight is required for speed adjustment) from cruising altitude/level to interception of a glide slope or to a minimum altitude specified for the initial or intermediate approach segment of a nonprecision instrument approach. The profile descent normally terminates at the final approach fix or where the glide slope or other appropriate minimum altitude is intercepted. [ARP4107]

profile drag The part of the airfoil drag that results from the skin friction on the airfoil surface and pressure drag caused by boundary-layer separation due to the shape of the airfoil. [ARP4386]

profile path An imaginary earth-referenced line in space connecting successive three-dimensional points through which flight is desired and/or controlled (altitude, latitude, and longitude). [AIR4102/9]

prognosis The forecast of future functional status and condition based on current and accumulated inputs. [ARP1587]

program *1*. In data processing, a series of instructions that tell the computer how to operate. *See also* routine and subroutine. *2*. Any series of actions proposed in order to achieve a certain result.

program, actuarial A statistical technique in which historical data are used to depict the life expectancy of an item with respect to age. [AIR4896]

program address counter *See* location counter.

program, block maintenance A program that divides major structural inspections and/or maintenance tasks into groups or blocks, which permit convenient, economical, and effective accomplishment of the inspections or tasks. [ARD50010]

program, continuous maintenance A type of complete maintenance program that is expected to assure continuous availability of the aircraft. Under such a program, the total maintenance effort is apportioned to each of the various and more frequent types of maintenance. NOTE: A complete overhaul at one point in time is not a part of a continuous maintenance plan. [ARD50010]

program control Describes a system in which a computer is used to direct an operation or process, automatically holding or making changes in the operation or process on the basis of a prescribed sequence of events.

program counter (PC) A register that contains the address of the next instruction to be executed. At the end of execution of each instruction, the program counter is incremented by 1, unless a jump is to be carried out, in which case the address of the jump label is entered.

program documentation The complete listing of the use, content, and installation of a program.

program, equalized maintenance A maintenance program whereby work packages are scheduled for accomplishment in such a manner that the required maintenance manpower will remain relatively constant. Portions of the heavier maintenance tasks are integrated into the lighter or lesser maintenance periods so that the workload fluctuations will be minimized. [ARD50010]

program evaluation and review technique (PERT) A management control tool for managing complex projects. According to PERT, project milestones are defined and interrelated; then, using a flowchart or computer, progress is measured against the milestones. Deviations from the integrated plan are used to trigger decisions or preplanned alternative actions to minimize adverse effects on the overall goal.

program execution time The time required for a real-time program to complete all calculations for a given time step. Real-time operation requires that the program execution time for a model element be smaller than the time step for that element. [ARP4148]

program generator Refers to computer software that translates simple statements into program codes.

programmable data distributor (PDD) For the telemetry frame synchronizer, an optional module that causes it to send certain predefined words from each telemetry frame out through a separate port.

programmable logic controller (PLC) *1.* A microcomputer-based control device used to replace relay logic. *2.* A solid-state control system that has a user-programmable memory for storage of instructions to implement specific functions, such as I/O control, logic, timing, counting, three-mode (PID) control, communication, and arithmetic, data, and file manipulation.

programmable read only memory (PROM) A hardware device that stores digital words which can be read by the computer but not modified

program, maintainability The planning, development, and implementation of the organized set of tasks directly related to the specification, assessment/prediction, and verification of the design characteristics of an item, making it possible to meet operational objectives with a minimum expenditure of maintenance and support effort. [AIR4896]

program, maintenance A program that defines a logical sequence of maintenance actions to be performed. When these actions are performed collectively, the desired maintenance standards will be achieved. [AIR4896]

programmer's console A man-machine interface consisting of various information entry/retrieval devices, arranged as a packaged unit. Used by the programmer of a computer control system for a manufacturing process, to monitor, modify, and control the internal behavior of the digital controller.

programming The design, writing, and testing of a program. *See* convex programming, dynamic programming, linear programming, mathematical programming, nonlinear programming, quadratic programming, macroprogramming, microprogramming, and multiprogramming.

programming language A language used to prepare computer programs.

progressive burning The burning of a propellant grain in which the reacting surface area or burning rate increases during combustion. The mass flow rate produced increases as the web increases. [ARP4386]

progressive geometry A propellant grain configured in such a manner that the exposed surface area increases as burning progresses. [AIR913]

projectile penetration *See* terminal ballistics.

projectiles *1*. Objects, especially missiles, fired, thrown, launched, or otherwise projected in any manner, such as bullets, guided rocket missiles, sounding rockets, or pilotless airplanes. *2*. Originally, objects such as bullets or artillery shells projected by applied external forces.

projection welding A welding process in which the arc is localized by projections, embossments, or intersections.

prolate spheroid Ellipsoid of revolutions, the longer axis of which is the axis of revolution.

PROM *See* programmable read only memory.

promoter A chemical that is itself a feeble catalyst, but that greatly increases the activity of a given catalyst. [AIR4844]

proof To test a component or system at its peak operating load or pressure. [AIR4844]

proof-of-concept testing The evaluation of the feasibility of new concepts during the development of a system. [ARP4107]

proof pressure *1*. The maximum total pressure required of a component or system without external leakage after which the component or system will meet all other specification requirements. [AIR4783] *2*. The static pressure for testing an assembly; a prescribed multiple of the nominal or operating pressure. [MA2005] *3*. The pressure above normal operating pressure that is used to non-destructively test a pressure container. [ARP906A] *4*. For a component or system, the pressure above nominal system pressure that, when applied at defined test conditions, does not lead to external leakage, permanent deformation, or any detrimental influence on function. [ARP4386]

proof speed The speed above maximum operating speed that is used to nondestructively test high-speed rotating components. [AS1606]

proof test For a shock absorber or other oil or air-oil unit, a high-pressure test over a specific period of time to insure against seal leakage or structural weakness. [AIR1489]

propagation Advancement of a wave through a medium. [ARP5089]

propagation constant Characteristic of a radio-frequency transmission line, a complex quality that indicates the effect of the line on the transmitted wave. The real part indicates the attenuation; the imaginary part, the phase shift. [ARP1931]

propagation delay The time between the instant that the control step reaches 50% of the final value and the instant the output reaches 50% of the final value. [ARP993]

propagation loss Refers to the reduction in amplitude of a radio telemetry signal due to natural laws of attenuation.

propellant *1*. Any agent that is used for consumption or combustion in rockets, and from which the rockets derive their thrust; for example, fuels, oxidizers, additives, catalysts, or any compounds or mixture of these. *2*. Specifically, fuels, oxidants, or a combination or mixture of fuels and oxidants used in propelling rockets. Propellants are commonly in either liquid or solid form.

propellant explosions Detonations of propellants as a result of motor malfunction.

propeller, inboard component The component of a dual-rotation or coaxial propeller that is mounted on the inboard shaft (nearest to the power source). [ARP355]

propeller meter An instrument for measuring the quantity of fluid flowing past a given point. The flowing stream turns a propeller-like device, and the number of revolutions is related directly to the volume of fluid passed.

propeller, outboard component The component of a duel-rotation or coaxial propeller that
is mounted on the outboard shaft (farthest from
the power source). [ARP355]

propeller turbine *See* turbopropeller engine.

propeller-turbine engine *See* turbopropeller
engine.

propeller wash *See* wake turbulence.

prop end The end of the engine to which the
propeller is attached; or the end from which
power is applied either directly or indirectly to
the propeller or other points of application.
Also known as the front of the engine.
[ARP169]

prop-fan technology Refers to technology utilizing a small-diameter, highly loaded, many-
bladed variable pitch advanced turboprop.

proportional band *1.* The total amount of
change in the controlled variable required for
the controller to move the controlled device
through its complete stroke; that is, the amount
that the control point of the system can be
changed by varying the controlled device over
its entire range. Also called throttling range.
[ARP89C] *2.* An expression of the gain of an
instrument; the wider the band, the lower the
gain.

proportional bandwidth (PBW) In FM telemetry, refers to the condition in which each
subcarrier is deviated a fixed percentage of
center frequency (and therefore an amount proportional to the center frequency) by data.

proportional control *1.* Control of an aircraft
whereby control surface deflection (or output)
is proportional to the movement of the remote
control (or input). [AIR1489] *2.* Type of control in which there is a definite value of controller output (and a definite position of the
final controlled device) for every value of the
error. The range of error that causes full range
of position of the controlled device is called
proportional band or throttling range. The proportional band is usually adjustable within the
controller. [ARP89C] *3.* A control mode in
which there is a continual linear relationship
between the deviation computer in the controller, the signal of the controller, and the position of the final control element.

proportional control (Type 0) A control system in which proportional forward and feedback control elements are used to provide an
output in response to the error signal. This is
referred to as a Type 0 control system.
[ARP4386]

proportional control action Corrective action
that is proportional to the error; that is, the
change of the manipulated variable is equal to
the gain of the proportional controller multiplied by the error (the activating signal).

proportional control mode *1.* A controller
mode in which the controller output is directly
proportional to the controlled variable error.
2. In a control system, a mode of action that
produces a value correction proportional to the
deviation of the controlled variable from the
set point.

proportional counter An instrument whose
primary element is a radiation counter tube or
chamber, operated in the range over which the
amplitude of each current pulse in the output
is proportional to the energy of the quantum of
radiation absorbed.

proportional, integral, and derivative (PID)
1. Describes a three-mode controller. *2.* Refers
to a control method in which the controller
output is proportional to the error, its time history, and the rate at which it is changing. The
error is the difference between the observed
and desired values of the variable that is under
control action. *3.* Proportional plus integral
plus derivative control, used in processes where
the controlled variable is affected by long lag
times.

proportional limit The greatest stress that a
material is capable of sustaining without deviation from proportionality of stress and strain.
[AIR4844]

proportional pitch In computer printers, a typeface in which each character is a different
width, such as a W versus an I.

proportional plus reset control *See* PI control.

proportional region In a radiation counter tube,
the range of operating voltage over which the
gas amplification factor is greater than 1, usually about 10^3 to 10^5; and the output current
pulse is proportional to the number of ions pro-

duced by the primary ionizing event, which in turn is proportional to the energy of the radiation quantum absorbed.

proportional speed floating control A variation of floating speed control in which the motor speed is proportional to the error. The speed is thus high for large errors and low for small errors. [ARP89C]

proportioning control or proportional control A control that has a unique stable position of the final control device for any given error. This may result in a difference between the control point and set point, known as droop. *See also* proportional band. [ARP89C]

proportioning probe Used in leak testing, a probe in which the ratio of air to tracer gas can be changed without changing the amount of flow transmitted to the detector.

propulsive efficiency The efficiency with which energy available for propulsion is converted into thrust by a rocket engine.

protected location A computer storage location that is reserved for special purposes, and in which data cannot be stored without undergoing a screening procedure to establish suitability for storage therein. May be indicated by a set guard bit.

protected wiring Any section of an electric wire or bundle in which an outer jacket is employed for mechanical or electrical protection. [ARP4404]

protecting tube A closed-end tube that surrounds the measuring junction of a thermocouple and protects it from physical damage, corrosion, or thermochemical interaction with the medium whose temperature is being measured. Also known as thermowell.

protective component A component, or an assembly of components, that is not liable to become defective, in service or in storage, in such a manner as to lower the intrinsic safety of the circuit. Such a component or assembly of components is considered as not subject to fault in that manner when tests of intrinsic safety are made.

protector A device that encloses a part or portion of a part to prevent mutilation. [ARP480A]

protein synthesis Process by which protein molecules are formed.

protocol A set of related rules describing specific processes or activities. [AIR4075]

protocol data unit (PDU) The form in which each of the seven OSI layers passes data to the layer below, after accepting data (SDUs) from the layer above, and adding its own header (PCI). Also the form in which each of the layers passes data to the layer above, after accepting data from the layer below, and stripping off its header.

proton-proton reactions Thermonuclear reactions in which two protons collide at very high velocities and combine to form deuterons. The resultant deuterons may capture other protons to form tritium and the latter may undergo proton capture to form helium. The proton-proton reactions are now believed to be the principal sources of energy within the sun and other stars of its class. A temperature of 5 million degrees Kelvin and high hydrogen (proton) concentrations are required for these reactions to proceed at rates compatible with energy emission by such stars.

protons Positively charged subatomic particles having a mass of 1.67252×10^{-24} g, slightly less than that of an electron.

protoplanets Transition objects formed during primeval cloud condensation into stellar systems (stars, planets, etc.). Protoplanets form the nucleus of planetary accretion.

proverse yaw A positive yawing moment produced by a positive rolling surface deflection. [ARP4386]

proving Determination of flowmeter performance by establishing the relationship between the volume actually passed through the meter and the volume indicated by the meter.

provisioning, initial The process by which an airline defines the range and depth of spare parts that are considered as necessary for the support of a forecast maintenance commitment covering the operation of new aircraft and/or aircraft items. [AIR4896]

proximate traffic Any traffic that is not generating a resolution advisory or traffic advisory, but is within 6 nautical miles horizontally and within ±1200 ft vertically. [AIR4102/10]

proximity detector A sensor that produces an electric signal when the distance from the sensor to another object is less than a predetermined value.

proximity effect (electricity) Refers to the redistribution of current in a conductor caused by the presence of another conductor.

proximity switch A device that senses the presence or absence of an object without physical contact and activates or deactivates an electrical circuit as a result.

PS *See* power setting parameters.

PSD *See* power spectral density.

pseudo fly-by-wire An FBW control system having a means to mechanically control or override the servoactuators. [ARP4386]

pseudo instruction *See* quasi instruction.

pseudoplastic Describes a material that exhibits flow (permanent deformation) at all values of shear stress, although in most cases the flow that occurs below some specific value (an apparent yield stress) is low and increases negligibly with increasing stress.

pseudopotentials Factors in a method for calculating energy bands in solids by the use of an approximation that includes the many body effect.

pseudo-random Satisfying one or more of the standard criteria for statistical randomness, but being produced by a definite calculation process.

pseudo-random number sequence A sequence of numbers, determined by some defined arithmetic process, that is satisfactorily random for a given purpose, for example, it satisfies one or more of the standard statistical tests for randomness. Such a sequence may approximate any one of several statistical distributions, such as uniform distribution or normal (Gaussian) distribution.

pseudo-variable A variable that requires manipulation prior to calculation or processing.

PSK *See* phase shift keying.

psophometer An instrument for measuring noise in electric circuits; its output is exactly one-half of the psophometric emf in a circuit when it is connected across a 600-ohm resistance in the circuit.

PSS *See* packet switching system.

psycholinguistics Study of linguistic behavior, such as conditioning by psychological factors, including the speaker's and listener's culturally determined categories of expression and comprehension.

psychrometer A device consisting of two thermometers, one of which is covered with a water-saturated wick, used for determining relative humidity. For a given set of wet-bulb and dry-bulb temperature readings, relative humidity is read from a chart. Also known as a wet-and-dry-bulb thermometer.

PT *See* procedure turn.

PTM (modulation) *See* pulse time modulation.

P$_{TVP}$ *See* true hydrocarbon vapor pressure.

puck A pad or cup that carries brake lining and is attached to the rotors and/or stators of the brake assembly. [AIR1489]

puckers Areas on prepreg materials where material has locally blistered from the separator film or release paper. [AIR4844]

puff A minor combustion explosion within the boiler furnace or setting.

pulled in filling Refers to an extra filling yarn dragged into the fabric along with a regular filling yarn and extending across a portion of the fabric. [AIR4844]

pull-off load The force required to pull the cable out of either the ferrule or cable end fitting. [AS4536]

pullout A maneuver whereby an aircraft is brought to (or toward) level flight from a diving attitude or a steep glide. [ARP4107]

pull-out force *1*. The axial force required to remove a terminated conductor from its attached contact or terminal. 2. The axial force required to remove a contact from its retention member. *See also* force contact retention. [ARP914A]

pull test *See* tensile test.

pullup A maneuver, in the vertical plane, whereby the aircraft is forced into a short climb, usually from approximately level flight. [ARP4107]

pulmonary stretch reflex *See* Hering-Breuer reflex.

pulp molding The process by which a resin-impregnated pulp material is preformed by application of a vacuum and subsequently is oven-cured or molded. [AIR4844]

pulsating current Unidirectional current whose magnitude alternately rises and falls in a regularly recurring pattern.

pulsating flow Irregular or repeating variations in fluid flow, often due to pressure variations in reciprocating pumps or compressors in the system.

pulsating pressure Pressure whose magnitude alternately rises and falls in a regularly recurring pattern, and whose variation exceeds 1% per second, or 5% per minute, of the scale on the measuring instrument.

pulsation Rapid fluctuation in furnace pressure.

pulsation dampening Refers to a device installed in a gas or liquid piping system to smooth out fluctuations due to pulsating flow and/or pressure.

pulse *1.* A significant and sudden change of short duration in the level of an electrical variable, usually voltage. *2.* A regular or intermittent variation in a normally constant quantity, characterized by a relatively rapid rise and subsequent decay within a finite time period.

pulse amplitude A general term indicating the magnitude of a pulse.

pulse amplitude modulation (PAM) The process (or the results of the process) in which a series of pulses is generated with amplitudes proportional to the measured signal samples.

pulse-averaging discriminator In an FM system, a subcarrier demodulator in which the width of each cycle of the subcarrier is used to derive a data output.

pulse band The range of controlled variable over which the final control device is pulsed to take corrective action; that is, the range of controlled variable that falls between a "maximum on" condition in one direction and a "maximum on" condition in the other direction. *See also* pulse modulation. [ARP89C]

pulse charging Rapid and efficient method for charging electric batteries.

pulse code *1.* A code in which sets of pulses have been assigned particular meanings. *2.* The binary representations of characters. *3.* A series of energy pulses, or a pulse train, modulated in accordance with a data signal. *4.* Generally, any data transmission scheme in which pulsed energy is used to encode the transmitted values.

pulse code modulation (PCM) *1.* Any modulation that involves a pulse code. *2.* The process of sampling a signal and encoding the height or amplitude of each sample into a series of uniform pulses. *3.* In telemetry, serial data transmission (generally a series of binary-coded words). *4.* A form of modulation in which the modulation signal is sampled, quantized, and coded so that each element of information consists of different types or numbers of pulses and spaces. [AS15531]

pulse count telemetering A method of transmitting information that involves an "off-on" switching signal whose number of signal pulses per unit time represents the transmitted value.

pulse decay time The time between the instant when the amplitude of a pulse begins to drop from a specified upper limit and the instant when it reaches a specified lower limit.

pulse discriminator A device that detects pulses having defined characteristics.

pulse Doppler radar A pulse radar system in which the Doppler effect is used for obtaining information about the target (not including simple resolution from fixed targets).

pulse duration The time interval between the first and last instances at which the instantaneous amplitude reaches a stated fraction of the peak pulse amplitude.

pulse duration modulation (PDM) *1.* A form of pulse-time modulation in which the duration of a pulse is varied. *2.* The process of sampling a signal and encoding each sample into a series of pulses whose duration or widths are proportional to the amplitude of the sample.

pulse duty factor The ratio of average pulse duration to average pulse spacing.

pulse-echo method An inspection method in which the presence and position of a discontinuity is indicated by the echo amplitude and time position. [ARP5089]

pulse-forming network Electrical circuitry used to generate high-voltage pulses of particular shapes, and to modify the shapes of pulses generated by other sources.

pulse frequency modulation (PFM) A form of pulse-time modulation in which the pulse repetition rate is the characteristic that is varied.

pulse height *See* pulse amplitude.

pulse-height discriminator An electronic circuit that selects and passes only those voltage pulses that exceed a given minimum amplitude.

pulse-height selector An electronic circuit that selects and passes only those voltage pulses whose peak amplitudes are within a specific range of values.

pulse input In process-control systems, a type of input used to measure pulse- or tachometer-type signals (speed, rpm, frequency, etc.).

pulse-interval modulation Modulation of a pulsed carrier wave in which the time interval between pulses is varied in accordance with the modulating signal.

pulsejet A jet-propulsion engine containing neither compressor nor turbine. Equipped with vanes in the front which open and shut; to create thrust it takes in air in rapid periodic bursts rather than continuously. [ARP4386]

pulse length A measure of the duration of a pulse, expressed in time or number of cycles. [ARP5089]

pulse mode A type of sequential circuit in which inputs are nonperiodic pulses, as opposed to logic levels used in conjunction with clock pulses (clock mode).

pulse modulation *1.* A method of varying the rate of corrective action as a function of error by moving the final control device in discrete steps. *See also* pulse band. [ARP89C] *2.* Modulation of a carrier wave by a pulsed modulating wave.

pulse-position modulation (PPM) A form of pulse-time modulation in which the position in time of a pulse varies in accordance with some attribute of the modulating wave.

pulse radar A type of radar, designed to facilitate range measurement, in which the transmitted energy is emitted in periodic short pulses.

pulse repeater An electronic device that receives pulses from one circuit and transmits them into another circuit, with or without altering the frequency and waveform of the pulses.

pulse repetition frequency The rate at which pulses are repeated in a periodic pulse train.

pulse repetition period The reciprocal of pulse repetition frequency.

pulse repetition rate *1.* The average rate at which pulses are repeated, whether or not the pulse train is periodic. *2.* The number of electric pulses per unit of time experienced by a point in a computer, usually the maximum, normal, or standard pulse rate.

pulse rise time The time between the instant when the amplitude of a pulse reaches some specified fraction of its peak value and the instant when it reaches some specified higher fraction; unless otherwise stated, the two fractions are taken to be 10% and 90%.

pulse spacing The average time interval between corresponding locations on the waveforms of consecutive pulses.

pulse spectrum The frequency distribution of a series of sine waves that can be combined to yield a given periodic pulse train.

pulse switch A switch that provides one pulse of electric current for each cycle of operation.

pulse telemetering Describes any system for transmitting information in terms of electric pulses that are independent of electrical variations in the transmission channel. Pulse telemetering systems can be classified as pulse duration, pulse count, or pulse code systems.

pulse-time modulation (PTM) Modulation of a pulsed carrier wave in which the time of occurrence of some specific point on each pulse waveform is varied from the unmodulated value in accordance with a modulating signal.

pulse transformer A type of transformer designed to convert an a-c input signal to a pulsed output signal.

pulse transmitter A device mounted on a meter to count the number of revolutions of a meter; the counting is done either electronically or optically. [AIR4783]

pulse width *See* pulse duration.

pulse width modulation *See* pulse duration modulation.

pulse width modulation control A control scheme whereby the controller output is a train of pulses of variable duration. The width of the "on" pulse is varied relative to the width of the "off" pulse as some function of the error. [ARP89C]

pultrusion Process of pulling continuous lengths of resin-impregnated fiber through a shaped, heated die to produce lengths of reinforced plastic.

pump *1.* A device for converting mechanical energy into fluid energy. [ARP4386] *2.* A machine that draws fluid into itself through an inlet and forces the fluid out through an exit port, often at higher pressure than at the inlet.

pump, fixed displacement *See* fixed displacement pump.

pump, jet (ejector) A device that raises the pressure of a substance by entraining it with a high-velocity jet of the same or a different substance, giving the resultant mixture a relatively high velocity. [ARP147C]

pump, mechanical volume control *See* mechanical volume control pump.

pump, pilot control *See* pilot control pump.

pump port A port receiving flow directly from the pump before entering the operating portion of the system. [ARP4386]

pump, pressure volume control *See* pressure volume control pump.

pump speed A measure of the capability of a pump to handle gases. Expressed in volumetric units, such as liters/second or cubic feet/minute. [AMS2769]

pump, variable delivery A pump whose delivery can be controlled independently of RPM by varying the output volume per cycle. [ARP4386]

punch *See* force.

purge *1.* To cause a liquid or gas to flow from an independent source into the impulse line(s). *2.* To introduce air into the furnace and the boiler flue passages in such volume and manner as to completely replace the air or gas-air mixture contained therein.

purge interlock A device so arranged that an air flow to the furnace above a minimum must exist for a definite time interval before the interlocking system will permit the automatic ignition torch to be placed in operation.

purge meter A device designed to measure the small flow rates of liquids and gases used for purging measurement piping.

purge post An acceptable method of scavenging the furnace and boiler passes to remove all combustible gases after flame failure controls have sensed pilot and main burner shutdown and safety shut-off valves are closed.

purge, pre-ignition An acceptable method of scavenging the furnace and boiler passes to remove all combustible gases before the ignition system can be energized.

purge-type liquid-level detector *See* air-bubbler liquid level detector.

purging Elimination of an undesirable gas or material from an enclosure by means of displacing the undesirable material with an acceptable gas or material.

push back Refers to the property of a braid or shield whereby it can be easily pushed back along the cable core. [ARP1931]

pushbroom sensor modes Spacecraft instrument arrangements in which large numbers of detectors comprising linear arrays are swept by the forward motion of the spacecraft to attain increased fidelity and high sensitivity in the data captured.

push-down list A list that is constructed and maintained so that the next item to be retrieved is the most recently stored item in the list; that is, last-in, first-out.

pusher A mechanism that exerts a pushing force on a specific part with respect to another. May be of the hydraulic, knocker, screw, lever, or cam type. *See also* driver and extractor. [ARP480A]

pusher propeller *1.* A propeller mounted aft of an engine. *2.* A propeller mounted behind the lateral axis or center of gravity of an aircraft. [ARP4107]

push-pull amplifiers Amplifiers in which there are two identical signal branch circuits so as to operate in phase opposition and with input and output connections each balanced to ground.

push-pull circuit breaker Circuit breakers that may be manually actuated by a "push" to close and a "pull" to open. [ARP1199A]

push-push actuator An actuator in which two linear pistons act on the opposite ends of a crank arm attached to a rotary output shaft. [ARP4386]

push-up list A list that is constructed and maintained so the next item to be retrieved and removed is the oldest item still in the list; that is, first-in, first-out.

P/V *See* converter.

PVC *See* polyvinylchloride.

PVD *See* plan-position indicator.

PV tracking Tracking in which a set point automatically tracks the process variable when the controller is in manual.

pycnometer A container of precisely known volume that is used to determine density of a liquid by weighting the filled container and dividing the weight by the known volume. Also spelled pyknometer.

pyknometer *See* pycnometer.

pylon *1.* A rigid structure protruding from the wing or fuselage to support an engine. *2.* A tower that marks a turning point in a race, or is used for precision flying exercises. [ARP4107]

pyranometer Actinometer that measures the combined intensity of incoming direct solar radiation and diffuse sky radiation. Consists of a recorder and a radiation-sensing element which is mounted so that it views the entire sky. Sometimes called solarimeter.

pyrgeometer An instrument for measuring radiation from the earth's surface into space.

pyrheliometer An instrument for measuring the intensity of direct solar radiation only.

pyroelectric detectors Detectors for visible, infrared, and ultraviolet radiation that rely on the absorption of radiation by pyroelectric materials. Heating of such materials by the absorbed radiation produces electric charges on opposite sides of a crystal, which can be measured to determine changes in the amount of radiation incident on the detector. These detectors usually also exhibit piezoelectric properties and may require isolation from acoustic or acceleration phenomena.

pyrogen A rocket-ignition system containing a solid-propellant grain as its main ignition material. [AIR913]

pyroheliometer An actinometer that measures the intensity of direct solar radiation, consisting of a radiation-sensing element enclosed in a casing, and a recorder unit. The casing is closed except for a small aperture, through which the direct solar rays enter.

pyrolysis *1.* Chemical decomposition by the action of heat. *2.* The thermal process by which organic precursor fiber materials, such as rayon, polyacrylonitrile, and pitch are chemically changed into carbon fiber by the action of heat in an inert atmosphere. [AIR4844]

pyrometer *1.* Instrument that measures high temperature, e.g., of molten lavas, by electrical or optical means. *2.* Any of a broad class of temperature-measuring instruments or devices. The term originally applied only to devices for measuring temperatures well above room temperature, but now it applies to devices for measuring temperatures in almost any range.

pyrotechnic Describes a mixture of chemicals designed to produce heat, light, noise, smoke, or gas. [AIR913]

Q

q *See* volumetric flow rate.

QAD *See* quick attach-detach.

QBC *See* queue control block.

QCD *See* quantum chromodynamics.

QEC *See* quick engine change.

Q factor *1.* A rating factor for electronic components such as coils, capacitors, and resonant circuits that is equal to reactance divided by resistance. *2.* In a periodically repeating mechanical, electrical, or electromagnetic process, the ratio of energy stored to energy dissipated per cycle.

Q flow *See* quota flow control.

Q meter A direct-reading instrument that measures the Q factor of an electric circuit at radio frequencies by determining the ratio of inductive reactance to resistance. Also known as quality-factor meter.

Q-switch An optical device that changes the Q (quality) factor of a laser cavity, typically raising it from a value below laser threshold to one well above threshold. This technique produces a short, intense pulse, known as a Q-switched pulse. Q-switches can be based on acousto-optic or electro-optic devices, rotating mirrors, frustrated internal reflection, or saturation of absorption in a dye.

quadracycle gear configuration *See* gear configuration, quadracycle.

quadrant detectors Detectors that are divided up into four angularly symmetric sectors or quadrants. The amounts of radiation incident on each quadrant can be compared to one another for applications such as making sure that a beam is centered on the detector.

quadrant-edged orifice An orifice having a rounded contour at the inlet edge to yield more constant and predictable discharge coefficient at low flow velocity (Reynolds number less than 10,000).

quadratic programming In operations research, a particular case of nonlinear programming in which the function to be maximized or minimized, and the constraints, are quadratic functions of the controllable variables. Contrast with convex programming, dynamic programming, linear programming, and mathematical programming.

quadrature *1.* Elongation of 90 deg, usually specified as east or west in accordance with the direction of the body from the sun. *2.* Describes the situation in which two periodic quantities differ by a quarter of a cycle.

quadrature voltage A voltage that differs by 90 deg time phase from the resolver output of fundamental frequency at maximum coupling. [ARP826]

quadrilateral mechanism A mechanical linkage composed of four elements or links, one of which may be the fixed base of the mechanism. Widely used in landing gear retraction and folding systems. Also called four-bar linkage. [AIR1489]

quadruplex Fourfold, as applied to a four-channel system. [ARP4386]

quadruplex system A control system containing four single paths so as to provide multiple failure capability, such as SFO/FS, DFO, or DFO/FS. [ARP4386]

quadrupole Describes a linear accelerator that has four longitudinal vanes in its resonating cavity, which are shaped to create RF electric fields that simultaneously accelerate, bunch, and focus the charged particle beam.

quadrupole mass spectrometer A type of mass spectrometer employing a filter that prevents certain ions from entering the detector. The filter consists of four conductive rods electrically connected in such a manner that, by varying the absolute potential applied to the rods, all ions except those possessing a specific mass-to-charge ratio are prevented from entering the detector.

quad-slope converter An integrating analog-to-digital converter that performs two cycles of dual slope conversion, one with zero input and one with the analog input being measured.

quad voter A voter that performs in a fashion to select one common output to represent the four signals or values.

qualification tests *1.* Tests required to qualify a particular design and size of fitting for use in a specific type and class of hydraulic or pneumatic tube system. [ARP899] *2.* Tests run on one or more units that are representative of the production article, to test the performance, endurance, environmental, and special features of a design to demonstrate specification compliance and suitability of the unit for production and use in service. [ARP906A]

qualified Competent, suited, or having met the requirements for a specific position or task.

qualified products list A list of commercial products that have been pretested and found to meet the requirements of a specification, especially a government specification. [AIR4844]

qualitative Describes inductive analytical approaches that are oriented toward relative, nonmeasurable, and subjective values. [ARP926A]

quality A measure of the degree to which a device conforms to specification and workmanship standards. [AIR4896]

quality assurance A set of systematic actions intended to provide confidence that a product or service will continually fulfill a defined need.

quality assurance system The combination of policy, organization, personnel, procedures, inspection, and test plans assuring that from receipt of materials to shipment of product, the quality of the product will not be compromised and the performance requirements will be met. [AS7200/1]

quality conformance tests Tests specified by the procuring agency to monitor and maintain continued quality. [ARP899]

quality control An aggregate of functions designed to insure adequate quality in manufactured products by initial critical study of engineering design, materials, processes, equipment, and workmanship; followed by periodic inspection and analysis.

quality-factor meter *See* Q meter.

quantitative Describes inductive or deductive analytical approaches that are oriented toward the use of numbers or symbols used to express a measurable quantity. [ARP926A]

quantity The value that marks frequency-distribution interval boundaries, determined by arranging a set of N observations in order of magnitude and marking off equal parts (N/P) of the total population P.

quantity meter A flowmeter in which the flow is separated into known isolated quantities, which are separately counted to determine the total volume passed through the meter.

quantity of light The product of the luminous flux and the time it is maintained. Unit is the lumen-hour. [ARP798]

quantization The subdivision of the range of values of a variable into a finite number of nonoverlapping, and not necessarily equal, subranges or intervals, each of which is represented by an assigned value within the subrange. For example, a person's age is quantized for most purposes with the quantum of one year.

quantization distortion Inherent distortion introduced when a range of values for a wave attribute is divided into a series of smaller subranges.

quantization level A particular subrange for a quantized wave, or its corresponding symbol.

quantize To convert information from an analog pulse (as from a multiplexer) into a digital representation of that pulse.

quantized pulse modulation Pulse modulation in which either the carrier wave or modulating wave is quantized.

quantum chromodynamics (QCD) A gauge theory describing the interaction between quarks and gluons.

quantum efficiency A measure of the efficiency of conversion or utilization of light or some other form of energy.

540

quantum electronics The branch of electronics that essentially deals with lasers and laser devices requiring quantum theory for their exact description.

quantum noise Noise due to the discrete nature of light, i.e., due to the quantization of light into photons.

quantum theory The theory first stated by Max Planck that all electromagnetic radiation is emitted and absorbed in quanta, each of magnitude hv, where h is the Planck constant and v is the frequency of the radiation.

quantum wells Effective potential wells created by a minimum in the conduction band or a maximum in the valence band that arises when a smaller bandgap semiconductor is sandwiched between a larger bandgap semiconductor.

quarantined To be located in a separate place awaiting some further action, for example, requalification test, inspection, repair information, spare parts, or disposal. [AIR4844]

quark parton model A theoretical model that summarizes our understanding of how protons and neutrons are made up of the fundamental subparticles called quarks.

quarter amplitude A process-control tuning criteria whereby the amplitude of the deviation (error) of the controlled variable, following a disturbance, is cyclic so that the amplitude of each peak is one quarter of the previous peak.

quarter-wave plate A polarization retarder that causes light of one linear polarization to be retarded by one quarter wavelength (90°) relative to the orthogonal polarization.

quartz A natural, transparent form of silica. May be marketed in its natural crystalline state, or crushed and remelted to form fused quartz.

quasi instruction Same as pseudo instruction.

quasi-isotropic laminate A laminate that is laid up symmetrically, with an equal number of plies at each 0°, 45°, and 90° angle.[AIR4844]

quasi-static unbalance A condition of unbalance in which the central principal axis intersects the shaft axis at a point other than the center of gravity. [ARP587]

quefrencies In cepstral analysis, the frequency of periodic ripples in a spectra of a signal that contains echoes. Quefrencies are expressed in cycles per hertz or in seconds.

quench cracks Propagating cracks that usually traverse an irregular and erratic course on the surface. [ARP4784]

quenching (atomic physics) Phenomenon whereby very strong electric fields cause the orbit of an electron or atom to precess rapidly so the average magnetic moment associated with its orbit angular momentum is reduced to zero.

query language 1. Command language used to search and retrieve information. 2. A means of getting information from a database without the need to write a program.

queue 1. Waiting line resulting from temporary delays in providing service. 2. In data processing, a waiting list of programs to be run next.

queue control block (QCB) A control block that is used to regulate the sequential use of a programmer-defined facility among requesting tasks.

queued access method Any access method that automatically synchronizes the transfer of data between the program using the access method and input/output devices, thereby eliminating delays for input/output operations.

queuing An ordered progression of items into and through a system or process, especially when there is waiting time at the point of entry.

queuing discipline Refers to the rules or priorities for queue formation within a system of "customers" and "servers," as well as the rules for arrival time and service time.

queuing theory A form of probability theory that is useful in studying delays or line-ups at servicing points.

Quevenne scale A specific gravity scale used in determining the density of milk. A difference of 1° Quevenne is equivalent to a difference of 0.001 in specific gravity, thus 20° Quevenne expresses a specific gravity of 1.020.

quick attach-detach (QAD) mounting flange A disconnect device by which the starter or accessory can be attached to the engine using simple tools on a single fastener. Usually incorporates a special flange that is pre-bolted to the mounting pad. [ARP906A]

quick charge A charging rate that ranges from 0.2 to 0.5 C rate. [ARP4386]

quick charge battery A nickel-cadmium battery that can be charged fully in 3 to 5 h by a constant-current charger, and is capable of continuous overcharge at this quick charge rate. [ARP4386]

quick disconnect A disconnect that can be uncoupled or coupled without the use of tools; may be self-sealing when disconnected. Commonly employed in hydraulic and pneumatic lines. *See also* coupling, quick disconnect. [ARP906A]

quick engine change (QEC) A package or kit of hardware items not included on the engine as delivered by the engine producer, but required in the build-up of the engine prior to installation in an aircraft. [ARP1587]

quick engine change kit Includes a nacelle or cowling set with engine-driven and required aircraft accessories or hardware; installed on the engine for faster engine removal and installation on the aircraft. [ARP741]

quick look *1.* Refers to a feature of NAS Stage A and ARTS which provides the controller with the capability to display full data blocks of tracked aircraft from other controller positions. [ARP4107] *2.* Essentially, an "instant replay" of data, generally at the same rate at which it was recorded.

quick return A device that makes the return stroke of a reciprocating machine element faster than the power stroke.

quiescent current A DC current that is present in each valve coil when using a differential coil connection. The polarity of the current in the coils is in opposition, so that no electrical control power exists. [ARP490]

quiescent flow A continuous leakage flow to return through a hydraulic control interface. [AIR1489]

quinoxalines A group of heterocyclic compounds consisting of a benzene ring condensed with a diazine ring.

quota flow control (Q flow) A flow-control procedure by which the central flow-control function (CFCF) restricts traffic to an ARTC center area having an impacted airport, thereby avoiding sector/area saturation. May be implemented to prevent air-traffic saturation in a sector/area serving an airport having weather or some facility problems that prevent it from accepting its normal flow of aircraft. [ARP4107]

R

R *See* Rankine.

racheting Discontinuous (jerky) movement or rotation of a dynamic display feature, caused by excessively large quantization steps in the translation or rotation of the particular feature. [ARP4256]

rack *1.* A framework, stand, or grating on or in which articles are placed. [ARP480A] *2.* A housing for electronic equipment which permits convenient removal of portions of the equipment. [ARP914A] *3.* A bar having radar beacons. *See* radar beacon.

rack-and-pinion fluid actuator An actuator in which the rotary motion is achieved by piston-driven linear racks engaging a rotary output gear. [ARP4386]

rack-and-pinion mechanical actuator A mechanical actuator generally consisting of reduction input gearing, outputting to a rotary-pinion gear that connects with a rack to provide linear output actuation force. [ARP4386]

rad *See* radian.

RADAR Acronym for radio detection and ranging (generally printed in lower case). Refers to a device that measures the time interval between transmission of radio pulses and reception of their echoes, and correlates the angular orientation of the radiated antenna beam or beams in azimuth and/or elevation, thereby providing information on range, azimuth, or elevation of objects in the path of the transmitted pulses. [ARP4107]

radar altimeter *See* radio altimeter.

radar altitude The altitude of an aircraft or spacecraft as determined by a radio altimeter; thus, the actual vertical distance from the terrain. [ARP4107]

radar approach control *See* RAPCON.

radar approach control facility *See* RAPCON facility.

radar astronomy The study of celestial bodies within the solar system by means of radiation originating on earth, but reflected from the body under observation.

radar beacon A beacon transmitting characteristic signals on radar frequency, permitting crafts to determine their bearings and sometimes the range of the beacons.

radar contact Used by air-traffic controllers to indicate that an aircraft is identified on the radar display and that radar service can be provided until radar identification is lost or radar service is terminated. When the pilot is informed of radar contact, he/she is no longer required to report over compulsory reporting points. [ARP4107]

radar-controlled departure The use of surveillance radar (ASR) vectors to establish an aircraft on the en route track, expediting the departure by using radar traffic separation standards. [ARP4107]

radar cross section The ratio of power returned in a radar echo to power received by the target reflecting the signal.

radar direction finder (RDF) *See* radio direction finder.

radar display *See* radarscope.

radar dome *See* radome.

radar flight following Observing the progress of a radar-identified aircraft whose primary navigation is being provided by the pilot. To accomplish this, the controller retains and correlates the aircraft identity with the appropriate target or target symbol displayed on the radar scope. [ARP4107]

radar handoff The action whereby radar identification of an aircraft is made known from one controller to another without interruption of the radar flight following. [ARP4107]

radar homing missiles Radar-following missiles designed to attack radar transmitters.

radar identification The process of ascertaining that an observed radar target is the radar return from the particular aircraft. [ARP4107]

radar-identified aircraft An aircraft whose position has been correlated with an observed target or symbol on the radar display. [ARP4107]

radar-monitored departure Use of ASR to monitor departing aircraft, with advisories given concerning other radar-observed traffic that might come into a conflict situation with the departing aircraft. [ARP4107]

radar monitoring *See* radar service.

radar navigational guidance *See* radar service.

radar networks A series of tracking stations, each of which can individually or jointly track a target by utilizing an interchange of radar information; used for multiradar tracking.

radar point out/point out Used between controllers to indicate radar handoff action in which the initiating controller plans to retain communications with an aircraft penetrating the other controller's airspace and additional coordination is required. [ARP4107]

radar range *1.* The distance from a radar to a target, as measured by the radar. *2.* The maximum distance at which a radar set is effective in detecting targets.

radar reflector Device capable of, or intended for, reflecting radar signals.

radar scanning The action or process of moving or directing a searching radar beam.

radarscope The viewing-screen portion of radar equipment (e.g., a cathode ray tube), upon which electronic pulses represent the distance and bearing of radar target returns. In modern radar systems, computer-generated displays represent the aircraft and necessary information about it, such as altitude or call sign. [ARP4107]

radar separation *See* radar service.

radar service A term that encompasses one or more of the following services, which are based on the use of radar, and can be provided by a controller to a pilot of a radar-identified aircraft: radar separation, radar navigational guidance, and radar monitoring. *Radar separation–* the use of radar to maintain spacing between aircraft in accordance with established minimums. *Radar navigational guidance–*radar vectoring of aircraft to provide course guidance. *Radar monitoring–*the radar flight following of an aircraft (whose primary navigation is being performed by the pilot) to observe and note deviations from its authorized flight path, airway, or route. Radar monitoring may include the monitoring of instrument approaches, as well as en route radar flight following. [ARP4107]

radar target Objects that reflect a sufficient amount of a radar signal to produce an echo signal on the radar screen.

radar vectoring The use of radar to provide navigational or traffic-separation guidance. In vectoring, the controller uses his/her radar display to ascertain the position and track of an aircraft with respect to the target of concern and gives the pilot the heading to fly in order to make the desired track or separation from other aircraft. [ARP4107]

radar weather echo intensity level Levels of the intensities of radar weather echoes established by the National Weather Service; established because radar echo intensities have a direct correlation with the degree of turbulence and other weather features associated with thunderstorms. These levels are: Level 1, WEAK; Level 2, MODERATE; Level 3, STRONG; Level 4, VERY STRONG; Level 5, INTENSE; and Level 6, EXTREME. *Level 1 (WEAK) and Level 2 (MODERATE)–*light to moderate turbulence possible, with lightning. *Level 3 (STRONG)–*severe turbulence possible, with lightning. *Level 4 (VERY STRONG)–*severe turbulence likely, with lightning. *Level 5 (INTENSE)–*severe turbulence, with lightning and organized wind gusts. Hail likely. *Level 6 (EXTREME)–*severe turbulence with large hail, lightning, and extensive wind gusts. [ARP4107]

radial A magnetic bearing extending from a VOR/VORTAC/TACAN navigation facility. [ARP4107]

radial engine An internal-combustion reciprocating engine, the cylinders of which are disposed radially about the crankshaft in circular rows of five or more cylinders to a row. The row(s) form a plane perpendicular to the crankshaft, which rotates while the cylinders remain stationary. *See also* rotary engine. [ARP4107]

radial piston motor A motor utilizing multiple pistons arranged radially around a shaft with an eccentric cam. [ARP4386]

radial play Refers to the maximum difference between the high and low readings of a dial indicator suitably arranged to measure the total radial travel of the shaft in its own bearings. [ARP667]

radial velocity In radar, the vector component of the velocity of a moving target that is directed away from or toward the ground station.

radian Metric unit for a plane angle.

radiance *1.* In radiometry, a measure of the intrinsic radiant intensity emitted by a radiator in a given direction. Equal to the irradiance (radiant flux density) produced by radiation from the source upon a unit surface area oriented normal to the line between source and receiver, divided by the solid angle subtended by the source at the receiving surface. It is assumed that the medium between the radiator and receiver is perfectly transparent, thus the radiance is independent of attenuation between source and receiver. *2.* Radiant flux per unit solid angle per unit of projected source area.

radiancy The rate of radiant energy emission from a unit area of a source in all the radial directions of the overspreading hemisphere.

radiant flux The rate of flow of radiant energy with respect to time.

radiant flux density *1.* Radiant flux per unit area. *2. See* irradiance.

radiant fluxmeter A device that measures the amount of radiant flux emitted or absorbed. Units are typically watts per unit area.

radiant intensity The energy emitted per unit time per unit solid angle along a specific linear direction.

radiation The propagation of energy through matter or space in the form of waves. [ARP5089]

radiation belts Envelopes of charged particles trapped in the magnetic field of a spatial body.

radiation counter Instrument used for detecting or measuring moving subatomic particles by a counting process.

radiation damage Generally, the deleterious effects of radiant energy (electromagnetic radiation or particulate radiation) on physical substances or biological tissues.

radiation error The difference between the gas temperature and the sensed temperature, caused by radiant heat transfer between the sensing element and surrounding areas. [AIR1900]

radiation loss In a boiler-unit heat balance, a comprehensive term used to account for the conduction, radiation, and convection heat losses from the settings to the ambient air.

radiation pressure Pressure exerted on any material body by electromagnetic radiation incident on it.

radiation pyrometer An instrument in which the radiant power emitted by a hot object is used in determining the temperature of the object.

radiation pyrometer temperature measurement system A temperature-measurement system that is located remotely from the medium whose temperature is to be measured, and whose indication depends on the wavelength and intensity of radiation from the medium to the measuring element of the pyrometer. [AIR1900]

radiation thermometer Any of several devices for determining the temperature of a body by measuring its emitted radiant energy, without physical contact between the sensor and the body.

radiation transport Refers to the study of radiation from emission to absorption.

radiation trapping Confinement of radiation with a magnetic field.

radiator 1. Any source of radiant energy, especially electromagnetic radiation. 2. A device that dissipates the heat from something, e.g., from water or oil, not necessarily by radiation only.

radical A very reactive chemical intermediate. [AIR4844]

radioactive half life The time it takes for the radioactivity of a specific nuclide to be reduced by one half.

radioactive tracer A radioactive nuclide used at small concentration to follow the progress of some physical, chemical, or biological process. Typically, a radioactive isotope is substituted for a small proportion of the same element normally present, and the movement and behavior of the elemental substance or of a radioactively "labeled" compound is observed with radiation-detection instruments.

radioactivity *1.* Spontaneous disintegration of atomic nuclei with emission of corpuscular or electromagnetic radiation. *2.* The number of spontaneous disintegrations per unit mass and per unit time of a given unstable (radioactive) element, usually measured in curies.

radio altimeter A device that measures the altitude of a craft above the terrain by measuring the elapsed time between the transmission of radio waves from the craft and the reception of the same waves reflected from the terrain. Also called a radar altimeter. [ARP4107]

radio assisted detection and ranging *See* radar.

radio astronomy The study of celestial objects through observation of radio-frequency waves emitted or reflected by these objects. Specifically, the study of celestial objects by measurement of the radiation emitted by them in the radio-frequency range of the electromagnetic spectrum.

radio beacon(s) (RBN) Transmitters, together with their associated equipment, that emit signals enabling the determination, by means of suitable receiving equipment, of direction, distance, or position with respect to the beacon. *See also* nondirectional beacon.

radio channel A band of frequencies wide enough to be used for radio communication.

radio control Remote control of a pilotless airplane, rocket, or spacecraft by means of radio signals that activate controlling devices.

radio detection and ranging *See* radar.

radio direction finder (RDF) A radio receiving set, together with its associated equipment, including a directional antenna that can be rotated. The position of the antenna when the peak signal is received indicates the direction to the transmitting station. [ARP4107]

radio-doppler Describes a technique whereby the radial component of relative velocity between an object and a fixed or moving point of observation is determined by measuring the difference in frequency between transmitted radio-frequency waves and reflected waves returning from the object.

radio frequencies (RF) Frequencies at which coherent electromagnetic radiation of energy is useful for communications purposes.

radio-frequency heating *See* electronic heating.

radio frequency interference (RFI) A type of electrical noise that can adversely affect electronic circuits.

radio frequency ion thrustor engines *See* RIT engines.

radio frequency radiation *See* radio waves.

radiograph A permanent visible image on a recording medium, produced by penetrating radiation passing through the material being tested. [ARP5089]

radiographic inspection The use of x-rays and/or nuclear radiation to detect discontinuities

radiographic interpretation The identification of subsurface discontinuities indicated on a radiograph. [ARP5089]

radiographic technique The selection of radiographic factors such as kilovoltage, milliamperage, type of film and screen, distance, and exposure time so as to render the best possible radiographic sensitivity. [ARP5089]

radiography A nondestructive testing method in which a source of x-rays or gamma rays is utilized to indicate the subsurface condition of an opaque material. [ARP5089]

radio horizons Loci or points at which direct rays from a radio transmitter become tangential to the earth's surface.

radio interferometer Interferometer operating at radio frequencies. Radio interferometers are used in radio astronomy and in satellite tracking.

radioisotope A radioactive isotope of a chemical element; a nonpreferred synonym for radionuclide.

radioisotope tracer flowmeter A device for determining flow rate by injecting a radioactive substance into the fluid stream. Two types are available, one based on peak timing and the other on the dilution method. In the peak timing device, flow rate is calculated from a measurement of the time it takes a peak concentration of tracer element to pass between two fixed points along the flow path. In the dilution method device, a known concentration of tracer is injected into the fluid stream and its concentration at a point downstream determined by analysis.

radio jets (astronomy) Jets of energetic particles occurring in radio galaxies and quasars, usually emitted from the nuclear (active) region of the extragalactic radio source.

radioluminescence Emission of light due to radioactive decay of a nuclide.

radio magnetic indicator (RMI) An aircraft navigational instrument coupled with a gyro compass or similar compass that indicates the direction of a selected NAVAID and bearing with respect to the heading of the aircraft. [ARP4107]

radio meteors Meteors that have been detected by the reflection of radio signals from their trails of relatively high ion density (ion columns).

radiometer *1.* An instrument for detecting and, usually, measuring radiant energy. *2.* An instrument that measures optical power in radiometric units (watts); it measures electromagnetic power linearly over its entire spectral range.

radiometric analysis A method of quantitative chemical analysis based on measuring the absolute disintegration rate of a radioactive substance of known specific activity.

radiometric correction An effort to correct the intensity range of an image.

radiometric rectification *See* radiometric correction.

radiometric resolution Refers to the sensitivity of a sensor to distinguish between gray levels.

radionuclide Any nuclide that undergoes spontaneous radioactive decay. *See also* radioisotope.

radio receiver *See* receiver.

radiosonde(s) A balloon-borne instrument for taking meteorological data. Consists of transducers for determining temperature, humidity and barometric pressure; a modulator for converting transducer outputs to radio signals; a selector switch for establishing transmission sequence; and a transmitter for generating a radio-frequency carrier wave.

radio spectra Frequencies of electromagnetic radiation usable for radio communication.

radio telescope Device for receiving, amplifying, and measuring the intensity of radio waves originating outside the earth's atmosphere or reflected from a body outside the atmosphere.

radio waves Waves produced by oscillation of an electric charge at a frequency useful for radio communication.

radius of gyration A quantity derived from the moment of inertia, I. If $I = K^2m$, where m is the mass, the quantity K is called the radius of gyration. Also called radius of inertia. [AIR1489]

radius of inertia *See* radius of gyration.

radius rod A rod or link in a mechanical system that serves to maintain a point in locus, equidistant from a given point. Used in landing gear retraction/extension system to predetermine and maintain the path of motion of the gear during the cycle. [AIR1489]

radix Also called the base number; the total number of distinct marks or symbols used in a numbering system. For example, since the decimal numbering system uses ten symbols (0, 1, 2, 3, 4, 5, 6, 7, 8, 9), the radix is ten; and since there are only two marks or symbols (0, 1) in the binary numbering system, the radix is two.

radix complement A number obtained by subtracting each digit of the given number from one less than the radix, then adding to the least significant digit, executing all required carries. In radix two (binary), the radix complement of a number is used to represent the negative number of the same magnitude.

radix notation A positional representation in which the significances of any two adjacent digit positions have an integral ratio called the radix of the less significant of the two positions. Permissible values of the digit in any position range from zero to one less than the radix of that position.

radix number The quantity of characters for use in each of the digital positions of a numbering system. In the more common number systems, the characters are some or all of the Arabic numerals. Unless otherwise indicated, the radix of any number is assumed to be 10. For positive identification of a radix 10 number, the radix is written in parentheses as a subscript to the expressed number. Synonymous with base and base number.

radix point The index that separates the integral and fractional digits of a quantity in the numbering system in which it is represented. Also called base point, binary point, decimal point, and others, depending on the numbering system;

radome(s) *1.* Contracted from radar dome; pronounced ray-domes. Dielectric housings for antennas. *2.* A weatherproof cover for a primary radar device, constructed of a material that is transparent to radio waves.

Raduga satellite A Soviet communications satellite in geostationary orbit for radio and TV transmission.

RAIL Acronym for runway alignment indicator light system.

railgun accelerator Linear d-c motor consisting of a pair of rigid, field-producing rails, and a movable conducting armature.

rails, optical Long, linear rods that are attached to an optical bench. Optical mounts can be affixed to the rail, creating an optical bench.

rain or high humidity Water forming ice or frost on an aircraft wing surface, when the temperature of the wing surface is at or below 0°C (32°F). [ARP4737]

rake A mechanical device with projecting tubes, prongs, etc., set in such a manner as to pick up temperature, pressure, etc. [ARP480A]

ram *1.* The moving portion in the head of a crimping tool. [ARP914A] *2. See* cylinder, actuating.

Raman shifter A device that alters the wavelength of light by inducing Raman shifts in the light passing through it. (Raman shifts are changes in photon energy caused by the transfer of vibrational energy to or from the molecule.)

ram extrusion An extrusion process in which the insulation material, in a fine powder form, is mixed with a lubricant and forced through a circular die without heating. [ARP1931]

ramjet engine Jet engine with no mechanical compressor; consists of specially shaped tubes or ducts open at both ends. The air necessary for combustion is shoved into the tubes or ducts and is compressed by the forward motion of the engine. After the air passes through a diffuser, it is mixed with fuel and burned; the exhaust gases issue in a jet from the rear opening. Ramjet engines cannot operate under static conditions.

ramp The area of an airport in which aircraft are loaded and unloaded. [ARP4107]

ramp encoder Describes an analog-to-digital conversion process whereby a binary counter is incremented during the generation of a ramp voltage; when the amplitude of the ramp voltage is equal to the amplitude of the voltage sample, the counter clock is inhibited. The counter contents therefore contain the binary equivalent of the sampled data.

ramp generator An electrical power supply that generates a voltage that increases at a constant rate. A plot of voltage vs. time shows a ramp-like waveform.

ramping A gradual, programmed increase/decrease in temperature or pressure to control cure or cooling. [AIR4844]

ramp response The total (transient plus steady-state) time response resulting from a sudden increase in the rate of change in the input from zero to some finite, constant value.

ramp weight The weight of an airplane before engine start; includes the takeoff weight plus a fuel allowance for engine start, taxi, runup, and takeoff ground roll to liftoff. *See also* takeoff weight. [ARP4107]

random access The process of obtaining data from, or placing data into, storage when there is no sequential relation governing the access time to successive storage location. Synonymous with direct access.

random effect A common shift in a group of measurements due to a random level change of a usually uncontrollable factor. [AIR4844]

random error(s) *1.* Errors that are not systematic, are not erratic, and are not mistakes. *2.* Precision or repeatability data that deviate from a mean value in accordance with the laws of chance.

randomized non-return-to-zero (RNRZ) Refers to a coding scheme that provides the greatest packing density on instrumentation tape with low possibility of DC components.

randomizer A hardware device used to inject a pseudo-random bit sequence into an NRZ wavetrain, thereby guaranteeing frequent data transitions so that the low-frequency component is not too low for transmission or recording. A "derandomizer" removes the sequence and restores data to the original form; a form of data enhancement.

random noise Oscillations whose instantaneous amplitudes occur, as a function of time, according to a normal (Gaussian) curve.

random number generator A special routine or hardware designed to produce a random number or series of random numbers according to specified limitations.

random numbers *1.* Expressions formed by sets of digits selected from a sequence of digits, in which each successive digit is equally likely to be any of the digits. *2.* A series of numbers obtained by chance.

random pattern Describes a winding with no fixed pattern. [AIR4844]

random RNAV routes *See* area navigation, random routes.

random routes of area navigation *See* area navigation, random routes.

random sampling Selecting a small number of items for inspection or testing (the sample) from a much larger number of items (the lot or population) in a manner that gives each item in the population an equal probability of being included in the sample.

random sequence In welding, refers to a technique of depositing a longitudinal weld bead in increments of random length and location.

random variables Variables that randomly assume different possible values. Mathematically, random variables are described by their probability distribution, which specifies the possible values of a random variable together with the probability associated (in an appropriate sense) with each value. Random variables are said to be continuous if their possible values extend over a continuum and discrete if their possible values are separated by finite intervals.

random walk The path followed by a particle that makes random scattering collisions with other particles in a gaseous or liquid medium.

random walk method In operations research, a variance-reducing method of problem analysis in which experimentation with probabilistic variables is traced to determine results of a significant nature. Uninteresting walks add only to the variance of the process and thus contribute nothing. An interesting walk, however, tends to lead toward a predictive solution.

range *1.* The maximum ultrasonic path length that can be displayed. [ARP5089] *2.* The difference between the maximum and minimum curve-fit values at one particular value of the independent variable. [AIR4979] *3.* For instrumentation, the set of values over which measurements can be made without changing the sensitivity of the instrument. *4.* The extent of a measuring, indicating, or recording scale. *5.* An area within defined boundaries or landmarks, used for testing vehicles, artillery, or missiles, or for other test purposes. *6.* The maximum distance that a vehicle, aircraft, or ship can travel without refueling. *7.* The distance between a weapon and a target. *8.* The maximum distance that a radio, radar, sonar, or television transmitter can send a signal without excessive attenuation.

rangeability *1.* Describes the relationship between the range and the minimum quantity that can be measured. *2.* For a meter, the ratio of the maximum flow rate to the minimum flow rate.

range check In data processing, a validation that data is within certain limits.

range errors Errors in radar range measurement due to the propagation of radio energy through a nonhomogeneous atmosphere. These errors occur because the velocity of radio wave propagation varies with the index of refraction and because ray travel is not in straight lines through actual atmospheres. The resulting range errors are generally insignificant.

range phase *See* mishap, maneuver.

range, wire The designation of wire/conductor sizes that a given conductor barrel, ferrule, grommet, or accessory will accommodate. [ARP914A]

ranging The measurement of distance by timing how long it takes a light, radio-frequency sound, or ultrasound pulse to make a round trip from the source to a distant object.

rank To arrange in an ascending or descending series according to importance.

Rankine Symbolized by R. An absolute temperature scale in which the zero point is defined as absolute zero (the point where all spontaneous molecular motion ceases), and the scale divisions are equal to the scale divisions in the Fahrenheit system. 0°F equals approximately 459.69°R.

Rankine cycle An ideal thermodynamic cycle consisting of heat addition at constant pressure, isentropic expansion, heat rejection at constant pressure, and isentropic compression. Used as an ideal standard for the performance of heat-engine and heat-pump installations operating with a condensable vapor as the working fluid, such as a steam power plant.

RAPCON Acronym for radar approach control.

RAPCON facility (radar approach control facility) A terminal ATC facility that uses radar and nonradar capabilities to provide approach control services to aircraft arriving, departing, or transiting airspace controlled by the facility. [ARP4107]

rapid quenching (metallurgy) Rapid cooling of molten metals or alloys to achieve maximum uniformity in the crystal structure.

rapid traverse A machine tool mechanism whereby the workpiece is quickly moved to a new position while the cutting tool is retracted.

rare gases Gases such as helium, neon, argon, krypton, xenon, and radon, all of whose shells of planetary electrons contain stable numbers of electrons, resulting in the atoms being almost completely chemically inactive. Also known as inert gases.

rare gas-halide lasers A class of lasers in which the inert gases are used as the amplifying medium.

RAST (recover, assist, secure, and traverse) A system that is installed in the helicopter landing area of small ships, and is used to guide the helicopter to the landing area by means of a tensioned cable. Once landed, the helicopter is automatically secured to the deck. Such a system further assists in traversing the helicopter between the landing area and the hangar. [AIR1489]

raster A set of lines that provide essentially uniform coverage of a given area; for example, the set of parallel lines on a television picture tube that are most easily seen when there is no picture.

raster display A display in which the entire screen is scanned sequentially. [ARP4032]

raster pitch The distance between corresponding points of adjacent active scan lines of a raster scan. [ARP1782]

raster scan *1.* A regular pattern of scanning lines. [ARP1782] *2.* A form of display similar to television, in which the signal is scanned backwards and forwards across the screen.

ratchet valve *See* lock valve.

rate action *1.* Refers to the derivative control mode. *2.* A control action that produces a corrective signal proportional to the rate at which the controlled variable is changing. Rate action produces a faster corrective action than proportional action alone.

rate control action *See* derivative control action.

rated average laboratory life With regard to lamps, the average life obtained when 50% of a statistically large group of the same lamps still survive in closely controlled laboratory life testing at their design voltage. [ARP881]

rated capacity *1.* The ampere-hours of discharge that can be removed from a fully charged battery at a specific constant discharge rate, a specified discharge temperature, and a given cutoff voltage. [AIR1898] 2. The manufacturer's stated capacity rating for mechanical equipment; for example, the maximum continuous capacity in pounds of steam per hour for which a boiler is designed.

rated conditions The inlet and outlet conditions at which a starter is designed to operate and at which performance calibration is desired. Usually expressed in terms of pressure, temperature, and flow rate; or current and volts. [ARP906A]

rated continuous power The maximum power that can be delivered by an actuation system for an indefinite period, without sustaining damage or reducing life. [ARP4386]

rated control pressure The specified control pressure corresponding to rated current with a specified supply pressure and load flow. Rated control pressure is normally specified for a blocked load (zero steady flow to the load). [ARP490]

rated current The specified input current (expressed in mA) of either polarity to produce rated flow. The particular coil connection (differential, series, or parallel) must be specified in conjunction with the rated current. NOTE: Rated current does not include null bias current. [ARP490]

rated discharge pressure Pressure against which a component performance rating is established. [AIR1749]

rated flow *1*. The specified control flow corresponding to rated current and specified load pressure drop. Rated flow is normally specified as the no-load flow. [ARP490] 2. For a component or system, specified flow at steady-state conditions. [ARP4386] *3*. For a piping system or process vessel, the design flow rate.

rated flow capacity Low flow corresponding to full displacement of the servovalve output stage with specified values of control pressure, supply pressure, and return pressure. [ARP490]

rated flow gain The ratio of rated flow to rated current, expressed in cis/mA, gpm/mA, or mL/s/mA. [ARP490]

rated hold current *See* minimum limit of ultimate trip.

rated horsepower The maximum or allowable power output of an engine, turbine, or other prime mover under normal, continuous operating conditions.

rated inflation pressure The specified inflation pressure corresponding to the static loaded radius of the tire at its rated load. [AS4833]

rated inlet pressure Pressure at the inlet fuel connections of a component, at which its performance rating is established. [AIR1749]

rated load *1*. The static linear impedance required to draw rated current under a steady state condition of rated units. [AS8011A] 2. A specified steady-state load applied to the actuator for determining rated velocity. Usually, rated load is an opposing load. [ARP4386] *3*. The maximum design load for a machine, structure, or vehicle.

rated power The power required at rated load and rated velocity; also called the power point. This point usually determines the maximum power required from the actuation system. [ARP4386]

rated pressure The nominal maximum input or operating pressure. Also called service pressure. [ARP4386]

rated pressure gain The ratio of rated control pressure and rated input current at a specified supply pressure, expressed in psi/mA or kPa/mA. This is the specified "nominal" pressure gain. [ARP490]

rated temperature The maximum temperature at which an electric component can operate for extended periods without loss of its basic properties. [ARP1931]

rated trip current *See* maximum limit of ultimate trip.

rated velocity *1*. The required actuation system output velocity when the specified rated load is applied to the actuation motion attach point. Rated velocity can be determined at any load position. [ARP4386] 2. The maximum velocity that a servoactuator is capable of under specified external load conditions and rated pressures. [ARP1281A]

rated voltage The voltage at which an electrical component can operate for extended periods of time without undue degradation. [ARP1931]

rate gyro A gyroscope that provides an output signal proportional to the turning rate about a given axis. [ARP419]

rate gyroscope A gyro wheel mounted in a single gimbal ring in such a manner that rotation about an axis perpendicular to both the gimbal axis and gyro axis produces a precessional torque proportional to the rotation rate.

rate limit The maximum velocity of the control system output. Usually, rate limit is defined under aiding load condition with saturation signals applied to the servoactuator. [ARP4386]

rate of blowdown A rate normally expressed as a percentage of the water fed.

rate-of-climb indicator A navigation instrument that indicates the rate at which an aircraft gains or loses altitude.

rate of flow change Time derivative of inspiratory or expiratory flow. [AIR1109]

rate response A relationship describing the output of a control system as a function of its input signal.

rate time In the action of a proportional-plus-rate or proportional-plus-reset-plus-rate controller, the time by which the rate action advances the proportional action on the controlled device.

rating The value of some characteristic of performance as specified in the model specification. *See also* load. [AS1607]

ratio controller *1.* A controller that maintains a predetermined ratio between two or more variables. *2.* A controller that maintains the magnitude of a controlled variable at a fixed ratio to another variable.

ratio meter A measuring instrument in which the pointer deflection is proportional to the ratio of the currents passing through two coils; used to measure the quotient of two electrical qualities.

ratio of specific heats Specific heat at constant pressure, divided by specific heat at constant volume.

ratio spectrofluorometer A type of instrument used in chemical assaying. Used to determine proportions of two compounds that are similar in bioassay and spectrophotometry but markedly different in fluorescence. Quantitative assays can be made either by proportionality or by use of a linearity curve.

ratio thermometer A device consisting essentially of two radiation thermometers in the same housing, the output of each thermometer having a separate wavelength response. The output of such a device is a ratio signal that is a function of temperature, but is relatively insensitive to target size; thus such a device is as accurate for small radiating bodies as it is for larger ones.

ratio-type telemeter A telemeter that translates data in terms of the relative phase relation between two electrical quantities, or their relative magnitudes.

RATO (rocket assisted takeoff) *See* jet-assisted takeoff.

raw data *1.* The sets of numbers that describe the body measurements made on each individual in a survey. [AIR5145] *2.* In data processing, information that has not been processed or analyzed by the computer.

rawinsondes Combinations of raob (radiosonde observation) and rawin (a method of winds-aloft observation); observations of temperature, pressure, relative humidity, and winds-aloft by means of radiosonde and radio direction-finding equipment or radar tracking.

raw material Material such as sheet, plate bar, forging, or castings. Usually identified by a heat or lot number, and usually tested destructively for acceptance. [AMS2770E]

ray acoustics *See* geometrical acoustics.

Rayleigh-Benard convection Describes the flow of a fluid that is contained between horizontal, thermally conducting plates and is heated from below. The Rayleigh number is proportional to the temperature difference between the plates.

Rayleigh disc A special form of acoustic radiometer used to measure particle velocities.

Rayleigh scattering Any scattering process produced by spherical particles whose radii are smaller than about one-tenth the wavelength of the scattered radiation.

Rayleigh waves *1.* Two-dimensional barotropic disturbances in a fluid having one or more discontinuities in the vorticity profile. *2.* Surface waves associated with the free boundary of a solid, such that a surface particle describes an ellipse whose major axis is normal to the surface and whose center is at the undisturbed surface. At maximum particle displacement away from the solid surface; the motion of the particle is opposite to that of the wave. *See* surface wave.

ray optics *See* geometrical optics.

ray tracing A procedure used in the graphical determination of the path followed by a single ray of radiant energy as it travels through media of varying indices of refraction.

RBN *See* radio beacon.

RCAG *See* remote communications air/ground facility.

RCLM Acronym for runway centerline marking.

RCLS *See* runway centerline light system.

RCM *See* reliability centered maintenance.

RCO *See* remote communications outlet.

RC oscillator Any oscillator whose frequency is determined by the interaction between resistors and capacitors in an electronic circuit.

RCR *See* runway condition reading.

RCTL *See* resistor-capacitor-transistor logic.

RDF *See* radio direction finder.

Re *See* Reynolds number.

reactance (X) *1.* Opposition offered to the flow of alternating current by inductance or capacitance of a component or circuit; the imaginary portion (j factor) of the impedance. [ARP1931] *2.* In an electrical circuit, a component that is due to the presence of capacitive or inductive elements (not resistive elements), opposing the flow of electric current.

reactance drop The voltage drop 90° out of phase with the current.

reaction A chemical transformation or change brought about by the interaction of two substances.

reaction bonding Chemical combining of ingredients to produce silicon nitride ceramics.

reaction injection molding A process for molding polyurethane, epoxy, and other liquid chemical systems. [AIR4844]

reaction products The substances formed in a chemical reaction—the desired items as well as the unwanted fumes, sludge, residues, etc.

reaction time *1.* The element of uptime needed to initiate a mission, measured from the time command is received. [AIR4896] *2.* In human engineering, the interval between the input signal (physiological) or a stimulus (psychophysiological) and the response elicited by the signal. *3.* An inherent perceptual limitation whereby an individual requires a specific amount of time for information processing before action is taken. [ARP4107]

reactive alerting system A system that senses and identifies the presence of windshear after the phenomenon is encountered. [AIR4102/11]

reactive diluent In epoxy formulations, a compound containing one or more epoxy groups that functions mainly to reduce the viscosity of the mixture. [AIR4844]

reactive volt-ampere meter *See* varmeter.

reactor *1.* A circuit element that introduces capacitative or inductive reactance. *2.* A vessel in which a chemical reaction takes place. *3.* An enclosed vessel in which a nuclear chain reaction takes place.

reactor cores In nuclear reactors, the regions containing the fissionable material.

reactor, endothermic A reactor that absorbs heat from the surroundings.

reactor, exothermic A reactor that generates heat.

reactor safety Refers to theoretical and experimental investigations of the behavior of reactor types and designs under various real or hypothetical accident conditions.

read *1.* To play back data from a tape, disk, or the like. *2.* To obtain the contents of computer memory.

readability *1.* The uncertainty in a temperature reading. [AMS2750] *2.* On an instrument, the smallest fraction of the scale that can be easily read—either by estimation or by use of a vernier.

readiness The probability that material will perform satisfactorily on demand. [ARP4386]

reading For an indicating instrument, the indicated value determined from the scale. For a recording instrument, the indicated value determined from the position of the index with respect to an appropriate indicating scale.

readout circuit The hydraulic to electrical system that permits the fine control of the impulse curve at the test manifold. [AIR1228]

ready time During a mission, the period of time that an item is available for operation, but is not required. [AIR4896]

real leak Any path through the wall of a vacuum chamber that allows a gas to pas through, however small the rate. [AMS2769]

real number Any number that can be represented by a point on a number line.

real-time *1.* Describes any measurement or inspection that can be interpreted as it is happening. [ARP5089] *2.* Pertaining to the actual time during which a physical process transpires. *3.* See response-critical.

real-time clock A clock that indicates the passage of actual time, in contrast to a fictitious time set up by the computer program; for example, the clock that indicates elapsed time in the flight of a missile, whereby a 60-second trajectory is computed in 200 actual milliseconds, or a 0.1 second interval is integrated in 100 actual microseconds.

real-time input Input data inserted into a system at the time of generation by another system.

reamer A special type of cutting device used to size close-tolerance holes. [ARP480A].

rear insertion-front release A type of connector in which contacts are inserted from the rear with the proper insertion tool; and released from the rear, with the removal tool inserted from the face of the connector. Requires demating of the plug and receptacle. [ARP914A]

rear insertion-rear release A type of connector in which contacts are both inserted and removed from the rear of the connector with the proper tools. Does not require demating of an electrical installation. [ARP914A]

rearward facing steps *See* backward facing steps.

reasonableness test A test that provides a means for detecting a gross error in calculation by comparing results against upper and lower limits representing an allowable reasonable range.

reassembly Assembling items that were removed during disassembly and closing the reassembled items. [AIR4896]

reassociation The recombination of the products of dissociation.

Reaumur scale A temperature scale in which 0° is the ice point and 80° is the steam point. NOTE: This scale is little used outside the brewing, winemaking, and distilling industries.

REB *See* relativistic electron beams.

reboiler *1.* The heat exchanger at the bottom of a distillation column. The reboiler generates vapors from the liquid that comes down the column; these vapors then ascend through the column. *2.* A closed heat exchanger in which one medium is used to heat a second. May be used to maintain a separation between the two mediums due to noncompatibility, or to prevent contamination between systems while transferring thermal energy.

rebound In a shock absorber, the phase of operation that occurs after the compression and storage of energy within the unit. In rebound, the stored energy forces the extension or rebound of the piston to release the energy. This sometimes causes the aircraft to bounce back into the air after initial impact, unless the rebound is snubbed properly. [AIR1489]

rebreather bag *See* bag, rebreather. [ARP171]

rebreather system A system in which the same volume of gas is breathed over and over again; the oxygen consumed is replaced from an external source, and the carbon dioxide and water vapor are removed by chemical means. [ARP171]

receiver *1.* A search unit or transducer element, used to receive ultrasonic energy from a test part. [ARP5089] *2.* An apparatus near the outlet of a compressor which collects excess oil or moisture in the compressed air, and which reduces or eliminates pressure pulsations and stores compressed air for later use. *3*. An electronic device that detects radio-wave signals and amplifies them into electrical signals of varying frequency and amplitude to drive an output device such as a loudspeaker. Also known as radio receiver.

receiver, fiber optic A fiber-optic detector (optical to electrical converter) plus signal-conditioning circuits. [ARD50024]

receiver gage A gage that is calibrated in engineering units and receives the output of a pneumatic transmitter.

receiving gage A fixed gage designed for inspecting several dimensions on a part, and for determining the relationships between dimensions.

receiving system *See* receive.

recharger An apparatus for refilling oxygen cylinders with oxygen. [ARP171]

recharging The restoring of discharged electric storage batteries to a charged condition by passing direct current through them in a direction opposite to that of the discharging current.

recipe A complete set of data and procedure that defines the control requirements of a particular product manufactured by a batch process. A recipe consists of a header, equipment requirements, procedure, and formula.

reciprocal linear dispersion *See* dispersion.

reciprocal transducer A transducer in which the output signal is proportional to the reciprocal of the level of the stimulus.

reciprocating engines *See* piston engines.

reciprocity theorem Any theorem expressing reciprocal relations for the behavior of some physical system, according to which input and output can be interchanged without altering the response of the system to a given excitation.

recirculation The reintroduction of part of a flowing fluid to repeat the cycle of circulation. *See also* circulation.

recline lock override Describes a design feature of a reclinable seat back whereby the seat back may be returned from any angle of recline to the normal upright position by the mere application of force in the direction of travel to the seat back. [ARP750]

recognition The psychological process whereby an observer interprets the visual or auditory

stimuli he/she receives from a distant object and thus forms a correct conclusion as to the exact nature of the object or sound.

recoil control The process of controlling the energy stored in a shock absorber after impact. The energy is dissipated in the extension phase usually by forcing hydraulic oil through orifices or passages. [AIR1489]

recoil snubber The valve, orifice(s), or snubbing device that controls impact forces within a shock absorber upon extension of the landing gear after takeoff or catapulting. [AIR1489]

recombination The reaction between an ion and one or more electrons, returning the ionized element or molecule to the neutral state.

recombination coefficient A measure of the specific rate at which oppositely charged ions join to form neutral particles; a measure of ion recombination.

recommended firing current (or energy) In an EED, the current (or energy) that must be applied to a bridge circuit to cause operation within a specified time. [AIR913]

recommended test current (or energy) In an EED, the current (or energy) that can be applied to a bridge circuit for extended periods without degrading the explosive material or firing the device. [AIR913]

recondition Refers to the work necessary to return an item to the highest standard specified in the relevant manual. [AIR4896]

reconditioning Performed periodically, a programmed electrical exercise whereby the capability of a battery that has electrically deteriorated while in service is restored to full energy and power delivery. Consists of a discharge/charge cycle(s); may also include water replacement and mechanical cleaning procedures. [AS8033]

reconfigurable Describes a fault-tolerant device or system in which continued functional operation, subsequent to a failure, is provided by rearranging or recombining the surviving control elements. Sometimes referred to as self-repairing. [ARP4386]

reconfiguration Process by which the operation of an HSRB is initialized using Beacon frames following the application of power, the presence of failures, or the detection of certain errors using Beacons. [AIR4289]

record *1.* A segment of a file consisting of an arbitrary number of words or characters. *2.* In data processing, a group of data that contains all the information about a single item.

recorded value The value of a measured variable as determined from the position of a trace or mark on chart paper; or as determined from a permanent or semipermanent effect on an alternative recording medium.

recorder An instrument that generates and displays a continuous graphic, acoustic, or magnetic record of a measured variable.

recording channel The circuit and associated apparatus needed to produce a single trace in a recording system that incorporates any means of producing multiple traces on a single chart, or multiple tracks on an alternative recording medium.

recording extensions Describes a recorder that is attached directly to the meter body, with the recorder pen positioned by the metering float through a magnetic coupling.

recording thermometer *See* thermograph.

recoverable Describes items that may be rehabilitated to a serviceable condition one or more times before scrapping. Rehabilitation is by rework or servicing, such as welding, refinishing, or recharging. [ARD50010]

recovery and soaking time Elapsed time between insertion of parts in a heating medium and start of soaking time. [AMS2771A]

recovery factor The ratio of the difference between indicated temperature and static temperature, to the difference between total temperature and static temperature. [ARP485]

recovery procedure A flight path control technique used to escape from an inadvertent encounter with windshear. [ARP4109]

recovery ratio The ratio of the indicated temperature to the total temperature. [ARP485]

recovery time In a radiation counter tube, the time that must elapse after detection of a photon of radiation before the instrument can deliver another pulse of substantially full response level upon interaction with another photon of ionizing radiation.

recovery voltage The voltage impressed across a protective device after the circuit has been interrupted—and after high-frequency transients have subsided. [ARP1199A]

recrystallization *1.* In metals, the change from one crystal structure to another, as occurs on heating or cooling through a critical temperature. *2.* The formation of a new strain-free grain structure from that existing in cold-worked metal, usually accomplished by heating.

recrystallization temperature The minimum temperature at which complete recrystallization of a cold-worked metal occurs in a specified time, usually one hour.

recrystallized beta grain Strain-free grains of beta phase that precipitate relatively little or no alpha phase during a decoration aging treatment of titanium. [AMS4897]

rectangular coordinates *See* Cartesian coordinates.

rectangular terminal A terminal with a rectangular-shaped tongue. [ARP914A]

rectennas Devices that convert microwave energy into direct-current power by utilizing a number of small diodes, each with its own diode rectifier.

rectifier(s) Static devices with an asymmetrical conduction characteristic, used to convert alternating current into direct current.

rectifier antennas *See* rectennas.

rectifier instrument An instrument for measuring certain quantities of an alternating-current circuit. With this instrument, the a-c input is converted to direct current by a rectifier, and the d-c signal is actually measured.

rectifier-type voltmeter An instrument that incorporates four semiconductor rectifier elements arranged in a square (full-wave bridge configuration), with the a-c input connected across one diagonal and a permanent-magnet moving-coil detector connected across the other diagonal. In this instrument, the d-c output is proportional to the average current or voltage over any given half-cycle of the a-c input. This makes the instrument useful for measuring nominal voltage or low-range (milliampere) current.

rectifying section In a distillation column, the section of trays above the feed plate. In this section, the vapor is enriched in the light components that are taken overhead.

recursion The property whereby a callable program is allowed to call itself.

recursive Describes a process that is inherently repetitive. The result of each repetition is usually dependent on the result of the previous repetition.

recycle synchronizer A method of telemetry subframe recognition in which a specific subcommutator word (or words) contains a unique signal that marks the recycling of the subcommutator.

red data Data that require safeguards. [AIR4271]

red dwarf stars Red stars of low luminosity, so designated by E. Hertzsprung. Red dwarf stars are commonly those main sequence stars fainter than an absolute magnitude of plus 1, and are the faintest and coolest of the dwarfs.

red giant stars Stars that have evolved to the point where hydrogen core burning has been completed, the helium core has become denser and hotter than originally, and the envelope has expanded to perhaps 100 times its initial size.

redout A temporary condition in which vision is obscured by a reddishness, or in which objects appear to have a reddish color. This condition, sometimes followed by unconsciousness (not considered a part of redout), is caused by blood rushing to the head. Redout is generally caused by the presence of negative G forces. [ARP4107]

Redox cells Cells for converting the energy of reactants to electrical energy. In a redox cell, an intermediate reductant in the form of liquid electrolyte reacts at the anode in a conventional manner and is regenerated by reaction with a primary fuel.

redox potential The electrochemical potential prevailing in a chemical reaction involving an exchange of electrons; the *red*uction-*ox*idation potential.

red plague A powdery, brown-red cuprous oxide deposit sometimes found on silver-coated copper conductors and shield braids. Red plague has a fungus-like appearance and will appear in random spots along the length of a conductor or shield. [ARP1931]

red tape *See* housekeeping.

reduced gravity A condition in which the acceleration acting on a body is less than normal gravity, between 0 and 1 G.

reducer A pipe fitting used to couple a pipe of one size to a pipe of a different size. The term "increaser" is also used—when the flow is from the smaller pipe to the larger.

reducing atmosphere An atmosphere that tends to: (a) promote the removal of oxygen from a chemical compound; or (b) promote the reduction of immersed materials.

reducing coupling A pipe fitting for connecting two pipes of different sizes.

reduction Removal of oxygen from a chemical compound.

reduction of area The difference between the original cross-sectional area of a tension-test specimen and the area of its smallest cross section, usually expressed as a percentage of the original area. [AIR4844]

redundancy *1.* The existence of more than one means for accomplishing a given function. Each means of accomplishing the function need not necessarily be identical. [ARD50013] *2.* In the transmission of information, the fraction of the gross information content of a message that can be eliminated without loss of essential information. *3.* Any circuitry or program instructions present solely to handle faults or errors and not necessary for normal system operation. *4.* A parallel or secondary system that takes over when the primary system fails so control can continue uninterrupted.

redundancy, active A type of redundancy whereby all redundant items are operating simultaneously, rather than being activated when needed. [ARD50010]

redundancy check An automatic or programmed check based on the systematic insertion of components or characters used especially for checking purposes.

redundancy management The portion of the system logic and control (hardware or software) that detects and isolates failures in a fault-tolerant system; and reconfigures the system after the failure is detected and isolated so as to maintain the same or a reduced level of operation. [ARP4386]

redundancy, standby A type of redundancy whereby the alternative means of performing the function is inoperative until needed, and is activated upon failure of the primary means of performing the function. [ARD50010]

redundant *1.* Describes a condition in which multiple independent means are incorporated to accomplish a given function. [ARP4754] *2.* Denotes the use of duplicate or alternate components for the purpose of improving mission safety or reliability. [ARP4386]

redundant actuator A multichannel actuator design in which the channels are identical, each receiving the same input, connecting the individual outputs to form a common, or single, output. [ARP4386]

redundant character A character that is specifically added to a group of characters to ensure conformity with certain rules that can be used to detect computer malfunction.

redundant coding The use of more than one symbol or symbol characteristic to convey the same information. [ARP4155]

redundant data bus The use of more than one data bus to provide more than one data path between the subsystems. [AS1773]

redundant design A technique of incorporating into a system two or more components that perform the same function(s), so that if one fails or malfunctions the other(s) will perform the necessary functions to enable the system to continue to operate safely. *See also* fail safe and fail operational. [ARP4107]

redundant sealing Supplemental sealing that serves as a backup for the primary seal, providing a form of insurance. [AIR4069]

reed A thin bar of metal, wood, or cane that is clamped at one end and set into transverse elastic vibration by a stimulating force such as wind pressure. Used to create sound in musical instruments; also acts as a frequency standard in some meters.

reel hoses Relatively long output hoses contained on reels for compact storage. [AIR4783]

re-engagement *1.* The act of the starter engaging the engine while the engine is at any speed condition between rest and starter cutoff. [ARP906A] *2. See* engagement.

re-enterable load module A load module that can be used concurrently by two or more tasks.

re-entrancy The property whereby a callable program is allowed to be called and executed before it has completed the execution from a previous call. The results of the previous call are not affected.

reentrant The property of a program that enables it to be interrupted at any point by another program, and then resume from the point at which it was interrupted.

re-entrant program A program that can be used for various tasks.

reentry The event occurring when a spacecraft or other object comes back into the sensible atmosphere after going to higher altitudes; or the action involved in this event.

reentry edge A negative contact angle between a fillet and its substrate, caused by improper tooling (fairing, feathering) of the edges of the fillet, or failure to tool at all. [AIR4069]

re-entry point The instruction at which a program is re-entered from a subroutine.

reentry trajectories Those parts of rocket trajectories that begin at reentry and end at target or at the surface.

refereed test A predetermined, destructive or nondestructive test performed by a regulatory body or a disinterested organization, often to fulfill a regulatory requirement. In some cases, the test may be performed by the regulated organization and merely be witnessed by an agent of the regulatory body.

referee fluid A substitute fluid used for testing and/or cleaning. [AIR4728]

reference accuracy In process instrumentation, a number or quantity that defines a limit that error will not exceed when a device is used under specified operating conditions (with error representing the difference between the measured value and the standard or ideal value).

reference address *See* base address.

reference block A block that is machined from material similar to the product to be inspected, and is used to produce a reflection of known characteristics. [AMS2633B]

reference coordinate system A coordinate system in which the structure is described with respect to loads and ply orientation. [AIR4844]

reference data base A collection of empirical or analytical data from which a model is derived and to which model output may be compared. [ARP4148]

reference dimension On a mechanical drawing, a dimension without tolerance that is given for information only and is not to be used in making or inspecting the part or assembly.

reference facility A test facility of known performance, traceable to the designated baseline facility, against which the customer's facility is compared. [ARP741]

reference input A signal from an independent reference source that is used as one of the inputs to an automatic controller.

reference junction *1.* The union of two dissimilar wires maintained at a known signal level. [ARP485] *2.* A device used to couple a thermocouple transducer to copper wires without introducing an error.

reference junction compensation A means of counteracting the effect of temperature variations of the reference junction, when allowed to vary within specified limits.

reference level *1.* The basis for comparison of an audio-frequency signal level given in decibels or volume units. *2. See* datum plane.

reference line A straight line on the profile of the basic rack, with reference to which the tooth dimensions are specified. [AS1560]

reference methods Methods that have been identified as being equivalent to those detailed in a specification, e.g., for the quantification of individual pollutant species, and that are recognized as acceptable alternative methods. [ARP4418]

reference operating conditions *See* operating conditions, reference.

reference plane *See* datum plane.

reference pressure The pressure against which another pressure can be compared or measured.

This pressure is arbitrary but agreed to and stipulated by those involved with its intended use. [ARP4386]

reference speed for final approach *See* V_{ref}.

reference standard A piece of material, part, or piece from a part, containing an artificial discontinuity of known size. Provides a means of producing a reflection of known characteristics; used to establish a measurement scale. [ARP5089]

refined metal Metal derived from raw materials in the form of virgin elements, master alloy, and/or revert melted in any combination, replenished as needed, fluxed for non-metallics, and degassed as necessary. [AMS5402B]

refinery gas The commercially noncondensible gas resulting from fractional distillation of crude oil, or the cracking of crude oil or petroleum distillates. Refinery gas is either burned at the refineries or supplied for mixing with city gas.

reflectance *1.* The ratio of the radiant flux reflected by a body to that incident upon it. *2.* The fraction of incident light that is reflected by a surface. Also called reflectivity.

reflected glare Glare resulting from specular reflections of high brightness in polished or glossy surfaces in the field of view. Usually associated with reflections from within a visual task or areas in close proximity to the region being viewed. [AIR1151]

reflected radiation *See* reflected waves.

reflected rays *See* reflected waves.

reflected waves *1.* Shock waves, expansion waves, or compression waves reflected by another wave incident upon a wall or other boundary. *2.* In electronics, radio waves reflected from a surface or object.

reflecting telescopes Telescopes that collect light by means of concave mirrors.

reflection *1.* An indication that has arisen as a result of an incident sound beam returning from the boundary of two materials of dissimilar acoustic impedance. [ARP5089] *2.* The process whereby a surface of discontinuity turns back a portion of the incident radiation into the medium through which the radiation approached.

reflection loss The part of a signal that is lost due to reflection of power at a line discontinuity. [ARP1931]

reflection nebulae Any celestial body having a hazy cloudy appearance whose brightness results from the scattering by dust particles of light from nearby stars.

reflectivity *See* reflectance.

reflectometer A photoelectric instrument for measuring the proportion of light reflected from a given surface.

reflector *1.* An interface at which an ultrasonic beam reflects. [ARP5089] *2.* A device whose chief use is to redirect the light of a lamp by reflection in a desired direction or directions. [ARP798]

reflector antenna Antenna consisting of a reflecting surface and a feed.

reflux The recycle stream that is returned to the top of a column. This stream supplies a liquid flow for the rectifying section, which enriches the vapor stream moving up the column. Material in the stream is condensate from the overhead condenser. Reflux closes the energy balance by removing heat introduced at the reboiler.

reflux drum *See* accumulator.

reflux ratio A quantity usually expressed as the ratio of the reflux flow to the distillate flow. Used primarily in column design.

reflux receiver *See* accumulator.

reflux-to-feed ratio *See* L/F.

refracted radiation *See* refracted waves.

refracted rays *See* refracted waves.

refracted wave(s) The resultant wave train produced when an incident wave crosses the boundary between its original medium and a second medium; in many cases, only a portion of the wave crosses the boundary, with the remainder being reflected from the boundary.

refracting telescopes Telescopes that collect light by means of a lens or system of lenses.

refraction The process whereby the direction of energy propagation is changed as the result of a change within the propagating medium; or as the energy passes through the interface representing a density discontinuity between two media. In the first instance, the rays undergo a smooth bending over a finite distance. In the second case, the index of refraction changes through an interfacial layer that is thin compared to the wavelength of the radiation; thus, the refraction is abrupt, essentially discontinuous.

refraction loss Reduction in amplitude or some other wave characteristic due to the refraction occurring in a nonuniform medium.

refractive index *1.* The ratio of the velocity of light in air to its velocity in the liquid under examination. *See also* refractivity. [AIR1116] *2.* The change in direction (angle) of a beam of light at the interface when progressing from one medium to another. *3.* The ratio of the phase velocity of a wave in free space to the phase velocity of the same wave in the specific medium.

refractivity The algebraic difference between a refractive index and unity.

refractometer An instrument for measuring the index of refraction of a liquid, gas, or solid. Measurement may be accomplished in any of several ways, including measuring the critical angle, measuring refraction produced by a prism, observing interference patterns in transmitted light, and measuring the dielectric constant of the substance.

refractor A device, usually of prismatic glass, that redirects the light of a lamp in desired directions, principally by refraction. [ARP798]

refractory Materials of construction capable of withstanding high temperatures in various industrial processes and operations. [AIR4844]

refractory baffle A baffle of refractory material.

refractory coating Pyrolytic material used for coating other materials exposed to high temperatures.

refractory lined fire-box boiler A horizontal fire-tube boiler, the front portion of which sets over a refractory or water-cooled refractory furnace. The rear of the boiler shell has an integral or separately connected section containing the first pass tubes, through which the products of combustion leave the furnace; they then return through the second-pass upper bank of tubes.

refractory metals Usually refers to alloys of high-melting point, hard-to-work metals, but can also refer to certain unalloyed elements.

refractory wall A wall made of refractory material.

refrigerant The medium of heat transfer in a refrigerating system. This medium picks up heat by evaporating at a low temperature and pressure, and gives up heat by condensing at a higher temperature and pressure. [ARP147C]

refrigerant charge *See* charge, refrigerant.

refrigeration cycle The complete course of operation of a refrigerant back to a starting point, evidenced by: a repeated series of thermodynamic processes, or flow through a series of apparatus, or a repeated series of mechanical operations. [ARP147C]

refrigeration system, air cycle A refrigeration cycle in which air is used as a refrigerant in an air cycle machine. [ARP147C]

refrigeration system, cascade A vapor cycle refrigeration system with two or more refrigerant circuits, each containing a compressor, condenser, and evaporator. The evaporator of one circuit cools the condenser of the other circuit. [ARP147C]

refrigeration system, compound A vapor refrigeration cycle using two or more stages of compression. [ARP147C]

refrigeration system, vapor cycle An assembly of connected components, usually consisting of an evaporator, expansion valve, compressor, condenser, and control elements through which a refrigerant is circulated for the purpose of extracting heat at a low temperature level and rejecting it at a high temperature level. [ARP147C]

refuel control panel Aircraft-mounted control panel used for controlling fueling and fuel distribution to aircraft fuel tanks. [AIR4783]

refueling A self-propelled vehicle with an onboard fuel supply, containing all necessary equipment, used to refuel an aircraft. [AIR4783]

refurbish (engine, module) To restore an engine or engine module to ensure that cost-effective operation is achieved. [AIR4896]

refusal speed The highest speed attained during takeoff acceleration from which the aircraft may be decelerated to a stop within the remaining runway length. [ARP4107]

refuse The solid portions of the products of combustion.

regain moisture content *See* dry basis.

regenerative cooling The cooling of a part of an engine by the fuel or propellant being delivered to the combustion chamber; specifically, the cooling of a rocket engine combustion chamber or nozzle by circulating the fuel and/or oxidizer around the part to be cooled.

regenerative feedback *See* positive feedback.

regenerator(s) *1.* Devices used in a thermodynamic process for capturing and returning to the process heat that would otherwise be lost. *2.* A repeater; that is, a device that detects a weak signal in a fiber-optic communication system, amplifies it, cleans it up, and retransmits it in optical form. *See* repeater.

register(s) *1.* A device used on or with a meter to register the volume of fluid passed through the meter. [AIR4783] *2.* Accurate matching or superimposition of two or more images. *3.* A subassembly of the burner on a furnace or oven that directs airflow into the combustion chamber. *4.* The component of a meter that counts the revolutions of a rotor or individual pulses of energy and indicates the number of counts detected. *5.* In data processing, the specific location of data in memory.

regrade A disposition of a nonconformity that: (a) determines that a product is not acceptable for it's original intended design, and (b) directs the product to be redesignated or modified for an alternate use. [ARD9000]

regression analysis A statistical technique for expressing functional relationships between quantities that can be measured or observed, and quantities that are to be predicated on a probabilistic basis. [ARP4293]

regressive burning A condition in which the mass flow produced by a propellant grain decreases as the web is consumed, due to decreasing area, decreasing burn rate, or both. [ARP4386]

regressive geometry A propellant grain configured in such a manner that the surface area decreases as burning progresses. [AIR913]

regrowth alpha Alpha that grows on preexisting (primary) alpha during cooling from some temperature high in the alpha-beta field. [AS1814]

regular reflection factor The ratio of the regularly reflected light to the incident light. [ARP798]

regular transmission Transmission in which the transmitted light is not diffused. In such transmission, the direction of the transmitted pencil of light has a definite geometrical relation to the corresponding incident pencil of light. [ARP798]

regular transmission factor The ratio of regularly transmitted light to the incident light. [ARP798]

regulated voltage range The range in adjustment of the line-to-line voltage as controlled by the regulator rheostat. [ARP1148A]

regulating transformer A transformer used to adjust the voltage and/or phase relation in steps, without interrupting the load. Generally consists of one or more windings connected in series with the load circuit, and one or more windings excited from the load circuit or from a separate source.

regulation *1*. Control of flow or of some other process variable. *2*. For a cold-cathode glow-discharge tube, the difference between maximum and minimum anode voltage drop over a range of anode current.

regulator *1*. A system for controlling the output voltage of an alternator. [ARP1148A] *2*. A pressure regulator. There are several different types: Type I, Type II, Type III, Class A, and Class B. *Type I*–provides an outlet flow or pressure that varies automatically with cabin pressure to ensure adequate delivery to the mask. *Type II*–may be manually adjusted to deliver the proper amount of pressure or flow for a selected cabin pressure. *Type III*–has a fixed outlet flow or pressure which has been pre-set to provide adequate flow up to indicated cabin pressure. *Class A*–designed to be cylinder mounted. *Class B*–intended for use in a line between the oxygen source and the dispensing unit or on an oxygen-cylinder valve.

[AS1197] *3*. A device for controlling pressure or flow in a process.

regulator, cabin pressure *1*. A pressure-regulator valve from a pressurized cabin. Regulates the pressure in that cabin by controlling the outflow of air from the cabin. *2*. A valve at the air supply to the cabin, regulating flow to the cabin in order to maintain prescribed pressure. [ARP147C]

regulator, cabin pressure auxiliary unit A unit that contains all of the parts of a cabin pressure regulator but the control elements, and is operated in parallel with and by cabin pressure from the control head of the master regulator. [ARP147C]

regulator, constant flow A regulator that delivers constantly a definite quantity of oxygen. This quantity varies with altitude and is adjusted either manually or automatically with an aneroid control. [ARP171]

regulator, demand A regulator that supplies gas when subjected to a slightly negative pressure as a result of each inspiration; flow normally ceases on exhalation. [ARP171]

regulator, diluter-demand A demand regulator with a device for diluting the oxygen with air. [ARP171]

regulator, gas pressure A spring-loaded, dead-weighted, or pressure-balanced device that will maintain the gas pressure to the burner supply line.

regulator, master cabin pressure A cabin-pressure regulator containing all the necessary control elements. [ARP147C]

regulator, oxygen-pressure A device that receives oxygen from a supply source and delivers it to a mask or other system component in suitable condition of pressure and flow. [ARP171]

regulator, pressure-demand A demand regulator that, by virtue of either mechanical or pneumatic loading of the diaphragm, delivers gas at a slightly positive pressure during each inspiration; and maintains a positive pressure, with no oxygen delivery, during expiration. [ARP171]

regulator sensing The means by which voltage is sensed and fed to the voltage regulator. [ARP1148A]

regulator, temperature A device that automatically controls the temperature of an air stream in a given region. [ARP147C]

regulatory control Maintaining the outputs of a process as close as possible to their respective setpoint values, despite the influences of setpoint changes and disturbances.

reheated steam Superheated steam that derived its superheat from a reheater.

reheat engines Turbojet or turbofan engines in which additional fuel is injected into the airflow—either bypass air or the exhaust air from the main combustion process—and burned there to raise the air temperature, the airflow volumetric flow rate, and therefore the engine thrust. [AS5116]

reheating The process of adding heat to steam to raise its temperature after it has done part of its intended work. This is usually done between the high-pressure and low-pressure sections of a compound turbine or engine.

REIL *See* runway end identification lights.

reinforced molding compound Compound supplied by raw-material producer in the form of ready-to-use materials, as distinguished from premix. [AIR4844]

reinforced plastics Molded, formed, filament-wound, tape-wrapped, or shaped plastic parts consisting of resins, to which reinforcing fibers, mats, fabrics, etc., have been added before the forming operation to provide some strength properties greatly superior to those of the base resin. [AIR4844]

reinforced reaction injection molding A reaction injection molding with a reinforcement added. [AIR4844]

reinforcement A strong material bonded into a matrix to improve its mechanical properties. [AIR4844]

reinforcing materials Fibers, filaments, fabrics, and other substances used for strengthening of matrices in composite materials.

Reissner-Nordstrom solution The unique solution of general relativity theory describing a nonrotating, charged black hole.

reject (suppression) Describes a control used for minimizing or eliminating low-amplitude signals so that larger signals are emphasized. [ARP5089]

rejected takeoff (RTO) The process of rejecting or aborting a takeoff of an aircraft because of engine or any other failure or command decision. *See* abort.

rejection Nonacceptance of an item because of its nonconformance to a specified requirement. [AIR4896]

rejection level That setting of the signal level above or below which all parts are rejectable; for example, the level in an automatic system at which objectionable parts will actuate the reject mechanism of the system. [ARP5089]

relative accuracy The maximum deviation from a straight line that passes through the end point of an ADC or a DAC transfer junction. Expressed as percent, ppm of the full-scale range, or in LSBs.

relative bearing *See* bearing.

relative density The ratio of the weight of any volume of a substance, at given temperatures, to the weight of an equal volume of standard substance. [AIR4783]

relative density bottle *See* specific gravity bottle.

relative gain An open-loop gain determined with all other manipulated variables constant, divided by the same gain determined with all other controlled variables constant.

relative humidity The ratio, expressed as a percentage, of the amount of water vapor (grams of water vapor per kilogram of dry air) actually present in the air, to the amount of water vapor that would be present if the air were saturated with respect to water at the same temperature and pressure. [AIR1335]

relative luminosity The ratio of measured luminosity at a particular wavelength to measured luminosity at the wavelength of maximum luminosity.

relative response The ratio of the response of a device or system under some specific condition to its response under stated reference conditions.

relative response of NVIS The response of a night-vision imaging system is in the spectral region between 450 and 930 nm. [ARP4392]

relative wind The direction and velocity with which the ambient air moves relative to an aircraft or airfoil. [ARP4107]

relativistic electron beams (REB) Beams of electrons traveling at approximately the speed of light.

relativistic particle A particle with a velocity so large that its relativistic mass exceeds its rest mass by an amount that is significant for the computation or other considerations at hand.

relativistic velocity A velocity sufficiently high that some properties of a particle at this velocity have values significantly different from those obtained when the particle is at rest.

relativity A principle postulating the equivalence of the description of the universe, in terms of physical laws, by various observers, or for various frames of reference.

relaxation The relief of stress with regard to time, temperature, and a specific strain. [ARP700]

relaxation method (mathematics) An iterative numerical method for solving elliptic partial differential equations, e.g., a Poisson equation.

relaxation oscillator A device that generates a periodic nonsinusoidal electrical signal by gradually storing electrical energy and then rapidly releasing it.

relaxation time *1.* The time required for a system, object, or fluid to recover to a specified condition or value after disturbance. *2.* The time taken by an exponentially decaying quantity to decrease in amplitude by a factor of $1/e$ or 0.3679.

relaxed static stability *1.* Refers to the use of automatic flight-control augmentation design to relax the conventional configuration constraints on the basic aerodynamic static stability of an aircraft. *2.* A lessening of the static stability of an aircraft to improve its maneuvering performance capability. [ARP4386]

relaxed stress During a stress-relaxation test, the initial stress minus the remaining stress at a given time. [AIR4844]

relay An electromechanical device in which a change in a relatively low-power electric signal controls the flow of electric current in one or more electric circuits generally not interconnected with the relay-control circuit.

relay-operated controller A control system or device in which the signal that operates the final control element or device is produced by supplementing the energy from the primary control element with energy from another source.

release agent *1.* A material that is applied in a thin film to the surface of a mold to keep the resin from bonding to the mold. [AIR4844] *2. See* parting agent.

released (signal) *1.* The logic 0 state of a bus signal line produced when no module asserts the signal associated with that line. *2.* The more positive of the two states of a bus signal line relative to the 0 volt logic reference. [AS4710]

release element In a catapult holdback, the member that initiates the release process. Two types are: (a) tension bar or frangible link (tension bar or ring), which fails at a predetermined load; and (b) hydromechanical device (repeatable release holdback bar), which releases at a given load and is reusable. [AIR1489]

release film An impermeable layer of film that does not bond to the resin being cured. [AIR4844]

release force The level to which force on the plunger must be reduced to allow the contacts to snap from the operated contact position to the normal contact position. [AIR4077]

release paper A sheet that serves as a protectant and/or carrier for an adhesive film or mass, and that is easily removed from the film or mass prior to use. [AIR4844]

release point The position of the plunger at which the contacts snap from the operated contact position to the normal contact position. [AIR4077]

release travel The distance through which the plunger moves when traveling from the release point to the free position. [AIR4077]

relevant Describes something that can occur or recur during the operational life of an item inventory. [AIR4896]

relevant indication An indication of a discontinuity that could be cause for rejection based on type, size, location, or distribution. [AMS2442]

reliability The probability that an item will perform its intended function for a specified interval under stated conditions. [ARD50010]

reliability analysis *See* analysis, reliability.

reliability, assessed The reliability of an item, determined within stated confidence limits, from tests or failure data on nominally identical items. Results can only be accumulated (combined) when all the conditions are similar. [ARD50013]

reliability assurance The exercise of positive and deliberate measures to provide confidence that a specified reliability will be obtained. [ARD50013]

reliability, basic *See* basic reliability.

reliability centered maintenance (RCM) A disciplined logic or methodology used to identify preventive maintenance tasks to realize the inherent reliability of equipment at a minimum expenditure of resources. [ARD50010]

reliability centered maintenance analysis The analysis process utilized to identify preventive maintenance requirements for aircraft and engine systems and equipment consistent with RCM principles. [AIR4896]

reliability demonstration *See* reliability qualification test.

reliability development/growth test A series of tests conducted to disclose deficiencies and to verify that corrective actions will prevent recurrence in the operational inventory. Also known as TAAF testing. [ARD50013]

reliability, dispatch The percentage of flights that depart without incurring a delay (technical) or cancellation (technical). [ARD50010]

reliability, enroute The probability of successfully completing a flight plan without incurring a failure that would cause deviation from flight plan. [ARD50010]

reliability, estimated A reliability factor that is postulated for a system, subsystem, or equipment under specified conditions of test or use. [ARP4386]

reliability goal The level of reliability desired of a design, often expressed as the reliability

design "objective" for development guidance. This is contrasted with the minimum acceptable reliability, expressed as a development requirement. [AIR4896]

reliability growth The positive improvement of the reliability of equipment through the systematic and permanent removal of failure mechanisms, regardless of their sources, by implementing corrective action. [ARD50010]

reliability growth management The systematic planning for reliability achievement as a function of time and other resources; controlling the ongoing rate of achievement by reallocation of resources based on comparisons between planned and assessed reliability values. [AIR4896]

reliability, inherent *See* inherent reliability.

reliability, mission *See* mission reliability.

reliability, operational The assessed reliability of an item based on field data. [ARD50010]

reliability, predicted The reliability of an item estimated from its design considerations, and from the reliability of its parts under the intended conditions of use. [AIR4896]

reliability prediction A method used to predict the reliability of an item at a future time. [ARP4386]

reliability qualification test A test conducted under specified conditions, by or on behalf of the government, using items representative of the approved production configuration, to determine compliance with specified reliability requirements, as a basis for production approval. Also known as a reliability demonstration or design approval test. [ARD50010]

reliability requirement A level of reliability expressed in an equipment specification as a design requirement, and supported with a reliability-acceptance test. [AIR4896]

reliability with repair *See* on line maintenance.

relic radiation Background radiation resulting from the primordial big bang.

relief Clearance around the cutting edge of a tool, provided by tapering or contouring the adjacent surfaces.

relief valve *1.* A valve that automatically releases pressure higher than its setting. [ARP4386] *2.* A modulating valve that limits supplied pressure by dumping from a system or compartment to a lower pressure region. [ARP986]

relink time The amount of time required to reconfigure a link when adding or removing individual sources or sinks. [AIR4911]

relocatable A program that can be moved about and located in any part of a system memory without affecting its execution.

relocatable coding Absolute coding containing relative addresses; when derelativized, may be loaded into any portion of the programmable memory of the computer and will execute the given action properly. The loader program normally performs the derelativization.

relocate In programming, to move a routine from one portion of storage to another, adjusting the necessary address references so that that routine, in its new location, can be executed.

relocation dictionary The part of an object or load module that identifies all relocatable address constants in the module.

reluctance Resistance of a substance to the passage of magnetic lines of force; the reciprocal of magnetic permeability. Also known as magnetic resistance.

reluctive pressure transducer A type of pressure sensor in which a moving armature attached to a pressure-sensitive element varies the reluctance of a magnetic circuit (either a permanent magnet or an electromagnet), thus producing an output current in a measuring coil.

REM *See* roentgen equivalent man.

remainder Refers to the basis element from which the alloy is made; assumed to be present in an amount approximately equal to the difference between 100% and the sum percentage of the alloying elements and listed impurities. [AMS2269E]

remanence The magnetic flux density that remains in a magnetic circuit after the removal of an applied magnetomotive force.

remedial maintenance The maintenance performed following equipment failure, as required, on an unscheduled basis. Contrast with preventive maintenance.

remote In data processing, refers to any devices that are not located near the main computer.

remote access Communication with a data-processing facility by one or more stations that are distant from that facility.

remote circuit breaker *1.* Circuit breaker that is not controllable by the crew during flight. *See* circuit breaker. [ARP4101/5] *2.* A circuit breaker composed of a solenoid-operated contractor in which the solenoid circuit is controlled by a current-sensitive element, plus a manual switching and trip-indicating device. [ARP4404]

remote communications air/ground facility (RCAG) An unmanned VHF/UHF transmitter/receiver facility that is used to expand ARTCC air/ground communications coverage, and to facilitate direct contact between pilots and controllers. [ARP4107]

remote communications outlet (RCO) An unmanned air/ground communications facility that is remotely controlled by air traffic personnel; may be UHF or VHF. Intended to extend the communication range of the air-traffic facility. [ARP4107]

remote control *1.* Control of an operation from a distance, especially by means of electricity or electronics. *2.* A controlling switch, lever, or other device used in remote control.

remote handling system The equipment that permits personnel to extend their manual capabilities into remote environments. [AIR4896]

remote manipulation Using electromechanical or hydromechanical equipment to enable a person to perform manual operations while remaining some distance from the work location. Usually used for handling radioactive or otherwise hazardous materials.

remote manipulator system Device used in space for deploying and retrieving payloads by remote control. Also used for space maintenance and/or servicing of satellites and other spacecraft.

remote measuring element A temperature-measuring device that is removed from the medium to be measured. [ARP485]

remote metering *See* telemetering.

remote processing unit (RPU) Field station containing input/output circuitry and the main processor. Remote processing units are used to measure analog and discrete inputs, convert these inputs to engineering units, perform analog and logical calculations (including control calculations) on these inputs, and provide both analog and discrete (digital) outputs.

remote sensing *1*. A means of providing constant voltage at the aircraft receptacles by sensing the voltage at the receptacle with three separate leads in the output cable. [ARP1148A] *2*. The sensing of remote phenomena by whatever means. *3*. Detecting, measuring, indicating, or recording information without actual contact between an instrument and the point of observation, for example, as in optical pyrometry.

remote terminals All terminals not operating as the bus controller or as a bus monitor. [AS15531]

removal, access The removal of an item for the sole purpose of allowing ease of approach to another component(s), where the manner of installation makes it impossible or impractical to do so otherwise. [AIR4896]

removal cannibalization The removal of an item to satisfy the needs of another aircraft or item. [AIR4896]

removal, condition analysis The removal of an item for the sole purpose of determining its serviceability state at specified time. [AIR4896]

removal, confirmed A removal wherein a failure or defect is found that substantiates the reason for removal. [ARD50010]

removal, deferred suspect The removal of an item wherein impending malfunction is suspected, but actual removal has been deferred to a convenient time and/or location for accomplishment. [AIR4896]

removal, engine predefined An engine removal occurring at or after achievement of a predefined engine running time. [AIR4896]

removal, engine premature An engine removal occurring prior to achievement of a predefined engine running time, set at the moment of engine installation

removal, justified Those removals for which a shop check reveals a defect. NOTE: The total removals of some unit may include items

removed for maintenance convenience, as well as those for which no problems are found, and those for which a shop check reveals a defect. The last group is the only one for which removals are justified. [ARP4386]

removal rate The number of removals of an item expressed in terms of a base period, usually per 1000 aircraft hours, 1000 items hours, 1000 engine hours; sometimes per 100 or 1000 departures. [ARD50010]

removal, scheduled *1*. Removal of an item when the hard time limit is reached. *2*. Removal of an "on-condition" item during a scheduled periodic inspection or test. [ARP4386]

removal, section The removal of an item in conjunction with the removal of a higher assembly. [AIR4896]

removal, service/performance evaluation The removal of an item for the purpose of evaluation. [AIR4896]

removal, shop check The removal of an item to specifically, functionally check the item within a workshop. [AIR4896]

removal, time stagger The removal of an item to reduce the probability or necessity of simultaneous removals in multi-item applications. [AIR4896]

removal, trouble shooting A removal wherein an item is removed and replaced by another item for the sole purpose of determining whether the malfunction persists. [AIR4896]

removal, unconfirmed The removal of an item wherein no defect or failure is found that substantiates the reason for removal, but another defect or failure may be found. [ARP4386]

removal, under investigation The removal of an item brought about by symptoms of operational problems/performance, but wherein the cause has not been identified. [AIR4896]

removal, unjustified The removal of an item wherein no defect or failure is found. [AIR4896]

removal, unscheduled The removal of an item brought about as a result of a known or suspect malfunction or defect—both at a time other than scheduled maintenance, inspection, or test. Includes: (a) all conditions-monitored (CM) items; (b) only those on-condition (OC) items removed prior to their scheduled check; and (c) only those hard-time (HT) items removed prior to their schedule. [ARP4386]

rendezvous *1.* Two or more objects meeting with zero relative velocity at a preconceived time and place. *2.* The point in space at which a rendezvous takes place, or is to take place.

REP *See* roentgen equivalent physical.

repairable item *See* item, repairable.

repair cycle The period that elapses from the time the item is removed in a repairable condition to the time it is returned to stock in a serviceable condition. [ARP4386]

repair cycle time The elapsed time from failure of an item until the item is repaired and restored to an RFI condition and returned to the operational site. [ARD50010]

repair, essential A repair that is necessary to ensure that the end item will fulfill its mission efficiently and safely. [ARP4386]

repair/overhaul, modular Refers to the application of maintenance procedures and techniques that concentrate attention on a defective subassembly or module and its repair or overhaul, rather than treating the complete assembly as an entity for all maintenance actions. [ARP4386]

repair part, high mortality A repair part with failures, anticipated or actual, of 30 or more per 100 end items per year. [ARP4386]

repair time The time spent replacing, repairing, or adjusting all items suspected to have been the cause of the malfunction, except those subsequently shown by interim test of the system not to have been the cause. [AIR4896]

repeatability error *See* precision error.

repeatability of LVDT The ability to reproduce the same output for repeated exact positioning of the armature. [ARP4386]

repeatable release holdback bar A type of catapult holdback that is designed to restrain the aircraft until a predetermined load is built up in the towing mechanism, then release the aircraft automatically, allowing the aircraft to be launched. The reusable release element may then be reset for the next launch. [AIR1489]

repeatback control unit *See* follow-up control unit.

repeater A device that amplifies or regenerates data signals in order to extend the distance between data stations. Also called a regenerator.

repetition rate The rate at which the individual pulses of acoustic energy are generated. [ARP5089]

replacement schedule The specified periods of time when operating equipment are to be replaced. [AIR4896]

replicated optics Optical components formed by transferring a master pattern to a roughly machined substrate, using an epoxy layer to form a final optical surface. The epoxy layer is then coated with a reflective layer to form the final component. This process allows mass production of complex surfaces much less expensively than conventional polishing techniques.

report generator A computer program that gives a less experienced user the ability to create reports from various files.

reporting point A geographical location in relation to which the position of an aircraft is reported. [ARP4107]

reproducibility The ability of an instrument to duplicate, with exactness, measurements of a given value. Usually expressed as a percent of span of the instrument. *See also* precision.

required time of arrival (RTA) Associated with 4D; represents the required ETA, or the ETA to be programmed, at a designated downstream waypoint or waypoints. [ARP1570]

rerun In data processing, to execute a program again.

resealing pressure The inlet pressure at which fluid no longer leaks past a relief valve after it is closed.

reseat pressure The differential pressure at which an open valve will revert to a specified flow rate condition when closed. [AIR1749]

reservation subfield Part of a control field that indicates the highest reservation made by a station with a pending message frame. [AIR4075]

reserve batteries Batteries that are chemically inert and unable to generate electrical energy until introduction of an external electrolyte. [AS4492]

reserved variable Any variable available only to specific programs in the system. Contrast with global variable.

reserved words In a language, certain words that may only appear in the context reserved for them.

reserve energy The capacity of a shock absorber to accept energy inputs over and above the design conditions, i.e., the reserve for overload or over design landing conditions. [AIR1489]

reserve, expiratory Difference in volume between normal expiration and maximum forced expiration. Volume average = 1.5 L = 1500 mL; volume range = 0.8 to 2.0 L. [ARP171]

reserve, inspiratory The air that can be inspired in addition to the tidal air by forcible inspiration. Volume average = 1.5 L = 1500 mL; volume range = 1.2 to 2.0 L. [ARP171]

reserve power An additional certificated rating above takeoff/go-around power. [ARP4102/5]

reserve static line A device that is connected to the main canopy and is capable of actuating the reserve parachute assembly following a breakaway from the main canopy. [AS8015]

reservoir A container for operating fluid supply. [ARP4386]

reservoir, pneumatic A pressure-storage chamber in which pneumatic pressure energy may be accumulated and from which pneumatic pressure energy may be withdrawn. [ARP4386]

reset *1.* The process for returning a system to an operational state after a failure has occurred and has been corrected or isolated. Some systems are capable of providing this function automatically. [ARP4386] *2.* An integrating control function. *3.* Any control characteristic that eliminates droop in a system. [ARP89C]

reset action *1.* A control action that produces a corrective signal proportional to the length of time the controlled variable has been away from the set point. *2.* An action that takes care of load changes. *3.* Integral control mode action.

reset control *See* integral control.

reset cycle To return a cycle index to its initial value.

resettable remote circuit breaker A circuit breaker that is resettable from the normal crew station. [ARP4101/5]

reset windup The development of saturation of the integral mode of a controller during times when control cannot be achieved; often causes the controlled variable to overshoot its setpoint when the obstacle to control is removed.

resident In data processing, describes a program that is permanently stored in the memory of the computer.

residual air The air remaining in the lungs after the most complete expiration. Also called residual volume. [ARP171]

residual elements, each, maximum The maximum amount that may be present of an individual element not mentioned specifically in the specified composition. [AMS2269E]

residual elements, total, maximum The sum percentage of the residual elements found. [AMS2269E]

residual error The error remaining after attempts at correction.

residual fuels Products remaining after the removal of some of the water and an appreciable percentage of the more volatile hydrocarbons from crude petroleum.

residual gas analysis The study of residual gases in vacuum systems using mass spectrometry. [AIR4844]

residual gases Those gases remaining after a furnace is evacuated to its initial vacuum level. [AMS2769]

residual strain The strain associated with residual stress. [AIR4844]

residual stress In structures, any stress in an unloaded body. These stresses arise from local yielding of the material due to machining, welding, quenching, or cold working. Also known as internal stress.

residual volume *See* residual air. [ARP171]

residue check A check of numerical data or arithmetic operations in which the number A is divided by n, and the remainder B accompanies A as a check digit.

resilience The ratio of energy returned on recovery from deformation, to the work input required to produce the deformation. [AIR4844]

resin An organic substance, of natural or synthetic origin, that is polymeric in structure.

resin applicator In filament winding, the device that deposits the liquid resin onto the reinforcement band. [AIR4844]

resin batch Resin mixed in one mixer in one operation; or resins blended together in one homogeneous mix with traceability to the individual component lots. [AIR4844]

resin content The amount of resin in a laminate, expressed as either a percentage of total weight or total volume. [AIR4844]

resin/fiber dust Nuisance dust composed of a mixture of resin and fiber formed from solid material by crushing, grinding, drilling, etc., of nonmetallic composites. [AIR4844]

resin, liquid *See* liquid resin.

resin matrix composites Composite materials in which a matrix of filaments and/or fibers of glass, metal, or other material is bound with a polymer or resin.

resinography The science of the morphology, structure, and related descriptive characteristics of resins, polymers, plastics, and their products, as correlated with their composition or conditions, and their properties or behaviors. [AIR4844]

resinoid Any of the class of thermosetting synthetic resins—either in their initial, temporarily fusible state, or in their final infusible state. [AIR4844]

resin pocket An apparent accumulation of excess resin in a small, localized section visible on cut edges of molded surfaces, or internal to the structure and nonvisible. [AIR4844]

resin-rich area Localized area filled with resin and lacking reinforcing material. [AIR4844]

resin ridge On the surface of a part, a sharp buildup consisting of only resin. [AIR4844]

resin-starved area Localized area of insufficient resin, usually identified by low gloss, dry spots, or fiber showing on the surface. [AIR4844]

resin system A mixture of resin and ingredients such as catalyst, initiator, and diluents required for the intended processing and final product. [AIR4844]

resin transfer molding A process whereby catalyzed, thermosetting resin is transferred or injected into an enclosed mold in which the fiber reinforcement has been placed. [AIR4844]

resistance *1.* The property of an electric circuit that determines, for a given current, the rate at which electric energy is converted into heat. [ARP1931] *2.* Impediment to gas flow, such as pressure drop or draft loss through a dust collector. Usually measured in inches water gage (w.g.).

resistance bulb A temperature-sensitive device composed of a material whose resistance increases in a reproducible manner as the temperature increases, with negligible hysteresis. [AIR1900]

resistance compensated ladder circuit A parallel circuit into which compensating resistors have been inserted so that the output voltage of the circuit will equal the arithmetic average of the voltages of the several individual measuring junctions. [ARP485]

resistance drop The voltage drop in phase with the current.

resistance magnetometer A device for measuring magnetic field strength by means of a change in electrical resistance of a material immersed in the magnetic field.

resistance meter Any instrument for measuring electrical resistance. Also called an ohmmeter or megger.

resistance spool A variable resistance used to adjust the resistance of the thermocouple measurement system to a predetermined value under given conditions. [ARP485]

resistance strain gage A fine wire or similar device whose electrical resistance changes in direct proportion to the amount of elastic strain it is subjected to.

resistance temperature detector A temperature-sensing device consisting of an electrical resistor that has a known characteristic of resistance versus temperature. [AS5116]

resistance thermometer A temperature-measuring device in which the sensing element is a resistor that has a known variation in electrical resistance with temperature.

resistance welding A group of welding processes in which heat is obtained from the resistance of the work to electrical current in a circuit of which the work is one part. Spot welding is an example of resistance welding.

resistive flowmeter A device for measuring liquid flow rates in which an electrical output signal proportional to flow rate is determined from the rise and fall of a conductive differential-pressure manometer fluid in contact with a resistance-rod assembly.

resistivity Electrical resistance per unit length and unit cross section.

resistor *1.* Passive fluidic element that, because of viscous losses, produces a pressure drop as a function of the flow through it, and has a transfer function of essentially real components (i.e., negligible phase shift) over the frequency range of interest. [ARP993A] *2.* An electrically conductive material shaped and constructed so that it offers a known resistance to the flow of electricity.

resistor-capacitor-transistor logic (RCTL) A type of computer circuit used to perform the "not or" logic function at speeds higher than can be achieved with RTL circuits; similar to RTL except that capacitors reduce switching time.

resistor compensated resolver A resolver that contains one or more resistors having external termination, intended for connection to an amplifier to bring the transformation ratio or phase shift of the resolver amplifier combination within close limits; or internal elements that compensate for changes in resolver characteristics with variation in temperature or other conditions. [ARP826]

resistor-transistor logic (RTL) A form of logic circuit that uses resistors and transistors and performs not or nor logic.

resizing tool A tool designed to shape or size to a specific dimension. *See also* anvil. [ARP480A]

resolution *1.* A measure of the ability to delineate picture detail on a CRT; or the smallest discernible or measurable detail in a visual presentation. [ARP1782] *2.* The ability of a film, a lens, a combination of film and lens, or a vidicon system to render barely distinguishable a standard pattern of black and white lines. *3.* In radar, the minimum angular separation at the antenna at which two targets can be distinguished (a function of beamwidth); or the minimum range at which two targets at the same azimuth can be separated (equal to one half the pulse height). *4.* For a gyro, a measure of response to small changes in input; the maximum value of the minimum input change that will cause a detectable change in the output for inputs greater than the threshold, expressed as a percent of one half the input range. *5.* The smallest increment of actuator output produced by a change in command input. [ARP4386] *6.* The lower limit capability of an instrument or analysis technique to differentiate between samples of differing concentrations. [ARP4418]

resolution advisory *1.* Aural and visual information given to pilots to avoid a potential collision. [AIR4102/10] *2.* According to the TCAS system, the display indication given to the crew recommending an escape maneuver to increase or maintain separation relative to an intruder aircraft. [ARP4153]

resolution, defect For a test system, the property whereby indications due to defects in a test specimen that are located in close proximity to each other may be separated. [ARP5089]

resolution of LVDT The smallest change in armature position that can be detected as a change in the output voltage. [ARP4386]

resolution sensitivity The smallest change in an input that produces a discernible response.

resolver Refers to any means for determining the mutually perpendicular components of a vector quantity.

resolver function error The difference between the actual fundamental in-phase output voltage and the theoretical voltage at any rotor displacement, expressed as a percentage of the actual fundamental voltage at +90 deg from the minimum voltage position of the winding under test.

resolving power *1.* A measure of the ability to respond to small changes in input. *2.* The ability of an optical device to separate the images of two objects very close together.

resolving time The minimum separation time between events that will enable a counting device to detect and respond to both events.

resonance *1.* The condition in which the frequency of the forced vibration is the same as the natural frequency of the body, resulting in abnormally large amplitudes of vibration. [ARP5089] 2. The phenomena whereby a free wave or oscillation of a system is amplified by a forced wave or oscillation of exactly equal period. The forced wave may arise from an impressed force upon the system or from a boundary condition. The growth of the resonant amplitude is characteristically linear in time. *3.* For a system in forced oscillation, the condition that exists when any change, however small, in the frequency of excitation causes a decrease in the response of the system.

resonance bridge An electrical network used in measuring inductance, capacitance, or frequency. Normally consists of four arms—one containing both inductance and capacitance and the other three containing only nonreactive resistances—and an adjustment device which balances the network by establishing resonance.

resonance fluorescence The emission of radiation by a gas or vapor as a result of excitation of atoms to a higher energy level by incident photons at the resonance frequency of the gas or vapor. Also called resonance radiation.

resonance lines Spectral lines that occur either as absorption or emission lines.

resonance radiation *See* resonance fluorescence.

resonant frequencies The wave frequency at which mechanical or electronic resonance is achieved.

resonator(s) *1.* In radio and radar applications, circuits that will resonate at a given frequency, or over a range of frequencies, when properly excited. *2.* Generally, a pair of mirrors located at either end of a laser medium, which cause light to bounce back and forth between them while passing through the laser medium.

resource Any facility of the computing system or operating system required by a job or task, including main storage, input/output devices, the central processing unit, data sets, and control processing programs.

resource access services Those low-level services that enable the system services to interact with physical resources. [AS4893]

resource manager Generally, any control program function responsible for the allocation of a resource.

respirable particles All airborne particles that lie in the respirable size range of 2 μm and smaller. [ARP4418]

respiration, types of Types of normal and abnormal breathing, classified as: apnea, bradypnea, Cheyne-Stokes respiration, costal breathing, dyspnea, eupnea, external respiration, hyperpnea, hyperventilation, hypopnea, hypoventilation, internal respiration, and tachypnea. *Apnea*–cessation of breathing. *Bradypnea*–decrease in subject's normal respiratory rate, or unusually slow breathing. *Cheyne-Stokes respiration*–abnormal breathing varying cyclically from very deep to very shallow, with varying periods of apnea. *Costal breathing*–respiration product solely by the intercostal muscles. *Dyspnea*–consciousness of air want; shortness of breath; labored breathing; subjective difficulty in breathing. *Eupnea*–normal breathing. *External respiration*–the interchange of gases between the blood and the atmosphere. *Hyperpnea*–increased ventilation, regardless of cause. *Hyperventilation*–an excessive rate and/or depth of breathing (excessive minute volume of respiration), with the subsequent "blow-off" of more than the normal amount of carbon dioxide. *Hypopnea*–decreased ventilation, regardless of cause. *Hypoventilation*–air exchange less than that required by metabolic needs, manifested by decreased rate and/or depth of breathing. *Internal respiration*–the interchange of gases between body-tissue cells and the blood. *Tachypnea*–increased respiratory rate. [ARP171]

respirator A device that delivers respirable gas to the airways of the user or recipient by intermittent positive pressure, with respiration either pressure-controlled, volume-controlled, or a combination of both. May be supplied by assisted (demand) or controlled respiration. *See also* resuscitator. [ARP171]

respiratory acidosis *See* acidosis, respiratory.

respiratory alkalosis *See* alkalosis, respiratory

respiratory cycle Time required for inspiration, expiration, and postexpiratory pause prior to next inspiration. [ARP171]

respiratory rate The number of respiratory cycles per minute. [ARP171]

respond in form Describes a remote terminal whose response to an illegal command consists of a response formatted as though it were a legal command. [AS4115]

response *1.* The propagation delay that occurs in response to the approximate step control of recommended amplitude. [ARP993] *2.* The motion (or other output) resulting from an excitation or input under specified conditions. Response characteristics, often presented graphically, give the response as a function of some independent variable, such as frequency or direction.

response-critical In process control, implies the need to react to random disturbances in time to prevent impairment of yield, or dangerous conditions. Real time is often used synonymously.

response, delayed *1.* The execution of a selected course of action so long after the decision was made that the selected course of action is no longer appropriate. *2.* An adaptive or goal-seeking response evoked a considerable time after the disappearance of its usual stimulus. [ARP4107]

response, poor Ineffective execution of a selected course of action due to cognitive or physical task saturation, an anomaly of attention, an anomaly of motivation, or lack of sufficient procedural knowledge. [ARP4107]

response pressure The pressure at which a function is initiated. [ARP4386]

response set A cognitive framework of expectations that predisposes a person to a certain course of action regardless of the environmental cues perceived. [ARP4107]

response time *1.* A measure of the time required to complete some part or all of a required action. The 63% point is commonly used, i.e., the time required for the cabin air temperature to reach 63% of the total change that results from a step change in the supply-air temperature. [ARP89] *2.* The time delay between the application of an input signal and the resulting output signal. [AIR1489]

responsiveness The ability of an instrument or control device to follow wide or rapid changes in the value of a dynamic measured variable.

restoration Maintenance tasks necessary to return an item to a specific standard. [AIR4896]

restoring torque gradient The rate of change, with respect to deflection, of the resultant of electric and mechanical torques tending to restore the moving element of an instrument to any position of equilibrium.

restricted area *See* special use airspace.

restricted burning grain A solid-propellant grain in which certain surfaces are restricted or inhibited to provide particular burning characteristics. [AIR913]

restrictor valve A valve whose function is to produce a relatively high pressure drop in a fluid circuit by means of a reduced flow area or orifice. [ARP4386]

resultants The sums of two or more vectors.

resuscitator *1.* A mechanical device that provides artificial respiration by alternatively blowing oxygen (or air) into a mask covering the mouth and nose and sucking it out gain; or by inflating the lungs to a preset pressure, then releasing pressure and allowing the elastic recoil of the thorax to expel the gas. *See* respirator. *2.* One who applies resuscitative measures, or operates a resuscitative device. [ARP171]

retainer A device designed to position and/or restrain the movement of one or more parts. [ARP480A]

retarded elastic-chamber gage A pressure gage whose sensitive element is an elastic chamber that moves freely only through the lower portion of its indicating range.

retarder A straight or helical strip inserted in a fire tube primarily to increase the turbulence.

retarding magnet A magnet used in a motor-type meter to limit rotor speed to a value proportional to the quantity being measured.

retest OK Refers to the subsequent passing of a previously failed test. [AIR4896]

retest phase A period of time following a formal maintainability demonstration test for repeat tests in the event of a reject decision, or for the purpose of collecting data on untested maintenance actions. [AIR4896]

reticle(s) A glass window onto which is etched or printed a pattern, typically for use in measurement or alignment. The simplest type of reticle is the crosshairs of an alignment telescope.

reticulated foam A three-dimensional, net-like, open-cell material, made of a flexible polyurethane compound. Useful for explosion suppression and slosh attenuation in aircraft fluid tanks. [AIR1664]

retirement for cause Primarily on aircraft, a procedure based on fracture mechanics which allows safe utilization of the full life capacities of each component.

retort processing One method for converting shale oil into oil similar to petroleum oils.

retract Refers to the port to which pressure is applied for retraction of an actuator such as a cylinder. [ARP4386]

retraction The process of retracting, lifting, or folding a landing gear within the confines of the mold lines of the air vehicle, to a stowed position, for high-speed, low-aerodynamic drag configuration of the air vehicle. [AIR1489]

retraction curve For a landing gear, a type of load-stroke curve that presents power requirements for the retraction cycle of operation. [AIR1489]

retraction geometry The geometric representation (mathematics) of the properties, measurement, and relationship of points, lines, angles, surfaces, and solids of component parts of a landing gear system in the retraction process. [AIR1489]

retractor A device for storing webbing in a torso-restraint system. [AS8043]

retractor plate *See* brake pressure plate.

retreading The methods of restoring a worn tire by renewing the tread area, or by renewing the tread area plus one or both sidewalls. [ARP4834]

retread level escalation The process used to verify that a population of retreaded tires is suitable for an additional service life. [ARP4834]

retreating blade On a rotary-wing aircraft in horizontal motion, any rotor blade or wing moving with the relative wind. *See also* advancing blade. [ARP4107]

retrofit Contracted from retroactive fit. A modification of an aircraft, aircraft component, or other object that duplicates a change or modification made in later models of the same type. [ARP4107]

retrofitting Modification of equipment to incorporate changes made in later production of similar equipment; the changes may be performed in the factory or in the field.

retroreflection Reflection in which the reflected rays return along paths parallel to those of their corresponding incident rays.

retroreflector(s) Class of optical instruments that cause reflected radiation to return along paths parallel to those of their corresponding incident rays.

retro-rocket A rocket fired in a direction opposite to the line of flight of the vehicle to which it is attached. [AIR913]

retrorocket engines Rocket engines fitted on or in spacecraft, satellites, or the like to produce thrust opposed to forward motion.

retrothrust Thrust used for a braking maneuver; reverse thrust.

return Refers to an outlet port having rated flow capacity for return of working fluid to reservoir. [ARP4386]

return flow oil burner A mechanical, atomizing oil burner in which part of the oil supplied to the atomizer is withdrawn and returned to storage or to the oil line supplying the atomizer.

return line patch A patch providing an indication of contaminant level or internal wear rate of the unit, other than that indicated by the outlet patches. [ARP1302]

return pressure (back pressure) Pressure caused by resistance to flow in the return line and/or by precharged reservoirs. [ARP4386]

reverberation *1.* The persistence of sound in an enclosed space, as a result of multiple reflections after the sound source has stopped. *2.* The sound that persists in an enclosed space, as a result of repeated reflection or scattering after the source of the sound has stopped.

reverberation chambers Chambers designed to eliminate outside noise for accurate acoustic measurement.

reverberation time The time in seconds that it takes for average sound-energy density to decrease to one-millionth of its original steady-state value after sound from the source has stopped.

reverberation time meter An instrument for determining reverberation time of an acoustic enclosure.

reverse-acting controller A controller in which the value of the output signal decreases as the value of the input (measured variable or controlled variable) increases.

reverse drawing Drawing, especially a deep-drawn part, a second time in a direction opposite to the original draw direction.

reverse helical winding A type of filament winding wherein, as the fiber-delivery arm traverses one circuit, a continuous helix is laid down, reversing direction at the polar ends. [AIR4844]

reverse impact test A test in which one side of a sheet of a material is struck by a pendulum or falling object and the reverse side is inspected for damage. [AIR4844]

reverse osmosis The application of pressure to stop or reverse the transport of solvent through a semipermeable membrane separating two solutions of different solute concentration. The applied pressure required to prevent the flow of solvent across a perfectly semipermeable membrane is called the osmotic pressure and is a characteristic of the solution.

reverse pitch The angle of a propeller blade that produces reverse thrust; a negative pitch angle on the propeller blade. [ARP4107]

reverse polarity *1.* Describes an electrical circuit in which the positive and negative electrodes have been interchanged. *2.* Describes an arc-welding circuit in which the electrode is electrically positive and the workpiece electrically negative.

reverse thrust The force developed by an air vehicle or a rocket engine in a reverse direction. Utilized for deceleration of the air vehicle in ground operations, such as landing rollout or rejected takeoff, to assist the wheel brakes and/or other systems of the air vehicle. [AIR1489]

reverse video A CRT screen display of dark characters on a light background; the opposite of the usual CRT screen display.

reversible adiabatic or isentropic expansion An expansion process in which no heat transfer takes place. This is the standard for determining energy extraction from expanded gas and represents the theoretical (100% efficient) reversible process. [ARP4386]

reversible drive A drive that, without input torque, can be back-driven from the output by a torque less than the maximum design load at the output. [ARP4386]

reversible transducer A transducer in which the transducer loss is independent of the direction of energy transmission through it.

reversing switch An electrical switch whose function is to reverse connections, on demand, of one part of the circuit.

reversing time The time required for a motor to reach 63.2% of the non-load speed upon the reversal of the control winding voltage, after initially running at the no-load speed in the opposite direction. [ARP667]

reversion The process for changing over control from an active to a standby channel, or from a primary to a secondary channel. [ARP4386]

revolution (motion) *See* revolving.

revolutions per minute (rpm) A standard unit of measure for rotational speed.

revolving Moving in a path about an axis, usually external to the body accomplishing the motion.

rework To restore an item to a condition exactly conforming to original design specifications. Usually applied to corrective action taken when an item has failed an inspection, but requires a relatively simple operation, such as replacing a part, to enable the item to pass an identical inspection.

REX A hardness The hardness of a sealant as measured by a REX A hardness gauge. [AS7200/1]

Reynolds number A dimensionless combination of parameters describing a fluid-flow field; the ratio of dynamic (inertia) forces to viscous forces. The product of fluid density, fluid velocity, a characteristic length (wing chord or pipe diameter), divided by the fluid absolute viscosity. Named after Osborne Reynolds.

Reynolds stress In the mathematical treatment of a viscous, incompressible, homogeneous fluid in turbulent motion, represents the transfer of momentum due to turbulent fluctuations.

RF *See* radio frequencies.

RF display A CRT signal display that is not rectified; displayed signals are both above and below the sweep or baseline. [ARP5089]

RFI *See* radio frequency interference.

RFI protector A device that protects a computer from strong radio or television transmissions.

rheocasting Use of partially solidified metal alloys (fractions solids) fed directly into a casting machine for forming into machine parts.

rheology The study of the deformation and flow of matter.

rheopectic substance A fluid whose apparent viscosity increases with time at any constant shear rate.

rheostat An adjustable variable resistor.

rhombic antenna An antenna composed of long wire radiators, which lie along the sides of a rhombus. The sides of the rhombus, the angle between the sides, the elevation, and the termination are proportioned to give the desired directivity. The antenna is usually terminated in an impedance.

rhomboids Parallelograms whose adjacent sides are not equal.

rib *1.* In a lifting surface, the structural frame that maintains the proper airfoil cross-sectional shape. *2.* A reinforcing member designed into a plastic part to provide lateral, horizontal, hoop, or other structural support. [AIR4844]

ribbon A fiber having essentially a rectangular cross-section, in which the width-to-thickness ratio is at least 4:1. [AIR4844]

ribbon cable A cable of individually insulated round conductors lying parallel and held together by means of films, adhesive, or woven textile yarn. *See also* flat cable. [ARP1931]

ribbon direction In honeycomb, the direction of the node bonds. [AIR4844]

ribbon-in-the-sky An artist's concept sketch, used in briefings and discussions, that provides a pictorial representation of the total flight path for a mission. The sketch includes three-dimensional data and notations of specific task objectives, for example, orbits and refuel. [ARP4107]

ribbon parachute Parachute whose canopy consists of an arrangement of closely spaced tapes. Ribbon parachutes have high porosity with attendant stability and slight opening shock.

riblets Longitudinal striations forming V-shaped grooves on aerodynamic and hydrodynamic surfaces. Riblets act to reduce large-scale disturbances near the boundary layer. The grooves are dimensional on the order of the wall vortices and turbulent dimensions.

Richardson number A nondimensional number arising in the study of shearing flows of a stratified fluid.

rich mixture A fuel-air mixture in which the ratio of fuel to air is higher than that required for efficient operation of the engine; generally with an air to fuel ratio from 12:1 to 15:1. *See also* lean mixture. [ARP4107]

rifled tube A tube that is helically grooved on the inner wall.

rigid coupling A coupling assembly that does not provide for any relative movement after assembly. [ARP4968]

rigidity Resistance of a body to instantaneous change of shape.

rigid pavement *See* pavement, rigid.

rigid plastics For purposes of general classification, a plastic that has a modulus of elasticity either in flexure or in tension greater than 690 MPa (100 ksi) at 23°C (70°F) and 50% relative humidity. [AIR4844]

rigid resin A resin having a modulus high enough to be of practical importance. [AIR4844]

rigid rotors (plasma physics) Ensembles of electrons moving in circular or nearly circular orbits at a constant angular frequency.

rigid thermocouple harness An assembly in which the lead supporting and protective structure is firm and nonpliant and requires application of appreciable force in order to obtain deflection. [ARP485]

rime icing/rime ice *See* ice, rime.

rim width On a wheel, the lateral dimension between the inside of the rim flanges which restrain the installed tire in a lateral direction; equal to the installed width between tire beads. [AIR1489]

ring *1*. A circular item(s) used for handling, assembling, fabricating, supporting, etc. of parts. [ARP480A] *2*. A sequential network topology in which each node is connected to exactly two nodes, and serves as a repeater when it is not sourcing data onto the network.

Ringelmann chart A series of four rectangular grids of black lines of varying widths printed on a white background, used as a criterion of blackness for determining smoke density.

ringing An oscillating transient in an output signal that occurs following a sudden rise or fall in the input signal.

ringing method A bonded-structure inspection method in which unbonds are indicated by increased amplitude ringing signals. [ARP5089]

ringing signals Closely spaced multiple signals that can be caused by multiple reflections in a thin material or continued vibration of a transducer element. [ARP5089]

ringing time *1*. The time that the mechanical vibrations of a transducer element continue after the electrical pulse has stopped. [ARP5089] *2*. In ultrasonic testing, the length of time that a piezoelectric crystal continues to vibrate after the ultrasonic pulse has been generated.

ring spring A type of spring used in shock-absorber design; contains circular rings with a double trapezoidal cross section. Ring springs are utilized in stacks and absorb energy by friction of the adjacent ring surfaces. [AIR1489]

ring-tongue terminal A terminal having a round-ended tongue with a hole to accommodate a screw or stud. [ARP914A]

rinse To remove liquid penetrant inspection materials from the surface of the test part by means of washing with water. [AMS2647A]

ripple The cyclic variation of voltage about the mean level of the d-c voltage during steady-state d-c electric-system operation. [AS1212]

ripstop Describes a mechanical design technique for achieving mechanical separation of hydraulic systems in which more than one pressure source exists. If material fracture occurs in one portion of a rip-stop design, the fracture cannot propagate from one hydraulic system containment to a second hydraulic system containment and cause loss of two hydraulic systems; thus, if two hydraulic supplies are to be used, they can never enter the same piece of material anywhere in a system designed to provide rip-stop. [ARP4386]

riser *See* feedhead.

rise rate, initial instantaneous pressure The pressure rise rate, measured in the first 1/4 s of pressure rise, expressed in psi/s, developed from the tangent to a pressure time curve. [ARP906A]

riser - R (height) The distance between the surface of the tread of one step and the surface of the tread of a step above or below, when measured perpendicularly between the tread surfaces. [ARP836A]

riser to tread ratio - R/T An arithmetic ratio of the height of one of the risers to the geometrical length of the tread; equal to the tangent of the angle of inclination of the stairway. [ARP836A]

rise time *1*. The time required for the leading edge of a pulse to rise from one-tenth of its final value to nine-tenths of its final value. Rise time is proportional to time constant. *See also* decay time. [AIR1489] *2*. The time required for the output voltage of a digital circuit to change from a logical low level (0) to a logical high level (1). *3*. In urethane foam molding, the time between the pouring of the urethane mix and the completion of foaming. [AIR4844]

risk *1*. Describes the frequency (probability) of an occurrence and the associated level of hazard. [ARP4754] *2*. As used in cost-effectiveness analysis and operational research, the product of the consequence of an outcome and its probability of occurrence. [ARP4293]

RIT engines Radio-frequency ion thrustors that generate thrust by converting electric energy into a reaction force through the use of an electromagnetic field.

riveter A manual or power-driven tool, designed to perform the complete heading operation when setting rivets. [ARP480A]

RLI *See* runway loading index.

RMI *See* radio magnetic indicator.

RMS or rms *See* root mean square.

RNAV *See* area navigation.

R-NAV (area navigation) A term created by ARINC; currently used interchangeably with L-NAV (lateral navigation). [ARP1570]

RNAV way point (W/P) A predetermined geographical position, used for route or instrument approach definition or progress reporting purposes, and defined relative to a VORTAC station position. [ARP4107]

RNRZ *See* randomized non-return-to-zero.

robot An intelligent, multi-purpose device, usually programmable, which carries out pick and place, assembly, or other manipulative operations.

robotics *1.* The study and development of reprogrammable devices that do multifunctional tasks, conventionally done by humans, using manipulative functions and/or sensory feedback. The extension of human capabilities to manipulate, repair, service, construct and/or manufacture in space or on the ground is of primary interest to the aerospace community. *2.* Artificial computer intelligence, as applied to the use of industrial robots.

robustness (mathematics) Insensitivity of systems to uncontrolled perturbations; independent of changes in environmental parameters as demonstrated mathematically.

rocket A missile containing combustibles, independent of atmospheric oxygen, which on being ignited liberate gases producing thrust. [AIR913]

rocket engines Reaction engines that contain or carry along all of the substances necessary for their operation or for the consumption or combustion of their fuel, not requiring any outside substance, and thus capable of operation in outer space.

rocket launchers Devices for launching rockets.

rocket lining In solid rockets, refers to the layers of inhibitors applied to the inner surface of the chamber holding the grain.

rocket nozzles The exhaust nozzles of rockets.

rocket propellants Agents used for consumption or combustion in rockets and from which the rockets derive their thrust. Includes fuels, oxidizers, additives, catalysts, or any compounds or mixtures of these.

rocket thrust The thrust of a rocket engine, usually expressed in pounds.

rocket vehicles Vehicles propelled by rocket engines, used to place satellites in orbit, place missiles on target, or carry passengers over rails as on rocket sleds.

rockoons High-altitude sounding systems that consist of small solid-propellant research rockets carried aloft by a large plastic balloons.

rockwell hardness A hardness value derived from the increase in depth of an impression as the load on an indenter is increased from a fixed minimum value to a higher value and then returned to the minimum value. [AIR4844]

rod-out The act of pushing a specially designed rod through a valve or opening to loosen deposits.

Roentgen A quantity of x-ray or gamma-ray radiation that produces, in air, ions carrying one electrostatic unit of electrical charge of either sign per 0.001293 gram of air.

roentgen equivalent man (REM) The unit of dose in radiation dosimetry; equal to the amount of radiation of any type that produces the same amount of biological damage in human beings as a dose of 1 roentgen of 200-kV x-rays.

roentgen equivalent physical (REP) A unit of radiation equal to the amount of radiation of any type that results in energy absorption of 93 ergs/g in soft tissue.

Roentgen rays *See* x-rays.

roll *1.* The act of rolling; rotational or oscillatory movement of an aircraft or similar body about a longitudinal axis through the body. 2. The amount of roll, i.e., the angle of roll.

roll axis The longitudinal (or X) axis of an aircraft about which roll occurs. [ARP4386]

roll control Control of the roll (or lateral) motion of a vehicle about the longitudinal axis. [ARP4386]

rolled joint A joint made by expanding a tube into a hole, using a roller expander.

rolled ring A ring produced from a cast stock or pre-formed biscuit that has been pierced and subsequently rolled into the final ring size. [AMS2355F]

roller A conical or cylindrical part that functions by rolling on a surface; for example, a roller adapter, forming roller, or spreading roller. [ARP480A]

rolling moment For an aircraft, a moment about the center of gravity that tends to cause the aircraft to roll about its longitudinal axis. [ARP4107]

rolling radius *See* tire rolling radius.

rolling resistance *See* tire rolling resistance.

rollout *1.* The process of landing an aircraft; the time from touchdown through deceleration and stop, i.e., roll-out time, roll-out distance etc; or rejected takeoff from brake application to stop. *2.* The first showing of a prototype. [AIR1489] *3.* An act or instance of recovering from a banked attitude. [ARP4107]

roll-over error For an analog-to-digital converter with bipolar input range, the output difference for inputs of equal magnitude but opposite polarity. Specified in counts or LSBs.

roll test For a wheel and/or tire, a type of test that consists of a series of landings, takeoffs, and taxis; or a continuous roll of the tire/wheel assembly against a rotating flywheel. Intended to provide an index of the anticipated service fatigue life of the wheel or tire. [AIR1489]

roll-yaw coupling A dynamic phenomenon whereby rapid rolling of an aircraft generates uncommanded yawing. [ARP4386]

Ronchi grating *See* optical grating.

Ronchi test An improvement of the Foucault knife-edge test for curved mirrors, in which the knife edge is replaced with a transmission grating with 15 to 80 lines per centimeter, and the

pinhole source is replaced with a slit or a section of the same grating.

roof prism *See* Amici prism.

room temperature A temperature in the range of 20 to 30°C (68 to 86°F). [AIR4844]

room-temperature curing adhesive An adhesive that sets within an hour at room temperature, and later reaches full strength without heating. [AIR4844]

room-temperature vulcanizing Vulcanization or curing at room temperature by chemical reaction. [AIR4844]

root *1.* The bottom of a screw thread. *2.* The points at which the fusion zone of a weld intersect the base metal.

root directory In MS-DOS, the primary directory of a floppy or hard disk that contains subdirectories.

root locus analysis A graphical procedure for determining all possible roots of the characteristic equation for a control system. [AIR1823]

root mean square (rms) A means of expressing AC voltage in terms of the DC voltage; equal to peak AC voltage divided by the square root of two. [ARP1931]

root-mean-square errors In statistics, the square root of the arithmetic mean of the squares of the deviations of the various items from the arithmetic mean of the whole.

root mean square value (RMS) The square root of the mean of the squares of sample values.

rope lay Refers to a method of stranding or cabling in which the members themselves are stranded groups. The groups (either conductor strands or wires in a cable) are combined in a concentric configuration to form the final conductor or cable. [ARP1931]

roping Periodic luminance modulation along a line, producing a rope-like appearance. [ARP4256]

rosette strain gage *See* strain rosette.

rosette-type strain gage A type of resistance strain gage having three individual gage elements arranged to measure strain in three different directions simultaneously; a typical arrangement has two elements oriented at 90°

to each other and the third at 45° to the first two.

rosin A resin obtained as a residue in the distillation of crude turpentine from the sap of the pine tree or from an extract of the stumps and other parts of the tree. [AIR4844]

rotable item *See* item, rotable.

rotables Repairable parts that can be exchanged between engines; that rotate between the first engine, the repair shop, stores, and the second engine. [ARP4294]

rotameter A variable-area, constant-head, indicating-type rate-of-flow volume meter in which fluid flows upward through a tapered tube, lifting a shaped plummet to a position where upward fluid force just balances the weight of the plummet.

rotary actuator *1.* An actuator that develops rotary motion and torque as outputs. The term "rotary actuator" implies that the actuator is capable of limited output rotation; if the device is capable of continuous rotation, it is referred to as a motor. [ARP4386] *2.* A device that converts electric energy into controlled rotary force; usually consists of an electric motor, a gear box, a control relay, and one or more limit switches.

rotary engine *1.* An engine in which the cylinders rotate around a stationary crankshaft. *See also* radial engine. [ARP4107] *2.* A positive-displacement engine consisting of a rotor and stator. The control volume that encloses the working fluid during the thermodynamic cycle moves in a generally circular motion, rather than a linear motion as in a piston engine.

rotary oil burner A burner in which atomization is accomplished by feeding oil to the inside of a rapidly rotating cup.

rotary valve A valve for the admission or exhaust of working fluid; includes a ported piston or disk that turns on its axis.

rotating *See* rotation.

rotating-cup viscometer A laboratory device for measuring viscosity in terms of the drag torque on a stationary element, such as a paddle or cylinder. The stationary element is immersed in a liquid contained in a cup that rotates at constant speed.

rotating meter *See* velocity-type flowmeter.

rotation The nose-up pitch rotation (pitch) of the aircraft by the pilot, increasing the angle of attack to generate additional lift for takeoff. *See also* V_r. [ARP4107]

rotational transition A change in the rotational state of a molecule. Rotational transitions involve less energy than either electronic or vibrational transitions, and typically correspond to wavelengths in the far infrared, longer than about 20 micrometers.

rotational viscometer A device for measuring the apparent viscosity of non-Newtonian fluids by determining the torque required to rotate a spindle in a container filled with the test fluid; in some instruments, the container may rotate while the spindle does not.

rotation speed *See* V_r.

rotor *1.* The rotating member of a turbine, electric motor, compressor, pump, or similar machine. *2.* Any rotating assembly of vanes or airfoils.

rotor angle For a resolver, the angular displacement of the rotor from the electrical zero position. Rotation in a counter-clockwise direction facing the shaft extension end is considered a positive angular increase. [ARP826]

rotor body interactions Aerodynamic interactions between a helicopter rotor and a body.

rotorcraft An aircraft that in all of its usual flight attitudes is supported in the air wholly or in part by a rotor or rotors; that is, by airfoils rotating or revolving about an axis. The term rotorcraft is commonly applied to a helicopter, autogyro, or the like, in which the sustaining airfoils rotate about a substantially vertical axis. [ARP4107]

rotorcraft-load combination The combination of a rotorcraft and an external load, including the means for attaching the external load. Rotorcraft-load combinations are designed as Class A, Class B, and Class C. *Class A*–a combination in which the external load cannot move freely, cannot be jettisoned, and does not extend below the landing gear. *Class B*–a combination in which the external load is jettisonable and is lifted free of land or water during the rotorcraft operation. *Class C*–a combination in which the external load is jettisonable and remains in contact with land or water during the rotorcraft operation. [ARP4107]

rotor-type vacuum gage A device for measuring low pressures, down to 10^{-7} torr, by sensing the deceleration of a rotor (usually a steel ball) levitated in a rotating magnetic field, the rotor being exposed directly to the evacuated space.

rotor wash *See* downwash and wake turbulence.

roughing pump A positive-displacement mechanical pump which may or may not be used in combination with a rotary-lobe blower. [AMS2769]

rough machined Metal part surfaces that have been machined to a semifinished configuration requiring further machining. [ARP1917]

roughness Refers to relatively finely spaced surface irregularities, the height, width, and direction of which establish the predominant surface pattern. [ARP5089]

roughness height rating Quantitative expression of the roughness of a surface. [ARP4784]

roughness width The distance in inches between the successive ridges that constitute the predominant pattern of the surface roughness. [AS291D]

roughness width cutoff (instrument cutoff) The unit length in inches over which the irregularities of the surface profile are to be averaged. The roughness width cutoff must always be greater than the roughness width in order to obtain a true roughness height rating. [AS291D]

round *See* round-off.

round-chart instrument A recording instrument whose output trace is written on a circular paper chart.

rounded orifice An orifice whose inlet side is rounded rather than sharp-edged.

rounding error The error resulting from rounding off a quantity by deleting the less significant digits and applying some rule of correction to the part retained; for example, 0.2751 can be rounded to 0.275 with a rounding error of 0.0001. Synonymous with round-off error.

roundness *See* circularity (roundness).

round-off Synonymous with round. *See also* rounding error and half-adjust.

round-off error *See* rounding error.

round robin testing Testing of specimens from the same sample by different laboratories and/or by different test methods or equipment. [AS7101]

route A defined path, consisting of one or more courses in a horizontal plane, which aircraft traverse over the surface of the earth. [ARP4107]

router A network device that interconnects two computer networks that have the same network architecture. A router requires OSI Level 1, 2, and 3 protocols. *See also* bridge.

routine A subdivision of a program consisting of two or more instructions that are functionally related; therefore, a program. *See also* program and subroutine.

roving A number of yarns, strands, tows, or ends collected into a parallel bundle with little or no twist. [AIR4844]

roving ball The supply package offered to the winder, consisting of a number of ends or strands wound to a given outside diameter onto a length of cardboard tube. [AIR4844]

roving cloth A textile fabric, coarse in nature, woven from rovings. [AIR4844]

row A group of cylinders arranged in a common plane at right angles to the axis of the crankshaft. [ARP169]

row nucleation The mechanism by which stress-induced crystallization is initiated, usually during fiber spinning or hot drawing. [AIR4844]

rpm *See* revolutions per minute.

RPU *See* remote processing unit.

RT Symbol for the radiographic method of nondestructive testing inspection. [ARP5089]

R/T *See* riser to tread ratio.

RTA *See* required time of arrival.

RTL *See* resistor-transistor logic.

RTO *See* rejected takeoff.

rudder A control surface, usually attached to the vertical stabilizer, which controls the yawing motion of the aircraft. [ARP4386]

rudder angle The acute angle between the chord of the rudder moved from a neutral position and the plane of symmetry of the aircraft. The angle is positive when the trailing edge of the rudder is to the left of neutral when viewed from behind. [ARP4386]

rudder control A control for producing a yawing moment on the airplane through the use of the rudder; a mechanism for moving the rudder. [ARP4386]

ruddevator/ruddervator One of a pair of control surfaces set in a V, each of which combines the functions of both a rudder and an elevator. *See also* stabilizer. [ARP4107]

ruled diffraction gratings Diffraction gratings in which the lines are mechanically ruled by a precision ruling engine. These gratings are normally replicated and the replicas are sold for most applications.

rule of mixtures General rule stating that when two materials are mixed together, the properties of the mixture are an average of the properties of the constituents according to the proportion of each in the mixture. [AIR4844]

Runge-Kutta method A method for the numerical solution of an ordinary differential equation.

run-in *See* break-in.

runner *1.* The secondary feed channel in an injection or transfer mold that runs from the inner end of the sprue to the cavity gate. *2.* The molding material in this secondary feed channel. [AIR4844]

running engagement *See* engagement, running.

running fit Any of a class of clearance fits that allow assembled parts to run freely, especially shafts within their bearings or pinned joints in linkages. *See also* sliding fit.

running re-engagement *See* engagement, running.

running torque The torque required to produce continuous nut rotation. Also called prevailing torque. [AS1895]

runout As applies to screw threads, refers to circular runout of the major cylinder (for external threads) or the minor cylinder (for internal threads) with respect to the pitch cylinder. [AS8879]

run-out of mounting diameters The maximum difference between the high and low readings

of a dial indicator suitably arranged to measure the eccentricity of the mounting diameters with respect to the rotational axis of the shaft. [ARP667]

run-out of smooth shaft Maximum difference between the high and low readings of a dial indicator suitably arranged to measure the eccentricity of the smooth shaft at a point 1/8 in. from its end when rotated in its own bearings with the body of the unit held stationary. [ARP667]

run-time failure A software event that occurs during the execution of an application program whenever the program attempts to perform an operation that is outside the realm of anticipated operation. [ARP4904]

runway A long and narrow prepared strip or piece of ground, usually hard-surfaced, for the takeoff and landing of aircraft. The term "runway" is rarely applied to the flight deck of an aircraft carrier; and it is applied to water area or strips only in a general sense, for example, "the natural runway provided by a river." *See also* water lane. [AIR1489]

runway centerline light system (RCLS) Flush centerline lights spaced at 50-ft intervals, beginning 75 ft from the landing threshold, and extending to within 75 ft of the opposite end of the runway. [ARP4107]

runway condition reading (RCR) Numerical decelerometer readings relayed by air-traffic controllers at USAF and certain civil bases for use by the pilot in determining runway braking action. *See also* braking action. [ARP4107]

runway edge light system Lights used to outline the edges of runways during periods of darkness and restricted visibility conditions. These light systems are classified according to the intensity or brightness that they are capable of producing, i.e., high-intensity runway lights (HIRL), medium-intensity runway lights (MIRL), and low-intensity runway lights (LIRL). [ARP4107]

runway end identification lights (REIL) Two synchronized flashing lights, one on each side of the runway threshold facing the approach area, which provide rapid and positive identification of the approach end of a particular runway. These lights are effective for: (a) identification of a runway surrounded by a preponderance of other lighting; (b) identification of a runway that lacks contrast with surrounding terrain; and (c) identification of a runway during reduced visibility. [ARP4107]

runway, grooved A runway that has lateral grooves cut in the surface to provide for water drainage, and the prevention of aircraft skids or hydroplaning. [AIR1489]

runway loading index (RLI) A factor that takes into account the runway load in terms of variations of pressure and single-wheel load. This factor can be given with reasonable accuracy by the expression: $RLI = \sqrt{PW}$, where P = tire pressure and W = single-wheel load. [AIR1489]

runway markings *See* airport marking aids.

runway profile descent An instrument flight rules (IFR) air-traffic control arrival procedure to a runway, published for pilot use in graphic and/or textual form. May be associated with a standard terminal arrival route (STAR) procedure. Runway profile descents provide routings, and may depict crossing altitudes, speed restrictions, and headings to be flown from the en route structure to the point where the pilot will receive clearance for and execute an instrument approach procedure. [ARP4107]

runway visibility value (RVV) The visibility determined for a particular runway by a transmissometer. A meter provides a continuous indication of the visibility (reported in miles or fractions of miles) for the runway. RVV is used in lieu of prevailing visibility in determining minimums for a particular runway. [ARP4107]

runway visual range (RVR) An instrumentally derived value, based on standard calibrations, that represents the horizontal distance a pilot will see down the runway from the approach end. Based on the sighting of either high-intensity runway lights or on the visual contrast of other targets, whichever yields the greater visual range. *See also* takeoff minimums. [ARP4107]

rupture *1.* A sudden cleavage or break resulting from physical stress. [ARP5089] *2.* In breaking strength or tensile strength tests, the point at which a material physically comes apart, as opposed to yield strength. [ARP1931]

rupture capacity The maximum current for which a device is designed to provide protection under fault condition. [AS1831] *See also* interrupting capacity.

rupture current In a circuit, a value of current that reflects the capabilities of the power source without the effects of the power controller. [AS1831]

rupture disc A diaphragm designed to burst at a predetermined pressure differential.

rupture disc device A nonreclosing pressure-relief device that relieves excessive static inlet pressure via a rupture disc.

rupture strength The stress in a material at failure, based on the ruptured cross-sectional area itself. [AIR4844]

rupture stress In a material or item, the tensile stress, based on the original cross section, at which failure occurs at a given temperature and time. [ARP700]

rut depth The distance between the original ground surface and rut-bottom elevations after a tire has passed over the surface and all rebound has been completed. [AIR1780]

Rutherford *1.* 10^{-6} radioactive disintegration per second. *2.* A quantity of a nuclide having an activity equal to one Rutherford.

RVR *See* runway visual range.

RVV *See* runway visibility value.

S

s *See* second.

S *1. See* siemens. *2. See* solubility, S.

S₁ *See* volume of air dissolved.

Sabin A unit of measure for sound absorption equivalent to one square foot of a perfectly absorptive surface.

sabotage Deliberate destructive action that may be directed against property, processes, systems, organizations, governments, or people, and that is intended to prevent a process, undermine a group, or interfere with progress toward a goal.

sabot projectile A projectile that has a device (sabot) fitted around or in back of it in a gun barrel or launching tube. The sabot supports or protects the projectile, or prevents the escape of gas ahead of the projectile; it separates from the projectile after launching.

saddle *1.* A device that supports or cradles a part by engaging the contour of the part. [ARP480A] *2.* A casting, fabricated chair, or member used for the purpose of support.

safe Denotes an acceptable level of risk, with relative freedom from and low possibility of personnel injury, fatality, damage to property, or loss of the function of critical equipment. [AIR4728]

safe and arm A device for interrupting (safing) and aligning (arming) an explosive train. [AIR913]

safe area An area in which explosive gas/air mixtures are not expected to be present so that special precautions for the construction and use of electrical apparatus are not required. Also known as a non-hazardous area.

safety *1.* Describes a state in which risk is lower than the boundary risk. The boundary risk is the upper limit of the acceptable risk; it is specific for a technical process or state. [ARP4754] *2.* The conservation of human life

and its effectiveness, and the prevention of damage to items, consistent with mission requirements. [ARD50010] *3.* To secure against loosening or rotating, for example, to safety a bolt by passing a restraining wire through its head. [ARP4107]

safety cable An inseparable assembly consisting of a length of cable with an end fitting affixed to one end. [AS4536]

safety cable assembly An assembly consisting of a ferrule affixed to a safety cable. [AS4536]

safety can A metal can with a special closure, used for storing, handling, and transporting flammable liquids.

safety coded functional significant item An FSI whose failure or secondary damage resulting from a failure has a direct adverse effect on operating safety. [AIR4896]

safety control *See* control, safety.

safety diaphragm A diaphragm, usually metal, that will rupture if excessive pressure develops. [ARP4386]

safety ground A connection between metal structures, cabinets, cases, etc. which is required to prevent electrical shock hazard to personnel.

safety hardener A curing agent that causes only a minimum toxic effect on the human body, either on contact with the skin or as concentrated vapor in the air. [AIR4844]

safety hoist A hoisting device that stops automatically when tension is released.

safety hook A lifting hook with a spring-loaded latch that prevents the lifting sling from accidentally slipping off.

safety of flight event An engine, engine component, or sub-system functional status or condition that could seriously jeopardize flight integrity. [ARP1587]

safety of flight warning Timely crew warning to permit in-flight correction of problems seriously affecting flight integrity or leading to catastrophic engine failure. The intent is to provide either real time crew warning or post-flight maintenance indication for all engine discrepancies that could cause an unsafe flight situation. [ARP1587]

safety plug A nonreclosable pressure-relief device containing a fusible element that melts at a predetermined temperature.

safety pressure A small positive pressure maintained in the oronasal mask cavity to prevent inboard leakage at specified peak rates of inspiratory flow. [ARP171]

safety relief device A device intended to prevent rupture of a pressure vessel under certain conditions of exposure. [ARP4386]

safety relief valve An automatic pressure-relieving device that is actuated by the pressure upstream of the valve and characterized by opening pop action, with further increase in lift with an increase in pressure over popping pressure.

safety shut down The action of shutting off all fuel and ignition energy to a burner by means of a safety control or controls such that restart cannot be accomplished without operator action.

safety stop *1*. On a hoisting engine, a device that automatically prevents the engine from overwinding or falling. *2*. A device that prevents mechanical over-travel on a piece of equipment.

safety valve A spring-loaded valve that automatically opens when pressure attains the valve setting. Used to prevent excessive pressure buildup in a boiler.

safing Actions that eliminate or control hazards. [AIR4728]

sag The "droop" of a sealant after it is applied to an overhead or vertical surface. [AIR4069]

sagging Run-off or flow-off of adhesive from an adherend surface due to application of excess or low-viscosity material. [AIR4844]

Sagnac effect A phase shift (and consequent measurable rotation rate) caused by nonreciprocity (different optical path lengths) of two counterpropagating light waves travel-ing in the same coil in a fiber-optic gyro or ring interferometer.

SALS Acronym for short-approach light system.

salt spray corrosion test A test that is used to determine the relative abilities of thin coatings of greases to prevent corrosion on steel surfaces in the presence of salt spray. [S-5C,40-55]

sample *1*. A small portion of a material or product intended to be representative of the whole. *2*. In statistics, the collection of measurements taken from a specified population. [AIR4844]

sample-and-hold per channel A method of sampling and holding analog data channels in advance of a normal multiplex sample-and-hold A/D encoding process.

sampled-data control A branch of automatic control theory that is concerned with the control of variables whose current values are not continuously available for comparison with the setpoint, but instead are sampled only at given intervals.

sampled data system A system that does not continuously update information; normally a digital/discrete system. Generally includes a data-hold device, inserted directly following the system, to preclude high-frequency components inherent in the sampled signal from disrupting the controlled parameter or controller operation. [ARP89]

sampled signal A signal that is updated only at given intervals by a new observation of the variable.

sample-hold A device that takes a "snapshot" of an analog signal so that it is held stationary for an A/D conversion.

sample/hold amplifier An amplifier that samples the input signal and holds the value for sufficient time before it is input to an analog to digital (A/D) converter.

sample inspection The monitoring and/or withdrawal of selected devices from service to establish realistic time limits. [ARP4386]

sample interval The time interval between measurements or observations of a variable.

sample mean The arithmetic average of the measurements in a sample; an estimator of the population mean. [AIR4844]

sample median The value of the middle observation—if the sample size is ordered from smallest to largest. [AIR4844]

sample plan The plan designed by a telemetry engineer to sample and encode data incrementally so that it may be accurately decoded and re-created.

sample size The number of units of product in a sample. [AIR4896]

sample standard deviation The square root of the sample variance. [AIR4844]

sample variance The sum of the square deviations from the sample mean, divided by n-1. [AIR4844]

sampling *1.* The collection of exhaust sample under controlled conditions for the purpose of analysis. [ARP1179A] *2.* Obtaining the values of a function for discrete, regularly, or irregularly spaced values of the independent variable. *3.* Selecting only part of a production lot or population for inspection, measurement, or testing.

sampling action A type of controller action whereby the value of the controlled variable is measured at intermittent intervals rather than continuously.

sampling controller A controller in which intermittently observed values of a signal, such as the setpoint signal, the actuating error signal, or the signal representing the controlled variable, are used to effect control action.

sampling error *See* precision error.

sampling rate For a given measurement, the number of samples per second in a time-division-multiplexed system. Typically, sampling rate is at least five times the highest data frequency of the measurement.

sampling theorem Theorem stating that equispaced data, with two or more points per cycle of highest frequency, allow reconstruction of band-limited functions.

sampling thermocouple An assembly that has the means for sampling gases at two or more points, and for mixing these at the measuring junction so as to give a useful representative value. [ARP485]

sandblasting *See* grit blasting.

sandwich braze A joining technique for reducing thermal stress in a brazed joint, in which a shim is placed between the opposing surfaces to act as a transition layer.

sandwich construction A technique for producing composite materials that consists of gluing hard outer sheets onto a center layer, usually a foamed or honeycomb material.

sandwich panel A panel consisting of two thin-face sheets bonded to a thick, lightweight honeycomb or foam core. [AIR4844]

SAP *See* service access point.

SAS *See* stability augmentation system.

satellite *1.* An attendant body that revolves about another body. *2.* A man-made object designed or expected to be launched as a satellite. [ARP4386]

satellite communication Use of communication satellites, passive reflecting belts of dipoles or needles, or reflecting orbiting balloons to extend the range of radio communication by returning signals to earth from the orbiting object, with or without amplification.

satellite defense *See* spacecraft defense.

satellites Any objects, man-made or natural, that orbit celestial bodies.

satellite surfaces The crust and soil of natural satellites.

satin Describes a type of finish having a satin or velvety appearance, specified for plastics or composites. [AIR4844]

satin finish A type of metal finish produced by scratch brushing a polished metal surface to produce a soft sheen.

SATS (short airfield for tactical support) A portable airfield design for use by one high-performance jet squadron. Consists of a surfaced runway 2210 feet long and 96 feet wide, turn-off areas at either end, a hot pad, and parking and maintenance. Also includes a CE1-3 catapult, two M-21 primary recovery systems, two Fresnel lens optical landings systems (FLOLS), and extensive field lighting and audio visual communications systems necessary to operate (launch and recovery) high-performance jet aircraft. A SATS field has the capability of being readily expanded to an expeditionary airfield. [AIR1489]

saturable-core magnetometer A magnetometer whose output is derived from changes in magnetic permeability of a ferromagnetic core as a function of the strength of the magnetic field being measured.

saturable-core reactor A device for introducing inductive reactance into a circuit, whereby the effective reactance can be varied by varying auxiliary direct-current excitation of a ferromagnetic core.

saturant *See* lacquer.

saturated air Air that contains the maximum amount of water vapor that it can hold at the given temperature and pressure.

saturated steam Steam at the temperature corresponding to its pressure.

saturated temperature The temperature at which evaporation occurs at a particular pressure.

saturated water Water at its boiling point.

saturation An equilibrium condition in which the net rate of absorption under prescribed conditions falls essentially to zero. [AIR4844]

saturation current *1.* In ionic conduction, the current obtained when the applied voltage is sufficient to collect all of the ions present. *2.* In an electromagnet, the excitation current required to produce magnetic saturation.

saturation level For an APC, the maximum counting rate of the electronic counting circuitry, specified by the manufacturer. The APC should be operated at the specified flow rate and at a particle concentration that yields counting rates below the saturation level. [ARP1192]

saturation voltage The minimum applied voltage that produces saturation current.

saturator A device, equipment, or person that saturates one material with another.

sawtooth wave A type of cyclic direct-current waveform in which voltage and current rise gradually from zero to a peak value, then drop instantaneously to zero and repeat the cycle.

Saybolt Furol viscosimeter An instrument that is similar to a Saybolt Universal viscosimeter, but with a larger diameter tube for measuring the viscosity of very thick oils. *See* Saybolt Universal viscosimeter.

Saybolt Universal viscosimeter An instrument for determining viscosity by measuring the time it takes an oil or other fluid to flow through a calibrated tube.

s-band In telemetry, the portion of the radio-frequency spectrum between 2200 and 2300 MHz.

S-basis Refers to the minimum value specified by the appropriate Federal, Military, Society of Automotive Engineers, American Society for Testing and Materials, or other recognized and approved specifications for the material. [AIR4844]

scab A surface defect on a casting or rolled metal product, consisting of a thin, flat piece of metal partly detached from the substrate.

scalar quantity Any quantity that can be described by magnitude alone—as opposed to a vector quantity, which can only be described by both magnitude and direction.

scale effect Any variation in the nature of the flow and in the force coefficients associated with a change in value of the Reynolds number, i.e., caused by change in size without change in shape.

scale factor *1.* The coefficients by which quantities in a problem are multiplied or divided in order to convert them so as to have them lie in a given range of magnitude, e.g., plus one to minus one. *2.* A constant multiplier that converts an instrument reading in scale divisions to a measured value in standard units. *3.* In analog computing, a proportionality factor that relates the value of a specified variable to the circuit characteristic that represents it in the computer. *4.* In digital computing, an arbitrary factor applied to some of the numerical quantities in the computer to adjust the position of the radix point; this is done so that the significant digits occupy specific positions.

scale height A measure of the relationship between density and temperature at any point in the atmosphere.

scale length The distance that the pointer of an indicating instrument, or the marking device of a recording instrument, travels in moving from one end of the instrument scale to the other, measured along the baseline of the scale divisions.

scale models Three-dimensional representations of objects or structures containing all parts in the same proportion as their true size.

scaler(s) A measuring-circuit or control-circuit component that produces one output pulse each time a specific number of input pulses have been received.

scale span The algebraic difference, measured in scale units, between the highest value that can be read from the scale and the lowest value.

scale-up To use data from an experimental model or pilot plant to design a larger (scaled-up) facility or device, usually of commercial size.

scaling Forming a thick layer of oxide on a metal, especially at high temperatures.

scaling circuit An electronic circuit that produces an output pulse whenever a predetermined number of input pulses has been received.

scaling factor A factor used in heat-exchange calculations to allow for reduced thermal conductivity across a tube or pipe wall due to scaling.

scalp To remove the surface layer of a billet, slab, or ingot, thereby removing surface defects that might persist through later operations.

scan *1*. Collection of data from process sensors by a computer for use in calculations, usually accomplished through a multiplexer. *2*. A single sweep of PC applications program operation. The scan operates the program logic based on I/O status, and then updates outputs and input status. The time required for this is called the scan time. *3*. To examine an area, volume, or portion of the electromagnetic spectrum, point by point, in an ordered manner.

scan linearity Degree of uniformity of deflection sensitivity at the face of a CRT. [ARP1782]

scanned image The representation of an image by successive scan lines of picture elements or PIXELS (PELS). Scanned images are often interchanged as facsimile data transmissions. [AS4159]

scanner(s) *1*. Radar mechanisms incorporating such things as rotatable antennas, radiators, motor drives, or mountings for directing a searching radar beam through space and imparting target information to an indicator. *2*. An instrument that automatically samples or interrogates the state of various processes, files, conditions, or physical states; and initiates action in accordance with the information obtained.

scanning In radar, refers to the motion of the antenna assembly when searching for targets.

scanning rate (or speed) The speed at which a computer can select and convert an analog-input variable.

scarf angle The angle of taper of a scarf joint, i.e., the taper ratio of length to thickness. [AIR4844]

scarf joint A joint made by cutting away similar angular segments on two adherents and bonding the adherents with the cut areas fitted together. [AIR4844]

scattering *1*. The process by which small particles suspended in a medium of a different index of refraction diffuse a portion of the incident radiation in all directions. In scattering, no energy transformation results, only a change in the spatial distribution of the radiation. *2*. A collision or other interaction that causes a moving particle or photon of electromagnetic energy to change direction.

scattering coefficients Measures of the attenuation due to scattering of radiation as it traverses a medium containing scattering particles.

scattering cross sections The hypothetical areas normal to the incident radiation that would geometrically intercept the total amount of radiation actually scattered by a scattering particle. Equivalently, the cross sectional areas of isotropic scatterers (spheres) that would scatter the same amount of radiation as the actual amount.

scattering functions The intensities of scattered radiation in a given direction per lumen of flux incident upon the scattering material.

scattering loss A reduction in the intensity of transmitted radiation due to internal scattering in the transmission medium, or to roughness of a reflecting or transmitting surface.

scatter plates (optics) Holograms of diffusing screens, used for scattering incident light by the process of diffraction.

scatter propagation Specifically, the long-range propagation of radio signals by scattering due to index of refraction inhomogeneities in the lower atmosphere.

scavenger *1.* A reactive metal added to molten metal to combine with and remove dissolved gases or other impurities. *2.* A chemical added to boiler water to remove oxygen.

scavenging *1.* Removing spent gases from the cylinder of an internal combustion engine and replacing them with a fresh charge or with air. *2.* Removing dissolved gases or other impurities from molten metal by reaction with an additive.

scenario An outline of a mission flight plan giving the particulars of each mission phase. *See also* design mission scenario and total mission scenario. [ARP4107]

Schach effect The effect that occurs when a slowly or nonrotating satellite is heated on its sunward side; the photons of thermal radiation carry away more momentum from the hot sunward side than the cold shadowed side, thereby given the satellite a certain net acceleration in the direction away from the sun.

scheduled maintenance Periodic prescribed inspection and/or serving of equipment accomplished on a calendar, mileage, or hours of operation basis. Included in preventive maintenance. [ARD50010]

schematic For a fluid unit or system of units, a flow diagram including all interconnections. [ARP4386]

Schering bridge A type of a-c bridge circuit that is particularly useful for measuring the combined capacitive and resistive qualities of insulating materials and high-quality capacitors.

schlieren photography A method of photography for flow patterns. This method takes advantage of the fact that light passing through a density gradient in a gas is refracted as though it were passing through a prism.

Schmitt trigger A bi-stable pulse generator in which an output pulse of constant amplitude exists only as the input voltage exceeds a certain DC value; the circuit can convert a slowly changing input waveform to an output waveform with sharp transitions.

scholander A simple gas analyzer for the chemical volumetric determination of oxygen and carbon dioxide in a gas; total inert gases present is determined by difference. [ARP171]

scintillation *1.* Generic term for rapid variations in apparent position, brightness, or color of a distant luminous object viewed through the atmosphere. *2.* A flash of light produced in a phosphor by an ionizing event. *3.* On a radar display, a rapid apparent displacement of the target from its mean position.

scintillation counter The combination of phosphor, photomultiplier tube, and associated circuits for counting scintillations.

scintillation spectrometer A scintillation system that is designed to separate and determine the energy distribution in heterogeneous radiation.

sclerometer An instrument for determining the hardness of a material by measuring the force needed to scratch or indent the surface with a diamond point.

scleroscope An instrument for determining the hardness of a material by measuring the height to which a standard steel ball rebounds when dropped from a standard height.

scoop An air inlet designed and located to deliver atmospheric air to an internal system by decelerating the air, thereby increasing the static pressure and producing flow. [ARP147C]

scoop proof Describes a design feature whereby exposed contacts of a connector cannot be touched or damaged by any portion of the mating connector. [ARP914A]

scooter gear *See* gear, scooter.

scoring Deep scratches on the surface of a metal.

scotch boiler A cylindrical steel shell with one or more cylindrical internal steel furnaces located (generally) in the lower portion, and with a bank or banks (passes) of tubes attached to both end closures.

scotchweld seal A faying-surface seal using a structural adhesive. [AIR4069]

Scotch yoke A type of four-bar linkage used to convert uniform rotation into simple harmonic motion. Also converts linear motion to rotary motion.

scouring Physical or chemical attack on internal surfaces of process equipment.

scraper ring A piston ring that scrapes oil from the cylinder wall to prevent it from being burnt.

scrap rate The number of items that are beyond the limits of economical repair, expressed as a quantity per unit of time. [ARD50010]

scratch An elongated surface discontinuity that is infinitely small in width compared to length. [AIR4844]

scratch hardness *1.* A measure of the resistance of minerals or metals to scratching. For minerals, scratch hardness is defined by comparison with ten selected minerals comprising the Mohs scale. *2.* A method of measuring metal hardness in which a cutting point is drawn across a metal surface under a specified pressure, and hardness is determined by the width of the resulting scratch.

screech High-frequency acoustic noise generated by engine exhaust-gas turbulence. [AIR1839]

screen *1.* A device made of meshed fabric, usually mounted on a frame, used to prevent ingestion of foreign objects into the air intake of jet engines. [ARP480A] *2.* In data processing, the plane surface of a CRT that is visible to the user. *3. See* shield.

screen analysis A method of determining the particle size distribution of any loose, flowing aggregate by sifting it through a series of standard screens with holes of various sizes and determining the proportion that passes each screen.

screened ground A wire or terminations provided with a means to adequately shield the inner conductor(s). [AIR4258]

screen effectiveness A measure of the capability of a screen to precipitate latent defects to failure. Sometimes used specifically to mean screening strength. [AIR4896]

screen grid An internal electron-tube element positioned between a control grid and an anode. The screen grid is usually maintained at a fixed positive potential so that electrostatic influence of the anode is reduced in the region between the screen grid and the cathode.

screening *1.* A process for inspecting items to remove those that are unsatisfactory or those that are likely to exhibit early failure. Inspec-tion includes visual examination, physical dimension measurement, and functional performance measurement under specified environmental conditions. [ARD50010] *2.* Separation of an aggregate mixture into two or more portions according to particle size, by passing the mixture through one or more standard screens. *3.* Removing solids from a liquid-solid mixture by means of a screen.

screening regimen A combination of stress screens applied to an equipment, identified in the order of application. [AIR4896]

screening strength *See* screen effectiveness.

screening test A test or combination of tests intended to remove unsatisfactory items or those likely to exhibit early failures. [AIR4896]

screen parameters In screening strength equations, parameters that relate to screening strength (e.g., vibration g-levels, temperature rate of change, and time duration). [AIR4896]

screw *1.* A cylindrical machine element with a helical groove cut into its surface. *2.* A type of fastener with a threaded, cylindrical or conical shank, and a slitted, recessed, flat, or rounded head which is usually larger in diameter than the shank.

screw conveyor A device for moving bulk material which consists of a helical blade or auger rotating within a stationary trough or casing. A screw elevator moves material in a vertical direction; a screw feeder moves material into a process unit. Also known as an auger conveyor, spiral conveyor, or worm conveyor.

screw pinch A cylindrical plasma equilibrium in which the axial and azimuthal components of the vacuum field are of the same size.

screw thread Any of several forms of helical ridges formed on or cut into the surface of a cylindrical or conical member. Standard thread designs are used to connect pipes to fittings or to construct threaded fasteners.

scrim A low-cost reinforcing fabric made from continuous-filament yarn in an open-mesh construction. [AIR4844]

scrubber A device for removing entrained dust or moisture from a process gas stream.

SCS *See* supervisory control system.

scuffing A form of mild adhesive wear generally exhibited as a dulling of the worn surface.

scuff patch/scuff pad An additional thickness of material designed to prevent tank-wall damage from any anticipated source. The material can be rubber-coated fabric or molded rubber applied to either the inside or outside surface of the tank. Also called a wear patch. [AIR1664]

scum A film of impurities on the surface of a liquid or a solid.

S/D *See* synchro-to-digital converter.

SDF *See* simplified directional facility.

SDLC *See* synchronous data link control.

seal *1.* To close a fuel tank or vessel to make it leakproof by the application of sealant to fasteners, seams, and any other possible leak path. *2.* Any device or system that creates a nonleaking union between two mechanical components.

sealant A material applied to a joint in paste or liquid form that hardens or cures in place, forming a seal against gas or liquid entry. [AIR4844]

seal chambers Enlarged pipe sections in measurement-impulse lines to: (a) provide a high area-to-volume displacement ratio to minimize error from hydrostatic head difference when using large volume displacement measuring elements; and (b) prevent loss of seal fluid by displacement into the process.

seal drain A drain provision to remove seal leakage oil from the accessory drive cavity, and/or accessory housing. [ARP906A]

sea level The level of the surface of the ocean; especially, the mean level halfway between high and low tide, used as a standard in reckoning land elevation or sea depths.

sealing voltage The voltage required to move the armature of a magnetic relay from the position where contacts first touch to its fully closed position.

seal, interfacial A seal provided at the interface of mated connectors to prevent the entry of fluids or contaminants across the interface. [ARP914A]

seal leg The piping from the instrument to the top elevation of the seal fluid in the impulse line.

seal, peripheral A seal provided around the periphery of connector inserts to prevent the entry of fluids or contaminants at the perimeter of mated connectors. [ARP914A]

seal plane All surfaces of a tank that establish fuel-seal continuity and are in immediate contact with fuel. These surfaces may be composed of structure, fastener, and/or sealing materials. [AIR4069]

seal weld A weld used primarily to obtain tightness and prevent leakage.

seam *1.* A surface crack resulting from a defect in casting or forging; or extraneous materials, stringer in the material, not homogeneous with the base metal. [MA2005] *2.* An open surface defect that is narrow and continuous, usually straight, running generally parallel to the nut axis. Seams are generally inherent in the bar from which the nut is made. [MA1568] *3.* An extended length weld. *4.* A mechanical joint, especially one made by folding edges of sheet metal together so that they interlock. *5.* A mark on ceramic or glass parts corresponding to the mold parting line.

seamless tubing Tubular products made by piercing and drawing a billet, or by extrusion.

seam welding A process for creating a weld between metal sheets; consists of a series of overlapping spot welds. Usually done by resistance welding but may be done by arc welding.

search unit A device for generating and/or receiving ultrasonic energy. [ARP5089]

season cracking Usually reserved for describing stress-corrosion cracking of copper or copper alloys in an environment that contains ammonium ions.

seat In a valve, the fixed area into which the moving part of the valve rests when it is closed to retain pressure and prevent flow.

seating, downstream Seating assisted by pressure differential across the closure component in the closed position; this moves the closure component slightly downstream into tighter contact with the seat-ring seal, which is supported by the body.

seating, upstream A seat on the upstream side of a ball, designed so that the pressure of the controlled fluid causes the seat to move toward the ball.

seat ring A part that is assembled in the valve body and may comprise part of the flow-control orifice. The seat ring may have special material properties and may provide the contact surface for the closure member.

seat, spring loaded A seat that is designed to exert a greater force at the point of closure component contact to improve the sealing characteristics, particularly at low pressure differential.

secant bulk modulus The total change in fluid pressure divided by the total change in fluid volume per unit of the initial volume under pressure. [AIR1116]

secant modulus Idealized Young's modulus derived from a secant drawn between the origin and any point on a nonlinear stress-strain curve. [AIR4844]

second Symbolized s. A unit of time in both the metric and English systems.

secondary air Combustion air introduced into a combustion chamber over the burner flame to provide excess air and ensure complete combustion.

secondary bonding The joining together, by the process of adhesive bonding, of two or more already-cured composite parts, where the only chemical or thermal reaction occurring is the curing of the adhesive itself. [AIR4844]

secondary calibration An electronic calibration or standardization performed using the reference system built into the instrument. [ARP1192]

secondary cell A cell that is capable of being recharged. [ARP4386]

secondary circuit The part of an electrical circuit that conducts current output from a transformer to perform the function of the circuit.

secondary combustion Combustion that occurs as a result of ignition at a point beyond the furnace. *See also* delayed combustion.

secondary cosmic rays Secondary emission in the atmosphere stimulated by primary cosmic rays.

secondary damage Damage resulting from a primary failure. [ARP1587]

secondary device Part of a flowmeter that receives a signal from the primary device and displays, records, and/or transmits it as a measure of the flow rate.

secondary electron An energetic electron set in motion by the transfer of momentum from primary electromagnetic or particulate radiation.

secondary emission Emission of subatomic particles or photons that is stimulated by primary radiation; for example, emission of particles and/or photons by particles whose electron configurations or nuclei have been disrupted by the impingement of cosmic rays.

secondary explosive A high explosive that is relatively insensitive to heat and shock, and is usually initiated by a primary explosive or by an exploding bridge wire. [AIR913]

secondary filter The filter most often installed immediately upstream of a fluid-system component, and downstream of the primary filter, to provide to that component positive protection from potential catastrophic failure caused by large particle ingestion. [AIR4057]

secondary hardening Hardening of certain alloy steels by precipitation hardening during tempering. Secondary hardening supplements hardening achieved by controlled cooling from above the critical temperature in a step that precedes tempering.

secondary isolation Isolation to the subunit component/part that must be replaced in the shop to return the subunit to serviceable condition. [AIR4896]

secondary loop The inner loop of a cascade system.

secondary power In an aircraft, all nonpropulsive power generation and transmission; includes electric, hydraulic, pneumatic, mechanical, and auxiliary power. Sometimes called flight vehicle power. *See* auxiliary power. [ARP906A]

secondary pressure control Pressure resulting from a valve or other such device which provides pressure control at a value greater than the primary pressure control level. [AIR4783]

secondary radar A radar technique or mode of operation in which the return signals are obtained from a beacon, transponder, or repeater carried by the target, as contrasted with primary radar in which the return signals are obtained by reflection from the target.

secondary seal A seal that alone cannot provide a dependable absolute seal. Sometimes used in conjunction with a primary seal. [AIR4069]

secondary steering system *See* classification of nosewheel steering systems.

secondary storage Those storage facilities that are not an integral part of the computer but are directly connected to and controlled by the computer, e.g., magnetic drum and magnetic tapes.

secondary structure In aircraft and aerospace applications, a structure that is not critical to flight safety but whose failure could cause significant problems. [AIR4844]

secondary treatment Treatment of boiler feed water; or internal treatment of boiler-water after primary treatment.

secondary winding The output winding of a transformer or similar electrical device.

second degree repair The repair of a damaged or nonoperating gas turbine engine, its accessories, or components to an acceptable operating condition. [AIR4896]

second-level address *See* indexed address.

sectional conveyor A belt conveyor that can be made shorter or longer by removing or adding interchangeable sections.

sectional header boiler A horizontal boiler of the longitudinal or cross-drum type, in which the tube bank is composed of multiple parallel sections, each section having a front and rear header connected by one or more vertical rows of generating tubes. The sections or groups of sections have a common steam drum.

sector gear A toothed machine component that looks like part of a gear wheel, containing the bearing and part of the rim and its teeth.

sector link In an elastic-chamber pressure gage, the connecting link between the elastic chamber and the sector gear that positions the pointer.

sediment *1.* In water, matter that can be removed from suspension by gravity or mechanical means. *2.* A non-combustible solid matter that settles out at the bottom of a liquid. A small percentage sediment is present in residual fuel oils.

sediment trap A device used for removing sediment from an instrument sensing line on a boiler.

see and avoid Refers to a visual procedure wherein pilots of aircraft flying in visual meteorological conditions (VMC), regardless of type of flight plan, are charged with the responsibility of observing the presence of other aircraft and maneuvering their aircraft as required to avoid collisions. [ARP4107]

Seebeck effect The establishment of an electric potential difference tending to produce a flow of current in a circuit of two dissimilar metals, the junctions of which are at different temperatures.

seed(s) *1.* Small bubbles of air or gaseous inclusions entrapped within the glass. [ARP924] *2.* A small single crystal of semiconductor material used to start the growth of a single large crystal from which semiconductor wafers are to be cut.

seepage Occurrence of extremely slight fluid at the surface of a component or part of a hydraulic system, normally due to "breathing" of seals under cyclic pressure load. No drops are allowed over an extended period of observation. [ARP4386]

segmented circle A system of visual indicators laid upon the ground, generally surrounding the wind sock or tetrahedron, designed to provide traffic-pattern information at airports that do not have operating control towers. [ARP4107]

segmented display A display in which the individual addressable display elements (segments) are of varying shape and/or orientation, each dedicated to the display of a specific type or specific types of symbolic or pictorial information. [ARP4256]

segments of an instrument approach procedure Any of four separate segments, depending on how the approach procedure is structured. The four possible segments are: (a) initial approach segment, (b) intermediate approach segment, (c) final approach segment, and (d) missed approach segment. *Initial approach segment*–the segment between the initial approach fix and the intermediate fix or the point where the aircraft is established on the intermediate course or final approach course. *Intermediate approach segment*–the segment between the intermediate fix or point and the final approach. *Final approach segment*–the segment between the final approach fix or point and the runway, airport, or missed approach point. *Missed approach segment*–the segment between the missed approach point, or point of arrival at decision height, and the missed approach fix at the prescribed altitude. [ARP4107]

segregation *1.* The maintenance of independence by means of a physical barrier between two hardware components. [ARP4761] 2. Keeping process streams apart. *3.* Nonuniform distribution of alloying elements and impurities in a cast metal microstructure. *4.* A series of close, parallel, narrow, and sharply defined wavy lines of color on the surface of a molded plastics part that differ in shade from surrounding areas and make it appear as if the components have separated. *5.* The tendency of refuse of varying compositions to deposit selectively in different parts of the unit.

seizure *See* freeze-up.

selective-ion electrode A type of pH electrode in which a metal-metal-salt combination is used as the measuring electrode; this makes the electrode particularly sensitive to solution activities of the anion in the metal salt.

selective plating Any of several methods of electrochemically depositing a metallic surface layer at only localized areas of a base metal, the remaining unplated areas being masked with a nonconductive material during the plating step.

selective reinforcement The addition of advanced composite materials to selected areas for local augmentation of strength or stiffness. [AIR4844]

selective surfaces Surfaces, often coated, of which the spectral optical properties, such as reflectance, absorptance, emittance, or transmittance vary significantly with wavelength. Such properties are of interest in solar energy applications.

selective system A system in which the protective device closest to the faulted circuit opens

and isolates that circuit without disturbing the remainder of the system. [ARP1199A]

selectivity The characteristic of an electronic receiver that determines the extent to which it can differentiate between a desired signal and electronic noise, or undesired signals of other frequencies.

selector *1.* A device for choosing objects or materials according to predetermined attributes. *2.* A device for starting or stopping a mechanism at predetermined positions or locations. *3.* A gearshift lever for operating an automatic transmission in a motor vehicle.

selector valve *1.* A control valve with more than one flow pattern; has a minimum of three working ports. [ARP4386] *2.* A cockpit-mounted valve that provides a manual/automatic selection of primary, emergency, or backup oxygen supply. [ARP171]

selenology That branch of astronomy that concerns the moon, its magnitude, motion, constitution, and the like. Selene is Greek for moon.

self absorption Attenuation of radiation due to absorption within the substance that emits the radiation.

self-adapting Pertaining to the ability of a system to change its performance characteristics in response to its environment.

self adaptive control systems Particular types of stability-augmentation systems that change the responses of given control inputs by constantly sampling responses and adjusting their gain, rather than having fixed or selective gain systems.

self-aligning torque *See* aligning torque.

self-balancing potentiometer *See* potentiometer.

self braking strut In a landing gear system, a folding brace or strut that carries its own actuator. [AIR1489]

self-check A means to passively verify the accuracy of output data. [ARD50024]

self-checking code *See* error detecting code.

self-cleaning Describes any device that is fitted with a mechanism that removes accumulated deposits from its interior without disassembly.

self-contained apparatus Apparatus that is not necessarily connected to equipment in the non-hazardous area, and thus is usually self-powered. NOTE: A self-contained apparatus is normally portable, e.g., walkie-talkie radios, but the term does not imply that it must be.

self-contained instrument An instrument in which all of the component parts are contained within a single case or enclosure, or are incorporated into a single assembly.

self diffusion (solid state) The spontaneous movement of an atom to a new site in a crystal of its own species.

self-discharge *1.* The spontaneous decomposition of battery materials from the charged to the discharged state. [ARP4386] *2.* The decrease in the state of charge of a cell, not resulting from an applied useful load, over a period of time, normally due to internal electrochemical losses and accelerated by higher temperature. [AS8033]

self-extinguishing The quality whereby a material stops burning once the source of a flame is removed.

self-extinguishing resin A resins formulation that will burn in the presence of a flame, but will extinguish itself within a specified time after the flame is removed. [AIR4844]

self-hardening steel *See* air-hardening steel.

self-ignition temperature The temperature of a material at which spontaneous combustion takes place when the temperature rises slowly. [AIR4844]

self-indicting fuse or limiter *1.* A type of fuse or limiter that incorporates as an integral part a device to visually denote severance of the fusible element. *2.* A limiter or fuse that visually displays the fusible element or incorporates a mechanical indicting device. [ARP1199A]

self-luminous display A display device that generates and emits light from the display face. [ARP4032]

self monitoring The capability of a component to monitor itself without additional sensors or other devices. [ARP4386]

self-organizing control A subclass of adaptive control in which the structure of the controller is modified to improve system performance by the use of a logic that is independent of the prior knowledge of the changing dynamics of the vehicle. [ARP4386]

self-organizing environment Refers to a class of equipment that may be characterized loosely as containing a variable network in which the elements are organized by the equipment itself, without external intervention, to meet criteria of successful operation.

self-quenched counter tube A type of radiation counter tube in which internal interactions inhibit the reignition of electron discharge.

self-repairing *See* reconfigurable.

self-sealing fastener A fastener that provides a fuel-tight seal without the need for sealant material or the use of a mechanical seal; for example, an interference fit fastener. [AIR4069]

self-sealing tank A tank that is constructed in such a way that an automatic sealing action occurs when penetrated by a projectile. The weight and thickness of construction is adjusted to the ballistic threat anticipated for the aircraft. [AIR1664]

self-skinning foam A urethane foam that, upon curing, produces a tough outer surface over a foam core. [AIR4844]

self sufficient Describes a starting system that is independent of all sources of energy external to the aircraft. [ARP906A]

self-tapping screw A threaded fastener with specially designed and hardened threads that form internal threads in a hole in sheet metal or soft materials as the screw is driven.

self-test *1.* A means by which a circuit or system acquires a real-time measure of its capacity to perform within design limits. [ARD50024] *2.* A test or series of tests that is performed by a device upon itself to show whether it is operating within design limits. This includes test programs on computers and ATE performing functional and diagnostic tests. [ARD50010]

self-vulcanizing Describes an adhesive that undergoes vulcanization without the application of heat. [AIR4844]

semantics The relationships between symbols and their meanings.

semaphore A task synchronization mechanism. *See* synchronization of parallel computational processes.

semi-articulated Describes a configuration that is a variation on the articulated shock strut configuration. The wheel is mounted on a lever, which pivots about the lower end of the piston of a telescoping shock absorber. The front end of the lever is restrained by a link to the outer cylinder of the shock absorber. [AIR1489]

semiautomatic controller A control device in which some of the basic functions are performed automatically.

semiautomatic mode That mode in which the operator is able to select and initiate a subset of the automatic mode tests. [ARP4904]

semiautomatic valve A valve incorporating automatic operation subject to external control. [ARP4386]

semi-conductor An electronic device such as an integrated circuit or transistor.

semiconductor devices Electronic devices in which the characteristic distinguishing electronic conduction takes place within semiconductors.

semiconductor diodes Two-electrode semiconductor devices in which the rectifying properties of junctions or point contacts are utilized.

semiconductors Electronic conductors, with resistivity in the range between metals and insulators, in which the electrical charge carrier concentration increases with increasing temperature over some temperature range. Certain semiconductors possess two types of carriers, namely, negative electrons and positive holes.

semiconductor strain gage A type of strain-measuring device particularly well suited for use in miniature transducer elements. Consists of a piezoresistive element that is either bonded to a force-collecting diaphragm or beam, or diffused into its surface.

semicrystalline In plastics, describes materials that exhibit localized crystallinity. [AIR4844]

semienclosed Having the ventilating openings in the frame protected with wire screen, expanded metal, or perforated covers. [ARP4404]

semi-graded index Describes an optical fiber with refractive index profile intermediate between step-index and graded-index. NOTE: Strictly speaking, this might be considered a type of graded-index fiber with refractive index profile somewhat steeper than normal.

sender *See* transmitter.

sense indicator An aircraft radio indicator that shows whether the aircraft is flying toward or away from an omnirange station. Sometimes called a to-from indicator. [ARP4107]

sense switch *See* alteration switch.

sensibility reciprocal A balance characteristic equal to the change in load required to vary the equilibrium position by one scale division at any load.

sensible heat Heat that causes a temperature change.

sensing element *1.* A device that measures the value of a variable by providing an output that can be utilized by the controller. [ARP89C] *2.* The portion of a device that is directly responsive to the value of the measured quantity; may include the case protecting the sensitive portion.

sensing level The liquid level at which the sensing element is actuated. [AIR1660A]

sensitive part A life-limited part, whose failure is not catastrophic but likely to seriously affect engine performance, reliability, or operating cost (durability critical). [AIR1872]

sensitive time For a cloud chamber, refers to the amount of time after expansion when the degree of supersaturation is sufficient to allow a track to form and be detected.

sensitivity *1.* For an explosive component, an expression of susceptibility to initiation by externally applied stimuli. [AIR913] *2.* In a sensing element, the ratio of change in output to a specified change in input. [ARP89C] *3.* Response of a mathematical model to variations of the input parameters. *4.* The least signal input capable of causing an output signal having desired characteristics. *5.* The degree to which a process characteristic can be influenced or changed by a small change in some physical or chemical stimulus.

sensitivity analysis Repetition of an analysis with different quantitative values for cost or operational assumptions or estimates, such as hit-kill probabilities, activity rates, or research and development costs, to determine their effects. Such results are then compared with the results of basic analysis. [ARP4293]

sensitivity of LVDT The nominal slope of the input/output curve. [ARP4386]

sensitization In stainless steels, a phenomenon whereby chromium at the grain boundaries is reduced by carbide precipitation to below the passive level. Sensitization makes the alloy susceptible to intergranular corrosion; it can be caused by improper welding, furnace brazing, milling, or heat treatment. [ARP4386]

sensitized stainless steel Any austenitic stainless steel having chromium carbide deposited at the grain boundaries. This deprives the base alloy of chromium, resulting in more rapid corrosion in aggressive media.

sensitometer An instrument for determining the sensitivity of light-sensitive materials.

sensitometry Measurement of the light-response characteristics of photographic film under specified conditions of exposure and development.

sensor *1.* A transducer, plus the means for formatting a signal that carries intelligence about the measurand, plus mechanical packaging suited to the intended application and environment. [ARD50024] *2.* A component that senses variables and produces a signal in a medium compatible with fluidic devices; for example, a temperature or angular-rate sensor. [ARP993A] *3.* The portion of a device that is directly responsive to the value of the measured quantity. May include the case protection and the sensitive portion. *4.* A device that produces a voltage or current output representative of some physical property being measured (e.g., speed, temperature, or flow). *5. See* primary element.

sensor, hybrid A sensor that has an electrical transducer and associated circuits, and that employs a fiber-optic signal transmission path. [ARD50024]

sensor, modulated, fiber optic A sensor that has a transducer that modulates an optical carrier signal supplied by its interface module. [ARD50024]

sensor output format The form of the signal or waveform supplied by the sensor; e.g., analog, pulse-frequency, or digital. [ARP1587]

sensor, self-luminous, fiber optic A sensor that has a transducer that derives optical power from the transducer environment, or internally generates optical power, the nature of which is determined by the measurand. [ARD50024]

sensor system A sensor, together with the necessary interconnection lines and the interface circuits. [ARD50024]

sensor/video interconnect subsystem The logical interconnect that provides the information transfer paths between sources and sinks. [ARD50012]

separation *1.* The breaking away of the boundary layer from the surface. *2.* The maintenance of independence by means of physical distance between two hardware components. [ARP4761] *3.* Spacing of aircraft to achieve their safe and orderly movement in flight and while landing and taking off. [ARP4107] *4.* An action that disunites a mixture of two phases into the individual phases. *5.* Partition of aggregates into two or more portions of different particle size, for example, by screening. *6.* The degree, in decibels, to which right and left channels of a stereophonic radio or sound system are isolated from each other. *7.* The parting of two connected members of a structure or system, as occurs at preplanned times after launching a multistage rocket. *8.* The removal of dust from a gas stream.

separation minima The minimum longitudinal, lateral, and vertical distances by which aircraft are spaced through the application of air-traffic control procedures. *See also* composite separation. [ARP4107]

separation velocity The velocity at which a space vehicle is moving when some part or section is separated from it; specifically, the velocity of an earth satellite at the time of separation from the carrier. [ARP4386]

separator *1.* A part of the accumulator (the piston) that separates the gas and the hydraulic fluid. [ARP4379] *2.* The movable or flexible member in a fluid container, such as an accumulator or reservoir, the function of which is to prevent intermixture of fluids, such as air and hydraulic fluid. [ARP4386] *3.* Layer(s) of insulating material (conventionally a fabric) that provide electronic separation between electrically conductive cell plates of opposing polarities. The separator is placed adjacent to the gas barrier and provides physical support for the gas-barrier material. [AS8033] *4.* Any machine for dividing a mixture of materials according to some attribute, such as size, density, or magnetic properties.

separator-filter A piece of process equipment that removes solids and entrained liquid from a fluid stream by passing the fluid both through a set of baffles or a coalescer and through a screen.

separator, water A device for removing free moisture from the air stream. [ARP147C]

separator, water, high pressure A heat-exchanger-type device that cools high-pressure air below its dew point to condense water vapor from it, then collects the condensate in the outlet ducting and provides a means for draining it. [ARP147C]

separator, water, low pressure A water separator located in the downstream low-pressure ducting of a refrigeration system. This separator agglomerates the water particles and separates them from evaporator exhaust air or air-cycle machine turbine exhaust air. [ARP147C]

septum Adhesive and prepreg (or a metal sheet) cured between two pieces of core. [ARP5089]

sequence A transaction comprising a number of ordered transfers performing one intended function. [AS4710]

sequence control A system of control in which a series of machine movements occurs in a desired order, the completion of one movement initiating the next; and in which the extent of the movements is not specified by numerical input data.

sequence monitor Refers to computer monitoring of the step-by-step actions that should be taken by the operator during a startup and/or shutdown of a power unit. As a minimum, the computer would check that certain milestones had been reached in the operation of the unit. The maximum coverage would have the computer check that each required step is performed, that the correct sequence is followed, and that every checked point falls within its prescribed limits. Should an incorrect action or result occur, the computer would record the fault and notify the operator. *See also* sequential control.

sequence of events *See* mishap, sequence of events.

sequencer A mechanical or electronic control device that not only initiates a series of events, but also causes the events to follow one another in an ordered progression.

sequence report The weather report transmitted hourly to all teletype stations, and available at all flight-service stations. [ARP4107]

sequencing In landing gear systems, the process of controlling gear retraction or extension in the proper order of succession. For example: (a) the gear control is placed in the gear "up" position; (b) heel doors open; (c) gear retracts within the wheel well and locks in the stowed position; (d) doors close and lock; and (e) system hydraulic pressure is deenergized, thus leaving the gear in position for high-speed flight. [AIR1489]

sequential control *1.* Control by completion of a series of one or more events. *2.* A mode of computer operation in which instructions are executed in consecutive order by ascending or descending addresses of storage locations, unless otherwise specified by a jump. *3.* A class of industrial process-control functions in which the objective of the control system is to sequence the process units through a series of distinct states. *See also* sequence monitor.

sequential sampling A method of inspection that involves testing an undetermined number of samples, one by one, until enough test results have been accumulated to allow an accept/reject decision to be made.

sequential simulator *See* fail all simulator.

serial For digital data, describes the presentation of data as a time-sequential bit stream, one bit after another.

serial computer A computer, some specified characteristic of which is serial; for example, a computer that manipulates all bits of a word serially. Contrast with multiprocessing.

serial data enhancement *See* enhancement, serial data.

serial-to-parallel converter In PCM telemetry, the circuitry that converts a serial bit stream into bit-parallel data outputs, each transfer representing one measurement.

series cascade action A type of control-system interaction whereby the output of each controller in a series (except the last one) serves as an input signal to the next controller.

series element *1.* Any of a number of two-terminal electronic elements that form a path from one node of a network to another, in such a way that only elements of the path terminate at intermediate nodes along the path. 2. Any of a number of two-terminal elements connected in such a way that any mesh containing one of the elements also contains the others.

series expansion In mathematics, a divergent series of terms whose sum is asymptotic or ascending.

series resistor A resistive element of the voltage circuit of an instrument which adapts the instrument to operate on some designated voltage.

series servo A servo so located in a control system that its output adds algebraically to that of a major input. This arrangement is commonly used with SAS actuators to superimpose control on primary commands. The series servo output will not cause motion at the major input. [ARP4386]

series wound motor A wound-field d-c motor in which the rotor conductors and stator field coil are connected in series. [ARP4386]

serious injury Any injury that: (a) requires hospitalization for more than 48 hours, commencing within 7 days from the date the injury was received; (b) results in a fracture of any bone (except simple fractures of fingers, toes, or nose); (c) causes severe hemorrhages, or severe nerve-, muscle-, or tendon-damage; (d) involves any internal organ; or (e) involves second- or third-degree burns, or any burns affecting more than 5 percent of the body surface. [ARP4107]

serrations *1.* Alteration of the inside surface of a conductor barrel to provide better gripping of the conductor; or on the outside of a connector housing to provide better gripping of the connector. 2. Protusions on the rear of a connector housing for positive orientation of accessories. [ARP914A]

serve A filament, or group of filaments, such as wires or fibers, helically wound around a central core, normally unbonded. [ARP1931]

service *1.* Identifies the primary type of aircraft lighting for which a lamp was designed. [ARP881] 2. The work performed for a subsystem, application, or end user. [AS4893]

serviceability Refers to the degree of ease or difficulty with which a system can be repaired. [AIR4896]

serviceable Describes the condition of an end item in which all requirement for repair, bench check, overhaul, or modification, as applicable, have been accomplished, making it capable of performing the function or requirements for which it was originally designed. An end item in which signs of previous use are apparent is not necessarily unserviceable. When appearance is not a primary consideration, and the condition of the item meets all safety and performance requirements, it is processed as serviceable. [ARD50010]

serviceable item *See* item, serviceable.

service access point (SAP) The connection point between a protocol in one OSI layer and a protocol in the layer above. SAPs provide a mechanism by which a message can be routed through the appropriate protocols as it is passed up through the OSI layers.

service bulletin The only document issued by the manufacturer to notify the airline of recommended modification, substitution of parts, special inspections/checks, reduction of existing life limits, or establishment of first time life limits and conversion from one engine model to another. [ARD50010]

service ceiling The height above sea level, under standard atmospheric conditions, at which a given airplane is unable to climb faster than 100 ft/min. [ARP4107]

service conditions The heat, cold, flexing, shock, impact, vibration, etc. that an adhesive or composite will be subjected to in service. [AIR4844]

service factor *1.* For a facility such as a chemical processing plant or electric generating station, the proportion of time the facility is operating; actual operating time in hours, divided by total elapsed time in hours, expressed as a percent. *2.* For electric motors, a factor describing how much a motor can be operated above rated current without damage. For example, an electric motor with a service factor of 1.15 can be operated up to 115% of rated current without damage.

service life *1.* The time in service up until a predetermined point is reached. This point is usually expressed as a maximum differential pressure across the filter element at rated flow and pressure. [AIR888] *2.* The total time, specified in hours of engine operation with specified mission profiles and mission mix, that a part can function with repair, provided that after each repair the part can function for a minimum of one engine repair interval. Random failure events, such as FOD, are not considered in determining service life. [ARD50010]

service limits An end-of-life condition under which the unit must be removed from the aircraft. [ARP4256]

service pressure *See* rated pressure.

service rating The maximum voltage or current at which a connector or electrical device is designed to function continuously at a specified temperature. [ARP914A]

service test A test of an item, system, material, or technique, conducted under simulated or actual operational conditions to determine whether the specified military requirements or characteristics are satisfied. [ARD50010]

servicing Performing any act needed to keep an item in operating condition, i.e., lubricating, fueling, oiling, cleaning, etc., but not including preventative maintenance of parts or corrective maintenance tasks. [ARP4386]

servicing point An access specifically provided to permit servicing (lubrication, filling, draining, charging, or cleaning) of equipment in the normal installed position. [ARD50010]

servo The actuating device that positions the aircraft control surfaces in response to a commanded signal from a controller, selector, or sensor. Also called a servomotor or servo actuator. [ARP419]

servoactuator *1.* Device consisting of the power modulator, the actuator, and the feedback element, typically packaged together as a single component. [ARP4386] *2.* An assembly of a controller, an actuator, and feedback element(s), typically packaged together as a single line replaceable unit (LRU). [ARP4386] *3. See* servo.

servoactuator flight control A mechanism capable of producing a controlled output position/motion of an aerodynamic control surface in response to a command input. [ARP1281A]

servo brake *1.* A motor-vehicle brake in which vehicle motion is used to increase the pressure on one of the brake shoes. *2.* A power-assisted braking device.

servocylinder A servo actuator having an integral follow-up mechanism that enables its final output to be made a function of the input signal to its control valve. [ARP4386]

servo engage/disengage Refers to the process of connecting or disconnecting the servo motor and gear train from the cable drum or sector through a mechanical or electrical clutch in the servo. [ARP419-57]

servomechanism(s) Control systems incorporating feedback in which one or more of the system signals represent mechanical motion.

servomotor *See* servo.

servo overpower The act of overpowering the servo by applying an appropriate pilot force at the control column or rudder pedals. [ARP419-57]

servo tab/servotab A tab that is directly actuated by the aircraft control system and that, when deflected, causes the control surface or other surface to which it is attached to be deflected or moved by the air forces acting on it. Sometimes called a Flettner tab or flying tab. [ARP4107]

servovalve A valve that is capable of proportional fluid control in response to an input. [ARP4386]

servovalve, electrohydraulic flow-control An electrical input, flow-control valve, that is capable of continuous control. Output stage flow is a direct function of input current. [ARP490]

set *1.* To place a storage device into a specified state, usually other than that denoting zero or blank. *2.* To place a binary cell into the state denoting one. *3.* In simulation theory, refers to entities with at least one common attribute. (The entities may also own any number of sets.) Sets may be arranged (topologically ordered) on a "first-in, first-out," "last-in, last-out," or ranked basis. *4.* In plastics processing, conversion of a liquid resin or adhesive into a solid material. *5.* Permanent strain in a metal or plastics material.

setpoint The position at which the control point setting mechanism is set. This is the same as the desired value of the controlled variable.

setpoint control A control technique in which the computer supplies a calculated setpoint to a conventional analog instrumentation control loop.

set pressure *1.* The pressure that a component is designed to provide for a defined operation. [ARP4386] *2.* The inlet pressure at which a safety-relief valve opens; usually a pressure established by specification or code.

set screw A small, headless machine screw used for holding a knob, gear, or collar on a shaft. Usually has a sharp or cupped point on one end and a slot or recessed socket on the other end.

setting In an adjustable or calibrated unit (such as a relief valve, pressure switch, or flow-control device), operating characteristics that result from an adjustment or setting. [ARP4386]

setting temperature The temperature to which an adhesive or assembly is subjected to set the adhesive. [AIR4844]

setting time The period of time during which an assembly is subjected to heat and/or pressure, to set the adhesive. [AIR4844]

settling Partial or complete separation of heavy materials from lighter ones by gravity.

settling time The time interval between the step change of an input signal and the instant when the resulting variation of the output signal does not deviate more than a specified tolerance from its steady-state value. Also known as correction time.

set up To harden, as in curing of a polymer resin. [AIR4844]

setup *1.* An arrangement of data or devices to solve a particular problem. *2.* Preliminary operations, such as control adjustments, installation of tooling, or filling of process fluid reservoirs, that prepare a manufacturing facility or piece of equipment to perform specific work.

severe weather avoidance plan (SWAP) A plan to minimize the effect of severe weather on traffic flows in impacted terminal and/or ARTCC areas. SWAP is normally implemented to provide the least disruption to the ATC system when flight through portions of airspace is difficult or impossible due to severe weather. [ARP4107]

severity The consequences of a failure mode; the worst potential consequence of a failure are considered, determined by the degree of injury, property damage, or system damage that could ultimately occur. [ARD50010]

sexadecimal number A number, usually of more than one figure, representing a sum in which the quantity represented by each figure is based on a radix of sixteen. Synonymous with hexadecimal number.

sextants Double reflecting instruments for measuring angles, primarily altitudes of celestial bodies.

S_f *See* volume of air dissolved.

SFA *See* single frequency approach.

SFO *See* single-fail operative.

S_g *See* volume of air evolved.

S-glass *1.* Structural glass, used as fiber reinforcement, designed to give high tensile strength. [AIR4844] *2. See* structural glass.

SGU *See* symbology generator unit.

shading Controlling the phase distribution and amplitude distribution of transducer action at the active face in order to control its directionality.

shadowgraph photography Photography in which steep density gradients in the flow about a body are made visible, the body itself being presented in silhouette.

shaft A cylindrical metal rod used to position, and sometimes to drive or be driven by, rotating parts such as gears, pulleys, or impellers, which transmit power and motion.

shaft balancing A method of reducing vibrations in rotating equipment by redistributing the mass to eliminate asymmetrical centrifugal forces.

shaft encoder A device for indicating the angular position of a cylindrical member. *See also* gray code and cyclic code.

shaft horsepower *1.* The horsepower delivered to the driving shaft of an engine, as measured by a torsion meter. *See also* brake horsepower. [ARP4107] *2.* The power input to a pump or compressor.

shakedown test A test of equipment, made during installation or prior to its initial production operation.

shake-table test A durability test in which a component or assembly is clamped to a table or platen and subjected to vibrations of predetermined frequencies and amplitudes.

shallow discontinuity A discontinuity that is open to the surface of a solid object, and that possesses little depth in proportion to the width of this opening. [ARP4784]

shape control The control of large, flexible platforms in orbit by means of actuators strategically located.

shape memory alloys Martensitic alloys (titanium-nickel) that exhibit shape-recovery characteristics through stress-induced transformation and reorientation. Reverse transformation during heating restores the original grain structure of the high-temperature phase.

shaping A machining process in which a reciprocating, single-point tool cuts a flat or simply-contoured surface.

shatter cones Distinctively striated, conical rock fragments along which fracturing has occurred, ranging in length from less than a centimeter to several meters. Generally found in nested or composite groups in rocks of crypto-explosion structures. Believed to be formed by shock waves generated by meteorite impact.

shattercrack *See* flake.

shaving *1.* Cutting a thin layer of material off the surface of a workpiece, for example, when bringing gear teeth to final shape. *2.* Trimming thin layers or burrs from forgings, stampings, or tubing to smooth parting lines, uneven edges, or flash.

shear *1.* A tool that cuts plate or sheet material by the action of two opposing blades that move along a plane approximately at right angles to the surface of the material being cut. *2.* A type of stress tending to separate solid material by moving the portions on opposite sides of a plane through the material in opposite directions.

shear crimping Buckling of compressive facing due to low core shear modulus. Usually causes the core to fail in shear at the crimp. [AIR4844]

shear edge The cutoff edge of a mold. [AIR4844]

shear fracture A mode of fracture resulting from translation along slip planes that are preferentially oriented in the direction of the shearing stress. [AIR4844]

shear lip A characteristic of ductile fractures whereby the final portion of the fracture separation occurs along the direction of principal shear stress. Such a characteristic is exhibited in the cup and cone fracture of a tensile-test specimen made of relatively ductile material.

shear modulus The ratio of shearing stress to shearing strain within the proportional limit of the material. [AIR4844]

shearography *1.* Developed for strain measurements, a process that provides a full-field video strain gauge, in real time, over large areas. *2.* An enhanced form of holography, which requires the part to be under stress. [AIR4844]

shear pin *1.* A pin or wire designed to hold parts in a fixed relative position until sufficient force to cut through the pin is applied to the assembly. *2.* A pin through the hub and shaft of a power-train member which is designed to fail in shear at a predetermined force, thereby protecting the mechanism from being overloaded.

shear rigidity The sandwich property that resists shear distortions. [AIR4844]

shear section A special section, usually in the starter or accessory output shaft, that is designed to shear or fail at a specified torque range. [ARP906A]

shear spinning A metal-forming process in which sheet metal or light plate is formed into a part having rotational symmetry. In this process, a tool is pressed against a rotating blank, deforming the metal in shear until it comes in contact with a shaped mandrel. The resulting part has a wall thinner than the original blank thickness.

shear stability The ability of a fluid to withstand shearing without permanent loss of viscosity. [AIR1116]

shear strain The tangent of the angular change, caused by a force between two lines originally perpendicular to each other through a point in a body. [AIR4844]

shear strength The maximum shear stress that a material is capable of sustaining. [AIR4844]

shear stress Component of any stress lying in the plane of the area where stress is measured. The existence of shear stress in fluid is evidence of viscosity.

shear test *1.* A test in which two narrow strips of aluminum are overlapped at one end by approximately 1 inch and bonded in this area by sealant; after cure, the specimen is mounted in a testing machine and pulled in opposite directions, creating shear forces in the sealant. The force required for the sealant to fail in shear is a measure of its shear strength. [AS7200/1] *2.* Any of various tests intended to measure the shear strength of a solid.

shear wave *1.* A type of wave in which the particle motion is perpendicular to the direction of propagation. [ARP5089] *2.* Wave motion in which the particle motion is perpendicular to the direction of propagation. [AMS2633B] *3.* In an elastic medium, a wave in which any element of the medium along the wave changes its shape without changing its volume.

sheet *See* sheeting.

sheeting *1.* Sheet made in continuous lengths and generally supplied in roll form. *2.* A synonym for sheet.

sheet metal A flat-rolled metal product, generally thinner than about 0.25 in.

sheet molding compound A composite of fibers; usually a polyester resin and pigments, fillers, and other additives that have been compounded and processed into sheet form to facilitate handling in the molding operation. [AIR4844]

shelf life The length of time that a material, substance, product, or reagent can be stored under specified environmental conditions and continue to meet all applicable specification requirements and/or remain suitable for its intended function. [AIR4844]

shell *1.* A thin metal cylinder. *2.* The outer wall of a tank or pressure vessel. *3.* A mold wall, made of sand and a thermosetting plastics material, used in certain casting processes. *4. See* housing, electrical

shell-and-tube heat exchanger A device for transferring heat from a hot fluid to a cooler one. In this device, one fluid passes through the inside of a bundle of parallel tubes, while the other fluid passes over the outside of the tubes, but inside the vessel shell; heat is transferred by conduction across the walls of the tubes.

sheltered area *See* area, sheltered.

shield *1.* A conductive layer of material (usually wire, foil, or tape) applied over a wire or cable to provide electrical isolation from external interference. Also known as a screen. [ARP1931] *2.* Any barrier to the passage of interference-causing electrostatic or electromagnetic fields. An electrostatic shield is formed by a conductive layer, like a foil, surrounding a cable core. *3.* An attenuating body that blocks radiation from reaching a specific location in space, or that allows only radiation of significantly reduced intensity to reach the specific location.

shielded cable One or more wires enclosed within a conductive shield to minimize the interference effects of internal or external circuits. *See also* shielded conductor. [ARP914A]

shielded conductor An insulated conductor encased in one or more conducting envelopes, usually made of woven wire mesh or metal foil. Similar products containing one or more insulated conductors are known as shielded cable, shielded conductor cable, or shielded wire.

shielded conductor cable *See* shielded conductor.

shielded metal arc welding (SMAW) Welding process in which metals are heated with an arc between a covered metal electrode and the work. Shielding is obtained from decomposition of the electrode covering. Pressure is not used and filler metal is obtained from the electrode.

shielded thermocouple assembly A thermocouple assembly in which the measuring junction is partially or completely enclosed by a housing, the primary purpose being to reduce the radiation error. [ARP485]

shielded wire *See* shielded conductor.

shield, electrical connector A device that is placed around the portion of a connector that is used for attaching wires or cables. Purpose is to shield against electromagnetic interference and/or protect the connector wires or cable from mechanical damage. [ARP914A]

shielding *1.* Surrounding an electronic circuit or signal-transmission cable with a ground plane, such as a foil or woven-metal sheath, so that capacitive coupling between the circuit and ground plane remains stable. *2.* Interposing a radiation-absorbing material between a source of ionizing radiation and personnel or equipment to reduce or eliminate radiation damage.

shielding attenuation The difference between power flowing in a path including the exterior surface of a shielded enclosure, and power flowing in a path including the interior surface of the shield and the internal conductors. [ARP1705]

shielding effectiveness *1.* For a given external source, the ratio of the electric or magnetic field strength at a point before and after the placement of the shield in question. [ARP1705] *2.* A value equal to 20 times the log to the base ten, of the ratio between the voltage impressed on a given nonshielded wire and that impressed on an identical wire with a shield, when these wires are placed in a fixed magnetic field. [ARP1931]

shielding increase The difference in shielding attenuation with a gasket in the seam of the test fixture and with metal-to-metal contact. [ARP1705]

shielding system Any set of conductors and loads surrounded by an electrically conductive shield. The system includes all conductive materials in the vicinity of the shield. [ARP1705]

shift register A data-storage location in which data is stored and moved (shifted) from one position to another in the register. This shifting is usually to perform some logic or arithmetic operation on the stored data.

shim A thin piece of material, usually metal, that is placed between two surfaces to compensate for slight variations in dimensions between two mating parts, and to bring about a proper alignment or fit.

shimmy *1.* Abnormal vibration or wobbling. *2.* A self-sustained vibration of a rolling wheel(s) about a real or theoretical swiveling axis. Sometimes classified as large-angle shimmy and small-angle shimmy. *Large-angle shimmy*—consists of more or less violent oscillations of the wheel assembly about its caster axis; the tire contact area is sliding bodily sideways at each swing and derives energy to maintain the oscillation from the forward motion of the vehicle. *Small angle or kinematic shimmy*—is a high-frequency oscillation within the adhesion range of the tire. [AIR1489]

shimmy damper *See* damper, shimmy.

shiva laser system High-energy, multi-arm, solid state (Nd doped ED-2 glass) infrared laser system used for laser-driven fusion experiments.

shock absorber *1.* A device for absorbing the energy from impact of a mass or force input. [AIR1489] *2.* Device for the dissipation of energy, used to modify the response of a mechanical system to applied shock. *3.* A component connected between a piece of equipment and its frame or support to damp out relative motion between them and reduce the effect of acceleration forces. Normally consists of a dashpot or a combination of a dashpot and a spring.

shock absorber, air-oil A shock absorber in which absorption of energy is accomplished by forcing oil through an orifice. This type of shock absorber also provides a chamber with gas (air, or more usually, dry nitrogen) under pressure for the purpose of supporting the static load on the unit. The gas pressure also causes extension of the shock absorber when the load is removed. [AIR1489]

shock absorber, constant A shock absorber in which a fixed or constant orifice is used in metering the fluid. [AIR1489]

shock absorber, dashpot *See* dashpot.

shock absorber, dual chamber A shock absorber that provides two separate chambers for gas charge and thus a two-stage spring rate as the shock absorber is compressed. [AIR1489]

shock absorber, frangible A shock absorber in which material failure is used as a shock-absorption element; for example: (a) buckling collapse of tubing; (b) fragmenting of tubing; (c) compression of honeycomb material; (d) compression of a fluid-filled device; or (e) compression of a solid thermoplastic. Such devices are "one-shot" devices and must be replaced after each use. [AIR1489]

shock absorber, friction A shock absorber in which friction is used as the energy-absorption mechanism; for example, a ring spring configuration. [AIR1489]

shock absorber, honeycomb A shock absorber in which a crushable honeycomb core material is used as the shock-absorbing element. [AIR1489]

shock absorber, leaf spring A shock absorber in which the deflection of a leaf spring is used for energy absorption capability. Common on light aircraft. [AIR1489]

shock absorber, liquid spring *See* liquid spring.

shock absorber, pneumatic A shock absorber in which air is used as the shock-absorbing medium. Unusual on present-day aircraft. [AIR1489]

shock absorber, rubber A shock absorber in which rubber is used as the shock-absorption medium. May be used with or without oil dampening. [AIR1489]

shock absorber, spring A shock absorber in which helical coil springs are used as shock-absorbing elements. May be used with or without oil damping. [AIR1489]

shock absorber, variable orifice A shock absorber in which a variable orifice meters the fluid that provides the shock-absorption medium. The resistance or load on the unit can therefore be programmed versus stroke. The most common method used for control is a metering pin of variable diameter operating through the orifice. *See* metering pin and metering tube. [AIR1489]

shock cord A strong, many-stranded, rubber cord encased in a braided-fabric sheath, used as a shock-absorber element on light planes or as a launching element for gliders. Also known as bungee cord. [AIR1489]

shock fronts Shock waves that are regarded as the forward surfaces of fluid regions having characteristics different from those of the region ahead of the wave; the front sides of shock waves.

shock motion A sudden, transient motion of large relative displacement.

shock mount A supporting structure that isolates sensitive equipment from the effects of mechanical shock or relatively high-amplitude vibrations.

shock resistance The ability to absorb mechanical shock without cracking, breaking, or excessively deforming.

shock spectra Plots of the maximum acceleration experienced by single-degree-of-freedom systems as a function of their own natural frequency in response to applied shocks.

shock stall A sudden reduction of lift on an airfoil, brought about at transonic speeds by airflow separation aft of a shock wave. [ARP4386]

shock strut A structural strut or link that embodies a shock-absorbing element. [AIR1489]

shock tubes Relatively long tubes or pipes in which very brief, high-speed gas flows are produced by the sudden release of gas at very high pressure into low-pressure portions of the tubes; the high-speed flows move into the region of low pressure behind shock waves.

shock tunnels Shock tubes used as wind tunnels.

shock wave(s) *1.* The intense compressive wave produced by the detonation of an explosive charge. [AIR913] *2.* Surfaces or sheets of discontinuity (i.e., abrupt changes in conditions) set up in a supersonic field of flow, through which the fluids undergo a finite decrease in velocity, accompanied by a marked increase in pressure, density, temperature, and entropy, as occurs, for example, in supersonic flows about bodies.

shoe *1.* In glass-fiber forming, a device for gathering filaments into a strand. [AIR4844] *2.* A renewable friction element whose contour fits that of a drum and stops it from turning when lateral pressure is applied. Also known as a brake shoe. *3.* A metal block used as a form or support during bending of tubing, wire, rod, or sheet metal.

shop replaceable assembly An item that is designated to be removed or replaced upon failure from a higher-level assembly in the shop (intermediate or depot maintenance activity), and is to be tested as a separate entity. [ARD50010]

shop replaceable unit An item that is designated to be removed or replaced upon failure from a higher-level assembly in the shop (intermediate or depot maintenance activity), and is to be tested as a separate entity. [ARD50010]

SHORAN Contracted from short range navigation; a precision electronic position-fixing system using a pulse transmitter and receiver and two transponder beacons at fixed points. [ARP4107]

shore Unit of durometer measurement. [AIR1558]

shore hardness *1.* A measure of the resistance of material to indentation by a spring-loaded indenter. The higher the number, the greater the resistance. [AIR4844] *2.* An instrument measure of the surface hardness of an insulation or jacket material. [ARP1931]

short airfield for tactical support *See* SATS.

short beam shear A flexural test of a specimen having a low test span-to-thickness ratio, such that failure is primarily in shear. [AIR4844]

short beam shear strength The interlaminar shear strength of a parallel-fiber-reinforced plastic material, as determined by three-point flexural loading of a short segment cut from a ring specimen. [AIR4844]

short circuit An abnormally low resistance between the positive and negative conductors. [AS1831]

short-circuit current(s) (fault current) *1.* The maximum current that a system can produce at the point of application of the protective device. [ARP1199A] *2.* The steady value of the input alternating currents that flow when the output direct-current terminals are short-circuited and rated line alternating voltage is applied to the line terminals.

short finish Gray appearance of a polished glass surface resulting from the polishing not being carried through to the point where all traces of the previous grinding or smoothing operation are removed. [ARP924]

short message count subfield Part of a control field that indicates the number of consecutive short messages, excluding the message frame, that have been transmitted following the claimed token subframe. [AIR4075]

shortness A form of brittleness in alloys, usually brought about by grain-boundary segregation. Shortness may be referred to as hot, red, or cold, depending on the temperature range in which it occurs.

short range clearance A clearance issued to a departing IFR flight which authorizes IFR flight to a specific fix short of the destination while air-traffic control facilities are coordinating and obtaining the complete clearance. [ARP4107]

short range navigation *See* SHORAN.

short range predictive alerting system A system that senses and identifies windshear shortly before an encounter, such that pilot action may precede the encounter. [AIR4102/11]

606

short run Failure of molten metal to completely fill the mold cavity.

short shot Injection of insufficient material to fill the mold. [AIR4844]

short term trending Tracking of engine, engine component, or sub-system operational degradation by noting data on a particular flight or ground check and comparing with a pre-established trend. [ARP1587]

short time duty A requirement of service that demands operation at a substantially constant load for a short and definitely specified time. [ARP1199A]

short time exposure Designates an exposure period of 120 min duration to any desired medium or condition. [AIR4844]

short-transverse Direction of maximum contraction of the metal during forging. [AMS6474]

shot *1.* Small spherical particles of a metal. *2.* Small, roughly spherical steel particles used in a blasting operation to remove scale from a metal surface. *3.* An explosive charge.

shot effect In an electron tube, random variation in electron emission from the cathode; or random variation in electron distribution between electrodes.

shoulder A portion of a cylindrical machine element, such as a shaft, screw, or flange, that is larger in diameter than the remainder.

shrinkage An internal cavity that has an irregular dendritic surface and is greater than 40% of the area of the shot particle area. [AMS2431/2A]

shrink fit A tight interference fit between mating parts, where the amount of interference varies almost directly with diameter. To form a shrink fit, the outer member is heated so that it expands; the parts are then assembled, allowing the outer member to cool and shrink onto the inner member. *See also* force fit.

shrink forming A process for forming metal parts in which a combination of mechanical force and shrinkage of a heated blank is used to achieve final shape.

shrink link In a landing-gear system, a link or rod element that, during retraction, provides a

kinematic constraint and shrinks or shortens a shock strut to reduce the required stowage volume within the aircraft or vehicle. [AIR1489]

shrink ring A heated ring or collar that is placed over an assembly and allowed to contract to hold the parts firmly in position.

shrink strut A strut that is shrunk or shortened before or during the retraction process to reduce the required stowage volume within the aircraft or vehicle. The shortening may be accomplished by means of a shrink link (kinematically); or hydraulically, electrically, mechanically, or by other methods. [AIR1489]

shroud *1.* A device used to limit airflow through a rotor during balancing operations. [ARP480A] *2.* A machine element used chiefly as a protective covering over other elements, especially in a rotating assembly.

shroud, insulation A part of a connector or device that provides physical protection to otherwise exposed contacts or terminals. *See also* insulation support. [ARP914A]

shroud line In a parachute, any one of a number of lines that attach the harness or load to the canopy. Also called a suspension line. [ARP4107]

shunt *1.* In an electric circuit, a low-resistance element connected in parallel with another portion of the circuit and used chiefly to carry most of the current flowing in the circuit. *See* circuit. *2.* To divert all or part of a process flow away from the main stream and into a secondary operation, holding area, or bypass.

shunt diode barrier A network consisting of a fuse or resistor provided to protect voltage-limiting shunt diodes, a current-limiting resistor, or other limiting components provided to limit current and voltage to intrinsically safe circuits.

shunt valve A valve that allows a fluid under pressure to escape into a passage that is of lower pressure or can accommodate higher flow rates than the normal passage.

shunt wound motor A brush-wound field d-c motor in which the rotor conductors and the stator field coil are connected in parallel to the d-c power source. [ARP4386]

shutdown Cessation of engine operation for any reason other than training or normal operating procedures. [ARD50010]

shut-down circuit An electronic, electrical, hydraulic or pneumatic circuit that provides controlled steps for turning off or closing down process equipment. Usually designed to automatically sequence shutdown actions and prevent equipment damage due to performing them out of sequence. May be used for normal or emergency situations.

shutdown, ground An engine shutdown that occurs at any time an aircraft is not airborne. [AIR4896]

shutdown, in-flight *See* in-flight shutdown.

shute Wires that run across the short way of the cloth as woven. Sometimes called filler or weft wires. [AIR888]

shutoff head The pressure developed by a centrifugal or axial-flow pump at its discharge when the discharge flow is zero.

shutoff valve *1.* A two-port, two-position valve that opens or closes a fluid passage. A shutoff valve may be reversible, operating equally well with pressure at either port; or nonreversible, performing satisfactorily only with pressure entering at one of its ports. [ARP4386] *2.* A device that is used to shut off or close down flow or pressure in a hydraulic or fluid system. Used in many landing-gear system applications, such as parking brakes or skid control. [AIR1489]

shuttle A back-and-forth motion in a machine, where the moving element continues to face one direction only.

shuttle derived vehicles New configuration resulting from the production and operation of the Space Shuttle.

shuttle engineering simulator Training equipment for crew members in mission operation procedures, including various approach maneuvers, braking, and final approach.

shuttle valve *1.* A three-port valve with one outlet and two inlet ports, which automatically connects the outlet port to the inlet port with the higher pressure and blocks the inlet port with the lower pressure. [ARP4386] *2.* A valve used in a brake or other fluid-power system, serving to shift from one power source to an alternate source or other similar applications. [AIR1489]

SI *See* Systeme International d'Unites.

SID Acronym for standard instrument departure (route). A preplanned instrument flight rule (IFR) air-traffic control departure procedure printed for pilot use in graphic and/or textual form. [ARP4107]

sidebands The frequency components on either side of center frequency which are generated when a carrier wave is modulated.

side brake A brake that is mounted at the side of the wheel. [AIR1489]

side-by-side actuator *1.* An actuator having two or more separate actuation elements in parallel, which provide a single output motion. [ARP4386] *2. See* parallel actuators.

side-draw product A product stream removed from the column midway between the top and bottom trays. If the side draw is above the feed tray, it is usually a vapor; if it is below the feed tray, it is usually a liquid.

side force coefficient A mathematical (numerical) measure of the physical force property that is constant in determination of side force on a tire. [AIR1489]

side friction coefficient *See* coefficient, side friction.

side milling Using a milling machine and a cutter with teeth on one or both sides to machine a vertical surface.

side rake The angle between a reference plane and the tool face of a single-point turning tool.

sidereal time Time based on the rotation of the earth relative to the vernal equinox.

side relief angle (SRF) In a cutting tool, the angle between a plane normal to the base and the flanks of the tool below the cutting edge.

side rod In a side-lever engine, one of the members linking the piston-rod crossheads to the side levers.

siderograph An instrument combining a clock and a navigation instrument which keeps a reference time equivalent to the time at 0° longitude (Greenwich meridian).

sideslip *1.* The motion of an aircraft having a component of velocity relative to the air along its pitch axis. Positive sideslip occurs when the vehicle moves to the right. [ARP4386] *2.* A slip in which the longitudinal axis of the aircraft remains parallel to the original flight path, but in which the actual flight path changes direction according to the steepness of the bank. A sideslip is used to make the aircraft move sideways through the air to counteract the drift that results from a crosswind. *See* slip. [ARP4107]

side-step maneuver A VFR maneuver accomplished by a pilot at the completion of an instrument approach to permit a straight-in landing on a parallel runway not more than 1200 ft to either side of the runway to which the instrument approach was conducted. Landing minimums for a side-step maneuver will be higher than those to the primary runway, but normally will be lower than the published circling minimums. [ARP4107]

sidetone Feature that enables a speaker to hear his/her own transmission in the headset. [ARP4107]

siemens Symbolized S. Metric unit of conductance.

sieve A meshed or perforated sheet, usually of metal, used for straining liquids, classifying particulate matter, or breaking up masses of loosely adherent or softly compacted solids.

sieve fraction The portion of a loose aggregate mass that passes through a standard sieve of given size number, but does not pass through the next finer standard sieve; usually expressed in weight percent.

sight glass A glass tube, or a glass-faced section of a process line, used for sighting liquid levels or taking manometer readings.

sighting tube A tube, usually made of a ceramic material, that is used primarily for directing the line of sight for an optical pyrometer into a hot chamber.

SIGMET Contracted from significant meteorological information. A weather advisory concerning weather significant to the safety of all aircraft. SIGMET advisories cover severe and extreme turbulence, severe icing, and widespread dust or sandstorms that reduce visibility to less than 3 miles. [ARP4107]

signal *1.* Vertical deflection from the baseline of an A-scan. [ARP5089] *2.* A measure or quantity of the medium used to communicate a condition, effect, or other desired intelligence from one point in the system to another. In an electrical system, this may be a voltage; in a hydraulic system, a pressure. A signal may be generated by a sensor, controller or selector, or other such reference device. [ARP419-57]

signal attenuation A reduction in the strength of an electrical signal.

signal common The reference point for all voltage signals in a system. Current flow into signal common is minimized to prevent IR drops, which induce inaccuracy in the signal common reference.

signal conditioning To process the form or mode of a signal so as to make it intelligible to or compatible with a given device, including such manipulation as pulse shaping, pulse clipping, digitizing, and linearizing.

signal distance The path length that a signal is required to traverse.

signal generator An instrument used in testing and calibrating other electronic instruments. The signal generator delivers an output waveform at an accurately calibrated frequency, anywhere in the audio to microwave range.

signal piping Piping that interconnects instruments, instrument devices, or bulkhead fittings.

signal simulator A hardware device that generates a signal similar in most respects to actual data from a test vehicle.

signal-to-noise ratio *1.* The ratio of the signal from the variable of interest to signals from variables that are of no interest. [ARP5089] *2.* A ratio that measures the comprehensibility of a data source or transmission link, usually expressed as the root mean square signal amplitude divided by the root mean square noise amplitude.

signal validation Refers to equipment "error" indications and program limit checks that detect faulty signals.

signature A signal or combination of data inputs that are characteristic of an individual engine, engine component, or subsystem, and that can be used to indicate functional status and condition. [ARP1587]

signature analysis For a process, identifying departures from the "reference signature" and recognizing the source of the departure. The reference signature is a particular noise spectrum or vibration spectrum that can be used to establish that a process is operating correctly.

significance In positional representation, the factor, dependent on the digit position, by which a digit is multiplied to obtain its additive contribution in the representation of a number. Synonymous with weight.

significant digit(s) *1.* Any digit that is necessary to define a value or quantity to the required level of accuracy. [AIR4844] *2.* A set of digits, usually from consecutive columns beginning with the most significant digit different from zero, and ending with the least significant digit, each of whose value is known and assumed relevant. For example, 2300.0 has five significant digits, whereas 2300 probably has two significant digits; 2301 has four significant digits; and 0.0023 has two significant digits.

significant event An event or condition that indicates a change in system status in an event-driven system; for example, a significant event is declared when an I/O operation completes. The following are considered to be significant events: I/O queuing, I/O request completion, a task request, a scheduled task execution, a mark time expiration, and a task exit. A declaration of a significant event indicates that the executive should review the eligibility of task execution since the event might unblock the execution of a higher-priority task.

significant item A collective term that includes functional significant items and structural significant items. [AIR4896]

significant meteorological information *See* SIGMET.

silica gel A form of colloidal silica that has the appearance of coarse sand and has many fine pores. Silica gel is extremely absorbent and is used as a catalytic material. [AIR4844]

silicon A basic material used to make semiconductors; has limited capacity for conductivity.

silicon-controlled rectifier (SCR) A semiconductor device used to provide stepless control of an electric power circuit without the necessity of load matching. Usually, two rectifiers are used in the circuit to provide full-wave control of the heater element, but in some instances, two SCRs are used in a single package, known as a triac.

silicone *1.* Polymeric materials in which the recurring chemical group contains silicon and oxygen atoms as links in the main chain. [ARP1931] *2.* A generic name for semiorganic polymers of certain organic radicals. Silicones can exist as fluids, resins, or elastomers, and are used in diverse materials, such as greases, rubbers, cosmetics, and adhesives.

siliconizing Producing a surface layer that is alloyed by diffusing silicon into the base metal at elevated temperature.

silicon-on-insulator semiconductors *See* SOI (semiconductors).

silky fracture *1.* A type of fracture surface appearance characterized by a fine texture, usually dull and nonreflective; typical of ductile fractures. *2. See* fibrous structure.

sill *See* crest.

silver hydrogen batteries Secondary batteries having silver and hydrogen electrodes. Silver hydrogen batteries have good energy density and cycle life.

silver solder A brazing alloy composed of silver, copper, and zinc which melts at a temperature below that of silver but above that of lead-tin solder.

similitude Something that resembles something else; for example, a process that has been scaled up from a laboratory or pilot plant operation to commercial size.

simmer Detectable leakage from a safety-relief valve at a pressure below the popping pressure.

simple balance A weighing device consisting of a bar resting on a knife edge and two pans, one suspended from each end of the bar. To determine precise weight, an unknown weight on one of the pans is approximately balanced by known weights placed in the other pan; a precise balance is obtained by sliding a very small weight along the bar until a pointer attached to the bar at the balance point indicates a null position.

simple harmonic motion A motion in which the displacement is a sinusoidal function of time.

simplex pump A reciprocating pump with only one power cylinder and one pumping cylinder.

simplified directional facility (SDF) A NAVAID used for nonprecision instrument approaches. The final approach course is similar to that of an ILS localizer except that the SDF course may be offset from the runway, generally not more than 3 deg; and the course may be wider than the localizer, resulting in a lower degree of accuracy. [ARP4107]

simplified short approach light system (SSALS) A light system that is similar to an approach light system, but the installation consists of fewer light fixtures. *See also* approach light system. [ARP4107]

simulate *1.* To artificially create behavior, environmental conditions, or operating conditions pertaining to one system by using another, different system. Usually done to accomplish testing, experimentation, or training that would be difficult or hazardous to accomplish with the real system. *2.* To represent physical phenomena by use of mathematical formulas.

simulation *1.* The construction of a working mathematical or physical model presenting similarity of properties or relationships with the natural or technological system under study. [ARP4293] *2.* The representation of certain features of the behavior of a physical or abstract system by the behavior of another system. For example, the representation of physical phenomena by means of operations performed by a computer; or the representation of operations of a computer by those of another computer. *3.* Using computers, electronic circuitry, models, or other imitative devices to gain knowledge about operations and interactions that take place in real physical systems.

simulation framework A simulation system capable of integrating and running two or more simulation algorithms in a single simulation environment.

sine bar An accurately constructed layout aid consisting essentially of a straight bar with cylindrical rests at each end. One end of the sine bar is placed on a surface plate or gage block and the other end on a stack of gage blocks equal to the sine of a desired angle to the surface plate or another reference plane.

sine galvanometer A magnetometer whose measuring element consists of a small magnet suspended in the center of a pair of Helmholtz coils. With the sine galvanometer, the magnitude of a magnetic field is determined from the position of the magnet when various known currents are passed through the coils.

sine wave A waveform in which the value of wave parameters, such as voltage and current in certain alternating-current circuits, varies directly as the sine of another variable, such as time.

sine-wave response *See* frequency response.

single acting Producing power or motion in one direction only.

single-axis tracking antenna A receiving antenna that tracks the transmitting station automatically in azimuth, but not in elevation.

single cascade action A type of control-system action whereby the input to the second of two automatic controllers is supplied by the first.

single channel actuator An actuator that contains a single torque- or force-generating device. [ARP4386]

single channel per carrier transmission Voice- and data-transmission system for satellite communication, featuring the use of a carrier frequency for each channel of communication.

single-circuit winding A widening in which the filament path makes a complete traverse of the chamber, after which the following traverse lies immediately adjacent to the previous one. [AIR4844]

single compression Describes a weave in which the shute wires are compressed tightly by the comb of the loom so that they are deformed in the machine direction; the shute wires are in contact with one another. [AIR888]

single ended actuator An actuator with a single cylinder and a piston that has only one rod extending to the atmosphere. [ARP4386]

single-ended amplifier An electronic amplifier in which each stage operates asymmetrically with respect to ground. Each stage contains one tube or amplifying transistor; or each contains two or more, connected in parallel.

single event upsets Radiation-induced errors in microelectronic circuits caused when charged particles (usually from the radiation belts or from cosmic rays) lose energy by ionizing the medium through which they pass, leaving behind a wave of electron-hole pairs.

single-fail operative (SFO) A quality whereby a control device or system can sustain any single failure and remain operative. Unless specifically stated, single-fail operative is understood to mean that no nominal loss of performance occurs after the failure. [ARP4386]

single failure point The failure of an item that would result in failure of the system, and is not compensated for by redundancy or alternative operational procedure. [AIR4896]

single frequency approach (SFA) A service provided to military, single-piloted turbojet aircraft which permits use of a single UHF frequency during approach for landing—so that pilots will not normally be required to change frequency from the beginning of the approach to touchdown. [ARP4107]

single harness reserve parachute assembly A certificated parachute assembly that is worn in conjunction with a main parachute assembly used for premeditated jumps. [AS8015]

single lap specimen In adhesive testing, a specimen made by bonding the overlapped edges of two sheets or strips of material, or by grooving a laminated assembly. [AIR4844]

single melted products The usable product of all steel melted at one time in a single furnace charge. [AMS2300G]

single-mode Describes an optical fiber that can carry only a single waveguide mode of light. Components such as connectors used with such fiber are also labeled single-mode.

single-phase meter An instrument for determining power factor in a single-phase alternating-current circuit; contains a fixed coil that carries the load current, and crossed coils connected to the load voltage. In the single-phase meter, the moving system is not restrained by a spring and therefore takes a position related

directly to the phase angle between voltage and current.

single phasing The tendency of a rotor to continue to rotate when one motor winding is opened and the other motor winding remains energized. [ARP667]

single point failure The failure of an item that would result in failure of the system, and is not compensated for by redundancy or alternative operational procedure. [ARD50010]

single point refueling Underwing refueling, commonly utilized for military aircraft in which there is only a single refueling connection. [AIR4783]

single pole A type of device, such as a switch, relay, or circuit breaker, that is capable of either opening or closing one electrical path.

single sampling A type of inspection whereby an entire lot or production run (population) is accepted or rejected based on results of inspecting a single group of items (sample) selected from the population.

single search unit A transducer or transducers contained within one element. [ARP5089]

single-sideband modulation A type of modulation whereby the spectrum of the modulated wave is translated in frequency by a specific amount, either with or without inversion.

single-speed floating control A type of control in which a reversible motor drives the final controlled device at a given speed in one direction when the temperature at the sensing element is above the set point, and reverses when the temperature is below the set point. A dead band in the controller causes the motor to remain stationary when there is no load change or other disturbance to the system. [ARP89C]

single spread Describes the application of adhesive to only one adherend of a joint. [AIR4844]

single-stage compressor A machine that raises pressure in a compressible fluid in a single pass through a single set of machine elements.

single-stage pump A machine that develops pressure to drive a relatively incompressible fluid through a system by passing the fluid through a single set of machine elements.

single-test specimen blank The volume of material from which only one test specimen can be machined. [AMS6474]

single tire applications Applications in which one tire/wheel assembly is provided to support the load carried by one landing gear strut. [AS4833]

single void combustion Combustion that occurs only in the void that is initially ignited, i.e., there is no propagation. [AIR4170]

single wheel gear configuration *See* gear configuration, single wheel.

single wheel load The share of aircraft (or vehicle) or gear load, as geometrically broken down and carried by a given wheel. [AIR1489]

sink *1.* The logical user of an information transfer. [AIR4911] 2. A reservoir into which material or energy is rejected.

sinkage The distance between the undisturbed soil surface and the bottom of the tire. [AIR1780]

sinkhead *See* feedhead.

sink mark A shallow depression or dimple on the surface of an injection-molded part due to collapsing of the surface following local internal shrinkage after the gate seals. [AIR4844]

sink speed *1.* The rate at which an aircraft loses altitude; especially the rate at which a heavier-than-air aircraft descends in a glide in still air under given conditions of equilibrium. 2. The vertical sink rate at which an aircraft (i.e., landing gear) contacts the ground. [AIR1489]

sintered filter *See* filter, sintered.

sintering Heating a powder metal compact at a temperature below the melting point to form diffusion bonds between the particles.

sintering temperature The temperature at which a given powdered compact will densify to a certain desired density, for example, 90% of the theoretical density during a certain heating period. [AIR4844]

sipes Lateral cuts in the tread of a tire. Under braking loads, the tread sections deflect, thus exposing the cut edges and increasing traction. [AIR1489]

siphon A tube, hose, or pipe for moving liquid from a higher to a lower elevation through a combination of gravity acting on liquid in the longer leg, and atmospheric pressure acting to keep the shorter leg filled.

siphoning The transfer of a liquid from a high to a lower level whereby atmospheric pressure forces the liquid up the shorter leg, while the weight of the liquid in the longer leg causes continuous downward flow.

SIS (semiconductors) Semiconductor devices consisting of an electrically insulating layer sandwiched between two semiconducting materials.

site of failure *See* locus of failure.

situational awareness Keeping track of the prioritized significant events and conditions in one's environment. [ARP4107]

size *1.* A specified value for some dimension that establishes the comparative bulk or magnitude of an object. 2. One of a set of standard dimensions used to select an object from among a group of similar objects to obtain a correct fit. *3.* In welding, the joint penetration of a groove weld, or the nugget diameter of a spot weld, or the length of the nominal legs of a fillet weld. *4.* A material such as casein, gum, starch, or wax used to treat the surface of leather, paper, or textiles.

size block *See* gage block.

size factor A factor used in tire design to compensate for variation in characteristics due to size of the tire. [AIR1489]

sizing *1.* The process of applying a size. *2.* Applying a material on a surface in order to fill pores and thus reduce the absorption of the subsequently applied coating. *3.* A term that is sometimes used incorrectly to describe a finish to improve adhesion. [AIR4844]

sizing content The percent of the total strand weight made up by the sizing; usually determined by burning off or dissolving the organic sizing. [AIR4844]

skew *1.* Having an oblique position in relation to a specific reference plane, reference direction, or physical object. 2. A tape motion characterized by an angular velocity between the gap center line and a line perpendicular to the tape center line.

skewing Describes a condition in which the warp and fill yarns are not at right angles to each other. [AIR4844]

skid *1.* A runner or slide used as an element of the landing gear of certain aircraft. *See also* wing skid and tail skid. [AIR1489] *2.* The act of sliding or slipping over a surface. [AIR1489] *3.* A term that is sometimes used to designate the portion of a tire tread that wears away in the normal service life of the tire, i.e., the skid depth. [AIR1489] *4.* On an object, a metal bar or runner that provides support or wear resistance when the object contacts a floor, runway, apron, or other flat areaway. *5.* A device placed under the wheel of a heavy, wheeled object, to prevent the wheel from turning while the object descends a steep hill.

skid control box A component of the skid control system containing the electrical and electronic components (black box). [AIR1489]

skid control system A group of interconnected components that interact to control excessive brake pressure, and thus prevent inadvertent tire skidding and contribute to shorter aircraft stopping distances. Also called antiskid system. [AIR1489]

skid control system operating environment The environment that the skid control system "feels" in performing its function. The skid-control operating environment includes the wheel-braking system, the airframe elastic structure (including struts), the wheel dynamic-loading tire elastic properties, runway friction levels available, runway roughness, and ambient and hydraulic fluid temperature. [AS483A]

skid control valve A component of the skid control system that controls or modulates brake pressure in response to an electrical (or mechanical) signal from the skid-control box. [AIR1489]

skid depth The depth dimension of the part of a tire tread that wears away in the normal service life of a tire. [AIR1489]

skid resistance The physical characteristic whereby a tire resists or opposes skidding. Skid resistance is affected by inflation pressure, tire shape, tread pattern, degree of wear, and other factors. [AIR1489]

skin *1.* Generally, a thin exterior covering. May be applied to the exterior walls of a building, the exterior covering of an airplane, a protective covering made of wood or plastics sheeting, or a thin layer on a mass of metal that differs in composition or some other attribute from the main mass of metal. *2. See* facings.

skin friction The drag or resistance caused by air against the outside of a moving aircraft or other object, especially at high speeds. [ARP4386]

skip-lot sampling A method of selecting samples for inspection by applying the principles of a continuous sampling plan to a continuing series of lots or batches, rather than to individual product lots. [AS1303]

skirt The extension of a motorcase from the tangency plane, used for interstage connections. Usually wound or laid up as an integral part of the case. [AIR4844]

skis Long, flat runners attached to the landing gear of an aircraft for takeoff, landing, and operation over snow or ice; or, in special cases, over water. [AIR1489]

skyhook balloons Large, free balloons having plastic envelopes, used especially for constant-level meteorological observations at very high altitudes.

sky waves In radio systems, radio energy that is received after having been reflected by the ionosphere.

slab A relatively thick piece of metal whose width is at least twice its thickness; generally used to describe a mill product intermediate between an ingot and a flat-rolled product, such as sheet or plate.

SLAM Acronym for scanning laser acoustic microscope.

slant-range distance The distance from an aircraft directly to an airfield, a navigational fix, or another aircraft that is at a different elevation. The slant range distance between two objects when they are at different elevations is greater than the horizontal distance. [ARP4107]

slash An elongated mark (slash, virgule) on a radar presentation screen indicating the radar beacon reply of an aircraft. [ARP4107]

slat *1.* A section of the wing leading edge that is free to move fore and aft on tracks. At high angles of attack, the local suction at the wing leading edge creates a chordwise force forward and displaces the slat from the contour of the designed wing leading edge. 2. A high-lift device designed to increase the maximum lift coefficient for low-speed flight. [ARP4107]

slave 1. A module that is selected by the bus master to participate in a message sequence. [AS4710] 2. A mechanical or electronic device that is under the control of a another device.

slave operated Pressure-operated, so as to position in a manner equivalent to rigid mechanical inter-lock, by means of a master or control unit. [ARP4386]

sleek A scratch having boundaries that appear polished. [ARP924]

sleeve 1. A hollow cylindrical member used to line a housing to impart different metallurgical properties to the rubbing surface than those inherent in the housing. 2. In a slide valve, the hollow cylindrical member that directly affects the flow pattern by means of its relative position to an internal slide or spool. [ARP4386]

sleeve bearing A cylindrical machine element that fits around a shaft and supports the shaft while it turns.

sleeve coupling A hollow cylinder that fits over the ends of two adjacent shafts or pipes to hold them together.

slenderness ratio 1. The unsupported effective length of a uniform column, divided by the least radius of gyration of the cross-sectional area. [AIR4844] 2. A geometric factor expressing the ratio of the length of a vehicle to its diameter. Sometimes referred to as fineness ratio. [ARP4386]

slewing 1. As applied to a gyro, the rotation of the spin axis caused by applying torque about the axis of rotation. 2. In radar, changing the scale on the display.

slew rate 1. The steady-state output velocity of the actuation system in response to a step input command. Slew rate is usually determined from the average velocity between 10 and 90% of load position (so as to avoid actuator acceleration and deceleration effects) when both static and dynamic loads are imposed at the servoactuator output. [ARP4386] 2. The maximum rate of change of an output signal from a device. 3. In a device or circuit, the limitation in the rate of change of output voltage, usually imposed by some basic circuit considerations.

slide(s) 1. In a slide valve, the moving member that directly affects the flow pattern, usually cylindrical and internal to a sleeve. *See also* spool. [ARP4386] 2. Any mechanism that moves with predominantly sliding motion. 3. The main reciprocating member of a mechanical press, which moves up and down in the press frame and carries the punch or upper die. 4. A flat-bottomed chute.

slide type damper *See* damper.

sliding-block linkage A mechanism for converting rotary motion into linear motion, or vice versa. Consists of a crank, a block that slides back and forth in a slot or on ways, and a link bar attached to the crank and block with pin joints.

sliding fit A type of clearance fit used to accurately locate parts that must assemble together without perceptible play (close-sliding fit); or used to allow assembled parts to move or turn easily, but not run freely (sliding fit). *See also* running fit.

sliding gear A gear set whose speed can be changed by sliding gears along their axes to put them in or out of mesh with other gears of different sizes.

sliding-vane rotary flowmeter A type of positive-displacement flowmeter in which radial vanes slide in or out to trap and release discrete volumes of the metered fluid, as a rotor containing the vanes revolves about a central cam surface that controls vane position.

slime 1. A soft, viscous, or semisolid surface layer, often resulting from corrosion or bacterial action, and often having a foul appearance or odor. *See also* sludge. 2. A mudlike deposit in the bottom of a chemical process or electroplating tank.

sling An item made of strap, chain, rope, webbing, or the like, specifically designed to hold securely something to be hoisted, lowered, carried, rotated, or suspended.

sling psychrometer A device for determining relative humidity. Consists of a wet-and-dry bulb thermometer mounted in a frame that can be whirled about, usually by means of a handle and short piece of chain or wire rope attached to the upper end of the frame.

slip *1.* The relative colinear displacement of the adherends on either side of the adhesive layer in the direction of the applied load. [AIR4844] *2.* A descent wherein one wing is lowered and the longitudinal axis of the aircraft is at an angle to the flight path. *See also* forward slip and sideslip. [ARP4107] *3.* A term commonly used to express leakage in positive-displacement flowmeters.

slip angle The angle at which a tensioned fiber will slide off a filament-wound dome. [AIR4844]

slip clutch A clutch that may be incorporated into a starter to limit the maximum torque (impact and steady) that can be transmitted by the starter to the accessory drive train or engine. [ARP906A]

slip flow Rarefied gas flow in the region between Knudsen numbers 0.01 and 0.1.

slip joint *1.* A telescoping joint between two parts in an assembly. *2.* A mechanical union that allows limited axial movement of one member, such as a pipe or duct, with respect to a mating member. *3.* In civil engineering, a type of contraction joint consisting of a tongue and groove. This type of joint allows independent movement between two members, such as wall sections, slabs, or precast structural units. *4.* A type of scarf joint used in flexible-bag molding in which plastics veneers are laid up so that their beveled edges overlap.

slippage *1.* Fluid leakage along the clearance between a reciprocating-pump piston and its bore. Also known as slippage loss. *2.* Movement that unintentionally displaces two solid surfaces in contact with each other. *3.* Movement of a gas phase through or past a gas-liquid interface, rather than movement that drives the interface forward. Especially applicable to certain phenomena in petroleum engineering.

slippage loss *See* slippage.

slipping turns A flight maneuver that consists of a turn in which the aircraft is allowed to slip. [ARP4107]

slip plane In ductile metals, a crystallographic plane along which dislocations move under local shear stresses to produce permanent plastic strain.

slip ratio *See* aircraft, slip ratio.

slip seal A seal between members designed to permit movement of either member by slipping or sliding.

slipstream The flow of air pushed back by a revolving propeller or rotor. *See also* wake turbulence. [ARP4107]

slit *1.* A long, narrow opening, often used for directing and shaping streams of radiation, fluids, or suspended particulates. *2.* To cut sheet metal, rubber, plastics, or fabric into sheet or strip stock of precise width using rotary cutters, knives, or shears.

sliver A number of staple or continuous-filament fibers aligned in a continuous strand without twist. [AIR4844]

slope control Electronically producing specific changes in a parameter with time, especially as applied to a method of varying welding current.

slot *1.* An air gap extending spanwise along an airfoil or between segments of an airfoil to allow additional airflow to pass across the upper surface of the airfoil. [ARP4386] *2.* Any of certain apertures in an airfoil to improve aerodynamic behavior. *3.* Any elongated opening in a machine part or structural member. *4.* A special socket in a PC designed to accept an additional circuit board.

slotted flap A trailing edge flap in which the hinge line is arranged so as to generate a slot between the flap and the wing when the flap is deflected. Air from under the wing flows through the slot to re-energize the airflow over the upper surface of the flap, thus increasing the lift. [ARP4386]

slotted nut A hexagonal nut with slots cut across the flats of the hexagon so that a lockwire or cotter pin can be used to prevent the nut from turning.

slotted tongue terminal A terminal with a bifurcated tongue that allows attachment to a screw or stud without removal of the mounting hardware. [ARP914A]

slow charge An overnight return of energy to a battery at 0.05 to 0.1 C rates. [ARP4386]

slow (time) code A spread-out time code, modulated in amplitude and width, suitable for display on a relatively low-speed chart recorder.

slow neutron A free (uncombined) neutron having a kinetic energy of about 100 eV or less. *See* also thermal neutron.

slub An abruptly thickened place in a yarn. [AIR4844]

sludge *1.* A soft water-formed sedimentary deposit that normally can be removed by blowing down. *2.* Fine sediment, such as may be found in the bottom of an oil crankcase or boiler drum. *See also* slime.

slug *1.* The fundamental unit of mass in the US customary system of units. A slug weighs 32.17 pounds. *2.* A large dose of chemical treatment applied internally and intermittently to a steam boiler. Also used sometimes instead of priming to denote a discharge of water out through a boiler steam outlet in relatively large intermittent amounts. *3.* A small, simply shaped piece of metal used as starting stock for forging, upsetting, or extrusion. *4.* The offal resulting from piercing a hole in sheet metal. *5.* Liquid that completely fills the internal passage of a tube for a short distance.

slugging Producing a substandard weld joint by adding a separate piece of material that is not completely fused into the joint.

sluice A waterway that is fitted with a vertical sliding gate for controlling the flow of water.

slump test A quality-control test for determining the consistency of concrete. The amount of slump is expressed as the decrease in height that occurs when a conical mold filled with wet concrete is inverted over a flat plate and then removed, leaving the concrete behind.

slurry *1.* A suspension of fine solids in a liquid which can be pumped or can flow freely in a channel. *2.* A semiliquid refractory material used to repair furnace linings. *3.* An emulsion of soluble oil and water, used as a cutting fluid in certain machining operations.

slurry preforming Method of preparing reinforced plastic preforms by wet-processing techniques similar to those used in the pulp-molding industry. [AIR4844]

slush drag The drag or resistance to forward motion of a tire(s) and the aircraft, generated as the aircraft moves through melting snow, slush, and water. [AIR1489]

small aircraft Aircraft of 12,500 lb or less, maximum certified takeoff weight. [ARP4107]

small scale integration Low density of integrated circuits per unit area.

smart actuator An electrohydraulic servo-actuator containing the feedback loop closure and some or all of the redundancy management electronics in a line replacement unit. [ARP4386]

smart pump A variable-displacement, pressure-compensated hydraulic pump, in which the displacement and the pressure and power setting may be controlled remotely; and in which transient cavitation protection, reduced starting torque, built-in test and health monitoring, and other optional features may be provided. [ARP4386]

smash In fabric, a place where a number of warp or filling yarns have been broken. [AIR4844]

SMAW *See* shielded metal arc welding.

smoke *1.* Small, gas-borne solid particles, including but not limited to black carbonaceous material from the burning of fuel, which in sufficient concentration create visible opacity. [ARP1179A] *2.* A dispersion of fine solid or liquid particles in a gas.

smoke point In a standard test, the maximum flame height that kerosene or jet fuel will burn at without smoking.

SN Stands for smoke number, a dimensionless term quantifying smoke emission. SN increases with smoke density and is rated on a scale from 0 to 100. SN is evaluated for a sample size of 0.0162 grams of exhaust gas per sq millimeter (0.0230 lb per sq in.) of filter area. [ARP1179A]

snap gage A device with two flat, parallel surfaces that are precisely spaced apart, used for checking one limit of tolerance on a diameter or length dimension. Sometimes, go-no-go tolerance limits are built into a single snap-gage frame to permit checking both high and low limits of tolerance at the same time.

snap ring A type of retaining fastener in the shape of the letter C, which is expanded across its diameter and allowed to snap back into a groove to hold parts in position, especially to keep them from sliding axially along a shaft.

snap roll A maneuver whereby an aircraft, by a quick movement of the controls, is made to complete a full rotation about its longitudinal axis, while maintaining an approximately level line of flight. [ARP4107]

snapshot dump A selective dynamic dump performed at various points in a machine run.

S-N diagram In fatigue testing, a plot of stress against the number of cycles to failure. [AIR4844]

sneak circuit analysis A procedure conducted to identify latent paths that cause occurrence of unwanted function or inhibit desired functions, assuming all components are functioning properly. [ARD50010]

snowflake *See* flake.

snubber A device that provides a resisting force to snub or damp an imposed load; for example, an arresting hook snubber, which resists the upward force imposed by the carrier deck or ground, prevents hook bounce, and maintains point contact with the deck. [AIR1489]

snubber valve Within a snubber, a device that provides resistance to fluid motion and thereby snubs or damps the force imposed. [AIR1489]

soaked sleeve A firesleeve that is so saturated with the fluid conveyed by the hose assembly to which it is mounted, that sweating or permeation of the fluid through the firesleeve results. A soaked sleeve is usually indicative of a damaged or faulty hose assembly or spillage of fluid from a nearby port. [ARP1658A]

soaking time The sum of the required time for the temperature of parts to be within the required range and the lag time. [AMS2771A]

soak test A test in which sealants are soaked in fluids as specified in specifications to determine resistance to these fluids. The conditions of the test are somewhat harsher than are found in actual use. [AS7200/1]

soap bubble solution A noncorrosive soap solution used in the detection of leaks. [AIR4069]

soap bubble test A leak test that is performed by applying a soap solution to the external surface or joints of a system under internal pressure, then observing the location, if any, at which bubbles form, indicating the existence of a gas leak.

soar To fly without propulsive power, as in a glider. This type of flying is called dynamic soaring, unless it is done on ascending air currents; then it is called up-current soaring. [ARP4107]

socket weld An external weld joining a plain-ended male portion and the corresponding socket; for example, a male valve inlet in a process line or vessel socket.

sodar Sound detection and ranging.

softening The act of reducing scale-forming calcium and magnesium impurities in water.

softening agent A substance, often an organic chemical, that is added to another substance to soften it.

softening range The range of temperatures over which a plastic changes from a rigid to a soft state. [AIR4844]

soft hammer A hammer whose head is made of annealed copper, leather, or plastic, thereby preventing it from damaging finished surfaces.

soft landing The act of landing on the surface of a planet or natural satellite without damage to any portion of the vehicle or payload, except possibly the landing gear.

software A set of programs, procedures, rules, and possibly associated documentation concerned with the operation of a computer system, for example, compilers, library routines, manuals, and circuit diagrams. Contrast with hardware.

software error failure A failure caused by an error in the computer program associated with the hardware. [ARD50010]

software, operational All software that is resident in an on-board system. [AIR4896]

software priority interrupt The programmed implementation of priority interrupt functions. *See* priority interrupt.

software, support All software that is used in the development, verification, validation, and modification of the operational software or related hardware. [AIR4896]

SOI (semiconductors) Semiconductor devices consisting of a silicon layer coupled to an electrically insulating layer.

solar activity Any type of variation in the appearance of energy output of the sun.

solar backscatter UV spectrometer A spaceborne spectrometer that measures solar-UV spectral irradiance incident on the earth and backscattered radiance from the earth, and thereby estimates the total atmospheric ozone content of the atmosphere and the attitude distribution of ozone.

solar blankets Large, high-temperature, low-mass solar arrays consisting of ultrathin silicon solar cells, interconnected, welded, and bonded to flexible substances.

solar blind A detector that contains filters to block sunlight, making it essentially "blind" to the sun. In most cases, this involves blocking wavelengths longer than approximately 300 nm, simulating the absorptive effects of upper-atmosphere ozone.

solar cells *1.* Cells constructed from semiconductor materials, materials whose resistivities are between those of metals (which conduct electricity well and have resistivities of less than 10^{-3} ohm-cm) and insulators (whose resistivities are in excess of 10^{12} ohm-cm). Common semiconductors are germanium, silicon, and cadmium; thus, some of the common solar cell materials are silicon and cadmium sulfide. [AIR744A] *2.* Photovoltaic cells that convert sunlight into electrical energy.

solar constant The rate at which solar radiation is received outside the earth's atmosphere on a surface normal to the incident radiation, and at the earth's mean distance from the sun.

solar cosmic rays Cosmic rays supposedly originating in the sun.

solar diameter Observable dimension of the sun.

solar dynamic power systems Electric power systems in which a solar-heated working fluid is used to drive a turboalternator. Primary applications are for space stations and spacecraft.

solar eclipses Obscurations of the light of the sun by the moon.

solar engine A device for converting thermal energy from the sun into electrical or mechanical energy; or for using thermal energy from the sun to run a refrigeration system.

solar faculae *See* faculae.

solarimeter *See* pyranometer.

solar neutrinos Neutral particles originating from nuclear reactions in the core of the sun.

solar oscillations Irregular oscillations in the solar atmosphere.

solar parallax The angle at the sun subtended by the equatorial diameter of the earth.

solar planetary interactions The interactions and subsequent effects caused by the interactions of solar activity and/or wind with a planet, its magnetic field, its atmosphere, or natural satellites.

solar plasma (radiation) *See* solar wind.

solar power Describes any of several methods whereby energy from the sun is used to perform useful work.

solar prominences Filamentlike protuberances from the chromosphere of the sun.

solar radiation The total electromagnetic radiation emitted by the sun.

solar radio bursts Sudden increases in the flux from the sun at radio frequencies.

solar radio emission Radiation at radio frequencies originating from the sun or its corona.

solar radio waves *See* solar radio emission.

solar simulators Devices that produce thermal energy that is equivalent in intensity and spectral distribution to that from the sun. Solar simulators are used in testing materials and space vehicles.

solar thermal propulsion Proposed energy source for spacecraft propulsion, created by passing hydrogen through a heat exchanger placed at the focal point of a large, parabolic-dish, solar-concentrator mirror.

solar wind Streams of plasma flowing approximately radially outward from the sun.

solder A joining alloy with a melting point below about 450°F, such as certain lead-base or tin-base alloys.

solder cup A solder-retaining cavity on the terminating end of solder contacts or terminals. [ARP914A]

solder eye A contact or terminal that has a hole at its terminating end, through which a conductor can be inserted prior to being soldered. [ARP914A]

solder glass A special glass that softens below about 900°F, and that is used to join two pieces of higher-melting glass without deforming them.

soldering A joining process in which an alloy, melting below 800°F, is used to form an electrical joint without alloying the base metals. [ARP1931]

soldering embrittlement Penetration by molten solder along grain boundaries of a metal with resulting loss of mechanical properties.

solderless wrap A technique whereby a solid conductor is mechanically winded to a terminal post having a series of edges. Also called wire wrap. [ARP914A]

solder short *See* bridging.

solder sleeve A heat-shrinkable tubing device containing a predetermined amount of solder and flux, used for environmental-resistant solder connections and shield termination. [ARP914A]

solenoid An item consisting of one or more coils surrounding an iron core, in which the coil(s) and the core are movable in relation to one another. The axial or rotary movement is a result of the magnetic flux of the coil. A solenoid is designed to convert electrical energy into mechanical energy. Does not include switch contacts. [ARP480A]

solenoid valve *1.* An electrically operated valve capable of two or more discrete operating positions, but not capable of proportional fluid control, except through pulse-width modulation techniques. [ARP4386] *2.* A shutoff valve whose position is determined by whether electric current is flowing through a coil surrounding a moving-iron valve stem. The valve may be normally open, in which case gas or liquid flows through it when electricity to the coil is turned off; or it may be normally closed, in which case gas or liquid flows only when electricity is turned on; or it may be three-way, in which gas or liquid flows in one path through the valve when electricity is off and in a different path when electricity is on.

soleplate A flat member in the frame of a machine on which a bearing can be mounted and, if necessary, adjusted slightly.

solettas Orbiting solar mirrors (reflectors).

solid coupling A device that is used to rigidly connect two shafts together, and is usually capable of transmitting full torque from one shaft to the other.

solid cryogen cooling Cooling with solidified cryogenic fluids.

solid damping *See* damping, structural.

solid die A one-piece tool with internal threads, used for cutting screw threads on rod stock or small-diameter pipe.

solid electrolytes Single crystals, certain alloys, alkaline metals, and other compact compounds used in galvanic cells (batteries).

solidification The change in state from liquid to solid in a material as its temperature passes through its melting temperature or melting range on cooling.

solid laminate A structurally reinforced, resin-impregnated composite cured to a solid state, containing no sandwich layers of honeycomb, plastic foam, or other material. [AIR4844]

solid propellant combustion The burning of solid propellants by rapid oxidation and production of expanding gases, heat, and light.

solid propellant gas generator A generator consisting of three components: (a) a self-sustaining combustible solid mixture, called the propellant; (b) an initiator to start the reaction upon command; and (c) a chamber that has the dual function of storing the above components and expelling the combustion gases at a rate determined by the propellant, its geometry and temperature, and the output or load orifice size. [AIR744A]

solid propellant rocket engines Rocket engines fueled with solid propellants. Such rocket engines consist essentially of a combustion chamber containing the propellant, and a nozzle for the exhaust jet—although they often contain other components, such as grids or liners.

solid propellants Specifically, a rocket propellant in the solid form, usually containing both fuel and oxidizer combined or mixed, and formed into a monolithic (not powdered or granulated) grain.

solids content In an adhesive, the percentage by weight of nonvolatile matter. [AIR4844]

solid state Describes an electronic device or circuit whose operation is controlled by some combination of electrical, magnetic, and optical phenomena within a circuit element consisting largely of a single piece of solid material, usually a crystalline semiconductor.

solid state devices Devices that utilize the electric, magnetic, and photic properties of the solid materials, e.g., binary magnetic cores or transistors.

solid state switch A nonmechanical switching device that changes output states upon electrical current/voltage or magnetic input. [AIR4258]

solid-state welding Describes any welding process that produces a permanent bond without exceeding the melting point of the base materials and without using a filler metal.

solo flight time Flight time during which a pilot is the only occupant of the aircraft. [ARP4107]

solubility coefficient The percentage of water or other fluid absorbed by a material at saturation at a given temperature. [AIR4844]

solubility parameter Symbolized by delta, a measure of the energy required to separate the molecules of a liquid. [AIR4844]

solubility, S The air solubility in volume percent, measured at 32°F and one atmosphere pressure, that will dissolve in a petroleum liquid when the air in equilibrium with a liquid is at a partial pressure of 760 mm of Hg. In the basic V/L formula, S is expressed in units of volume of air, measured at 32°F and one atmosphere pressure, dissolved in 100 volumes of fuel at 60°F. [AIR1326]

soluble oil An oil-based fluid that can form a stable emulsion or colloidal suspension with water. Used principally as a cutting fluid or coolant.

solute The dissolved material. [AIR4844]

solution A liquid, such as boiler water, containing dissolved substances.

solution heat treating Heating to a sufficiently high temperature, followed by water quench-

ing, to hold one or more elements in solid solution for subsequent precipitation hardening. [AMS2728]

solution heat treatment Heating an alloy into a temperature range at which the principal alloying element(s) become dissolved in a single solid phase, then cooling the material rapidly enough to prevent precipitation of secondary phases.

solvation The process of swelling, gelling, or dissolving of a material by a solvent; for resins, the solvent can be plasticized.

solvent A substance used for dissolving and/or cleaning materials during reinforced plastics operations. *See also* thinner. [AIR4844]

solvent activated adhesive A dry-film adhesive that is rendered tacky by the application of a solvent just prior to use. [AIR4844]

solvent adhesive An adhesive having a volatile organic liquid as a vehicle. [AIR4844]

solvent retention Refers to the occurrence of solvent residues in chemical or material end products or intermediates.

solvent welding/solvent cementing Process used for the mass production of strong and reliable joints for thermoplastic materials. [AIR4844]

somatogravic illusion *See* illusions, vestibular.

somatogyral illusion *See* illusions, vestibular.

sonar A method or system, analogous to radar used under water, in which high-frequency sound waves are emitted so as to be reflected back from objects and used to detect the objects of interest.

sonic barrier Popular term for the large increase in drag encountered when the speed of an aircraft or missile approaches the speed of sound in air; the speed at which this occurs is somewhat indefinite, and depends on altitude and general atmospheric conditions.

sonic booms Noises created by shock waves that emanate from aircraft or other objects traveling at or above sonic velocity.

sonic fatigue stresses The fluctuating stresses induced in structure and fastener assemblies by oscillating sound waves that may be produced by pulse, ram or turbo jet, and rocket motors. [ARP700]

sonic opacity A characteristic of a medium, such as a medium containing a large quantity of particles or small bubbles, whereby sound or ultrasound is reflected randomly from the discontinuities, rather than being transmitted through the medium.

sonic speed The speed of sound in the specific medium of concern. *See also* acoustic velocity.

sonic thermocouple assembly A thermocouple assembly designed so that the gas can be made to flow over the measuring junction at Mach 1, resulting in maximum heat transfer to the junction. [ARP485]

sonic vibration Refers to the dynamics of mechanical vibration waveforms in the frequency range up to 20kHz. [ARP1587]

sonobuoy A combination of a passive sonar set and a radio transmitter, mounted in a buoy that can be dropped by parachute from an aircraft. Underwater sounds, such as those from a submarine, are picked up by the sonar set and transmitted by radio to a receiver on the aircraft or a ship. The source of underwater sound can be determined by triangulation using several sonobuoys dropped in a known pattern, and comparing time-delay data from their signals by computer analysis.

sonograph An instrument for recording sound or seismic vibration patterns.

sorbates Gas taken up by sorbents.

sorbents Materials that take up gas by sorption.

sorption The taking up of gas by absorption, adsorption, chemisorption, or any combination of these processes.

sortie *1.* An operational flight by one aircraft. [ARD50010] *2.* Specifically refers to flight on a combat mission, but also commonly used to identify any flight, from takeoff to landing, that is part of a larger mission which may contain several flights. A flight from takeoff to landing that remains in the traffic pattern is not considered a sortie. [ARP4107]

sound A pressure wave in an elastic or compressible material which exists in the form of alterations in pressure, stress, particle displacement or particle velocity; usually restricted to waves whose frequency is in the range of human hearing.

sound analyzer A device for measuring the band pressure level at various sound frequencies.

sound detection and ranging *See* sodar.

sound fields Regions containing sound waves.

sound fixing and ranging A method for acoustically tracking submerged bodies or floats utilizing fixed hydrophones.

sound generators *See* acoustic generator.

sounding Any penetration of the natural environment for scientific observation, usually by sounding rockets or balloons.

sounding rockets Rockets designed primarily for routine upper-air observation (as opposed to research) in the lower 250,000 feet of the atmosphere, especially that portion inaccessible to balloons, i.e., above 100,000 feet.

sound intensity At a point, the average rate of sound energy transmitted in a specified direction through a unit area normal to this direction at the point considered.

sound-level meter An electronic instrument for measuring noise or sound levels in either decibels or volume units.

sound-powered telephone A type of telephone usually used for emergency communications over short distances. Electric current for transmitting the signal is generated by the speaker's voice in a specially designed microphone, and no external source of power is required.

sound pressure At a point, the total instantaneous pressure at that point in the presence of a sound wave, minus the static pressure at that point.

sound pressure level (SPL) The intensity of a sound wave; in decibels, this is equal to $20 \log (P_s/P_r)$, where P_s is the pressure produced by the sound and P_r is a stated reference pressure.

sound reproduction Any process for detecting sound at one location and time and regenerating the same sound at the same location and time, and/or at another location and time, with any desired intensity.

sound velocity *See* acoustic velocity.

sound waves Mechanical disturbances advancing with finite velocity through an elastic medium and consisting of longitudinal displacements of the medium, i.e., consisting of compressional and rarefactional displacements parallel to the direction of advance of the disturbance. Sound waves are small-amplitude adiabatic oscillations.

source *1.* The origin of radiation, e.g., an x-ray tube or radioisotope. [ARP5089] *2.* The logical producer of an information transfer, whether data, control, or status. [AIR4911]

source code *1.* The program instructions written in high-level languages or assembly languages. The program must be translated into object code before it can be executed by the computer. *2.* Software generated by a programmer in assembly language, generally with comments, headings, and other annotation.

source-film distance The distance between the focal spot of an x-ray tube or radiation source and the film; generally expressed in inches. [ARP5089]

source impedance *See* output impedance.

source voltage The voltage available at the generator terminals. [AS1831]

southern sky That portion of the celestial sphere between the celestial equator and the celestial south pole, and generally visible from areas in the earth's southern hemisphere.

space *1.* The region above the earth's atmosphere, defined in various contexts as beginning at from 1000,000 to 250,000 feet altitude. *2.* In data processing, a unit of storage that is empty.

space based radar Radar systems installed on large space structures.

spaceborne experiments A collective term designating the various experiments performed or planned in orbiting spacecraft and usually involving physical phenomena in space environments.

space capsules Containers used for carrying out experiments in space.

space charge *1.* The electric charge carried by a cloud or stream of electrons or ions in a vacuum or a region of low gas pressure, when the charge is sufficient to produce local changes in the potential distribution. *2.* The net electric charge within a given volume.

space commercialization Refers to for-profit activities in space or prefatory to space activity.

spacecraft Devices, manned and unmanned, that are designed to be placed into an orbit about the earth or into a trajectory to another celestial body.

spacecraft availability The percent of on-orbit time that a spacecraft is capable of performing the specified missions. [AIR4896]

spacecraft charging Electric charge induction on the surface of a spacecraft by magnetospheric plasmas or other ion sources.

spacecraft defense All measures designed to destroy attacking enemy vehicles, including missiles, while in space; or measures designed to nullify or reduce effectiveness of such attack.

spacecraft docking *See* docking.

spacecraft survivability The ability of a spacecraft to survive adverse conditions, including reentry problems.

space observations (from earth) Surveillance of extraterrestrial phenomena from the earth's surface.

space operation center A proposed NASA space station to be assembled in space, designed for conducting space-based operations, such as satellite servicing, orbit transfer vehicle launch and recovery, and assembly of large space structures. Onboard capabilities could include space manufacturing and research experiments. When fully assembled, the space operation center will be larger in size than the Space Shuttle.

space perception The ability to estimate depth or distance between points in the field of vision.

space plasmas Concentrations of free electrons and protons in the ionosphere, plasmasphere, and beyond.

space platforms Gimbal-mounted platforms equipped with gyros and accelerometers for maintaining a desired orientation in inertial space, independent of spacecraft motion.

space polar coordinates *See* polar coordinates.

space processing Forming and fabrication techniques aboard a spacecraft in a weightless or low-gravity environment, and involving improved chemical and/or physical procedures for the creation of new or better products.

space radiation *See* extraterrestrial radiation.

space segment availability The percent of time a space segment is operational

Space Shuttle ascent stage Shuttle takeoff configuration comprising the orbiter, solid rocket boosters, and external tank.

Space Shuttle main engine Liquid-propellant propulsion system in which fuel drawn from external tanks is used to provide power for the orbiter to attain orbital speed.

space simulator Device used to simulate one or more parameters of the space environment, used for testing space systems or components. Specifically, a closed chamber capable of approximating the vacuum and normal environments of space.

space suits Pressure suits for wear in space or at very low ambient pressures within the atmosphere, designed to permit the wearer to leave the protection of a pressurized cabin.

space technology The systematic application of science and engineering to the exploration and exploitation of outer space.

spacetennas The transmitting antennas of a solar-power satellite transmission system, which directs the high-power beam from space to a focus on the rectennas on earth.

space transportation system flights Revised collective designation for all Space Shuttle flights.

spall To detach material from a surface in the form of thin chips whose major dimensions are in a plane approximately parallel to the surface.

span *1.* A structural dimension measured in a straight line between two specific extremities, such as the ends of a beam or two columnar supports. *2.* The dimension of the wings of an aircraft, from tip to tip, measured in a straight line. *3.* The difference between maximum and minimum calibrated measurement values; for example, an instrument with a calibrated range of 20–120 has a span of 100.

span adjustment *See* adjustment, span.

span drift The time-related change in response of the analyzer in repetition of a span gas measurement under identical conditions of flow and concentration. [ARP1256]

span error The difference between actual span and ideal span, usually expressed as a percent of ideal span.

span gas A calibration gas to be used for routine verification and adjustment of analyzer response. [ARP1256]

spanloader aircraft Advanced distributed-load cargo aircraft configurations in which the payloads are distributed across the span of the wing, to provide a close match between aerodynamic and inertial loading, for minimal bending stresses.

spanner An attachment for a sextant that establishes an artificial horizon.

spar A spanwise structural member in a wing or other lifting surface.

spare, preferred An item that is recommended as the current preferred spare for replenishment purposes. [AIR4896]

spare, ready An item that has been assembled to the point that it can be attached to an aircraft without further buildup or assembly other than that necessary to fit it to its unique position. [AIR4896]

spares Reparable components or assemblies used for maintenance replacement purposes in major end items of equipment. [ARD50010]

spares float The quantity of individual items held for the purpose of providing replacements for those removed from aircraft for overhaul, repair, or rectification. [AIR4896]

spares, provisional Those spare parts that are procured under certain special procedures at a certain point in the system acquisition cycle. [ARP4386]

spares range The total number of different items individually identified by part number or description. This excludes any reference to quantities of individual items. [AIR4896]

spark arrester *1.* A device that reduces or prevents electric sparks at a point where a power circuit is opened or closed, such as at a circuit breaker or knife switch. *2.* A device that prevents airborne embers from escaping from a chimney.

spark duration The length of time, usually expressed in microseconds, required to dissipate the total energy of any one spark discharge occurring between the electrodes of a spark igniter. [AIR784]

spark energy The energy (joules) released between electrodes of a spark igniter. [AIR784]

spark igniter configuration A low-tension-type of spark igniter in which a material classified as a semi-conductor is used to bridge the gap from the center electrode to the outer or ground shell. [ARP846]

sparking voltage For a particular shunted surface gap spark igniter design, the lowest voltage applied to the igniter that will initiate a spark. [ARP846]

spark rate The number of spark discharges per unit time occurring at a spark igniter under a given set of conditions.

spark recorder A type of recorder whereby sparks passing between a metal pointer and an electrically grounded plate periodically burn small holes in recording paper as it moves slowly across the face of the plate. Sparks are produced at regular intervals by a circuit powering an induction coil, and the varying lateral position of the moving pointer creates the trace.

spark test A test used to locate pinholes in wire or cable insulation by the application of an electrical potential across the insulation while it is passing through an ionized field. [ARP1931]

spatial Relating to space; concerning geometric position and three-dimensional volume.

spatial coherence The coherence of light over an area of the wavefront of a beam; where the beam hits the surface.

spatial disorientation (Type I) Unrecognized incorrect orientation in space. May result from an illusion, an anomaly of attention, or an anomaly of motivation, but is not accompanied by discomfort or confusion because it is not noticed. Also referred to as spatial misorientation. [ARP4107]

spatial disorientation (Type II) Recognized incorrect orientation in space typified by a discrepancy between sensory information and cognitive expectancy. The illusory sensory source may be visual, kinesthetic, or vestibular, and the effect of the cognitive conflict may range from mild discomfort or confusion to incapacitation. [ARP4107]

spatial marching Techniques for solving partial differential equations that move along in a space direction.

spatial misorientation *See* spatial disorientation (Type I).

spatial resolution The precision with which an optical instrument can produce separable images of different points on an object.

spatial unorientation Lack of knowledge as to orientation in space due to the inability to detect orienting cues, as in a rapidly spinning or tumbling aircraft. In this case, the lack of orientation is recognized, but there are neither usable orienting cues nor a cognitive expectancy of true orientation. [ARP4107]

spatter *1.* In evaporated coating processes, a term used to denote the condition resulting when large particles of coating material condense on the glass surface, and adhere there. [ARP924] *2.* Particles of molten metal expelled during a welding operation, adhering to an adjacent surface.

special application Denotes removers and developers that are qualified for use only with a specific penetrant. [AMS2644]

special inspection *1.* Any inspection resulting from failure indication, overstress or severe weather flight occurrence, fleet data, similar failures, etc. [AIR4896] *2. See* inspection, aircraft.

specially orthotropic ply An orthotropic ply in which the loads are in the direction of the principal plane of elastic symmetry. [AIR4844]

special observation Refers to a category of aviation weather observations taken to report significant changes in one or more of the observed elements since the last preceding record or special observation. [ARP4107]

special purpose test equipment Equipment that can only be used to test a specific prime equipment. [AIR4896]

special sortie A sortie that is singularly urgent or is a measure of capability. [ARP4107]

special sortie, actual A sortie that is objectively urgent or is a measure of capability, such as combat, medical evacuation, weather evacuation, or search and rescue. [ARP4107]

special sortie, perceived A sortie that is subjectively perceived to be urgent, or is a measure of capability, such as a checkride, an operational readiness inspection (ORI), or higher headquarters exercise. [ARP4107]

special test equipment (STE) Equipment developed for the principal purpose of maintaining quality assurance of end items development and production. Some STE may be used for depot repair. [ARD50010]

special use airspace Airspace of defined dimensions, identified by an area on the surface of the earth, to which activities must be confined because of their nature, and/or wherein limitations may be imposed on aircraft operations that are not part of those activities. Special use airspace includes such areas as military operating areas, prohibited areas, and restricted areas. [ARP4107]

special VFR conditions Weather conditions in a control zone that are less than the basic VFR weather minima, but in which some aircraft are permitted flight under visual flight rules. [ARP4107]

special VFR operations Aircraft operating in accordance with clearances within control zones in weather conditions less than the basic VFR weather minima. Such operations must be requested by the pilot and approved by ATC. [ARP4107]

specific acoustic impedance The complex ratio of sound pressure to particle velocity at a given point within the medium.

specific acoustic reactance The imaginary component of specific acoustic impedance.

specific acoustic resistance The real component of specific acoustic impedance.

specific address *See* absolute address.

specific adhesion Adhesion between surfaces that are held together by valence forces of the same type as those that give rise to cohesion. [AIR4844]

specification *1.* A collection of requirements that, when taken together, constitute the criteria that define the functions and attributes of a system or an item. [ARP4754] *2.* A detailed, precise description of a weapon system, its hardware, software, geometry, or other design parameters. [ARP4107] *3.* A list of requirements that must be met when making a material, part, component, or assembly; installing it in a system; or testing its attributes or functions.

specification range The concentration range of a constituent in a solution required by the applicable process specification or work instruction. [ARP4992]

specific code *See* absolute code.

specific fuel consumption The fuel flow rate, usually lb/h, required per unit of power or thrust, usually horsepower or lb of thrust, produced by an engine. For rockets, *see* specific impulse.

specific gravity (sp gr) The density of any material divided by that of water at a standard temperature. [AIR4844]

specific gravity bottle A small flask used to determine density. The precise weight of the bottle is determined when empty, when filled with a reference liquid such as water, and when filled with a liquid of unknown density. Also known as a density bottle or relative density bottle.

specific gravity, gas The density of a gas compared to the density of air.

specific gravity, liquid The density of a liquid compared to the density of water.

specific heat (sp ht) *1.* The quantity of heat, expressed in Btu, required to raise the temperature of 1 lb of a substance 1°F. *2.* The ratio of the thermal capacity of a substance to that of water. The specific heat at constant pressure of a gas is designated c_p. The specific heat at constant volume of a gas is designated c_v. The ratio of the two (c_p/c_v), is called the ratio of specific heats, k.

specific heat ratio The ratio of the specific heat of a material at constant pressure and the specific heat at constant volume. *See* specific heat. [AIR1116]

specific humidity The weight of water vapor in a gas-water vapor mixture per unit weight of dry gas.

specific impulse In rocket engines, the fuel flow rate, usually in lb/s, required per unit of thrust produced, usually lb. Specific impulse is thus expressed in units of seconds.

specific properties Refers to material properties divided by the material density. [AIR4844]

specimen A piece or portion of a sample or other material taken to be tested. In the case of adhesives, a specimen is made up to an ASTM or other standard using metal or composite adherends bonded together in order to test the adhesive and not the adherends. [AIR4844]

specimen blank The volume of material that encompasses one or more single-test specimen blanks. [AMS6474]

speckle holography An imaging technique whereby a speckle pattern results when a diffusely reflecting surface is illuminated by a laser and interference occurs between the fields passing through the various portions of the lens aperture. Information about the motion of an object can be obtained from the imaged fringes resulting from the translation of two speckle patterns.

speckle interferometry An imaging process whereby the interference between two mutually coherent, but randomly speckled, fields of two lens-formed images from laser illuminated, diffusely reflecting surfaces results in a pattern on the image plane of an interferometer.

spectral absorption *See* absorption spectra.

spectral analysis *1.* A frequency decomposition of the analog input signals. *2.* Identification of the frequency spectrum.

spectral density In PCM-coded data, the amount of a signal level at each frequency or portion of the spectrum.

spectral emissivity The ratio, at a specified wavelength, of thermal radiation emitted from a non-blackbody, to that emitted from a blackbody at the same temperature.

spectral lines *See* line spectra.

spectral response *See* spectral sensitivity.

spectral sensitivity *1.* In electronics, radiant sensitivity considered as a function of wavelength. *2.* In physics, the response of a device or material to monochromatic light as a function of wavelength. Also known as spectral response.

spectral shift control A type of reactor moderator control in which the neutron spectrum is intentionally changed.

spectrofluorometer An instrument for determining chemical concentration by fluorometric analysis using two monochromators: one to analyze the wavelength of strongest emission, and the other to select the wavelength of best excitation in the sample.

spectroheliograph Instrument for taking photographs (spectroheliograms) of the image of the sun in monochromatic light. The wavelength of light chosen for this purpose corresponds to one of the Fraunhofer lines, usually the light of hydrogen or ionized calcium.

spectrometer A spectroscope that includes an angular scale for measurement of the angular deviation and wavelengths of the components of the spectrum.

spectrophotometric titration Instrumented titration in which the end point is determined by measuring a change in absorbed radiation with a spectrophotometer.

spectrophotovoltaics The enhancement of solar-cell productivity by concentrating and subdividing the sunlight spectrum and focusing on specific spectrum-efficient solar cells.

spectroradiometer An instrument that measures power as a function of wavelength.

spectroreflectometer A device that measures the reflectance of a surface as a function of wavelength.

spectroscope A device that spreads out the spectrum for analysis. The simplest type is a prism or diffraction grating, which spreads out the spectrum on a piece of paper or ground glass.

spectroscopic analysis Identification of chemical elements by characteristic emission and absorption of light rays.

spectrum An array of the frequency component amplitudes of a signal, arranged in order of frequency.

spectrum analyzer *1.* An instrument for measuring the distribution of energy among the frequencies emitted by a pulse magnetron. *2.* An electronic instrument for analyzing the output, amplitude, and frequency of audio- or radio-frequency generators or amplifiers under normal or abnormal operating conditions.

spectrum display unit *1.* An adjunct to a radio receiver that displays the radio spectrum in and on each side of the carrier being received. *2.* On a telemetry receiver, a device that displays the spectrum at and on both sides of the frequency to which the receiver is tuned.

specular reflection Reflection in which the reflected radiation is not diffused; reflection as from a mirror.

specular transmission density The value of photographic density obtained when only the normal component of transmitted flux is measured for source illumination, the rays of which are perpendicular to the plane of the film.

speed *See* velocity.

speed brakes Control surfaces that are deployed to cause an aerodynamic drag on the aircraft and hence reduce its speed. *See also* ground spoilers and drag brake. [ARP4386]

speed climb/speed descent Vertical modes that maintain an assigned speed (or Mach), usually with the elevator or stabilizer; the throttle setting may be fixed or variable and may be applied independently. Such modes usually can be flown independent of waypoints. [ARP1570]

speed, critical engine failure For a multiengine aircraft, that speed at which, during takeoff, the aircraft can fail the critical engine and: (a) elect to continue takeoff and just clear the ground at the end of the runway; or (b) elect to abort the takeoff and just stop the aircraft within the limit of the remaining runway. [AIR1489]

speed, dynamic hydroplaning The speed at which dynamic hydroplaning of a tire or complete air vehicle occurs. [AIR1489]

speed, engine idle The minimum normal operating speed of the engine. [ARP906A]

speed, engine lite-off The rotor speed at which combustion is initiated. [ARP906A]

speed, engine self-sustaining The engine rotor speed from which the engine is capable of accelerating to idle speed, within specified engine limits, without the assistance of a starter. [ARP906A]

speed, free running *See* no load speed.

speed, landing gear operational limit The maximum airspeed of the aircraft at which the landing gear is designed to retract and/or extend. [AIR1489]

speed, landing gear structural design The maximum airspeed of the aircraft at which the landing gear may be exposed to the airstream without danger of structural damage. [AIR1489]

speed management–time control (4D) The functions that provide guidance for adjusting speed and/or flight path so as to conform with air-traffic flow control arrival time requirements at a designated downstream waypoint or waypoints. The requirements would normally also include a specified altitude and an arrival speed. An assigned speed at the final (or designated) waypoints is also usually implied. (4D stands for four-dimensional, i.e., latitude, longitude, altitude, and time.) Such command signals may be directed to the auto flight system as well as pilot displays. [ARP1570]

speed, no load *See* no load speed.

speed, overrunning The speed at which the engine side of an engaging mechanism overruns the starter side. [ARP906A]

speed, pad Engine rotor speed as related to the accessory pad. [ARP906A]

speed rating The maximum ground speed to which a tire has been tested in accordance with the specification. [AS4833]

speed reducer A gear train for transmitting power from a motor to the machinery it drives, at a rotational speed less than that of the motor.

speed, rotor The speed of the engine rotor or starter turbine. [ARP906A]

speed, runaway In electrical starter-generator systems, the speed to which the starter-generator will go without load at maximum applied voltage. [ARP906A]

speed, starter cut-off The speed up to which the starter provides starting assistance, and at which the power source of the starter is switched off. [ARP906A]

spent fuels Nuclear reactor fuels irradiated to the extent that they no longer can effectively sustain a chain reaction.

spent liquor The liquid effluent from the pulping stage of papermaking. Consists of wood chemicals, such as lignin, and partly reacted digestion chemicals (caustic, sulfite or sulfate, depending on which pulping process was used).

sp gr *See* specific gravity.

spherical aberration A lens defect that makes rays from the peripheral part of the lens focus at a different point than do rays from the central portion of the lens; this produces an image lacking in contrast.

spherical bearings An integral unit in which a spherically shaped (ball) inner race is free to rotate and misalign within the confines of the outer race. [AIR1594]

spherical candlepower (mean) For a lamp, the average candlepower of the lamp in all directions in space; equal to the total luminous flux of the lamp in lumens divided by 4π. [ARP798]

spherical coordinates A system of curvilinear coordinates in which the position of a point in space is designated by its distance from the origin or pole (the radius vector); the angle phi between the radius vector and a vertically directed polar axis (the cone angle or coaltitude); and the angle theta between the plane of the phi and a fixed meridian plane through the polar axis (the polar angle or longitude).

spherical plasmas Confined circular plasmas.

spherical wave A wave whose equiphase surfaces form a series of concentric spheres.

spheroids Ellipsoids; figures resembling spheres.

Spheromaks Toroidal fusion reactors.

spherometer A device for measuring the spherical curvature of a surface.

spherulitic-graphite cast iron *See* ductile iron.

sp ht *See* specific heat.

spicules Bright spikes extending into the chromosphere of the sun from below.

SPIFR Acronym for single pilot IFR. Generally used to describe a general aviation aircraft operated in instrument flight conditions with only one pilot at the controls. [ARP4107]

spikes Variations from the surge level or from the controlled steady-state level of a characteristic that reaches its greatest amplitude in an extremely short time. Results from very high-frequency currents of complex waveform produced when loads are switched. Transients so generated usually consist of a train of spikes. [AS1212]

spin A maneuver, either deliberate or inadvertent, of a stalled airplane whereby the airplane descends in a helical path at an angle of attack greater than the angle of maximum lift. The nose of the aircraft in a spin is usually, though not necessarily, pointed sharply downwards. In a normal spin, the longitudinal axis of the aircraft inclines downward at an angle greater than 45 deg. *See also* flat spin and inverted spin. [ARP4107]

spindle An element of a landing gear that provides an axis for larger revolving parts. [AIR1489]

spin down *1*. The process of angular deceleration of the wheels of an aircraft after takeoff. *2*. The decrease in angular velocity of a wheel (increase in slip ratio) as measured by the skid-control system. [AIR1489]

spin glass A magnetic alloy in which the concentration of magnetic atoms is such that below a certain temperature their magnetic moments are no longer able to fluctuate thermally in time, but are still directed at random—in loose analogy to the atoms of ordinary glass.

spinning *1*. Production of plastics filament by extrusion through a spinneret. *2*. Forming sheet metal into rotationally symmetrical shapes such as bowls or cones by pressing a round-ended tool against the flat stock and forcing it to conform to the shape of a rotating mandrel.

spinning solid upper stage Space shuttle upper stage designed for launching of satellites not requiring the full capacity of the interim upper stage; does not require inertial guidance system or three-axis stabilization.

spin rocket A small rocket that imparts spin to the airframe of a missile. [ARP4386]

spin stabilization As applied to a spacecraft, directional stability obtained by the action of gyroscopic forces which result from spinning the body about its axis of symmetry.

spin up *1*. The process of accelerating the wheel of an aircraft from zero to high angular velocity in a very small increment of time at touchdown and contact with the runway, thereby inducing high drag loads in the landing gear structure. *2*. Increase in angular velocity of a wheel (decrease in slip ratio) as measured by the skid-control system. [AIR1489]

spiral *1*. A maneuver or performance, especially of an airplane, in which the craft ascends or descends in a helical (corkscrew) path. A spiral is distinguished from a spin in that the angle of attack is within the normal range of flight angles. 2. The flight path of an aircraft in a spiral. [ARP4107]

spiral bevel gear A bevel gear with curved, oblique teeth, which provide for gradual engagement and bring more teeth into contact with one another at any given time than would occur with an equivalent straight bevel gear.

spiral conveyor *See* screw conveyor.

spiral divergence mode A spiral mode corresponding to unstable lateral motion. The flight of an aircraft in this mode consists of a truly banked turn of every-decreasing radius. [ARP4386]

spiral flow test A test for determining the flow characteristics of thermoplastic resins by measuring the length and weight of resin flowing along a spiral cavity.

spiral gear A helical gear that transmits power from a driving shaft to a nonparallel driven shaft.

spiral welded pipe Pipe made by forming steel plate into long helical strips, fitting the strips together, and welding the spiral seams.

spiral wrap A helical wrap of material over a core. Also called a serve. [ARP1931]

SPL *See* sound pressure level.

splash An intermediate tool made using a fiber-filled synthetic plaster material. [AIR4844]

splashing Cutting into or slicing of core. [AIR4844]

splash lubrication A method of lubricating a piston engine whereby the connecting-rod bearings are dipped into troughs filled with oil, splashing it onto other engine parts.

splay A fanlike surface defect near the gate on a part. [AIR4844]

splice *1*. To connect two pieces, forming a single longer piece, as in connecting the ends of wire, rope, or tubing. In a splice, the connection may be made by any of several methods, including weaving and welding, and may be made with or without a connector. 2. A permanent junction between two optical fiber ends. This junction can be a mechanical splice, formed by gluing or otherwise attaching the ends together mechanically; or a fusion splice, formed by melting the ends together.

splice housing A housing designed to protect a splice in an optical fiber from damage, such as from the application of stress on the fiber.

splice plate A piece of flat-rolled stock used to connect the webs or flanges of two girders together.

spline *1*. A design method to lock or key two rotating members together. Splines may be square, involuted, or crown. [ARP4386] 2. One of a set of axial keyways or gearlike ridges on the end of a shaft or the interior of a hub. In use, the splined shaft fits into a mating splined hub to transmit rotational power and motion, while permitting limited axial play between the two members.

splined coupling A coupling device consisting of a female splined tube attached to a shaft, coupled through a male spline member that is attached to a second shaft. [ARP4386]

splintering A combination of cracking and delamination of the outer skin. [AIR4844]

split-beam colorimeter An instrument for determining the difference in radiation absorption by a sample at two wavelengths in the visible or ultraviolet region.

split-beam ultraviolet analyzer An instrument for monitoring the concentration of a specific chemical substance in a process stream or coating by measuring the amount of ultraviolet light absorbed at one wavelength and comparing it to the amount absorbed at a reference wavelength that is only weakly absorbed by the sample. Also known as a dual beam analyzer.

split charge rate A charging method in which a battery is charged at a high rate, then the charge rate is automatically reduced, the battery being charged at a lower charge rate as it approaches full charge. [ARP4386]

split core Core cell walls that are ruptured or split. [AIR4844]

split flap A kind of flap in which the upper surface of the wing trailing edge is fixed and only the lower surface deflects downward. [ARP4386]

split S A flight maneuver consisting of a half snap roll followed by a pullout, accomplishing a 180 deg change in direction accompanied by a loss of altitude. [ARP4107]

splitter Plates that are spaced in an elbow of a duct and are so disposed as to guide the flow of fluid through the elbow with uniform distribution, minimizing pressure drop.

splitter vanes In a gas conduit, a set of curved, parallel strips of metal placed along the flow direction to guide gas flow around a sharp bend in the conduit.

split winding A winding divided into two equal parts, allowing it to be externally connected in series or in parallel. [ARP667]

spoiler *1.* An aerodynamic surface used to disturb (spoil) the usually smooth airflow across a vehicle surface. [ARP4386] *2.* A plate, series of plates, comb, tube, bar, or other device that projects into the airstream about a body to break up or spoil the smoothness of the flow, especially such a device that projects from the upper surface of an airfoil, giving an increased drag and a decreased lift. Spoilers are normally movable and consist of two basic types: the flap spoiler, which is hinged along one edge and lies flush with the airfoil or body when not in use; and the retractable spoiler, which retracts edgewise into the body. [ARP4107]

spoilers, ground *See* ground spoilers.

spoke A bar, rod, or wire connecting the hub of a wheel to its rim.

spokeshave A small tool for planing concave or convex surfaces.

sponge metal Any metal mass produced by decomposition or chemical reduction of a compound at a temperature below the melting temperature of the metal.

sponging In sealants, a phenomenon that is sometimes observed after fuel soak and temperature cycling. The sealant swells and a cross-section reveals many voids of various sizes. Sponging can also be caused by other factors, but the occurrence is rare. [AIR4069]

spontaneous combustion Ignition of combustible materials following slow oxidation without the application of high temperature from an external source.

spontaneous ignition temperature A measure of the flammability characteristics of a fluid. [AIR1116]

spool *1.* The internal member of a cylindrical slide valve which directly affects the flow pattern through its relative position to a surrounding sleeve. [ARP4386] *2.* The drum of a hoist. *3.* The movable member of a slide-type hydraulic valve. *4.* A reel or drum for winding up thread or wire. *5.* A relatively short transition member for making a welded connection between two lengths of pipe. Also known as a spool piece.

spooldown The process of rotor deceleration from part or full power following a blowout or fuel cutoff. [ARP906A]

spooldown start An engine airstart initiated during rotor spooldown, prior to reaching equilibrium windmilling speed. [ARP906A]

spooling The technique by which output to low-speed devices is placed into queues on faster devices to await transmission to the slower devices.

spool piece *See* spool.

spot The small luminescent area of the screen surface instantaneously excited by the impact of the electron beam. [ARP1782]

spot check A type of random inspection in which only a very small percentage of total production is checked to verify that a process remains within its control limits.

spot drilling Drilling a small, shallow hole in a surface to act as a centering guide in a subsequent machining operation.

spot facing Producing a flat, machined surface concentric with a drilled hole to serve as a seat for a washer or bolthead, or to allow for flush mounting of mating parts.

spot welding A form of resistance welding whereby a weld nugget is produced along the interface between two pieces of metal, usually sheet metal, by passing electric current across the joint; the joint is clamped between two small-diameter electrodes or between an electrode and an anvil or plate.

spray A mechanically produced dispersion of liquid drops in a gas stream; the larger the drops, the greater must be the gas velocity to keep the drops from separating out by gravity.

spray angle The angle included between the sides of the cone formed by liquid fuel discharged from mechanical, rotary atomizers, and by some forms of steam or air atomizers.

sprayed metal molds Molds made by spraying molten metal onto a master until a shell of predetermined thickness is achieved. [AIR4844]

sprayer plate A metal plate used to atomize the fuel in the atomizer of an oil burner.

spray nozzle A nozzle from which a liquid fuel is discharged in the form of a spray.

spray painting A process in which compressed air atomizes paint and carries the resulting spray to the surface to be painted.

spray tower A duct through which liquid particles descend countercurrent to a column of gas. A fine spray is used when the object is to concentrate the liquid; a coarse spray when the object is to clean the gas by entrainment of the solid particles in the liquid droplets.

spray-up Describes a technique in which a spray gun is used as an application tool. [AIR4844]

spread The quantity of adhesive per unit joint area applied to an adherend, usually expressed in pounds of adhesive per thousand square feet of joint area. [AIR4844]

spreader A device used to expand or open a part to facilitate assembly or disassembly of the part with respect to its final position. *See also* expander. [ARP480A]

spread or diffusing Describes surfaces (and media) that break up incident light and distribute it as though the surface were incandescent, uniformly bright in all directions or approximately so. Examples are rough plaster, white glass, white plastic. [ARP798]

spread reflection Reflection of electromagnetic radiation from a rough surface with large irregularities.

spread spectrum transmission Communications technique whereby many different signal waveforms are transmitted in a wide band; power is spread thinly over the band so that narrow-band radios can operate within the band without interference.

spring A machine element whose chief purpose is to store mechanical energy or to induce mechanical force through elastic deformation of the material of which it is made. The element may be shaped in the form of a plate, leaf, flat-wound helix, coil, or washer; it may be made of almost any relatively hard metal or alloy; and it may be stressed in tension, compression, bending, or torsion. In most spring designs, the amount of deflection is directly proportional to applied load; if the load is released, the element returns to its normal, unstressed shape or position.

springback *1.* In landing gear systems, a structural reaction that follows spin up. The energy stored in the gear structure by spin up causes a cyclic reaction to produce forward drag forces in the system. [AIR1489] *2.* Movement of a part in the direction of recovering original size or shape upon release of elastic stress. *3.* In flash, upset, or pressure welding, the amount of deflection in the welding machine due to the upsetting pressure.

spring clip *1.* A U-shaped fastener, used to attach a leaf spring to an axle. *2.* A fastener that grips a part by elastic force; used chiefly in electrical connections.

spring constant The number of pounds required to compress a spring or specimen 25 mm in a prescribed test procedure. [AIR4844]

spring coupling A flexible coupling with resilient parts.

spring hook A hook-shaped device with a spring-loaded member spanning the gap to form an eye. The spring-loaded member allows a bight of rope or cable to be quickly inserted into the eye and prevents the rope from slipping off the hook unless the member is deliberately depressed toward the center of the eye.

spring load A load that varies proportionally with load position. Such a load may be caused by aerodynamic forces on a movable surface, or by thrust deflection reaction forces on a movable nozzle element; and it can be either opposing or aiding, depending on output position and motion direction. [ARP4386]

spring rate The force required to deflect a member a specified amount. [AIR1489]

spring steel Carbon or low-alloy steel that is cold-worked or heat treated to give it the high yield strength normally required in springs; if it is a heat-treatable composition, the springs may be formed prior to heat treatment (hardening).

spring temper A level of hardness and strength for nonferrous alloys and some ferrous alloys corresponding approximately to a cold-worked state, two-thirds of the way from full hard to extra spring temper.

sprocket A tooth on the periphery of a wheel or spool for engaging the links of a chain, or the perforations in computer paper, or motion picture film, or some other similar device, so that the chain or paper or film can be driven without slippage and/or will traverse the wheel without lateral movement.

sprocket chain A flat chain, usually with pinned links, that meshes with the teeth of a sprocket for transmitting motion and mechanical power from one sprocket (the driving sprocket) to another (the driven sprocket).

sprocket hole Any of a series of perforations along the edge of motion picture film, paper tape, computer paper or continuous stationery which engage the teeth of a sprocket wheel or spool so that the material can be driven through a mechanical device, such as a camera, projector, printer, or recording instrument.

sprue A single hole through which molding compounds are injected directly into the mold cavity. [AIR4844]

sprung arch An arch in the form of a segment of a circle, supported by skew blocks at the two ends.

sprung mass The portion of a landing gear that is attached to fixed air-vehicle structure, as opposed to the unsprung mass, (tires, wheels, brakes, axle, etc.), separated from the air vehicle by means of the shock absorber. [AIR1489]

spun roving A heavy, low-cost glass or aramid fiber strand consisting of filaments that are continuous but doubled back on themselves. [AIR4844]

spur gear A toothed wheel whose teeth run parallel to the axis of the hub.

spurious error Error due to instrument malfunction or to human "goof-ups."

sputtering Dislocation of surface atoms of a material as a result of bombardment with high-energy atomic particles.

sputter-ion pump *See* getter-ion pump.

squama A scale or structure resembling a scale.

square mesh In wire cloth or textile fabric, a weave in which the number of wires or threads per inch is the same both with the weave and in the cross-weave direction.

square mil The area of a square, one mil by one mil. [ARP1931]

square thread A machine thread with a square cross section, in which the widths of land and groove are each equal to one-half the pitch.

square wave(s) *1.* Oscillations, the amplitudes of which show periodic discontinuities between two values, remaining constant between jumps. *2.* In radar, pulses initiated by a rapid rise to peak power, maintained at a constant peak power over the finite pulse length, and terminated by rapid decrease from peak power. *3.* A wave in which the dependent variable assumes one fixed value for one-half of the wave period, then assumes a second fixed value for the other half, with negligible time of transition between the two fixed values at each transition point.

square wells The impurity potential areas that bound an electron or hole in semiconducting crystals such as silicon.

squash dimension *See* apparent bond width.

squat switch An electrical switch that is activated by virtue of presence of a load on the landing gear. This signal from such a switch provides intelligence to many aircraft systems as to whether the aircraft is airborne. [AIR1489]

squawk (mode, code, function) Request to a flight-crew person to activate specific modes/codes/functions on the aircraft transponder; for example, "Squawk three/alpha, two one zero five, low." [ARP4107]

squeal *See* brake squeal.

squeezed states (quantum theory) Single-mode minimum uncertainty states for which the fluctuations in one quadrature phase of the field are smaller than would occur for a coherent state.

squeeze films Thin viscoelastic fluid films squeezed between two usually planar structures to serve as sealants, load dampers, lubricants, etc.

squeeze-out time *1.* Refers to a test performed on faying surface sealants at standard conditions. The squeeze-out time is essentially the time from the start of mixing of a two-part kit (or the thawing of a frozen tube of sealant) to the tightening of two surfaces with sealant between them. *2.* The period of time that still allows partially cured sealant to be squeezed out to a preselected thickness between two surfaces. [AS7200/1]

squeeze roll One of two opposing rollers designed to exert pressure on a material passing between them.

squeeze time In resistance welding, the time from initial application of pressure until welding current begins to flow.

squib *1.* Generally, any of various small-size pyrotechnic or explosive devices containing deflagrating materials. [AIR913] *2.* Various small explosive devices. *3.* Explosive device used in the ignition of a rocket.

squid (detectors) Superconducting quantum interference device magnetometers.

squirrel cage motor An induction motor in which the rotor consists of inserted or cast parallel conductors, electrically shorted at each end. [ARP4386]

squirter system An ultrasonic scan system, usually a through-transmission system, in which sound is coupled to the part through water jets. [ARP5089]

SRF *See* side relief angle.

SSALS *See* simplified short approach light system.

SSALSR Acronym for simplified short approach light system with runway alignment indicator lights.

SST *See* synchronous system trap.

stabilator Slab horizontal tail used as single primary control.

stability *1.* In meteorology, a state in which the vertical distribution of temperature is such that a parcel of air will resist displacement from its initial level. [ARP4107] *2.* In aerodynamics, the inherent flight characteristics of an aircraft tending to restore it to its original condition when disturbed by an unbalancing force or moment. *See also* stability, inherent; and sta-

bility, aerodynamic. [ARP4107] *3.* The ability of an explosive material to retain its original properties without degradation when exposed to various environmental conditions over a period of time. [AIR913] *4.* The relative ability of a substance to retain its mechanical, physical, and chemical properties during service. *5.* The relative ability of a chemical to resist decomposition during storage.

stability, aerodynamic The stability of a body with respect to aerodynamic forces. [ARP4107]

stability assessment The procedure by which destabilizing effects are accounted for in an engine. [ARP1420]

stability augmentation Maintenance of aircraft stability in flight by means of automatic control devices that supplement a pilot's manipulation of the aircraft controls. The automatic controls are used to modify inherent aircraft handling problems.

stability augmentation system (SAS) A function of the flight control system, including sensors and actuators, performing in such a manner so as to augment the basic dynamic stability of the aircraft. When considered as an entity, the SAS is essentially a closed-loop regulator control system. The SAS generally has limited authority; to prevent undesirable coupling between the SAS signals and the pilot inputs, SAS signals are normally introduced by a series servo that does not cause stick motion or forces. [ARP4386]

stability, directional *See* directional stability.

stability, dynamic *See* dynamic stability.

stability, inherent Stability of an aircraft due solely to the disposition and arrangement of its fixed parts; that is, the characteristic that causes an aircraft, when displaced, to return to its normal attitude of flight without the use of controls or the interposition of any mechanical devices. [ARP4107]

stability, lateral The characteristic of an aircraft whereby it remains stable or regains stability when caused to roll or sideslip. [ARP4107]

stability, longitudinal The characteristic of an aircraft whereby it rights itself or retains stability with respect to vertical displacement of its nose and tail about the center of lift (i.e., pitching motion). [ARP4107]

stability matrix The eigenvalues of a system about an operating point. [AIR1823]

stability, negative dynamic The characteristic of an aircraft whereby the amplitude of an oscillatory motion increases with time. [ARP4107]

stability, negative static The characteristic of an aircraft whereby, when disturbed from equilibrium, it continues to change attitude in the direction of disturbance. Also called static instability. [ARP4107]

stability, neutral static The characteristic of an aircraft whereby, when disturbed from equilibrium, it neither continues in the direction of displacement nor returns to its original attitude. A neutrally stable aircraft, if once disturbed from a state of steady flight, will not return to its original flight attitude but may seek any new flight attitude and state of steady flight. Dynamically, such an airplane is neither stable nor unstable. [ARP4107]

stability of a linear system The characteristic of a linear system whereby, having been displaced from its steady state by an external disturbance, it comes back to that steady state when the disturbance has ceased.

stability, positive dynamic The characteristic of an aircraft whereby the amplitude of an oscillatory motion decreases with time. [ARP4107]

stability, positive static The characteristic of an aircraft whereby, when disturbed, it returns to its previous attitude of equilibrium. [ARP4107]

stability, static *See* static stability.

stabilization In carbon-fiber forming, the process used to render the carbon-fiber precursor infusible prior to carbonization. [AIR4844]

stabilized core Honeycomb cores in which the cells have been filled with a specified reinforcing material for the purpose of supporting the cell walls during machining. [AIR4844]

stabilized honeycomb compressive strength The compressive strength of honeycomb material for which the plane surface of the test

specimen has been stabilized, with either a plastic resin or by the attachment of facings. [AIR4844]

stabilizer(s) *1.* Structural devices used to reduce the lateral deflection of vehicles, applied within the envelope of the vehicle. [ARP1328] *2.* An airfoil or combination of airfoils, considered a single unit, whose principal function is to maintain stable flight of an aircraft or missile. *3.* Any chemical added to a formulation for the chief purpose of maintaining mechanical or chemical stability throughout the useful life of the substance.

stabilizing treatment Any of various treatments—mechanical or thermal—intended to promote dimensional or microstructural stability in a metal or alloy.

stable element Any device, such as a gyroscope, used to maintain a stable spatial position of devices such as instrumentation or ordnance mounted in a ship or aircraft.

stacking A lamination sequence in which the warp surface of one ply is laid against the fill surface of the preceding ply. [AIR4844]

stacking sequence A description of a laminate that details the ply orientations and their sequence in the laminate. [AIR4844]

stadimeter Instrument for determining the distance to an object of known dimension by measuring the angle subtended at the observer by the object; graduated directly in distance.

stage *1.* A hydraulic amplifier used in a servovalve. Servovalves may be single-stage, two-stage, three-stage, etc. [ARP490] *2.* In electronics, that portion of a circuit between the control tap of one tube or transistor and the control tap of another.

staggered-intermittent fillet welding Welding a T-joint on both sides of the tee in such a manner that the weld bead is segmented, with the segments on either side being opposite gaps between segments on the opposing side.

staging Heating a premixed resin system, such as in a prepreg, until the chemical reaction starts, but stopping the reaction before the gel point is reached. [AIR4844]

stagnation *1.* Describes a thermocouple assembly that is designed to measure a temperature that approximates the total temperature value. [ARP485] 2. The condition of being free from movement or lacking circulation.

stagnation point In a field of flow about a body, a point at which the fluid particles have zero velocity with respect to the body.

stagnation pressure A theoretical pressure that could be developed if a flowing fluid could be brought to rest without loss of energy (isentropically).

stagnation temperature 1. The temperature created when a gas at high velocity is brought to rest. [ARP4386] 2. The temperature that would be attained if all of the kinetic energy of a moving stream of fluid were converted to heat.

stainless alloy Any member of a large and complex group of alloys containing iron, at least 5% chromium, and often other alloying elements, and whose principal characteristic is resistance to atmospheric corrosion or rusting. Also known as stainless steel.

stainless steel *See* stainless alloy.

stair length The distance, measured horizontally and parallel to the tread surfaces, between the face of the nose of the lowermost tread to the face of the nose of the uppermost tread. This is a geometric dimension only and does not include allowance for approach at the lower end, intermediate platforms in excess of the normal tread length, or lengths of the upper platforms. [ARP836A]

stair stepping Refers to discrete steps occurring along edges of a line or symbol caused by quantization phenomena. [ARP4256]

stall 1. A condition in which the airflow separates from the airfoil surface. The result of a stall is a loss or severe reduction in the force or lift effect of the airfoil. [ARP4107] 2. A flow breakdown at one or more compressor blades. [ARP1420]

stalling speed *See* V_s.

stall load The steady-state force or torque from the actuator at zero velocity, when the controller has a saturated input; the load that the servoactuator cannot overpower. [ARP4386]

stall speed Speed at which an aircraft will stall under given flight conditions. [ARP4107]

stall torque *See* torque, stall.

standard air Dry air weighing 0.075 lb per cu ft at sea level (29.92 in. barometric pressure) and 70°F.

standard atmosphere *See* atmosphere, standard. [ARP171]

standard atmospheric pressure 1. Mean atmospheric pressure related to sea level. [ARP4386] 2. A standard unit of atmospheric pressure; the pressure exerted by a 760 mm column of mercury at standard gravity (980.665 cm/s or 9.8066 cm/s^2), at temperature 0° Centigrade. 3. 760 mm of mercury, or 29.9213 in of mercury, or 1013.25 millibars. [ARP4107]

standard cell A reference cell for electromotive force.

standard components In the EASY computer program, subroutines that define dynamic performance relations of ECS components and controls, dynamic performance relations of general basic controls, and miscellaneous analytical functions. [AIR1823]

standard conditions Temperature of 59°F and pressure of 14.7 psia. [ARP906A]

standard control A measuring or controlling device used to provide assurance that allowable process variation has not been exceeded. [AS7200/1]

standard deviation 1. A measure of the spread or dispersion of the distribution of a random variable, mathematically expressed as the positive square root of the second moment about the mean. [ARP4386] 2. The square root of the average squares of deviations from an arithmetic mean. [ARP1192]

standard error of estimate The residual scatter of points about a curve fit; calculated by statistical methods from the analysis of data. [AIR4979]

standard fit Any fit between mating parts whose allowance and tolerance have been standardized.

standard flue gas Gas weighing 0.078 lb per cu ft at sea level (barometric pressure of 29.92 in. of mercury) and 70°F.

standard gage A highly accurate gage used only as a reference standard for checking or calibrating working gages.

standard instrument arrival route *See* STAR.

standard instrument departure *See* SID.

standardization *1.* The act or process of reducing something to, or comparing it with, a standard; a measure of uniformity. *2.* A special case of calibration whereby a known input is applied to a device or system for the purpose of verifying the output or adjusting the output to a desired level or scale factor.

standardize *See* normalize.

standard leak A controlled, finite amount of tracer gas allowed to enter a leak detector during adjustment and calibration.

standard parts drawing Document specifying all the requirements for insert. [AS3506]

standard patch A sample patch used as a comparison with patches taken from patch tests run on production units. [ARP1302]

standard pitch The geometric pitch of a propeller taken at two-thirds of its radius. *See* geometric pitch. [ARP4107]

standard rate turn A turn of 3 deg per second. [ARP4107]

standard, reference *See* reference standard.

standard sensor interface A universal interface circuit capable of interchangeably servicing the sensing elements for many types of measurand. [ARD50024]

standard sphere gap The maximum distance between the surfaces of two metal spheres, measured along a line connecting their centers, at which spark-over occurs when a dynamically variable voltage is applied across the spheres under standard atmospheric conditions. The standard sphere gap value is a measure of the crest value of an alternating-current voltage.

standard terminal arrival *See* STAR. [ARP4107]

standard, transfer *See* transfer standard.

standard volume For a gas, defined as the volume at 25°C (77°F) and 1001.32 kPa (29.921 in. Hg Abs.). [ARP1179A]

standard wire rope Wire rope made of six wire strands laid around a sisal core.

standby Describes the normal status of a channel in a fault detection-correction fault-tolerant system when that channel may be switched into control in the event of a failure of a normally active operating channel. [ARP4386]

standby charge A low overcharge current rate, on the order of 0.01 to 0.03 C, applied continuously to a vented-cell battery to maintain its capacity in a ready-to-discharge state. Often called trickle charging. [ARP4386]

standing wave *1.* The wave that builds up in front of and behind the contact area of a tire when it is operated at high speeds, and the centrifugal forces, internal stresses, external loads, and deflections imposed result in a departure from a round shape. This condition represents a dangerous mode of operation for the tire. [AIR1489] *2.* Periodic wave having a fixed distribution in space which is the result of interference of progressive waves of the same frequency and kind. Such waves are characterized by the existence of nodes or partial nodes and antinodes that are fixed in space. *3.* A wave in which the ratio of the instantaneous value of any of the dependent wave functions at one point on the wave, to its instantaneous value at any other point does not vary with time.

standing wave meter An instrument for measuring the standing-wave ratio in a radio-frequency transmission line.

standpipe A vertical tube filled with a liquid such as water.

staple fibers Fibers of spinnable length, manufactured directly or by cutting continuous filaments to short lengths. [AIR4844]

star *1.* Self-luminous celestial body, exclusive of nebulas, comets, and meteors; a sun seen in the heavens. Distinguished from a planet or natural satellite that shines by reflected light. *2.* Describes a wiring technique whereby devices are interconnected via a central hub or wiring closet.

STAR Contracted from standard instrument arrival route. A preplanned instrument flight rule (IFR) air-traffic control arrival procedure published for pilot use in graphic or textual form. STARs provide transitions from the en route structure to an outer fix or an instrument approach fix/arrival waypoint in the terminal area. Also called standard terminal arrival. [ARP4107]

star clusters Groups of stars physically close together.

star coupler A coupler in which many fibers are brought together to a single optical element, where their signals are mixed. The mixed signals are then transmitted back through all the fibers.

star formation The collapse under gravity of molecular clouds of interstellar matter to form clusters of protostars, and the continuing collapse of the protostars to form main-sequence stars.

star formation rate The rate at which stars are formed within a specified region or galaxy; sometimes expressed as the number of solar masses per year.

star grain A solid-propellant grain with an internal star-shaped cross section. [AIR913]

Stark effect Refers to the broadening or splitting of a spectral line observed when a luminous gas is acted upon by a strong electric field.

star network A set of three or more branches in an electronic network, where one terminal of each branch is connected at a common node.

starspots Temporary disturbed areas in the stellar photosphere that appear dark because they are colder than the surrounding areas.

start cycle The events that take place between start initiation and the point of engine idle. [ARP906A]

starter *1.* An electric motor and gear used to turn the crankshaft of an internal combustion engine until its operation becomes self-sustaining. *2.* In some chemical processes, a reactive mixture used to initiate a reaction between less reactive chemicals. Also known as starting mix.

starter-assisted start An airstart in which the aircraft starter is engaged either during spooldown, during windmilling, or from zero rpm high-pressure rotor speed. [ARP906A]

starter drive coupling The elements of the starter that engage the engine, and through which torque to accelerate the engine is transmitted. [ARP906A]

starter envelope The external three-dimensional shape of the starter; or space allowance for the starter. [ARP906A]

starter, fixed displacement A hydraulic starter in which there is a constant flow of fluid per revolution. [ARP906A]

starter/generator An electrical starter that will function as a starting motor when external electrical energy is applied, and will also function as an electric generator when driven by the engine. [ARP906A]

starter, hydraulic *See* hydraulic starter.

starter power transform The relationship between starter power input, expressed in terms such as voltage and current, or pressure and flow rate, and starter power output, expressed in terms of torque and speed. Usually presented graphically on torque vs. speed axes with input axes overlaid. [ARP906A]

starter/pump, hydraulic A hydraulic starter that will function as a motor when external hydraulic energy is applied, and will also function as a hydraulic pump when driven by the engine. [ARP906A]

starter, variable displacement A starter whose displacement during the starting cycle is automatically controlled, usually to limit the fluid flow rate to a predetermined maximum. [ARP4386]

starting mix *See* starter.

starting resistance The force needed to produce an oil film in a set of journal bearings supporting a shaft when the shaft first begins to turn.

starting torque Torque required to initiate engine flywheel rotation. [AIR4152]

starting voltage The voltage of rated frequency applied to the control-voltage winding necessary to start the rotor turning at no-load conditions, with rated voltage and frequency on the fixed-voltage winding and energizing-voltage winding. [ARP667]

star trackers Telescopic instruments on rockets or other flight-borne vehicles which lock onto a celestial body and give guidance reference to the vehicles during flight.

star tracking *See* star trackers.

start switch A switch or other device used to initiate a start cycle. [ARP906A]

start time, actual The measured length of engine starting time, from initiation of the start cycle to engine idle speed. [ARP906A]

start time, calculated The calculated time of engine starting time, from initiation of the start to engine idle speed, analytically obtained using graphical or tabulated data, or equations describing engine and starter characteristics. [ARP906A]

starved area In a plastic part, an area that has an insufficient amount of resin to wet out the reinforcement completely. [AIR4844]

starved joint An adhesive joint that has been deprived of the proper film thickness of adhesive due to insufficient adhesive spreading, or to the application of excessive pressure during the lamination process. [AIR4844]

state A description of the process in terms of its measured variables; or a description of the condition of a circuit or device, as in "logic state 1."

state changes Conditions involving one or more bits changing from 0 to 1, or from 1 to 0. [ARP1834]

state equations *See* equations of state.

state-space model *See* piece-wise linear model.

state-variable model *See* piece-wise linear model.

state variables *1.* Variables that represent energy storage, such as rotor speeds and metal temperatures. [AIR4548] *2.* Variables that completely describe the behavior of a dynamic system. [AIR1823] *3.* The output(s) of the memory element(s) of a sequential circuit.

static *1.* At rest, or at rest relative to a solid surface. *2.* Describes a structural test with application of a single increasing load. *3.* Communication interference due to discharge of static electricity.

static accuracy The degree to which the controlled temperature coincides with the specified or selected temperature after all transients have decayed. Static accuracy is usually specified as a deviation from nominal. [ARP89C]

static charge The electric charge produced by the relative motion of a nonconducting material over a nonconducting plastic material. [AIR4844]

static connection A pipe tap on a manifold used to connect process pressure to an instrument.

static discharge The vertical distance from the centerline of pump suction to the point of free discharge level. [AIR4783]

static discharge head The vertical distance from the centerline of the pump discharge to the free discharge fuel level. [AIR4783]

static efficiency For a fan, the mechanical efficiency multiplied by the ratio of static pressure differential to the total pressure differential, from fan inlet to fan outlet.

static fatigue Failure of a part under continued static load. [AIR4844]

static firing The firing of a rocket engine in a hold-down position to measure thrust and accomplish other tests.

static generation Net electrical charge generated within a system caused by friction between different substances. [AIR4783]

static ground line S The ground line with the aircraft at rest, in the basic mission takeoff configuration. [ARP1538]

static head Pressure, measured in height, of the side wall of a tank, pipe/plumbing, or a system. [AIR4783]

static-head liquid-level meter A pressure-sensing device, such as a gage, so connected in the piping system that any dynamic pressures in the system cancel each other, and only the pressure difference due to liquid head above the gage position is registered.

static information Information that remains constant during the time of relevant task performance. [ARP4155]

static instability *See* stability, negative static.

static load The load that is imposed on a member when in a static state or 1 g condition. [AIR1489]

static loaded radius For a tire, the perpendicular distance between the axle centerline and the surface against which the tire is loaded, when supporting its rated load, while inflated to its rated inflation pressure. [AS4833]

static model Set of equations or physical laws to determine a balance of systems at rest. *See also* steady-state model.

static modulus The ratio of stress to strain under static conditions. [AIR4844]

static port An opening used as a source of ambient (static) pressure in the pitot-static system. One static port can generally be found on each side of an aircraft, in an area where there is usually no dynamic (positive or negative) pressure due to the motion of the aircraft through the air. Static air pressure is used to determine altitude and vertical velocity of an aircraft; and, when compared with dynamic or impact pressure from the pitot tube, it is used in determining airspeed. *See also* pitot-static tube and static tube. [ARP4107]

static position The state of deflection or compression of a landing gear when the aircraft is in a static or 1g condition. Usually, further compression of the strut (taxi stroke) is available for dynamic taxi conditions. Decrease of the 1g static load permits extension of the strut (landing stroke region) to the free flight or full-extended position. [AIR1489]

static pressure *1.* The pressure measured by a gauge attached to the side wall of a pipe; or the total pressure minus the dynamic pressure. [AIR4783] *2.* The local pressure in a fluid that has no element due to velocity of the fluid. [ARP4386] *3.* Pressure exerted by a gas at rest; or pressure measured when the relative velocity between a moving stream and a pressure-measuring device is zero.

static pressure gage An indicating instrument for measuring pressure.

static pressure tube *See* static tube.

static seal *See* gasket.

static stability The property of a physical system whereby it maintains constancy in its static and dynamic responses despite changes in its internal conditions and variations in its environment. Compare with dynamic stability.

static stability, negative *See* stability, negative static.

static stability, neutral *See* stability, neutral static.

static stability, positive *See* stability, positive static.

static steering torque The torque required to turn the nose wheel(s) of an aircraft as it sits static on the runway surface (without the benefit of engine thrust or forward motion of the aircraft). [AIR1489]

static stores Digital registers in telemetry devices that hold set-up instructions from the computer.

static stress A stress in which the force is constant or slowly increasing with time, for example, test of failure without shock. [AIR4844]

static suction head Pressure, measured in height, along the side wall of the flow inlet of the pump, with the pump not working and the delivery valve closed. [AIR4783]

static suction lift Differential height measured between the fuel surface level and flow inlet of the pump, with the pump not working and the delivery valve closed. [AIR4783]

static temperature *1.* A measurable property that is directly proportional to the mean kinetic energy of the particles. [AIR1900] *2.* The temperature of a fluid as measured under conditions of zero relative velocity between the fluid and the temperature-sensitive element, or as measured under conditions that compensate for any relative motion.

static test *1.* Refers to any measurement taken in a normally dynamic system under static conditions; for example, a pressure test of a hydraulic system under no-flow conditions. *2.* Specifically, a test to verify structural characteristics of a rocket, or to determine rocket-engine thrust, while a rocket is in a stationary or hold-down position. *3.* Structural strength test of an aircraft or missile in which flight loads are simulated statically by weights or hydraulic jacks.

static torque The torque that a brake is capable of resisting in a static (non rotating) state, as opposed to dynamic torque. [AIR1489]

static tube *1.* A tube vented to the atmosphere, used in measuring ambient (static) air pressure for comparison with impact air pressure to determine airspeed. *See also* pitot-static tube and static port. [ARP4107] *2.* A device used to measure static pressure in a stream of fluid. Normally, a static tube consists of a perforated, tapered tube, with a branch tube for connecting it to a manometer. A related device, called a static pressure tube, consists of a smooth tube with a rounded nose that has radial holes in the tube behind the nose.

static unbalance A condition of unbalance in which the central principal axis is displaced only parallel to the shaft axis. A quantitative measure of static unbalance is given by the resultant of the two dynamic unbalance vectors. [ARP588A]

static weighing A weighing method in which the net mass of liquid collected is deduced from tare (empty tank) and gross (full tank) weighings made, respectively, before the flow is diverted into the weighing tank and after it is diverted to the by-pass.

station A device capable of originating to or utilizing data from the HSDB (high-speed data bus). [AIR4271]

stationary orbit A circular orbit around a planet in the equatorial plane, having a rotational period equal to that of the plant. For earth, the stationary orbit is about 26,000 miles (approximately 42,000 kilometers) in radius. A body moving in a stable stationary orbit appears fixed in the sky relative to an observer on the surface of the planet. [ARP4386]

stationary wave A standing wave in which the energy flux is zero at all points on the wave.

stationkeeping The sequence of maneuvers that maintains a vehicle in predetermined orbit.

station lines Reference lines on loft and print, used for locating and dimensioning purposes. [AIR4844]

statistical error *1.* Generally, any error in measurement resulting from statistically predictable variations in measurement system response. *2.* Specifically, an error in radiation-counter response resulting from the random time distribution of photon-detection events.

statistical quality control Any method for controlling the attributes of a product or controlling the characteristics of a process that is based on statistical methods of inspection.

stator(s) In machinery, parts or assemblies that remain stationary with respect to rotating or moving parts; for example, assemblies such as the field frames of electric motors or generators; or the stationary casings and blades surrounding axial flow compressor rotors or turbine wheels.

statoscope *1.* A barometer for recording small changes in atmospheric pressure. *2.* An instrument for indicating small changes in altitude of an aircraft.

status message segment A status word and associated data words, if any. [AS4113]

statute mile A unit of distance equal to 5280 feet.

stay A tensile stress member used to hold material or other members rigidly in position. *See also* drag brace.

staybolt A bolt threaded through or welded at each end, into two spaced sheets of a firebox or box header to support the flat surfaces against internal pressure.

STC *See* supplemental type certificate.

STE *See* special test equipment.

steady flow A flow in which the flow rate in a measuring section does not vary significantly with time.

steady state Describes a substance or system whose local physical and chemical properties do not vary with time.

steady-state deviation *1.* The system deviation after transients have expired. *2. See* offset.

steady-state distortion The spatial distortion measured by low-response pressure instrumentation. [ARP1420]

steady state flow *See* equilibrium flow.

steady-state model A mathematical model that represents a process at equilibrium (infinite time) conditions. *See also* static model.

steady-state optimization A method of optimizing some criterion function of a process, usually using a steady-state model of the process. Linear programming is frequently used as the optimization method, with a function approximating the profit of the process being a typical optimizing criterion. Contrast with dynamic optimization.

steady state response to steering The stable operating condition of a tire under lateral load inputs from the steering system. [AIR1489]

steady-state vibration Within a vibrating system, a condition in which the velocity of each moving particle can be described by a periodic function.

steady state wind A wind that produces a constant force. [ARP1328]

steam *1.* The vapor phase of water substantially unmixed with other gases. *2. See* water vapor.

steam atomizing oil burner A burner for firing oil atomized by steam. May be of the inside- or outside-mixing type.

steam attemperation Reducing the temperature of superheated steam by injecting water into the flow or passing the steam through a submerged pipe.

steam binding A restriction in circulation due to a steam pocket or a rapid steam formation.

steam cock A valve for admitting or releasing steam.

steam cure To hasten the curing cycle of concrete or mortar by the use of heated water vapor, at either atmospheric or higher pressure.

steam dryer A device for removing water droplets from steam. *See also* steam scrubber.

steam-free water Water containing no steam bubbles.

steam gage A device for measuring pressure in a steam system.

steam generating unit A unit to which water, fuel, and air are supplied, and in which steam is generated. Consists of a boiler furnace and fuel-burning equipment; may also include water walls, a superheater, a reheater, an economizer, an air heater, or any combination thereof.

steam jacket A casing around the cylinders and heads of a steam engine, or around some other mechanism or space, to keep the surfaces hot and dry.

steam-jet blower A device in which the energy of steam flowing through a nozzle or nozzles is used to induce a flow of air to be supplied for combustion.

steam purity The degree of contamination, usually expressed in ppm.

steam quality The percent by weight of vapor in a steam and water mixture.

steam scrubber A series of screens, wires, or plates through which steam is passed to remove entrained moisture.

steam separator A device for removing the entrained water from steam.

steam trace A technique for preventing freezing in a pipe or tubing line by means of an adjacent steam line; usually 1/4-in. to 1/2-in. copper tubing. *See also* heat tracing.

steam tracing Refers to an arrangement for heating a process line or instrument-air line to keep liquids from freezing or condensing. Often, a piece of pipe or tubing carrying live steam is simply run alongside, or is coiled around, the line to be heated.

steam trap A device that automatically collects condensate in a steam line and drains it away.

steel Any alloy of iron with up to 2% carbon which may or may not contain other alloying elements to enhance strength or other properties.

steerable antennas Directional antennas whose major lobe can be readily shifted in direction.

steer damper A device that provides power for steering (usually on a nose gear), as well as damping to resist shimmy. [AIR1489]

steering angle The angle through which the landing gear or wheel(s) may be steered. Usually expressed as degrees left and right (one-half the total angle). [AIR1489]

steering bars *See* command bars.

steering control system *See* system, steering control.

steering rate The rate of change of nosewheel steering angle in degrees (or radians) per second. [ARP1595]

steering ratio The relationship between input control movement and output nosewheel angle change. [ARP1595]

steering torque Steering unit output torque about strut centerline. [AIR1752]

Stefan-Boltzmann law One of the radiation laws; states that the amount of energy radiated per unit time from a unit surface area of an ideal black body is proportional to the fourth power of the absolute temperature of the black body.

stellar activity A general term encompassing stellar phenomena such as stellar flares, starspot activity, magnetic activity, and nuclear fusion.

stellarators Experimental thermonuclear devices whereby containment in a magnetic field is achieved by closing the field upon itself, thus allowing the particles to perform endless spiral motion.

stellar color The particular wavelengths of optical radiation emitted by a star.

stellar cores The central portion of the interior of stars.

stellar coronas Ionized regions about stars, formed by x-rays emitted during stellar flares.

stellar flares Ejections of material from stars in eruptions that last from a few minutes to an hour or more.

stellar interiors The subsurface portions of stars.

stellar magnitude A measure of the relative brightness of a star. Stellar magnitudes are expressed in a variety of ways, according to the method or process of observation or determination.

stellar mass accretion Process by which a star accumulates matter as it moves through dense clouds of interstellar gas.

stellar oscillations Irregular fluctuations of stellar atmospheres.

stellar parallax The subtended angle at a star formed by the mean radius of the earth's orbit; indicates distance to a star.

stellar systems Gravitationally bound groups of stars.

stem rotation A phenomenon that occurs in linear-motion valves when the hydraulic forces from the process fluid cause the closure component to rotate about the stem axis.

stenosis Narrowing of a tube. [ARP171]

step bearing A bearing that supports the lower end of a vertical shaft. Also known as pivot bearing.

step brazing Making a series of brazed joints in a single assembly by sequentially making up individual joints and heating each one at a lower temperature than the previous joint, in order to maintain the joint integrity of previous joints. This process requires a lower-melting brazing alloy for each successive joint in the assembly.

step change The change from one value to another in a single increment, in negligible time.

stepdown fix A fix permitting additional descent within a segment of an instrument

approach procedure by identifying a point at which a controlling obstacle has been safely overflown. [ARP4107]

step gage *1.* A plug gage consisting of a series of cylindrical gages of increasing diameter mounted on the same axis. *2.* A gage for measuring the height of a step or shoulder; consists of a gage body and a sliding blade.

step-growth-polymerization A chemical reaction in which polymers are formed by the stepwise intermolecular addition of molecules through reactive groups. [AIR4844]

step-index fiber An optical fiber in which there is a discontinuous change in refractive index at the boundary between fiber core and cladding. Such fibers have a large numerical aperture (light-accepting angle), and are simple to connect; but have lower bandwidth than other types of optical fibers.

stepped stud insert An insert that is designed to retain a stud, with the nut-end thread being one thread size smaller in diameter than the stud-end thread. [AS1229A]

stepped wedge A device that is used, with appropriate parameters on each step, for the inspection of parts having great variations in thickness or complex geometry. [ARP5089]

stepper motor A motor whose rotor moves in discrete angular steps determined by the number of poles of stator and rotor located around the circumference of the motor air gap. [ARP4386]

stepping motor A motor that is useful for low-torque applications and suitable for computer interfacing; pulse input results in a precise rotary step, typically 0.8° per pulse or 1.6° per pulse. A stepping motor is often operated in open-loop mode.

step recovery diodes Varactors in which forward voltage injects carriers across the junction, but before the carriers can combine, the voltage reverses and carriers return to their origin in a group. The result is an abrupt cessation of reverse current and a harmonic-rich waveform.

step response *1.* The time response of the actuation system output following a step-command input; the time required to reach a particular percentage of the final output, together with limits on the percentage overshoot. [ARP4386] 2. The time response of a device or process when subjected to an instantaneous change in input from one steady-state value to another.

step stress test A test consisting of several stress levels applied to a sample sequentially, for periods of equal duration. [AIR4896]

step turn A maneuver used to put a float plane in planing configuration prior to entering an active sea lane for takeoff. [ARP4107]

step wedge A reference standard used to calibrate an ultrasonic instrument to accurately display the depth of a flaw or surface. The step wedge must be made of material acoustically similar to that being inspected. [ARP5089]

step width - W Dimensional distance from one side of a step to the other side. [ARP836A]

step width - W$_1$ The width of the step surface as measured along the nose of the step. [ARP836A]

steradian The solid angle subtended at the center of a sphere by an area on the surface equal to a square with sides of length equal to the radius of the sphere.

stereophonics Reproducing or reinforcing sound by using two or more audio channels, so that the sound gives three-dimensional sensations similar to those given by the sound sources.

stereo route A routinely used route of flight established by users and ARTCCs, identified by a coded name; for example, ALPHA 2. Such routes minimize flight-plan handling and communications. [ARP4107]

stick gage A vertical rod or stick, with a graduated scale or markings, that is fixed in an open tank or vessel so that liquid-level changes can be observed directly.

stick shaker A device that induces vibration felt in the pilot's yoke, warning of an approach to a stall; usually set to activate approximately 10 knots above the stall speed. [ARP4107]

stick shaker speed *See* V$_{ss}$.

stiction Force required to initiate relative movement between surfaces in contact.

stiffener A plate, angle, channel, or similar structural element attached to a slender beam or column to prevent it from buckling by increasing its stiffness.

stiffness *1.* A measure of modulus, the relationship of load and deformation; the ratio between the applied stress and resulting strain. [AIR4844] 2. As applied to a servoactuator, the ability of the output to resist motion when subjected to static and dynamic external loading. Measured in units of output motion/external force. [ARP1281A] *3.* Force or torque per unit displacement. [ARP4386]

stiffness to ground The combined stiffness of all spring rates that determines the spatial reference of the load mass. This term is frequency-dependent due to servoloop characteristics. Stiffness to ground, in conjunction with the load inertia, determines the load natural frequency. [ARP4386]

Stilb A unit of luminescence equal to one candela per cm^2; rarely used, as the candela per m^2 is preferred.

stilling basin An area ahead of the weir plate large enough to pond the liquid so that the liquid approaches the weir plate at low velocity. Also called weir pond.

stimulate To cause an occurrence or action artificially, rather than waiting for it to occur naturally, as to stimulate an event.

Stirling cycle A theoretical heat-engine cycle in which heat is added at constant volume, followed by isothermal expansion with heat addition. The heat is then rejected at constant volume, followed by isothermal compression with heat rejection.

stitch bonding A method of making wire connections on an integrated circuit board using impulse welding or heat and pressure to bond a connecting wire at two or more points, while feeding the wire through a hole in the welding electrode.

stitch welding Making a welded seam using a series of spot welds that do not overlap.

stochastic Pertaining to direct solution by trial-and-error, usually without a step-by-step approach, and involving analysis and evaluation of progress made, as in a heuristic approach to trial-and-error methods. In a stochastic approach to a problem solution, intuitive conjecture or speculation is used to select a possible solution, which is then tested against known evidence, observations, or measurements. Intervening or intermediate steps toward a solution are omitted. *See* heuristic.

stochastic process Ordered set of observations in one or more dimensions, each being considered as a sample of one item from a probability distribution.

stoichiometric conditions In chemical reactions, the point at which equilibrium is reached, as calculated from the atomic weights of the elements taking part in the reaction; stoichiometric equilibrium is rarely achieved in real chemical systems; rather, empirically reproducible equivalence points are used to closely approximate stoichiometric conditions.

stoichiometric mixture The optimum ratio of air and fuel required to provide the maximum possible combustion pressure. [AIR4170]

Stokes A unit of kinematic viscosity (dynamic viscosity divided by sample density); the centistoke is more commonly used.

STOL Acronym for short takeoff and landing; describes an aircraft that has the capability of operating from a short runway in accordance with applicable airworthiness and operating regulations. *See also* VTOL. [ARP4107]

stone guard A coarse mesh strainer normally located upstream of a hydrant valve or similar installation to prevent stones and other foreign objects from entering the valve, where they could cause damage or otherwise prevent correct operation. [AIR4783]

stop altitude squawk Command used by ATC to request an aircraft to turn off the automatic altitude-reporting feature of its transponder. Used when the verbally reported altitude varies 300 ft or more from the automatic altitude report. [ARP4107]

stop cock A small valve for roughly controlling or shutting off the flow of fluid in a pipe.

stop module A mechanical device that limits rotational displacement between predetermined values. [ARP4386]

stop nut *1.* A nut positioned on an adjusting screw to restrict its travel. *2.* A nut with an insert made of a compressible material; this keeps the nut tight without requiring a lock washer.

stops Metal pieces inserted between die halves, used to control the thickness of a press-molded part. [AIR4844]

stop squawk Command used by ATC to tell the pilot to turn off specified functions of the aircraft transponder. [ARP4107]

stopway An area beyond the takeoff runway, no narrower than the runway and centered upon the extended centerline of the runway, able to support the airplane during an aborted takeoff without causing structural damage to the airplane; designated by the airport authorities for use in decelerating the airplane during the aborted takeoff. [ARP4107]

storage buffer *1.* A synchronizing element between two different forms of storage, usually between internal and external storage. *2.* An input device in which information is assembled from external or secondary storage and stored, ready for transfer to internal storage. *3.* An output device into which information is copied from internal storage and held for transfer to secondary or external storage. Computation continues while transfers between buffer storage and secondary or internal storage (or vice versa) take place.

storage calorifier *See* cylinder.

storage cell An elementary unit of storage, for example, a binary cell or a decimal cell.

storage cycle A periodic sequence of events occurring when information is transferred to or from the storage device of a computer. This sequence includes storing, sensing, and regeneration.

storage device A device into which data can be inserted, in which it can be retained, and from which it can be retrieved.

storage dump *See* dump.

storage life The length of time an item can be stored under specified conditions and still meet specified requirements. [ARP4386]

storage stability The resistance of a fluid to change during long periods of storage or disuse. [AIR1116]

stored energy The energy (joules) stored in the tank or storage capacitor of a capacitor discharge system; or in the inductance coil of an inductive discharge system. [AIR784]

stowage bucket/box A device designed to accept the stowage of a nozzle or coupler while the unit is not in service. [AIR4783]

STPD Abbreviation for standard temperature and pressure, dry. Conditions comprising a temperature of 0°C (32°F) and one atmosphere absolute pressure (101.3 kPa or 760 mm of Hg), dry basis (partial pressure of water vapor equals zero). [AS1303]

straight beam A vibrating pulse wave train traveling normal to the test surface. [ARP5089]

straightening vanes Horizontal vanes inside a fluid conduit or pipe to reduce turbulent flow ahead of an orifice or venturi meter.

straight fittings Parts such as unions, machined out of bar stock, connecting to a port or tube-to-tube. [MA2005]

straight-in approach, IFR An instrument approach in which final approach is begun without first having executed a procedure turn. [ARP4107]

straight-in approach, VFR Entry into the traffic pattern by interception of the extended runway centerline (final approach course) without executing any other portion of the traffic pattern. [ARP4107]

straight-in landing A landing made on a runway aligned within 30 deg of the final approach course, following completion of an instrument approach. [ARP4107]

straight-in landing minimums *See* landing minimums/IFR landing minimums.

straight polarity Describes arc welding in which the electrode is connected to the negative terminal of the power supply.

strain Elastic deformation due to stress. [AIR4844]

strain aging A change in properties of a metal or alloy that occurs at room temperature or slightly elevated temperature following cold working.

strainer A screen or porous medium positioned in a flowing stream of fluid (such as a water intake) to separate out harmful objects or particles before the fluid enters process equipment.

strain foil A type of strain gage made by photoetching a resistance element out of thin foil.

strain gage(s) *1.* Instruments used to measure the strain or distortion in a member or test specimen (such as a structural part) subjected to a force. *2.* A device that can be attached to a surface, usually with an adhesive, to indicate strain magnitude in a given direction, the strain being determined by changes in electrical resistance of fine wire. May be used to measure strain due to static or dynamic applied loading, in tension and/or compression, depending on design of the gage, bonding technique, and type of instrumentation used to determine resistance changes in the strain element. *3.* A high-resistance, fine-wire or thin-foil grid used in a measuring bridge circuit. When the grid is securely bonded to a specimen, its resistance changes as the specimen is stressed. Such devices are used in many forms of transducers. *4.* A transducer that converts information about the deformation of solid objects, called the strain, into a change in resistance.

strain hardening The increase in tensile and yield strengths, and the corresponding reduction in ductility, associated with plastic deformation of a metal at temperatures below its recrystallization range.

strain in contact area Within a tire carcass, the internal strain in the contact area that is deflected from a normal state. [AIR1489]

strain relaxation Reduction in internal strain over time. [AIR4844]

strain relief Refers to a technique involving devices or methods of termination or installation to reduce the transmission of mechanical stresses to the conductor termination. [ARP914A]

strain rosette An assembly of two or more strain gages used for determining biaxial stress patterns. Also known as rosette strain gage.

strain sensitivity A characteristic of a conductor that describes its change in resistance in relation to a corresponding change in length; can be calculated as $\Delta R/R$ divided by $\Delta L/L$. When referring to a specific strain-gage material, strain sensitivity is commonly known as the gage factor.

strake *1.* A long but shallow surface normal to the skin and aligned with local airflow; an extremely low-aspect-ratio fin. *2.* One row from stem to stern of single plates cladding marine aircraft.

strand *1.* A group of wires helically wound around a core wire in a left-hand direction or a right-hand direction. [AS4536] *2.* A rod or filament of metal or electrically conductive material. [AS1198]

strand cast heat Describes a process for casting the product of a single furnace load in a continuous cast fashion; ingots are not involved. [MAM2304]

strand count The number of strands in a plied yarn. [AIR4844]

strand integrity The degree to which the individual filaments making up a strand or end are held together by the applied sizing. [AIR4844]

strand size The size of a strand; the number corresponding to the AWG size of the strand. [AS1198]

strand tensile test A tensile test of a single resin-impregnated strand of any fiber. [AIR4844]

strange attractors In theoretical physics, abstract geometrical objects representing motion that is bounded but not periodic. The detailed behavior of strange attractors is sensitive to external perturbations, but their overall qualitative behavior is stable. Strange attractors are of particular interest in the study of turbulence.

strapping interval *See* interval.

strategic message A clearance or flight plan message concerned with the overall mission of the flight. [ARP4791]

strategy, frame synchronizer The procedure defined by an operator to emphasize rapid acquisition, or to emphasize accuracy of acquisition, or any point between those extremes.

stratification Non-homogeneity existing transversely in a gas stream.

stratiform clouds Layered clouds of extensive horizontal development. [AIR4367]

stratosphere The upper portion of the atmosphere between approximately 35,000 ft (11 km) and 100,000 ft (30 km) altitude, in which the temperature changes but little with altitude, and clouds of water vapor do not normally form. [ARP147C]

stratosphere radiation Any infrared radiation involved in the complex infrared exchange continually proceeding within the stratosphere.

stratospheric warming A temperature rise in the global stratosphere.

stratus A low-level principal stratiform cloud with a uniform gray base at altitudes usually below 2000 m (6500 ft). [AIR4367]

stray current corrosion Galvanic corrosion of a metal or alloy induced by electrical leakage currents passing between a structure and its service environment.

stray light Light emitted by components of the lighting assembly directly into the crew station area. Stray light is not visible to crew-station members, but contributes to the ambient light level; normally associated with flood-lit integrally-lighted displays. [ARP1161]

streak cameras Cameras for measuring radiation pulses by deflection of an electron beam.

streak photography The process of taking a time-exposure photograph of a tracer particle in a fluid. The photograph reveals the motion of each tracer particle in the form of a streak which may be interpreted as a velocity vector.

streamline flow A type of fluid flow in which flow lines within the bulk of the fluid remain relatively constant with time. *See also* laminar flow.

streamlining Contouring the exterior shape of a body to reduce drag due to relative motion between the body and a surrounding fluid.

stream tube In the characterization of fluid flow, an imaginary tube whose wall is generated by streamlines passing through a closed curve.

street elbow A pipe elbow with an external thread at one end and an internal thread at the other end.

strength The maximum stress that a material is capable of sustaining. [AIR4844]

strength weld A weld capable of withstanding a design stress.

stress The force per unit area of a body that tends to produce a deformation. [ARP4107]

stress amplitude One-half the algebraic difference between the maximum stress and minimum stress in one cycle of repeated variable loading.

stress analysis In the design of a system or equipment, the evaluation of stress conditions existing in parts when loads are applied. [AIR4896]

stress concentration *1.* A localization of high stresses due to the presence of shoulders, grooves, holes, keyways, threads, etc. which result in a modification of the simple stress distribution. [ARP700] *2.* In structures, a localized area of high stress.

stress concentration factor A measure of the severity of stress concentration; expressed as the ratio of the maximum local stress to the nominal simple stress, the latter of which is determined irrespective of stress concentration. [ARP700]

stress corrosion Preferential attack of areas under stress in a corrosive environment, where such an environment alone would not have caused corrosion. [AIR4844]

stress-corrosion cracking Deep cracking in a metal part due to the synergistic action of tensile stress and a corrosive environment, causing failure in less time than could be predicted by simply adding the effects of stress and the corrosive environment together. The tensile stress may be a residual or applied stress, and the corrosive environment need not be severe, but only must contain a specific ion that the material is sensitive to.

stress crack External or internal crack in a plastic caused by tensile stresses less than that of its short-time mechanical strength, frequently accelerated by the environment to which the plastic is exposed. [AIR4844]

stress cycles A variation of stress with time, repeated periodically and identically.

stress intensity factors In materials, load-induced variables in tension, compression and/or shear which are conducive to crack initiation and propagation and fatigue fracture.

stress raiser A discontinuity or change in contour that induces a local increase in stress in a structural member.

stress ratio The ratio of the minimum stress to the maximum stress in one cycle, considering tensile stresses as positive, compressive stresses as negative. [ARP700]

stress relaxation The decrease in stress under sustained, constant strain. [AIR4844]

stress relieving Heating to a suitable temperature, holding long enough to reduce residual stress, and then cooling slowly enough to avoid inducing new residual stresses.

stress, residual *See* residual stress.

stress, rupture *See* rupture stress.

stress-strain Stiffness at a given strain. [AIR4844]

stress-strain curve Simultaneous readings of load and deformation, converted to stress and strain, plotted as ordinates and abscissae, respectively, to obtain a stress-strain diagram. [AIR4844]

stress-strain relationship Relationship between the stress or load on a structure, structural member, or specimen, and the strain or deformation that follows.

stress tensors Complete sets of stress components in a solid or fluid medium.

stretcher leveling Removing warp and distortion in a piece of metal by gripping it at both ends and subjecting it to tension loading at stresses higher than the yield strength.

stretch forming Shaping a piece of sheet metal or plastics sheet by applying tension and then wrapping the sheet around a die form; may be performed cold or the sheet may be heated first. Also known as wrap forming.

striae (cords) Apparent streaks or veins in glass which are the result of minor variations in the index of refraction within the body of the glass. [ARP924]

striation technique A method of making sound waves in air visible; relies on the ability of individual sound waves to refract light.

strike *1.* A thin electroplated film to be followed by other plated coatings. *2.* A plating solution of high covering power and low efficiency, used for electroplating very thin metallic films. *3.* A local crater or remelted zone caused by accidental contact between a welding electrode and the surface of a metal object. Also known as arc strike.

string A linear sequence of entities, such as characters or physical elements.

stringer *1.* A solid, nonmetallic impurity in the parent metal, often the result of an inclusion that has been stretched during a rolling process. [AS3071A] *2.* Slender, lightweight, lengthwise fill-in structural member in a rocket body, or the like, serving to reinforce and give shape to the skin.

string manipulation The handling of string data by various methods, generally in terms of bits, characters, and sub-strings.

string-shadow instrument An indicating instrument in which the measured value is indicated by means of the shadow of a filamentary conductor whose position in an electric or magnetic field depends on the magnitude of the quantity being measured.

stringy alpha Platelet alpha that has been elongated and distorted by nondirectional metal working, but not broken up or recrystallized. Also called wormy alpha. [AS1814]

strip chart A hardware device that records analog data (generally, six or eight channels) on a continuous chart.

strip-chart recorder Any instrument that produces a trace or series of data points, using one or more pens or a print wheel, on a grid printed on a continuous roll of paper that is moved at a uniform rate of travel in a direction perpendicular to the motion of the indicating mechanism of the instrument. The resulting trace is a graph of the measured variable as a function of time.

stripper *1.* A tool or chemical used to remove insulation material from wire or cable. [ARP914A] *2.* A distillation column that has no rectifying section. In such a column, the feed enters at the top, and there is no other reflux.

stripping The process of removing insulation material from a wire or cable. [ARP1931]

stripping section That section of a distillation column below the feed. The stripping section strips the light components from the liquid moving down the column.

strip terminal A contact or terminal supplied in some means of continuous form, for use in automatic or semiautomatic crimping machines. [ARP914A]

strobe pulse A pulse of light whose duration is less than the period of a recurring event or periodic function, and which can be used to render a specific event or characteristic visible so it can be closely observed.

stroboscope A device for intermittently viewing or illuminating moving bodies so that they appear to be motionless, either by placing an intermittent shutter between the object and an observer or by repeatedly flashing a brilliant light on the object. In this manner, a vibrating or rotating object can be made to appear stationary by adjusting the frequency of the stroboscope; the indicated frequency of the stroboscope is equal to the vibrational or rotational frequency of the object.

stroboscopic flicker As applied to an electronic display image, an appearance of flickering or jumping when the image has a short rise and fall time and there is relative motion between the observer and the display. [ARP4256]

stroboscopic tachometer A stroboscopic lamp with a variable-flashing-rate control circuit that enables the frequency to be adjusted until a rotating object appears to stand still. The frequency is read from a calibrated dial, and represents either the fundamental rotational speed in cycles per unit time or one of its harmonics; sometimes, a patterned disk centered on the axis of rotation is used to make it easier to determine fundamental frequency.

stroke The linear extent of movement of a reciprocating mechanical part.

stroke-written symbology (vector-written) A type of system that allows the electron beams to create symbology by moving directly from one location to another on the CRT screen; the speed of the beam may be varied, or the symbol may be retraced to increase luminance. [ARP1782]

strong interactions (field theory) One of the fundamental interactions of elementary particles, primarily responsible for nuclear forces and other interactions among hadrons.

strongly coupled plasmas Highly compressed and collisional plasmas with electron densities on the order of 10^{24} per cubic centimeter or more. The mean kinetic and potential energies of particles in the plasma are typically of the same order of magnitude.

Strouhal number A nondimensional number occurring in the study of periodic or quasiperiodic variations in the wake of objects immersed in the fluid stream. Defined as: $S = fh/V$, where f is frequency, V is velocity, and h is reference length.

structural adhesive Adhesive used for transferring required loads between adherents exposed to service environments typical for the structure involved. [AIR4844]

structural analysis Determination of the stresses and strains in a structural member due to combined gravitational and applied service loading.

structural assembly One or more structural elements that together provide a basic structural function. [AIR4896]

structural bond A bond that joins basic load-bearing parts of an assembly. The load may be either static or dynamic. [AIR4844]

structural damage Damage that may affect the structural integrity. [ARP5089]

structural detail The lowest functional level in an aircraft structure. [AIR4896]

structural element Two or more structural details that together form an identified manufacturer's assembly part. [AIR4896]

structural function The mode of action of aircraft structure; includes acceptance and transfer of specified loads in items and provides consistently adequate aircraft response and flight characteristics. [AIR4896]

structural glass/S-glass A magnesia/alumina/silicate glass reinforcement providing high strength. [AIR4844]

structural sandwich construction A laminar construction comprising a combination of alternating, dissimilar, simple or composite materials assembled and intimately fixed in relation to one another so as to use the properties of each to attain specific structural advantages for the whole assembly. [AIR4844]

structural significant item A structural detail, element, or assembly that is judged significant because the consequence of its failure could be a reduction in aircraft/engine residual strength or function to the extent that safety or mission is adversely impacted. [AIR4896]

strut *1.* A slender structural element that provides axial force, usually compression, to support another component or piece of structure.

Commonly used to attach wheels to a vehicle, support airplane wings, and hold doors open. *2. See* cylinder, actuating.

strut inclination Forward angle of strut centerline to the vertical. [AIR1752]

STS *1. See* entry guidance. *2. See* payload delivery. *3. See* power modules. *4. See* turnaround.

stub connection Connection designated for attachment of the stub cable to the coupler. [AS4117]

stud *1.* A post used for connecting conductors or terminals; may be threaded, serrated, or plain. [ARP914A] *2.* A projecting pin serving as a support or means of attachment.

stud arc welding (SW) Welding in which metals are heated with an arc between a metal stud, or similar part, and the work. Once the surfaces to be joined are properly heated, they are brought together under pressure.

stud hole A hole or opening in the tongue of a terminal lug that is intended to accommodate a screw or stud. [ARP914A]

stud-type board A terminal board used for connecting conductors or terminals by means of binding posts or stud terminations. *See* terminal board. [ARP914A]

stud welding Producing a joint between the end of a rod-shaped fastener and a metal surface, usually by drawing an arc briefly between the two members, then forcing the end of the fastener into a small weld puddle produced on the metal surface.

stuffing *See* packing.

stuffing box A cavity around a rod or shaft that penetrates a pump casing, valve body, or other portion of a pressure boundary, which can be filled with packing material and compressed to form a leak-tight seal while still permitting axial or rotary motion of the shaft.

subassembly Two or more parts that form a portion of an assembly or component replaceable as a whole, but having a part or parts that are individually replaceable. [ARD50010]

subcarrier A carrier applied as a modulating wave to another carrier or an intermediate subcarrier.

subcarrier band A band (of frequencies) associated with a given subcarrier and specified in terms of maximum subcarrier deviation.

subcarrier channel The channel required to convey telemetry information involving a subcarrier band.

subcarrier discriminator In FM telemetry, the device that is tuned to select a specific subcarrier and demodulate it to recover the data.

subcarrier oscillator The basic subcarrier frequency generator, whose output frequency is used as the transmission or carrier medium of desired signal information. In telemetry, the desired signal information is most often used to frequency modulate the subcarrier for transmission.

subcommutation Commutation of a number of channels, with the output applied to an individual channel of the primary commutator. Subcommutation is synchronous if its rate is a submultiple of that of the primary commutator.

subcommutation frame In PCM systems, a recurring integral number of subcommutator words including a single subcommutation frame synchronization word. The number of words in a subcommutation frame is equal to an integral number of primary commutator frames. The length of a subcommutation frame is equal to the total number of words or bits generated as a direct output of the subcommutator.

subconscious level of awareness See awareness, subconscious level.

subcooler, vapor cycle A heat exchanger in which liquid flowing to the expansion valve is subcooled; this is often done by gas leaving the evaporator. [ARP147C]

subcooling The process of cooling a liquid below its condensing temperature for a particular saturated liquid pressure. [ARP147C]

subframe A multiplex generated at a slower rate than a frame, and input to the frame through one of the channels.

subgiant stars Celestial bodies whose position on the Hertzsprung-Russell (H-R) diagram is intermediate between that of the main-sequence stars and normal giants of the same spectral type.

subgravity See reduced gravity.

subharmonic A sinusoidal function whose frequency is a submultiple of some other periodic function to which it is related.

subjective brightness The attribute of any light sensation giving rise to the percept of luminous intensity, including the whole range of qualities of being bright, light, brilliant, dim, or dark. [AIR512B]

subjective fatigue See fatigue, subjective.

sublimation The transition of a substance directly from the solid state to the vapor state, or vice versa, without passing through the intermediate liquid state.

submerged-arc welding An electric-arc welding process in which the arc between a bare-wire welding electrode and the workpiece is completely covered by granular flux during welding.

submersible/immersible So constructed that it will operate successfully when submerged in water under specified conditions of pressure and time. [ARP4404]

submultiplexing See block switching.

subnetwork One of the interoperable communications networks, such as HF, VHF, satellite, or Mode S secondary radar, that will support aeronautical data communications. [ARP4791]

sub-optimization The process of fulfilling or optimizing some chosen objective that is an integral part of a broader objective. Usually the broad objective and lower-level objective are different.

subroutine A set of instructions necessary to direct a computer to carry out a well-defined mathematical or logical operation; a subunit of a routine, usually coded in such a manner that it can be treated as a black box by the routine using it. A single routine may simultaneously be both a subroutine with respect to another routine and a master routine with respect to a third. Usually control is transferred to a single subroutine from more than one place in the master routine, and the reason for using the subroutine is to avoid having to repeat the same sequence of instructions in different places in the master routine. See also program and routine.

subsatellite An object designed to be carried into orbit inside an artificial earth satellite, but later ejected to serve a particular purpose. [ARP4386]

subscale Subsurface oxides formed by reaction of a metal with oxygen that diffuses into the interior of the section rather than combining with metal in the surface layer.

subset *1.* In data processing, any set of items that relate to a larger set. *2.* A subscriber apparatus in a communications network.

subsieve analysis Determination of particle-size distribution in a powdered material, none of which is retained on a standard 44-micrometer sieve.

subsonic *1.* In aerodynamics, describes speeds less than the speed of sound. [ARP4107] *2.* For an aircraft, describes any speed from hovering (zero) up to about 85% of the speed of sound in the atmosphere at ambient temperature.

subsonic flow Flow of a fluid, such as air over an airfoil, at speeds less than acoustic velocity.

substantial damage Damage or failure that adversely affects the structural strength, performance, or flight characteristics of the aircraft, and which normally would require major repair or replacement of the affected component. [ARP4107]

substitute item An item that possesses such functional and physical characteristics as to be capable of being exchanged for another only under specified conditions or in particular applications, without alteration of the item itself or of adjoining items. [AIR4896]

substitutional element An alloying element with an atomic size and other features similar to the titanium atom, which can replace or substitute for the titanium atoms in the lattice and form a significant region of solid solution in the phase diagram. Such elements, which are used in alloying titanium, include but are not limited to aluminum, vanadium, molybdenum, chromium, iron, tin, and zirconium. [AS1814]

substitution equipment An alternative type aircraft assigned to replace the aircraft originally assigned, when the originally assigned aircraft encounters a technical problem. [AIR4896]

substrate A material upon whose surface an adhesive or resin is spread for any purpose, such as bonding or coating. [ARP5089]

sub-subcommutation Commutation of a number of channels, with the output applied to an individual channel of a subcommutator. Unique identification must be provided for sub-subcommutation frame synchronization.

subsystem *1.* The device or functional unit receiving data transfer service from the data bus. [AS1773] *2.* A major functional portion of a system that contributes to operational completeness of the system. [ARD50010]

subthreshold *See* threshold.

successive approximation A type of analog-to-digital conversion in which the unknown input is compared with sums of accurately known binary fractions of full scale, starting with the largest, and rejecting any that changes the state of the comparator. At the end of conversion, the output of the converter is a digital representation of the ratio of the input to full scale by a fractional binary code.

suction head *See* suction lift.

suction lift The pressure, in feet of fluid, that a pump must induce on the suction side to raise the fluid from the level in the supply well to the level of the pump. Also known as suction head.

suction line A tube, pipe, or conduit that leads fluid from a reservoir or intake system to the intake port of a pump or compressor.

suction pressure For values below atmospheric pressure, atmospheric pressure minus absolute pressure. [ARP4386]

sudden ionospheric disturbances Complex combinations of sudden changes in the conditions of the ionosphere and the effects of these changes.

summation action A type of control-system action in which the actuating signal is the algebraic sum of two or more controller output signals; or in which the actuating signal depends on a feedback signal that is the algebraic sum of two or more controller output signals.

summing point Any point in a control system where an algebraic summation of two or more control-loop variables is performed. [ARP4386]

sunseeker Two-axis device actuated by servos and controlled by photocells to keep instruments pointed toward the sun despite rolling and tumbling of an aerospace vehicle in which the instruments are carried. [ARP4386]

supercharger A high-speed impeller driven by the engine or its exhaust gases to increase manifold pressure and thus enhance the performance of the engine.

supercommutation *1.* Commutation at a rate higher than once per commutator cycle, accomplished by connecting a single data input source to equally spaced contacts of the commutator (cross-patching); corresponding cross-patching is required at the decommutator. *2. See* interval.

supercompressibility The extent to which behavior of a gas departs from Boyle's law.

superconducting quantum interferometers. *See* squid (detectors).

superconductivity Near-zero electrical resistance exhibited by some metals. With such metals, extremely powerful currents and magnetic fields are possible.

superconductor A compound capable of exhibiting superconductivity, that is, an abrupt and large increase in electrical conductivity as the temperature of the material approaches absolute zero.

superfines The portion of a metal powder whose particle size is less than 10 micrometers.

superfinishing Producing a finely honed surface by rubbing a metal with abrasive stones.

superform Refers to a patented double-metallic diaphragm process in which the unique superplastic deformation properties of aluminum cauls are used to form and consolidate thermoplastic composite parts. [AIR4844]

superheat To raise the temperature of steam above its saturation temperature.

superheated steam Steam at a higher temperature than its saturation temperature.

superheated vapor *See* vapor, superheated.

superheater A nest of tubes in the upper part of a steam boiler whose function is to raise the steam temperature above saturation temperature.

superlattices Crystals grown by depositing semiconductors in layers whose thickness is measured in atoms.

superluminescent diode A compromise between a diode laser and LED; operated at the high drive currents characteristic of diode lasers, but lacks the cavity-mirror feedback mechanisms that produce stimulated emission. Used when high power output is desired, but coherent emission is not.

supernatant liquor The liquid above settled solids, as in a gravity separator.

superplastic forming A strain-rate-sensitive metal-forming process utilizing the characteristics of materials exhibiting high elongation-to-failure. [AIR4844]

superplasticity The unusual ability of some metals and alloys to elongate uniformly by several thousand percent at elevated temperatures without separating.

superpressure balloons Meteorological balloons consisting of nonextensible envelopes designed to withstand higher internal pressure differentials than external ones. Such balloons will maintain constant elevations until sufficient gas diffuses from them to cause a change in buoyancy.

superrotation The generally more rapid relative motions found in the very tenuous regions of the atmosphere, at heights around 300 km. The density of the atmosphere decreases rapidly with height and more than 95% of the mass of the atmosphere is contained within the troposphere and lower stratosphere; these regions of the atmosphere rotate faster on average than the underlying solid earth.

supersede Describes the condition when a new item will be used, with the old item being eventually forced out of use; i.e., the new item supersedes the old item. [AIR4896]

supersonic *1.* A generic term roughly designating a speed that exceeds the speed of sound in a given fluid medium. *2.* For an aircraft, any speed that exceeds Mach 1, which is about 650 to 750 mph depending on atmospheric conditions and altitude.

supersonic compressors Compressors by which supersonic velocity is imparted to the fluid relative to the rotor blades and/or the stator blades, producing oblique shock waves over the blades to obtain a high pressure rise.

supersonic diffusers Diffusers designed to reduce the velocity and increase the pressure of fluid moving at supersonic velocities.

supersonic flow In aerodynamics, flow of a fluid over a body at speeds greater than the acoustic velocity, in which the shock waves start at the surface of the body.

supersonic nozzles Converging diverging nozzles designed to accelerate a fluid to supersonic speed.

supervisory control *1.* Implies a type of control in which controller output or computer program output is used as an input to other controllers, e.g., generation of setpoints in cascaded control systems. Also known as digitally-directed analog (DDA) control (which is distinct from direct digital control). *2.* An analog system of control in which controller setpoints can be adjusted remotely, usually by a supervisory computer.

supervisory control system (SCS) Remote setpoint information to single-loop analog controllers provided by a digital computer.

supervisory pressure A motivating factor stemming from a person's need to meet perceived supervisory expectations. [ARP4107]

supervisory program A program used in supervisory control.

supplemental type certificate (STC) An FAA approval of a major change in an aircraft, engine, or subsystem that is not great enough to require a new application for a type certificate. The STC supplements the type certificate data sheet of the aircraft to include the approved modification. [ARP4107]

supply pressure In a hydraulic or pneumatic system, the output pressure from the primary source of pressure; this is subsequently regulated or controlled to provide desired system functions.

support A device or structure that serves to hold in position and/or act as a proper foundation by bearing the weight or stress of another part or parts. [ARP480A]

supportability The degree to which system design characteristics and planned logistics resources, including manpower, meet system availability requirements. [ARP4293]

support and test equipment One of the nine principle elements of ILS; consists of tools, metrology and calibration equipment, performance-monitoring and fault-isolation equipment, maintenance stands, and handling devices required to support the operation and maintenance of systems. Items are categorized as peculiar (to the system under development) and common (commercially available), or currently in the defense inventory. Includes equipment categorized as ground support equipment (GSE) or aerospace ground equipment (AGE). [ARD50010]

supported adhesive film An adhesive supplied in a sheet or in a film form with an incorporated carrier that remains in the bond when the adhesive is supplied and used. [AIR4844]

support equipment Equipment required to support the operation and maintenance of the aircraft and all of its airborne equipment. [ARD50010]

supporting electrode In a spectroscopic apparatus, an electrode, other than a self electrode, that is designed to hold the analytical sample on or inside it.

support system A programming system used to support the normal translating functions of machine-oriented, procedural-oriented, and problem-oriented language processors.

support tube The portion of a thermocouple assembly that is immersed in the medium to be measured and provides support for the measuring junction and lead wires. [ARP485]

suppressed range An instrument range that does not include zero. The degree of suppression is expressed by the ratio of the value at the lower end of the scale to the span.

suppressed weir A rectangular weir in which the width of the approach channel is equal to the crest width, i.e., there are no end contractions.

suppressed-zero instrument Any indicating or recording instrument whose zero (no-load) indicator position is offscale, below the lower limit of travel for the pointer or marking device.

suprathreshold *See* threshold.

surface *1.* For a part, the area that continues uninterrupted until it adjoins a fillet, corner, or another individual surface. [AS291D] *2.* The exterior skin of a solid body, considered to have zero thickness.

surface activation The chemical process of making a surface more receptive to bonding to a coating or an encapsulating material. [AIR4844]

surface-active agent *See* surfactant.

surface analyzer An instrument that measures irregularities in the surface of a body by moving a stylus across the surface in a predetermined pattern and producing a trace showing minute differences in height above a reference plane magnified as much as 50,000 times.

surface blowoff Removal of water, foam, etc. from the surface at the water level in a boiler; or the equipment for such removal.

surface combustion The non-luminous burning of a combustible gaseous mixture close to the surface of a hot, porous, refractory material through which it has passed.

surface condenser Any of several condenser designs for inducing a change of state from gas to liquid by allowing the gas phase to come in contact with a surface, such as a plate or tube, that is cooled on the opposite side, usually by being in direct contact with flowing cooled water.

surface density Any amount distributed over a surface, expressed as amount per unit area of surface.

surface effect ships Vessels that are based on the ground effect principle and have submerged rigid sidewalls (sealants).

surface filter Porous materials that retain contaminants primarily on the influent face. In a surface filter, the filtration holes for particle retention are on the same plane within the filtration media. [AIR888]

surface filter element A filter element that removes particles from a fluid system by direct interception. [ARP4386]

surface finish The roughness of a surface after finishing, measured either by comparing its appearance with a set of standards of different patterns and lusters, or by measuring the height of surface irregularities with a profilometer or surface analyzer.

surface flaws Irregularities of any sort that occur at only one place, or at relatively infrequent and widely varying random intervals on a surface. A flaw may be a scratch, ridge, hole, peak, crack, check, etc. Unless otherwise specified, the effect of flaws is not be included in roughness height measurements. *See also* surface irregularities. [AS291D]

surface gage A scribing tool in an adjustable stand that is used to check or lay out heights above a reference plane.

surface hardening Any of several processes for producing a surface layer on steel that is harder and more wear-resistant than the softer, tougher core. Usually involves some kind of heat treatment, and may or may not involve changing the chemical composition of the surface layer. *See also* case hardening.

surface irregularities Deviations from the nominal surface, as roughness, waviness, or flaws. *Roughness*–relatively finely spaced irregularities, the height, width, shape, and direction of which establish the predominant surface pattern. *Waviness*–irregularities of the nominal surface evidenced by recurrent forms of waves. Waviness may be caused by factors such as machining deflections, vibrations, heat treatment, or warping strains. *Flaws*–irregularities of any sort that occur at only one place or at relatively infrequent and widely varying random intervals on a surface. A flaw may be a scratch, ridge, hole, peak, crack, check, etc. Unless otherwise specified, the effect of flaws is not included in the roughness height measurement. [AS291D]

surface, nominal *See* nominal surface.

surface plate A table, usually made of granite or steel at least 2 ft square, that has a very accurate flat plane surface; used primarily in inspection and layout work as a reference plane for determining heights.

surface preparation Physical and/or chemical preparation of an adherent to make it suitable for adhesive bonding. [AIR4844]

surfacer Material applied on the toolside surface of a part to fill in resin starvation, porosity, or roughness, to maintain contour or aerodynamic smoothness. [AIR4844]

surface resistivity The ratio of the potential gradient parallel to the current along the surface of a material, to the current per unit width of the surface. [ARP1931]

surface roughness Relatively finely spaced irregularities the height, width, shape and direction of which establish the predominant surface pattern. *See also* surface irregularities. [AS291D]

surface, roughness height rating *See* roughness height rating.

surface, roughness width *See* roughness width.

surface tension The contractive force in the surface film of a liquid which tends to make the liquid occupy the least possible volume. [AIR4844]

surface transfer impedance When a current is caused to flow on the outside of a cable shield, the ratio of the induced voltage resulting along the inside of the shield, to the driving current. [ARP1931]

surface treating Any of several processes for altering properties of a metal surface, making it more receptive to ink, paint, electroplating, adhesives, or other coatings; or making it more resistant to weathering or chemical attack.

surface treatment *1.* A material applied to fibrous material during the forming operation or in a subsequent process. *2.* The preparation of surfaces to be bonded together. [AIR4844]

surface wave A type of wave in which the molecules vibrate in an elliptical motion to a depth of one wavelength in the carrier material. *See also* Rayleigh waves. [ARP5089]

surface, waviness Irregularities of the nominal surface evidenced by recurrent forms of waves. Waviness may be caused by factors such as machining deflections, vibration, heat treatment, or warping strains. *See also* surface irregularities. [AS291D]

surface, waviness height value *See* waviness height value.

surface, waviness width value *See* waviness width value.

surfacing Depositing filler metal on the surface of a part by welding or thermal spraying.

surfacing mat A thin mat of fine fibers used primarily to produce a smooth surface on an organic matrix composite. [AIR4844]

surfactant Surface active agent, usually liquid or particulate, that alters surface tension or performs other tasks at boundaries between dissimilar materials.

surge *1.* A variation from the controlled steady-state level of a characteristic, resulting from the inherent regulation of the electric-power supply system and remedial action by the regulator. [AS1212] *2.* A transient rise in power, pressure, etc., such as a brief rise in return pressure in a hydraulic system. *See* transient. [AIR1489] *3.* A response of the entire engine that is characterized by a flow stoppage or reversal in the compression system. [ARP1420] *4.* In compressors, an unstable operating regime in which internal oscillations persist.

surge damping valve A valve whose function is to reduce surge pressures during intermittent flow. [ARP4386]

surge line The locus of surge points on a compressor map. [ARP1420]

surge margin The pressure ratio range, at a constant corrected airflow, through which a compressor may be operated between its operating point and surge line without surge. [ARP1420]

surge point On a compressor map, the point at the limit of stable operation, defined by the maximum pressure ratio at a given airflow. [ARP1420]

surge pressure A sudden rise in fuel pressure, usually resulting from system start-up, shutdown, or the operation of valves, cylinders, or other components connected to the same system. [AIR1749]

surge-pressure control Suppression accomplished by controlling the flow rate of liquid into the operator chamber in a fashion to control the time-rate-of-change of poppet position with respect to the seat. [AIR1660A]

surge-pressure relief Suppression accomplished by limiting the pressure in the operator chamber of the shutoff valve. [AIR1660A]

surge protector A device positioned between a computer and a power outlet designed to absorb power bursts that could damage the computer. *See also* power line protector.

surge tank *1.* A standpipe or storage reservoir in a downstream channel or conduit to absorb sudden rises in pressure and to prevent starving the conduit during sudden drops in pressure. *2.* An open tank connected to the top of a surge line which maintains steady loading on a pump.

surveillance approach An instrument approach in which the air-traffic controller issues instructions for pilot compliance based on aircraft position in relation to the final approach course (azimuth) and the distance (range) from the end of the runway as displayed on the controller's radar scope. [ARP4107]

survivability A measure of the degree of capability of a system to withstand a hostile environment without suffering an abortive impairment of its ability to accomplish its designated function. [ARP4386]

survivor curve A type of reliability curve that shows the average percent of total production of a given model or type of machine still in service after various lengths of service life in hours.

survivor locator light A device consisting of parts and components that may include, but is not limited to, a light source, a battery pack, interconnecting electrical wires, rigging/activation means, and integral attachment/security means. [AS4492]

susceptibility meter An instrument for determining magnetic susceptibility at low magnitudes.

susceptometer A device for measuring the magnetic susceptibility of ferromagnetic, paramagnetic, or diamagnetic materials.

suspend A halt in the process of attaching the aircraft to the catapult, resulting from the need to ascertain whether the launch can be safely accomplished. [AIR1489]

suspended arch An arch in which the refractory blocks or shapes are suspended by metallic hangers.

suspended solids Undissolved solids in boiler water.

suspension *1.* A dispersion of a solid in a liquid. [AIR4844] *2.* A fine wire or coil spring that supports the moving element of a meter or other instrument. *3.* A system of springs, shock absorbers, and other devices that support the chassis of a motor vehicle on its running gear. *4.* A system of springs or other devices that support an instrument or sensitive electronic equipment on a frame, and reduce the intensity of mechanical shock or vibration transmitted through the frame to the instrument.

suspension line *See* shroud line.

sustained flight vehicle A powered vehicle deriving most of its lift from aerodynamic forces, but augmented by centrifugal force at altitudes considered arbitrarily to be at 150,000 ft (45,720 m). [ARP4386]

sustainer charge A component of an ignition train that maintains the operating pressure until thermal equilibrium is obtained. [ARP4386]

sustainer grain A propellant or pyrotechnic grain used in a pressure cartridge or igniter to sustain burning. [AIR913]

sustainer rocket A rocket engine used as a sustainer, especially on an orbital glider or orbiting spacecraft that dips into the atmosphere at its perigee. [ARP4386]

SW *See* stud arc welding.

swaging (swedging) *1.* The mechanical reshaping of barrels; an obsolete term for crimping. [ARP914A] *2.* Any of several methods of tapering or reducing the diameter of a rod or tube, most commonly involving hammering, forging, or squeezing between simple concave dies.

SWAP *See* severe weather avoidance plan.

swath width The width of the area covered by an imaging sensor, determined by the geometry of the instrument.

S waves Waves in an elastic media which cause an element of the medium to change its shape without a change in volume. Mathematically, S waves are waves whose velocity field has zero divergence.

sweat *1.* The condensation of moisture from a warm, saturated atmosphere on a cooler surface. *2.* A slight weep in a boiler joint, but not in sufficient amount to form drops.

sweat cooling A process by which a body having a porous surface is cooled by forced flow of coolant through the surface from the interior.

swedging *See* swaging.

sweep The uniform and repeated movement of an electron beam across the CRT. [ARP5089]

sweep delay *1.* A delay in time, after the initial pulse, of starting the sweep presentation. *2.* The control used for adjusting the time of starting the sweep presentation. [ARP5089]

swell A sudden increase in the volume of steam in a water-steam mixture below the water level.

swept wing Visibly obvious backward inclination of an airfoil from root to tip, so that the leading edge meets relative wind obliquely.

swinging load A load that changes at relatively short intervals.

swing joint A connection between two pipes that allows them to be repositioned with respect to each other.

swing pipe A discharge pipe whose inlet end can be raised or lowered within a tank.

switch, air flow proving A device installed in an air stream which senses air flow or loss thereof and electrically transmits the resulting impulses to the flame-failure circuit.

switch, high pressure A device that monitors liquid, steam, or gas pressure, and is arranged to open and/or close contacts when the pressure value is exceeded.

switching pressure The pressure at which a system or a component is activated, deactivated, or reversed. [ARP4386]

switching time In a switching device, the time interval between the reference time, or time at which the leading edge of a switching or driving pulse occurs, and the last instant at which the instantaneous output response reaches a stated fraction of its peak value.

switch loading - switch refueling Refueling with a grade of fuel different than the residual fuel occupying the tank. [AIR4170]

switch, low pressure A device to monitor liquid, steam, or gas pressure, and arranged to open and/or close contacts when the pressure drops below the set value.

switch, oil temperature limit A device to monitor the temperature of oil between preset limits, and arranged to open and/or close contacts should improper oil temperature be detected.

switch, pressure A device that opens or closes an electrical circuit at a give fluid pressure. [ARP4386]

swivel *1.* A rotatable fluid connection. [ARP4386] *2.* A mechanical device that can move freely about a pin joint.

swivel angle *1.* The angle through which a gear or wheel is free to swivel or caster. [AIR1489] *2.* A transient variation in the current and/or potential at some point in the circuit. *3.* An upheaval of liquid in a process system, which may result in carryover of liquid into vapor lines. *4.* An unstable pressure buildup in a process system.

swivel fitting A fitting, directionally adjustable without lateral movement, usually retained by a flange, and not free to rotate in service. [ARP4386]

swivel joint A fluid connection that is free to rotate in service, usually under pressure. [ARP4386]

symbol An identifiable display element characterized by shape, size, structure, location, brightness, and color; any of these characteristics may be static or dynamic. One or more of these characteristics is used to encode information. [ARP4155]

symbology The use of one or more symbols that make up a format to portray/define information. [ARP4107]

symbology generator unit (SGU) Component that provides symbology information to the image source. [ARP4102/8]

symbol-task pairing Refers to the relationship between symbol attributes, information content, and the task to be accomplished by the end-user. [ARP4155]

symmetrical laminate A composite laminate in which the sequence of plies below the laminate midplane is a mirror image of the stacking sequence above the midplane. [AIR4844]

symmetrical transducer A transducer in which all possible termination pairs may be interchanged without affecting transducer function.

symmetry breaking *See* broken symmetry.

symmetry/symmetrical *1.* Corresponding to form and arrangement of parts on opposite sides of a boundary, such as a plane or line, or around a point or axis. *2.* An arrangement with balanced or harmonious proportions. *3.* The degree of equality between the normal flow gain of one polarity and that of the reversed polarity. Symmetry is measured as the difference in normal flow gain of each polarity, expressed as a percent of the greater. [ARP490]

sympathetic detonation (ignition) The explosion of a second charge or device caused by nearby detonation (ignition) of another. [AIR913]

synchronization Keeping one part of a process in a fixed relationship to another part of the process; for example, keeping the advance of strip stock through a stamping press timed to move ahead only when the press has raised the die free of the work.

synchronization of parallel computational processes Controlling the execution of parallel computational processes so as to maintain some desired sequential relationship between programmed actions within the processes. *See also* semaphore.

synchronization pattern A sequence of ones and zeros that signals the start of each frame of PCM data; the pattern generally chosen is a pseudo-random sequence, one that is unlikely to occur randomly in data.

synchronize To lock one element of a system into step with another. Usually refers to locking a receiver to a transmitter, but can also refer to locking the data terminal equipment bit rate to the data-set frequency.

synchronous As applied to two or more circuits, motors, machines, or other devices, electrically or mechanically in phase or in step.

synchronous data link control (SDLC) A type of data link protocol.

synchronous platforms Space platforms whose rotation is synchronized with that of earth. Also known as geostationary platforms.

synchronous satellites Equatorial west-to-east satellites orbiting the earth at an altitude of approximately 35,900 kilometers; at this altitude, they make one revolution in 24 hours, synchronous with the earth's rotation. Also known as geostationary satellites.

synchronous system trap (SST) A system condition that occurs as a result of an error or fault within the executing tasks.

synchroscope An instrument for indicating whether two periodic quantities are synchronous. This is done by means of a rotating pointer or cathode-ray oscilloscope; the position of the pointer or pattern on the

oscilloscope tube indicates the instantaneous phase difference between the two quantities.

synchro-to-digital converter (S/D) An electronic device for converting the analog output signal of a rotary transformer (synchro) into a digital word for further processing.

synchrotrons Devices for accelerating particles, ordinarily electrons, in a circular orbit in an increasing magnetic field by means of an alternating field applied in a synchronism with the orbital motion.

synergetic curve A curve plotted for the ascent of an aerospace vehicle, determined to give the missile or other vehicle an optimum economy in fuel with an optimum velocity. [ARP4386]

synergism An action whereby the total effect of two components or agents is greater than the individual effects of the components when simply added together; for instance, in stress-corrosion cracking, cracks form and propagate deep into a material in a much shorter time and at a much lower stress than could be predicted from known effects of stress and the corrosive environment.

syntax *1.* The structure of expressions in a language. *2.* The rules governing the structure of a language.

synthetic aperture radar Active microwave sensors providing all-weather, high-resolution imagery.

synthetic apertures In radar technology, the simulations of large antennas by correcting the phase and magnitude of the return signals from smaller antennas, permitting the use of lower frequencies for airborne radars.

synthetic quench A water solution of polyalkylene glycol or other synthetic material, used for quenching when minimum distortion or low residual stresses are desired. [AMS2771A]

syntony Describes the condition of two or more oscillating circuits having the same resonant frequency.

system administrative time All system downtime other than active maintenance time and logistic time. [AIR4896]

system analysis The examination of an activity, procedure, method, technique, or business to determine behavioral relationships, or what must be accomplished and how.

system, arresting A system comprising the gear and apparatus designed to arrest the aircraft in its landing roll. *See also* arresting gear. [AIR1489]

systematic error *1.* An error in a set of measurements or control that can be predicted from scientific principles. *2.* Error that cannot be reduced by increasing the number of measurements if the equipment and conditions remain unchanged. *3.* Any constant or reproducible error introduced into a measured or controlled value due to failure to control or compensate for a specific side effect.

system availability The percent of time that a space system is capable of or is performing the specified missions. [AIR4896]

system, brake control *See* brake control system.

system characteristic equation The denominator of the system transfer function. [AIR1823]

system check A check on the overall performance of the system, usually not made by built-in computer check circuits, e.g., control totals, hash totals, and record counts.

system downtime The time interval between the commencement of work on a system malfunction and the time when the system has been repaired and/or checked by the maintenance man, and no further maintenance activity is executed. [AIR4896]

system effectiveness The probability that a system can successfully meet an operational demand within a given time when operated under specified conditions. [ARD50010]

Systeme Internationale d'Unites (SI) The current International System of Units.

system/equipment The item under analysis, be it a complete system, or any portion thereof being procured. [AIR4896]

system error In a control system, the difference between the value of the ultimately controlled variable and its ideal value.

system final test time The time spent confirming that a system is in satisfactory operating condition following maintenance. [AIR4896]

system, gear positioning A system comprising the gear and apparatus for positioning the landing gear in its respective operational positions. Includes uplocks, downlocks, actuator, sequencing, and indicating subsystems. [AIR1489]

system generated electromagnetic pulses Electromagnetic fields generated by the emission of a large electronic current from a metallic body in space, caused by the incidence on the body's surface of strong ionizing radiation pulses (usually x-ray) from space.

system integration The combining of subsystems, each with numerous interfaces for the input and output of data, and each with specified functions vital to the planned success of the main system.

system, landing impact A system comprising the gear and apparatus for absorbing and dissipating the energy and loads imposed on the air vehicle by impact with the ground. [AIR1489]

system reliability The probability that a specific system will complete a specific mission without failure of mission-essential system functions. [ARD50013]

system resonance The closed-loop resonant frequency observed with complete system operation. The system resonance is lower than the load natural frequency due to dynamic effects of system control elements, damping, and backlash. [ARP4386]

system R&M parameter A measure of reliability or maintainability in which the units of measurement are directly related to operational readiness, mission success, maintenance manpower cost, or logistic support cost. [AIR4896]

system safety assessment A systematic, comprehensive evaluation of the implemented system to show that the relevant safety requirements are met. [ARP4754]

system simulation The simulation of any dynamic system.

system, steering control A system comprising the gear and apparatus for steering the aircraft in ground operations. Includes an input control (wheel or rudder pedals), mechanical or electrical connection to power control valves, a power unit, and other elements. [AIR1489]

system synchronization A means by which transmitting and receiving stations establish a time reference. [AIR4271]

system test Determining performance characteristics of an integrated, interconnected assemblage of equipment under conditions that evaluate its ability to perform as intended and that verify suitability of its interconnections.

T

T *See* tesla.

T₁ *See* tread depth.

T₂ *See* effective tread depth.

TAAF testing *See* reliability development/growth test.

tab A small auxiliary control surface attached to a larger control surface such as an aileron, rudder, or elevator, usually at the trailing edge. Deflection of the tab causes the larger surface to deflect in the opposite direction. [ARP4107]

TACAN Contracted from tactical air navigation, an ultra-high-frequency electronic air navigation aid that provides suitably equipped aircraft with a continuous indication of bearing and distance to the TACAN station. [ARP4107]

tachometer An instrument designed to indicate the revolutions per minute of an engine crankshaft. [ARP4107]

tachometer generator A generator that is designed to be energized from a single-phase voltage source and to deliver a generated output voltage essentially proportional to speed and energizing voltage. (In practice, residual voltages at zero speed exist.) The frequency of this output voltage is the same as the fundamental energizing voltage frequency. [ARP667]

tack *1.* Stickiness of an adhesive or filament-reinforced resin prepreg material. *2.* The property of an adhesive whereby it is able to form a bond of measurable strength immediately after the adhesive and the adherent are brought into contact under low pressure. [AIR4844]

tack-free A condition in which a plastic material can be dented with an inert object without sticking to it. [AIR4844]

tack free time The time required at standard conditions for a curing sealant to lose its surface tackiness, as determined by placing a small piece of polyethylene film on its surface and then peeling the film away. The sealant is tack free when no sealant is removed by the film. [AIR4069]

tack weld Any small, isolated arc weld, especially one that does not bear load but rather merely holds two pieces in a fixed relationship.

tactical air navigation *See* TACAN.

tactical message A clearance message relating to a situation of a local nature. [ARP4791]

tag number Instrument loop identification number.

tail assemblies *1.* The rear part of a body, as of an aircraft or a rocket. Also referred to as tails. *2.* The tail surfaces of an aircraft or rocket.

tail bumper A landing gear installation at the aft end of the fuselage or tail of the aircraft to protect it from contact with the ground; may be a simple skid, a shock-absorbing unit, or a complete retractable installation. [AIR1489]

tail gear *See* gear, tail.

tails (assemblies) *See* tail assemblies.

tail skid A tail bumper installation that consists of a simple skid or runner. *See* skid. [AIR1489]

tail tire An auxiliary tire that is mounted on the rear of the aircraft. [AS4833]

tail wheel A wheel at the tail, used to support the tail section on the ground and/or protect that portion of the aircraft during takeoff and landing. A tail wheel may be retractable, steerable, fixed, castering, etc. [AIR1489]

takeoff *1.* The action of a rocket vehicle departing from its launch pad. *2.* The action of an aircraft as it becomes airborne.

takeoff, aborted A takeoff that has to be discontinued for any reason. [AIR4896]

takeoff decision speed *See* V_1.

takeoff minimums The minimum weather conditions for aircraft to take off; generally 1 mile visibility or RVR 50 for one- and two-engine aircraft, and 1/2-mile visibility or RVR 24 for aircraft with three or more engines. *See* runway visual range (RVR). [ARP4107]

takeoff phase *See* mishap, maneuver.

takeoff power *1.* The power certificated for takeoff or go-around. [ARP4102/5] *2.* With respect to reciprocating engines, the brake horsepower that is developed under standard sea-level conditions, under the maximum conditions of crankshaft rotational speed and engine manifold pressure approved for the normal takeoff, and limited in continuous use to the period of time shown in the approved engine specification. *3.* With respect to turbine engines, the pounds of thrust that are developed under static conditions at a specified altitude and atmospheric temperature, under the maximum conditions of rotor shaft rotational speed and gas temperature approved for the normal takeoff, and limited in continuous use to the period of time shown in the approved engine specification. [ARP4107]

takeoff roll The movement of an aircraft along the runway from the start or takeoff to rotation and liftoff. [ARP4107]

takeoff safety speed *See* V_2.

takeoff thrust The equivalent of takeoff power for a turbojet or turbofan engine.

takeoff weight The weight of an aircraft at liftoff. *See also* ramp weight. [ARP4107]

talcum powder (red) A powder used in determining leak exit location; the powder turns bright red in fuel. [AIR4069]

tan delta *1.* The ratio of the loss modulus to the storage modulus, measured in compression, tension or flexure, or shear. *2.* The ratio of the out-of-phase components of the dielectric constant to the in-phase component of the dielectric constant. [AIR4844]

tandem actuators Two or more coaxial actuators that are mechanically connected to move together. A tandem actuator usually has two pistons on the same rod, carried in dual actuator cylinder housings. Separate cylinders (with a common piston rod) can be used to give partial rip-stop protection. [ARP4386]

tandem gear configuration *See* gear configuration, tandem.

tandem networks An arrangement of two-terminal-pair networks in which the output terminals of one network are directly connected to the input terminals of the other network.

tangent bulk modulus The change of fluid pressure with respect to volume change. May be calculated isothermally or adiabatically. [AIR1116]

tangent galvanometer A galvanometer consisting of a small compass mounted horizontally in the center of a large vertical coil of wire; the current through the coil is proportional to the tangent of the angle that the compass needle makes with its rest (no current) position.

tangent modulus The slope of the line at a predetermined point of a static stress-strain curve, expressed in force per unit area per unit strain. [AIR4844]

tank, barrier A film or material applied to the tank inner liner that resists fluid permeation. Used on most non-self-sealing and self-sealing fluid tanks containing aviation fuels. [AIR1664]

tank building form A male shape duplicating the shape of the finished tank. The various plies of rubber-coated fabric or spray-on polyurethane are fabricated over this shape. The form maintains the tank shape during vulcanization, and is removed after vulcanization. [AIR1664]

tank circuit A resonant electronic circuit that consists of a capacitor and an inductor connected in parallel.

tank, convolution Generally, a recess projecting into the tank surface to clear structural ribs, formers, longerons, etc. Convolutions permit utilization of the maximum volume of the structure for fluid storage. [AIR1664]

tank, crash-resistant A variation of either a non-self-sealing tank or self-sealing flexible tank fabricated of very strong elastomer-coated fabric and high-strength fittings. Such a tank is intended to retain fluid without leakage under impact-survivable crash conditions. [AIR1664]

tank, fiber-lock fitting A fitting design developed under the auspices of the US Army, specifically for crash-resistant tanks. In such a tank, numerous individual fibers are fabricated readily into the tank wall as means of retaining the metal fitting in the tank. [AIR1664]

tank, fit check fixture A fixture accurately simulating the size and shape of the aircraft flexible tank cavity. Used as a quality-assurance acceptance test tool. [AIR1664]

tank fitting, flange Plies of rubber-coated fabric or molded rubber used as a means of attaching fitting subassemblies to a flexible tank wall. Flanges are bonded to the metal rings of the fitting and also to the tank wall and become an integral part of the tank assembly during vulcanization. [AIR1664]

tank fittings, mounted Subassemblies fabricated into the tank wall for the purpose of: (a) attaching the tank to structure; (b) providing a sealing surface for and a means of attaching equipment items to the tank; (c) providing for the attachment and sealing of vent lines and fluid flow lines to the tank. [AIR1664]

tank, hanger fittings *See* hanger fittings.

tank, hydrodynamic ram *See* hydrodynamic ram.

tank, lace line *See* lace line.

tank, liner (inner liner) *See* liner.

tank, molded in place seal *See* molded in place seal.

tank, nipple fitting *See* nipple fitting.

tank, non-self-sealing *See* nonself-sealing tank.

tank-selector valve A valve used to select one of the separate gas tanks from which fuel is to be drawn; usually located in the flight station or cockpit of an aircraft. [ARP4107]

tank test *See* wet dielectric test.

tap *1.* An internal threading tool with longitudinally separated cutting edges designed to form specific threads. [ARP480A] *2.* A threaded plug, in which the threads are of accurate form and dimensions, and have cutting edges that form internal threads in a hole as the plug is screwed into the hole. *3.* A small hole in the wall of a pipe or process vessel, usually threaded, where an instrument, control device, or sampling device is attached. *4.* A coupler in which part of the light carried by one fiber is split off and inserted into another fiber; essentially the same as a tee coupler. *5.* To withdraw a quantity of molten metal from a

refining or remelting furnace.

tape A ribbon made of plastic, metal, paper, or other flexible material suitable for data recording by means of electromagnetic imprinting, punching or embossing patterns, or printing.

tape-and-plumb-bob liquid-level gage *See* plumb-bob gage.

tape-controlled machine A machine tool operated automatically by means of control signals read off a length of magnetic or punched paper tape.

tape drive A device that moves magnetic tape past a head that can "read" the tape.

tape formatter A device, including buffers and controls, for recording ordered data on magnetic tape in gapped form, in a format recognized by a computer.

tape header data Several recorded characters at the beginning of a magnetic tape, used to identify the content of the tape.

taper A dimensional feature whereby thickness, height, diameter, or some other measurement varies linearly with distance along a given axis.

tapered-roller bearing A roller bearing having tapered rollers that run in conical races; can support both radial and thrust loads.

tapered-tube rotameter A type of variable-area flowmeter in which a float that has greater density than the fluid rides inside a tapered tube in such a manner that fluid flowing upward through the tapered section carries the float with it until the upward force exerted by the flowing fluid just balances the downward force due to float weight; as the float rides upward, the annular area around it becomes larger and force on the float decreases. If the tube is made of glass, it can be graduated so that flow rate is read directly by observing float position; otherwise flow rate must be determined from an indirect indication of float position.

tapered waveguide A waveguide section having a continuous change in cross section.

taper pin An electric solderless terminal used to connect a wire to a special receptacle. A taper pin has a round, tapered shank on one end, and a wire barrel on the other end that is crimped to the wire. Formed and solid pins are available. [ARP592]

taper pin receptacles Electric terminals having round, tapered holes that mate with taper pins. Formed and solid receptacles are available. [ARP592]

taper pin terminal blocks Electric connector blocks made from insulating material, and containing taper pin receptacles. [ARP592]

tape search A hardware process by which an operator or computer can cause instrumentation tape to be searched automatically for specific start and stop times for data reduction.

tape speed compensation signal A signal recorded on instrumentation tape along with the data (preferably on the same track as the data), used to correct electrically for tape speed errors during playback.

tape transport *See* transport, tape.

tape-type liquid-level gage A liquid-level gage consisting of a tape wound around a drum that is attached to a pointer or other level indicator, with one end of the tape attached to a float and the other counterweighted to keep the tape taut.

tape wrap A spirally or longitudinally applied insulating tape, wrapped around a conductor or cable, either insulated or uninsulated, and used as an insulation or mechanical barrier. [ARP1931]

taphole A hole in the side or bottom of a furnace or ladle for draining off molten metal.

tappet An oscillating part, such as a lever, that is operated by a cam or push rod, and is used to tap or push another machine element, such as a valve.

tappet rod A pivot rod carrying one or more tappets and acting as a fulcrum for their motion.

tap test Refers to the practice of tapping a part with a solid object and listening for acoustic changes which may indicate flaws. [ARP5089]

tare weight In any weighing operation, the residual weight of any containers, scale components, or residue, which is included in total indicated weight and must be subtracted to determine weight of the live load.

tare weight (filter) The initial weight as received, after equilibration to ambient room conditions. [ARP785]

target *1.* A goal or standard against which some quantity, such as productivity, is compared. *2.* A point of aim or object to be observed by visual means, electromagnetic imaging, radar, sonar, or similar noncontact method.

target acquisition The process of optically, manually, mechanically, or electronically orienting tracking systems in the direction and range to lock on a target.

target computer A computing system on which a real-time program derived from the model is to execute. [ARP4148]

target drone An unmanned aircraft or a missile used as a target for testing interception equipment and procedures. [ARP4386]

target flowmeter A device for measuring fluid-flow rates by means of the drag force exerted on a sharp-edged disk centered in a circular flowpath, the drag force being due to differential pressure created by fluid flowing through the annulus. Usually, the disk is mounted on a bar whose axis coincides with the tube axis, and drag force is measured by a secondary device attached to the bar.

target masking Technique used in vision contrast discrimination involving the ratio of the luminance of a target (object) to the luminance of the background, especially when light and dark adaptation are factors.

target penetration *See* terminal ballistics.

target-type flowmeter An instrument for measuring fluid flow in which the fluid exerts force on a small circular disc suspended in the center of the flow conduit by means of a pivoted bar. The force exerted on the bar by a force-balance transmitter to counteract the fluid force on the target is an indication of the flow rate.

TAS *See* true airspeed.

TAT *See* total air temperature.

taut-band ammeter An instrument for measuring electric current in which a moving coil is mounted on a taut metal band held rigidly at the ends; when current flows through the coil, it deflects within the gap of a permanent magnet, twisting the metal band. The magnitude of current is indicated by a pointer attached to the coil when the torque exerted by magnetic-field interaction is balanced by restoring torque in the twisted band.

taxi To move an aircraft about on the ground, water, or other surface (such as a carrier deck), under its own power, except during the takeoff and landing rolls. [ARP4107]

taxi phase *See* mishap, maneuver.

taxistrip *See* taxiway.

taxi stroke The stroke of a shock absorber that is used mainly during taxi operations. Generally the portion from "static" toward "full compressed." [AIR1489]

taxiway Also referred to as taxistrip. A strip or area of the airfield set aside for taxi operations, from parking areas to the runway or interconnecting runways. Not intended to be used for takeoffs and landings. [AIR1489]

TCA *See* controlled airspace, terminal control area.

TCAS Partial acronym (pronounced Tee-CAS) for traffic alert and collision avoidance system. [ARP4153] An avionics system designed to present real-time information about the close proximity of other aircraft in order to facilitate the pilot's responses and to avoid mid-air collisions. [ARP4107]

TCAS-I The first version of TCAS; provides the pilot with vertical (altitude) information about other aircraft in close proximity. [ARP4107]

TCAS-II Proposed second-generation TCAS that would give pilots left-right guidance information in addition to vertical (altitude) information about other aircraft in close proximity to their own aircraft, thereby enabling them to take evasive measures to avoid mid-air collisions. [ARP4107]

TCH *See* threshold crossing height.

TDM *See* time division multiplex.

TDMA *See* time division multiplex access.

TDZE *See* touchdown zone elevation.

TDZL *See* touchdown zone lights.

team player *See* internalized unit values.

teardown inspection Inspection of the unit detail parts usually imposed as a part of the qualification testing of a starter or at the completion of a test program. [ARP906A]

tear-down time The amount of time required to disassemble a machine set-up following a production run and prior to setting up the jigs and fixtures for the next order.

teardrop procedure turn An instrument approach procedure turn that starts from over the approach facility on an outbound course that is displaced to the left (to the right if a nonstandard approach) of the reciprocal of the inbound course, and is followed by a turn to the right (left for nonstandard) to intercept the inbound course. [ARP4107]

tearing modes (plasmas) Explosive reconnections of energetic particle accelerations at high voltages in the magnetosphere during substorms.

tears Sharp breaks or fissures in the surface of a hose cover, generally caused by strain and service conditions. [ARP1658A]

tear strength *1.* The tearing point measured in pounds per linear inch or SI equivalent. [AIR1558] *2.* Force required to initiate or continue a tear in a material under specified conditions. [ARP1931]

TED *See* transferred electron devices.

tee coupler *1.* A fiber-optic coupler in which the three fiber ends are joined together, and a signal transmitted from one fiber is split between the other two. *2. See* tap.

tee joint A junction, such as in piping or a weldment, where a branch member is connected at one end to a cross member running at right angles to the branch.

teeming Pouring molten metal into an ingot mold. Most often used with reference to steel production.

teldata The common parallel interface for telemetry data, as from a frame synchronizer to a buffered data channel.

telecommunications Pertaining to the transmission of signals over long distances, such as by telegraph, radio, or television.

telemetering *1.* The transmission of a measurement over long distances, usually by electromagnetic means. *2.* Using radio waves, wires, or other means to transmit instrument readings to a remote location. Also known as remote metering or telemetry.

telemetry *1.* The science of measuring quantities, transmitting the results to a distant station, and interpreting, indicating, and/or recording the quantities measured. *2. See* telemetering.

telemetry front end (TFE) Hardware devices that accept multiplexed data and time, establish synchronization, convert to parallel data, and provide timing pulses, status, and the like for computer entry.

telemetry input channel A device that prepares telemetry data for input to a real-time computer.

telephotometry The body of principles and techniques concerned with measuring atmospheric extinction using various types of telephotometers.

teleran An aircraft navigation system that combines radar position information with a television image. With such a system, ground plan position indicator, map, and weather information are displayed together in the aircraft.

telescoping gage An adjustable gage for measuring inside dimensions such as hole diameters. Consists of a spring-loaded member that extends until it touches both sides of a hole; this is locked in place to prevent further extension as it is withdrawn from the hole. An outside micrometer or vernier calipers is used to measure the length of the locked gage member.

telltale A marker on the outside of a tank that indicates water level on the inside of the tank.

telluric currents Large-scale surges of electric charges within the earth's crust, associated with disturbances of the ionosphere. Also known as earth currents.

telluric lines In a solar spectrum, absorption lines produced by constituents of the atmosphere of the earth itself rather than by gases in the outer solar atmosphere, such as those responsible for the Fraunhofer lines.

TELSET The common parallel interface for telemetry set-up, as from a buffered data channel.

temper *1.* The relative hardness and strength of flat-rolled steel or stainless steel that cannot be further hardened by heat treatment. *2.* The relative hardness and strength of nonferrous alloys, produced by mechanical and/or thermal treatment, and characterized by a specific structure, range of mechanical properties, or reduction of area during cold working. *3.* In the production of casting molds, to moisten mold sand with water. *4.* In the heat treatment of ferrous alloys, to reheat after hardening for the purpose of decreasing hardness and increasing toughness without undergoing a eutectoid phase change. *5.* In tool steels, an imprecise shop term sometimes used to denote carbon content. *6.* In glass manufacture, to anneal or toughen by heating below the softening temperature. *7.* To moisten and mix clay, mortar, or plaster to a consistency suitable for use. *8.* A master alloy added to tin to make the finest pewter. *9. See* draw.

temperature *1.* The thermal state of a body considered with reference to its ability to communicate heat to other bodies. [AIR1900] *2.* The intensity of heat as measured on some definite temperature scale, by means of any of various types of thermometers. *3.* In statistical mechanics, a measure of translational molecular kinetic energy (with three degrees of freedom). *4.* In thermodynamics, the integrating factor of the differential equation referred to as the first law of thermodynamics.

temperature, ambient The static temperature of the medium surrounding any unit under consideration. [ARP171]

temperature, autogenous ignition Temperature at or above which a self-sustaining combustion of a critical volume of fluid will take place without the presence of an external source of ignition, such as a spark or flame. [ARP699D]

temperature coefficient *1.* A nonlinear value expressed as the allowed change in sensitivity as a percentage of the nominal sensitivity, when averaged over a specified temperature range. [ARP4386] *2.* The rate of change of some physical property (e.g., electrical resistivity) with temperature; the coefficient may be constant or nearly constant, or it may vary itself with temperature.

temperature coefficient of resistivity The amount of resistance change of a material per degree of temperature rise. [ARP1931]

temperature compensating circuit breaker A circuit breaker in which means may be inherent or otherwise provided for partially or completely neutralizing the effect that the ambient temperature may have on its tripping characteristics. [ARP1199A]

temperature compensation Refers to any construction or arrangement that makes a measurement device or system substantially unaffected by changes in ambient temperature.

temperature, corrected The dry air rated temperature available for cooling the heat load after the corrections have been made for removal of water in a low-pressure water separator, or reheating by a high-pressure water separator. [ARP147C]

temperature dependence For a material, refers to a characteristic that is dependent on changes in the ambient temperature.

temperature, design The selected normal operating temperature. [ARP699D]

temperature, dew point *See* dew point temperature.

temperature, differential *1.* In a duct, the difference between the internal temperature and a reference temperature. *2.* In a flowing system, the difference in temperature between two points due to temperature drop. [ARP699D]

temperature, dry air rated Equivalent turbine discharge air temperature that would be obtained if the supply air humidity were low enough to prevent the condensation of moisture. [ARP147C]

temperature, dry bulb *See* dry-bulb temperature.

temperature, dynamic The difference between the total temperature and the static temperature of a stream of compressible fluid. [ARP147C]

temperature, effective An arbitrary index that combines into a single value the effect of air temperature, humidity, and air movement on the sensation of warmness or coldness felt by the human body. The numerical value is that of still, saturated air that will induce the identical sensation. [ARP147C]

temperature, envelope The range of temperatures to which a part is subjected in normal operation. [ARP699D]

temperature error An error in an instrument reading caused by a difference between the ambient temperature of the instrument and some desired standard temperature.

temperature indicating system The part of the temperature-measurement system extending physically from the engine-airframe interface up to and including the aircraft cockpit indicator. [AIR1900]

temperature indicator A device that responds to a signal from a temperature-measuring system to provide a reproducible indication of the signal level. [ARP485]

temperature, maximum operating The maximum temperature that a duct will experience under all possible operating conditions. [ARP699D]

temperature measurement The observation of the relative heat energy present as it affects engine performance and/or life. [AIR1900]

temperature measurement system A system consisting of one or more measuring elements to quantitatively measure temperature by heat transfer. Includes an indicator and a means of transmitting the resultant single output to the indicator, and if applicable, to the engine control device. [AIR1900]

temperature measuring element The element or portion of a temperature-measuring system that yields a measurable signal that is reproducible as a function of temperature. [AIR1900]

temperature range of the fluid Specified range of fluid temperature that should not be exceeded for satisfactory operation of a system. [ARP4386]

temperature rate rise Temperature change in an individual thermocouple, divided by the elapsed time between the two measurements. [AIR4844]

temperature rise, ram The positive increment of temperature achieved by reducing the relative velocity between the aircraft and the ambient air. Ram temperature rise is equivalent to the dynamic temperature rise when the relative velocity is zero. [ARP147C]

temperature sensing system The part of the temperature-measurement system including the temperature-measuring element, and extending physically from the temperature-measuring element to the engine-airframe interface and, if applicable, to the engine control. [AIR1900]

temperature sensitivity In phase-speed-sensitive output voltages, the changes of phase shift and zero speed voltages due only to changes in ambient temperature within a specified temperature range. [ARP667]

temperature, static *See* static temperature.

temperature, total *See* total temperature.

temperature, turbine adiabatic discharge The turbine discharge temperature, based on a reversible adiabatic expansion of dry air. [ARP147C]

temperature, wet bulb *1*. The temperature at which liquid or solid water, by evaporating into air, can bring the air to saturation adiabatically. *2*. Without qualification, the temperature indicated by a wet bulb psychrometer constructed and used according to specifications. [ARP147C]

tempering *1*. Heating a martensitic structure developed by hardening to an intermediate temperature to develop a specific combination of strength and ductility. [AMS2759B] *2*. Heating hardened ferrous alloys below the transformation temperature to reduce hardness and improve toughness. *3*. Adding moisture to molding sand, clay, mortar, or plaster. *4*. Heating glass below its softening temperature.

tempering air Air at a lower temperature added to a stream of pre-heated air to modify its temperature.

tempilstick A crayon made of a material having a sharp reaction at a specific temperature. Such a crayon, sensitive to a specific temperature, is used to mark the surface of a metal to be heated; when the mark changes color, it provides confirmation that the intended temperature was reached or exceeded.

temporal coherence The coherence of light over time. Light is temporally coherent when the phase change during an interval T remains constant regardless of when the interval is measured.

temporal distribution The statistical distribution based on time of phenomena, occurrences, or events.

temporal resolution The precision with which an optical instrument or a system differentiates between time intervals.

tenacity In yarn manufacture, the strength of a yarn or filament of a given size. [AIR4844]

tensile bar *1*. A compression- or injection-molded specimen of specified dimensions used to determine the tensile properties of a material. *2. See* tensile specimen. [AIR4844]

tensile specimen A bar, rod, or wire of specified dimensions used in a tensile test. Also known as tensile bar or test specimen.

tensile strength *1*. The stress computed at maximum load to cause fracture and based on the original cross section area of the material. [ARP700] *2*. The resistance to lengthwise stress, measured by the greatest load in weight per unit area, pulling in the direction of length, that a given substance can bear without tearing apart. [ARP1931]

tensile stress Force per unit cross-sectional area applied to elongate a material. [ARP1931]

tensile test A method of determining mechanical properties of a material by loading a machined, cast, or molded specimen of specified cross-sectional dimensions in uniaxial tension until it breaks. Used principally to determine tensile strength, yield strength, ductility, and modulus of elasticity. Also known as pull test.

tension The partial pressure of a gas, either in a mixture of gases or in a liquid. [ARP171]

tensioning The operation prior to firing the catapult, during which a specified tension force is applied to the bridle, pendant, or launch bar by moving the catapult tow fitting forward. [AIR1489]

tension system A duct system in which the fluid column or longitudinal forces due to internal pressure are not transmitted to the supporting structure. [ARP699D]

tensors Arrays of functions that obey certain laws of transformation. A one-row or one-column tensor array is a vector.

tenth-scale vessel A filament-wound material test vessel based on a one-tenth subscale of the prototype. [AIR4844]

ten to the minus ninth Written 10^{-9} and meaning one-billionth. Used in the design of a system to define the probability of failure; that is, a failure probability of one in 1,000,000,000 flight hours (events). The likelihood of such a failure is expressed as "extremely improbable." [ARP4107]

ten to the minus sixth Written 10⁻⁶ and meaning one-millionth. Used in the design of a system to define the probability of failure; that is, a failure probability of one in 1,000,000 flight hours (events). The likelihood of such a failure is expressed as "improbable." [ARP4107]

terminal *1.* Airport buildings through which all passengers and cargo traffic passes. These buildings link passengers and cargo to the aircraft. *2.* Refers to the final portion of the flight path of a missile. *3.* The electronic module necessary to interface the data bus with the subsystem and the subsystem with the data bus. [AS1773] *4.* A device attached to the end of a conductor to provide both electrical and mechanical connection to a post, stud, chassis, or another terminal. [ARP1931] *5.* The contacts of a fuse or limiter, either blade or ferrule, for connecting the current-responsive element of the fuse or limiter to the fuse clips. [ARP1199A] *6.* An I/O device that includes a keyboard and a display mechanism; used as the primary communication device between a computer system and a person. A terminal can be "dumb," with no processing capability; or "intelligent," when some processing capability is included within it. *7.* The connection points where the field wiring is brought to the I/O modules.

terminal area Generally, airspace in which approach-control service or airport traffic-control service is provided. [ARP4107]

terminal ballistics A branch of ballistics dealing with the motion and behavior of projectiles at the termination of their flight, or in striking and penetrating a target.

terminal block An assembly containing connection provisions to facilitate the connection of one or more conductors. [ARP914A]

terminal board A structural component that provides one or more electrical terminals that are electrically insulated from the chassis or mounting, and, almost always, from each other. Also called terminal strip. *See also* stud-type board.

terminal control area (TCA) *See* controlled airspace, terminal control area.

terminal pair A set of two associated terminals, so arranged as to be accessible for connecting a pair of associated leads.

terminal plate A conductive bushing bar or commoning bar (link, jumper bar). [ARP914A]

terminal radar service area (TRSA) Airspace surrounding designated airports wherein ATC provides radar vectoring, sequencing, and separation on a full-time basis for all IFR and participating VFR aircraft. Service in a TRSA is called Stage III service. [ARP4107]

terminal strip *See* terminal board.

terminal velocity The maximum velocity attainable, especially by a free-falling body, under given conditions.

terminal very high frequency omnidirectional range station (T-VOR) A very high-frequency terminal omnirange station located on or near an airport, and used as an approach aid. [ARP4107]

termination point The point at which the cable-end fitting or ferrule attaches to the cable. [AS4536]

terminator The shielded resistor used to terminate the bus in its characteristic impedance. [AS4117]

termini Hardware to physically constrain and position an optical fiber at a signal-path disconnect point; the optical equivalent of an electrical pin or socket contact. [ARD50020]

terrain clearance indicator An instrument for measuring absolute altitude. Also known as absolute altimeter.

terrain following (TF) Describes the flight of a military aircraft maintaining a constant altitude above the terrain or the highest obstruction. The altitude of the aircraft will change with the varying terrain and/or obstruction(s). [ARP4107]

terrestrial magnetism *See* geomagnetism.

tertiary air Air for combustion supplied to the furnace to supplement the primary and secondary air.

tertiary isolation Isolation at the lowest replacement level of a subunit or a component/part belonging to a subunit, to return the item in question to serviceable condition. [AIR4896]

tertiary steering system *See* classification of nosewheel steering systems.

tesla Symbolized T. Metric unit for magnetic flux density.

test A procedure or action taken to determine, under real or simulated conditions, the capabilities, limitations, characteristics, effectiveness, reliability or suitability of a material, device, system, or method. [ARD50010]

testability A design characteristic that allows the status of a unit to be confidently determined in a timely fashion. [AIR4896]

testability measurement A measurement of the ease and adequacy of testing derived as a result of demonstration or failed history analysis. [AIR4896]

testability prediction A prediction of the ease and adequacy of testing developed through use of models or schematics. [AIR4896]

test, acceptance *See* acceptance test.

test, analyze and fix A test concept designed to provide for positive reliability growth of equipment under simulated mission environments by removing all failure mechanisms, through failure analysis and positive corrective actions, until the required reliability level is achieved. [ARD50010]

test block A volume of material within which a specimen blank is embedded to adjust the cooling rate of the specimen blank during heat treatment. [AMS6474]

test chamber An environmental chamber capable of controlling the test ambient temperatures within the parameters of the test document. [AIR1228]

test circuit The hydraulic system that provides the impulse at the test manifold. [AIR1228]

test, confidence *See* confidence test.

test, design verification *See* design verification tests.

tester A portable, self-contained unit designed to test the function of a component or system. Includes connecting adapters, harnesses, hoses, etc. [ARP480A]

test facility/test cell An area in which a gas turbine engine is operated to determine its performance and other information as required by a given test. [ARP741]

test firing The firing of a rocket engine, either live or static, with the purpose of making controlled observations of the engine or an engine component.

test flight Any flight made to check any item or the aircraft. [AIR4896]

test, follow-on operational *See* follow-on operational test.

test, functional The testing of installed aircraft/engines, accessories, and equipage to determine proper functioning, particularly with respect to the applicable system. [AIR4896]

test gage A pressure gage specially built for test service or other types of work requiring a high degree of accuracy and repeatability.

testing development Refers to a series of tests conducted to disclose deficiencies and to verify that corrective actions will prevent recurrence in the operational inventory. [AIR4896]

testing, nondestructive Testing of a nature that does not impair the usability of the item. [AIR4896]

testing, quantitative Testing that monitors or measures the specific quantity, level, or amplitude of a characteristic to evaluate the operation of an item. [AIR4896]

testing without replacement Life-test sampling without replacement; a life-test procedure whereby failed units are not replaced. [ARD50010]

testing with replacement Life-test sampling with replacement; a life-test procedure whereby the life test is continued with each failed unit of product replaced by a new one drawn at random from the same lot as soon as the failure occurred. [ARD50010]

test measurement and diagnostic equipment Any system or device used to evaluate the condition of an item to identify or isolate any actual or potential failures. [ARD50010]

test, nondestructive *See* nondestructive testing.

test pattern generators Image-processing software.

test point *1.* A means of safe access to permit a measurement that will facilitate maintenance, repair, calibration, alignment, or monitoring. *2.* A jack or similar fitting to which a test probe is attached for measuring or observing a circuit parameter or waveform. [AIR4896]

test ports Calibration connection points on the manifold, between the manifold block valves and the instrument.

test pressure of a vessel The minimum pressure at which a pneumatic pressure vessel must be tested as prescribed in the specification for compressed gas cylinders by the DOT or CTC. [ARP4386]

test program A set of directed commands, instructions, and data designed to automatically accomplish a test function. [AIR4896]

test program instruction An item of technical data that contains information that cannot be displayed by the ATE and is required to accomplish maintenance of the particular unit under test. [AIR4896]

test program set The combination of test program, test program instruction, and interconnection device that together allow a test system to run the gamut of tests necessary to check and/or diagnose a system, equipment, or module. [AIR4896]

test, qualification *See* qualification tests.

test specimen The configuration of test material in which testing to determine properties is performed. [AMS6474]

test surface The test part surface through which the ultrasonic energy used for inspection initially enters the test part. [ARP5089]

test to failure Testing conducted on one or more items until a predetermined number of failures have been observed. [ARD50010]

tetrahedron A device normally located on uncontrolled airports and used as a landing-direction indicator; the small end of a tetrahedron points in the direction of landing. At controlled airports, the tetrahedron, if installed, should be disregarded because tower instructions supersede the indicator. [ARP4107]

TF *See* terrain following.

TFE *See* telemetry front end.

theodolite *1.* Optical instrument that consists of a sighting telescope, mounted so that it is free to rotate around horizontal and vertical axes, and graduated scales on which the angle of rotation may be measured. *2. See* transit.

theoretical acceleration at stall A figure of merit derived from the stall torque to rotor inertia ratio which indicates how rapidly a motor will accelerate from stall. [ARP667]

theoretical air The amount of air required to completely burn a given amount of a combustible material.

theoretical cutoff frequency Disregarding any dissipation effects, the characteristic frequency at which the image attenuation constant of a transducer changes from zero to a positive value, or vice versa.

theoretical draft The draft that would be available at the base of a stack if there were no friction or acceleration losses in the stack.

theoretical plate A hypothetical device for bringing two streams of material into such perfect contact that they leave the device in equilibrium with each other.

therm A unit of heat applied especially to gas. One therm equals 100,000 Btu.

thermal accommodation coefficients *See* accommodation coefficient.

thermal aging Exposure to a given thermal condition or a series of conditions for prescribed periods of time. [ARP1931]

thermal-agitation voltage The electrical potential difference induced in circuits by the agitated motion of electrons in the circuit conductors.

thermal analysis Determining transformation temperatures and other characteristics of materials or physical systems by making detailed observations of time-temperature curves obtained during controlled heating and cooling.

thermal-arrest calorimeter A device for measuring heats of fusion in which a sample is frozen under vacuum at subzero temperatures and thermal measurements are taken as the calorimeter warms to room temperature.

thermal barrier The zone of speed at which friction heat generated by rapid passage of an object through the atmosphere exceeds endurance compatible with the function of the object. [ARP4386]

thermal bulb A device for measuring temperature in which the liquid in a bulb expands and contracts with changes in temperature, causing a Bourdon-tube element to elastically deform, thereby moving a pointer in direct relation to the temperature at the bulb.

thermal comfort The condition at which satisfaction with the thermal environment is expressed, and which is measured by such factors as air temperature, relative humidity, and air velocity.

thermal conductivity A measure of the quantity of heat that will flow in a unit time through a unit area and thickness having a difference in temperature between its faces. [AIR1116]

thermal conductivity gage A device for measuring pressure in a high-vacuum system by observing changes in thermal conductivity of an electrically heated wire that is exposed to the low-pressure gas in the system.

thermal contraction Contraction caused by a decrease in temperature. [AIR4844]

thermal converter A device consisting of one or more thermoelectric junctions in contact with or integral with an electric heater. The output of the thermoelectric component is directly related to the current flowing in the electric heater.

thermal cutout A device for protecting a circuit or electrical device from excessive current. Consists of a heater element and a replaceable fusible link which melts and opens the circuit when too much current flows through the heater element.

thermal decomposition The braking apart of complex molecules into simpler units by the application of heat.

thermal degradation Impairment of properties caused by exposure to heat.

thermal delay timer A timing device relying on the movement of a heated bimetal to actuate a set of contacts.

thermal detector *See* bolometer.

thermal diffusion Spontaneous movement of solvent atoms or molecules to establish a concentration gradient as a direct result of the influence of a temperature gradient.

thermal efficiency *See* thermodynamic efficiency.

thermal electromotive force The electromotive force developed across the free ends of a

bimetallic couple when heat is applied to a physical junction between the opposite ends of the couple. Also known as thermal emf.

thermal emf *1.* The electrical potential generated in a conductor or circuit due to thermal effects, usually differences in temperature between one part of the circuit and another. *2. See* thermal electromotive force.

thermal emission The process by which a body emits electromagnetic radiation as a consequence of its temperature only.

thermal endurance For a material or system of materials, the time at a selected temperature to deteriorate to some predetermined level of electrical, mechanical, or chemical performance under prescribed conditions of test. [AIR4844]

thermal energy *1.* Energy that flows between bodies because of a difference in temperature. *2. See* heat.

thermal expansion A physical phenomenon whereby raising the temperature of a body causes it to change dimensions (usually increasing) in a manner characteristic of the material of construction.

thermal expansion molding A process in which elastomeric tooling details are constrained within a rigid frame to generate consolidation pressure by thermal expansion during the curing cycle of the autoclave molding process. [AIR4844]

thermal fatigue Failure induced by clinical varying temperature gradients that create cyclic thermal stress and strain in the material. Because the tensile and compressive stresses occur in very localized areas, their calculation requires detailed knowledge of temperature distribution. [AIR1872]

thermal instability In a fluid, the conditions of temperature gradient, thermal conductivity, and viscosity that lead to the onset of convection.

thermal instrument Any instrument that measures a physical quantity by relating it to the heating effect of an electric current, such as in a hot-wire instrument.

thermal neutron(s) *1.* Neutrons in thermal equilibrium with the medium in which they exist. *2.* A free (uncombined) neutron having a kinetic energy approximately equivalent to the kinetic energy of its surroundings. *See also* slow neutron.

thermal noise The noise at radio frequency caused by thermal agitation in a dissipative body.

thermal pollution Environmental temperature rise due to waste heat disposal.

thermal power plant A facility or system for converting thermal energy into electric power.

thermal processing Any process in which metals are exposed to controlled heating, soaking, or cooling. [AMS2750]

thermal radiation Electromagnetic radiation that is absorbed by a grey or black body from a source at a higher temperature than the absorbing body.

thermal rating The maximum or minimum temperature at which a material will perform its design function without undue degradation. [ARP1931]

thermal relief valve A valve designed to bypass only the additional volume caused by thermal expansion of the fluid. [ARP4386]

thermal runaway A condition in which a battery on constant-potential charge at elevated temperature will destroy itself through internal heat generation caused by high overcharge currents during constant-potential charging. [ARP4386]

thermal shock The development of a steep temperature gradient and accompanying high stresses within a structure.

thermal spraying A method of coating a substrate by introducing finely divided refractory powder or droplets of atomized metal wire into a high-temperature plasma stream from a special torch, which propels the coating material against the substrate.

thermal stress *1.* In a laminate, stress caused by the different expansion in different directions of plies at different orientations. *2.* In metals or other materials, the stress that is developed if materials having different coeffficients of thermal expansion are bolted or otherwise joined together and heated. *3.* The stress developed if a block of material is restrained from expanding while being heated; or is restrained from contracting while being cooled. [AIR4844]

thermal stress cracking Crazing and cracking of some thermoplastic resins, resulting from overexposure to elevated temperatures. [AIR4844]

thermal transducer Any device that converts thermal energy into electric power or other useful measuring medium. An example is a thermocouple.

thermal-type flowmeter An apparatus in which heat is injected into a flowing fluid stream and flow rate is determined from the rate of heat dissipation; either the rise in temperature at some point downstream of the heater, or the amount of thermal or electrical energy required to maintain the heater at a constant temperature, is measured.

thermal-type liquid-level meter Any of several devices that indicate the position of liquid level in a vessel by means of a thermally activated property, such as an abrupt change in temperature, evaporation or condensation effects, or thermal expansion effects.

thermal variable A characteristic of a material or system that depends on its thermal energy; for example, temperature, thermal expansion, calorific value, specific heat or enthalpy.

thermionic emission Direct ejection of electrons as a result of heating a material; occurs because the heating raises the energy of the electrons beyond the binding energy that holds them to the material.

thermionics The study of the emission of electrons by heat.

thermionic tube An electron tube in which at least one of the electrodes is heated to induce electron or ion emission.

thermistor(s) *1.* A temperature-sensitive device composed of a material whose resistance decreases in a reproducible manner as the temperature increases, with negligible hysteresis. [AIR1900] *2.* Electron devices employing the temperature-dependent change of resistivity of a semiconductor. *3.* A temperature transducer constructed from semiconductor material, in which the temperature is converted into a resistance, usually with negative slope and highly nonlinear. Usual applications are as a nonlinear circuit element (either alone or in combination with a heater); as a temperature compensator in a measurement circuit; or as a temperature-measurement element.

thermites Fire-hazardous mixtures of ferric oxide and powdered aluminum. Upon ignition with a magnesium ribbon, such mixtures reach temperatures up to 4000°F (sufficient to soften steel).

thermit welding A fusion-welding process in which a mixture of finely divided iron oxide and aluminum particles is ignited, reducing the iron oxide and producing a molten ferrous alloy that is then cast in a mold built up around the joint to be welded.

thermoammeter A device used chiefly to measure radio-frequency (RF) currents. In this device, the RF current is run through a wire of appropriate size, on which is mounted a thermocouple. The output of the thermocouple is proportional to the temperature of the wire, which is a function of the RF current passing through the wire. Also called a hot-wire ammeter.

thermochemistry A branch of chemistry that treats the relations of heat and chemical changes.

thermocouple *1*. A temperature-sensing device that consists of two conductors of dissimilar electrothermal characteristics. The conductors are joined together at one end (hot junction), where the junction is placed in contact with the item whose temperature is to be measured. At the other end (cold junction), a voltmeter is used to measure the electrical voltage that is produced by the difference in temperature between the two junctions. [AS5116] *2*. A junction of two dissimilar metals in which the energy is changed directly into electrical energy. [ARP480A]

thermocouple assembly That part of the thermocouple sensing system including all components from the measuring junction to the electrical connector, or to the harness junction box, in the case of an integral harness. [AIR1900]

thermocouple assembly extension lead An assembly that physically supports, protects, and insulates the conductors, extending from the thermocouple assembly to a point within the thermocouple harness. [ARP485]

thermocouple boss A device that provides a means of attaching a thermocouple assembly. [ARP485]

thermocouple cable A two-conductor cable, each conductor employing a dissimilar metal, made up specifically for temperature measurements. [ARP1931]

thermocouple calibration The plot of the EMF output of a temperature-measuring junction versus known temperatures. [AIR1900]

thermocouple calibration accuracy The ratio, expressed in percentage, of the difference between the EMF output of a temperature-measuring junction at a given temperature and the specified EMF for that temperature, to the specified EMF. [AIR1900]

thermocouple harness An assembly that physically supports, protects, and insulates the conductors, extending from the thermocouple assemblies to the thermocouple output harness connector. [ARP485]

thermocouple harness circuitry The complete electrical path designed to conduct the output EMFs generated by the individual temperature-measuring junctions into a single useable value. [ARP485]

thermocouple harness extension lead An assembly that physically supports, protects, and insulates the conductors, extending from the thermocouple harness output connector to the engine-airframe disconnect and/or engine control. [ARP485]

thermocouple indicating system That part of the thermocouple-type measurement system extending physically from the engine-airframe disconnect to and including the cockpit indicator. [AIR1900]

thermocouple instrument An electrothermic instrument having a direct-current mechanism, such as a permanent-magnet moving coil, that is driven by the output of one or more thermojunctions heated directly or indirectly by electric current.

thermocouple lead That portion of the thermoelements from the electrical connector to the measuring junction. [ARP485]

thermocouple sensing system That part of the thermocouple-type measurement system including the thermocouple assembly, and extending physically from the thermocouple assembly to the engine-airframe disconnect; and, if applicable, to the engine control. [AIR1900]

thermocouple-type temperature measurement system A thermocouple and associated supporting structure and electrical insulation up to and including the indicator; and, if applicable, to the control device. [AIR1900]

thermocouple vacuum gage A device for measuring pressures in the range of about 0.005 to 0.02 torr by means of current generated by a thermocouple welded to the midpoint of a small heating element exposed to the vacuum chamber. Alternatively, the instrument current may be generated in a specially constructed thermopile that serves as both heater and thermoelectric element.

thermodynamic efficiency In thermodynamics, the ratio of the work done by a heat engine to the total heat supplied by the heat source.

thermodynamic equilibrium In statistical mechanics, a very general result stating that if a system is in equilibrium, all processes that can exchange energy must be exactly balanced by the reverse process so that there is no net exchange of energy.

thermodynamics A branch of engineering concerned primarily with systems and processes for converting thermal energy into mechanical energy, or mechanical energy into thermal energy. In aircraft these processes are used for: (a) generation of mechanical power, (b) refrigeration (reversed power cycles), and (c) compression of air for control of cabin atmosphere, and for power- and heat-distribution systems. [AIR1168/9]

thermoelasticity Rubberlike elasticity resulting from an increase in temperature. [AIR4844]

thermoelectric cooling A method of cooling a chamber based on the Peltier effect. In this method, an electric current is circulated in a thermocouple whose cold junction is coupled to the chamber; the hot junction dissipates heat to the environment. Also known as thermoelectric refrigeration.

thermoelectric heating A method of heating that involves a device similar to one used for thermoelectric cooling, except that the direction of current is reversed in the circuit. *See* thermoelectric cooling.

thermoelectric hygrometer A condensation-type hygrometer in which the mirror element is chilled thermoelectrically.

thermoelectric refrigeration *See* thermoelectric cooling.

thermoelectric series A tabulation of metals and alloys, arranged in order according to the magnitude and sign of their characteristic thermal emf.

thermoelectric thermometer A thermometer in which a thermocouple or thermocouple array is in direct contact with the body whose temperature is to be measured. The temperature reading is given in relation to the reference junction, whose temperature is known or automatically compensated for.

thermoforming A method of forming sheet plastic by heating it then pulling it over a contoured mold surface.

thermograd probe An instrument that records temperature versus depth as it is lowered to the ocean floor. Used to record the flow of heat through the ocean floor.

thermograph An instrument for recording air temperature. Also known as a recording thermometer.

thermographic inspection A method for optically viewing the temperature distribution on skins or components by measuring the radiation produced at infrared wavelengths. [ARP5089]

thermography *1.* An inspection method that remotely detects surface temperature variations. Can be used for fluid ingression detection on nonmetallic composite parts. [ARP5089] *2.* Technique employing heat transfer transients. *3.* Refers to either of two methods, contact thermography or projection thermography; both of which are used for measuring surface temperature using thermoluminescent materials.

thermogravimetric analysis The study of the mass of a material under various conditions of temperature and pressure. [AIR4844]

thermojunction Either of the two locations at which the conductors of a thermocouple are in electrical contact. One, the measuring junction, is in thermal contact with the body whose temperature is being determined; the other, the reference junction, is generally held at some known or controlled temperature.

thermomechanical treatment Combination of material-forming processes with heat treatments in order to obtain specific material properties.

thermometer *1.* An immersion- or contact-type device whose operation is based on thermal equilibrium between a sensitive measuring element and the medium to be measured. [AIR1900] *2.* An instrument for measuring temperature, usually involving a change in a physical property such as density, or electrical resistance of a temperature-sensitive material.

thermomigration A technique for doping semiconductors in which exact amounts of known impurities are made to migrate from the cool side of a wafer of pure semiconductor material to the hotter side when the wafer is heated in an oven.

thermophone An electroacoustic transducer that produces sound waves when a conductor whose temperature varies in response to a varying electric current causes air adjacent to the conductor to expand and contract.

thermophoresis A process whereby particles migrate in a gas under the influence of forces created by a temperature gradient.

thermopile(s) An array of thermocouples used for measuring temperature or radiant energy, or for converting radiant energy into power. The thermocouples may be connected in series to give a higher-voltage output, or in parallel to give a higher-current output.

thermoplastic A classification of resin that can be readily softened and resoftened by heating. [ARP1931]

thermoplastic films Materials with a linear macromolecular structure that will repeatedly soften when heated and harden when cooled.

thermoplastic resin An organic solid that will repeatedly soften when heated and harden when cooled; examples include styrene, acrylics, polyethylene, vinyl, and nylon.

thermoregulation A mechanism by which mammals and birds balance heat gain and loss in order to maintain a constant body temperature.

thermoregulator A highly accurate or highly sensitive thermostat; for example, a mercury-in-glass thermometer with sealed-in electrodes, which turns an electric circuit on and off as the level of mercury rises and falls past the position of the electrodes.

thermoset Refers to a plastic that, when cured by application of heat or chemical means, changes into a substantially infusible and insoluble material. [AIR4844]

thermosetting A classification of resin that cures by chemical reaction when heated, and when cured cannot be softened by heating. [ARP1931]

thermosetting resin An organic solid that sets up (solidifies) under heat and pressure, and cannot be softened and remolded readily; examples include phenolic, epoxy, melamine, and urea.

thermostat A temperature-sensitive element used as the prime actuator of a temperature regulator. [ARP147C]

thermostatic bulb *See* bulb, thermostatic.

thermostatic switch An electric switch whose contacts open and close in response to the amount of heat received by conduction or convection from the device whose power it regulates.

thermowell *See* protecting tube.

THETA *See* observed mean time between failure.

THETA 1 *See* lower test MTBF.

THETA D *See* demonstrated MTBF interval.

thickener Equipment for removing free liquid from a slurry or other liquid-solid mixture to give a solid or semisolid mass without using filtration or evaporation; usually, the process involves centrifuging or gravity settling.

thick film metallization Conductive metallization applied to an insulating substrate by either: (a) silk screening an admixture of finely divided gold powder and base metal oxides suspended in an organic vehicle, and firing in air to produce adherence; or (b) silk screening an admixture of finely divided base metal powder and base metal oxides suspended in an organic vehicle, firing in a controlled atmosphere to produce adherence, and electroplating a gold overlayer. [AS1346]

thickness *1.* The minimum dimension of the heaviest section of a part. [AMS2759B] *2.* A physical dimension usually considered to represent the shortest of the three principal measurement axes. In flat products such as sheet or plate, the thickness is the distance between top and bottom surfaces measured along an axis mutually perpendicular to them.

thickness gage A device for measuring thickness of sheet material; may involve physical gaging, but more often involves methods such as radiation absorption or ultrasonics.

thickness ratio A geometric factor expressing the ratio of the maximum thickness of an airfoil section to its chord. [ARP4386]

thin film metallization Conductive metallization applied to an insulating substrate by evaporating, sputtering, or plating metallic layers, or by an equivalent process. [AS1346]

thin-film potentiometer A potentiometer in which the conductive element is a thin film of a cermet (metal mix), conductive plastic, or deposited metal. Usually useful when a stepless output is desired.

thin-film strain gage A strain gage in which the gage is produced by depositing an insulating layer (usually a ceramic) onto the structural element, then depositing a metal gage element onto the insulation layer by sputtering or vacuum deposition through a mask that defines the strain gage configuration. Thin-film techniques are used almost exclusively for transducer applications; for greatest sensitivity, four bridge elements, wiring between the elements, balance components, and temperature-compensation components are deposited simultaneously.

thin-layer chromatography (TLC) A form of chromatography in which a small amount of sample is placed at one end of thin sorbent layer deposited on a metal, glass, or plastic plate. After washing a solvent through the sorbent bed by capillary action, the individual components of the sample may be detected by visual means, ultraviolet analysis, radiochemistry, or other suitable technique—or by a combination of techniques. Thin-layer chromatography is a rapid and inexpensive method for screening and selecting solvent-stationary-phase systems for liquid chromatographic analysis.

thinner(s) An organic liquid added to a mixture such as paint to reduce its viscosity and make it more free-flowing. *See also* solvent.

third degree repair Repair encompassing the same gas turbine engine repair capability as the second-degree repair, except that certain functions that require high maintenance man-hours and are of low incident rate are excluded. [AIR4896]

thixotropic substance A substance whose flow properties depend on both shear stress and agitation. At a given shear stress, flow increases with increasing time of agitation; when agitation stops, internal shear stress exhibits hysteresis.

thixotropy The property of a nonsag material whereby it can be moved (stirred or extruded) with less force than would be required for a Newtonian fluid. Non-Newtonian pseudoplastic materials (such as the nonsag sealants) stand like whipped cream without seeking their own level, but flow easily from a sealant gun under relatively low pressure. [AIR4069]

thread *1.* A continuous helical rib used to provide interconnection by twisting of the ribbed member into a mating ribbed member; used extensively in pipe-and-fitting connections and in threaded fasteners. *2.* A thin, single-strand or twisted filament of natural or synthetic fibers, plastics, metal, glass, or ceramic.

threaded coupling *See* coupling, threaded.

threat Describes traffic that has satisfied the threat detection logic and requires a resolution advisory. [AIR4102/10]

threat aircraft An intruder that has been determined by the CAS threat-detecting logic to warrant either a caution or a warning alert. [ARP4153]

threat evolution The evolution of the potential harm of an approaching aircraft or other objects.

three axis stabilization Maintenance of a stable platform in a desired three-axis orientation in inertial space by utilizing gyros and accelerometers, independent of vehicle motion.

three body problem In classical celestial mechanics, the problem that treats the motion of a small body, usually with negligible mass, relative to and under the gravitational influence of two other finite point masses.

three-mode control *See* PID control.

three-mode controller *See* PID controller.

three-quarters hard A temper of nonferrous alloys and some ferrous alloys that corresponds approximately to a hardness and tensile strength midway between those of half hard and full hard tempers.

three term control Proportional integral derivative control.

three-way servoactuator A servoactuator with a single control port connecting to a three-way servovalve. The control port is connected to the large actuator area and system pressure is usually applied to the small area. A three-way linear servoactuator is usually a single-ended actuator. [ARP4386]

three-way valve A multiorifice flow-control valve with a supply port, return port, and one control port arranged so that valve action in one direction opens the supply port to the control port, and reversed valve action opens the control port to the return port. [ARP4386]

threshold *1.* The minimum signal amplitude necessary to trigger an electronic alarm. [ARP5089] *2.* The beginning of that portion of the runway usable for landing. *3.* An inherent perceptual limitation that requires stimuli to be presented within a certain range of intensity and duration to ensure perception by the organism. The different types of stimuli are subthreshold stimuli, suprathreshold stimuli, and transthreshold stimuli. *Subthreshold stimuli*–stimuli that are presented below a detectable range or duration and thus are not perceived. *Suprathreshold stimuli*–stimuli that are presented above a detectable range or duration and thus are perceived. *Transthreshold stimuli*–stimuli that are presented within a detectable range and duration and thus are perceived. [ARP4107]

threshold age *See* age, threshold.

threshold crossing height (TCH) The theoretical height above the runway threshold at which the glide slope antenna of the aircraft would be if the aircraft maintains the trajectory established by the mean ILS glidepath. [ARP4107]

threshold sensitivity The lowest value of a measured quantity that a given instrument or controller responds to effectively.

threshold voltage In solid state devices, the threshold energy necessary to remove an electron from the bound position to the conduction band.

throat *1.* The narrowest portion of a constricted duct, as in a diffuser, or a venturi tube. *2.* The shortest distance from the root of a fillet weld to its face. *3.* In a machine such as a resistance spot welder or arbor press, the distance from the centerline of the electrodes or punch to the nearest point of interference with the frame or other machine component. The throat establishes the maximum distance from the edge of flat work that the machine can perform an operation.

throat-injection amplifier An amplifier in which auxiliary flow at a nozzle throat is used as a control signal to modulate the output flow. The pressure level of the control signal may either be above or below local throat pressure to result in a positive or negative (suction) quiescent control flow. [ARP993A]

thrombus A clot of blood formed within the heart or blood vessels, and fixed to the vascular or heart wall. If a thrombus becomes detached in whole or in part, it then becomes a thrombotic embolus. [ARP171]

throttle burst Instantaneous advancement of the throttle from any given setting to the full-open position. [ARP4107]

throttle chop Instantaneous retarding of the throttle from any given position to the full-closed position. [ARP4107]

throttle control A function or mode of operation that provides power or throttled control in response to pitch attitude, glide slope, or airspeed signals. [ARP419-57]

throttle valve A device for regulating flow of a fluid by alternatively opening up or closing down a restriction in a passage or inlet.

throttling Pertains to an irreversible adiabatic process that consists of lowering pressure by expansion without doing work. [ARP147C]

throttling calorimeter An instrument that determines the moisture content of steam by admitting steam to a well-insulated expansion chamber through an orifice, and then measuring steam temperature; moisture content is found by referring to steam tables.

throttling range *See* proportional band.

through bulkhead initiator An initiator that carries an ignition command through the wall of a pressure vessel. [ARP4386]

throughput Digital information transfer rate. [AIR4548]

throughput rate The net rate at which data can be received by a device, manipulated as specified, and output to some other specified device.

through stay A brace used in fire-tube boilers between the heads or tube sheets.

through transmission method An inspection method in which ultrasonic energy is generated by one search unit and received by another at the opposite surface of the test part. [ARP5089]

throwing power The relative ability of an electroplating solution or electrophoretic paint to cover irregularly shaped parts with a uniform coating. Contrast with covering power.

throwout The device for disengaging the driving and driven plates of a motor vehicle clutch.

thrust *1.* The forward-directed force (positive thrust) developed by a jet engine or rocket, or by the rotation of a propeller. *See also* reverse thrust. [ARP4107] *2.* Generically, the force that any body exerts on another body; both can be stationary, both can be in motion, or one can be stationary and the other in motion. *3.* The resultant force in the direction of motion due to the components of pressure forces in excess of ambient atmospheric pressure, acting on all inner surfaces of the vehicle propulsion system parallel to the direction of motion. [ARP4386]

thrust augmentation The increasing of the thrust of an engine or power plant, especially of a jet engine, and usually for a short period of time, over the thrust normally developed.

thrust bearing(s) *1.* A bearing that supports axial load on a shaft and prevents the shaft from moving in an axial direction. *2.* Bearings designed to carry thrust loads. [AIR1594]

thrust collar A device that creates an axial force proportional to the load on the output. [ARP4386]

thrust distribution The location of areas of upward thrust (lift) on wings, airfoils, etc.

thrust reverser A mechanism to control and reverse the aircraft engine propulsive thrust. [ARP4386]

thrust vectoring Refers to an attitude-control design whereby the direction of the main thruster of the vehicle—and thus its velocity vector—can be changed. [ARP4386]

thumbwheel A multiple-position switch driven by a notched disc that can be rotated by thumb action. Frequently used to enter numerical information into a PC from a machine location.

tie down *1.* The process of bridling or mooring an aircraft to the ground or deck to protect it from movement due to winds or deck roll. *See* mooring. *2.* The chain, cable, or strap used to secure the aircraft (ground support equipment). [AIR1489]

tie rod An item, long in relation to its cross-section, that connects and maintains relative position between two parts. [ARP480A]

tiger start A type of start in which the power control is set at part power or above at fuel manifold pressurization, instead of at the usual idle detent. [ARP906A]

tight *1.* Having inadequate clearance, or barest minimum clearance, between moving parts. *2.* Describes a pressurized or vacuum system that is free from leaks. *3.* A class of fit having slight negative allowance, requiring light to moderate levels of force to assemble mating parts together.

tight burr A burr that is closely compacted and binding in the periphery of a part without any loose ends, and is within the dimensional limits of the part. [MA3375]

TIG welding *See* gas tungsten-arc welding.

tile A preformed refractory; usually refers to shapes other than standard brick.

tiltmeter Instrument used to measure small changes in the tilt of the earth's surface, usually in relation to a liquid-level surface or to the rest position of a pendulum.

tilt rotor aircraft A type of convertible aircraft that takes off, hovers, and lands as a helicopter, but is converted into a fixed-wing aircraft, by the 90-degree tilting of its rotor or rotors for use as a propeller, for forward flight.

tilt-switch level detector A relatively simple device for detecting high level in a bulk solids container by means of a free-hanging sensor that produces a switch action when the rising level of bulk material tilts the sensor from its normal vertical position.

time, active The time during which an item is in the operational inventory. [AIR4896]

time, active maintenance *See* active maintenance time.

time, active technician The time expended by the technicians in active performance of a maintenance task. [AIR4896]

time, adjustment and calibration The time for recalibration returning, etc., when adjustments are necessary, either to compensate for performance degradation, or to compensate for differences between the operating characteristics of the replacement item and those of the original item. [ARD50010]

time, alert That element of uptime during which an item is assumed to be in specified operating conditions, and is awaiting a command to perform its intended mission. [AIR4896]

time, awaiting maintenance That time during which a vehicle is not operationally ready because of maintenance (NORM) and no maintenance work is as yet being performed on either the vehicle or its related equipment. [ARP4386]

time base error In instrumentation tape recording and playback, the data error that results from a difference between tape recording speed and tape playback speed.

time between overhaul The maximum time that an item is permitted to operate between overhauls. [ARD50010]

time, block The time an aircraft is underway, covering the time from pulling the chocks at the parked position until parked in the destination. Block time is entered on the aircraft log sheets and is used in computing airline statistical information. It includes taxi time, waiting on taxiways and at the end of runways, takeoff run, flight time, landing roll, and taxi-in. It excludes running of engines and/or systems, while aircraft are parked, or for strictly maintenance check/test purposes. [ARP4386]

time, checkout *See* checkout time.

time, cleanup The portion of total maintenance time following reassembly and checkout of an item, in which tools, test equipment, and material not required for operation are removed from the equipment operating area. [ARD50010]

time code A serial BCD code, superimposed on a carrier so that it can be recorded on instrumentation tape, to annotate the time of day at which all data were recorded.

time code translator A hardware device to accept the serial time code (as from a separate track of an instrumentation tape), recognize synchronization, and prepare the time of day in parallel format for computer entry.

time constant *1.* The time required for a unit to accelerate from 0 to 63.2% of its final no-load speed when rated voltage with proper phase relationship is applied to both motor windings and energizing voltage winding. [ARP667] *2.* Generally, the time required for an instrument to indicate a given percentage of the final reading result from an input signal; the relaxation time of an instrument. In the case of instruments such as thermometers, whose response to step changes in an applied signal is exponential in character, the time constant is equal to the time required for the instrument to indicate 63.2% of the total change, that is, when the transient error is reduced to 1/e of the original signal change. Also called lag coefficient. [AIR1489]

time, corrective maintenance That part of the maintenance time during which corrective maintenance is performed on an item. [AIR4896]

time critical warning A warning condition in which the time to respond is extremely limited and the response to the alert is the most important action that the crew can take at that specific time; for example, ground proximity, windshear, or collision avoidance. [ARP4153]

time-delay Indicates that there is a purposely introduced delayed action, the delay decreasing as the magnitude of the current increases. [ARP1199A]

time-delay fuse A fuse whose total clearing time is deliberately delayed in the overload current range. [ARP1199A]

time division multiple access Radio transmission method in which each station of a satellite communication network is assigned a time schedule for transmission (in lieu of frequency division); a multi-element antenna with an adaptive null steering array eliminates interference.

time division multiplex (TDM) A system for transmitting information about two or more quantities (measurands) over a common channel, in which available time intervals are divided among the measurands to form a composite pulse train. In such a system, information may be transmitted by variation of pulse duration (PDM), pulse amplitude (PAM), pulse position (PPM), or by a pulse code (PCM).

time division multiplex access (TDMA) A type of protocol for access to a communication system network.

time-division multiplexing *1.* A system for transmitting information about two or more quantities (measurands) over a common channel by dividing available time intervals among the measurands to form a composite pulse train. *2.* A digital technique for combining two or more signals into a single stream of data by interleaving bits from each signal; for example, bit one might be from signal one, bit two from signal two, etc.

time, down *See* downtime.

time error The difference between the current ETA and the required time of arrival (RTA) at any designated downstream waypoint or waypoints. [ARP1570]

time, fault correction Time required to perform corrective maintenance on a failed item, in place and on the equipment; or to replace the failed item with a replacement item drawn from storage. [ARD50010]

time, fault detection *See* fault detection time.

time, fault location *See* fault location time.

time, final test The portion of active repair time required after completion of maintenance, adjustments, and calibration to verify by measurement that equipment performance is within specified or otherwise previously established limits. [ARD50010]

time, flight The airborne time elapsed from "wheels off" at takeoff to "touchdown" at landing. [ARP4386]

time gate A transducer that gives an output signal only during specific time intervals.

time, inactive The period of time during which an item is in reserve. [AIR4896]

time, in house lead The maximum number of calendar days from request to completion for an activity that is to be accomplished within an operator's own facility. [AIR4896]

time, initial provisioning lead The maximum number of calendar days quoted by the supplier to cover the period of time from receipt of a customer's initial provisioning order to shipment. [ARD50010]

time-inverse A time-current relationship whereby the protective device opening time decreases as the current increases. [ARP1199A]

time lag The total time between the application of a signal to a measuring instrument and the full indication of that signal within the uncertainty of the instrument. [AIR1489]

time limited dispatch A means of allowing aircraft dispatch for a limited time with certain faults present. [ARP4754]

time, logistic *See* logistic time.

time, logistics delay The portion of downtime during which repair is delayed solely because of the necessity to wait for a replacement part or other subdivisions of the system. [ARP4386]

time, maintenance An element of down time that excludes modification and delay time. [ARD50010]

time marching Refers to techniques for solving a problem with partial differential equations that have a time derivation.

time, mission *See* mission time.

time, modification *See* modification time.

time, not operating *See* not operating time.

time, operating The time period during which the equipment is performing its intended function. [ARP4386]

timer *1.* A device that automatically starts or stops a machine function, or series of functions, depending on either the time of day or elapsed time from an arbitrary starting point. *2.* An instrument that measures elapsed time from some arbitrary starting point. *3.* A device that fires the ignition spark in an internal combustion engine at a preset point in the engine cycle. *4.* A device that opens or closes a set of contacts, and automatically returns them to their original position after a preset time interval has elapsed. Also known as interval timer.

time, reaction *See* reaction time.

time, reorder lead The maximum number of calendar days required by the supplier to cover the period of time from receipt of a customer's stock replenishment order to shipment of a normal replenishment quantity. [AIR4896]

time response The variation of an output variable of an element or a system, produced by a specified variation of one of the input variables.

time-schedule controller A controller in which the setpoint (or reference input signal) automatically adheres to a predetermined time schedule.

time series A discrete or continuous sequence of quantitative data assigned to specific moments in time, usually studied with respect to their distribution in time.

time sharing *1.* Describes a computer system in which CPU time and system resources are shared with a number of tasks or jobs under the direction of a scheduling formula or plan. *2.* The use of a device for two or more purposes during the same overall time interval, accomplished by interspersing component actions in time. *3.* Participation in available computer time by multiple users, via terminals. Characteristically, the response time is such that the computer seems dedicated to each user.

time signals Accurate signals marking specified times or time intervals; used primarily for determining errors of timepieces. Such signals are usually sent from an observatory by radio or telegraph.

time since installation Time accumulated since a given item was last installed. [AIR4896]

time since overhaul Time accumulated since a given item was last overhauled. [AIR4896]

time slicing A method of scheduling programs in a multiprogramming environment in which specific fixed time periods or "time slices" are assigned to each task in a cyclic fashion.

time, supply delay That element of delay time during which a needed replacement item is being obtained. [ARD50010]

time, task elapsed The calendar time from the commencement to the completion of a defined task. [ARD50010]

time, testing The time required to determine whether designated characteristics of a system are within specified values. [ARP4386]

time to go (TTG) *1.* The time in min (or min:sec) to fly the great circle course from present position to the next waypoint. *2.* The cumulative time between present position and a designated downstream waypoint or waypoints. [ARP1570]

time, total The operating time an item has accumulated since new. [AIR4896]

time, turnaround *1.* The time needed to repair, service, or checkout an aircraft for recommitment to operational service. *2.* The total number of calendar days required to complete a specified task from receipt of an item by the maintenance facility to availability for issue, shipment, or reuse, as appropriate. [AIR4896]

time, up That element of active time during which an item is in condition to perform its required functions. [ARD50010]

time-variant distortion The spatial distortion measured by high-response total-pressure instrumentation. [ARP1420]

tinning Coating with a thin layer of molten solder or tin to prevent corrosion or prepare a connection for soldering.

tipback angle *1.* For a tricycle gear aircraft, the angle between a vertical line through the center of gravity (CG) of the aircraft (normal to the static ground line) and a line that passes through the CG and the ground contact point of the main landing gear. *2.* The angle to which an aircraft may rotate when the CG is vertically above the pivot point of the main landing gear and tipback is impending. The pivot point is the axle on a single axle configuration, or the bogie pivot axis on a multi-axle configuration. *3.* Generally considered to be the minimum angle achievable as determined by the combination of possible CG locations and landing gear positions corresponding to those locations. [AIR1489]

tip path/tip path plane The circular plane described by the rotating tip (the furthermost rotating part from the axis of rotation) of a propeller or rotor. [ARP4107]

tip speed The (rotational) velocity of that part of a propeller or rotor blade nearest the perimeter of its rotation. Tip speed (in miles per hour) = $2\pi \times 1.894 \times 10^{-4}$ r $\times$ RPH, where r = the length of the blade in feet from the center of rotation to the tip, and RPH is rotational speed of the propeller or rotor in revolutions per hour. [ARP4107]

tip stall A type of stall at the wingtips of an airplane caused by a loss of lift resulting from a reduction in circulation at the tips. [ARP4386]

tip vanes Wing-mounted rotor tips, the spans of which are oriented approximately parallel to the local free stream to increase the capture area and power output of the rotor.

tip vortex A vortex flowing from the tip of a wing, owing to the flow of air around the tip from the high-pressure region below the surface to the low-pressure region above it. [ARP4107]

tire *1.* A solid or air-filled covering for a wheel, typically of rubber and/or fabric, fitted around the rim of the wheel, to absorb shock and provide traction. *2.* A hoop of iron or heavy rubber fitted about the rim of a wheel. [AIR1489]

tire, anti-shimmy A tire with a channel-tread cross section that provides contact points with the ground in two separate areas outboard (both sides) from the plane of symmetry of the tire. Often used for tail wheel tires or other applications to resist a tendency to shimmy. [AIR1489]

tire aspect ratio The ratio of the section height (outside radius minus bead seat radius) to the section width. [AIR1489]

tire blister A thin swelling of the tread or outer ply of a tire due to air entrapped between plies of the carcass. [AIR1489]

tire bead *See* bead.

tire centrifugal growth *See* centrifugal growth.

tire, chafer *See* chafer strip.

tire, chevron cuts A pattern of cuts in a tire tread resembling chevrons or Vs, due to a tearing of the surface rubber under traction. [AIR1489]

tire, chine A tire whose cross section has a flare near the shoulder, similar to the chine of a boat, for the purpose of modifying the normal spray or splash pattern as the tire runs through water. Used for engine inlet protection or other purposes. [AIR1489]

tire cornering force *See* cornering force.

tire cornering power *1.* In an aircraft, the power developed by a tire in producing turning capability. *2.* Resistance to lateral movement in a rolling tire. [AIR1489]

tire, cut protected A tire with provisions for protection from cutting. May consist of inclusions (wire, fabric, etc.) in the tread or other design features. *See also* tire, cut resistant. [AIR1489]

tire-cut resistant A tire that has provisions for cut resistance. *See also* tire, cut protected. [AIR1489]

tire deflection The difference between loaded and unloaded tire section heights, where tire section height is the distance between the wheel bead seat and the tire tread against a rigid surface. [AIR1780]

tire dynamic rating The dynamic load for which a tire is rated, such as load on a nose wheel tire induced by aircraft braking or a dynamic landing load (as opposed to static load). [AIR1489]

tire, expandable *See* expandable tire. [AIR1489]

tire, fabric reinforced tread A tire whose tread is reinforced with layers of fabric to provide greater strength and resistance to centrifugal force at high rotational speeds. [AIR1489]

tire footprint, gross The total area within the boundary of a tire footprint. *See also* contact area. [AIR1489]

tire footprint, net The total area within the boundary of a tire footprint less the ineffective area that might be a result of tread pattern, pressure inefficiency in a low aspect ratio tire, etc. [AIR1489]

tire, groove cracking The tendency of a tire to crack at the juncture of tread ribs to the basic carcass, or in the grooves of the tread. [AIR1489]

tire growth Increase in size of a tire for a number of reasons including: (a) pressure growth, (b) age growth, and (c) centrifugal growth. [AIR1489]

tire, ice grip A tire configuration that includes special provisions for service on ice, such as steel wire in the tread to provide traction or grip on ice. [AIR1489]

tire loaded radius Generally given at the rated load of the tire; the normal dimension from the axle centerline to the surface against which the tire is loaded. Generally considered as a static condition, as opposed to rolling radius, which is a dynamic condition. [AIR1489]

tire ply One of the layers of fabric in the tire carcass. [AIR1489]

tire ply rating A number that represents a strength rating of the tire; may or may not be the same as the number of plies or fabric layers of the carcass. [AIR1489]

tire, pneumatic A tire that contains a gas (e.g., air or dry nitrogen) as a shock-absorbing and load-carrying medium. [AIR1489]

tire rated pressure The service pressure at which a tire is rated to carry a given load at a specified deflection. [AIR1489]

tire rib undercutting The tendency of a tire to develop cuts at the base of a rib-type tread due to bending of the tread from side loads, and thus undercutting of the tread. [AIR1489]

tire rolling and loaded radius The instantaneous distance from the axle center to the tire-ground interface. This distance should be used in calculation of drag forces; however, in calculating the airplane synchronous velocity, it is recommended that rolling radius be used. [AS483A]

tire rolling radius The normal dimension from the axle centerline to the surface on which the tire is rolling. This represents a dynamic condition, as opposed to loaded radius, which represents a static condition. [AIR1489]

tire rolling resistance The resistance a tire exhibits to pure rolling movement; generally a function of the air pressure, the radial load, the tire construction, and other factors. [AIR1489]

tire self aligning torque *See* aligning torque.

tire sipes *See* sipes.

tire skid depth *See* skid depth.

tire slip The difference between aircraft velocity and tire-tread velocity, divided by aircraft velocity. [AIR1780]

tire slip angle The difference in angle between actual steered wheel angle and the direction in which the wheel is traveling on the surface on which it is rolling. [AIR1489]

tire slip ratio In a braked wheel, the ratio of the difference in actual surface speed and the surface speed of a free-rolling wheel, to the actual surface speed of a free-rolling wheel. [AIR1489]

tire, solid rubber A tire configuration that utilizes a solid rubber cross section. Sometimes used as tailwheels for light aircraft. [AIR1489]

tire standing wave *See* standing wave.

tire static rating The static load for which a tire is rated; generally given at a specified service pressure and deflection. [AIR1489]

tire to runway coefficient The coefficient of friction that a tire will develop against the runway surface; a function of tire design and construction, runway construction, climatic conditions, etc. [AIR1489]

tire traction force *1.* The tractive force that a tire will develop. *2.* Adhesive friction of a wheel (tire) to accelerate a load or provide braking. [AIR1489]

tire tread The outside diameter layer of a tire, which contacts the surface on which the tire rolls and provides a specified depth for wear in service. [AIR1489]

tire tread chunking A throwing off of the tread layer or sections of the tread of a tire due to abuse, improper construction, or high centrifugal forces in high-speed operation. [AIR1489]

tire, tubeless A tire that is designed to retain air without use of an inner tube. Must be used with a wheel that is also designed to retain air in the rim section. [AIR1489]

tire, tube type A tire that is designed to be used with an inner tube within for the purpose of air retention, as opposed to a tubeless tire, which is designed to operate without a tube. [AIR1489]

tire undertread The layer of the tire carcass directly under the tread; may contain shredded wire, fabric, or other construction for reinforcement. [AIR1489]

tire vent holes Small holes, usually pierced, in the outer side walls of a tire (part way through only) for the purpose of relieving trapped air within the carcass plies and for prevention of blisters. [AIR1489]

tire wear indicator A means of indicating the amount of wear that a tire has used (or how much remains). May be in the form of colored fabric plies in the tread or carcass, molded dimples, or recesses in the tread. [AIR1489]

titration curve A plot in which pH is the ordinate and units of reagent added per unit of sample is the abscissa.

TKE *See* track angle error.

TLC *See* thin-layer chromatography.

T network A network consisting of three branches, one terminal of each branch being connected at a common node, and the remaining terminals being connected, respectively, to an input junction, an output junction, and a common input and output junction.

TOD *See* top of descent.

toe The junction between the face of a weld and the adjacent base metal.

toe crack A crack in a weldment that runs into the base metal from the toe of the weld.

to-from indicator *See* sense indicator.

toggle *1.* Pertaining to a manually operated on-off switch, i.e., a two-position switch. *2.* Pertaining to flip-flop, see-saw, or bistable action. *3.* A pinned lever that can be used to amplify forces.

toggle action A mechanism that exerts pressure developed by the application of force on a knee joint; used as a method of closing presses and applying pressure at the same time. [AIR4844]

toggle circuit breaker A circuit breaker that has a toggle actuating means. [ARP1199A]

toggle switch A manually operated electric switch with a small projecting knob or arm that may be placed in either of two positions, "on" or "off," and will remain in that position until changed.

tokamak devices Experimental toroidal magnetic confinement devices in which toroidal current runs through a plasma in order to produce fusion-reactor-like plasma conditions. The name is a Russian acronym for toroidal magnetic current.

token ending delimiter field A field that indicates the ending boundary of the token. [AIR4075]

token status subfield Part of a control field that indicates the beginning of a token frame. [AIR4075]

tol *See* tolerance.

tolerance (tol) *1.* Permissible variation in the dimension of a part. *2.* Permissible deviation from a specified value; may be expressed in measurement units or percent.

tolerance limits The extreme upper and lower boundaries of a specified range; computed from the nominal value and its tolerance.

tomography Technique of making radiographs of plane sections of a body or an object. Purpose is to show detail in a predetermined plane of the body, while blurring the images of structures in other planes. Also called planigraphy.

ton *1.* A weight measurement of which there are three types, the short ton, the long ton, and the metric ton. *Short ton*–2,000 lb (avoirdupois). *Long ton*–2,240 lb (avoirdupois). *Metric ton*–1,000 kg. *2.* A unit volume of sea freight equal to 40 cu ft. *3.* A unit of measurement for refrigerating capacity equal to 200 Btu/min, or about 3517 W; derived from the capacity equal to the rate of heat extraction needed to produce a short ton of ice having a latent heat of fusion of 144 Btu/lb from water at the same temperature for 24 h.

tool steel Any of various steel compositions containing sufficient carbon and alloying elements to permit hardening to a level suitable for use in cutting tools, dies, molds, shear blades, metalforming rolls, and other tooling applications.

top coat A material applied as a thin coating over the surface of applied sealant to protect it from the possible deleterious effects of fuel. [AIR4069]

top dead center The position of a piston and its connecting rod when at the extreme outer end of its stroke.

top of descent (TOD) The point at which the main descent from en route cruise is initiated. [ARP1570]

topology *1.* The arrangement or layout of elements comprising a system. [AIR4289] *2.* The logical interconnection between devices. Local area networks typically use either a broadcast topology (bus), in which all stations receive all messages; or a sequential topology (ring), in which each station receives messages from the station before them and transmits (repeats) messages to the station after them. Wide area networks typically use a mesh topology, in which each station is connected to one or more other stations and acts as a bridge to pass messages through the network.

topping charge A reduced rate charge that completes (tops) the charge on a cell, and that can be continued in overcharge without damaging the cell. [ARP4386]

topworks A nonstandard term for actuator.

tor *See* torr.

torching *1.* The burning of fuel at the end of an exhaust pipe or stack of a reciprocating aircraft engine, the result of an excessive richness in the fuel-air mixture. [ARP4107] *2.* The rapid burning of combustible material deposited on or near boiler-unit heating surfaces.

torn sleeve A firesleeve that evidences surface tear, or that has been stretched in two or fissured to the degree that it will no longer provide adequate fire protection to the hose assembly. [ARP1658A]

toroidal wheels Doughnut-shaped wheels designed particularly for vehicles used in soft, granular soil (planetary surfaces).

torque About an axis, the product of a force and the distance of its line of action from the axis. Steering torque, for example, is the torque available to rotate the steered wheel(s) about the spindle or steering axis. [AIR1489]

torque, accelerating The net torque (algebraic summation of applied torque, accessory torque, and engine-starting torque) that produces acceleration of the combined engine, accessory, and starter, as typified by the starting cycle. [ARP906A]

torque, accessory The torque attributable to the accessories driven by the engine which must be overcome during the start cycle. [ARP906A]

torque, aerodynamic The portion of the engine drag torque that results from the engine compressor pumping. [ARP906A]

torque amplifier A two-shaft device that supplies power to rotate the output shaft, maintaining corresponding position between output and input shafts, but without imposing any additional torque on the input shaft.

torque, applied Starter output torque applied to the engine. [ARP906A]

torque arms A contrivance consisting of two (usually metal) arms hinged to one another at their ends (apex), the other ends being hinged to two separate parts, between which rotation would be possible except for the constraint provided by the torque arms. Reciprocation of the two separate parts is permitted by virtue of the hinges of the torque arms. [AIR1489]

torque, breakaway Torque required to induce engine rotor rotation. [ARP906A]

torque-coil magnetometer An instrument for measuring properties of a magnetic field, the output of which is related to the torque developed in a coil that can turn within the field being measured.

torque converter *1.* A hydraulic device consisting of a pump runner and turbine in a close coupled package. The torque converter is capable of torque multiplication and can be utilized in a turbine engine starting system, in combination with an APU. [ARP906A] *2.* Device for changing the torque speed or mechanical advantage between an input shaft and an output shaft.

torque efficiency The ratio of the measured brake torque integral to the integral of straight lines between peak torques taken from torque-time history instrumentation. [AIR1489]

torque, engine drag The sum total of torque loads developed by the engine that resist engine rotation; absolute value is considered to be negative. [ARP906A]

torque, engine starting The sum total of torque developed by the engine during the start cycle; absolute value is negative when resisting acceleration and positive when assisting acceleration. [ARP906A]

torque/force summing drive A multichannel summing arrangement that sums the torque or force outputs of multiple actuators in which individual actuators are constrained to have the same position. [ARP4386]

torque, friction That portion of engine drag starting torque required to overcome the friction of mechanical and/or fluid components. [ARP906A]

torque, impact The highest transient torque that may occur when the starter first engages the engine during a start or during re-engagement, and that exceeds that torque nominally generated by the starter. [ARP906A]

torque limiting brake A mechanical device that permits free flow of torque below a predetermined level. Once that level is exceeded, the excess torque is routed to ground through a brake; or the torque is disrupted in slippage, as in a slip clutch. [ARP4386]

torque motor An electromagnetic device that provides a rotary output displacement directly proportional to input current. [ARP4386]

torque, rated Starter torque developed at rated input and output conditions. [ARP906A]

torque, stall That torque developed by the starter at zero output speed. [ARP906A]

torque, starter The torque produced by the starter during the start cycle; a function of starter, speed, and starter power supply. [ARP906A]

torque, steady state That torque developed by the starter or engine while operating with its output held to a constant speed. [ARP906A]

torque-to-inertia or force-to-mass ratio The ratio of actuator stall load capability to rotary actuator inertia or linear actuator mass. This

ratio determines the maximum actuator output acceleration. [ARP4386]

torque, transient That torque developed by the starter or engine while operating with its drive accelerating. [ARP906A]

torque tube A tubular shaft utilized to carry mechanical torque from one mechanical element to another, thereby transmitting mechanical power by rotational means. [ARP4386]

torque-tube flowmeter A device for measuring liquid flow through a pipe. In this device, differential pressure due to the flow operates a bellows, the motion of which is transmitted to a recorder arm by means of a flexible torque tube.

torque-type viscometer An instrument for measuring the viscosity of Newtonian fluids, non-Newtonian fluids, and suspensions by determining the torque needed to rotate a vertical paddle or cylinder submerged in the fluid.

torque, valve The maximum actuating moment required at a given fluid pressure (usually rated pressure) to move the valve mechanism from one position to another. [ARP4386]

torr Also spelled tor. A unit of pressure equal to the pressure exerted by a column of mercury 1 mm high at 0°C.

torsiometer An instrument consisting of angular scales mounted around a rotating shaft to determine the amount of twist in the loaded shaft, and thereby determine the power transmitted. Also known as torsionmeter.

torsion Twisting stress. [AIR4844]

torsional elastic moment The total torque moment that a tire is capable of producing through the footprint in contact with a surface. [AIR1489]

torsional modulus In a material, the ratio of the torsion stress to the strain over the range for which this value is constant. [AIR4844]

torsional pendulum A device for performing dynamic mechanical analysis, in which the sample is deformed torsionally and allowed to oscillate in free vibration. [AIR4844]

torsional rigidity Refers to the resistance of a fiber to twisting. [AIR4844]

torsional spring rate The stiffness of a mechanical drive train, expressed in terms of torque per angular displacement. [ARP906A]

torsional stress The shear stress on a transverse cross section caused by a twisting action. [AIR4844]

torsion balance An instrument for measuring minute magnetic, electrostatic, or gravitational forces by means of the rotational deflection of a horizontal bar suspended on a torsion wire whose other end is fixed.

torsion bar A type of spring that flexes by twisting about its axis, rather than by bending.

torsion galvanometer A galvanometer whose reading is determined by the angle through which a moving system must be rotated to bring it to its zero position while under the influence of a specific force between the fixed and moving systems.

torsion hygrometer An instrument for measuring humidity in which a substance sensitive to humidity is twisted or spiraled under tension in such a manner that changes in length of the sensitive element will rotate a pointer in direct relation to atmospheric humidity.

torsionmeter *See* torsiometer.

total absorption spectrometer An instrument that measures the total amount of x-rays absorbed by a sample and compares it to the amount absorbed by a reference sample; the sample may be solid, liquid, or gas.

total adjusted error The maximum output deviation from the ideal expected values. Expressed as LSBs or percent of full-scale range at a fixed reference voltage.

total air The total quantity of air supplied to the fuel and products of combustion. Percent total air is the ratio of total air to theoretical air, expressed as percent.

total air temperature (TAT) The temperature of an airstream brought to rest isentropically. [AIR4367]

total case depth The maximum depth below the surface at which a Rockwell C hardness, or equivalent, of five HRC units above core hardness is achieved. [AMS2759/8]

total catch efficiency The total water catch on a body, divided by the amount of water contained in the volume of air swept out by the frontal area of the body. [AIR1168/9]

total clearing time The total time measured from the beginning of the specified overcurrent condition until the interruption of the circuit. For a fuse or limiter, the total clearing time is equal to the sum of the melting time and the arcing time. [ARP1199A]

total control of flight Automated control of aircraft state variables and flight path to meet mission task objectives responding to real time changes in data regarding objectives. [ARP4386]

total dynamic suction head The static head less the friction loss in the suction piping. [AIR4783]

total dynamic suction lift The total static suction lift plus the friction loss in the suction piping. [AIR4783]

total electrical power The sum of the instantaneous electrical control power and the electrical quiescent power, expressed in mW. [ARP490]

total emissivity The ratio of the total amount of thermal radiation emitted by a non-blackbody to the total amount emitted by a blackbody at the same temperature.

total energy systems Energy systems that supply both electrical and heat requirements.

total harmonic content For a complex wave, the total rms value remaining when the fundamental component is removed. [ARP1212]

total head Pressure, measured in height; the sum of dynamic delivery head, plus static head, plus differential height between source and discharge of flow. [AIR4783]

total hydrocarbons The total of hydrocarbon compounds of all classes and molecular weights. [AIR1533]

totalizer The nonresettable set of numbers on a meter register reflecting the cumulative quantity that has passed through the meter. [AIR4783]

totally enclosed *See* enclosed, totally.

total mission scenario(s) All portions of all missions or sorties that the aircraft will accomplish. Typically, many portions are repetitious, such as takeoff, climb, and cruise segments accomplished under similar environmental conditions. All mission scenarios are designed to assure that all aircraft capabilities are defined. [ARP4107]

total null The residual voltage measured with a vacuum tube voltmeter, indicating the average value of the voltage wave in terms of the rms value of an equivalent sine wave, when the fundamental in-phase output voltage obtained at the minimum voltage positions is zero. [ARP826]

total pressure The sum of the static and dynamic pressures at a location. [ARP4386]

total radiation pyrometer *See* wideband radiation thermometer.

total range The portion of a system of units that is between the upper and lower scale limits of an instrument, and therefore defines the values of the measured quantity that can be indicated or recorded.

total shielding effectiveness The difference of an electromagnetic amplitude emanating from a source measured internal to an enclosure, and that from the same source measured in free space. [ARP1173]

total solids concentration The weight of dissolved and suspended impurities in a unit weight of boiler water, usually expressed in ppm.

total system downtime The time interval between the reporting of a system malfunction and the time when the system has been repaired and/or checked by the maintenance man, and no further maintenance activity is executed. [AIR4896]

total temperature The temperature indicated by an error-free instrument having a fixed position in the gas stream; the sum of the static temperature and the temperature rise due to the conversion of kinetic energy to heat, as the compression occurs at the sensing element. [ARP485]

total travel The distance from the plunger free position to the full overtravel point. [AIR4077]

total void Sum of plan voids and component voids. [AIR4170]

total water catch The total mount of water (or ice) on the aircraft surface; the integrated value of the local catch. [AIR1168/9]

touch and go/touch and go landing A landing in which the aircraft touches down on the runway, but does not come to a stop before taking off again. [AIR1489]

touchdown The first point of contact of an air vehicle with the landing surface in the landing process. [AIR1489]

touchdown protection A skid-control system feature that provides a full brake pressure release upon extending the landing gear prior to arming the system at touchdown. [AIR1489]

touchdown zone The first 3000 ft of the runway, beginning at the threshold. This area is used in determining the touchdown zone elevation (TDZE) in the development of straight-in landing minimums for instrument approaches. [ARP4107]

touchdown zone elevation (TDZE) The highest elevation, expressed in feet above MSL, in the first 3000 ft of the landing surface. TDZE is indicated on the instrument approach procedure chart when straight-in landing minimums are authorized. [ARP4107]

touchdown zone lights (TDZL or TZL) Two rows of transverse light bars located symmetrically about the runway centerline in the runway touchdown zone. The TDZL system starts 100 ft from the landing threshold and extends to 3000 ft from the threshold or the midpoint of the runway, whichever is the lesser. [ARP4107]

touch feedback In a manipulator, a type of interaction in which servos provide force feedback to the manipulator fingers, providing a sense of resistance so the operator does not crush the object.

toughness *1.* The property of a material whereby it absorbs work. *2.* The actual work per unit volume or unit mass of material that is required to rupture the material. [AIR4844]

tow An untwisted bundle of continuous filaments. [AIR4844]

tow bar A device or assembly for the purpose of connecting an air vehicle to a tow tractor, and serving to carry the loads imposed in moving the vehicle in the towing process. [AIR1489]

tower en route control service/tower-to-tower control service The control of IFR en route traffic within delegated airspace between two or more adjacent approach control facilities. This service is designed to expedite traffic and reduce control requirements and pilot communication requirements. [ARP4107]

tow fitting A structural fitting attached to the landing gear or airframe structure that attaches to the tow bar for the aircraft towing process. [AIR1489]

towing The process of moving an air vehicle by an external power source provided by a tow tractor or vehicle, connected to the air vehicle by means of a tow bar. [AIR1489]

towing tank *See* model basin.

Townsend discharge A type of direct-current discharge between two electrodes immersed in a gas; requires electron emission from the cathode.

T-peel strength The average load per unit width of bond line required to produce progressive separation of two bonded, flexible adherends under specific conditions. [AIR4844]

traceability The characteristic by which requirements at one level of a design may be related to requirements at another level. [ARP4754]

trace element Any element occurring in the specified alloy systems that was not intentionally added as an alloying addition, and that may have a deleterious effect on mechanical properties. [AMS2281]

trace icing *See* icing, trace.

tracer A colored thread or filament visible in the insulation on an electrical wire so that the wire can be easily identified or traced between connections.

tracer gas A gas used in connection with a leak-detecting instrument to find minute openings in a sealed vacuum system.

track *1.* The actual flight path of an aircraft over the surface of the earth. [ARP4107] *2.* To follow the movement of an object; for example, by continually repositioning a telescope or radar set so its line of slight is always on the object. *3.* In data processing, a specific area on any storage medium that can be read by drive heads.

track angle error (TKE) The angle between the actual ground track of an aircraft and the desired ground track; or the angular difference between ground track angle and desired track angle. Track angle error is "left" when the actual track angle is less than the desired track angle, and "right" when the actual track angle is greater than the desired track angle. [ARP1570]

tracked vehicle Land vehicle equipped with continuous roller belts over cogged wheels for moving over rough terrain.

tracking *1.* The formation of a conducting path across the surface of an insulating material by current discharge or leakage. [AIR4844] *2.* In the nose-gear launch catapult-hookup process, the movement of the aircraft under its own power from entry into the aft section approach ramp, to the hookup condition. [AIR1489]

tracking accuracy The accuracy with which a channel output or combination of channel outputs match one another and/or the commanded output or outputs. [ARP4386]

tracking error In lateral mechanical recording equipment, the angle between the vibration axis of the pickup and a plane that is both perpendicular to the record surface and tangent to the unmodulated recording groove at the point where the needle rides in the groove.

tracking filter *1.* A narrow bandpass filter, the center frequency of which is controlled to follow the frequency of a varying parameter of interest. [AS8054] *2.* Electron device for attenuating unwanted signals while passing desired signals; uses phase-lock techniques that reduce the effective bandwidth of the circuit and eliminate amplitude variations.

tracking problem The problem of controlling a system so that the output follows a given path.

tracking radar A radar used for following a target.

tracking stations Stations set up to track objects moving through the atmosphere or space, usually by means of radio or radar.

tracking system Any device that continually repositions a mechanism or instrument to follow the movement of a target object.

track, landing gear test A track installation designed and built for the purpose of testing landing gears, with the means provided for imposing loads simulating use on an aircraft. [AIR1489]

track, tape The path traversed by one head during the record or playback process.

track type landing gear A landing gear that has endless belts running on bogies and rollers (as in a caterpillar tractor), instead of wheels. [AIR1489]

tractor Describes a propeller mounted forward of the engine; or an engine mounted forward of the main supporting surface. [ARP4107]

tradeoff The procedure of trading a degree of one attribute to gain a degree of another attribute, e.g., a degree of reliability might be sacrificed to obtain a greater degree of performance. [AIR4896]

traffic *1.* A term used by a controller to transfer radar identification of an aircraft to another controller for the purpose of coordinating separation action. Traffic is normally issued: (a) in response to a handoff or point out; (b) in anticipation of a handoff or point out; or (c) in conjunction with a request for control of an aircraft. 2. A term used by ATC to refer to one or more aircraft. [ARP4107]

trafficability cone penetrometer A probe-type instrument designed to measure soil strength. [AIR1780]

traffic advisories Advisories issued to alert a pilot to other known or observed air traffic that may be in such proximity to the position or intended route of flight of his/her aircraft as to warrant his/her attention. [ARP4107]

traffic advisory In the TCAS system, refers to a display indication that there is a traffic situation that could subsequently require a resolution advisory. No suggested maneuver is contained in the indication. [ARP4153]

traffic control Control of vehicular traffic, such as control of priority highway lanes, stoplight control, rapid-transit train control, or air-traffic control.

traffic information *See* traffic advisories.

traffic in sight Term used by pilots to inform a controller that traffic previously issued as a traf-

fic advisory from the controller is in sight. [ARP4107]

traffic pattern The traffic flow that is prescribed for aircraft landing at, taxiing on, or taking off from an airport. The components of a typical traffic pattern are an upwind leg, a crosswind leg, a downwind leg, a base leg, and a final approach. [ARP4107]

trail angle, arresting hook The angle at which the arresting hook trails in a free-flight configuration ready for touchdown. Usually given as the angle between a line normal to the aircraft horizontal reference line and through the hook pivot, and a line through the hook pivot and extending through the throat of the hook point. [AIR1489]

trail, geometric The offset of a tire contact area centroid behind the caster axis, resulting from the caster axis being canted forward (wheel(s) mounted on the centerline of the caster axis). [AIR1489]

trailing edge *1.* The second transition of a pulse. 2. The downstream edge of a wing or airfoil.

trailing vortex A vortex trailing from a wing or other lifting body; or a vortex trailing from a wing tip. [ARP4107]

trail, mechanical *See* mechanical trail.

trail, pneumatic The offset of a tire contact area centroid behind the caster axis, resulting from aft movement of the tire contact centroid as it rolls. Additive to geometric and/or mechanical trail. [AIR1489]

train *See* line.

trajectories In general, paths traced by bodies moving as a result of an externally applied force, considered in three dimensions.

trajectory control The process of guiding or directing a vehicle so that its trajectory will follow a predetermined curve. [ARP4386]

transceiver A device that provides an electrical interface to a physical medium. Typically used to connect a node to a baseband network. *See also* modem and transmitter receivers.

transconductance The change in plate current divided by the change in control-grid voltage that causes it, when the plate voltage and all other voltages are kept constant.

transducer *1.* A device capable of being actuated by energy from one or more transmission systems or media and of supplying related energy to one or more other transmission systems or media. [AIR1489] *2.* A hardware sensing device that measures a physical phenomenon (e.g., pressure, temperature, or position) and outputs a calibrated signal. [ARP1587] *3.* A component that converts a signal from one medium to an equivalent signal in a second medium, one of which is compatible with fluidic devices. [ARP993A] *See also* primary element and transmitter.

transducer dissipation loss The ratio of the power delivered by a specified source to a transducer connected to a specified load, to the power available from the transducer connected to the same source.

transducer, fiber optic The portion of a fiber-optic coupled sensing system that responds to the measurand based on some defined physical law. [ARD50024]

transducer gain The ratio of the power a transducer delivers to a specified load under specified operating conditions, to the available power of the specified source. Usually expressed in terms of decibels. [AIR1489]

transducer loss The reciprocal of transducer gain.

transfer *1.* Compensation for differences in signal amplitude from equivalent reflectors in a test part and the reference standard used in an inspection. [ARP5089] *2.* The conveyance of control from one mode to another by means of instructions or signals. *3.* The conveyance of data from one place to another. Transfer is sometimes taken to refer specifically to movement between different storage media. *4.* An instruction for transfer. *5.* An instruction that provides the ability to break the normal sequential flow of control. Synonymous with jump.

transfer admittance *1.* The current density induced along one surface of a gasketted panel joint, divided by unit normal electric field at the other surface. [ARP1705] *2.* In an electrical transducer, the complex ratio of current at a second pair of terminals, to the emf across a given pair of terminals, at a specified frequency, both pairs being terminated in a specified manner.

transfer chamber In plastics molding, an intermediate chamber or vessel for softening a thermosetting resin with heat and pressure before admitting it to the mold for final curing.

transfer characteristic The current at one electrode expressed as a function of voltage at another electrode, with all other voltages held constant. This relationship is usually shown graphically.

transfer constant A transducer rating. Consists of a complex number equal to one-half the natural logarithm of the complex ratio of the product of voltage and current entering the transducer to that leaving the transducer when the transducer is connected to its image impedance. The real part of the transfer constant is the image attenuation constant and the imaginary part is the image phase constant. Transfer constants can also be determined for pressure and volume flow rate or force and velocity, instead of voltage and current. Also known as transfer factor.

transfer ellipse Path followed by a body moving from one elliptical orbit to another. Transfer ellipses that intersect departure and arrival orbits at large angles are most expensive in energy requirements. [ARP4386]

transfer factor *See* transfer constant.

transfer function *1.* The ratio of the Laplace transform of the output to the Laplace transform of the input. [AIR1823] *2.* A mathematical expression that describes the relationship between physical conditions at two different points in time or space in a given system; may also describe the role played by the intervening time or space. *3.* The response of an element of a process-control loop, showing how the output of the device is determined by the input.

transfer impedance *1.* The potential difference induced along one surface of a gasketted panel joint, divided by unit current flowing across the other surface. [ARP1705] *2.* The complex ratio of applied a-c voltage, force, or pressure at one point in a transducer, to the a-c current, velocity, or volume velocity at another point in the same transducer, with all inputs and outputs being connected to the system in some specified manner.

transfer instruction *See* branch instruction.

transfer lag *See* capacity lag.

transfer molding Method of molding thermosetting materials in which the plastic is first softened by heat and pressure in a transfer chamber and then forced by high pressure through suitable sprues, runners, and gates into the closed mold for final shaping and curing. [AIR4844]

transfer of control *See* branch.

transfer operation An operation that moves information from one storage location or one storage medium to another, e.g., read, record, copy, transmit, or exchange.

transfer orbits In interplanetary travel, elliptical trajectories tangent to the orbits of both the departure planet and the target planet.

transfer ratio The ratio of the number of turns in the secondary winding to the number of turns in the primary winding.

transferred electron devices (TED) Electronic equipment utilizing diodes exhibiting negative conductance and susceptance.

transferring controller/facility A controller/facility transferring control of an aircraft to another controller/facility. [ARP4107]

transfer standard A laboratory instrument that is used to calibrate working standards. The transfer standard is periodically calibrated against the laboratory standard. [AIR1678]

transfer switch A switch that controls whether a given conductor is connected to one circuit or to another.

transfer time *1.* The time interval from the removal of a valve or component from one conditioning chamber to the insertion of the valve or component into another conditioning chamber; or the time interval from the removal of a valve or component from the conditioning chamber to the initiation of testing. [AS1607] *2.* The time interval between the instant the transfer of data to or from storage commences and the instant it is completed.

transfer vector A transfer table used to communicate between two or more programs. This

table is fixed in relation to the program it serves, and provides a communication linkage between that program and any remaining subprograms.

transform To change the form of data according to specific rules.

transformation ratio The ratio of the no-load maximum fundamental secondary voltage to the fundamental supply voltage applied to the primary. [ARP826]

transformation ratio unbalance A quantity that is determined by noting the maximum difference in the numerical value of transformation ratio of each output winding as each input winding is excited. [ARP826]

transformation ratio variation The change in a numerical value of any particular transformation ratio relative to ambient temperature, input voltage level, or excitation frequency. Expressed as a transformation ratio difference relative to the value of transformation ratio under nominal excitation voltage, nominal excitation frequency, and a specified temperature. [ARP826]

transformation temperature The temperature at which a phase change occurs in a crystalline solid; sometimes applied to the upper or lower limit of a transformation range.

transformed beta A local or continuous structure composed of decomposition products arising either by martensitic or by nucleation and growth processes during cooling from either above the beta transus or some temperature high in the alpha-beta phase field. A transformed beta structure typically consists of alpha platelet, which may or may not be separated by beta phase. [AS1814]

transformer An electrical device in which electromagnetic induction is used to transfer power from one electric circuit to another at the same frequency, usually increasing voltage and decreasing current (or vice versa) in the process.

transformer voltage divider An inductive-type voltage divider used in some a-c bridge circuits to provide high accuracy, much as a KVVD is used in some d-c bridge circuits.

transgranular corrosion A slow mode of failure resulting from the combined action of stress and aggressive environment. In this mode of failure, the path of failure runs through the grains, producing branched cracking.

transient *1.* A temporary deviation of an aircraft from its normal attitude, produced by momentarily overpowering the automatic controls or by momentarily injecting a signal to cause a deviation. [ARP419-57] *2.* A dynamic condition or characteristic (e.g., power level, voltage, magnetic field strength, force, or pressure) that is not periodically repeated; often implies an anomalous, temporary departure from a steady-state condition, the latter being either constant or cyclic. *See also* surge. *3.* Pertaining to rapid change.

transient analyzer An electronic device used to capture a record of a transient event for later analysis.

transient digitizer A device that records a transient analog waveform and converts the information it has collected into digital form.

transient distortion Inability of equipment to reproduce very brief signals.

transient fatigue *See* fatigue, acute/transient.

transient overshoot An excursion beyond the final steady-state value of the output as the result of a step-input change; usually referred to as the first excursion, and expressed as a percentage of the steady-state output step.

transient overvoltage In a signal or supply line of a device, the occurrence of a momentary excursion in voltage that exceeds the maximum rated conditions specified for the device.

transient reproducibility factor The ratio, expressed in percent, between the indicated standard response time and the standard response time predetermined for the configuration tested. [ARP485]

transient response For a system responding to an input, a response that is nonsinusoidal, may change in magnitude abruptly, and may be of a long or short duration. [ARP1281]

transient time The elapsed time period between two successive mission elements. [ARP1352]

transilluminated systems Integrally lit displays in which the light source is behind the display and the light passes through the lit portions of the display. [ARP1161]

transistor A three-terminal solid-state semiconductor device that can be used as an ampli-fier, a switch, a detector, or wherever a three-terminal device with gain or switching action is required.

transistor/transistor logic (TTL) A type of digital circuitry.

transit A surveying instrument having a telescope mounted for measuring both horizontal and vertical angles. Also known as a transit theodolite or theodolite.

transition *1.* Generally, describes the change from one phase of flight or flight condition to another; for example, transition from en route flight to approach, or transition from instrument flight to visual flight. *2.* A published procedure used to connect the basic SID to one of several en route airways/jet routes (SID transition); or a published procedure used to connect one of several en route airways/jet routes to the basic STAR (STAR transition). [ARP4107] *3.* The switching from one state to another (for example, positive voltage to negative) in a serial transmission.

transitional flow Flow between laminar and turbulent flow; generally between a pipe Reynolds number 2000 and 7000.

transition area *See* controlled airspace, transition area.

transition, first-order *See* first-order transition.

transition frequency *See* crossover frequency.

transition loss The ratio of signal power delivered to a portion of a transmission system following a discontinuity, after insertion of an ideal transducer, to the signal power delivered to the same portion prior to insertion of the ideal transducer.

transition points In aerodynamics, the points of change from laminar to turbulent flow.

transition pressure The pressure at which phase transition occurs.

transition temperature *1.* An arbitrarily defined temperature within the temperature range in which metal fracture characteristics (determined usually by notched tests) are changing rapidly, such as from primarily fibrous (shear) to primarily crystalline (cleavage) fracture. *2.* An arbitrarily defined temperature in a range in which the ductility of a material changes rapidly with temperature.

transit theodolite *See* transit.

transit time The time it takes for a particle such as an electron or atom to move from one point to another in a system or enclosure.

translator *1.* A program whose input is a sequence of statements in some language, and whose output is an equivalent sequence of statements in another language. *2.* A translating device.

transmission factor The ratio of the light transmitted to the incident light. [ARP798]

transmission line A continuous conductor or other pathway capable of transmitting electromagnetic power from one location to another while maintaining the power within a system of material boundaries.

transmission loss In a transmission system, a reduction in the magnitude of some characteristic of a signal between two stated points.

transmission network The physical interconnect that provides the information transfer paths between physical entities. [AIR4911]

transmission network interface The physical interface between a source or sink and the transmission network. [ARD50012]

transmissometer *1.* An instrument system that measures and indicates the transmissivity of light in the atmosphere. The transmissometer compares the actual transmissivity of the light along a known baseline with the total possible, and computes either: (a) the visibility, or (b) the visual range. Since the instrument is located along a runway, it is the source of runway visibility and runway visual range data. [ARP4107] *2.* An instrument for measuring the extinction coefficient of the atmosphere, and for determining visual range. Also known as a hazemeter or transmittance meter.

transmittance *1.* The ratio of the radiant flux transmitted by a medium or a body to the incident flux. *2.* For a body that is wholly or partly transparent to the particular wavelength(s) involved, the ratio of transmitted electromagnetic energy to incident electromagnetic energy impinging on the body.

transmittance meter *See* transmissometer.

transmittance of combiner Percentage of white light from an external source passing through the combiner, measured at or near the DEP. [ARP4102/8]

transmitter *1.* A device that consists of an emitter plus its driving and controlling circuitry. Also called sender. [ARD50024] *2.* A device that translates the low-level output of a transmission to a site where it can be further processed. *3.* In process control, a transmitter mounted together with a sensor or transducer in a single package, designed to be used at or near the point of measurement. *See* transducer. *4.* A light source (LED or diode laser) in combination with the electronic circuitry to drive it.

transmitter receivers Combinations of transmitters and receivers in single housings, with some components being used by both units. *See also* modem and transceiver.

transmutation A nuclear reaction that changes a nuclide of one element into a nuclide of a different element.

transonic For an aircraft, a flight regime at such a speed that the flow over the aircraft is supersonic in some places and subsonic in others. [ARP4386] *See also* transonic flow.

transonic flow In aerodynamics, flow of a fluid over a body in the range just above and just below the acoustic velocity. *See also* transonic.

transonic speed The speed of a body, relative to the surrounding fluid, at which the flow on the body is in some places subsonic and in other places supersonic. [ARP4107]

transparent *1.* The quality of a substance whereby light, some other form of electromagnetic radiation, or particulate radiation is permitted to pass through it. *2.* In data processing, a programming routine that allows other programs to operate identically regardless of whether the transparent instructions are installed or not installed.

transparent glass Glass having no apparent diffusing properties. Varieties of such glass are referred to as flint, crown, crystal, or clear. [ARP798]

transpiration The passage of gas or liquid through a porous solid (usually under conditions of molecular flow).

transponder *1.* The airborne radar beacon receiver/transmitter portion of the air-traffic control radar beacon system (ATCRBS) which automatically receives radio signals from interrogators on the ground, and selectively replies with a specific reply pulse or pulse group only to those interrogations being received on the mode to which it is set to respond. [ARP4107] 2. Combined receiver and transmitter whose function is to transmit signals automatically when triggered by an interrogator.

transponder codes *See* codes (transponder codes).

transportability A measure of the ability to reuse computer programs on an industrial computer.

transport delay The time from the initiation of an input signal until the first discernible change in the output caused by the input signal. [AIR1489]

transport, tape A hardware device that moves magnetic tape past heads for recording or playback.

transthreshold *See* threshold.

transverse *1.* Perpendicular to longitudinal. *2.* In a uni-directional composite or prepreg tape, in the material plane and perpendicular to the fibers. [AIR4844]

transverse electric wave A type of electromagnetic wave in which the electric field vector everywhere is perpendicular to the direction of propagation in a homogeneous isotropic medium.

transverse electromagnetic wave A type of electromagnetic wave in which the electric field vector and the magnetic field vector are everywhere perpendicular to the direction of propagation in a homogeneous isotropic medium.

transverse loads Loads that are not perpendicular to the facings. [AIR4844]

transversely isotropic Describes a material exhibiting a special case of orthotropy, in which properties are identical in two orthotropic dimensions, but not the third. [AIR4844]

transverse oscillation Oscillation in which the direction of motion of the particles is perpendicular to the direction of advance of the oscillatory motion; in contrast with longitudinal oscillation, in which the direction of motion of the particles is the same as the direction of advance of the oscillatory motion.

transverse strain The linear strain in a plane perpendicular to the loading axis of the specimen. [AIR4844]

transverse waves Waves in which the direction of displacement at each point of the medium is parallel to the wave front.

trap *1.* Conditional jump to a known location, automatically activated by hardware or software; the location from which the jump occurred is recorded. A trap is often used as a temporary measure to determine the source of a computer bug. 2. A device on the intake, or high-vacuum side, of a diffusion pump to reduce backflow of oil or mercury vapors from the pumping medium into the evacuated chamber.

trapped-air process A method of forming closed, blow-molded plastics objects, in which sliding machine elements pinch off the top of the object after blowing to form a sealed, inflated product.

trapped fuel Any fuel in a fuel-delivery system, such as the fuel system of an internal combustion engine, that is not contained in the tanks.

trapped vortexes Air flow that is in rotary motion, but is trapped relative to leading edge vortex separation, increasing not only lift but also drag. Trapped vortexes result in thrust and reduced drag.

travel cycle Travel of the closure component from its closed position to the rated travel opening and and back to the closed position.

traveler A piece of documentation that travels with material in-process, identifying it, indicating the steps of processing necessary, the inspections required, the approvals as initialed by compounders and inspectors. (Compounders indicate steps completed. Inspectors confirm quality standards have not been compromised.) [AS7200/1]

traveling wave A wave in which the ratio of the instantaneous value for any component of the wave field at one point to the instantaneous value at any other point varies with time. A traveling wave also has the property of transmitting energy from one point to another along its direction of propagation.

traveling wave tubes Electron tubes in which streams of electrons interact continuously or repeatedly with guided electromagnetic waves moving substantially in synchronism with them, and in such a way that there is a net transfer of energy from the streams to the waves. Also known as a helix tube.

travel time The time required for one-half of a travel cycle at specified conditions.

traverse *1.* To swivel a gun, antenna, tracking device or similar mechanism in a horizontal plane. *2.* A survey consisting of a set of connecting lines of known length which meet each other at specific angles.

tray In a distillation column, a horizontal plate that temporarily holds a pool of descending liquid until it flows into a vertical "downcomer" and onto the next tray. Each tray has openings to permit passage of ascending vapors.

tread *1.* The lateral dimension between gears mounted in line (laterally) on an aircraft; generally specified dimensionally. Measured to the centerline of the wheel, for a single wheel gear; or to the centroid of the wheel pattern, for a multi-wheel gear. [AIR1489] *2.* The outer surface of a wheel or tire which contacts the roadway or rails.

tread depth T_1 Theoretical dimension from one stair nosing to the adjacent stair nosing. [ARP836A]

tread depth, tire The thickness of depth of tread that may be worn away before a tire becomes unserviceable. [AIR1489]

treadle A bar or machine element that is pivoted at one end and connected to one or more other machine elements so that when it is stepped on, power and/or motion are transmitted to the other elements.

treated water Water that has been chemically treated to make it suitable for boiler feed.

tree A decoder, the diagrammatic representation of which resembles the branching of a tree.

trend analysis A technique whereby deviation of recorded data and signature characteristics with respect to time are utilized to diagnose and prognosticate a malfunction or failure. [ARP1587]

trending A technique in which deviation of recorded data from a baseline or signature characteristic is presented with respect to a time scale. [AIR1873]

triac Semiconductor switching element. Commonly used in a-c output modules for PCs.

trial-for-ignition The period of time during which the programming flame failure controls permit the burner fuel valves to be open before the flame-sensing device is required to detect the flame.

trial for main flame ignition A timed interval during which, with the ignition means proved, the main valve is permitted to remain open. If the main burner is not ignited during this period, the main valve and ignition means are cut off; a safety switch lockout follows.

trial for pilot ignition A timed interval during which the pilot valve is held open and an attempt is made to ignite and prove it. If the presence of the pilot is proved at the termination of the interval, the main valve is energized; if not, the pilot and ignition are cut off, followed by a safety lock-out.

triangular wings *See* delta wing(s).

triangulation Determining position by laying out lines of sight to three celestial bodies or landmarks, widely spaced around the horizon; if properly corrected for time of observation, for the speed and heading of the ship or aircraft, and for current or wind, the lines will meet at a point or will form a small triangle on a map that indicates position at the time of observation.

triaxial A cable construction having three coincident axes, such as conductor, first shield, and second shield, all insulated from one another. [ARP1931]

triboelectrical The separation of electrical charges of opposite sign by processes such as friction between two solid bodies. [AS5116]

tribology Science of friction, wear, and lubrication.

triboluminescence The emission of light caused by application of mechanical energy to a solid.

trickle charging *See* standby charge.

tricycle gear configuration *See* gear configuration, tricycle.

trifilter hydrophotometer An instrument for measuring the transparency of water at three wavelengths, using red, green, and blue optical filters.

trim Describes an aircraft in which the control surfaces are positioned in such a way that the aircraft attitude remains constant about all axes. An airplane may be trimmed for level flight, climb, or dive. [ARP4386]

trim, anti-cavitation A combination of control valve trim that by its geometry reduces the tendency of the controlled liquid to cavitate.

trim, anti-noise A combination of control valve trim that by its geometry reduces the noise generated by fluid flowing through the valve.

trim, balanced Control valve trim designed to minimize the net static and dynamic fluid-flow forces acting on the trim.

trim condition data Data that indicate an out-of-trim engine condition and implement a corrective action to return to scheduled trim. [ARP1587]

trim indicator An instrument used to indicate the relative amount of servo effort being used to maintain the aircraft trim. [ARP419-57]

trim, reduced Control valve trim that has a flow area smaller than the full flow area for that valve.

trim, soft seated Valve trim with an elastomeric, plastic, or other readily deformable material, used either in the closure component or seat ring to provide shutoff with minimal actuator forces.

triode An electron tube containing three electrodes: an anode, a cathode, and a control electrode or grid.

trip To release a catch or to free a mechanism.

trip-free circuit breaker A circuit breaker designed so that its pole(s) cannot be maintained closed when carrying overload currents that automatically trip it to the open position; and so that none of its poles recloses while the operating mechanism is maintained in the closed position. [ARP1199A]

triple point A temperature at which all three phases of a pure substance—solid, liquid, and gas—are in mutual equilibrium.

triple-tandem actuator A tandem actuator having three separate actuation sections. [ARP4386]

triple wheel gear configuration *See* gear configuration, triple wheel.

triplex Threefold, as in a triplex valve or a triplex actuator. [ARP4386]

triplex system A fault-tolerant control system containing three signal paths. [ARP4386]

trisonic wind tunnels Wind tunnels designed for subsonic, transonic, and supersonic flows.

tristable control *See* bistable and tristable control.

tri-state A type of logic device that has a high impedance state in addition to a high- and low-level output state; the high impedance state effectively disconnects the output of the device from the circuit. This type of logic device is thus useful in the design of bus-oriented systems.

tri states A three-way switch in which the three states are 1, 0, and a neutral state (effectively disconnected).

tristimulus values The amounts of the three primary color stimuli required to give, by additive mixture, a color match with the color stimulus considered. [ARP1782]

tri-tandem gear configuration *See* gear configuration, tri-tandem.

tritium An isotope of hydrogen having atomic weight of 3 (one proton and two neutrons in the nucleus).

tri-twin tandem gear configuration *See* gear configuration, tri-twin tandem.

trochotron A multiple-electrode electron tube that generates an output signal proportional to an input signal by charging elements in sequence with an electron beam manipulated by a magnetic field.

trombe walls Structures with passive solar collectors in the walls.

tropopause The boundary between the troposphere and the stratosphere, usually characterized by an abrupt change of lapse rate. The change is in the direction of increased atmospheric stability from regions below to regions above the tropopause; its height varies from

15 to 20 kilometers in the tropics, to about 10 kilometers in polar regions.

troposphere That portion of the atmosphere from the earth's surface to the stratosphere; that is, the lowest 10 to 20 kilometers of the atmosphere. The troposphere is characterized by decreasing temperature with height, appreciable vertical wind motion, appreciable water vapor content, and weather. Dynamically, the troposphere can be divided into the following layers: surface boundary layer, Ekman layer, and free atmosphere.

tropospheric waves Radio waves that are propagated by reflection from a place of abrupt change in the dielectric constant, or the gradient of dielectric constant, in the troposphere.

trouble-shoot To search for the cause of a malfunction or erroneous problem behavior, in order to remove the malfunction or erroneus problem behavior. *See also* debug.

TRSA *See* terminal radar service area.

truck On a multi-wheel gear, an assembly of more than one axle, wheels, tires, connecting structural beam, and other equipment. *See also* bogie. [AIR1489]

true airspeed (TAS) Indicated airspeed adjusted for error from the installation of the sensing equipment, the compressibility of the air, and the density of the air. [ARP4107]

true bearing *See* bearing.

true bond width On a bonded device, the maximum width of the beam lead in the bonded area directly over the conductor film metallization. [AS1346]

true complement *See* complement.

true concentric stranding Stranding that is composed of a central wire, surrounded by one or more layers of helically laid wires, with a reversed direction of lay, and an increased length of lay, for each successive layer. [ARP1931]

true hydrocarbon vapor pressure (P_{TVP}) Physical data, in psia, obtained from TVP vs. temperatures curves for various air-free hydrocarbon fuel blends at a vapor-liquid ratio of zero. [AIR1326]

true mass flow A measurement that is a direct measurement of mass, and is independent of the properties and the state of the fluid.

true north The direction of the earth's north terrestrial pole, i.e., the northern extremity of the earth's rotational axis. Also called geographic north. [ARP4107]

true ratio For an instrument transformer, a characteristic equal to root-mean-square primary current (or voltage), divided by root-mean-square secondary current (or voltage), determined under specified conditions.

true strain For a body subjected to an axial force, the natural logarithm of the ratio of gauge length at the moment of observation to the original gauge length. [AIR4844]

true stress The stress along the axis, calculated on the actual cross section at the time of observation, instead of the original cross-sectional area. [AIR4844]

true value The actual value of the parameter being measured. [AIR1678]

truncate *1.* To terminate a computational process in accordance with some rule, e.g., to end the evaluation of a power series at a specified term. *2.* To drop digits of a number of terms of a series, thus lessening precision; for example, the number 3.14159265 is truncated to five figures in 3.1415, whereas one may round off to 3.1416.

trunk *1. See* bus. *2. See process* I/O bus.

trunnion mounted actuator An actuator in which the body is mounted through a trunnion to the actuator support. The output rod can articulate in one axis only. [ARP4386]

truth table A table that describes a logic function by listing all possible combinations of input values and indicating, for each combination, the true output values.

TTG *See* time to go.

TTL *See* transistor/transistor logic.

tube *1.* A hollow product that is long in relation to its cross section, and has uniform wall thickness except as affected by corner radii. [ARP1917] *2.* A hollow device, round, square or any other cross section, used to convey fluid gases and/or semi-solids. [ARP480A] *3.* An evacuated glass-enveloped device used in electronic equipment to modify operating characteristics of a signal. Also known as an electron tube.

tube hole A hole in a drum, header, or tube sheet to accommodate a tube.

tube lasers Stimulated emission devices activated with shock tubes.

tube plug A solid plug driven into the end of a tube.

tubercle A localized scab of corrosion products covering an area of corrosive attack.

tube seat That part of a tube hole with which the tube makes contact.

tube sheet A perforated plate for mounting an array of tubes so that fluid on one side of the plate is admitted to the interior of the tubes and is kept separate from fluid on the outside of the tubes, as in a shell-and-tube heat exchanger.

tube turbining Passing a power-driven rotary device through a length of tubing to clean its interior surface.

tubular terminal A terminal manufactured from tubing rather than flat stock. [ARP914A]

tubular-type collector A collector utilizing a number of essentially straight-walled cyclone tubes in parallel.

tuck-down *See* pitchunder.

tumbling *1.* Loss of control in a two-frame free gyroscope due to a slowing of the wheel. *2.* A process for smoothing and polishing small parts by placing them in a barrel with wooden pegs, sawdust, and abrasives, or with metal slugs, and rotating the barrel about its axis until the desired surface smoothness and luster is obtained.

tumbling motion An attitude situation in which the vehicle continues on its flight, but turns end over end about its center of mass.

tunable lasers Stimulated emission devices with selectable frequency output.

tuner In a telemetry receiver, the input circuitry that selects and amplifies the desired frequency band.

tungsten inert-gas welding *See* gas tungsten-arc welding.

tuning In algorithms or analog controllers, the adjustment of control constants to produce the desired control effect.

turbidity The optical obstruction to the passing of a ray of light through a body of water, caused by finely divided suspended matter.

turbine A machine for converting thermal energy in a flowing stream of fluid into rotary mechanical power by expanding the working fluid through one or more sets of vanes on the periphery of a rotor.

turbine blades The blades of a turbine wheel.

turbine cooling system An air-cycle refrigeration system in which a reduction in air temperature is accomplished across an air expansion turbine. [ARP147C]

turbine engine Engine incorporating a turbine as a principal component, especially a gas turbine engine.

turbine meter A volumetric flow-measuring device in which the rotation of a turbine-type element is used to determine flow rate.

turbine wheels Multivaned wheels or rotors, especially in gas turbine engines, rotated by the impulse from or reaction to a fluid passing across the vanes.

turboblower An axial-flow or centrifugal compressor.

turbofan A turbojet engine in which additional propulsive thrust is gained by extending a portion of the compressor or turbine blades outside the inner engine case. The extended blades propel bypass air, which flows along the engine axis but between the inner and outer engine casing. This air is not combusted, but does provide additional thrust (30 to 40 percent), caused by the propulsive effect imparted to it by the extended compressor blading. [ARP4107]

turbofan engine A turbojet engine that draws in more air than the combustion process will use, compresses the bypass air slightly, and discharges it together with the core engine exhaust. [AS5116]

turbojet engine A jet engine incorporating a turbine-driven air compressor to take in and compress the air for the combustion of fuel, the gases of combustion being used both to rotate the turbine and to create a thrust-producing jet. [ARP4107]

turbojet en route descent A procedure for effecting the descent of jet aircraft from an en route altitude to the final approach without execution of the maneuvers prescribed in a published high-altitude instrument approach procedure. Purpose is to expedite the movement of air traffic. [ARP4107]

turbomechanical actuators Types of actuation systems utilizing a high-speed turbine powered by a warm gas generator. In such systems, turbine output is coupled to the load by some type of mechanical motion conversion. [ARP4386]

turboprop *See* turbopropeller engine.

turbopropeller engine An aircraft engine of the gas-turbine type in which the turbine power is used to drive both a compressor and a propeller. This type of engine usually delivers jet thrust in addition to its propeller thrust. Often called a turboprop or turboprop engine; sometimes called a propeller turbine or propeller-turbine engine. [ARP4107]

turboprop engine *1.* A turboshaft engine that is designed to drive an aircraft propeller. [AS5116] 2. *See* turbopropeller engine.

turboshaft engine A gas turbine engine that primarily delivers its output power to turn an output shaft. [AS5116]

turbosupercharger A gas-turbine-driven air compressor used to increase air-intake pressure of a reciprocating internal-combustion engine.

turbulence A state of flow in which the instantaneous velocities exhibit irregular and apparently random fluctuations so that in practice only statistical properties can be recognized and subjected to analysis.

turbulence amplifier An amplifier in which control of the laminar-to-turbulent transition of a power jet is used to modulate the output. [ARP993A]

turbulent boundary layer The layer in which the Reynolds stresses are much larger than the viscous stresses. When the Reynolds number is sufficiently high, there is a turbulent layer adjacent to the laminar boundary layer.

turbulent burner A burner in which fuel and air are mixed and discharged into the furnace in such a manner as to produce turbulent flow from the burner.

turbulent flow Fluid motion in which random motions of parts of the fluid are superimposed upon a simple pattern of flow. All or nearly all fluid flow displays some degree of turbulence.

The opposite is laminar flow. *See* laminar flow. *See also* viscous flow. [AIR1489]

turn angle The angle between the plane of the tire and the velocity vector at the vertical axis of the wheel. [AIR1780]

turnaround (STS) The intervals between flights of the shuttle orbiters.

turnaround inspection *See* inspection, aircraft.

turnaround time *See* time, turnaround.

turn back, air (technical) The return of an aircraft to the airport of origin as a result of the malfunction or suspected malfunction of any item on the aircraft. [AIR4896]

turn back, ground (technical) The situation that occurs when an aircraft leaves the block and returns for a technical reason before starting its takeoff roll. [AIR4896]

turndown The ratio of the maximum plant-design flow rate to the minimum plant-design flow rate.

turner fluorometer A type of uv fluorometer in which primary filters pass only uv radiation to excite the sample, and secondary filters pass only visible light to the photomultiplier tube. In the turner fluorometer, the intensity of emitted light is proportional to sample concentration—even when exciting and measured light are not at optimum wavelengths.

turning clearance radius The dimensional radius required for clearance for turning an aircraft; generally specified. May be determined by wing-tip clearance, tail clearance, nose clearance, or other type of clearance. [AIR1489]

turning radius Usually specified as the minimum radius about which an aircraft will turn; must be specified. Examples include nose-wheel turning radius and outside main-gear turning radius. [AIR1489]

turnover angle The angle formed between a vertical line through the center of gravity of the aircraft, and a line through the center of gravity of the aircraft and normal to a line between the turnover points (for example, nose gear and outside main gear). In a turn, the turnover angle is the roll angle at which turnover is impending. [AIR1489]

turnover frequency *See* crossover frequency.

turnstile antennas Antennas composed of two dipole antennas, normal to each other, with their axes intersecting at their midpoints. Usually, the currents are equal and in phase quadrature.

T-VOR Partial acronym for terminal very high-frequency omnidirectional range.

Twaddle scale A specific gravity scale that attempts to simplify measurement of liquid densities greater than that of water, for example, the densities of industrial liquors. In the Twaddle scale, the range of density from 1.000 to 2.000 is divided into 200 equal parts, so that one degree Twaddle equals a difference in specific gravity of 0.005; on this scale, 40° Twaddle indicates a specific gravity of 1.200.

twilled binding Describes a weave in which each shute wire passes successively over two and under two warp wires, and each warp wire passes successively over two and under two shute wires. [AIR888]

twin Two portions of a crystal having a definite crystallographic relationship; one may be regarded as the "parent," the other as the "twin." The orientation of the twin is either a mirror image of the orientation of the parent about a "twinning plane" or an orientation that can be derived by rotating the twin portion about a "twinning axis." [AS1814]

twin tricycle gear configuration *See* gear configuration, twin tricycle.

twin wheel gear configuaration *See* gear configuration, twin wheel.

twisted pair A cable composed of two insulated conductors, twisted together, without common covering. [ARP1931]

two body problem The problem in classical celestial mechanics that treats the relative motion of two point masses under their mutual gravitational attraction.

two-film technique In radiography, a procedure whereby two films of different relative speeds are used simultaneously to radiograph both the thick and the thin sections of an item. [ARP5089]

two-position action A type of control-system action that involves positioning the final control device in either of two fixed positions, without permitting it to stop at any intermediate position.

two's complement *1.* Refers to a method of representing negative numbers in binary; formed by taking the radix complement of a positive number. *2.* Refers to a form of binary arithmetic used in most computers to perform both addition and subtraction with the same circuitry, where the representation of the numbers determines the operation to be performed.

two-sided sampling plan Any statistical quality control method whereby acceptability of a production lot is determined against both upper and lower limits.

two-stage A filter separator containing two kinds or types of replaceable elements. [AIR4783]

two-stroke cycle In a reciprocating internal combustion engine, an engine cycle that requires two strokes of the piston to complete.

two-way restrictor valve A restrictor valve that restricts flow in either direction, usually through a fixed orifice. [ARP4386]

two-way valve An orifice flow-control component with supply and one control port arranged so that action is in one direction only, from supply to control port. [ARP490]

Tyndall effect First observed by Sir John Tyndall, a physical phenomenon whereby particles suspended in a fluid can be seen readily if illuminated by strong light and viewed from the side, even though they cannot be seen when viewed from the front in the same light beam. The Tyndall effect is the basis for nephelometry, which involves measurement of the intensity of side-reflected light, and is commonly used in such applications as analyzing for trace amounts of silver in solution; determining the concentration of small amounts of calcium in titanium alloys; measuring bacterial growth rates; and controlling the clarity of beverages, potable water, and effluent discharges.

type As used with respect to the certification, ratings, privileges, and limitations of aircraft, a specific make and basic model of aircraft, including modifications thereto that do not change its handling or flight characteristics.

Type 0 control system *See* proportional control (Type 0).

type of hydraulic system A classification standard for a military aircraft hydraulic systems based on minimum and maximum allowable fluid temperatures, defined in ISO 6771 as Type I, Type II, and Type III. *Type I*: –54 to +71°C (-65 to +160°F). *Type II*: –54 to +135°C (-65 to +275°F). *Type III*: –54 to +232°C (-65 to +450°F). NOTE: The upper limit of Type III was reduced from +232°C (+450°F) by the US in recognition of hydraulic fluid limitations. [ARP4386]

TZL *See* touchdown zone lights.

U

UCS diagram *See* uniform-chromaticity-scale diagram.

UDMH Acronym for unsymmetrical dimethyl-hydrazine.

UHF *See* ultra high frequency.

ullage The empty volume of a propellant tank that is not occupied by fuel or oxidizer. [AIR913]

ultimate analysis *See* analysis, ultimate.

ultimate cycle method *See* Ziegler-Nichols method.

ultimate elongation The elongation at rupture. [AIR4844]

ultimate pressure *1*. The lowest pressure reached in an empty, clean, degassed furnace after pumping for a reasonable length of time at ambient temperature. [AMS2769] *2. See* burst pressure, minimum. [ARP4386]

ultimate tensile strength The ultimate or final stress sustained by a specimen in a tension test. [AIR4844]

ultra high frequency (UHF) The frequency band between 300 and 3000 MHz; the band of radio frequencies used for military air/ground voice communications. NOTE: In some instances, frequencies as low as 225 MHz may still be called UHF. [ARP4107]

ultralight aircraft An aircraft for one person, weighing less than 254 pounds, with a top speed of 55 knots and a maximum stalling speed of 24 knots.

ultrasonic Pertaining to mechanical vibrations having a frequency greater than approximately 20,000 Hz. [ARP5089]

ultrasonically-assisted machining A machining method in which ultrasonic vibrations are imparted to a tool to make it cut with better quality or speed than would be possible using the same machining process, but without vibrating the tool.

ultrasonic atomizer A type of atomizer that produces uniform droplets at low feed rates by flowing liquid over a surface that is vibrating at ultrasonic frequency.

ultrasonic bonding A method of joining two solid materials by subjecting a joint under moderate clamping pressure to vibratory shearing action at ultrasonic frequencies until a permanent bond is achieved. May be used on both soft metals and thermoplastics.

ultrasonic cleaning Removing soil from the surface of a part by the combined action of ultrasonic vibrations and a chemical solvent, usually with the part immersed.

ultrasonic coagulation A process that uses ultrasonic energy to bond small particles together, forming an aggregated mass.

ultrasonic delay line A constrained pathway for propagating sound so that the transit time along the pathway becomes a fixed time delay for another signal.

ultrasonic densimeters Density-measuring instruments utilizing ultrasonic devices (sensors).

ultrasonic density sensor A device for determining the density of a liquid or semisolid from the attenuation of ultrasound beams passing through the liquid or semisolid. A typical application involves immersing an ultrasonic transducer in fully agitated lime slurry, thus avoiding the coating and clogging that occurs with other devices.

ultrasonic detector Any of several devices for detecting ultrasound waves and measuring one or more wave attributes.

ultrasonic drilling A method of producing holes of almost any desired shape in very hard materials, such as tungsten carbide or gemstones, by causing a suitable tool that is pressed against the workpiece to vibrate axially under the driving force of an ultrasonic transducer.

ultrasonic flowmeter A device for measuring flow rates across fluid streams by either Doppler-effect measurements or time-of-transit determination. In both types of these flow measurement devices, displacement of the portion of the flowing stream carrying the sound waves is determined and flow rate calculated from the effect on soundwave characteristics.

ultrasonic frequency For compression waves resembling sound, any frequency that is above the audible range; that is, above about 15–20 kHz.

ultrasonic generator A device for producing compression waves of ultrasonic frequencies.

ultrasonic level detector Any of several devices that use either time-of-transit or intensity attenuation of an ultrasonic beam to determine the position of the upper surface in a body of confined liquid or bulk solids.

ultrasonic light diffraction The formation of optical diffraction patterns when light passing through a longitudinal ultrasound field is refracted by interaction with the sound waves.

ultrasonic machining A machining method in which an abrasive slurry is driven against a workpiece by a tool vibrating axially at high frequency; used to cut an exact shape in the workpiece surface.

ultrasonic material dispersion Using ultrasound waves to break up one component of a mixture and disperse it in another to create a suspension or emulsion.

ultrasonics The technology associated with the production and utilization of sound having a frequency higher than about 15 kHz.

ultrasonic stroboscope A device for producing pulsed light by using ultrasound to modulate a light beam.

ultrasonic testing (UT) A nondestructive testing method in which high-frequency sound waves are projected into a solid to detect and locate flaws, to measure thickness, or to detect structural differences.

ultrasonic thickness gage Any of several devices in which either resonance or pulse-echo techniques are used to determine the thickness of metal parts, for example, sheet or plate thickness, or pipe-wall thickness. Such devices may also be used to determine coating thickness in applications in which a suitable reflection can be obtained from the coating-substrate interface.

ultrasonic transducer A device for converting high-frequency electric impulses into mechanical vibrations, or vice versa, usually through the use of a magnetostrictive or piezoelectric material.

ultrasonic vibration Refers to the dynamics of mechanical vibration waveforms in the frequency ranges of more than 20kHz. [ARP1587]

ultrasonic welding A method of joining thermoplastic composites in which vibrations at a fixed frequency are imparted to a properly designed set of parts. In this method, the rapid agitation of the joint area under pressure creates frictional heat, melting the plastic in a fraction of a second. [AIR4844]

ultrasonoscope An instrument for displaying an echosonogram on an oscilloscope; sometimes provides auxiliary output to a chart recorder.

ultraviolet Refers to the zone of invisible radiation, beyond the violet end of the spectrum of visible radiation. [AIR4844]

ultraviolet astronomy Use of special optical instruments for the observation of astronomical phenomena in the ultraviolet spectrum.

ultraviolet degradation Degradation caused by long-time exposure of a material to sunlight or other ultraviolet rays. [ARP1931]

ultraviolet light *See* ultraviolet radiation.

ultraviolet radiation Electromagnetic radiation having wavelengths shorter than visible light and longer than low-frequency x-rays; that is, wavelengths of about 14 to 400 nanometers.

ultraviolet spectrophotometry Determination of the concentration of various compounds in a water solution or gas stream based on characteristic absorption of ultraviolet rays. Ultraviolet (UV) absorption patterns are not as distinctive "fingerprints" as their infrared (IR) counterparts; but, in many cases, UV absorption patterns are more selective and sensitive for use in process-control applications.

ultraviolet stabilizer Any chemical compound that, when admixed with a resin, selectively absorbs ultraviolet rays. [AIR4844]

ultraviolet telescopes Optical telescopes designed to collect ultraviolet light, of wavelengths not capable of passing through earth's atmosphere; as such, they must be used in space.

umbilical connector *See* connector, umbilical.

umbilical cord A cable fitted to a missile with a quick disconnect plug, through which missile equipment is controlled and tested while the vehicle is still attached to launching equipment. [ARP4386]

umbras *1.* The darkest parts of shadows in which light is completely cut off by intervening objects. Lighter parts surrounding the umbras, in which the light is only partly cut off, are called penumbras. *2.* The darker central portions of sun spots, surrounded by lighter penumbra.

Umkehr effect Due to the presence of the ozone layer, an anomaly of the relative zenith intensities of scattered sunlight at certain wavelengths in the ultraviolet as the sun approaches the horizon.

unaccounted-for loss The portion of a boiler heat balance representing the difference between 100 percent and the sum of the heat absorbed by the unit and all the classified losses, expressed as percent.

unarming Removal of normal skid-control function upon reading a low predetermined wheel velocity threshold corresponding to a low aircraft forward velocity. [AIR1489]

unavailability cost The sum, over the life of a system, of the cost incurred in providing replacement facilities to maintain the specified requirement in cases where the primary source is unavailable. [ARP4293]

unbalance The condition that exists in a rotor when vibratory force or motion is imparted to its bearings as a result of centrifugal forces. [ARP587]

unbalanced (to ground) Opposite of balanced (to ground); describes cable pairs that can be susceptible to noise and crosstalk and can cause crosstalk to other pairs.

unbond Within a bonded interface between two adherents, an area where the intended bonding action failed to take place, or where two layers of prepreg in a cured component do not adhere to each other. [AIR4844]

unbonded beam lead length On a bonded device, the distance from the edge of the dielectric layer to the apparent inner edge of the bonded area. [AS1346]

unbonded strain gage A type of wire strain gage, sometimes used in transducer applications, with which strain is determined from elastic tension developed across the gage between mechanical end connections.

unburned combustible The combustible portion of the fuel that is not completely oxidized.

unburned combustible loss *See* combustible loss.

uncertainty *1.* A state in which there is no objective basis for assigning numerical probability weights to the different possible outcomes, or there is no way to describe the possible outcomes. [ARP4293] *2.* The interval within which the true value of a measured quantity is expected to lie with a stated probability.

uncertainty limit The expected error limit of a result for a given coverage. [AIR1678]

uncompensated integrating tachometer generator A generator characterized by a relatively high output-to-null ratio; as such, it is employed in high-gain rate and computing servomechanism applications. Such applications require a high degree of linearity over the range of speeds of the generator. [ARP667]

uncontrolled airspace The portion of the airspace that has not been designated as continental control area, control area, control zone, terminal control area, or transition area, and within which ATC has neither the authority nor the responsibility for exercising control over air traffic. [ARP4107]

uncontrolled reentry (spacecraft) The descent into a denser atmosphere of a spacecraft in an elliptical orbit, due to aerodynamic drag and other perturbation forces. In uncontrolled reentry, the gradually increasing deceleration causes some kinetic energy to be converted into atmospheric heat; the centrifugal force decreases and gravity pulls the spacecraft further into the atmosphere; and the spacecraft eventually burns.

uncouple To disengage a screwed, pinned, or latched connection.

uncoupled modes Modes of vibration that can exist in systems concurrently with, and independently of, other modes.

undamped natural frequency For a landing gear or mechanical system, the frequency of free vibration resulting from only elastic and inertial forces of the system. [AIR1489]

underbead crack A crack in the heat-affected zone of a weldment that does not extend to the base metal surface.

undercarriage The landing gear of an aircraft. *See* landing gear. [AIR1489]

undercure For a molded article, a condition resulting from the allowance of too little time and/or temperature or pressure for adequate hardening of the molding. [AIR4844]

undercut A protuberance or indentation that impedes the withdrawal of a molded part from a two-piece, rigid mold. [AIR4844]

underfill A condition whereby the face of a weld is lower than the position of an adjacent base metal surface.

underflow Pertaining to the condition that arises when a machine computation yields a nonzero result that is smaller than the smallest nonzero quantity that the intended unit of storage is capable of storing.

underlap A lap condition that results in an increased slope of the normal flow curve in the null region. [ARP490]

undershoot Movement of an aircraft whereby it stops short of the commanded reference altitude, attitude, or flight path, before settling out. [ARP419-57]

understressing Repeatedly stressing a part at a level below the fatigue limit or below the maximum service stress for the purpose of improving fatigue properties.

under surface blowing Use of jets blowing on the underside of airfoils for variations in pressure distribution.

under the hood Indicates that the pilot is using a hood to restrict visibility outside the cockpit while simulating instrument flight. An appropriately rated pilot is required in the other control seat while this operation is being conducted. [ARP4107]

undertread *See* tire undertread.

underwing refueling The refueling of an aircraft system by means of a nozzle that is attached to one or more fuel-tight connections positioned on the underside of the aircraft wing, sometimes on the fuselage, within the wheel well of a military aircraft, or at another location generally accessible from the ground or suitable refueling vehicle. [AIR4783]

undetected failure A failure that is not identifiable until a second and detectable failure has occurred. [ARP4386]

unequal area actuator An actuator having different effective piston area for each direction of motion. An unequal area actuator may be either single-ended or double-ended. [ARP4386]

unfired pressure vessel A vessel designed to withstand internal pressure, that is neither subjected to heat from products of combustion nor an integral part of a fired pressure vessel system.

unfired torque The torque required to rotate an engine during the start cycle before light off has occurred; the sum of compressor pumping torque, bearing friction, and engine accessory drag. [ARP906A]

uniaxial load Describes a condition in which a material is stressed in only one direction along the axis or centerline of component parts. [AIR4844]

Unicom A non-government facility that may provide airport information at certain airports. [ARP4107]

unidirectional Having filament orientation in the lengthwise direction only. [AIR4844]

unidirectional concentric stranding A stranding in which each successive layer has a different lay length, thereby retaining a circular form without migration of strands from one layer to another. [ARP1931]

unidirectional laminate A reinforced plastic laminate in which substantially all of the fibers are oriented in the same direction. [AIR4844]

unidirectional lay A variation of concentric lay in which all of the helical layers of strands comprising the concentric conductor have the same direction of lay. This type of construction includes normal unidirectional lay, in which each successive layer has a greater lay length than the preceding layer; and unidirectional equal lay (unilay), generally limited to 19 strands, in which all helical layers have the same length of lay. *See also* concentric lay. [AS1198]

unidirectional pulse A wave pulse in which intended deviations from the normally constant values occur in only one direction.

unidirectional stranding A term denoting a stranded conductor in which all layers have the same direction of lay. [ARP1931]

unified field theory Any theory that attempts to express gravitational theory and electromagnetic theory within a single unified framework; usually an attempt to generalize Einstein's general theory of gravitation alone to a theory of gravitation and classical electromagnetism.

unified screw thread A system of standard 60° V threads that are classified as coarse (UNC), fine (UNF), and extra fine (UNEF) to provide different levels of strength and clamping power.

uniform-chromaticity-scale diagram (UCS diagram) A chromaticity diagram in which the coordinate scales are chosen with the intention of making equal intervals represent, as nearly as possible, equal steps of discrimination for colors of the same luminance at all parts of the diagram. [ARP1782]

uniform corrosion Chemical reaction or dissolution of a metal characterized by uniform receding of the surface.

uniform quality A requirement whereby all specified mechanical and metallurgical properties and conditions of the material must not exhibit variations exceeding the commonly recognized industry standards. [ARP1917]

uniform screen An optical display, such as a monochrome CRT, that has a continuous light-emitting surface, uninterrupted by any physical structure. [ARP1782]

unilateral tolerance A method of dimensioning in which either the upper or lower limit of the allowable range is given as the stated size or location, and the permissible variation is given as a positive or negative tolerance from that size, but not both. The decision as to whether the upper or lower limit is given as the stated dimension depends on the critical value for the dimension, and is chosen so that the tolerance is always away from the critical value and toward a less critical condition.

unilateral transducer A transducer that produces output waves related to input waves when connected in one direction, but that cannot produce such waves when input and output connections are reversed.

unilay *See* unidirectional lay.

unilay strand A conductor constructed with a central core surrounded by more than one layer of helically-laid wires, with all layers having a common length and direction of lay. [ARP1931]

uninterruptible power source The use of resident batteries in a device—to phase in when external power is interrupted.

uninterruptible power supply (UPS) A type of power supply that can provide electrical power even when line power is lost.

union A threaded assembly for joining the ends of two pipes or tubes, in which neither can be rotated to complete the joint. Usually consists of flanged members that are threaded or soldered onto the pipe ends, and a ring member that surrounds the flange edges and makes a leak-tight seal.

unipolar data Data-signal transmission that includes only two possible states, in contrast with bipolar data, which includes two active states and an inactive state. Unipolar data transmission is typical of fiber-optical signaling, whereby optical power is typically modulated between the state of 1 (high light level) to off (low light level). [AS1773]

unit *1.* A quantity that serves as a standard of measurement. Units for angles are degrees, minutes, and seconds; for potential, volts rms; for impedance, ohms; for current, amperes rms; and for temperature, degrees Celsius. [ARP826] *2.* The complete power package, for example, the prime mover, alternator, and all associated equipment and systems. [ARP1148A] *3.* An assembly or any combination of parts, subassemblies, and assemblies mounted together, normally capable of independent operation in a variety of situations. [ARD50013] *4.* A device having a special function.

unit sensitivity The specific amount that a measured quantity must rise or fall to cause a pointer or other indicating element to move one scale division on a specific instrument.

unit under test (UUT) A component tested on the ATE. [ARP4904]

universal instrument *See* altazimuth.

universal joint A linkage for transmitting rotational motion and power between two shafts whose axes do not coincide. Such a linkage is used when the axis of one shaft must be allowed to pivot through a small angle with respect to the other during operation.

universal load cell *See* bidirectional load cell.

universal output transformer An output transformer for an electronic device that has several taps on its secondary winding so that it can be connected to almost any loudspeaker system when the proper tap is chosen.

universal ratio set (URS) An arrangement of variable resistors, used as a highly accurate, continuously adjustable arm of a Wheatstone bridge, with a resolution of 0.001 ohm and a total range of 2111.110 ohms. A URS is particularly well suited for measuring unknown resistances in terms of highly accurate fixed-decade standards within a 10:1 ratio of the unknown resistances.

universal time Time defined by the rotational motion of the earth and determined from the apparent diurnal motions that reflect this rotation. Because of variations in the rate of rotation, universal time is not rigorously uniform. Also called Greenwich mean time (GMT).

unlink time The amount of time required to remove a link. [ARD50012]

unloaded operating inflation pressure The calculated inflation pressure for the operating load of a tire based on the load rating of the tire at rated pressure. [AS4833]

unmate To disengage, disconnect, or uncouple mated connectors. [ARP914A]

unsafe condition Any condition within an aircraft that jeopardizes the safety of the aircraft or the personnel aboard. [AS1212]

unsaturated compound Any compound that has more than one bond between two adjacent atoms, usually carbon atoms, and is capable of adding other atoms at that site to reduce it to a single bond. [AIR4844]

unsaturation (chemistry) In an organic compound, a state in which the atomic bonds of the chain or ring of the compound are not completely satisfied (not saturated). Unsaturation usually results in a double bond (as in the olefins) or a triple bond (as in the acetylenes).

unscheduled landing A landing at other than the scheduled destination, for mechanical or operational reasons. [ARP4107]

unscheduled maintenance Any maintenance other than scheduled preventative maintenance; for example, corrective maintenance required as a result of a problem uncovered during scheduled preventative maintenance. [ARD50010]

unscheduled support costs The cost of unscheduled maintenance work arising because reliability or maintainability or availability, or any combination thereof, fails to meet the specification. [ARP4293]

unsprung mass The portions of the landing gear (wheel, tires, axles, lower part of shock absorber, etc.) that are supported or partially supported by the ground, as opposed to the spring mass, which is separated by the shock absorber. *See also* sprung mass. [AIR1489]

unsprung weight The portion of the gross weight of a vehicle that is made up of the wheels, axles, and various other components not supported by the springs of the vehicle.

unsteady flow A flow in which the flow rate fluctuates randomly with time, with the mean value not being constant.

unsymmetric laminate A laminate having an arbitrary stacking sequence without midplane symmetry. [AIR4844]

up-converters Parametric amplifiers in which the output signal frequencies are greater than the input signal frequencies.

update time Time between updates of a model with sensor data, environmental, or other information; the interval within which model execution typically must be completed. [AIR4548]

upgrade To increase the value or quality of an operating system or commercial product by incorporating changes in design or manufacture without changing its basic function.

uplink Message or data transmitted by the ground network to an aircraft by address (registration mark or flight number). [ARP4102/13]

uplinking In telecommunication systems, the transmission of signals from ground terminals to satellites.

uplift The amount of fuel input into an aircraft at a single refueling operation. [AIR4783]

uplock A locking device to hold a landing gear in the retracted or stowed position. [AIR1489]

upper atmosphere Generally, the atmosphere above the troposphere.

upper surface blowing Use of jet blowing on the upper surface of airfoils to create variations in pressure distribution.

upper torso restraint The portion of a torso-restraint system intended to restrain movement of the chest and shoulder region; commonly referred to as a shoulder harness. [AS8043]

UPS *See* uninterruptible power supply.

upset To cause a local increase in diameter or other cross-sectional dimension by applying an axial deforming force to a piece of rod or wire, as in the production of heads on nails or screws.

upsetting *See* cold heading.

uptake A conduit for exhaust gases, connecting the outlet of a furnace or firebox to a chimney or stack.

up time *1.* The time during which equipment is either producing work or is available for productive work. Contrast with downtime. *2. See* time, up.

uptime ratio The percentage of time that operational equipment is able to satisfy mission demands. [AIR4896]

upward compatibility In data processing, the ability of a computer device or program to function on newer models.

URS *See* universal ratio set.

usable life The maximum number of hours allowed to elapse between the time an adhesive is sampled for quality assurance evaluation testing and the time the adhesive is cured. [AIR4844]

USASCII Stands for US Standard Code for Information Exchange. The standard code used for information exchange among data processing systems, communications systems, and associated equipment. This code uses a character set consisting of 7-bit coded characters (8 bits including parity check), with control characters and graphic characters. *See also* ASCII.

use factor *1.* A factor for adjusting base failure rate to specific use environment and packaging configurations other than those applicable to ground-based systems. [ARP926A] *2.* The ratio of hours in operation to the total hours in that period.

useful life The length of time that a population of items is expected to operate with a constant failure rate. This excludes infant mortality and wearout periods, hence it is the total operating time between debugging and wearout. [ARP4386]

user *1.* The group that operates the ATE. [ARP4904] *2.* A company or body of users that receives standards from external standards-making organizations and uses the standards via their computer system. [AS4159]

user-computer interface *See* man-computer interface.

user interface *1.* The logical interface between a source or sink and the sensor/video interconnect subsystem. [AIR4911] *2.* The means by which a program communicates with an operator.

user-selectable A function descriptor that requires that the system provide the operator with a predefined set of options, within the ATE hardware constraints, for that function. [ARP4904]

user-specifiable A function descriptor that requires that the system allow the operator to specify a particular parameter, within the ATE hardware constraints, for that function. [ARP4904]

UT Symbol for the ultrasonic method of nondestructive testing inspection. [ARP5089]

utility *1.* Any general-purpose computer program included in an operating system to perform common functions. *2.* Any of the systems in a process plant or manufacturing facility that are not directly involved in production. May include any or all of the following: steam, water, refrigeration, heating, compressed air, electric power, instrumentation, waste treatment, and effluent systems.

utilization, aircraft The average daily flying hours for one in-service aircraft. [AIR4896]

utilization equipment An individual unit, set, or a complete system to which electrical power is applied. [AS1212]

utilization rate The planned or actual number of life units expended, or missions attempted, during a stated interval of calendar time. [ARP4386]

utilization systems Includes all of the many electrically operated devices or systems, the load circuits from the power-using equipment to its load bus, and whatever circuit breakers or protective devices may be employed. [ARP4404]

U-tube manometer A device for measuring gage pressure or differential pressure by means of a U-shaped transparent tube partly filled with a liquid, commonly water. With this type of gage, a small pressure above or below atmospheric pressure is measured by connecting one leg of the U to the pressurized space and observing the height of liquid while the other leg is open to the atmosphere. Similarly, a small differential pressure is measured by connecting both legs to pressurized space, for example, high- and low-pressure regions across an orifice or venturi.

UUT diagnostic fault isolation test Refers to a sequence of UUT individual tests, the results of which identify failed components in the UUT. [ARP4904]

UUT individual test A sequence of operations that verify either a single UUT performance requirement or a set of similar UUT performance requirements. [ARP4904]

UUT performance verification test A sequence of UUT individual tests, the results of which verify that the UUT is fully operational and meets all of its performance requirements. [ARP4904]

V

V *1.* Velocity; the speed of an aircraft, expressed in knots or statute miles per hour. [ARP4107] *2.* Volume; the measured sample size. [ARP1179A] *3. See* volt.

V₁ Takeoff decision speed (formerly denoted as critical engine failure speed); the speed below which an aircraft can lose an engine and still stop on the remaining runway. [ARP4107]

V₂ Takeoff safety speed; the normal takeoff speed for a multiengine aircraft. If an aircraft loses an engine after V_1, the pilot should continue to accelerate to this airspeed and attempt to take off. [ARP4107]

Vₐ Design maneuvering speed; the maximum speed at which application of full aerodynamic control will not overstress the aircraft. V_a generally decreases as the gross weight of the aircraft decreases. [ARP4107]

vacuum A given space filled with gas at pressures below atmospheric pressure.

vacuum bag The plastic or rubber layer used to cover a part so that a vacuum can be drawn. [AIR4844]

vacuum bag molding A process whereby a sheet of flexible transparent material, plus bleeder cloth and release film, are placed under the lay-up on the mold and sealed at the edges. [AIR4844]

vacuum bag sealers The sealing tape or putty used on the periphery of a tool to seal a bag to the tool. [AIR4844]

vacuum brake A type of power-assisted vehicle brake in which the released position is maintained by maintaining a pressure below atmospheric pressure in the actuating cylinder; and the actuated position is obtained by admitting air at atmospheric pressure to one side of the cylinder.

vacuum breaker A device used in a water supply line to relieve a vacuum and prevent backflow. Also known as a backflow preventer.

vacuum degassing Removing dissolved or trapped gases in a metal by melting or heating the metal under high vacuum.

vacuum deposition *See* vacuum plating.

vacuum filtration A process for separating solids from a suspension or slurry by admitting the suspension or slurry to a filter at atmospheric pressure (or higher) and drawing a vacuum on the outlet side to assist the liquid in passing through the filter element.

vacuum forming A method of forming sheet plastics by clamping the sheet to a stationary frame, then heating it and drawing it into a mold by pulling a vacuum in the space between the sheet and mold.

vacuum fusion A laboratory technique for determining dissolved gas content of metals by melting the metals in vacuum and measuring the amount of hydrogen, oxygen, and sometimes nitrogen released during melting. This process can be used on most metals, except reactive elements, such as alkali and alkaline-earth metals.

vacuum gage Any of several devices for measuring pressures below ambient atmospheric pressure.

vacuum-gage control circuit An electric circuit that: (a) energizes the tube of an electrically operated vacuum gage; (b) controls and measures gage currents or voltages; and sometimes (c) supplies and regulates power that degasses tube elements.

vacuum-gage tube An enclosed portion of a pressure-measuring system connected to an evacuated chamber or system. The essential component of a vacuum gage tube is a pressure-sensing element, but it also includes the envelope and any support structure, plus the means for connecting the gage to the evacuated space.

vacuum hot pressing A method of processing materials at elevated temperatures and consolidation pressures, and low atmospheric pressures. [AIR4844]

vacuum photodiode A vacuum tube in which light incident on a photoemissive surface (cathode) frees electrons, which are collected by the positively biased anode.

vacuum plating A process for producing a thin film of metal on a solid substrate by depositing a vaporized compound on the work surface, or by reacting a vapor with the surface, in an evacuated chamber. Also known as vacuum deposition or vapor deposition.

vacuum pressure The pressure under the vacuum bag. [AIR4844]

vacuum pump A device similar to a compressor, the inlet of which is attached to a chamber to remove noncondensible gases such as air, and maintain the chamber at a pressure below atmospheric.

vacuum system(s) Chambers having walls capable of withstanding atmospheric pressure and having an opening through which the gas can be removed through a pipe or manifold to a pumping system. The pumping system may or may not be considered as part of the vacuum system.

vacuum tube *1.* A device used in an electronic circuit to amplify d-c, audio, or microwave frequencies; or to rectify radio-frequency signals. Consists of an arrangement of metal emitters, grids, and plates enclosed in a thin, evacuated glass envelope, with a molded plastic base containing pin connectors that are attached to the tube internals. *2. See* valve.

validated adhesive An adhesive that has satisfactorily passed its evaluation tests and has a current usable life. [AIR4844]

validation The process of demonstrating, through testing in the real environment, or an environment as real as possible, that a system satisfies the user's requirements. [ARP1834]

validity Correctness—especially the degree of closeness by which iterated results approach the correct result.

validity check A check based on known limits or on given information or computer results,

e.g., a calendar month will not be numbered greater than 12, and a week does not have more than 168 hours.

valid status message segment A condition that occurs within a bus controller when it determines that a status message segment contains only a clear status and valid data words; that the data words are contiguous to the status word and to each other; and that the response time is as specified. [AS4113]

valid word A word that conforms to the following minimum criteria: (a) the word begins with a valid sync field; (b) the bits are in a valid Manchester II code; (c) the information field has 16 bits plus parity; (d) the word parity is odd. [AS4113]

valve *1.* A device for directing, regulating, or stopping flow, or for regulating pressure in a fluid system, usually through the operation of one or more movable members. [ARP4386] *2.* Among the British, denotes a vacuum tube (because vacuum tubes operate with a stream of electrons through valve action).

valve, aneroid-operated A valve that is operated automatically, at the desired altitude, by the expansion of an aneroid. [ARP171]

valve, automatic *See* automatic valve.

valve, ball *See* ball valve.

valve, blow-off *See* blow-off valve.

valve, brake *See* brake valve.

valve, build-up and vent On a cryogenic converter, a two-position (building-up and vent), manually operated valve. In the building-up position, the valve allows pressure to build-up to a preset value, then allows excess pressure to vent to atmosphere; in the vent position, contents are vented to atmosphere with no pressure build-up.

valve, check *See* check valve.

valve, compensated exhalation An expiratory valve used on pressure-demand masks, in which the exhalation resistance is controlled by, and usually is slightly greater than, the pressure at which gas is delivered from a pressure-demand regulator to the mask. [ARP171]

valve, control A valve in which flow pattern or rate is controlled by external means. [ARP4386]

valve, controllable check *See* controllable check valve.

valve, diaphragm type *See* diaphragm valve.

valve, directional control *See* directional control valve.

valve, dump *See* dump valve.

valve, engine start *See* engine start valve.

valve, expansion A regulator that operates to control the fluid flow rate, classified as constant pressure valves or constant superheat valves. *Constant pressure*–a regulator that operates to control the fluid flow rate to maintain a constant evaporator pressure in a vapor refrigeration cycle. *Constant superheat*–a regulator that operates to control the fluid flow rate to maintain nearly constant temperature difference between evaporator outlet vapor and the saturation temperature corresponding to the pressure of the vapor at the system design point. [ARP147C]

valve, expiratory A valve, usually located in the mask, for permitting escape of expired air from the mask. [ARP171]

valve flame tests A requirement imposed on an aircraft valve that depends on whether the valve must be fire-proof or fire-resistant. Fire-proof valves are usually mounted on the engine side of a firewall or form part of the firewall. Fuel-actuated valves also fall within this category. Flame-resistant valves are valves mounted in similar regions where fires could occur, but the environment is not as severe. [ARP986]

valve, floating ball A valve with a full ball positioned within it; the full ball contacts either of two seat rings and is free to move toward the seat ring opposite the pressure source when in the closed position to effect shutoff.

valve flow gain The increment of valve flow per change in valve input variable at no load across the output. [ARP4386]

valve, flow regulating (flow regulator) *See* flow regulating valve.

valve follower A linkage that transmits motion from a cam to the push rod of a valve, especially in an internal combustion engine.

valve, fuel control An automatically or manually operated device consisting essentially of a regulating valve and an operating mechanism. This device is used to regulate fuel flow and is usually in addition to the safety shut-off valve.

valve, globe *See* globe valve.

valve, lock *See* lock valve.

valve, manual A valve controlled by hand, as opposed to an aneroid-operated valve. [ARP171]

valve, manual gas shutoff A manually operated valve in a gas line for the purpose of completely turning on or shutting off the gas supply.

valve, manual oil shutoff A manually operated valve in the oil line for the purpose of completely turning on or shutting off the oil supply to the burner.

valve, manual reset safety shut-off A manually opened, electrically latched, electrically operated safety shut-off valve designed to automatically shut off fuel when de-energized.

valve, motor driven reset safety shut-off An electrically operated safety shut-off valve designed to automatically shut off fuel flow upon being de-energized. Opened and reset automatically by integral motor device only.

valve, overpressure relief A valve designed to relieve pressure at a predetermined level. Provides safety from explosive failure. [AIR1489]

valve position The position of the valve mechanism, which determines the flow pattern. [ARP4386]

valve pressure drop The sum of the differential pressures across the control orifices of the output stage, expressed in psi (or kPa); equal to the supply pressure, minus the return pressure, minus the load pressure drop. [ARP490]

valve pressure gain Ratio of the change in controlled pressure to the corresponding change in a controlling variable at no flow. [ARP4386]

valve, pressure-operated A valve that allows gas to flow through or from a system when the pressure upstream of the valve reaches a preset value. [ARP171]

valve, pressure reducing (pressure reducer) *See* pressure reducing valve.

valve, pressure regulating (unloader valve) *See* pressure regulating valve.

valve, pressure relief A valve that limits maximum pressure in a circuit by releasing excess to return on the basis of differential pressure between pressure and return ports. A pressure relief valve must be able to accommodate full rated flow of the line size for which it is constructed. [ARP4386]

valve, pressure relief modulated check A valve acting to limit pressure-induced flow to a preset magnitude to equalize pressures on both sides of the valve. [ARP171]

valve, priority *See* priority valve.

valve, purge A valve used to eliminate noncondensable gases from a refrigeration system. [ARP147C]

valve, relief *See* relief valve.

valve, restrictor *See* restrictor valve.

valve, restrictor, adjustable *See* adjustable restrictor valve.

valve, restrictor, one-way (restrictor check valve) A restrictor valve that permits free flow in the reverse direction. [ARP4386]

valve, restrictor, two-way *See* two-way restrictor valve.

valve, selector *See* selector valve.

valve, semiautomatic *See* semiautomatic valve.

valve, servo A directional control valve that infinitely modulates flow or pressure as a function of its input signal. [ARP4386]

valve, servo, electro-hydraulic *See* servovalve, electrohydraulic flow control.

valve, shutoff *See* shutoff valve.

valve, shuttle *See* shuttle valve.

valve, surge damping (surge damper) A valve whose function is to reduce surge pressures during intermittent flow. [ARP4386]

valve, thermal relief *See* thermal relief valve.

Van de Graaf generator An electrostatic device in which a system of belts is used to generate electric charges and carry them to an insulated electrode, which becomes charged to a high potential.

van der Waals forces *See* dispersion forces.

vane *1.* A flat or curved machine element attached to a hub or rotor which is acted upon by a flowing stream of fluid to produce rotary motion. *2.* A fixed or adjustable plate inserted in a gas or air stream, used to change the direction of flow.

vane actuator An actuator in which fluid pressure is applied to either side of a vane or vanes

that are integral with or geared to a rotary output shaft. [ARP4386]

vane control In the inlet of a fan, a set of movable vanes to provide regulation of air flow.

vane guide A set of stationary vanes to govern direction, velocity, and distribution of air or gas flow.

vane motor A motor having a number of vanes attached to a shaft or an outer housing which are eccentric with respect to one another. [ARP4386]

vane pump A variable- or fixed-displacement pump having a number of cavities. In such a pump, the rotation varies the volume of fluid entrapped between vanes, which creates a pumping action through positive volume displacement. [ARP4386]

vane starter An engine-starting device incorporating a positive-displacement vane-type air/gas motor, driven by cold or heated compressed air or other gas. May include speed-increasing or -reduction gears, and usually includes a jaw-type engaging mechanism or an overrunning clutch. [ARP906A]

vapor(s) *1.* Gases whose temperatures are below their critical temperatures, so that they can be condensed to the liquid or solid state by increase of pressure alone. *2.* The gaseous product of evaporation.

vapor area, flammable Any area in an aircraft where flammable fluids and vapors may collect due to leaky lines or fittings in the area, or due to seepage from some other area. [ARP4404]

vapor barrier A sheet or coating of low gas permeability that is applied to a structural wall to prevent condensation and absorption of moisture.

vapor degreasing A cleaning process in which the hot vapors of a chlorinated solvent are used to remove soils, especially oil, grease, waxes, fingerprints etc. [AIR4844]

vapor deposition *See* vacuum plating.

vapor-filled thermometer A type of filled-system thermometer with which temperature is determined from the vapor pressure developed from partial vaporization of a volatile liquid contained within the system.

vapor generator A container of liquid, other than water, that is vaporized by the absorption of heat.

vaporimeter *1.* An apparatus in which the volatility of an oil is estimated by heating it in a current of air. *2.* An instrument used to determine alcohol content of a substance by measuring the vapor pressure of the substance.

vaporization The change from liquid or solid phase to the vapor phase. Also called volatilization.

vaporization cooling A method of cooling hot electronic equipment by spraying it with a volatile, nonflammable liquid of high dielectric strength; the liquid absorbs heat from the electronic equipment, vaporizes, and carries the heat to enclosure walls or to a radiator or heat exchanger. Also known as evaporative cooling.

vapor liquid ratio, V/L The equilibrium ratio of volume of vapor (actually air and fuel vapor) to volume of liquid, both at the same temperature. [AIR1326]

vapor-liquid-solid process A process utilizing vapor feed gases and a liquid catalyst, and producing solid crystalline whisker growth. [AIR4844]

vapor phase epitaxy A crystal growth process whereby an element or a compound is deposited in a thin layer on a slice of substrate single crystal material by the vapor-phase technique.

vapor pressure For a given temperature, the pressure at which a liquid is in equilibrium with its vapor. As a liquid is heated, its vapor pressure increases until it equals the pressure above the liquid; at this point the liquid begins to vaporize.

vapor pressure element A temperature-sensitive device whose signal is proportional to the pressure of vapor in coexistence with its liquid phase and is independent of the specific volume. [AIR1900]

vapor pressure thermometer A temperature transducer in which the pressure of vapor in a closed system of gas and liquid is a function of temperature.

vapor, superheated Vapor or gas existing at a temperature higher than the saturated temperature corresponding to the existing pressure. [ARP147C]

vaportight So enclosed that vapor will not enter the enclosure. [ARP4404]

var A unit of measure for reactive power; calculated by taking the product of voltage, current, and the sine of the phase angle.

variable The symbolic representation of a logical storage location that can contain a value that changes during a discrete processing operation. *See also* measurand.

variable cambered wing A wing so constructed that its camber can be adjusted or shaped by use of a control mechanism. [ARP4386]

variable displacement pump A pumping unit in which the displaced volume is varied by a control mechanism; the output can be controlled independently of revolutions per minute by varying the output volume per cycle. [ARP4386]

variable frequency variable voltage a-c control An a-c motor controlled by varying the frequency and voltage of the a-c power supplied to it. The power is regulated to maintain a constant voltage to frequency ratio. [ARP4386]

variable-inductance accelerometer An instrument for measuring instantaneous acceleration of a body. Consists of a differential transformer with a center coil excited from an external a-c signal, the magnitude of which is proportional to displacement of a ferromagnetic core mass suspended on springs in the center of the three coils.

variable-inductance pickup A transducer that converts mechanical oscillations into audio-frequency electrical signals by varying the inductance of an internal coil.

variable-length record format A file format in which records are not necessarily the same length.

variable lift devices Means of varying the lift of a basic airfoil through the use of various structures; for example, flaps, slots, and slats. [ARP4107]

variable pitch propeller *See* controllable pitch propeller.

variable position valve A valve that controls flow by selective throttling of the flow passage. Intermediate areas may be infinitely or incrementally selectable. [ARP986]

variable pulse width d-c control A d-c motor controlled by modulating d-c power into a pulse train and varying the individual pulse widths. The pulse train frequency is usually held constant. [ARP4386]

variable ratio drive A motion transmission that has the capability of varying output velocity and torque with constant velocity and torque input. [ARP4386]

variable-reluctance pickup A transducer that converts mechanical oscillations into audio-frequency electrical signals by varying the reluctance of an internal magnetic circuit.

variable reluctance proximity sensor A device that senses the position (presence) of an actuating object by means of the voltage generated across the terminals of a coil surrounding a pole piece extending from one end of a permanent magnet. In this device, coil voltage is proportional to the rate of change of magnetic flux as the actuating object passes through the field near the pole piece.

variable reluctance tachometer A type of tachometer designed to measure rotational speeds of 10,000 to 50,000 rpm by detecting electrical pulses generated as an actuating element integral with the rotating body repeatedly passes through the magnetic field of a variable-reluctance sensor. The pulses are amplified and rectified, then used to control direct current to a milli-ammeter, which is calibrated directly in rpm.

variable-resistance accelerometer An instrument that measures acceleration by determining the change in electrical resistance in a measuring element, such as a strain gage or slide wire, whose dimensions are changed mechanically under the influence of acceleration.

variable-resistance pickup A transducer that converts mechanical oscillations into audio-frequency electrical signals by varying the electrical resistance of an internal circuit.

variable-riser stairway A stairway designed to allow variation of the riser to tread ratio, and thus variation of the angle of inclination, to suit varying elevations. [ARP836A]

variable stream control engines Advanced, moderate bypass-ratio turbofan configurations in which duct burner thrust augmentation and coannular nozzles are used for jet noise reduction.

variable voltage d-c control A mechanically commutated d-c motor controlled by varying the voltage of the d-c power supplied to it. [ARP4386]

variable word-length Having the property that a machine word may have a variable number of characters. May be applied either to a single entry whose information content may be changed from time to time, or to a group of functionally similar entries whose corresponding components are of different lengths.

variation limit, under minimum or over maximum A variation limit whereby an individual determination for a specified element may vary under or over the specified composition limit. [AMS2269E]

variometer *1.* Instrument for comparing magnetic forces, especially of the earth's magnetic field. *2.* A form of variable inductance, consisting of two coils connected in a series and arranged one inside the other. The inner coil is equipped to rotate and, thereby, vary the mutual inductance between coils. The variometer was principally used in the early days of radio communications, but has found continued usefulness in electronics.

varistors Two electrode semiconductor devices having a voltage-dependent nonlinear resistance.

varmeter An instrument for measuring the electric power drawn by a reactive circuit. Also known as reactive volt-ampere meter.

varying duty A requirement of service that demands operations at loads, and for intervals of time, both of which may be subject to wide variations. [ARP1199A]

VASI Stands for visual approach slope indicator. An airport lighting facility providing vertical visual approach slope guidance to aircraft during approach to landing. The VASI radiates a directional pattern of high-intensity red and white focused light beams, which indicate to the pilot whether he/she is "on path" (if red/white), "above path" (if white/white), or "below path" (if red/red). Some airports serving large aircraft have three-bar VASIs which provide two visual glide paths to the same runway. [ARP4107]

vasoconstrictor An agent that causes narrowing of the blood vessels. [ARP171]

vasodilator An agent that causes dilation of the blood vessels. [ARP171]

VATOL aircraft Abbreviation for vertical attitude takeoff and landing aircraft.

V_b The design speed for maximum gust intensity; maximum turbulence penetration speed. [ARP4107]

V-band coupling A sheet-metal band clamp with sheet-metal retainer sections welded or riveted to the band. [AIR869]

V_c In designing the aircraft, the design cruising speed used in calculation of structural strength; at this speed, the structure will sustain 50 ft/s gusts. [ARP4107]

VCO See voltage controlled oscillators.

V-couplings The family of V-band couplings and V-retainer couplings. [AIR869]

V_d The design diving speed to which the aircraft is carried in official certification tests; this is the speed chosen by the designer, at which the aircraft must be flown to ascertain that no adverse flight conditions exist (at that speed). [ARP4107]

V_{df} Maximum demonstrated flight diving speed. Generally demonstrated only by test pilots. See also V_{ne}. [ARP4107]

VDU See video display unit.

vection illusion See illusions, vection.

vectopluviometer A rain gage, or a circular array of four or more rain gages, used to measure the direction and inclination of falling rain.

vector(s) Quantities such as force, velocity, or acceleration that have both magnitude and direction at each point in space, as opposed to scalars that have magnitude only. Such a quantity may be represented geometrically by an arrow of length proportional to its magnitude, pointing in the assigned direction.

vectoring See radar vectoring.

vector voltmeter A two-channel, high-frequency sampling voltmeter that can be connected to two input signals of the same frequency to measure not only their voltages but also the phase angle between them.

vegetative index Linear combinations of spectral band responses in digital count, reflectance factor, or voltage to determine the vigor, greenness, and/or biomass of the vegetation. Observations of the spectral band responses can be made by satellite-borne, aircraft-borne, truck-mounted, or hand-held spectrometers.

vehicle 1. A body such as an aircraft or rocket designed to carry a payload aloft. 2. A self-propelled machine for transporting goods or personnel. 3. A solvent or other carrier for the resins and pigments in paint, lacquer, shellac, or varnish.

vehicle instability Deflection of a vehicle due to a wind force, causing unsafe working conditions. [ARP1328]

vehicle pivot point The point of the vehicle that is in contact with the ground, on the side of the vehicle opposite to that where the wind force is applied, and farthest from the point where the wind is applied. [ARP1328]

vehicle tip point Maximum vehicle instability when the vehicle center of gravity has been rotated by a wind force to a point directly above the vehicle pivot point. [ARP1328]

vehicle velocity The velocity of the origin of the body axis, usually the center of gravity, relative to the air, unaffected by the aerodynamic field of the vehicle. [ARP4386]

veil An ultra-thin mat, similar to a surface mat, often composed of organic fibers as well as glass fibers. [AIR4844]

veiling brightness Brightness superimposed on the retinal image, which reduces its contrast. This veiling effect, produced by bright sources or areas in the visual field, results in decreased visual performance and visibility. [AIR1151]

velocimeter An instrument for measuring the speed of sound in gases, liquids, or solids.

velocity 1. The distance an ultrasonic wave travels in unit time. [ARP5089] 2. Denotes the maximum design system operating velocity—the takeoff speed under maximum gross weight, standard hot-day, 8000 ft (2432m) altitude, or as defined by the SPS. [AS483A] 3. Rate of motion. Referred to as linear speed when it the rate of motion in a straight line; and angular speed when it is the change of direction per unit time. *4.* The rate of change of a position vector with respect to time at any given point in space; the first derivative of distance with respect to time.

velocity coupling The response of a burning propellant surface to the local velocity; includes both mean flow and acoustic velocity (both being parallel to the burning surface).

velocity gain The proportional change of actuator velocity with error signal. This parameter is usually specified for no-load conditions and is usually measured under open-loop conditions. [ARP4386]

velocity head The pressure, measured in height of fluid column, needed to create a fluid velocity. Numerically, velocity head is the square of the velocity divided by twice the acceleration of gravity.

velocity meter A flowmeter that measures rate of flow of a fluid by determining the rotational speed of a vaned rotor inserted into the flowing stream; the vanes may or may not occupy the entire cross section of the flowpath.

velocity of approach A factor (F) determined by the ratio (m) of the valve orifice area to the inlet pipe area.

velocity of propagation In cable measurements, the transmission speed of an electric signal down a length of cable compared to its speed in free space; a function of dielectric constant. [ARP1931]

velocity outlet *See* gasper air outlet.

velocity pressure A measure of the kinetic energy of a fluid.

velocity-type flowmeter A flow-measurement device in which the fluid flow causes a wheel or turbine impeller to turn, producing a volume-time readout. Also known as current meter or rotating meter.

velocity versus force or torque characteristic (speed versus force or torque characteristic) The curves of steady-state velocity vs. steady-state force or torque when the actuator is driven by the power modulator with constant values of signal input. [ARP4386]

velocity, vibration A quantity expressed in units of velocity, e.g., inches per second or millimeters per second. Velocity is the first integral of acceleration with respect to time. [AIR1839]

Venn diagram A graphical representation in which sets are represented by closed regions that may bear all kinds of relations to one another, such as being partially overlapped, completely separated from one another, or contained totally one within another. All members of a set are considered to lie within or be contained within the closed region representing the set. The Venn diagram is used to facilitate the determination of whether several sets include or exclude the same members.

vent *1.* Auxiliary port used to establish a reference pressure in a particular region of the amplifier; analogous to an electrical ground potential. [ARP993] *2.* Any opening or passage that allows gases to escape from a confined space to prevent the buildup of pressure or the accumulation of hazardous or unwanted vapors.

vent cap The check valve component on top of a cell which regulates the maximum internal pressure that the cell jar will hold, while preventing the entrance of atmospheric gases under conditions of fluctuating ambient pressures. The vent cap is removable to allow water replacement in the electrolyte. [AS8033]

vent cloth A layer or layers of open-weave cloth used to provide a path for the vacuum to "reach" the area over a laminate being cured, so that volatiles and air can be removed. [AIR4844]

vented amplifier An amplifier that utilizes auxiliary ports to establish a reference pressure in a particular region of the amplifier geometry. This is in contrast with a closed amplifier, which has no communication with an independent reference. [ARP993A]

venting In autoclave curing of a part or assembly, refers to turning off the vacuum source and venting the vacuum bag to the atmosphere. [AIR4844]

ventricular fibrillation *See* fibrillation, ventricular.

venturi In a pipe, tube, or flume, a constriction consisting of a tapered inlet; a short, straight, constricted throat; and a gradually tapered outlet. Fluid velocity is greater and pressure is lower in the throat area than in the main conduit upstream or downstream of the venturi.

A venturi can be used to measure flow rate or to draw another fluid from a branch into the main fluid stream.

venturi meter A type of flowmeter that measures flow rate by determining the pressure drop through a venturi constriction.

venturi tube(s) Short tubes of smaller diameter in the middle than at the ends. When fluids flow through such tubes, the pressure decreases as the diameters become smaller, the amount of decrease being proportional to the speed of flow and the amount of restriction.

vent valve A device installed on the upper surface of a vehicle tank to allow air to escape during loading or enter during discharge. [AIR4783]

verification The evaluation of an implementation of requirements to determine that they have been met. [ARP4754]

verify To determine whether a transcription of data or other operation has been accomplished accurately.

Verneuil process Method of single-crystal growth in which powder is dropped through an oxy-hydrogen flame, falling molten on crystal seed.

vernier A short auxiliary scale that slides along a main instrument scale and permits accurate interpolation of fractional parts of the least division on the main scale.

vernier engines Rocket engines of small thrust, used primarily to obtain a fine adjustment in the velocity and trajectory of a rocket vehicle just after the thrust cutoff of the last sustainer engine; and used secondarily to add thrust to a booster or sustainer engine.

vertical axis Also referred to as the normal axis. An imaginary line passing through the center of gravity and lying in the plane of symmetry of an aircraft; this line is perpendicular to both the longitudinal axis and the lateral axis. Angular movement about the vertical axis is called yaw. [ARP4107]

vertical axis freedom As applied to a horizontal balancing machine bearing carriage or bearing housing, freedom to rotate by a few degrees about the vertical axis through the center of the support. [ARP587]

vertical bearing (V/B) The angle in the vertical plane between the horizon and a line from the present position in space to the next three-dimensional defined waypoint. [ARP1570]

vertical bleed Removal of volatiles and excess resin through a perforated release sheet into a bleeder cloth over the whole area of a part. [AIR4844]

vertical boiler A fire-tube boiler consisting of a cylindrical shell, with tubes connected between the top head and the tube sheet forming the top of the internal furnace. The products of combustion pass from the furnace directly through the vertical tubes.

vertical bond voids Absence of a bond or bonding material between the vertical edges of two core sections or core and inserts. [AIR4844]

vertical deviation The vertical distance or altitude between the existing flight altitude and the programmed path or profile altitude. [ARP1570]

vertical firing Describes a burner so arranged that air and fuel are discharged into the furnace, in practically a vertical direction.

vertical gyro *See* attitude gyro.

vertical junction solar cells Solar cells made from wafers on which narrow grooves are formed using a preferential KOH etch. The grooved region is radiation tolerant.

vertical motion simulators Vibration machines that produce mechanical oscillations parallel to the vertical axis.

vertical navigation (VNAV) Those functions that provide guidance signals to control an aircraft during climb or descent. [ARP1570]

vertical orientation The attitude of an object in reference to a plane parallel to the direction of gravity (determined with a plumbline).

vertical orifice installation An orifice installation used in a vertical pipeline. *See* orifice plate, orifice run, and meter run.

vertical path The climb and descent modes that control the vertical trajectory to specific earth-referenced paths or profiles including one or more three-dimensional waypoints. [AIR4102/9]

vertical profile (VPROF) *See* profile. *See also* path profile and V-path.

vertical situation display (VSD) A display that represents the situation of the aircraft on an imaginary vertical plane ahead of the aircraft. The basic dimensions of a VSD are azimuth, elevation, and attitude with respect to roll. Lateral translation of display elements signifies a change in aircraft heading; vertical translation of display elements represents a change in pitch or vertical flight path; and revolution of the display elements denotes rotation of the aircraft about the roll axis. [ARP4107]

vertical speed (VS) The rate of change of altitude. Usually refers to barometric altitude rate, but may also refer to radar altitude rate or inertial vertical velocity. [ARP1570]

vertical takeoff and landing aircraft (VTOL aircraft) *See* VTOL.

vertical track distance The vertical distance or height between existing altitude and the programmed path or profile altitude. [AIR4102/9]

vertigo The sensation by a person that the outer world is revolving about him (objective vertigo); or that he himself is moving in space (subjective vertigo). NOTE: The term "vertigo" is frequently used erroneously as a synonym for dizziness or giddiness, indicating an unpleasant sensation of disturbed relations to surrounding objects in space.

very fast-acting fuse A fuse that opens a circuit without deliberate time-delay, and whose short-circuit opening time is faster than a normal-opening fuse. [ARP1199A]

very high frequency (VHF) The frequency band between 30 and 300 MHz. [ARP4107]

very high frequency omnidirectional range (VOR) A ground-based electronic navigational aid transmitting very high-frequency navigation signals, 360 deg in azimuth, oriented from magnetic north. The VOR periodically identifies itself by Morse code and may have an additional voice identification feature. VOR is used as the basis for navigation in the National Airspace System. [ARP4107]

very high speed integrated circuits *See* VHSIC (circuits).

very large scale integration (VLSI) A very complex integrated circuit that contains ten thousand or more individual devices, such as basic logic gates and transistors, placed on a single semiconductor chip.

very long base interferometry (VLBI) The simultaneous observation of radio sources by two radio telescopes spaced very far apart to enhance angular resolution. The signals are recorded on magnetic tapes and combined electronically on a computer.

vestibular illusions *See* illusions, vestibular.

v_f The design maximum speed with the wing flaps in a prescribed extended position. [ARP4107]

V/F (boilup-to-feed ratio) A quantity used to analyze the operation of a distillation column.

V_{fe} The highest permissible speed at which wing flaps may be actuated. [ARP4107]

VFR *See* visual flight rules.

VFR aircraft/VFR flight An aircraft conducting flight in accordance with visual flight rules. [ARP4107]

VFR conditions Weather conditions equal to or better than the minimum for flight under visual flight rules. [ARP4107]

VFR conditions on-top/VFR-on-top ATC authorization for an IFR aircraft to operate in VFR conditions at any appropriate VFR altitude (as specified in FAR and as restricted by ATC), at or above the minimum en route altitude (MEA), that is at least 1000 ft above a cloud layer. [ARP4107]

VFR over-the-top Special permission for the operation of an aircraft over the top of an area of poor weather conditions under VFR when the aircraft is not being operated on an IFR flight plan. [ARP4107]

V_h The maximum level flight airspeed with maximum continuous power applied. [ARP4107]

VHF *See* very high frequency.

VHSIC (circuits) Stands for very high speed integrated circuits. Chips being developed by a DOD program to provide high-speed MIL spec VLSI devices for use in military systems.

vibrating density sensor Any of several devices in which a change in natural oscillating frequency of a device element (e.g., a cylinder, single tube, twin tube, U-tube, or vane) is detected and related to the density of process fluid flowing through the system.

vibrating quartz-crystal moisture sensor A device for detecting the presence of moisture in a sample gas stream by dividing the stream into two portions, one of which is dried, then alternately passing the two streams across the face of a hygroscopically sensitized quartz crystal. The wet and dry vibrational frequencies of the quartz crystal are continuously monitored and compared to the frequency of an uncoated, sealed reference crystal.

vibrating-reed electrometer An instrument in which a vibrating capacitor is used to measure small electrical charges, often in combination with an ionization chamber.

vibrating-reed tachometer A tachometer consisting of an extended series of reeds of various lengths mounted on the same base. The tachometer is placed on a vibrating surface, such as the enclosure of rotating equipment, and the frequency determined by observing which of the reeds is vibrating at its natural frequency.

vibration *1.* Motion due to a continuous change in the magnitude of a given force that reverses its direction with time. *2.* Motion of an oscillating body during one complete cycle; two oscillations.

vibration acceleration The response of a mass to an applied force; usually stated as a ratio (g) of the acceleration with respect to the acceleration due to gravity (G). [AIR1839]

vibration damping Any method of converting mechanical vibrational energy into heat.

vibration displacement A quantity expressed in units of displacement, e.g., inches, millimeters, or mils. Displacement is the second integral of acceleration with respect to time. [AIR1839]

vibration isolators Resilient supports that tend to isolate systems from steady-state excitation.

vibration meter A device for measuring vibrational displacement, velocity, and acceleration. Consists of a suitable pickup, electronic amplification circuits, and an output meter.

vibration mode In a system undergoing vibration, a characteristic pattern assumed by the system whereby the motion of every particle is simple harmonic with the same frequency.

vibration-type level detector A device for detecting the level of solids in a bin or hopper, in which a tuning fork driven by a piezoelectric crystal vibrates freely when the level is below the sensor position and is inhibited from vibrating when bulk material surrounds the sensor.

vibration velocity *See* velocity, vibration.

vibratory separation A technique for separating or classifying particulate solids using screens that are subjected to vibratory or oscillating motion.

vibrograph An instrument for making an oscillographic recording of the amplitude and frequency of a mechanical vibration, such as by producing a trace on paper or film using a moving stylus.

vibrometer A device for measuring the amplitude of a mechanical vibration. Also known as a vibration meter.

vibronic isolation Refers to systems that minimize the transfer of vibrations from the floor and surrounding environment to the surface of an optical table or other equipment mounted on them.

vibronic transition A simultaneous change in both the vibrational and electronic energy state of a molecule, with the amount of energy involved being similar to that for electronic transitions.

Vickers hardness *See* diamond-pyramid hardness.

Victor airway Phonetic designation of a VOR airway; for example, Victor 123. [ARP4107]

video display unit (VDU) Any one of several types of shared human interface devices that use digital video technology.

video landmark acquisition and tracking Shuttle era system for earth-feature identification, acquisition, and tracking.

video map An electronically displayed map on the radar display which may depict data such as airports, heliports, runway centerline extensions, hospital emergency landing areas, NAVAIDs and fixes, reporting points, airway/route centerlines, boundaries, handoff points, special-use tracks, obstructions, prominent geographic features, map alignment indicators, range accuracy marks, and minimum vectoring altitudes. [ARP4107]

video signals Signals with a bandwidth of over 20 kHz.

viewable screen size (useful screen size) The dimensions of the useable light-emitting area of the CRT faceplate, projected through the glass onto a plane perpendicular to the Z-axis of the faceplate. [ARP1782]

view effects Effects of change in angular size of field of view on receptors of radiation.

viewing envelope The envelope in space, defined by horizontal and vertical angles measured from the perpendicular to the display face, within which a viewer can perceive the symbology presented by the display. [ARP1782]

virtual address space A set of memory addresses that are mapped into physical memory addresses by the paging or relocation hardware located where the program is executed.

virtual leak A gradual release of gas by desorption from the interior walls of a vacuum system in a manner that cannot be accurately predicted. The effect of a virtual leak on system operation resembles that of an irregularly variable physical leak.

viscoelastic damping The absorption of oscillatory motions by materials that are viscous while exhibiting certain elastic properties.

viscoelastic flow *See* viscoelasticity.

viscoelasticity A property involving a combination of elastic and viscous behavior. A viscoelastic material is considered to combine the features of a perfectly elastic solid and a perfect fluid. [AIR4844]

viscometer An instrument that measures the viscosity of a fluid.

viscometer gage An instrument that determines pressure in a vacuum system by measuring the viscosity of residual gases.

viscosity *1.* A property of a fluid relating to shearing stress and continued resistance to flow. The viscosity of sealants is measured at standard conditions by a device that has a spindle mounted on a gauge. [AS7200/1] *2.* A measure of the internal friction of a fluid or its resistance to flow.

viscosity index A measure of the change of viscosity of a fluid with temperature. [AIR1116]

viscous damping *1.* The dissipation of energy that occurs when a particle in a vibrating system is resisted by a force that has a magnitude proportional to the magnitude of the velocity of the particle, and direction opposite to the direction of the particle. *2.* A method of converting mechanical vibration energy into heat by means of a piston attached to a vibrating object which moves against the resistance of a fluid (usually a liquid or air) confined in a cylinder, or a bellows attached to a stationary support.

viscous-drag-type density meter A type of meter for determining gas density by comparing the drag force on linked impellers driven by flow of a standard gas and the test gas. In such a meter, the balance point is a function of gas density, thus such a meter can be calibrated to read directly in density units.

viscous flow The flow of a fluid through a duct under conditions in which the mean free path is very small in comparison with the smallest dimensions of a transverse section of the duct. This flow may be either laminar or turbulent. *See* laminar flow and turbulent flow.

viscous fluids Fluids whose molecular viscosity is sufficiently large to make the viscous forces a significant part of the total force field in the fluid.

viscous friction load A load opposing motion and proportional to load velocity. [ARP4386]

visibility The ability, as determined by atmospheric conditions and expressed in units of distance, to see and identify prominent unlighted objects by day and prominent lighted objects by night. The term "visibility," as reported in statute miles, hundreds of feet, or meters, normally refers to horizontal visibility. [ARP4107]

visibility meter An instrument for directly or indirectly determining visual range in the earth's atmosphere.

visible infrared spin scan radiometer A radiometer used for satellite sounding of the atmosphere.

visible radiation *See* light.

visible spectrum *1.* The range of wavelengths of visible radiation. *2.* A display or graph of the intensity of visible radiation emitted or absorbed by a material as a function of wavelength or some related parameter.

visual approach An approach whereby an aircraft on an IFR flight plan, operating in VFR conditions under the control of an air-traffic control facility and having an air traffic control authorization, may proceed to the airport of destination in VFR conditions. [ARP4107]

visual approach slope indicator *See* VASI.

visual flight rules (VFR) *1.* Rules that govern the procedures for conducting flight under visual conditions. *2.* As used in the US, indicates weather conditions that are equal to or greater than minimum VFR requirements. [ARP4107]

visual illusions *See* illusions, visual.

visual line width The total width of a display line as seen by the human eye, with or without the use of a magnifier. [ARP4067]

visual meteorological conditions (VMC) Meteorological conditions expressed in terms of visibility, distance from cloud, and ceiling equal to or better than specified minimums. [ARP4107]

visual photometry A subjective approach to the problem of photometry, in which the human eye is used as the sensing instrument; to be distinguished from photoelectric photometry.

visual separation A means employed by ATC to separate aircraft in terminal areas. There are two ways to effect this separation: (a) the tower controller may see the aircraft involved and issue instructions, as necessary, to ensure that the aircraft avoid each other; or (b) a pilot may see the other aircraft involved and upon instructions from the controller provide his/her own separation by maneuvering the aircraft, as necessary, to avoid it. [ARP4107]

vitreous enamel *1.* A coating applied to metal by covering the surface with powdered alkali borosilicate glass frit and fusing it onto the surface by firing at a temperature of 800 to 1600°F (425 to 875°C). Also known as porcelain enamel. *2. See* enamel.

V/L *See* vapor liquid ratio.

VLBI *See* very long base interferometry.

V/L calculation *See* calculation, V/L.

V$_{le}$ The maximum safe speed for flight with the landing gear in the extended position. [ARP4107]

V$_{lo}$ Maximum speed for safe extension or retraction of landing gear (landing gear operating speed). [ARP4107]

V$_{lof}$ Liftoff speed; the velocity at which an aircraft separates from the ground during takeoff. [ARP4107]

VLSI *See* very large scale integration.

VMC *See* visual meteorological conditions. [ARP4107]

V$_{mca}$ Minimum speed at which a multiengine aircraft is controllable in flight with the critical engine inoperative and the remaining engine(s) at takeoff power. It is not required that an aircraft be able to climb or hold its altitude under these conditions, only that it be able to maintain its heading. [ARP4107]

V$_{mcg}$ The minimum speed at which a multiengine aircraft is controllable with the critical engine inoperative during the takeoff roll. [ARP4107]

V$_{mo}$ Maximum operating limit speed. Limiting airspeed for turboprop and jet aircraft based on 80 percent of the flight-demonstrated diving speed. Flight at or beyond this speed has a high probability of causing serious structural damage to the aircraft. Similar to V$_{ne}$. [ARP4107]

V$_{mu}$ Minimum unstick speed; that is, the calibrated airspeed at and above which the aircraft can safely lift off the ground and continue flight. [ARP4107]

VNAV *See* vertical navigation.

V$_{ne}$ Never-exceed speed; the airspeed that is 90 percent of the maximum flight-demonstrated diving speed. Flight at or beyond this airspeed entails a high risk of structural damage to the aircraft, especially if it is poorly rigged or a slight bit of turbulence is encountered. Use of this term is limited to piston-powered aircraft. Similar to V$_{mo}$. [ARP4107]

V$_{no}$ The maximum structural cruising speed; the maximum speed for normal operation. [ARP4107]

vocabulary A list of operating codes or instructions available to the programmer for writing the program for a given problem, for a specific computer, or for a specific language.

voice control Using the voice to activate devices that respond or operate by means of speech recognition.

voice print An acoustic spectrograph that can be used to analyze sound patterns, especially the harmonic patterns that distinguish one person's voice from another's.

void *1.* Discontinuity in which there is a physical separation between opposite walls. [ARP5089] *2.* Any opening, small crack, or crevice occurring at the juncture of structural members (such as chambers, reliefs, joggles, butt joints, or fasteners.) [AIR4069]

void content Volume percentage of voids, usually less than 1% in a properly cured composite. [AIR4844]

void seal A seal used to fill holes, joggles, channels, and often other voids caused by the build-up of structure in a fuel tank. The void seal provides continuity of sealing where fillet seals are interrupted by such structure gaps. [AIR4069]

vol *See* volume.

volatile *1.* In reference to a liquid, having appreciable vapor pressure at room or slightly elevated temperature. *2.* In reference to a computer, having memory devices that do not retain information if the power is interrupted.

volatile content The percent of volatiles that are present in a plastic or an impregnated reinforcement. [AIR4844]

volatile loss Weight loss by evaporation. [AIR4844]

volatile matter Those products given off by a material as gas or vapor, determined by definite prescribed methods.

volatile memory Memory whose contents are lost when the power is switched off.

volatile storage *1.* A storage device in which stored data are lost when the applied power is removed, for example, an acoustic delay line. *2.* A storage area for information subject to dynamic change.

volatility *1.* A property of a substance that has a low vapor pressure or high boiling point. Low volatility is a desirable characteristic of a hydraulic fluid. [AIR1116] *2.* The tendency of a liquid fuel to evaporate or change to a vapor. [AIR4783]

volatilization *See* vaporization.

volt Symbolized by V. A unit of electromotive force; when applied to a conductor of one ohm resistance, one volt will produce a current of one ampere.

voltage *See* electric potential.

voltage amplification The ratio of the voltage of an output signal to the voltage of the corresponding input signal.

voltage breakdown test. Test to determine the maximum voltage that can be withstood by an insulted wire before electrical current leakage through the insulation. [ARP1931]

voltage, common mode (CMV) *See* common mode voltage.

voltage controlled oscillators (VCO) An oscillator whose frequency of oscillation can be varied by changing an applied voltage.

voltage divider An electronic network that consists of multiple impedance elements connected in series. For a given voltage impressed across the network, one or more lower output voltages can be obtained by tapping across one or more node pairs in the network.

voltage drop The amount of voltage loss from original input in a conductor of given size and length. [ARP1931]

voltage gradient The output voltage expressed as a function of input angle over the specified linear range. [ARP826]

voltage limit In a charge-controlled battery, the limit of voltage beyond which battery potential is not permitted to rise. [ARP4386]

voltage modulation The cyclic variation and/ or random variation about the mean level of the a-c peak voltage during steady-state electric-system operation, such as caused by voltage regulation and speed variations. The modulation envelope is formed by a continuous curve connecting the successive peaks of the basic voltage wave. [AS1212]

voltage modulation frequency characteristics The component frequencies that make up the modulation envelope wave form. [AS1212]

726

voltage-range multiplier A separate device installed externally to an instrument so that the voltage range of the instrument can be extended beyond the upper limit of its scale; consists principally of a special type of series resistance or impedance element.

voltage rating The maximum alternating current and/or the direct-current voltage at which a protective device is designed to operate. [ARP1199A]

voltage regulation The band that the output voltage stays within, except during transients. [ARP1148A]

voltage regulation steady state The band that the output voltage stays within with a fixed load. [ARP1148A]

voltage standing-wave ratio For a waveguide, the ratio of the amplitude of the electric field at a voltage minimum, to the amplitude at an adjacent voltage maximum.

voltage transient recovery The time required for the output voltage to recover to and remain within the prescribed limits after full-load application or removal. [ARP1148A]

voltage transients The maximum momentary deviation of the output voltage from the voltage regulation band. [ARP1148A]

voltage-type telemeter A system for transmitting information to a remote location using the amplitude of a single voltage as the telemeter signal.

volt-ampere meter An instrument for measuring apparent power (the product of voltage and current) in an a-c power circuit. In high-power applications, the scale is usually graduated in kilovolt-amperes.

voltmeter An instrument for determining the magnitude of an electrical potential. Generally constructed as a moving-coil instrument having high internal series resistance; if the high internal resistance is replaced with a low-resistance shunt connected in parallel with the instrument terminals, it can function as an ammeter.

voltmeter-ammeter An instrument consisting of a voltmeter and an ammeter in the same housing, but with separate electrical connections.

volt-ohm-milliammeter A test instrument having different ranges for measuring voltage, resistance, and current flow (in the milliampere range) in electrical or electronic circuits. Also known as a circuit analyzer, multimeter, or multiple-purpose meter.

volume (vol) *1.* The magnitude of a complex audio-frequency current measured in standard volume units on a graduated scale. *2.* The three-dimensional space occupied by an object.

volume control A device or system that regulates or varies the output-signal amplitude of an electronic circuit, such as for varying the loudness of reproduced sound.

volume dynamics The time-dependent response of the thermodynamic properties within a confined volume. Simulation of volume dynamics is usually associated with high-frequency phenomena, such as surge, or long pipes and ducts. [AIR4548]

volume flow rate Flow rate calculated using the area of the full closed conduit and the average velocity, in the form, $Q = V \times A$, to arrive at the total quantity of flow.

volume fraction Fraction of a constituent material based on its volume. [AIR4844]

volume indicator A standard instrument for indicating the magnitude of a complex wave, such as an electronic signal for reproducing speech or music. The magnitude of the complex wave, in volume units, is equal to the number of decibels above a reference level established by connecting the instrument across a 600-ohm resistor that is dissipating 1 mW of power at 100 Hz.

volume median droplet size In an icing cloud, the droplet diameter, in microns, compared to which the droplets comprising one-half of the water in the cloud are smaller. [AIR1168/9]

volume meter Any flowmeter in which actual flow of a fluid is determined by measuring a characteristic associated with the flow.

volume of air The number of cubic feet of air per minute expressed at fan outlet conditions.

volume of air dissolved (S$_1$) Volume of air initially dissolved in a fuel, measured at 32°F and one atmosphere pressure in the vapor phase, per 100 volumes of the fuel at 60°F. [AIR1326]

volume of air dissolved (S$_f$) Volume of air measured at 32°F and one atmosphere pressure that a fuel will dissolve under the lower or final partial pressure of air in the vapor phase, per 100 volumes of the fuel at 60°F. [AIR1326]

volume of air evolved (S$_g$) Calculated from the values of volume of air dissolved, S$_1$ and S$_f$. Equals S$_1$-S$_f$; measured at 32°F and one atmosphere pressure in the vapor phase per 100 volumes of the fuel at 60°F. [AIR1326]

volume resistance Between two electrodes in contact with or embedded in a specimen, the ratio of the direct voltage applied to the electrodes, to the portion of the current between the electrodes that is distributed through the volume of the specimen. [AIR4844]

volume resistivity The ratio of the potential gradient parallel to the current in the fluid, to that of the current density. Measured in ohm-centimeters. [AIR1116]

volumetric efficiency *See* efficiency, volumetric.

volumetric flow *See* flow, volumetric.

volumetric flow rate (q) The volume of fluid moving through a pipe or channel within a given period of time.

volute A spiral casing for a centrifugal pump or fan; it allows the speed developed at the rotor vanes to be converted to pressure without hydraulic shock.

VOR See very-high-frequency omnidirectional range.

VORTAC Partial acronym for VHF omnidirectional range/tactical air navigation. A combination of VOR (very-high-frequency omnidirectional range) and TACAN (tactical air navigation) ground-based radio navigation facilities. [ARP4107]

vortex Fluid in rotational motion (possessing vorticity), e.g. streamed behind a wingtip or across the leading edge of a slender delta.

vortex advisory system Display system that compares measured on-minute-average wind magnitudes and direction with the wind-rose criterion to predict wake vorticity, and to indicate to the air-traffic controller (by means of a red or green light) when the interarrival spacings for landings may be reduced to the 3-nautical mile limit.

vortex alleviation The alteration of airfoil configurations to change the airflow patterns directly behind the wings, to eliminate or inhibit the vertical motion that directly affects the aircraft immediately following, during closely spaced landings.

vortex amplifier An amplifier in which the pressure drop across a controlled vortex is used for modulating the output. [ARP993A]

vortex avoidance Schemes that involve airborne or ground-based equipment to track, monitor, and/or predict vortex behavior that might affect the approach and landing operations.

vortex filaments The fine-scale structure of turbulent flow; the small non-energy containing eddies convected at mean freestream velocities.

vortex flaps Leading edge flap designs for highly swept wings, in which the leading edge tabs are counter reflected, causing vortices to form on the flap. The trapped vortices cause significantly improved wind-flow characteristics.

vortex flow *See* vortices.

vortex generator A small blade perpendicular to the skin of an aircraft or another body, set at an angle of attack to produce a streamwise wingtip vortex, thereby stirring the boundary layer. Usually used to increase the relative speed of the boundary layer and keep it attached to the surface.

vortex shedding Periodic separation of a fluid flowing past an unstreamlined body; a phenomenon that occurs when fluid flows past an obstruction. In such a case, the shear layer near the obstruction has a high velocity gradient, making it inherently unstable; at some point downstream of the immediate vicinity of the obstruction, the shear layer breaks down into well-defined vortices, which are captured by the flowing stream and carried further downstream.

vortex streets Two parallel rows of alternately placed vortices along the wake of an obstacle in a fluid of moderate Reynolds number.

vortex traps *See* trapped vortexes.

vortex-type flowmeter A device in which the differential-pressure variations associated with formation and shedding of vortices in a stream of fluid flowing past a standard flow obstruction (usually a circular element with a T-shaped cross section) are used to actuate a sealed detector at a frequency proportional to vortex shedding; this, in turn, provides an output signal directly related to flow rate.

vortices In fluids, circulations drawing their energy from flows of much larger scale, and brought about by pressure irregularities. *See also* eddy.

vorticity equations Dynamic equations for the rate of change of the vorticity of a parcel, obtained by taking the curl of the vector equation of motion.

voter A logic element or device that selects one signal to represent the output (of the voter) from three or more signals input to the device. [ARP4386]

V/P *See* converter.

V-Path, path descent (climb), profile descent (climb) Terms used to designate modes that control the vertical trajectory to a specific earth-referenced path or profile including one or more three-dimensional waypoints (latitude, longitude, and altitude). The path or profile may be linear or non-linear. Usually it is controlled by elevator or stabilizer, and the speed by throttles or drag; however, other techniques (e.g., "backside") may be used. Such terms are currently used mainly in reference to descents, but climbs are not excluded. *See also* path profile and profile. [ARP1570]

VPROF *See* profile. *See also* path profile and V-path.

V_r Rotation speed. The airspeed during take-off roll, at which the aircraft is rotated up in pitch to attain its takeoff attitude. [ARP4107]

V_{ref} Reference speed for final approach; generally $1.3 \times V_s$. [ARP4107]

V-retainer coupling Coupling in which a V-retainer is an integral part of a coupling segment. V-retainer couplings have no straps; the latch components are attached to lugs on the retainer segments. [AIR869]

V_s Stalling speed; the minimum steady flight speed at which the airplane is controllable. [ARP4107]

VSD *See* vertical situation display.

V_{so} Power-off stalling speed or minimum steady speed at which an airplane in the landing configuration is controllable. [ARP4107]

V_{ss} Stick shaker speed. That velocity at which the pilot's control stick (column) is caused to shake, warning the pilot of the nearness to stall speed. [ARP4107]

V/STOL (vertical/short takeoff and land) Refers to a type of aircraft designed to make a vertical or short run takeoff and landing. [AIR1489]

V/STOL aircraft A hybrid form of heavier-than-air aircraft that is capable, by virtue of one or more horizontal rotors or units acting as rotors, of taking off, hovering, and landing as, or in a fashion similar to, a helicopter. Once aloft, and moving forward, such an aircraft is capable, by means of a mechanical conversion of one sort or another, of flying as a fixed-wing aircraft, especially in its higher speed ranges.

V/STOL forward operating facility A portable airfield capable of providing support by STOL fixed-wing aircraft, as well as helicopters. [AIR1489]

VTOL Partial acronym (pronounced VEE-TOL) for vertical takeoff and landing (aircraft). Aircraft capable of vertical climbs and descents, and of using very short runways or small areas for takeoff and landings. *See also* STOL. [ARP4107]

VTOL forward landing site A portable airfield of minimum size designed for operations dependent on logistic or tactical support by helicopters. [AIR1489]

V_x Best angle of climb speed; the airspeed that delivers the greatest gain of altitude for a given horizontal distance. [ARP4107]

V_{xse} Best angle of climb speed with one engine inoperative. [ARP4107]

V_y Best rate of climb speed; the airspeed that delivers the greatest gain in altitude for a given period of time. [ARP4107]

V_{yse} Best rate of climb speed with one engine inoperative. [ARP4107]

W *1.* The calculated mass of the measured sample volume. [ARP1179A] *2. See* step width - W. *3. See* watt.

W₁ *See* step width - W_1.

wafer *1.* A reinforcement for motorcase port openings. [AIR4844] *2.* A thin disc of a solid substance. *3.* A thin part or component, such as a filter element.

wafer-type temperature detector A type of resistance-thermometer element designed with fine insulated wire of copper, nickel, or platinum sandwiched between protective sheets of insulating material, inside a sealed wafer-type enclosure. This type of temperature detector combines small mass with good thermal contact to give fast response times.

waist The center portion of a vessel, tank, or container which is smaller in cross section than adjacent sections.

wake turbulence Phenomenon resulting from the passage of an aircraft through the atmosphere; includes vortices, thrust stream turbulence, jet blast, jet wash, propeller wash, and rotor wash, both on the ground and in the air. [ARP4107]

walking beam landing gear A type of bogie landing gear configuration in which a forward and aft axle are carried on separate levers pinned to a structural post. A shock-absorber linkage is attached to the levers, and at the upper end to a swinging link or walking beam that is pivoted to the structural post. [AIR1489]

wall attachment amplifier An amplifier in which output is modulated by controlling the attachment of a free jet to a wall (Coanda effect). Usually employed as a digital amplifier. [ARP993A]

wall box A structure in a wall of a steam generator through which apparatus, such as sootblowers, extend into the setting.

wall thickness The thickness of a layer of applied insulation. [ARP1931]

WAN (wide area network) *See* network.

wandering sequence A welding technique in which increments of a weld bead are deposited along the seam randomly with regard to both increment length and location.

warheads Originally, the parts of the missile carrying the explosive, chemical, or other charge intended to damage the enemy; by extension, sometimes used synonymously with payload or nose cone.

warm gas Gas at 1000°F (540°C) to 2500°F (1400°C), typically available from the decomposition of a liquid monopropellant, or from burning liquid propellants at nonstoichiometric conditions. [ARP4386]

warm setting adhesives Intermediate-temperature-curing adhesives. [AIR4844]

warning alert Emergency operational or aircraft system conditions that require immediate corrective or compensatory action by the crew. [ARP4153]

warning indicating system A system that indicates to the pilot or crew member that a hazardous condition requiring immediate action exists, such as fire warning. [ARP1088]

warning, landing gear A device, mechanism, or system to provide a warning to the pilot when the landing gear is not in the landing configuration (extended) under specified conditions (e.g., aircraft altitude or engine power condition); may provide visual or aural warning. [AIR1489]

warp The yarn running lengthwise in a woven fabric. [AIR4844]

warp clock A composite fabrication and engineering symbol used as reference for aligning the warp yarns or tows in the desired direction. [AIR4844]

warp direction The direction of the warp yarns or tows in a fabric or tape. [AIR4844]

warp surface The surface of a fabric that has a majority of warp fibers woven above the fill fibers. [AIR4844]

warranty, failure free *See* failure free warranty.

wash *1.* Airflow or turbulence induced by a propeller, rotor, jet engine, or airfoil. [ARP4107] *2.* In castings, a surface defect caused by heat from the metal rising in the mold, thereby inducing expansion and shear of interface sand in the cope cavity. *3.* A coating applied to the face of a mold prior to casting. Also known as mold wash. *4.* To remove cuttings or debris from a hole during drilling by introducing a liquid stream into the borehole and flushing it out.

washer *1.* An intermediate object with a ratio of thickness to outside diameter of less than 25%; used to provide a fixed dimension between two or more parts. [ARP480A] *2.* A ring-shaped component used to distribute the holding force of a fastener, insulate or cushion a nut or bolthead from its bearing surface, lock a nut in place, or improve tightness of a bolted joint.

waste fuel Any byproduct fuel that is waste from a manufacturing process.

waste heat In non-combustible gases, sensible heat.

waste lubrication A method of delivering oil or other lubricant to a bearing surface by wicking action using cloth waste to absorb and transfer the lubricant.

waste treatment The processing of waste materials (liquid and solid) with chemicals, high temperature, chopping, grinding, and filtering equipment; or with bacterial action, dryers, and separators, for conversion to useful products.

watchdog In control systems, a combination of hardware and software that acts as an interlock scheme, disconnecting the output of the system from the process in event of system malfunction.

watchdog timer An electronic internal timer that will generate priority interrupt unless periodically recycled by a computer. Used to detect program stall or hardware failure conditions.

water absorption Ratio of the weight of water absorbed by a material to the weight of the dry material. [AIR4844]

water-break-free surface A surface that maintains a continuous water film for a period of at least 30 s after having been spray or immersion rinsed in clean water at a temperature below 100°F. [AS1576]

water break test A test in which water is applied to a prepared surface and should remain in a continuous film over the whole area for at least 30 s. [AIR4844]

water calorimeter A device for measuring radio-frequency power by determining the rise in temperature of a known volume of water by which the radio-frequency power is absorbed.

water column (w.c.) A vertical tubular member connected at its top to the steam and at its bottom to the water space of a boiler. The water gage, gage cocks, high- and low-level alarms, and fuel cutoff may be connected to the water column.

water content The percentage of water contamination of a sample of water-washable penetrant or lipophilic emulsifier taken from the process tanks. [AMS2647A]

water cooling Using a stationary or flowing volume of water to absorb heat and disperse it or carry it away.

water delay column A hollow column, filled with water and attached to a search unit, which causes a time delay between the initial pulse and front surface signal. [ARP5089]

water-flow pyrheliometer A device for determining intensity of solar radiation in which the radiation sensor is a blackened water calorimeter. With this device, radiation intensity is calculated from the rise in temperature of water flowing through the calorimeter at a constant rate.

water gage (w.g.) The gage glass and its fittings for attachment.

water gas Gaseous fuel consisting primarily of carbon monoxide and hydrogen, made by the interaction of steam and incandescent carbon.

water hammer *1*. A sudden increase in pressure of water due to an instantaneous conversion of momentum to pressure. *2*. A series of shocks, sounding like hammer blows, caused by suddenly reducing fluid-flow velocity in a pipe.

water heating The heating of water by any means, including solar technology.

water jacket A casing around a pipe, process vessel, or operating mechanism for circulating cooling water.

water jet Water emitted from a nozzle under high pressure. [AIR4844]

water lane A lane or strip of water marked or set aside and maintained for the takeoff and landing of seaplanes. *See also* runway. [AIR1489]

water level The elevation of the surface of the water in a boiler.

water line A horizontal reference plane used for identifying vertical locations within the aircraft. [AS1426]

water path In ultrasonic testing, in a water column or immersion test setup, the distance from the search unit to the workpiece.

water probe A device inserted into the sump area of a filter separator which detects the presence of water at a predetermined level. [AIR4783]

waterproof Impervious to water. Compare with water resistant.

waterproof grease A viscous lubricant that does not dissolve in water, and that resists being washed out of bearings or other moving parts.

waterproofing agent A substance used to treat textiles, paper, wood, and other porous or absorbent materials to make them shed water, rather than allow the water to penetrate.

water resistant Slow to absorb water or to allow water to penetrate. Water resistance is often expressed as a maximum allowable immersion time. Compare with waterproof.

water rudder A small auxiliary rudder attached to the rear portion of a float or hull of a seaplane for the purpose of aiding directional control while the seaplane is on the water. [ARP4107]

water ski *See* hydro ski.

water slug valve A valve used to stop the flow of fuel in a system when signaled by a water-detection system that there is water in the fuel. [AIR4783]

water sump control A float-type device that, by sensing the difference in density of fuel and water, signals the refueling system to activate the deadman system to stop flow, sound an alarm, or cause the sump to automatically drain the water. [AIR4783]

water tube In a boiler, a tube having the water and steam on the inside and heat applied to the outside.

water tube boiler A boiler in which the tubes contain water and steam, the heat being applied to the outside surface.

water vapor *1*. Water (H_2O) in gaseous form; also called aqueous vapor. *2*. Steam, usually of low absolute pressure.

water vapor pressure The partial pressure exerted by the water vapor present in a gas. [ARP171]

watt (W) Metric unit of power. The rate of doing work, or the power expended, equal to 10^7 ergs/s, 3.4192 Btu/h, or 44.27 footpounds/min.

watt-hour A measure of energy or work accomplished. [AIR1898]

watt-hour meter An integrating meter that automatically registers the integral of active power in a circuit with respect to time, usually providing a readout in kW-h.

wattmeter An instrument for measuring the magnitude of the active power in an electric circuit. Such instruments are provided with a scale usually graduated in either watts, kilowatts, or megawatts. If the scale is graduated in kilowatts or megawatts, the instrument is usually designated as a kilowattmeter or a megawattmeters.

wave Variation of a physical attribute of a solid, liquid or gaseous medium in such a manner that some of its parameters vary with time at any position in the medium, while at any instant of time the parameters vary with position.

wave analyzer An electronic instrument for measuring magnitude and frequency of the various sinusoidal components of a complex electrical signal.

wave filter A transducer that separates waves by introducing relatively small insertion loss into waves of one or more frequency bands, while introducing relatively large insertion loss into waves of other frequencies.

waveform digitizer A device that generates a digital signal corresponding to an analog waveform that it receives.

waveforms The graphical representations of waves, showing variation of amplitude with time.

wave front Of a wave propagating in a bulk medium, any continuous surface where the wave has the same phase at any given instant in time.

wave gage A device for measuring the height of waves on the ocean or a large lake, and for measuring the period between successive waves.

waveguide An elongated volume of air or other dielectric used in guided transmission of electromagnetic waves. Usually consists of a circular or rectangular tube of dimensions chosen for efficient propagation of a specific frequency or frequencies. The tube walls may be electrically conductive, or they may constitute surfaces where permittivity and/or permeability are discontinuous.

waveguide lasers Pump sources for deuterium oxide lasers.

wave impedance In an electromagnetic wave, the ratio of the transverse electric field to the transverse magnetic field.

wave interference A pattern of varying wave amplitude caused by superimposing one wave on another in the same medium.

wavelength In any periodic wave, the distance from any point on the wave to a point having the same phases on the next succeeding cycle; the wavelength (λ) equals the phase velocity (n) divided by the frequency (f).

wavelength-division multiplexing Combination of two or more signals so they can be transmitted over a common optical path, usually through a single optical fiber, by a technique in which the signals are generated by light sources having different wavelengths.

wavelength meter An instrument that measures the wavelength of a laser beam or other monochromatic source of light.

wavemeter An instrument for determining the wavelength of an a-c or high-frequency signal, either directly or indirectly, usually by measuring the frequency and converting to wavelength. Also called frequency meter.

wave motor A power-conversion device for producing mechanical power from the lifting power of sea waves.

wave normal A unit vector perpendicular to a wave front and having a positive component coincident with the direction of propagation; in an isotropic medium, the wave normal lies along the direction of the propagation.

wave radiation Radiation made up of oscillating electric and magnetic fields and propagated with speed of light.

wave soldering A soldering technique used extensively to bond electronic components to printed circuit boards. In this technique, the soldering is precisely controlled by moving the assemblies across a flowing wave of solder in a molten soldering bath. This technique also minimizes heating of the assemblies, and thus avoids one of the causes of early failure in electronic components. Also known as flow soldering.

wave tail The trailing portion of a signal-wave envelope, in time or distance, in which the rms wave amplitude decreases from its steady-state value to the end of the wave train.

wave train A succession of ultrasonic waves arising from the same source, having the same characteristics, and propagating along the same path. [ARP5089]

waviness height value A physical measurement in inches that represents the maximum height of the waves—from wave peak to wave valley. [AS291D]

waviness width value A physical measurement in inches that represents the width of the waves—from wave peak to wave peak. [AS291D]

waypoint (WP) *See* area navigation, RNAV waypoint.

Wb *See* weber.

w.c. *See* water column.

WD *See* wind direction.

weak interactions (field theory) One class of the fundamental interactions among elementary particles, responsible for beta decay of nuclei, and for the decay of elementary particles with lifetimes greater than about 10^{-10} sec, such as muons, K mesons, and lambda hypersons. Also known as beta interactions.

weapon replacement assembly *See* line replaceable unit.

weapons delivery Total requirements for locating the target, establishing the release conditions, and maintaining to the target (if required); includes the detection, recognition, and acquisition of the target and the weapons release, as well as guidance.

weapons replaceable assembly (WRA) A generic term that includes all of the replaceable packages of an avionic equipment, pod, or system as installed in an aircraft weapon system, with the exception of cables, mounts, and fuse boxes or circuit breakers. [ARD50010]

weapon system A weapon and those components/parts required for its operation. [ARD50010]

weapon system reliability The probability that a given weapon system that is initially available will successfully complete its designed mission/function. [AIR4896]

wear life The total accumulated operating time of a flight-control actuator, usually measured in hours from the initial pre-installation, including calibration, to the time that the actuator fails to meet specified performance requirements because of wear. [ARP4386]

wear limit of involute splines *See* involute splines, wear limit.

wear patch *See* scuff patch/scuff pad.

wear track On a test ring, the width of the wear mark in millimeters. [AIR1794]

weathercocking/weathervaning The tendency of an aircraft to head into the wind. [ARP4107]

weathercock stability *See* directional stability.

weathering *1.* Exposure of plastics to the outdoor environment. [AIR4844] *2.* The surface deterioration of a hose cover during outdoor exposure, such as checking, cracking, crazing, or chalking. [ARP1658A]

weatherometer A test apparatus used to estimate the resistance of materials and finishes to deterioration when exposed to climatic conditions. This apparatus subjects test surfaces to accelerated weathering conditions, such as concentrated ultraviolet light, humidity, water spray, and salt fog.

weatherproof Capable of being exposed to an outdoors environment for an extended period of time without substantial degradation.

weather resistance The relative ability of a material or coating to withstand the effects of wind, rain, snow, and sun on its color, luster, and integrity.

web *1.* In a grain of propellant, the minimum distance that can burn through as measured perpendicular to the burning surface; the distance between perforations in a grain. [AIR913] *2.* The vertical plate connecting upper and lower flanges of a rail or girder. *3.* In a twist drill or reamer, the central portion of the tool body. *4.* A thin section of a casting or forging connecting two regions of substantially greater cross section.

webbing A narrow fabric woven with continuous filling yarns and finished selvages. [AS8043]

weber (Wb) Metric unit for magnetic flux.

wedge lock fitting A fitting design developed under the auspices of the US Army, specifically for use in crash-resistant fuel tanks; so named because it uses a wedging action to retain the metal fitting rings to the flexible tank materials. [AIR1664]

weep Refers to a minute leak in a boiler joint that forms droplets (or tears) of water very slowly.

weepage Occurrence of slight fluid at the surface of sliding parts, such as piston rods, pump shafts, and valve spools, due to wiping of fluid from the wetted surface by seals or scrapers. [ARP4386]

weeping Slow leakage manifested by the appearance of a fluid on a surface. [AIR4844]

weft *1.* The transverse threads or fibers in a woven fabric. Also known as woof. [AIR4844] *2. See* pick.

weft wires *See* shute.

Weibel instability For collisionless plasmas, an instability characterized by the unstable growth of transverse electromagnetic waves and large magnetic field fluctuation, brought about by an anisotropic distribution of electronic velocities.

weight *See* significance.

weight, carrier landing design gross The maximum weight at which an aircraft is designed to be landed aboard an aircraft carrier. [AIR1489]

weight, catapulting design gross The maximum weight at which an aircraft is designed to be catapulted from an aircraft carrier. [AIR1489]

weight, first landing The maximum weight at which an aircraft is designed to be landed after takeoff in a fully loaded condition. [AIR1489]

weight flow *See* flow, weight.

weighting Artificial adjustment of a measurement to account for factors peculiar to conditions prevailing at the time the measurement was taken.

weight, landing design gross The weight at which an aircraft is designed to be landed for a specified rate of sink velocity. [AIR1489]

weight, landplane landing design gross Applies to land-based aircraft and to field landing of carrier-based and amphibious types; the weight at which the aircraft is designed to be landed at a specified rate of sink velocity. [AIR1489]

weightlessness A condition in which no acceleration, either gravity or other force, can be detected by an observer within the system in question. Also called zero gravity.

weight, maximum design gross The weight of the aircraft with maximum internal and maximum external load for which provision is required. [AIR1489]

weight, maximum landing design gross Applicable to land-based operations; the maximum design gross weight of the aircraft less certain permissible reductions, such as: (a) assist takeoff fuel, (b) droppable fuel and tanks, (c) dumpable fuel, (d) other items expended during or immediately after takeoff as routine takeoff procedure. [AIR1489]

weight, maximum takeoff design gross The maximum taxi design gross weight, less weight for fuel required in taxi-out. [AIR1489]

weight, maximum taxi design gross The maximum weight at which the aircraft is designed to be taxied; not necessarily the same as maximum design gross weight. [AIR1489]

weir An open-channel flow-measurement device analogous to an orifice plate-flow constriction.

weir pond *See* stilling basin.

weld area For fusion welds, the area adjacent to a weld joint, within which the specified weld is to be made. [AS4488]

weld-breaking theory (fretting) *See* fretting, weld-breaking theory.

welded strain gage A type of foil strain gage, specially designed to be attached to a metal substrate by spot welding; used almost exclusively in stress analysis.

welding Producing a coherent bond between two similar or dissimilar metals by heating the joint, with or without pressure, and with or without filler metal, to a temperature at or above the melting points of the metal(s).

welding force *See* electrode force.

welding ground *See* work lead.

weld line Visible on a finished part, the mark made by the meeting of two flow fronts of plastic material during molding. [AIR4844]

weld mark *See* flow line.

weldment A structure or assembly whose parts are joined together by welding.

weld metal The metal in the fusion zone of a welded joint.

weld repair Welding required to make a part conform to the engineering drawing after the initial welding has been accomplished. [AS4488]

well *1.* A compartment, bay, or volume within an air vehicle for stowage of the retracted landing gear wheel assembly, strut, and/or other associated equipment. [AIR1489] *2.* A pressure-tight tube or similarly shaped chamber, closed at one end, and usually having external threads so that it can be screwed into a tapped hole in a pressure vessel; used to form a pressure-tight means of inserting a thermocouple or other temperature-measuring element. Also known as a thermowell. *3.* A chamber at the bottom of a condenser, vacuum filter, or other vessel where liquid droplets collect. Also known as hot well.

well-type manometer A type of double-leg, glass-tube manometer in which one leg is substantially smaller than the other; the large-diameter leg acts as a reservoir, the liquid level of which does not change appreciably with changes in pressure.

Westphal balance An instrument for determining the density of solids and liquids by direct reading, using a balance with movable weights. To measure liquid density, a plummet of known weight and volume is suspended in the liquid; to measure solid density, a sample of the solid is suspended in a liquid of known density.

wet-and-dry bulb thermometer *See* psychrometer.

wet back Baffle provided in a firetube boiler that is completely water cooled; joins the furnace to the second pass to direct the products of combustion.

wet-back boiler A baffle provided in a firetube boiler or water leg construction; covers the rear end of the furnace and tubes, and is completely water cooled. The products of combustion leaving the furnace are turned in this area and enter the tube bank.

wet basis In industrial measurement, the more common basis for expressing moisture content. In the wet basis method, moisture is determined as the quantity present per unit weight or volume of wet material. In contrast, the textile industry uses dry basis or regain moisture content as the measurement standard.

wet-bulb temperature The lowest temperature that a water-wetted body will attain when exposed to an air current; this is the temperature of adiabatic saturation.

wet-bulb thermometer A thermometer in which the bulb is covered with a piece of fabric, such as muslin or cambric, that is saturated with water. Most often used as an element in a psychrometer.

wet cell A cell utilizing liquid electrolyte. [ARP4386]

wet classifier A device for separating solids in a liquid-solid mixture into fractions by making use of the difference in settling rates between small and large particles. A wet classifier may be used on a moving stream with appropriate spiral trays, or in a tank or pond that is not agitated.

wet-coupled bondtester A bondtester that operates on the resonance/impedance principle, generally at frequencies above 100 kHz. Requires a liquid couplant. [ARP5089]

wet dielectric test A voltage dielectric test whereby the specimen to be tested is submerged in a liquid, and a voltage potential placed between the conductor and the liquid. Also called a tank test. [ARP1931]

wet flexural strength The flexural strength after water immersion, usually after boiling the specimen for 2 h in water. [AIR4844]

wet installation Refers to a bolted joint in which sealant is applied to the head and shank of the fastener, thereby providing a seal, after assembly, between the fastener and the elements being joined. [AIR4844]

wet installed fasteners Fasteners that are coated on the shank and under the head with a curing-type sealant to provide a corrosion barrier and a secondary seal. [AIR4069]

wet lay-up A method of making or repairing a reinforced product by applying the resin system as a liquid when the reinforcement is put in place. [AIR4844]

wet leg The liquid-filled low-side impulse line in a differential pressure-level-measuring system.

wetness *1.* Designates the percentage of water in steam. *2.* Describes the presence of a water film on heating surface interiors.

wet-out Describes an impregnated roving or yarn in which substantially all voids between the sized strands and filaments are filled with resin. [AIR4844]

wet-out rate The time required for a resin to fill the interstices of a reinforcement material and wet the surface of the reinforcement fibers, usually determined by optical or light-transmission means. [AIR4844]

wet spinning The production of synthetic and man-made filaments by extruding the chemical solution through spinnerets into a chemical bath where they coagulate.

wet strength The strength of an adhesive joint determined immediately after removal from a liquid in which it has been immersed under specified conditions of time, temperature, and pressure. [AIR4844]

wetting The spreading, and sometimes absorption, of a fluid on or into a surface. [AIR4844]

wet-type differential pressure meter A design of differential-pressure instrument in which pressure difference is determined across an intermediate liquid in the instrument, by means of either an inverted-bell or float-type indicating mechanism.

wet winding In filament winding, a process of winding glass on a mandrel in which the strand is impregnated with resin just before contact with the mandrel. [AIR4844]

w.g. *See* water gage.

Wheatstone bridge A four-arm resistance bridge, usually having three fixed resistances and one variable resistance.

wheelbarrowing A pilot-induced condition resulting when the nose wheel, rather than the main gear, is forced to support an abnormal share of the aircraft weight. This condition is aggravated by a crosswind and often results in loss of directional control. [ARP4107]

wheelbase The distance from the front gear axle to the aft gear axle(s) of an aircraft. If the aft gear axle is of multi-wheel or bogie configuration, the wheelbase is the dimension to the pivot point of the assembly. [AIR1489]

wheel bead seat The portion of the rim of the wheel upon which the tire bead rests. The bead seat diameter is a basic dimension for a wheel. [AIR1489]

wheel, castering *See* castering wheel.

wheel, convertible A wheel that is designed to be used either with a tube-type tire or a tubeless tire. [AIR1489]

wheel, demountable flange A wheel that is designed to permit removal of one rim flange to allow tire installation and removal. [AIR1489]

wheel diagram A method of representing a time-multiplexed format in the manner of a mechanical commutator or rotary switch.

wheel, drop center divided A split-type wheel with a drop center rim to aid in tire installation and removal. [AIR1489]

wheel flange A protruding rim, edge, etc., as on a wheel, used to strengthen the wheel and hold the tire in place. *See also* wheel rim flange. [AIR1489]

wheel, main Those wheels of an aircraft that carry the main portion of the load. In a standard tricycle configuration, the main wheels are normally located just aft of the center of gravity of the air vehicle, and carry 85–90% of the total aircraft weight. A main wheel is different from a nose wheel in that it is designed to house a brake(s) assembly. [AIR1489]

wheel, nonfrangible A wheel configuration designed to provides a means of continued load support and rolling capability in case of failure of the tire(s). [AIR1489]

wheel, nose *See* nose wheel.

wheel, outrigger A wheel that is located in an outboard location, such as a wing tip gear, and carries little or no load. Such a wheel provides stability for the air vehicle in ground operations. [AIR1489]

wheel rim flange The flange on the outboard edge of a wheel rim which holds the tire in place laterally. *See also* wheel flange. [AIR1489]

wheels, co-rotating Multiple (usually two) wheels rotating about a common axis and coupled together to rotate as a unit. Sometimes employed as an antishimmy measure in landing-gear design. [AIR1489]

wheels, coupled *See* wheels, co-rotating.

wheel search unit Ultrasonic device that couples ultrasonic energy to a test part through the rolling contact area of a wheel containing a liquid and one or more transducer elements. [ARP5089]

wheel speed transducer A mechanism that generates an electrical signal with voltage or frequency proportional to wheel velocity. [AIR1489]

wheel, split A wheel configuration designed to provide a means of separating the wheel into two halves or parts for tire installation and removal. [AIR1489]

wheel, sprung A type of wheel designed to contain a shock absorber inside; sometimes combined with a rigid-structure landing gear. [AIR1489]

wheel, tail *See* tail wheel.

whip antennas Thin, flexible monopole antennas.

whisker A short, single-crystal fiber or filament used as reinforcement in a matrix. [AIR4844]

whispering gallery modes Electromagnetic (or elastic) waves that differ in frequency by more than an order of magnitude.

whistlers Radio-frequency electromagnetic signals generated by some lightning discharges.

white holes (astronomy) Time-reversed black holes, expanding sources with growing intensity and photon energy.

white light A mixture of colors of visible light that appears white to the eye. A mixture of the three primary colors is sufficient to produce white light.

white light meter A meter that will measure white light in foot candles or lumens. [AMS2647A]

white noise Ideally, a noise in which power is distributed uniformly over all frequencies, with a mean noise power-per-unit bandwidth. Ideal white noise is not attainable, however, so bandwidth restrictions must be applied.

white radiation *See* Bremsstrahlung.

Whitworth screw thread A British standard screw thread characterized by a 55° "V" form with rounded crests and roots.

wicking *1.* Distribution of a liquid by capillary action into the pores of a porous solid or between the fibers of a material such as cloth. *2.* Flow of solder under the insulation on wire, especially stranded wire.

wide angle diffusion Diffusion in which light is scattered over a wide angle. [ARP798]

wide-area network In data processing, the interconnection of computers that can be miles apart, in contrast to computers that are interconnected within one building.

wideband radiation thermometer A low-cost pyrometer that responds to a wide spectrum of the total radiation emitted by a target object. Depending on the lens or window material used, the instrument responds to wavelengths from 0.3 μm to between 2.5 and 20 μm, which usually represents a significant fraction of the total radiation emitted. Also known as broadband pyrometer or total radiation pyrometer.

wide range metering The measurement of flow rates over a wide range of values at a defined accuracy.

Widmanstatten structure *See* basketweave.

Wien bridge A type of a-c bridge circuit in which one leg containing a variable resistance and a variable capacitance is used to achieve balance with a leg containing a resistance and capacitance in parallel. In addition to applications of this type, the Wien bridge is also used to determine the frequency of an unknown a-c excitation signal, this being the more usual application.

wiggler magnets Components used in the production of coherent x-rays by the pumping of a gas with synchrotron radiation in combination with low-energy photon beams.

wind Movement of air that causes a force to be imposed on surfaces of aircraft ground-support equipment. [ARP1328]

wind angle The angular measure in degrees between the direction parallel to the filaments and an established reference. [AIR4844]

windbox A chamber below the grate or surrounding a burner, through which air under pressure is supplied for combustion of the fuel.

windbox pressure The static pressure in the windbox of a burner or stoker.

wind direction (WD) The clockwise angle from true north to the wind velocity vector. [ARP1570]

wind gust A temporary increased wind force that exceeds the steady-state wind force. [ARP1328]

winding compensated resolver A resolver that contains one or more windings magnetically coupled to the primary, with external terminations for providing a feedback voltage to a booster amplifier which energizes the primary.

This design usually provides close control on transformation ratio and phase shift, and reduces effects of changes in temperature, frequency, applied voltage, or impedance loading to very low levels. [ARP826]

winding pattern The total number of individual circuits required for a winding path to begin repeating by laying down immediately adjacent to the initial circuit. [AIR4844]

winding tension In filament winding or tape wrapping, the amount of tension on the reinforcement as it makes contact with the mandrel. [AIR4844]

windmill With reference to a propeller or rotor, to rotate without engine power, as from the force of the relative wind, or from prior existing power. [ARP4107]

windmilling Steady-state engine rotation caused solely by air being forced through the engine by motion of flight (ram air). [ARP906A]

windmill start An airstart initiated after the engine has reached equilibrium windmilling speed. [ARP906A]

wind-on-nose (WON) Refers to the wind component acting along the longitudinal axis of the aircraft. A tailwind is positive and is sometimes designated as S by some navigators (acting on the stern). A headwind is negative and is sometimes designated as N (acting on the nose). [ARP1570]

window *1.* An interval of time during which conditions are favorable for taking some predetermined action, such as launching a space vehicle. *2.* A defect in thermoplastic sheet or film, similar to a fisheye, but generally larger. *3.* An aperture for passage of magnetic or particulate radiation. *4.* An energy range or frequency range that is relatively transparent to the passage of waves. *5.* In data processing, an area of a computer screen that can be temporarily opened to run a second program without disturbing the original program.

wind propagation The use of current, present position, instantaneous winds for all downstream leg ETEs (estimated times en route) and ETAs (estimated times of arrival). The ETEs and ETAs are continuously revised as the present position winds change. The accuracy of this method is increased if the wind is resolved into directional components corresponding to the downstream leg direction of flight, as opposed to simply using present position ground speed for all downstream legs. [ARP1570]

wind shear A change in the wind speed and/or wind direction in a very short distance, resulting in a tearing or shearing effect. Wind shear can exist in a horizontal or vertical direction, and occasionally in both directions. [ARP4107]

windshear advisory An alert set at a windshear level which requires crew awareness and may require crew action. [AIR4102/11]

windshear cautionary alert An alert set at a windshear level which requires pilot awareness and may require pilot action. [ARP4109]

windshear warning alert An alert set at a windshear level which requires immediate corrective or compensatory action by the pilot. [ARP4109]

windshield combiner Refers to the use of an area of the windshield suitably treated to enable it to function as a combiner. [ARP4102/8]

wind speed The velocity of the wind with respect to a point on the earth's surface. [ARP1570]

wind tunnel(s) Tubelike structures or passages, sometimes continuous, together with their adjuncts, in which high-speed movements of air or other gases are produced (e.g., by fans), and within which objects such as engines or aircraft, airfoils, or rockets (or models of these objects), are placed to investigate the airflow about them and the aerodynamic forces acting upon them.

wind turbines Machines that convert wind energy into electricity.

windup Saturation of the integral mode of a controller developing during times when control cannot be achieved, causing the controlled variable to overshoot its setpoint when the obstacle to control is removed.

wing The primary lifting surface of an airplane. [ARP4386]

wing camber For a wing, a design characteristic describing the variation of airfoil camber from wing root to wing tip.

winglets In aerospace engineering, small, nearly vertical, winglike surfaces mounted rearward above the wing tips to reduce the drag coefficients of lifting conditions.

wing load(ing) Gross weight of an aircraft divided by the area of the wing. [ARP4107]

wing nacelle configurations Aerodynamic configurations involving various arrangements of wings and nacelles (over-the-wing, etc.).

wing nut An internally threaded fastener having radial projections that allow it to be tightened or loosened by finger pressure.

wingover A flight maneuver or stunt in which the aircraft is put into a steep climbing turn until almost stalled, after which the nose is allowed to drop while the turn is continued, until normal flight is attained in a direction opposite to the original heading. [ARP4107]

wing root The base of a wing where it joins the fuselage. [ARP4107]

wing skid A skid or runner located in an outboard location near the wing tip to serve as protection to the air vehicle structure in case of tip-over or instability. *See* skid. [AIR1489]

wingtip gear A landing-gear assembly located at or near the wingtip, usually to provide stability for ground operations when the main load-carrying gear assemblies are located near the centerline of the aircraft. [AIR1489]

wiped joint A type of soldered joint in which molten filler metal is applied, then distributed between the faying surfaces by sliding mechanical motion.

wiping action The interaction of two electrical contacts that come in contact through the sliding of their surfaces against each other. *See also* contact wipe. [ARP914A]

wire *1.* A single metallic conductor of solid, stranded, or tensile construction designed to carry current in an electric circuit, but without a metallic covering, sheath, or shield. [ARP1931] *2.* A very long, thin length of metal, usually of circular cross section, and usually made by drawing through a die.

wire cloth Screening made of wires crimped or woven together.

wired program computer A computer in which the instructions that specify the operations to be performed are specified by the placement and interconnection of wires. The wires are usually held by a removable control panel, allowing flexibility of operation; however, there are also permanently wired machines, referred to as fixed program computers.

wiredrawing Reducing the cross section of wire or rod by pulling it through a die.

wire gage A device for measuring the diameter of wire or thickness of sheet metal.

wire insulation A flexible covering for electrical wire that has relatively high dielectric strength to prevent inadvertently grounding the conductor, and that may have specific mechanical or chemical attributes to protect the conductor from damage or environmental effects in service.

wire range *See* range, wire.

wire rope A flexible length of stranded metal wire suitable for carrying substantial axial tension loads. Most often used to haul or hoist loads, or to act as a tensioned stationary support member in a structure, framework, or truss.

wire size A numerical designation for a conductor, usually expressed in terms of American Wire Gage (AWG), based on the approximate circular mil area of the conductor. [ARP1931]

wire throw-out In braided hose, refers to a broken end or ends in the wire reinforcements protruding from the surface of the braid. [ARP1658A]

wire wrap *See* solderless wrap.

wiring *1.* Wires, cables, groups, harnesses, and bundles, and their terminations, associated hardware, and support, installed in the vehicle. [ARP1931] *2.* The act of fabricating and installing wiring in the vehicle. [ARP1931]

Wobbe index The ratio of the heat of combustion of a gas to its specific gravity. For light hydrocarbon gases, the Wobbe index is almost a linear function of the specific gravity of the gas.

wobble switch Any of several designs of momentary or limit switches that are actuated by physical contact of an object with an extended wire, rod, or cable projecting from the switch body. In most instances, wobble switches can be actuated by an object approaching from any direction in a plane perpendicular to the axis of the actuator.

Wolf-Rayet (W-R) stars Very luminous, very hot (as high as 50,000K) stars whose spectra have broad emission lines (mainly He I and He II), which are presumed to originate from material ejected from the stars at very high velocities. Some W-R spectra show emission lines due to carbon (CWC stars); others show emission lines due to nitrogen (WN stars).

Wollaston wire Extremely fine platinum wire used in electroscopes, microfuses, and hot-wire instruments. Made by coating platinum wire with silver, drawing the sheathed wire, and dissolving the silver away with acid.

WON *See* wind-on-nose.

Wood's glass A type of glass that is relatively opaque to visible light, but relatively transparent to ultraviolet rays.

woof *See* weft. [AIR4844]

word *1.* A sequence of 16 bits plus sync and parity. There are three types of words: command, status, and data. [AS1773] *2.* A collection of bits (8, 12, 16, or 32) that represents the basic instruction or data in the computer.

word processing Software and computer hardware that can manipulate text in a variety of ways, such as formatting, moving, or replacing.

word selector A hardware device that can select a given word or words whenever they occur in the format, and present each to the user as a display and/or analog output.

word time In a storage device that provides serial access to storage locations, the time interval between the appearance of corresponding parts of successive words.

work-around An action required to complete a process run, even though all equipment is not working satisfactorily. A work-around may involve running part of the process on manual, or jumping out of an interlock until maintenance can be scheduled.

work breakdown structure A product-oriented family tree, composed of hardware, services, and data, that completely describes a program or project; results from the identification of acquisition tasks during the development and production of a system or equipment. [ARP4293]

work curve A graph in which the pull-out force, indent force, and relative conductivity of a crimp joint are plotted as a function of various depths of crimping. [ARP914A]

work function An electrical potential difference corresponding to the amount of work that must be done to remove an electron from the surface of a metal.

work hardening Increase in resistance to further deformation with continuing distortion. [AIR4844]

working fluid(s) Fluids (gas or liquid) used as the medium for the transfer of energy from one part of a system to another.

working life The period of time during which a liquid resin or adhesive, after mixing with catalyst, solvent, or other compounding ingredients, remains usable. [AIR4844]

working load The maximum service load that an individual structural member, or an entire structure, is designed to carry.

working pressure The maximum allowable operating pressure for an internally pressurized vessel, tank, or piping system, usually defined by applying the ASME Boiler and Pressure Vessel Code or the API piping code.

working standard *1.* An instrument that is calibrated in a laboratory against an interlab or transfer standard, and is used as a standard in calibrating measuring instruments. [AIR1678] *2.* Any standardized luminous source for daily use in photometry. [ARP798]

working temperature The temperature of the fluid immediately upstream of a primary device.

working voltage *See* service rating.

working zone The portion of the enclosed volume of a piece of thermal-processing equipment occupied by parts or raw material during the soaking portion of a thermal treatment. [AMS2750]

work lead The electrical conductor that connects the workpiece to a welding machine or other source of welding power. Also known as ground lead or welding ground.

work life *See* assembly time.

work softening The phenomena whereby the yield strength of a metal drops when it has been strained or cold-worked at low temperature and subsequently strained at an elevated temperature, causing the dislocations to become unstable.

World Federation The joining together of three international regions: (a) The Americas (Canadian MAP Interest Group and US MAP/TOP Users Group) and Western Pacific (Australian MAP Interest Group), (b) Asia (Japan MAP Users Group), and (c) Europe (European MAP Users Group).

worm conveyor *See* screw conveyor.

wormy alpha *See* stringy alpha.

woven fabric composite A major form of advanced composite in which the fiber constituent consists of woven fabric. [AIR4844]

woven roving A heavy glass-fiber fabric made by weaving roving or yarn bundles. [AIR4844]

wow In instrumentation tape recording and playback, high-frequency tape speed variations.

W/P *See* area navigation, RNAV waypoint.

WRA *See* line replaceable unit.

wrap forming *See* stretch forming.

wrapper sheet *1.* In a firebox or locomotive boiler, the outside plate enclosing the firebox. *2.* The thinner sheet in the shell of a two-thickness boiler drum.

wringing fit A type of interference fit having zero to slightly negative allowance.

wrought alloy A metallic material that has been plastically deformed, hot or cold, after casting to produce its final shape or an intermediate semifinished product.

W-R stars *See* Wolf-Rayet (W-R) stars.

wye A pipe fitting similar to a tee, but in which the branch is at a 45°angle to the run.

wye-delta bridge A d-c bridge arrangement in which resistors no larger than one megohm are used in wye configurations to simulate high-accuracy, high-stability resistors of one gigohm (10^9 ohms) or greater. Wye delta bridges are used to accurately determine the value of very large resistances or to calibrate a bridge.

X

X *See* reactance.

X-axis In composite laminates, an axis in the plane of the laminate which is used as the 0° reference for designating the angle of a lamina. [AIR4844]

xenon chloride lasers Rare-gas-halide lasers using XeCl as the active material.

xenon fluoride lasers Lasers using XeF as the active material.

x-ray binaries Bright galactic x-ray sources consisting of a compact star (neutron star or black hole) accreting matter from a close companion star.

x-ray diffraction analyzer Any of several devices for detecting the positions of monochromatic x-rays diffracted from characteristic scattering planes of a crystalline material; used primarily in detecting and characterizing phases in crystalline solids.

x-ray diffractometer An instrument used in x-ray crystallography to measure the diffracted angle and intensity of x-radiation reflected from a powdered, polycrystalline, or single-crystal specimen.

x-ray emission analyzer An apparatus for determining the elements present in an unknown sample (usually a solid) by bombarding it with electrons and using x-ray diffraction techniques to determine the wavelengths of characteristic x-rays emitted from the sample. With this apparatus, wavelength is used to identify the specific atomic species responsible for the emission, and relative intensity at each strong emission line can be used to quantitatively or semiquantitatively determine composition.

x-ray fluorescence analyzer An apparatus for analyzing the composition of materials (solid, liquid, or gas) by exciting them with strong x-rays and determining the wavelengths and intensities of secondary x-ray emissions.

x-ray goniometer An instrument for measuring the angle between incident and refracted beams of radiation in x-ray analysis.

x-ray imagery Reproduction of an object by means of focusing penetrating electromagnetic radiation (with wavelengths ranging from 10^{-6} to 10^2 nanometers) coming from the object or reflected by the object. Analogous to infrared imagery, radar imagery, and microwave imagery using, respectively, IR, radar and microwave frequencies.

x-ray microscope An apparatus for producing greatly enlarged images by projection, using x-rays from a special ultra-fine-focus x-ray tube, which acts essentially as a point source of radiation.

x-ray monochromator A device for producing an x-ray beam having a narrow range of wavelengths. Usually consists of a single crystal of a selected substance mounted in a holder that can be adjusted to give proper orientation.

x-ray NDI Nondestructive inspection using x-rays. [AIR4844]

x-rays Short-wavelength electromagnetic radiation, having a wavelength shorter than about 15 nanometers, usually produced by bombarding a metal target with a stream of high-energy electrons. The wavelengths of x-rays are in the same range as gamma rays, longer than cosmic rays, but shorter than ultraviolet; like gamma rays, x-rays are very penetrating and can damage human tissues, induce ionization, and expose photographic films. Also called Roentgen rays.

x-ray stars Stars with strong emission in the x-ray portion of the electromagnetic spectrum.

x-ray thickness gage A device used to continually measure the thickness of moving cold-rolled sheet or strip during the rolling process. Consists of an x-ray source on one side of the strip and a detector on the other. With this device, thickness is proportional to the loss in intensity as the x-ray beam passes through the moving material.

x-ray tubes Vacuum tubes designed to produce x-rays by accelerating electrons to a high velocity by means of an electrostatic field, then suddenly stopping them by collision with a target.

X-value A term sometimes used to designate the inductive or capacitative reactance of an a-c electrical device or circuit.

x-wing rotors A new VTOL concept utilizing the stopped rotor x-wing aircraft.

XY-plane In composite laminates, the reference plane parallel to the plane of the laminate. [AIR4844]

XY plotter A device used in conjunction with a computer to plot coordinate points in the form of a graph.

XY recorder A recorder for automatically drawing a graph of the relationship between two experimental variables. In this type of recorder, the position of a pen or stylus at any given instant is determined by signals from two different transducers; this drives the pen-positioning mechanism in two directions at right angles to each other.

Y

Y *See* expansion factor.

yagi antennas Directional antennas used on some types of radar and radio equipment, consisting of an array of elemental, single-wire dipole antennas and reflectors.

Yang-Mills fields Types of fields based on Yang-Mills Theory.

Yang-Mills theory Mathematical theory for describing interactions among elementary particles, based on the idea of gauge invariance under a non-Abelian group.

yaw The lateral rotational or oscillatory movement of an aircraft about the vertical axis. [ARP4107]

yaw axis The vertical or Z axis of an aircraft, about which yawing occurs. [ARP4386]

yawed angle (angle of yaw) The amount of angular movement about a vertical axis. In a tire, the yawed angle is the angle between the plane of symmetry of the wheel and the direction line of roll of the wheel/tire assembly. [AIR1489]

yawed rolling The process of rolling in a yawed mode of operation. [AIR1489]

yawing moments Moments that tend to rotate aircraft, airfoils, rockets, or spacecraft about a vertical axis.

yawing response to steer angle The tire response (yawed rolling angle) for a given input steering signal (angle). [AIR1489]

Y-axis In composite laminates, the axis in the plane of the laminate that is perpendicular to the X-axis. [AIR4844]

y factor (air) The ratio of the isentropic temperature rise to the total absolute temperature at the inlet, given as a function of pressure ratio across the compressor. [ARP147C]

yield The quantity of a substance produced in a chemical reaction or other process from a specific amount of incoming material.

yield load Applied structural load that produces stress equal to the yield stress. For mechani-

cally fastened joints, a fitting factor (usually 1.15) is often included.

yield point In a material, the first stress, less than the maximum attainable stress, at which the strain increases at a higher rate than the stress. [AIR4844]

yield strength *1.* The stress at which a material exhibits a specified deviation from proportionality of stress and strain. [ARP700] *2.* The lowest stress at which a material undergoes plastic deformation; below this stress the material is elastic; above it, viscous. [ARP1931]

yield strength-offset On a stress-strain curve, the distance along the strain coordinate between the initial portion of the curve and a parallel line intersecting the curve at a value of stress used as a measure of yield strength. Yield-strength offset is used for materials that have no obvious yield point; a value of 0.2% of the strain is normally used. [ARP700]

yield stress The force per unit area at the onset of plastic deformation, as determined in a standard mechanical-property test, such as a uniaxial tension test.

yoke *1.* The cockpit control in an aircraft, used to manually input pitch and roll commands. *2.* A C-shaped transducer holder used to align transducers during through-transmission testing. [ARP5089] *3.* A clamp or structure that attaches to two or more other parts to hold them in place. [ARP480A] *4.* A slotted crosshead used in some steam engines instead of a connecting rod. *5.* The framework surrounding the rotor of a d-c generator or motor, which supports the field coils and provides magnetic linkage between them.

Young's modulus The ratio of normal stress to corresponding strain for tensile or compressive stresses less than the proportional limit of the material. [AIR4844]

Z

Z *See* compressibility factor.

zap In data processing, a slang word meaning to erase or wipe-out data.

Z-axis In composite laminates, the reference axis normal to the plane of the laminate. [AIR4844]

ZEL (zero launch) The launch of a rocket or aircraft by a zero length launcher. [AIR1489]

zenith That point of the celestial sphere vertically overhead. The point 180 deg from the zenith is called the nadir.

zeolites Any of a group of hydrated silicates of aluminum with alkali metals, used for their molecular sieve properties. [ARP171]

zero a device To erase all of the data stored on a volume and reinitialize the format of the volume.

zero adjuster A mechanism for repositioning the pointer on an instrument so that the instrument reading is zero when the value of the measured quantity is zero.

zero adjustment *See* adjustment, zero.

zero bias A positive or negative adjustment to instrument zero to cause the measurement to read as desired.

zero bleed Refers to a laminate fabrication procedure that does not allow loss of resin during cure. [AIR4844]

zero code error A measure of the difference between the ideal (0.5 LSB) and the actual differential analog input level required to produce the first positive LSB code to transition (00...00 to 00...01).

zero defects Refers to a management program that encourages perfect performance in a manufacturing operation, and usually provides workers with rewards and incentives for achieving perfection.

zero drift Time-related deviation of analyzer output from zero setpoint when it is operating on gas free of the component to be measured. Zero drift is not to be confused with interference. [ARP1256]

zero elevation Biasing the zero output signal to raise the zero to a higher starting point. Usually used in liquid-level measurement for starting measurement above the vessel connection point.

zero-fuel weight The maximum permissible weight of an aircraft in flight when there is no fuel in the wing tanks to reduce the structural stresses in the wing structure.

zero gas A gas to be used in establishing the zero (or no-response) adjustment of an analyzer. [ARP1256]

zero governor A regulating device that is normally adjusted to deliver gas at atmospheric pressure within its flow rating.

zero grade gas A gas that is used to establish the zero (or no response) adjustment of an instrument, and that is free from the component to be measured by the instrument. [ARP4418]

zero gravity *See* weightlessness.

zero lap The valve lap condition that results in the meeting of the straight line extensions of the normal flow curve. [ARP4386]

zero launch *See* ZEL.

zero length launcher A launcher that holds a missile or vehicle in position and releases it simultaneously at points at which a buildup of thrust (normally rocket thrust) is sufficient to take the missile or vehicle directly into the air, without need for a takeoff run and without imposing a pitch rate upon release. [AIR1489]

zero level A reference level used for comparing signal intensities in electronic or sound-reproduction systems. In electronics, the zero level is usually taken as 0.006 W of power; in sound reproduction, it is usually taken as the threshold of hearing.

zero lift Angle of attack at which an airofoil generates neither positive nor negative lift, i.e., no force normal to the airstream.

zero-lift angle of attack The angle between the chord plane of an airfoil and the relative wind at which the airfoil produces no lift. [ARP4107]

zero point energy Kinetic energy retained by molecules of a substance at a temperature of absolute zero.

zero shift A shift in the instrument-calibrated span evidenced by a change in the zero value. Usually caused by temperature changes, overrange, or vibration of the instrument.

zero suppression *1.* The elimination of non-significant zeros in a numeral. *2.* Biasing the zero output signal to produce the desired measurement. Used in level measurement to counteract the zero elevation caused by a wet-leg.

zero time The time at which starting-switch actuation occurs. [AS1606]

zero-zero Refers to a condition of no effective visibility in either a horizontal or a vertical direction, as in zero-zero weather. [ARP4107]

zero/zero out The procedure of adjusting a measuring instrument to the proper output value for a zero-measurement signal.

zeta pinch Type of plasma pinch produced by an electric current applied axially to a plasma cylinder in a controlled fusion reactor.

Ziegler-Nichols method A method of determining optimum controller settings when tuning a process-control loop, based on finding the proportional gain that causes instability in a closed loop. Also called the ultimate cycle method.

zincblende Zinc sulfide, ZnS; a cubic crystal.

zinc-bromide batteries Electric cells in which, during charge, zinc is plated on the anode and bromine is evolved at the cathode; the bromine is transferred to an external chamber for mixing and storing with an organic liquid complexing oil. During discharge, the zinc is oxidized at the anode and the complexed bromine is reduced at the cathode.

zinc chlorides Reaction products of hydrochloric acid and zinc; white crystals, soluble in water and alcohol, with a melting point of 290°C.

zinc-chlorine batteries Candidate electric cells under development for electric vehicles.

zinc plating An electroplating coating of zinc on a steel surface which provides corrosion protection in a manner similar to galvanizing.

zinc rich coating (ZRC) A single-component zinc-rich coating that can be applied by brush, spray, or dip, and dries to a gray matte finish. ZRC is accepted by Underwriters Laboratories, Inc. as being equivalent to hot-dip galvanizing.

Zo *See* characteristic impedance.

zonal circulation *See* zonal flow (meteorology).

zonal flow (meteorology) The flow of air along a latitude circle; more specifically, the latitudinal (east or west) of existing flow. Also called zonal circulation.

zone *1.* A portion of internal storage allocated for a particular function or purpose. *2.* The international method for specifying the probability that a location is made hazardous by the presence, or potential presence, of flammable concentrations of gases or vapors.

Zone 0 A location in which a concentration of a flammable gas or vapor mixture is continuously present, or is present for long periods of time.

Zone 1 A location in which a concentration of a flammable gas or vapor mixture is likely to occur in normal operation.

Zone 2 A location in which a concentration of a flammable gas or vapor mixture is unlikely to occur in normal operation; and if it does occur, will exist only for a short period of time.

zone bit *1.* One of the two leftmost bits in a binary coding system in which six bits are used to define each character. *2.* Any bit in a group of bit positions used to indicate the classification of the group; for example, numeral, alpha character, special sign, or command.

zone control A method of controlling temperature or some other process characteristic by dividing a physical area or process flowpath into several regions or zones, and independently controlling the process characteristic in each zone.

zones Part of area classification; formerly called divisions. Areas in which the probability of the presence and concentration of a potentially explosive mixture is similar.

zooplankton The aggregate of passively floating or drifting animal organisms in aquatic ecosystems.

ZRC *See* zinc rich coating.

Z-value Sometimes used to designate the impedance of a device or circuit, which is the vector sum of resistance and reactance (X-value).

zyglo method A technique for liquid-penetrant testing to detect surface flaws in a metal; uses a special penetrant that fluoresces when viewed under ultraviolet radiation.